D1690174

CHEMICAL SYNTHESIS OF
NUCLEOSIDE ANALOGUES

CHEMICAL SYNTHESIS OF NUCLEOSIDE ANALOGUES

Edited by

Pedro Merino
Universidad de Zaragoza
Zaragoza, Aragon, Spain

WILEY

A JOHN WILEY & SONS, INC., PUBLICATION

Copyright © 2013 by John Wiley & Sons, Inc. All rights reserved.

Published by John Wiley & Sons, Inc., Hoboken, New Jersey.
Published simultaneously in Canada.

No part of this publication may be reproduced, stored in a retrieval system, or transmitted in any form or by any means, electronic, mechanical, photocopying, recording, scanning, or otherwise, except as permitted under Section 107 or 108 of the 1976 United States Copyright Act, without either the prior written permission of the Publisher, or authorization through payment of the appropriate per-copy fee to the Copyright Clearance Center, Inc., 222 Rosewood Drive, Danvers, MA 01923, (978) 750-8400, fax (978) 750-4470, or on the web at www.copyright.com. Requests to the Publisher for permission should be addressed to the Permissions Department, John Wiley & Sons, Inc., 111 River Street, Hoboken, NJ 07030, (201) 748-6011, fax (201) 748-6008, or online at http://www.wiley.com/go/permission.

Limit of Liability/Disclaimer of Warranty: While the publisher and author have used their best efforts in preparing this book, they make no representations or warranties with respect to the accuracy or completeness of the contents of this book and specifically disclaim any implied warranties of merchantability or fitness for a particular purpose. No warranty may be created or extended by sales representatives or written sales materials. The advice and strategies contained herein may not be suitable for your situation. You should consult with a professional where appropriate. Neither the publisher nor author shall be liable for any loss of profit or any other commercial damages, including but not limited to special, incidental, consequential, or other damages.

For general information on our other products and services or for technical support, please contact our Customer Care Department within the United States at (800) 762-2974, outside the United States at (317) 572-3993 or fax (317) 572-4002.

Wiley also publishes its books in a variety of electronic formats. Some content that appears in print may not be available in electronic formats. For more information about Wiley products, visit our web site at www.wiley.com.

Library of Congress Cataloging-in-Publication Data:

Chemical synthesis of nucleoside analogues / edited by Pedro Merino.
   p. ; cm.
Includes bibliographical references and index.
ISBN 978-1-118-00751-8 (cloth)
I. Merino, Pedro, 1962–
[DNLM: 1. Nucleosides–chemical synthesis. 2. Nucleosides–pharmacology. 3. Anti-Infective Agents–chemical synthesis. 4. Anti-Infective Agents–pharmacology. QU 58]
616.9′1061–dc23

2012040271

To my sons Pedro, Javier, and Ignacio

CONTENTS

Contributors		ix
Foreword		xiii
Preface		xv
1	**Biocatalytic Methodologies for Selective Modified Nucleosides** Miguel Ferrero, Susana Fernández, and Vicente Gotor	1
2	**Nucleosides Modified at the Base Moiety** O. Sari, V. Roy, and L. A. Agrofoglio	49
3	**Chemical Synthesis of Acyclic Nucleosides** Hai-Ming Guo, Shan Wu, Hong-Ying Niu, Ge Song, and Gui-Rong Qu	103
4	**Phosphonated Nucleoside Analogues** Roberto Romeo, Caterina Carnovale, Antonio Rescifina, and Maria Assunta Chiacchio	163
5	**Chemical Syntheses of Nucleoside Triphosphates** Lina Weinschenk and Chris Meier	209
6	**Mononucleotide Prodrug Synthetic Strategies** Suzanne Peyrottes and Christian Périgaud	229
7	**Synthesis of C-Nucleosides** Omar Boutureira, M. Isabel Matheu, Yolanda Díaz, and Sergio Castillón	263
8	**Methodologies for the Synthesis of Isomeric Nucleosides and Nucleotides of Antiviral Significance** Maurice Okello and Vasu Nair	317
9	**Synthesis of Conformationally Constrained Nucleoside Analogues** Esma Maougal, Jean-Marc Escudier, Christophe Len, Didier Dubreuil, and Jacques Lebreton	345

10	**Synthesis of 3′-Spiro-Substituted Nucleosides: Chemistry of TSAO Nucleoside Derivatives** *María-José Camarasa, Sonsoles Velázquez, Ana San-Félix, and María-Jesús Pérez-Pérez*	427
11	**L-Nucleosides** *Daniela Perrone and Massimo L. Capobianco*	473
12	**Chemical Synthesis of Carbocyclic Analogues of Nucleosides** *E. Leclerc*	535
13	**Uncommon Three-, Four-, and Six-Membered Nucleosides** *E. Groaz and P. Herdewijn*	605
14	**Recent Advances in Synthesis and Biological Activity of 4′-Thionucleosides** *Varughese A. Mulamoottil, Mahesh S. Majik, Girish Chandra, and Lak Shin Jeong*	655
15	**Recent Advances in the Chemical Synthesis of Aza-Nucleosides** *Tomás Tejero, Ignacio Delso, and Pedro Merino*	699
16	**Stereoselective Methods in the Synthesis of Bioactive Oxathiolane and Dioxolane Nucleosides** *D. D'Alonzo and A. Guaragna*	727
17	**Isoxazolidinyl Nucleosides** *Ugo Chiacchio, Antonino Corsaro, Salvatore Giofrè, and Giovanni Romeo*	781
18	**Synthetic Studies on Antifungal Peptidyl Nucleoside Antibiotics** *Apurba Datta*	819
19	**Chemical Synthesis of Conformationally Constrained PNA Monomers** *Pedro Merino and Rosa Matute*	847
Index		881

CONTRIBUTORS

L. A. AGROFOGLIO, Institut de Chimie Organique et Analytique, Université d'Orléans, Orléans, France

OMAR BOUTUREIRA, Departamento de Química Analítica i Química Orgànica, Universitat Rovira i Virgili, Tarragona, Spain

MARÍA-JOSÉ CAMARASA, Instituto de Química Médica, CSIC, Madrid, Spain

MASSIMO L. CAPOBIANCO, Istituto per la Sintesi Organica e la Fotoreattività, Consiglio Nazionale delle Ricerche, Bologna, Italy

CATERINA CARNOVALE, Dipartimento Farmaco-Chimico, Università di Messina, Messina, Italy

SERGIO CASTILLÓN, Departamento de Química Analítica i Química Orgànica, Universitat Rovira i Virgili, Tarragona, Spain

GIRISH CHANDRA, Department of Bioinspired Science and Laboratory of Medicinal Chemistry, College of Pharmacy, Ewha Womans University, Seoul, Korea

MARIA ASSUNTA CHIACCHIO, Dipartimento di Scienze del Farmaco, Università di Catania, Catania, Italy

UGO CHIACCHIO, Dipartimento di Scienze del Farmaco, Università di Catania, Catania, Italy

ANTONINO CORSARO, Dipartimento di Scienze del Farmaco, Università di Catania, Catania, Italy

D. D'ALONZO, Dipartimento di Scienze Chimiche, Università degli Studi di Napoli Federico II, Napoli, Italy

APURBA DATTA, Department of Medicinal Chemistry, University of Kansas, Lawrence, Kansas, USA

IGNACIO DELSO, Servicio de Resonancia Magnética Nuclear, Centro de Química y Materiales de Aragón, Universidad de Zaragoza, CSIC, Zaragoza, Aragón, Spain

YOLANDA DÍAZ, Departamento de Química Analítica i Química Orgànica, Universitat Rovira i Virgili, Tarragona, Spain

DIDIER DUBREUIL, Université de Nantes, Faculté des Sciences et des Techniques, Nantes, France

JEAN-MARC ESCUDIER, Université Paul Sabatier, Laboratoire de Synthèse et Physicochimie des Molécules d'Intérêt Biologique, Toulouse, France

SUSANA FERNÁNDEZ, Departamento de Química Orgánica e Inorgánica y Instituto Universitario de Biotecnología de Asturias, Universidad de Oviedo, Oviedo, Asturias, Spain

MIGUEL FERRERO, Departamento de Química Orgánica e Inorgánica y Instituto Universitario de Biotecnología de Asturias, Universidad de Oviedo, Oviedo, Asturias, Spain

SALVATORE GIOFRÈ, Dipartimento Farmaco-Chimico, Università di Messina, Messina, Italy

VICENTE GOTOR, Departamento de Química Orgánica e Inorgánica y Instituto Universitario de Biotecnología de Asturias, Universidad de Oviedo, Oviedo, Asturias, Spain

E. GROAZ, Laboratory for Medicinal Chemistry, Rega Institute for Medical Research, Leuven, Belgium

A. GUARAGNA, Dipartimento di Chimica Organica e Biochimica, Università di Napoli Federico II, Napoli, Italy

HAI-MING GUO, College of Chemistry and Environmental Science, Key Laboratory of Green Chemical Media and Reactions of Ministry of Education, Henan Normal University, Xinxiang, China

P. HERDEWIJN, Laboratory for Medicinal Chemistry, Rega Institute for Medical Research, Leuven, Belgium

LAK SHIN JEONG, Department of Bioinspired Science and Laboratory of Medicinal Chemistry, College of Pharmacy, Ewha Womans University, Seoul, Korea

JACQUES LEBRETON, Université de Nantes, Faculté des Sciences et des Techniques, Nantes, France

E. LECLERC, Institut Charles Gerhardt, Ecole Nationale Supérieure de Chimie de Montpellier, Montpellier, France

CHRISTOPHE LEN, Université de Technologie de Compiègne, Ecole Supérieure de Chimie Organique et Minérale, Transformaciones Intégrées de la Matière Renouvelable, Centre de Recherche Royallieu, Compiègne, France

MAHESH S. MAJIK, Department of Bioinspired Science and Laboratory of Medicinal Chemistry, College of Pharmacy, Ewha Womans University, Seoul, Korea

ESMA MAOUGAL, Université de Nantes, Faculté des Sciences et des Techniques, Nantes, France

M. ISABEL MATHEU, Departamento de Química Analítica i Química Orgànica, Universitat Rovira i Virgili, Tarragona, Spain

ROSA MATUTE, Department of Chemical Engineering and Environment Technologies, University of Zaragoza, Zaragoza, Aragón, Spain

CHRIS MEIER, Department of Chemistry, Faculty of Sciences, University of Hamburg, Hamburg, Germany

PEDRO MERINO, Departamento de Síntesis y Estructura de Biomoléculas, y Departamento de Química Orgànica, Instituto de Sintesis y Catalisis Homogenea, Universidad de Zaragoza, CSIC, Zaragoza, Aragón, Spain

VARUGHESE A. MULAMOOTTIL, Department of Bioinspired Science and Laboratory of Medicinal Chemistry, College of Pharmacy, Ewha Womans University, Seoul, Korea

VASU NAIR, Center for Drug Discovery and College of Pharmacy, University of Georgia, Athens, Georgia, USA

HONG-YING NIU, College of Chemistry and Environmental Science, Key Laboratory of Green Chemical Media and Reactions of Ministry of Education, Henan Normal University, Xinxiang, China

MAURICE OKELLO, Center for Drug Discovery and College of Pharmacy, University of Georgia, Athens, Georgia, USA

MARÍA-JESÚS PÉREZ-PÉREZ, Instituto de Química Médica, CSIC, Madrid, Spain

CHRISTIAN PÉRIGAUD, Institut des Biomolécules Max Mousseron, Université Montpellier 2, Montpellier, France

DANIELA PERRONE, Dipartimento di Scienze Chimiche e Farmaceutiche, Università di Ferrara, Ferrara, Italy

SUZANNE PEYROTTES, Institut des Biomolécules Max Mousseron, Université Montpellier 2, Montpellier, France

GUI-RONG QU, College of Chemistry and Environmental Science, Key Laboratory of Green Chemical Media and Reactions of Ministry of Education, Henan Normal University, Xinxiang, China

ANTONIO RESCIFINA, Dipartimento di Scienze del Farmaco, Università di Catania, Catania, Italy

GIOVANNI ROMEO, Dipartimento Farmaco-Chimico, Università di Messina, Messina, Italy

ROBERTO ROMEO, Dipartimento Farmaco-Chimico, Università di Messina, Messina, Italy

V. ROY, Institut de Chimie Organique et Analytique, Université d'Orléans, Orléans, France

ANA SAN-FÉLIX, Instituto de Química Médica, CSIC, Madrid, Spain

O. SARI, Institut de Chimie Organique et Analytique, Université d'Orléans, Orléans, France

GE SONG, College of Chemistry and Environmental Science, Key Laboratory of Green Chemical Media and Reactions of Ministry of Education, Henan Normal University, Xinxiang, China

TOMÁS TEJERO, Departamento de Síntesis y Estructura de Biomoléculas, Instituto de Sintesis y Catalisis Homogenea, Universidad de Zaragoza, CSIC, Zaragoza, Aragón, Spain

SONSOLES VELÁZQUEZ, Instituto de Química Médica, CSIC, Madrid, Spain

LINA WEINSCHENK, Department of Chemistry, Faculty of Sciences, University of Hamburg, Hamburg, Germany

SHAN WU, College of Chemistry and Environmental Science, Key Laboratory of Green Chemical Media and Reactions of Ministry of Education, Henan Normal University, Xinxiang, China

FOREWORD

Nucleosides and nucleotides are implicated in all aspects of cellular life—metabolic regulation, catalysis, energy supply, and the storage of genetic information—through the nucleic acids. This broad and important role suggests that the chemistry of this class of compounds represents a foremost research topic in bioorganic and medicinal chemistry. Since the two final decades of the twentieth century it has been clear that the synthesis of modified nucleosides and nucleoside analogues would be the key to understanding the cellular functions and mechanisms involved in a variety of diseases. In fact, research has resulted in several important successes, as many anticancer and antiviral drugs have been produced as nucleoside analogues, in addition to several other compounds employed as diagnostic tools.

The present volume provides an analysis of the chemical syntheses of nucleoside analogues that address systematically almost all the diverse classes of nucleoside analogues, giving a picture of the state of the art and of the new directions that research in this field is undertaking. The selection of topics in this volume correctly reflects the dynamism of investigations in progress in the area of nucleosides and nucleotides and draws attention to the expanding opportunities for medicinal chemistry. The scientific credentials of the editor and the contributors, all well-recognized international experts in the field, assure readers of the quality of the work.

This important volume will be of interest to a wide audience of multidisciplinary researchers in organic, bioorganic, and medicinal chemistry, and will be especially helpful for inexpert researchers who are entering the field with fresh enthusiasm to put forth new ideas.

ALBERTO BRANDI

University of Florence

PREFACE

Nucleosides are the building blocks required in constructing nucleic acids. As we learned in school, billions of combinations are possible using only four different nucleosides, and thus the genetic code is generated, carrying all our past and future features, behavior, and illnesses. But nucleosides also take part in many other biological processes, and they can be found as discrete molecules not closely related to nucleic acids. This is the case for cyclic AMP, a second messenger involved in many biological processes, and for the well-known ATP, a coenzyme used by cells for intracellular energy transfer, or S-adenosylmethionine, the "natural" methylating agent employed for a great variety of living beings and required for cellular growth.

In all cases, either forming a part of nucleic acids or as discrete entities, nucleosides are composed of a sugar residue of furanose (either D-ribose or 2-deoxy-D-ribose) and a heterocyclic base (adenine, guanine, thymine, cytosine, uracil, and the more recently discovered methylcytosine). Only combinations of these components are found in nature, with very few exceptions. However, for many years synthetically modified nucleosides have proven their valuable utility as therapeutic drugs. Among other uses they have been employed as antiviral and antitumoral agents with great success, although they often cause important and undesirable side effects. For this reason it is still a challenge to develop new nucleoside analogues that can be used as therapeutic drugs while minimizing side effects. The only access to nucleoside analogues is chemical synthesis, and a great effort is still necessary to develop new routes to a wide variety of modified nucleosides. Indeed, such modifications can assist in several ways, including modification of the heterocyclic base and changes at the furanose ring. The former is somewhat limited, because to achieve some biological activity the ability to form hydrogen bonds must be retained. Accordingly, the number of heterocycles that can be used as alternative bases is limited even though there is a considerable catalogue of possibilities. Modifications at the furanose ring are more versatile. It is possible to eliminate hydroxyl groups (deoxynucleosides), to change their configuration or even to exchange other substituents, as in the case of the well-known anti-HIV agent AZT. It is also possible to replace the furanose ring by a different heterocyclic ring (aza-nucleosides, thionucleosides, and dioxolanyl, oxathiolanyl, and isoxazolidinyl nucleosides are good examples), by a carbocyclic ring (carba-nucleosides), and by an acyclic chain (acyclic nucleosides). These modifications have important effects on the conformation of nucleosides that can also be achieved by introducing conformational restrictions, as in the case of constrained nucleosides and spironucleosides. Additional modifications consist of replacing the C–N nucleosidic bond by a hydrolytically stable C–C bond as in the case of C-nucleosides, moving the base moiety at the furanose ring to a different position (isonucleosides) and synthesizing the enantiomers of natural nucleosides (L-nucleosides). All these modifications, should, however, preserve a relative disposition between the base moiety and the hydroxymethyl group, which is crucial for biological activity; very few exceptions are found in which such a group does not seem to be necessary. The hydroxymethyl group is required for further phosphorylation of the nucleoside analogue to form the corresponding nucleotide, which is the biologically active form. This phosphorylation should be carried out by kinases, a family of enzymes that is usually very selective, and therefore is not always easy to achieve for modified compounds. Thus, the preparation of phosphonated analogues and pronucleotides capable of generating phosphorylated analogues in situ, is of great importance. In addition, nucleoside analogues can form a part of more complex structures, as is true of nucleoside antibiotics. They can also be used as monomers in the construction of novel peptide nucleic acids, promising molecules for the treatment of autoimmune diseases and allergies and for use as molecular probes.

These transformations are somewhat different and require different synthetic approaches. In this book a collection of various possibilities are presented that have been studied by the authors in accessing the great variety of nucleoside analogues. Without their effort it would not be possible to have at our disposal a huge arsenal of nucleoside analogues to be tested as new therapeutic drugs. Chemical synthesis is the invaluable tool that makes possible our opportunity to fight against such important illnesses as antiviral and bacterial infections, malaria, cancer, and autoimmune diseases. We sincerely thank all synthetic chemists who are dedicating their professional lives to this mission, often without the recognition by society that they deserve.

Pedro Merino

Universidad de Zaragoza

1 Biocatalytic Methodologies for Selective Modified Nucleosides

MIGUEL FERRERO, SUSANA FERNÁNDEZ, and VICENTE GOTOR

Departamento de Química Orgánica e Inorgánica y Instituto Universitario de Biotecnología de Asturias, Universidad de Oviedo, Oviedo, Asturias, Spain

1. INTRODUCTION

Nucleosides are fundamental building blocks of biological systems that are widely used as therapeutic agents to treat cancer, fungal, bacterial, and viral infections.[1] Since the latter 1980s, nucleoside analogues have been investigated with renewed urgency in the search for agents effective against the human immunodeficiency virus (HIV) and to use as a more effective treatment for other viral infections. This has resulted in an explosion of synthetic activity in the field of nucleosides, and consequently, extensive modifications have been made to both the heterocyclic base and the sugar moiety to avoid the drawbacks shown by nucleosides or analogues in certain applications.

The intense search for clinically useful nucleoside derivatives has resulted in a wealth of new approaches to their synthesis, and most important, their enantioselective synthesis. Thus, especially for organic chemists, biocatalytic methods have been recognized as practical procedures in the nucleoside area.[2] For the manipulation of protecting groups, the use of biocatalysts in organic synthesis has become an attractive alternative to conventional chemical methods, due to their simple feasibility and high efficiency. In general, these catalysts are inexpensive and satisfy increasingly stringent environmental constraints. Due to these advantages, biocatalyzed reactions are playing an increasing role primarily in the preparation of nonracemic chiral biologically active compounds not only in the laboratory but also in industrial production, in which enzyme-catalyzed chemical transformations are in great demand in the pharmaceutical and chemical industries.[3] In addition, enzyme-catalyzed reactions are less hazardous, less polluting, and less energy consuming than are conventional chemistry-based methods.

Chemical Synthesis of Nucleoside Analogues, First Edition. Edited by Pedro Merino.
© 2013 John Wiley & Sons, Inc. Published 2013 by John Wiley & Sons, Inc.

The synthetic potential of enzymes related to nucleoside synthesis has been applied profusely, especially since the introduction of organic solvent methodology. It is our aim in this chapter to cover the literature of the last decade or so relative to nucleosides with selected examples because of special significance. Our desire is to show a range of examples that cover nucleoside analogue syntheses through enzymatic procedures and to summarize and offer an easily accessible visual reference review. Due to the vastness of the bibliographic material related to nucleosides, we do not cover other enzymatic processes, such as preparation of nucleoside antibiotics using microorganisms,[4] nucleoside syntheses mediated by glycosyl transfer,[5] or halogenation enzymes.[6]

Most enzymatic reactions, just like those included here, are performed by a small number of biocatalysts. With the passing of time, their nomenclature has changed in an effort to unify criteria and to refer to a given enzyme by only one name. Table 1 lists the enzymes mentioned in this review, sorted alphabetically. These

Table 1. Enzymes Commonly Used in Biocatalytic Processes Shown in This Review

Accepted Name (Abbreviation)
   Other Denominations

Adenosine deaminase (ADA)
Adenylate deaminase (AMPDA)
   $5'$-adenylic acid deaminase, AMP deaminase
***Candida antarctica* lipase A (CAL-A)**
   lipase A
***Candida antarctica* lipase B (CAL-B)**
   Novozym-435, SP-435, lipase B
***Candida rugosa* lipase (CRL)**
   Candida cylindracea lipase (CCL)
ChiroCLEC BL
Lipase M (from *Mucor javanicus*)
Lipozyme
   Mucor miehei lipase, Lipozyme IM
***Pseudomanas cepacia* lipase (PSL)**
   Pseudomonas sp. lipase, *Pseudomonas fluorescens* lipase (PFL), PCL, lipase P, lipase PS, LPS, amano PS, amano lipase PS
   Burkholderia cepacia
Pig liver esterase (PLE)
Porcine pancreas lipase (PPL)
Subtilisin
***Thermomyces lanuginosa* lipase (TL IM)**
   Lipozyme TL IM

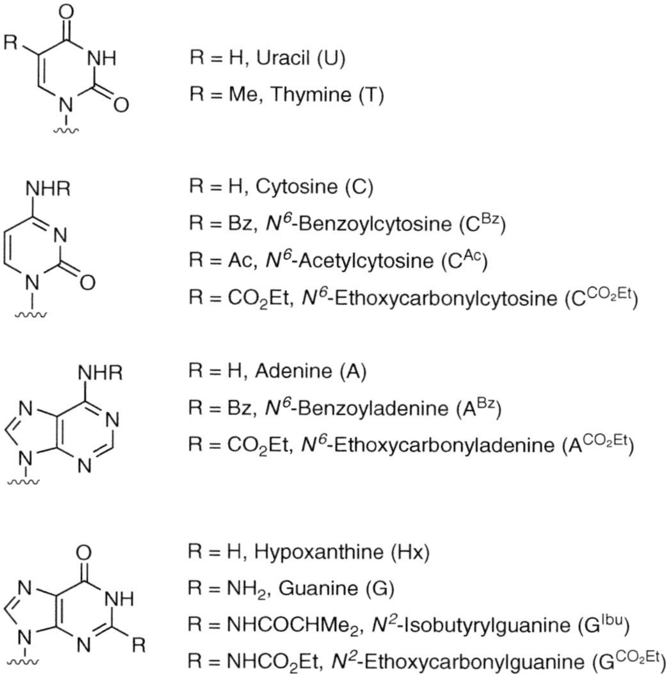

R = H, Uracil (U)
R = Me, Thymine (T)

R = H, Cytosine (C)
R = Bz, N^6-Benzoylcytosine (C^{Bz})
R = Ac, N^6-Acetylcytosine (C^{Ac})
R = CO_2Et, N^6-Ethoxycarbonylcytosine (C^{CO_2Et})

R = H, Adenine (A)
R = Bz, N^6-Benzoyladenine (A^{Bz})
R = CO_2Et, N^6-Ethoxycarbonyladenine (A^{CO_2Et})

R = H, Hypoxanthine (Hx)
R = NH_2, Guanine (G)
R = $NHCOCHMe_2$, N^2-Isobutyrylguanine (G^{ibu})
R = $NHCO_2Et$, N^2-Ethoxycarbonylguanine (G^{CO_2Et})

Figure 1. Pyrimidine and purine bases, their more common protected versions used in this review, and their abbreviations.

are cited as in the original papers to facilitate checking the original work, together with their corresponding new denominations.

To simplify the schemes, Figure 1 collects the common abbreviations of nucleoside bases, their protected version used in this chapter, and their structures.

2. TRANSFORMATIONS ON THE SUGAR MOIETY

Modification of nucleosides via enzymatic acylation has been one of the most extensively used methodologies over recent years, since in some cases a simple acylation of one of the hydroxyl groups in a nucleoside can result in an increase in their biological activity compared with that of the unmodified derivative.[7]

2.1. Enzymatic Acylation/Hydrolysis

An interesting family of nucleoside analogues is that of the amino sugar nucleosides, since they possess anticancer, antibacterial, and antimetabolic activities.[8] Gotor, Ferrero, and co-workers have synthesized, through short and convenient syntheses, pyrimidine 3′,5′-diamino analogues of thymidine (T),[9] 2′-deoxyuridine (dU),[9] 2′-deoxycytydine (dC^{Bz}),[10] (*E*)-5-(2-bromovinyl)-2′-deoxyuridine (BVDU, Brivudin),[11] and the purine 3′,5′-diamino analogue of 2′-deoxyadenosine (dA^{Bz}).[10] Regioselective protection of one of the primary amino groups situated in the 3′- or 5′-position is a very difficult task, since traditional chemical methods do not distinguish between them, and moreover, there are other reactive points on the molecule, such as the nitrogen atoms on the bases. They report the regioselective enzymatic acylation of the amino groups in the sugar moiety of pyrimidine and purine 3′,5′-diaminonucleosides.[9,12] This enzymatic strategy made it possible for the first time to regioselectively synthesize $N^{3'}$- or $N^{5'}$-acylated pyrimidine and purine 3′,5′-diamino nucleoside derivatives by means of a very simple and convenient procedure using immobilized *Pseudomonas cepacia* lipase (PSL-C) or *Candida antartica* lipase B (CAL-B) as a biocatalyst, respectively (Scheme 1).

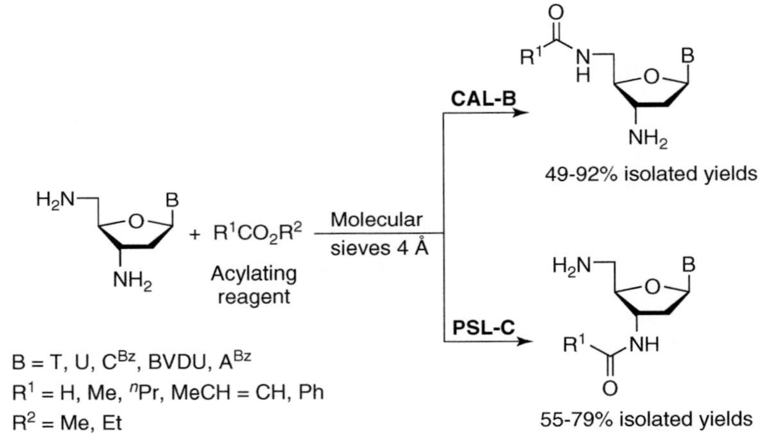

Scheme 1.

Although oxime esters are good acylating agents in regioselective enzymatic acylations of nucleosides,[13] nonactivated esters such as alkyl esters are used since amines are much more nucleophilic than alcohols, and they react nonenzymatically with oxime esters. To confer versatility to this enzymatic reaction, other acyl moieties besides acetyl, such as formyl, alkyl, alkenyl, or aryl, are introduced.

An efficient new approach to the synthesis of oligonucleotides via a solution-phase *H*-phosphonate coupling method has been reported.[14] It is particularly suitable when multikilogram quantities of oligonucleotides are required and is an alternative method of choice to traditional solid-phase synthesis. The key building blocks for solution-phase oligonucleotide synthesis are 3′- and/or 5′-protected nucleosidic monomers. Among the limited protecting groups available, the levulinyl group is frequently chosen to protect the 3′- and/or 5′-hydroxyl of the nucleosides, since this group is stable to coupling conditions and can be cleaved selectively without affecting other protecting groups in the molecule. Until recently, the preparation of these building blocks has been carried out through several tedious chemical protection and deprotection steps.

To avoid this classical approach, Gotor, Ferrero, and co-workers report,[15] for the first time, a short and convenient synthesis of 3′- and 5′-O-levulinyl-2′-deoxynucleosides from the corresponding 3′,5′-di-O-levulinyl derivatives by regioselective enzymatic hydrolysis, avoiding several tedious chemical protection–deprotection steps (Scheme 2).

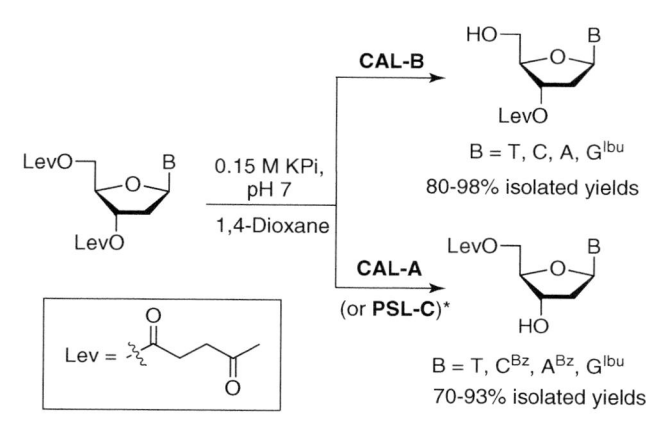

*PSL-C was used for di-Lev-dG^{Ibu} since CAL-A did not catalyze the hydrolysis.

Scheme 2.

Thus, CAL-B selectively hydrolyzes the 5′-levulinate esters, furnishing 3′-O-levulinyl-2′-deoxynucleosides in >80% isolated yields. On the other hand, PSL-C and *Candida antarctica* lipase A (CAL-A) exhibit the opposite selectivity toward hydrolysis at the 3′-position, affording 5′-O-levulinyl derivatives in >70% yields.

A similar hydrolysis procedure has been extended successfully to the synthesis of 3′- and 5′-O-levulinyl-protected 2′-O-alkylribonucleosides (Scheme 3). This work demonstrates for the first time application of commercial CAL-B and PSL-C toward regioselective hydrolysis of levulinyl esters with excellent selectivity and yields. It is noteworthy that protected cytidine and adenosine base derivatives are not adequate substrates for enzymatic hydrolysis with CAL-B, whereas PSL-C is able to accommodate protected bases during selective hydrolysis.

In addition, they also report improved synthesis of dilevulinyl esters using a polymer-bound carbodiimide as a replacement for dicyclohexylcarbodiimide (DCC), thus considerably simplifying the workup for esterification reactions (Scheme 4).

Another efficient synthesis of 3′- and 5′-O-levulinyl-2′-deoxy- and 2′-O-alkylribonucleosides is described from parent nucleosides using enzyme-catalyzed regioselective acylation in organic solvents (Scheme 5).[16]

Several lipases are screened in combination with acetonoxime levulinate as an acylating agent. PSL-C is selected for acylation of the 3′-hydroxyl group in nucleosides, furnishing 3′-O-levulinylated products in excellent yields. Similarly, CAL-B provides 5′-O-levulinyl nucleosides in high yields. Base-protected cytidine and

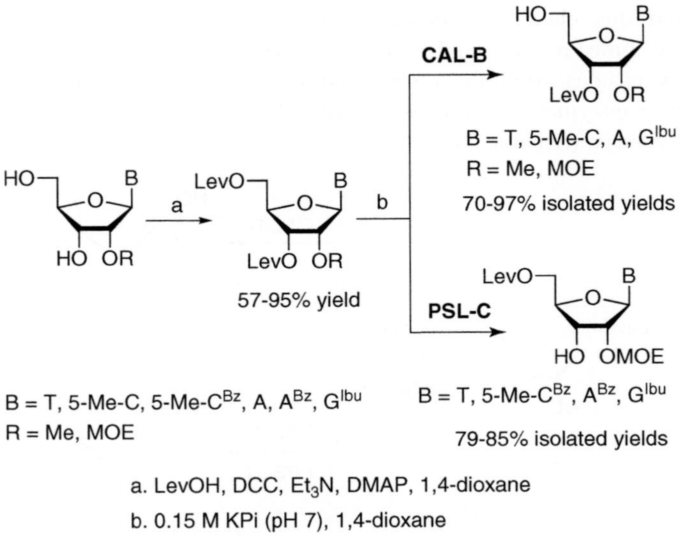

B = T, 5-Me-C, A, G^{Ibu}
R = Me, MOE
70-97% isolated yields

57-95% yield

B = T, 5-Me-C^{Bz}, A^{Bz}, G^{Ibu}
79-85% isolated yields

B = T, 5-Me-C, 5-Me-C^{Bz}, A, A^{Bz}, G^{Ibu}
R = Me, MOE

a. LevOH, DCC, Et$_3$N, DMAP, 1,4-dioxane
b. 0.15 M KPi (pH 7), 1,4-dioxane

Scheme 3.

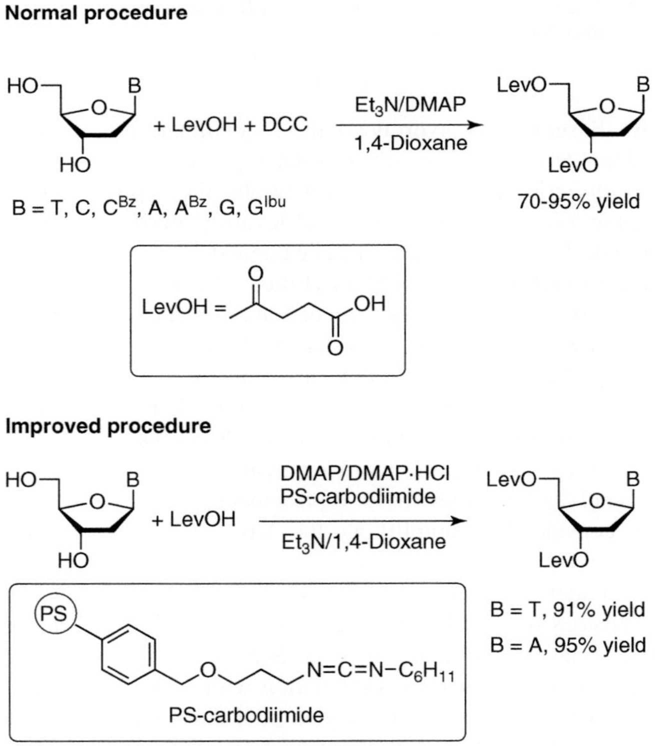

Normal procedure

B = T, C, C^{Bz}, A, A^{Bz}, G, G^{Ibu}

70-95% yield

LevOH = (structure: levulinic acid)

Improved procedure

PS-carbodiimide

B = T, 91% yield
B = A, 95% yield

Scheme 4.

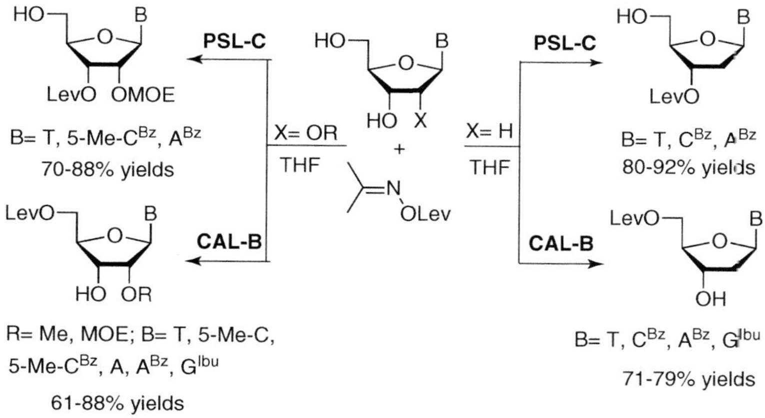

Scheme 5.

adenosine analogues are good substrates for lipase-mediated acylations. Thus, N-benzoyl-3'-O-levulinyl cytidine and adenosine derivatives are obtained in good yield, overcoming the limitations of the original hydrolysis protocol.[15] The new and improved method is shorter than the one described earlier, allowing greater atom efficiency and lowering the cost of enzyme and reagents via recycling. To demonstrate the industrial utility of this method, 3'-O-levulinyl thymidine and N^2-isobutyryl-5'-O-levulinyl-2'-deoxyguanosine are synthesized on a 25-g scale.[17] Additionally, PSL-C is reused to make the processes even more economical.[18]

The crotonyl group is present in different biological active compounds, such as antitumor agents COTC [2-crotonyloxymethyl-($4R,5R,6R$)-4,5,6-trihydroxy-2-cyclohexenone][19] and COMC (2-crotonyloxymethyl-2-cyclohexenone).[20] The activity of this type of derivative can be ascribed to the presence of the α,β-unsaturated ester, which can undergo Michael-type additions of nucleophiles within an enzyme. However, introduction of this moiety on nucleosides has rarely been studied.[21] Previously, 3'-amino-5'-crotonylamino-3',5'-dideoxythymidine was synthesized,[9] and preliminary biological studies have shown that it inhibits the in vitro replication of HIV-1 and HIV-2.[22] Due to the fact that this compound cannot be 5'-phosphorylated, it may suffer a Michael-type addition from a specific enzyme. Moreover, the presence of this moiety on nucleosides would afford excellent starting compounds for the synthesis of β-amino acid analogues of potential interest.[23]

Nevertheless, the synthesis of O-crotonyl derivatives is not trivial because under normal conditions, using a base-catalyzed process with crotonyl chloride, mixtures of desired compounds and β/γ-unsaturated analogues are obtained, due to deconjugation of the double bond.

Regioselective syntheses of several O-crotonyl-2'-deoxynucleoside derivatives using biocatalytic methodology has been reported[24] (Scheme 6).

While CAL-B affords 5'-O-acylated compounds, PSL-C provides the 3'-O-crotonylated analogues. Since classical chemical approaches did not work

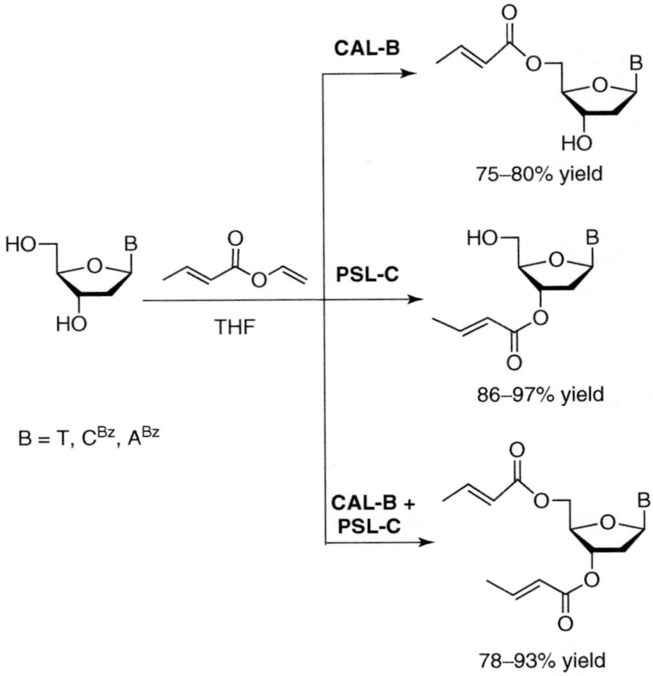

Scheme 6.

appropriately due to side isomerization reactions, a mixture of both lipases is used to achieve a useful synthetic route toward diacylated nucleosides.

Benzoylation remains one of the methods used most frequently for the protection of hydroxyl and amino functions in nucleoside and nucleotide chemistry.[25] The selective manipulation of hydroxyl over amino groups of nucleobases is an important reaction in oligonucleotide synthesis. The classical method of benzoylation of the hydroxyl group in nucleosides with benzoyl chloride or benzoic anhydride provides nonselective reactions. Other mild benzoylating reagents are reported for this purpose; however, lower selectivity has been observed toward the acylation of different hydroxyl functions.

To overcome these problems, a mild and efficient procedure for the selective benzoylation of 2′-deoxynucleosides through direct enzymatic acylation with vinyl benzoate, a commercially available reagent, is reported (Scheme 7).[26] CAL-B is selected, due to its well-demonstrated selectivity in the transesterification of the 5′-hydroxyl group and ability to recycle.

To demonstrate the suitability of the reaction for industrial applications, both the acylating agent and the enzyme were reused for subsequent reactions. The recycled CAL-B maintained total selectivity toward acylation of the 5′-OH, with the exception of the longer reaction rate. On the other hand, benzoylation with recycled vinyl benzoate gives results identical to those obtained using fresh acylating agent.

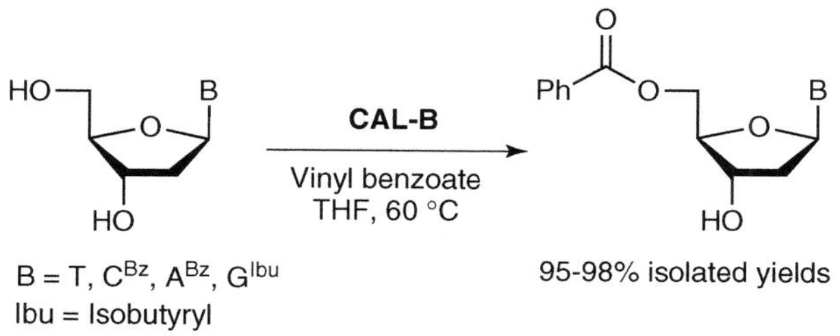

Scheme 7.

Experiments on the large-scale acylation of nucleosides are carried out on a 5- and 25-g scale of the starting material. Excellent results are obtained with CAL-B catalyzing the acylation process, with total selectivity furnishing 5′-O-benzoylated derivatives in quantitative yields.

Ionic liquid-containing systems offer new opportunities for the enzymatic acylation of nucleosides.[27] In PSL-C-mediated benzoylation of floxuridine and its analogues (Scheme 8), enzyme performance, including enzyme activity and 3′-regioselectivity, are enhanced significantly using $[C_4MIm]PF_6$-containing systems, up to excellent conversions (>99%) and good-to-excellent 3′-regioselectivity (81–99%). It is observed that enzyme performance depends not only on the anion of ionic liquid (IL), but also on the cation, and that a proper combination of the cation and anion is critical to allow the enzyme to exhibit excellent performances. The optimal IL content in IL-containing systems is 5% v/v.

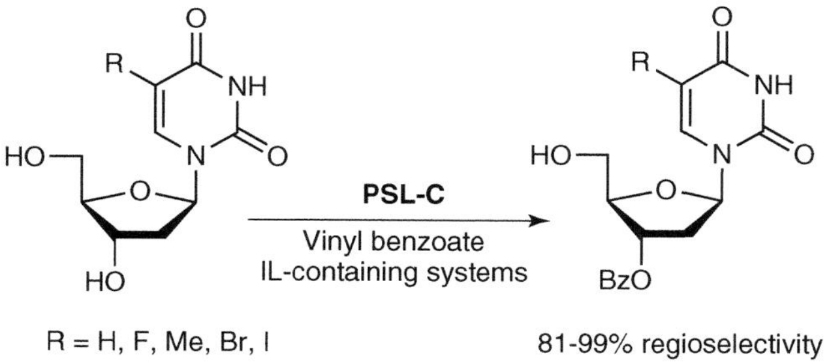

Scheme 8.

Highly regioselective acylations at 3′ or 5′ of fluorouridine, floxuridine, 6-azauridine, and their derivatives are performed using PSL-C or CAL-B/lipase from *Thermomyces lanuginosa* (TL IM), respectively (Scheme 9).[28]

The effects of some crucial factors on the enzymatic processes are examined. The optimum reaction medium, molar ratio of nucleoside to vinyl ester, reaction temperature, and enzyme dosage are investigated. A great variety of acyl donors are tested, from alkyl to aryl or alkenyl or haloalkyl chains on the vinyl ester.

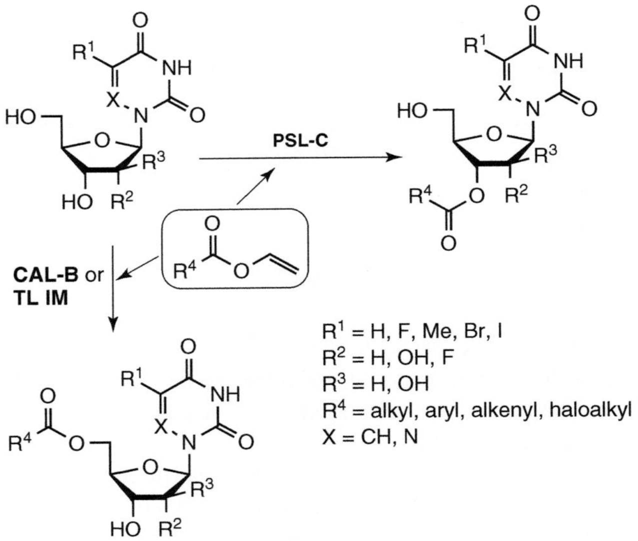

Scheme 9.

R^1 = H, F, Me, Br, I
R^2 = H, OH, F
R^3 = H, OH
R^4 = alkyl, aryl, alkenyl, haloalkyl
X = CH, N

Efficient protocols for the selective synthesis of polymeric prodrugs of ribavirin, ddI, cytarabine, acyclovir, 5-fluorouridine, or aracytosine are developed using different biocatalysts: CAL-B, Lipozyme, or PSL-C (Scheme 10).[29]

Transesterifications are performed using vinyl carboxylates ranging from 4 to 10 carbon atoms. A series of analogues are prepared, with high acylation regioselectivity at 5′-OH when CAL-B or Lipozyme biocatalysts are used. However, PSL-C in DMSO selectively acylates the amino group. The influence of reaction parameters, including enzyme, solvents, molar ratio of substrates, reaction time, carbon length of the acyl donor, and reaction temperature, are investigated in detail.

Synthesis of lobucavir prodrug requires regioselective coupling of one of the two hydroxyl groups of lobucavir with valine (Scheme 11).[30]

Either hydroxyl group of lobucavir could be selectively aminoacylated with valine by using enzymatic reactions. One of them is obtained in 83–87% yield by selective hydrolysis of di-O-valinyl derivative with lipase M from *Mucor javanicus* or lipase from *Candida cylindracea* (CCL). The final active intermediate for lobucavir prodrug is prepared by transesterification of lobucavir using ChiroCLEC BL (61% yield), or more selectively by using PSL-C (84% yield).

Ribavirin is a powerful antiviral agent used to treat hepatitis C. Although this therapy is very effective in eradicating hepatitis C virus, it has several side effects. It was suggested that the administration of ribavirin in the form of a prodrug might improve its pharmacokinetic profile and reduce side effects. Indeed, a series of preclinical evaluations demonstrated that the bioavailability and variability of the alanine ester of ribavirin are improved compared to those of ribavirin. To satisfy the prodrug requirements to be used in toxicological studies, formulation development,

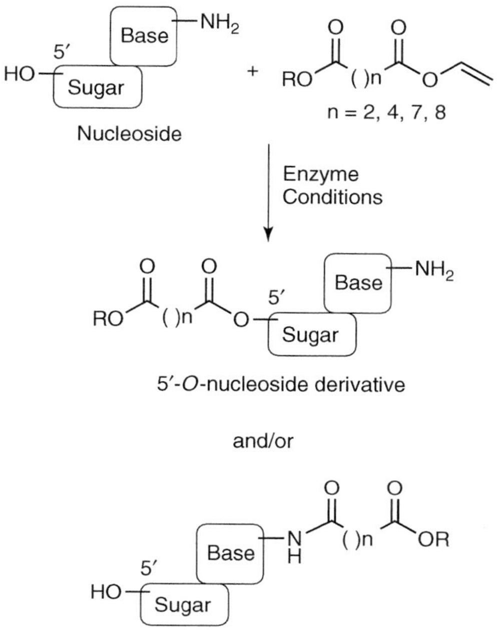

Scheme 10.

a. Ribavirin, **CAL-B**, acetone, 50 °C
b. ddI, **Lipozyme**, acetone
c. Cytarabine, **CAL-B** or **Lipozyme**, acetone
d. Acyclovir, **PSL-C**, acetone or DMSO
e. 5-Fluorouridine, **CAL-B**, THF
f. Ribavirin or *ara*-cytosine, **CAL-B**, acetone

and early clinical trials, efficient synthesis of 5′-*O*-alanylribavirin has been reported (Scheme 12).[31]

The final ester is synthesized via CAL-B-catalyzed acylation of ribavirin with the oxime ester of L-Cbz-Ala in anhydrous THF. The reaction was highly regioselective, resulting in the exclusive acylation of the 5′-hydroxyl. The process is also scaled-up on a pilot-plant scale to produce 82 kg of final ester in three batches in an average isolated yield of 82%.

The protection of hydroxyl groups as esters is one of the oldest and most frequently used strategies in the synthesis of nucleosides. Acetyl and benzoyl protecting groups are prized because they can be removed by alkaline hydrolysis without cleaving the glycosidic bond in nucleosides.[25b] Protecting groups that can be removed under milder acid conditions or even under neutral conditions are of considerable value. The merits of acetal groups such as THP and THF lies in their stability under a variety of conditions, such as basic media, alkyl lithiums, metal hydrides, Grignard reagents, oxidative reagents, and alkylating or acylating reagents and cleavage under mild acidic conditions or heating.[32]

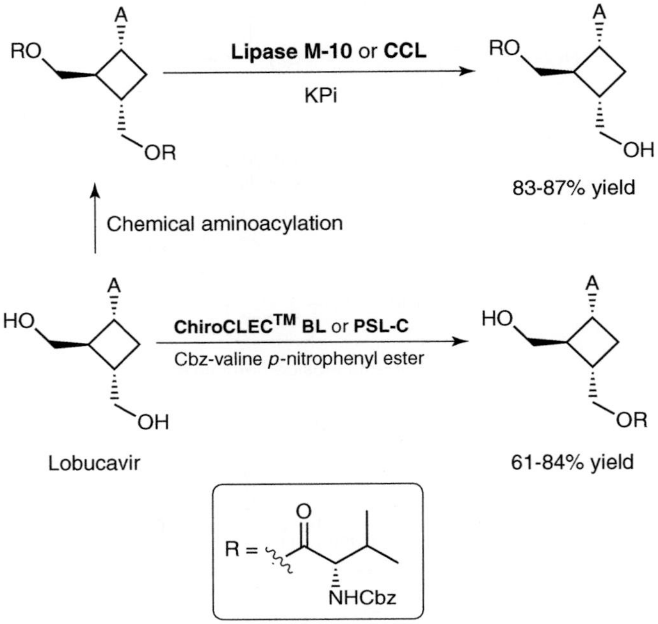

Scheme 11.

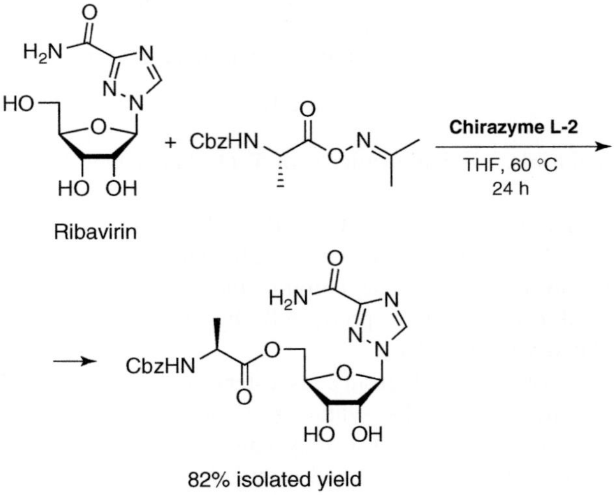

Scheme 12.

Numerous methods have been reported for the tetrahydropyranylation of alcohols.[33] However, protocols for the synthesis of 3'-O-THP and 3'-O-THF ethers of 2'-deoxynucleosides are not efficient. Often, dimethoxytrityl (DMTr) and silyl protecting groups are employed as blocking groups for the 5'-hydroxyl group during the synthesis of 3'-O-acetal. These reagents have scale-up limitations, due to the higher cost and corrosive nature of the silyl chloride.

A protocol is developed[34] based on the regioselective synthesis of 5'-benzoyl derivatives of 2'-deoxynucleosides catalyzed by CAL-B. It is a simple and convenient synthetic strategy for the preparation of tetrahydropyranyl, 4-methoxytetrahydropyranyl, and tetrahydrofuranyl ethers of 2'-deoxynucleosides, which are useful building blocks for nucleic acid chemistry and thus a significant improvement over reported methods (Scheme 13).

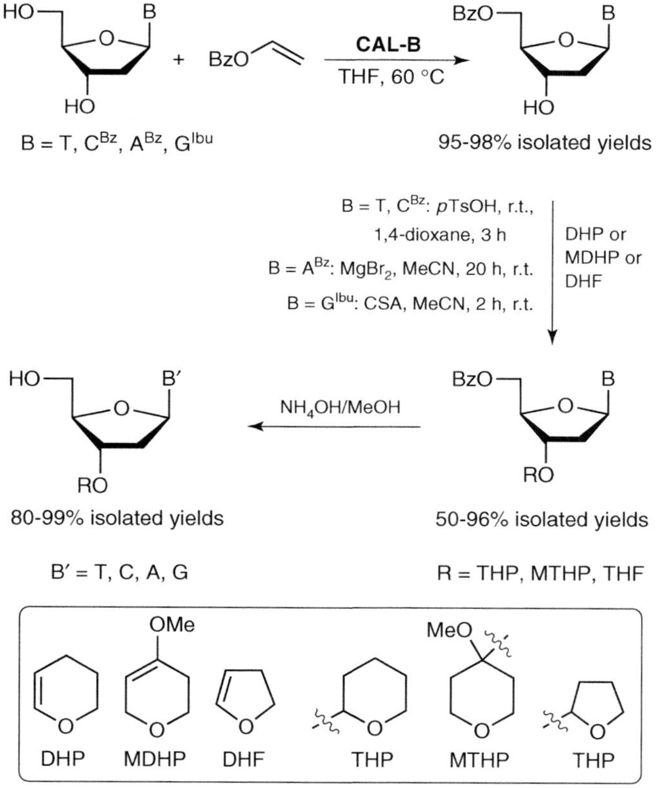

Scheme 13.

After enzymatic benzoylation, tetrahydropyranylation, and tetrahydrofuranylation at the 3'-hydroxyl group of 2'-deoxynucleosides with p-toluenesulfonic acid, $MgBr_2$, or camphorsulfonic acid as catalysts, deprotection of the 5'-O-benzoyl group furnishes 3'-O-acetal-protected 2'-deoxynucleosides. The three-step process enables large-scale synthesis of protected nucleosides.

2.2. Resolutions of Racemic Mixtures

Until recently, only nucleosides possessing the natural β-D-configuration have been studied as chemotherapeutic agents, due to their easy access. However, the discovery of lamivudine (L-2′,3′-dideoxy-3′-thiacytidine, 3TC), the first compound with the unnatural β-L-configuration approved by the U.S. Food and Drug Administration (FDA) for use in combination therapy against human immunodeficiency virus type 1 (HIV-1) and hepatitis B virus (HBV), has sparked tremendous interest in the synthesis of β-L-nucleosides.[35] As a result, several L-nucleosides are undergoing clinical trials as potential antiviral or antitumor agents.[36] Favorable features of L-nucleosides include lower toxicity while maintaining antiviral activity comparable and sometimes greater than that of their D-counterparts, and higher metabolic stability. In addition, the L-series nucleosides are also of interest as precursors to nuclease-stable L-oligonucleotides.[37] The tremendous therapeutic potential of L-nucleosides has stimulated interest in their synthesis. Formation of a mixture of D- and L-nucleoside is a common occurrence during the synthesis of these compounds. This results in a challenging separation of the racemic mixtures.

One of theses examples is the new method[38] for the synthesis of 2′,3′-dideoxynucleoside analogues (Scheme 14). Thus, electrochemical activation of 2-substituted furans, followed by coupling with a pyrimidine or purine base, gives planar furyl nucleosides as key intermediates, which are hydrogenated *cis*-selectively to give the corresponding β-2′,3′-dideoxynucleosides as racemic mixtures. An enzymatic kinetic resolution gives rise to β-D- and β-L-configured derivatives in high optical purity. This is exemplified by the synthesis of β-D- and β-L-3′-deoxythymidine.

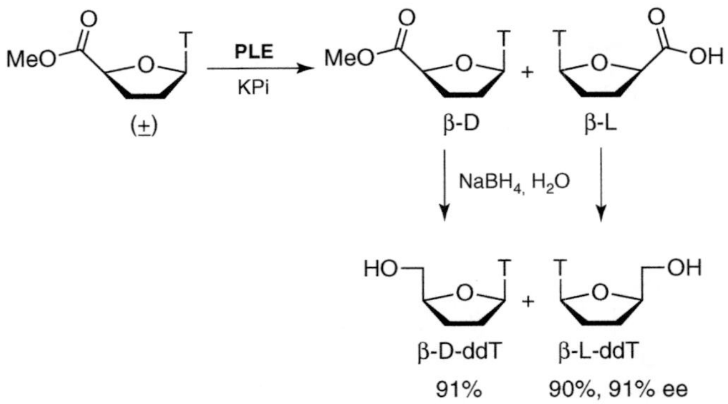

Scheme 14.

The bicyclo[3.1.0]hexane scaffold can lock the conformation of a carbocyclic nucleoside into one of the two antipodal (north or south) conformations typical of conventional nucleosides that normally exist in a rapid two-state equilibrium in solution. A general practical method to access bicyclo[3.1.0]hexane pseudosugar for the north antipode via a lipase-catalyzed double-acetylation reaction has been reported[39] which successfully resolves a diol precursor into enantiomerically pure (+)-diacetate and (−)-monoacetate (Scheme 15).

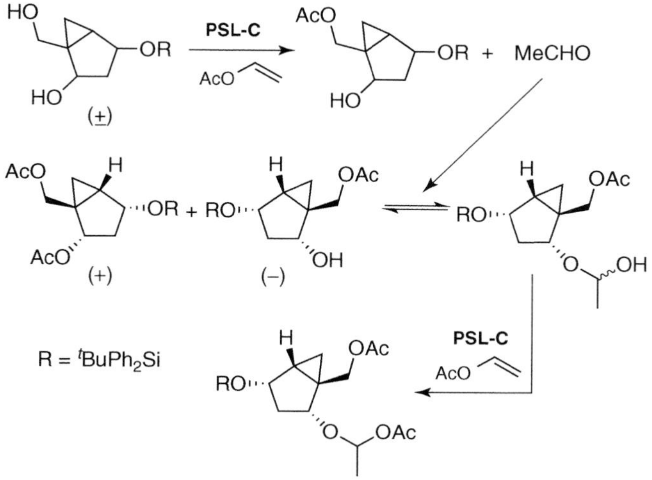

Scheme 15.

The former diacetate is converted to the conformationally locked (north)-carbocyclic guanosine derivative. The most attractive features of this synthesis are that a relatively complex synthon is obtained from simple and inexpensive starting materials and that the resulting racemic mixture is resolved successfully by PSL-C. During the lipase-catalyzed resolution, the presence of an unusual acetal-forming reaction that consumes small amounts of the unreactive (−)-monoacetate is detected. This side reaction is also enzyme-catalyzed and is triggered by the by-product acetaldehyde generated during the reaction.

The preparative-scale chemoenzymatic resolution of novel (R)- and (S)-spiro[2.3]hexane nucleosides has been described (Scheme 16).[40] The key step

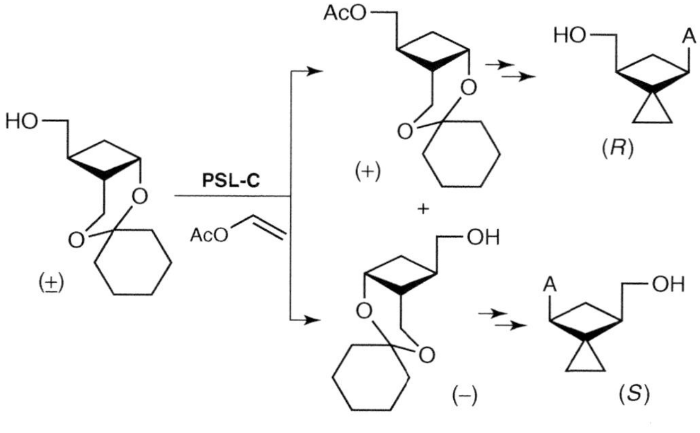

Scheme 16.

involves the PSL-C-catalyzed resolution of racemic compound, synthesized in seven steps, starting from diethoxyketene and diethyl fumarate, to give (+)-acetate and (−)-alcohol.

The (+)-acetate and (−)-acetate corresponding compounds are converted to (R)- and (S)-9-(6-hydroxymethylspiro[2.3]hexane)-4-adenine, respectively, through condensation with 6-chloropurine, followed by ammonolysis that readily provides novel spiro[2.3]hexane nucleosides in high optical purity. Among the several lipases studied, PSL-C gives the highest optical and chemical yield on a multigram scale.

4′-C-Ethynyl-2′-deoxynucleosides belong to a novel class of nucleoside analogues endowed with potent activity against a wide spectrum of HIV viruses, including a variety of resistant clones. Although favorable selectivity indices are reported for several of these analogues, some concern still exists regarding the 3′-OH group and its role in cellular toxicity. To address this problem, the 3′-OH group in 4′-C-ethynyl-2′-deoxycytidine is removed. This compound is chosen because of its combined high potency and low selectivity index. Removal of the 3′-OH is not straightforward; it requires a different synthetic approach from the one used to synthesize the parent compound.[41] Starting with glycidyl-4-methoxyphenyl ether, the target 4′-C-ethynyl-2′,3′-dideoxycytidine analogue in its racemic form is obtained after 13 steps. Then the lipase-catalyzed resolution of the latter racemic mixture gives β-D-dideoxyribo and β-L-dideoxyribo (Scheme 17), which can be separated, being the key step in achieving synthesis of the corresponding 5′-triphosphates to use in the biological assays.

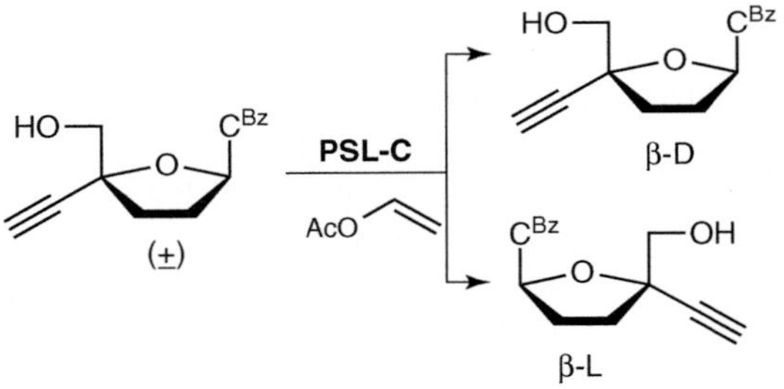

Scheme 17.

Gotor, Ferrero, and co-workers report for the first time a practical synthesis of β-L-3′- and β-L-5′-O-levulinyl-2′-deoxynucleosides through enzymatic acylation and/or hydrolysis processes.[42] The opposite selectivity during acylation exhibited by PSL-C with β-D- and β-L-nucleosides furnishes acylated compounds that have different R_f values. As a consequence, isolation of both products is achieved by simple column chromatography, allowing the parallel kinetic resolution of D/L nucleosides (Scheme 18).

They extend the procedure to β-D/L-mixtures of ribonucleosides[43] due to their applications in the synthesis of therapeutic aptamers.[44] Therefore, they investigate the enzymatic acylation of β-D- and β-L-ribonucleosides with CAL-B and PSL-C and its application in the resolution of racemic mixtures. PSL-C also

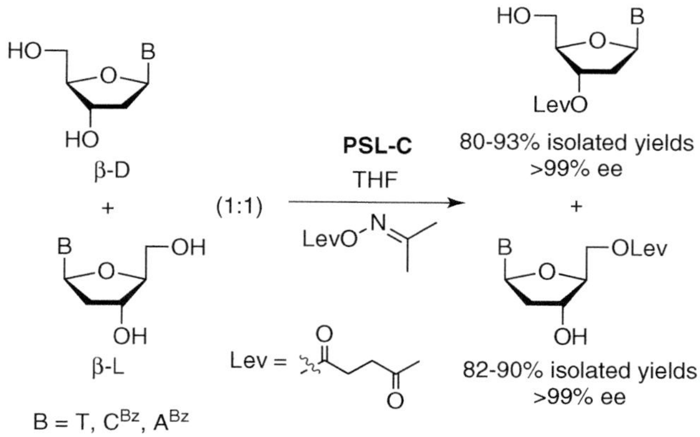

Scheme 18.

exhibits total selectivity toward the acylation of the 5′-hydroxyl group of β-L-uridine, whereas a mixture of acylated derivatives is obtained during bioacylation of β-D-uridine. Thus, a double-sequential parallel kinetic resolution catalyzed by PSL-C and CAL-B is developed to separate the β-D/L-uridine racemic mixture.

Among the approaches tested, the route specified in Scheme 19 provides the most convenient strategy. Treatment of β-D/L-uridine with acetonoxime levulinate

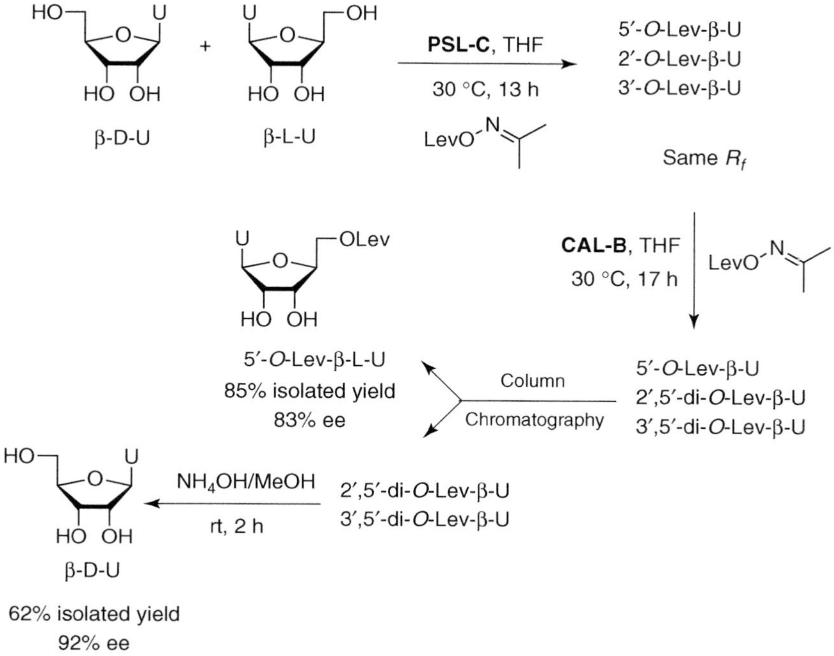

Scheme 19.

in the presence of PSL-C at 30 °C in THF furnishes multiple inseparable monoacylated products with the same R_f. They attribute the 5′-O-levulinylation of the L-isomer due to the expected selectivity observed with PSL-C, whereas secondary hydroxyl groups of the D-isomer are acylated in a nonselective manner with PSL-C. To accomplish the separation of these products, they carry out a second acylation step catalyzed by CAL-B. Since this lipase is regioselective toward the 5′-hydroxyl group of the D-nucleosides, 2′- and 3′-O-acyl derivatives are transformed into 2′,5′- and 3′,5′-di-O-levulinyl nucleosides while the 5′-O-levulinyl compound remained unreacted. The diacylated products and 5′-O-Lev-β-L-U are easily separated by column chromatography. Subsequently, the diacyl derivatives are treated with aqueous ammonium in MeOH to give the corresponding nucleoside β-D-U, which is obtained with a satisfactory 92% ee, determined by transformation into 5′-O-Lev-β-D-U. The chiral HPLC analysis of 5′-O-Lev-β-L-U reveals 83% ee during acylation with PSL-C.

An efficient biocatalytic procedure is described[45] to obtain chiral nonracemic N,O-nucleoside derivatives consisting of a lipase-catalyzed resolution of the corresponding racemates in organic solvent. Despite the low enantioselectivity shown by the lipases investigated, single enantiomers of (+)-thymine and (+)-cytosine derivatives are obtained in high enantiomer purity by developing a lipase-catalyzed double-sequential kinetic resolution route (Scheme 20).

Enantioselective esterification of thymine and cytosine derivatives is investigated by comparing the efficiency of different lipases and acyl donors. Since esterifications occur with low enantioselectivity ($E \leq 14$) with all the lipases considered, for preparative purposes, a double-sequential kinetic resolution is achieved using Lipozyme IM as the best catalyst. This approach enables to obtain all the enantiomers with >95% ee.

2.3. Asymmetrizations and Desymmetrizations

The selectivity of CAL-B is used in the manipulation of a diastereotopic furanose diol as the key step in the synthesis of a novel bicyclo-3-amino-3-deoxyfuranose derivative, which is an important intermediate for the synthesis of bicyclic analogues of AZT (Scheme 21).[46] The asymmetrization of the diol has been achieved with the preferred formation of a monoacylated product with 100% diastereoselectivity. Further, this approach also has advantages with respect to high stereoselectivity and better yields obtained through ecofriendly routes.

The same group also studies the inverse enzymatic process, which is the enzymatic hydrolysis reaction.[47] Lipozyme TL IM catalyzes the deacylation of 4-C-acyloxymethyl-3,5-di-O-acyl-1,2-O-(1-methylethylidene)-β-L-threopentofuranose to form 3,5-di-O-acyl-4-C-hydroxymethyl-1,2-O-(1-methylethylidene)-α-D-xylopentofuranose in a highly selective and efficient manner (Scheme 22).

The rate of lipase-catalyzed deacylation of tributanoyl furanose is 2.3 times faster than the rate of deacylation of the triacetyl furanose derivative. To confirm the structure of the lipase-catalyzed deacylated product, it is converted to a bicyclic sugar derivative which can be used for the synthesis of bicyclic nucleosides of

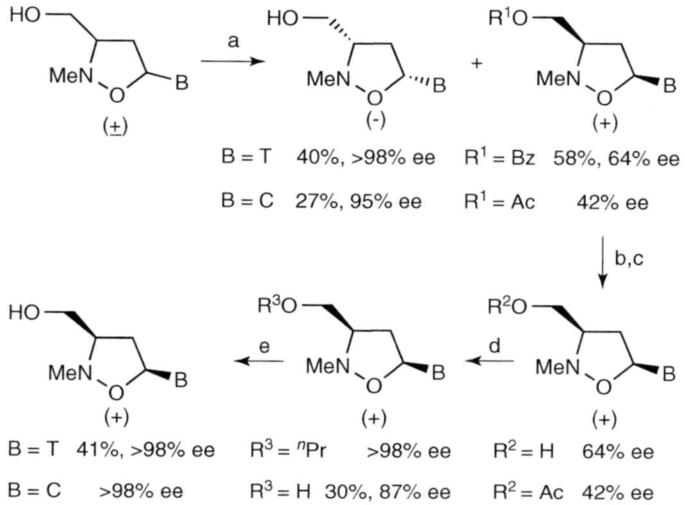

Scheme 20.

a. **Lipozyme IM**, vinyl benzoate (B = T); vinyl acetate (B = C)
b. Preparative chromatography (B = T, C)
c. Chemical hydrolysis (B = T); none (B = C)
d. **Lipozyme IM**, vinyl propionate (B= T); **Lipozyme IM**, ^{n}BuOH (B = C)
e. Chemical hydrolysis (B = T); Crystallization (B = C)

Scheme 21.

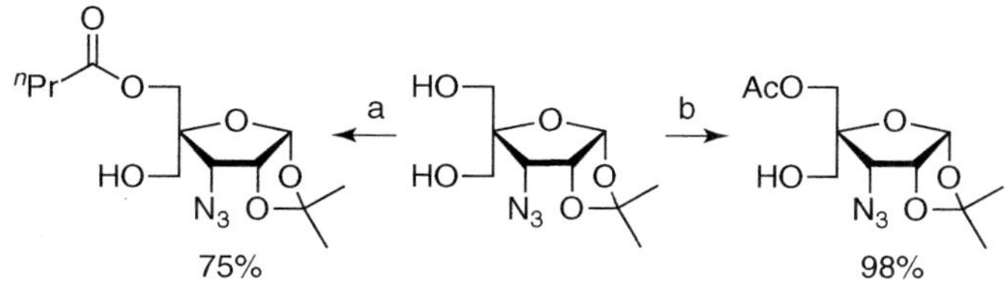

a. **CAL-B**, toluene, 2,2,2-trifluoroethyl butyrate
b. **CAL-B**, toluene, vinyl acetate

importance in the development of novel antisense and antigene oligonucleotides. Further, it is established that the monohydroxy product of the lipase-catalyzed reaction is the result of selective deacylation of the 4-C-acyloxymethyl moiety in the substrate and not of any acyl migration process.

A CAL-B catalytic methodology is developed[48] for selective deacylation of one of the acyloxy functions involving a primary hydroxyl group over the other acyloxy

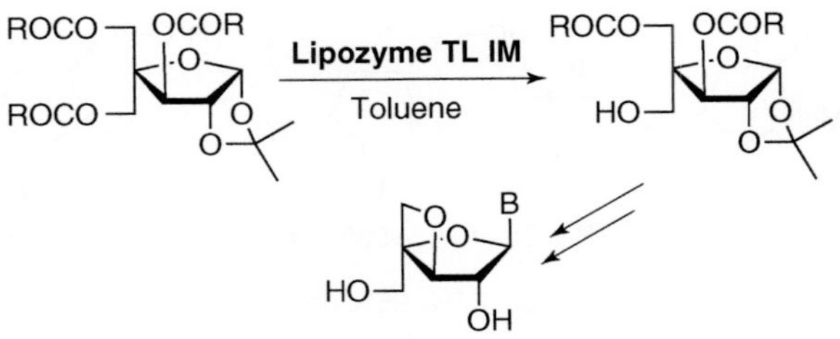

Scheme 22.

functions involving primary and secondary alcohol groups in 4′-C-acyloxymethyl-2′,3′,5′-tri-O-acyl-β-D-xylofuranosylnucleosides (Scheme 23).

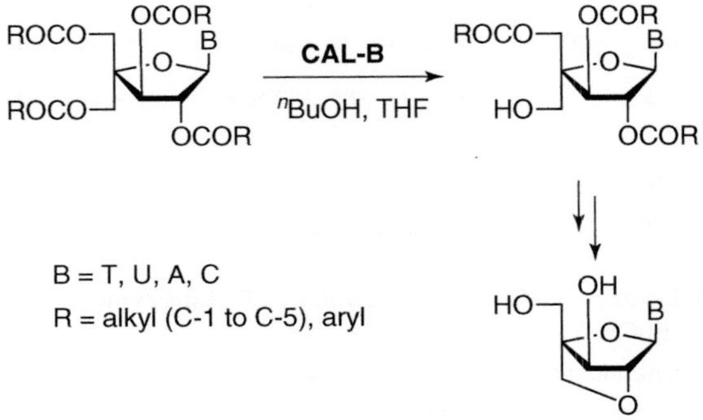

B = T, U, A, C
R = alkyl (C-1 to C-5), aryl

Scheme 23.

Optimization of the biocatalytic reaction reveals that tetra-O-butanoyl-β-D-xylofuranosyl nucleosides are the best substrates for the enzyme. The possibility of acyl migration during enzymatic deacylation reactions is ruled out by carrying out biocatalytic deacylation reactions on mixed esters of 4′-C-hydroxymethyl-2′,3′,5′-tri-O-acetyl-β-D-xylofuranosylnucleosides. The methodology developed is used for the efficient synthesis of xylo-LNA monomers T, U, A, and C in good yields.

The enzymatic desymmetrization of methylenecyclopropane diol or its corresponding diacetate derivative, generated from a [2 + 1]-cycloaddition between dioxepin and methylchlorocarbene, has been described[49] (Scheme 24).

After screening five commercial lipases, the two enantiomers of acetic acid 2-hydroxymethyl-3-methylenecyclopropylmethyl ester are obtained in high yields and excellent enantioselectivities using *Pseudomonas fluorescens* lipase (PFL) or porcine pancreas lipase (PPL) in organic solvent. The stereostructure of the desymmetrization products is established by x-ray analysis. Using these enantiopure building blocks, a synthesis of novel nucleoside analogues is also reported.

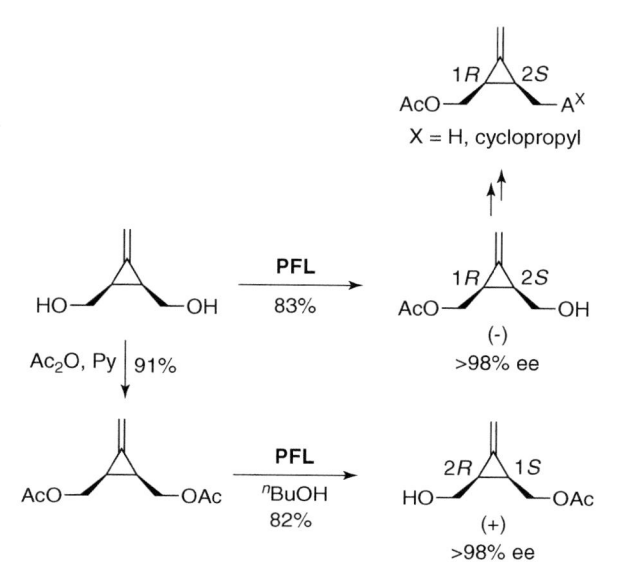

Scheme 24.

2.4. Separation of Mixtures

A majority of nucleosides or oligonucleotides with therapeutic applications are β-nucleoside analogues or oligonucleotides assembled from β-nucleosides. The principal reason for the use of β-nucleosides lies in their availability from natural sources and well-established synthetic routes. On the contrary, the use of α-nucleosides in therapeutics has been limited because they are unnatural and difficult to synthesize in a pure state. Despite these limitations, α-nucleosides continue to be of general interest because of their unique properties. Furthermore, the α-oligonucleotides and their derivatives have been found to be attractive agents for diagnostic applications via the antisense or antigene mode of action. These oligonuclotides have shown good hybridization properties toward DNA or RNA with stability against nuclease degradation.

Due to the ongoing interest in the utilization of α-nucleosides for therapeutic applications, the development of efficient methods for the synthesis of α-nucleoside monomers is an important endeavor. Several methods have been reported in the literature for the synthesis of α-nucleosides. However, these methods often result in a mixture of α- and β-anomers that are tedious to separate. The separation of synthetically derived α/β-nucleosides remains an underexplored area of research.

Several lipase-catalyzed methodologies have been described for the separation of α/β-anomeric nucleosides.[13b,50] One of them is the selective acylation of α-thymidine using acetonoxime butyrate catalyzed by PSL and CAL-B. Additional examples in the literature are the lipase-catalyzed diastereoselective deacetylation of an anomeric mixture of peracetylated α/β-thymidine and the separation of α/β-L-2′,3′-dideoxynucleosides using cytidine deaminase (CDA). All of them are around 20 years old and are not described here.

However, an efficient and high-yield protocol for the synthesis and separation of 3′- and/or 5′-protected α-2′-deoxynucleosides is described[51] through regioselective acylation–deacylation processes catalyzed by enzymes, which allow the separation of α/β-nucleosides frequently obtained during the synthesis of nucleosides utilizing glycosylation protocols.[52] Thus, PSL-C is highly chemo- and regioselective toward the 3′-position of β-2′-deoxynucleoside derivatives, whereas PSL-C displays opposite selectivity toward the 5′-position for the corresponding α-anomer (Scheme 25).

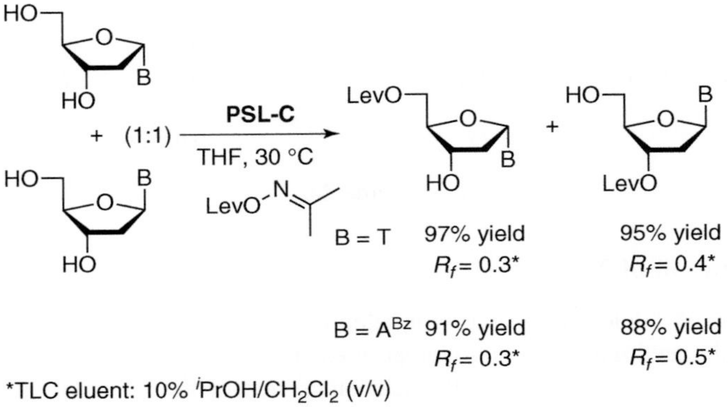

Scheme 25.

The usefulness of this protocol is clearly demonstrated by the separation of α/β-nucleosides for two industrial projects. The first project is the successful application of this protocol to a convenient separation of α/β-mixture of thymidine derivatives from an industrial waste stream (Scheme 26).

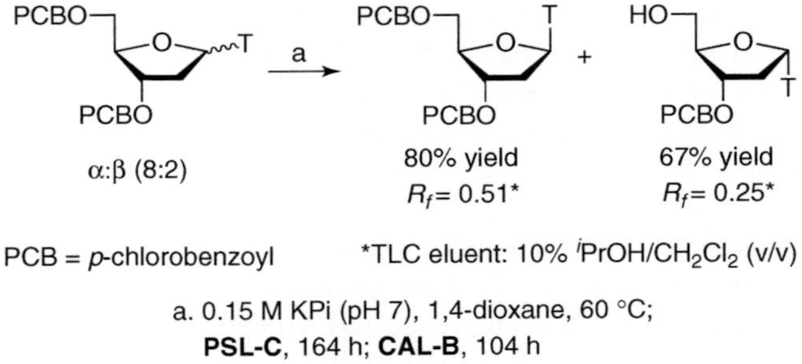

Scheme 26.

In particular, the isolation of α-thymidine from industrial waste is of paramount importance considering the commercial value[53] of this product, now accessible in two simple steps. The waste mother liquor represents approximately 5% of the total β-thymidine production via the glycosylation procedure. This protocol permits the environmentally benign biodegradation of industrial waste and transforms it into two valuable products: (1) α-thymidine, which is the key raw material for the synthesis of α-nucleosides and oligonucleotide analogues as therapeutics, and (2) β-thymidine, which is a starting material for the synthesis of AZT.

Furthermore, this technique is also applied for the separation of anomeric mixture of 2'-deoxy-2'-fluoro-α/β-arabinonucleosides (FANA), which are useful building blocks for the antisense constructs (Scheme 27). It is found that PSL-C has different substrate specificity for α- and β-FANA-nucleosides, where both are hydrolyzed at the 5'-hydroxyl groups with comparable rates.

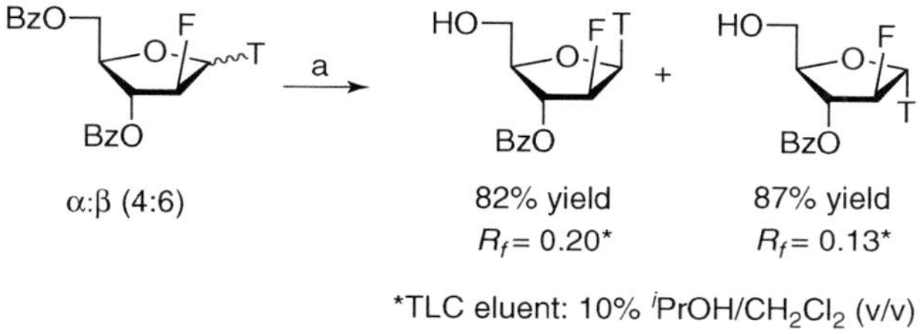

*TLC eluent: 10% ^{i}PrOH/CH$_2$Cl$_2$ (v/v)

a. 0.15 M KPi (pH 7), 1,4-dioxane, 60 °C; **PSL-C**, 165 h

Scheme 27.

CAL-B immobilized on lewatite selectively acylates the primary hydroxyl group of the furanosyl nucleoside in a mixture of 1-(α-D-arabinofuranosyl)thymine and 1-(α-D-arabinopyranosyl)thymine (Scheme 28).[54]

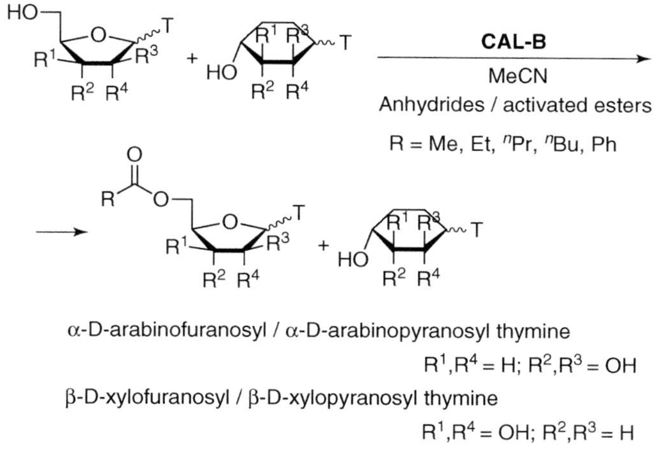

α-D-arabinofuranosyl / α-D-arabinopyranosyl thymine
$R^1, R^4 = H; R^2, R^3 = OH$

β-D-xylofuranosyl / β-D-xylopyranosyl thymine
$R^1, R^4 = OH; R^2, R^3 = H$

Scheme 28.

This selective biocatalytic acylation of furanosyl nucleoside allows easy separation of arabinofuranosyl thymine from an inseparable mixture with arabinopyranosyl thymine. The primary hydroxyl-selective acylation methodology of arabinonucleoside is also used successfully for the separation of 1-(β-D-xylofuranosyl)thymine and 1-(β-D-xylopyranosyl)thymine from a mixture of the two, demonstrating the generality of the enzymatic methodology for the separation of furanosyl and pyranosyl nucleosides.

In recent years, several oligonucleotide-based drugs have entered human clinical trials for the treatment of a variety of viral, infectious, and cancer-related diseases.[55] Among these, 2'-O-methylribonucleotides have been used extensively as a second-generation chemistry to elicit high nuclease resistance, cellular uptake, and improved binding affinity for the RNA target. Additionally, methylribonucleotides have found applications in studying pre-mRNA splicing, examining the structures of spliceosomes, and preparing nuclease-resistant hammerhead ribozymes. Also, 2'-O-alkylribonucleosides are present in RNAs as minor components.

The importance of methylribonucleotides, which constitute a significant portion of the raw material cost in the preparation of oligonucleotides, has stimulated a widespread effort toward the synthesis of 2'-O-methylated nucleosides as key building blocks using several synthetic approaches. Although some of these strategies have been implemented successfully for the production of pyrimidine-containing 2'-O-methyl nucleosides, these protocols are much less efficient for purine analogues.[56] In the latter case, the problems derive from the inherent reactivity of purine bases toward alkylation and the need for selective protection of the 3'- and/or 5'-hydroxyl groups.

The increasing demand for 2'-O-methylribonucleosides, due to its incorporation in oligonucleotide-based therapeutics and the interesting profile of 3'-O-methylribonucleosides, were motivations to develop an approach where both products could be harvested from a single process that is environmentally safe and scalable (Scheme 29).[57]

Thus, efficient separation of a 1:1 mixture of 2'/3'-O-methyladenosine regioisomers is carried out by selective enzymatic acylation using PSL-C in combination with acetonoxime levulinate as an acyl donor. The 3'-hydroxyl group of 2'-O-methyladenosine is acylated with high selectivity (ca. 70%), whereas an equal amount of 3'-O-methyladenosine in the same solution resulted in minor acylation of 5'-hydroxyl group (ca. 8%). The differential behavior of both regioisomers toward enzymatic acylation allows a separation protocol. Upon extraction of the acylated products, the 3'-O-methyladenosine is isolated in 81% yield and 97% purity from the aqueous layer. Hydrolysis of acylated products in organic layer furnished 2'-O-methyladenosine in 67% yield and 99% purity.

The separation process is applied successfully to the crude reaction mixture of methylated products (ca. 3:1 of 2'/3'-O-methyladenosine) on a 5-g scale. Also reported is the use of methyl p-toluenesulfonate as a safe reagent for the 2'-O-methylation of adenosine (Scheme 30). Thus, an economical methodology enabling the synthesis and efficient separation of the isomeric mixture of

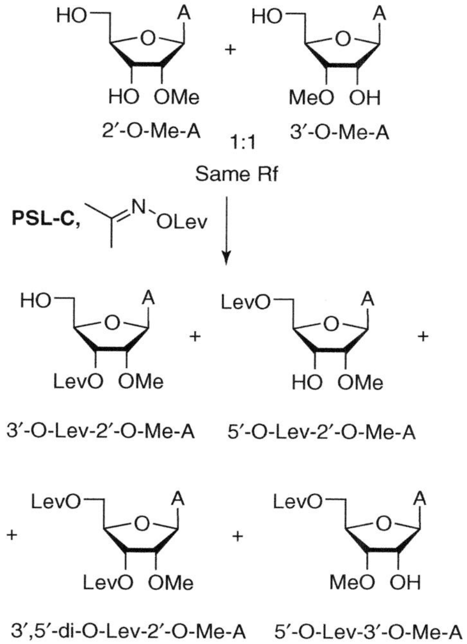

Scheme 29.

2'/3'-O-methyladenosine formed during the direct methylation of inexpensive adenosine has been described for the first time.

2.5. Preparation of Nucleoside Derivatives

As modified oligonucleotides have become a major field of investigation for chemists, methods for their suitable protection and deprotection for the synthesis of nucleoside monomers have become equally important. Among the plethora of synthetic tools available to chemists, application of biocatalysts in organic chemistry has become one of the most attractive alternatives to conventional chemical methods.

One of these monomers consists of 3'-O-dimethoxytrityl-protected nucleosides, which are valuable building blocks for the assembly of oligonucleotides. Use of the DMTr group for protection of the primary 5'-hydroxyl group in nucleoside and oligonucleotide synthesis is well established.[58] By contrast, efficient methods for the preparation of 3'-O-DMTr-protected nucleosides are not available. To perform an inverse (5'→3') oligodeoxyribonucleotide synthesis,[59] large quantities of 3'-O-DMTr-protected nucleosides are needed. Interestingly, the recently approved drug Macugen,[60] a vascular endothelial growth factor antagonist, possesses a 3'-thymidine residue that is inverted.[61] The incorporation of inverted residue in Macugen is accomplished via 3'-O-DMTr-protected thymidine.

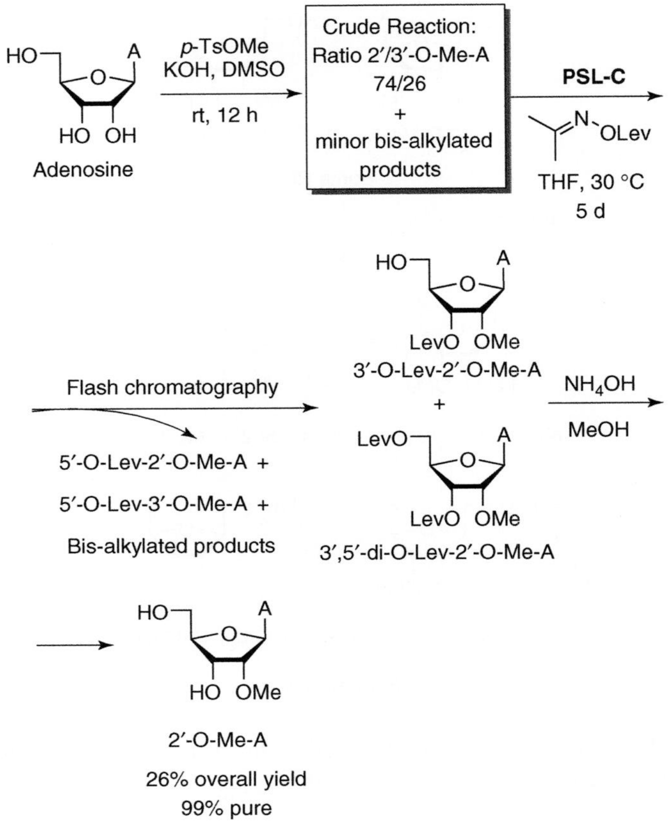

Scheme 30.

The overall approach for the syntheses of 3′-O-DMTr derivatives is outlined in Scheme 31, which is an easy, efficient, and scalable chemoenzymatic strategy for the synthesis of 3′-O-dimethoxytrityl-2′-deoxynucleosides. A key feature of this approach is the regioselective synthesis of 5′-O-levulinyl-2′-deoxynucleosides through enzymatic acylation in the presence of CAL-B. In addition, it was observed that the deblocking of the levulinyl group from the 5′-position is perfectly compatible with conventional base-protecting groups. To demonstrate the scalability of this method, 3′-O-dimethoxytritylthymidine is synthesized on a 25-g scale.

Chemically modified nucleosides are widely used as therapeutic agents to treat cancer and fungal, bacterial, and viral infections.[62] Similarly, carbohydrates are of paramount importance in intercellular recognition, bacterial and viral infection processes, and inflammation events, making them an attractive target for drug development.[63] Consequently, a conjugate of these two classes of molecules may offer an avenue to the design and development of novel therapeutics with improved biological functions. The use of a phosphate ester linkage for conjugation is widely practiced because of its natural occurrence in cellular physiology and as a backbone

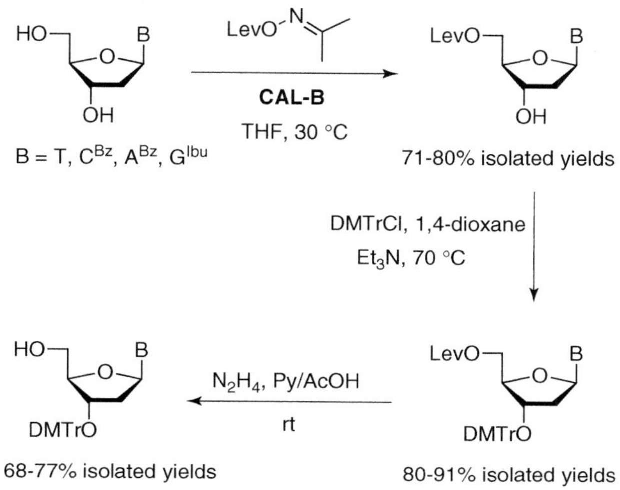

Scheme 31.

for DNA and RNA molecules. The culmination of these attributes motivated the use of natural phosphate linkage to conjugate a biologically active nucleoside with a carbohydrate unit, anticipating that it could modulate the therapeutic effects of known nucleosides in a favorable manner.

Several modified nucleosides (AZT, d4T, 3TC, ddI, ddC, and abacavir) have been approved by the FDA as anti-HIV drugs. Although these drugs are effective for the treatment of HIV, some limitations, such as toxicity, short half-life, and dependence on cellular enzymes, preclude their wider use. To overcome these limitations and to improve the therapeutic potential of these nucleosides, several phosphate-based prodrug strategies have been developed.[64] The prodrug approach appears to be a promising way to improve the anti-HIV activity of the approved nucleosides, to reduce their cellular toxicity, to enhance cellular uptake, and to prevent viral resistance.[65]

Among the various protocols available for conjugation, phosphoramidite chemistry offers the best yields. Tetraacetylphosphoramidite (Scheme 32)[66] is envisioned as the key building block for synthesis of the target nucleoside–carbohydrate prodrugs, for several reasons. First, the phosphoramidite chemistry offers excellent yield during solution-phase coupling of various nucleosides. Second, phosphoramidite compounds are reasonably stable and easy to handle during synthesis. Third, the glucose–amidite would be an ideal molecule offering a versatile conjugation unit for attachment of glucose to a wide range of nucleosides. Thus, the synthesis of glucosyl phophoramidite is accomplished in a few steps via a regioselective enzymatic hydrolysis of 1,2,3,4,6-penta-O-acetyl-α-D-glucopyranose.

Commercial *Candida rugosa* lipase (CRL) was found to be an efficient catalyst for both regio- and chemoselective deacetylation of the primary hydroxyl group in the peracetylated α-D-glucose, furnishing the 6-OH derivative in excellent yield after crystallization from diethyl ether/n-hexane (4:1). Synthesis

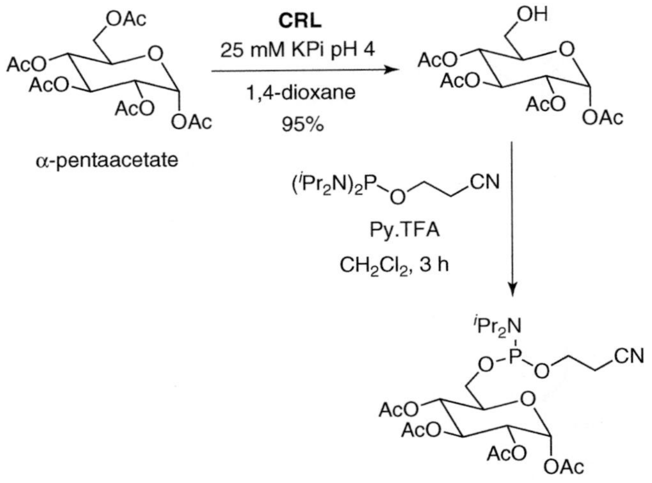

Scheme 32.

of the final phosphoramidite is carried out by phosphitylation of corresponding tetraacetyl alcohol with 2-cyanoethyl-N,N,N',N'-tetraisopropylphosphoramidite and pyridinium trifluoroacetate as an activator in high yield.

The conjugation[67] of glucosyl phosphoramidite is done using standard amidite coupling conditions. Phosphoramidite is first activated with $1H$-tetrazole, and the resulting intermediate was oxidized in situ using an iodine/pyridine/water solution to give the phosphotriester derivative in 39–49% isolated yields after flash chromatography. Following this procedure, d4U, d4T, ddI, IdU, ddA, virazole, ara-A, and ara-C glucose conjugates are synthesized (Scheme 33). Several of the nucleoside precursors shown in Scheme 33 are prepared using biocatalytic methodology.

Abasic sites are one of the most common cellular DNA damage areas in the genome. They result from cleavage of the glycosidic bond, which can occur spontaneously under physiological conditions or during the intermediate steps of base excision repair when DNA is exposed to various endogenous and exogenous damaging agents.[68] Therefore, studies of repair of DNA are important for maintaining the integrity of the genome.[69] Invivo, the abasic sites are believed to be repaired by the action of cellular endonucleases.[70]

Several oligonucleotide analogues containing abasic site have been synthesized to determine the structural basis of these lesions, which is responsible for their distinct biochemical effects.[71] In addition, abasic sugars were introduced into therapeutic siRNAs to inhibit degradation by nucleases.[72]

The widespread use of furanoid glycals as key intermediates for the synthesis of interesting biological compounds[73] has aroused considerable interest in scalable and cost-efficient procedures for their preparation. Most synthetic methods employed 2-deoxy-D-ribose as a starting material and imply loss of atoms.[74] Clearly, these protocols are not atom-efficient, as a good portion of the molecule is sacrificed to

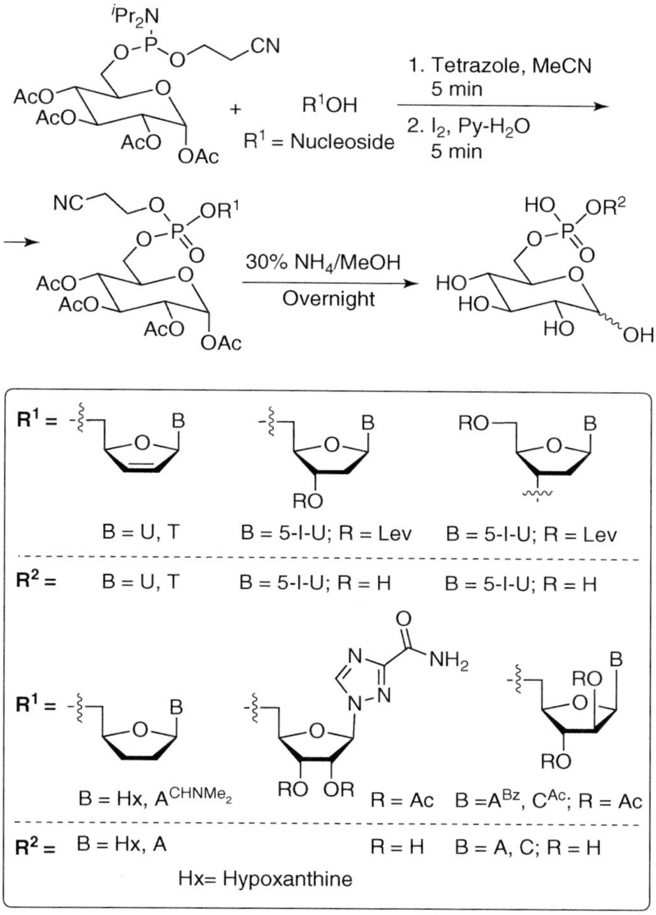

Scheme 33.

obtain 1,2-dideoxy-D-ribose. To overcome these limitations, a green protocol for the preparation of 3- and 5-O-DMTr-1,2-dideoxy-D-ribose is reported (Scheme 34).[75]

CAL-B and PSL-C catalyze the regioselective hydrolysis of di-O-toluoyl-1,2-dideoxy-D-ribose, furnishing the 3-O-toluoyl derivative in quantitative yield. Meanwhile, both lipases exhibit complementary behavior in the acylation reaction of 1,2-dideoxy-D-ribose, providing the opposite regioisomer exclusively. The presence of electron-donating substituents in the *para*-position of the benzoate vinyl ester decreases the rate of enzymatic acylation. However, reaction with the electron-withdrawing group in the acylating agent proceeds at a faster rate. The protocol described offers a fast, reliable, and scalable route to orthogonally protected carbohydrate building blocks in excellent overall yields following an atom-efficient and low environmental impact protocol without sacrificing the stereo- and regio-control requirements of modern carbohydrate chemistry. Specially, the 3-O-DMTr

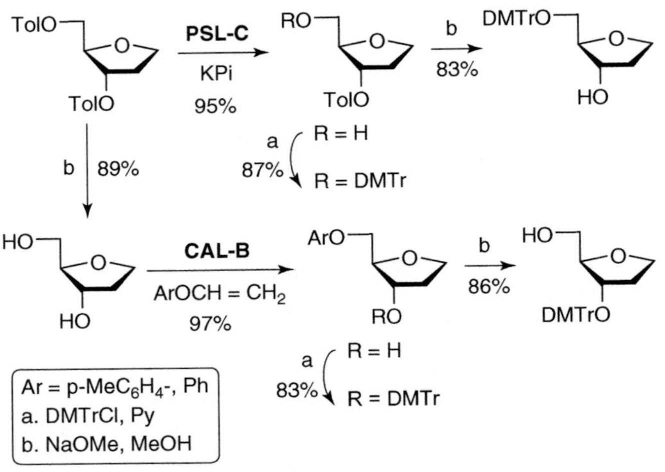

Scheme 34.

precursor is very useful when reverse amidite is required for therapeutic oligonucleotides synthesis. In addition, the enzyme has been reused to make the process more economical for industrial applications.

In the synthesis of modified nucleosides, esters are currently used to protect ribose hydroxyl groups. As their subsequent removal is carried out by treatment with ammonia or methoxide, reactive groups present either in the ribose or in the base may interfere. Thus, a mild procedure for removal of acyl groups is an alternative route of synthetic value. Although enzymatic techniques appear to be very attractive for such a purpose, due to their well-known mildness and efficiency,[76] little attention has been paid to hydrolase-catalyzed deacylation of ribonucleosides.

In this context, Iglesias, Iribarren, and co-workers[77] have investigated two different enzymatic procedures to achieve full deacetylation of nucleosides: lipase-catalyzed alcoholysis and hydrolysis (Scheme 35).

Simple and mild enzymatic transformations have been described for full removal of acetates as well as regioselective conditions for the production of compounds, in good yields, derived from 2′,3′,5′-tri-O-acetyluridine, 2′,3′,5′-tri-O-acetyl-2′-C-methyluridine, base-labile 6-chloro-2′,3′,5′-tri-O-acetylpurine riboside, 2-amino-6-chloro-2′,3′,5′-tri-O-acetylpurine riboside, 3′,5′-di-O-acetyl-2′-deoxynucleosides, 2′,3′,5′-tri-O-acetylcytidine, 4-N-acetyl-2′,3′,5′-tri-O-acetylcytidine, and a set of 3-deazauridine and 6-azauridine peracylated derivatives. Among the enzymes tested, CAL-B is the biocatalyst that gives the best results. In some cases, the alcohol employed in the biotransformation affects the rate of the enzymatic reaction and the yield of the products obtained, but in all cases only one regioisomer is formed. It is of note that this mild and simple enzymatic technique represents a convenient procedure for the removal of acetyl groups from base labile halogenated nucleosides.

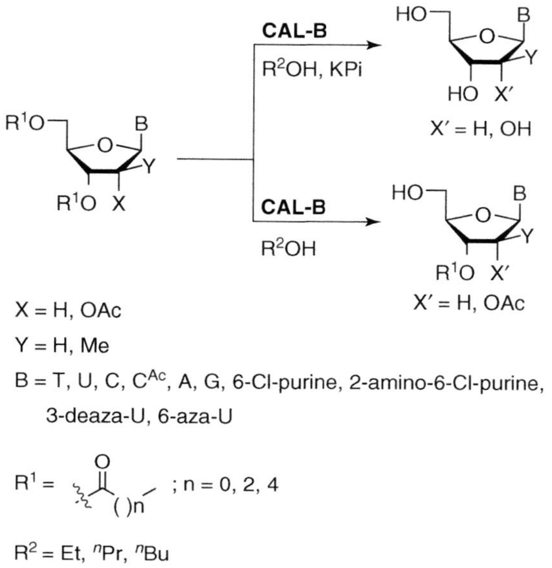

Scheme 35.

X = H, OAc
Y = H, Me
B = T, U, C, C^{Ac}, A, G, 6-Cl-purine, 2-amino-6-Cl-purine, 3-deaza-U, 6-aza-U

R^1 = acyl (CH$_2$)$_n$; n = 0, 2, 4

R^2 = Et, ^{n}Pr, ^{n}Bu

2.6. Alkoxycarbonylations and Hydrolysis

Lipases are one of the most used biocatalysts, due to their versatility in accepting a wide range of nucleophiles (alcohols, amines, thiols, water, etc.) and carbonylating agents (esters, anhydrides, carbonates, etc.).[76] The enzymatic alkoxycarbonylation reaction has been studied very little,[78] despite the biological relevance shown by carbonates and carbamates.[79] It is noteworthy that through these processes it is possible to introduce functionalities selectively, which act as protected or activated groups in alcohols and amines. In the last case, reactions are irreversible since carbamates are not substrates to lipases.

In the treatment of acquired immunodeficiency syndrome (AIDS),[80] many nucleosides have been recognized as potent and selective inhibitors of the replication of HIV. Although zidovudine (AZT) was the first compound approved by the FDA, it is still one of the most potent agents active against HIV. Stavudine (d4T) showed a clinical benefit superior to that of zidovudine.[81] On the other hand, serious side effects are associated with the administration of AZT and d4T and often require cessation of treatment. The search for new combinations of compounds with improved selectivity, lipophilicity, and efficiency that could overcome problems of resistance as well as toxicity is a field of great interest.

Based on this strategy, a new series of homo- and heterodimers of AZT and d4T, which contain carbonate and carbamate linkages, has been prepared.[82] Several arguments support this work: (1) As long as the linkage between the nucleosides (AZT and d4T) is not hydrolyzed extracellularly, the delivery and bioavailability might be enhanced, depending on the lipophilic character of these new models; (2) some synergetic effects on the inhibition of HIV replication could be expected; and (3) depending on the nature of the chemical bond between the two nucleosides, intracellular hydrolysis could regenerate the two nucleosides in the cytoplasm. Owing to these facts, the dimers cited could be considered as prodrugs.

To prepare homo- and heterodimer carbonates, vinyl and oxime esters derived from AZT and d4T are used (Scheme 36). AZT and d4T are converted into 5'-O-alkoxycarbonylated derivatives by reaction with acetone O-[(vinyloxy)carbonyl]oxime acting as an alkoxycarbonylating reagent and CAL-B as a biocatalyst. In addition, 5'-O-acetonoximecarbonylated nucleosides are obtained as minor compounds. Vinyl carbonates as well as oxime carbonates can be used as intermediates (90% and 92% isolated combined yields, respectively) in the next step when they react with AZT or d4T.

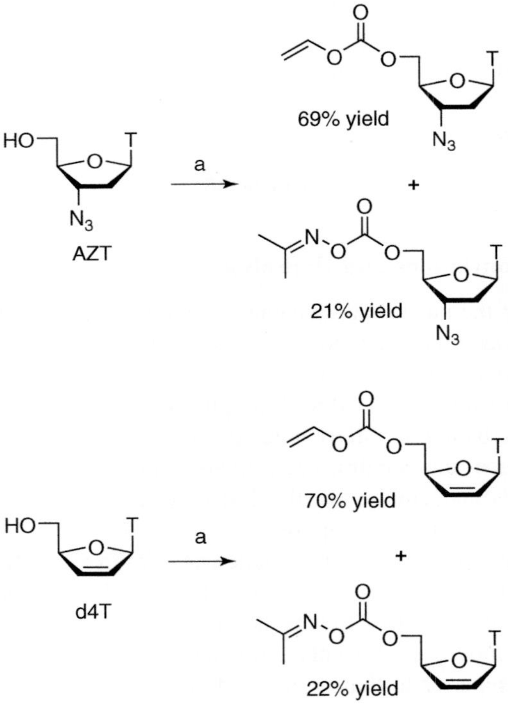

T = thymin-1-yl

a. **CAL-B**, $CH_2 = CHOCO_2N = CMe_2$

Scheme 36.

Then AZT reacts with NaH in THF at 0 °C, giving an alkoxide which subsequently reacts with activated nucleosides (Scheme 37). After workup and

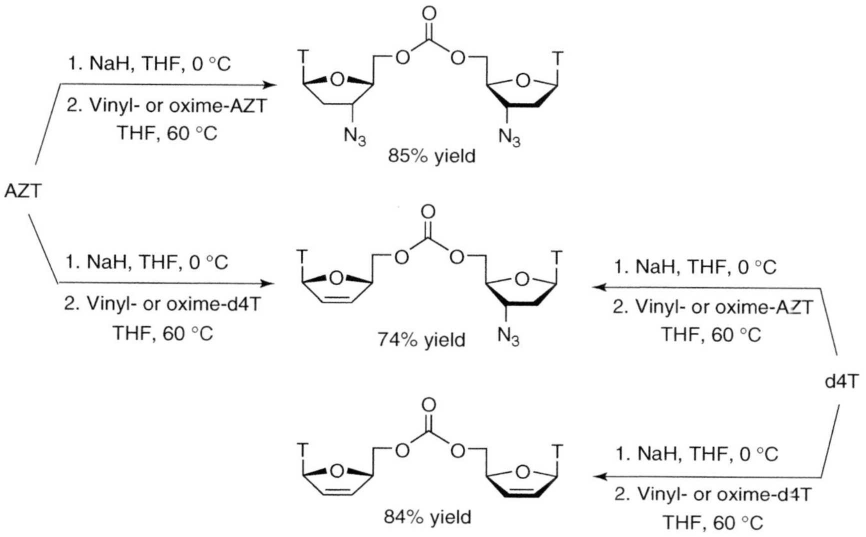

Scheme 37.

purification on silica gel chromatography column, the homodimer AZT–AZT is isolated in an 85% yield. The heterodimer AZT–d4T is prepared by reaction of the alkoxide from either AZT or d4T. The process is carried out by reaction of the AZT alkoxide with activated derivatives of d4T or by reaction of d4T alkoxide with activated derivatives of AZT, in all cases in an isolated yield of 74%. Finally, the homodimer d4T–d4T is prepared following the same procedure with an isolated yield of 84%.

5′-O-Carbamate-AZT and 5′-O-carbamate-d4T are obtained by condensation of 5′-O-vinylcarbonate–AZT (or 5′-O-oximecarbonate–AZT) or 5′-O-vinylcarbonate–d4T (or 5′-O-oximecarbonate–d4T) with 1,4-diaminobutane in THF at 60 °C, respectively (Scheme 38). Then those amino carbamates are condensed with activated carbonates derived from AZT and d4T, or with a mixture of both, giving place to homodimer dicarbamate AZT–AZT in an 83% isolated yield, heterodimer dicarbamate AZT–d4T in an 80% yield, and homodimer d4T–d4T in an 81% yields.

Ferrero and Gotor have a program devoted to a study of the synthesis of nucleoside[83] and nucleotide[84] derivatives prepared through a chemoenzymatic procedure. Both of these 3′- and 5′-alkylidencarbazoylnucleoside analogues are promising precursors for novel types of therapeutic nucleoside derivatives. The first step is the enzymatic alkoxycarbonylation reaction, which takes place with high regioselectivity and yield toward the 3′- or 5′-position, depending on the conditions. The carbonate nucleosides then react with hydrazine to give carbazoyl nucleoside analogues, which through reaction with aldehydes provide the corresponding alkylidencarbazoyl nucleoside derivatives. When applied to adenosine, this general procedure failed because of the low solubility of this nucleoside in the solvents

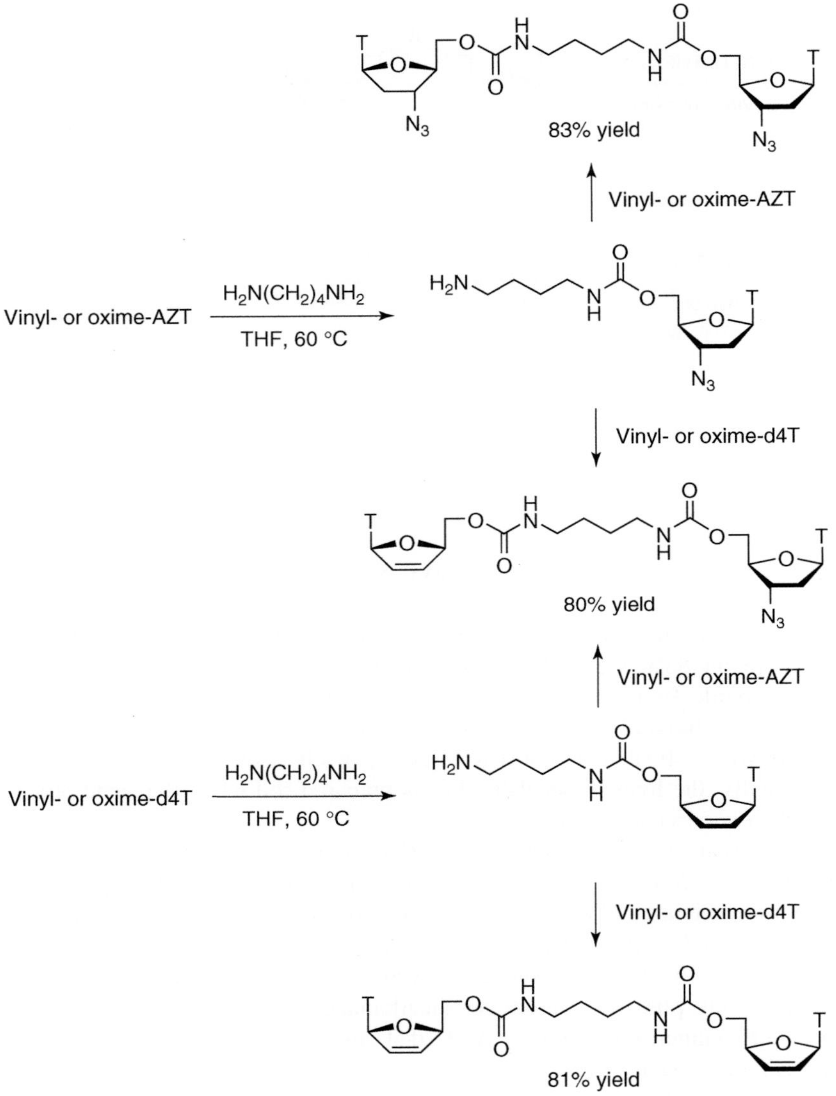

Scheme 38.

required for the enzymatic process. As a consequence, this is a limiting step in the preparation of the corresponding nucleoside derivatives (Scheme 39).

To address this limitation,[85] a commercially available 2′,3′-protected adenosine derivative is used. This alternative pathway led to an excellent yield of 5′-alkylidencarbazoyl adenosine analogues, through the formation of vinylcarbonate and/or oximecarbonate of 2′,3′-protected adenosine, hydrazinolysis of which give vinylcarbonate, which is transformed in carbazoyl nucleoside. This finally

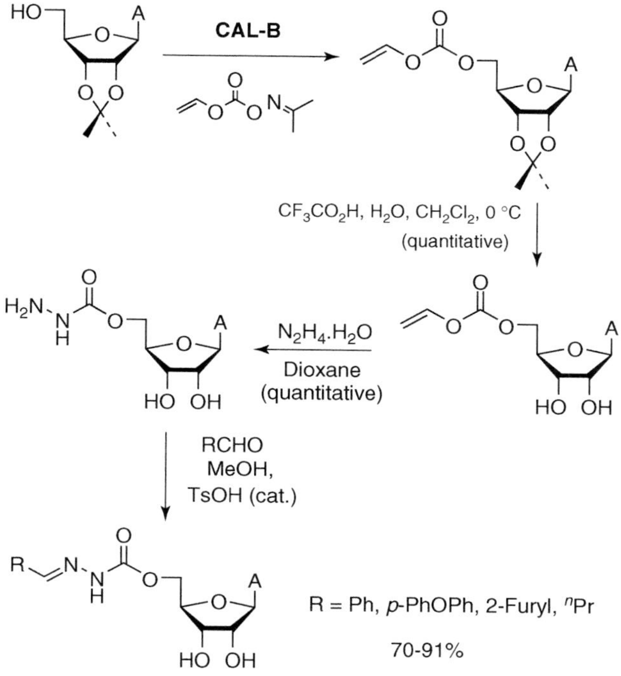

Scheme 39.

reacted with different aldehydes to yield 5′-alkylidencarbazoyl adenosine analogues.

In recent years, the search for new nucleoside derivatives using enzymatic clean, simple, and efficient methodology has received a great deal of attention from organic chemists. In this context, oxime carbonates have been used with lipases for direct and selective protection of natural nucleosides.

This regioselective process would be more difficult in the case of two primary amines, due to the major nucleophilic character of the amino group compared to the hydroxyl group. The first regioselective enzymatic alkoxycarbonylation of primary amino groups[86] is reported in pyrimidine 3′,5′-diaminonucleoside derivatives to synthesize novel pyrimidine aminonucleosides regioselectively protected as carbamates. CAL-B catalyzes this reaction with nonactivated homocarbonates, allowing the selective synthesis of several N-5′-carbamates, including (E)-5-(2-bromovinyl)-2′-deoxyuridine (BVDU) analogues, with moderate-to-high yields, whereas PSL-C affords mixtures of alkoxycarbonylated regioisomers (Scheme 40).

Since direct enzymatic methodology did not allow for N-3′-alkoxycarbonylated nucleoside derivatives in good yields, an orthogonal protection scheme is designed (Scheme 41). The strategy starts with protection of the 5′-NH$_2$ group as allyloxycarbamate. Treatment of the later compounds with benzyloxycarbonyl chloride at room temperature affords dicarbamate nucleosides with high efficiency.

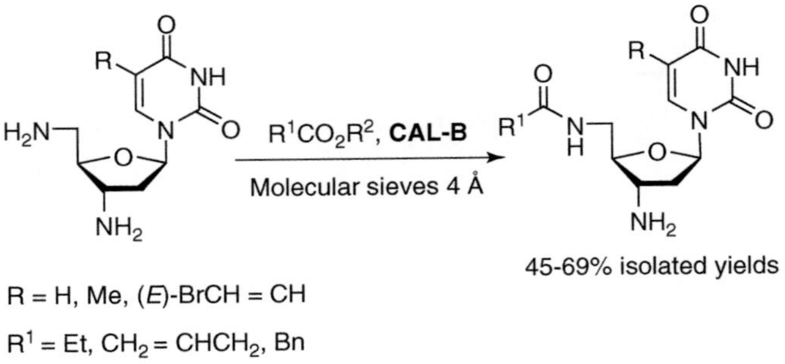
R¹CO₂R², CAL-B
Molecular sieves 4 Å
45-69% isolated yields
R = H, Me, (E)-BrCH = CH
R¹ = Et, CH₂ = CHCH₂, Bn

Scheme 40.

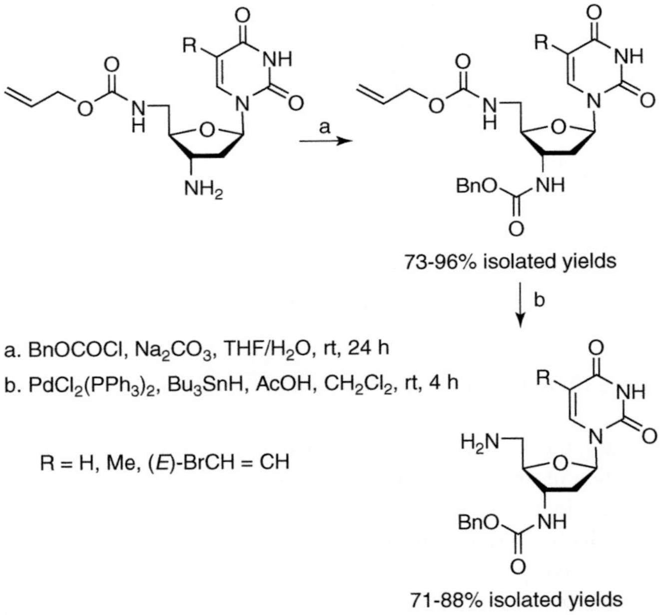
a
73-96% isolated yields
b
a. BnOCOCl, Na₂CO₃, THF/H₂O, rt, 24 h
b. PdCl₂(PPh₃)₂, Bu₃SnH, AcOH, CH₂Cl₂, rt, 4 h
R = H, Me, (E)-BrCH = CH
71-88% isolated yields

Scheme 41.

Then, selective deprotection of the allyloxycarbonyl group gives place to *N*3′-Cbz-protected derivatives through a short and efficient chemoenzymatic route.

Carbonates display a wide range of applications in synthetic, polymer, and medicinal chemistry.[87] Whereas in the former field, carbonates are employed as protective groups for diols and poliols, in the latter the lipophilic nature of carbonates can afford prodrugs of pharmacological active compounds. Therefore, the use of carbonates in nucleoside chemistry offers an interesting potential. However, the use of alkyl carbonates as protective groups of nucleosides has been very restricted, because their removal requires a strong basic medium, frequently not compatible with the complex structure of such molecules. To avoid this limitation, the report[88] on the removal of carbonates by hydrolases in nucleoside alkyl carbonates is of great interest (Scheme 42).

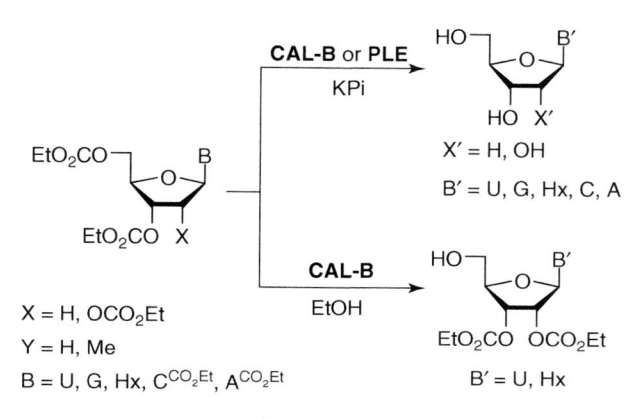

Scheme 42.

A set of mono-, di-, tri-, and tetraalkoxycarbonylated nucleosides is tested to assess their enzymatic hydrolysis. All the alkoxycarbonyl groups of the substrates, from both carbonate and carbamate functions, are quantitatively hydrolyzed using pig liver esterase (PLE) at pH 7 and 60 °C, regardless of the nucleoside base. Quantitative full alkoxycarbonyl group removal is also reached by CAL-B under mild conditions, but in this case, longer reaction times are required. Thus, PLE appears as a useful catalyst for the mild and quantitative deprotection of nucleoside carbonates and carbamates in the synthesis of modified nucleosides.

3. BIOTRANSFORMATIONS THAT MODIFY THE BASE

A systematic study of the role of the sugar ring in the process of recognition and binding of nucleosides, nucleotides, and oligonucleotides to their target enzymes has been undertaken.[89] The picture that is emerging from these studies shows that the majority of enzymes appear to have strict conformational requirements for substrate binding with the furanose ring in a well-defined shape. In particular, methanocarbanucleosides built on a rigid bicyclo[3.1.0]hexane template have been instrumental in defining the role of sugar puckering in nucleosides and nucleotides by stabilizing the active receptor-bound conformation and thereby identifying the biologically favored sugar conformer.

The synthetic approach to these conformationally locked carbocyclic nucleosides is not simple and relies on the availability of the chiral synthon that constitutes the sugar moiety (Scheme 43).[90] The overall yields of amino and diamino derivatives obtained from two extremely simple and cheap starting materials, such as

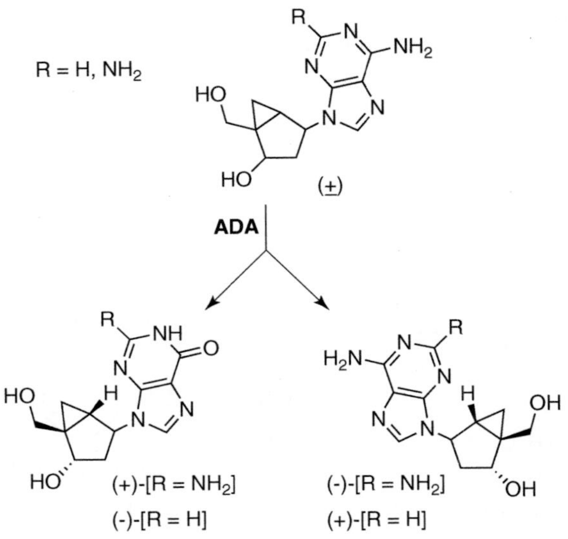

Scheme 43.

acrolein and ethyl acetoacetate, are 15% and 19%, respectively. This is remarkable considering the complexity of the resulting structures with four asymmetric carbons.

The key step on the syntheses is the enzymatic resolution of an aqueous solution of bicyclonucleosides with adenosine deaminase (ADA). The two purine analogues built on the complex bicyclo[3.1.0]hexane platform are resolved efficiently with remarkable enantioselectivity by ADA.

To study the ribosomal peptidyl transfer, puromycin analogues are of interest in which adenine has been replaced by hypoxanthine. Inosine puromycin analogues from 3′-azidodeoxyadenosine derivatives for the quantitative transformation of the N-heterocycle are synthesized using adenylate deaminase (AMPDA) (Scheme 44).[91] The amino acid coupling was carried out under Staudinger–Vilarrasa conditions in 94% yield starting with protected azide and in 82% yield using unprotected azide: thus, in the presence of two hydroxyls and a lactam function.

A chiral synthesis of a series of hexahydroisobenzofuran (HIBF) nucleosides is accomplished via glycosylation of a stereo-defined (*syn*-isomer) sugar motif with the appropriate silylated bases.[92] All nucleoside analogues are obtained in 52–71% yield as a mixture of α- and β-anomeric product, increasing the breadth of the novel nucleosides available for screening. The structure of the novel bicyclic HIBF nucleosides is established by a single-crystal x-ray structure of the β-HIBF thymine analogue. Furthermore, the sugar conformation for these nucleosides is established as N-type. Among the novel HIBF nucleosides synthesized, 25 compounds have been tested as an inhibitor of HIV-1 in human peripheral blood mononuclear (PBM) cells, and seven were found to be active (EC_{50} = 12.3 – 36.2 μM). Six of these

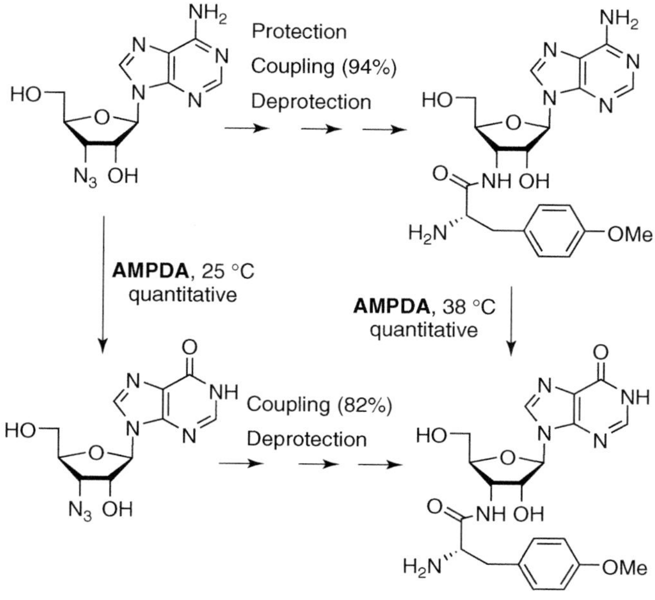

Scheme 44.

compounds are purine analogues, with β-HIBF inosine analogue being the most potent (EC_{50} = 12.3 μM) among all compounds tested. The striking resemblance between didanosine (ddI) and β-HIBF inosine analogue may explain the potent anti-HIV activity.

This inosine compound is obtained via enzymatic deamination of the corresponding adenosine analogue (Scheme 45). This reaction is slower than the deamination of natural adenosine reported and afforded only a 50% yield of the desired product after 72 h. Authors attribute this sluggish reactivity to the lipophilic nature of the HIBF nucleoside. Nonetheless, it is remarkable to observe that the presence of the cyclohexene ring is not incompatible with recognition by the enzyme. Recognition as a substrate for adenosine deaminase further confirms that the 5′-hydroxyl group was accessible to the active site in enzyme for its transformation.[93]

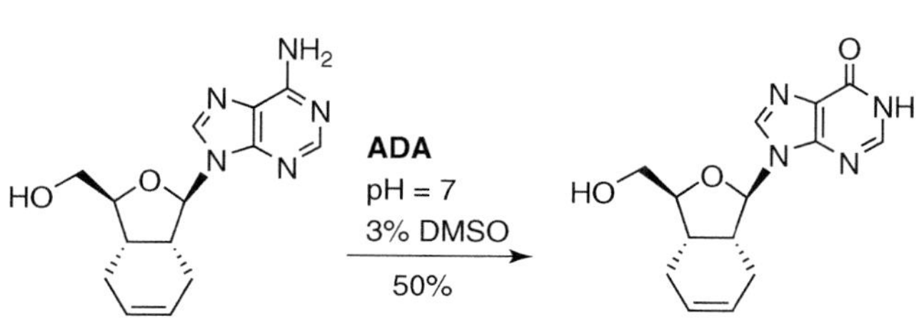

Scheme 45.

4. MOLECULAR MODELING

Due to the cleanness, simplicity, and efficiency of enzymatic reactions, they are often the best route to complex molecules such as nucleoside analogues, which are drug candidates for several diseases. Modifications of the sugar moiety are important sources of new compounds, with promising chemotherapeutic properties. One sugar modification is the lipase-catalyzed regioselective acylation of natural 2'-deoxy- and ribonucleosides with vinyl and oxime esters, and 3',5'-diamino-2',3',5'-trideoxynucleosides using nonactivated esters, as described previously in several examples in this chapter. One of the most useful lipases, CAL-B, shows high regioselectivity for the functional group at the ribose 5'-position for many different derivatives. The molecular basis of this regioselectivity is analyzed in depth[94] for CAL-B-catalyzed regioselective acylation of natural thymidine with oxime esters, and also regioselective acylation of an analogue, 3',5'-diamino-3',5'-dideoxythymidine, with nonactivated esters. In both cases, acylation favors the less hindered 5'-position up to 80-fold over the 3'-position. Computer modeling of phosphonate transition-state analogues for acylation of thymidine suggests that CAL-B favors acylation of the 5'-position because this orientation allows the thymine ring to bind in a hydrophobic pocket and forms stronger key hydrogen bonds as compared to acylation of the 3'-position.

On the other hand, computer modeling of phosphonamidate analogues of the transition states for acylation of either the 3'- or 5'-amino groups in 3',5'-diamino-3',5'-dideoxythymidine shows similar orientations and hydrogen bonds and thus does not explain the high regioselectivity. However, computer modeling of inverse structures, where the acyl chain binds in the nucleophile pocket, and vice versa, does rationalize the regioselectivity observed. The inverse structures fit the 5'-intermediate thymine ring, but not the 3'-intermediate thymine ring, in the hydrophobic pocket and form a weak new hydrogen bond between the carbonyl O-2 of the thymine and the nucleophile amine only for the 5'-intermediate. A water molecule may transfer a proton from the ammonium group to the active-site histidine. As a test of this inverse orientation, the acylation of thymine and 3',5'-diamino-3',5'-dideoxythymidine has been compared with butyryl acyl donors and with the isosteric methoxyacetyl acyl donors. Both acyl donors reacted at equal rates for thymidine, but the methoxyacetyl acyl donor reacted four times faster than the butyryl acyl donor for 3',5'-diamino-3',5'-dideoxythymidine. This faster rate is consistent with an inverse orientation for 3',5'-diamino-3',5'-dideoxythymidine, where the ether oxygen of the methoxyacetyl group can form a hydrogen bond similar to that of the nucleophilic amine. This combination of modeling and experiments suggests such lipase-catalyzed reactions of apparently close substrate analogues, as alcohols and amines may follow different pathways. The alternative mechanism for amines is shown in Scheme 46.

As it has been shown, hydrolases, and more specifically, lipases, can regioselectively acylate primary hydroxyl groups in the presence of secondary hydroxyl groups in nucleosides. For example, CAL-B catalyzes the selective enzymatic acylation of the primary hydroxyl in several 2'-deoxynucleosides. In special cases,

Scheme 46.

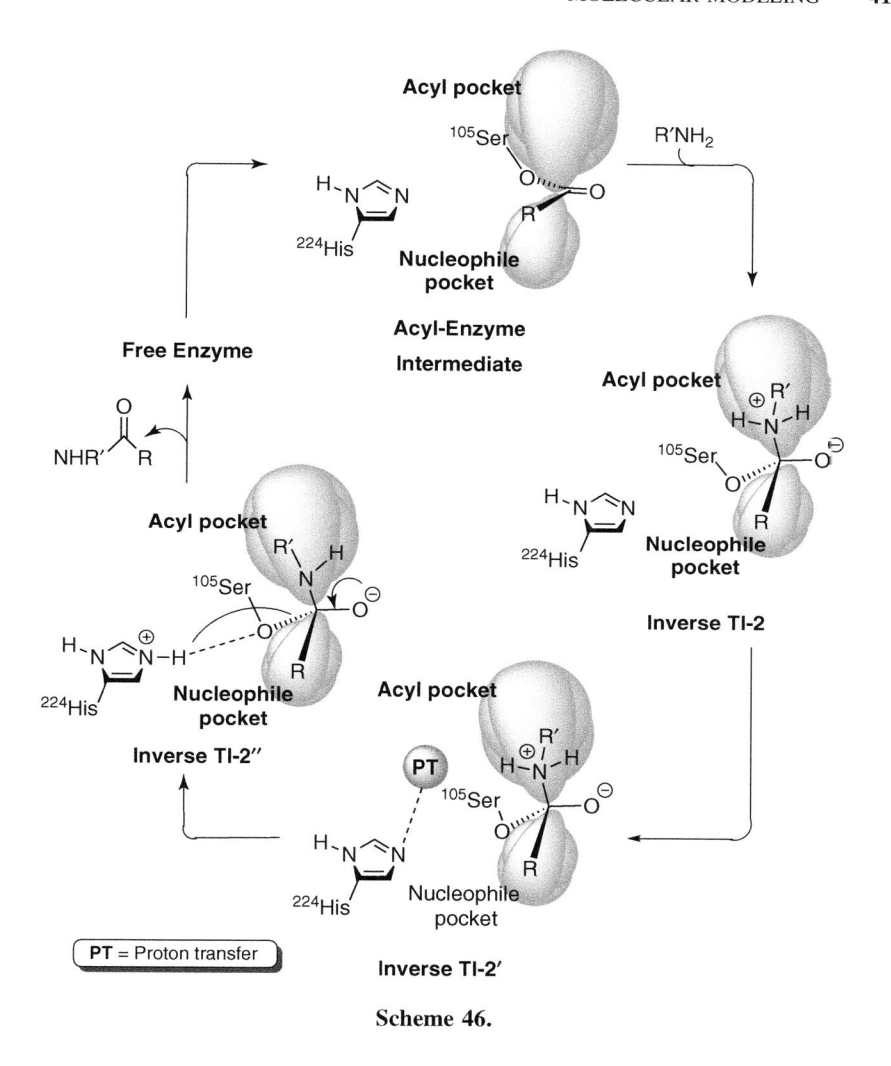

chemical methods can favor reactions at a secondary over a primary hydroxyl group. Also, PSL-C catalyzes the regioselective acylation of the more hindered secondary 3'-position in β-D-2'-deoxynucleosides, as has been shown. The explanation of this "abnormal" behavior is the focus of a paper[95] that identifies the molecular basis of this selectivity using computer-aided molecular modeling.

Surprisingly, PSL-C favors acylation of the secondary hydroxyl at the 3'-position over the primary hydroxyl at the 5'-position in 2'-deoxynucleosides by up to >98:1. Molecular modeling found catalytically productive tetrahedral intermediate analogues for both orientations. However, acylation of 3'-hydroxyl places the thymine base in the alternative hydrophobic pocket of PSL-C's substrate-binding site, where it can hydrogen-bond to the side-chain hydroxyls of Tyr23 and Tyr29 and the main-chain carbonyl of Leu17. Conversely, acylation of the 5'-hydroxyl leaves the thymine base in the solvent, where there is no favorable binding to the enzyme. It is proposed that these remote stabilizing interactions between the thymine base and the PSL-C's substrate-binding site stabilize the 3′-acylation transition state and thus account for the unusual regioselectivity.

5. CONCLUDING REMARKS

Chemoenzymatic approaches to the synthesis of nucleosides have been demonstrated to be a powerful tool to generate a great variety of natural and nonnatural analogue structures with a wide structural diversity. The number of reports describing enzyme-mediated hydrolysis, protections, and resolutions has been increasing in the past decade, not only applied to laboratory scale but also in industrial production. This is due to the favorable features of biocatalytic systems, especially the selectivity. With the development of modern biotechnology, we predict that enzymes will become available more cheaply and will enable industrial preparation of nucleoside derivatives via biotransformations. As the area continues to grow, new enzymatic procedures for the synthesis of many natural and unnatural nucleosides will continue to be developed. It is believed that more effective synthetic strategies based on combined chemical and enzymatic methods will have to be developed to tackle the new generation of problems associated with nucleosides, particularly those involved in modifications in both the heterocyclic base and the sugar moiety, to avoid the drawbacks shown in certain applications, such as agents effective against viral infections or cancer, or applications related antisense oligonucleotide preparation. Moreover, biocatalysts are ecologically beneficial natural catalysts that offer the opportunity to carry out highly chemo-, stereo-, and regioselective synthesis of nucleosides, which could not be performed by classical chemical methodologies. Recently, molecular modeling is helping to explain the selectivity shown by several biocatalysts. The aim of this tool is to predict which catalyst would be best suited for a particular substrate, thus avoiding investing so much time and money in screening.

Acknowledgments

The authors wish to express their gratitude to the Spanish Ministerio de Educación y Ciencia (MEC) (projects MEC-CTQ-2007-61126 and MICINN-12-CTQ2011-24237).

ABBREVIATIONS

A	adenine
AIDS	acquired immunodeficiency syndrome
ara-A	arabinoadenosine
ara-C	arabinocytosine
AZT	zidovudine
C	cytosine

Cbz	benzyloxycarbonyl
DCC	dicyclohexylcarbodiimide
ddA	2′,3′-dideoxyadenosine
ddC	2′,3′-dideoxycytosine
ddI	2′,3′-dideoxyinosine
d4T	2′,3′-dideoxy-2′,3′-didehydrothymidine (stavudine)
d4U	2′,3′-dideoxy-2′,3′-didehydrouridine
DHF	dihydrofuran
DHP	dihydropyran
DMSO	dimethylsulfoxide
DMTr	dimethoxytrityl
FDA	U.S. Food and Drug Administration
G	guanine
HBV	hepatitis B virus
HIBF	hexahydroisobenzofuran
HIV	human immunodeficiency virus
IdU	5-yodo-2′-deoxyuridine
IL	ionic liquid
MDHP	methoxydihydropyran
MTHP	methoxytetrahydropyran
PBM	peripheral blood mononuclear
Py	pyridine
T	thymine
THF	tetrahydrofuran
THP	tetrahydropyran
3TC	lamivudine

REFERENCES

1. (a) De Clercq, E. *Curr. Opin. Microbiol.* **2005**, *8*, 552–560. (b) *Frontiers in Nucleosides and Nucleic Acids*; Schinazi, R. F.; Liotta, D. C., Eds.; IHL Press: Tucker, GA, **2004**. (c) Rachakonda, S.; Cartee, L. *Curr. Med. Chem.* **2004**, *11*, 775–793. (d) De Clercq, E. *J. Clin. Vir.* **2004**, *30*, 115–133. (e) *Recent Advances in Nucleosides: Chemistry and Chemotherapy*; Chu, C. K., Ed.; Elsevier Science: New York, **2002**.
2. (a) Ferrero, M.; Gotor, V. *Chem. Rev.* **2000**, *100*, 4319–4347. (b) Ferrero, M.; Gotor, V. *Monatsh. Chem.* **2000**, *131*, 585–616.
3. (a) *Industrial Enzymes: Structure, Functions and Applications*; Polaina, J.; MacCabe, A. P., Eds.; Springer-Verlag: Dordrecht, The Netherlands, **2007**. (b) *Biocatalysis in the Pharmaceutical and Biotechnology Industries;* Patel, R. M., Ed.; CRC Press, Taylor & Francis Group: Boca Raton, FL, **2007**. (c) Straathof, A. J. J.; Panke, S.; Schmid, A. *Curr. Opin. Biotechnol.* **2002**, *13*, 548–556.
4. Utagawa, T. *J. Mol. Catal. B: Enzymatic* **1999**, *6*, 215–222.
5. (a) Mikhailopulo, I. A. *Curr. Org. Chem.* **2007**, *11*, 317–333. (b) Lewkowicz. E. S.; Iribarren, A. M. *Curr. Org. Chem.* **2006**, *10*, 1197–1215.
6. Li, N.; Smith, T. J.; Zong, M.-H. *Biotechnol. Adv.* **2010**, *28*, 348–366.

7. (a) Hamamura, E. K.; Prystasz, M.; Verheyden, J. P. H.; Moffat, J. G.; Yamaguchi, K.; Uchida, N.; Sato, K.; Nomura, A.; Shiratori, O.; Takese, S.; Katagiri, K. *J. Med. Chem.* **1976**, *19*, 654–662. (b) Hamamura, E. K.; Prystasz, M.; Verheyden, J. P. H.; Moffat, J. G.; Yamaguchi, K.; Uchida, N.; Sato, K.; Nomura, A.; Shiratori, O.; Takese, S.; Katagiri, K. *J. Med. Chem.* **1976**, *19*, 667–674.
8. Suhadolnik, R. J. In *Nucleoside Antibiotics*; Wiley-Interscience: New York, **1970**.
9. Lavandera, I.; Fernández, S.; Ferrero, M.; Gotor, V. *J. Org. Chem.* **2001**, *66*, 4079–4082.
10. Lavandera, I.; Fernández, S.; Ferrero, M.; Gotor, V. *Tetrahedron* **2003**, *59*, 5449–5456.
11. Lavandera, I.; Fernández, S.; Ferrero, M.; Gotor, V. *Nucleosides Nucleotides Nucleic Acids* **2003**, *22*, 833–836.
12. Lavandera, I.; Fernández, S.; Ferrero, M.; De Clercq, E.; Gotor, V. *Nucleosides Nucleotides Nucleic Acids* **2003**, *22*, 1939–1952.
13. (a) Moris, F.; Gotor, V. *J. Org. Chem.* **1993**, *58*, 653–660. (b) Moris, F.; Gotor, V. *Tetrahedron* **1993**, *49*, 10089–10098. (c) Gotor, V.; Moris, F. *Synthesis* **1992**, 626–628.
14. Reese, C. B.; Yan, H. *J. Chem. Soc., Perkin Trans. 1* **2002**, 2619–2633.
15. García, J.; Fernández, S.; Ferrero, M.; Sanghvi, Y. S.; Gotor, V. *J. Org. Chem.* **2002**, *67*, 4513–419.
16. García, J.; Fernández, S.; Ferrero, M.; Sanghvi, Y. S.; Gotor, V. *Nucleosides Nucleotides Nucleic Acids* **2003**, *22*, 1455–1457.
17. García, J.; Fernández, S.; Ferrero, M.; Sanghvi, Y. S.; Gotor, V. *Tetrahedron: Asymmetry* **2003**, *14*, 3533–3540.
18. Lavandera, I.; García, J.; Fernández, S.; Ferrero, M.; Gotor, V.; Sanghvi, Y. S. In *Current Protocols in Nucleic Acid Chemistry*; Egli, M.; Herdewijn, P.; Matsuda, A.; Sanghvi, Y. S., Eds.; Wiley: Hoboken, NJ, **2011**; Chapter 2, pp. 2.11.1–2.11.36.
19. Aghil, O.; Bibby, M. C.; Carrington, S. J.; Double, J.; Douglas, K. T.; Phillips, R. M.; Shing, T. K. M. *Anticancer Drug Des.* **1992**, *7*, 67–82.
20. (a) Takeuchi, T.; Chimura, H.; Hamada, M.; Umezawa, H.; Yoshioka, O.; Oguchi, N.; Takahashi, Y.; Matsuda, A. *J. Antibiot.* **1975**, *28*, 737–742. (b) Chimura, H.; Nakamura, H.; Takita, T.; Takeuchi, T.; Umezawa, H.; Kato, K.; Saito, S.; Tomisawa, T.; Iitaka, Y. *J. Antibiot.* **1975**, *28*, 743–748.
21. It has been introduced as a protecting group in thymidine derivatives: Arentzen, R.; Reese, C. B. *J. Chem. Soc., Chem. Commun.* **1977**, 270–272.
22. Biological assays have been performed by Erik de Clercq in Belgium, unpublished results
23. Guibourdenche, C.; Podlech, J.; Seebach, D. *Liebigs Ann.* **1996**, 1121–1129.
24. Díaz-Rodríguez, A.; Fernández, S.; Lavandera, I.; Ferrero, M.; Gotor, V. *Tetrahedron Lett.* **2005**, *46*, 5835–5838.
25. (a) For comprehensive literature, see *Current Protocols in Nucleic Acid Chemistry*; Egli, M.; Herdewijn, P.; Matsuda, A.; Sanghvi, Y. S., Eds.; Wiley: Hoboken, NJ, **2011**; Chapter 2. (b) *Protective Groups in Organic Synthesis*; 4th ed.; Wuts, P. G. M.; Green, T. W., Eds.; Wiley, Hoboken, NJ, **2006**.(c) *Protecting Groups*; Kocienski, P. J., Ed.; Georg Thieme Verlag: Stuttgart, Germany, **2004**; pp. 464–475.
26. García, J.; Fernández, S.; Ferrero, M.; Sanghvi, Y. S.; Gotor, V. *Tetrahedron Lett.* **2004**, *45*, 1709–1712.
27. Li, N.; Ma, D.; Zong, M.-H. *J. Biotechnol.* **2008**, *133*, 103–109.

28. (a) Wang, Z.-Y.; Li, N.; Zong, M.-H. *J. Mol. Catal. B: Enzymatic* **2009**, *59*, 212–219. (b) Li, N.; Zong, M.-H.; Ma, D. *Tetrahedron* **2009**, *65*, 1063–1068. (c) Wang, Z.-Y.; Zong, M.-H. *Biotechnol. Prog.* **2009**, *25*, 784–791. (d) Li, N.; Zong, M.-H.; Ma, D. *Eur. J. Org. Chem.* **2008**, 5375–5378. (e) Li, N.; Zong, M.-H.; Liu, X.-M.; Ma, D. *J. Mol. Catal. B: Enzymatic* **2007**, *47*, 6–12. (f) Wang, H.; Zong, M.-H.; Wu, H.; Lou, W.-Y. *J. Biotechnol.* **2007**, *129*, 689–695.

29. (a) Qian, X.; Liu, B.; Wu, Q.; Lv, D.; Lin, X.-F. *Bioorg. Med. Chem.* **2008**, *16*, 5181–5188. (b) Wu, Q.; Xia, A.; Lin, X.-F. *J. Mol. Catal. B: Enzymatic* **2008**, *54*, 76–82. (c) Li, X.; Wu, Q.; Lv, D.-S.; Lin, X.-F. *Bioorg. Med. Chem.* **2006**, *14*, 3377–3382. (d) Sun, X.-F.; Wu, Q.; Wang, N.; Cai, Y.; Lin, X.-F. *Biotechnol. Lett.* **2005**, *27*, 113–117. (e) Liu, B.-K.; Wang, N.; Wu, Q.; Xie, C.-Y.; Lin, X.-F. *Biotechnol. Lett.* **2005**, *27*, 717–720. (f) Wang, N.; Chen, Z. C.; Lu, D. S.; Lin, X.-F. *Bioorg. Med. Chem. Lett.* **2005**, *15*, 4064–4067.

30. Hanson, R. L.; Shi, Z.; Brzozowski, D. B.; Barnerjee, A.; Kissick, T. P.; Singh, J.; Pullockaran, A. J.; North, J. T.; Fan, J.; Howell, J.; Durand, S. C.; Montana, M. A.; Kronenthal, D. R.; Mueller, R. H.; Patel, R. N. *Bioorg. Med. Chem.* **2000**, *8*, 2681–2687.

31. Tamarez, M.; Morgan, B.; Wong, G. S. K.; Tong, W.; Bennett, F.; Lovey, R.; McCormick, J. L.; Zaks, A. *Org. Process Res. Dev.* **2003**, *7*, 951–953.

32. Parham, W. E.; Anderson, E. L. *J. Am. Chem. Soc.* **1948**, *70*, 4187–4189.

33. Sartori, G.; Ballini, R.; Bigi, F.; Bosica, G.; Maggi, R.; Righi, P. *Chem. Rev.* **2004**, *104*, 199–250.

34. Rodríguez-Pérez, T.; Fernández, S.; Martínez-Montero, S.; González-García, T.; Sanghvi, Y. S.; Gotor, V.; Ferrero, M. *Eur. J. Org. Chem.* **2010**, 1736–1744.

35. (a) Chang, C.-N.; Doong, S.-L.; Zhou, J. H.; Beach, J. W.; Jeong, L. S.; Chu, C. K.; Tasi, C.-H.; Cheng, Y.-C. *J. Biol. Chem.* **1992**, *267*, 13939–13942. (b) Doong, S.-L.; Tasi, C.-H.; Schinazi, R. F.; Liotta, D. C.; Cheng, Y.-C. *Proc. Natl. Acad. Sci. USA* **1991**, *88*, 8495–8499.

36. For recent reviews: (a) Mathé, C.; Gosselin, G. *Antiviral Res.* 2006, 71, 276–281. (b) Sommadossi, J.-P. *In* Recent Advances in Nucleosides; Chu, C. K., Ed.; Elsevier: Amsterdam, **2002**; pp. 417–432. (c) Damaraju, V. L.; Bouffard, D. Y.; Wong, C. K. W.; Clarke, M. L.; Mackey, J. R.; Leblond, L.; Cass, C. E.; Grey, M.; Gourdeau, H. *BMC Cancer* **2007**, *7*, 121. (d) Swords, R.; Giles, F. *Hematology* **2007**, *12*, 219–227. (e) Gumina, G.; Chong, Y.; Chu, C. K. In *Cancer Drug Discovery and Development: Deoxynucleoside Analogs in Cancer Therapy*; Peters, G. J., Ed.; Humana Press: Totowa, NJ, **2006**; pp. 173–198.

37. Nolte, A.; Klussmann, S.; Bald, R.; Erdmann, V. A.; Fürste, J. P. *Nat. Biotechnol.* **1996**, *14*, 1116–1119.

38. Albert, M.; De Souza, D.; Feiertag, P.; Hönig, H. *Org. Lett.* **2002**, *4*, 3251–3254.

39. Yoshimura, Y.; Moon, H. R.; Choi, Y.; Marquez, V. E. *J. Org. Chem.* **2002**, *67*, 5938–5945.

40. Bondada, L.; Gumina, G.; Nair, R.; Ning, X. H.; Schinazi, R. F.; Chu, C. K. *Org. Lett.* **2004**, *6*, 2531–2534.

41. Siddiqui, M. A.; Hughes, S. H.; Boyer, P. L.; Mitsuya, H.; Van, Q. N.; George, C.; Sarafinanos, S. G.; Marquez, V. E. . *J Med. Chem.* **2004**, *47*, 5041–5048.

42. García, J.; Fernández, S.; Ferrero, M.; Sanghvi, Y. S.; Gotor, V. *Org. Lett.* **2004**, *6*, 3759–3762.

43. Martínez-Montero, S.; Fernández, S.; Sanghvi, Y. S.; Gotor, V.; Ferrero, M. . *J Org. Chem.* **2010**, *75*, 6605–6613.
44. Eulberg, D.; Jarosch, F.; Vonhoff, S.; Klussmann, S. In The Aptamer Handbook; Klussmann, S., Ed.; Wiley-VCH: Weinheim, Germany, **2006**; Chapter 18, pp. 417–439.
45. Carnovale, C.; Iannazzo, D.; Nicolosi, G.; Piperno, A.; Sanfilippo, C. *Tetrahedron: Asymmetry* **2009**, *20*, 425–429.
46. Prasad, A. K.; Roy, S.; Kumar, R.; Kalra, N.; Wengel, J.; Olsen, C. E.; Cholli, A. L.; Samuelson, L. A.; Kumar, J.; Watterson, A. C.; Gross, R. A.; Parmar, V. S. *Tetrahedron* **2003**, *59*, 1333–1338.
47. Prasad, A. K.; Kalra, N.; Yadav, Y.; Singh, S. K.; Sharma, S. K.; Patkar, S.; Lage, L.; Olsen, C. E.; Wengel, J.; Parmar, V. S. *Org. Biomol. Chem.* **2007**, *5*, 3524–3530.
48. Singh, S. K.; Sharma, V. K.; Bohra, K.; Olsen, C. E.; Prasad, A. K. *J. Org. Chem.* **2011**, *76*, 7556–7562.
49. Obame, G.; Pellissier, H.; Vanthuyne, N.; Bongui, J.-B.; Audran, G. *Tetrahedron Lett.* **2011**, *52*, 1082–1085.
50. (a) Damkjaer, D. L.; Petersen, M.; Wengel, J. *Nucleosides Nucleotides* **1994**, *13*, 1801–1807. (b) Van Draanen, N. A.; Koszalka, G. W. *Nucleosides Nucleotides* **1994**, *13*, 1679–1693.
51. García, J.; Díaz-Rodríguez, A.; Fernández, S.; Sanghvi, Y. S.; Ferrero, M.; Gotor, V. *J. Org. Chem.* **2006**, *71*, 9765–9771.
52. *Handbook of Nucleoside Synthesis*; Vorbrüggen, H.; Ruh-Pohlenz, C., Eds.; Wiley: New York, **2001**, and references cited therein.
53. The price of α-thymidine from Aldrich is 766 €/g.
54. Maity, J.; Shakya, G.; Singh, S. K.; Ravikumar, V. T.; Parmar, V. S.; Prasad, A. K. *J. Org. Chem.* **2008**, *73*, 5629–5632.
55. For selected recent reviews in this area, see (a) Sohail, M. *Drug Discov. Today* **2001**, *6*, 1260–1261. (b) Dean, N. M. *Curr. Opin. Biotechnol.* **2001**, *12*, 622–625. (c) Crooke, S. T. *Oncogene* **2000**, *19*, 6651–6659. (d) Green, D. W.; Roh, H.; Pippin, J.; Drebin, J. A. *J. Am. Coll. Surg.* **2000**, *191*, 93–105. (e) Koller, E.; Gaarde, W. A.; Monia, B. P. *Trends Pharmacol. Sci.* **2000**, *21*, 142–148. (f) Sanghvi, Y. S. In: *Comprehensive Natural Products Chemistry*; Barton, D. H. R.; Nakanishi, K., Eds-in-Chief; Pergamon Press: New York, **1999**; Vol. 7 (Kool, E. T., Ed.), p. 285.
56. For recent accounts on 2′-O-methylation of guanosine and adenosine, see (a) Kore, A. R.; Parmar, G.; Reddy, S. *Nucleosides Nucleotides Nucleic Acids* **2006**, *25*, 307–314. (b) Chow, S.; Wen, K.; Sanghvi, Y. S.; Theodorakis, E. A. *Bioorg. Med. Chem. Lett.* **2003**, *13*, 1631–1634. (c) Wen, K.; Chow, S.; Sanghvi, Y. S.; Theodorakis, E. A. *J. Org. Chem.* **2002**, *67*, 7887–7889.
57. Martínez-Montero, S.; Fernández, S.; Rodríguez-Pérez, T.; Sanghvi, Y. S.; Wen, K.; Gotor, V.; Ferrero, M. *Eur. J. Org. Chem.* **2009**, 3265–3271.
58. (a) Chattopadhyaya, J. B.; Reese, C. B. *J. Chem. Soc., Chem. Commun.* **1978**, 639–640. (b) Barnett, W. E.; Needham, L. L.; Powell, R. W. *Tetrahedron* **1972**, *28*, 419–424.
59. (a) D'Onofrio, J.; Montesarchio, D.; De Napoli, L.; Di Fabio, G. *Org. Lett.* **2005**, *7*, 4927–4930. (b) Wagner, T.; Pfleiderer, W. *Helv. Chim. Acta* **2000**, *83*, 2023–2035. (c) Esipov, D. S.; Esipova, O. V.; Korobko, V. G. *Nucleosides Nucleotides* **1998**, *17*, 1697–1704. (d) Robles, J.; Pedroso, E.; Grandas, A. *Nucleic Acids Res.* **1995**, *23*, 4151–4161.

60. Macugen is a drug developed and marketed by Eyetech Pharmaceuticals, Inc. and Pfizer, Inc.
61. Nimjee S. M.; Rusconi C. P.; Sullenger B. A. *Annu. Rev. Med.* **2005**, *56*, 555–583.
62. (a) De Clercq, E. *Curr. Opin. Microbiol.* **2005**, *8*, 552–560. (b) Schinazi, R. F.; Liotta, D. C. *In* Frontiers in Nucleoside and Nucleic Acids; IHL Press: Tucker, GA, **2004**. (c) Rachakonda, S.; Cartee, L. *Curr. Med. Chem.* **2004**, *11*, 775–793. (d) De Clercq, E. *J. Clin. Vir.* **2004**, *30*, 115–133. (e) Chu, C. K. In Recent Advances in Nucleosides: *Chemistry and Chemotherapy*; Elsevier Science: New York, 2002.
63. (a) Dove, A. *Nat. Biotechnol.* **2001**, *19*, 913–917. (b) Dwek, R. A. *Chem. Rev.* **1996**, *96*, 683–720. (c) Varki, A. *Glycobiology* **1993**, *3*, 97–130.
64. Peterson, L. W., McKenna, C. E. *Expert Opin. Drug Deliv.* **2009**, *6*, 405–420.
65. Herdewijn, P. In *Current Protocols in Nucleic Acid Chemistry*; Beaucage, S. L.; Bergstrom, D. E.; Glick, G. D.; Jones, R. A., Eds.;Wiley: Hoboken, NJ, **2007**; pp. 15.0.1–15.0.4.
66. Rodríguez-Pérez, T.; Lavandera, I.; Fernández, S.; Sanghvi, Y. S.; Ferrero, M.; Gotor, V. *Eur. J. Org. Chem.* **2007**, 2769–2778.
67. Rodríguez-Pérez, T.; Fernández, S.; Sanghvi, Y. S.; Detorio, M.; Schinazi, R. F.; Gotor, V.; Ferrero, M. *Bioconjug. Chem.* **2010**, *21*, 2239–2249.
68. Friedberg, E. C.; Walker, G. C.; Siede, W.; Wood, R. D.; Schultz R. A.; Ellenberger, T. *In* DNA Repair and Mutagenesis; 2nd ed.; ASM Press: Washington, DC, **2006**.
69. For a minireview, see Boiteux, S.; Guillet, M. *DNA Repair* **2004**, *3*, 1–12.
70. (a) Xu, Y. J.; Kim E. Y.; Demple, B. *J. Biol. Chem.* **1998**, *273*, 28837–28844. (b) Masuda, Y.; Bennett R. A.; Demple, B. *J. Biol. Chem.* **1998**, *273*, 30352–30359.
71. (a) Huang, H.; Greenberg, M. M. *J. Org. Chem.* **2008**, *73*, 2695–2703. (b) Chen, J.; Dupradeau, F.-Y.; Case, D. A.; Turner, C. J.; Stubbe, J. *Nucleic Acids Res.* **2008**, *36*, 253–262. (c) Takeshita, M.; Chang, C.-N.; Johnson, F.; Will, S.; Grollman, A. P. *J. Biol. Chem.* **1987**, *262*, 10171–10179.
72. Morrissey, D. V.; Blanchard, K.; Shaw, L.; Jensen, K.; Lockridge, J. A.; Dickinson, B.; McSwiggen, J. A.; Vargeese, C.; Bowman, K.; Shaffer, C. S.; Polisky, B. A.; Zinnen, S. *Hepatology* **2005**, *41*, 1349–1355.
73. (a) Lin, Y.-I.; Bitha, P.; Sakya, S. M.; Lang, S. A., Jr.; Yang, Y.; Weiss, W. J.; Petersen, P. J.; Bush, K.; Testa, R. T. *Bioorg. Med. Chem. Lett.* **1997**, *7*, 3063–3068. (b) Lin, Y.-I.; Bitha, P.; Sakya, S. M.; Strohmeyer, T. W.; Li, Z.; Lee, V. J.; Lang, S. A., Jr.; Yang, Y.; Bhachech, N.; Weiss, W. J.; Petersen, P. J.; Jacobus, N. V.; Bush, K.; Testa, R. T.; Tally, F. P. *Bioorg. Med. Chem. Lett.*, **1997**, *7*, 1671–1676. (c) Chen, J. J.; Walker, J. A.; Liu, W.; Wise, D. S.; Townsend, L. B. *Tetrahedron Lett.* **1995**, *36*, 8363–8366. (d) Ravikumar, V. T.; Wyrzykiewicz, T. K.; Cole, D. L. *Nucleosides Nucleotides Nucleic Acids* **1994**, *13*, 1695–1706.
74. (a) Chenault, H. K.; Mandes, R. F. *Tetrahedron* **1997**, *53*, 11033–11038. (b) Plavec, J.; Tong, W.; Chattopadhyaya, J. *J. Am. Chem. Soc.* **1993**, *115*, 9734–9746. (c) Iyer, R. P.; Bogdan, U.; Boal, J.; Storm, C.; Egan, W.; Matsukura, M.; Zon, S.; Broder, G.; Wilk, A.; Koziolkiewicz, M.; Stec, W. J. *Nucleic Acids Res.* **1990**, *18*, 2855–2859. (d) Eritja, R.; Walker, P. A.; Randall, S. K.; Goodman, M. F.; Kaplan, B. E. *Nucleosides Nucleotides Nucleic Acids* **1987**, *6*, 803–814.
75. Martínez-Montero, S.; Fernández, S.; Sanghvi, Y. S.; Gotor, V.; Ferrero, M. *Org. Biomol. Chem.* **2011**, *9*, 5960–5966.

76. (a) Faber, K. In *Biotransformations in Organic Chemistry*; 5th ed.; Springer-Verlag, Berlin, 2004. (b) Bornscheuer, U. T.; Kazlauskas, R. J. *In* Hydrolases in Organic Synthesis; Wiley-VCH: Weinheim, Germany, **1999**.
77. (a) Zinni, M. A.; Iglesias, L. E.; Iribarren, A. M. *J. Mol. Catal. B: Enzymatic.* **2007**, *47*, 86–90. (b) Zinni, M. A.; Rodríguez, S. D.; Pontiggia, R. M.; Montserrat, J. M.; Iglesias, L. E.; Iribarren, A. M. *J. Mol. Catal. B: Enzymatic* **2004**, *29*, 129–132.(c) Roncaglia, D. I.; Schmidt, A. M.; Iglesias, L. E.; Iribarren, A. M. *Biotechnol. Lett.* **2001**, *23*, 1439–1443.(d) Iglesias, L. E.; Zinni, M. A.; Gallo, M.; Iribarren, A. M. *Biotechnol. Lett.* **2000**, *22*, 361–365.
78. Gotor, V. *Bioorg. Med. Chem. Lett.* **1999**, *7*, 2189–2197.
79. Parrish, J. P.; Salvatore, R. N.; Jung, K. W. *Tetrahedron* **2000**, *56*, 8207–8237.
80. Barré-Sinoussi, F.; Chermann, J. C.; Rey, F.; Nugeyre, M. T.; Chamaret, S.; Gruest, J.; Dauguet, C.; Axler-Blin, C.; Vézinet-Brun, F.; Rouzioux, C.; Rozenbaum, W.; Montagnier, L. *Science* **1983**, *220*, 868–871.
81. (a) De Clercq, E. *Nat. Rev. Drug Discov.* **2002**, *1*, 13–25. (b) De Clercq, E. *J. Med. Chem.* **1995**, *38*, 2491–2517.
82. (a) Taourirte, M.; Mohamed, L. A.; Rochdi, A.; Vasseur, J.-J.; Fernández, S.; Ferrero, M.; Gotor, V.; Pannecouque, C.; De Clercq, E.; Lazrek, H. B. *Nucleosides Nucleotides Nucleic Acids* **2004**, *23*, 701–714. (b) Taourirte, M.; Lazrek, H. B.; Vasseur, J.-J.; Fernández, S.; Ferrero, M.; Gotor, V. *Nucleosides Nucleotides Nucleic Acids* **2001**, *20*, 959–962.
83. Magdalena, J.; Fernández, S.; Ferrero, M.; Gotor, V. *J. Org. Chem.* **1998**, *63*, 8873–8879.
84. Magdalena, J.; Fernández, S.; Ferrero, M.; Gotor, V. *Tetrahedron Lett.* **1999**, *40*, 1787–1790.
85. Magdalena, J.; Fernández, S.; Ferrero, M.; Gotor, V. *Nucleosides Nucleotides Nucleic Acids* **2002**, *21*, 55–64.
86. Lavandera, I.; Fernández, S.; Ferrero, M.; Gotor, V. *J. Org. Chem.* **2004**, *69*, 1748–1751.
87. Parrish, J. P.; Salvatore, R. N.; Jung, K. W. *Tetrahedron* **2000**, *56*, 8207–8237.
88. (a) Capello, M.; Imanishi, L.; Iglesias, L. E.; Iribarren, A. M. *Biotechnol. Lett.* **2007**, *29*, 1217–1220. (b) Capello, M.; González, M.; Rodríguez, S. D.; Iglesias, L. E.; Iribarren, A. M. *J. Mol. Catal. B: Enzymatic* **2005**, *36*, 36–39.
89. Prota, A.; Vogt, J.; Perozzo, R.; Pilger, B.; Wurth, C.; Marquez, V. E.; Russ, P.; Schultz, G. E.; Folkers, G.; Scapozza, L. Biochemistry **2000**, *39*, 9597–9603, and references cited therein.
90. Moon, H. R.; Ford, H., Jr.; Marquez, V. E. *Org. Lett.* **2000**, *2*, 3793–3796.
91. Charafeddine, A.; Chapuis, H.; Strazewski, P. *Org. Lett.* **2007**, *9*, 2787–2790.
92. Díaz-Rodríguez, A.; Sanghvi, Y. S.; Fernández, S.; Schinazi, R. F.; Theodorakis, E. A.; Ferrero, M.; Gotor, V. *Org. Biomol. Chem.* **2009**, *7*, 1415–1423.
93. Gupta, M., Nair, V. *Coll. Czech. Chem. Commun.* **2006**, *71*, 769–787.
94. Lavandera, I.; Fernández, S.; Magdalena, J.; Ferrero, M.; Kazlauskas, R. J.; Gotor, V. *ChemBioChem.* **2005**, *6*, 1381–1390.
95. Lavandera, I.; Fernández, S.; Magdalena, J.; Ferrero, M.; Grewal, H.; Savile, C. K.; Kazlauskas, R. J.; Gotor, V. *ChemBioChem.* **2006**, *7*, 693–698.

2 Nucleosides Modified at the Base Moiety

O. SARI, V. ROY, and L. A. AGROFOGLIO

Institut de Chimie Organique et Analytique, Université d'Orléans, Orléans, France

1. INTRODUCTION

For four decades, nucleoside mimetics have attracted considerable attention in synthetic chemistry, owing to their antiviral activities against HIV, HBV, HSV, and others[1] Actually, with more than 40 approved molecules, antiviral chemotherapy consists of a huge panel of powerful nucleoside analogues. However, owing to numerous limitations, such as toxicity and mutation of viruses, it has become necessary to develop new original molecules. In this field, the commonly known active compounds belong to the categories of sugar- or base-modified nucleosides. The intense search for clinically useful nucleoside derivatives has resulted in a wealth of new approaches to their syntheses. These strategies have involved several formal modifications of naturally occurring nucleosides, especially alteration of the base moiety. Since the discovery of the first successful antiviral drug, 5-iodouracil (IDU),[2] in 1958, interest has diversified toward compounds in which the heterocyclic components of the nucleoside have departed significantly from the natural form. Some of the compounds made exemplify structures containing unusual substituents such as ribavirin (**1**), BVDU (**2**), pseudouridine (**3**), and showdomcin (**4**) (Figure 1).

These novel types of nucleosides have been found variously to have anticancer, antiviral, and antibacterial activity. A variety of strategies have been devised to design effective, selective, and nontoxic antiviral agents based on nucleoside analogues. The common approach thus has been to vary the chemical structure of the natural nucleosides systematically in the hope that they would still be capable of enzymatic phosphorylation in the host but that the activated forms would be able to block only the viral metabolic processes (Figure 2).

Because this topic has been reviewed extensively in recent years, we decided to provide only a few highlights of the fascinating world of nucleoside analogues. The reader is referred to earlier reviews in the literature on base-modified nucleosides,

Chemical Synthesis of Nucleoside Analogues, First Edition. Edited by Pedro Merino.
© 2013 John Wiley & Sons, Inc. Published 2013 by John Wiley & Sons, Inc.

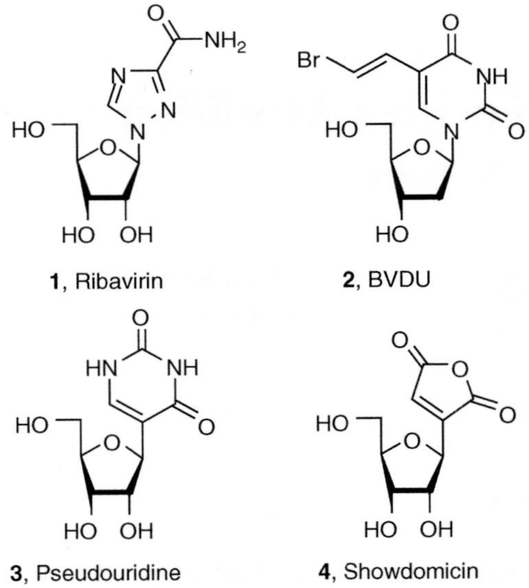

1, Ribavirin

2, BVDU

3, Pseudouridine

4, Showdomicin

Figure 1.

especially regarding C-nucleosides[3] and "basic" pyrimidine functionalization,[4] which are not covered herein, as well as to recent contributions in *Modified Nucleosides*, edited by Piet Herdewijn.[5]

2. RUTHENIUM-CATALYZED C–H BOND ACTIVATION FOR NUCLEOSIDE SYNTHESES

Novel reactions that can functionalize carbon–hydrogen bonds selectively are of intense interest to the chemical community because they offer new strategic approaches for synthesis. In this perspective, the introduction of new functionalities directly through selective transformation of ubiquitous C–H bonds, instead of conducting transformations on preexisting functional groups, have been a major breakthrough in modern chemistry. Thus, metal-mediated C–H activation reactions have been an area of rapid growth for the past decade, as an alternative to the well-known cross-coupling reactions involving organometallics. The C–H activation catalyzed by Pd for the creation of a C–C bond has emerged as a promising field for new organic transformation, and readers are referred to excellent reports on that subject[6] (Scheme 1).

Although several C–H arylations of arenes and heterocycles have been achieved using transition metals (e.g., Rh, Ru, or Pd), their applications to natural nucleobases and azole analogues are quite limited. However, from a chemical point of view it has been reported that directed C–H activation is favored by α-chelating a heteroatom that is conjugated to the aryl rings; in fact, many recent examples

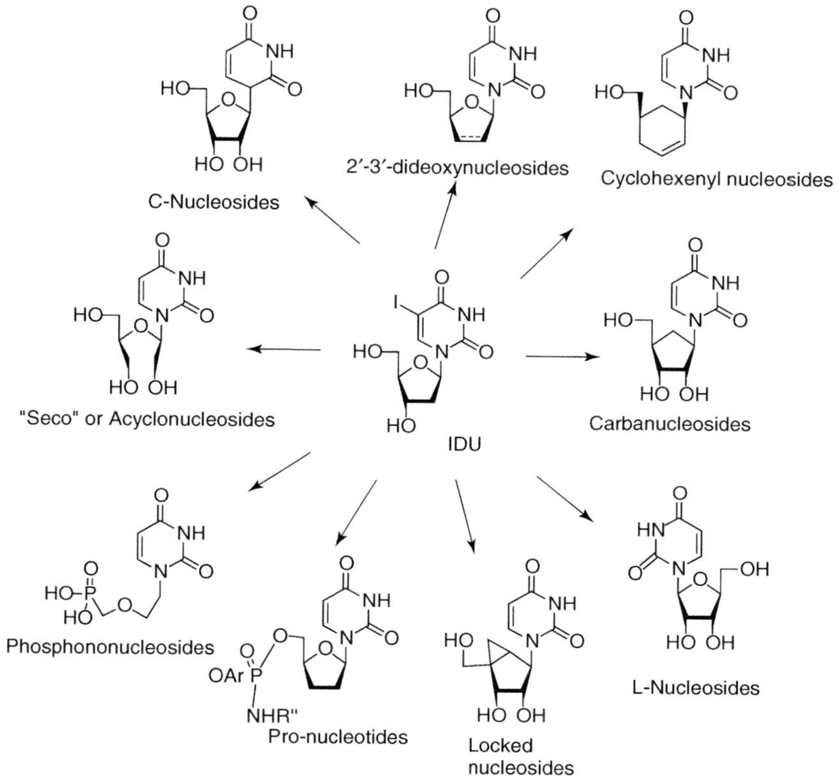

Figure 2. Various families of bioactive nucleosides.

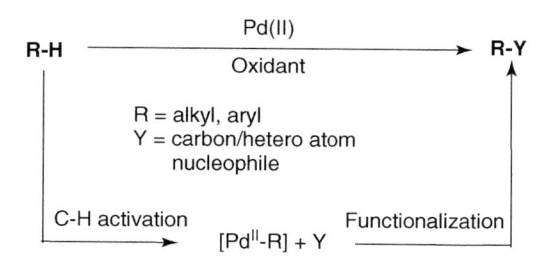

Scheme 1. Palladium(II)-catalyzed functionalization of C–H bonds. (From ref. 5a.)

of pyridine- and carbonyl-directed arylation of sp^2 and sp^3 C–H bonds have been reported (Figure 3).

Thus, the nitrogen atoms of purines, pyrimidines, imidazoles, or other azole moieties can direct C–H bond activation, drawing the catalyst proximal to the reaction center. So in the following we summarize the most recent development in

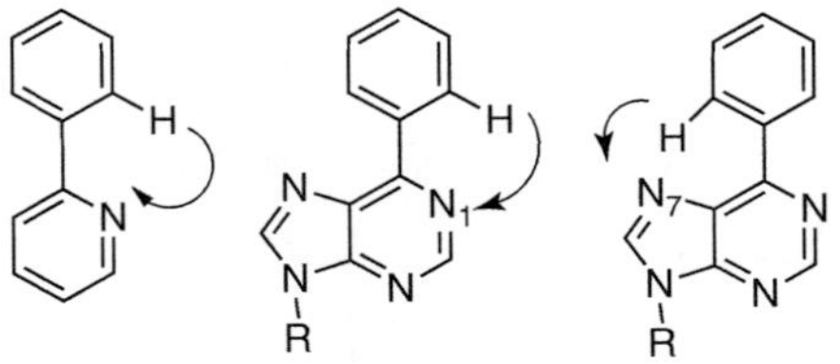

Figure 3. N-directed C–H bond activation in 2-phenylpyridines.

C–H activation reaction in the context of nucleobase/nucleoside functionalization. In 2002, Ellman et al.[7] proposed the first C–H functionalization of a purine moiety **5** by a Rh-catalyzed alkylation with 3,3-dimethylbut-1-ene (**6**) in the presence of a phosphine ligand and a Lewis acid to afford compound **7** in 89% yield (Scheme 2). Once the possibility of functionalizing imidazole/benzimidazole C–H bonds was established, alternative methodologies have been proposed.

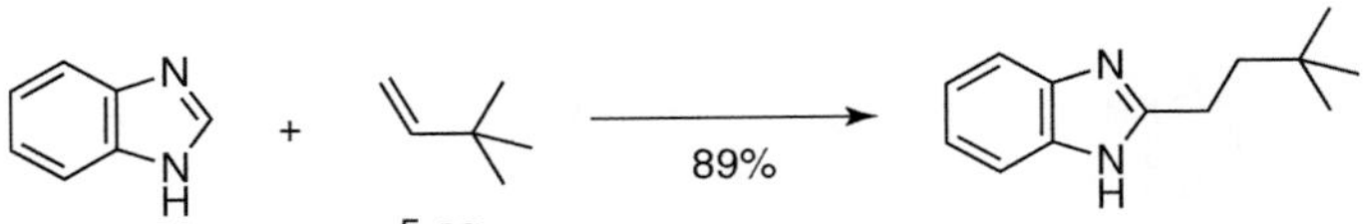

Scheme 2. *Reagents and conditions:* [RhCl(coe)$_2$]$_2$ (5 mol%), PCy$_3$ (7.5 mol%), MgBr$_2$ (5 mol%), THF, 150°C, 89%.

Later, Viel et al.[8] described a regioselective preparation of 1,5-diaryl-1*H*-imidazoles by palladium-catalyzed direct C5 arylation of *N*-arylated precursor **8** using 4-iodoanisole (**9**) (Scheme 3).

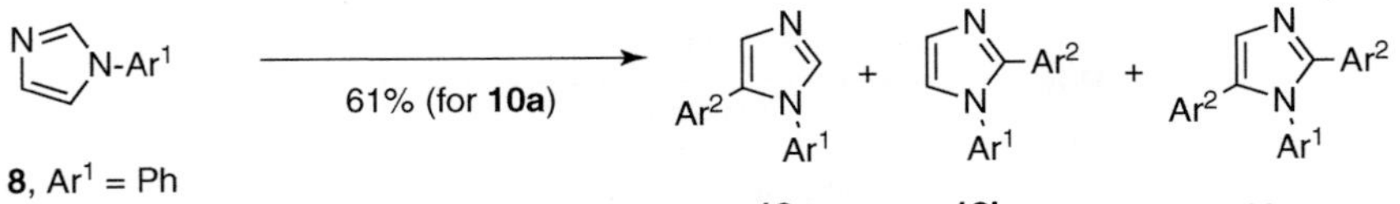

8, Ar^1 = Ph
9, Ar^2 = 4-MeOPh

Scheme 3. *Reagents and conditions:* Ar^2-I, Pd(OAc)$_2$ (5 mol%), AsPh$_3$ (10 mol%), CsF (2.0 equiv), DMF, 140°C, 61% (for **10a**).

The C5-arylated product **10a** was obtained in 61% yield along with undesired C2 arylation **10b** and C2–C5-diarylation product **10c**. Otherwise, slight modification of the previous conditions by adding copper iodide to the reaction made it possible to obtain C2 arylation product **12** exclusively without the need of a ligand or base in 76% yield[9] (Scheme 4). This procedure was also extended to benzimidazole ring **14** with an 81% yield.

Later in 2006, Hocek et al.[10] described the first palladium-catalyzed C8-arylation reaction of purine. Hence, coupling of N^9-benzylated purine **15** with *p*-tolyliodide

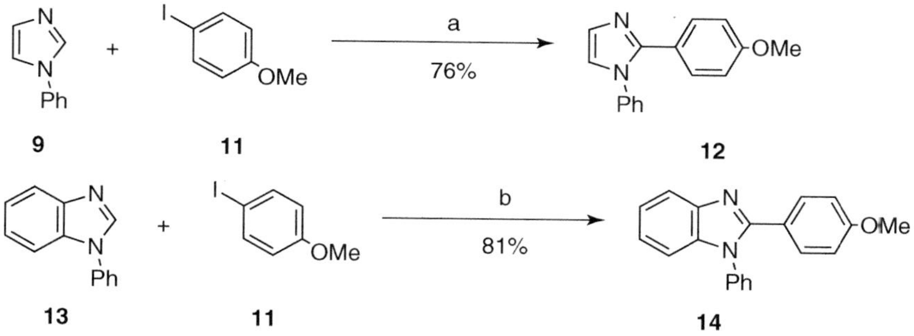

Scheme 4. *Reagents and conditions:* (a) Pd(OAc)$_2$ (5 mol%), CuI (2.0 equiv), DMF, 140°C, 26 h, 76%; (b) Pd(OAc)$_2$ (5 mol%), CuI (2.0 equiv), DMF, 140°C, 48 h, 81%.

under optimized conditions using palladium acetate (5 mol%), CuI (3 equiv), and cesium carbonate (2.5 equiv) gave the desired C8-arylated product **16** in 95% yield (Scheme 5). Interestingly, when the reaction was conducted under air, formation of dimer 8,8′-bispurine and arylation to the *ortho*-position of the phenyl group at position 6 was observed.

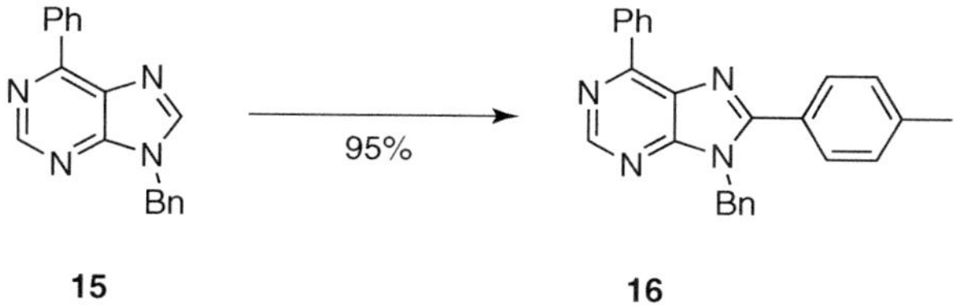

Scheme 5. *Reagents and conditions:* 4-iodotoluene (2 equiv), Pd(OAc)$_2$ (5 mol%), Cs$_2$CO$_3$ (2.5 equiv), CuI (3.0 equiv), DMF, 160°C, 60 h, 95%.

Moreover, this optimized procedure has also been applied to the regioselective synthesis of 2,6,8-trisubstituted purines **18a–c** bearing three different C-substituents following a consecutive three-step palladium-catalyzed pathway (Scheme 6).

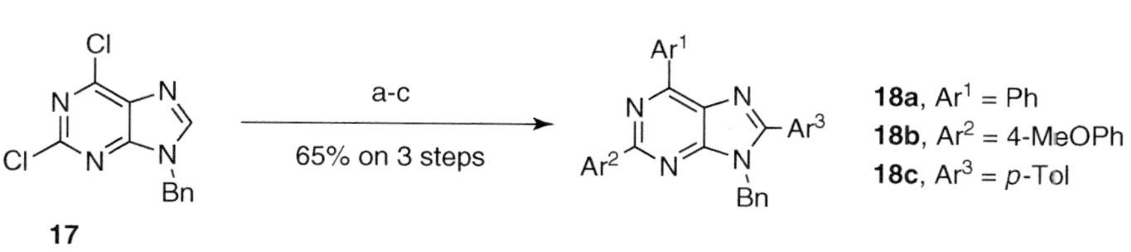

18a, Ar^1 = Ph
18b, Ar^2 = 4-MeOPh
18c, Ar^3 = *p*-Tol

Scheme 6. *Reagents and conditions:* (a) Ar^1-B(OH)$_2$, Pd(Ph$_3$P)$_4$; (b) Ar^2-B(OH)$_2$, Pd(Ph$_3$P)$_4$; (c) Ar^3-I, Pd(OAc)$_2$, CuI, 65% in three steps.

Soon after, Alami and co-workers described the direct C8-arylation of free-(NH$_2$)-adenines catalyzed by Pd(OH)$_2$/C (Pearlman's catalyst) with a stoichiometric amount of copper iodide under ligandless conditions.[11] Using optimized conditions, the coupling between N^9-benzylated purine **19** with 4-iodoanisole under microwave irradiation at 160°C for 15 min gave the expected purine derivative **20** in 90% yield (Scheme 7).

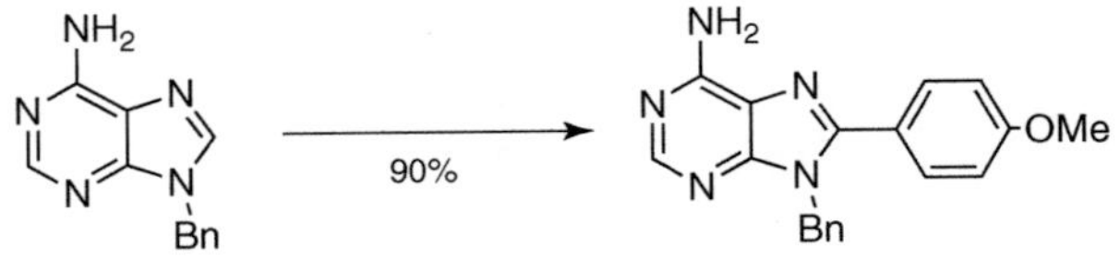

Scheme 7. *Reagents and conditions:* 4-iodoanisole (1.5 equiv), Pd(OH)$_2$/C (5 mol%), CuI (1 equiv), Cs$_2$CO$_3$ (2 equiv), NMP, 160°C, 0.25 h, MW, 90%.

Also, C–H arylation of free-NH$_2$ adenines[12] and caffein[13] (Scheme 8) using less reactive aryl chloride was reported to be an effective coupling partner.

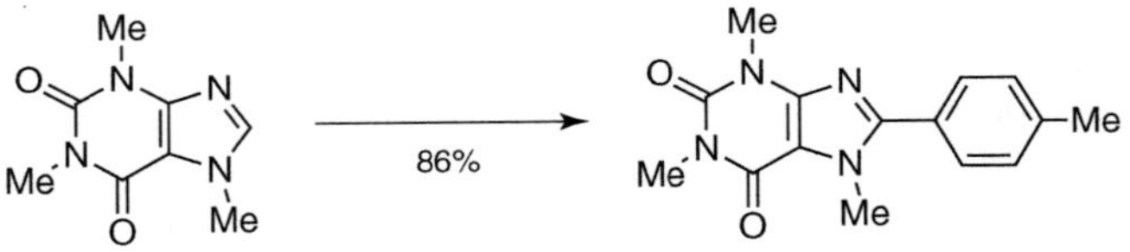

Scheme 8. *Reagents and conditions:* 4-chlorotoluene (1.5 equiv), Pd(OAc)$_2$ (5 mol%), *n*-BuAd$_2$P (10 mol%), K$_3$PO$_4$ (2 equiv), NMP, 125°C, 86%.

Some examples of direct C8 alkenylation of purines were also reported by Alami and co-workers.[14] Thus, Pd(acac)$_2$-catalyzed coupling of **21** or **19** with β-(*E*)-bromostyrene afforded alkenylated products **23** and **24**, respectively, in 86% and 66% yield as unique *E*-isomers (Scheme 9).

Subsequently, Piguel and co-workers proposed an alternative procedure for direct C8-alkenylation of 6-benzylsulfanyl-9-benzylpurine **25**.[15] Following the alkenylation conditions depicted in Scheme 10, the authors obtained the desired alkenylated purine **26** in better yield at a lower temperature (74%, 120°C). Then the resulting purine derivative was readily converted in two steps to N^6,N^6-disubstituted adenines **27** by oxidation of the sulfanyl group followed by an SNAr with appropriate amines.

In 2010, Hocek and co-workers[16] published efficient intramolecular direct C–H arylation reactions for the synthesis of fused purines such as purino[8,9-*f*]phenanthridine **30** and 5,6-dihydropurino[8,9-*a*]isoquinoline **34** (Scheme 11). Whereas a double C–H arylation pathway involving purine **28** and diiodobenzene **29** gave a low yield (35%) of **30**, the alternative tandem Suzuki/C–H arylation process in **31** with boronic acid **32** afforded the fused purine **30** in 82%

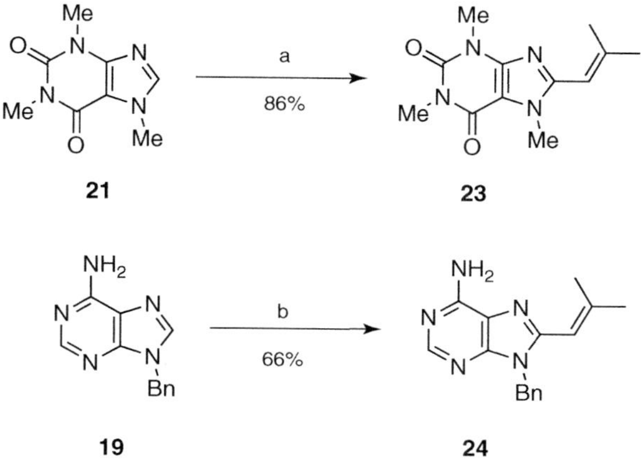

Scheme 9. *Reagents and conditions:* (a) β-(E)-bromostyrene (1.2 equiv), Pd(acac)$_2$ (2.5 mol%), CuI (20 mol%), P(o-tolyl)$_3$ (20 mol%), t-BuOLi (2 equiv), THF, 130°C, 2 h, 86%; (b) β-(E)-bromostyrene (1.2 equiv), Pd(acac)$_2$ (2.5 mol%), CuI (1 equiv), P(o-tclyl)$_3$ (5 mol%), t-BuOLi (2 equiv), NMP, 130°C, 2 h, 66%.

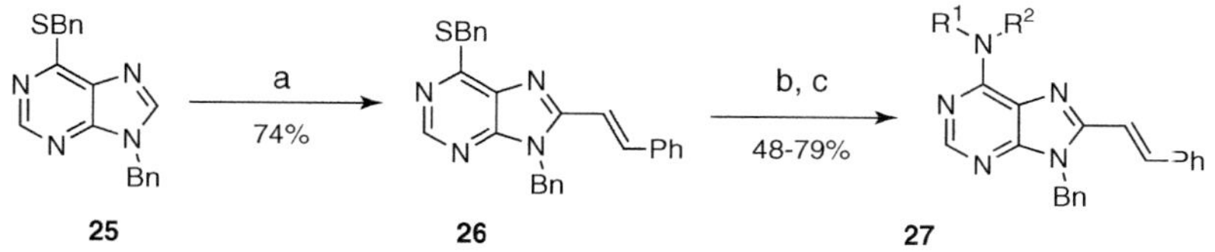

Scheme 10. *Reagents and conditions:* (a) E-styryl bromide (2 equiv), Pd(OAc)$_2$ (5 mol%), CuI (10 mol%), phenanthroline (20 mol%), t-BuOLi (2 equiv), 1,4-dioxane, 120°C, 30 min, MW, 74%; (b) oxidation; (c) R^{1}R^{2}NH, 48–79% in two steps.

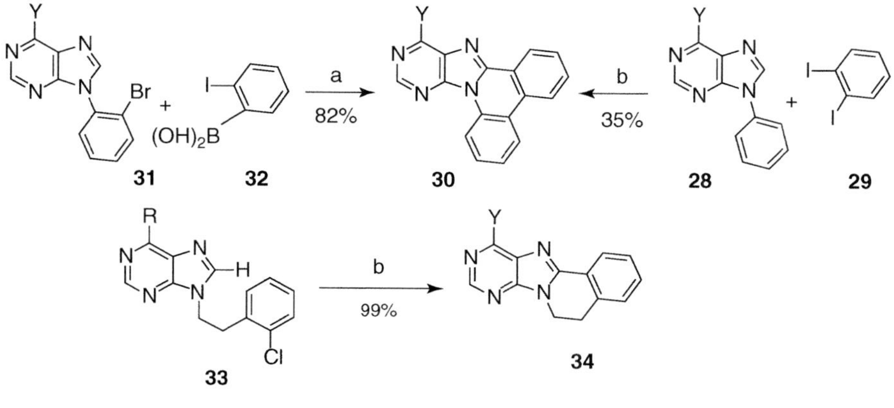

Scheme 11. *Reagents and conditions:* (a) Pd(OAc)$_2$ (5 mol%), TBAB (1 equiv), KOAc (4 equiv), DMF, 140°C, 25 h, 35%; (b) Pd(OAc)$_2$ (5 mol%), P(Cy)$_3$·HBF$_4$ (10 mol%), K$_2$CO$_3$ (2.5 equiv), DMF, 150°C, 17–20 h, 82% (for 30) and 99% (for **34**).

over two steps. 5,6-Dihydropurino[8,9-a]isoquinoline **34** was also obtained by intramolecular direct C–H arylation using N-alkylated precursor **33** in 99% yield.

A combinatorial approach has been described by Vaňková et al.[17] for direct C8-arylation of purines immobilized on Wang resin (Scheme 12). Thus, the reaction of supported 2,6,8-trisubstituted purines **35** with various aryl halides using a classical palladium acetate–copper iodide–piperidine catalytic system in anhydrous DMF followed by resin cleavage yielded up to 16 different products **36** in 7 to 70% yield. Note that a lower yield was observed when the reaction was conducted on free N^9.

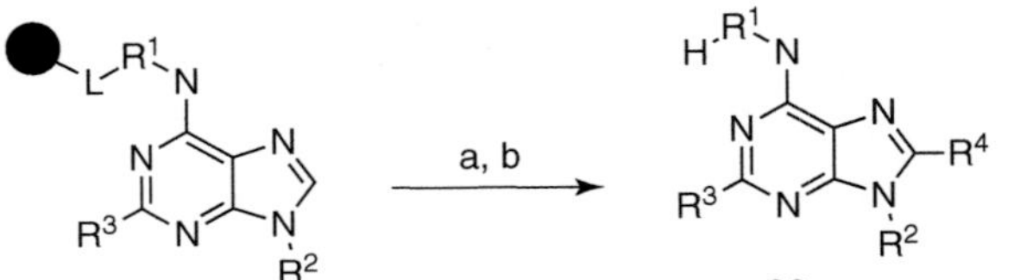

Scheme 12. *Reagents and conditions:* (a) ArX, CuI, Pd(OAc)$_2$, piperidine, anh. DMF, 115–135°C, 24 h; (b) 50% TFA, DCM, rt, 1 h.

More recently, Guo et al.[18] proposed the first nickel-catalyzed direct C8-arylation of purine **37** using Grignard reagent **38**. While employing palladium complexes as Pd(OAc)$_2$ or Pd(dppf)Cl$_2$ gave only trace amounts of the arylated purine **39** desired, use of Ni(dppp)Cl$_2$ together with DCE as an oxidant afforded **39** in an 89% yield (Scheme 13).

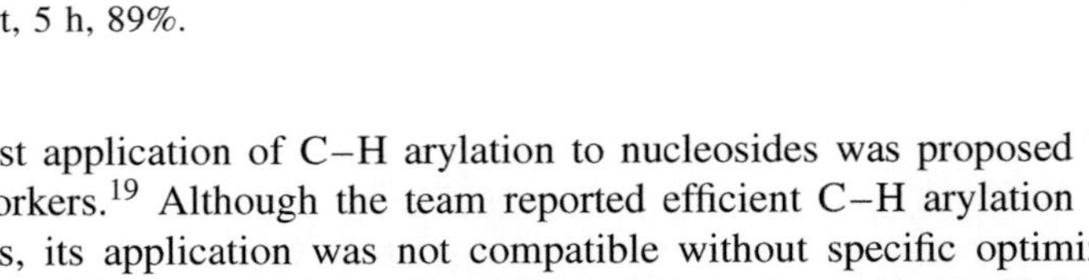

Scheme 13. *Reagents and conditions:* Ni(dppp)Cl$_2$ (30 mol%), 1,2-dichloroethane (3 equiv), N$_2$, THF, rt, 5 h, 89%.

The first application of C–H arylation to nucleosides was proposed by Hocek and co-workers.[19] Although the team reported efficient C–H arylation on purine derivatives, its application was not compatible without specific optimization for labile nucleosides, due to the harsh conditions required to achieve efficient conversions. Thus, adenosine (**40**) was subjected to direct C8-arylation with *p*-tolyliodide using palladium acetate, copper iodide, and piperidine in DMF at 150°C to afford C8-arylated adenine **41** in 68% yield (Scheme 14). The key feature of this reaction was the use of piperidine. In fact, the authors supposed that the dimethylamine generated by long-term heating of DMF could influence the rate of the reaction.

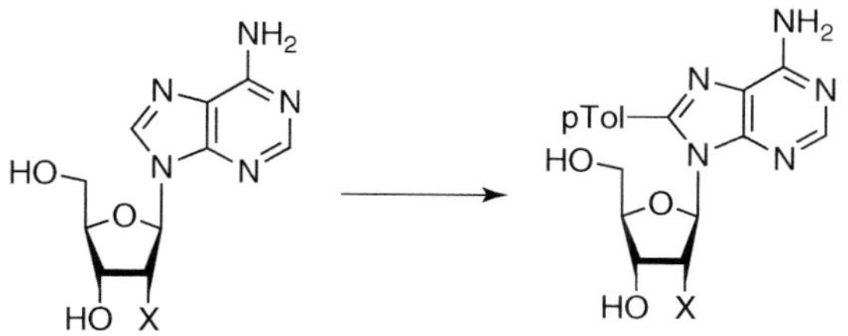

Scheme 14. *Reagents and conditions:* 4-iodotoluene, Pd(OAc)$_2$, CuI, piperidine, DMF, 150°C, 5h.

Thus, the addition of piperidine notably accelerated the conversion, which allowed the reaction temperature to be reduced. Finally, the reaction was also conducted on labile 2′-deoxyadenosine using a lower reaction temperature (125°C) to afford corresponding arylated purine **42** in 31% yield. A common side product for both reactions was the *N*-6,8-diarylated nucleoside as the product of copper-catalyzed *N*-arylation.

An example of intramolecular C2-arylation of benzimidazole was reported by Barbaro et al.[20] As depicted on Scheme 15, the *N*-alkylated benzimidazole derivative **43** was involved in an intramolecular C2-arylation using copper iodide and LiO/*t*-Bu as a base without the need of a transition metal catalyst to afford, almost quantitatively, the expected derivative **44**.

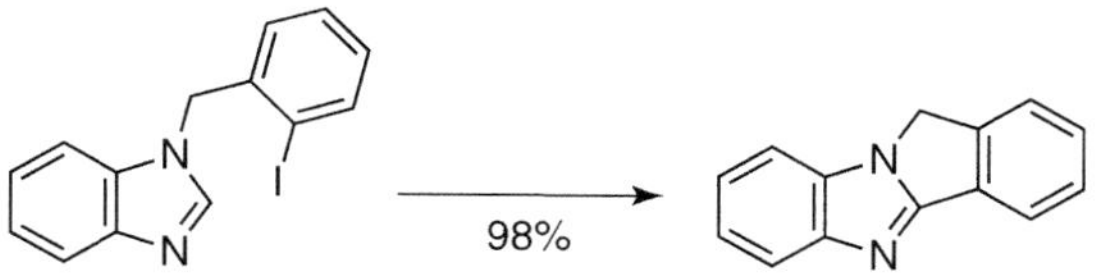

Scheme 15. *Reagents and conditions:* CuI (10 mol%), LiO/*t*-Bu (3 equiv), *o*-xylene, 150°C, 98%.

In 2009, Hocek and co-workers[21] proposed the first direct arylation on pyrimidines. Following three different procedures, authors were able to arylate position C5 or C6 more or less efficiently using iodobenzene (and various substituted aromatic halides) as a coupling partner. Thus, when the reaction was performed on **45** in the presence of Pd(OAc)$_2$, P(perFPh)$_3$, and cesium carbonate, C5-arylated pyrimidine **46** was obtained in 53% yield as a major product along with 9% of C6-arylated derivative **47** (Scheme 16). Interestingly, the addition of copper iodide switched the regioselectivity at position C6 to give **47** in 73% yield along with 5% of **46**. Finally, removal of the phoshine ligand gave C6-arylation exclusively but with decreased yield (35%).

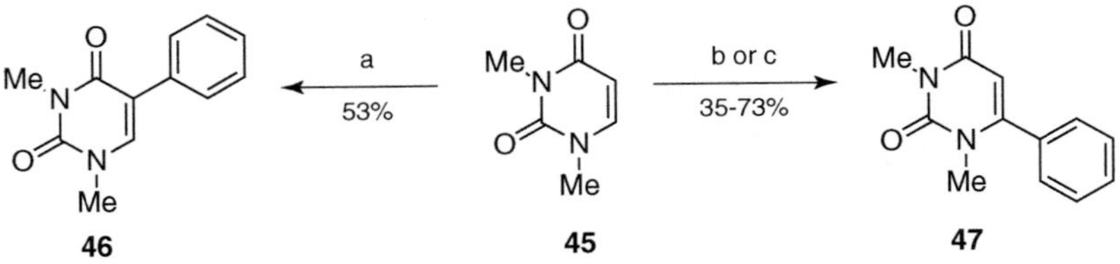

Scheme 16. *Reagents and conditions:* (a) iodobenzene (2 equiv), Pd(OAc)$_2$ (5 mol%), P(perFPh)$_3$ (10 mol%), Cs$_2$CO$_3$ (2.5 equiv), DMF, 160°C, 50 h, 53%; (b) iodobenzene (2 equiv), Pd(OAc)$_2$ (5 mol%), P(perFPh)$_3$ (10 mol%), CuI (3 equiv), DMF, 160°C, 50 h, 73%; (c) iodobenzene (2 equiv), Pd(OAc)$_2$ (5 mol%), CuI (3 equiv), DMF, 160°C, 50 h, 35%.

Soon after, Kim and co-workers[22] proposed alternative procedures for direct arylation of 1,3-dimethyluracil **45**. These procedures exhibited improved regioselectivity and better yields in addition to compatibility with aryl halide bearing an electron-withdrawing substituent. Thus, the coupling of 1,3-dimethyluracil with bromobenzene catalyzed by Pd(OAc)$_2$ under the presence of Ph$_3$P, K$_2$CO$_3$, and PivOH gave C5-arylated pyrimidine derivative **46** in 79% yield along with 10% of C6-arylation product **47** (Scheme 17). The authors assumed that use of a relatively poorly coordinating carboxylate as the counterion could allow the dissociation of ArPdBr into more electrophilic arylpalladium species ArPd$^+$[OCOR]$^-$.[23] Finally, when the reaction was conducted in benzene at reflux using Pd(TFA)$_2$, Ag(OAc)$_2$, and PivOH, C6-arylation product **47** was obtained in 85% yield with a trace amount of C5-arylated derivative **46** (<6%).

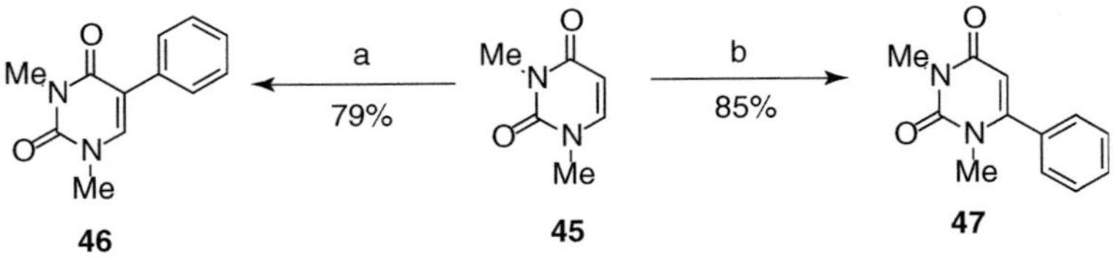

Scheme 17. *Reagents and conditions:* (a) bromobenzene, Pd(OAc)$_2$ (10 mol%), Ph$_3$P (3 equiv), K$_2$CO$_3$ (3 equiv), PivOH (30 mol%), DMF, 130°C, 12 h, 79%; (b) benzene (60 equiv), Pd(TFA)$_2$ (5 mol%), AgOAc (3 equiv), PivOH (6 equiv), reflux, 20 h, 85%.

In 2009, Chattopadhyay et al.[24] published the first palladium-catalyzed intramolecular C–H arylation of pyrimidines for the synthesis of new benzannulated pyridopyrimidines. Hence, the C5-functionalized 1,3-dimethylpyrimidine **48** was involved in a C–H arylation reaction at position C6 using palladium acetate, tetrabutylammonium bromide, and potassium acetate in DMF at 140°C to give annulated pyridopyrimidine **49** in 91% yield. The same procedure was applied to perform arylation at position C5 using C6-functionalized 1,3-dimethylpyrimidine **50** to afford C5-arylated derivative **51** in 95% yield (Scheme 18).

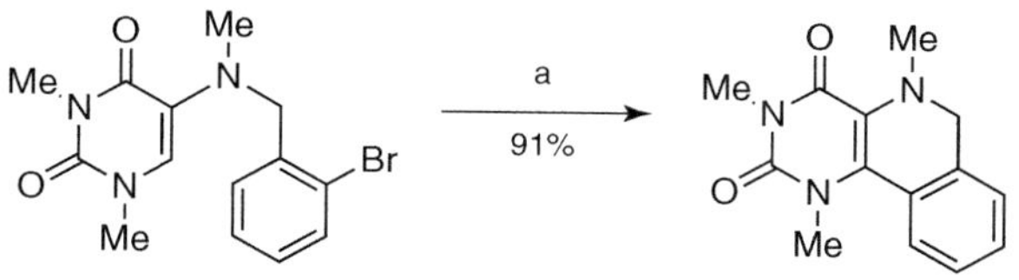

Scheme 18. *Reagents and conditions:* (a) Pd(OAc)$_2$ (10 mol%), KOAc, TBAB, DMF, 140°C, 24 h, 91% (for **49**) and 95% (for **51**).

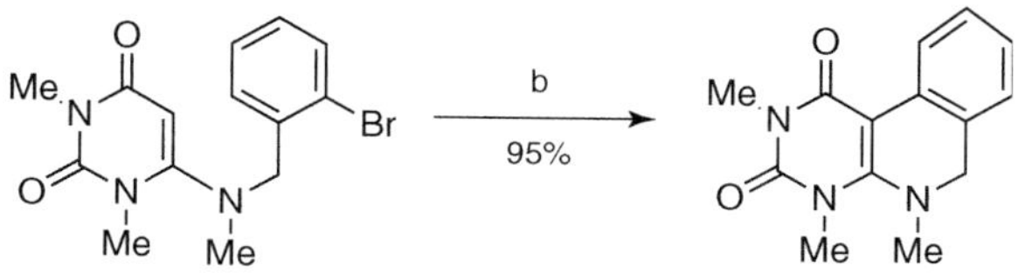

Direct alkynylation of some oxazole derivatives via a C–H activation reaction was also reported by Piguel and co-workers.[25] As depicted in Scheme 19, the coupling between 5-phenyloxazole **52** and bromophenylacetylene **53** under optimized conditions with CuBr·SMe$_2$, DPEPhos, and LiO/t-Bu in dioxane at 120°C rapidly afforded the alkynylated oxazole derivative **54** in 89% yield.

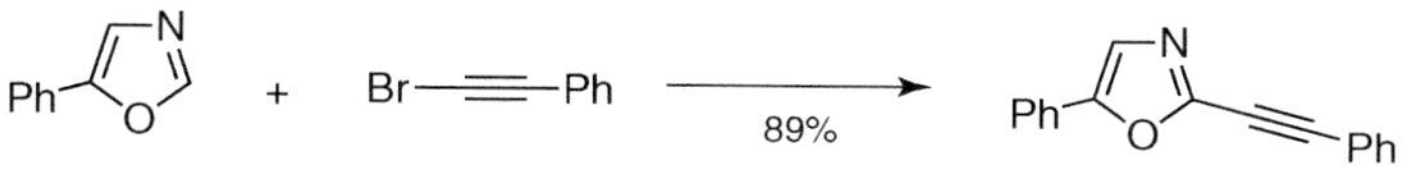

Scheme 19. *Reagents and conditions:* CuBr·SMe$_2$ (15 mol%), DPEPhos (15 mol%), LiO/t-Bu (2 equiv), 1,4-dioxane, 120°C, 1 h, 89%.

Furthermore, authors described another approach to alkynylation azoles using 1,1-dibromoalkenes as coupling partners under similar reaction conditions.[26] Notably, Bessellèvre Piguel and used this modified procedure for direct alkynylation of N-benzyl-6-chloropurine **55** to afford the desired alkynylated purine derivative **57** in 35% yield (Scheme 20). Alternatively, some nickel/copper-catalyzed direct alkenylation of similar azole derivatives was reported by Miura and co-workers.[27]

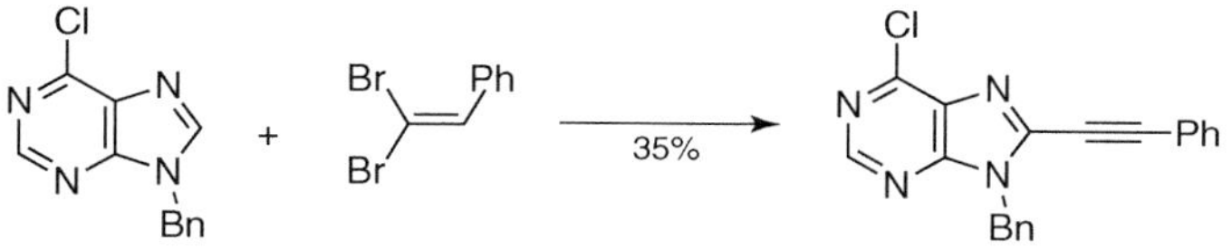

Scheme 20. *Reagents and conditions:* CuBr.SMe$_2$ (5 mol%), DPEPhos (10 mol%), LiO/t-Bu (6 equiv), 1,4-dioxane, 120°C, 2 h, 35%.

An example of efficient direct C2-arylation of indole was described by Lebrasseur and Larrosa.[28] In fact, the coupling at room temperature of N-metylindole **58** and iodobenzene **59** catalyzed by Pd(OAc)$_2$ gave the arylated indole derivative **60** in 92% yield (Scheme 21). Interestingly, the use of silver carboxylate, generated in situ by the use of Ag$_2$O and *ortho*-nitrobenzoic acid, was found to be essential. Actually, silver(I) salts are generally used to abstract halide anions from transition metal complexes to increase their electrophilicity.[29]

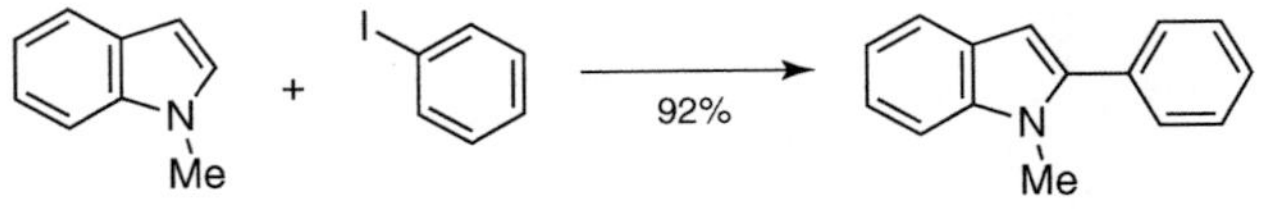

Scheme 21. *Reagents and conditions:* Pd(OAc)$_2$ (5 mol%), Ag$_2$O (0.75 equiv), *o*-nitrobenzoic acid, DMF, 25°C, 15 h, 92%.

Furthermore, Waser et al. reported the direct C3-alkynylation of indole and pyrrole heterocycles.[30] Thus, as depicted on Scheme 22, the gold chloride–catalyzed coupling between NH-free indole **61** and benziodoxolone-derived reagent **62** gave the C3-alkynylated indole **63** in 86% yield, whereas both palladium and copper catalysis were ineffective.

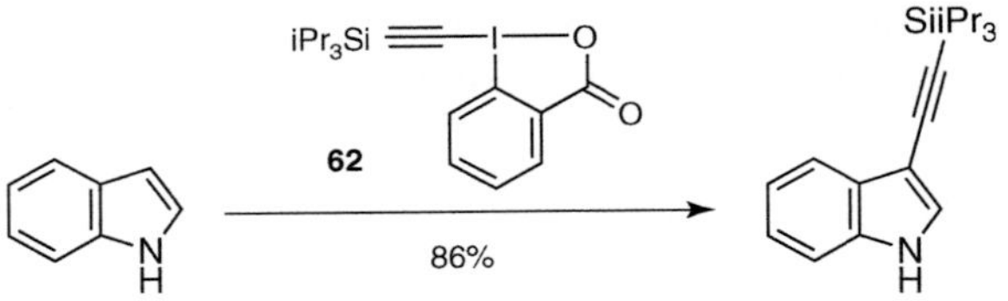

Scheme 22. *Reagents and conditions:* AuCl (5 mol%), Et$_2$O, 23°C, 12 h, 86%.

Finally, recently, Lakshman et al. reported the direct arylation of 6-phenylpurine and 6-arylpurine nucleosides by ruthenium-catalyzed C–H bond activation.[31] As discussed previously, the nitrogens of purine itself can direct the C–H bond activation. Starting from more complex 2′-deoxyribonucleoside substrates, various 6-aryl purine nucleosides (**64**) were submitted to C–H bond activation and arylation with aryl iodides and aryl bromides, with 10 mol% of the Ru catalyst (Scheme 23). With either aryl iodides or aryl bromides, mono (**65**) and diarylated (**66**) products were obtained. A possible mechanism has been proposed involving the purinyl N^1 atom, and the monoarylated products could reenter the catalytic cycle, resulting in diarylated compounds. The exact role of purinyl N^7 has yet to be determined.

3. METALLO-CATALYZED SYNTHESES AND MODIFICATIONS OF TRIAZOLO NUCLEOSIDES

Based on the discovery of U.S. Food and Drug Administration (FDA)–approved ribavirin (**1**) and the anti-HIV-1 reverse transcriptase inhibitor TSAO (**67**), the

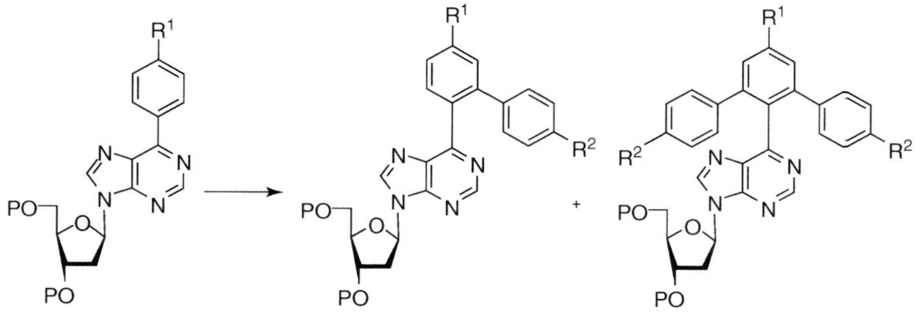

64 **65** **66**

R^1 = H, F, OPh, OMe
R^2 = Me, MeO, Me(C=O), H
P = H or protecting group

Scheme 23. *Reagents and conditions:* aryl halides (2 equiv), $[RuCl_2(benzene)]_2$ (5 mol%), Ph_3P (40 mol%), K_2CO_3 (3 equiv), NMP, 120°C.

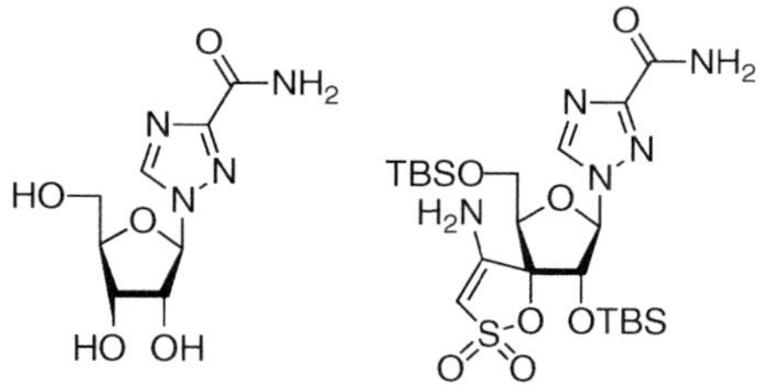

1, Ribavirin **67**, TSAO

Figure 4. Chemical structures of ribavirin and TSAO.

five-membered triazole moiety attracted increased interest in the development of nucleoside analogues (Figure 4). In recent years, metallo-catalyzed reactions have been used as efficient and versatile routes to improve the emergence of a functionalized triazole moiety, including cross-coupling or regioselective organic reactions.

3.1. Pd(0)-Catalyzed Functionalization of a 1,2,4-Triazole Ring

3.1.1. Under Suzuki–Miyaura Conditions. Novel substituted 1,2,4-triazolyls in acyclo- and ribonucleoside series were developed by Peng's team for their antiviral or anticancer properties. Various modifications of the triazol ring at the C5- or C3-position via Suzuki, Sonogashira, or Huisgen reactions were used. Wan et al.[32] optimized the Suzuki coupling reaction between 5-bromotriazole nucleoside **68** and various aromatic groups using $Pd(Ph_3P)_4$ as a catalyst in the presence of K_2CO_3 under conventional heating and microwave irradiation conditions, which increase the yields of **69** and reduce the reaction time (Scheme 24).

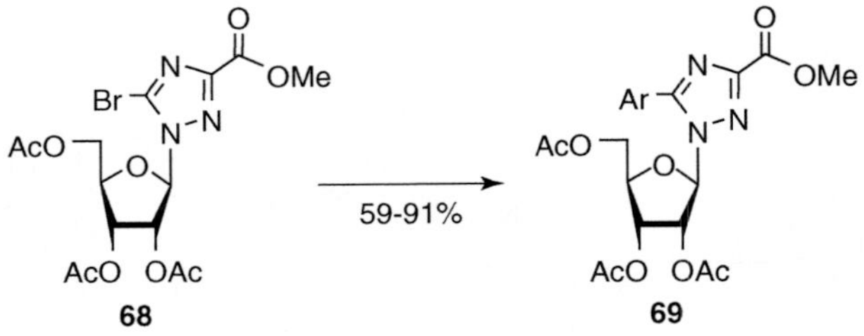

Scheme 24. *Reagents and conditions:* Ar-B(OH)$_2$, Pd(Ph$_3$P)$_4$, K$_2$CO$_3$, toluene, MW, 150°C, 15 min, 59–91%.

3.1.2. Under Sonogashira Conditions.

Following the same approach, this team also investigated the synthesis of modified nucleosides bearing an ethynyl moiety through a Pd(0) Sonogashira coupling reaction at the C3- or C5-position[33] of acyclic[34] and ribo[35] nucleosides. Their investigations were promoted by microwave irradiation to afford the desired nucleoside analogues **71** and **73**, respectively, in good to excellent yields (Scheme 25).

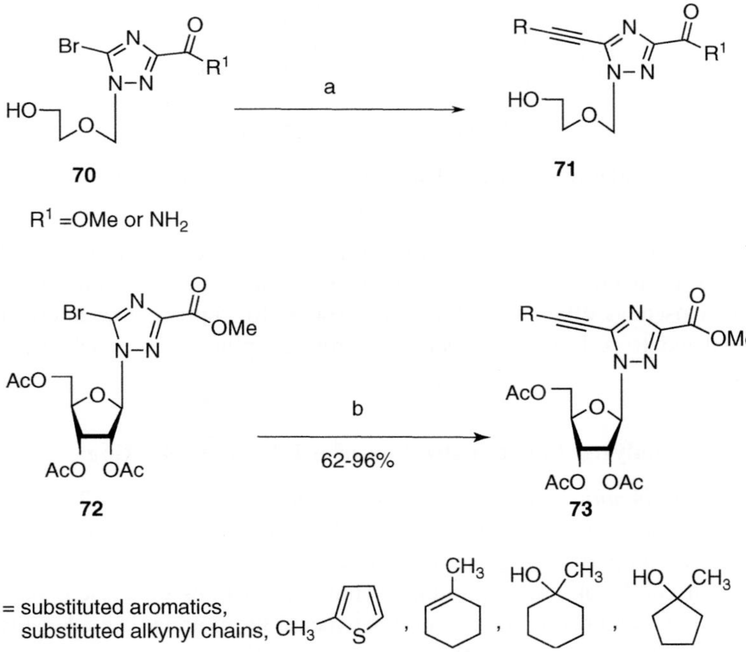

R^1 = OMe or NH$_2$

R = substituted aromatics, substituted alkynyl chains, CH$_3$

Scheme 25. *Reagents and conditions:* (a) alkyne, Pd(Ph$_3$P)$_4$, CuI, Li$_2$CO$_3$, dioxane/H$_2$O (3:1), MW, 100°C, 25 min; (b) alkyne, Pd(Ph$_3$P)$_4$, CuI, Et$_3$N, CH$_3$CN, MW, 100°C, 30 min.

3.2. Copper/Ruthenium-Catalyzed Azide–Alkyne Cycloadditions to a 1,2,3-Triazole Ring

Recent advances in the powerful linking Huisgen azide–alkyne cycloaddition reaction[36] through the use of copper (CuAAC)[37] or ruthenium (RuAAC),[38] which catalyzed mainly 1,4 or 1,5 regioisomer formation, respectively, followed the mechanism proposed by Fokin et al.[39] (Scheme 26). The formation of 1,4-substituted-1,2,3-triazoles at the 1'-position of the sugar moiety have attracted a significant interest, as reported recently by Amblard et al.[40]

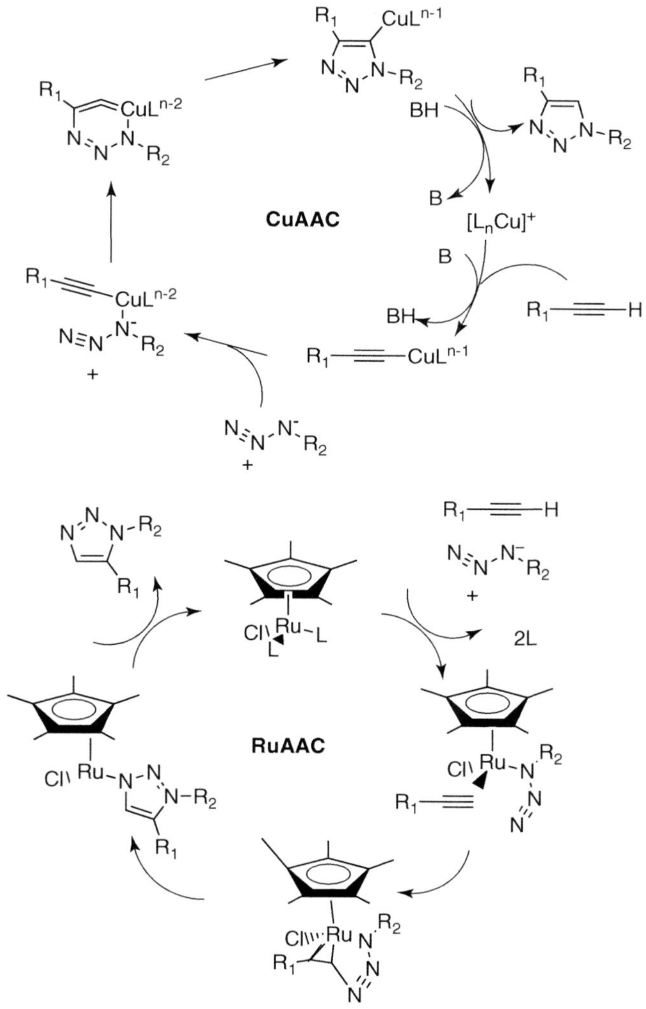

Scheme 26. (From Ref. 39a.)

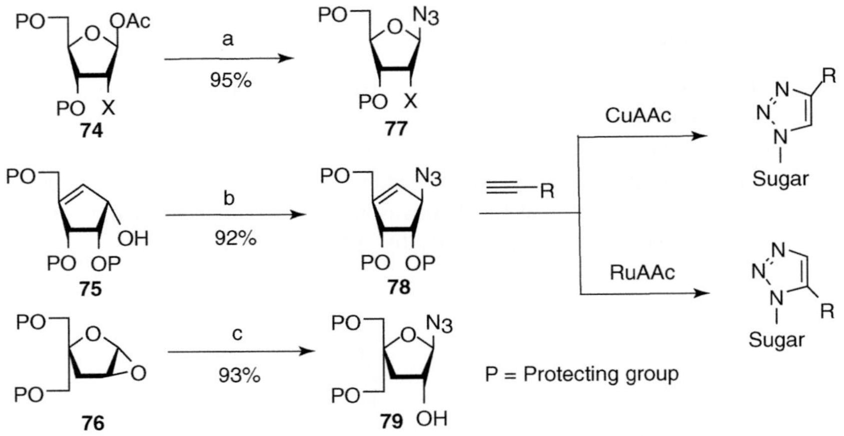

Scheme 27. *Reagents and conditions:* (a) TMSN$_3$, BF$_3$.Et$_2$O, CH$_2$Cl$_2$, 95%; (b) (i) MsCl, Et$_3$N, CH$_2$Cl$_2$, 0°C, 2 h; (ii) NaN$_3$, DMF, 80°C, 24 h, 92%; (c) NaN$_3$, MeOH/H$_2$O, reflux, 4 h, 93%.

3.2.1. Cycloaddition from a Corresponding Azido Sugar.

Huisgen's azide–alkyne cycloaddition required, first, the functionalization of a sugar moiety at the anomeric position by an azido or alkyne group. Several teams have chosen to start from a substituted 1-azido sugar moiety, such as ribofuranosyl **74** or the cyclopentenyl analogues **75** or **76**,[41] which were synthesized as shown in Scheme 27.

Beginning in 2006, Benhida's group has investigated the azide–alkyne cycloaddition of both the azido-2-deoxyribose and -ribose series for their anticancer properties. Starting with compounds **80** and **81**, the reaction proceeds in 24 h with a stoichiometric amount of CuI and 87% yield, whereas only 5 min was required under microwave irradiation with only 5% catalyst loading on silica gel and a 95% yield (Scheme 28).[42] Thus, the influence of microwave activation was investigated and showed notable acceleration of the reaction with a lower catalyst loading. Given the efficiency of microwaves on Huisgen's 1,3-dipolar cycloaddition for protected 1′-azido-2′-deoxyribose, several teams have reported the synthesis of nucleoside analogues bearing a triazol moiety.

More recently and using the same type of substrate, Agrofoglio et al.[43] used both the CuAAC 1,4- and 1,5-regioselective RuAAc cycloaddition reactions developed in Sharpless's group to synthesize the 1,5-substituted-1,2,3-triazolo ribonucleoside analogues **85**. The microwave activation was particularly interesting because the reaction was not only faster (from 6 h to 5 min) but could be carried out under uncertain conditions (e.g., in the presence of water). During our investigation, we also observed that the RuAAC gave mainly 1,5-regioisomers, with, however, 3 to 7% 1,4-regioisomers (Scheme 29).

Starting from protected 1′-azidoribose and 2′-deoxyribose **86a,b**, Emolart'ev et al. have also reported that cycloaddition with 2-ethynylfuro[2,3-*b*]pyrazines

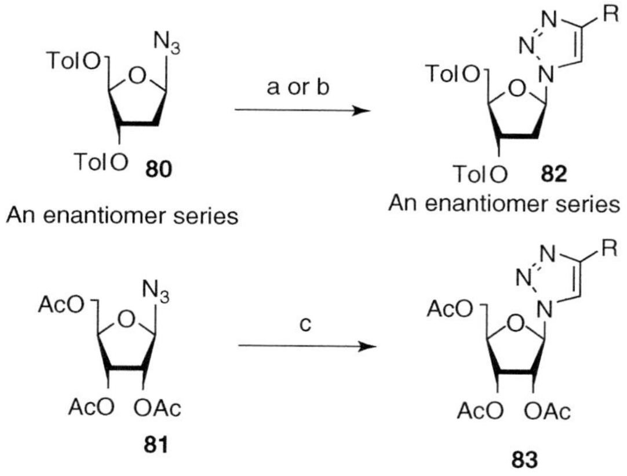

Scheme 28. *Reagents and conditions:* (a) alkyne, CuI, DIPEA, toluene, thermal heating, reflux, 24 h, 87%; (b) alkyne, CuI, DIPEA, SiO$_2$, MW (95–115°C), 1.5–3 min, 95%; (c) alkyne, CuI, DIEA, SiO$_2$, MW (95–115°C), 1–2 min, 85–95%.

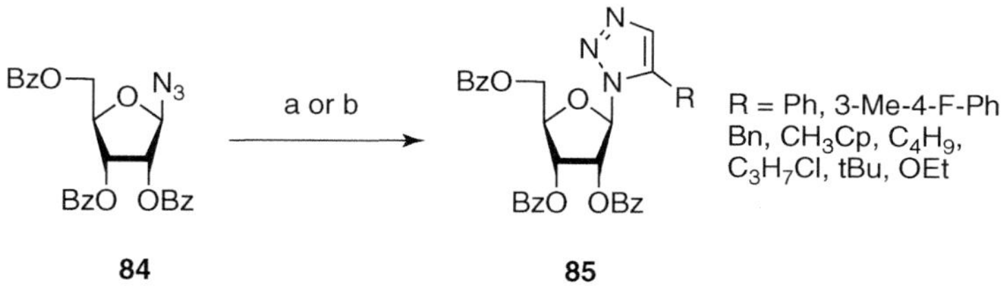

Scheme 29. *Reagents and conditions:* (a) alkyne, Cp*RuCl(Ph$_3$P)$_2$, THF, thermal heating, 50°C, 6 h, 54–83%; (b) alkyne, Cp*RuCl(Ph$_3$P)$_2$, THF, MW, 100°C, 5 min, 87–95%.

87 could be sped up dramatically under microwave irradiation to give **88** (Scheme 30).[44]

At the same time, two teams have developed a three-component route to obtaining 4,5-disubstituted triazolyl nucleosides **89** and **90**. They report an efficient method starting from 1-azido nucleoside **81** and alkyne in the presence of CuI/NBS[45] and proceeding to **89**, or CuI/I$_2$/cerium ammonium nitrate (CAN)[46] and *N*,*N*-diisopropylethylamine (DIPEA) in THF, which generate an in situ electrophile and Cu$^+$-catalyzed click reaction, to **90**. They extend their procedure to different electrophilic additions, opening the way for further C5-triazole substitutions **92** (Scheme 31).

Exploring new copper catalysts, Agrofoglio and co-workers have shown that Huisgen's 1,3-cycloaddition could be realized efficiently between azido carba-sugar **93** and different alkynes under microwave irradiation with other sources of copper(I) to afford **94** in good to excellent yield and a short reaction time, with organic complexes [Cu(CH$_3$CN)$_4$]PF$_6$ and IMesCuBr (Scheme 32).[47]

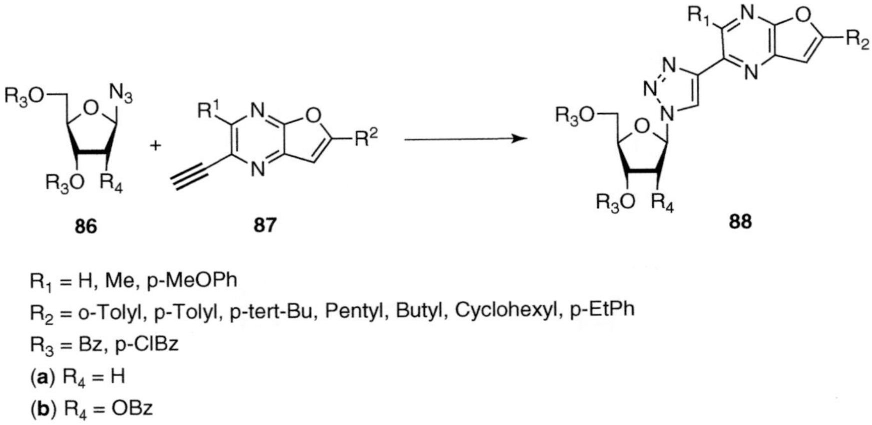

R_1 = H, Me, p-MeOPh
R_2 = o-Tolyl, p-Tolyl, p-tert-Bu, Pentyl, Butyl, Cyclohexyl, p-EtPh
R_3 = Bz, p-ClBz
(a) R_4 = H
(b) R_4 = OBz

Scheme 30. *Reagents and conditions:* Cu(0)/CuSO$_4$, amine ligand, THF/i-PrOH/water, MW, 90°C, 5–10 min, 54–91%.

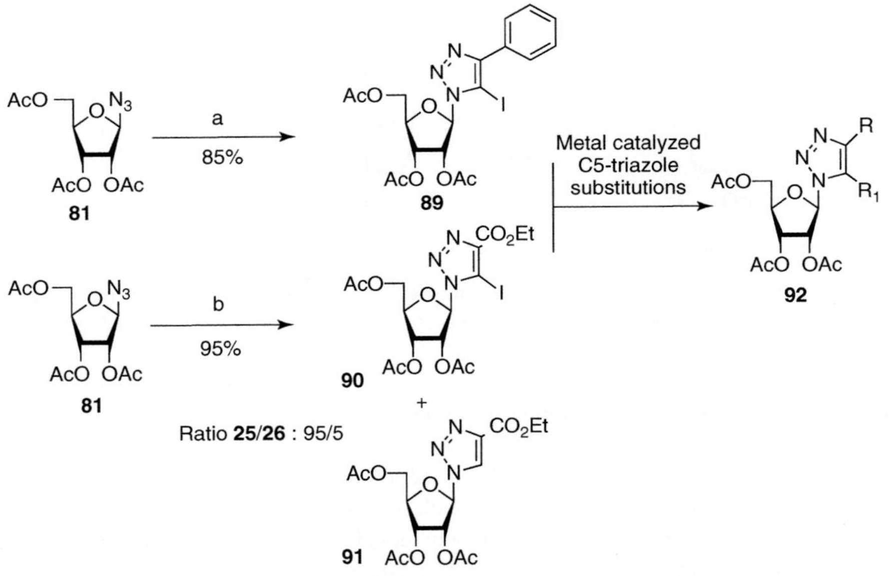

Scheme 31. *Reagents and conditions:* (a) phenylacetylene, CuI, NBS, DIPEA, THF, rt, 5 h, 85%; (b) ethyl propiolate, CuI, I$_2$, CAN, DIPEA, THF, rt, 5 h, 95%.

Novel ribavirin analogues were synthesized by Uziel et al.[48] starting from *C*-ribosylpropiolate **95** through Huisgen's 1,3-dipolar cycloaddition under catalysis. They have shown that the choice among four surfactants depends on the degree of substitution of the ribosylalkynes. However, the use of neutral surfactant (OGH) was effective in any degree of alkyne substitution. In the presence of OGH, they obtained a mixture of cyclized compounds **96** and **97** in 24 h in 67% to compare

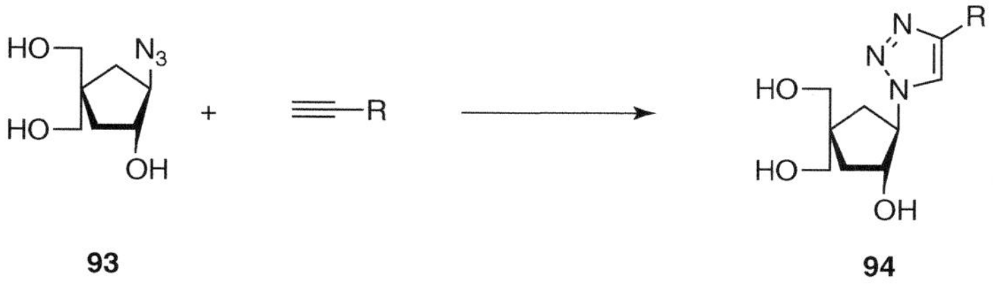

R	Cu(0)/CuSO$_4$	[Cu(CH$_3$CN)$_4$]PF$_6$	IMesCuBr
Ph	98%, 2 min	98%, 15 min	94%, 9 h
C$_5$H$_{11}$	98%, 1 h	60%, 10 h	93%, 2 h
CH$_2$OAc	95%, immediate	92%, 2 min	93%, 5 min
CH$_2$CH$_2$OH	95%, 2 min	92%, 2 h	96%, 15 min

Scheme 32. *Reagents and conditions:* copper catalyst, *t*-BuOH/H$_2$O, MW, 125°C.

with the use of SDS, which provides 58%. Ribavirin analogues **98** are obtained following a conventional deprotection procedure (Scheme 33).

3.2.2. Base-Modified Nucleoside-Bearing Triazolyl Unit.
In the part of their program targeting the discovery of a molecule showing antiviral activity against tobacco mosaic virus, Li et al. developed a series of bitriazolyl nucleosides. 3- and 5-Azidotriazole nucleosides (acyclic and ribonucleoside) **99** were subjected to the Cu(I)-catalyzed Huisgen reaction in the presence of various substituted alkynes.[49] In the case of cycloaddition reaction with 3-azidotriazole nucleoside, the desired compounds **100** were obtained in good yield (Scheme 34).

However, when starting from the electron-deficient 5-azidotriazole nucleoside **101**, two major unexpected derivatives, the amine **102** and the amide **104** (in the ribofuranosyl series), were isolated under mild Cu(II)–ascorbate reaction conditions. They observed that conventional reaction conditions (CuSO$_4$/Na ascorbate) could reduce azido compounds to their respective amines. This usually "clean" reaction, known to yield a single product, leads to the formation of numerous side products, which cause some difficulty in purifying the desired acyclic nucleosides **103** (Scheme 35).

Some analogues of 2′-deoxyuridine bearing triazole moiety at the C5-position were synthesized by several teams for different applications. Park et al. described some hydrogel properties of this type of structure.[50] Nielsen et al. reported the effect of stacking triazole and aromatic groups in the major grooves of nucleic acid duplex stability.[51] Two other teams have obtained small libraries of nucleosides with a 1,2,3-triazole scaffold for their anticancer[52] or antiviral activity. Agrofoglio et al.[53] described the synthesis of two series of C5-(1,4 and 1,5-disubstituted-1,2,3-trazolo)-nucleoside analogues **106** through CuAAC and RuAAC azide–alkyne cycloaddition reactions, which possess antiviral activity (Scheme 36).

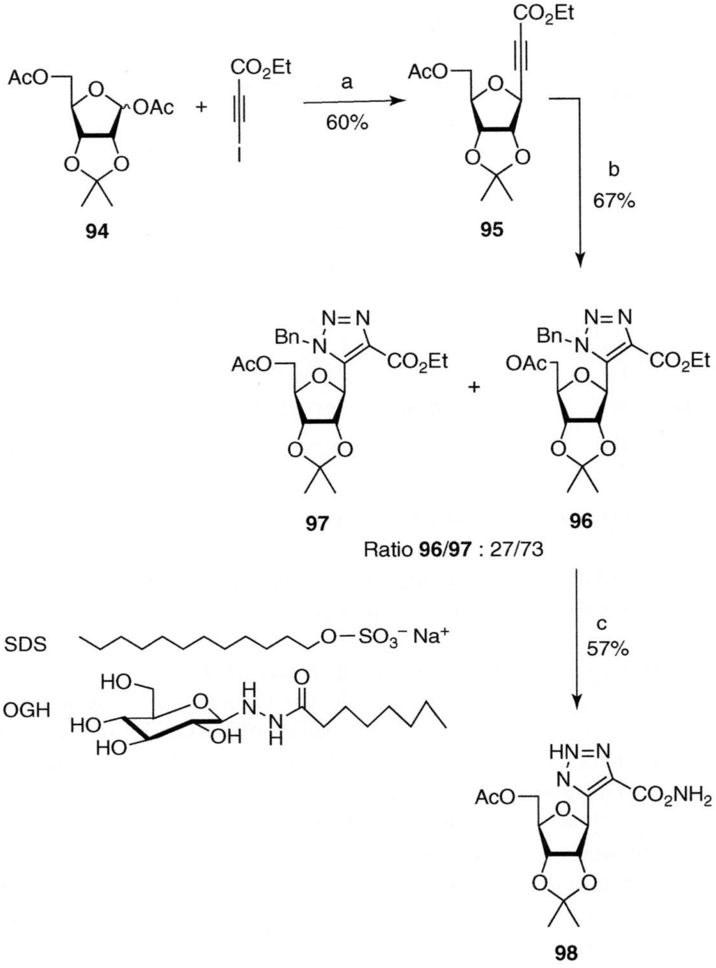

Scheme 33. *Reagents and conditions:* (a) In(0), CH_2Cl_2, reflux, 60%; (b) benzyl azide, H_2O, OGH, 24 h, 67%; (c) (i) NH_3, MeOH; (ii) Dowex 50W×8, MeOH/H_2O; (iii) H_2, Pd/C, MeOH, 57%.

As part of their selective A_3 adenosine receptor agonists and antagonist research program, Cosyn et al.[54] developed a series of 2-(1,2,3-triazolyl)adenosine derivatives, **110** and **111**. Starting from 2-iodo-N^6-methyladenosine **107**, they obtained as intermediates the 2-ethynyl- (**108**) and 2-azido- (**109**) adenosine derivatives through a [Pd]- and a [Cu]-catalyzed reaction, respectively. Both structures were then subjected to Huisgen's 1,3-dipolar cycloaddition in the presence of $CuSO_4$ and sodium ascorbate, to provide a 1,4-disubstituted triazole moiety (**110** and **111**). Due to the presence of Cu(I), which catalyzed both azidation and then cycloaddition, they

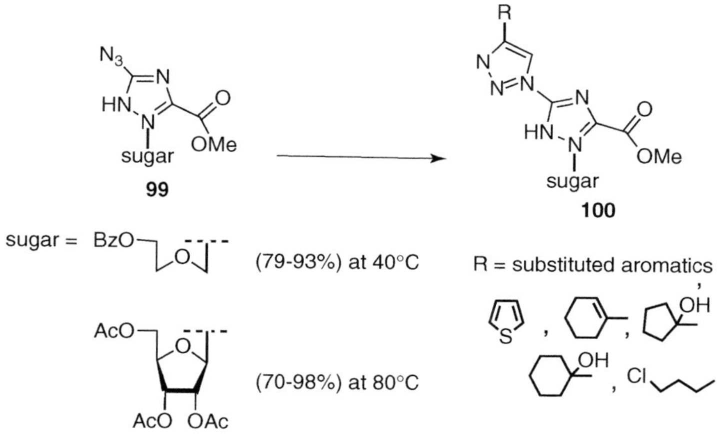

Scheme 34. *Reagents and conditions:* alkyne, CuSO$_4$·5H$_2$O, sodium ascorbate, THF/H$_2$O.

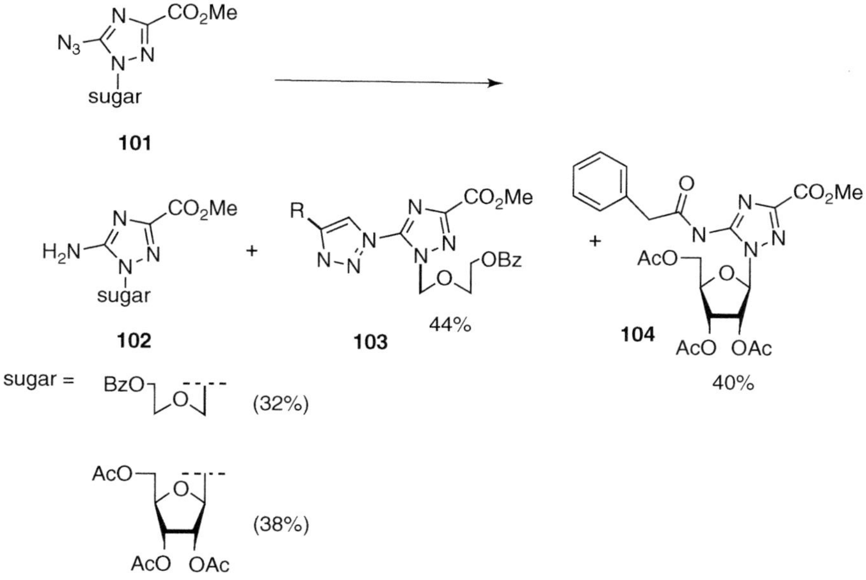

Scheme 35. *Reagents and conditions:* Phenylacetylene, CuSO$_4$·5H$_2$O, sodium ascorbate, THF/H$_2$O.

first attempted a one-step conversion from compound **107** to **112**. Unfortunately, compound **109** (which exists as an equilibrium of two forms, **109a** and **109b**) was isolated as a main product, with a small amount of the desired compound **112**, perhaps, due to the shift in equilibrium observed with in the tautomeric fused tetrazole by NMR during the preparation of compound **112** (Scheme 37).

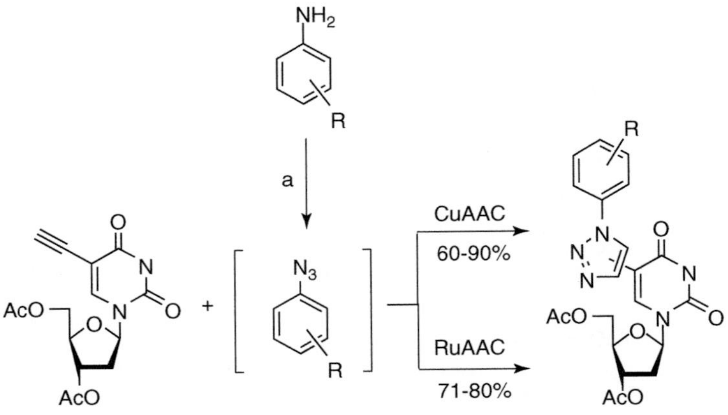

Scheme 36. *Reagents and conditions:* (a) *t*-BuONO, TMSN$_3$, CH$_3$CN, 2 h, 0°C to rt.

4. SELECTED NUCLEOBASE-MODIFIED NUCLEOSIDES

4.1. Ribavirin Analogues

Azole nucleosides are a class of antimetabolites related structurally to 5-amino-1-β-D-ribofuranosyl imidazole-4-carboxamide (AICAR), whose 5'-monophosphate is a key intermediate in purine biosynthesis. Important members of this class endowed with immunosuppressive, antitumor, and antiviral activity are discussed below (Figure 5.)

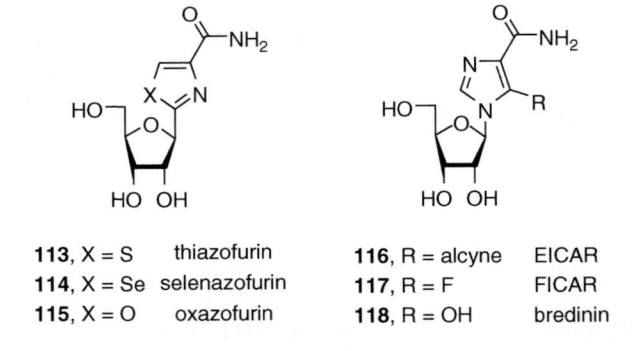

113, X = S   thiazofurin
114, X = Se   selenazofurin
115, X = O   oxazofurin

116, R = alcyne   EICAR
117, R = F   FICAR
118, R = OH   bredinin

Figure 5.

Chen et al. synthesized various triazole nucleoside derivatives (Figure 6.)[55] They speculated that the triazole-3-carboxamide ring was essential for activity against HIV-1, HSV-2, guinea pig–like virus, and murine leukemia viruses. Indeed, a comparison of two modified nucleosides revealed the necessity of having a carboxamide group.

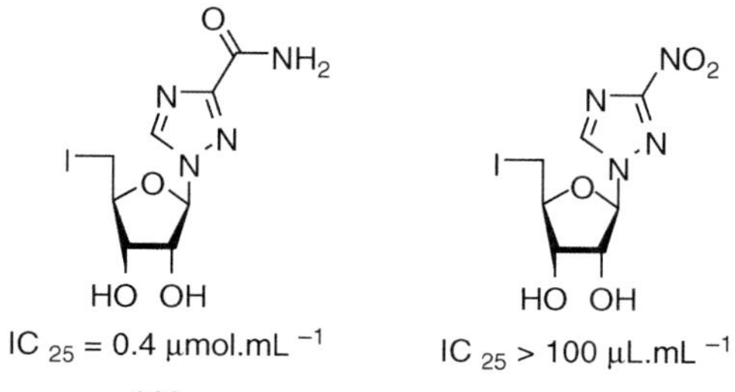

Figure 6.

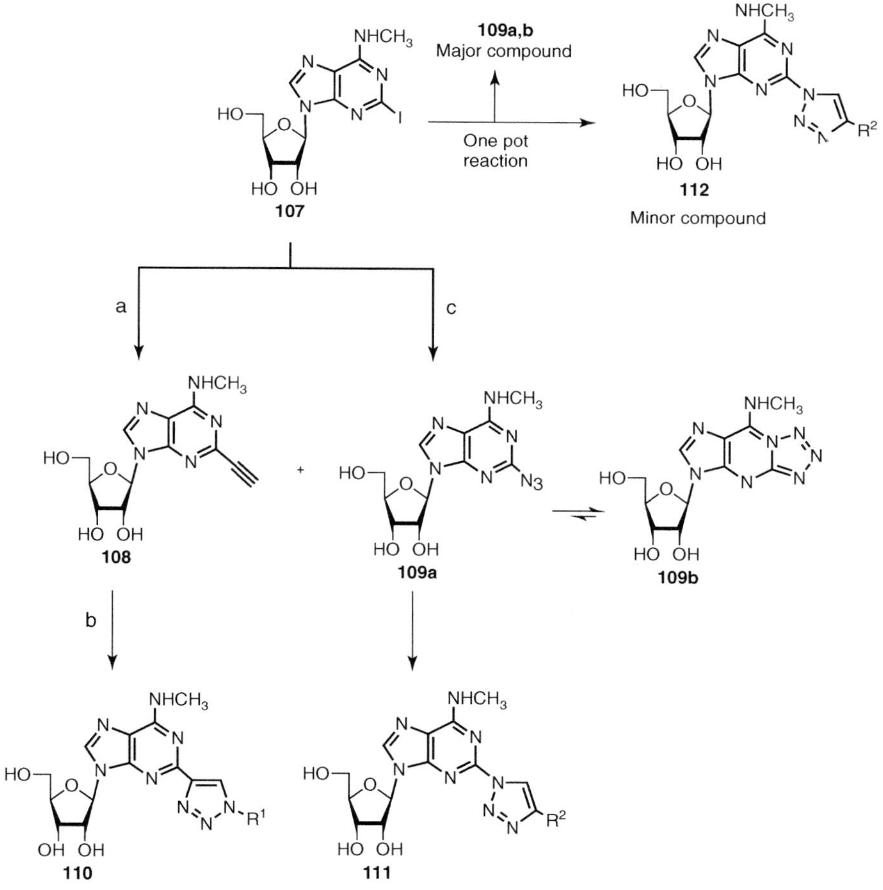

119: IC$_{25}$ = 0.4 μmol.mL^{-1}

120: IC$_{25}$ > 100 μL.mL^{-1}

Scheme 37. *Reagents and conditions:* (a) (i) trimethylsilylacetylene, CuI, PdCl$_2$(Ph$_3$P)$_2$, DMF; (ii) NH$_3$/MeOH (7 N); (b) azide, CuSO$_4$·5H$_2$O, sodium ascorbate, H$_2$O/*t*-BuOH, rt; (c) CuSO$_4$·5H$_2$O, sodium ascorbate, L-proline, Na$_2$CO$_3$, NaN$_3$, H$_2$O/*t*-BuOH, 60°C.

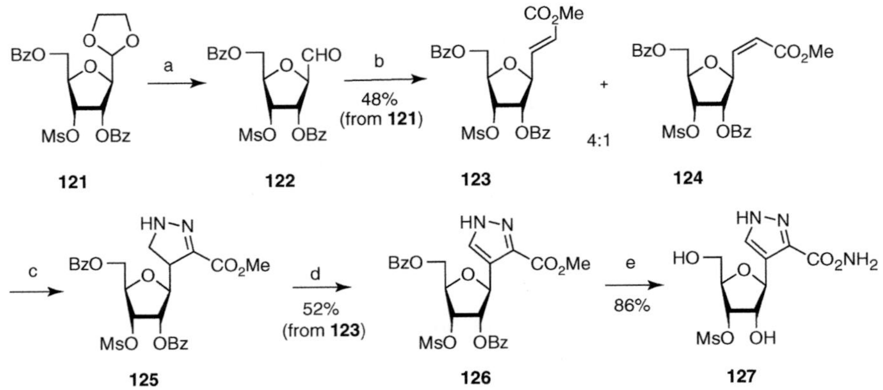

Scheme 38. *Reagents and conditions:* (a) 4:1 TFA–6M HCl, 4°C, 24 h; (b) Ph$_3$P=CHCOOMe, CH$_2$Cl$_2$, rt, 24 h, 48% from **121**; (c) CH$_2$N$_2$, Et$_2$O, 0°C, 5 h; (d) Cl$_2$/CCl$_4$, CH$_2$Cl$_2$, rt, 3 h, 52% from **123**; (e) NH$_3$, MeOH, rt, 7 days, 86%.

Popsavin et al. reported the synthesis and the biological evaluation of a new pyrazole-related *C*-nucleoside **127** (Scheme 38).[56]

Hydrolytic removal of the dioxolane protective group of **121** was achieved with a 4:1 mixture of trifluoroacetic acid and 6 M hydrochloric acid at 4°C. Due to its instability the crude **122** was treated immediately with carbomethoxymethylene–triphenylphosphorane in dichloromethane to afford a 4:1 mixture of the corresponding *E* and *Z* unsaturated esters **123** and **124**, respectively, in 48% overall yield. Both of these esters were readily separated by column chromatography and converted to the *C*-nucleoside by 1,3-dipolar cycloaddition with an excess of diazomethane in ether at 0°C to yield pyrazoline **125** as the only reaction product. Without further purification, the intermediate **125** was treated with a saturated solution of chlorine in carbon tetrachloride to give the pyrazole derivative **126**. Finally, treatment of **126** with methanolic ammonia afforded the *O*-deprotected *C*-nucleoside **127**, ready for biological testing.

The same group also investigated the synthesis of tetrazole-related *C*-nucleoside **133** (Scheme 39). Starting from **128**, hydrolytic removal of the dioxolane protective group of **128** was followed by subsequent treatment of the liberated aldehyde **129** with hydroxylamine hydrochloride, in the presence of sodium acetate in ethanol.

The mixture of *E*- and *Z*-oximes **130** was treated with mesyl chloride in pyridine to afford the corresponding nitrile **131** in an overall yield of 63%. The intermediate **131** was then reacted with an excess of sodium azide in the presence of ammonium chloride, to give the corresponding tetrazole derivative **132** in good yield. Moreover, the action of sodium methoxide on **132** furnished the *O*-deprotected *C*-nucleoside **133** in 72% yield.

Veronese et al. developed a powerful synthesis of pyrazole *C*-nucleoside via a tin(IV) chloride–promoted key step reaction of β-D-ribofuranosyl cyanide with β-dicarbonyl compounds **137a–c** (Scheme 40).[57] The reaction of **134** with methyl acetoacetate afforded the β-D-ribofuranosyl enaminoketo ester **135** in good yield.

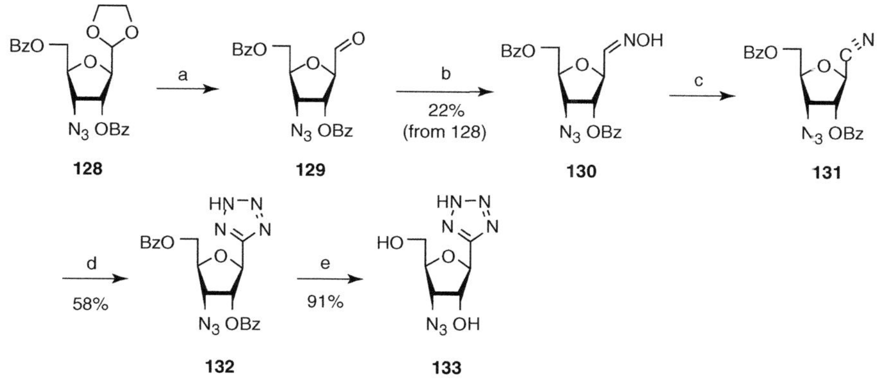

Scheme 39. *Reagents and conditions:* (a) 4:1 TFA–6 M HCl, 4°C, 96 h; (b) NH$_2$OH·HCl, NaOAc, EtOH, rt, 2 h; (c) MsCl, pyridine, −15°C to rt, 2.5 h, 22% from **128**; (d) NaN$_3$, NH$_4$Cl, DMF, 100°C, 2.5 h; (e) NaOMe, MeOH, rt, 2.5 h.

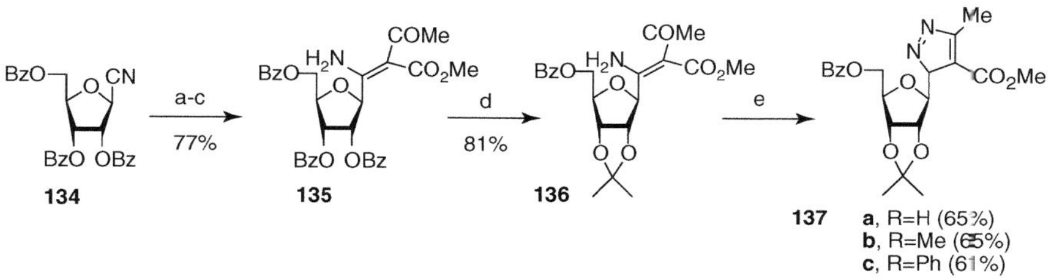

Scheme 40. *Reagents and conditions:* (a) MeCOCH$_2$CO$_2$Me, SnCl$_4$, 85%; (b) K$_2$CO$_3$, MeOH, rt, 45 min, 90%; (c) Me$_2$C(OMe)$_2$, TsOH, acetone, rt, 2 h; (d) i) K$_2$CO$_3$, MeOH, then 2,2-dimethoxypropane, acetone, H$^+$; (ii) TBDMSCl, imidazole, DMF, rt, 12 h, 81%; (e) RNHNH$_2$, EtOH, reflux, 2 h.

The tri-*O*-benzoyl enaminoketo ester **135** was deprotected by treatment with potassium carbonate in methanol at room temperature to give the β-ribofuranosyl enaminoketo ester, which was treated directly with 2,2-dimethoxypropane in acetone to afford, after silylation of primary alcohol, the derivative **136** in an 81% overall yield (from **135**). The reaction of **136** with hydrazine was then carried out using absolute ethanol, in which the least epimerization was found. Under these conditions, the β-D-ribofuranosyl pyrazole **137a** was obtained in 65% yield. Under similar conditions, **136** was reacted with methyl and phenylhydrazines to afford derivatives **137b** and **137c**, in 65% and 61% yield, respectively.

4.2. Thio or Seleno Derivatives

A number of studies have reported the syntheses and anticancer activity of ribonucleoside analogues containing a thio-substituted five-membered heterocyclic base. Due to their nucleophilicity and reversible redox properties, sulfur atoms play

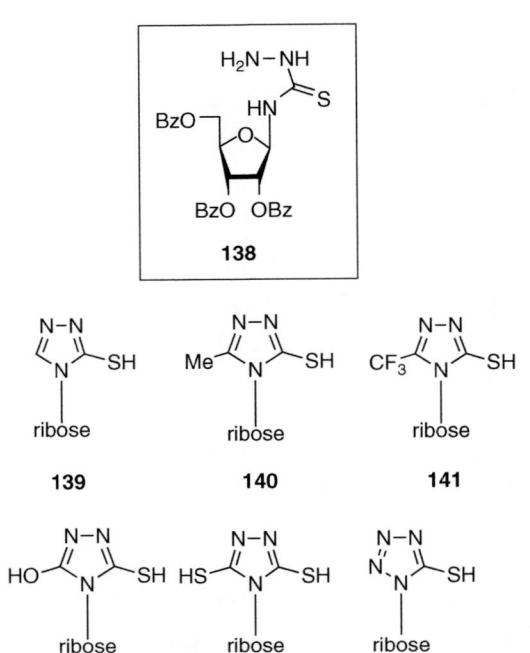

Scheme 41. *Reagents and conditions:* **139**–HCOOH, (MeCO)$_2$O, DMAP, 50°C, 30 min, 65%; **140**–MeC(OMe)$_3$, CH$_2$Cl$_2$, reflux, 10 h, 65%; **141**–(CF$_3$CO)$_2$O, CH$_2$Cl$_2$, rt, 1 h, 55%; **142**–Im$_2$O, THF, rt, 4 h, 75%; **143**–CS$_2$, reflux, 4 h, 64%; **144**–NaNO$_2$, MeCOOH, CH$_2$Cl$_2$, rt, 10 min, 83%.

an important role in the life processes. Nucleoside analogues containing a sulfur atom in the sugar moiety or the heterocyclic moiety have been synthesized, and some of them have been found to be active against cancer cells or viruses (Scheme 41).[58] Compound **138**, the central intermediate in the synthesis, was obtained by N-substitution of silver thiocyanate on 1-bromo-2,3,5-tri-*O*-benzoyl-β-D-ribofuranose followed by hydrazinolysis.[59] Condensation of **138** with various acids, esters, amides, or anhydrides in anhydrous solvent gave corresponding products, which were debenzoylated by sodium methoxide to give the final products **139–144**, respectively.

Compounds of the second type were synthesized via both condensation and fusion procedures, producing two positional isomers, **148** and **149** (Scheme 42).

In 1997, various groups synthesized selenophenfurin, an inosine 5′-monophosphate analogue of selenazofurin which acts as a dehydrogenase inhibitor. This selenium analogue of thiazofurin is a widely studied agent with a diverse array of biological effects. These include potent antitumor and antiviral activity, as well as efficacy as a maturation-inducing agent.[60] Selenazofurin is 5- to 10-fold more potent than thiazofurin in several antitumor screens and in vitro

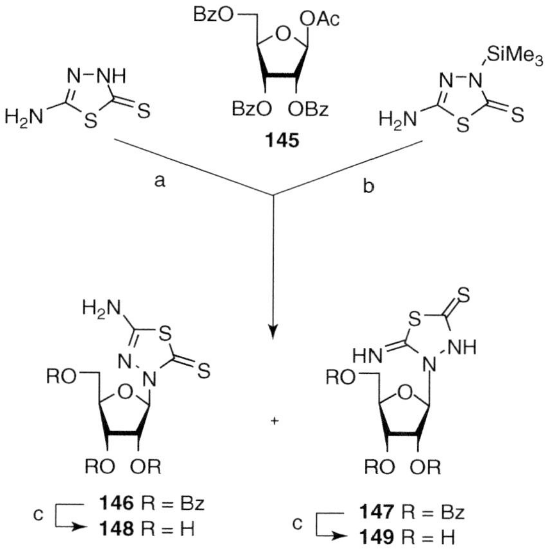

Scheme 42. *Reagents and conditions:* (a) *p*-TSA, 120–130°C, 20 mmHg, 30 min, 38%; (b) SnCl$_4$, (CH$_2$Cl)$_2$, rt, 8 h, 52%; (c) NaOCH$_3$, MeOH, rt, 4 h.

studies.[61] This team also synthesized thiophenfurin, a *C*-nucleoside isostere of thiazofurin in which the thiazole ring is replaced by a thiophene heterocycle.[62] Like thiazofurin, thiophenfurin was found to be active as an antitumor agent both in vitro and in vivo. This finding prompted the researchers to extend the study of structure relationships in this type of *C*-nucleoside to selenophenfurin, the selenophene analogue of selenazofurin.[63]

The ethyl selenophen-3-carboxylate **156** was obtained by reacting the 3-carboxylic acid **155** with SOCl$_2$ and ethanol. Compound **155** was obtained in four steps. Tetra-iodoselenophene **152**, prepared by reaction of **151** with mercuric acetate in glacial acetic acid and then with iodine, was treated with zinc powder and acetic acid under reflux to give 3-iodoselenophen **153**. Compound **153** was converted into nitrile **154** by cyanation with trimethylsilyl cyanide in anhydrous triethylamine in the presence of tetrakis(triphenylphosphine)palladium(0). Hydrolysis of **154** with hydrochloric acid under reflux gave the acid **155** (Scheme 43).

The reaction of **156** with 1,2,3,5-tetra-*O*-acetyl-β-D-ribofuranose in 1,2-dichloroethane in the presence of SnCl$_4$ gave 2- and 5-glycosylated regioisomers as a mixture of α- and β-anomers, **157**. The mixture was treated with a catalytic amount of sodium methoxide in ethanol to give deprotected ethyl esters (**158**). Treatment of **158** with ammonium hydroxide gave selenophenfurin **159** (Scheme 44). Selenophenfurin and its parent compounds were tested for inhibitory properties against IMPDH from human myelogenous leukemia K562 cells in culture. The selenophenfurin was found to have the highest activity in enhancing

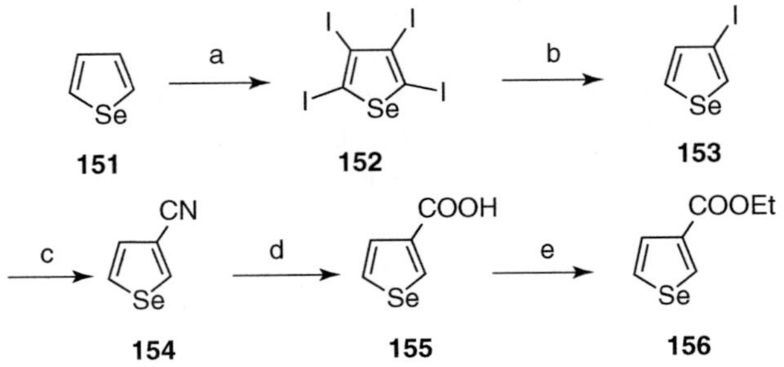

Scheme 43. *Reagents and conditions:* (a) Hg(OCOCH$_3$)$_2$, H$^+$, I$_2$; (b) Zn/AcOH; (c) Me$_3$SiCN, (Ph$_3$P)$_4$Pd(0), Et$_3$N; (d) HCl; (e) SOCl$_2$, EtOH; (f) SnCl$_4$, (CH$_2$Cl)$_2$; (g) EtONa, EtOH; (h) 30% NH$_4$OH.

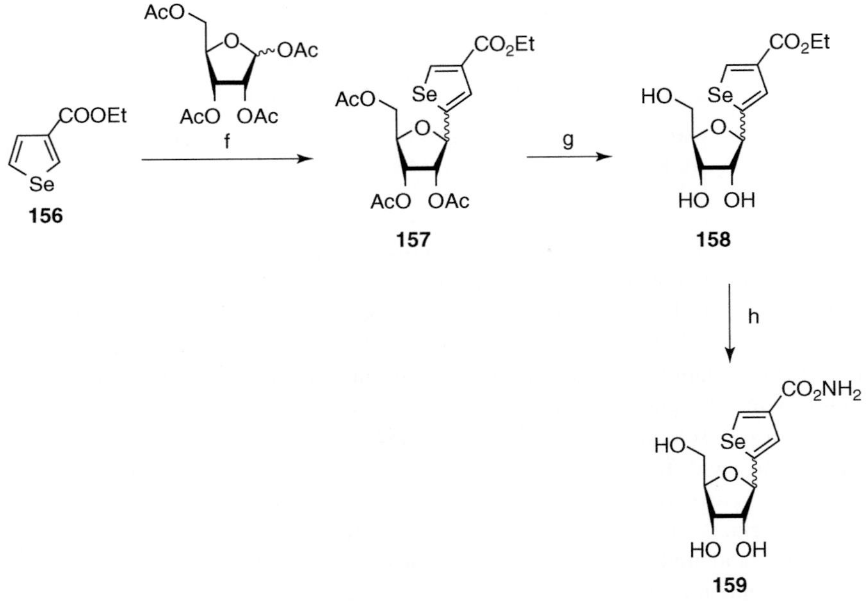

Scheme 44. *Reagents and conditions:* (f) SnCl$_4$, (CH$_2$Cl)$_2$; (g) EtONa, EtOH, and column chromatography; (h) 30% NH$_4$OH.

the IMP level and decreasing the guanine nucleoside pools. No other purine and pyrimidine nucleotide concentrations were tested.

4.3. Fleximers

In recent years, a new class of shape-modified nucleosides has been introduced. The purine heterobases of adenosine and guanosine have been split into their imidazole and pyrimidine components, thereby introducing flexibility while retaining the elements necessary for recognition. As a consequence, these novel "fleximers" should find use as bioprobes for investigating enzyme–coenzyme binding sites as well as nucleic acid and protein interactions.[64] It was also shown that substitution of the purine nucleobase found in the natural nucleosides with a more flexible two-ring heterocyclic system strongly increased the population of *anti*-conformation around the glycosidic bond.[65]

As shown in Scheme 45, 4, 5-dibromoimidazole **160** was coupled to the commercially available tetraacetate-protected ribose using bis(trimethylsilyl)acetamide and trimethylsilyltriflate. Removal of the labile acetate groups and subsequent conversion to more stable benzyl ethers, accomplished using a modified benzylation procedure, gave **161** in a 60% overall yield (three steps). Grignard treatment of **161** with ethylmagnesium bromide and dimethylformamide produced aldehyde **162**, which was then converted to oxime, which by dehydration gave the nitrile **163** (in 72% overall yield). Displacement of the bromide at C4 with thioglycolamide yielded **164** (50%). Closure to the thiophene ring was quite facile and yielded **165** (90%). Subsequent treatment of **165** with a 1:1 mixture of acetic anhydride and triethyl orthoformate facilitated closure of the pyrimidine ring, which was subjected to thiocarbonylation and immediate methylation to achieve the desired **166** (80%). Ammonolysis of **166** yielded tricyclic nucleoside **167** (75%), which was then refluxed with Raney nickel to provide the protected fleximer (74%) and after removal of protected groups, the desired adenosine fleximer **168** (91%).

Guanosine fleximer was realized from sequential treatment of **169** with sodium hydroxide and then heating with carbon disulfide, followed by the addition of hydrogen peroxide, and finally, ammonia provided tricyclic **170** (77%) (Scheme 46). Again, using Raney nickel, the thiophene ring of **170** was cleaved successfully (76%) and, following deprotection, gave guanosine fleximer **171**.

Taha et al. reported the synthesis of acyclic fleximer nucleosides in order to determine the influence on biological evaluation of the 1,2,3-triazolyl-(4 or 5)-yl methyl moiety as a spacer between the 4-substituted pyrazolo[3,4-*d*]pyrimidines and the (2,3-dihydroxy-1-propoxy)methyl chain (Scheme 47).[66]

The 4-methylthio-1-propargyl-*1H*-pyrazolo-[3,4-*d*]-pyrimidine **172** was reacted with **173**, synthesized in five steps[67] under 1,3-dipolar cycloaddition reaction in anhydrous toluene under reflux, to afford as a major product the regioisomer-1,4 (**174**) in 78% overall yield. The benzoyl and acetyl groups were subsequently removed from the protected acyclic nucleoside **174** by treatment with a solution of methanol saturated with ammonia in a sealed reacting vessel at 25°C, affording the required nucleoside **175** in quantitative yield. Further classical functionalization can be carried out with various nucleophiles (such as ammonia, primary amines, sodium methanolate, or hydroxide), affording acyclic fleximer **176**.

More recently, Seley-Radtke[65b] reported that a series of 5′-nor carbocyclic "reverse" flexible nucleosides or fleximers have been designed in which the nucleobase scaffold resembles a "split" purine as well as a substituted pyrimidine (Figure 7).

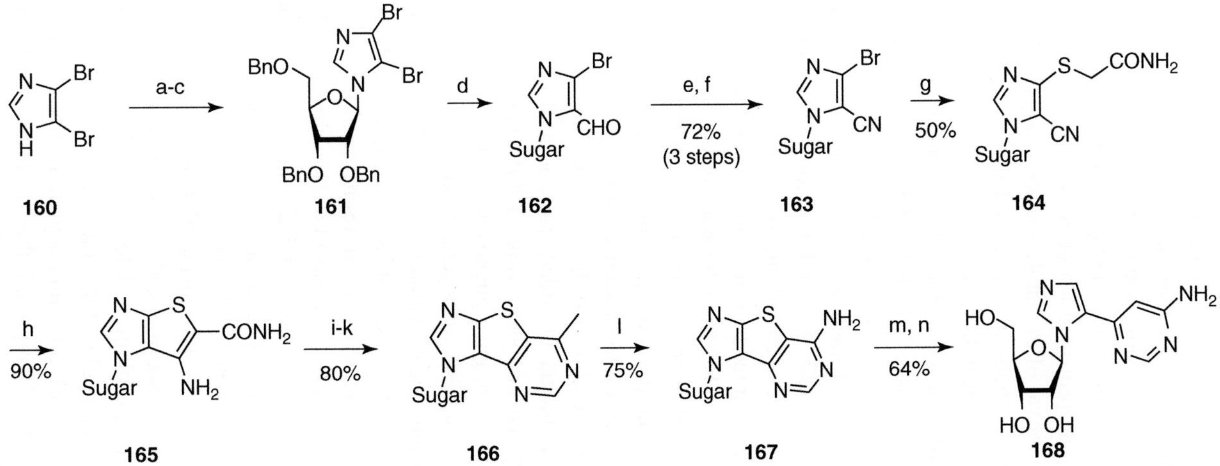

Scheme 45. *Reagents and conditions*: (a) 1,2,3,5-tetra-*O*-acetyl-β-D-ribofuranose, BSA, MeCN, and then TMSOTf; (b) NH₃, EtOH; (c) NaH, BnBr, Bu₄NI; (d) EtMgBr, anhydrous DMF; (e) NH₂OH·HCl, NaHCO₃; (f) Ac₂O; (g) NH₂COCH₂SH, K₂CO₃; (h) NaOEt, EtOH; (i) CH(OEt)₃, Ac₂O (1:1); (j) P₂S₅, pyridine; (k) MeI, K₂CO₃; (l) NH₃, MeOH; (m) Raney Ni, MeOH; (n) BF₃·OEt₂, EtSH, CH₂Cl₂.

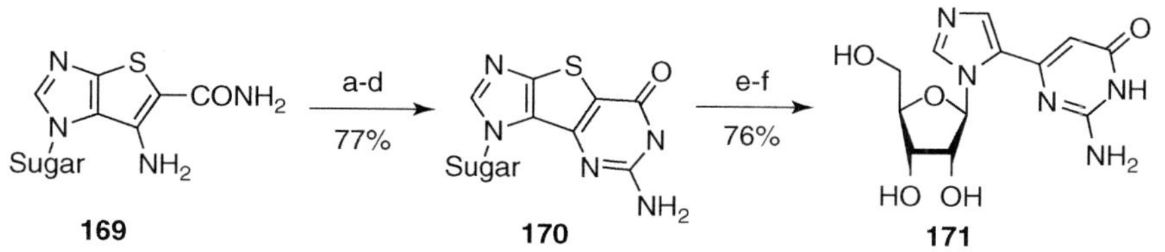

Scheme 46. *Reagents and conditions:* (a) NaOH, MeOH; (b) CS$_2$, heat; (c) H$_2$O$_2$; (d) NH$_3$; (e) Raney Ni, MeOH; (f) BF$_3$·Et$_2$O, EtSH, CH$_2$Cl$_2$.

4.4. Selected Modified Uracil Analogues

4.4.1. Deaza Analogues.
The 3-deazapyrimidine nucleosides have been shown to have a wide scope of biological, biochemical, and medicinal applications. They have functioned effectively as antivirals, tumor growth inhibitors, antibacterial agents,[68] and RNA viral suppressors.[69] The chemical, biochemical, and pharmacological properties of 3-deazauridine have been widely evaluated.[70,71] The most common synthesis[72] of 3-deaza-2'-deoxycytidine begins with construction of the pyrimidine heterocycle using 1-methoxy-1-butene-3-yne, a material that represents an unattractive synthetic target, owing to its explosive nature. Searles and McLaughlin opted for a new route (Scheme 48). It begins with the 2-chloropyridine **178** and relies on the ability to introduce the requisite exocyclic amino group through nitration or reduction (to **179**) followed by hydrolysis of the chloro substituent to generate the 2-pyridone ring system (**180**).[73] **180** was then introduced onto a sugar moiety (to **182**) and served as a monomer (**184**) of oligo-DNA.

More recently, Camplo et al. developed the synthesis of 3-hydroxy-2-methylpyridin-4-one dideoxy nucleoside derivatives (**185**, **186**) (Figure 8).[74] In fact, 3-hydroxypyridin-4-ones are currently one of the main candidates for the development of orally active iron chelators. Recently, a clear synergism in HIV-1 inhibition was observed by combining iron chelators with the anti-HIV nucleoside analogue ddI, and it was suggested that in combination with existing antivirals, iron chelation could have a beneficial effect on HIV disease.[75]

Many papers have been published on *C*-nucleosides (including the deazanucleoside analogues) in the past decade. The reader is referred to a recent review of Stambasky et al. concerning this type of base-modified nucleoside.[76]

4.4.2. Pyridino Derivatives.
In the past three decades, a tremendous effort has been directed to the search for nucleoside anticancer and antiviral drugs. In addition to other nucleoside analogues, a large number of sangivamycin and toyocamycin analogues have been synthesized and evaluated for anticancer activities since these two compounds were isolated four decades ago.[77,78] Recently, Wang et al. synthesized and evaluated a series of pyridipyrimidine nucleosides in search of selective inhibitors of cancer cells (Scheme 49).[79] The substituted pyridine **187** was treated with trimethylsilyl iodide to remove the methyl group to give **188**, which was

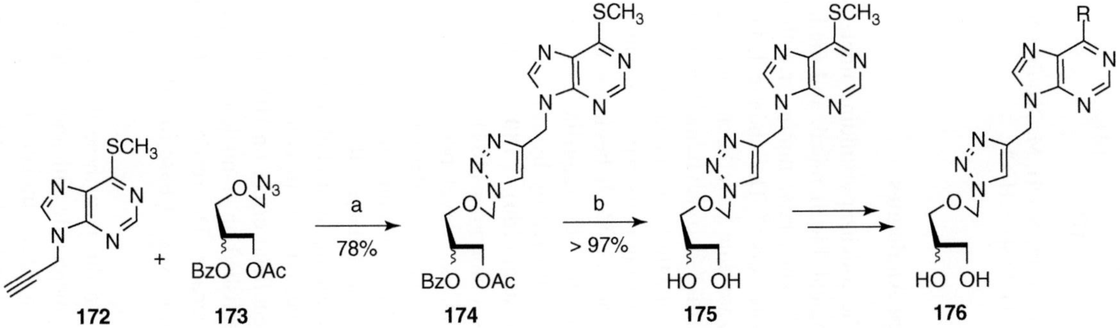

Scheme 47. *Reagents and conditions:* (a) toluene, reflux; (b) NH$_3$/MeOH, rt.

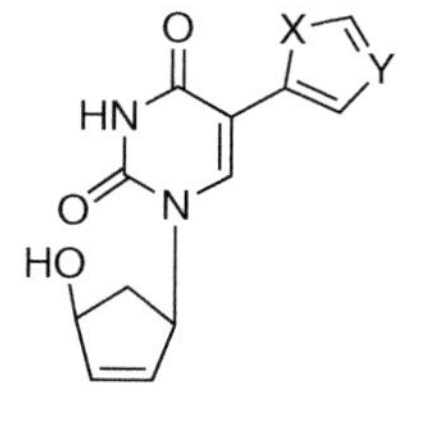

177a, X=S, Y=CH
  b, X=O, Y=CH
  c, X=S, Y=N

Figure 7. 2,3′-Dideoxy reverse carbocyclic targets.

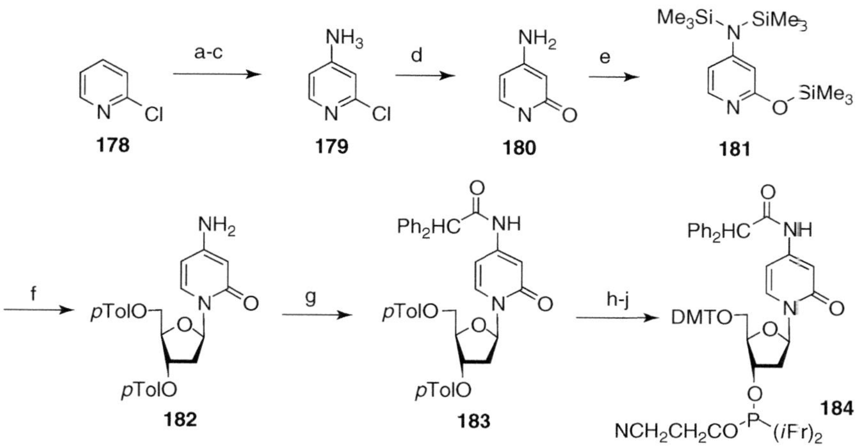

Scheme 48. *Reagents and conditions:* (a) H$_2$O$_2$; (b) HNO$_3$/H$_2$SO$_4$; (c) Zn/AcOH; (d) NaOH/MeOH; (e) BSA. (f) SnCl$_4$; (g) Ph$_2$CHCOCl. pyridine; (h) NaOMe/MeOH; (i) DMTCl, pyridine; (j) CNCH$_2$CH$_2$OP(Cl)-*i*-Pr$_2$/DIEA.

refluxed with formamidine acetate in ethoxyethanol to effect a cyclization. 4-Amino-5-oxopyrido[2,3-*d*]pyrimidine **189** was obtained in very good yield.

4.4.3. Diazino Derivatives.
As part of an ongoing project leading to the synthesis of anti-HIV polynucleotides, Broom's team became interested in the synthesis of 3-(β-D-ribofuranosyl)isoguanine, (Scheme 50).[80]

Glycosylation of silylated 4-amino-2-oxo-1,26-dihydroxypteridine **190** gave predominantly the N^1-glycosylated pteridine nucleoside **191**. Nitrozation of **191** in acetic acid below 10°C for 4 h gave the 5-nitroso derivative **192** in quantitative yield. Quantitative reduction of nitroso to amine was achieved by the treatment of **192** with sodium hydrosulfite and water in DMF at 50°C for 4 h to yield **193**. Reaction of **193** with glyoxal in DMF at room temperature for 1 h gave a 3:2 mixture of 4-amino-2-oxo-1-(2,3,5-tri-*O*-benzoyl-β-D-ribofuranosyl)-1,2-dihydropteridine **194** and 4-amino-2-oxo-3-(2,3,5-tri-*O*-benzoyl-β-D-ribofuranosyl)-2,3-dihydropteridine

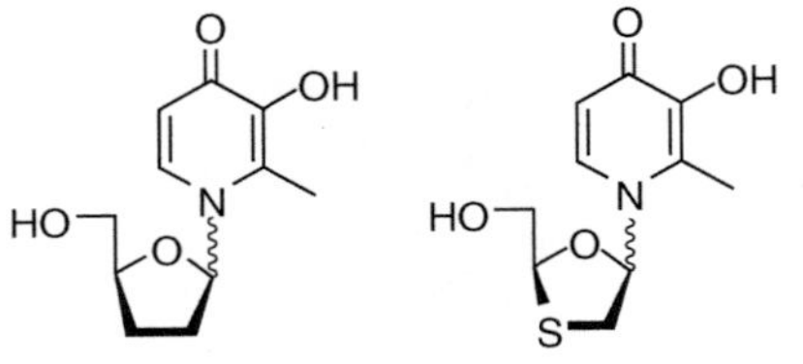

185 **186**

Figure 8. Two 3-hydroxy-2-methylpyridin-4-one dideoxy nucleosides.

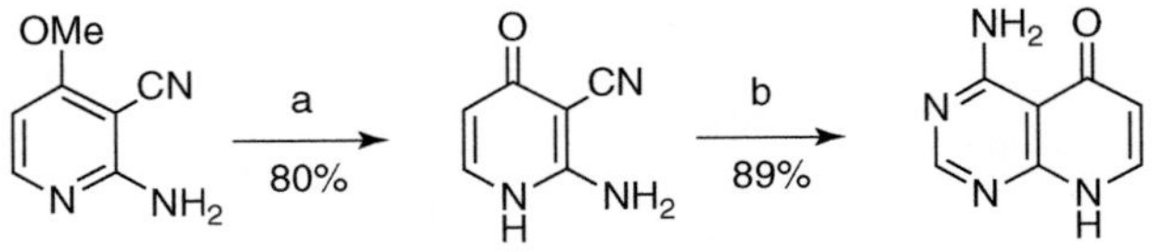

187 **188** **189**

Scheme 49. *Reagents and conditions:* (a) Me_3SiH, MeCN, reflux, 20 h; (b) $HC(=NH)NH_2 \cdot AcOH$, 2-ethoxyethanol, reflux, 30 h.

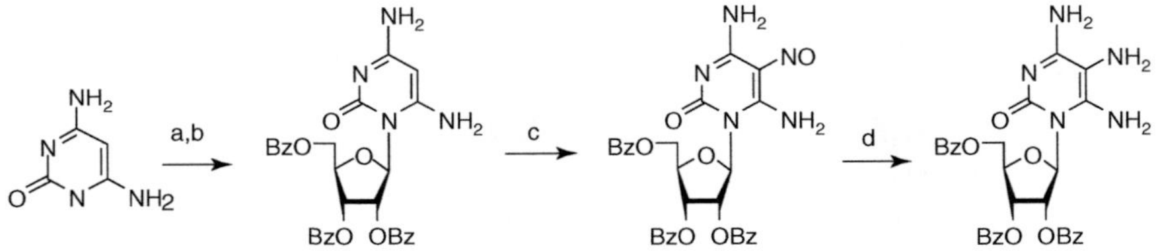

190 **191** **192** **193**

Scheme 50. *Reagents and conditions:* (a) HDMS, TMSCl cat., reflux, 6 h; (b) 1-O-acetyl-2′,3′,5′-(tri-O-benzoyl)-β-ribofuranose, Sn(IV)Cl, rt, 24 h, 88%; (c) $NaNO_2$, HOAc, H_2O, <10°C, 4 h, >95%; (d) Na_2SO_4, DMF, H_2O, 50°C, 4 h, >95%.

195, respectively (Scheme 51). Treatment of **193** with the Vilsmeier reagent at room temperature gave predominantly 6-amino-3-(2,3,5-tri-O-benzoyl-β-D-ribofuranosyl)purin-2-one **196** in 85% yield.

All the previous molecules have been deprotected by treatment with sodium methanolate in methanol or with sodium hydroxide in a mixture of water and methanol. Rollin et al. recently developed a five-step convergent process based on carbohydrate-derived oxazolidinethiones.[81] Condensation of 1,3-oxazoline derivative **197** with various anthranilic acid derivatives in dry *tert*-butanol offers an efficient access to sugar-derived quinazolinone **198** in mostly good yields (Scheme 52). To validate their approach to base-modified nucleosides, they investigated the ring cleavage of quinazolines **198**. The 2,2′-anhydronucleoside **198** was submitted to acidic conditions to afford **199** with 74% yield after further hydrogenolysis of benzyl groups.

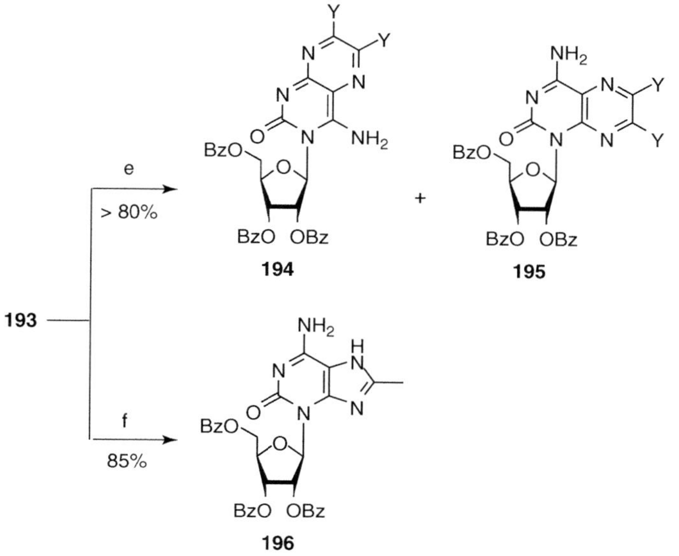

Scheme 51. *Reagents and conditions:* (e) Y–CO–CO–Y, DMF; (f) X–CO–NMe$_2$, FOCl$_3$.

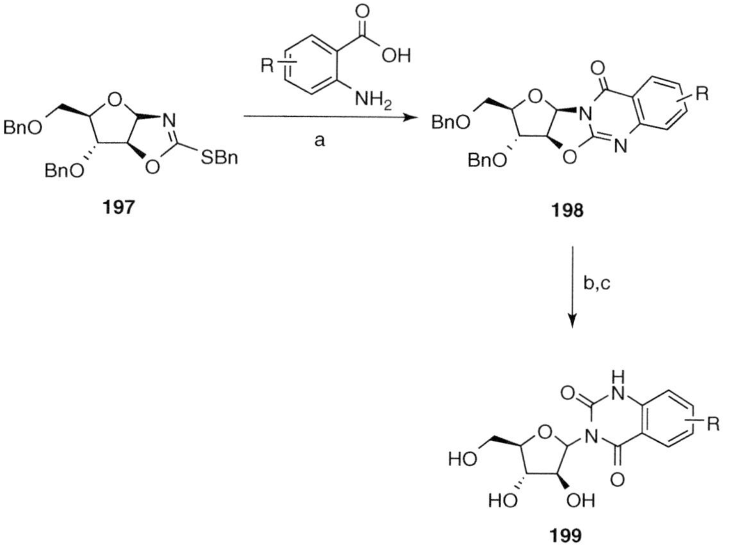

Scheme 52. *Reagents and conditions:* (a) anthranilic acid, *t*-BuOH, molecular 4-Å, 65–92%; (b) NaOH 5%, EtOH or 1 N HCl, EtOH; (c) Pd/C, H$_2$, MeOH.

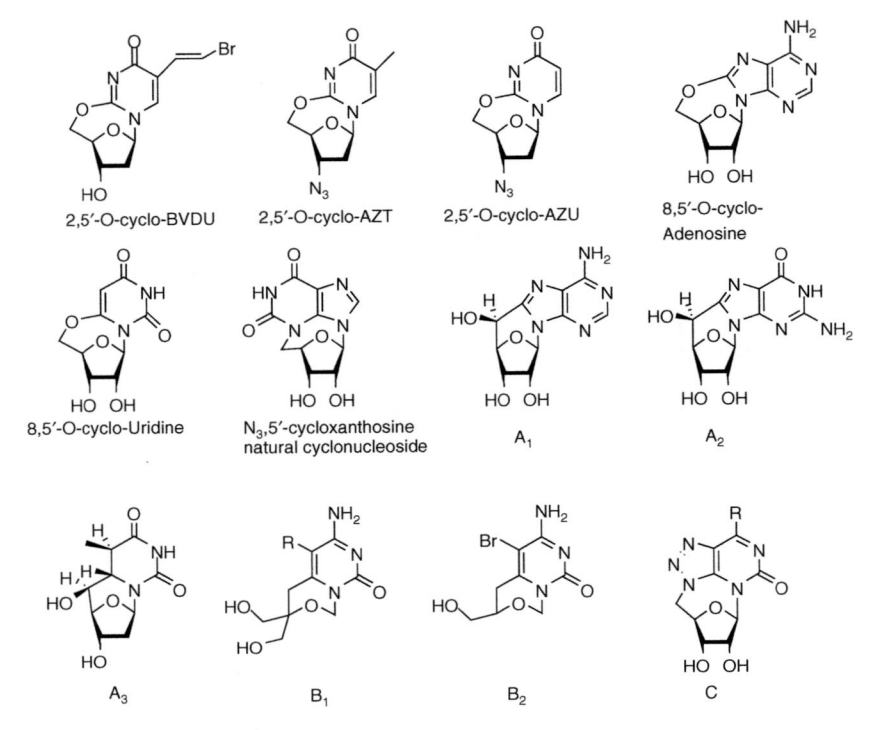

Figure 9. Structural diversity of bioactive cyclonucleosides.

4.5. Cyclonucleosides

In 2005, $N^3,5'$-cycloxanthosine was isolated from marine *Eryus* sp. sponge.[82] This was the first natural occurrence of a cyclonucleoside reviewed since that by Agrofoglio et al.[83] The preparation of C-bridged cyclonucleosides has also been reviewed by Len et al.[84] This discovery is of great importance, as natural nucleosides from marine sponges (spongothymidine and spongouridine isolated from a Caribbean sponge)[85] have inspired the development of antileukemic and antiviral agents, ara-C and ara-A, respectively. In fact, $N^3,5'$-cycloxanthosine was patented[86] in 2004 as a member of a family of synthetic nucleosides with antiviral properties, (Figure 9).

To exemplify this chemistry, the first new example of the Mitsunobu reaction in the synthesis of an *S*-bridge is reported (Scheme 53). The Mitsunobu reaction is widely used in the synthesis of pyrimidine cyclonucleosides possessing $-O-$,[87] $-S-$,[88] or $-N-$[89] linkages and has found many applications in the synthesis of purine cyclonucleosides. During research on the synthesis of acyclic nucleoside and nucleotide analogues, Janeba et al.[90] reported the synthesis of acyclic analogues of 8-cyclonucleosides and 8-cyclonucleotudes possessing a *S*-anhydro linkage, obtained from 1-substituted acyclic 8-bromoadenine derivatives (Scheme 1). 8-Bromo-1-propyladenine **200**[91] and its nucleotide analogue **201**[92] were transformed into their appropriate 8-mercapto derivatives **202** and **203** by

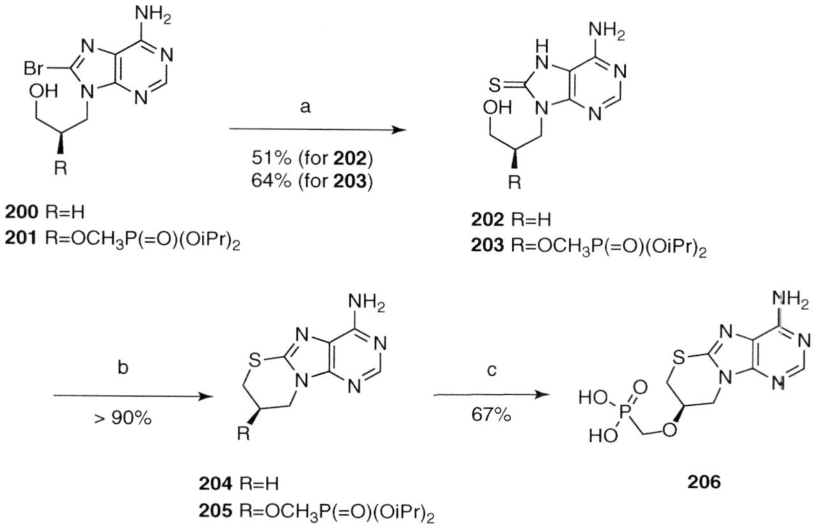

Scheme 53. *Reagents and conditions:* (a) thiourea, EtOH, reflux, 15 h; (b) Ph₃P, DEAD, DMF, −100°C to rt, overnight (procedure repeated twice), > 90%; (c) TMSBr, CH₂CN, rt, overnight, 67%.

refluxing with thiourea in ethanol. To avoid unnecessary cyclization to $N^3,5'$-cyclonucleosides, an intramolecular Mitsunobu reaction was used to form an of S-linkage. After treatment with triphenylphosphine, diethyl azodicarboxylate S-bridged purine cyclonucleoside analogues **204** and **205** were obtained in a high yield. Deprotection of a phosphonate group in nucleotide **205** by trimethysilyl bromide led to free phosphonic acid **206**.

A carbonate method developed by Hampton and Nichol[93] found many applications in the synthesis of 2,2'-pyrimidine cylonucleosides.[94] It was also used successfully in the cyclization of bicyclic nucleoside analogues resembling pyrimidine nucleosides, which could also be treated as analogues of purine nucleosides a possessing a modified base (Scheme 54).

When treated with diphenyl carbonate and sodium bicarbonate in DMF, 1-(β-D-ribofuranosyl)-2,4-quinazolinedione (**207**)[95] and 6,7-diphenyl-1-(β-D-ribofuranosyl)lumazine (**209**)[96] gave appropriate 2,2'-O-anhydro-1-(β-D-arabinofuranosyl)-4-quinazolone (**208**) and 2,2'-O-anhydro-1-(β-D-arabinofuranosyl)-6,7-diphenyllumazine (**210**), respectively, with high and moderate yields.

A diphenyl carbonate method applied in the synthesis of purine cyclonucleosides made possible the synthesis of compounds with various types of linkages; however, in initial reports this method turned out to be unsuitable for the synthesis of O-bridged purine cyclonucleosides, giving none of the product sought[97] and no traces of the product.[98] On the contrary, Shugar et al.[99] reported that 8-(α-hydroxyisopropyl)adenosine (**212**) is unexpectedly

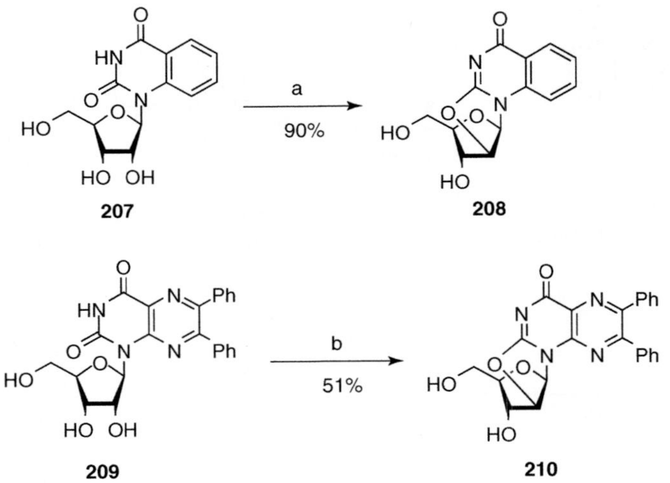

Scheme 54. *Reagents and conditions:* (a) diphenyl carbonate, NaHCO$_3$, DMF, 150°C, 30 min, 90%; (b) diphenyl carbonate, NaHCO$_3$, DMF, 155–160°C, 30 min, 51%.

transformed into cyclic derivative via 2′,3′-cyclic carbonate. Adenosine (**211**) underwent photochemical addition of isopropanol with di-*tert*-butyl peroxide as an initiator, giving 8-(α-hydroxyisopropyl)adenosine (**212**) as the main product and 2,8-di(α-hydroxyisopropyl)adenosine (**213**) as a minor one (Scheme 55). When

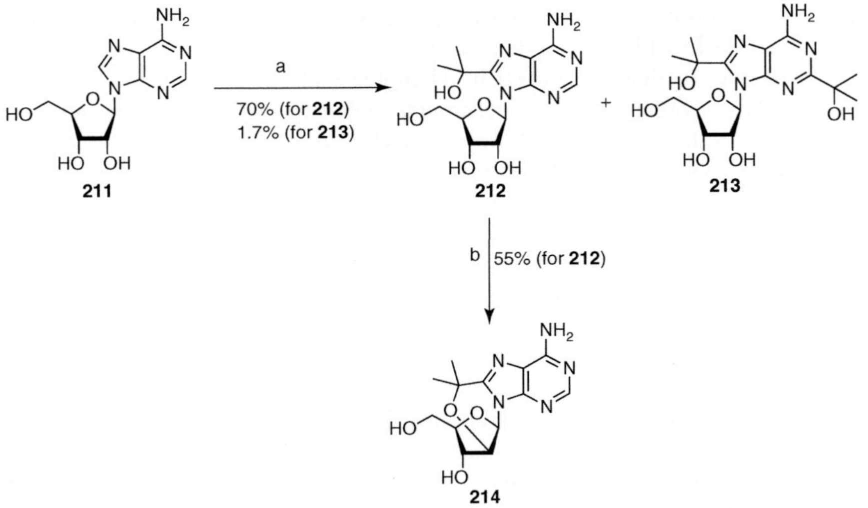

Scheme 55. *Reagents and conditions:* (a) di-*t*-BuOOH, 80%, *i*-PrOH, irradiation, 60 h; (b) diphenyl carbonate, NaHCO$_3$, DMF, 140–145°C, 30 min, 55%.

compound **212** was heated with diphenyl carbonate and sodium bicarbonate in DMF, cyclonucleoside **214** was obtained in moderate yield.

Synthesis of pyrimidine cyclonucleosides via 2′,3′-O-sulfinate intermediates generated by treatment with thionyl chloride in pyridine was widely explored for classical[100] and carbocyclic[101] nucleoside analogues. Recently, a similar method utilizing 3′,5′-sulfinyl derivatives was developed for pyrimidine nucleosides.[102] For some reason, this method was less well explored than that for purine nucleosides. Robba et al.[103] reported that treatment of thieno[3,2-d]pyrimidine nucleoside **215** with thionyl chloride in pyridine led to 2′,3′-sulfinyl derivative **216**, which treated in situ with mesyl chloride in pyridine gave 2,2′-anhydro-1-(5′-chloro-5′-deoxy-3′-O-mesyl-β-D-arabinofuranosyl)thieno[3,2-d]pyrimidin-4-one (**217**) (Scheme 56).

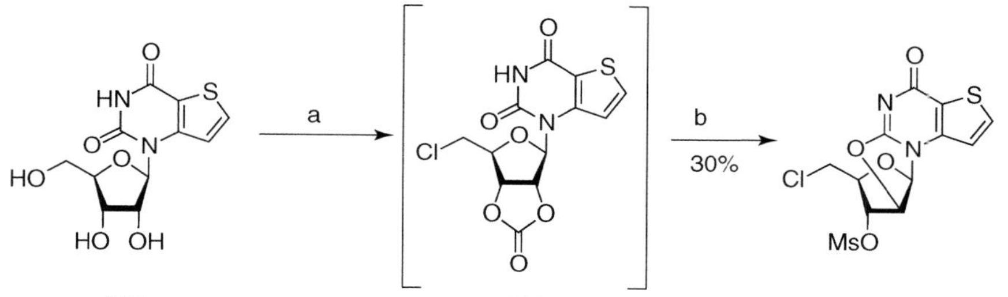

Scheme 56. *Reagents and conditions:* (a) SOCl$_2$, pyridine, CH$_3$CN, reflux, 1 h; (b) MsCl, pyridine, 0°C for 2 h, rt for 1 h, 30%.

Ahrem et al.[104] reported that when treated with thionyl chloride in a pyridine–acetonitrile mixture, 8-bromoadenosine-5′-carboxylic acid ethyl ester (**218**), smoothly gave 2′,3′-O-sulfinyl derivative (**219**), transformed into 8,2′-S-bridged nucleoside **220** by heating with thiourea in butanol (Scheme 57). Treatment with acetic anhydride in pyridine led to 3′-O-acetyl derivative **221**, which underwent elimination of acetic acid by treatment with DBU, giving unsaturated 8,2′-S-bridged nucleoside **222**. Prolongation of the reaction time led to slow decomposition of **222** and cleavage of anhydro linkage, giving furane nucleoside **223**. Attempts at formation of 8,2′-O-bridged nucleoside by heating of **219** with sodium acetate in an acetic acid/acetic anhydride mixture led exclusively to formation of furan nucleoside **224**.[105]

Maki et al. reported[106] that treatment of 2′,3′-O-isopropylideneadenosine (**225**) with N-bromosuccinimide (NBS) at acetic acid at room temperature led to formation of 8,5′-O-cycloadenosine (**226**) in almost quantitative yield (Scheme 58). No other products were detected by TLC analysis of the reaction mixture. Additionally, the employment of N-chlorosuccinimide (NCS) or N-iodosuccinimide (NIS) in place of NBS also allowed this conversion without side reactions. In the case of guanosine, protection of the N^6-benzoyl group was required, as reaction of N-unprotected guanosine led predominantly to 8-bromoguanosine with 60% yield, while the nucleoside sought was obtained as a minor product with 18% yield.

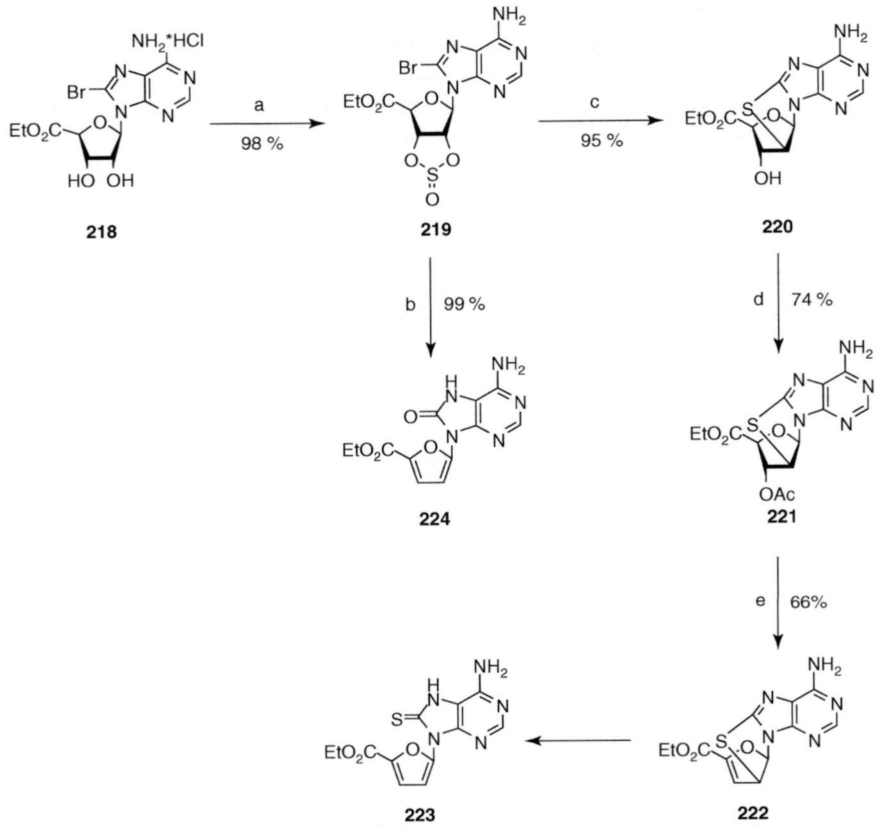

Scheme 57. *Reagents and conditions:* (a) SOCl$_2$, pyridine, CH$_3$CN, 0°C, 3.5 h, 98%; (b) NaOAc, AcOH, Ac$_2$O, reflux, 1 h, 99%; (c) BuOH, thiourea, reflux, 5 h, 95%; (d) Ac$_2$O, pyridine, 10 h, rt, 74%; (e) DBU, pyridine, 2 h, rt, 66%.

Protection of the amine group as a benzamide dramatically increased the yield of cyclonucleoside to 95%, while there was only a trace of the 8-bromo derivative.

When the reaction with NBS was carried out in dry acetonitrile or DMF, the 8-bromo-2′,3′-*O*-isopropylideneadenosine was obtained at high yield without the formation of **226**. These results can be explained by the initial attack of a halogenium ion at the imidazole ring nitrogen N^7 to generate a purinyl cation **A**. The subsequent intramolecular capture of the transient cationic species **A** by the 5′-hydroxy group on the 2′,3′-*O*-isopropylidene-protected ribofuranosyl ring could yield the intermediate **B**, which is driven to the final product as a result of the elimination of hydrogen halide.

Recently, Agrofoglio et al. investigated the synthesis of C5-substituted 6,5′-*O*-cyclonucleosides **227–240** (Figure 10).[107] The key step involved a bromine–lithium exchange at C5 followed by an exchange with an alkyl or alkenyl group. Palladium-mediated or palladium-catalyzed reactions, used widely

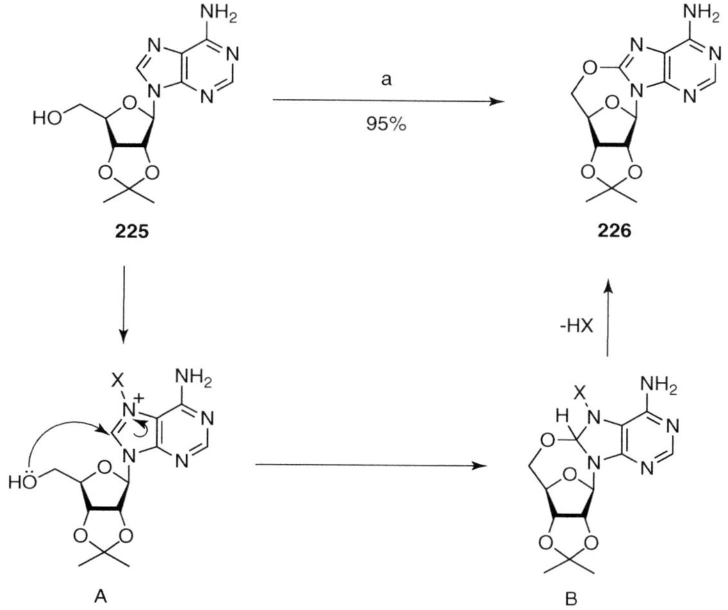

Scheme 58. *Reagents and conditions:* (a) NBS, AcOH, rt, 1 day, 95%.

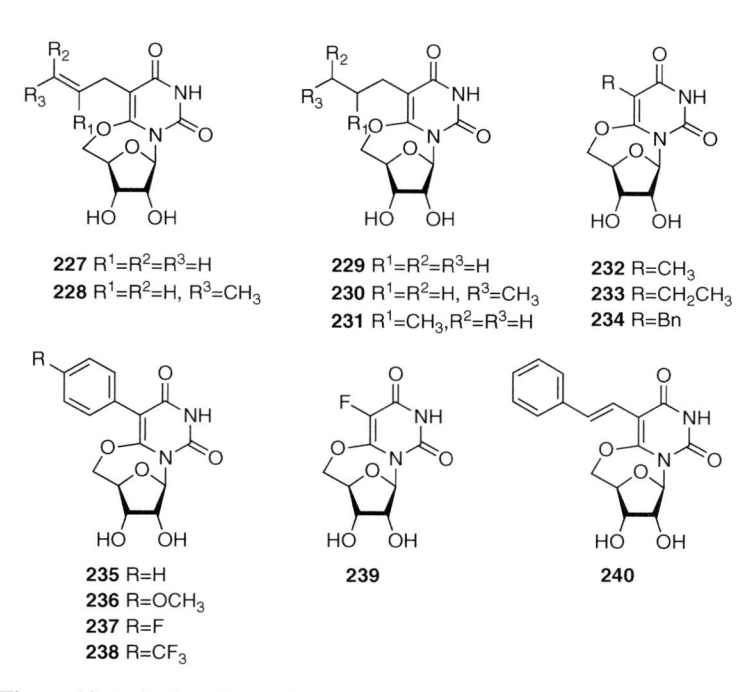

227 $R^1=R^2=R^3=H$
228 $R^1=R^2=H, R^3=CH_3$

229 $R^1=R^2=R^3=H$
230 $R^1=R^2=H, R^3=CH_3$
231 $R^1=CH_3, R^2=R^3=H$

232 $R=CH_3$
233 $R=CH_2CH_3$
234 $R=Bn$

235 R=H
236 $R=OCH_3$
237 R=F
238 $R=CF_3$

239

240

Figure 10. Palladium(0)-mediated syntheses of pyrimidine cyclonucleosides.

in the synthesis of new nucleoside analogues,[108] were also explored by our group[109] for the modification at C5 of pyrimidine cyclonucleosides **235–240**.

4.6. Miscellaneous 5-Substituted Nucleosides

Finally, numerous modifications have been made at the C5-position of pyrimidine nucleosides, due to their great significance in biochemistry and biotechnology. C5 pyrimidine nucleosides are synthesized by an organopalladium coupling reaction. Readers are referred to the comprehensive review on that subject by Agrofoglio et al. (Figure 11).[4]

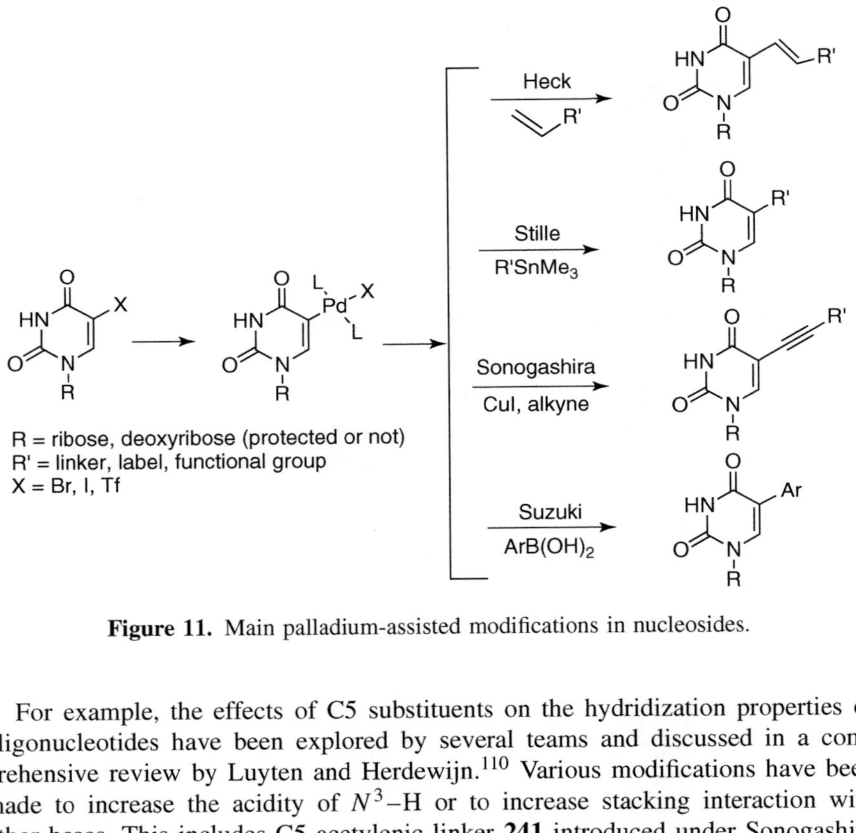

Figure 11. Main palladium-assisted modifications in nucleosides.

For example, the effects of C5 substituents on the hydridization properties of oligonucleotides have been explored by several teams and discussed in a comprehensive review by Luyten and Herdewijn.[110] Various modifications have been made to increase the acidity of N^3–H or to increase stacking interaction with other bases. This includes C5 acetylenic linker **241** introduced under Sonogashira cross-coupling (Figure 12).[111]

As part of this research, considerable work has been done to synthesize amino-containing C5 side chains **242**, which in many cases do enhance stabilization through stacking of the C5 amide group with neighboring bases, including an amido moiety and a guanidinium side chain, (Figure 13).[112]

Duplex and triplex formation of mixed pyrimidine oligonucleotides with stacking of phenyltriazole moities of nucleosides **246** (or **247**) in the major groove has

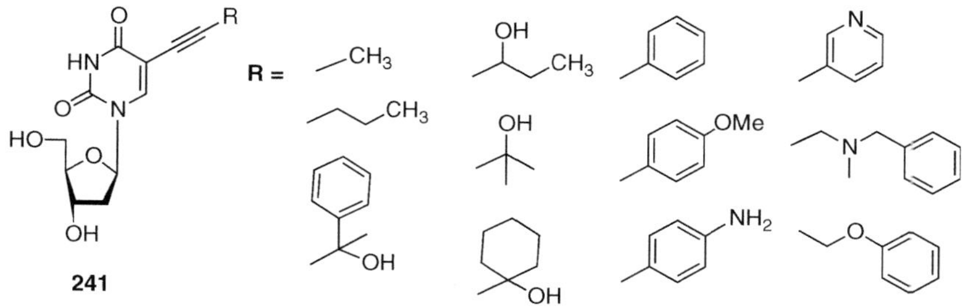

Figure 12. Some examples of C5 substituents that stabilize DNA duplexes.

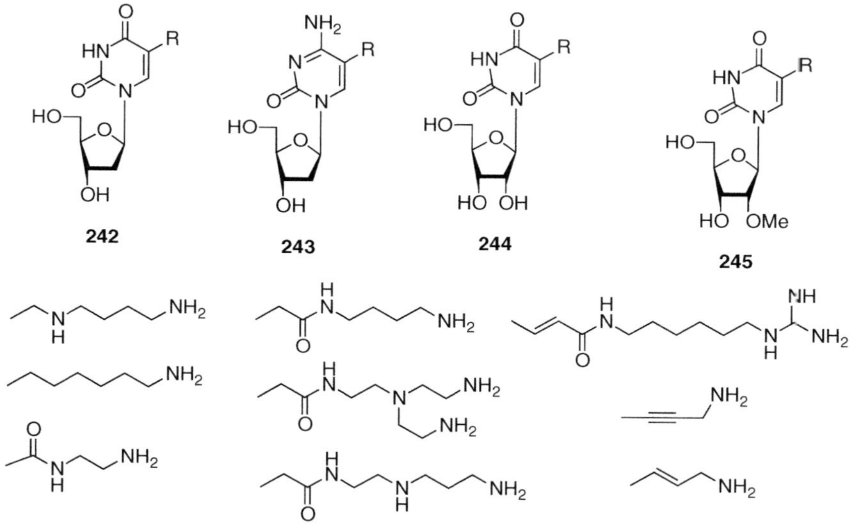

Figure 13. Some examples of amino-containing C5 substituents that stabilize DNA duplexes.

recently been reported (Figure 14). For example, the 5-(1-phenyl-1,2,3-triazol-4-yl)-2′-deoxycytine **247** was synthesized from a CuAAC procedure and incorporated into mixed pyrimidine oligonucleotide sequences.[113]

Several groups have explored the repertoire of acceptable functional groups, including sulfur,[114] at C5 of pyrimidine nucleosides for their incorporation into DNA. For example, Eaton and co-workers prepared several UTP derivatives (Figure 15) as substrates for T7 RNA-polymerase-mediated synthesis of modified RNA.[115]

Fluorescent nucleoside analogues **249** bearing an aromatic hydrocarbon (e.g., naphthalene, phenanthrene, pyrene) (Figure 16) as a nucleobase surrogate is an important class of compounds. Such compounds absorb light in the ultraviolet region around 345 nm, and emission occurs in the violet-blue region of the spectra.[116]

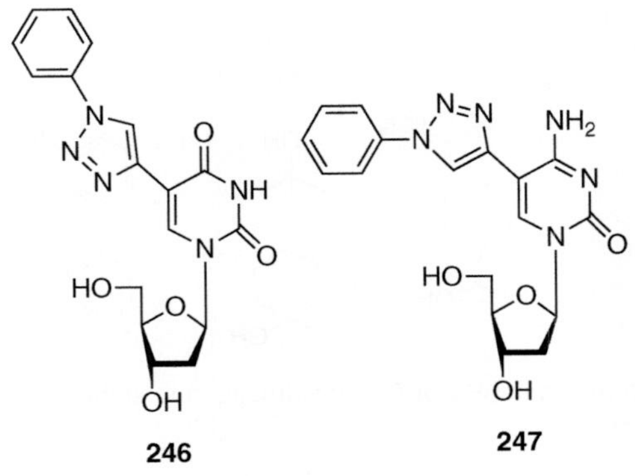

Figure 14. 5-(Phenyltriazole)pyrimidine nucleosides.

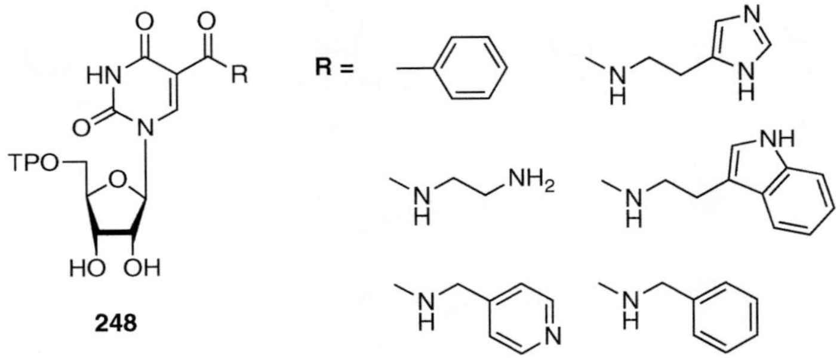

Figure 15. Examples of modified nucleoside triphosphate bearing a fluorescent probe at C5.

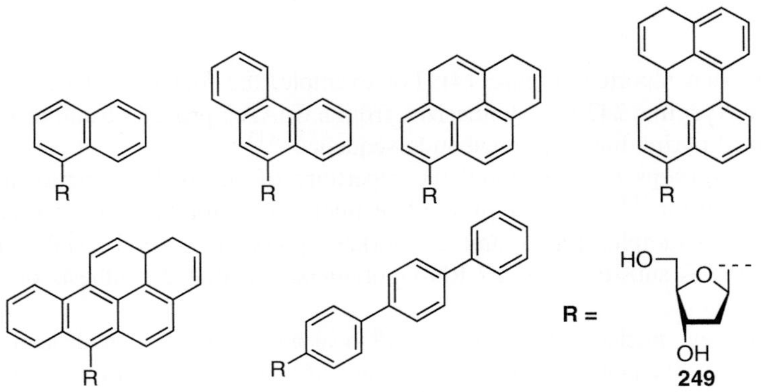

Figure 16. Fluorescent nucleoside analogues with polycyclic aromatic hydrocarbons.

However, several other classes of fluorescent nucleosides have been reported, such as polythiophene derivatives, coumarin nucleosides, and Nile Red nucleosides; Seela and co-workers employed an ethynyl linker at position 7 of deaza-adenosine to develop a family of compounds with enhanced fluorescent properties.[117] The fluorescent nucleosides have been described in a comprehensive review by Wilhelmsson.[118]

The heterocycles have also been modified in the frame of the concept of nonpolar isosteres of the natural DNA bases (Figure 17). The purpose of such compounds was to use them as probes to base pairing by hydrogen bonding. Structures **250–253** (which lack NH and O hydrogen-bonding groups) use benzene or indole (or benzimidazole) as a framework for pyrimidines and purines, respectively.[119]

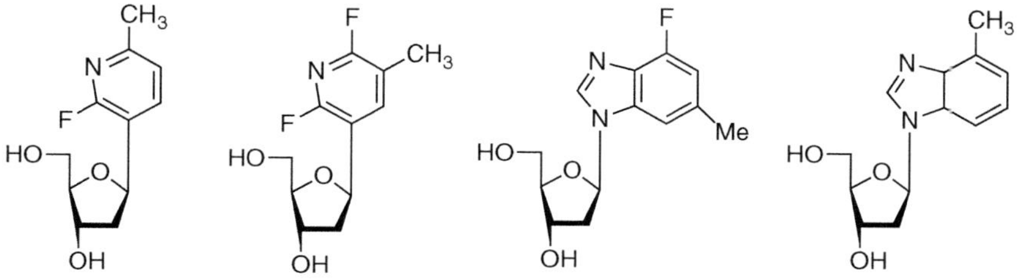

Figure 17. Examples of nonpolar nucleoside isosteres.

Finally, an unusual mechanism of thymidylate biosynthesis has recently been discovered in organisms containing the *thyX* gene; this gene codes for a flavin-dependent thymidylate synthase (FDTS). Following the recent trapping of an intermediate in the reaction catalyzed by FDTS[120] and because several human pathogens depend on FDTS for DNA biosynthesis, it makes an attractive target for antibiotic drugs.[121] Thus, recently, Herdewijn et al. reported the synthesis of 5-substituted 2'-deoxyuridine monophosphate analogues as inhibitors of FDTS in *Mycobacterium tuberculosis*.[122] Various functional groups were introduced at the C5-position of pyrimidine 2'-deoxynucleosides **254–259** from an initial Pd(0)-catalyzed cross-coupling (under Suzuki–Miyaura, Sonogashira, or Heck conditions) (Figure 18).

5. CONCLUDING REMARKS

During recent years, new nucleosides have been approved to treat important viral diseases. Many other nucleosides are used in biochemistry and biotechnology, where, for example, they are incorporated into oligonucleotides as molecular tools for various applications or for probing base-pair recognition by polymerases. Modified nucleobases can also be capable of forming a Watson–Crick type of base pair. In the near future, more accurate compounds will be required for both medical applications and for chemically oriented methods of gene detection. Thus, nucleosides are very important not only because of their biological properties as

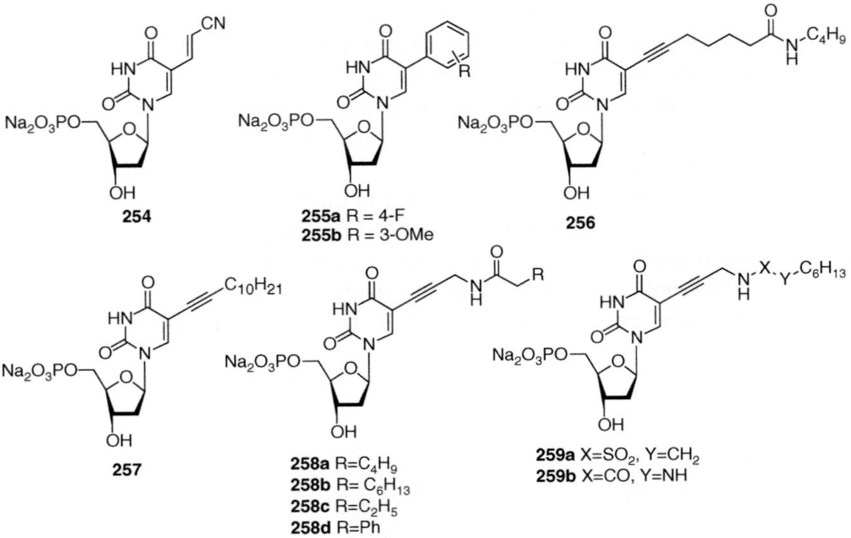

Figure 18. Structures of synthesized ThyX-inhibitors.

antiviral, antitumor, and antibiotic agents but also because they have induced or take advantage of new developments in both heterocycle and nucleoside chemistry. As the field of transition metal catalysis has matured, new areas have emerged that challenge the frontiers of synthetic organic chemistry and allow the creation of an unlimited variety of modified nucleosides. The best is yet to come!

ABBREVIATIONS

acac	acetylacetonate
AICAR	5-amino-1-β-D-ribofuranosyl imidazole-4-carboxamide
ara-A	arabinoadenosine
ara-C	arabinocytidine
BSA	bis(trimethylsilyl)acetamide
BVDU	5-(2-bromovinyl)-arabinouracil
CAN	cerium ammonium nitrate
coe	cyclooctene
CuAAC	copper-catalyzed azide-alkyne cycloadditions
Cy	cyclohexyl
DBU	1,8-Diazabicyclo[5.4.0]undec-7-ene
DCM	dichloromethane
ddI	2′,3′-dideoxyinosine
DIPEA	N,N-diisopropylethylamine
DMF	dimethylformamide
DNA	deoxyribonucleic acid

DPEPhos	bisdiphenylphosphinophenyl ether
dppp	1,3-bis(diphenylphosphino)propane
FDA	U.S. Food and Drug Administration
FDTS	flavin-dependent thymidylate synthase
HBV	hepatitis B virus
HIV	human immunodeficiency virus
HSV	herpes simplex virus
IDU	5-iodo-2′-deoxyuridine
MW	microwave
NBS	N-bromosuccinimide
NCS	N-chlorosuccinimide
NIS	N-iodosuccinimide
NMP	N-methylmorpholine
NMR	nuclear magnetic resonance
Piv	pivaloyl
RNA	ribonucleic acid
RuAAC	ruthenium-catalyzed azide–alkyne cycloaddition
SNAr	nucleophilic aromatic substitution
TBAB	tetra-n-butylammonium bromide
TFA	trifluoroacetic acid
THF	tetrahydrofuran
Tol	p-tolyl
TSAO	[2′,5′-bis-O-($tert$-butyldimethylsilyl)-β-D-ribofuranosyl]-3′-spiro-5″-(4″-amino-1″,2″-oxathiole-2″,2″-dioxide)nucleoside

REFERENCES

1. (a) Yokohama, M.; Momotake, A. *Synthesis* **1999**, 1541–1554. (b) Borthwick, A. D.; Biggadike, K. *Tetrahedron* **1992**, *48*, 571–774. (c) Agrofoglio, L. A.; Suhas, E.; Farese, A.; Condom, R.; Challand, S. R.; Earl, R. A.; Guedj, R. *Tetrahedron* **1994**, *50*, 10611–10838. (d) Akella, L. B.; Vince, R. *Tetrahedron* **1996**, *52*, 2789–2794. (e) Ferrero, M.; Gotor, V. *Chem. Rev.* **2000**, *100*, 4319–4348. (f) Zemlicka, J. *Pharmacol. Ther.* **2000**, *85*, 251–266. (g) Agrofoglio, L. A.; Challand, S. R. Acyclic, Carbocyclic and L-Nucleosides; Kluwer Academic Publishers: Dordrecht, **1998**. (h) Ichikawa, E.; Kato, K. *Synthesis* **2002**, 1–28. (i) Ichikawa, E.; Kato, K. *Curr. Med. Chem.* **2001**, *8*, 385–423. (j) Huryn, D. M.; Okabe, M. *Chem. Rev.* **1992**, *92*, 1745–1768.
2. Prusoff, W. H. *Biochim. Biophys. Acta.* **1959**, *32*, 295–296.
3. Wu, Q.; Simons, C. *Synthesis* **2004**, 1533–1553.
4. Agrofoglio, L. A.; Gillaizeau, I.; Saito, Y. *Chem. Rev.* **2003**, *103*, 1875–1916.
5. Herdewijn, P., Ed. *Modified Nucleosides in Biochemistry*, Biotechnology and Medicine; Wiley-VCH: Weinheim, Germany, **2008**.
6. For reviews on C–H activation chemistry, see (a) Chen, X.; Engle, K. M.; Wang, D.-H.; Yu, J.-Q. *Angew. Chem. Int. Ed.* **2009**, *48*, 5094–5115. (b) Crabtree, R. H. *Chem. Rev.* **1985**, *85*, 245–269. (c) Shilov, A. E.; Shul'pin, G. B. *Chem. Rev.* **1997**, *97*, 2879–2932. (d) Stahl, S. S.; Labinger, J. A.; Bercaw, J. E. *Angew. Chem. Int. Ed.*

1998, *37*, 2180–2192. (e) Bergman, R. G. *Nature* **2007**, *446*, 391–393. (f) Godula, K.; Sames, D. *Science* **2006**, *312*, 67–72. (g) Labinger, J. A.; Bercaw, J. E. *Nature* **2002**, *417*, 507–514. (h) Jia, C.; Kitamura, T.; Fujiwara, Y. *Acc. Chem. Res.* **2001**, *34*, 633–639. (i) Dyker, G. *Angew. Chem. Int. Ed.* **1999**, *38*, 1698–1712. (j) Pfeffer, M.; Ritleng, V. *Chem. Rev.* **2002**, *102*, 1731–1769. (k) Ackermann, L.; Vicente, R.; Kapdi, A. R. *Angew. Chem. Int. Ed.* **2009**, *48*, 9792–9826.

7. Tan, K. L.; Bergman, R. G.; Ellman, J. A. *J. Am. Chem. Soc.* **2002**, *124*, 13964–13965.
8. Bellina, F.; Cauteruccio, S.; Mannina, L.; Rossi, R.; Viel, S. *J Org. Chem.* **2005**, *70*, 3997–4005.
9. Bellina, F.; Cauteruccio, S.; Rossi, R. *Eur. J. Org. Chem.* **2006**, *11*, 1379–1382.
10. Čerňa, I.; Pohl, R.; Klepetářová, B.; Hocek, M. *Org. Lett.* **2006**, *8*, 5389–5392.
11. Sahnoun, S.; Messaoudi, S.; Peyrat, J.-F.; Brion, J.-D.; Alami, M. *Tetrahedron Lett.* **2008**, *49*, 7279–7283.
12. Sahnoun, S.; Messaoudi, S.; Brion, J.-D.; Alami, M. *Org. Biomol. Chem.* **2009**, *7*, 4271–4278.
13. Chiong, H. A.; Daugulis, O. *Org. Lett.* **2007**, *9*, 1449–1451.
14. Sahnoun, S.; Messaoudi, S.; Brion, J.-D.; Alami, M. *Eur. J. Org. Chem.* **2010**, *12*, 6097–6102.
15. Vabre, R.; Chevot, F.; Legraverend, M.; Piguel, S. *J. Org. Chem.* **2011**, *76*, 9542–9547.
16. Čerňa, I.; Pohl, R.; Klepetářová, B.; Hocek, M. *J. Org. Chem.* **2010**, *75*, 2302–2308.
17. Vaňková, B.; Krchňák, V.; Soural, M.; Hlaváč, J. *ACS Comb. Sci.* **2011**, *13*, 496–500.
18. Qu, G.-R.; Xin, P.-Y.; Niu, H.-Y.; Wang, D.-C.; Ding, R.-F.; Guo, H.-M. *Chem. Commun.* **2011**, 11140–11142.
19. Cerna, I.; Pohl, R.; Hocek, M. *Chem. Commun.* **2007**, 4729–4730.
20. Barbero, N.; SanMartin, R.; Dominguez, E. *Org. Biomol. Chem.* **2010**, *8*, 841–845.
21. Čerňová, M.; Pohl, R.; Hocek, M. *Eur. J. Org. Chem.* **2009**, *11*, 3698–3701.
22. Kim, K. H.; Lee, H. S.; Kim, J. N. *Tetrahedron Lett.* **2011**, *52*, 6228–6233.
23. Lebrasseur, N.; Larrosa, I. *J. Am. Chem. Soc.* **2008**, *130*, 2926–2927.
24. Majumdar, K. C.; Sinha, B.; Maji, P. K.; Chattopadhyay, S. K. *Tetrahedron* **2009**, *65*, 2751–2756.
25. Besselièvre, F.; Piguel, S. *Angew. Chem. Int. Ed.* **2009**, *48*, 9553–9556.
26. Pacheco Berciano, B.; Lebrequier, S.; Besselièvre, F. O.; Piguel, S. *Org. Lett.* **2010**, *12*, 4038–4041.
27. Matsuyama, N.; Hirano, K.; Satoh, T.; Miura, M. *Org. Lett.* **2009**, *11*, 4156–4159.
28. Lebrasseur, N.; Larrosa, I. *J. Am. Chem. Soc.* **2008**, *130*, 2926–2927.
29. Liston, D. J.; Lee, Y. J.; Scheidt, W. R.; Reed, C. A. *J. Am. Chem. Soc.* **1989**, *111*, 6643–6648.
30. Brand, J. P.; Charpentier, J.; Waser, J. *Angew. Chem. Int. Ed.* **2009**, *48*, 9346–9349.
31. Lakshman, M. K.; Deb, A. C.; Chamala, R. R.; Pradhan, P.; Pratap, R. *Angew. Chem. Int. Ed.* **2011**, *50*, 11400–11404.
32. Wan, J.; Zhu, R.; Xia, Y.; Qu, F.; Wu, Q.; Yang, G.; Neyts, J.; Peng, L. *Tetrahedron Lett.* **2006**, *47*, 6727–6731.
33. Xia, Y.; Liu, Y.; Wan, J.; Wang, M.; Rocchi, P.; Qu, F.; Iovanna, J. L.; Peng, L. *J. Med. Chem.* **2009**, *52*, 6083–6096.

34. Zhu, R.; Wang, M.; Xia, Y.; Qu, F.; Neyts, J.; Peng, L. *Bioorg. Med. Chem. Lett.* **2008**, *18*, 3321–3327.
35. Wan, J.; Xia, Y.; Liu, Y.; Wang, M.; Rocchi, P.; Yao, J.; Qu, F.; Neyts, J.; Iovanna, J. L.; Peng, L. *J. Med. Chem.* **2009**, *52*, 1144–1155.
36. Huisgen, R. *Angew. Chem. Int. Ed. Engl.* **1963**, *2*, 565–598.
37. (a) Rostovtsev, V. V.; Green, L. G.; Fokin, V. V.; Sharpless, K. B. *Angew. Chem. Int. Ed. Engl.* **2002**, *41*, 2596–2599. (b) Tornoe, C. W.; Christensen, C.; Meldal, M. *J. Org. Chem.* **2002**, *67*, 3057–3064.
38. Zhang, L.; Chen, X.; Xue, P.; Sun, H. H. Y.; Williams, I. D.; Sharpless, K. B.; Fokin, V. V.; Jia, G. *J. Am. Chem. Soc.* **2005**, *127*, 15998–15999.
39. (a) Rodionov, V. O.; Fokin, V. V.; Finn, M. G. *Angew. Chem. Int. Ed.* **2005**, *44*, 2210–2215. (b) Himo, F.; Lovell, T.; Hilgraf, R.; Rostovtsev, V. V.; Noodleman, L.; Sharpless, B. K.; Fokin, V. K. *J. Am. Chem. Soc.* **2005**, *127*, 210–216. (c) Boren, B. C.; Narayan, S.; Zhang, L.; Zhao, H.; Lin, Z.; Jia, G.; Fokin, V. V. *J. Am. Chem. Soc.* **2008**, *130*, 8923–8930.
40. Amblard, F.; Cho, J. H.; Schinazi, R. F. *Chem. Rev.* **2009**, *109*, 4207–4220.
41. Cho, J. H.; Bernard, D. L.; Sidwell, R. W.; Kern, E. R.; Chu, C. K. *J. Med. Chem.* **2006**, *49*, 1140–1148.
42. (a) Guerguez, R.; Bougrin, K.; El Akri, K.; Benhida, R. *Tetrahedron Lett.* **2006**, *47*, 4807–4811; (b) El Akri, K.; Bougrin, K.; Balzarini, J.; Faraj, A.; Benhida, R. *Bioorg. Med. Chem. Lett.* **2007**, *17*, 6656–6659.
43. Pradere, U.; Roy, V.; McBrayer, T. R.; Schinazi, R. F.; Agrofoglio, L. A. *Tetrahedron* **2008**, *64*, 9044–9051.
44. Ermolat'ev, D. S.; Pravinchandra Mehta, V.; Van der Eycken, E. V. *QSAR Comb. Sci.* **2007**, *26*, 1266–1273.
45. Li, L.; Zhang, G.; Zhu, A.; Zhang, L. *J. Org. Chem.* **2008**, *73*, 3630–3633.
46. Malnuit, V.; Duca, M.; Manout, A.; Bougrin, K.; Benhida, R. *Synlett* **2009**, *13*, 2123–2128.
47. (a) Broggi, J.; Joubert, N.; Aucagne, V.; Zevaco, T.; Berteina-Raboin, S.; Nolan, S. P.; Agrofoglio, L. A. *Nucleosides Nucleoides. Nucleic Acids* **2007**, *26*, 779–783. (b) Broggi, J.; Joubert, N.; Aucagne, V.; Berteina-Raboin, S.; Diez-Gonzales, S.; Nolan, S.; Topalis, D.; Deville-Bonne, D.; Balzarini, J.; Neyts, J.; Andrei, G.; Snoeck, R.; Agrofoglio, L. A. *Nucleosodes Nucleoties Nucleic Acids* **2007**, *26*, 1391–1394. (c) Broggi, J.; Díez-González, S.; Petersen, J. L.; Berteina-Raboin, S.; Nolan, S.; Agrofoglio, L. A. *Synthesis* **2008**, 141–148.
48. Youcef, R. A.; Dos Santos, M.; Roussel, S.; Baltaze, J.-P.; Lubin-germain, N.; Uziel, J. *J. Org. Chem.* **2009**, *74*, 4318–4323.
49. Li, W.; Xia, Y.; Fan, Z.; Qu, F.; Wu, Q.; Peng, L. *Tetrahedron Lett.* **2008**, *49*, 2804–2809.
50. Park, S. M.; Lee, Y. S.; Kim, B. H. *Chem. Commun.* **2003**, 2912–2913.
51. Kocalka, P.; Andersen, N. K.; Jensen, F.; Nielsen, P. *ChemBioChem* **2007**, *8*, 2016–2116.
52. Lee, Y. S.; Park, S. M.; Kim, H. M.; Park, S. K.; Lee, K.; Lee, C. W.; Kim, B. H. *Bioorg. Med. Chem. Lett.* **2009**, *19*, 4688–4691.
53. Montagu, A.; Roy, V.; Balzarini, Y.; Snoeck, R.; Andrei, G.; Agrofoglio, L. *Eur. J. Med. Chem.* **2011**, *46*, 778–786.

54. Cosyn, L.; Palaniappan, K. K; Kim, S.-K.; Duong, H. T.; Gao, Z.-G. ; Jacobson, K. A.; Calenbergh S. V. *J. Med. Chem.* **2006**, *49*, 7373–7383.
55. Li, Z.; Chen, S.; Jiang, N.; Cui, G. *Nucleosides Nucleotides Nucleic Acids* **2003**, *22*, 419–435.
56. Popsavin, M.; Torović, L.; Spaić, S.; Stankov, S.; Kapor, A.; Tomić, Z.; Popsavin, V. *Tetrahedron* **2002**, *58*, 569–580.
57. Manferdini, M.; Morelli, C. F.; Veronese, A. C. *Tetrahedron* **2002**, *58*, 1005–1013.
58. For examples, see: (a) Xiang, Y.; Teng, Q.; Chu, C. K. *Tetrahedron Lett.* **1995**, *36*, 3781–3784. (b) Van Praanen, N. A.; Freeman, G. A.; Short, S. A.; Harvey, R.; Jansen, R.; Szczech, G.; Koszalka, G. W. *J. Med. Chem.* **1996**, *39*, 538–542. (c) Rahim, S. G.; Trivedi, N.; Bogunovic-Batchelor, M. V.; Hardy, G. W.; Mills, G.; Selway, J. W. T.; Snwoden, W.; Littler, E.; Coe, P. L.; Basnak, I.; Walker, R. T. *J. Med. Chem.* **1996**, *39*, 789–795. (d) Saluja, S.; Zou, R.; Drach, J. C.; Townsend, L. B. *J. Med. Chem.* **1996**, *39*, 881–891. (e) Nishizono, N.; Koike, N.; Yamagata, Y.; Fujii, S.; Matsuda, A. *Tetrahedron Lett.* **1996**, *37*, 925–928.
59. Hohnjec, M.; Kobe, J.; Valcavi, U.; Japelj, M. *J. Heterocycl. Chem.* **1975**, *12*, 909–912.
60. (a) Srivastava, P. C.; Robins, R. K. *J. Med. Chem.* **1983**, *26*, 445–448. (b) Streeter, D.; Robins, R. K. *Biochem. Biophys. Res. Commun.* **1983**, *115*, 544–550. (c) Kirsi, J. J.; North, J.; McKernan, P. A.; Murray, B. K.; Canonico, P. G.; Huggins, J.; Srivastava, P. C.; Robins, R. K. *Antimicrob. Agents Chemother.* **1983**, *24*, 353–361. (d) Goldstein, B. M.; Leary, J. F.; Farley, B. A.; Marquez, V. E.; Rowley, P. T. *Blood* **1991**, *78*, 593–598.
61. Jayaram, H. N.; Dion, R. L.; Glazer, R. I.; Johns, D. G.; Robins, R. K.; Srivastsava, P. C.; Cooney, D. A. *Biochem. Pharmacol.* **1982**, *31*, 2371–2380.
62. Franchetti, P.; Cappellacci, L.; Grifantini, M.; Barzi, A.; Nocentini, G.; Yang, H.; O'Connor, A.; Jayaram, H. N.; Carrell, C.; Goldstein, B. M. *J. Med. Chem.* **1995**, *38*, 3829–3837.
63. Franchetti, P.; Cappellacci, L.; Sheikha, G. A.; Jayaram, H. N.; Gurudutt, V. V.; Sint, T.; Schneider, B. P.; Jones, W. D.; Goldstein, B. M.; Perra, G.; De Montis, A.; Loi, A. G.; La Colla, P.; Grifantini, M. *J. Med. Chem.* **1997**, *40*, 1731–1737.
64. Seley, K. L.; Zhang, L.; Hagos, A. *Org. Lett.* **2001**, *3*, 3209–3210.
65. (a) Polak, M.; Seley, K. L.; Plavec, J. *J. Am. Chem. Soc.* **2004**, *126*, 8159–8166. (b) Zimmermann, S. C.; Sadler, J. M.; Andrei, G.; Snoeck, R.; Balzarini, J.; Seley-Radtke, K. L. *Med. Chem. Commun.* **2011**, *2*, 650–654.
66. Moukha-Chafiq, O.; Taha, M. L.; Lazrek, H. B.; Vasseur, J.-J.; Pannecouque, C.; Witvrouw, M.; De Clercq, E. *Farmaco* **2002**, *57*, 27–32.
67. Taha, M. L.; Lazreck, H. B. *Bull. Soc. Chim. Belg.* **1997**, *106*, 163–168.
68. Bloch, A.; Gutschman, G.; Currie, G. L.; Robbins, R. K.; Robins, M. J. *J. Med. Chem.* **1973**, *16*, 294–297.
69. Khare, G. P.; Sidwell, R. W.; Huffman, J. H.; Tolman, R. L.; Robins, R. K. *Proc. Soc. Exp. Biol. Med.* **1972**, *40*, 990–994.
70. Devadas, B.; Rogers, T. E.; Gray, S. H. *Synth. Commun.* **1995**, *25*, 3199–3210.
71. Momparler, R. L.; Momparler, L. F. *Cancer Chemother. Pharmacol.* **1989**, *25*, 51–54.
72. (a) Currie, B. L.; Robins, R. K.; Robins, M. J. *J. Heterocycl. Chem.* **1970**, *7*, 323–327. (b) Cook, D.; Day, R. T.; Robins, R. K. *J. Heterocycl. Chem.* **1977**, *14*, 1295–1298.
73. Searls, T.; McLaughlin, L. W. *Tetrahedron* **1999**, *55*, 11985–12270.

74. Barral, K.; Hider, R. C.; Balzarini, J.; Neyts, J.; De Clercq, E.; Camplo, M. *Bioorg. Med. Chem. Lett.* **2003**, *13*, 4371–4374.
75. (a) Georgiou, N. A.; van der Bruggen, T.; Hider, R. C.; Marx, J.; van Asbeck, B. S. *Eur. J. Clin. Invest.* **2002**, *32*, 91–96. (b) Georgiou, N. A.; van der Bruggen, T.; Oudshoorn, M.; Nottet, H.; Marx, J.; van Asbeck, B. S. *J. Clin. Virol.* **2001**, *20*, 141–147.
76. Stambasky, J.; Hocek, M.; Kocovsky, P. *Chem. Rev.* **2009**, *109*, 6729–6764.
77. Revankar, G. R.; Robins, R. K. In Chemistry of Nucleosides and Nucleotides; Townsend, L. B., Ed.; Plenum: New York, **1991**; Vol. 2, pp. 161–398.
78. (a) Bobek, M.; Bloch, A. *Nucleosides Nucleotides* **1994**, *13*, 429–435. (b) Krawczyk, S. H.; Nassiri, M. R.; Kucera, L. S.; Kern, E. R.; Ptak, R. G.; Wotring, L. L.; Drach, J. C.; Townsend, L. B. *J. Med. Chem.* **1995**, *38*, 4106–4114. (c) Krawczyk, S. H.; Renau, T. E.; Nassiri, M. R.; Westerman, A.; Wotring, L. L.; Drach, J. C.; Townsend, L. B. *J. Med. Chem.* **1995**, *38*, 4115–4119. (d) Finch, R. A.; Revankar, G. R.; Chan, P. K. *Anticancer Drug Des.* **1997**, *12*, 205–215. (e) Loomis, C. R.; Bell, R. M *J. Biol. Chem.* **1988**, *263*, 1682–1692.
79. Girardet, J.-L.; Gunic, E.; Esler, C.; Cieslak, D.; Pietrzkowski, Z.; Wang, G. *J. Med. Chem.* **2000**, *43*, 3704–3713.
80. Rajeev, K. G.; Broom, A. D. *Org. Lett.* **2000**, *2*, 3595–3598.
81. (a) Girniene, J.; Gueyrard, D.; Tatibouët, A.; Sackus, A.; Rollin, P. *Tetrahedron Lett.* **2001**, *42*, 2977–2980. (b) Girniene, J.; Apremont, G.; Tatibouët, A.; Sackus, A.; Rollin, P. *Tetrahedron* **2004**, *60*, 2609–2619.
82. Capon, R. J.; Trotter N. S. *J. Nat. Prod.* **2005**, *68*, 1689–1691.
83. Mieczkowski, A.; Roy, V.; Agrofoglio, L. A. *Chem. Rev.* **2010**, *110*, 1828–1856.
84. Len, C.; Mondon, M.; Lebreton, J. *Tetrahedron* **2008**, *64*, 7453–7475.
85. (a) Bergmann, W.; Feeney, R. J. *J. Am. Chem. Soc.* **1950**, *72*, 2809–2810. (b) Bergmann, W.; Feeney, R. J. *J. Org. Chem.* **1951**, *16*, 981–987. (c) Newman, D. J.; Cragg, G. M. *J. Nat. Prod.* **2004**, *67*, 1216–1238.
86. Wang, P.; Stuyver, L. J.; Watanabe, K. A.; Hassan, A.; Chun, B.-K.; Hollecker, L. WO Patent 2004013300, **2004**.
87. (a) Meena, S. M.; Pierce, K.; Szostak, J. W.; McLaughlin, L. W. *Org. Lett.* **2007**, *9*, 1161–1163. (b) Gaynor J. W.; Bentley J.; Cosstick R. *Nat. Protocols* **2007**, *2*, 3122–3135. (c) Sabbagh, G.; Fettes K. J.; Gosain R.; O'Neil I. A.; Cosstick, R. *Nucleic Acids Res.* **2004**, *32*, 495–501. (d) Perbost, M.; Hoshiko, T.; Morvan, F.; Swayze, E.; Griffey, R. H.; Sanghvi, Y. S. *J. Org. Chem.* **1995**, *60*, 5150–5156. (e) Seela, F.; Worner, K.; Rosemeyer, H. *Helv. Chim. Acta 77*, **1995**, 883–896. (f) Kimura, J.; Yagi, K.; Suzuki, H.; Mitsunobu, O. *Bull. Chem. Soc. Jpn.* **1980**, *53*, 3670–3677. (g) Watanabe, K. A.; Chu, C. K.; Reichman, U.; Fox, J. J. In Nucleic Acid Chemistry, vol. 1 (ed.Townsend, L. B.), Wiley, New York, **1978**, 343–345. (h) Kimura, J.; Fujisawa, Y.; Sawada T.; Mitsunobu, O. *Chem. Lett.* **1974**, 691–692. (i) Wada, M.; Mitsunobu, O. *Tetrahedron Lett.* **1972**, 1279–1282. (j) Shibuya, S.; Kuninaka, A.; Yoshino, H. *Chem. Pharm. Bull.* **1974**, *22*, 719–721.
88. Chen, L.-C.; Su, T.-L.; Pankiewicz, K. W.; Watanabe, K. A. *Nucleosides Nucleotides* **1989**, *8*, 1179–1188.
89. Shibuya, S.; Ueda, T. *Chem. Pharm. Bull.* **1980**, *28*, 939–946.
90. Janeba, Z; Holy, A.; Masojidkova, M. *Collect. Czech. Chem. Commun.* **2000**, *65*, 1698–1712.

91. Holý, A.; Kohoutova, J.; Merta, A.; Votruba, I. *Collect. Czech. Chem. Commun.* **1986**, *51*, 459–477.
92. Janeba, Z; Holy, A.; Masojidkova, M. *Collect. Czech. Chem. Commun.* **2000**, *65*, 1126–1144.
93. Hampton, A.; Nichol, A. W. *Biochemistry* **1966**, *5*, 2076–2082.
94. (a) Ogilvie, K. K.; Iwacha, D. J. *Can. J. Chem.* **1974**, *52*, 1787–1797. (b) Cook, A. F. *J. Med. Chem.* **20**, **1977**, 344–348.
95. Stout, M. G.; Robins, R. K. *J. Org. Chem.* **1968**, *33*, 1219–1225.
96. Hutzenlaub, W.; Kobayashi, K.; Pfleiderer, W. *Chem. Ber.* **1968**, *33*, 3217–3227.
97. Olgivie, K. K.; Slotin, L. *J. Org. Chem.* **1971**, *36*, 2556–2558.
98. Ikehara, M.; Tezuka, S. *Tetrahedron Lett.* **1972**, *13*, 1169–1170.
99. Dudycz, L.; Stolarski, R.; Pless, R.; Shugar, D. *Z. Naturforsch.* **1979**, *34c*, 359–373.
100. (a) Sowa, T.; Tsunoda, K. *Bull. Chem. Soc. Jpn.* **1975**, *48*, 505–507. (b) Krecmerova, M.; Hrebabecky, H.; Holý, A. *Collect. Czech. Chem. Commun.* **1996**, *61*, 645–655. (c) Krecmerova, M.; Hrebabecky, H.; Holy, A. *Collect. Czech. Chem. Commun.* *61*, **1996**, 627–644.
101. (a) Shealy, Y. F.; O'Dell, C. A. *J. Pharm. Sci.* **1979**, *68*, 668–670. (b) Shealy, Y. F.; O'Dell C. A. *J. Heterocycl. Chem.* **1980**, *17*, 353–358.
102. (a) Takatsuki, K.; Yamamoto, M.; Ohgushi, S.; Kohmoto, S.; Kishikawa, K.; Yamashita, H. *Tetrahedron Lett.* **2004**, *45*, 137–140. (b) Takatsuki, K.; Ohgushi, S.; Kohmoto, S.; Kishikawa, K.; Yamamoto, M. *Nucleosides Nucleotides* **2006**, *25*, 719–734.
103. Fossey, C.; Laduree, D.; Robba, M. *J. Heterocycl. Chem.* **1995**, *32*, 627–635.
104. Ahrem, A. A.; Ermolenko, T. M.; Tumoschiuk, B. A. *Zh. Org. Khim.* **1985**, *XXI*, 2185–2190.
105. Ahrem, A. A.; Ermolenko, T. M.; Tumoschiuk, B. A. *Zh. Org. Khim.* **1985**, *XXI*, 2190–2195.
106. Maki, Y.; Sako, M.; Saito, T.; Hirota, K. *Heterocycles* **1985**, *27*, 347–350.
107. Mieczkowski, A.; Peltier, P.; Zevaco, T.; Agrofoglio, L. A. *Tetrahedron* **2009**, *65*, 4053–4059.
108. Agrofoglio, L. A.; Gillaizeau, I.; Saito, Y. *Chem. Rev.* **2003**, *103*, 1875–1916.
109. Mieczkowski, A.; Roy, V.; Blu, J.; Agrofoglio, L. A. *Tetrahedron* **2009**, *65*, 9791–9796.
110. Luyten, I.; Herdewijn, P. *Eur. J. Med. Chem.* **1998**, *33*, 515–576.
111. Kottysch, T.; Ahlborn, C.; Brotzel, F.; Ritcher, C. *Chem. Eur. J.* **2004**, *10*, 4017–4028.
112. (a) Ozaki, H.; Mine, M.; Shinozuka, K.; Sawai, H. *Nucleosides Nucleotides Nucleic Acids* **2004**, *23*, 339–346. (b) Ito, T.; Ueno, Y.; Komatsu, Y.; Matsuda, A. *Nucleic Acids Res.* **2003**, *31*, 2514–2523. (c) Ozaki, H.; Mine, M.; Ogawa Y.; Sawai H. *Bioorg. Chem.* **2001**, *29*, 187–197.
113. Andersen, N. K.; Døssing, H.; Jensen, F.; Vester, B.; Nielsen, P. *J. Org. Chem.* **2011**, *76*, 6177–6187.
114. Held, H. A.; Benner, S. A. *Nucleic Acids Res.* **2002**, *30*, 3857–3869. (b) Roychowdhury, A.; Illangkoon, H.; Hendrickson, C. L.; Benner, S. A. *Org. Lett.* **2004**, *6*, 489–492.
115. Vaught, J. D.; Dewey, T.; Eaton, B. E. *J. Am. Chem. Soc.* **2004**, *126*, 11231–11237.

116. (a) Ren, R.; Chaudhuri, N. C.; Paris, P. L.; Rumney, S. IV; Kool, E. T. *J. Am. Chem. Soc.* **1996**, *118*, 7671–7678. (b) Strässler, C.; Davis, N. E.; Kool, E. T. *Helv. Chim. Acta* **1999**, *82*, 2160–2171. (c) Wilson, J. N.; Kool, E. T. *Org. Biomol. Chem.* **2006**, *4*, 4265–4274. (d) Gao, J. M.; Strassler, C.; Tahmassebi, D.; Kool, E. T. *J. Am. Chem. Soc.* **2002**, *124*, 11590–11591. (e) Segal, M.; Fischer, B. *Org. Biomol. Chem.* **2012**, *10*, 1571–1580. (f) Hurley, D. J.; Seaman, S. E.; Mazura, J. C.; Tor, Y. *Org. Lett.* **2002**, *14*, 2305–2308.

117. Seela, F.; Zulauf, M.; Sauer, M.; Deimel, M. *Helv. Chim. Acta* **2000**, *83*, 910–927.

118. Wilhelmsson, L. M. Q. *Rev. Biophys.* **2010**, *43*, 159–183.

119. (a) Schweitzer, B. A.; Kool, E. T. *J. Org. Chem.* **1994**, *59*, 7238–7242. (b) Guckian, K. M.; Morales, J. C.; Kool, E. T. *J. Org. Chem.* **1998**, *63*, 9652–9656.

120. Mishanima, T. V.; Koehn, E. M.; Conrad, J. A.; Palfey, B. A.; Lesley, S. A.; Kohen, A. *J. Am. Chem. Soc.* **2012**, *134*, 4442–4448.

121. (a) Koehn, E. M.; Fleischmann, T.; Conrad, J. A.; Palfey, B. A.; Lesley, S. A.; Mathews, I. I. *Nature* **2009**, *458*, 919–923. (b) Myllykallio, H.; Lipowski, G.; Leduc, D.; Filee, J.; Forterre, P.; Liebl, U. *Science* **2002**, *297*, 105–107.

122. Kögler, M.; Vanderhoydonck, B.; De Jonghe, S.; Rozenski, J.; Van Belle, K.; Herman, J.; Louat, T.; Parchina, A.; Sibley, C.; Lescrinier, E.; Herdewijn, P. *J. Med. Chem.* **2011**, *54*, 4847–4862.

3 Chemical Synthesis of Acyclic Nucleosides

HAI-MING GUO, SHAN WU, HONG-YING NIU, GE SONG, and GUI-RONG QU

College of Chemistry and Environmental Science, Key Laboratory of Green Chemical Media and Reactions of Ministry of Education, Henan Normal University, Xinxiang, China

1. INTRODUCTION

Human infectious diseases caused by viruses, increase with time because viruses reside in living body cells during reproduction and propagation. Viruses are difficult to wipe out and tend to mutate, attacking the body repeatedly and spreading diseases. Just like natural nucleosides, nucleoside compounds are made up of bases and ribose moieties. They have varying degrees of structural resemblance to natural nucleosides. Therefore, they can interfere with or directly attack the biosynthesis of proteins and nucleic acids. Many nucleoside analogues can inhibit the activity of DNA polymerase, thymidine kinase, or reverse transcriptase to terminate or restrain the extension and synthesis of viral DNA strands, so nucleosides play an important role as antiviruses. At present, most are universally recognized as potential antiviral drugs.

Researchers involved in the development of nucleoside drugs have found that these drugs have the disadvantages of strong toxic side effects with prolonged use and drug tolerance. Acyclic nucleosides, an important group of nucleoside drugs, are low in mammalian toxicity and low in drug tolerance. Since acyclovir (ACV) **1** was first reported to be a potent antiherpes drug in 1977–1978,[1] more and more chemists have been interested in searching intensively for clinically effective acyclic nucleosides, owing to their potential inhibitory potency to target such pathogenic agents as human immunodeficiency virus (HIV), herpes simplex virus (HSV), hepatitis B virus (HBV), hepatitis C virus (HCV), varicella zoster virus (VZV) and cytomegalovirus (CMV).[2,3]

Acyclic nucleoside is a nucleoside compound with acyclic chains in side a ribose ring. According to their bases, they are classified as purine acyclic nucleosides (Figure 1: **1–3**) and pyrimidine acyclic nucleosides (**4**). Based on the difference in

---

Chemical Synthesis of Nucleoside Analogues, First Edition. Edited by Pedro Merino.
© 2013 John Wiley & Sons, Inc. Published 2013 by John Wiley & Sons, Inc.

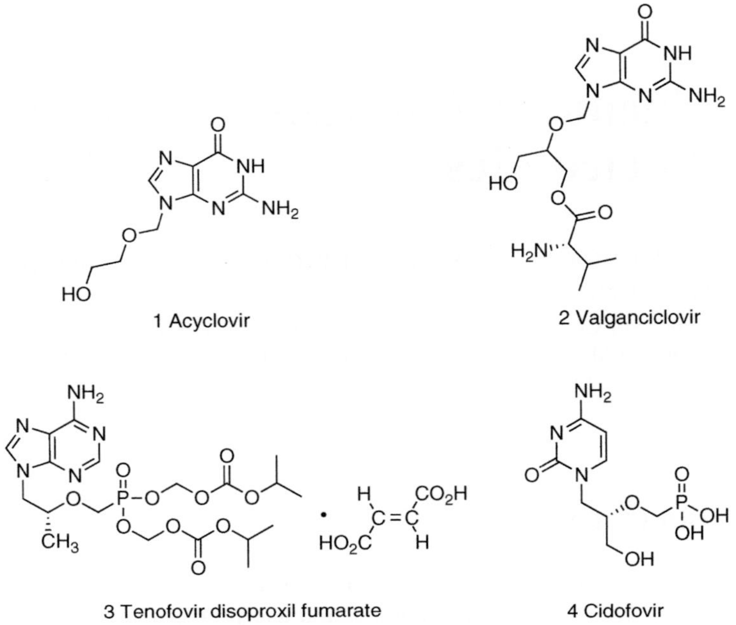

Figure 1. Representative acyclic nucleoside drugs.

their acyclic chains, they are classed as achiral acyclic nucleosides (**1**) and chiral acyclic nucleosides (**2–4**). The acyclic chains that link to basic groups are different in a myriad of ways. Representative acyclic nucleoside drugs include Zovirax topical (a brand name for acyclovir topical), Valcyte (a brand name for valganciclovir **2**), Valtrex (valacyclovir hydrochloride), Viread (tenofovir disoproxil fumarate **3**), Atripla (a fixed-dose combination tablet containing efavirenz, emtricitabine, and tenofovir disoproxil fumarate **3**), and Truvada (a fixed-dose combination of tenofovir and emtricitabine), which were among the top 200 brand-name drugs by retail dollars in 2008.

Although modifications can be obtained in different ways, methods such as alkylation, Mitsunobu reaction, Michael addition, and ring cleavage reaction seem to be most common. Aza-Michael addition plays an important part in building a C–N bond in organic chemistry, especially in the synthesis of acyclic nucleoside compounds. In 2005, He et al.[4] reported that they used the aza-Michael addition reaction to synthesize chiral acyclic nucleosides. Additionally, since the microwave was discovered to be a way of providing unconventional energy, it has been used widely in organic chemistry. For example, in 2007, Qu et al.[5] synthesized a series of acyclic nucleosides utilizing the microwave.

It is unfortunate that the present cultural documentary falls short of a systematic and practical summary. Therefore, our goal is to describe the main synthetical methods and detailed experimental procedures for most representative clinical drugs and their analogues. It is well known that structure determines the nature of most drugs.

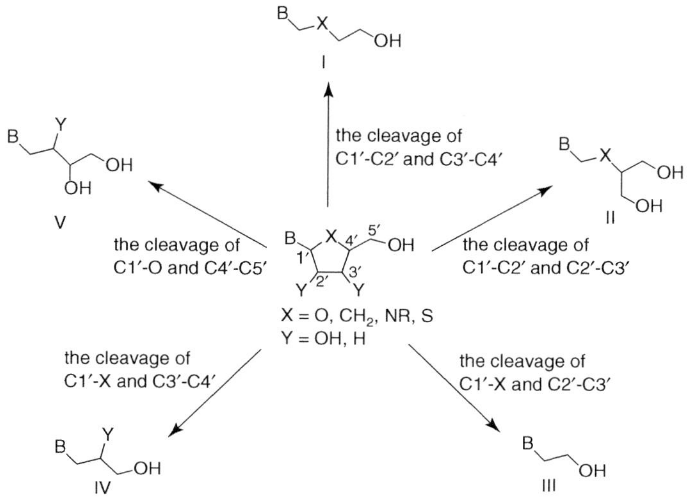

Figure 2. Types of acyclic nucleosides.

In this chapter we divide these compounds chiefly into five types according to the various acyclic side chains, because they are very similar to the various breakages of the sugar ring. To simplify description in the following passages, we call them types I, II, III, IV and V. In the structure of side chains, type I is very similar to products with breaking the C1′–C2′ and C3′–C4′ bonds in sugar rings. Type II and compounds resulting from breaking the C1′–C2′ and C2′–C3′ bonds in sugar rings are alike in the side chains. From the same perspective, type III is similar to the products of breaking the C1′–X and C2′–C3′ bonds in sugar rings. In the structure of side chains, type IV is very similar to the products of breaking the C1′–X and C3′–C4′ bonds in sugar rings. Type V and products with cleavage of C1′–O and C4′–C5′ bonds have similar side chains (Figure 2). When we tried to give a recapitulative name to each type, we encountered difficulties because of the multiplex changes in the side chains. Through comprehensive consideration, we decided to use the names of the most common drugs with a simple structure in each type. In these types, the side chains of acyclovir, ganciclovir, adefovir, cidofovir, and buciclovir have the simplest and most original structures; hence we call them acyclic nucleosides with acyclovir, ganciclovir, adefovir, cidofovir, and buciclovir analogue side chains.

2. SYNTHESIS OF ACYCLIC NUCLEOSIDES

2.1. Synthesis of Acyclic Nucleosides with Acyclovir Side-Chain Analogues (Type I)

From the type I structure we can see that there is no chiral center in this type. Each compound differs from the others depending on several distinct heteroatoms and whether it contains side chains (Figure 3). This type contains

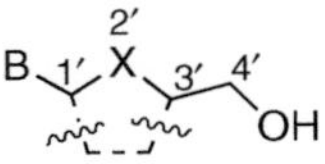

Figure 3. Structure of acyclovir side-chain analogues (type I).

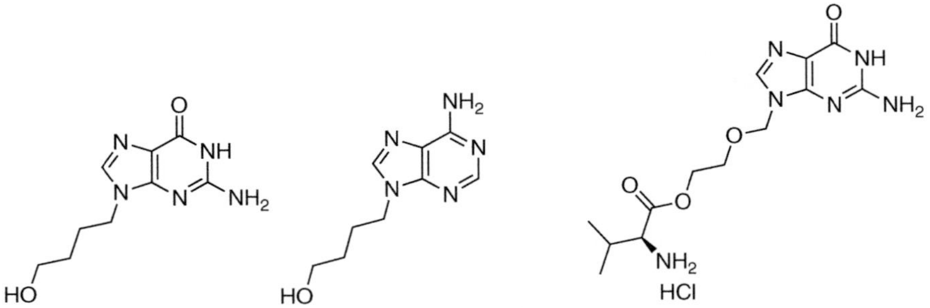

X = O, CH$_2$, NR, S

I

5 HBG
6 4-(6-amino-9H-purin-9-yl)butan-1-ol
7 Valacyclovir

Figure 4. Typical drugs in type I.

lots of clinical drugs and their derivatives, such as acyclovir (ACV, 9-[(2-hydroxyethoxy)methyl]guanine) **1**, HBG [9-(4-hydroxybutyl)guanine] **5**, 4-(6-amino-9H-purin-9-yl)butan-1-ol **6**, and valaciclovir (VCV) **7** (Figure 4).

2.1.1. With a 2′-O Side Chain (IA).

Acyclovir, the first acyclic nucleoside to be developed and listed, was developed and marketed by Burroughs Wellcome Company. It is a broad-spectrum highly active hypotoxic drug that has special characteristics used in treating the infections caused by herpes varicella zoster virus, cytomegalovirus (CMV), and others. Its inventor, Gertrude Belle Elion, won the Nobel Prize in Physiology or Medicine in 1988 owing partially to this achievement. There are many reports on its synthetic methods, in which the synthetic route starting from guanine is said to be best because guanine is easily obtained. In 1979, Barrio et al.[6] also reported a direct route to preparing 2-hydroxyethoxymethyl analogues of guanine, adenine, and cytosine. Chu and co-workers[7] synthesized a series of 5-bromo-, 5-iodo-, and 5-fluorouridines, thymidine, and cytidine, which were modified to 2′-halo-, azido-, and amino derivatives in 1981. In 1986, Kelley and Schaeffer[8] synthesized ACV1 in five steps starting from 7-formamido-5-methylthiofurazano[3,4-d]-pyrimidine.

Kumar and co-workers[9] overcame the long-standing problem of regioselective alkylation of guanine in 1999. They discovered the role of acids as catalysts in the coupling of N^2,N^9-diacetylguanine (DAG) with 2-oxa-1,4-butane-dioldiacetate

(OBDDA) and synthesized N^2-acetyl-9-(2-acetoxyethoxymethyl)guanine and N^2-acetyl-7-(2-acetoxyethoxymethyl)guanine in high regioselectivity.

In 1982, Robins and Hatfield[10] reported that they used several materials—2-amino-6-chloro-9-[(2-hydroxyethoxy)methyl]purine, 9-[(2-acetoxyethoxy) methyl]-2,6-dichloropurine and 2-amino-6-chloropurine—to synthesize ACV1. They also synthesized a series of N-[(2-hydroxyethoxy)methyl]heterocycles. These compounds came from (2-acetoxyethoxy)methyl bromide **8**, which is derived from 1,3-dioxolane reacting with acetyl bromide, coupled with trimethylsilylated bases (Scheme 1). Kelley et al.[11] and Schroeder et al.[12] reported that they also researched the synthetic method of this type of compound.

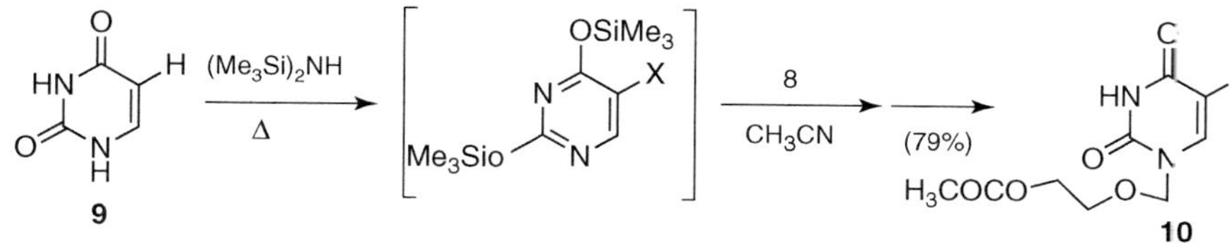

Scheme 1. Synthesis of 1-[(2-acetoxyethoxy)methyl]uracil **10**.

Experimental Procedure

1-[(2-Acetoxyethoxy)methyl]uracil (10). To a suspension of 168 mg (1.5 mmol) of uracil **9** in 5 mL of hexamethyldisilazane was added a drop of chlorotrimethylsilane, and the stirred mixture was heated at reflux with exclusion of moisture until a clear solution was obtained. Excess silylating reagent was removed in vacuo with protection against moisture. The residual clear oil was dissolved in 15 mL of dry acetonitrile and cooled to 0°C. A solution of 197 mg (1 mmol) of (2-acetoxyethoxy)methyl bromide **8** in 5 mL of dry acetonitrile was added slowly with stirring. The solution was allowed to be stirred for 2 h while warming to room temperature, at which time TLC indicated that the reaction was complete. Volatile materials were evaporated in vacuo. The resulting yellow oil was chromatographed on a column (3 cm in diameter) of 20 g of silica gel using 2% MeOH/CHCl$_3$, for elution. Fractions containing **10** were combined and evaporated. The residue was crystallized from CHCl$_3$ with diffusion of Et$_2$O to give 181 mg (79%) of pure **10**.

In 1991, Clausen and co-workers[13] used a simple method, cyclodesulfurization of 5-[(thiocarbamoyl)amino]imidazole-4-carboxamide **11b** under alkaline conditions, to synthesize acyclic nucleoside analogues **14** from 5-aminoimidazole-4-carboxamide **11a** (AICA) (Scheme 2).

A U. S. patent application[14] uses condensation of the hydroxyl group of acyclovir **1** and N-carbobenzoxy-L-valine and draws off a carbobenzoxy group through Pd/C hydrogenation to synthetize a first-generation nucleoside antivirus prodrug, valaciclovir **7** (VCV), which has improved oral bioavailability.

Researchers continue to develop more and more prodrugs of acyclovir **1** to achieve the very best treatment effect. Recently, Gomes and co-workers[15] synthesized water-miscible dipeptide ester prodrugs of acyclovir (ACV) **1** (Scheme 3)

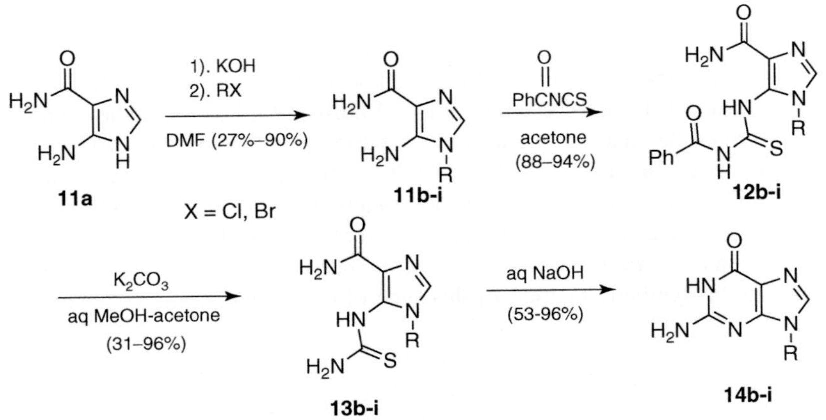

a: R = H
b: R = 1-b-D-ribofuanosyl
c: R = methyl
d: R = 1-ethyl
e: R = propyl
f: R = phenylmethyl
g: R = [2-(acethyloxy)ethoxy]methyl*
h: R = (2-hydroxyethoxy)methyl*
i: R = (1,3-dihydroxy-2-propoxy)methyl
* 12g → 13h

Scheme 2. Synthesis of acyclic nucleoside analogues **14**. *Reagents and conditions: (the last step)* method A: heavy-metal salt (Cu^{2+}, Ag^+, Hg^{2+}) in aq. NaOH; method B: 35% H_2O_2/Na_2WO_4 in aq. NaOH.

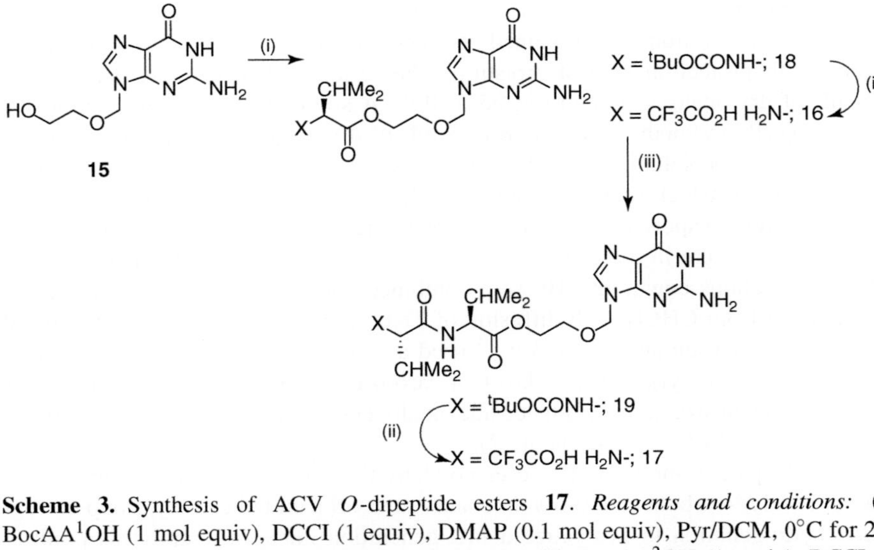

Scheme 3. Synthesis of ACV *O*-dipeptide esters **17**. *Reagents and conditions:* (i) BocAA^{1}OH (1 mol equiv), DCCI (1 equiv), DMAP (0.1 mol equiv), Pyr/DCM, 0°C for 2h, then for rt for 48 h; (ii) neat TFA, DCM, rt, 30 min; (iii) BocAA^{2}OH (1 mol equiv), DCCI (1 equiv), DMAP (0.1 mol equiv), TEA (1.2 mol equiv), DCM, 0°C for 2 h to rt for 48 h.

and worked over their chemical durability, kinetics, cytotoxicity, and bioactivity. The results showed that linking a Val group to ACV was very important to advance chemical durability and to adjust the qualities of the prodrug. Further search for dipeptide esters of ACV as possibly applicable in clinical ACV prodrugs is certainly worthwhile.

Experimental Procedure

Preparation of Compounds 17. Acyclovir **1** (1 equiv) was suspended in dry pyridine and the suspension put under magnetic stirring in a water–ice–acetone bath. The relevant BocAA$_1$OH (1 equiv) and DMAP (0.1 equiv), dissolved previously in dry DCM, were then added to the mixture, after which a suspension of DCCI (1 equiv) in dry DCM was added slowly dropwise. The mixture was stirred at 0°C for 2 h, then at room temperature for a further 24 h. The suspension was again led to 0°C by immersion in a water–ice–acetone bath and, again, DCCI (1 equiv) in DCM was slowly added dropwise, after which the reaction was allowed to proceed at room temperature for a further 24 h. The crude mixture was evaporated to dryness, resuspended in warm acetone, and left overnight at 4°C. The N,N'-dicyclohexylurea (DIU) fraction precipitated was removed by suction filtration and the filtrate evaporated to dryness and submitted to column chromatography on silica.

In 2009, a group of workers[16] synthesized a series of 3,9-dihydro-9-oxo-5H-imidazo[1,2-a]purine nucleosides **21** (called simply tricyclic nucleosides). Beauchamp et al.[17] reported a type of tricyclic nucleoside compound in 1985. Although these compounds did not appear to exhibit any activity against the virus, they attracted immense attention. In 1991, Boryski et al.[18] reported that they synthesized a type of straight-line tricyclic nucleoside. At the same time, more and more chemists joined in this field and made great progress.

It is well known that tricyclic nucleoside is synthesized from the condensation of guanosine **20** with phenylacyl bromide (Scheme 4). The data indicated that tricyclic derivatives of acyclovir (ACV) **22** and ganciclovir (GCV or DHPG) **23** (Figure 5) demonstrate similar bioactivities against herpes simplex virus (HSV).[19] Because of this activity, Golankiewicz and Ostrowski researched the synthetic method and activities against HIV and HCV. Through their research, they reported the instability of some compounds in an aqueous medium for the first time.

Linking a group to the side line, the compound of this category can produce a chiral center. It is apparent that there are two major routes in the synthesis of chiral acyclonucleosides and analogues. The first way, a chiral side chain with multistep modification introduced to purine or pyrimidine bases, involves an aza-Michael addition reaction, alkylation, a Mitsunobu reaction, and the ring-opening reaction of a pyranoid ring or propylene oxide. The second route is based on using an asymmetric catalytic method to prompt a reaction between a low-cost achiral side chain and bases. This route makes up for the deficiency in the first method. The chiral side chains are quite difficult to obtain or need lengthy processes; thus, the second way opens up an extremely productive and very meaningful protocol. It is a pity that until now, there have been only three examples of asymmetric catalytic

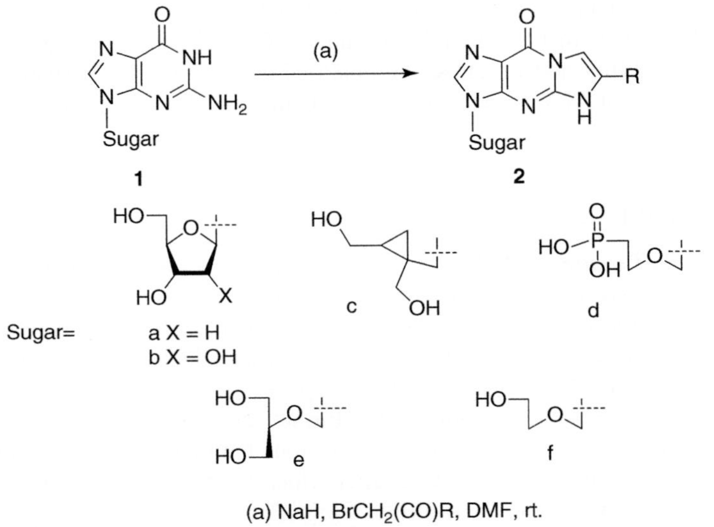

(a) NaH, BrCH₂(CO)R, DMF, rt.

Scheme 4. Synthesis of tricyclic nucleosides.

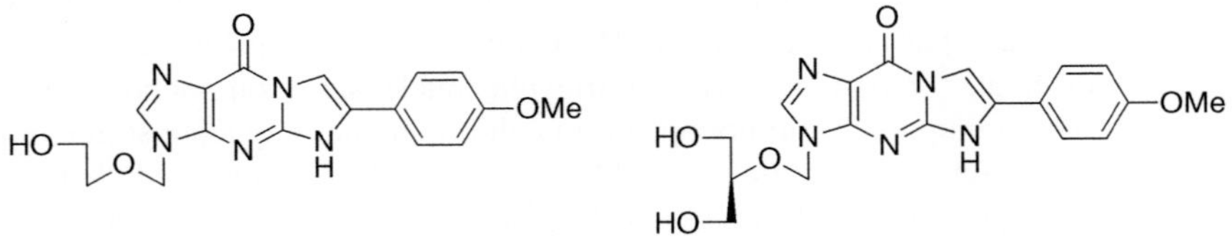

Figure 5. Structure of tricyclic derivatives of ACV and GCV.

ways methods involving chiral acyclic nucleosides. In the following sections we provide a detailed description. First we discuss a case that uses the first route to synthesize chrial acyclic nucleosides.

In 2002, Len and co-workers[20] completed asymmetric synthesis of both enantiomers of 1-{[(2-hydroxy-1-phenyl)ethoxy]methyl}thymine **25** and 9-{[(2-hydroxy-1-phenyl)ethoxy]methyl}guanine **26** (Scheme 5), which are the analogues of d4T **24** (Figure 6) and acyclovir **1**, respectively. In 1994, Chu and co-workers[21] produced compound **25** and its analogues in racemic form as intermediates to preparing HEPT[22] and EBPU[23] analogues. Len researched a much more convenient way to synthesize the title compound from (R)- and (S)-1-phenylethan-1,2-diol in five steps.

In 2002, Sufrin and co-workers[24] synthesized and evaluated a series of derivatives of 5′-([(Z)-4-amino-2-butenyl]methylamino)-5′-deoxyadenosine (Schemes 6 and 7), which are the analogues of cis-5′-deoxy-5′-(4-amino-2-butenyl)methyl aminoadenosine (AbeAdo) **34** (Figure 7).

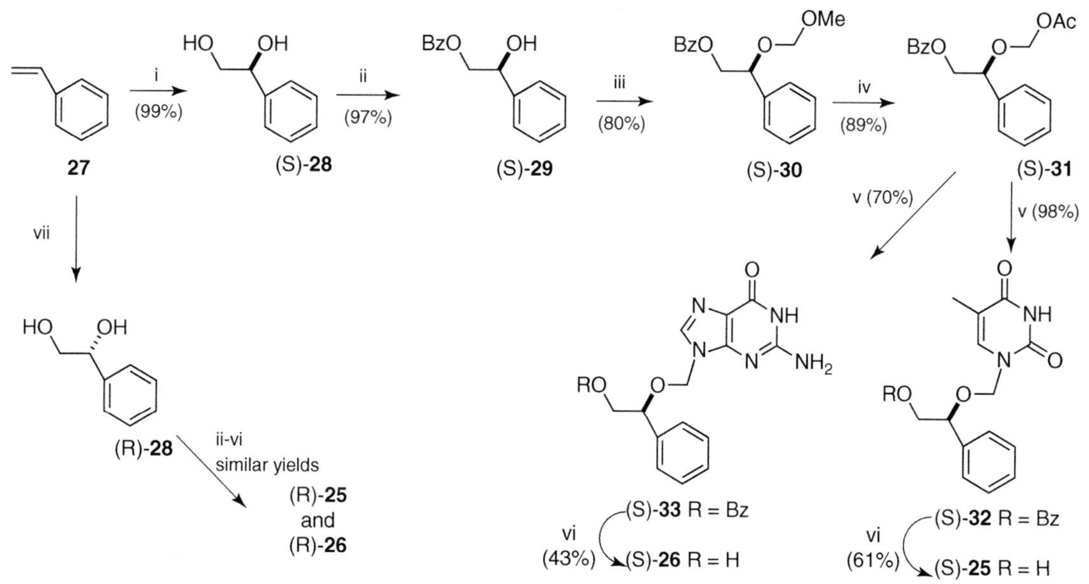

Scheme 5. Synthesis of d4T analogues. *Reagents and conditions*: (i) AD-mix, α, *t*-BuOH, H$_2$O; (ii) BzCl, pyridine; (iii) dimethoxymethane, P$_2$O$_5$, chloroform; (iv) acetic anhydride, BF$_3$-Et$_2$O; (v) silylated base, dibenzo-18-crown-6-ether, KI, acetonitrile, toluene; (vi) K$_2$CO$_3$, MeOH; (vii) AD-mix β, *t*-BuOH, H$_2$O.

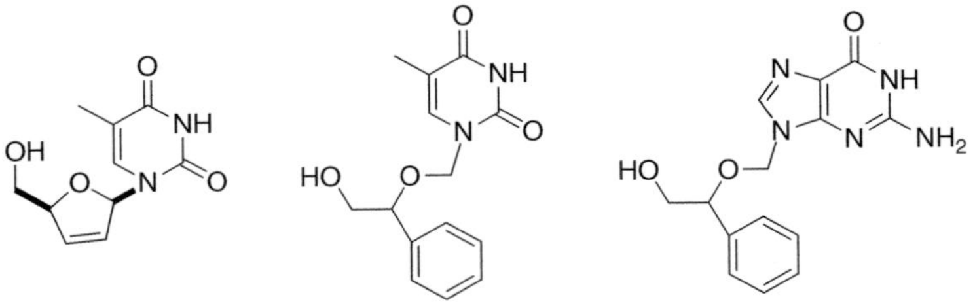

24 d4T **25** **26**

Figure 6. Structure of d4T and its analogues.

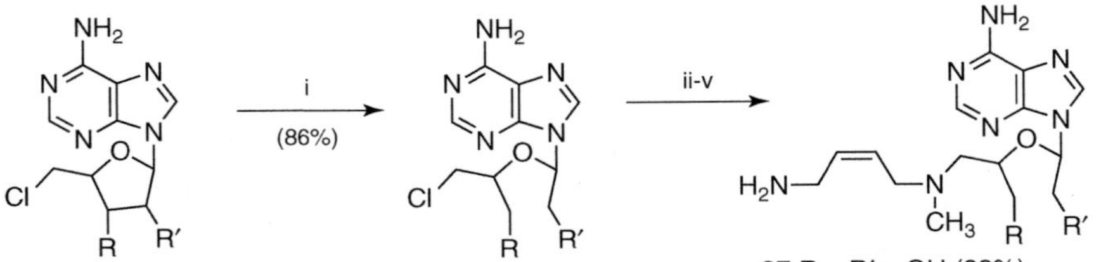

35 R = R′ = OH

36 R = R′ = OH

37 R = R′ = OH (88%)
38 R = R′ = OC(O)CH3 (76%)

Scheme 6. Synthesis of acyclic nucleoside analogues **37** and **38**. *Reagents and conditions:* (i) $NaIO_4/NaBH_3$; (ii) methylamine; (iii) *cis-tert*-butoxycarbonyl-4-chloro-2-butenyl-1-amine; (iv) acetic anhydride/DMAP; (v) trifluoroacetic acid.

2.1.2. With a 2′-C Side Chain (IB). Compared to other types, up to now there have been few reports in the literature about side chains without heteroatoms. Kjellberg [25] and Kjellberg and Liljenberg [26] researched a method of 9-(4-hydroxy-butyl)guanine (HBG) **4** synthesis in which the reactivity and regioselectivity of alkyl halides coupling with guanine were studied. Diacetoxyglyoxal-N^2-acetylguanine adduct **45** and diisobutyroxyglyoxal-N^2-acetylguanine guanine adducts **46** were alkylated with 4-bromobutyl acetate under an alkaline environment (Scheme 8).

In 1988, Howson et al.[27] synthesized and evaluated ATP analogues as P2X receptor antagonists, and in 2009, Cristalli et al.[28] synthesized a novel family of adenine-based acyclic nucleosides with a mono-, di-, triphosphorylated four-carbon side chain in the 9-position (Schemes 9 and 10) and found that some acyclic nucleotides weaken the activation of $P2X_3$ receptors, which are considered related to an increase in swelling and lingering pain. They synthesized 2-substituted 9-hydroxybutyladenine and then modified the side line. Janeba and co-workers[29] also reported a detailed synthesis of 9-(4-hydroxybutyl)adenine and 9-(4-hydroxybutyl)hypoxanthine involving deamination of 9-(4-hydroxybutyl)adenine with isoamyl nitrite.

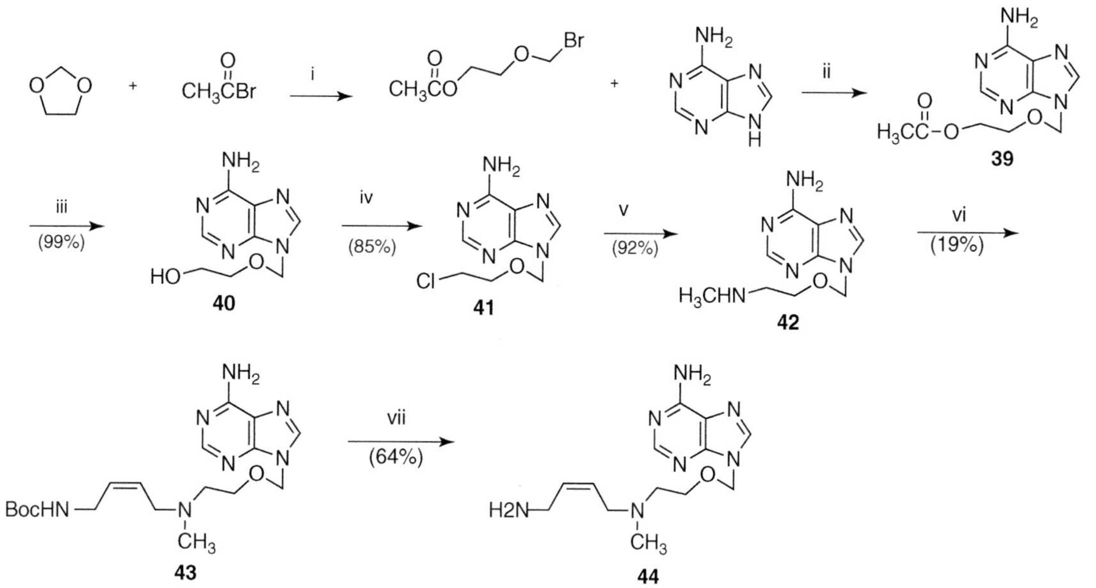

Scheme 7. Synthesis of acyclic nucleoside analogues **44**. *Reagents and conditions*: (i) stir at rt; (ii) NaH; (iii) sodium methoxide; (iv) thionyl chloride; (v) methylamine; (vi) *cis-tert*-butoxycarbonyl-4-chloro-2-butenyl-1-amine; (vii) H_2SO_4.

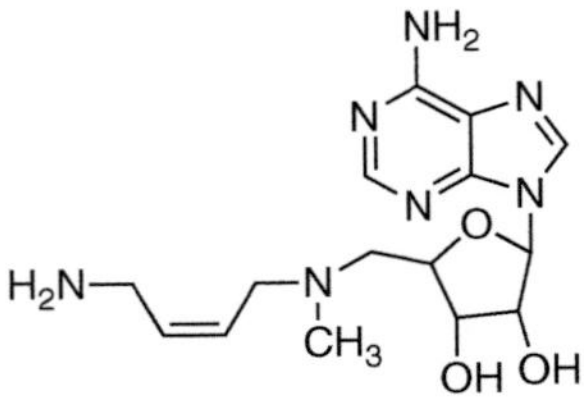
NH2
H2N
N
CH3
OH OH

34

Figure 7. Structure of AbeAdo (MDL73811).

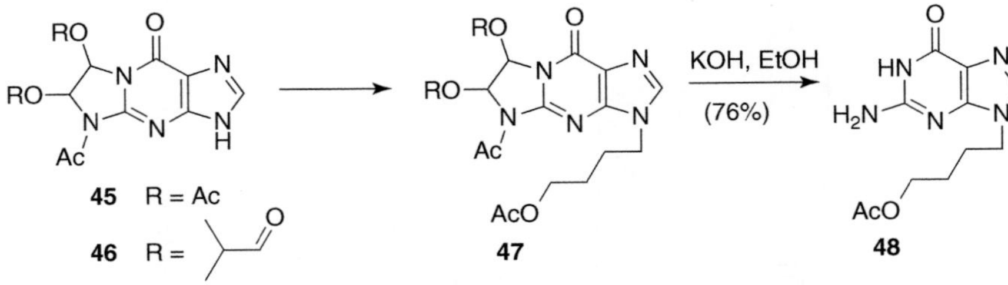
45 R = Ac
46 R =
47
KOH, EtOH
(76%)
48

Scheme 8. Synthesis of 9-(4-hydroxybutyl)guanine **48**.

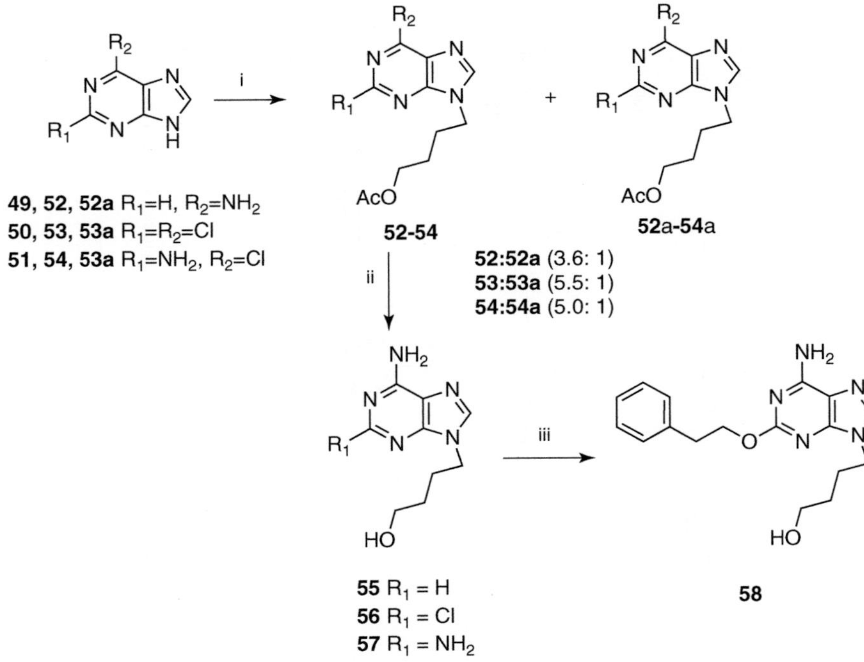
i
ii
iii
49, 52, 52a R1=H, R2=NH2
50, 53, 53a R1=R2=Cl
51, 54, 53a R1=NH2, R2=Cl
52-54
52a-54a
52:52a (3.6: 1)
53:53a (5.5: 1)
54:54a (5.0: 1)
55 R1 = H
56 R1 = Cl
57 R1 = NH2
58

Scheme 9. Synthesis of 2-substituted 9-hydroxybutyladenines **56–58**, the R1 in compound 55 is H. *Reagents and conditions:* (i) DMF, $Br(CH_2)_4OCOCH_3$, rt, 16 h; (ii) NH_3/CH_3OH, 120°C, 16 h; (iii) $C_6H_5CH_2CH_2OH$, NaOH, 80°C, 4 h.

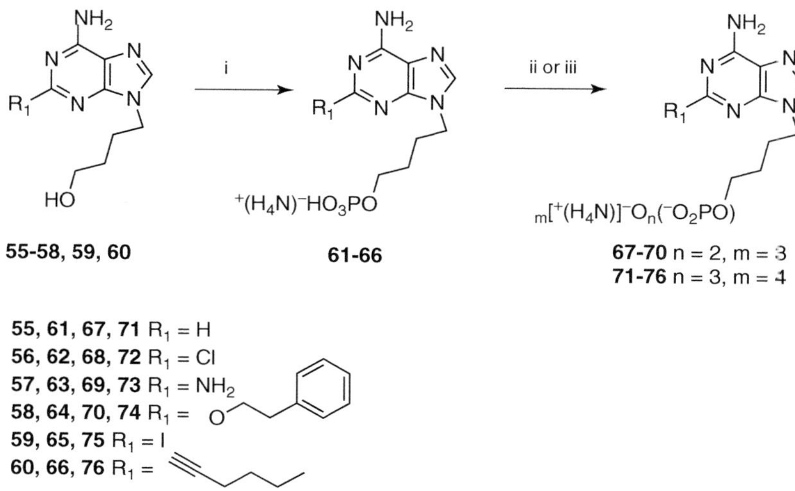

Scheme 10. Synthesis of 2′-substituted 9-hydroxybutyladenine mono-(**61–66**), di-(**67–70**), and triphosphates (**71–76**). *Reagents and conditions:* (i) POCl$_3$, (CH$_3$O)$_3$PO, rt, 3 h; (ii) CDI, tri-*n*-butylammonium phosphate, DMF, rt, 16 h; (iii) CDI, bis(tri-*n*-butylammonium)pyrophosphate, DMF, rt, 16 h.

Experimental Procedures

General Procedure for the Preparation of Compounds 52–54 and 52a–54a. To a solution of the suitable purine **49–51** (10 mmol) in dry DMF (13.0 mL), K$_2$CO$_3$ (1.66 g, 12.0 mmol) was added. After stirring for 5 min, 4-bromobutylacetate (1.57 mL, 11.0 mmol) was added and the reaction was left at room temperature for 16 h. Reaction completion was checked by TLC eluted with CHCl$_3$–CH$_3$OH (9:1): the *N*9-isomers showed R_f values ranging from 0.2 to 0.6, and the *N*7-isomers showed R_f values, of 0.1–0.5, respectively. The mixture was concentrated under vacuum and chromatograrphed over a flash silica gel column using the appropriate eluent to obtain compounds **52–54** and **52a–54a** as white solids.

General Procedure for the Synthesis of Monophosphates 61–66. Compounds **55–58**, **59**, and **60** (1.44 mmol) were in turn dissolved in trimethyl phosphate (3.0 mL), and then 4 equiv of POCl$_3$ (539 μL, 5.78 mmol) was added at 0°C. The solution was stirred at room temperature for 3 h, and then H$_2$O (3.0 mL) was added to the reaction maintained at 0°C, and the solution was neutralized by adding triethylamine. The reactions were monitored by TLC, using precoated TLC plates with silica gel 60 F-254 (Merck) and *i*-C$_3$H$_7$OH/H$_2$O/NH$_4$OH (5.5:1.0:3.5) as a mobile phase; monophosphate derivatives showed R_f values of 0.4–0.5. The nucleotides were purified by means of ion-exchange chromatography on a Sephadex DEAE A-25 (Fluka) column (HCO$_3^-$ form) equilibrated with H$_2$O and eluted with a linear gradient of H$_2$O/0.5 M NH$_4$HCO$_3$.

In the same year, Sauer and co-workers[30] synthesized a series of uracil nucleotide analogues with an acyclic chain as a P2Y$_2$ receptor, one of the P2Y receptor antagonists (Scheme 11). Some P2Y receptor antagonists, such as clopidogrel[31] and cangrelor,[32] are used as antithrombotic drugs or in clinical trials.

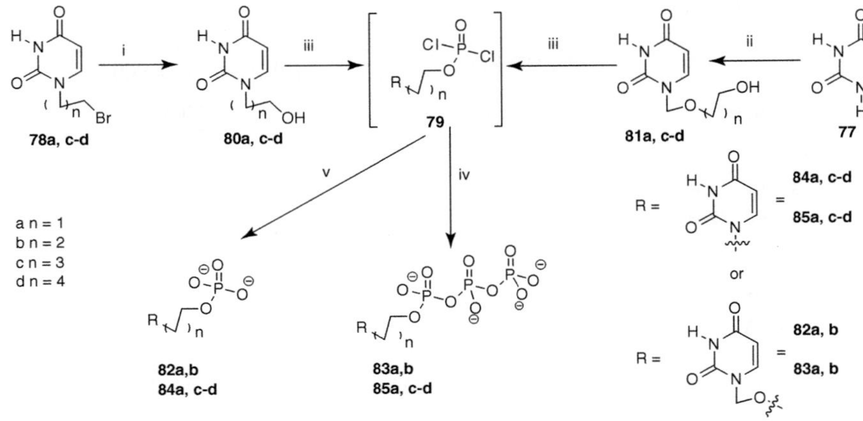

Scheme 11. Synthesis of acyclic nucleoside analogs **80a,c–d** and **81a,b** and acyclic nucleotide analogs **82a,b**, **84a,c-d** (monophosphates) and **83a,b**, **85a,c-d** (triphosphates). *Reagents and conditions:* (i) 1N NaOH, 48h, 50°C, 45-70%; (ii) (1), bis(trimethylsilyl)acetonitrile, 3h, rt; (2), KI, Me$_3$SiI, 1,3-dioxolane or 1,3-dioxane, 24h, rt; **81a** (25%), **81b** (35%); (iii) POCl$_3$, OP(OMe)$_3$, 1h, 0-4°C; (iv) (1), (BuNH+)$_4$P$_2$O$_7$, Bu$_3$N, DMF, 1h, 0-4°C; (2), (Et$_3$NH)HCO$_3$, 25-30%; (v) (Et$_3$NH)HCO$_3$; **82a,b** (50-60%), **84a,c-d** (10-15%).

2.1.3. With a 2′-N Side Chain (IC).

Acyclic nucleosides that possess a nitrogen atom at the 2′-position in the side chain are called acyclic aza-nucleosides. At present there is in sufficient study on the synthesis and biological activity of acyclic aza-nucleosides. In 2006, Koszytkowska-Stawińska et al.[33] synthesized a series of acyclic aza-nucleosides from *N*-(pivaloyloxymethyl)sulfonamides through a one-pot base silylation/nucleoside coupling process[34] (Scheme 12) and studied their biological activity.

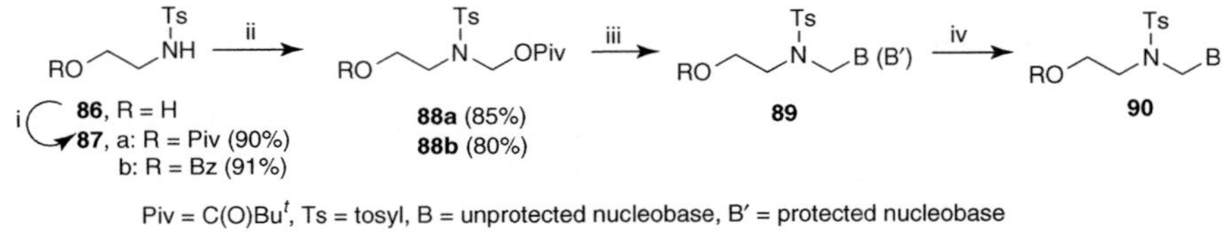

Piv = C(O)Bu^t, Ts = tosyl, B = unprotected nucleobase, B′ = protected nucleobase

Scheme 12. Synthesis of acyclic aza-nucleosides **90**. *Reagents and conditions:* (i) BzCl or PivCl, Py, rt, 1 day; (ii) PivOCH$_2$Cl, K$_2$CO$_3$, DMF, rt, 5 days; (iii) B(TMS)$_n$ or B′ (TMS)$_n$, Lewis acid; (iv) NH$_3$ aq./MeOH.

In 2009, these researchers[35] synthesized and evaluated acyclic 2′-aza-nucleosides with a phosphonomethoxy group in the side chain through a Vorbrüggen-type reaction of diethyl{2-[N-(pivaloyloxymethyl)-N-(p-toluenesulfonyl)amino]ethoxymethyl}phosphonate and pyrimidine (Scheme 13).

2.1.4. With a 2′-S Side Chain (ID).

In 1979, Schaeffer and Raleigh reported a patent[36] in which there is the synthesis of 2′-amido-9-(2-acyloxyethoxymethyl) hypoxanthines, including 9-(2′-hydroxyethylthiomethyl)adenine **95** (Figure 8).

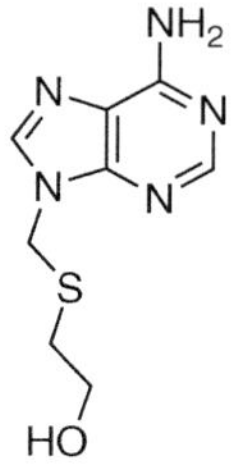

Figure 8. Structure of 9-(2′-hydroxyethylthiomethyl)adenine.

2.1.5. Others.

In 1989, Scheiner et al.[37] reported that they synthesized (R, S)-1-[1-(2-hydroxyethoxy)-3-azidopropyl]thymine, an acyclic derivative of 3′-azido-3′-deoxythymidin (AZT) **96** (Figure 9), through a Michael-type addition reaction (Schemes 14 and 15).

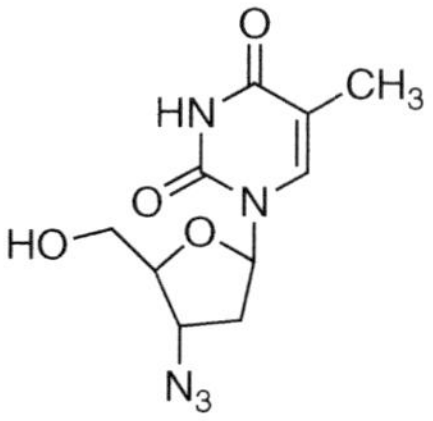

Figure 9. Structure of AZT.

Experimental Procedures

(R,S)-Methyl 3-Thymin-1-yl-3-[2-(trityloxy)ethoxy]propanoate (99). A suspension of thymine **98** (1.50 g, 11.9 mmol), **97** (18.5 g, 47.6 mmol), DBU (100 μL, 0.7 mmol) in acetonitrile (100 mL) and dimethyl sulfoxide (10 mL) was stirred at

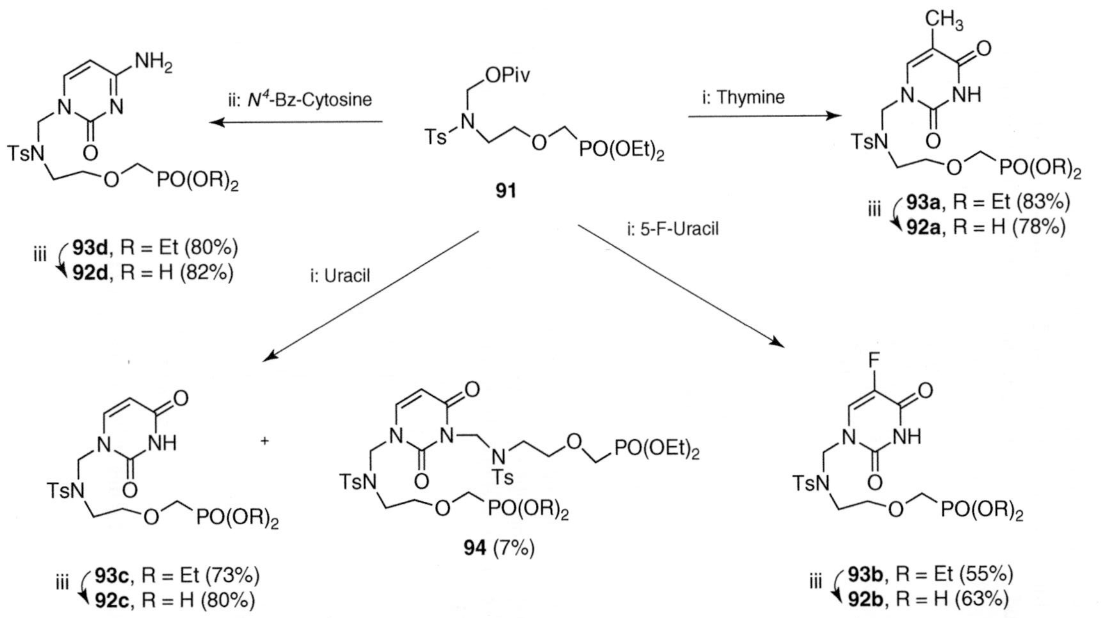

Scheme 13. Synthesis of acyclic 2′-aza-nucleosides with a phosphonomethoxy group. *Reagents and conditions:* (i) (1) nucleobase (thymine, 5-F-uracil, or uracil), BSA, CH$_3$CN, rt, 1 h; (2) **91**, SnCl$_4$/CH$_2$Cl$_2$, CH$_3$CN, rt, 2 days; (ii) (1) N^4-Bz-cytosine, BSA, CH$_3$CN, rt, 1 h; (2) **91**, SnCl$_4$/CH$_2$Cl$_2$, CH$_3$CN, rt, 2 days; (3) NH$_4$OH, MeOH, rt, 20 h; (iii) (1) TMSBr, CH$_3$CN, rt, 20 h; (2) acetone, H$_2$O, rt, 2 h.

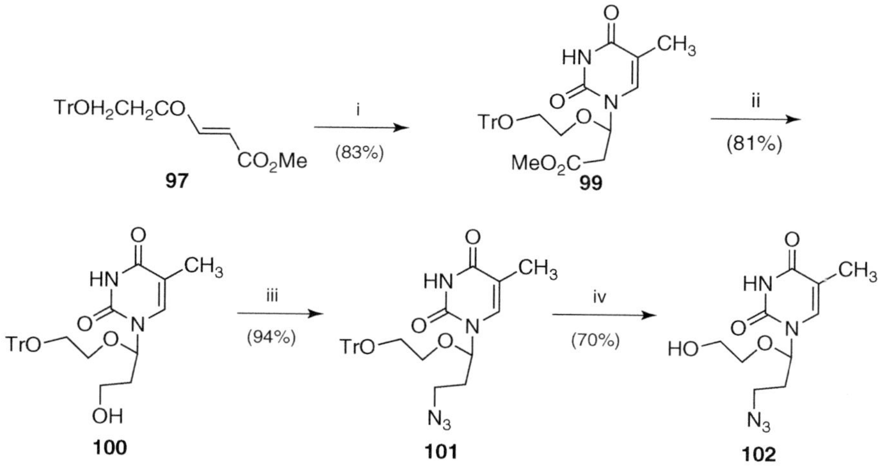

Scheme 14. Synthesis of 2'-O side-chain acyclic nucleoside derivative of AZT. *Reagents and conditions:* (i) thymine **98**; (ii) LiAlH$_4$; (iii) MsCl, then LiN$_3$; (iv) AcOH, H$_2$O.

room temperature for 3 weeks, by which time a solution was obtained. Acetic acid (0.5 mL) was added and solvent removed under reduced pressure. The resulting semisolid was chromatographed over silica gel (150 g). Elution with petroleum ether/EtOAc (2:1) afforded 5.08 g (83%) of a white solid. Recrystallizations from EtOAc/petroleum ether (2:1) gave **99**, mp 172–173°C. NMR (CDCl$_3$): δ 9.55 (s br, 1, NH), 7.30 [m, 16, ArH, C(6)H], 6.12 [t, 1, J = 6.3 Hz, C(1')H], 3.68 (s, 3, OCH$_3$), 3.7–3.4, 3.4–3.1 (ms, 2,2, CH$_2$CH$_2$), 2.83 [d, 2, J = 6.3 Hz, C(2')H], 1.90 [s, 3, C(5)CH$_3$]. *Anal.* (C$_{30}$H$_{30}$N$_2$O$_6$) C, H, N.

(R, S)-Methyl 3-[(2-Hydroxyethyl)thio]-3-thymin-1-yl-propanoate (105). A mixture of **104** (0.420 g, 2.00 mmol), 2-mercaptoethanol (1.0 mL, 15 mmol), DBU (15 μL, 0.1 mmol), and THF (5 mL) was stirred overnight at room temperature. Glacial HOAc (3 drops) was added to the clear solution and the volatile material removed under reduced pressure. The amorphous residue was taken up in boiling EtOAc, diluted with petroleum ether, and refrigerated. Crystals of **105** were collected (0.477 g, 83%), mp 127–129°C; recrystallized from EtOAc, mp 130–131°C. NMR (DMSO-d_6): δ 11.2 (s, 1, NH), 7.73 [s, 1, C(6)H], 6.00 [t, 1, J = 7 Hz, C(1')H], 3.65 (s, 3, OCH^3), 3.53 [t, 2, J = 6 Hz, CH$_2$O], 3.08 [d, 2, J = 7 Hz, C(2')H], 2.60 (t, 2, J = 6 Hz, CH$_2$S), 1.92 [s, 3,C(5)CH$_3$]. *Anal.* (C$_{11}$H$_{16}$N$_2$O$_5$S) C, H, N, S.

In 1993, Hendry and co-workers[38] synthesized a new potent inhibitor of methylthioadenosine (MTA) nucleosidase, 9-(4'-methylthiobutyl)adenine **107** (Figure 10). MTA nucleosidase is a principal enzyme in methionine and adenine salvage.

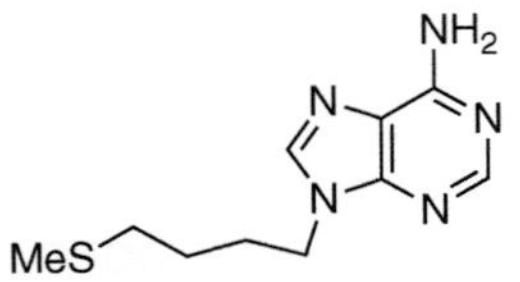

Figure 10. Structure of 9-(4'-methylthiobutyl)adenine **107**.

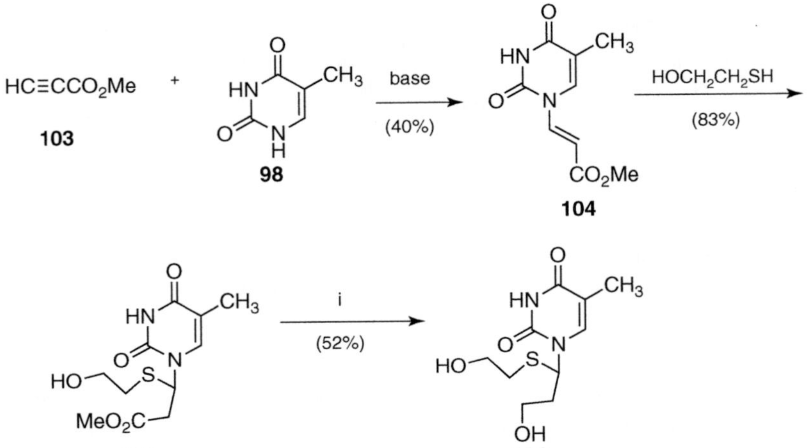

Scheme 15. Synthesis of 2'-S side-chain acyclic nucleoside derivative of AZT. *Reagents and conditions:* (i) LiAlH$_4$.

Experimental Procedure

9-(4'-Methylthiobutyl)adenine (107). Triphenylphosphine (1.71 g, 6.52 mmol) and DEAD (1.02 cm^3, 1.13 g, 6.48 mmol) were added to a solution of 6-chloropurine (1.01 g, 6.53 mmol) in THF (50 cm^3). After 5 min, 4-methylthiobutan-1-ol (0.61 cm^3, 0.61 g, 5.04 mmol) was added dropwise and stirred at room temperature for 26 h. The solvent was removed under reduced pressure and the residue was purified by silica gel chromatography using light petroleum/ethyl acetate (2:1) as eluent, to give 6-chloro-9-(4'-methylthiobutyl)purine as an oil (0.88 g, 69%), R_f = 0.61 CH$_2$Cl$_2$/MeOH (9:1); v_{max}(film)/cm 1595 and 1561 (purine); δ_H(CDCl$_3$) 8.72 (s, 1H, ArH), 8.12 (s, 1H, ArH), 4.31 (t, 2H, J = 7 Hz, CH$_2$–N), 2.52 (t, 2H, J = 7 Hz, SCH$_2$), 2.11–1.98 (m, 5H, CH$_2$CH$_2$–N and CH$_3$S) and 1.69–1.56 (m, 2H, SCH$_2$CH$_2$). Found: [M + H]$^+$, 257.0628; C$_{10}$H$_{13}$ClN$_4$S requires [M + H] 257.0628.

An excess of liquid ammonia was added to this chloropurine (0.14 g, 0.55 mmol) in a Teflon tube. This was then fitted into a stainless steel vessel, sealed, and left overnight at room temperature. After 18 h the pressure was released and ammonia was allowed to evaporate off. The residue was purified by silica gel chromatography using dichloromethane/methanol (29:1) as eluent to give the title compound **107** as a white solid (0.10 g, 75%, $R_f = 0.35$ CH_2Cl_2/MeOH (9:1); mp 121–123°C; υ_{max}(KBr)/cm 3275 (NH_2), 1678 and 1603 (purine); δ_H (CD_3OD) 8.20 (s, 1H, ArH), 8.11 (s, 1H, ArH), 4.24 (t, 2H, J 7, CH_2Ad), 2.50 (t, 2H, J = 7 Hz, SCH_2), 2.06–1.91 (m, 5H, CH_2CH_2Ad and CH_3S), and 1.65–1.52 (m, 2H, SCH_2CH_2); δ_C(CD_3OD) 157.3 (ArC), 153.7 (ArCH), 150.7 (ArC), 142.7 (ArCH), 120.1 (ArC), 44.5 (CH_2Ad), 34.3 (SCH_2), 30.1 (CH_2CH_2Ad), 27.0 (SCH_2–CH_2), and 15.2 (CH_3S). Found: [M + H]$^+$, 238.1126. $C_{10}H_{15}N_5S$ requires [M + H] 238.1126. Found: C, 50.5; H, 6.2; N, 29.1; S, 13.1. $C_{10}H_{15}N_5S$ requires C, 50.6; H, 6.4; N, 29.5; S, 13.5%.

In 2002, Ewing and co-workers[39] synthesized d4T **24** via ring-closure metathesis for the first time, as well as several unsaturated acyclic nucleosides: 1-{(R)-1-[((S)-1-hydroxybut-3-en-2-yl)oxy]allyl}pyrimidine-2,4(1H,3H)-dione **115**, 1-{(R)-1-[((S)-1-hydroxybut-3-en-2-yl)oxy]allyl}-5-methylpyrimidine-2,4(1H,3H)-di-one **116**, and 4-amino-1-{(R)-1-[((S)-1-hydroxybut-3-en-2-yl)oxy]allyl}pyrimidin-2 (1H)-one **117** (Scheme 16).

In 2008, Lequeux et al.[40] synthesized fluorophosphonylated acyclic nucleotides via a ring-opening reaction of cyclic sulfates (Scheme 17). The introduction of fluorine will change the antiviral activity at a certain excess.

2.2. Synthesis of Acyclic Nucleosides with Ganciclovir Side-Chain Analogues (Type II)

At the superficial level, ganciclovir side-chain analogues have no chiral center, but they can change into chiral compounds if the hydroxyl group is protected (Figure 11).

2.2.1. With a 2′-O Side Chain (IIA).
There are many representative drugs in this type, such as ganciclovir (DHPG)[41] **125** (Figure 12) and valganciclovir[42] **2** (Figure 1). Ganciclovir **125**, developed by the American Syntex Company, is the first drug ratified to guard against cytomegalovirus (CMV), but it has high levels of toxicity. The antiviral activity and mechanism action of valganciclovir **2**, the prodrug of DHPG **125**, has many affinities with **125**. Valganciclovir has not less than 10 times the bioavailability of ACV **1**.

In 1983, Kendall O. Smith et al.[43] synthesized a series of type II compounds: 9-[2-hydroxy-1-(hydroxymethyl)ethoxylmethyl]adenine/isoguanine/xanthine and their derivants (Scheme 18), such as one or two hydroxyls protected by Si/Ts or substituted by amino/nitrine. Follow-up studies[44–49] report different research points, but all use similar methods to synthesize this type of compound.

Experimental Procedure

9-{[2-Benzyloxy-1-(benzyloxymethyl)ethoxy]methyl}-6-chloropurine (128).
6-Chloropurine (9 g, 58.5 mmol) was dissolved in DMF (50 mL) and triethylamine (7 g) was added. The solution was cooled in an ice–salt bath and compound **126** (58.5 mmol in 50 mL of DMF) was added. Triethylamine

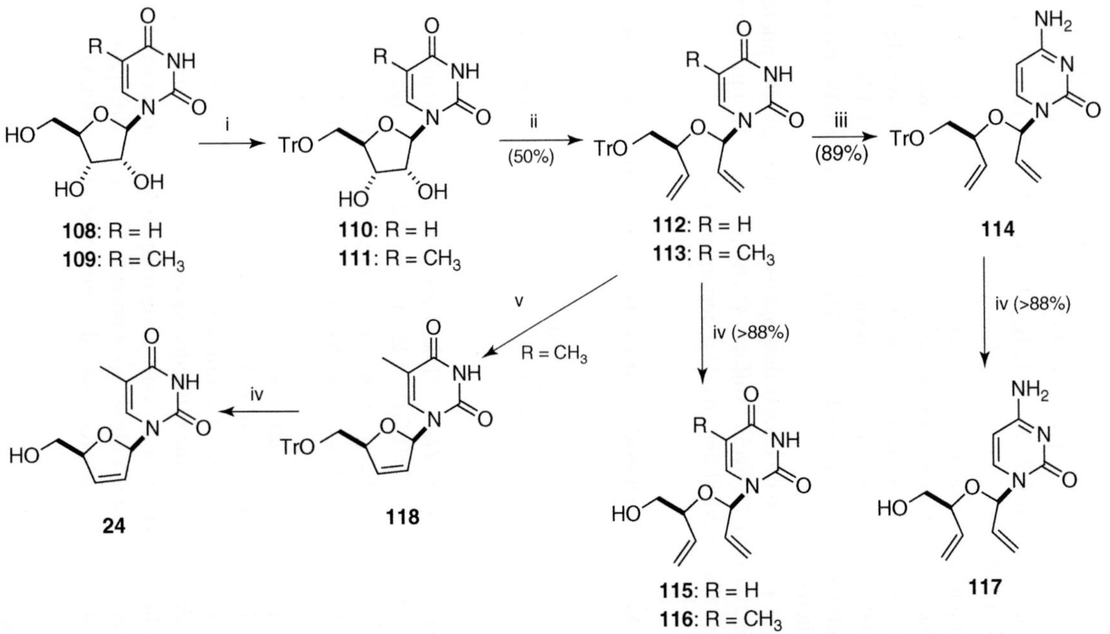

Scheme 16. Synthesis of d4T analogues. *Reagents and conditions:* (i) TrCl, Py.; (ii) (1) NaIO$_4$, EtOH/H$_2$O, (2) Ph$_3$PCH$_3$Br, *t*-BuOK, toluene; (iii) (1) 1,2,4-triazole, POCl$_3$, CH$_3$CN, Et$_3$N, (2) aq. 30% NH$_3$ in 1,4-dioxane; (iv) AcOH 80%; (v) Grubbs' reagent, CH$_2$Cl$_2$.

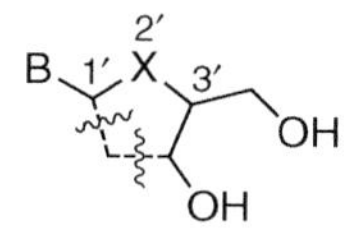

X = O, CH_2, NR,S

II

Figure 11. Structure of ganciclovir side-chain analogues (type II).

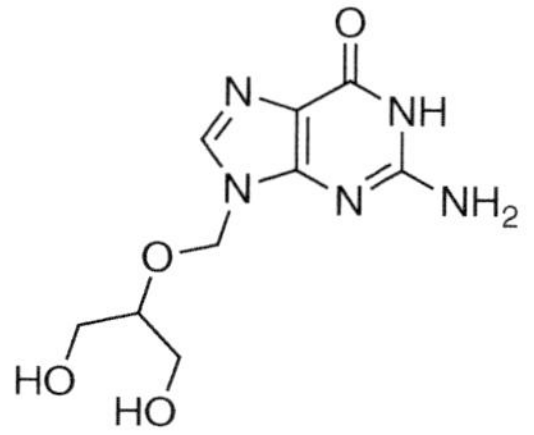

125 Ganciclovir

Figure 12. Structure of DHPG 125.

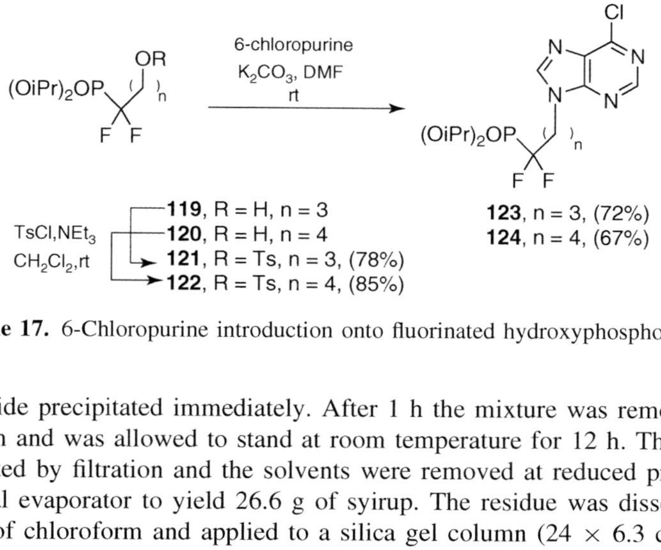

Scheme 17. 6-Chloropurine introduction onto fluorinated hydroxyphosphonates.

hydrochloride precipitated immediately. After 1 h the mixture was removed from the ice bath and was allowed to stand at room temperature for 12 h. The solution was collected by filtration and the solvents were removed at reduced pressure on a horizontal evaporator to yield 26.6 g of syirup. The residue was dissolved in a minimum of chloroform and applied to a silica gel column (24 × 6.3 cm) eluted with 1% methanol in chloroform. The yield of compound **128** was 14.8 g (60%) and that of compound **127** was 6.1 g (23%).

Compound **128** showed λ_{max}: 264 nm (pH 1), 264 nm (H_2O), and 264 nm (pH 13). The NMR spectrum in $CDCl_3$, showed signals at δ (ppm) of 3.53 (d, 4H, J = 5.5 Hz, $-CH_2CHCH_2-$), 4.03 (m, 1H, –CH–), 4.45 (s, 4H, 2 × $PhCH_2-$), 5.88 (s, 2H, $-OCH_2N-$), 7.33 (m, 10H, 2 × Ph–), 8.23 (s, 1H, H-2), and 8.78 (s, 1H, H-8).

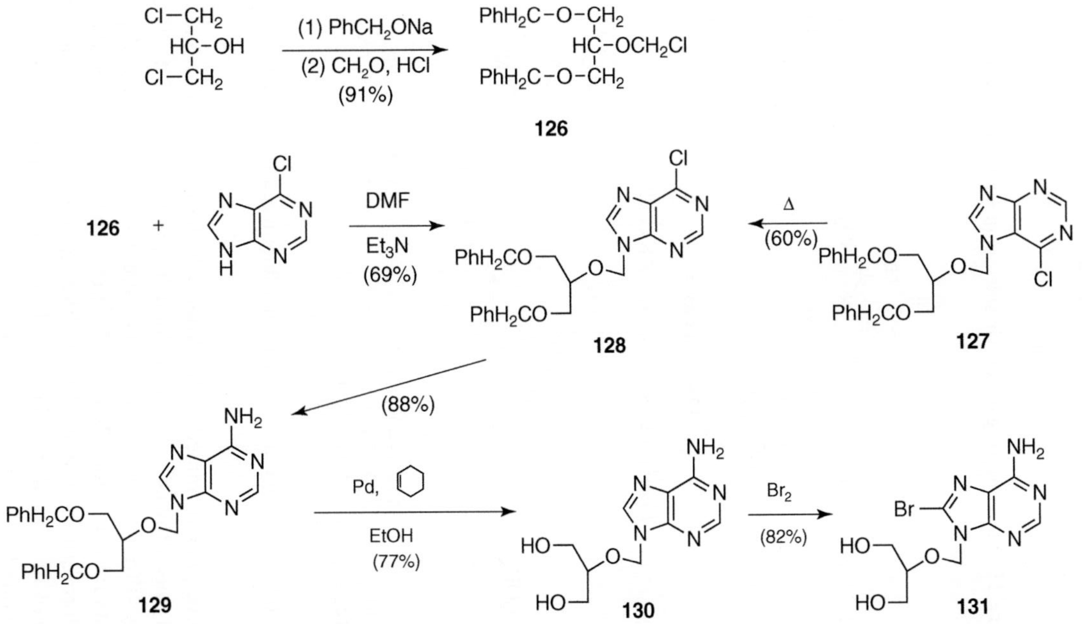

Scheme 18. Synthesis of **128** and its derivants.

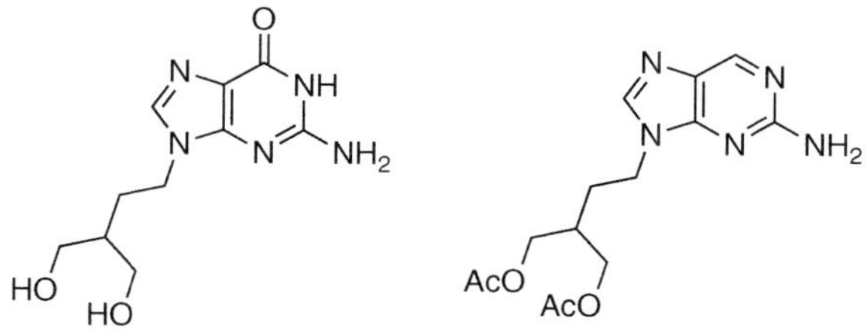

Figure 13. Structure of PCV and FCV.

2.2.2. With a 2′-C Side Chain (IIB). Penciclovir[50] (PCV) **132** is a new drug that has activity against herpes varicella zoster virus developed by SKB. Its prodrug famciclovir[51] (FCV) **133** was also developed by SKB as a substitute for orally indigestible PCV (Figure 13).

In 2003, Wang et al.[52] used a convenient route to synthesize ganciclovir (GCV or DHPG) **125** (Figure 12), PCV **132** (Scheme 19), 9-[(3′-fluoro-1′-hydroxy-2′-propoxy)methyl]guanine (FHPG) **150**, and 9-(4′-fluoro-3′-hydroxymethylbutyl)guanine (FHBG) **151** (Scheme 20).

In 1997, Hocková and Holý [53] synthesized four new "abbreviated" NAD$^+$ analogues, 1-[2-(adenin-9-yl)-3-hydroxypropyl]-3-carbamoylpyridinium chloride **158** and 1-[4-(adenin-9-yl)-2-(hydroxymethyl)butyl]-3-carbamoylpyridinium chloride **159**, using the Zincke reaction and several acyclic nucleosides, including carboxyl, hydroxyl, and amino groups (Scheme 21). NAD (nicotinamide adenine dinucleotide) is a transmission-electron coenzyme that is found frequently in cell metabolism reaction.

Experimental Procedure

9-[2-(2,2-Dimethyl-1,3-dioxan-5-yl)ethyl]adenine (153). A stirred mixture of adenine (3.0 g, 22 mmol), cesium carbonate (3.6 g, 11 mmol), and dimethylformamide (30 mL) was heated at 120°C for 2 h. After addition of 2,2-dimethyl-5-[2-(*p*-toluenesulfonyloxy)ethyl]-1,3-dioxane **152**[54] (10 g, 32 mmol) in dimethylformamide (5 mL), heating at 120°C was continued for 4 h. The reaction mixture was taken down, codistilled with water, and the product was isolated by preparative TLC on silica gel (15% methanol in chloroform). Yield: 0.9 g (15%), mp 145–147°C. ^{1}H-NMR spectrum [(CD$_3$)$_2$SO$_4$]: 8.16 s, 1H (H-2); 8.14 s, 1H (H-8); 7.20 br s, 2H (NH$_2$); 4.15 t, 2H, $J(1′,2′) = 7.1$ Hz (H-1′); 3.76 dd, 2H, $J(3′,4a′) = J(3′,5a′) = 5.4$ Hz, $J_g = 12.0$ Hz (H-4a′ and H-5a′); 3.53 dd, 2H, $J(3′,4b′) = J(3′,5b′) = 8.8$ Hz, $J_g = 12.0$ Hz (H-4b′ and H-5b′); 1.75 q, 2 H, $J(1′,2′) = J(3′,2′) = 7.1$ Hz (H-2′); 1.55 m, 1H (H-3′); 1.32 s and 1.25 s, 2 × 3 H (CH$_3$). For C$_{13}$H$_{19}$N$_5$O$_2$ (277.3):calc: 56.30% C, 6.91% H, 25.25% N; found: 55.98% C, 7.12% H, 24.89% N. Mass spectrum (FAB), *m/z* (rel. %): 278 (100) [M + H].

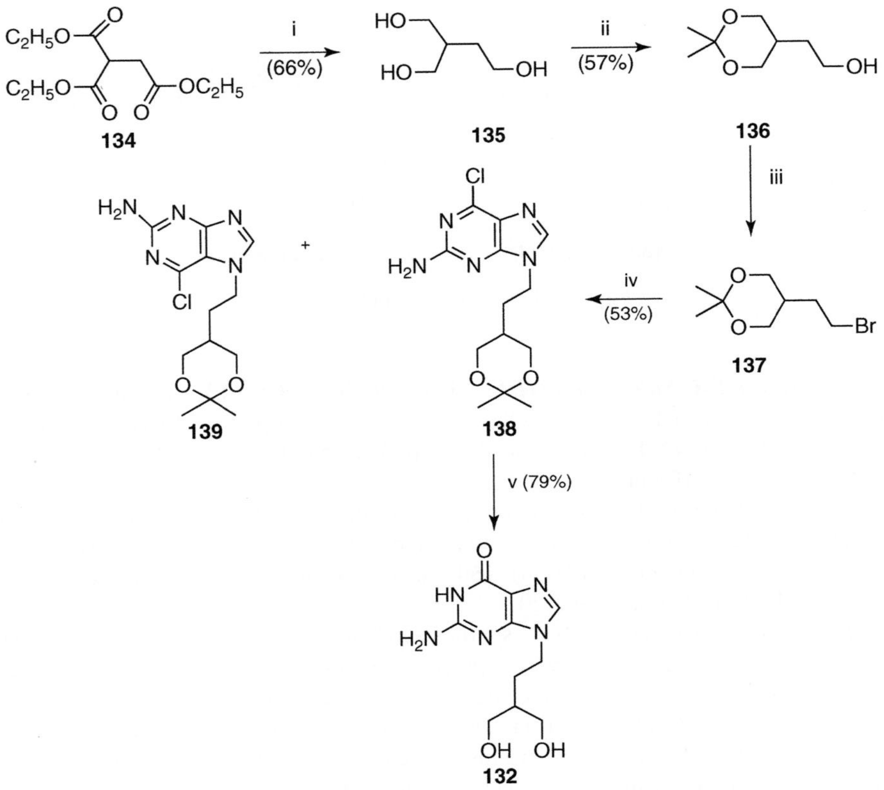

Scheme 19. Synthesis of PCV. *Reagents and conditions:* (i) NaBH$_4$, *t*-BuOH; (ii) 2,2-dimethoxypropane, *p*-toluenesulfonic acid, THF; (iii) CBr$_4$, Ph$_3$P, THF, 0°C; (iv) 2-amino-6-chloropurine, K$_2$CO$_3$, DMF; (v) 2 M HCl.

2.2.3. With a 2′-N Side Chain (IIC).

In 2010, Koszytkowska-Stawińska[55] reported that he used *N*-[1,3-di(pivaloyloxy)prop-2-yl]-*N*-(pivaloyloxymethyl)-acetamide **163** and a silylated nucleobase to synthesize *O*′-pivaloyl diesters of *N*′-acetylnucleosides (Scheme 22).

In 2006, Sas et al.[56] synthesized aza-analogues of ganciclovir **125** (Figure 12) via coupling of silylated nucleosides with *N*-(pivaloyloxymethyl)amides or sulfonamides in the presence of Lewis acid (Scheme 23). Differing from the example above, these analogues are protected by a mesyl or tosyl group. But the last coupling step is similar to that of the example above. In 2004, Koszytkowska-Stawińska and Sas[57] also synthesized this type of compound using a similar method.

2.2.4. With a 2′-S Side Chain (IID).

In 1985, Martin et al.[58] synthesized the thio analogue (thio-DHPG) of 9-[(1′,3′-dihydroxy-2′-propoxy)methyl]guanine (DHPG) through condensation of acetoxymethyl sulfide and diacetylguanine following

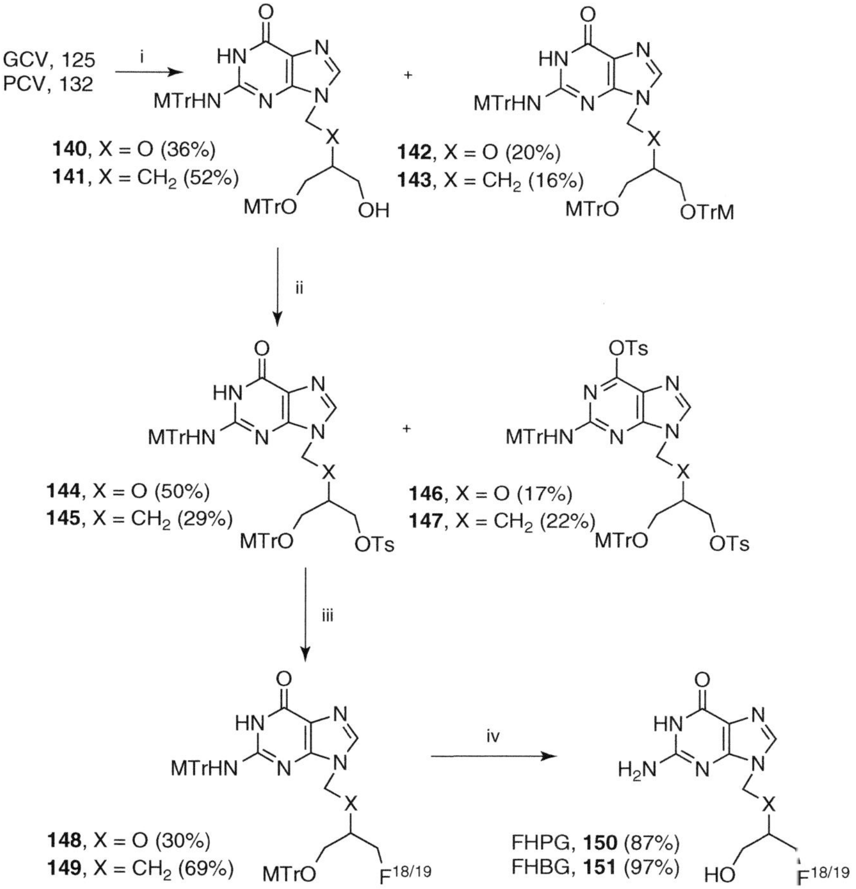

Scheme 20. Synthesis of FHPG, FHBG, [^{18}F]FHPG, and [^{18}F]FHBG. *Reagents and conditions:* (i) MTrCl, DMAP, Et$_3$N, DMF; (ii) TsCl, pyridine; (iii) [$^{18/19}$F]K, Kryptofix 2.2.2, CH$_3$CN; (iv) 1 M HCl, MeOH.

deprotection (Scheme 24). They found that some compounds have activity similar to that of ganciclovir against HSV-1 but less effect on HSV-2 and CMV.

Experimental Procedure

N^2-Acetyl-9-[[1′, 3′-bis(benzyloxy)-2′-propylthio]methyl]guanine (181). Hydrogen chloride gas (dried through concentrated H$_2$SO$_4$) was bubbled into a stirred mixture of paraformaldehyde (1.3 g, 43.3 mmol) and **178** (6.7 g, 23.2 mmol) in 1,2-dichloroethane (60 mL) at 0°C for 1.5 h. The resulting solution was dried over MgSO$_4$ and evaporated to dryness. The resulting oil was treated with sodium acetate (3.8 g, 46.3 mmol) in DMF (10 mL) at room temperature for 2 h. The resulting suspension was filtered. The filtrate was diluted with dichloromethane, washed with

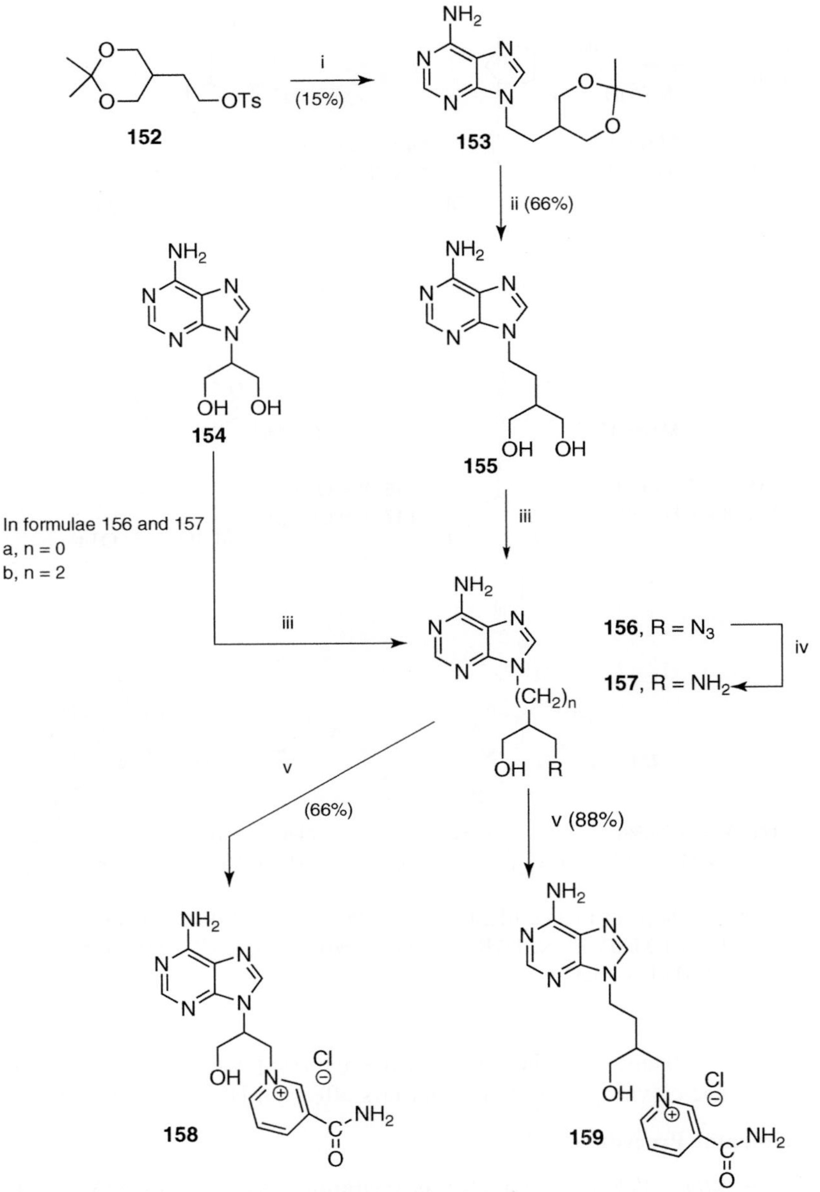

Scheme 21. Synthesis of **158** and **159**. *Reagents and conditions:* (i) adenine, Cs_2CO_3, DMF; (ii) 0.25 M H_2SO_4; (iii) (1), TsCl, pyridine; (2), NaN_3, DMF; (iv) Pd/C, H_2, MeOH; (v) 3-carbamoyl-1-(2,4-dinitrophenyl)pyridinium chloride, MeOH.

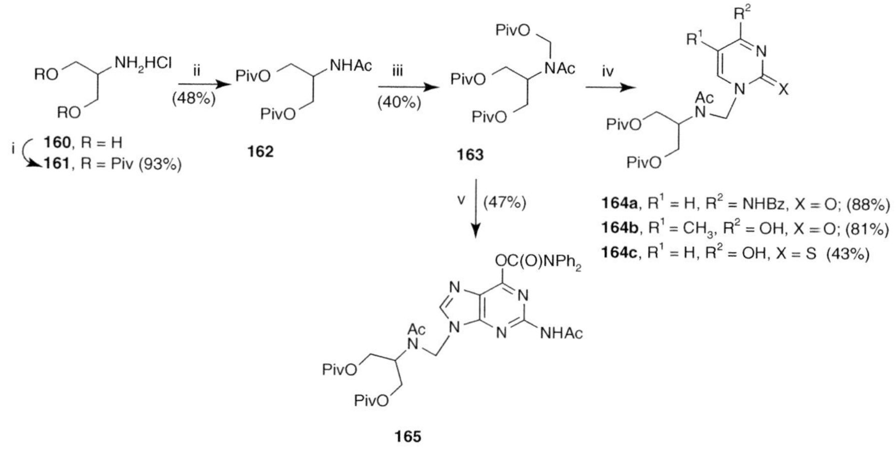

Scheme 22. Synthesis of **164** and **165**. *Reagents and conditions:* (i) PivCl, 90°C 4 h; (ii) Ac$_2$O, pyridine, rt, 1 day; (iii) ClCH$_2$OPiv, NaH, DMF, 4 days; (iv) (1), nucleobase *(N^4-benzoylcytosine, thymine, or 2-thiouracil)*, BSA, acetonitrile, rt 1 h; (2), **163**, TMSOTf, acetonitrile, rt, 2 days; (v) (1), *N^2-acetyl-O^6-(diphenylcarbamoyl)guanine*, BSA, CH$_2$Cl$_2$, 80°C, 15 min; (2), **163**, TMSOTf, toluene, 80°C, 1 h.

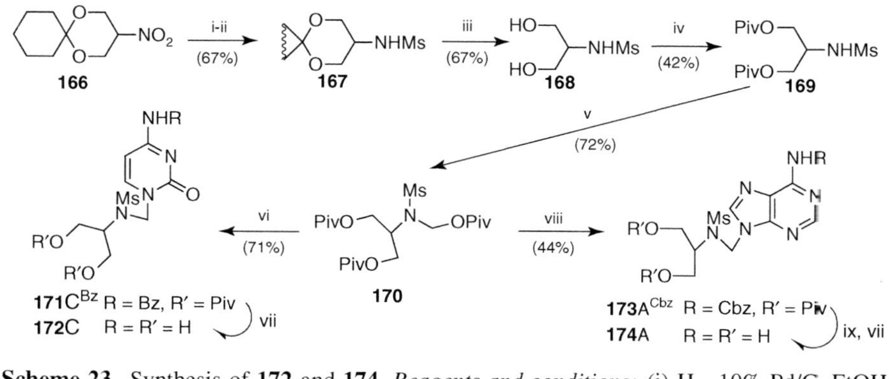

Scheme 23. Synthesis of **172** and **174**. *Reagents and conditions:* (i) H$_2$, 10% Pd/C, EtOH, rt, 60 bar, 24 h, quantitative; (ii) MeSO$_2$Cl (MsCl), pyridine, CH$_2$Cl$_2$, rt; (iii) MeOH, Dowex 50 (H$^+$), rt, 48 h; (iv) PivCl, pyridine, rt, 24 h; (v) NaH, DMF, ClCH$_2$OPiv, 72 h; (vi) *N^4-Bz-cytosine(C^{Bz})*, BSA, TMSOTf, MeCN, rt, 24 h; (vii) NH$_4$OH$_{coned}$, MeOH, sealed tube, 70°C, 1 day; (viii) *N^6-Cbz-adenine(A^{Cbz})*, BSA, SnCl$_4$, MeCN, 24 h; and (ix) H$_2$ (balloon), 10% Pd/C, MeOH, rt, 1 day.

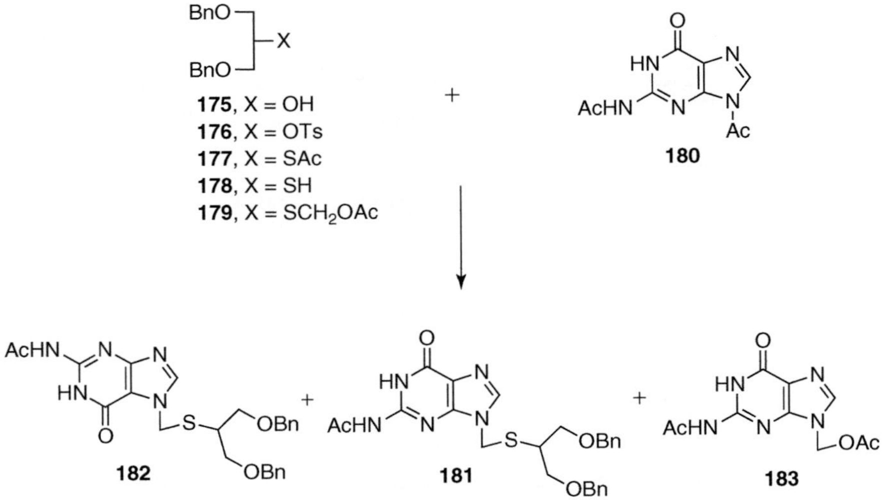

130   CHEMICAL SYNTHESIS OF ACYCLIC NUCLEOSIDES

175, X = OH
176, X = OTs
177, X = SAc
178, X = SH
179, X = SCH$_2$OAc

Scheme 24. Synthesis of thio analogues of DHPG.

saturated Na$_2$CO$_3$ and 10% HCl, dried over MgSO$_4$, and evaporated to give **179** as a pale yellow oil. A mixture of **179**, diacetylguanine **180** (5.4 g, 22.9 mmol), bis(4-nitrophenyl)phosphate (100 mg), and sulfolane (10 mL) was heated with stirring at 100°C for 6 h and then at room temperature for 18 h. The resulting mixture was diluted with dichloromethane and filtered through Celite. To remove sulfolane and by-product **183**, the filtrate was filtered through a short silica gel column, eluting with a gradient of dichloromethane to 5% methanol/dichloromethane. Selected fractions were pooled and chromatographed (3% methanol/dichloromethane) to give, in order of elution, 1.0 g (9%) of **182** and 1.2 g (11%) of **181** as white solids.

2.3. Synthesis of Acyclic Nucleosides with Adefovir Side-Chain Analogues (Type III)

The modifications of them concentrate on replacing the hydroxyl group, adding to C1'/C2' or changing of bases (Figure 14).

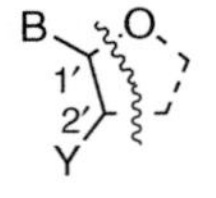

Y = OH, H

III

Figure 14. Structure of adefovir side-chain analogues (type III).

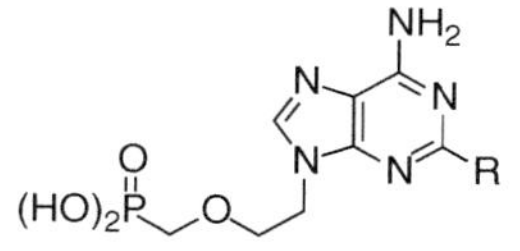

184 PMEA, R = H
185 PMEDAP, R = NH_2

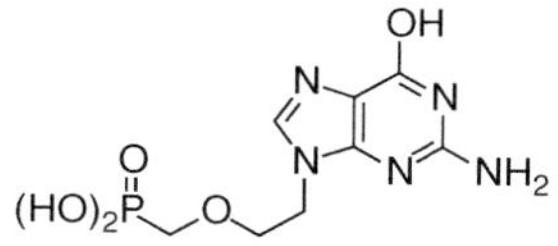

186 PMEG

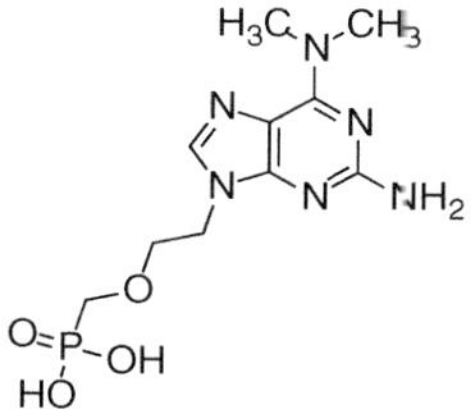

187 6-Me_2PMEDAP

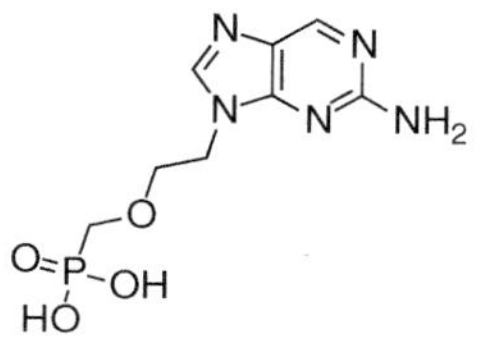

188 PMEMAP

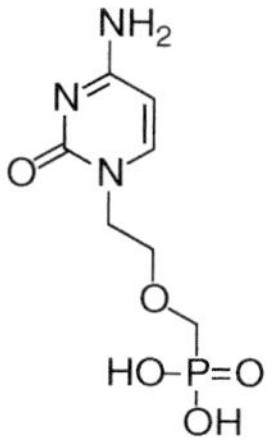

189 PMEC

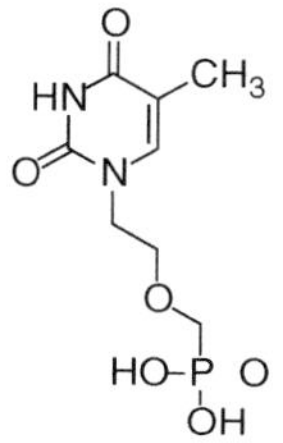

190 PMET

Figure 15. Some typical drugs in type IIIA.

2.3.1. Without a Substituent at C1′ and C2′ (III A). Typical drugs in this species are adefovir[59] (PMEA) **184**, PMEDAP[60] **185**, PMEG **186**, 6-Me_2PMEDAP **187**, PMEMAP **188**, PMEC **189**, and PMET **190** (Figure 15). PMEDAP and PMEG own more antiviral potency than PMEA. PMEDAP, especially, is immensely bioactive against retroviruses.[61]

In 1986, De Clercq[62] first suggisted the antivirus potential of adefovir (PMEA). Its prodrug, adefovir dipivoxil, developed by the Gilead Sciences company, was authorized to treat hepatitis B by the U.S. Food and Drug Administration (FDA) in 2002. Hadziyannis et al.[63] studied and collected a mass of clinical data to verify that the HBV DNA of e-antigen-negative hepatitis B patients was reduced. Meanwhile, Marcellin et al.[64] confirmed a reduction in the number of e-antigen-positive hepatitis B patients. Adefovir dipivoxil (ADV) thus became the first new drug authorized by the FDA for the treatment of chronic hepatitis B. Adefovir dipivoxil changes rapidly to adefovir in vivo, confirmed as the key to active suppression of the copying of HBV DNA chains.

In 1985, Holý et al.[65] synthesized a series of 8-bromoadenine derivatives. In this essay, they synthesized 9-(2′–hydroxyethyl)adenine **191** (Figure 16), the

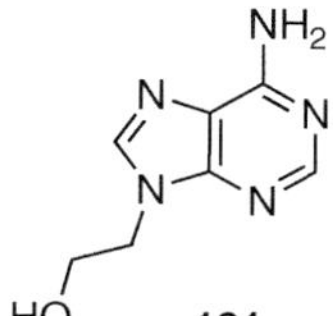

191

Figure 16. Structure of 9-(2′-hydroxyethyl)adenine.

simplest structure of this type, as starting material for research on the mechanism of formation.

Experimental Procedure

9-(2'-Hydroxyethyl)adenine (191). A stirred mixture of adenine (14 g, 0.1 mol), ethylene carbonate (10 g, 0.18 mol), and dimethylformamide (400 mL) was taken to the boil (calcium chloride protecting tube), and solid sodium hydroxide (0.15 g) was added. The mixture was refluxed under stirring for 2 h (bath temperature 150°C) and evaporated at 60°C/2 kPa. The residue was taken up in boiling ethanol (800 mL) and the extract was filtered through Celite and set aside overnight in a refrigerator. The crystallized product was collected on a filter, washed with ethanol and ether, and dried. The mother liquor was taken down in vacuo, the residue was crystallized from ethanol and processed as above; the total yield of compound Va, mp 241–242°C, was 13.2 g (71%). For C_7H_9NSO (179.2): calculated: 46.91% C, 5.06% H, 39.09% N; found: 47.06% C, 5.07% H, 39.04% N.

In 1996, Kim and co-workers[66] coupled the tosylate side chain with nucleosides to synthesize 9-[(2'-hydroxy-1-phosphonylethoxyl)ethyl]guanine, 1'-[(2'-hydroxy-1'-phosphonylethoxy)ethyl]cytosine, and 9-[(2'-hydroxy-1'-phosphonyle-thoxyl)ethyl]adenine, which are structurally closely related to PMEA **184** (Scheme 25).

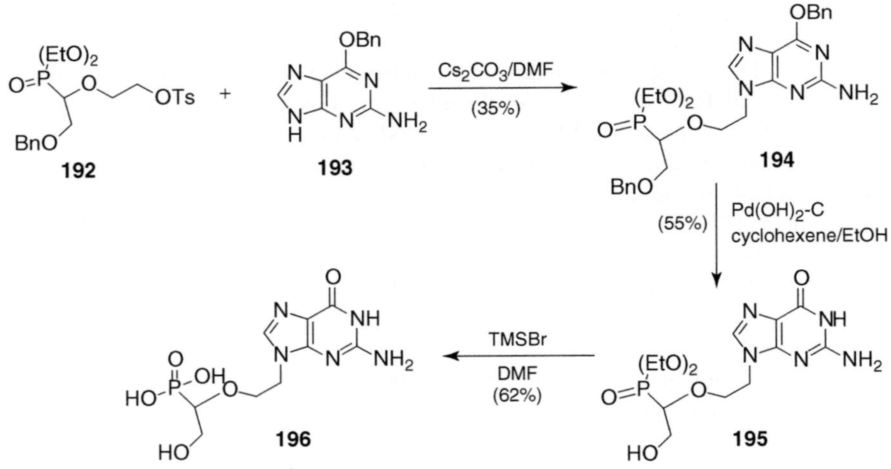

Scheme 25. Synthesis of 9-[(2'-hydroxy-1'-phosphonylethoxyl)ethyl]guanine.

In 1999, Holý et al.[67] reported that they used prepared di(2-propyl)-2-chloroethoxymethylphosphonate coupling with a pyrimidine base and a purine base to synthesize pyrimidine and purine *N*-[2-(2-phosphonomethox)ethyl] nucleotide analogues and then to made a series of derivatives (Scheme 26). Then,

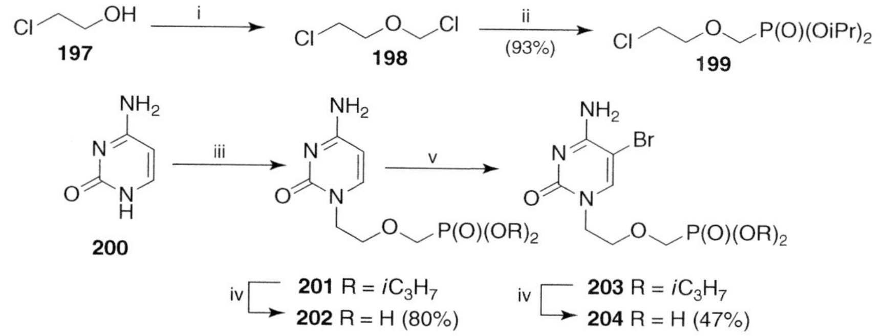

Scheme 26. Synthesis of **204**. *Reagents and conditions:* (i) $(CH_2O)_3$, HCl; (ii) $P(O\text{-}i\text{-}Pr)_3$, 10°C; (iii) **199**, Cs_2CO_3; (iv) TMSBr, CH_3CN, H_2O; (v) Br_2, DMF, CCl_4.

in 2000, they[68] used this method again to make a series of new modification of nucleosides.

Experimental Procedure

Bis(2'-propyl) 1'-[2'-(Phosphonomethoxy)ethyl]cytosine (201). A mixture of cytosine (4.4 g, 40 mmol) and cesium carbonate (6.6 g, 20 mmol) in DMF (100 mL) was stirred at 100°C for 1 h with exclusion of moisture, and after the addition of bis(2-propyl) 2-chloroethoxymethylphosphonate **199** (16 g, 50 mmol) the mixture was heated under stirring at 100°C for an additional 12 h. The mixture was filtered while hot and taken to dryness at 50°C/13 Pa. The residue in methanol (200 mL) was treated with silica gel (100 g), and the slurry was evaporated in vacuo. This material was applied on a column of silica gel (500 mL) in chloroform, and the column was eluted with chloroform and then with a chloroform/methanol mixture (4:1). The product was collected by filtration from an ether suspension and dried: yield 6.25 g (47.3%) of compound 6, mp 151°C. Anal. ($C_7H_{11}N_6O_4P$) C, H, N, P. ^{1}H-NMR [$(CD_3)_2SO$]: δ 7.47 d, 1H (H-6), $J(6,5) = 7.1$ HZ; 5.61 d, 1H (H-5), $J(5,6) = 7.1$ HZ; 4.55 dsept, 2H (p-OCH), $J(CH,CH_3) = 6.1$ HZ, $J(P,CH) = 7.1$ HZ; 3.80 t, 2H (H-1'), $J(1', 2') = 5.0$HZ; 3.74 d, 2H (PCH_2), $J(P,CH) = 8.6$ HZ; 3.67 t, 2H (H-2'), $J(2',1') = 5.0$ HZ; 1.22 and 1.20 2 × d, 6H(4 × CH_3), $J(CH_3,CH) = 6.1$ HZ.

In 2006, Shatila and Bouhadir[69] reported that they synthesized nucleic bases substituted with a diallylaminoethyl group using two methods and compared each protocol as to yield, expense, and difficulty (Scheme 27). According to the experimental data, they found that synthesis of cytosine derivatives achieved similar yields following both routes, but the adenine derivatives were formed in lower yields when using the first method because of the ion-exchange chromatography step.

In 1995, Chu et al.[70] reported that they synthesized a series of 1,3-dioxolane nucleosides by coupling 2-methyl-1,3-dioxolane with silylated pyrimidine in the presence of TMSOTf (Scheme 28). Under certain base conditions, 1,3-dioxolane

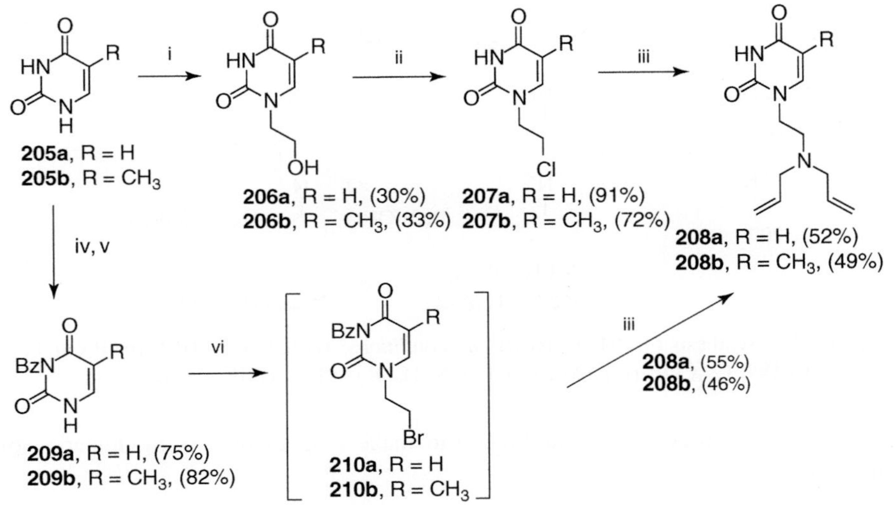

Scheme 27. Synthesis of nucleosides with a diallylaminoethyl side chain. *Reagents and conditions:* (i) ethylene carbonate, NaOH, dioxane; (ii) SOCl$_2$, pyridine; (iii) diallylamine, dioxane; (iv) PhCOCl, pyridine, MeCN; (v) K$_2$CO$_3$, dioxane; (vi) BrCH$_2$CH$_2$OH, DIAD, Ph$_3$P, dioxane.

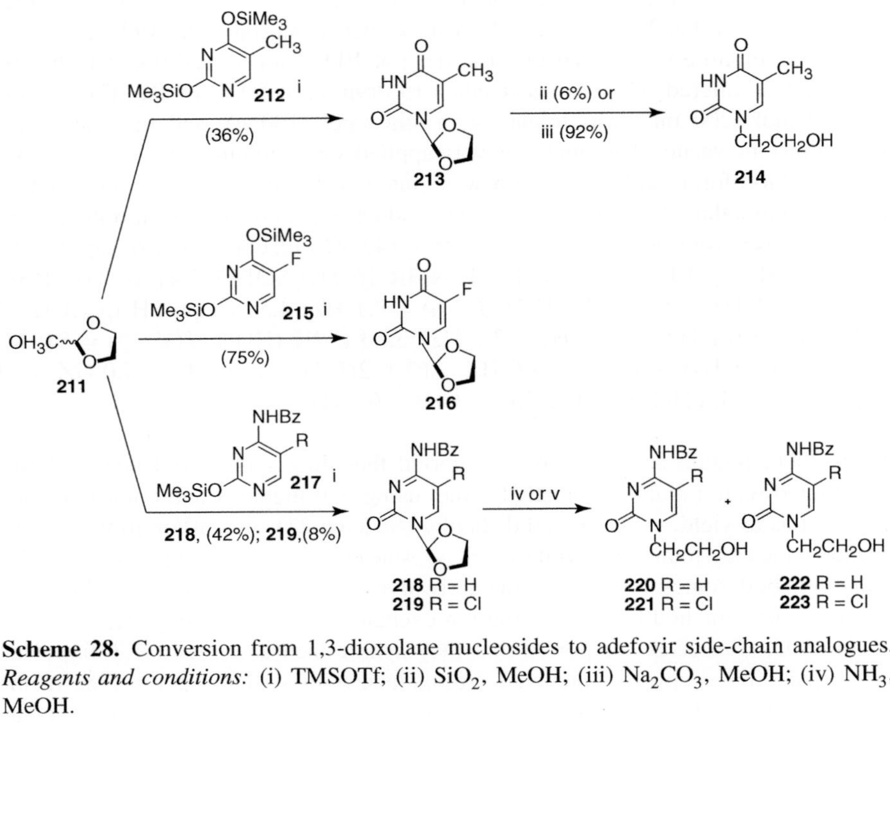

Scheme 28. Conversion from 1,3-dioxolane nucleosides to adefovir side-chain analogues. *Reagents and conditions:* (i) TMSOTf; (ii) SiO$_2$, MeOH; (iii) Na$_2$CO$_3$, MeOH; (iv) NH$_3$, MeOH.

nucleosides could decompose and then rearrange to acyclic nucleosides. According to the experimental results, they found that 5-fluorouracil nucleosides are the most stable.

Experimental Procedure

1-(1,3-Dioxolan-2-yl)thymine (213) and 1-(2-Hydroxyethyl)thymine (214). A mixture of thymine (2.0 g, 16 mmol) and ammonium sulfate (30 mg) in hexamethyldisilazane (HMDS) (50 mL) was refluxed for 4 h under an argon atmosphere. The clear solution was allowed to cool to room temperature and the HMDS was removed under reduced pressure under anhydrous conditions. Dry CH_3CN (70 mL) was added to the silylated thymine base followed by the addition of **211** (1.5 g, 14.4 mmol) in 5 mL of dry CH_3CN. This suspension was cooled in an ice–water bath to 5°C and treated with TMSOTf (3.0 mL, 16 mmol). The reaction mixture was stirred at room temperature for 48 h under argon and then poured into a saturated aqueous $NaHCO_3$ solution (20 mL) and ethyl acetate (50 mL). The organic layer was washed with H_2O (20 mL) and brine (20 mL), and dried (Na_2SO_4). The solvents were removed under reduced pressure, and the residue was purified as a silica gel column (hexane/EtOAc, 3:1) to give **213** (1.0 g, 36%) and **214** (136 mg, 5.5%). Compound **213** was recrystallized from EtOA/hexane to yield white crystals, mp 168–170°C: UV (H_2O) λ_{max} = (pH 11) 271.5 nm (ε = 2800), λ_{max} = (pH 7) 273.0 (ε = 4175), λ_{max} (pH 2) 271.0 (ε = 5345). ^1H = NMR (DMSO-d_6): δ 11.3 (s, lH, D_2O exchangeable), 8.22, (s, lH), 7.52 (s, lH), 4.30 (t, J = 5.2 Hz, 2H), 3.91 (t, J = 5.2 Hz, 2H), 1.74 (s, 3H). ^{13}C-NMR (DMSO-d_6): δ 164.05, 161.66, 150.73, 141.41, 108.20, 60.60, 56.95; MS m/e 198 (M)$^+$. Anal. Calcd for $C_8H_{10}N_2O_4$: C, 48.44; H, 5.04; N, 14.14; found: C, 48.39; H, 5.10; N, 14.26. Compound **214** was recrystallized from MeOH/Et_2O, mp 180–181°C: ^1H-NMR (DMSO-d_6): δ 11.22 (s, lH, D_2O exchangeable), 7.44, (s, lH), 4.90 (t, J = 5.3 Hz, D_2O exchangeable), 3.68 (t, J = 4.9 Hz, 2H), 3.57 (m, 2H), 1.75 (s, 3H); ^{13}C-NMR (DMSO- d_6): δ 164.43, 150.90, 142.46, 107.55, 62.70, 58.56; MS m/e 170 (M)$^+$. Anal. Calcd for $C_7H_{10}N_2O_3$: C, 49.36; H, 5.88; N, 16.45; found: C, 49.31; H, 5.88; N, 16.36.

In 2009, Hocková and co-workers[71] synthesized two series of branched 9-[2'-(2'-phosphonoethoxy)ethyl]purines to research their inhibiton of plasmodium falciparum hypoxanthine–guanine–xanthine phosphoribosyltransferase (pfHGXPRT) and human HGPRT (Scheme 29).

Recently, Hocková et al.[72] improved their method to synthesize a series of β-branched acyclic nucleoside phosphonates with the phosphonoethoxyethyl (PEE) moiety.

As noted earlier, we know that tricyclic nucleoside has great research value. In 2006, Holý and co-workers[73] synthesized a series of 9-substituted [2-(3H-imidazo[1,2-a]purin-3-yl)ethoxy]methylphosphonic and 4-substituted [2-(1H-imidazo[2,1-b]purin-1-yl)ethoxy]methylphosphonic acids which were tricyclic etheno analogues of PMEDAP **185** and PMEG **186** (Scheme 30)

In 2010, Holý and co-workers[74] synthesized a new type of ANP with a arylphosphonate moiety in the side chain using a microwave-assisted method (Scheme 31).

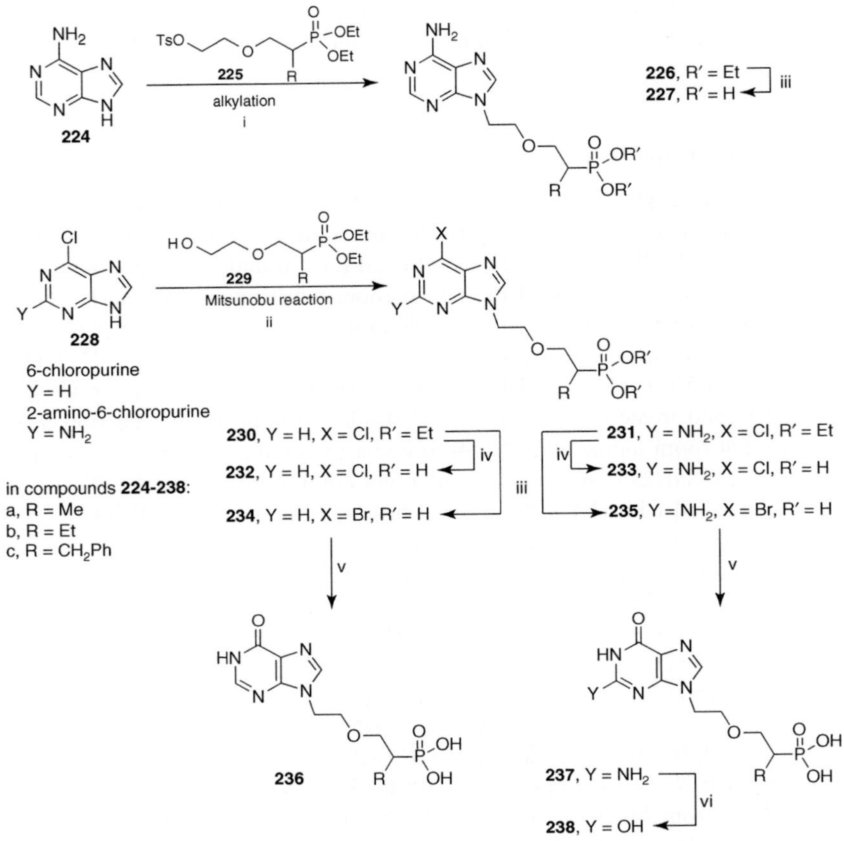

Scheme 29. Synthesis of **236** and **238**. *Reagents and conditions:* (i) Cs$_2$CO$_3$, DMF; (ii) Ph$_3$P, DIAD, HF; (iii) CH$_3$CN, MeSiBr; (iv) CH$_3$CN, 2,6-lutidine, MeSiBr; (v) TFA, H$_2$O; (vi) NaNO$_2$, HCl, H$_2$O.

2.3.2. With Substituent at C1′ or C2′ (III B).

Tenofovir [PMPA, (R)-9-(2′-phosphatemethoxylpropyl)adenine] **255** (Figure 17), which has been approved to treat HIV patients, is a new anti-HIV nucleotide drug. PMPA shows double activity against HIV and HBV in vitro. The oral bioavailability of PMPA is low, so tenofovir disoproxil fumarate **3** (tenofovir DF) has been designed as a sustained-release agent.

In 2010, Boddy and co-workers[75] synthesized tenofovir DF **3** via hydrolysis of tenofovir diethyl ester at low temperature (Scheme 32).

Wiemer et al.[76] developed a synthetic method of producing acycleoside and acycleotide analogues from common amino acids. They used D- or L-alanine as a starting material to assemble (R)- and (S)-2-amino-1-propanol, which were used to synthesize (R)- and (S)-9-(1′-methyl-2′-phosphonomethoxyethyl)adenine (Scheme 33).

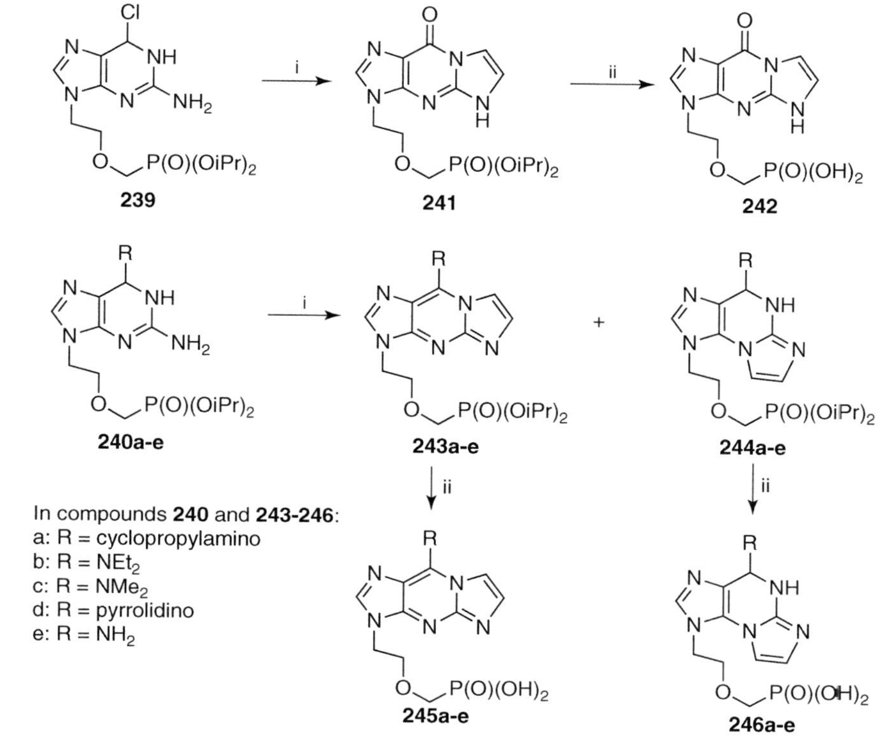

Scheme 30. Synthesis of etheno-bridged PMEDAP and PMEG derivatives. *Reagents and conditions:* (i) 1 M aq. ClCH$_2$CHO, water, dioxane, 70°C, 8 h; (ii) Me$_3$SiBr, acetonitrile, rt, 24 h.

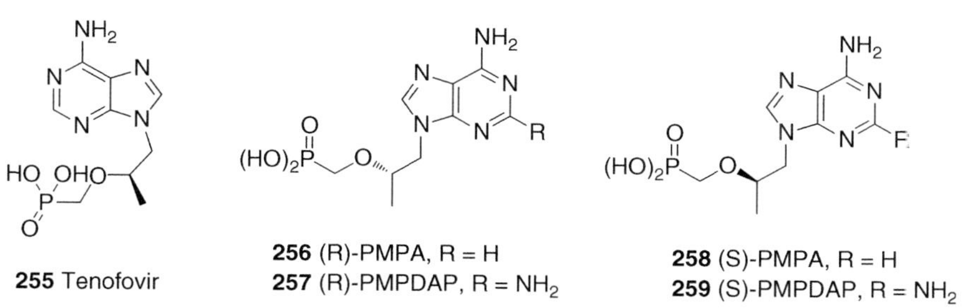

Figure 17. Typical type III B drugs.

Experimental Procedures

(2R,3R)-2-[(5-Amino-6-chloropyrimidin-4-yl)amino]-1,3-butanediol **(270)**. A solution of diol **269** (2.70 g, 11.3 mmol) in ethanol (30 mL) was treated with 10% Pd-C (0.36 g, 0.34 mmol), triethylamine (0.56 g, 5.5 mmol), and triethylsilane

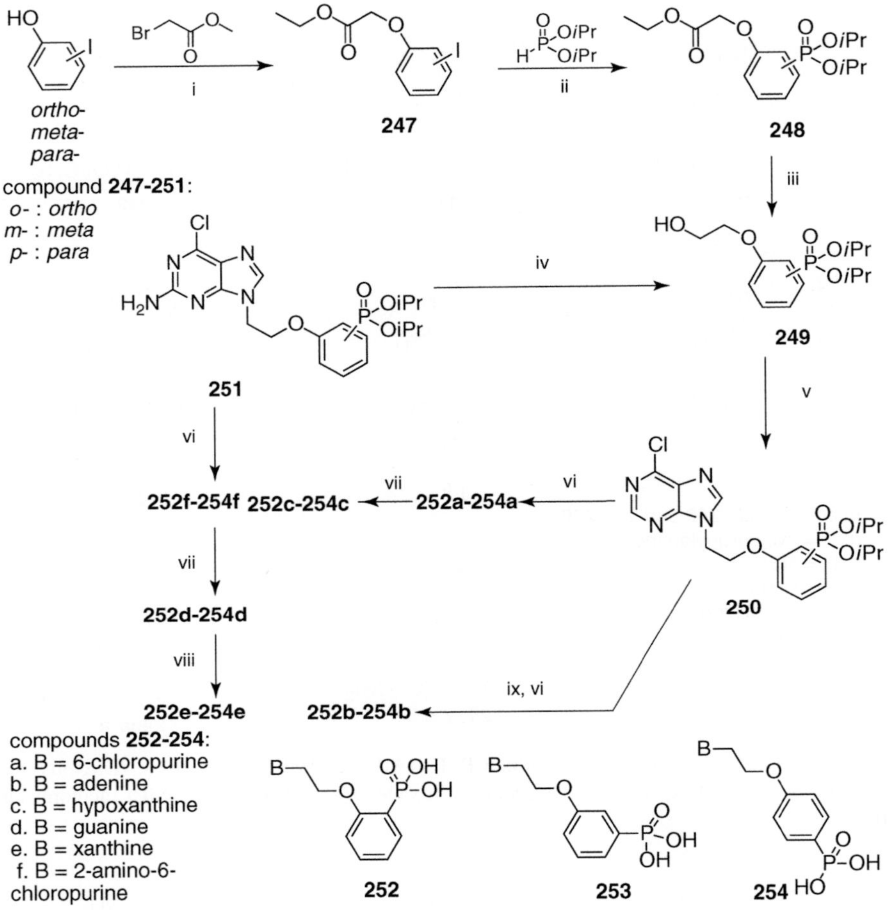

Scheme 31. Synthesis of ANPs with an arylphosphonate moiety in the side chain. *Reagents and conditions:* (i) K$_2$CO$_3$, DMF; (ii) Pd(Ph$_3$P)$_4$, Et$_3$N, THF, MW; (iii) BH$_3$·THF; (iv) 2-amino-6-chloropurine, DIAD, Ph$_3$P, THF; (v) 6-chloropurine, DIAD, Ph$_3$P, THF; (vi) BrSiMe$_3$, MeCN; (vii) TFA; (viii) NaNO$_2$, HCl; (ix) NH$_3$, MeOH.

(4.01 g, 34.5 mmol). The reaction mixture was heated at reflux until the starting material could not be detected by TLC (ca. 3 h). After the solution cooled to room temperature, the Pd-C was removed by filtration and the filtrate was concentrated to afford a clear oil. This oil was treated with 5-amino-4,6-dichloropyrimidine (1.82 g, 11.1 mmol) and sodium bicarbonate (1.15 g, 13.7 mmol) in 1-butanol (20 mL) and heated at reflux. After 48 h, the solution was allowed to cool to room temperature and concentrated in vacuo. The residue was dissolved in ethyl acetate and filtered through a short plug of silica gel to provide the pure product **270** (1.82 g, 69%). ^{1}H-NMR (DMSO): δ 7.69 (s, 1H), 6.29 (d, J = 8.2 Hz, 1H,

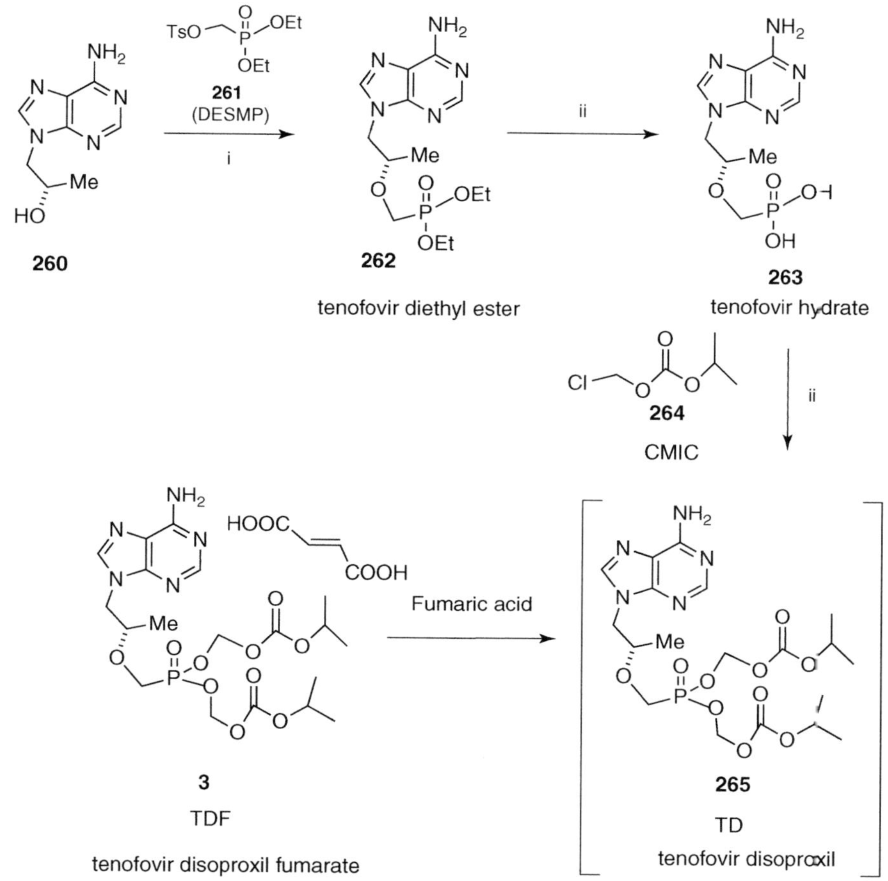

Scheme 32. Synthesis of TDF. *Reagents and conditions:* (i) DMF. THF, Mg(O-*t*-Eu)$_2$; (ii) Me$_3$SiBr, H$_2$O, 100–10°C, acetone wash and recrystallization; (iii) Et$_3$N, DMF.

exch.), 5.11 (s, 2H, exch. with D$_2$O), 4.62 (m, 2H, exch. with D$_2$O), 4.12 (m, 1H), 3.98 (m, 1H), 3.56 (m, upon D$_2$O addition: dd, J = 10.9, and 6.1 Hz, 1H), 3.45 (m, upon D$_2$O addition: dd, J = 10.9, and 6.3 Hz, 1H), 1.03 (d, J = 5.4 Hz, 3H).^{13}C-NMR (DMSO): δ 152.5, 145.5, 137.0, 123.4, 64.5, 60.2, 57.4, 20.0; ESI HRMS calcd for C$_8$H$_{13}$N$_4$O$_2$Cl: [M + H]$^+$−233.0805; found: 233.0800.

(1′R,2′R)-9-(1′-Hydroxymethyl-2′-hydroxypropyl)-6-chloropurine **(271)**. A solution of pyrimidine **270** (0.81 g, 3.5 mmol) in trimethylorthoformate (20 mL) was treated with HCl (0.3 mL, 9.8 mmol). The resulting solution was stirred at room temperature for 5 h and then concentrated in vacuo. The residue was crystallized from acetone to afford the analytically pure product **271** (0.74 g, 88%). ^{1}H-NMR (DMSO): δ 8.76 (s, 1H), 8.65 (s, 1H), 5.08 (d, J = 4.7 Hz, 1H,

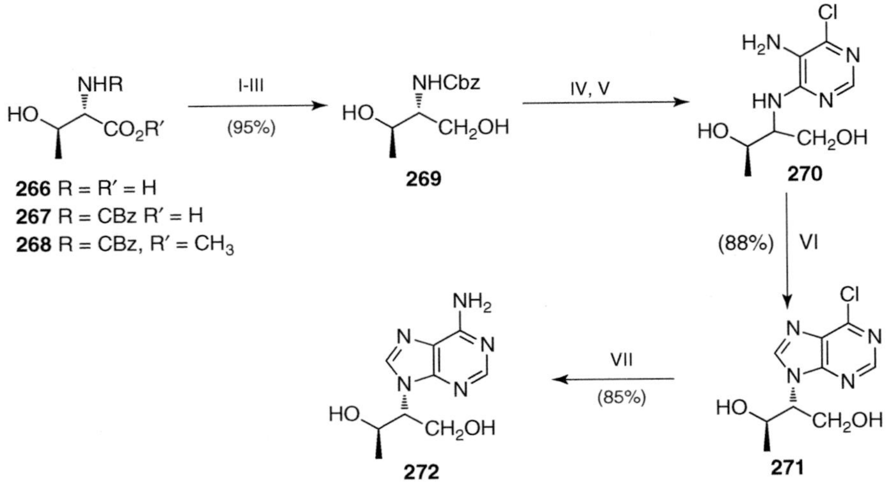

Scheme 33. Synthesis of (2R,3R)-2-(6-amino-9H-purin-9-yl)butane-1,3-diol. *Reagents and conditions:* (i) CbzCl; (ii), SOCl$_2$, MeOH; (iii) LiBH$_4$; (iv) Pd-C, Et$_3$SiH; (v) 5-amino-4,6-dichloropyrimidine, n-BuOH; (vi) HC(OMe)$_3$; (vii) NH$_3$, MeOH.

exch. with D$_2$O), 4.98 (dd, J = 5.5 Hz, 1H, exch. with D$_2$O), 4.53 (m, 1H), 4.22 (m, 1H), 3.96 (m, 1H), 3.87 (m,1H), 1.01 (d, J = 6.2 Hz, 3H). ^{13}C-NMR (DMSO): δ 152.9, 151.2, 148.7, 147.1, 130.5, 64.4, 63.6, 60.4, 20.7. *Anal.* Calcd for C$_9$H$_{11}$N$_4$O$_2$Cl: C, 44.55; H, 4.57; N, 23.09; Cl,14.61; found: C, 44.96; H, 4.64; N, 22.69; Cl, 14.28.

In 2006, Barral and co-workers[77] did research on 9-[2'-(boranophosphonomethoxy)ethyl]adenine **277a** and (R)-9-[2'-(boranophosphonomethoxy)propyl]adenine **277b**, with a borane group replacing one unbridging oxygen atom, that focused on the synthetic method, antiviral evaluation, and stability (Scheme 34).

Experimental Procedure

9-[2'-(Boranophosphonomethoxy)ethyl]adenine (277a). Compound **276a** (200 mg, 0.78 mmol) was dried over phosphorus pentoxide under vacuum for 5 h and then dissolved in anhydrous tetrahydrofuran (10 mL) that had been flushed with argon. BSA (960 μL, 3.89 mmol) was added by syringe, and the solution was stirred for about 1h at room temperature, under an argon atmosphere. DIPEA · BH$_3$ (270 μL, 1.56 mmol) was added, the solution was stirred for 1 h, and concentrated ammonium hydroxide (30%) in methanol (10 mL, 1:1 v/v) was added to the solution. After the solvents were evaporated under reduced pressure, the residue was purified by reversed-phase column chromatography (linear gradient 0-100% buffer B). Product fractions were collected and evaporated to dryness. An excess of triethylammonium bicarbonate was removed by repeated

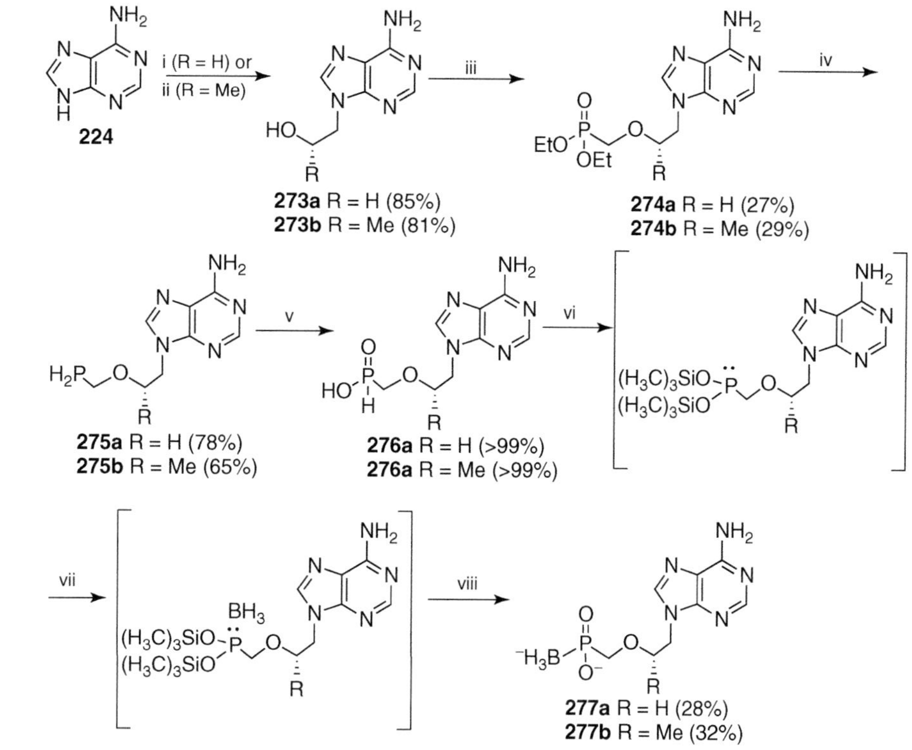

Scheme 34. Synthesis of **277**. *Reagents and conditions*: (i) 2-bromoethyl benzoate, NaH, DMF, 60°C, 16 h, then sat NH$_3$/MeOH, 14 h; (ii) (R)-propylene carbonate, NaOH, DMF 140°C, 16 h; (iii) diethyl {[(p-toluenesulfonyl)oxy]methyl}phosphonate, sodium *tert*-butoxide, DMF, rt, 72 h; (iv) TMSCl, LiAlH$_4$, THF, −78°C, rt, 2 h; (v) H$_2$O$_2$, H$_2$O/THF, rt, 1 h; (vi) BSA, THF, rt, 1 h; (vii) BH$_3$ · DIPEA, THF, rt, 1 h; (viii) NH$_4$OH/MeOH (1:1 v/v).

freeze-drying with deionized water to give **277a** (58 mg, 28%) as a white powder. HPLC purity: 98%. ^{1}H-NMR (DMSO-d_6): δ 8.16 (s, 1H, H-8), 8.08 (s, 1H, H-2), 7.12 (b s, 2H, NH$_2$), 4.22 (t, J = 5.2 Hz, 2H, CH$_2$N), 3.79 (t, J = 5.2 Hz, 2H, CH$_2$O), 3.33 (m, 2H, CH$_2$P), 0.5 to 0.10 (q, J = 88.0 Hz, 3H, BH$_3$). ^{31}P-NMR (DMSO-d_6): δ 83.04 (q, J_{PB} = 117.0 Hz). ^{11}B-NMR (DMSO-d_6): δ −31.38 (d, J_{PB} = 116.0 Hz). MS (GT, FAB$^-$): 270 (M)$^-$, 515 (M−BH$_3$)$^-$. HRMS (FAB): calcd for C$_8$H$_{14}$N$_5$O$_3$PB:(M)$^-$, 270.0927; found, 270.0904.

In 2001, Holý et al.[78] synthesized a series of N^6-substituted adenine and 2,6-diaminopurine analogues of 9-[2′-(phosphonomethoxy)ethyl] (PME) **243−324**, 9-[(R)-2′-(phosphonomethoxy)propyl] [(R)-PMP] **325−350** and (S)-PMP **351−362** (Scheme 35).

Experimental Procedure

6-Amino-9-[2′-(phosphonomethoxy)ethyl]purines (243−270). A primary or secondary amine (5 mL) was added to a solution of compound **236** (1.9 g, 5 mmol) in ethanol (50 mL) and the solution was refluxed for 6−8 h. The course of the reaction was monitored by TLC in systems S1 and S2. After completion, the mixture was taken down in vacuo and the residue was co-distilled twice with ethanol (20 mL each). The residue was treated with aqueous methanol (1:4, 50 mL) and Dowex 50 × 8 (H$^+$ form) until it reached dissolution and persistent acid reaction. This suspension was applied to a column (150 mL) of the same ion exchanger equilibrated in aqueous methanol (1:4), and the column was washed with the same eluent. The elution was continued until the UV absorption of the eluate (254 nm) dropped to the original value. The column was then washed with 2.5% ammonia in aqueous methanol (1:4), and the UV-absorbing ammonia fraction was collected. After evaporation in vacuo, the residue of compound **242** was co-distilled with ethanol (2 × 25 mL) and dried overnight in vacuo over phosphorus pentoxide. Acetonitrile (40 mL) and Me$_3$SiBr (4 mL) were added and the solution was left to stand overnight at ambient temperature. The mixture was evaporated in vacuo, co-distilled with acetonitrile (20 mL), and water (50 mL) was added. After 10 min, the acid solution was alkalinized with concentrated aqueous ammonia and evaporated. The residue in water (20−30 mL) was applied to a column (100 mL) of Dowex 50 × 8 (H$^+$ form) equilibrated in water and the desalting procedure described was repeated. The residue from the ammonia eluate was dissolved in water (20 mL) by adding concentrated aqueous ammonia to pH 10−10.5, and the solution was applied to a column (100 mL) of Dowex 1 × 2 (acetate form) washed with 0.02 M acetic acid. The column was washed first with 0.02 M acetic acid until the UV absorption dropped and then with a linear gradient of acetic acid (0.02−0.30 M, 1 L each). The main UV-absorbing fraction was taken down to dryness and the residue was co-distilled with water (3 × 25 mL). The product was obtained by recrystallization of the residue from water. After standing overnight in a refrigerator, the product was collected, washed with ethanol and ether, and dried.

Recently, Hostetler et al.[79] synthesized octadecyloxyethyl 9-(S)-[3′-methoxy-2′-(phosphonomethoxy)propyl]adenine [ODE-(S)-HPMPA], in which a methoxy

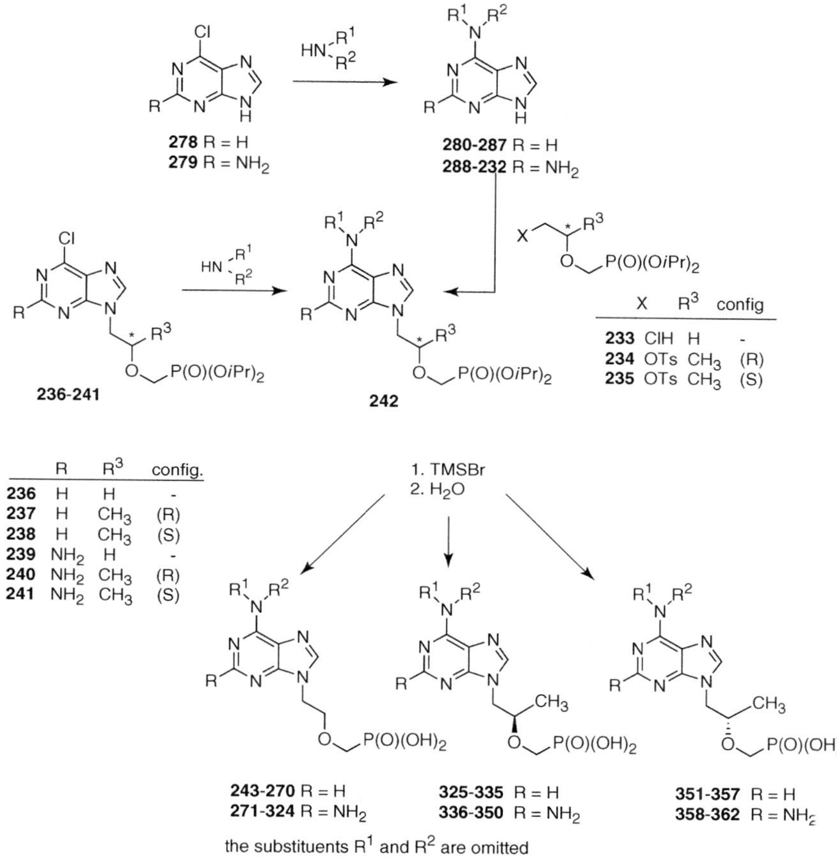

Scheme 35. Synthesis of N^6-substituted adenine analogues of PME, (R)-PMP, and (S)-PMP.

group replaces the acyclic side-chain hydroxyl of ODE-(S)-HPMPA[80] with anti-HCV activity but with toxicity in vitro and in mice. The modification reduces toxicity but retains good antiviral activity against HCV.

Janeba and co-workers[81] found a convenient way to synthesize and test the activity of N^9-[3-fluoro-2-(phosphonomethoxy)propyl] (FPMP) analogues of purine bases (Scheme 36). Replacement of the hydroxyl group by fluorine gives acyclic nucleoside phosphonates (ANPs) that have good activity against retrovirus, which compensates for the lack of anti-DNA virus activity.

Experimental Procedure

(S)-6-Chloro-9-{2'-[(diisopropoxyphosphoryl)methoxy]-3'-fluoropropyl}-purine (369b). Microwave-assisted: Sodium hydride (56 mg, 1.4 mmol) was added to a solution of 6-chloropurine (199 mg, 1.29 mmol) in DMF (5 mL) and

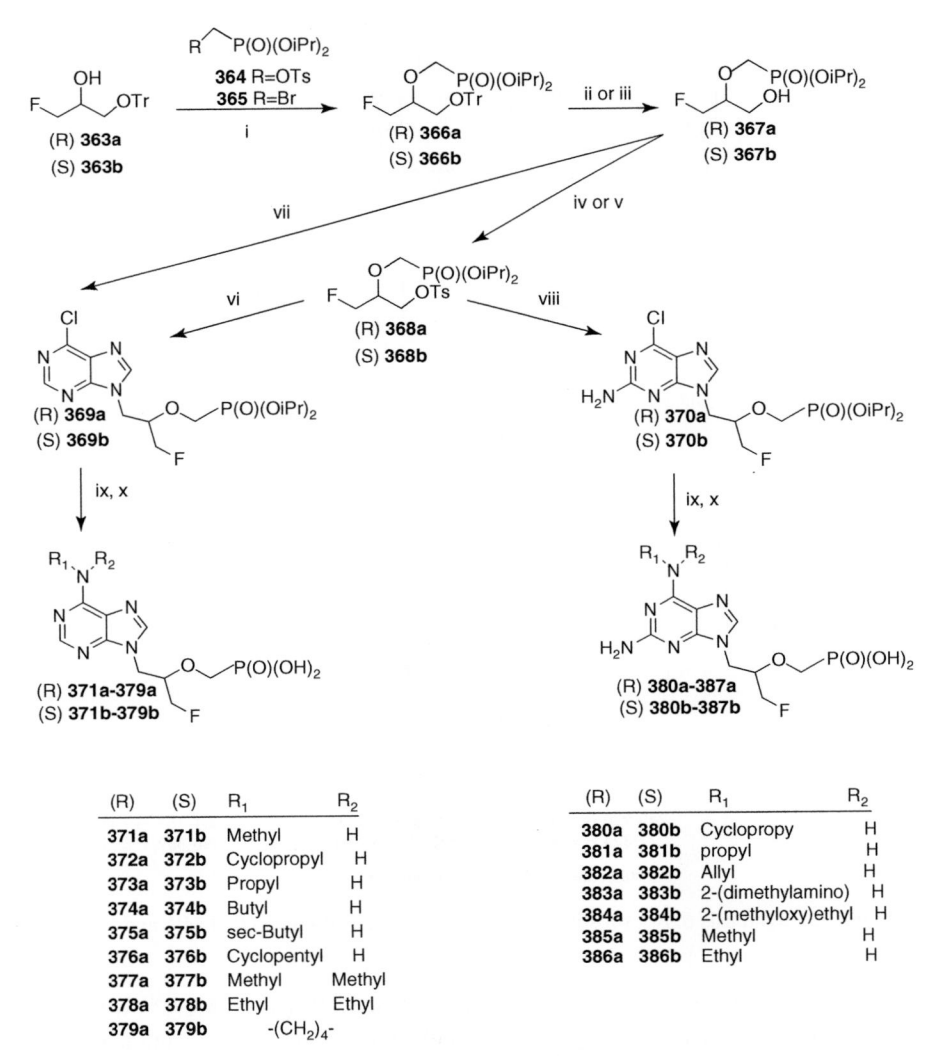

Scheme 36. Synthesis of FPMP purine analogues. *Reagents and conditions:* (i) NaH, DMF, −20°C to rt; (ii) 80% CH_3COOH, 90°C; (iii) Dowex D50W×8 (H^+ form), aq. MeOH, reflux; (iv) TsCl, Et_3N, CH_2Cl_2, 0°C to rt; (v) TsCl, Py, DMAP, 0°C; (vi) 6-chloropurine, NaH, DMF, 60°C (conventional) or 120°C (microwave); (vii) 6-chloropurine, Ph_3P, DIAD, THF, rt to 60°C; (viii) 2-amino-6-chloropurine, NaH, DMF, 90°C; (ix) amine, CH_3CN, 70 or 80°C ;(x) TMSBr, CH_3CN, rt.

(R)	(S)	R_1	R_2
371a	371b	Methyl	H
372a	372b	Cyclopropyl	H
373a	373b	Propyl	H
374a	374b	Butyl	H
375a	375b	sec-Butyl	H
376a	376b	Cyclopentyl	H
377a	377b	Methyl	Methyl
378a	378b	Ethyl	Ethyl
379a	379b	-$(CH_2)_4$-	

(R)	(S)	R_1	R_2
380a	380b	Cyclopropy	H
381a	381b	propyl	H
382a	382b	Allyl	H
383a	383b	2-(dimethylamino)	H
384a	384b	2-(methyloxy)ethyl	H
385a	385b	Methyl	H
386a	386b	Ethyl	H

the mixture was stirred for 15 min at room temperature. A solution of tosylate **368b** (0.5 g, 1.17 mmol) in DMF (2 mL) was added, then the reaction vial was sealed with a Teflon septum and microwave-irradiated at 120°C for 30 min. After cooling to room temperature, the solvent was removed in vacuo and the residue was

chromatographed. Compound **369b** as white foam; $[\alpha]_D$ −23.5 μ (c 0.4, CHCl$_3$).
^{1}H-NMR: δ 1.25 and 1.29 and 1.31 and 1.33 (4 × d, 4 × 3H, J_{CH_3-CH} = 5.2 Hz, CH$_3$-i-Pr), 3.76 (dd, 1H, J_{gem} = 13.9 Hz, J_{CH_2-P} = 8.7 Hz, C4$_b$), 4.16 (m, 1H, 2), 3.92 (dd, 1H, J_{gem} = 13.9 Hz, J_{CH_2-P} = 8.5, C4$_a$), 4.47 (dd, 1H, J_{gem} = 14.7 Hz, J_{1b-2} = 7.1 Hz, C1$_b$), 4.48 (ddd, 1H, J_{gem} = 10.5 Hz, J_{3b-2} = 4.7 Hz, J_{3b-F} = 46.7 Hz, C3$_b$), 4.62 (ddd, 1H, J_{gem} = 10.5 Hz, J_{3a-2} = 4.0 Hz, J_{3a-F} = 47.1 Hz, C3$_a$), 4.70 (m, 2H, CH-i-Pr),4.63 (ddd, 1H, J_{gem} = 14.7 Hz, J_{1a-2} = 3.8 Hz, J_{1a-F} = 0.9 Hz, C1$_a$), 8.35 (s,1H, Pu8), 8.75 (s, 1H, Pu2).^{13}C-NMR: 23.86 (m, CH$_3$-i-Pr), 43.97(d, J_{1-F} = 7.6 Hz, C1), 65.12 (d, J_{C-P} = 168.2 Hz, C4), 71.29 (m, CH-i-Pr), 77.76 (dd, J_{2-F} = 19.7 Hz, J_{2-P} = 9.1 Hz, C2), 81.37 (d, J_{3-F} = 174.3 Hz, C3), 131.26 (Pu5), 146.35 (Pu8), 150.94 (Pu6), 151.83 (Pu4), 151.86 (Pu2).^{19}F-NMR: δ −227.70 (td, J_{F-3a} = J_{F-3b} = 46.7 Hz, J_{F-2} = 18.6 Hz); MS (ESI) m/z: 409 [M + H]$^+$. *Anal*. Calcd (C$_{15}$H$_{23}$ClFN$_4$O$_4$P): C, 44.07; H, 5.67; Cl, 8.67; F, 4.65; N, 13.71; P, 7.58; found: C, 44.22; H, 5.86; N, 13.56; F, 4.59; P, 7.73.

2.4. Synthesis of Acyclic Nucleosides with Cidofovir Side-Chain Analogues (Type IV)

This type is obtained by breaking the C1'-O and C3'-C4' bonds in the sugar ring (Figure 18). In 1996, the FDA ratified parenteral cidofovir **4**, which possesses a very strong inhibitory effect on cytomegalovirus (CMV), developed by the Gilead company under the trade name Vistide. The disadvantage of cidofovir **4** is its renal toxicity.

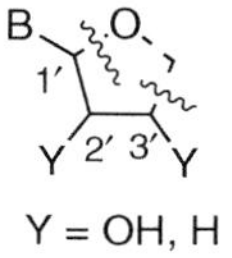

Figure 18. Structure of cidofovir side-chain analogues (type IV).

Sources have reported various methods of synthesizing cidofovir **4**. Bronson et al.[82] and Brodfuehrer et al.[83] reported using (*R*)-glycidol as a starting material for synthesis. Vemishetti et al. reported in a patent that they used a condensation of benzoyl cytosine and chiral triphenylmethoxylmethylepoxy ethane followed by deprotection, hydrolysis, and a second deprotection to get cidofovir **4**. We will look at just one method.

In 1994, Brodfuehrer and co-workers reported having synthesized (*S*)-HPMPC **4** in high yield via coupling with *R*-glycidol and cytosine in DMF (Scheme 37).

In 1988, Holý et al.[84] prepared several derivatives of 9-(*S*)-(3'-hydroxy-2'-phosphonylmethoxypropyl)adenine (HPMPA) from racemic or (*S*)-*N*-(2',3'-dihydroxypropyl) analogues. In this report, 1'-(*S*)-(2', 3'-dihydroxypropyl)cytosine **406a** and 1'-(*S*)-(2',3'-dihydroxypropyl)-5-methylcytosine **406b** were synthesized via a condensation reaction (Scheme 38).

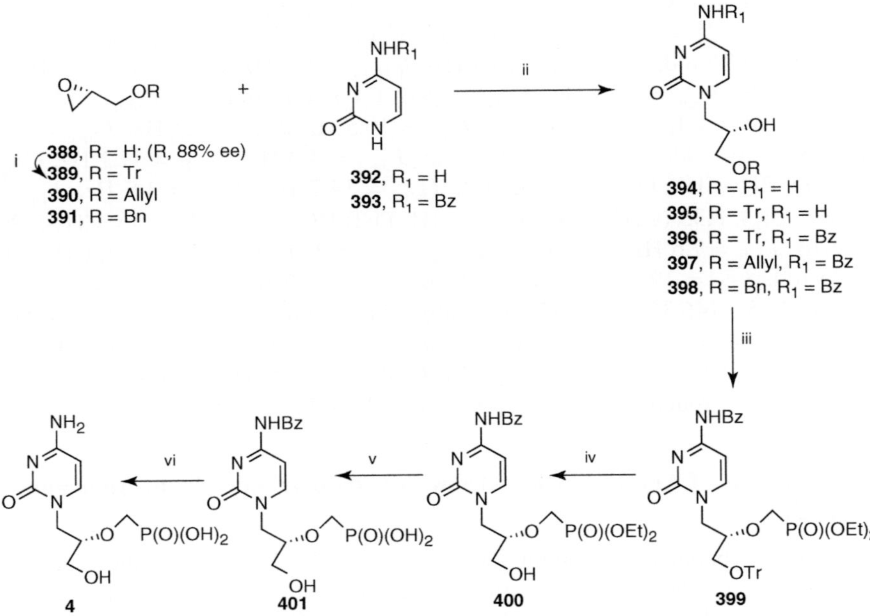

Scheme 37. Synthesis of (S)-HPMPC **4**. *Reagents and conditions:* (i) TrCl, TEA, CH$_2$Cl$_2$, rt, 3 h; (ii) K$_2$CO$_3$, DMF, 72°C, 5 h or NaH (0.22 equiv), DMF, 105°C, 5 h; (iii) TsOCH$_2$P(O)(OEt)$_2$, NaH (3 equiv), DMF, 0°C, 6 h; (iv) HCl$_2$, CH$_2$Cl$_2$, 0–5°C, 10 min, 55%; (v) TMSBr, CH$_2$Cl$_2$, rt, 18 h; (vi) conc. NH$_4$OH, rt, 4 h, 78%.

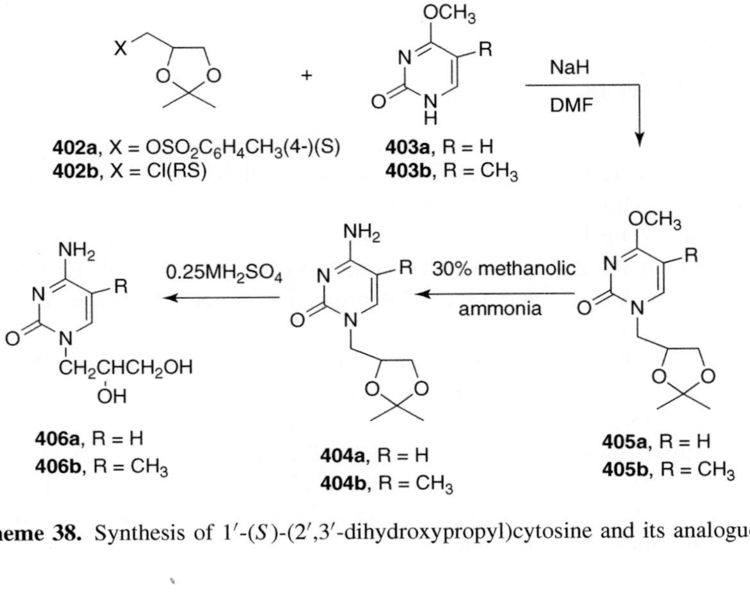

Scheme 38. Synthesis of 1′-(S)-(2′,3′-dihydroxypropyl)cytosine and its analogues.

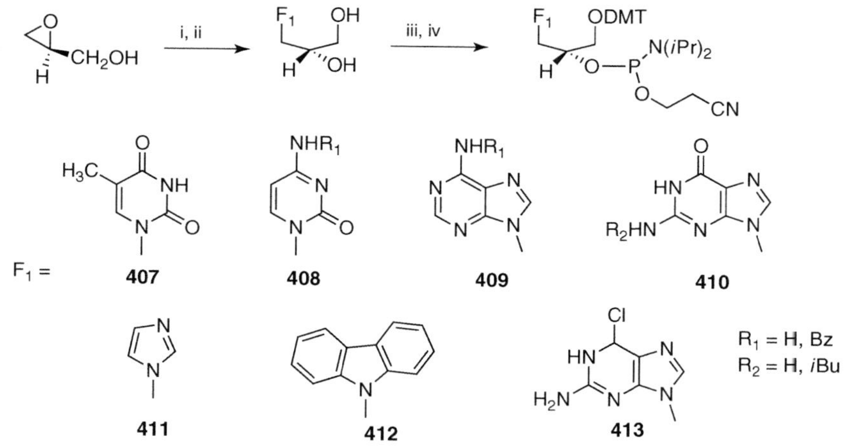

Scheme 39. Synthesis of acyclic nucleoside analogues. *Reagents and conditions:* (i) F₁H/potassium carbonate/DMF; (ii) flash chromatography or trituration with hot methanol; (iii) dimethoxytrityl chloride–pyridine; (iv) 2-cyanoethyl- N,N-diisopropylchlorophosporamidite, diisopropyl ethylamine in anh. acetonitrile.

In 1996, Acevedo and Andrews[85] used heterocyclic bases and R-(+)-glycidol to synthesize propane-2,3-diols or acyclic nucleosides (Scheme 39). Compared to the last example, this is similar but easy.

In 1971, Takemoto et al.[86] reported that they synthesized a series of purine and pyrimidine base derivatives with an N-(2,3-dihydroxypropyl) side chain via the reaction of bases with glycidol or glycerol α-chlorohydrine in DMF containing a catalytic amount of potassium carbonate (Scheme 40).

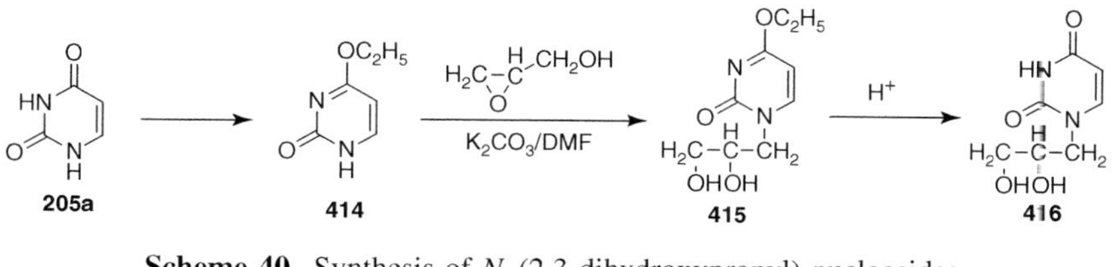

Scheme 40. Synthesis of N-(2,3-dihydroxypropyl) nucleosides.

Experimental Procedure

1-(4-Ethoxy-1,2-dihydro-2-keto-1-pyrimidyl)-2,3-propanediol (415). Dimethylformamide solution (15 mL) containing 4-ethoxy-1,2-dihydro-2-ketopyrimidine (1.4 g, 0.01 mol), glycidol (0.8 g, 0.011 mol) and a trace of anhydrous potassium carbonate was stirred at 60–65°C for 5 h. The solution was then evaporated to dryness under reduced pressure. The residue was chromatographed over 50 g of silica gel. Elution with a benzene–ethanol solution (3:1 v/v %) gave 0.8 g of the propanediol derivative. The product was recrystallized from benzene–ethanol to give white rods, mp 112.5–113.5°C. NMR (DMSO-d_6):τ 2.25, 4.15 (2d, 2, J = 7 Hz, ur. C_5 and C_6-H), 5.65 (br s, 2, pro. $C_{2,3}$-OH), 5.75 (q, 2, J = 7 Hz, ur. C_4-CH_2, 6.0–6.5 (m, 3, pro. $C_{2,3}$-H), 6.7 (d, 2, J = 5 Hz, pro. C_1-H), and 8.75 (t, 3, J = 7 Hz, ur. C_4-CH_3); UV, λ_{max} 274 mμ; IR(potassium bromide) υ 3150. 3300 (OH) and 1040,1110 (primary and secondary alcohol C–O). *Anal*. Calcd for $C_9H_{14}N_2O_4$: C, 50.46; H, 6.59; N, 13.08; found: C, 50.42; H, 6.72; N, 13.45.

In 2006, Horhota et al.[87] synthesized (S)-glycerol nucleoside triphosphates (gNTPs) and analyzed their activities to enzymatic polymerization (Figure 19). Glycerol nucleic acid (GNA) with a plain acyclic nucleotide frame is a possible progenitor of RNA,[88] and threose nucleoside triphosphates (tNTPs) also demonstrate remarkable research value (Scheme 41).

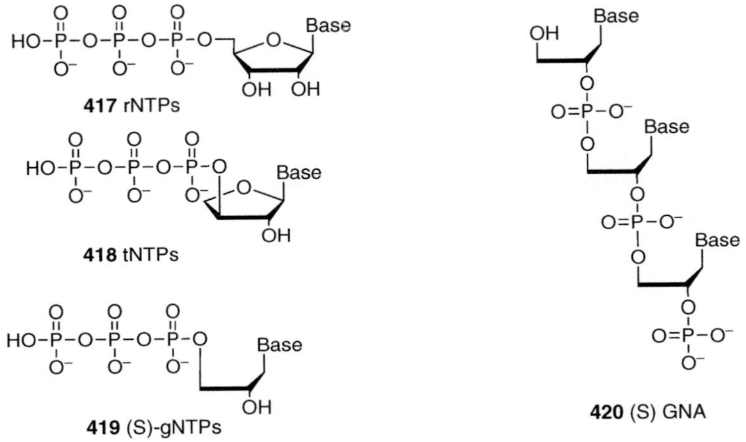

Figure 19. Structure of rNTPs and its analogues.

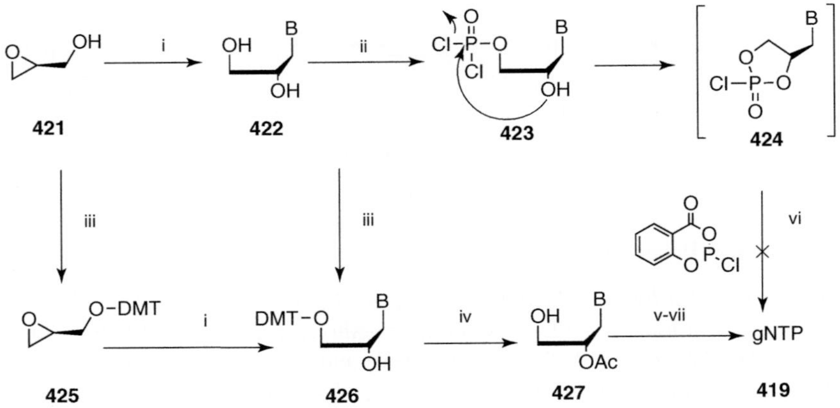

Scheme 41. Synthesis of gNTPs. *Reagents and conditions:* (i) K_2CO_3, nucleobase; (ii) $POCl_3$; (iii) DMT-Cl, pyridine; (iv) Ac_2O, pyridine, then H^+; (v) PPi; (vi) I_2, pyridine, H_2O; (vii) NH_4OH.

In 1999, Vilarrasa et al.[89] synthesized acyclic nucleosides via a ring-opening reaction (Scheme 42).

2.5. Synthesis of Acyclic ucleosides with Buciclovir Side-Chain Analogues (Type V)

Typical type V compounds are buciclovir[90] **433** and D-eritadenine[91] **434** (Figures 20 and 21), which have antiviral activity with chiral carbons in the side chain.

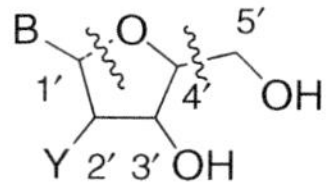

Figure 20. Structure of buciclovir side-chain analogues (type V).

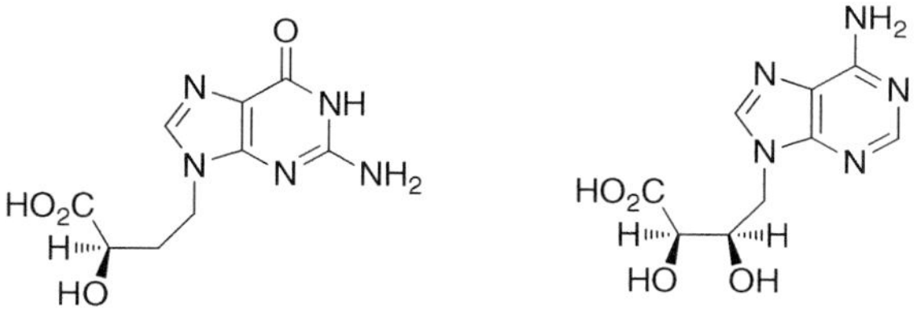

433 Buciclovir   **434** D-eritadenine

Figure 21. Structure of buciclovir and D-eritadenine.

In 1997, Evans et al.[92] synthesized 2'-phosphonomethyl(3',4'-dihydroxylbutyl) nucleotides of guanine **443** (Scheme 44) that have (2'S, 3'S) absolute configuration from L-ascorbic acid **435**. This is a retrosynthesis analysis (Scheme 43).

In 1993, Hirota et al.[93] developed a novel method to synthesize 9-D-ribitylpurines by breaking the C1'–O4' bond of purine in the presence of diisobutylaluminum hydride (DIBAL), and then using 9-D-ribitylpurines as a chiral pool to synthesize L-eritadenine, (2'S,3'R)-9-(2',3',4'-trihydroxybutyl)purine, and (2'S, 3'R)-9-(2',3',4'-trihydroxybutyl)adenine.[94] In developing their research, they in 1998 found an effective route to synthesizing purine acyclic nucleosides with chiral carbons in the side chain using 9-D-ribitylpurines **444a,b** as starting materials[95] (Scheme 45).

In 2004, Zakirova and co-workers[96] reported finding a novel route to the synthesis of chiral acyclic nucleoside and nucleotide analogues from D(−)- or L(+)-riboses **450** (Scheme 46).

In 1997, Hirota et al.[97] synthesized 9-D-ribitylpurines via cleavage of the C1'–O4' bond in the ribose ring of purine nucleosides (Scheme 47).

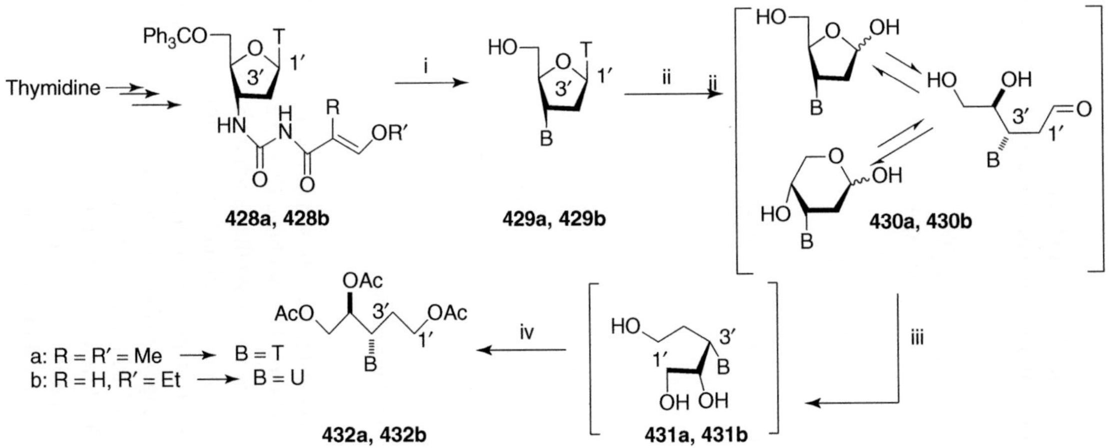

Scheme 42. Synthesis of **432**. *Reagents and conditions:* (i) 1 M H$_2$SO$_4$/dioxan 1:1, heat, 4 h; (ii) 4 M H$_2$SO$_4$/dioxane 1:1, heat, 20–24 h; (iii) NaBH$_4$, MeOH, 0°C, 20 min; (iv) Ac$_2$O, pyridine, rt, 7 h, 34% overall yield of **432a** from **428a**, 21% overall yield of **432b** from **428b**.

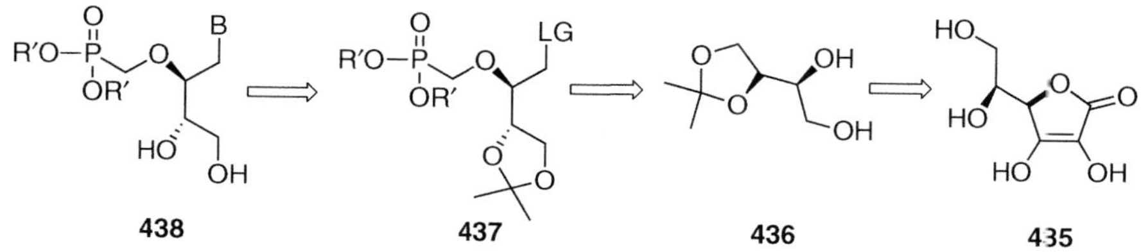

Scheme 43. Retrosynthesis analysis of 2′-phosphonomethyl(3′,4′-dihydroxylbutyl) nucleotides of guanine.

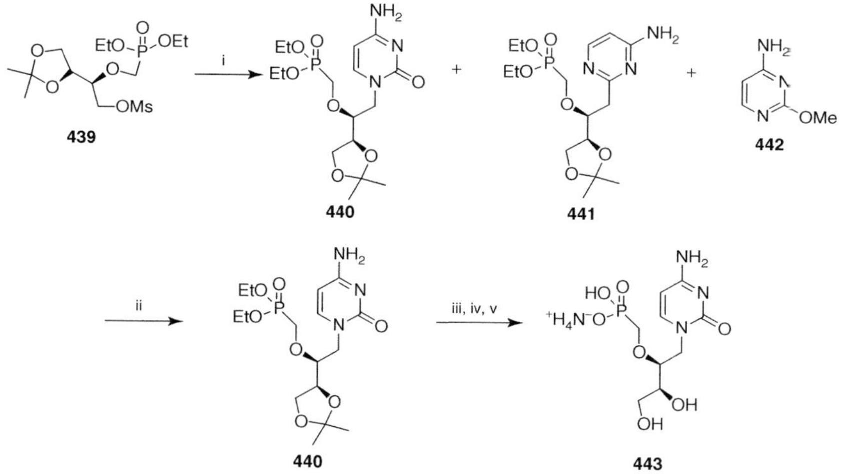

Scheme 44. Synthesis of **443**. *Reagents and conditions:* (i) cytosine, Cs$_2$CO$_3$, DMF, 90°C, 3 h; (ii) flash chromatography; (iii) 2 N HCl, MeOH, rt, 3 h, 52%; (iv) TMSBr, acetonitrile, rt, 16 h; (v) HPLC purification, 40% over (iv) and (v).

2.6. New Progress

With the development of research on acyclic nucleosides and nucleotides, more and more compounds that do not belong to any of the five major types have been synthesized. We list just a few typical examples.

In 2010, Otmar and co-workers[98] reported that they synthesized 8-aza-7,9-dideazaxanthine acyclic nucleosides and found that they inhibit thymidine phosphorylase (Scheme 48).

In 2009, Lequeux et al.[99] prepared some acyclic nucleosides with a difluoromethylphosphonate moiety and a triazole group in the side chain from difluorophosphonylated azides and propargylated nucleobases through a copper-catalyzed Huisgen 1,3-dipolar cycloaddition reaction. Then, in 2010,

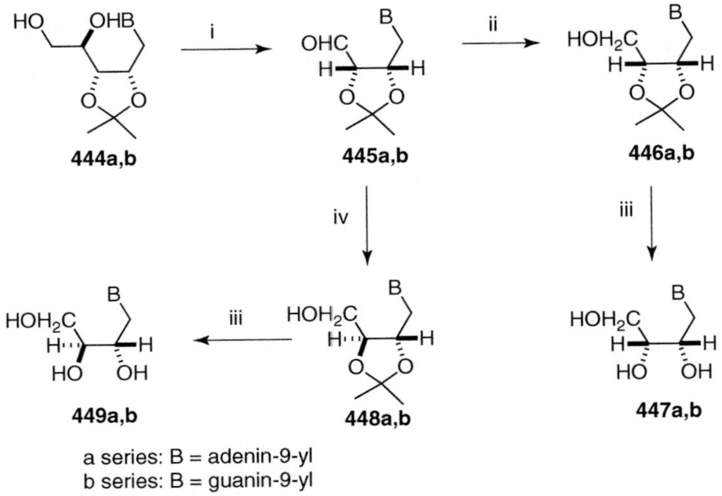

152 CHEMICAL SYNTHESIS OF ACYCLIC NUCLEOSIDES

Scheme 45. Synthesis of **449**. *Reagents and conditions:* (i) for **445a**, aq. NaIO$_4$; for **445b**, NaIO$_4$, AcOH–AcONa buffer (pH 4); (ii) aq. NaBH, pH 7–8; (iii) 80% AcOH; (iv) for **448a**, NaOMe, MeOH, then aq. NaBH$_4$; for compound **448b**, K$_2$CO$_3$, MeOH, then aq. NaBH$_4$.

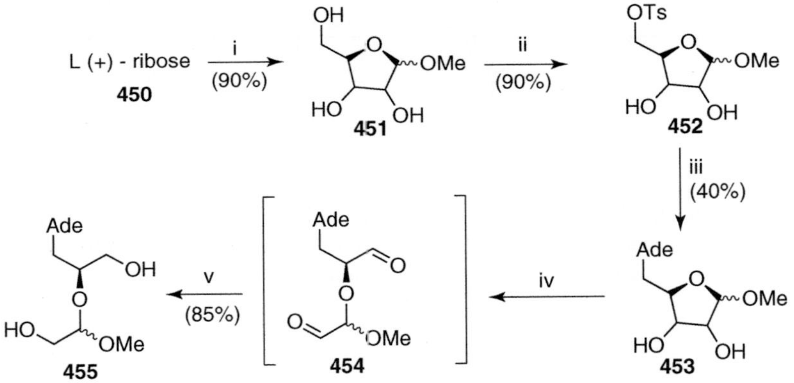

Scheme 46. Synthesis of **455**. *Reagents and conditions:* (i) AcCl/MeOH, rt, 16 h; (ii) TsCl/Py, 4°C, 16 h; (iii) Ade$^-$/DMF, 80°C, 5 days; (iv) NaIO$_4$, 4°C, 1 h; (v) NaBH$_4$, 4°C, 3 h.

Smietana and co-workers[100] synthesized triazoloacyclic nucleoside phosphonates via copper(I)-catalyzed azide–alkyne 1,3-dipolar cycloaddition (Scheme 49).

In 2009, Stanley and Hartwig[101] found highly regio- and enantioselective *N*-allylations of benzimidazoles, imidazoles, and purines via iridium catalysis. Then Guo and co-workers[102] synthesized chiral acyclic nucleosides with an amino side chain from nucleoside bases and protected L-serine **270** via the Mitsunobu coupling reaction (Scheme 50).

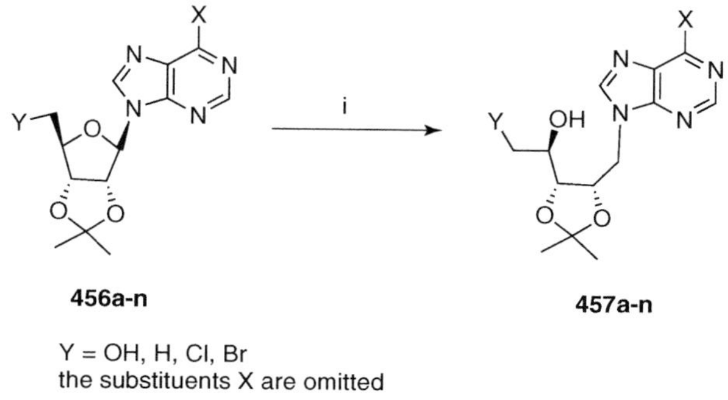

Scheme 47. DIBAL-H reduction of 6-substituted purine nucleosides and purine 5′-deoxynucleosides. *Reagent and conditions:* (i) DIBAL-H (5 equiv), THF 25°C, 24h.

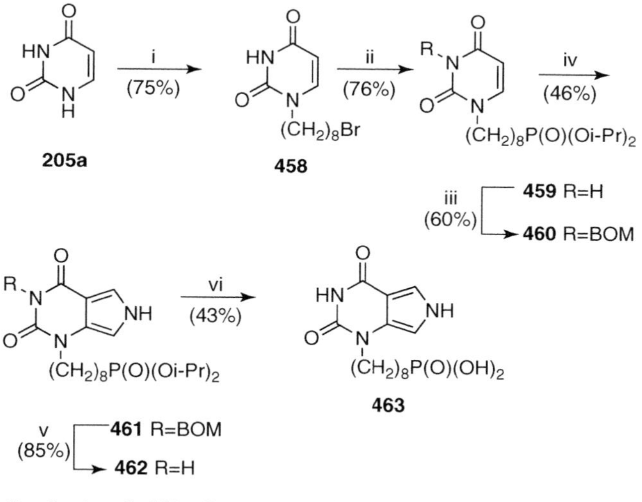

Scheme 48. Synthesis of **463**. *Reagents and conditions:* (i) BSA, 1,8-dibromooctane, CH_3CN, 75°C, 3 h; (ii) P(O-i-Pr)$_3$, 160°C, 6 h; (iii) NaH, BOM-Cl, DMF, 75°C, 5 h; (iv) TosMIC, DMSO/dioxane (1:4), 75°C, 6 h; (v) H$_2$/Pd, MeOH, overnight; (vi) (1) Me$_3$SiBr, CH_3CN, overnight; (2) NH$_3$/H$_2$O.

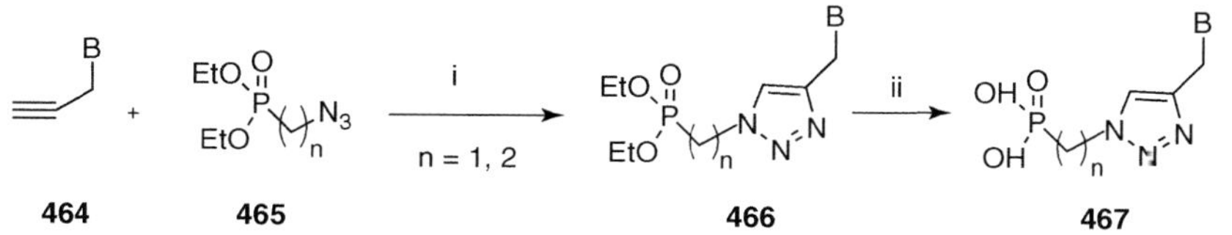

Scheme 49. Triazoloacyclic nucleoside phosphonates synthesized through a CuAAC reaction. *Reagents and conditions:* (i) CuSO$_4$ (5 mol%), sodium ascorbate (10 mol%), dioxane, H$_2$O; (ii) Me$_3$SiBr, DMF.

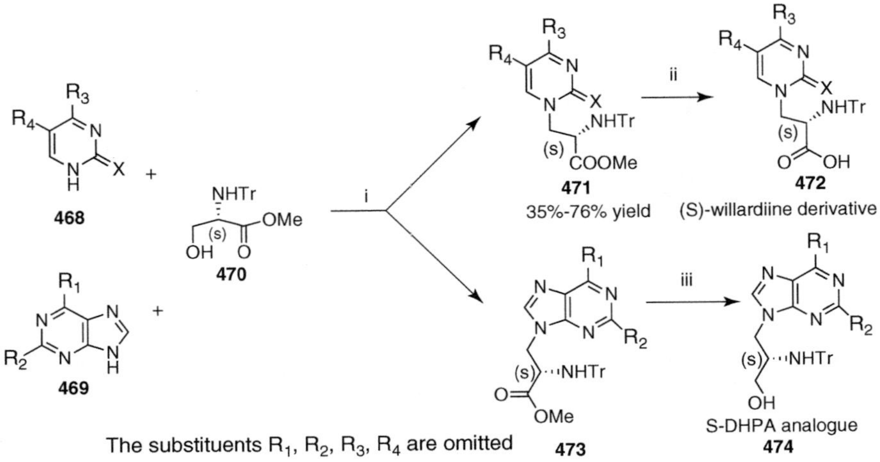

Scheme 50. Synthesis of chiral acyclic nucleosides with an amino side chain. *Reagents and conditions:* (i) Ph$_3$P, DIAD; (ii) LiOH; (iii) LiAlH$_4$, THF.

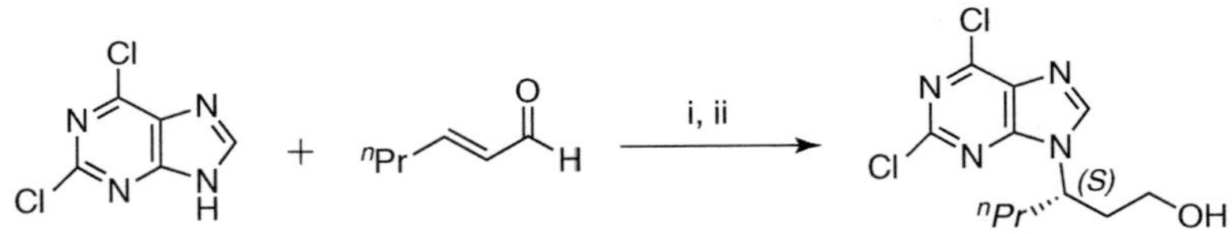

Scheme 51. Synthesis of chiral acyclic nucleosides via organocatalytic aza-Michael addition reaction. *Reagents and conditions:* (i) **475** (in figure 22) (10 mol%), PhCO$_2$H (10 mol%), toluene; (ii) NaBH$_4$, MeOH.

Recently, Guo et al.[103] synthesized chiral acyclic nucleosides and nucleotides via an organocatalytic aza-Michael addition reaction (Scheme 51), which is the third report on the synthesis of chiral acyclonucleosides from achiral materials. First, they optimized the reaction conditions and found the catalyst **475** to be best (Figure 22). Then they prepared S-DHPA and R-PMPA analogues. In The first, Gandelman and Jacobsen[104] reported the conjugate addition of purine to α,β-unsaturated enones or imides catalyzed by [(salen)Al] complexes, which reduced further could produce chiral acyclic nucleoside analogues. The second is Stanley and Hartwig's work, which we described earlier.

Experimental Procedure

Preparation of (S)-3-(2,6-dichloro-9H-purin-9-yl)hexan-1-ol (478). To a sample vial equipped with a magnetic stirring bar were added the catalyst **475** (0.01 mmol, 5.97 mg), **477** (20 μL, 0.15 mmol), toluene (1 mL), and benzoic acid (0.01 mmol, 1.22 mg). The mixture was stirred for 30 min at ambient temperature and then **476**

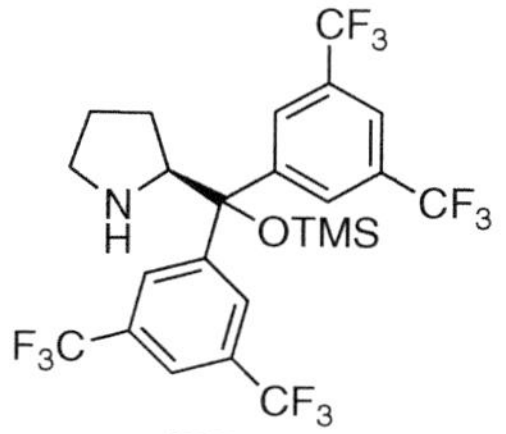

475

Figure 22. Structure of catalyst.

(0.1 mmol, 18.8 mg) was added at −30°C. The reaction was complete after 24 h, as monitored by TLC, and the resulting mixture was diluted with precooled MeOH (1.5 mL) and reduced with NaBH$_4$ (0.2 mmol, 7.6 mg). The reaction mixture was quenched with saturated NH$_4$Cl and extracted with ethyl acetate (3 × 10 mL). The organic layers were collected, dried over anhydrous Na$_2$SO$_4$, and concentrated under vacuum. The resulting residue was purified by column chromatography over silica gel (ethyl acetate/petroleum ether) to give the desired product **478**. White solid. mp 105–107°C; $[\alpha]_D^{20}$ = −8.0 (c = 0.4, CHCl$_3$). ^{1}H-NMR (400 MHz, CDCl$_3$): δ = 8.12 (s, 1H), 4.77–4.84 (m, 1H), 3.66–3.70 (m, 1H), 3.29–3.34 (m, 1H), 2.17–2.23 (m, 2H), 2.10–2.15 (m, 1H), 1.91–2.00 (m, 1H), 1.79 (s, 1H), 1.24–1.33 (m, 1H), 1.14–1.20 (m, 1H), 0.93 ppm (t, J = 7.6 Hz, 3H). ^{13}C-NMR (100 MHz, CDCl$_3$): δ = 153.2, 152.6, 151.7, 145.2, 130.9, 58.5, 54.3, 36.9, 36.2, 19.4, 13.5 ppm; HRMS: calcd for C$_{11}$H$_{15}$C$_{l2}$N$_4$O:[M + H]$^+$ 289.0623; found: 289.0623; the *ee* was determined by HPLC using a Chiralpak AD column [hexane/i-PrOH (97:3)]; flow rate 1.0 mL/min; major = 38.3 min, minor = 34.8 min (96% ee).

Over the past decades, microwave-promoted techniques, which are rapid and efficient, have attracted interest in the synthetic chemistry field. In 2007, Guo and co-workers[5] synthesized a series of acyclic nucleosides using microwave-promoted Michale addition in neat water (Table 1).

3. OUTLOOK

Today, more and more chemists are paying attention to acyclic nucleoside and nucleotide drugs because of their immense research value. But their use has some problems, such as toxic side effects and drug tolerance. The development of and research on acyclic nucleoside drugs have reduced their toxicity to a certain degree. At the present time, we do not understand the relation between drug structure and mechanism of action. Therefore, we need to find the relation and then use computer simulation to design new types of drugs. Many faults, rigid reaction conditions, low yields, and by-products complicate experimental synthesis of acyclic nucleosides. Therefore, founding new and green routes is very important. Looking to the future, there is widespread promise. We hope that this chapter will provide useful information for those involved in the synthesis of acyclic nucleosides.

Table 1. Michael Addition of Uracil Derivatives to Acrylonitrile[a]

Entry	R	X	Product	Yield (%)
1	H	O	481a	82
2	Cl	O	481b	78
3	I	O	481c	72
4	Me	O	481d	82
5	H	S	481e	74

[a] Reaction conditions: nucleobases (2 mmol), **480** (6 mmol), H_2O (5 mL), Et_3N (6 mmol), MWI 250 W (100°C). Isolated yields based on nucleobases.

Acknowledgments

The authors wish to express their gratitude to the National Nature Science Foundation of China (grants 20802016, 21072047, and 21172059), the Program for New Century Excellent Talents in the University of Ministry of Education (NCET-09-0122), the Excellent Youth Foundation of the Henan Scientific Committee (No. 114100510012), the Program for Changjiang Scholars and Innovative University Research Team (IRT1061), the National Students Innovation Experiment Program, and the Excellent Youth Program the Program for Innovative Research Team in University of Henan Province (2012IRTSTHN006), of Henan Normal University.

ABBREVIATIONS

A	adenine
Ac	acetyl
AcOH	acetic acid
ACV	acyclovir
AICA	5-aminoimidazole-4-carboxamide
ANPs	acyclic nucleoside phosphonates
aq.	aqueous
AZT	3'-azido-3'-deoxythymidine
B	base
Bn	benzyl
Boc	t-butyloxycarbonyl
BSA	N,O-bis(trimethylsilyl)acetamide
Bz	benzoyl
C	cytosine

ABBREVIATIONS

cat.	catalyst
Cbz	carbobenzoxy
CDI	N,N'-carbonyldiimidazole
CMV	cytomegalovirus
d4T	$2',3'$-dideohydro-$3'$-deoxythymidine
DAG	N^2,N^9-diacetylguanine
DCCI	N,N'-dicyclohexylcarbodiimide
DHPG	9-[(1,3-dihydroxy-2-propoxy)methyl]guanine/ganciclovir
DIBAL	diisobutylaluminium hydride
DIU	N,N'-dicyclohexylurea
DMA	dimethylamine
DMAP	4-(N,N-dimethylamino)pyridine
DMF	dimethylformamide
DMSO	dimethyl sulfoxide
DNA	deoxyribonucleic acid
FCV	famciclovir
FDA	U. S. Food and Drug Administration
FHBG	9-(4-fluoro-3-hydroxymethylbutyl)guanine
FHPG	9-[(3-fluoro-1-hydroxy-2-propoxy)methyl]guanine
FPMP	N^9-[3-fluoro-2-(phosphonomethoxy)propyl]
G	guanine
GCV	9-[(1,3-dihydroxy-2-propoxy)methyl]guanine/ganciclovir
gNTPs	glycerol nucleoside triphosphates
HBG	9-(4-hydroxybutyl)guanine
HBV	hepatitis B virus
HCV	hepatitis C virus
HIV	human immunodeficiency virus
HMDS	hexamethyl disilazane
HSV	herpes simplex virus
i-Bu	isobutyryl
Ms	methanesulfonyl
MWI	microwave irradiation
NAD	nicotinamide adenine dinucleotide
OBDDA	2-oxa-1,4-butane-dioldiacetate
ODE-(S)-MPMPA	octadecyloxyethyl 9-(S)-[3-methoxy-2-(phosphonomethoxy)propyl]adenine
PCV	penciclovir
PEE	phosphonoethoxyethyl
PfHGXPRT	plasmodium falciparum hypoxanthine–guanine–xanthine phosphoribosyltransferase
PMEA	adefovir
PMPA	tenofovir
Py	pyridine
T	thymine
TDF	tenofovir disoproxil fumarate

TLC	thin-layer chromatography
TMSCl	trimethylsilyl chloride
TMSOTf	trimethylsilyl trifluoromethanesulfonate
tNTPs	threose nucleoside triphosphates
Tr	trityl
Ts	tosyl
VCV	valaciclovir
VZV	varicella zoster virus

REFERENCES

1. (a) Schaeffer, H. J.; Beauchamp, L.; De Miranda, P.; Elion, G. B.; Bauer, D. J.; Collins, P. *Nature (London)* **1978**, *272*, 583. (b) Elion, G. B.; Furman, P. A.; Fyfe, J. A.; De Miranda, P.; Beauchamp, L.; Schaeffer, H. *Proc. Natl. Acad. Sci. USA* **1977**, *74*, 5716.
2. De Clercq, E.; Holý, A. *Nat. Rev. Drug Discov.* **2005**, *4*, 928.
3. Bonate, P. L.; Arthaud, L.; Cantrell, W. R.; Stephenson, K.; Secrist, J. A., III; Weitman, S. *Nat. Rev. Drug Discov.* **2006**, *5*, 855.
4. He, L.; Liu, Y. M.; Zhang, W.; Li, M. Q.; Chen, H. *Tetrahedron* **2005**, *61*, 8505.
5. Guo, H.-M.; Qu, G.-R.; Zhang, Z.-G.; Geng, M.-W.; Xia, R.; Zhao, L. *Synlett* **2007**, *5*, 721.
6. Barrio, J. R.; Bryant, J. D.; Keyser, G. E. *J. Med. Chem.* **1980**, *23*, 572.
7. Abrams, H. M.; Ho, L.; Chu, S. H. *J. Heterocycl. Chem.* **1981**, *18*, 947.
8. Kelley, J. L.; Schaeffer, H. J. *J. Heterocycl. Chem.* **1986**, *23*, 271.
9. Singh, D.; Wani, M. J.; Kumar, A. *J. Org. Chem.* **1999**, *64*, 4665.
10. Robins, M. J.; Hatfield, P. W. *Can. J. Chem.* **1982**, *60*, 547.
11. Kelley, J. L.; Kelsey, J. E.; Hall, W. R.; Krochmal, M. P.; Schaeffer, H. J. *J. Med. Chem.* **1981**, *24*, 753.
12. Schroeder, A. C.; Hughes, R. G.; Bloch, A. *J. Med. Chem.* **1981**, *24*, 1078.
13. Alhede, B.; Clausen, F. P.; Juhl-Christensen, J.; McCluskey, K. K.; Preikschat, H. F. *J. Org. Chem.* **1991**, *56*, 2139.
14. Beauchamp, L. M.; Raleigh, N. C. Therapeutic valine esters of acyclovir and pharmaceutically acceptable salts thereof. U.S. Pat. Appl. 4957924. **1990**.
15. Santos, C. R.; Capela, R.; Pereira, C. S.; Valente, E.; Gouveia, L.; Pannecouque, C.; De Clercq, E.; Moreira, R.; Gomes, P. *Eur. J. Med. Chem.* **2009**, *44*, 2339.
16. Amblard, F.; Fromentin, E.; Detorio, M.; Obikhod, A.; Rapp, K. L.; McBrayer, T. R.; Whitaker, T.; Coats, S. J. *Eur. J. Med. Chem.* **2009**, *44*, 3845.
17. Beauchamp, L. M.; Dolmatch, B. L.; Schaeffer, H. J.; et al. *J. Med. Chem.* **1985**, *28*(8), 982.
18. Boryski, J.; Golankiewicz, B.; De Clercq, E. *J. Med. Chem.* **1991**, *348*, 2380.
19. Golankiewicz, B.; Ostrowski, T. *Antiviral Res.* **2006**, *71*, 134.
20. Ewing, D. F.; Glaçon, V.; Mackenzie, G.; Len, C. *Tetrahedron Lett.* **2002**, *43*, 989.
21. Pan, B. C.; Chen, Z. H.; Piras, G.; Dutschman, G. E.; Rowe, E. C.; Cheng, Y. C.; Chu, S. H. *J. Heterocycl. Chem.* **1994**, *31*, 177.

22. Miyasaka, T.; Tanaka, H.; Baba, M.; Hayakawa, H.; Walker, T. R.; Balzarini, J.; De Clercq, E. *J. Med. Chem.* **1989**, *32*, 2507.
23. Tanaka, H.; Baba, M.; Hayakawa, H.; Sakamati, T.; Miyasaka, T.; Ubasawa, M.; Takashima, H.; Sekiya, K.; Nitta, I.; Shigeta, S.; Walker, T. R.; Balzarini, J.; De Clercq, E. *J. Med. Chem.* **1991**, *34*, 349.
24. Marasco, C. J.; Kramer, D. L.; Miller, J.; Porter, C. W.; Bacchi, C. J.; Rattendi, D.; Kucera, L.; Iyer, N.; Bernacki, R.; Pera, P.; Sufrin, J. R. *J. Med. Chem.* **2002**, *45*, 5112.
25. Kjellberg, J. *J. Heterocycl. Chem.* **1986**, *23*, 625.
26. Kjellberg, J.; Liljenberg, M. *Tetrahedron Lett.* **1986**, *27*, 877.
27. Howson, W.; Taylor, E. M.; Parsons, M. E.; Novelli, R.; Wilczynska, M. A.; Harris, D. T. *Eur. J. Med. Chem.* **1988**, *23*, 433.
28. Volpini, R.; Mishra, R. C.; Kachare, D. D.; Ben, D. D.; Lambertucci, C.; Antonini, I.; Vittori, S.; Marucci, G.; Sokolova, E.; Nistri, A.; Cristalli, G. *J. Med. Chem.* **2009**, *52*, 4596.
29. Janeba, Z.; Holy, A.; Votavova, H.; Masojidkove, M. *Collect. Czech. Chem. Commun.* **1996**, *61*, 442.
30. Sauer, R.; El-Tayeb, A.; Kaulich, M.; Müller, C. E. *Bioorg. Med. Chem.* **2009**, *17*, 5071.
31. Escolar, G.; Heras, M. *Drugs Today* **2000**, *36*, 187.
32. Ingall, A. H.; Dixon, J.; Bailey, A.; Coombs, M. E.; Cox, D.; McInally, I. J.; Hunt, S. F.; Kindon, N. D.; Teobald, B. J.; Willis, P. A.; Humphries, R. G.; Leff, P.; Clegg, J. A.; Smith, J. A.; Tomlinson, W. *J. Med. Chem.* **1999**, *42*, 213.
33. Koszytkowska-Stawińska, M.; Kaleta, K.; Sas, W.; De Clercq, E. *Nucleosides Nucleotides Nucleic Acids* **2007**, *26*, 51.
34. Koszytkowska-Stawińska, M.; Sas, W. *Tetrahedron Lett.* **2004**, *45*, 5437.
35. Koszytkowska-Stawińska, M.; De Clercq, E.; Balzarini, J. *Bioorg. Med. Chem.* **2009**, *17*, 3756.
36. Schaeffer, H. J. Raleigh, N. C.2-amido-9-(2-acyloxyethoxymethyl)hypoxanthines. U.S. Pat. Appl. 4146715. **1979**.
37. Scheiner, P.; Geer, A.; Bucknor, A.-M.; Imbach, J.-L.; Schinazi, R. F. *J. Med. Chem.* **1989**, *32*, 73.
38. Hendry, D.; Hutchinson, E. J.; Roberts, S. M.; Dunn, S. M.; Bryant, J. A. *J. Chem. Soc., Perkin Trans. 1*, **1993**, 1109.
39. Ewing, D. F; Glaçon, V.; Mackenzie, G.; Len, C. *Tetrahedron Lett.* **2002**, *43*, 3503.
40. Diab, S. A.; Sene, A.; Pfund, E.; Lequeux, T. *Org. Lett.* **2008**, *10*, 3895.
41. (a) Field, A. K.; Davies, M. E.; De Witt, C.; et al. *Proc. Natl. Acad. Sci. USA* **1983**, *80*, 4139. (b) Markham, A.; Faulds, D. *Drugs* **1994**, *48*, 455.
42. (a) Paya, C.; Humar, A.; Dominguez, E.; et al. *Am. J. Transplant.* **2004**, *4*, 611. (b) Daniel, F.; Martin, M. D.; et al. *N. Engl. J. Med.*, **2002**, *346*, 1119.
43. Ogilvie, K. K.; Nguyen-Ba, N.; Radatus, B. K.; Cheriyan, U. O.; Rizk Hanna, H. *Can. J. Chem.* **1984**, *62*, 241.
44. Hakimelahi, G. H.; Khalafi-Nezhad, A. *Helv. Chim. Acta* **1989**, *72*, 1495.
45. Ogilvie, K. K.; Hamilton, R. G.; Gillen, M. F.; Radatus, B. K. *Can. J. Chem.* **1984**, *62*, 16.

46. Ogilvie, K. K.; Cheriyan, U. O.; Radatus, B. K. *Can. J. Chem.* **1982**, *60*, 3005.
47. Martin, J. C.; Jeffrey, G. A.; McGee, D. P. C.; Tippie, M. A.; Smee, D. F.; Matthews, T. R.; Verheyden, J. P. H. *J. Med. Chem.* **1985**, *28*, 358.
48. Beauchamp, L. M.; Serling, B. L.; Kelsey, J. E.; Biron, K. K.; Collins, P.; Selway, J.; Lin, J.-C.; Schaeffer, H. J. *J. Med. Chem.* **1988**, *31*, 144.
49. Kumar, R.; Semaine, W.; Johar, M.; Tyrrell, D. L.; Agrawal, B. *J. Med. Chem.* **2006**, *49*, 3693.
50. Harnden, M. R.; Jarvest, R. L.; Bacon, T. H.; et al. *J. Med. Chem.*, **1987**, *30*, 1636.
51. Pue, M. A.; Pratt, S. K.; Fairless, A. J.; et al. *J. Antimicrob. Chemother.* **1994**, *33*, 119.
52. Wang, J.-Q.; Zheng, Q.-H.; Fei, X.; Mock, B. H.; Hutchins, G. D. *Bioorg. Med. Chem. Lett.* **2003**, *13*, 3933.
53. Hocková, D.; Holý, A. *Collect. Czech. Chem. Commun.* **1997**, *62*, 948.
54. Harnden, M. R.; Jarvest, R. L. *Tetrahedron Lett.* **1985**, *26*, 4265.
55. Koszytkowska-Stawińska, M. *Nucleoside Nucleotide Nucleic Acids* **2010**, *29*, 768.
56. Koszytkowska-Stawińska, M.; Sas, W.; De Clercq, E. *Tetrahedron* **2006**, *62*, 10325.
57. Koszytkowska-Stawińska, M.; Sas, W. *Tetrahedron Lett.* **2004**, *45*, 5437.
58. McGee, D. P. C.; Martin, J. C.; Smee, D. F.; Matthews, T. R.; Verheyden, J. P. H. *J. Med. Chem.* **1985**, *28*, 1242.
59. (a) Naesens, L.; Balzarini, J.; De Clercq, E. *Rev. Med. Virol.* **1994**, *4*, 147. (b) Gong, Y.-F.; Marshall, D. R.; Srinivas, R. V. *Antimicrob. Agents Chemother.* **1994**, *38*, 1683. (c) Calio, R.; Villani, N.; Balestra, E. *Antiviral Res.* **1994**, *23*, 77.
60. (a) Balzarini, J.; Naesens, L.; Herdewijn P.; et al. *Proc. Natl. Acad. Sci. USA* **1989**, *86*(1), 332. (b) Gangemi, J. D.; Cozens, R. M.; De Clercq, E.; et al. *Antimicrob. Agents Chemother.* **1989**, *33*(11), 1864. (c) Balzarini, J.; Hao, Z.; Herdewijn, P.; et al. *Proc. Natl. Acad. Sci. USA* **1991**, *88*(4), 1499. (d) Balzarini, J.; Perno, C. F.; Schols D.; et al. *Biochem. Biophys. Res. Commun.* **1991**, *178*(1), 329.(e) Naesens, L.; Neyts, J.; Balzarini, J.; et al. *J. Med. Virol.* **1993**, *39*(2), 167. (f) Naesens, L.; Balzarini, J.; De Clercq, E. *Rev. Med. Virol.* **1994**, *4*, 147. (g) Naesens, L.; Neyts, J.; Rosenberg I.; et al. *Eur. J. Clin. Microbiol. Infect. Dis.* **1989**, *8*, 1043. (h) Šolínová V.; Kašička V.; Sázelová, P.; et al. *Electrophoresis* **2009**, *30*, 2245.
61. (a) Naesens, L.; Balzarini, J.; Rosenberg, I.; Holý, A.; De Clercq, E. *Eur. J. Clin. Microbiol. Infect. Dis.* **1989**, *8*, 1043. (b) Naesens, L.; Balzarini, J.; De Clercq, E. *Antiviral Res.* **1991**, *16*, 53. (c) Naesens, L.; Neyts, J.; Balzarini, J.; Holý, A.; Rosenberg, I.; De Clercq, E. *J. Med. Virol.* **1993**, *39*, 167. (d) Vahlenkamp, T. W.; De Ronde, A.; Balzarini, J.; Naesens, L.; De Clercq, E.; van Eijk, M. J.; Horzinek, M. C.; Egberink, H. F. *Antimicrob. Agents Chemother.* **1995**, *39*, 746.
62. De Clercq, E. *Biochem Pharmacol.* **1987**, 2567.
63. Hadziyannis, S. J.; Tassopoulos, N. C.; Heathcote, E. J.; et al. *N. Engl. J. Med.* **2003**, 800.
64. Marcellin, P.; Chang, T. T.; Lim, S. G.; et al. *N. Engl. J. Med.* **2003** (*348*), 808.
65. Holý, A.; Kohoutová, J.; Merta, A.; Votruba, I. *Collect. Czech. Chem. Commun.* **1986**, *51*, 459.
66. Kim, D.-K.; Gam, J.; Kim, K. H. *J. Heterocycl. Chem.* **1996**, *33*, 1865.
67. Holý, A.; Günter, J.; Dvořáková, H.; Masojídková, M.; Andrei, G.; Snoeck, R.; Balzarini, J.; De Clercq, E. *J. Med. Chem.* **1999**, *42*, 2064.

68. Meszárosová, K.; Holý, A.; Masojídková, M. *Collect. Czech. Chem. Commun.* **2000**, *65*, 1109.
69. Shatila, R. S.; Bouhadir, K. H. *Tetrahedron Lett.* **2006**, *47*, 1767.
70. Liang, C. Y.; Lee, D. W.; Gary Newton, M.; Chu, C. K. *J. Org. Chem.* **1995**, *60*, 1546.
71. Hocková, D.; Holý, A.; Masojídková, M.; Keough, D. T.; de Jersey, J.; Guddat, L. W. *Bioorg. Med. Chem.* **2009**, *17*, 6218.
72. Hocková, D.; Holý, A.; Andrei, G.; Snoeck, R.; Balzarini, J. *Bioorg. Med. Chem.* **2011**, *19*, 4445.
73. Hořejší, K.; Andrei, G.; De Clercq, E.; Snoeck, R.; Pohl, R.; Holý, A. *Bioorg Med. Chem.* **2006**, *14*, 8057.
74. Hocková, D.; Dračínský, M.; Holý, A. *Eur. J. Org. Chem.* **2010**, 2885.
75. Houghton, S. R.; Melton, J.; Fortunak, J.; Brown Ripin, D. H.; Boddy, C. N. *Tetrahedron* **2010**, *66*, 8137.
76. Jeffery, A. L.; Kim, J.-H.; Wiemer, D. F. *Tetrahedron* **2000**, *56*, 5077.
77. Barral, K.; Priet, S.; Sire, J.; Neyts, J.; Balzarini, J.; Canard, B.; Alvarez, K. *J. Med. Chem.* **2006**, *49*, 7799.
78. Holý, A.; Votruba, I.; Tlouštová, E.; Masojídková, M. *Collect. Czech. Chem. Commun.* **2001**, *66*, 1545.
79. Valiaeva, N.; Wyles, D. L.; Schooley, R. T.; Hwu, J. B.; Beadle, J. R.; Prichard, M. N.; Hostetler, K. Y. *Bioorg. Med. Chem.* **2011**, *19*, 4616.
80. Wyles, D. L.; Kaihara, K. A.; Korba, B. E.; Schooley, R. T.; Beadle, J. R.; Hostetler, K. Y. *Antimicrob. Agents Chemother.* **2009**, *53*, 2660.
81. Baszczyňski, O.; Jansa, P.; Dračínský, M.; Klepetářová, B.; Holý, A.; Votruba, I.; de Clercq, E.; Balzarini, J.; Janeba, Z. *Bioorg. Med. Chem.* **2011**, *19*, 2114.
82. Bronson, J. J.; Ferrara, L. M.; Howell, H. G.; et al. *Nucleosides Nucleotides* **1990**, *9*(6), 745.
83. Brodfuehrer, P. R.; Howell, H. G.; Sapino, C.; Vemishetti, P. *Tetrahedron Lett.* **1994**, *35*, 3243.
84. Holý, A.; Rosenberg, I.; Dvořáková, H. *Collect. Czech. Chem. Commun.* **1989**, *54*, 2470.
85. Acevedo, O. L.; Andrews, R. S. *Tetrahedron Lett.* **1996**, *37*, 3931.
86. Ueda, N.; Kawabata, T.; Takemoto, K. *J. Heterocycl. Chem.* **1971**, *8*, 827.
87. Horhota, A. T.; Szostak, J. W.; McLaughlin, L. W. *Org. Lett.* **2006**, *8*, 5345.
88. (a) Zhang, L.; Peritz, A.; Meggers, E. *J. Am. Chem. Soc.* **2005**, *127*, 4174. (b) Zhang, L.; Pertiz, A. E.; Carroll, P. J.; Meggers, E. *Synthesis* **2006**, 645.
89. Costa, A. M.; Faja, M.; Vilarrasa, J. *Tetrahedron* **1999**, *55*, 6635.
90. Larsson, A.; Öberg, B.; Alenius, S.; Hagberg, C.-E.; Johansson, N.-G.; Lindborg, B.; Stening, G. *Antimicrob. Agents Chemother.* **1983**, *23*, 664.
91. (a) Rokujo, T.; Kikuchi, H.; Tensho, A.; Tsukitani, Y.; Takenawa, T.; Yoshida, K.; Kamiya, T. *Life Sci.* **1970**, *9*, 379. (b) Chibata, I.; Okamura, K.; Takeyama, S.; Kotera, K. *Experientia* **1969**, *25*, 1237. (c) Kamiya, T.; Saito, Y.; Hashimoto, M.; Seki, H. *Tetrahedron Lett.* **1969**, 4729. (d) Votruba, I.; Holý, A. *Collect. Czech. Chem. Commun.* **1982**, *47*, 167. (e) Holý, A.; Votruba, I.; De Clercq, E. *Collect. Czech. Chem. Commun.* **1982**, *47*, 1392.

92. Evans, C. A.; Hewgill, R. T.; Mansour, T. S. *Tetrahedron: Asymmetry* **1997**, *8*, 2299.
93. Kitade, Y.; Hirota, K.; Maki, Y. *Tetrahedron Lett.* **1993**, *34*, 4835.
94. Hirota, K.; Monguchi, Y.; Sajiki, H.; Kitade, Y. *Synlett* **1997**, *6*, 697.
95. Hirota, K.; Monguchi, Y.; Sajiki, H.; Sako, M.; Kitade, Y. *J. Chem. Soc., Perkin Trans.1* **1998**, 941.
96. Zakirova, N. F.; Shipitsyn, A. V.; Belanov, E. F.; Jasko, M. V. *Bioorg. Med. Chem. Lett.* **2004**, *14*, 3357.
97. Hirota, K.; Monguchi, Y.; Kitade, Y.; Sajiki, H. *Tetrahedron* **1997**, *53*, 16683.
98. Marak, D.; Otmar, M.; Votruba, I.; Dracinsky, M.; Krecmerova, M. *Bioorg. Med. Chem. Lett.* **2011**, *21*, 652.
99. Diab, S. A.; Hienzch, A.; Lebargy, C.; Guillarme, S.; Pfund, E.; Lequeux, T. *Org. Biomol. Chem.* **2009**, *7*, 4481.
100. Elayadi, H.; Smietana, M.; Pannecouque, C.; Leyssen, P.; Neyts, J.; Vasseur, J. J.; Lazrek, H. B. *Bioorg. Med. Chem. Lett.* **2010**, *20*, 7365.
101. Stanley, L. M.; Hartwig, J. F. *J. Am. Chem. Soc.* **2009**, *131*, 8971.
102. Guo, H. M.; Wu, Y. Y.; Niu, H. Y.; Wang, D. C.; Qu, G. R. *J. Org. Chem.* **2010**, *75*, 3863.
103. Guo, H. M.; Yuan, T. F.; Niu, H. Y.; Liu, J. Y.; Mao, R. Z.; Li, D. Y.; Qu, G. R., *Chem. Eur. J.* **2011**, *17*, 4095.
104. Gandelman, M.; Jacobsen, E. N. *Angew. Chem. Int. Ed.* **2005**, *44*, 2393.

4 Phosphonated Nucleoside Analogues

ROBERTO ROMEO and CATERINA CARNOVALE
Dipartimento Farmaco-Chimico, Università di Messina, Messina, Italy
ANTONIO RESCIFINA and MARIA ASSUNTA CHIACCHIO
Dipartimento di Scienze del Farmaco, Università di Catania, Catania, Italy

1. INTRODUCTION

Nucleoside analogues, where extensive modifications have been made to a heterocyclic base and/or sugar moiety to avoid the disadvantages due primarily to enzymatic degradation, constitute a highly successful group of anticancer[1] and antiviral drugs.[2] Several classes of nucleoside analogues have been designed and synthesized to increase the resistance to enzymatic degradation and/or to reduce toxicity and the cross-resistance problems.[3] Some nucleoside analogues are incorporated with unconventional nucleobases (e.g., ribavirin, where a triazole carboxyamide base mimics either adenine or guanine).[4] Other analogues, known as C-*nucleosides* (i.e., pyrazo-, tiazo-, selenazo-, and oxazofurin) are characterized by the replacement of the acid-labile C–N glycosidic bond by a stable C–C bond.[5–8]

Moreover, different types of modifications have been carried out through the insertion of heteroatoms or replacement of the furanose oxygen with other atoms (heterocyclic nucleosides).[9,10] In the same context, substitution of the ribose ring by an acyclic system gives rise to acyclic nucleosides,[11,12] while displacement of the oxygen atom by a methylene unit gives rise to a series of carbocyclic nucleosides.[13]

All these compounds act as reverse transcriptase inhibitors (NRTIs). NRTIs approved in current therapy, which resemble both endogenous purine and pyrimidine nucleosides, include abacavir (ABC), lamivudine (3TC), didanosine (ddI), zidovudine (ZDV, AZT), stavudine (d4T), zalcitabine (ddC), and tenofovir disoproxyl (PMPA) (Figure 1).[14]

To be active, once they have entered the cell, nucleoside analogues have to be phosphorylated by a combination of human intracellular enzymes, through three consecutive phosphorylation steps, before they can interact in their active

---

Chemical Synthesis of Nucleoside Analogues, First Edition. Edited by Pedro Merino.
© 2013 John Wiley & Sons, Inc. Published 2013 by John Wiley & Sons, Inc.

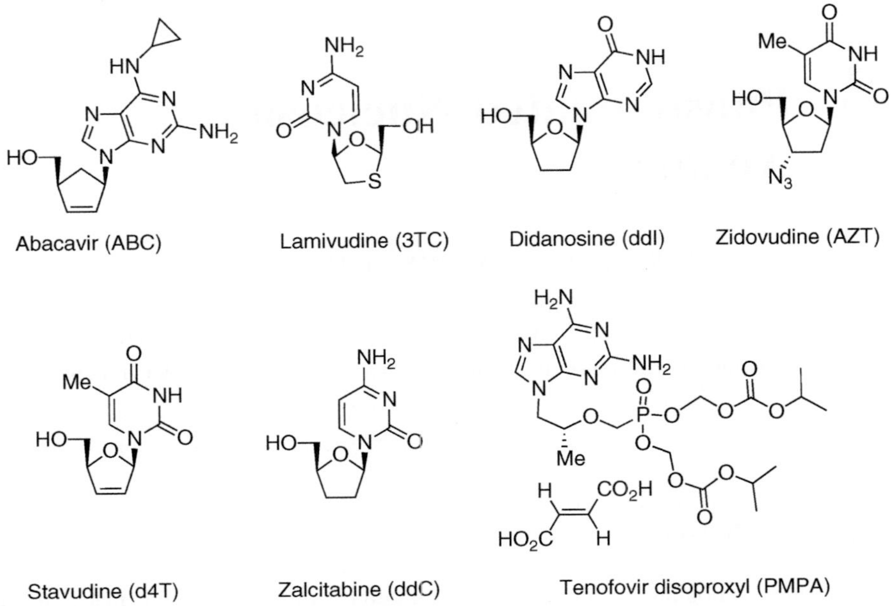

Figure 1.

triphosphate form with their target enzyme, the viral DNA polymerase.[15] The therapeutic effect depends on the rate of intracellular phosphorylation. In their triphosphate form, the compounds compete with the normal substrates [2′-deoxynucleoside 5′-triphosphates (dNTPs)] for binding sites on reverse transcriptase. When incorporated into nascent viral DNA, they may act as chain terminators, thus preventing further chain elongation.

Of crucial importance in the phosphorylation process is the monophosphorylation step, which is mediated by a specific virus-encoded thymidine kinase (TK) (for HSV and VZV) or a specific virus-encoded (UL97) protein kinase (PK) (for CMV).[16] Once the compounds have been monophosphorylated, the cellular kinases (i.e., guanosine monophosphate kinase GMP and nucleoside diphosphate kinase NDP) will afford further phosphorylation to the di- and triphosphate stages, respectively.[16–18] The first phosphorylation is usually the rate-limiting step for most nucleosides.

Many efforts have been made to overcome this issue and improve the therapeutic properties of nucleosides by shortening this cascade and bypassing at least the first phosphorylation step. To improve the cellular permeability and enhance the anti-HIV activity of nucleoside analogues, two main synthetic strategies have been exploited. The first is based on the mononucleotide prodrug approach, where the polar monophosphate unit of the nucleoside, masked by different groups, such as phosphoramidate, bis-S-acyl thioethyl esters (bis-SATE), bis-pivaloxymethyl (bis-POM), cyclo-Saligenyl, S-pivaloyl-2-thioethyl (t-BuSATE) and phenyl, S-acyl thioethyl mixed phosphate esters (mix-SATE), undergoes transient esterase-labile phosphate protection.[19] The drug-design rationale on which this approach is based is that these lipophilic nucleoside phosphotriesters are able to bypass the first monophosphorylation step catalyzed by dCK or TK1 and to deliver, by hydrolysis and/or enzymatic cleavage, the corresponding 5′-mononucleotide inside the cells.[20] The prodrug approach has been shown to be effective for both antiviral and anticancer applications. The possibilities offered and the success of this technology are well documented in the literature.

The second strategy involves the design of monophosphate analogues where the phosphate moiety is changed to an isosteric and isoelectronic phosphonate unit.[21] A phosphonated nucleoside, where the phosphonate group is attached to the acyclic nucleoside moiety through a stable P–C bond, shows an advantage, over its phosphate counterpart in being more stable metabolically and chemically. These 5′-mononucleotide mimics are able to overcome the instability of mononucleotides toward phosphodiesterases and to enhance cellular uptake by bypassing the initial enzymatic phosphorylation step.[22] Furthermore, within cells, they must be phosphorylated by cellular nucleotide kinases to the corresponding diphosphates and then triphosphates to exert biological activity.

The various types of nucleoside phosphonate analogues reported in the literature can be classified according to the following scheme: (a) carbocyclic phosphonated nucleosides; (b) heterocyclic phosphonated nucleosides; (c) pyranose phosphonated nucleosides; and (d) acyclic phosphonated nucleosides.

The prodrug approach is described in another section of the book. In this chapter, attention is directed to the carbocyclic and acyclic phosphonate nucleoside analogues, covering the literature up to 2011.

2. PHOSPHONATED CARBOCYCLIC NUCLEOSIDES

Carbocyclic nucleosides are structural analogues of natural and synthetic nucleosides in which the endocyclic oxygen atom of common nucleosides is replaced by a methylene group. Due to the absence of the labile glycosidic bond, these analogues are chemically and enzymatically more stable than are natural nucleosides.

Relevant members of this family of compounds are shown in Figure 2. The naturally occurring carbocyclic nucleosides aristeromycin and neplanocine A exhibit powerful antitumor and antiviral activities; carbovir and its prodrug abacavir, and carbocyclic-ddA, are synthetic derivatives that show high activity against HIV and hepatitis B virus, respectively.[23] Depending on the size of the ring, phosphonated cyclopropyl, cyclobutyl, cyclopentyl, and cyclohexyl carbanucleosides are reported in the literature.

2.1. Phosphonated Cyclopropyl Nucleosides

Phosphonated cyclopropyl nucleosides can be classified into different groups depending on the relative position of the nucleobase and the phosphonic group (Figure 3).

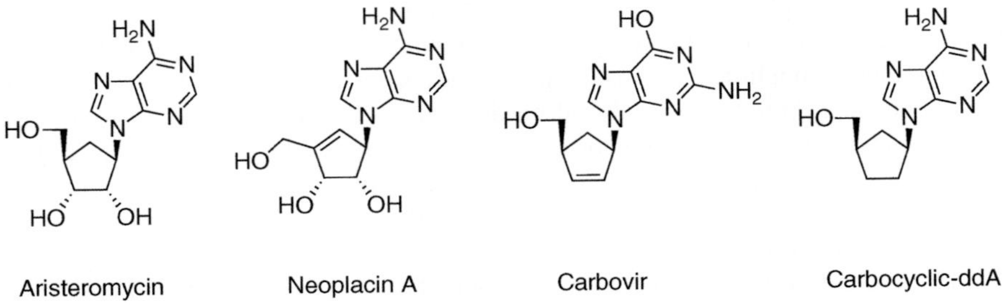

Figure 2. Relevant phosphonated carbocyclic nucleosides.

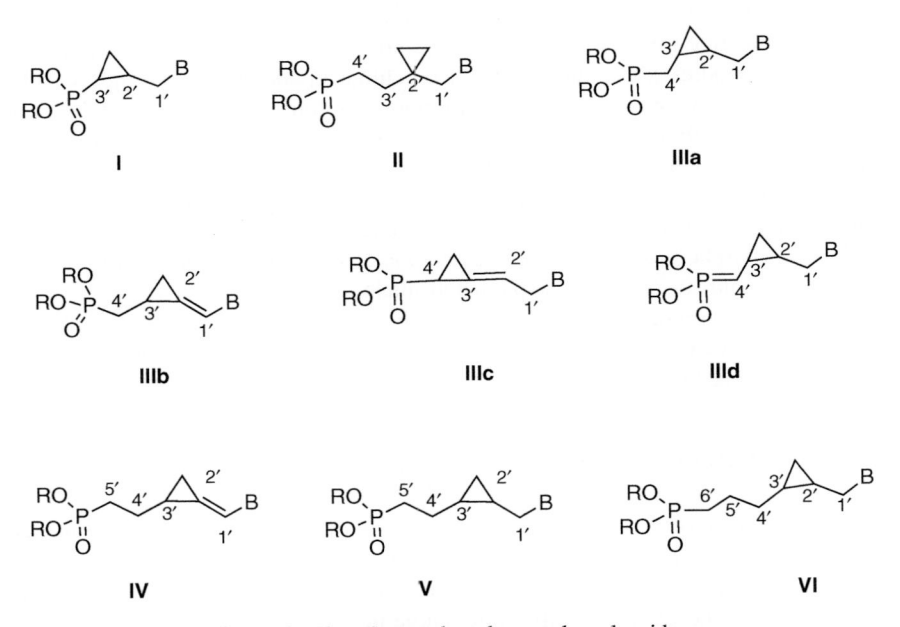

Figure 3. Phosphonated cyclopropyl nucleosides.

Compounds of Structure I. Cyclopropyl phosphonated derivatives containing adenine and 6-chloropurine nucleobases **6** and **7** have been prepared starting from alkylphosphonates **1** that have been lithiated and reacted with allyl bromide to give, after epoxydation with MCPBA, the γ,δ-epoxyalkanephosphonates **3**.[24] Subsequent treatment with LDA (R = CH$_2$Ph) or BuLi (R = H, CH$_3$, Ph) at −78°C afforded (hydroxymethyl)cyclopropylphosphonates **5a–d** (67–84% yield), in a *trans* configuration, via a ring opening/ring closure reaction. The target-modified nucleosides **6a** and **7b** were then synthesized by the replacement of the hydroxymethyl group, through a coupling reaction, with adenine or 6-chloropurine, under Mitsunobu conditions, in 42% and 46% yield, respectively.

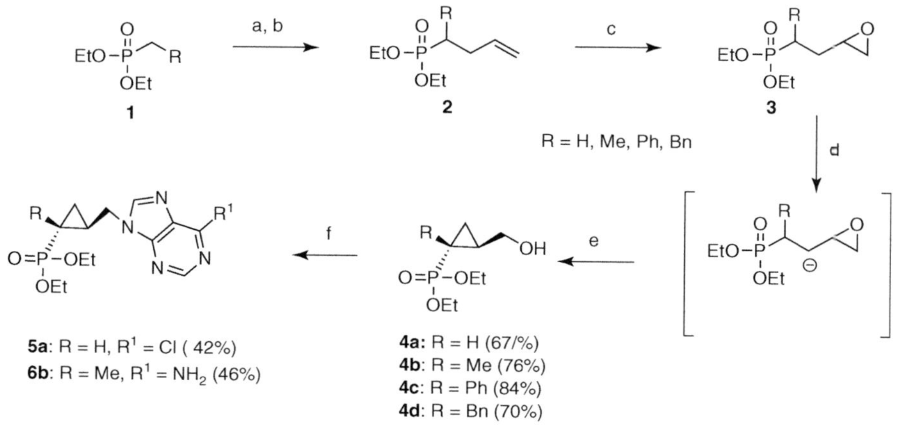

Scheme 1. *Reagents and conditions:* (a) base, −78°C, THF; (b) allylbromide; (c) MCBA, rt; (d,e) base, THF, −78°C to rt; (f) 6- chloropurine or adenine, Ph₃P, THF −10°C, DEAD, rt, 10 h.

Experimental Procedure

Preparative Procedure for Diethyl [(1SR,2RS)-2-(6-amino-9H-purin-9-yl)cyclopropyl]phosphonate (7a). A mixture of diethyl[(1SR,2RS)-2-(hydroxymethyl) cyclopropyl]phosphonate **4a** (166 mg, 0.80 mmol), 6-chloropurine (150 mg, 0.97 mmol), and triphenylphosphine (505 mg, 1.93 mmol) in anhydrous THF, cooled to −10°C, was treated with diethyl azodicarboxylate (280 mg, 1.59 mmol). The mixture was stirred at room temperature for 10 h and the solvent was then removed under reduced pressure to give, after chromatography purification with acetone, **7a** (109 mg, 034 mmol). ^{1}H-NMR, (300 MHz, CDCl₃): δ 8.27 (s, 1H), 7.87 (s, 1H), 4.25 (ddd, 1H, J = 14.45, 6,23, and 2.07 Hz), 3.30−3.17 (m, 2H), 3.06, (br s, 1H), 1.97−1.91 (m, 1H), 1.61−1.55 (m, 1H), 4.00−3.92 (m, 3H), 3. 87−3.66 (m, 2H), 1.81−1.77 (m, 1H), 1.23−1.00 (m, 6H), 0.99−0.93 (m, 2H). ^{13}C-NMR (75 MHz, CDCl₃): δ 155.70, 152.92, 149.90, 139.68, 61.86 (dd, J = 11.10, and 6.08 Hz), 46.52 (d, J = 3.75 Hz), 17.14 (d, J = 3.90 Hz), 12.20 (dd, J = 11.63, and 6.08 Hz), 12.00, 9.37 (dd, J = 11.18, and 4.80 Hz).

The enantiopure synthesis of nucleosides **6a** and **7a** has been reported by Midura et al.[25] In particular, the enantiopure (*S*)-(+)-(1-diethoxyphosphoryl)vinyl *p*-tolylsulfoxide **8** was transformed to a mixture of cyclopropanes **10a–d** in the relative ratio 36:3:12:1, through a cyclopropanation reaction performed with dimethyl(ethoxycarbonylmethyl)sulfonium bromide **9** in the presence of DBU in toluene (95% yield). The major stereoisomer **10a** thus obtained was converted to enantiopure nucleosides **6a** and **7a** in three steps: (1) reaction of **10a** with MeMgI to give **11a** as a single steroisomer (85%); (2) reaction of **11a** with LiBH₄ to give **12a** (90%); and (3) reaction of **12**a with the corresponding purine nucleobases by a Mitsunobu reaction (Scheme 2).

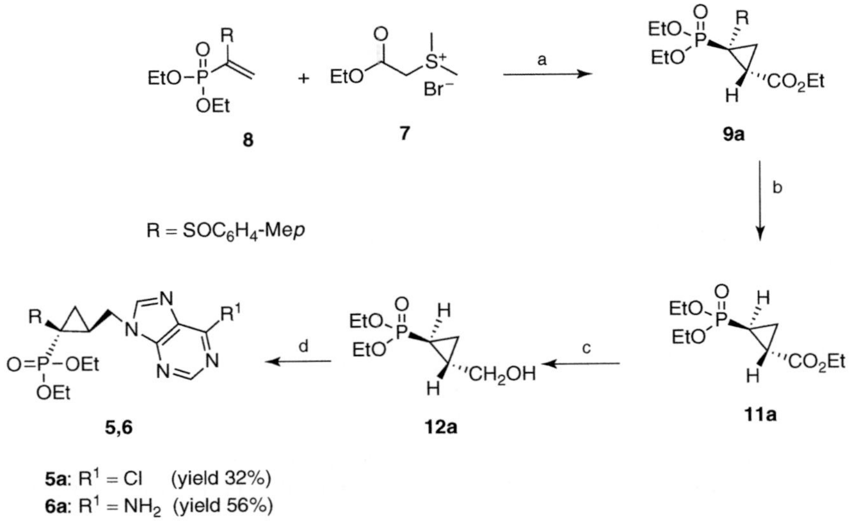

Scheme 2. *Reagents and conditions:* (a) DBU, toluene, heat; (b) MeMgl, Et$_2$O, $-30°$C; (c) LiBH$_4$, MeOH rt, 10 h; (d) 6-chloropurine or adenine, Ph$_3$P, THF, $-10°$C, DEAD, rt, 10 h.

Compounds of Structure II. The 9-[1-(phosphonomethoxycyclopropyl)methyl] guanine (PMCG **18**), a compound representative of the class of phosphonated nucleosides containing a cyclopropyl moiety at the 2′-position (type II), is endowed with selective anti-HBV activity (EC$_{50}$ = 0.5 μM). This compound has been synthesized by Choi et al.[26] The process involves as a key reaction a titanium-mediated Kulinkovich cyclopropanation. The synthetic route is achieved in seven steps, starting with ethyl 2-{[*tert*-butyl(diphenyl)silyl]oxy}acetate **13**, which was converted into the cyclopropanol derivative **14** in 80% yield by reaction with ethylmagnesium bromide and titanium(IV) isopropoxide. Compound **14** was then reacted with diisopropylbromomethylphosphonate and transformed to diisopropyl{[1-({[*tert*-butyl(diphenyl)silyl]oxy}methyl)cyclopropyl]oxy}-methylphosphonate **15** (70%), from which the target phosphonated **18** was obtained by reactions with TBAF, mesylchloride, 2-amino-6-chloro-9*H*-purine, trimethylsilylbromide, and hydrolysis (Scheme 3).

Compound **19**, containing the 2-amino-9*H*-purin-9-yl group, was prepared from **17** via hydrogenation followed by hydrolysis and converted to the orally available Dipivoxil **20** by etherification with chloromethyl pivalate in 39% yield (Scheme 3, route b).

Compounds of Structure IIIa. Diethyl[{(1′*S*,2′*S*)-2-[(2-amino-6-oxo-1,6-dihydro-9*H*-purin-9-yl)methyl]cyclopropyl}difluoromethyl]phosphonate (**26**) and its enantiomer [27] (type IIIa structures) were prepared by using as the key intermediate the optically active (+)- and (−)-diethyl{difluoro[(1*S*,2*S*)-2-(hydroxymethyl)cyclopropyl]methyl}phosphonate **22**. These enantiomers were synthesized with high

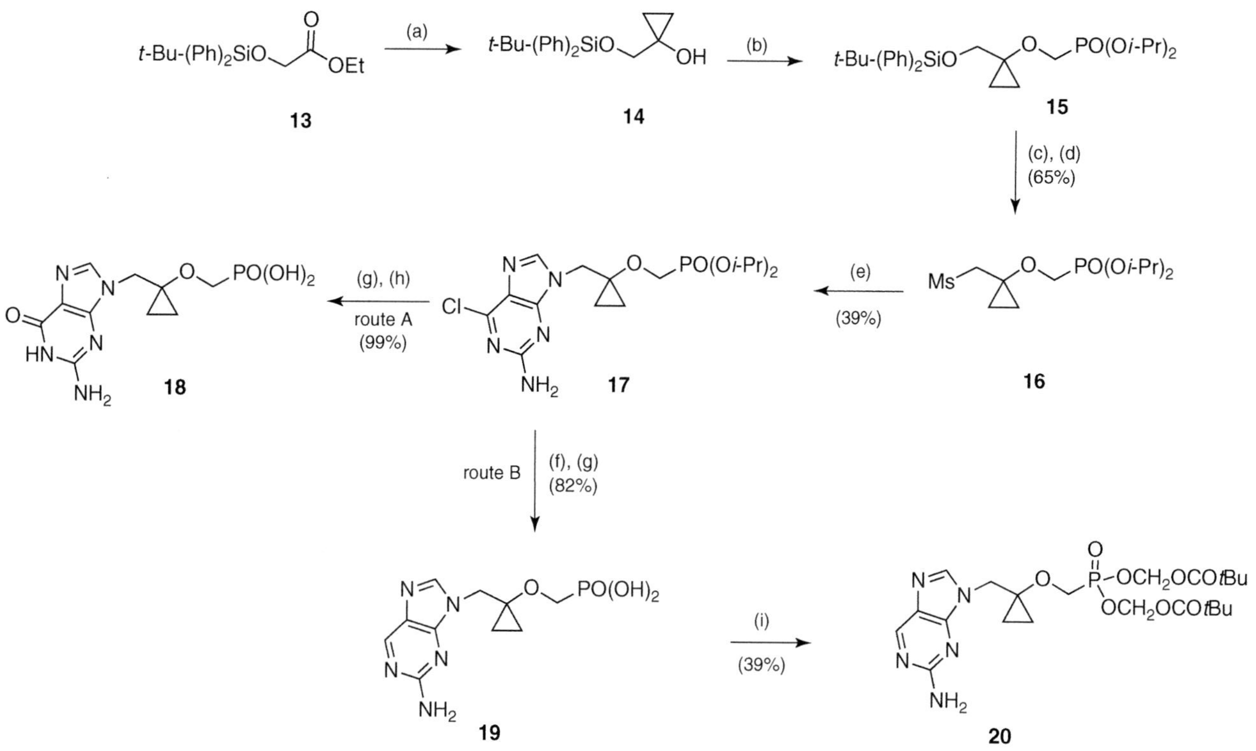

Scheme 3. *Reagents and conditions:* (a) CH$_3$CH$_2$MgBr, Ti(O-*i*-Pr)$_4$ (0.25 equiv), THF, 0–25°C, 10 h; (b) BrCH$_2$P(O)(O-*i*-Pr)$_2$, LiO/*t*-Bu, LiI (cat.), DMF, THF, 60°C, 4 h; (c) NH$_4$F, MeOH, reflux, 10 h; (d) MsCl, TEA, MDC, 0–25°C; (e) 6-chloroguanine, N$_a$H, DMF, 80°C, 4 h; (f) H$_2$, 5% Pd on C, THF, 1 atm, 18 h; (g) TMSBr, MDC, reflux, 18 h; (h) 2 N HCl, reflux, 6 h; (i) chloromethyl pivalate, TEA, 1-methyl-2-pyrrolidinone, 25°C, 48 h.

enantiomeric purity by cyclopropanation reaction of α,α-difluoroallylphosphonate **21** with an excess of diazomethane in the presence of Pd(OAc)$_2$, followed by deprotection with p-TsOH in methanol, and enzymatic double resolution of the corresponding alcohol (±)-**22** thus obtained. The condensation of (+)-**22** with −6-chloropurine or 2-amino-6-chloropurine via tosylate gave the purine derivatives (+)-**24** and (+)-**25** in 55% and 43% yield, respectively. Transformation of (+)-**25** into the final product (+)-**26** was achieved by reaction with TMSBr and aqueous acid hydrolysis (Scheme 4). (−)-**22** was converted to the enantiomer (−)-**26** in the same way as for (+)-**26**.

The inhibitory activity of (+)-**26** and (−)-**26** toward purine nucleoside phosphorylase (PNP) was evaluated compared with that of 9-(difluorophosphono-pentyl)guanine and found to be 2400-fold less than that of the acyclic derivative. Difluoro {(1SR,2SR)-2[(1SR)-1-(6-oxo-1,6-dihydro-9H-purin-9-yl)ethyl]cyclopropyl}methylphosphonic acid **37a** and related analogues were prepared as "multisubstrate analogue" inhibitors for purine nucleoside phopshorylase (PNP).[28] Alcohols **27** and **28** were converted by Swern oxidation into aldehydes **29** and **30**, alkylated with various Grignard reagents and oxidized by the Jones reagent to give the corresponding ketone derivatives **31a–d**. These compounds were transformed stereoselectively in the alcohols **32a–d** by reduction with K-Selectride. Finally, conversion of alcohols **32** into the target nucleotide analogues **37** was accomplished by condensation with 6-chloropurine under Mitsunobu conditions, followed by treatment with bromotrimethylsilane and H$_2$O. The diastereomeric nucleotide analogue **35** was also prepared from **33a**, via **34**, in a manner similar to the transformation of **32a** in **33a** via Mitsunobu inversion (Scheme 5).

Compounds of Structure IIIb. Phosphonates of type IIIb have been prepared by involving as the starting compound the (methylenecyclopropyl)methyl methanesulfonate **38**.[29] This compound was treated with NaH and diisopropyl phosphite to give the corresponding phosphonate **39** in 84% yield. Addition of bromine followed by alkylation–elimination gave a mixture of E/Z-isomers **41** + **42** (69%) via the intermediate **40**. Therefore, the E/Z-isomers **41/42** were dealkylated with Me$_3$SiI at −40°C to afford a mixture of E/Z-isomers **43** + **44** which have been separated by chromatography on Dowex I (formate) using formic acid as the eluent (Scheme 6). Nucleoside analogues containing guanosine and cytosine were prepared in a similar way.

Compounds of Structure IIIc. The synthesis of diisopropyl ethylidencyclopropyl nucleoside phosphonates **53a–d** belonging to class IIIc has been described by Zemlicka et al.[30] The synthetic scheme involves the formation of diisopropyl (Z)- and (E)-(2-hydroxyethylidene)-1-cyclopropylphosphonate **51** and **52**, which were prepared in 10 steps starting from the diisopropyl (E)-2-hydroxymethyl-1-cyclopropylphosphonate **45**. These intermediates were independently transformed to adenine and 2-amino-6-chloropurine derivatives **53a–d** via nucleophilic bromuration and amination. Acid hydrolysis of compounds **53a–d** provided the adenine and guanine analogues **54a, 54b, 55a,** and **55b** (Scheme 7). Phosphonates **54b** and **55b** are potent inhibitors of replication of Epstein-Barr virus (EBV).

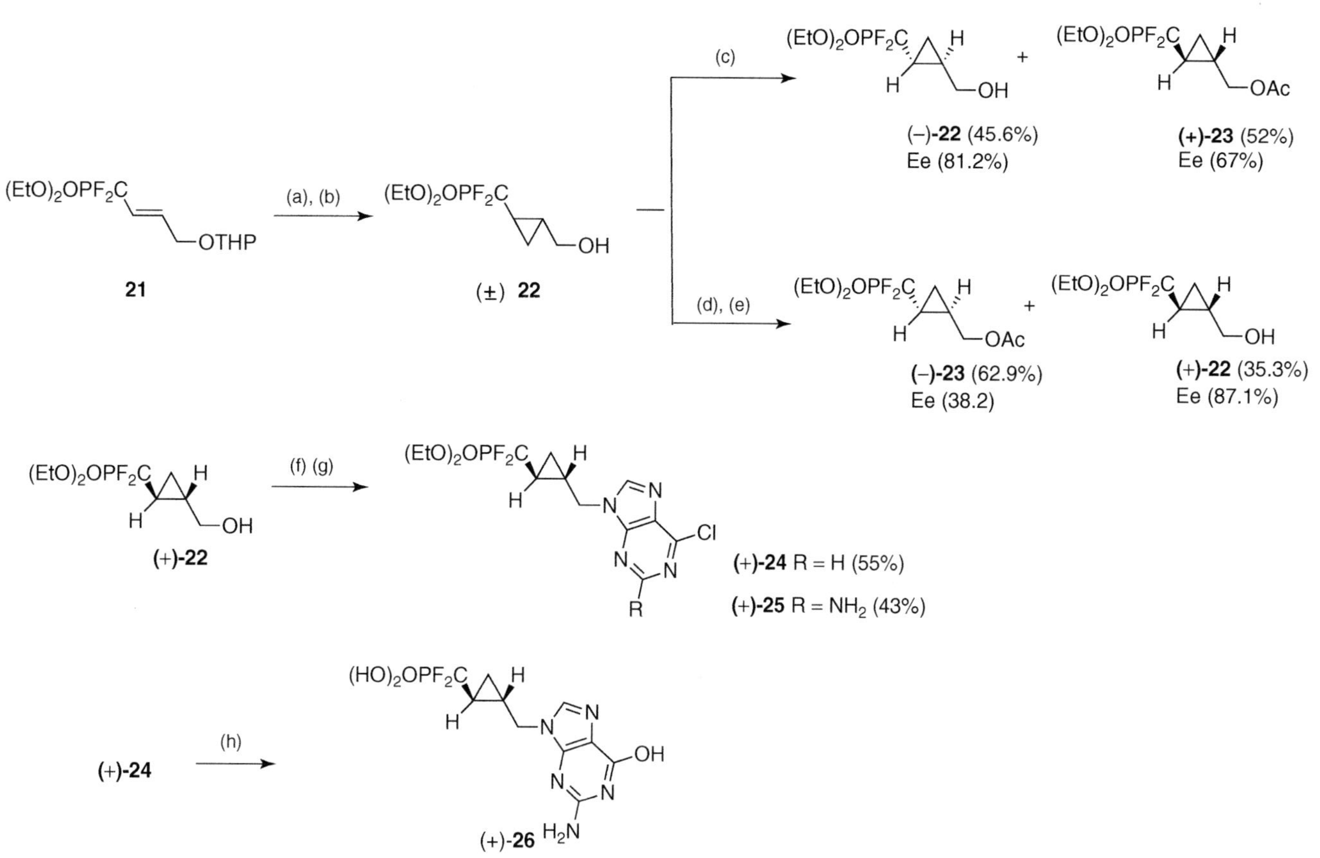

Scheme 4. *Reagents and conditions:* (a) CH$_2$N$_2$/Pd(OAc)2, (b) *p*-TsOH, MeOH, rt, 5 h; (c) vinyl acetate, PLE, THF, 37°C, 32 h; (d) Ac$_2$O, Py, rt, 3 h; (e) PPL (50 mg), 0.1 M phosphate buffer, *i*-Pr$_2$O; (f) TsCl, Et$_3$N; (g) purine derivatives, K$_2$CO$_3$, DMF; (h) TMSBr, MeOH, HCl.

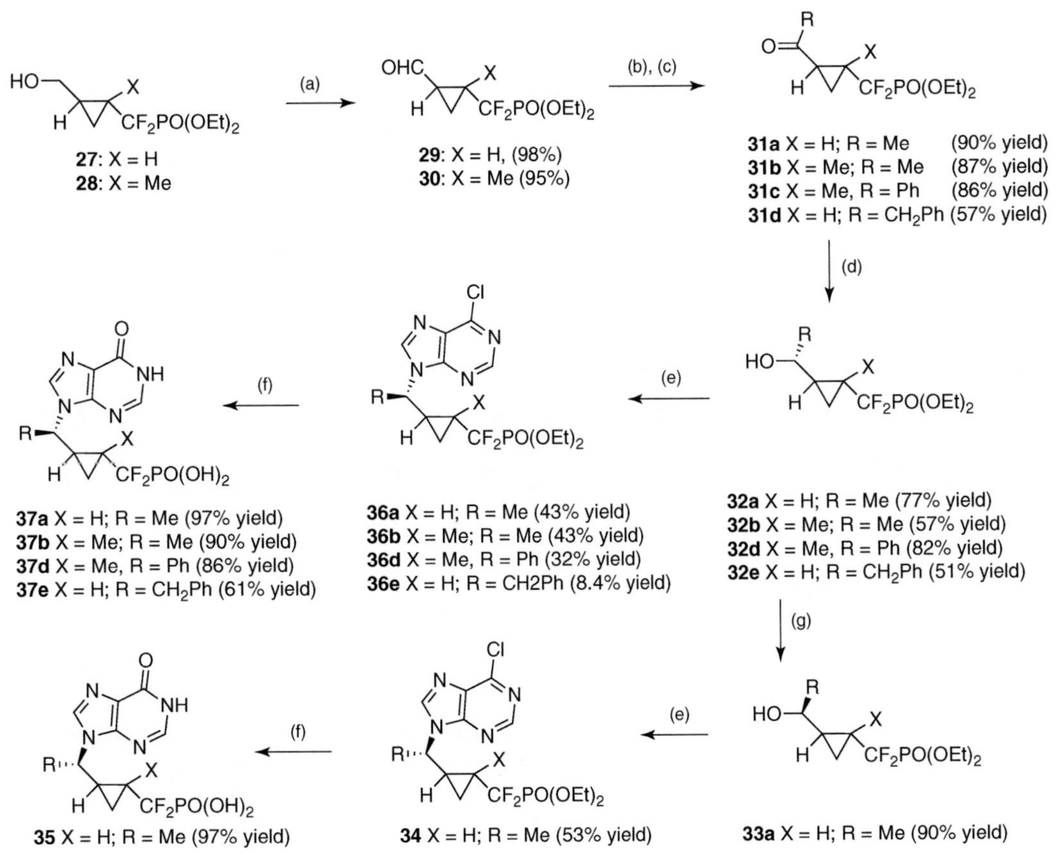

Scheme 5. *Reagents and conditions*: (a) oxalyl chloride, DMSO, −78°C; (b) MeMgX, PhMgX, or PhCH$_2$MgX, 0°C; (c) Jones' reagent 0°C; (d) K-Selectride −78°C; (e) 6-chloropurine, Ph$_3$P, DEAD, THF, rt, 14 h; (f) Me$_3$SiBr, rt, 37 h, H$_2$O, 48 h; (g) PhCO$_2$H, PH$_3$P, DEAD, rt, 15 h, NaOH, MeOH.

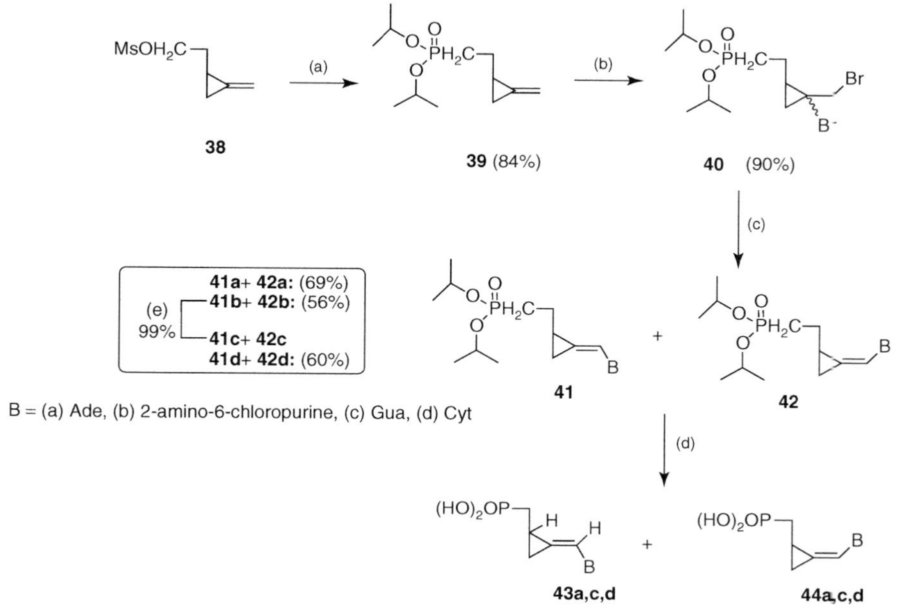

Scheme 6. *Reagents and conditions:* (a) diisopropyl phosphite, NaH, THF, 0°C, 2 h, then at −40°C 19 was added and stirred at rt for 24 h; (b) pyridine, HBr$_3$, CH$_2$Cl$_2$, 0°C; (c) adenine, 2-amino-6-chloropurine or cytosine, K$_2$CO$_3$, DMF, 110°C, 18 h; (d) MeSi$_3$Br or Me$_3$SiI, DMF; (e) 80% formic acid.

Compounds of Structure IIId. Reversed methylenecyclopropane analogues of antiviral phosphonates **63a**, **64a**, **63b**, and **64b**, of structure IIId, have been synthesized by Li and Zemlicka.[31] Thus, 1-bromo-1-bromomethylcyclopropane **56** was converted to the bromocyclopropyl phosphonate **57** by Michaelis–Arbuzov reaction with triisopropyl phosphate. β-Elimination, followed by nucleobase introduction performed with a Mitsunobu reaction and hydrolysis, furnishes the target phosphonates **63** and **64** (Scheme 8).

Compounds of Structure IV. Phosphonates **71** and **72** of structure IV, homologues of **43** and **44** (see Scheme 6), have been synthesized starting from dibromocyclopropane **65**.[29,32] Thus, **65** was converted to intermediate **67** by reaction with diisopropyl methyl phosphonate and subsequent β-elimination. Compound **67** was transformed to vicinal dibromide **68**, from which, following the sequence described for the series of lower homologues, the target compounds **71** and **72** were obtained (Scheme 9). The cytosine derivatives **69/70** and **71/72** were obtained as inseparable mixtures. All the target phosphonates were found inactive against herpes viruses as well as against hepatitis B virus and HIV-1.

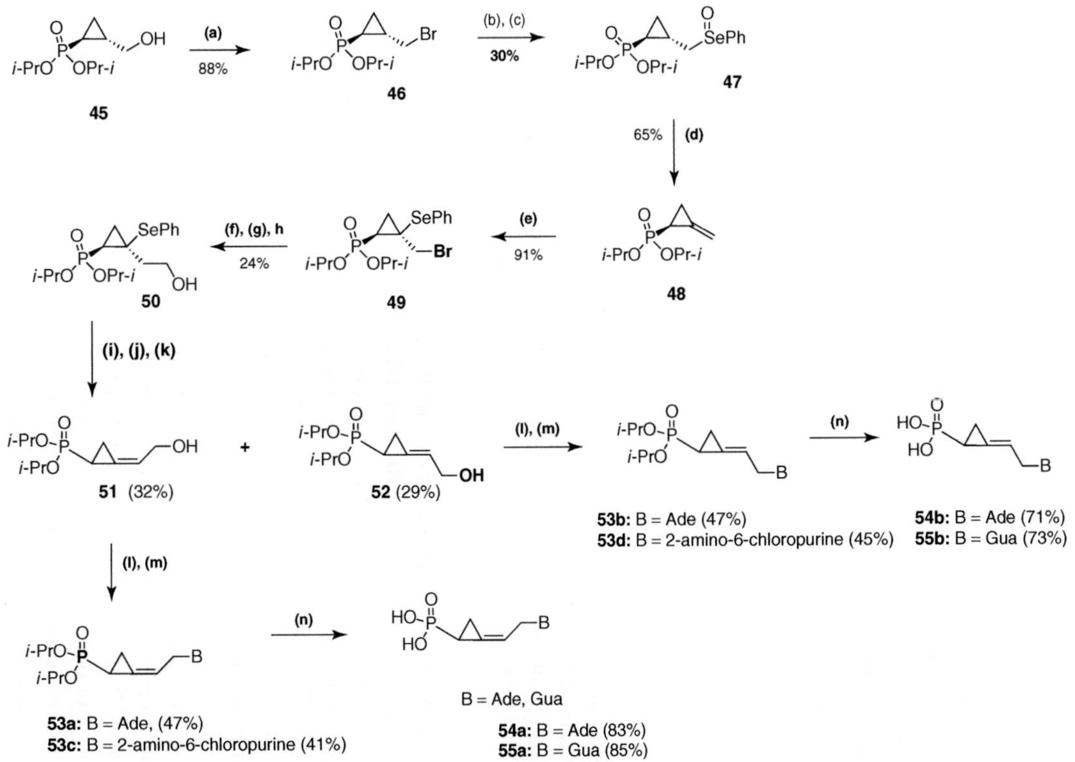

Scheme 7. *Reagents and conditions*: (a) Ph$_3$P, Br$_2$, CH$_2$Cl$_2$; (b) (PhSe)$_2$, EtOH; (c) H$_2$O$_2$, THF; (d) (*i*-Pr)$_2$NEt, PhMe; (e) (PhSe)$_2$, NBS, CH$_2$Cl$_2$; (f) Me$_3$SiCN, TBAF, MeCN; (g) HCl(g), MeOH; (h) LiBH$_4$, THF; (i) Ac$_2$O, Py; (j) H$_2$O$_2$, THF, K$_2$CO$_3$, MeOH/H$_2$O; (l) Ph$_3$P, CBr$_4$, CH$_2$Cl$_2$; (m) adenine or 2-amino-6-chloropurine; (n) HCl.

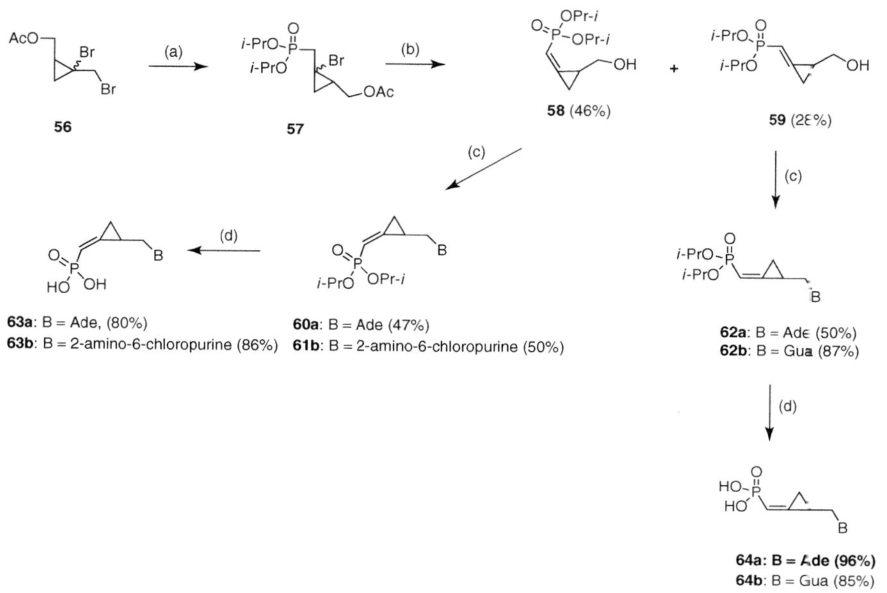

Scheme 8. *Reagents and conditions:* (a) triisopropyl phosphite, 120°C, 48 h; (b) NaOH (2%), rt, 40 min; (c) Ph$_3$P, adenine or 2-amino-6-chloropurine, DEAD, 0°C, rt, 12 h.

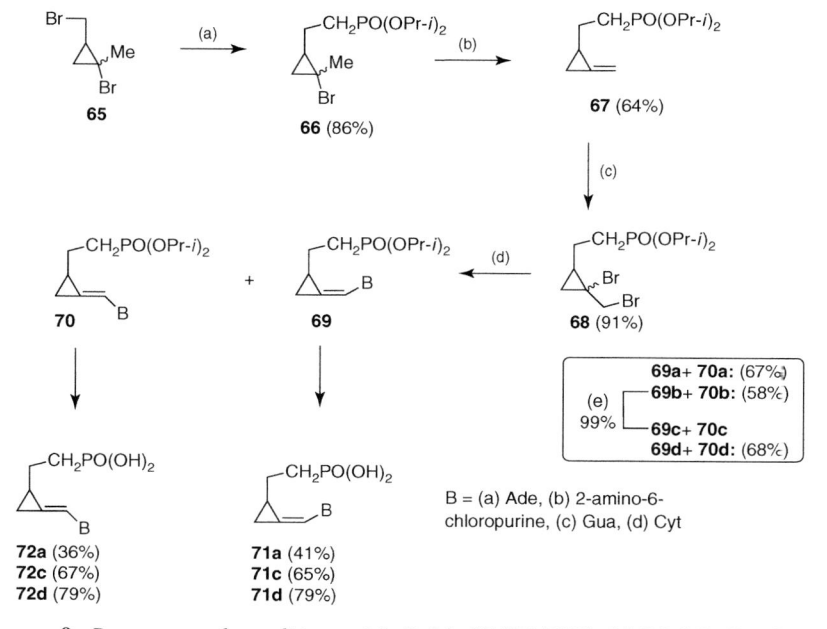

Scheme 9. *Reagents and conditions:* (a) BuLi, HMPA/THF, MePO(*i*-PrO); (b) NaH, THF/DMF, heat; (c) HBr$_3$, pyridine, CH$_2$Cl$_2$; (d) adenine, 2-amino-6-chloropurine or cytosine, K$_2$CO$_3$, or Cs$_2$CO$_3$, DMF, 110°C; (e) Me$_3$SiBr, DMF; (f) 80% formic acid.

Compounds of Structure V. Starting from (*E*)-allyl-α,α-difluorophosphonate **73**, a racemic mixture of {2-[2-(2-amino-6-oxo-6,9-dihydro-1*H*-purin-9-yl)ethyl]cyclopropyl}(difluoro)methylphosphonic acid **77a** and difluoro{2-[2-(6-oxo-6,9-dihydro-1*H*-purin-9-yl)ethyl]cyclopropyl}methylphosphonic acid **77b** were prepared for evaluation of their PNP inhibitory activity.[33] In particular, compound **73** was cyclopropanted to **74** with diazomethane in the presence of palladium diacetate. After removal of the tetrahydropyranyl group, **77a,b** were obtained via **75** and **36a,b** (Scheme 10).

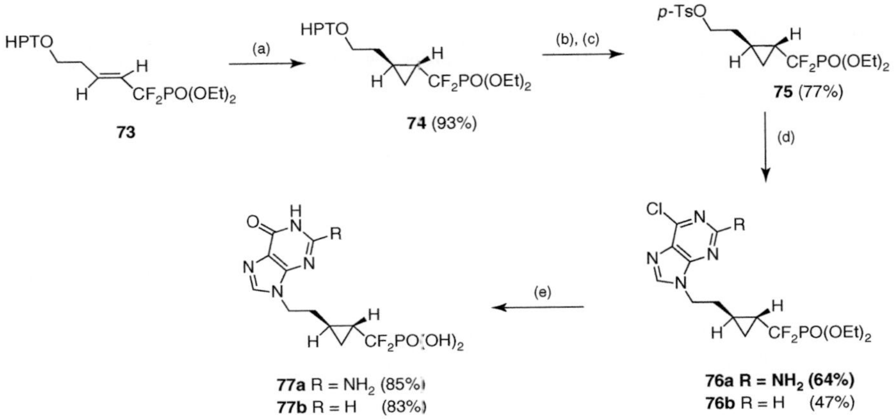

Scheme 10. *Reagents and conditions:* (a) CH_2N_2, Pd(OAc)$_2$, ether; *p*-TsOH, MeOH; (c) *p*-TsCl, Et$_3$N, CH$_2$Cl$_2$; (d) 2-amino-6-chloropurine or 6-chloropurine, K$_2$CO$_3$, DMF, H$_2$O.

Compound **77b**, containing the hypoxantine residue, showed IC$_{50}$ and K_i values toward a PNP of 70 nM and 8.8 nM, respectively. The result indicates that with respect to acyclic derivatives, the binding affinity (K_i), but also the inhibitory potency (IC$_{50}$) toward *Cellulomonas* sp. PNP is significantly improved by introduction of the cyclopropane ring. Phosphonated analogues of the antiviral cyclopropane nucleoside A-5021, **92**, were synthesized from (1*SR*,7*RS*)-3,5-dioxa-4,4-diphenylbicyclo[5.1.0]octane-1-methanol **78** by a 10-step process.[34] Thus, compound **78** was transformed by a transacetalization reaction, performed with a catalytic amount of *p*-TsOH, into the spiro compound **79** in 43% yield. The reaction proceeded with the introduction of a diethoxyphosphonyl group to give compound **81**, achieved by Swern oxidation followed by a Horner–Wittig reaction. Catalytic hydrogenation of **81** furnished **82**, which with benzoyl chloride gave a mixture of **83** and **84** in 47% yield and in about a 2:1 ratio (Scheme 11).

The phosphonate analogue of A-5021 **92** was finally obtained from compound **84**, by a three-step sequence involving (1) bromiation, (2) amination, and (3) hydrolysis. The adenine analogue **87** was prepared analogously, using, as a base, adenine instead of 2-amino-6-benzyloxypurine, while the regiosiomers **89** and **91** were synthesized from **83** via **88** and **90** using the same method (Scheme 12). In contrast

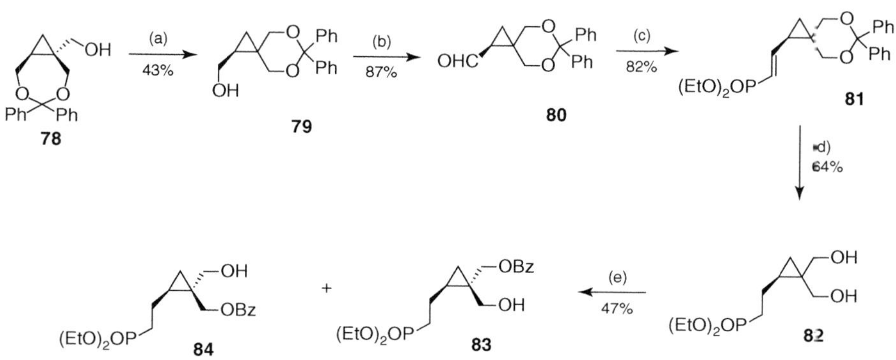

Scheme 11. *Reagents and conditions:* (a) cat. *p*-TsOH, THF, rt, 12 h; (b) DMSO, (COCl)$_2$, CH$_2$Cl$_2$, −50°C; (c) tetraethyl methylenediphosphonate, BuLi, −78°C, THF; (d) H$_2$, cat. Pd/C, EtOH–AcOH, rt; (e) PhCOCl, Et$_3$N, THF, rt.

to the potent antiherpetic activity of A-5021, all these derivatives do not show any antiviral activity.

Compounds of Structure VI. Methyl branched cyclopropyl phosphonates of structure VI have been synthesized by Kim et al.[35] The cyclopropyl moiety was constructed employing the Simmons–Smith reaction as a key step. Thus, the allylic alcohol **93** was reacted with ZnEt$_2$ and CH$_2$I$_2$ to form the cyclopropyl alcohol **94**, which was converted in compound **95**, treating it with diisopropyl bromomethylphosphonate. Compound **95** was desilylated with TBAF, mesylated, and then easily converted into the phosphonated nucleosides **96a–d** by reaction with suitable nucleobases (Scheme 13). The phosphonic acid nucleosides **97a–d**, obtained by hydrolysis of corresponding phosphonated with trimethylsilylbromide, have been evaluated for their antiviral activities. Unfortunately, all compounds proved to be inactive against HIV-1, HSV-1, HCMV, and CoxB3, except **96b**, which shows low activity against HIV-1 and CoxB3 (EC$_{50}$ 55.7 and 43.5 mg/mL, respectively). Analogously, phenyl branched nucleosides have been prepared starting from (*E*)-4-(*tert*-butyldimethylsilyloxy)-3-phenylbut-2-en-1-ol.[36]

In the same context, phosphonated difluorocyclopropyl analogue **109**, which exhibits in vitro anti-HIV-activity similar to that of PMEA (EC$_{50}$ = 0 2.4 μmol), and its isomer **113**, have recently been prepared.[37] The synthetic route starts from **98** or **99**, which were acetylated to give compounds **100** and **101** and then subjected to a difluorocyclopropanation reaction to afford the cyclopropanes **102** and **103**, respectively. Treatment of **102** and **103** with sodium methoxide in MeOH forms the intermediates **104** and **105** (Scheme 14).

Compounds **104** and **105** were, independently, converted to phosphonated nucleosides **109** and **113**, respectively, by reaction with diethylphosphonomethyl triflate, followed by (1) desilylation, (2) bromination, (3) amination, and (4) hydrolysis (Scheme 15).

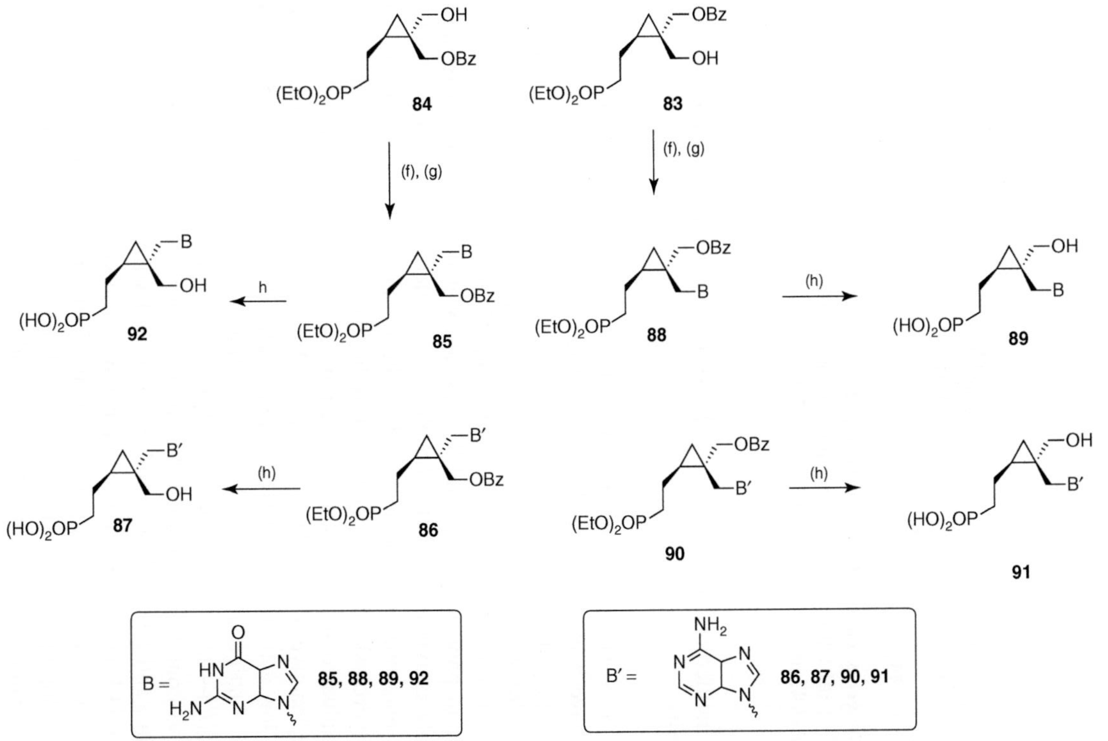

Scheme 12. *Reagents and conditions:* (a) TBDMSCl, imidazole, CH$_2$Cl$_2$; (b) CH$_2$I$_2$, Et$_2$Zn, CH$_2$Cl$_2$; (c) diisopropyl bromomethylphosphonate, LiO*t*-Bu, LiI, DMF; (d) TBAF, THF; (e) MsCl, TEA, CH$_2$Cl$_2$; (f) adenine, cytosine, thymine, or uracyl, K$_2$CO$_3$, 18-C-6, DMF; (g) Me$_3$SiBr, CH$_2$Cl$_2$.

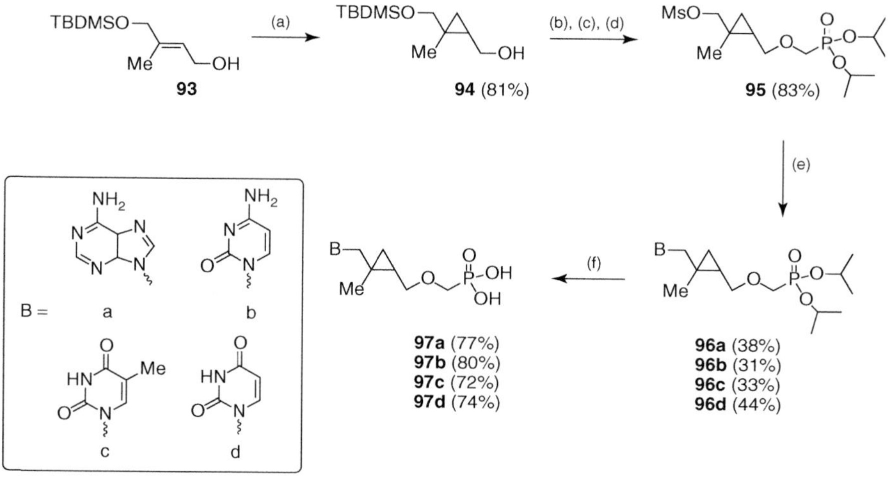

Scheme 13. *Reagents and conditions:* (a) CH$_2$I$_2$, Et$_2$Zn, CH$_2$Cl$_2$; (b) diisopropyl bromomethylphosphonate, LiO/t-Bu, LiI, DMF; (c) TBAF, THF; (d) MsCl, TEA, CH$_2$Cl$_2$; (e) adenine, cytosine, thymine, or uracil, K$_2$CO$_3$, 18-C-6, DMF; (f) Me$_3$SiBr, CH$_2$C$_{-2}$.

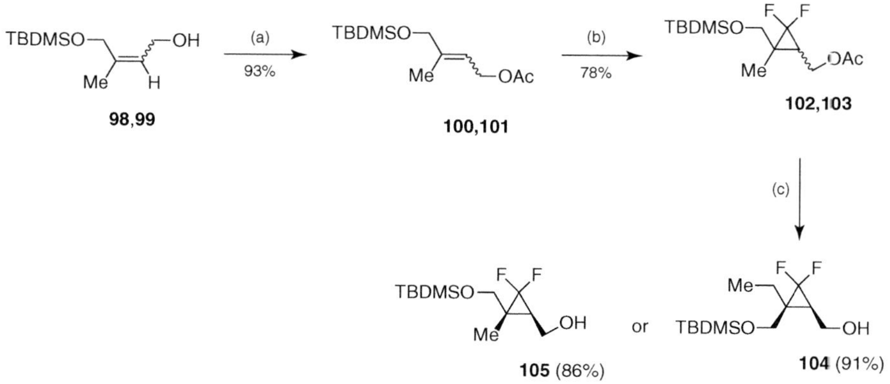

Scheme 14. *Reagents and conditions:* (a) Ac$_2$O, DMAP, Py, rt; (b) sodium chlorodifluoroacetate, diglyme, 190°C, reflux; (c) NaOMe, MeOH, 0°C.

Various unsubstituted cyclopropyl phosphonates of structure VI have been prepared by Hong et al., starting from *cis*-4-(*t*-butyldimethylsilanyloxy)but-2-en-ol **114** (Scheme 16),[38] applying the same methodology reported by Kim et al.[35] Only the adenine derivative **122** exhibited moderate activity against HMCV (EC$_{50}$ = 22.8 μg/mL), without significant toxicities to the host cell.

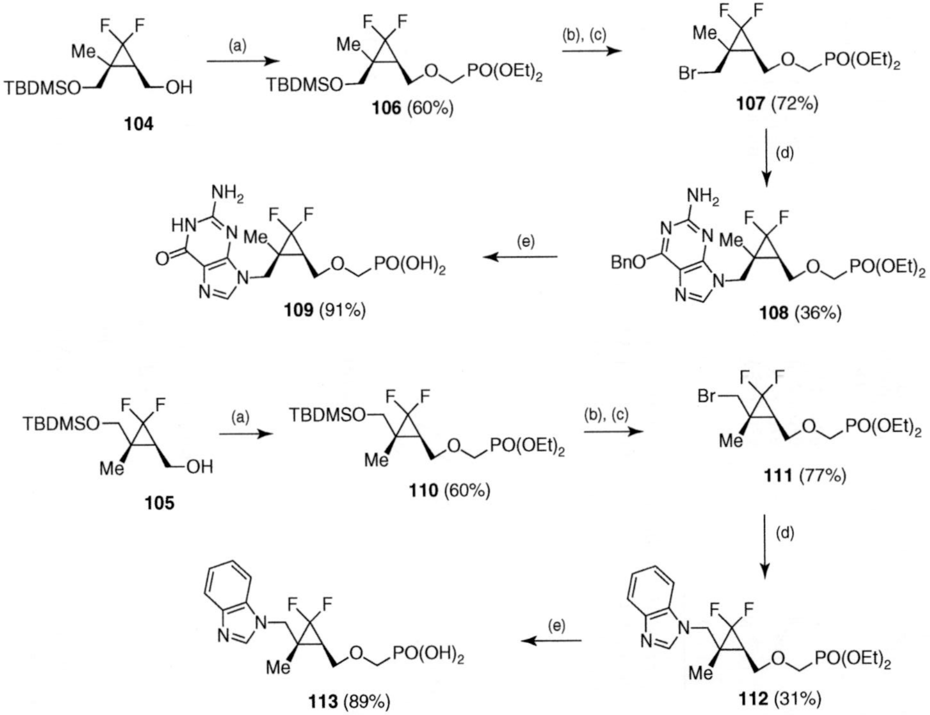

Scheme 15. *Reagents and conditions*: (a) (EtO)$_2$POCH$_2$OTf, LiO/-*t*-Bu, THF; (b) TBAF, THF; (c) NBS, Ph$_3$P, CH$_2$Cl$_2$; (d) 2-amino-6-benzoyloxypurine, K$_2$CO$_3$, DMF; (e) Me$_3$SiBr, DMF, rt, aq. HCl, rt, NaOH, H$_2$O.

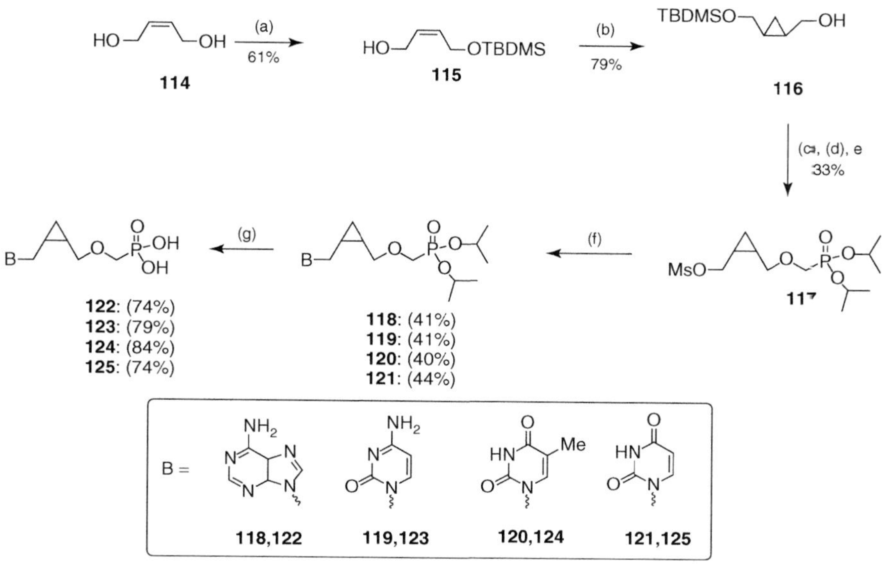

Scheme 16. *Reagents and conditions:* (a) TBDMSCl, imidazole, CH_2Cl_2; (b) CH_2I_2, Et_2Zn, CH_2Cl_2; (c) diisopropyl bromomethylphosphonate, LiO/t-Bu, LiI, DMF; (d) TBAF, THF; (e) MsCl, TEA, CH_2Cl_2; (f) adenine cytosine, thymine, or uracyl, K_2CO_3, 18-C-6, DMF; (g) Me_3SiBr, CH_2Cl_2.

2.2. Phosphonated Cyclobutyl Nucleosides

A unique example of cyclobutyl nucleoside analogues has been reported in the literature. Choi et al.[26] have obtained the derivative **124** in very low yield (1%) when {1-[(diisopropylphosphoryl)methoxy]cyclopropyl}methyl methane sulfonate **123** was coupled with 6-chloroguanine. The formation of this product was rationalized through a rearrangement in which **123** solvolyzes to generate the cyclopropylmethyl cation, which undergoes a ring enlargement to cyclobuane cation, followed by reaction with 6-chloroguanine anion (Scheme 17).

2.3. Phosphonated Cyclopentyl Nucleosides

2.3.1. 1,2-Substituted Cyclopentane Derivatives.

The synthesis of cyclopentane derivatives **130** and **131** is outlined in Scheme 17. Cyclopentane epoxide **125** has been reacted with the lithium salt of 6-O-benzylguanine to afford the N^9-alkylated *trans*-alcohol **126** in 43% yield and its N^7-isomer in 22% yield (Scheme 18). The N^9 derivative **126** was protected at the amino group by a monomethoxytrityl group (**127**) and left to react with the anion of diethyl(p-toluenesulfonyloxy)methylphosphonate, generated by treatment with sodium alkoxide, to afford the phosphonate diester **128**. Sequential deprotection of the guanine and phosphonate moiety gave the *trans*-cyclopentane derivative **130**.

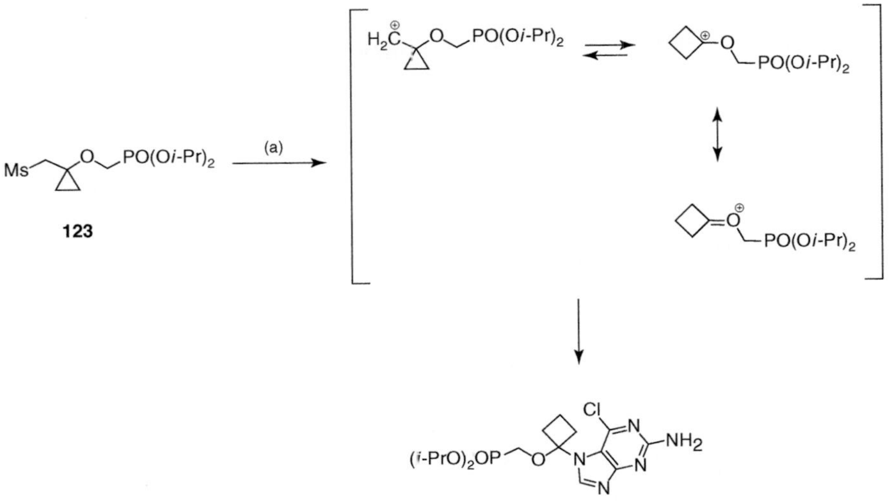

Scheme 17. *Reagents and conditions:* (a) 6-chloroguanine, NaH, DMF, 80°C, 4 h.

The *cis* derivative **131** was prepared starting from *cis*-alcohol obtained by the Mitsunobu reaction of **126**, followed by the same reactions described for the transformation of **126** to **130**, in 36% overall yield. The compounds obtained were then tested for their antiviral activity against herpes simplex virus type 2 and human immunodediecncy virus. Only *cis*-cyclopentane **131** showed moderate activity ($IC_{50} = 20$ μg/mL) against HIV; the *trans*-isomer **130** was inactive.[39]

2.3.2. 1,3-Substituted Cyclopentane Derivatives. Reaction of 6-oxabicyclo[3.1.0]hex-2-ene **132** with 2-amino-6-chloropurine in the presence of a catalytic amount of tetrakis(triphenylphosphine)palladium (10 mmol%) gave the alcohol **133** in 91% yield. This compound was transformed into the methoxy derivative **134** in 86% yield by reaction with K_2CO_3 and then converted into the diester **135**, in 25% yield, by reaction with NaH followed by addition of diethyl(*p*-tolylsulfonyloxymethyl)phosphonate. Reaction of **135** with trimethylsilyl iodide, followed by hydrolysis with an aqueous solution of ammonium hydrogen carbonate, and hydrogenation, using palladium on carbon as a catalyst, gave the cyclopentane derivative **137** (90%) via compound **136** (25%).[39,40] The unsaturated derivative **136** displayed modest potency against HIV ($IC_{50} = 66$ μg/mL), while the corresponding saturated derivative **137** was inactive (Scheme 19).

The reaction of **132** with thymine gave two products, and the major product **138** (25% yield) was transformed into the phosphonate **141** according to a similar procedure (Scheme 20).

The saturated phosphonate **144** was obtained in moderate yield not from hydrogenation of **141** but starting from **138** and according to the sequence of reactions reported in Scheme 21.

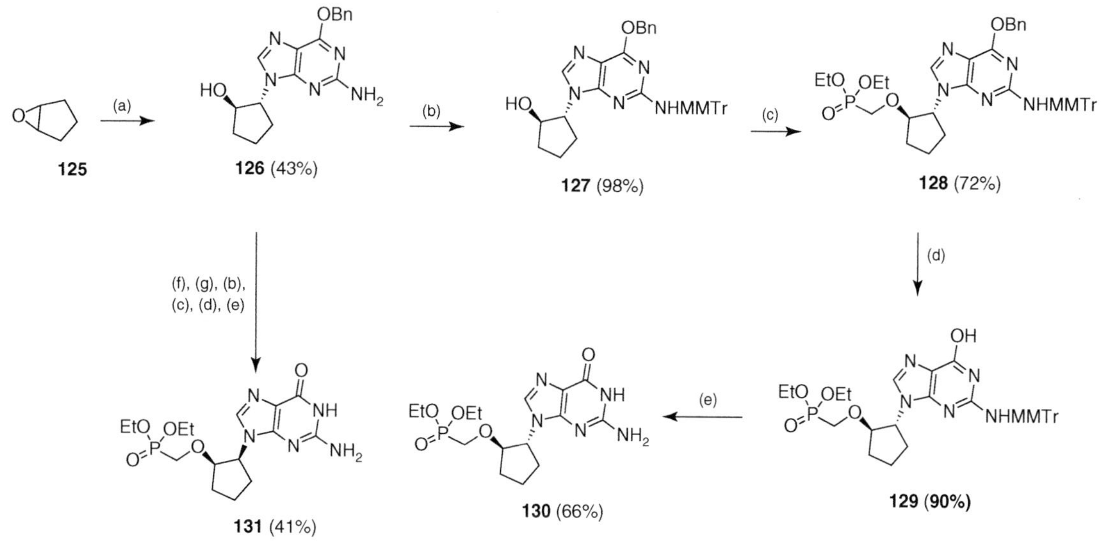

Scheme 18. *Reagents and conditions*: (a) 6-*O*-benzylguanine, LiH, DMF, 60°C; (b) Et$_3$N, (4-MeO-C$_6$H$_4$)Ph$_2$CCl, 4-dimethylaminopyridine, CH$_2$Cl$_2$, rt; (c) NaH, DMF, 80°C, (EtO)$_2$POCH$_2$OTs, rt; (d) 20% Pd(OH)$_2$-C, ethanol/cyclohexene, 1:1, then 80% AcOH, 60°C; (e) Me$_3$SiI, DMF, rt.

Bn = -CH$_2$Ph
MMTr = (4-MeO-C$_6$H$_4$)Ph$_2$C-

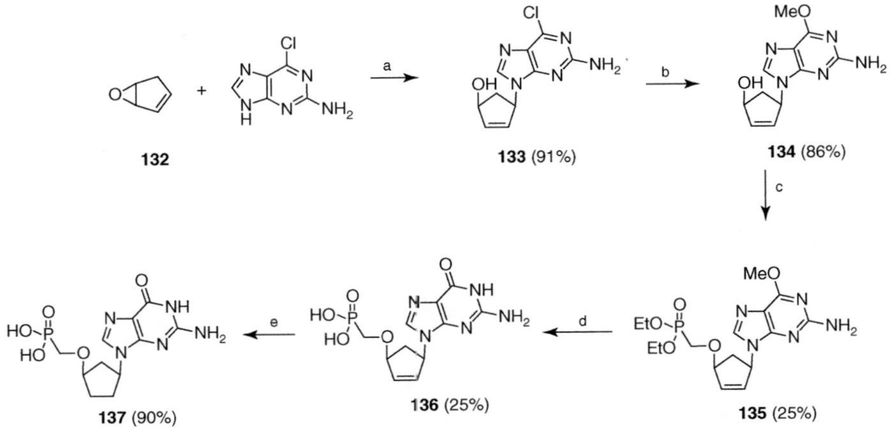

Scheme 19. *Reagents and conditions:* (a) (Ph$_3$P)$_4$Pd, DMSO-THF, 0°C to rt; (b) K$_2$CO$_3$ · MeOH, heat; (c) NaH, THF, 0.5 h, then *p*-MeC$_6$H$_4$SO$_3$CH$_2$PO(OEt)$_2$ in THF, rt; (d) Me$_3$SiI, DMF, rt, then NH$_4$HCO$_3$ · H$_2$O; (e) Pd-C, H$_2$O, H$_2$, rt, 6 h.

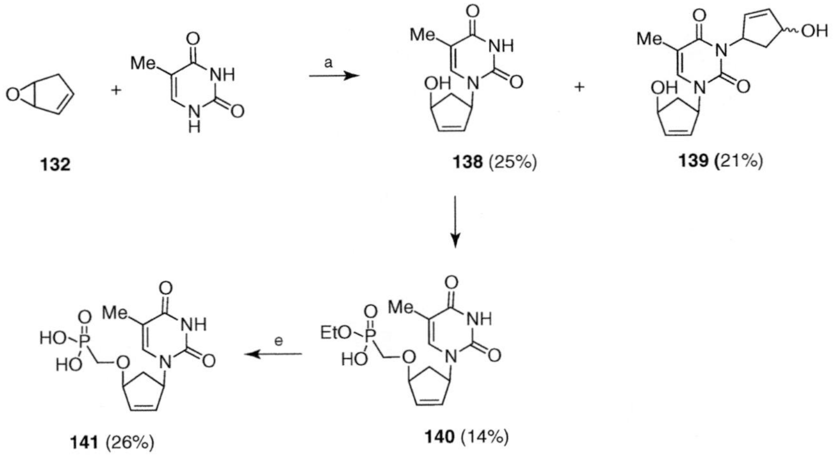

Scheme 20. *Reagents and conditions:* (a) (Ph$_3$P)$_4$Pd, DMSO-THF, 0°C to rt; (b) NaH, THF, 0.5 h, then *p*-MeC$_6$H$_4$SO$_3$CH$_2$PO(OEt)$_2$, in THF, rt; (d) Me$_3$SiBr, DMF, 0°C, then NH$_4$HCO$_3$ · H$_2$O.

It is interesting to note that the conversion of phosphonates **137** and **144** into the corresponding diphosphorylphosphonates **145** and **146** (Figure 4) affords potent inhibitors of human immunodeficiency virus reverse transcriptase (IC$_{50}$ = 0.06–0.11 µmol/dm^3).

To increase the lipophilic nature of the side chain, the same authors have prepared, as outlined in Scheme 22, the derivative **149** starting from **134**. Unfortunately, this compound showed no activity against the reverse transcriptase HIV,

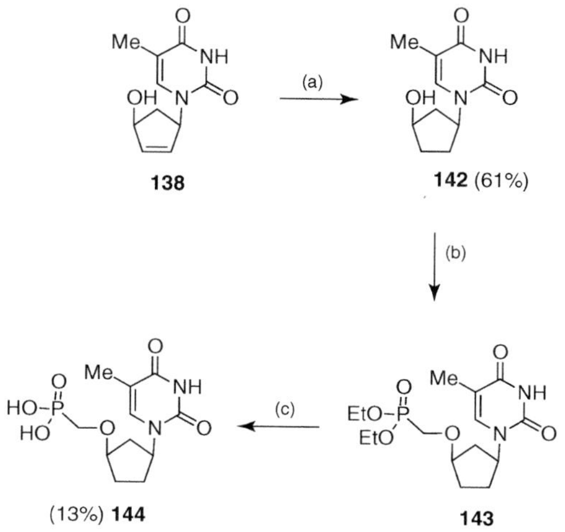

Scheme 21. *Reagents and conditions:* (a) TBDMSCl, imidazole, DMF, Pd-C, H$_2$, EtOH, rt, TBAF; (b) NaH, DMF, rt, then *p*-MeC$_6$H$_4$SO$_3$CH$_2$PO(OEt)$_2$ in THF, rt; (c) Me$_3$SiBr, DMF, 0°C to rt, then NH$_4$HCO$_3$ · H$_2$O, 0°C to rt.

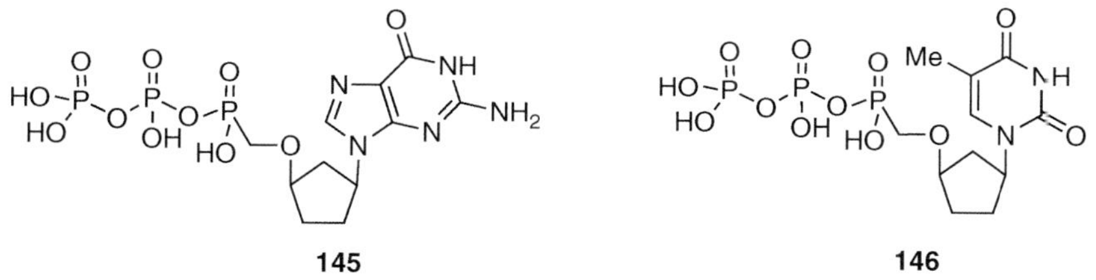

Figure 4.

showing that the greater length of the side chain is responsible for the lack of activity of the compound.

Phosphonated derivatives of carbocyclic 2′-deoxyguanosine (CDG, **154**) and 6-methylthio analogues **158** have been synthesized by Montgomery et al. to study their cytotoxic activity against different human cancer cell lines. These compounds were prepared through standard procedures (Schemes 23 and 24), involving as the key intermediate the (1α,2β,4β)-4-amino-2-(benzyloxy)cyclopentanol **150**. In particular, compound **150** was converted in seven steps into phosphonate **154** and in four steps into phosphonate **158**.[41]

A flexible synthesis of phosphonated (±)-cyclopentane phosphonic acid nucleoside analogues has been described by Pryde and co-workers.[42] The key step involves the ring opening of *trans*- or *cis*- epoxy alcohols **159** and **160**, with either a nucleoside base or a selenyl anion, to access the target-modified nucleosides **162**, **167**, **172**, and **173**. In particular, compound **159** was converted into

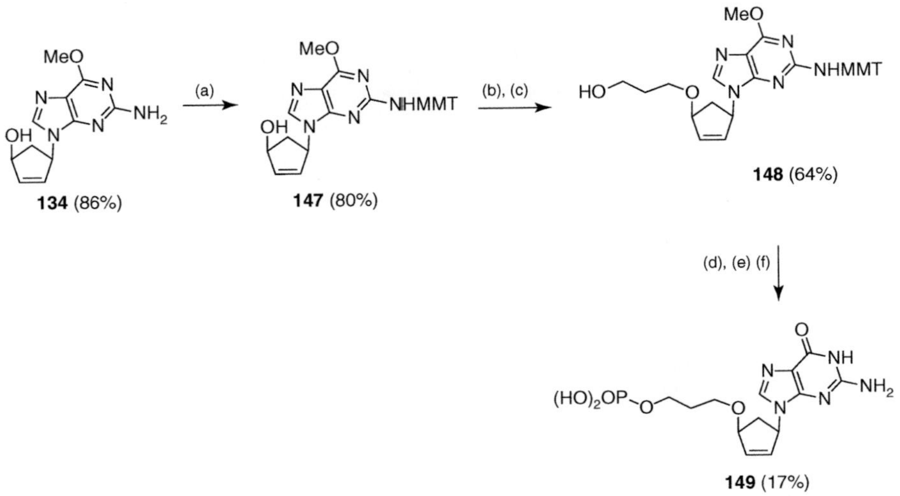

Scheme 22. *Reagents and conditions:* (a) MMTCl, Et$_3$N, DMAP, CH$_2$Cl, 0°C; (b) NaH, DMF, then Br(CH$_2$)$_3$OTBDMS; (c) TBAF, THF, rt; (d) NaH, THF, then (EtO)$_2$POCH$_2$OTs; (e) AcOH, heat; (f) Me$_3$SiI, DMF, rt, then NH$_4$HCO$_3$ · H$_2$O.

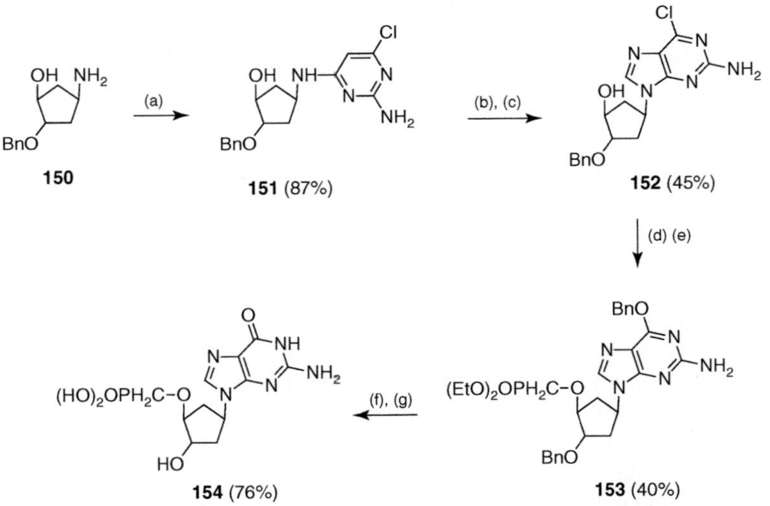

Scheme 23. *Reagents and conditions:* (a) 2-amino-4,6-dichloropyridine, N,N-diisopropylethylamine, BuOH, rt, 48 h; (b) 4-Chlorobenzendiazonium chloride, NaOAc, AcOH, H$_2$O, rt, 20 h; (c) EtOH, H$_2$O, AcOH, Zn, 90°C, 5 h, diethoxymethyl acetate, 80°C, 24 h; (d) BnOH, NaH, 20 h; (e) NaH, THF, then (EtO)$_2$POCH$_2$OTs, 18 h; (f) H$_2$Pd/C, rt, 20 h; (g) Me$_3$SiI, DMF, rt, then NH$_4$HCO$_3$ · H$_2$O.

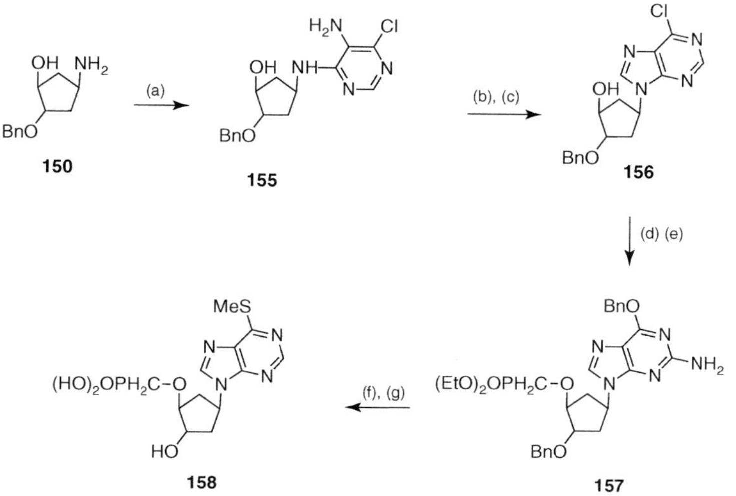

Scheme 24. *Reagents and conditions:* (a) 5-amino-4,6-dichloropyridine, N,N-diisopropylethylamine, BuOH, reflux, 22 h; (b) 4-chlorobenzendiazonium chloride, NaOAc, AcOH, H$_2$O, rt, 20 h; (c) EtOH, H$_2$O, AcOH, Zn, 90°C, 5 h, diethoxymethyl acetate, 80°C, 24 h; (d) BnOH, NaH, 20 h; (e) NaH, THF, then (EtO)$_2$POCH$_2$OTs, 18 h; (f) H$_2$Pd/C, rt, 20 h; (g) Me$_3$SiI, DMF, rt, then NH$_4$HCO$_3 \cdot$ H$_2$O.

the corresponding iodide and then into the diisopropyl phosphonate ester **161** via an Arbuzov reaction. Reaction of **161** with adenine followed by TMSBr-mediated deprotection gave the corresponding phosphonic acid **162** in 16% overall yield (Scheme 25).

The carbocyclic derivatives **167**, **168**, **172**, and **173** were, instead, prepared starting from *cis*-epoxy alcohol **160** (Schemes 25 and 26). Thus, **160** was reacted with NaSePh, generated *in situ* with diphenyldiselenide and NaBH$_4$; the intermediate seleno-alcohol thus obtained was protected as its acetate ester, and then converted to allyl acetate **164**, by oxidation–elimination of selenide function and TBAF treatment. The transformation of **164** in **165** was performed readily via tosylation and an Arbuzov reaction. The phosphonate ester **165** was converted in the cyclopentene phosphonic acid **167** via **166** by a nucleosidation reaction with uracil activated with tetrakis-(triphenylphosphine) palladium, followed by hydrolysis with TMSBr (Scheme 25).

Derivatives **168**, **172**, and **173** were synthesized involving as the starting compound the phosphonate ester **166**. In particular, **166** was hydrogenated under Pd catalysis and deprotected to give the analogous cyclopentane **168** (Scheme 26). In the meantime, **166** was used to obtain the derivatives **172** and **173**. Thus, **166** was converted to *cis*-cyclic osmate **169** by reaction with OsO$_4$ and NMO, which in turn is converted in a mixture of *cis*- and *trans*-derivatives **170** and **171**. Deprotection of each isomer separately gave the final phosphonic acids **172** and **173** (Scheme 26).

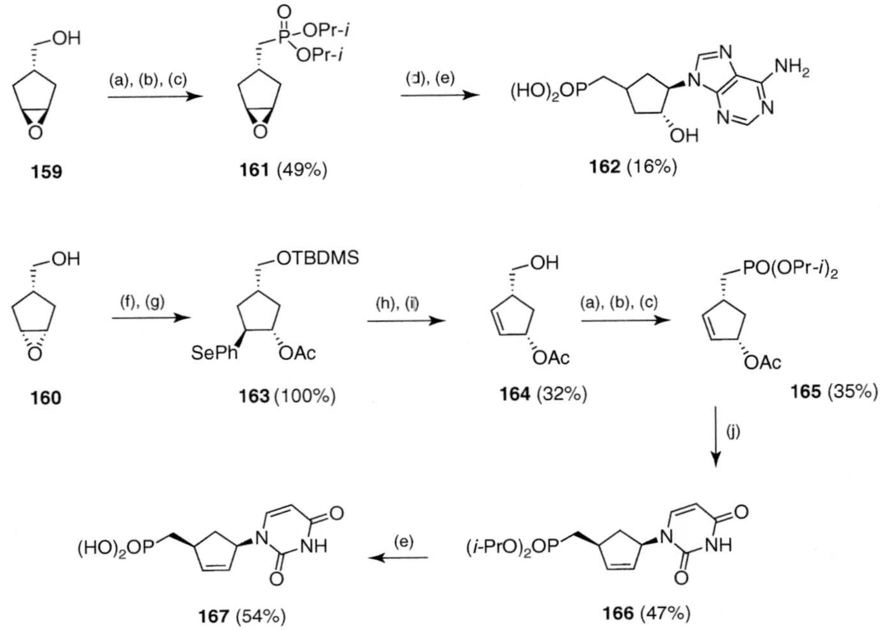

Scheme 25. *Reagents and conditions:* (a) TsCl, Et$_3$N, DMPA, CH$_2$Cl$_2$, 0°C, 16–48 h; (b) NaI, butanone, 5 h, reflux; (c) P(OPr-*i*), 120°C, 30 h; (d) adenine, Cs$_2$CO$_3$, DMF, Kryptofix, 120°C, 9 days; (e) TMSBr, MeCN, rt, 16 h; (f) TBDMSCl, imidazole, CH$_2$Cl$_2$, rt, 2 h; (g) (PhSe)$_2$, NaBH$_4$–EtOh, 0°C, 1 h, rt, 16 h; (h) H$_2$O$_2$, Hunig's base, 1-butanol, CH$_2$Cl$_2$, 50°C, 1 h; (i) TBAF, rt, 1 h; (j) uracil, Pd(Ph$_3$P)$_4$, DMSO, LiH, 50°C, 0.5 h.

Recently, different C-branched carbocyclic nucleoside phopshonates **177** and **178** bearing a 4′-methyl, a 4′-vinyl, and a 4′-ethynyl group have been synthesized.[43] The synthetic scheme starts from the commercially available desymmetrized cyclopentendiol **174** (Scheme 27). This compound was first reacted with 6-chloropurine under Mitsunobu conditions to provide **175**, deacetylated, and oxidized with Dess–Martin reagent to enone intermediate **176**. **176** was alkylated with cerium reagents and converted into compounds **177** and **178** by standard reactions. Furthermore, compound **178** was transformed in its diphosphonate **179** and, finally, in **180** through Lindlar hydrogenation. The biological assays performed on these compounds have shown that compound **178** is a potent inhibitor of HIV reverse transcriptase with an IC$_{50}$ of 0.10 µM.

In the same context, Hong et al. reported the synthesis of novel 4′-ethyl-5′-norcarbocyclic adenosine phosphonic acid analogues **186**, **187**, **188**, and **190**.[44] The synthetic route involves preparation of the intermediate (1*SR*,4*SR*)-4-ethyl-4-(4-methoxybenzyloxy)cyclopent-2-enol **185b**, obtained from propionaldehyde **181** using reiterative Grignard additions and ring-closing methatesis as key reactions (Scheme 28).

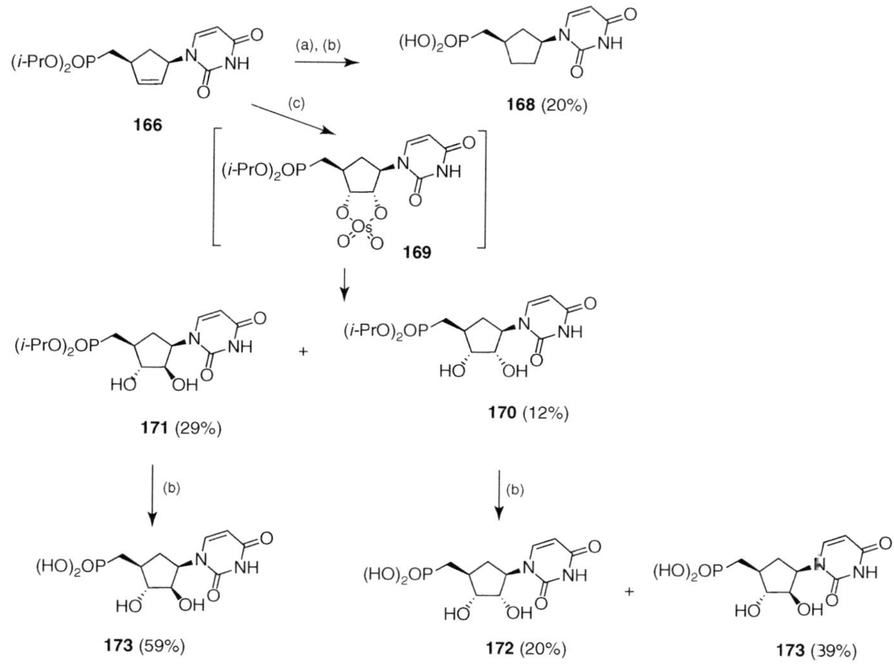

Scheme 26. *Reagents and conditions:* (a) TsCl, Et$_3$N, DMPA, CH$_2$Cl$_2$, 0°C, 16–48 h; (b) NaI, butanone, 5 h, reflux; (c) P(OPr-*i*), 120°C, 30 h; (d) adenine, Cs$_2$CO$_3$, DMF, Kryptofix, 120°C, 9 days; (e) TMSBr, MeCN, rt, 16 h; (f) TBDMSCl, imidazole, CH$_2$Cl$_2$, rt, 2 h; (g) (PhSe)$_2$, NaBH$_4$, EtOh, 0°C, 1 h, rt, 16 h; (h) H$_2$O$_2$, Hunig's base, 1-butanol, CH$_2$Cl$_2$, 50°C, 1 h; (i) TABF, rt, 1 h; (j) uracil, Pd(Ph$_3$P$_4$), DMSO, LiH, 50°C, 0.5 h.

Compound **185b** was then converted into the target phosphonate nucleosides **186**, **187**, **188**, and **190** according to Scheme 29, using simple methods already reported in the literature.

Compounds **186** and **187** have been shown to possess moderate antiviral activity against HIV-1 with EC$_{50}$ values of 55 and 21 μM, respectively, while the other nucleotide analogs **188** and **190** did not show any activity up to 100 μM.

6′-Fluoro-6′-methyl-5′-nor-adenosine nucleoside phosphonic acid **195** and its SATE prodrug **196** have been prepared from the key fluorinated alcohol **192**, obtained through a selective ring opening of epoxide **191**. Coupling of **192** with N^6-bis-Boc-adenine under a Mitsunobu reaction, followed by phosphonation and deprotection, gave the nucleoside phosphonic acid **195**, from which the final *t*-Bu-SATE prodrug **196** was obtained by reaction with AS-2-hydroxyethyl-2,2-dimethylpropanethioate in the presence of 1-(2-mesitylenesulfonyl)-3-nitro-1*H*,1,2,4-triazole (Scheme 30).[45]

The phosphonic nucleoside analogue **195** and its prodrug **196** showed antiviral activity against HIV-1 with EC$_{50}$ values of 62 and 16.7 μM, respectively.

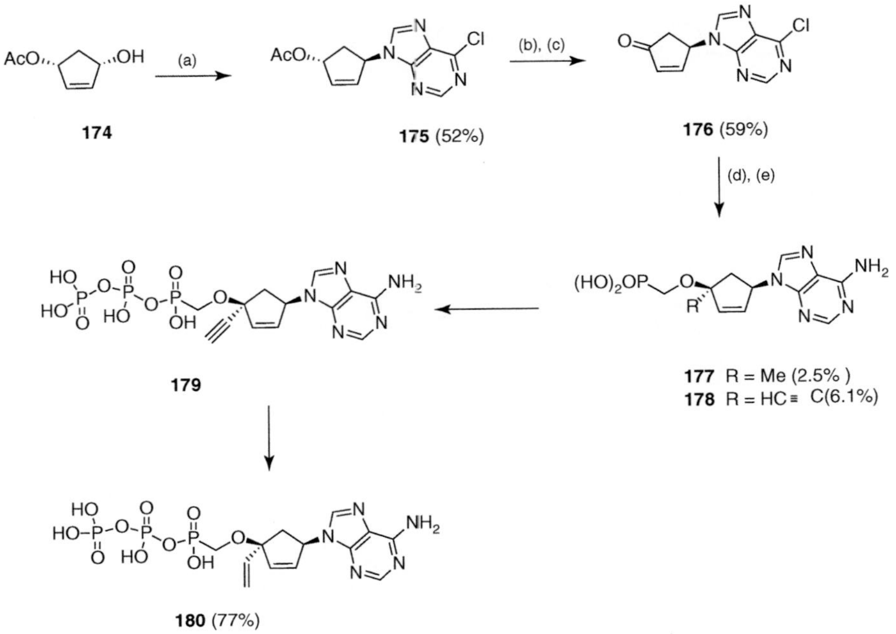

Scheme 27. *Reagents and conditions:* (a) 6-chloropurine, Ph₃P, DIAD, dioxane; (b) MeOH/NH₃ · H₂O 8:1:1; (c) Dess–Martin reagent; (d) CeCl₃, MeMgBr, or TMSC≡CMgBr, THF, −78°C; (e) diisopropyl bromomethylphosphonate, LiO/t-Bu, THF, 50°C; (f) NH₃, MeOH, 50°C; (g) TMSBr, 2,6-lutidine, DMF, 50°C, then MeOH, NH₄OH; (h) literature conditions; (j) Lindlar catalyst, quinoline, H₂, H₂O.

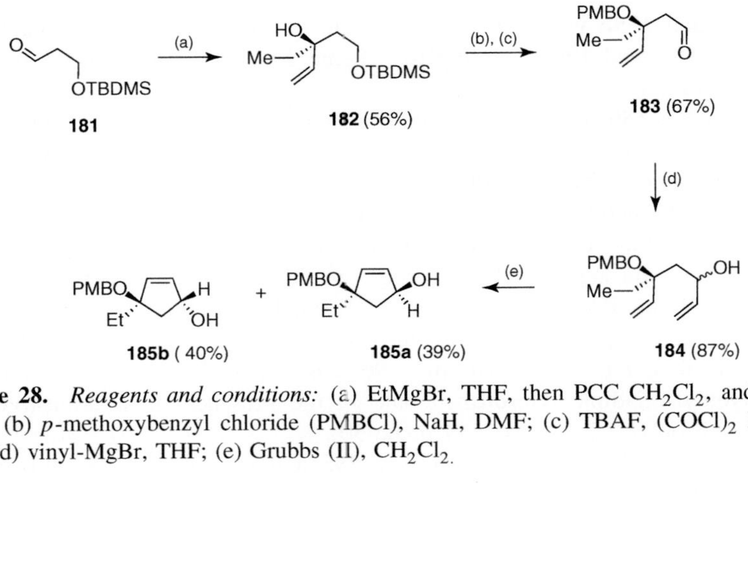

Scheme 28. *Reagents and conditions:* (a) EtMgBr, THF, then PCC CH₂Cl₂, and vinyl-MgBr; (b) *p*-methoxybenzyl chloride (PMBCl), NaH, DMF; (c) TBAF, (COCl)₂ DMSO, TEA; (d) vinyl-MgBr, THF; (e) Grubbs (II), CH₂Cl₂.

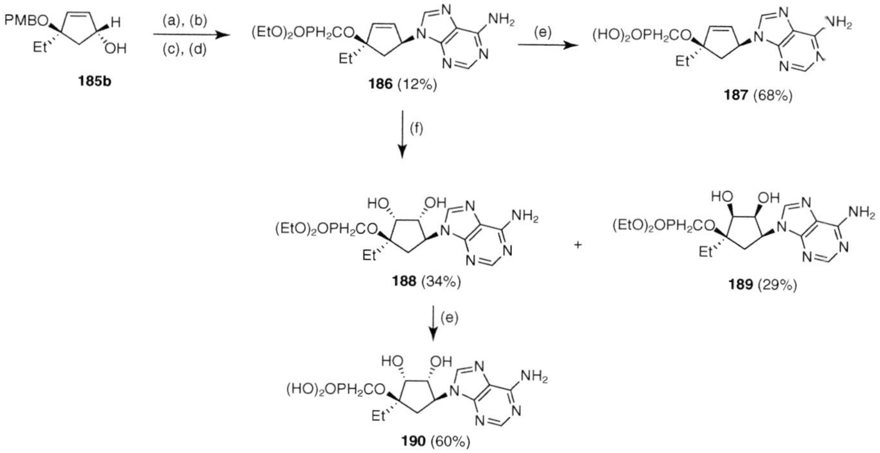

Scheme 29. *Reagents and conditions:* (a) 6-chloropurine, DIAD, THF; (b) DDQ, CH$_2$Cl$_2$/H$_2$O, RT; (c) (EtO)$_2$POCH$_2$OTf, LiO/t-Bu; (d) NH$_3$/MeOH, 70°C; (e) TMSBr, 2,6-lutidine, MeCN; (f) OsO$_4$, NMO, acetone/t-BuOH/H$_2$O.

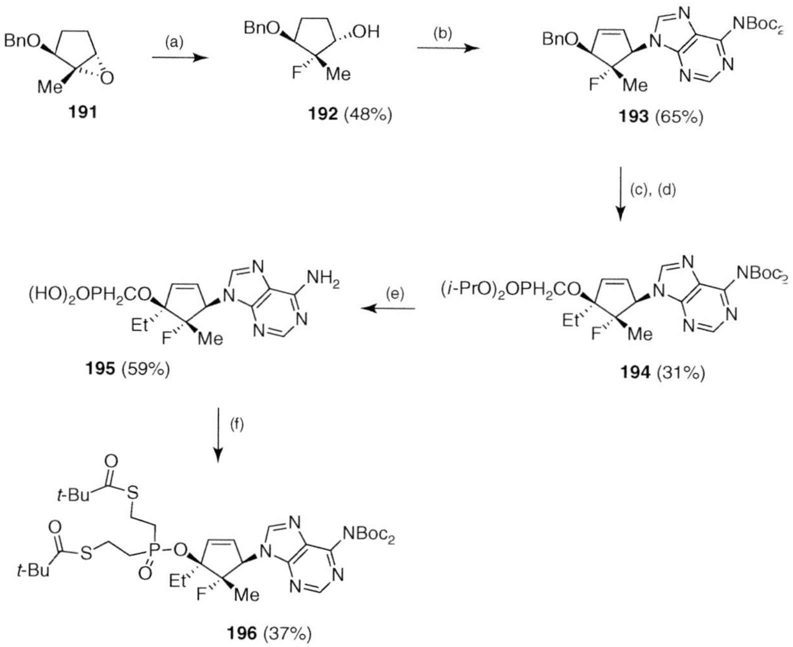

Scheme 30. *Reagents and conditions:* (a) 47% HF, (NH$_4$)$_2$SiF$_8$, CsF; (b) Ph$_3$F, N^6-bis-BOC-adenine, DIAD; (c) Pa(OH)$_2$, cyclohexene, MeOH, heat; (d) (*i*-PrO)$_2$POCH$_2$Br, LiO/t-Bu, LiI, DMF, 60°C, 3 h; (e) TMSBr, MeCN, 60°C, 12 h; (f) Bu$_3$N, MeOH, *S*-2-hydroxyethyl 2,2-dimethylpropanethioate, 1-(2-mesitylenesulfonyl)-3-nitro-1*H*-1,2,4-triazole, rt, 6 h.

The increased anti-HIV activity for the neutral phosphodiester **196** is the result of increased cellular uptake followed by intracellular release of the parent phosphonic acid.

2.4. Acyclic Nucleoside Phosphonates

Acyclic nucleoside phosphonates (ANPs) constitute a class of modified nucleotide analogues where the sugar ring has been replaced by a functionalized acyclic system linking the nucleobase and the phosphonic acid group. The presence of a phosphonate (P–C) instead of a phosphoester (P–O) linkage confers on these compounds metabolic stability and resistance to phosphatases. Following two subsequent phosphorylation steps, ANPs are converted into the corresponding triphosphates and interfere with nucleic acid biosynthesis, acting as DNA chain terminators. The various types of acyclic nucleoside phosphonates reported in this chapter are shown in Figure 5.

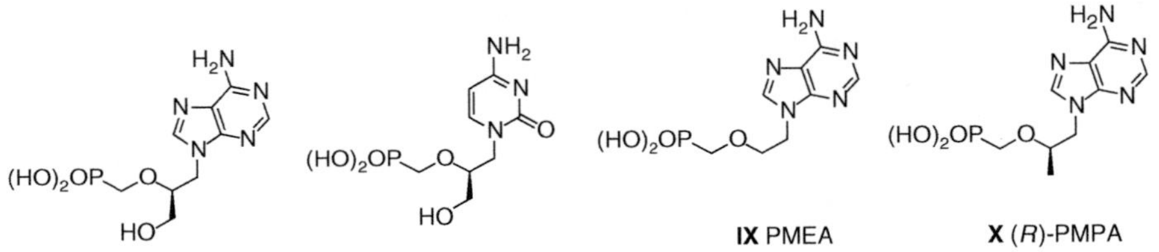

VII (S)-HPMPA   **VIII** (S)-HPMPC   **IX** PMEA   **X** (R)-PMPA

Figure 5. Acyclic nucleoside phosphonates.

Compounds of Structure VII. The first nucleoside phosphonate that opened the era of acyclic nucleotide analogues was the (S)-9-(3-hydroxy-2-phosphonylmethoxypropyl)adenine (S)-HPMPA **206**.[46–48] This compound exhibits strong activity against a variety of DNA viruses, including HSV-1, HSV-2, VZV, CMV, poxviruses, adenoviruses, and HBV. A recent synthesis of (S)-HPMPA **206** and its enantiomer has been described by Zakirova et al., starting from L(+)- or D(−)-ribose, respectively.[49] In particular, L(+)-ribose **197** was methylated to α,β-methylribofuranoside **198**, converted to 5-tosyl(methyl)riboside **199**, and coupled with adenine under anhydrous conditions in the presence of NaH, to afford the L-9-(α,β-1-methyl-5-ribofuranosyl)adenine **200** as a key intermediate. This product was oxidized to dialdehyde **201** with sodium periodate, reduced with NaBH$_4$ to give the optically active (S)-9-(3-hydroxy-2-oxy-[(R,S)-1-methoxy-2-hydroxyethyl)propyl]adenine **202**, which by treatment with 60% formic acid afforded the diol **203**. The reaction of **203** with iodomethylphosphonic acid, in the presence of DDC, resulted in the formation of **204**, which underwent an intramolecular cyclization to **205**, from which **206** has been isolated, when the reaction mixture was treated with water (Scheme 31). The corresponding enantiomer (R)-HPMPA was obtained in a similar way from D(-)-ribose.

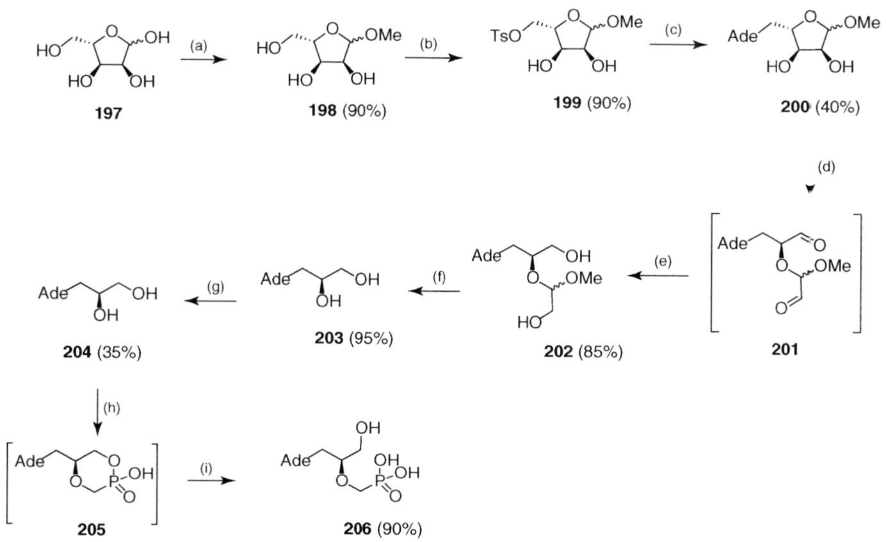

Scheme 31. *Reagents and conditions:* (a) AcCl/MeOH, RT, 16 h; (b) TsCl/Py, 4°C, 16 h; (c) adenine, NaH, DMF, 8°C, 5 days; (d) NaIO$_4$, 4°C, 1 h; (e) NaBH$_4$, 4°C, 3 h; (f) H$^+$, 6 h; (g) ICH$_2$PO(OH)$_2$/DCC/Py, 16 h; (h) NaH/DMF, 6 h; (i) galcial AcOH, 3 h.

Alkoxyalkyl esters of (S)-HPMPA, as amphiphilic prodrugs, have been prepared and evaluated biologically. The activity against CMV and adenovirus is increased markedly when **206** is transformed into the hexadecyloxypropyl (HDP) **207** or octadecyloxyethyl (ODE) **208** esters. It is noteworthy that while (S)-HPMPA is virtually inactive, HDP and ODE esters are active against HIV-1 in the nanomolar range, and also show activity against HCV replication at a 50% effective concentration of about 1 μM.[21e] The synthesis of ODE-(S)-HPMPA and HDP-(S)-HPMPA has been described by Beadle and Hostetler.[50] and is reported in Scheme 32.

Experimental Procedures

General Procedure for the Synthesis of Alkoxyalkyl Toluenesulfonyloxymethylphosphonates 210 and 211. Bromotrimethylsilane (27 g, 175 mmol) was added to a solution of diethyl toluenesulfonyloxymethylphosphonate **209** (9.5 g, 29.5 mmol) in dichloromethane anhydrous (150 mL). The mixture was stirred at room temperature under a nitrogen atmosphere for 18 h and then concentrated under vacuum to remove the solvent and the excess TMSBr. To the mixture redissolved in dichloromethane (150 mL) and cooled at 0°C, N,N-DMF (0.5 mL) and a solution of oxalyl chloride (22 g, 175 mmol) in CH$_2$Cl$_2$ (50 mL), dropwise, over 30 min, were added. After stirring for additional 5 h, the mixture was evaporated to an oil which was then redissolved in Et$_2$O (100 mL). A solution of 3-hexadecyloxy-1-propanol or 2-octadecyloxy-1-ethanol (21.5 mmol) and pyridine (10 mL) in Et$_2$O

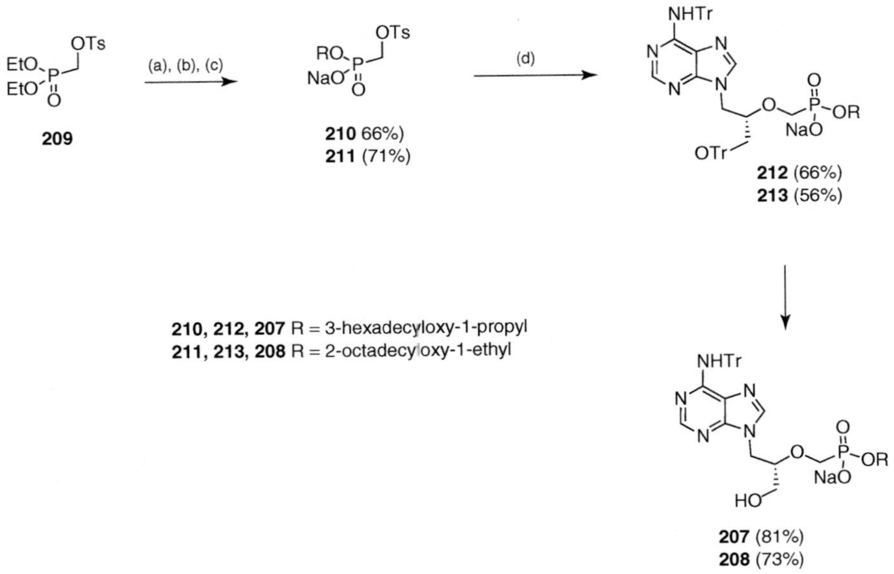

Scheme 32. *Reagents and conditions:* (a) TMSBr, CH$_2$Cl$_2$; (b) oxalyl chloride, *N*,*N*-DMF, (0.5 mL), 0°C, 5.5 h; (c) 3-hexadecyloxy-1-propanol or 2-octadecyloxy-1-ethanol, Et$_2$O, Py, 3 h, aq. NaHCO$_3$, 1 H; (d) (*S*)-9-[3-trityloxy-2-hydroxypropyl]-*N*^{6}-trityladenine, NaH, Et$_3$N (solvent), 50°C, 12 h; (e) 80% aq. AcOH, 60°C, 1 h.

210, 212, 207 R = 3-hexadecyloxy-1-propyl
211, 213, 208 R = 2-octadecyloxy-1-ethyl

(50 mL) was added, and stirring was continued for 3 h. The reaction mixture was added to cold, saturated NaHCO$_3$ and stirred vigorously for 1 h. After the completion of hydrolysis, the organic layer was separated, dried over MgSO$_4$, and evaporated under vacuum to give the crude esters, which were purified by flash chromatography (elution solvent: 15% EtOH/CH$_2$Cl$_2$).

3-(Hexadecyloxy)propyl Toluenesulfonyloxymethylphosphonate Sodium Salt (210).
^{1}H-NMR (CDCl$_3$): δ 0.88 (t,3H), 1.25 (br s, 26H), 1.54 (m, 2H), 1.83 (qt, 2H), 2.46 (s, 3H), 3.38 (t, 2H), 3.47(t, 2H), 3.91 (dt, 2H), 4.02 (d, 2H), 7.37 (d, 2H), 7.80 (d, 2H); MS (ES) *m/z* 571.32 [M + Na]$^+$.

2-(Octadecyloxy)ethyl Toluenesulfonyloxymethylphosphonate Sodium Salt (211).
^{1}H-NMR (CDCl$_3$): δ 0.88 (t, 3H), 1.26 (br s, 30H), 1.54 (m, 2H), 2.60(s, 3H), 3.51 (t, 2H), 3.65 (t, 2H), 4.03 (dt, 2H), 4.16 (d, 2H), 7.35 (d, 2H), 7.78 (d, 2H); MS (ES) *m/z* 585 [M + Na]$^+$.

General Procedure for Alkylation of (S)-9-[3-Trityloxy-2-hydroxypropyl]-N^6-trityladenine with 210 and 211. Sodium hydride (24 mg, 1.0 mmol) was added to a stirred solution of (*S*)-9-[3-trityloxy-2-hydroxypropyl]-*N*^{6}-trityladenine (640 mg, 0.62 mmol) in dry triethylamine (10 mL). After 15 min, the appropriate alkoxyalkyl toluenesulfonyloxymethylphosphonate (210, 211 0.65 mmol) was added, and the reaction mixture was heated at 50°C and kept there overnight. After cooling, the mixture was quenched with saturated NaCl/H$_2$O (50 mL) and extracted with ethyl acetate (3 × 15 mL). The organic layer, after drying over MgSO$_4$, was evaporated. The residue was purified by flash chromatography, where the products were eluted with 10% EtOH/CH$_2$Cl$_2$.

3-(Hexadecyloxy)propyl-(S)-9-[3-trityloxy-2-(phosphonomethoxy)propyl]-N^6-trityladenine (212). ^{1}H-NMR (CDCl$_3$): δ 0.88 (t, 3H), 1.33 (br s, 26H), 1.46 (m, 2H), 1.77 (qt, 2H), 3.37–3.8 (m, 9H), 3.86 (m, 2H), 4.03 (m, 2H), 7.2–7.5 (br m, 30H), 7.82 (s, 1H), 8.20 (s, 1H). MS (ES) m/z 1071 [M + H]$^+$.

2-(Octadecyloxy)ethyl-(S)-9-[3-trityloxy-2-(phosphonomethoxy)propyl]-N^6-trityladenine (213). ^{1}H-NMR (CDCl$_3$): δ 0.90 (t, 3H), 1.33 (br s, 30H), 1.41 (m, 2H), 3.3–3.7 (m, 7H), 3.8 (m, 2H), 4.21 (m, 2H), 7.2–7.6 (m, 30H), 7.93 (s, 1H), 8.10 (s, 1H). MS (ES) m/z 1085 [M + H]$^+$.

General Procedure for Deprotection and Isolation of (S)HPMPA Alkoxyalkyl Esters 207 and 208. Derivatives **212** and **213** were suspended in 80% aqueous acetic acid (20 mL) and heated to 60°C for 1 h. After cooling and evaporation of the solvent, the crude mixtures were purified by flash chromatography. Elution with 30% MeOH/CH$_2$Cl$_2$ provided compounds **207** and **208** as white powdery solids.

3-(Hexadecyloxy)propyl-(S)-9-[3-hydroxy-2-(phosphonomethoxy)propyl]adenine, Sodium Salt (HDP-HPMPA) (207). ^{1}H-NMR (CD$_3$OD): δ 0.84 (t, 3H), 1.20 (br s, 18H), 1.23 (m, 2H), 1.63 (qt, 2H), 3.20–3.55 (m, 11H), 3.67 (m, 2H), 4.05–4.34 (pair dd, 4H), 8.13 (s, 1H), 8.17 (s, 1H). ^{31}P-NMR δ 15.18. MS (ES) m/z 586 [M + H]$^+$.

2-(Octadecyloxy)ethyl-(S)-9-[3-hydroxy-2-(phosphonomethoxy)propyl]adenine, Sodium Salt (ODE-HPMPA) (208). ^{1}H-NMR (CD$_3$OD): δ 0.85 (t, 3H), 1.22 (br s, 30H), 1.43 (m, 2H), 1.89 (t, 2H), 3.35–3.80 (m, 11H), 4.00–4.15 (m, 2H), 8.11 (s, 1H), 8.15 (s, 1H). ^{31}P-NMR δ 15.06. MS (ES) m/z 600 [M + H]$^+$.

(S)-1-(3-Hydroxy-2-phosphonylmethoxypropyl)cytosine **223** (HPMPC, cidofovir) was first reported in 1987.[51] A valuable synthetic route that allows the preparation of multigram quantities of this compound has been described by Bronson et al.[52] (Scheme 33). The synthetic scheme develops in eight steps starting from (R)-2,3-O-isopropylideneglycerol **214**. The key step of the synthetic route is the coupling reaction of cytosine with (R)-3-O-benzyl-2-O-[diethylphosphonyl)methyl]-1-O-(methylsulfonyl)glycerol **217**.

The antiviral activity spectrum of cidofovir is similar to that of HPMPA; the compound is active against virtually all DNA viruses, including polyoma-, papilloma-, adeno-, herpes-, and poxviruses.[2b] Similar to (S)-HPMPA, its adenine counterpart, alkoxyalkyl esters of cidofovir have been designed as prodrugs, to increase the oral bioavailability and reduce the toxicity of cidofovir. The ODE and HDP derivatives **228** and **229** are the most interesting compounds; in particular,

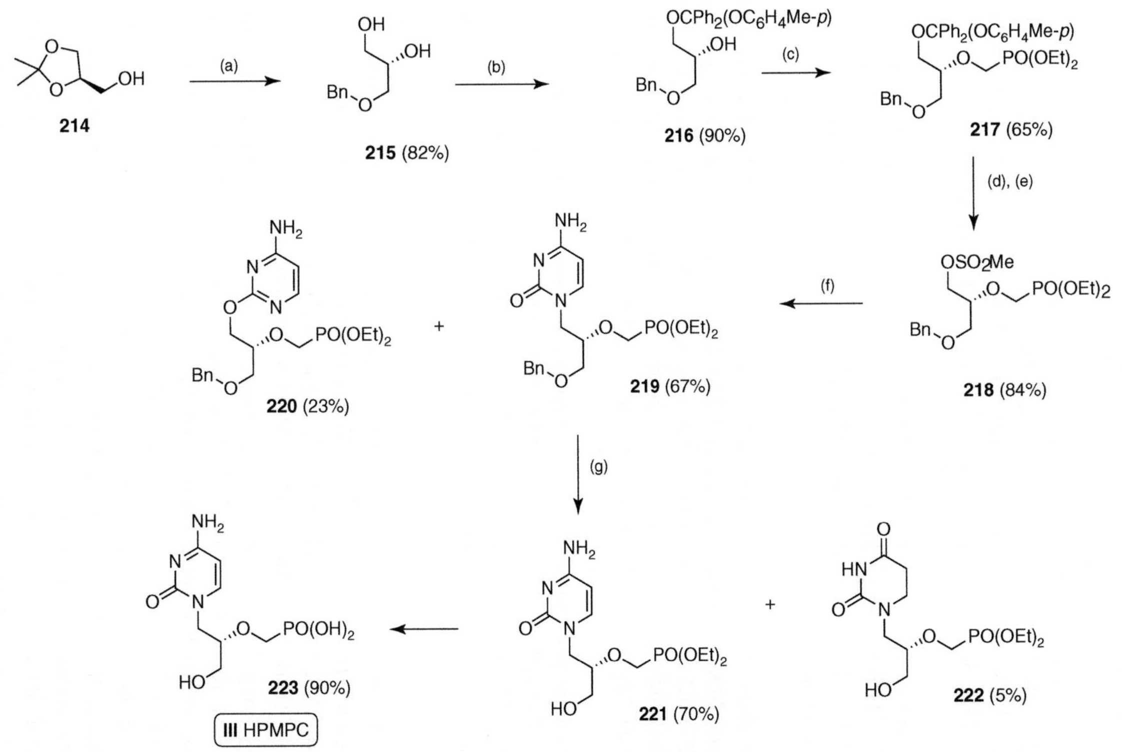

Scheme 33. *Reagents and conditions*: (a) BnBr, benzyltriethylammonium bromide, aq. NaOH, 90–95°C, 15 h, then aq. H$_2$SO$_4$, 90°C, 5 h; (b) (*p*-methoxyphenyl)diphenylmethyl chloride, DMAP, Et$_3$N, 20 min at 0°C, then 16 h at rt; (c) NaH, diethyl[(tosyloxy)methyl]phosphonate, 0°C, then 14 h at rt; (d) aq. AcOH at 100°C, 20 min; (e) Et$_3$N, MeSO$_2$Cl at 0°C, then 16 h at rt; (f) cytosine, CsCO$_3$, 90°C, 2.5 h; (g) 20% Pd(OH)$_2$/C, H$_2$, cyclohexene/EtOH; (h) TMSBr, MeCN, 14 h at rt.

HDP (S)-HPMPC (CMX001) is currently being developed for use in prophylactic and preemptive therapy of dsDNA viral infections.[53] The synthetic scheme for these prodrugs starts from the cyclic form of cidofovir **224**, which undergoes a nucleophilic substitution by the corresponding haloderivative (Scheme 34).[54]

225 n = 2, m = 17 (33%)
226 n = 3, m = 15 (31%)
227 n = 3, m = 17 (41%)

228 n = 2, m = 17 (41%)
229 n = 3, m = 15 (31%)
230 n = 3, m = 17 (72%)

231 R = H
232 R = Et

233 (S)-HPMP-5-azaC

234 (S)-HPMP-5-azaC

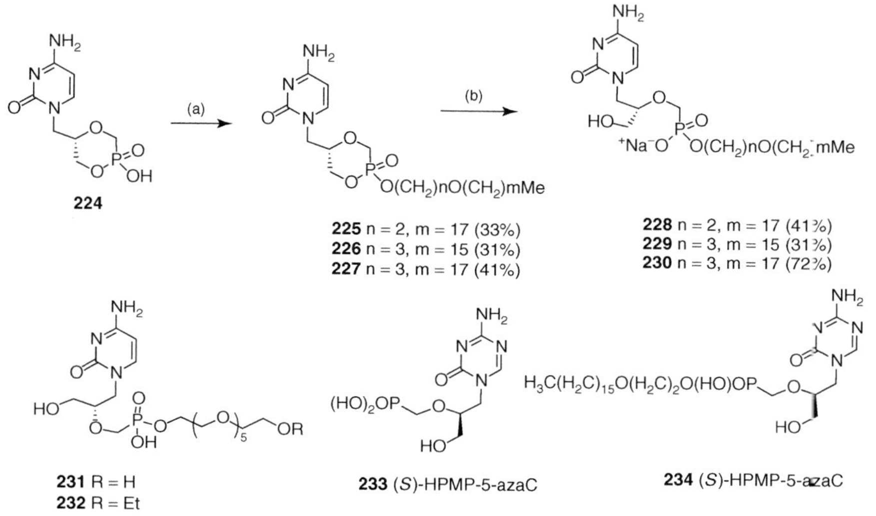

Scheme 34. *Reagents and conditions:* (a) ODE-br, ODP-br, or HDP-Br, N,N-DMF, 80°C, 6 h; (b) 0.5 M NaOH, 1.5 h, AcOH.

Two new prodrugs of (S)-HPMPC, **231** and **232**, bearing an hexaethylene glycol promoiety, have recently been synthesized.[54] **231** and **232** inhibit the replication of different herpes viruses and poxvirus vaccinia virus with an activity equivalent to that of cidofovir. A new acyclic nucleoside, the (S)-HPMP-5-aza-C **233**, originates from replacement of the pyrimidine ring in (S)-HPMPC by a triazine system. This compound shows activity against DNA viruses comparable to that of the parent compound.[2d] The insertion of an alkoxyalkyl group, as in its hexadecyloxyethyl (HDE) ester **234**, leads to enhancement of the antiviral activity against infections sensitive to cidofovir, in particular CMV, HSV, HPV, and adeno- and poxvirus infections.[2e]

Compounds of Structure IX. The antiretroviral properties of 9-(2-phosphonylmethoxyethyl)adenine **237** (PMEA, adefovir) were first reported by De Clercq et al.[47]; the anti-HIV activity has been documented further by Pawels et al.[55] Two synthetic routes, both based on the use of dialkyl *p*-toluenesulfonyloxymethanephosphonates have been reported.[56,57] Schultze et al. have developed an efficient, kilogram-scale synthesis of PMEA, according to an alkoxide-alkylation process starting from 2-(6-amino-9*H*-purin-9-yl)ethanol **233** (Scheme 35).[58]

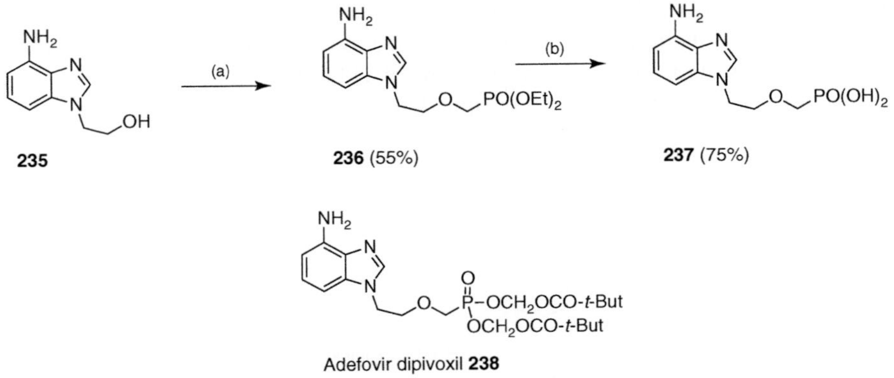

Scheme 35. *Reagents and conditions:* (a) OLi/*t*-Bu, diethyl *p*-toluenesulfonyloxy-methanephosphonate, 35°C, 1 h, aq. AcOH; (b) TMSBr, MeCN.

The low oral bioavailability of PMEA prompted the synthesis of an oral prodrug, the bis(pivaloyloxy-methyl)ester **240**,[59,60] which has been licensed successfully at a dose of 10 mg per day for the treatment of HBV.[61] In the same context, a set of amphiphilic prodrugs of PMEA have been prepared and evaluated in vitro (Scheme 36).[54] The phosphonate group was masked with a hexaethylene glycol unit or hydroxylated decyl or decyloxyethyl chain. A loss in the antiviral activities of hexaethylene glycol esters and hydroxylated decyl or decyloxyethyl esters has been observed, while the (5-methyl-2-oxo-1,3-dioxolen-4-yl)methyl ester of PMEA **242** showed significant activity against HIV and herpesviruses.

Two novel analogues of adefovir have been reported[62]: These compounds are characterized by the presence of a carbinol moiety that replaces the oxygen atom of the phosphonomethoxyethyl chain. Neither (*R*)- nor (*S*)-β-hydroxyphosphonate derivatives (Figure 6) revealed significant antiviral activity against a wide variety of DNA and RNA viruses.

By applying the strategy used for the synthesis of cidofovir, Bronson et al. have prepared **246**, PMEC, the cytosine analogue of adefovir, starting from 2-[(diethyoxyphosphoryl)methoxy]ethyl methanesulfonate **243** (Scheme 37).[52] PMEC was not active against any of the virus strains tested.

Compounds of Structure X. In 1991, De Clerq and Holý described the 9-(2*R*, *S*)-3-fluoro-2-phosphonylmethoxypropyl derivatives of adenine and 2,6-diaminopurine (FPMA), **247** and (FPMPDAP) **248** as potent and selective antiretroviral agents.[63] Later, in 1993, they reported the unfluorinated derivatives PMPA **249** and PMPDAP **250**, and showed that the anti-HIV activity of this class of compounds resided in the (*R*)-enantiomers (Figure 7).[64]

Although (*R*)-PMPDAP is about 10-fold more potent than (*R*)-MPA against HIV and HBV,[65] only the latter (the well-known tenofovir) has been selected as a drug and actually plays a key role in the control of HIV infections.

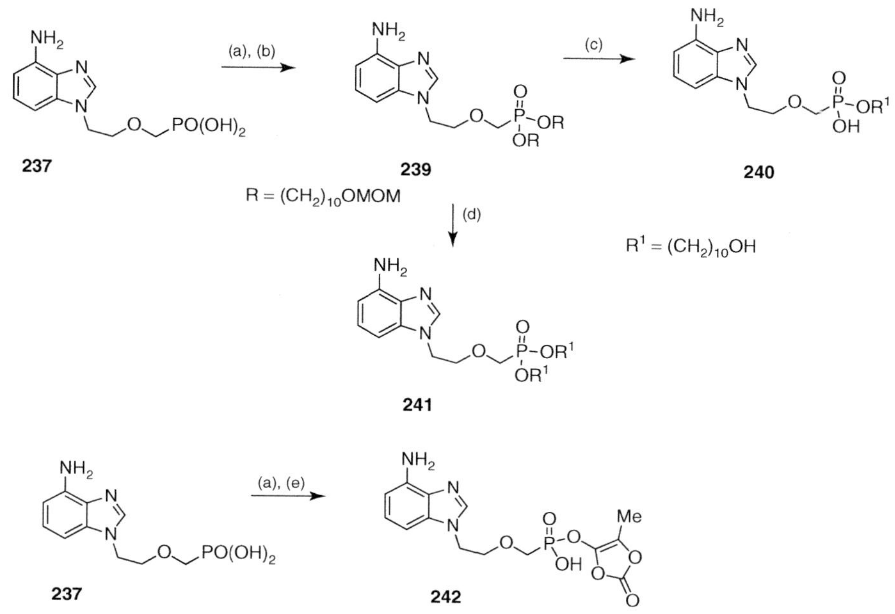

Scheme 36. *Reagents and conditions:* (a) Bu$_4$NOH, MEOH; (b) Br(CH$_2$)$_{10}$OMOM, DMF, 100°C; (c) LiN$_3$, DMF, 100°C; (d) HCl, MeOH, 65°C; (e) (5-methyl-2-oxo-1,3-dioxolen-4-yl)methylbromide, DMF.

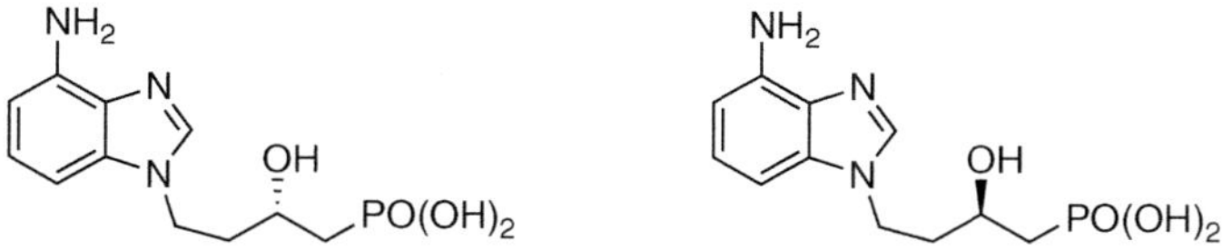

Figure 6. (S)- and (R)-β-hydroxyphosphonate acyclonucleosides as analogues of adefovir.

A kilogram-scale synthesis of tenofovir **249** has been reported according to a three-step sequence[58]: (1) condensation of adenine with (R)-propylene carbonate **251**; (2) alkylation of the corresponding (R)-9-(2-hydroxypropyl)adenine **252** with diethyl *p*-toluenesulfonyloxymethanephosphonate using lithium *t*-butoxide, and (3) hydrolysis of ester **253** with bromotrimethylsilane (Scheme 38).

Because of its poor oral bioavailability, a set of prodrugs of (R)-PMPA have been designed (Fig. 8). The fumarate salt of bis(isopropoxycarbonyloxymethyl) ester of (R)-9-[2-(phosphonomethoxy)propyl]adenine (tenofovir disoproxil fumarate, TDF) is actually in clinical use for the treatment of HIV and HBV infections.[66,67] GS-7340, the alanyl phenyl ester phosphonoamidate of tenofovir, showed 1000-fold enhanced potency against HIV-1 with respect to tenofovir[68]: this compound enhances the delivery of tenofovir to the lymphatic tissues. A third prodrug, in

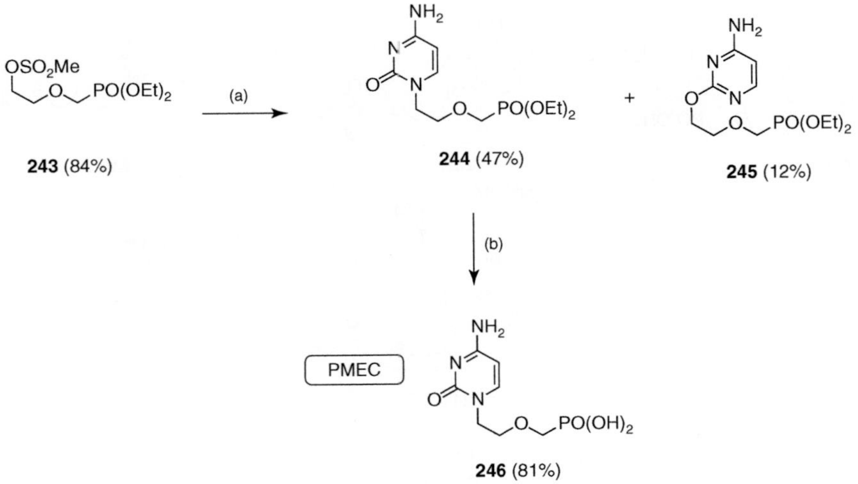

Scheme 37. *Reagents and conditions:* (a) cytosine, CsCO$_3$, 90°C, 2.5 h; (b) 20% Pd(OH)$_2$/C, H$_2$, cyclohexene/EtOH; (h) TMSBr DMF, rt, 14 h.

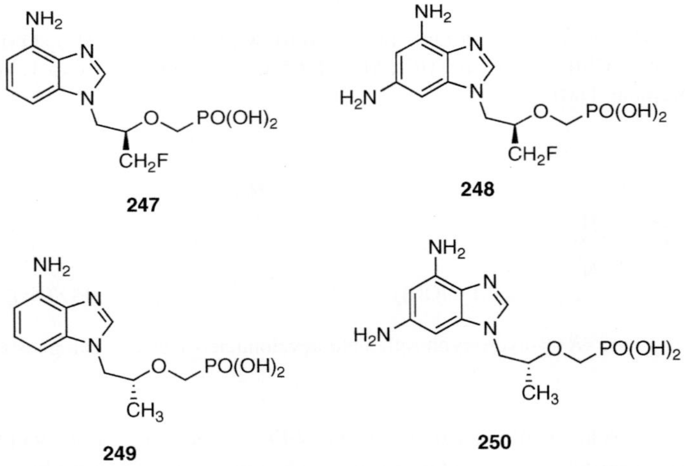

Figure 7. Structures of (R)-FPMA, (R)-FPMDAP, (R)-PMPA, and (R)-PMPDAP.

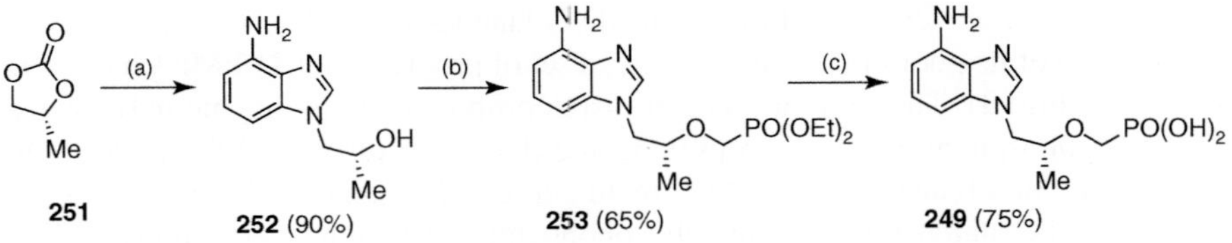

Scheme 38. *Reagents and conditions:* (a) adenine, NaOH, DMF, 140°C, 20 h; OLi/t-Bu, diethyl p-toluenesulfonyloxymethanephosphonate, 35°C, 1 h, aq. AcOH; (c) TMSBr, MeCN.

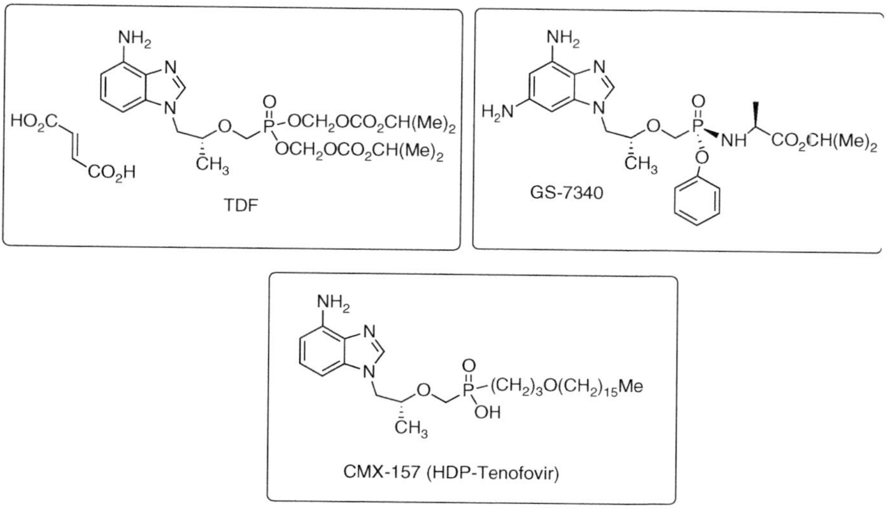

Figure 8. Structures of tenofovir diisoproxil fumarate TDF, GS-7340, and CMX-157.

the preclinical development phase for the treatment of HIV, is CMX-157 (hexadecyloxypropyl tenofovir), which shows potency in vitro against resistant HIV strains.[69]

2.5. 2,4-Diaminopyrimidine Nucleosides

A recent class of acyclic nucleoside phosphonates is composed of DAPys, called 2,4-diaminopyrimidine derivatives, where the disubstituted pyrimidine ring is linked through an oxygen bridge to either the 2-phosphonylmethoxyethyl or 2-phosphonomethoxypropyl moiety. Some of these compounds have shown promising antiviral activity. In particular, PMEO-DAPy **254**, 5-Br-PMEC-DAPy **255**, 5-Me-PMEO-DAPy **256**, and (*R*)-PMPO-DAPy **257** have shown activity against HIV-1, HIV-2, and HBV retroviruses, while (*R*)-HPMPO-DAPy **258** has shown activity against herpes-, adeno, pox-, and papillomaviruses (Figure 9).[70–72]

The synthesis of 2,4-diamino-6-[2-(phosphonomethoxy)ethoxy]pyridine (PMEO-DAPy) **254** is reported in Scheme 39 according to a procedure developed by Holý et al.[70]

The diisopropyl 2-chloroethoxymethyphosphonate **259** has been reacted with 2,4-diamino-6-hydroxypyrimidine in the presence of Cs_2CO_3 to give a mixture of diesters **260** and **261** in 17.2% and 26% yield, respectively. The **260** isolated was then transformed to the free phosphonic acid **254** (78.3%) by conventional reaction with TMSBr.

5-Br-PMEO-DAPy **255** and 5-Me-PMEO-DAPy **256** have been prepared by Hockova, Holy, et al. according to Scheme 40.[72a] In particular, compound **255** has been prepared starting from derivative **260**, an intermediate of the synthesis of PMEO-DAPy, by reaction with bromine followed by reaction with TMSBr

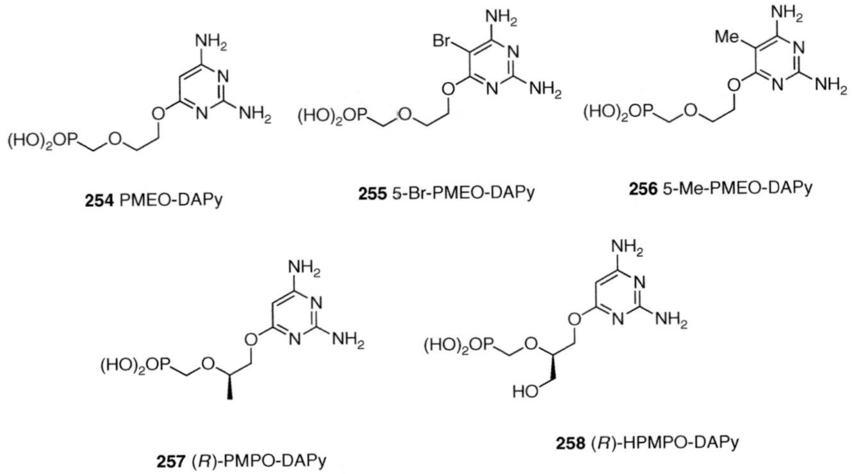

254 PMEO-DAPy **255** 5-Br-PMEO-DAPy **256** 5-Me-PMEO-DAPy

257 (*R*)-PMPO-DAPy **258** (*R*)-HPMPO-DAPy

Figure 9. 2, 4-Diaminopyrimidine nucleosides.

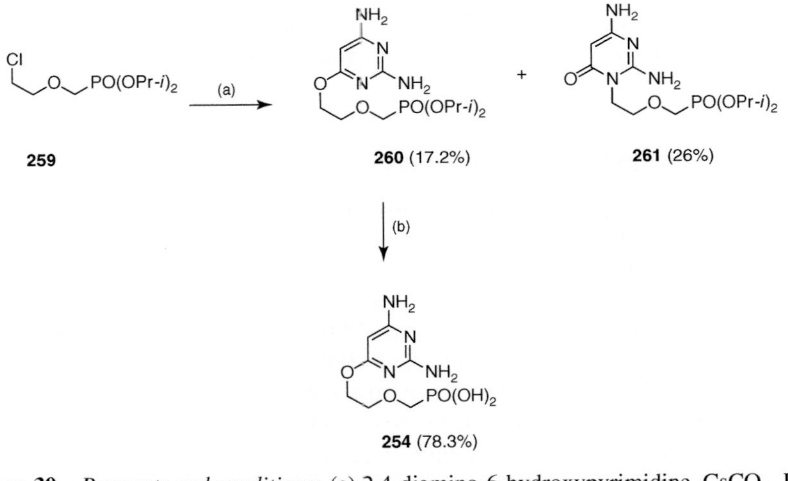

259 **260** (17.2%) **261** (26%) **254** (78.3%)

Scheme 39. *Reagents and conditions:* (a) 2,4-diamino-6-hydroxypyrimidine, $CsCO_3$, DMF, 80°C, 30 min, then 100°C, 16 h; (b) TMSBr, MECN rt, 14 h, aq. NH_3.

(Scheme 40). 5-Me-PMEO-DAPy **256** has been synthesized by a coupling reaction of **262** performed with Me_3Al and $Pd(PPh_3)_4$ and subsequent hydrolysis of **263** and **264**.

(*R*)-PMPO-DAPy **257** has been synthesized by reactions similar to those depicted in Scheme 39, involving as starting material the (*R*)-[2-(diisopropyl-phosphoryl)methoxypropyl tosylate **265** (Scheme 41). Interestingly, (*R*)-PMPO-DAPy has shown pronounced antiherpes and antiretroviral activity, comparable to

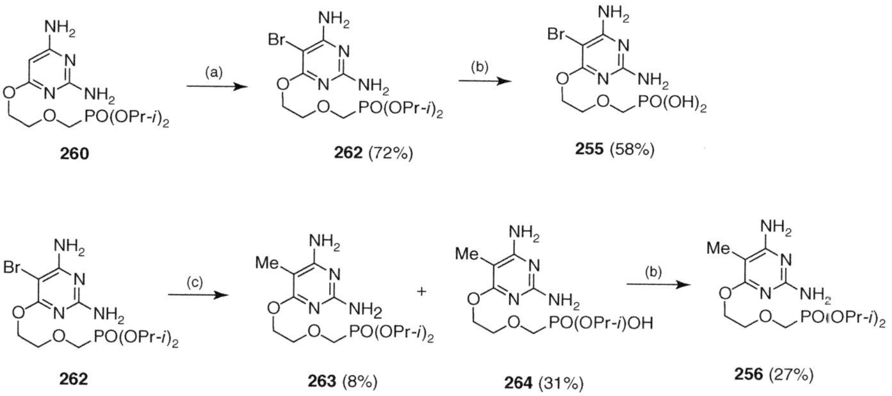

Scheme 40. *Reagents and conditions:* (a) Br$_2$, DMF/CCl$_4$, CsCO$_3$, DMF, 80°C, 30 min, then 100°C, 16 h; (b) TMSBr, MECN rt, 14 h, aq. NH$_3$; (c) AlMe$_3$, Pd(Ph$_3$P)$_4$, THF.

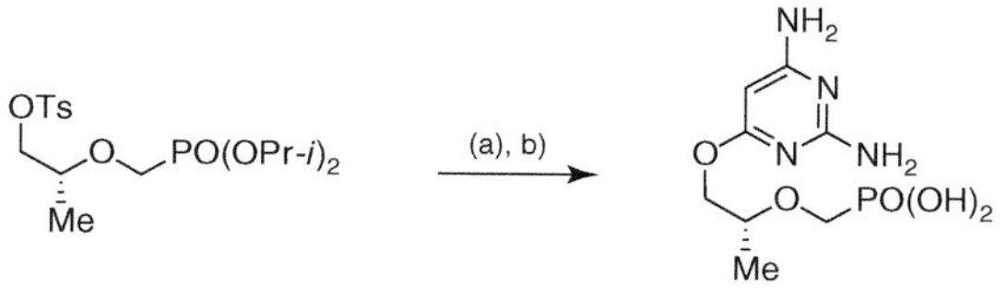

Scheme 41. *Reagents and conditions:* (a) 2,4-diamino-6-hydroxypyrimidine, DBU, DMF, 90–100°C, 24 h; (b) TMSBr, MECN rt, 14 h, aq. NH$_3$.

that of **254**, whereas the (S)-enantiomer, which has been prepared similarly to **257** from (S)-[2-(diisopropylphosphoryl)methoxypropyl tosylate, was virtually devoid of antiviral activity.

Finally, compound **258** has been prepared according to Scheme 42 in five steps, involving a sequence of reactions that starts from compound **214**.[73]

3. CONCLUDING REMARKS

Nucleoside analogues, in which the phosphoric ester group [(HO)$_2$(O)P–O–] is replaced by the corresponding phosphonomethyl group [(HO)$_2$(O)P–C–], represent an important class of compounds endowed with potent antiviral activity. This modification confers on the molecule particular stability toward enzymatic degradation by phosphatases, making it possible to overcome the first limiting phosphorylation step. In this context, acyclic phosphonated nucleosides (ANPs) have emerged as lead compounds. Their development has resulted in three approved drugs, and research on ANPs continues to provide new active derivatives.[21d,e,54] To achieve

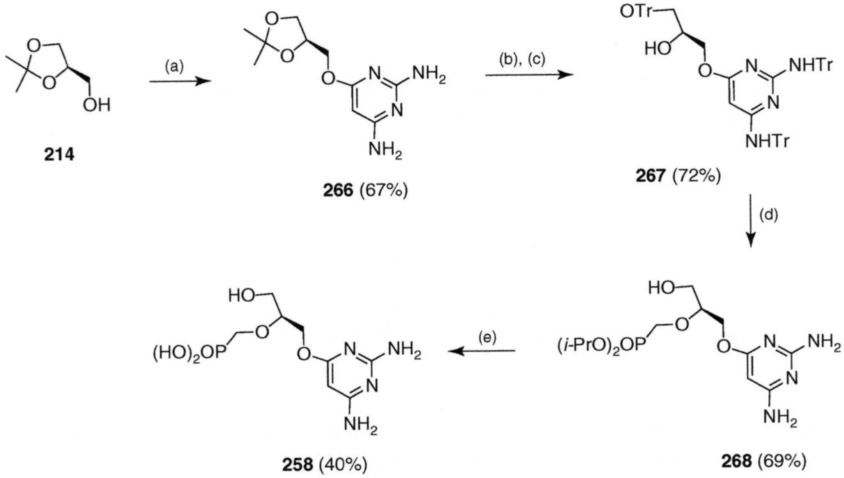

Scheme 42. *Reagents and conditions:* (a) NaH, THF, 2,4-diamino-6-chloropyrimidine, 50°C, 12 h; (b) H$_2$SO$_4$, rt, 12 h; (c) trityl chloride, DMAP, Py, 50°C, 15 h; (d) diisopropyl *p*-toluenesulfonyloxymethylphosphonate, NaH, rt, 3 days; (e) TMSBr, MeCN, 12 h.

better oral uptake, the synthesis of prodrugs is often required. The increasing availability of prodrug options suggests the exploration of phosphonic acids in the search for new, efficient antiviral agents.

REFERENCES

1. (a) Parker, W. B. *Chem. Rev.* **2009**, *109*, 2880–2893; (b) Liu, G.; Pranssen, E.; Fitch, E. F.; Warnew, E. *J.Clin. Oncol.* **1997**, *15*, 110–115; (c) Hishitsuka, H.; Shimma, N. In *Modified Nucleosides in Biochemistry, Biotechnology and Medicine*, Herdewijn, P., Ed., Wiley, Hoboken, NJ, **2008**, pp. 587–600; (d) Takahashi, T.; Shimuzu, M.; Akinaga, S. *Cancer Chemother. Pharmacol.* **2002**, *50*, 1930–1201; (e) Thottassery, J. V.; Westbrook, L.; Someya, H.; Parker, W. B. *Mol. Cancer Ther.* **2006**, *5*, 400–410; (f) Miura, S.; Izuta, S. *Current Drug Targets*, **2004**, *5*, 191–195; (g) Klopfer, A.; Hasenjager, A.; Belka, C.; Schulze-Osthoff, K.; Dorken, B.; Daniel, P. T. *Oncogene* **2004**, *23*, 9408–9418.
2. (a) De Clercq, E. *Rev. Med. Virol.* **2009**, *19*, 287–299; (b) De Clercq, E. *Biochem. Pharmacol.* **2007**, *73*, 911–922; (c) Cihlar, T.; LaFlamme, G.; Fisher, R.; Carey, A. C.; Vela, J. E.; Mackman, R.; Ray, A. S. *Antimicrob. Agents Chemother.* **2009**, *53*, 150–156; (d) Choo, H.; Beadle, J. R.; Kern, E. R.; Prichard, M. N.; Keith, K. A.; Hartlina, C. B.; Trahan, J.; Aldern, K. A.; Korba, B. E.; Hostetler, K. Y. *Antimicrob. Agents Chemother.* **2007**, *51*, 611–615; (e) Krcmerova, M.; Holý, A.; Piskala, A.; Masojidkova, M.; Andrei, G.; Naesens, L.; Neyts, J.; Balzarini, J.; De Clercq, E.; Snoeck, R. *J.Med. Chem.* **2007**, *50*, 1069–1077; (f) Lebeau, I.; Andrei, G.; Krecmerova, M.; De Clercq, E.; Holý, A.; Snoeck, R. *Antimicrob. Agents Chemother.* **2007**, *51*, 2268–2273.

3. (a) Vanek, V.; Budesinsky, M.; Rinnova, M.; Rosemberg, I. *Tetrahedron* **2009**, *65*, 862–876; (b) Kumamoto, H.; Topalis, D.; Broggi, J.; Pradere, U.; Roi, V.; Berteina-Raboin, S.; Nolan, S. P.; Deville-Bonne, D.; Andrei, G.; Snoeck, R.; Garin, D.; Grance, G. M.; Agrofoglio, L. A. *Tetrahedron*, **2008**, *64*, 3517–3526; (c) Vrbkova, S.; Dracinsky, M.; Holý, A. *Tetrahedron* **2007**, *63*, 11391–11398.

4. Wedemeyer, H.; Hardtke, S.; Cornberg, M. *Chemother. J.* **2012**, *21*, 1–7.

5. Elgemeie, G. H.; Zaghary, W. A.; Amin, K. M.; Nasr, T. M. *Nucleosides Nucleotides Nucleic Acids* **2005**, *24*, 1227–1247.

6. (a) Mironiuk-Puchalska, E.; Koszytkowska-Stawinska, M.; Sas, W.; De Clercq, E.; Naesens, L. *Nucleosides Nucleotides Nucleic Acids* **2012**, *31*, 72–84; (b) Merino, P.; Tejero, T.; Unzurrunzaga, F. J.; Franco, S.; Chiacchio, U.; Saita, M. G.; Iannazzo, D.; Piperno, A.; Romeo, G. *Tetrahedron Asymmetry*, **2005**, *16*, 3865–3876; (c) Chiacchio, U.; Rescifina, A.; Saita, M. G.; Iannazzo, D.; Romeo, G.; Mates, J. A.; Tejero, T.; Merino, P. *J.Org. Chem.* **2005**, *70*, 8991–9001.

7. Franchetti, P.; Cappellacci, L.; Perlini, P.; Jayaram, H. N.; Butler, A.; Schneider, B. P.; Collart, F. R.; Huberman, E.; Grifantini, M. *J. Med. Chem.* **1998**, *41*, 1702–1707.

8. Franchetti, P.; Cappellacci, L.; Marchetti, S.; Martini, C.; Costa, B.; Varani, K.; Borea, P. A.; Grifantini, M. *Bioorg. Med. Chem.* **2000**, *8*, 2367–2373.

9. Romeo, G.; Chiacchio, U.; Corsaro, A.; Merino, P. *Chem. Rev.* **2010**, *110*, 3337–3370.

10. Merino, P. *Curr. Med. Chem.* **2006**, *13*, 539–545.

11. (a) Hirota, K.; Monguchi, Y.; Sajiki, H. In *Recent Adv. Nucleosides*, Chu, C. K., Ed., **2002**, 57–70; (b) El Ashry, E. S. H.; Rashed, N. *Curr. Org. Chem.* **2000**, *4*, 609–651.

12. Littler, E.; Zhou, X.-X. In *Compr. Med. Chem. II*, Taylor, J. B.; Triggle, D. J., Eds., **2006**, *7*, 295–327.

13. (a) Wang, J.; Rwal, R. K.; Chu, C. K. In *Med. Chem. Nucleic Acids*, Zhang, L.-H.; Xi, Z.; Chattopadhyaya, J., Eds., **2011**, 1–100; (b) Gundersen, L. *Targets Heterocycl. Syst.* **2008**, *12*, 85–119.

14. Sharma, P. L.; Nurpeisov, V.; Hernandez-Santiago, B.; Beltran, T.; Schinazi, R. F. *Curr. Top. Med. Chem.* **2004**, *4*, 895–919.

15. (a) De Clercq, E.; Neyts, J.; Hand B. *Exp. Pharmacol.* **2009**, *189*, 53–84; (b) Berdis, A. J. *Biochemistry*, **2008**, *47*, 8253–8260.

16. Hutter M. C.; Helms V. *ChemBioChem* **2002**, *3*, 643–651.

17. (a) Schneider, B.; Sarfati, R.; Deville-Bonne, D.; Veron, M. *J.Bioenerg. Biomembr.* **2000**, *32*, 317–324; (b) Stein, D. S.; Moore, K. H. P. *Pharmacotherapy* **2001**, 2137–2146; (c) Lascu, I.; Gonin, P. *J.Bioenerg. Biomembr.* **2000**, *32*, 237–246.

18. Miller, W.H. Miller,R. L. *J. Biol. Chem.* **1980**, *255*, 7204–7207; *Biochem. Pharmacol.* **1982**, 31, 3879–3884.

19. (a) Hecker, S. J.; Erion, M. D.*J. Med.Chem.* **2008**, *51*, 2328–2345; (b) Schultz, C. *Bioorg. Med. Chem.* **2003**, *11*, 885–898; (c) Mackman, R. L., Cihlar, T. *Annu. Rep. Med. Chem.* **2004**, *39*, 305–321; (d) He, G.-X.; Krise, J. P.; Oliyai, R.. In *Prodrugs: Challenges and Rewards*, Springer-Verlag, New York, **2007**, pp. 223–264; (e) Ariza, M. E. *Drug Des. Rev.* **2005**, *2*, 273–387; (f) Congiatu, C.; McGuigan, C.; Jiang. W. G.; Davies, G.; Mason, M. D. *Nucleosides Nucleotides Nucleic Acids*, **2005**, *24*, 485–489.

20. (a) McGuigan, C.; Harris, S. A.; Daluge, S. M.; Gudmundsson, K. S.; McLean E. W.; Burnette, T. C.; Marr, H.; Hazen, R. *J. Med. Chem.* **2005**, *48*, 3504–3515; (b) Perrone, P.; Luoni, G. M.; Kelleher, M. R.; Daverio, F.; Angell; A.; Mulready, S.; Congiatu,

C.; Rajyaguru, S.; Martin, J. A.; Leveque, V.; Le Pogam, S.; Najera, I.; Klumpp, K.; Smith, D. B.; McGuigan, C. *J. Med. Chem.* **2007**, *50*, 1840−1849; (c) McGuigan, C.; Derudes, M.; Bugert, J. J.; Andrei, G.; Snoecke, R.; Balzarini, J. *Bioorg. Med. Chem. Lett.* **2008**, *18*, 4364−4367.

21. (a) De Clercq, E.; Holý, A. *Nat. Res. Drug Discov.* **2005**, *4*, 928−940; (b) Deville-Bonne, D.; El Amri, C.; Meyer, P.; Chen, Y. X.; Agrofoglio, L. A.; Janin, J. *Antiviral Res.* **2010**, *86*, 101−120; (c) De Clercq, E. *Med. Res. Rev.* **2009**, *29*, 571−610; (d) De Clercq, E. *Antiviral Res.* **2010**, *85*, 19−24; (e) De Clercq, E. *Biochem. Pharmacol.* **2011**, *82*, 99−109.

22. (a) Gallier, F.; Péyrottes, S.; Périgaud C. *Tetrahedron*, **2009**, *65*, 6039−6046; (b) Meurillon, M.; Gallier, F.; Peyrottes, S.; Périgaud, C. *Eur. J. Org. Chem.* **2007**, 925−933; (c) Gallier, F.; Alexandre, J. A. C.; El Amri, C.; Deville-Bonne, D.; Peyrotts, S.; Périgaud, C. *Chem. Med. Chem.* **2011**, *6*, 1094−1106.

23. Wang, P.; Schinazi, R. Y.; Chu, C. K. *Bioorg. Med. Chem. Lett.* **1988**, *8*, 1585−1588.

24. Hah J. H.; Gil, J. M.; Oh, D. Y. *Tetrahedron Lett.* **1999**, *40*, 8235−8238.

25. Midura, W. H.; Krysiak, J. A.; Mikolajcvzyk, M. *Tetrahedron Asymmetry* **2003**, *14*, 1245−1249.

26. Choi, J.-R.; Cho, D.-G.; Roh, K. Y.; Hwang, J.-T.; Ahn, S.; Jang, H. S.; Cho, W.-Y.; Kim, K. W.; Cho, Y.-G.; Kim, J.; Kim, Y.-Z. *J.Med. Chem.* **2004**, *47*, 2864−2869.

27. Yokpmatsu, T.; Sato, M.; Abe, H.; Suemune, K.; Matsumoto, K.; Kihara, T.; Soeda, S.; Shimeno, H.; Shibuya, S. *Tetrahedron* **1997**, *53*, 11297−11306.

28. Yokomatsu, T.; Yamagishi, T.; Suemune, K.; Abe, H.; Kihara, T.; Soeda, S.; Shimeno, H.; Shibuya, S. *Tetrahedron* **2000**, *56*, 7099−7108.

29. Guan, H.-G.; Qiu, Y.-L.; Ksebati, M. B.; Kern, E. R.; Zemlicka, J. *Tetrahedron* **2002**, *58*, 6047−6059.

30. Yan, Z.; Zhou, S.; Kern, E. R.; Zemlicka, J. *Tetrahedron* **2006**, *62*, 2608−2615.

31. Li, C.; Zemlicka, J. *Nucleosides Nucleotides Nucleic Acids* **2007**, *26*, 111−120.

32. Lai, M.-T.; Liu, L.-D.; Liu, H.-W. *J. Am. Chem. Soc.* **1991**, *113*, 7388−7397.

33. Yokomatsu, T.; Abe, H.; Sato, M.; Suemune, K.; Kihara, T.; Soeda, S.; Shimeno, H.; Shibuya, S. *Bioorg. Med. Chem.* **1998**, *6*, 2495−2505.

34. Onishi, T.; Sekiyama, T.; Tsuji, T. *Nucleosides Nucleotides Nucleic Acids* **2005**, *24*, 1187−1197.

35. Kim, J. W.; Ko, O. H.; Hong, J. H. *Arch. Pharm. Res.* **2005**, *28*, 745−749.

36. Oh, C. H.; Hong, J. H. *Arch. Pharm.* **2006**, *339*, 505−512.

37. Li, H.; Yoo, J. C.; Hong, J. H. *Nucleosides Nucleotides Nucleic Acids* **2011**, *30*, 945−960.

38. Kim, A.; Hong, J. H.; Oh, C. H. *Nucleosides Nucleotides Nucleic Acids* **2006**, *25*, 1399−1406.

39. Bronson, J. J.; Ferrara, L. M.; Martin, J. C.; Mansuri, M. M. *Bioorg. Med. Chem. Lett.* **1992**, *2*, 685−690.

40. Coe, M. D.; Roberts, S. M.; Storer, R. *J. Chem. Soc. Perkin Trans. 1* **1992**, 2695−2704.

41. Elliot, R. D.; Rener, G. A.; Riorda, J. M.; Secrist, J. A., III; Bennett, L. L., Jr.; Parker, W. B.; Montgomery, J. A. *J.Med. Chem.* **1994**, *37*, 739−744.

42. Wainwright, P.; Maddaford, A.; Bissel, R.; Fisher, R.; Leese, D.; Lund, A.; Runcie, K.; Dragovich, P. S.; Gonzales, J.; Kung, P.-P.; Middleton, D. S.; Pryde, D. C.; Stephenson, P. T.; Sutton, S. C. *SynLett* **2005**, *5*, 765−768.

43. Boojamra, C. G.; Parrish, J. P.; Sperandio, D.; Gao, Y.; Petrakovsky, O. V.; Lee, S. K.; Markevitch, D. Y. *Bioorg. Med. Chem.* **2009**, *17*, 1739–1746.
44. Yoo, J. C.; Lee, W.; Hong, J. H. *Bull. Korean Chem. Soc.* **2010**, *31*, 3348–3352.
45. Yoo, L. L.; Baik, Y. C.; Lee, W.; Hong, J. E. *Bull. Korean Chem. Soc.* **2010**, *31*, 2514–2518.
46. De Clercq, E.; Holý, A. *J.Med. Chem.* **1985**, *28*, 282–287.
47. De Clercq, E.; Holý, A.; Rosenberg, T.; Sakuma, T.; Balzarini, J.; Maudgal, P. C. *Nature* **1986**, *323*, 464–467.
48. Rosenberg, T.; Holý, A. *Nucleic Acids Symp. Ser.* **1987**, *18*, 33–36.
49. Zakirova, B. F.; Shipitsyn, A. V.; Belanov, E. F.; Jasko, M. V. *Bioorg. Med. Chem. Lett.* **2004**, *14*, 3357–3360.
50. (a) Beadle, J. R.; Hostetler, K. Y. *PCT Int. Appl. 2005087788*, 22 **Sep 2005**, p. 44; (b) Beadle, J. R.; Wan, W. B.; Ciesla, S. L.; Keith, K. A.; Hartline, C.; Kern, E. R.; Hostetler, K. Y. *J.Med. Chem.* **2006**, *49*, 2010–2015.
51. (a) De Clercq, E.; Sakuma, T.; Baba, M.; Pauwels, R.; Balzarini, J.; Rosenberg, T.; Holý, A. *Antiviral Res.* **1987**, *8*, 261–267; (b) Snoeck, R.; Sakuma, T.; De Clercq, E.; Rosenberg, T.; Holý, A. *Antimicrob. Agents Chemother.* **1988**, *32*, 1839–1842.
52. Bronson, J. J.; Ghazzouly, I.; Hitchcock, M. J. M.; Webb, R. R.; Martin, J. C. *J.Med. Chem.* **1989**, *32*, 1457–1463.
53. Beadle, J. R.; Hartline, C.; Aldern, K. A.; Rodriguez, N.; Harden, E.; Kerr, E. R.; Hostetler, K. I. *Antimicrob. Agents Chemother.* **2002**, *46*, 2381–2386.
54. Tichy, T.; Andrei, G.; Dracinsky, M.; Holý, A.; Balzarini, J.; Snoeck, R.; Krecmerova, M.; *Bioorg. Med. Chem.* **2011**, *19*, 3527–3539.
55. Pawels, R.; Balzarini, J.; Schols, D.; Baba, M.; Desmyter, J. *Antimicrob. Agents Chemother.* **1988**, *32*, 1025–1030.
56. Holý, A.; Dvorakova, H.; Masojidkova, M. *Collect. Czech. Chem. Commun.* **1995**, *60*, 1390–1393.
57. Holý, A.; Masojidkova, M. *Collect. Czech. Chem. Commun.* **1995**, *60*, 1196–1199.
58. Schultze, L. M.; Louie, M. S.; Postich, M. J.; Prisbe, E. J.; Rohloff, J. C.; Yu, R. H. *Tetrahedron Lett.* **1998**, *39*, 1853–1856.
59. Starrett, J. E.; Tortolani, D. R.; Hitchcock, M. J.; Martin, J. C.; Mansuri, M. M. *Antiviral Res.* **1992**, *19*, 267–273.
60. Cundy, K. C.; Fishback, J. A.; Shaw, J. P.; Lee, M. L.; Soike, K. F.; Visor, G. C. *Pharm. Res.* **1994**, *11*, 839–843.
61. (a) Hadziyannis, S. J.; Tassopoulos, N. C.; Heathcote, E. J.; Chang, T. T.; Kitis, G.; Rizzetto, M.; Marcelin, P.; Seng, G.; Goodman, Z.; Wulfsohn, M.; Xiong, S.; Fry, J.; Brosgart, C. L. N. *Engl. J. Med.* **2003**, *348*, 800–807; (b) Marcellin, P.; Chang, T. T.; Ling, S. G.; Tong, M. G.; Sievert, W.; Shiffman, M. L.; Jeffers, L.; Goodman, Z.; Wulfsohn, M. S.; Xiong, S.; Fry, J.; Brosgart, C. L. N. *Engl. J. Med.* **2003**, *348*, 808–816.
62. Kasthuri, M.; Chaloin, L.; Périgaud, C.; Peyrotttes, S. *Tetrahedron Asymmetry* **2011**, *22*, 1505–1511.
63. Balzarini, J.; Holý, A.; Jindrich, J.; Dvorakova, H.; Hao, Z.; Snoeck, R.; Herdewijn, P.; Jhons, D. G.; De Clercq, E. *Proc. Natl. Acad. Sci. USA* **1991**, *88*, 4961–4965.
64. Balzarini, J.; Holý, A.; Jindrich, J.; Naesens, L.; Snoeck, R.; Schols, D.; De Clercq, E. *Antimicrob. Agents Chemother.* **1993**, *37*, 332–338.

65. Heijtink, R. A.; Kruining, J.; de Wilde, G. A.; Balzarini, J.; De Clercq, E.; Schalm, S. V. *Antimicrob. Agents Chemother.* **1994**, *38*, 2180−2182.
66. Robbins, B. L.; Srinivas, R. V.; Kim, C.; Bischofberger, N.; Fridland, A. *Antimicrob. Agents Chemother.* **1998**, *42*, 612−617.
67. Naesens, L.; Bischofberger, N.; Augustijns, P.; Annaert, P.; Van den Mooter, G.; Arimilli, M. N.; Murty, N.; Kim, C. U.; De Clercq, E. *Antimicrob. Agents Chemother.* **1998**, *42*, 1568−1573.
68. Lee, W. A.; He, G. X.; Eisenberg, E.; Cihlar, T.; Swaminathan, S.; Mulato, A.; Cundy, K. C. *Antimicrob. Agents Chemother.* **2005**, *49*, 1898−1906.
69. Lanier, E. R.; Ptak, R. G.; Lampert, B. M.; Keilholz, L.; Hartman, T.; Buckheit, R. W., Jr.; Mankowski, M. K.; Osterling, M. C.; Almond, M. R.; Painter, G. R. *Antimicrob. Agents Chemother.* **2010**, *54*, 2901−2909.
70. Holý, A.; Votruba, I.; Masojidkova, M.; Andrei, G.; Snoeck, R.; Naesens, L.; De Clercq, E.; Balzarini, J. *J. Med. Chem.* **2002**, *45*, 1918−1829.
71. De Clercq, E.; Andrei, G.; Balzarini, J.; Leyssen, P.; Naesens, L.; Neyts, J.; Pannecouque, C.; Snoeck, R.; Ying, C.; Hockova, D.; Holý, A. *Nucleosides Nucleotides Nucleic Acids* **2005**, *34*, 331−341.
72. (a) Hochkova, D.; Holý, A.; Masojidkova, M.; Andrei, G.; Snoeck, R.; De Clercq, E.; Balzarini, J. *J.Med. Chem.* **2003**, *46*, 5064−5073; (b) Balzarini, J.; Pannecouque, C.; De Clercq, E.; Aquaro, S.; Perno, C.-F.; Egberink, H.; Holý, A. *Antimicrob. Agents Chemother.* **2002**, *46*, 2185−2193.
73. Balzarini, J.; De Clercq, E.; Holý, A. Patent **2003**/002580 A1, 22 Jan **2003**, p. 83.

5 Chemical Syntheses of Nucleoside Triphosphates

LINA WEINSCHENK and CHRIS MEIER

Department of Chemistry, Faculty of Sciences, University of Hamburg, Hamburg, Germany

1. INTRODUCTION

Nucleoside triphosphates play an essential role in biological systems. Besides the important role of adenosine triphosphate as the primary energy source in many systems, naturally occurring deoxyribo- and ribonucleoside triphosphates are the modules for the biosyntheses of DNA and RNA. Therefore, chemically synthesized natural NTPs have value not only for the structural investigation of nucleic acids of crucial interest. Modified NTPs have important therapeutic and diagnostics applications; the corresponding nucleoside analogues are used clinically, for example, in antiviral therapy. There, triphosphates is the ultimate active form that acts as a chain termination agent. Thus, syntheses of these nucleoside triphosphates are needed for biochemical and pharmacological studies. Various biochemical methods exist that give the natural triphosphates in high yields. But nonnatural nucleosides are bad substrates for enzymes. Therefore, chemical methods for the syntheses of NTPs are of great importance. But finding a universal method that is compatible with a huge range of nucleoside analogs is still a challenge. Some methods are well suited for some nucleosides and give high yields, but the yields decrease significantly when applied to other nucleosides.

There are many challenges in the chemical synthesis of NTPs: (1) demanding reaction conditions, (2) sensitivity of the reactions against water, (3) lability of the target molecules towards hydrolysis, and (4) problematic isolation and purification. In most cases protocols involve the conversion of an ionic compound with a more lipophilic structure. Finding a solvent that leads to a homogeneous reaction solution is therefore often difficult. In addition, anhydrous reaction conditions are of crucial importance to avoid the generation of by-products by nucleophilic side reactions with water. Furthermore, nucleoside triphosphates are labile compounds under certain conditions: Under basic as well as acidic conditions, hydrolysis of the

---

Chemical Synthesis of Nucleoside Analogues, First Edition. Edited by Pedro Merino.
© 2013 John Wiley & Sons, Inc. Published 2013 by John Wiley & Sons, Inc.

phosphate anhydride bonds occurs. If the protonated form lowers the pH in non-buffered media, it also causes a storage problem. Final isolation of the NTPs poses another hurdle. On the one hand, chromatography of highly charged and thus polar compounds is always a challenge, and on the other hand, less phosphorylated nucleosides or other phosphate-bearing molecules as by-products have similar characteristics. Although advances in the field of chromatography are marked, isolation and purification is often the limiting step in the synthesis of nucleoside triphosphates.

Two reviews on this topic were published in 2000 and 2005.[1,2] Here, we focus primarily on new methods for the syntheses of nucleoside triphosphates— without, of course, disregarding the most successful established methods.

2. SYNTHESIS OF NUCLEOSIDE TRIPHOSPHATE

Scheme 1 shows the two general approaches to the synthesis of NTPs: (1) a nucleophilic attack of the phosphate unit on an activated nucleosidic or nucleotidic unit (route A), and (2) activation of the phosphate unit followed by a nucleophilic attack by the nucleosidic or nucleotidic unit (route B). The most important prerequisite for the used starting materials is, of course, easy access. Syntheses that follow route A are discussed in Sections 2.1 and 2.2, and those that follow route B are presented in Section 2.3.

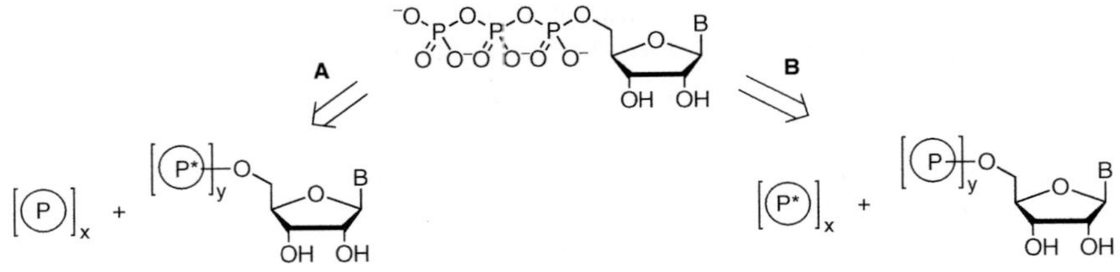

Scheme 1. Possibilities for the synthesis of nucleoside triphosphates (P* = activated phosphate unit; $x = 1, 2, 3$; $y = 0, 1, 2$, with $x + y = 3$).

2.1. Synthesis by Nucleophilic Attack of a Pyrophosphate on an Activated Nucleoside Monophosphate

The most popular method for the synthesis of nucleoside triphosphates is the activation of a nucleoside monophosphate followed by nucleophilic attack of a pyrophosphate salt. Here, the activated nucleoside monophosphates have to be prepared first. The most frequently used are dichlorophosphates, phosphites, phosphoramidates, and H-phosphonates.

2.1.1. Syntheses Involving Nucleoside Dichlorophosphates.
The $5'$-phosphorylation of nucleosides to generate nucleotides was first reported, in 1967, by Yoshikawa et al.[3] By treatment of $2',3'$-protected or unprotected nucleosides with phosphoryl chloride in trimethyl phosphate and further hydrolysis, monophosphates were obtained in very good yields, in the range 88 to 98% (Scheme 2).

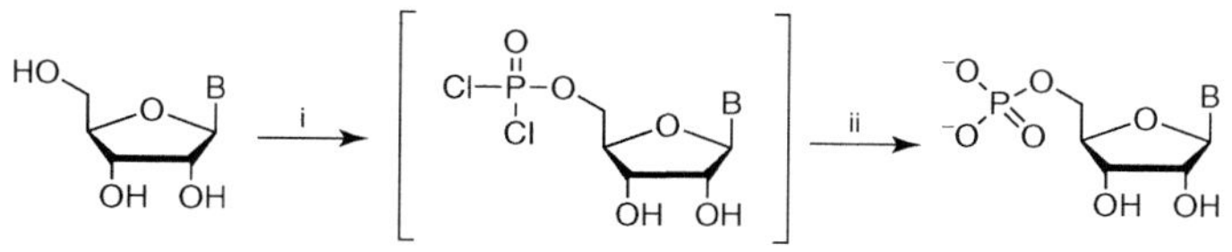

Scheme 2. *Reagents and conditions:* (i) 3 POCl$_3$, 1 H$_2$O, (MeO)$_3$PO, 0°C, 6 h; (ii) hydrolysis.

Based on this protocol for the syntheses of nucleotides, Ludwig developed a method for the synthesis of nucleoside triphosphates.[4,5] After generating the dichlorophosphate **1** following Yoshikawa's procedure, the addition of bis(tri-*n*-butylammonium)pyrophosphate in dry DMF followed to obtain intermediate **2**. In the presence of a tertiary amine base such as tributyl amine, high yields were obtained. After only 1 min the reaction mixture was quenched by the addition of triethylammonium carbonate, which finally led to the nucleoside triphosphate after the usual workup. Scheme 3 shows the protocol for the syntheses of ATP **3**, which was obtained in an excellent yield of 85% after chromatography on a DE-32 (HCO$_3^-$) ion-exchange column. In contrast to Yoshikawa, protocol, the first step was carried out under anhydrous conditions.

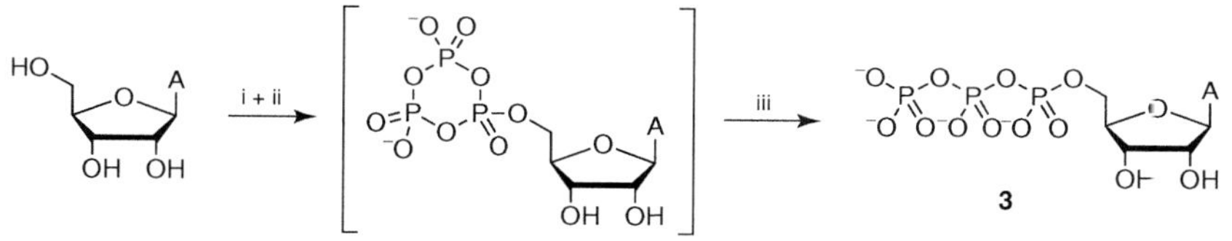

Scheme 3. *Reagents and conditions:* (i) POCl$_3$, (MeO)$_3$PO, 0°C, 1.5 h; (ii) [(HNBu$_3$)$_2$] H$_2$P$_2$O$_7$, NBu$_3$, DMF, 25°C, 1 min; (iii) (Et$_3$NH)HCO$_3$, H$_2$O, 0°C, 3 h.

In addition to the synthesis of ATP, conversions to 2'-dATP and 3'-dATP (cordycepintriphosphate) as well as 5-[(*E*)- and (*Z*)-bromovinyl]-2'-deoxyuridine and 5-[(*E*)-bromovinyl]-2'-uridine-5'-triphosphate were successful in yields up to 79%.[5] It is important to notice that this "one-pot, three-step" synthesis does not require protecting groups at the nucleoside.

A similar approach was published in 1981 by Ruth and Cheng,[6] the difference being that phosphoryl chloride was added in two portions and the reaction was carried out at −10°C. Although the yields were found to be between 8 and 46% only, they were able to synthesize triphosphates of seven different nucleoside analogs. The method of Ludwig was also applied to generate analogues of nucleoside triphosphates as 5'-*O*-(1-thiotriphosphates) by the use of thiophosphoryl chloride in the first step[7] or nucleoside β,γ-imidotriphosphates by the use of imidodiphosphate in the second step.[5] By the addition of proton sponge [e.g, 1,8-bis(dimethylamino)naphthalene], the 5'-triphosphorylation of acid-labile nucleoside analogues was also possible, although in moderate yields.[8]

Despite the advantages that the reaction is easily carried out and that the reagents required are cheap, the Ludwig procedure has some disadvantages. The reaction times depend strongly on the nucleosides used and not in all cases of nucleoside analogs is triphosphate formed.[6] In addition, the yields are often poor because of the formation of a set of by-products in the first step.[9] The formation of by-products was reduced by the optimization introduced by Gillerman and Fischer.[10] By reducing the temperature from 0°C to −15°C and by prolonging the reaction time in the second step, they eliminated most of the by-products except those of the corresponding nucleotides, which could be removed by liquid chromatography. Chemical yields were between 51 and 74%.

An approach by Broom for the syntheses of nucleoside diphosphates based on the one-pot procedure of Yoshikawa et al.[3] and Ludwig[4,5] led to another possibility for the syntheses of nucleoside triphosphates via dichlorophosphates **1**.[11] The replacement of tributylammonium pyrophosphate (Ludwig's approach)[4,5] by tributylammonium phosphate should give the 5′-diphosphates. However, the diphosphate expected was not the major product, but the nucleoside triphosphate. The proposed reaction sequence is shown in Scheme 4.

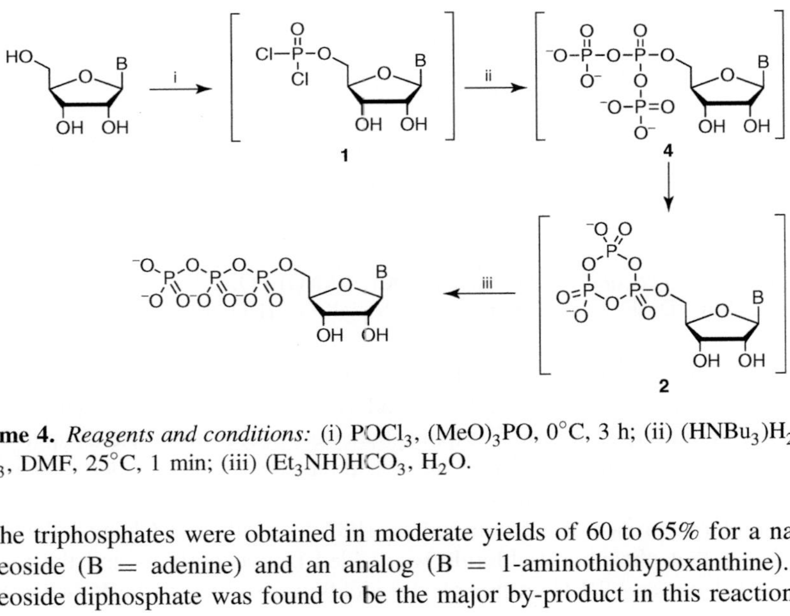

Scheme 4. *Reagents and conditions:* (i) POCl$_3$, (MeO)$_3$PO, 0°C, 3 h; (ii) (HNBu$_3$)H$_2$PO$_4$, NBu$_3$, DMF, 25°C, 1 min; (iii) (Et$_3$NH)HCO$_3$, H$_2$O.

The triphosphates were obtained in moderate yields of 60 to 65% for a natural nucleoside (B = adenine) and an analog (B = 1-aminothiohypoxanthine). The nucleoside diphosphate was found to be the major by-product in this reaction.

2.1.2. Syntheses Involving Phosphites. Another approach that involves the formation of cyclic triphosphate intermediate **2** was published by Ludwig and Eckstein in 1989.[12] The first phosphate was introduced by reaction of the nucleoside with a salicyl phosphorochloridite **5**. Although the selectivity for the 5′-position is higher than for the other hydroxy groups, protecting groups at the sugar moiety are essential for avoiding the formation of by-products. They proposed that first a nucleophilic attack involves the carboxyl group as the leaving group, followed by a second intramolecular nucleophilic attack in which the phenol moiety leaves and a cyclic triphosphite **7** is formed. Further oxidation with iodine in aqueous media led to a triphosphate (Scheme 5). The presence of intermediate **7** was observed in ^{31}P-NMR studies. Oxidation of the cyclic triphosphite **7** with sulfur offers a possibility for the synthesis of 5'-α-thiophosphates.

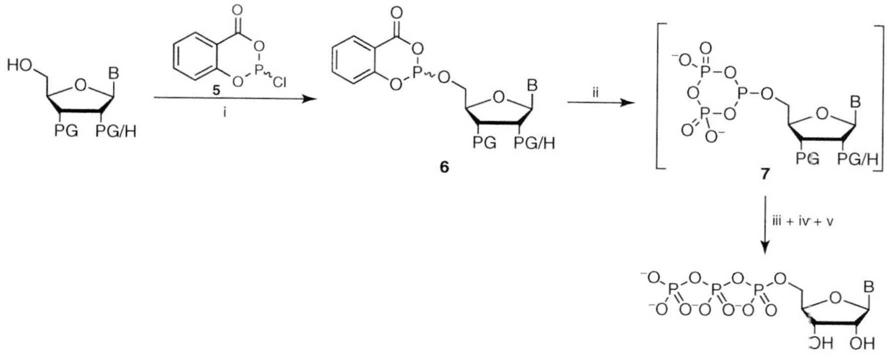

Scheme 5. *Reagents and conditions:* (i) pyridine/dioxane or pyridine/DMF, 10 min; (ii) [(HNBu$_3$)$_2$]H$_2$P$_2$O$_7$, NBu$_3$, DMF, 10 min; (iii) 1% iodine in pyridine/water, 15 min; (iv) 5% aq. solution of NaHSO$_3$; (v) NH$_3$, 1 h.

By this rapid one-pot reaction, all eight natural nucleoside triphosphates as well as 3'-azido-3'-deoxy- and 3'-fluoro-3'-deoxythymidine triphosphate were synthesized in approximately 65% yields. However, one-pot reactions always have the problem of the generation of a variety of by-products. In the past, variants involving solid-phase reactions of this method have also been published.[13,14] An interesting approach was reported by Krupp et al. in 1992.[14] Attaching the 3'-hydroxy group of 2'-*O*-methylribonucleosides or deoxyribonucleosides to an amino-functionalized solid support via a succinate linker made it possible to remove by-products simple by washing under an argon atmosphere. Although a final chromatographic step was necessary, the products were obtained in yields of 60 to 65%. This method represents a fast and easy way to handle synthesis on a solid support.

2.1.3. Syntheses Involving Phosphoramidates.

Phosphoramidates are common molecules for the activation of phosphate groups. They are easy to synthesize, relatively stable, and afford a highly activated center for a nucleophilic attack by a phosphate group after former acidic activation. For this reason, many protocols for the syntheses of nucleoside (poly)phosphates are published involving phosphoramidates.

An important contribution to the field of nucleotide synthesis was the phosphoromorpholidate method reported by Moffat.[15] This method was based on the protocol of Chambers and Khorana,[16] which includes the reaction of a nucleoside-5'-phosphate with a large excess of inorganic pyrophosphate in the presence of dicyclohexylcarbodiimide (DCC) to give nucleoside di- and triphosphates.

However, the low solubility of the compounds and the generation of a mixture of nucleoside di-, tri-, and higher polyphosphates results in low yields. Although it was possible to eliminate the problems arising from low solubility by using anhydrous pyridine as a solvent and the tributylammonium salts of the phosphoric acid, the final purifications by charcoal filtration and ion-exchange chromatography led to a significant loss of nucleotidic material. The method of Moffat used morpholidates as **8** instead of the nucleoside-5′-phosphate.[15] But using the same reaction conditions, with anhydrous pyridine as solvent, an interesting behavior was noticed. At first the generation of nucleoside triphosphate was observed followed by its disappearance and simultaneous generation of the corresponding diphosphate. Later, the reaction was carried out using nucleoside-5′-phosphormorpholidate **8** and tributylammonium pyrophosphate in dimethyl sulfoxide (Scheme 6).

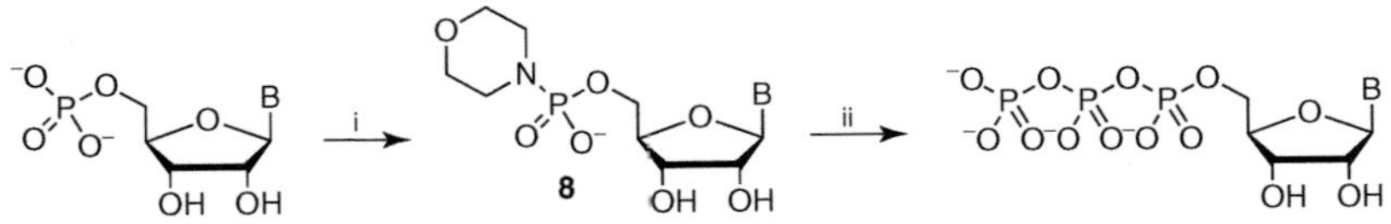

Scheme 6. *Reagents and conditions:* (i) DCC, morpholine, *t*-BuOH/H_2O, reflux; (ii) [$(HNBu_3)_2$]$H_2P_2O_7$, DMSO dried over a molecular sieve for 2–4 days.

It is crucial that the reaction be carried out under rigorously dry conditions. Otherwise, a huge amount of nucleotide is generated as a major by-product together with several others. However, when dimethyl sulfoxide was dried by storage over molecular sieve for several days prior to the reaction, the triphosphates of dA, dC, dG, T, C, and 6-azauridine were isolated in good yields of 73–80%.[15] The synthesis of the nucleoside-5′-phosphormorpholidate **8** starting from nucleoside-5′-phosphate was published by Moffat and Khorana in yields of 89–95% (Scheme 6).[17]

A drawback to this method is, of course, the previous synthesis of the nucleoside-5′-phosphate. Therefore, in 1975, van Boom and co-workers published an variation of the morpholidate method for the synthesis of nucleoside triphosphates, starting from the nucleoside.[18] The phosphorylation of some modified ribonucleosides of adenosine with 2,2,2-tribromoethyl phosphoromorpholinochloridate generated the phosphoramidate **9** in yields of 75–80%, which was converted to the phosphoromorpholidate by selective removal of the 2,2,2-tribromoethyl protecting group (Scheme 7). Without isolation, the unprotected phosphoromorpholidate was reacted

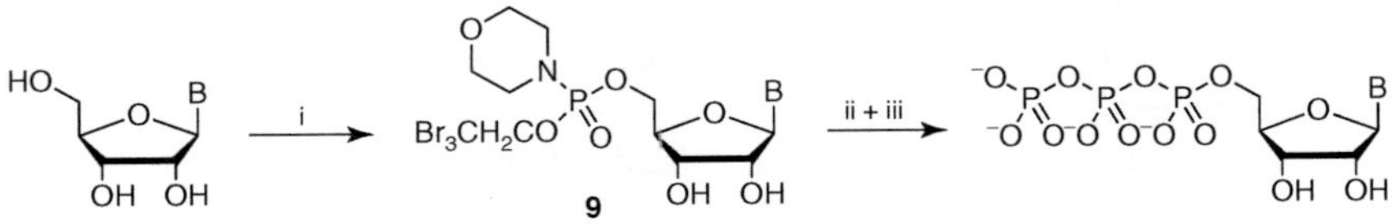

Scheme 7. *Reagents and conditions:* (i) 2,2,2-tribromoethyl phosphoromorpholinochloridate, pyridine, 20°C, 48 h; (ii) Cu/Zn, DMF, 20°C, 20 min; (iii) [$(HNBu_3)_2$]$H_2P_2O_7$, DMF, 20°C, 3 h.

with tributylammonium pyrophosphate in dimethylformamide to give the desired nucleoside triphosphates in 81–85% yield.

2,2,2-Tribromoethyl phosphoromorpholinonucleoside **9** was prepared easily by the reaction of 2,2,2-bromoethyl phosphorodichloridate with morpholine and a nucleoside in dry diethyl ether. The product was isolated by recrystallization in 80% yield.[18]

Another approach, reported by Hoard and Ott, employed phosphorimidazolidate as the activated nucleoside monophosphate in a one-pot reaction.[19] Phosphorimidazolidates **10** were easily prepared under mild conditions from the nucleoside-5'-phosphate and 1,1'-carbonyldiimidazole. After unreacted 1,1'-carbonyldiimidazole was trapped with methanol, an excess of tributylammonium pyrophosphate was added to the reaction mixture to generate the nucleoside triphosphate within 24 h (Scheme 8).

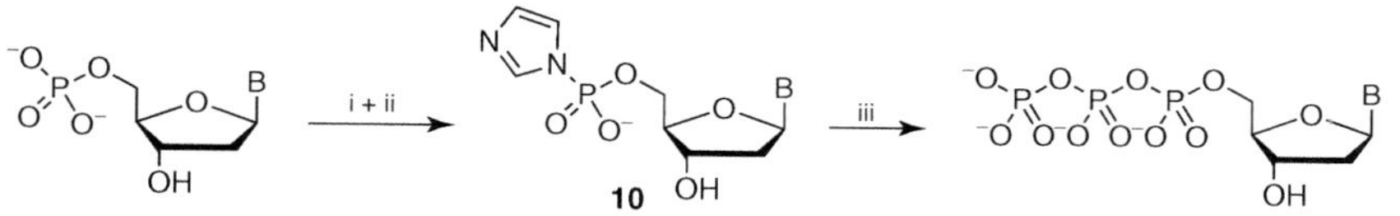

Scheme 8. *Reagents and conditions:* (i) 1,1'-carbonyldiimidazole, pyridine, DMF, rt, 30 min to 4 h; (ii) methanol, rt, 30 min; (iii) [(HNBu$_3$)$_2$]H$_2$P$_2$O$_7$, DMF, rt, 1 day.

After purification, various natural mono- and oligonucleotides in the range of 20–70% were obtained. However, often a complex mixture of by-products is also formed, which is difficult to remove. Therefore, the solid-phase synthesis recently developed by Peyrottes and co-workers is of interest.[20] The use of poly(ethylene glycol) (PEG) as a soluble support linked to the base offered the possibility of synthesizing the triphosphates of C, dC, and ara-C starting from the nucleosides. After attachment of the succinyl-linked PEG **11** in good chemical yield, a 5'-phosphorylation adapted by the synthesis in solution of Yoshikawa et al.[3] followed and gave the monophosphates **12** in 86–92% yield (Scheme 9). Then, the method of Hoard and Ott[19] described above was used to obtain the triphosphates after final cleavage from the support. The method allowed an increase in the yields and a decrease in the overall process time by simplification of the purification steps.

A completely new strategy involving phosphoramidates was published in 2004 by Borch et al.[21] The strategy employs the generation of highly active pyrrolidinium phosphoramidates **14** as a result of the in situ catalytic hydrogenolysis of the starting phosphoramidates **13** building the phosphoramidate zwitterion intermediates **14** (Scheme 10).

The pyrrolidinium phosphoramidates **14** reacted rapidly by a nucleophilic attack with a pyrophosphate salt to yield the nucleoside triphosphate. The advantages of the reaction are short reaction times, facile purification steps, and good yields of 55–77%. The hydrogenolysis, including the formation of **14**, is finished within 1 to 2 h, and just a filtration of the catalyst is needed. The next reaction using tris(tetra-*n*-butylammonium)hydrogen pyrophosphate was quantitative after only 10 to 30 min and needed just an anion-exchange chromatography for isolation of

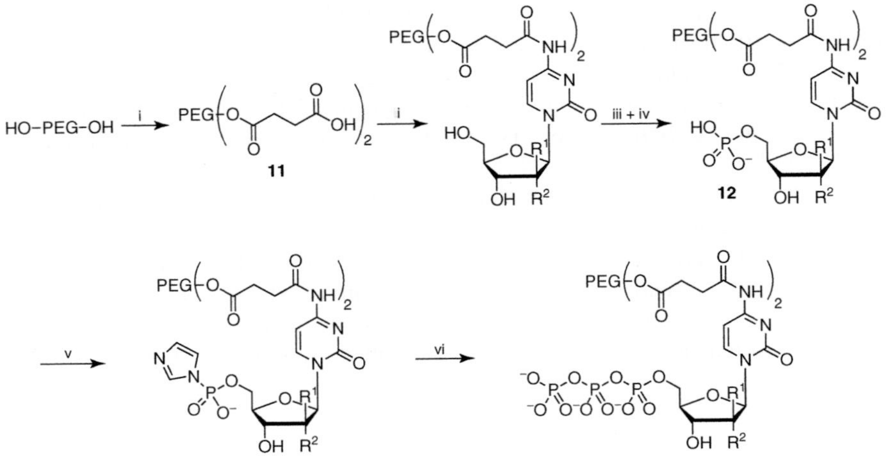

Scheme 9. *Reagents and conditions:* (i) succinic anhydride, DMAP, pyridine, rt, 2 days; (ii) nucleoside, DCC, HOBT, CH_2Cl_2, DMF, 60°C, 6 h;(iii) $POCl_3$, $PO(OEt)_3$/60°C or CH_3CN/0°C, 1 h; (iv) TEAB (1 M), pH 7.5; (v) 1,1′-carbonyldiimidazole, DMF, rt, 3 h; (vi) $[(HNBu_3)_2]H_2P_2O_7$, DMF, rt, 24 h.

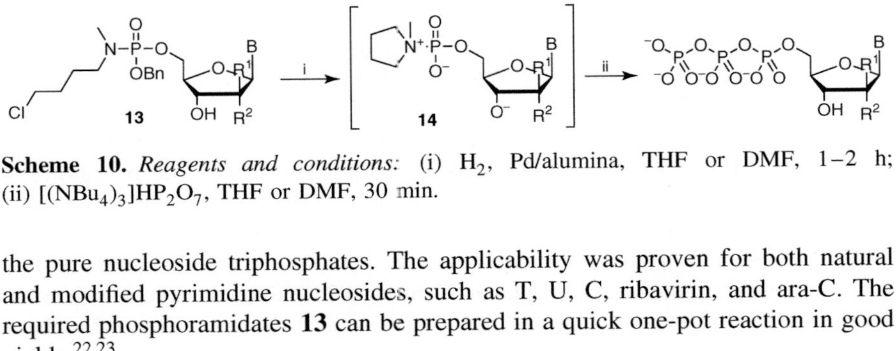

Scheme 10. *Reagents and conditions:* (i) H_2, Pd/alumina, THF or DMF, 1–2 h; (ii) $[(NBu_4)_3]HP_2O_7$, THF or DMF, 30 min.

the pure nucleoside triphosphates. The applicability was proven for both natural and modified pyrimidine nucleosides, such as T, U, C, ribavirin, and ara-C. The required phosphoramidates **13** can be prepared in a quick one-pot reaction in good yields.[22,23]

2.1.4. Syntheses Involving H-Phosphonates.

Recently, *H*-phosphonates gained interest as a precursor for the synthesis of nucleoside triphosphates. *H*-Phosphonates are relatively stable and can be generated in high yields. Therefore, these compounds offer an easy-to-handle class of molecules that can be converted in a variety of reactions to highly activated phosphates. A one-pot method for the generation of nucleoside triphosphates starting from nucleoside-5′-*H*-phosphonates **15** via pyridinium phosphoramidates was recently published by Peterson et al.[24] The reaction, shown in Scheme 11, involves treatment of 5′-*H*-phosphonates **15** with an excess of TMSCl and pyridine in DMF and further oxidation with iodine as the first step.

The mechanism proposed involves the formation of silyl-*H*-phosphonate or bis(silyl)phosphite, followed by oxidation and generation of the reactive pyridinium

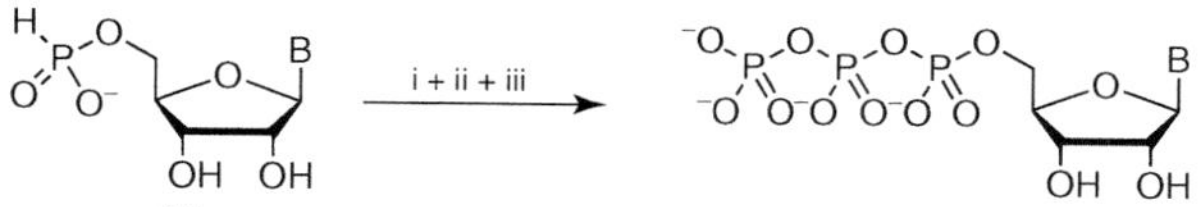

Scheme 11. *Reagents and conditions:* (i) 5–8 equiv. TMSCl, 25 equiv. pyridine, DMF, 5 min; (ii) iodine, DMF, 2 min; (iii) [(NBu$_4$)$_3$]HP$_2$O$_7$, 30 min.

phosphoramidate. A nucleophilic attack of tris(tetra-*n*-butylammonium)hydrogen pyrophosphate led to the nucleoside triphosphate. The procedure was applied to all natural nucleosides and the nucleoside analogues ribavirin and 6-methylpurine ribonucleoside, and the corresponding triphosphates were obtained in moderate yields of 26–41% after purification by gel chromatography followed by preparative reversed-phase HPLC. Formation of dinucleoside di- and triphosphates as a result of partial hydrolysis may be responsible for these low yields. However, the method is of interest because of the easy and fast reaction; the total reaction time is less than 45 min. In addition, the *H*-phosphonate precursor **16** can be synthesized in a rapid phosphitylation of 2′,3′-*O*-isopropylidene-protected nucleosides.[25] Further deprotection generated the *H*-phosphonates **15** in yields of 71–85% (Scheme 12).[26]

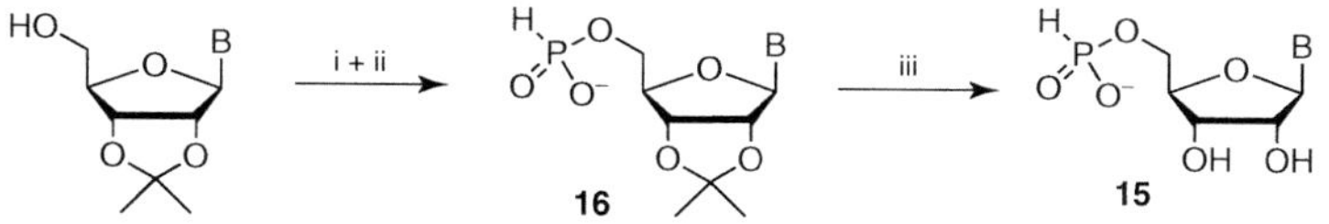

Scheme 12. *Reagents and conditions:* (i) PCl$_3$/CH$_2$Cl$_2$ or salicyl phosphorchloridite/pyridine/dioxane, −20–22°C, 5 h; (ii) H$_2$O, 10 min; (iii) aq. trifluoroacetic acid, 45 min.

More recently, Morvan et al. reported on a synthesis of different DNA and RNA triphosphates on a solid support.[26] Here, the *H*-phosphonate **18** was generated by the reaction of controlled pore glass–supported oligonucleotide **17** with a solution of diphenyl phosphite in pyridine, followed by hydrolysis. In a further two-step activation–phosphorylation procedure, *H*-phosphonate **18** was oxidized to an activated 5′-phosphorimidazolidate **19** by carbon tetrachloride in the presence of imidazole and *N,O*-bistrimethylsilylacetamide. Triphosphates **20** were obtained by subsequent reaction of an excess of (tri-*n*-butylammonium)pyrophosphate with **19** (Scheme 13). The procedure was described for a variety of 5′-triphosphates and oligonucleotides with different lengths and sequences containing different 5′-terminal nucleotides and with and without internal sugar-backbone modifications in yields ranging from 26 to 76%.

2.1.5. Syntheses Using Other Intermediates for Activation of Nucleoside-5′-Phosphates.

Originally used as a protecting group for the syntheses of oligonucleotides,[27] 8-quinoyl nucleoside 5′-phosphates **21** also found application

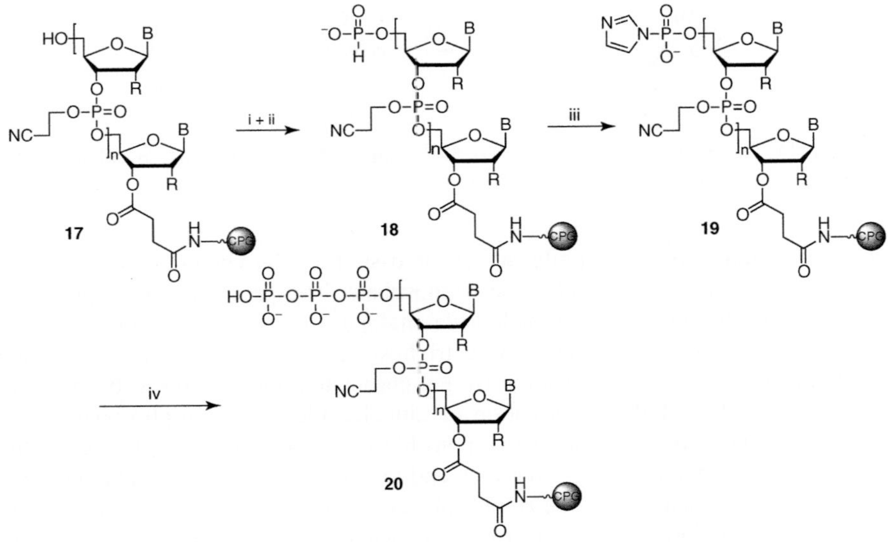

Scheme 13. *Reagents and conditions:* (i) diphenyl phosphite, pyridine, rt, 30 min; (ii) 100 mM (EtNH$_3$)$^+$ HCO$_3^-$ pH 8.0, rt, 2 h; (iii) imidazole, N,O-bistrimethylsilylacetamide, CCl$_4$, CH$_3$CN, triethylamine, rt, 5 h; (iv) [(NHBu$_3$)$_3$]HP$_2$O$_7$, DMF, rt for 17 h or MW/60°C for 40 min.

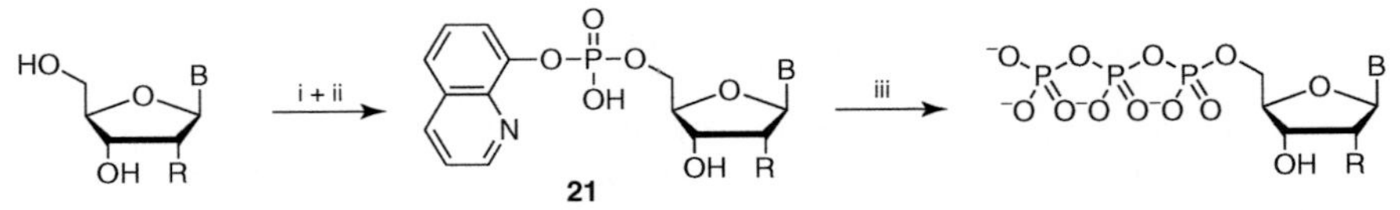

Scheme 14. *Reagents and conditions:* (i) 8-quinoyl phosphate, triphenylphosphine, 2,2'-dipyridyl diselenide, pyridine, rt, 6 h; (ii) H$_2$O, rt, 2 h; (iii) [(NHBu$_3$)$_2$]H$_2$P$_2$O$_7$, CuCl$_2$, DMSO, rt, 6 h.

for the syntheses of nucleoside 5'-di- and triphosphates.[28] The procedure is shown in Scheme 14. By reaction of 8-quinoyl phosphate with a nucleoside in the presence of triphenylphosphine and 2,2'-dipyridyl diselenide in dry pyridine, 8-quinoyl nucleoside 5'-phosphates **21** were obtained in yields of 70–87% after workup and chromatography. Deprotection of an 8-quinoyl group was successful by cupric chloride in aqueous dimethyl sulfoxide. When copper (II) chloride was used for the cleavage of the quinoyl group, a reaction with phosphoric acid, or rather, pyrophosphoric acid in dry DMSO, di- or triphosphates were obtained, respectively.

The method was applied to all natural ribonucleosides as well as thymidine. After treatment with Dowex and chromatography, the triphosphates were obtained in yields of 76–83%. Needless to say, the absence of water is absolutely necessary for the reaction.

Based on the *cyclo*Sal technique, developed originally as a prodrug concept,[29] we recently published a method for the synthesis of nucleoside (poly)phosphates.[30] *cyclo*Saligenyl-nucleoside monophosphates **22** are cyclic phosphate ester derivatives in which a saligenol and a nucleoside monophosphate are doubly esterified. The most important feature of this structure is the chemical differentiability of the various phosphate ester bonds that allow selective release of the nucleotide. The cleavage was initiated by nucleophilic attack of a phosphate or pyrophosphate salt on the neutral phosphate triester leading to an intermediate benzyl phosphate diester. Subsequently, the phosphate diester formed intermediately is cleaved spontaneously to yield the nucleoside di- and triphosphate, respectively, and a saligenol. An advantage of this method is avoidance of the synthesis of the monophosphate. Protected nucleosides were converted with *cyclo*Sal-phosphorochloridites **23** and, subsequently, oxidation with oxone to the corresponding *cyclo*Sal-nucleotides **22** in yields of 45–90%. Reaction with tetra-*n*-butylammonium pyrophosphate led to the corresponding triphosphates in 40–80% yield after removing the protecting groups, ion exchange, and RP-18 chromatography (Scheme 15).

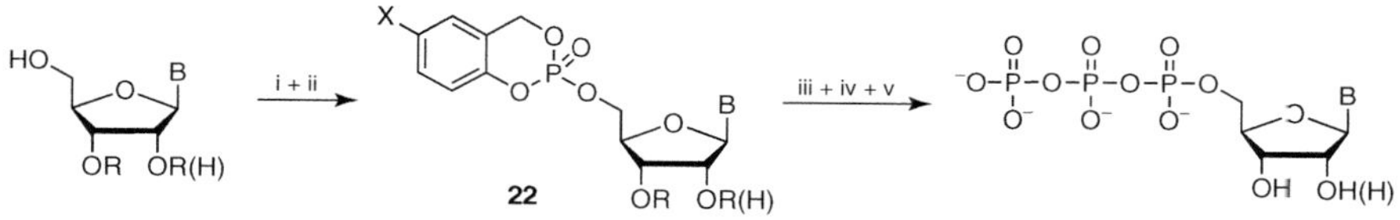

Scheme 15. *Reagents and conditions:* (i) 5-nitro- or 5-chloro-*cyclo*Sal-phosphorochloridite **22**, CH_3CN or DMF/THF, $-20\ °C$ to rt, 2–3 h; (ii) oxone in water, $-10°C$ to rt, 15 min for X = NO_2 or *t*-butyl hydroperoxide (5.5 M), $-20°C$, 1 h for X = Cl; (iii) $[(NBu_4)_3HP_2O_7]$, DMF, rt, 3–5 h; (iv) $MeOH/H_2O$, NEt_3 7:3:1, rt, 16 h; (v) ion exchange (NH_4^+).

The reactivity of the *cyclo*Sal-nucleotides **22** depends on the substituent in the aromatic system. The electrophilicity of the phosphorus atom and therefore the hydrolysis behavior is tunable by different substituent designs. Using *cyclo*Sal-nucleotides as prodrugs, slow hydrolysis is needed to ensure that the active agent reaches the intracellular target (e.g., the reverse transcriptase). Therefore. donor substituents such as the methyl group were used. In contrast, for use as chemical reagents, short reaction times are needed and therefore acceptor substituents such as the chloro or nitro-residue were used. Phosphorochloridites **23** were prepared in 70–80% yield from the corresponding saligenols with phosphorotrichloride, while the alcohols were prepared by reduction of 5-NO_2-salicyl aldehyde[31] or 5-Cl-salicylic acid,[32] respectively. The method was used for the synthesis of the natural ribo- and deoxyribonucleosides adenosine, guanosine, cytosine, thymidine, and uracil as well as for the analogues 5-[(*E*)-bromovinyl]uracil (BVdU) and carba-thymidine (carba-dT). Variations in the yields of 40–83% were probably a result of differences in the chromatographic properties. Counterions were replaced by NH_4^+ or Et_3NH^+ ions prior to ion-exchange chromatography to improve the chromatographic behavior. But removing the pyrophosphate salt was still difficult in some cases.

To solve this problem, we recently reported on a method that combines *cyclo*Sal-active esters **22** with a solid-phase strategy.[33] The general protocol is illustrated in Scheme 16.

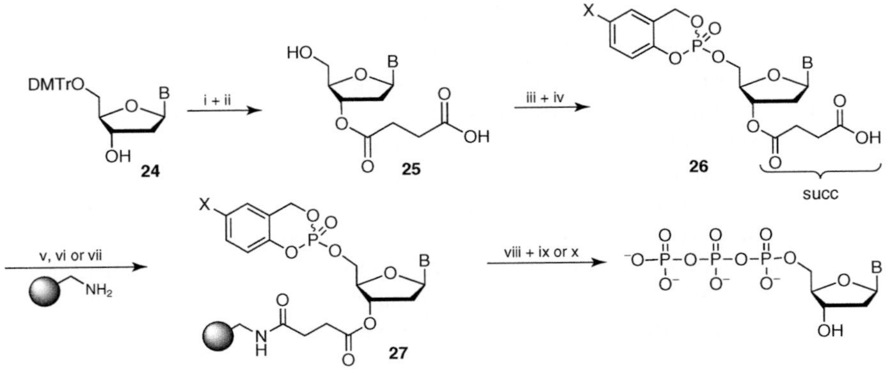

Scheme 16. *Reagents and conditions:* (i) DBU, succinic anhydride, CH_2Cl_2, rt, 45–120 min, CH_3COOH; (ii) TFA, CH_2Cl_2, rt, 8–120 min; (iii) 5-methylsulfonyl- or 5-chloro-*cyclo*Sal-phosphorchloridite 22, CH_3CN or DMF, DIPEA, $-35°C$ to rt, 2–3 h. (iv) oxone in water, $-10°C$, 15 min; (v) HOBt, DIC, DMF, rt, 16 h–2 days; (vi) TBTU, 4-ethylmorpholine, DMF, rt, 2 days; (vii) 3% TCA in CH_2Cl_2, rt, 24 h; (viii) $[(n\text{-Bu})_4N]_2H_2P_2O_7$ DMF, rt, 16 h; (ix) 25% aq. NH_3, 50°C, 2 h; (x) $CH_3OH/H_2O/Et_3N$ (7:3:1), rt, 20 h.

The procedure includes three steps. First, succinyl-linked acceptor-substituted-*cyclo*Sal-nucleotides **26** were attached to aminomethyl polystyrene as an insoluble solid support. Second, a nucleophile was reacted with the immobilized *cyclo*Sal-triesters **27**, which led to the formation of the target molecules still bound to the solid support. Finally, the cleavage of the products from the support was carried out under basic conditions.

The method was used for the di- and triphosphate synthesis of dC, U, dA, and the analog BVdU. Starting with 5'-*O*-DMTr-protected nucleoside **24**, the linker was attached followed by deprotection and further conversion to succinyl-*cyclo*Sal-nucleotides **26**. It is important to say that in case of deoxynucleosides there is no need for protecting groups at the 2'-hydroxy group because of use of the linker residue at this position. In the case of ribonucleoside uridine, the 2'- or 3'-position was first acetylated before the compound was attached to the linker. For the *cyclo*Sal moiety a chloro- or methylsulfonyl substituent was used as an acceptor. The synthesis was successful in good purity after extraction and gave succinyl-*cyclo*Sal-nucleotides **26** in yields of 56–83% after purification by silica gel chromatography. For the immobilization, 2'/3'-*O*-succinyl-*cyclo*Sal-nucleotides were anchored to aminomethyl polystyrene through an amide bond. Using a known procedure,[30] the immobilized 2'/3'-*O*-succinyl-*cyclo*Sal-nucleotides **27** were converted to nucleoside triphosphates in high purity. By simple washing of the polymer-bound target molecules, phosphate salts and the saligenol released were removed completely. This is a significant advantage, in contrast to the solution synthesis. After cleavage from the solid support and, in the case of ribonucleosides, of the acetyl-protecting group, purities of 78–90% were obtained. For increasing purity the target compounds were converted into their Et_3NH^+ salts and purified by a rapid chromatography on RP-18 silica gel.

Using *cyclo*Sal-nucleotides as an activated nucleotide moiety in reactions with other phosphate nucleophiles led to other structures, such as NDP- or NMP-sugars (here the 1-deprotonated peracetylated glycosides were used) and dinucleoside-(poly)phosphates or nucleotide analogs such as nucleoside-5'-β-γ-methylenetriphosphates (Scheme 17).[33]

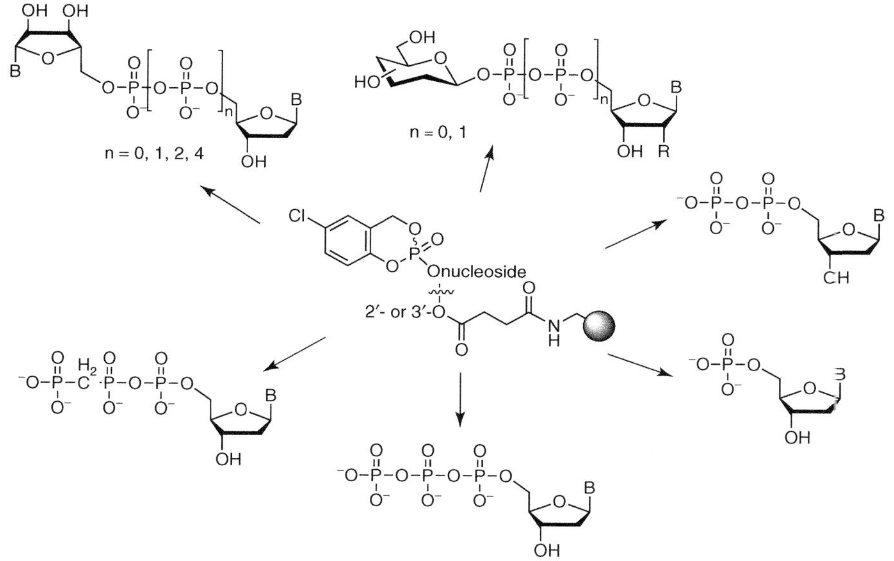

Scheme 17. Possibilities of *cyclo*Sal-nucleotides.

2.2. Synthesis by Nucleophilic Attack of a Triphosphate on an Activated Nucleoside

In addition to a wide application of reactions of pyrophosphate with an activated nucleoside monophosphate, Dixit and Poulter published the synthesis of ATP **3** by conversion of 5'-*O*-tosylated adenosine **28** with tetra-*n*-butylammonium triphosphate salt (Scheme 18).[34]

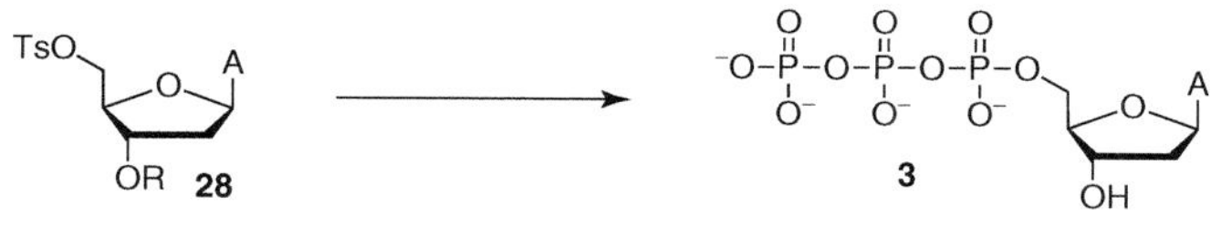

Scheme 18. *Reagents and conditions:* $[(n\text{-}Bu_4N)_4]HP_3O_{10}$, CH_3CN, 25°C, 48 h

The nucleophilic triphosphate salt was prepared by the titration of an aqueous solution of triphosphoric acid with tetra-n-butylammonium hydroxide. Reaction of the lyophilized salt with 5′-O-tosylate adenosine **28** gave ATP **3** in a good yield of 72% after flash chromatography on cellulose. A drawback is the long reaction time of 48 h. But the authors showed for the corresponding formation of the nucleoside diphosphate that the reaction time is reduced to 2 h by using 5′-O-tosylate-2′,3′-O-isopropylidene-adenosine as a starting material for the conversion with the pyrophosphate salt. In this case the yield increased from 74% to 93%.[34] They assumed that the isopropylidene group induces a preference for a more reactive conformation at C5′ or prevents unproductive hydrogen bonding between the pyrophosphate and the 2′- and 3′-hydroxyl groups. In addition to pyro- and triphosphate, methylenediphosphonate was used and also gave a good yield (72%).[34] All these reactions were conducted only with adenosine, but it shows the possibilities of the conversion of an activated nucleoside with nucleophilic phosphate salts.

2.3. Synthesis by Nucleophilic Attack of an Activated Mono- or Triphosphate on a Nucleoside/Nucleoside Diphosphate

Instead of the activation of the nucleoside, there are a few publications that describe a nucleophilic attack of a nucleoside or nucleotide on an activated phosphorus compound.

Sekine et al. reported on a zinc(II) chloride–catalyzed reaction of methyl phosphorimidazolidate **29** with the bis(tetrabutylammonium) salt of guanosine diphosphate **30** to yield P^1-methyl-P^3-(5′-guanosyl)triphosphate **31** in 86%.[35] The reaction is shown in Scheme 19.

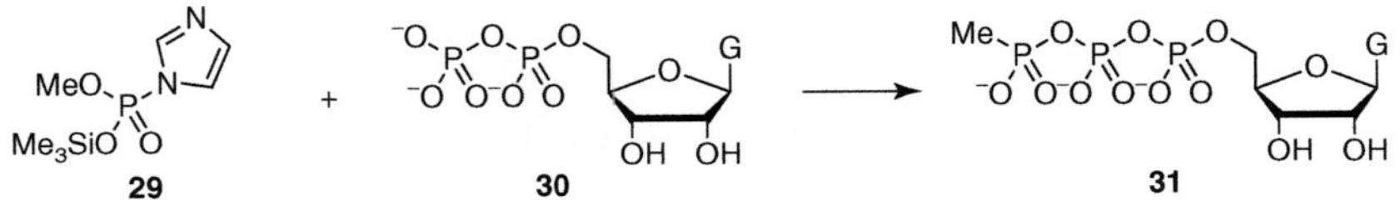

Scheme 19. *Reagents and conditions:* ZnCl$_2$, DMF, 25°C, 4 h.

Of course, the procedure was developed for P^1-methyl triphosphates. A synthesis of nucleoside triphosphates that needs the previous synthesis of the corresponding diphosphate is obviously not useful. But this method represents a possibility for the selective synthesis of methylated nucleoside triphosphates or other NTPs with modifications at the P^1-position.

A further solid-support method was described by Ahmadibeni and Parang and involves the nucleophilic attack of the nucleoside on an activated triphosphate.[36] A phosphoramidite **32** attached by a 4-acetoxybenzyl linker to an aminomethyl polystyrene resin reacted after oxidation and deprotection in a $1H$-tetrazole-mediated coupling with an unprotected nucleoside to yield the corresponding triphosphate (Scheme 20).

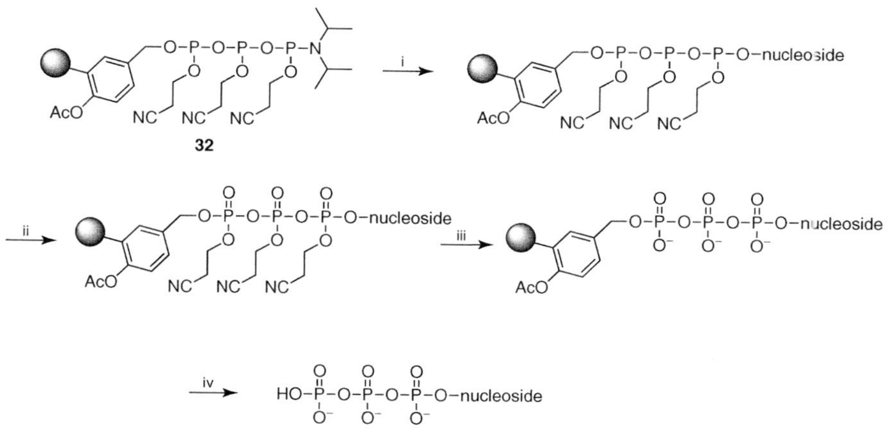

Scheme 20. *Reagents and conditions:* (i) nucleoside, 1*H*-tetrazole, THF/DMSO, rt, 28 h; (ii) *t*-BuOOH in decane, THF, rt, 1 h; (iii) DBU, THF, rt, 48 h; (iv) TFA/DCM/H_2O, rt, 25 min.

The reaction was successful for thymidine, uridine, adenosine, and 3′-azido-3′-deoxythymidine in an overall yield of 60–79% with a purity of 78–91%. The phosphitylating agent **32** can be prepared by a four-step reaction in an excellent yield of 94% and further coupling to the resin-attached linker (Scheme 21).

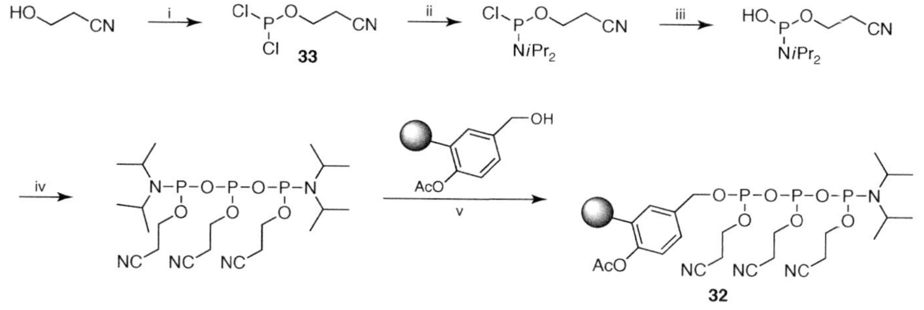

Scheme 21. *Reagents and conditions:* (i) PCl_3, CH_3CN, rt, 10 min; (ii) *i*-Pr_2NH, CH_3CN, rt, 30 min; (iii) H_2O, CH_3CN, rt, 10 min; (iv) 0.5 equiv **33**, CH_3CN, rt, 45 min; (v) CH_3CN/THF, 1*H*-tetrazole, rt, 24 h.

The benefit of solid-phase reactions is, of course, the simple washing and filtration steps after each reaction step. Only the target molecules have to be purified using C_{18} SepPak.

Very recently, Huang et al. published another approach that includes an activated triphosphate as a phosphitylating reagent.[37,38] Based on the studies of Ludwig and Eckstein,[12] a mild in situ phosphitylating reagent **34** was formed by the reaction of salicyl phosphorochloridite 5 with tributylammonium pyrophosphate in a first step. Nucleophilic attack of an unprotected nucleoside and oxidation followed by hydrolysis gave the corresponding triphosphate (Scheme 22).

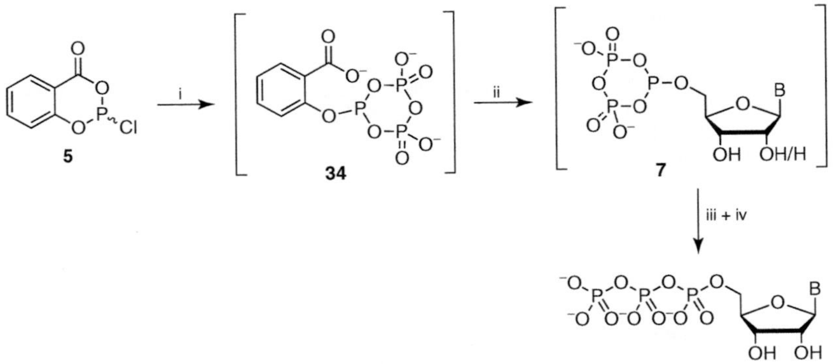

Scheme 22. *Reagents and conditions:* (i) $HNBu_3$ H_2PO_4, tributylamine, DMF or DMF/DMSO 1:1, 25°C, 30 min; (ii) nucleoside, DMF, or DMF/DMSO 1:1, 25°C, 1.5 h; (iii) I_2/pyridine/water; (iv) water.

The reaction was carried out for all natural ribo-[37] and deoxyribonucleosides[38] and ethanodeoxyadenosine,[38] but the triphosphates were isolated in low yields, from 19% to 46%. However, this rapid one-pot reaction offers the possibility for a fast synthesis starting directly with the nucleoside, without previous protection or deprotection reactions. Of course, the procedure also provides opportunities for different oxidizing agents, as described before for the method of Ludwig and Eckstein[12] to gain 5′-(α-P-thiotriphosphates), 5′-(α-P-boranotriphosphates), or 5′-(α-P-selenotriphosphates).

3. CONCLUDING REMARKS

Since the first chemical synthesis, isolation, and characterization of ATP by Baddiley, Michelson, and Todd in 1949,[39] the development of new methods for the synthesis of NTPs gain more and more interest. Just in the last decade many new approaches in this field were published. Some of the procedures have short reaction times, are easy to handle, and give satisfying yields for some nucleosides, but there is no universal method for all of them. That is, of course, a problem for biochemical and medicinal use, where a fast and noncomplicated access to the triphosphates is required. In addition, all solution syntheses need extensive final cleaning steps by chromatography. In our opinion, in the future, solid-phase strategies should be taken into greater consideration. The major amount of by-product and excess phosphate salt can be removed by simple washing steps. Furthermore, the use of commercially available solid-phase-linked reactive precursors may offer easy access to the important class of nucleoside triphosphate for many scientists.

Acknowledgment

The authors thank the University of Hamburg and the Deutsche Forschungsgemeinschaft for continuous and generous support.

ABBREVIATIONS

A	adenine or adenosine
ara-C	arabinofuranosyl cytidine
ATP	adenosine triphosphate
B	pyrimidine or purine base
BvdU	5-(E)-bromovinyl
C	cytosine
carba-dT	carba-thymidine
*cyclo*Sal	*cyclo*Saligenyl
dA	deoxyadenosine
DBU	diazabicyclo undecen
dC	deoxycytosine
DCC	N,N-dicyclohexyl carbodiimide
dG	deoxyguanosine
DIC	N,N-diisopropyl carbodiimide
DIPEA	diisopropylethylamine
DMAP	dimethyl aminopyridine
DMF	dimethylformamide
DMSO	dimethyl sulfoxide
DMTr	dimethyl trityl
DNA	deoxyribonucleic acid
HOBT	1-hydroxybenzotriazole
HPLC	high-performance liquid chromatography
MW	microwave
NDP	nucleoside diphosphate
NMP	nucleoside monophosphate
NTP	nucleoside triphosphate
PCR	polymerase chain reaction
PEG	poly(ethylene glycol)
RNA	ribonucleic acid
RP	reversed phase
T	thymidine
TBTU	2-($1H$-benzotriazole-1-yl)-1,1,3,3-tetramethyluranium tetrafluoroborate
TCA	trichloroacetic acid
TFA	trifluoroacetic acid
THF	tetrahydrofurane
TMS	trimethylsilyl
U	uridine

REFERENCES

1. Vaghefi, M. Nucleoside Triphosphates and Their Analogs, Taylor & Francis, Boca Raton, FL, **2005**, pp. 122.
2. Burgess, K.; Cook, D. *Chem. Rev.* **2000**, *100*, 2047–2059.
3. (a) Yoshikawa, M.; Kato, T.; Takenishi, T. *Tetrahedron Lett.* **1967**, *50*, 5065–5068.
   (b) Yoshikawa, M.; Kato, T.; Takenishi, T. *Bull. Chem. Soc. Jpn.* **1969**, *42*, 3505–3508.
4. Ludwig, J. *Acta Biochim. Biophys. Acad. Sci. Hung.* **1981**, *16*, 131–133.
5. Ludwig, J. *Bioact. Mol.* **1987**, *3*, 201–204.
6. Ruth, J. L.; Cheng, Y.-C. *Mol. Pharmacol.* **1981**, *20*, 415–422.
7. Eckstein, F. *Annu. Rev. Biochem.* **1985**, *54*, 367–402.
8. Kovács T. Z.; Ötvös, L. S. *Tetrahedron Lett.* **1988**, *29*, 4525–4528.
9. Knoblauch, B. H. A.; Müller, C. E.; Järlebark, L. Lawoko, G.; Kottke, T.; Wikstrom, M. A.; Heilbronn, E. *Eur. J. Med. Chem.* **1999**, 34, 809–824.
10. Gillerman, I.; Fischer, B. *Nucleosides Nucleotides Nucleic Acids* **2010**, *29*, 245–256.
11. Mishra, N. C.; Broom, A. D. *J. Chem. Soc. Chem. Commun.* **1991**, 1276–1277.
12. Ludwig, J.; Eckstein, F. *J. Org. Chem.* **1989**, *54*, 631–635.
13. Schoetzau, T.; Holletz, T.; Cech, D. *J. Chem. Soc. Chem. Commun.* **1996**, *54*, 387–388.
14. Gaur, R. K.; Sproat, B.S.; Krupp, G. *Tetrahedron Lett.* **1992**, *33*, 3301–3304.
15. Moffat, J. G. *Can. J. Chem.* **1964**, *42*, 599–604.
16. Chambers, R. W.; Khorana, H. G. *J. Am. Chem. Soc.* **1957**, *79*, 3747–3755.
17. Moffat, J. G.; Khorana, H. G. *J. Am. Chem. Soc.* **1961**, *83*, 649–658.
18. van Boom, J. H.; Crea, R.; Luyten, W. C.; Vink, A. B. *Tetrahedron Lett.* **1975**, *32*, 2779–2782.
19. Hoard, D. E.; Ott, D. G. *J. Am. Chem. Soc.* **1965**, *87*, 1785–8788.
20. Crauste, C.; Périgaud, C.; Peyrottes, S. *J. Org. Chem.* **2009**, *74*, 9165–9172.
21. Wu, W.; Freel Meyers, C. L.; Borch, R. F. *Org. Lett.* **2004**, *6*, 2257–2260.
22. Freel Meyers, C. L.; Hong, L.; Joswig, C.; Borch, R. F. *J. Med. Chem.* **2000**, *43*, 4313–4318.
23. Tobias, S. C.; Borch, R. F. *J. Med. Chem.* **2001**, *44*, 4475–4480.
24. Sun, Q.; Edathil, J. P.; Wu, R.; Smidansky, E. D.; Cameron, C. E.; Peterson, B. R. *Org. Lett.* **2008**, *10*, 1703–1706.
25. van der Marel, G. A. *Tetrahedron Lett.* **1986**, *27*, 2661–2664.
26. Zlatev, I.; Lavergne, T.; Debart, F.; Vasseur, J.-J.; Manoharan, M.; Morvan, F. *Org. Lett.* **2010**, *12*, 2190–2193.
27. Takaku, H.; Shimada, Y. *Chem. Soc. Jpn., Chem. Lett.* **1975**, 873–874.
28. Takaku, H.; Konishi, T.; Hata, T. *Chem. Soc. Jpn., Chem. Lett.* **1977**, 655–658.
29. (a) Meier, C. *Eur. J. Org. Chem.* **2006**, 1081–1102. (b) Ducho, C.; Görbig, U.; Jessel, S.; Gisch, N.; Balzarini, J.; Meier, C. *J. Med. Chem.* **2007**, *50*, 1335–1346. (c) Gisch, N.; Balzarini, J.; Meier, C. *J. Med. Chem.* **2007**, *50*, 1658–1667. (d) Jessen, H. J.; Balzarini, J.; Meier, C. *J. Med. Chem.* **2008**, *51*, 6592–6598. (e) Gisch, N.; Balzarini, J.; Meier, C. *J. Med. Chem.* **2008**, *51*, 6752–6760. (f) Gisch, N.; Pertenbreiter, F.; Balzarini, J.; Meier, C. *J. Med. Chem.* **2008**, *51*, 8815–8123.
30. Warnecke, S.; Meier, C. *J. Org. Chem.* **2009**, *74*, 3024–3030.

31. Arenz, C.; Giannis, A. *Eur. J. Org. Chem.* **2001**, 137–140.
32. Yoon, N. M.; Pak, C. S.; Brown, H. C.; Krishnamurthy, S.; Stocky, T. P. *J. Org. Chem.* **1973**, *38*, 2786–2792.
33. Tonn, V. C.; Meier, C. *Chem. Eur. J.* **2011**, *17*, 9823–9842.
34. Dixit, V. M.; Poulter, C. D. *Tetrahedron Lett.* **1984**, *25*, 4055–4048.
35. Kadokura, M.; Wada, T.; Urashima, C.; Sekine, M. *Tetrahedron Lett.* **1997**, *38*, 8359–8362.
36. Ahmadibeni, Y.; Parang, K. *Org. Lett.* **2005**, *7*, 5589–5592.
37. Caton-Williams, J.; Lin, L.; Smith, M.; Huang, Z. *R Soc. Chem., Chem. Commun.* **2011**, *47*, 8142–8144.
38. Caton-Williams, J.; Smith, M.; Carrasco, N.; Huang, Z. *Org. Lett.* **2011**, *13*, 4156–4159.
39. (a) Baddiley, J.; Michelson, A. M.; Todd, A. R. *J. Chem. Soc.* **1949**, 582–586.
    (b) Michelson, A. M.; Todd, A. R. *J. Chem. Soc.* **1949**, 2487–2490.

6 Mononucleotide Prodrug Synthetic Strategies

SUZANNE PEYROTTES and CHRISTIAN PÉRIGAUD

Institut des Biomolécules Max Mousseron, Université Montpellier 2, Montpellier, France

1. INTRODUCTION

During the past two decades, an impressive number of nucleoside analogues and their related prodrugs were developed for the treatment of infectious diseases and cancers, and new derivatives are still under clinical development. Once inside the cells, nucleoside analogues must be phosphorylated to their corresponding 5'-triphosphate forms, which interfere with nucleic acid biosynthesis as viral and/or cellular DNA and RNA polymerase inhibitors or as chain terminators. During this pharmacological process, few steps are often considered to be limiting, such as the absorption of nucleoside analogues, and their phosphorylation by nucleoside kinases. To overcome some of these limitations, various prodrug approaches have been used to improve oral absorption (lipophilic derivatives should cross cell membranes easily by passive diffusion) and/or to bypass the monophosphorylation step. Herein we focus on prodrug approaches that allow the intracellular delivery of nucleoside 5'-monophosphates (5'-mononucleotides), also called pronucleotide approaches.

These strategies (concept and biological data) have been well documented in the literature.[1] The aim of this chapter is to highlight recent synthetic pathways for pronucleotide preparation, primarily those reported during the past decade. These synthetic pathways may involve PIII and/or PV chemistry and are described according to the structure of the final pronucleotide (e.g., phosphotriester, phosphodiester, phosphoramidate), most of them illustrated in Figure 1. Because some of these compounds are prepared and evaluated as a diastereomeric mixture, and because the two diastereomers have been shown to exhibit different biological activity, we also report recent attempts for their diastereoselective synthesis.

Chemical Synthesis of Nucleoside Analogues, First Edition. Edited by Pedro Merino.
© 2013 John Wiley & Sons, Inc. Published 2013 by John Wiley & Sons, Inc.

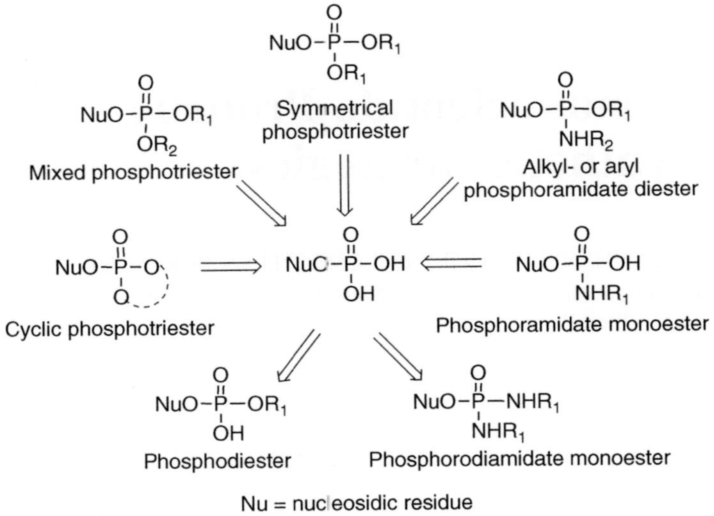

Figure 1. Generic structures of mononucleotide prodrugs.

2. PHOSPHOTRIESTER PRODRUGS OF NUCLEOSIDE ANALOGUES

Neutral phosphotriesters were one of the first series of nucleotide derivatives designed two decades ago, with the aim to increase the lipophilicity of the nucleoside 5′-monophosphate by masking the negative charges of the phosphate moiety with simple alkyl and/or aryl groups.[2] Thus, they should be taken up easily by cells through passive diffusion. However, due to their chemical stability under physiological conditions and to the fact that eukaryotic phosphotriesterase activity has yet to be reported, they are no longer considered to be effective prodrugs.

To retain the phosphotriester skeleton as pronucleotides, various protecting groups of the nucleoside 5′-monophosphates have been proposed. In this respect, the nature of the pro-moiety should allow conversion of the phosphotriester derivative to a phosphodiester intermediate, either by a nonenzymatic or an enzymatic process, and then the final release of the corresponding 5′-mononucleotide through phosphodiesterase activity, as an example.

2.1. Symmetrical Phosphotriesters

The term *symmetrical phosphotriester* refers to derivatives where the phosphorus atom is substituted by two identical biolabile protecting groups. Such pronucleotides have been synthesized by coupling the nucleoside (eventually protected) with various reagents, including both PV and PIII species (Figure 2): (1) the esterification of a preformed 5′-mononucleotide, (2) the preparation of a disubstituted chlorophosphate and further coupling of the nucleoside, (3) the activation of an *H*-phosphonate

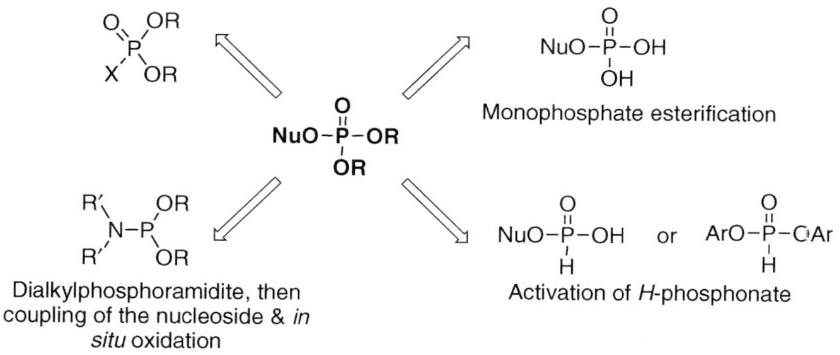

Figure 2. Main strategies for symmetrical phosphotriester synthesis.

intermediate, and (4) the use of di(alkyl)phosphoramidite, which is coupled with the nucleoside and in situ oxidation of the resulting phosphite triester.

2.1.1. Esterification of a Nucleoside 5′-Monophosphate.

The esterification procedure was one of the first developed, involving activation of the nucleoside 5′-monophosphate with an excess of 2,4,6-triisopropylbenzenesulfonyl chloride (TPSCl)[3] and then subsequent addition of the desired alcohol.[4] The yield of the reaction is highly dependent on the nature of the alcohol engaged (primary or secondary hydroxyl group and/or the presence of a bulky substituent).[4b,5]

Another method has also proved to be convenient for the synthesis of bis(acyloxymethyl)phosphotriester. Briefly, the nucleoside 5′-monophosphate is treated by a large excess of halogenomethyleneoxy ester (a few reagents are commercially available, such as the chloromethylpivalate, POM-Cl) in the presence of triethylamine,[6] and the desired phosphotriester derivatives were isolated in modest yields.

2.1.2. Use of Chloro- or Bis(alkyl)phosphate Intermediates.

The reactivity of chlorophosphate or bis(chloro)phosphate has been used extensively for the preparation of symmetrical and mixed (see Section 2.2.) phosphotriester derivatives. For example, Hwang and Cole[7] have described the use of bis(POM)phosphoryl chloride as a convenient reagent to obtain the corresponding bis(POM)phosphotriester derivatives (Figure 3) in modest yield (ca. 50%).

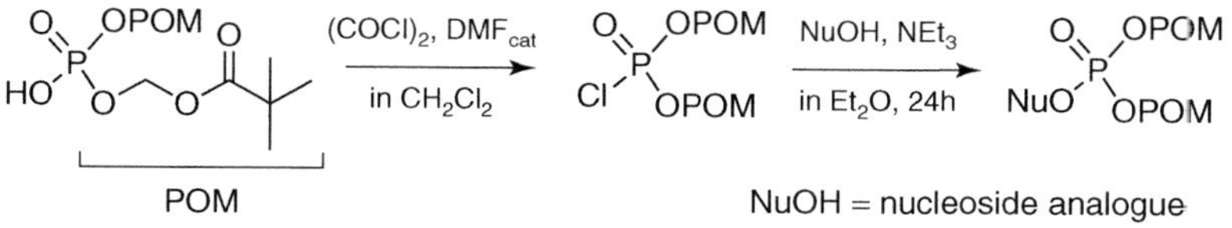

Figure 3. Synthesis of bis(POM)phosphotriester derivatives.

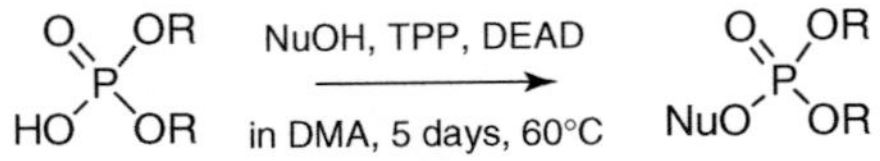

Figure 4. Synthesis of phosphotriester derivatives using the Mitsunobu reaction.

The bis(POM) reagent was prepared by transesterification of trimethylphosphate with POM-Cl and subsequent hydrolysis to the corresponding phosphodiester. The latter was converted into the chlorophosphate with oxalyl chloride and a catalytic amount of dimethylformamide. Interestingly, bis(alkyl)phosphate may be coupled with a nucleoside analogue under Mitsunobu reaction conditions (Figure 4; e.g., with triphenylphosphine (TPP) and diethylazodicarboxylate (DEAD)), affording the corresponding nucleoside phosphotriesters in modest yield (<50%).[8]

2.1.3. Activation of an H-Phosphonate Monoester.

Somehow, *H*-phosphonate chemistry has received less attention to the preparation of phosphotriester derivatives than that of phosphoramidate diesters (see Section 4.1). Nevertheless, it has been used successfully by the team of Kraszewski[4b] to obtain bis(aryl) nucleoside phosphotriesters. Thus, the nucleoside *H*-phosphonate monoester is activated by DPCD (diphenyl chlorophosphate), coupled with the first aryl moiety, and in situ oxidized in the presence of iodine. The resulting aryl nucleoside phosphodiester is then treated by TPSCl in the presence of *N*-methylimidazole (*N*-MeIm) to obtain the corresponding bis(aryl)phosphotriester, which may incorporate similar or different aryl substituents (see Figure 9).

2.1.4. Use of a Dialkylphosphoramidite Intermediate.

The phosphoramidite approach is probably the most convenient one to synthesize nucleoside phosphotriesters, and it has been used extensively by us[4a,9] and others[1c,10] during the past 20 years. Briefly, the phosphitylation reagent [i.e., bis(alkyl)-*N*,*N*-diisopropylphosphoramidite] is prepared initially from the commercially available dichloro-*N*,*N*-diisopropylphosphoramidite and the appropriate alcohol, corresponding to the precursor of the pro-moiety (Figure 5). Then, coupling of the nucleoside analogue (eventually protected on the other reactive groups) is

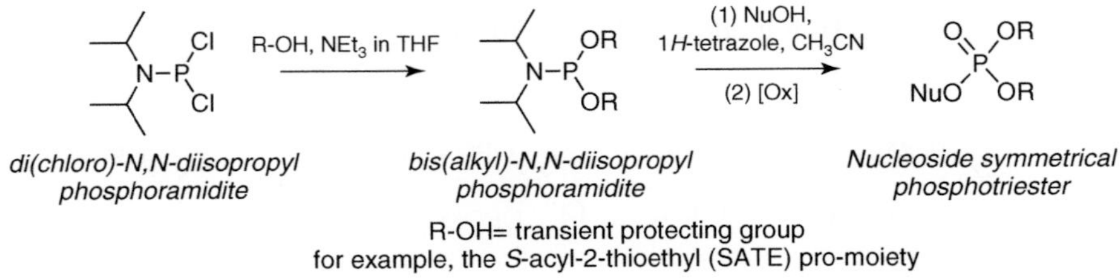

R-OH= transient protecting group
for example, the *S*-acyl-2-thioethyl (SATE) pro-moiety

Figure 5. Synthesis of phosphotriester derivative using PIII intermediates.

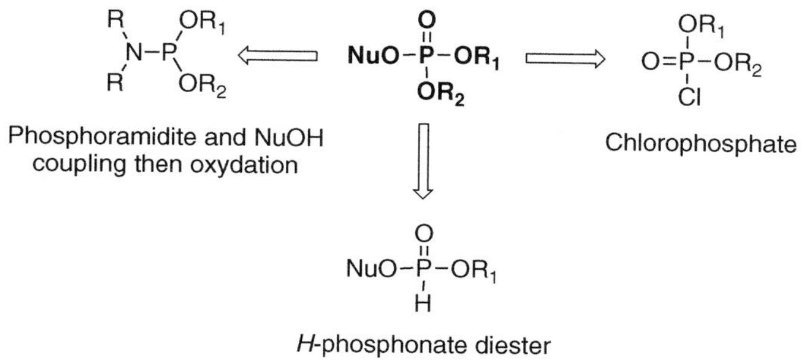

Figure 6. Main strategies for mixed phosphotriester synthesis.

carried out in the presence of $1H$-tetrazole followed by in situ oxidation using 3-chloroperbenzoic acid (m-CPBA)[10a] or $tert$-butylhydroperoxide (t-BuOOH).[1c]

2.2. Mixed Phosphotriesters

Mixed phosphotriester derivatives incorporating two different substituents as biolabile phosphate-protecting groups have been designed to overcome some of the limitations encountered during the development of symmetrical analogues. As an example of S-acyl-2-thioethyl (SATE) arylphosphotriester derivatives,[11] the first hydrolytic step is mediated by esterases, resulting in the formation of an arylphosphodiester intermediate which should be a substrate for phosphodiesterase activity, leading to the selective release of the nucleoside 5′-monophosphate. Such mixed phosphotriesters are usually obtained via a phosphoramidite or a chlorophosphate intermediate (Figure 6). The use of H-phosphonate chemistry has only been reported.[4b]

Using PV chemistry, phenyl(SATE)phosphotriester derivatives were obtained in a one-pot procedure[12] involving the coupling of commercial phenyldichlorophosphate with various SATE alcohols to give the corresponding phenyl(SATE)phosphorochloridates, which were used directly without further purification (Figure 7). This last intermediate was reacted with the nucleoside analogue in the presence of N-MeIm and afforded the desired mixed phosphotriesters in good yields (60 to 75% over the two steps).

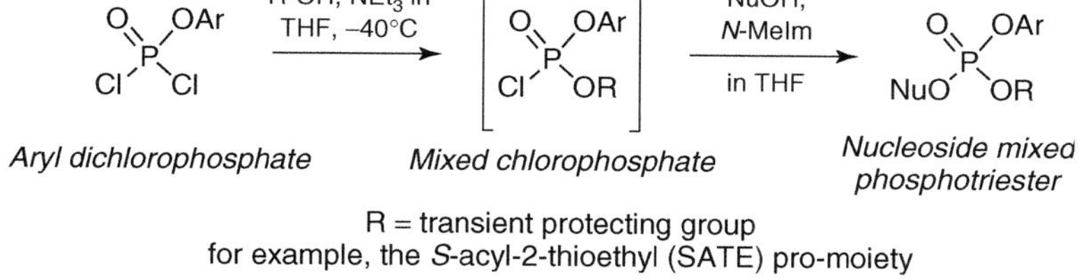

R = transient protecting group
for example, the S-acyl-2-thioethyl (SATE) pro-moiety

Figure 7. Synthesis of aryl(SATE)phosphotriester derivatives using PV intermediates.

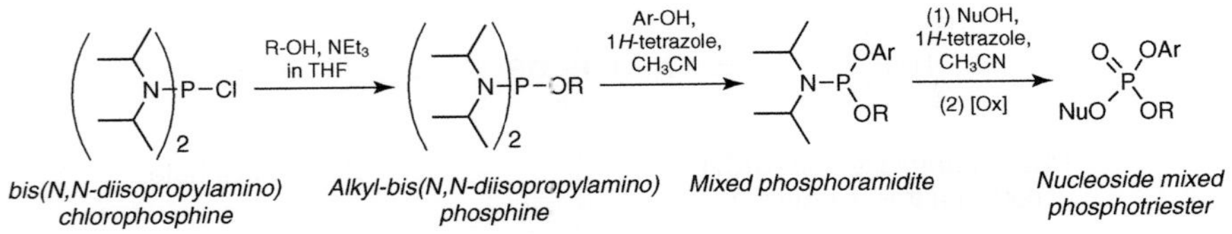

Figure 8. Synthesis of aryl(SATE)phosphotriester derivatives using PIII intermediates.

When the aryldichlorophosphate is not readily available, the phosphoramidite approach may be useful (Figure 8). Thus, the mixed phosphitylating agent was prepared from commercial bis(diisopropylamino)chlorophosphine with subsequent coupling of the SATE chain and then the aryl moiety in the presence of 1H-tetrazole. In the second step, the reaction of the phosphoramidite intermediate with the nucleoside analogue was carried out, followed by in situ oxidation of the resulting phosphite triester using t-BuOOH.[13]

Interestingly, Petersen et al.[14] have broadened the scope of the "mixed SATE pronucleotide concept" in designing a dual prodrug incorporating a nucleoside analogue, a SATE chain, and a nonnucleoside reverse transcriptase inhibitor (NNRTI) as an aryl substituent. The SATE-bis(N,N-diisopropylamino)phosphine was used as the starting material and was reacted successfully with the nucleoside analogue in the presence of a base (often, triethylamine) and then NNRTI, using diisopropylammonium tetrazolide as an activating agent, followed by oxidation with t-BuOOH to give the target compound in about 40% overall yield over the three steps.

Synthesis of nucleoside di(aryl)phosphotriesters was also accomplished starting from a nucleoside H-phosphonate monoester (Figure 9).[4b] This latter was coupled with a first aryl residue in the presence of DPCP (or pivaloyl chloride, PivCl), giving rise to formation of the corresponding aryl nucleoside H-phosphonate diester. Then, two procedures may be applied: iodine-promoted oxidative coupling of the second phenol derivative or its oxidation, leading to an aryl nucleoside phosphodiester, subsequently converted into the desired di(aryl)phosphotriester by coupling the second aryl substituent in the presence of TPSCl and N-MeIm.

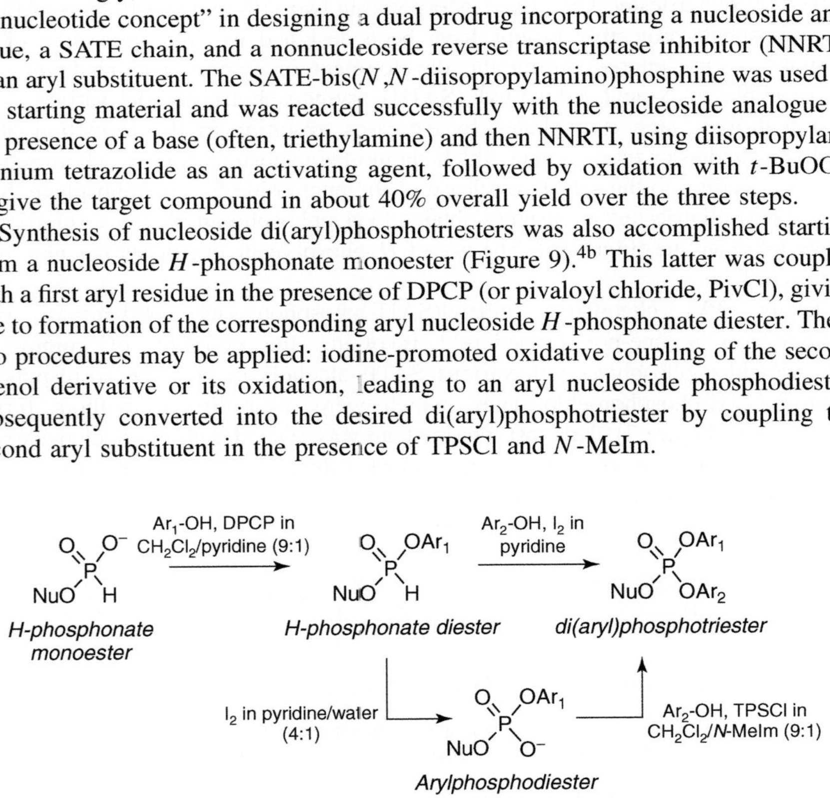

Figure 9. Synthetic pathways to di(aryl)phosphotriester derivatives.

2.3. Cyclic Phosphotriesters

2.3.1. Cyclo-Sal Pronucleotides.
The cyclosaligenyl (cyclo-Sal) phosphotriester approach has been developed extensively by Meier et al.[15] and was designed initially to release the nucleoside 5'-monophosphate selectively by a pH-driven chemical tandem hydrolysis mechanism. The synthesis of cyclo-Sal pronucleotides has been carried out using either PIII reagents (cyclic chlorophosphite or phosphoramidite) or PV intermediates (cyclic phosphorochloridate or nucleoside phosphorodichloridate), the latter approach often leading to lower yields (Figure 10). All pathways required preparation of the desired salicyl alcohols from the corresponding aldehydes, esters, or carboxylic acids.

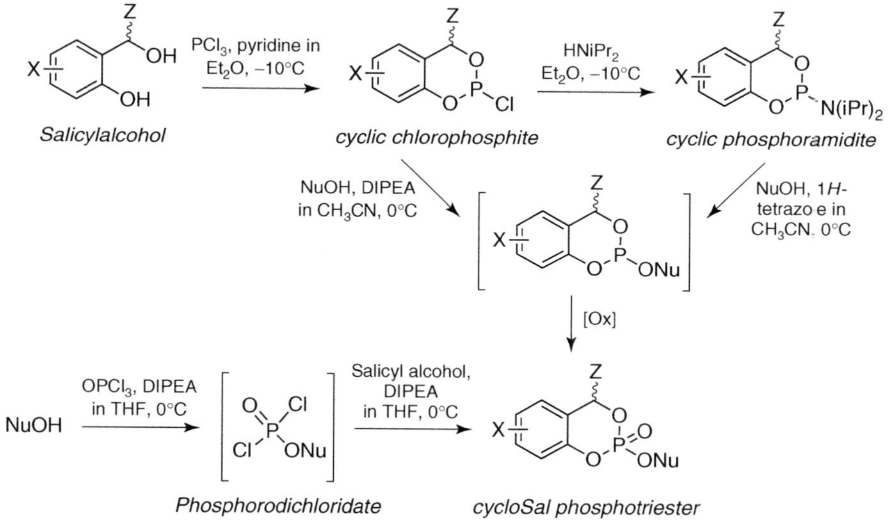

Figure 10. Synthetic pathways to cyclo-Sal phosphotriester derivatives.

Then the salicyl alcohol or its analogue is treated by phosphorus trichloride in the presence of pyridine to give the cyclic chlorophosphite.[16] This intermediate may be coupled directly with the nucleoside analogue in the presence of Hünig's base, and the resulting phosphite triester was oxidized in situ.[16c,17] Alternatively, and depending of the nature of the nucleoside analogue, the cyclic chlorophosphite was reacted with diisopropylamine to lead to the corresponding cyclo-Sal phosphoramidite, which was coupled with the nucleoside in the presence of a coupling agent such as 1H-tetrazole, imidazolium triflate, or pyridinium chloride.[18] Use of this last protocol, involving a phosphoramidite intermediate, seems to result in improved yields.

The synthetic pathway involving PV chemistry is a two-step one-pot reaction. First phosphorylation of the nucleoside analogue was performed using phosphorus oxychloride, and then the cyclo-Sal moiety was introduced.[19] This reaction was carried out in the presence of Hunig's base, in tetrahydrofuran or acetonitrile as solvent, and the phosphotriester derivative desired was usually obtained in modest to low yields (<40%). The last alternative is based on the use of a selected Cyclo-Sal phosphorochloridate (obtained previously by treatment of the salicyl alcohol with phosphorus oxychloride) coupled with the nucleoside in pyridine at a low temperature.[20]

2.3.2. HepDirect Pronucleotides.

Another type of six-membered ring containing pronucleotide (namely, HepDirect prodrugs) is based on a 1,3-cyclopropylphosphotriester skeleton and has been designed by Erion et al.[21] These compounds were prepared by coupling the nucleoside analogue of interest either to a phosphoramidite intermediate (PIII chemistry) or to a 4-nitrophenylphosphotriester reagent (PV chemistry). In either case, the commercially available (or previously synthesized) 1-aryl-1,3-propanediol is used as a starting material (Figure 11).

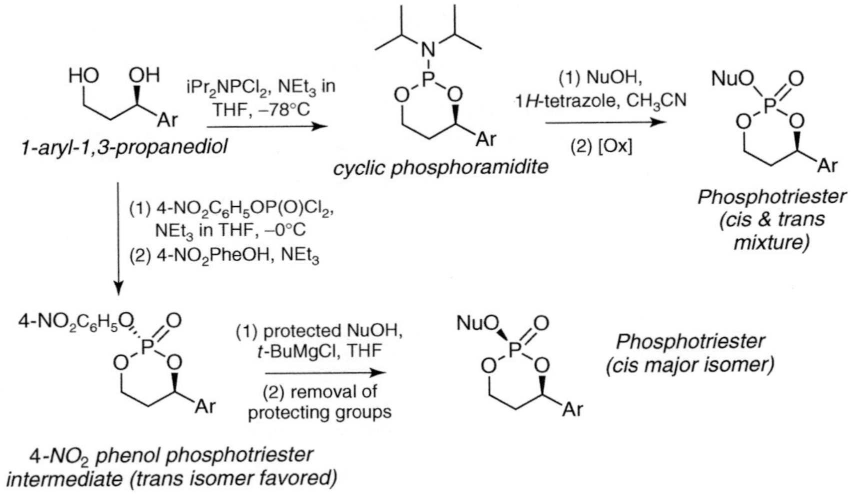

Figure 11. Two synthetic pathways to HepDirect pronucleotides.

On the one hand, the diol was reacted with diisopropylphosphoramidous dichloride and the resulting cyclic phosphoramidite was coupled directly with the suitable nucleoside analogue, eventually unprotected, in the presence of 1H-tetrazole. Then the phosphite triester was in situ oxidized with t-BuOOH to lead to the corresponding nucleoside phosphotriester.[22]

On the other hand, the diol was coupled with 4-nitrophenylphosphorodichloridate to generate the PV reagent.[23] The nucleophilic displacement of the 4-nitrophenyl group by the 5′-hydroxyl function of the protected nucleoside analogue was accomplished in the presence of a magnesium salt such as tert-butyl magnesium chloride (t-BuMgCl).[24]

These two protocols appeared not to be equivalent in terms of stereoselectivity. The cis- and trans-isomers refer to the relative stereochemistry of the activating aryl group and the nucleoside. The phosphoramidite procedure gave rise to a mixture (often not separable) of *cis*- and *trans*-phosphoesters, which is a disadvantage because only the *cis* form of such derivatives was shown to be hydrolyzed efficiently by the targeted enzymatic system (CYP3A4) to release the nucleoside 5′-monophosphate. Thus, the second protocol is the method of choice because it is stereospecific for the preparation of *cis*-pronucleotide but required the use of *O*-protected nucleosides. Consequently, this last synthetic pathway involved more steps and required the availability of sufficient quantities of starting material, especially the nucleoside analogue.

3. PHOSPHODIESTER PRODRUGS OF NUCLEOSIDE ANALOGUES

Phosphodiesters are unlikely to cross cell membranes, due to the presence of a negative charge on the phosphorus moiety. Despite this, several groups have synthesized and studied the biological properties of nucleotide–lipid or nucleotide–steroid conjugates, based on the premise that the lipophilic biolabile counterpart may facilitate cell penetration and/or targeted delivery of the nucleotides to tissues able to absorb lipids.

3.1. Phospholipid Conjugates

Phospholipid nucleoside conjugates mainly incorporate a glycerol scaffold. They were prepared using either PV or *H*-phosphonate chemistry, the nucleoside analogue being most often coupled in the last step (Figure 12). Thus, the primary (positions 1 and 3) or secondary (position 2) alcohol function of the glycerol (substituted by two lipidic chains preliminary coupled via either ether or ester bonds) was reacted with phosphorus oxychloride in the presence of pyridine[25] or triethylamine[26] and afforded the phospholipid intermediate. The latter was conjugated with the 5′-hydroxyl function of the desired nucleoside analogue (eventually bearing protecting groups on the other reactive groups) using either TPSCl[25,27] or dicyclohexylcarbodiimide (DCC)[26] in pyridine.

The second alternative involved preparation of the *H*-phosphonate monoester derivative of the lipid by treating the glycerol intermediate with

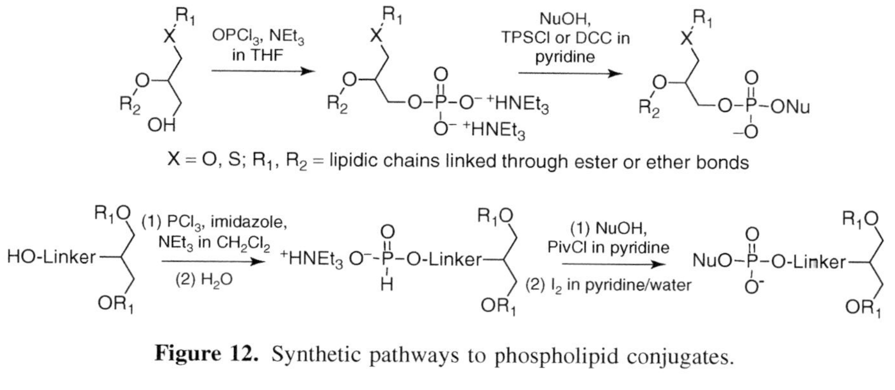

Figure 12. Synthetic pathways to phospholipid conjugates.

tris(imidazolyl)phosphine, generated in situ from PCl_3 and imidazole in the presence of triethylamine, and hydrolyzed by the addition of water. The resulting phospholipid intermediate was activated with PivCl and coupled with the nucleoside analogue to lead to an H-phosphonate diester, which was oxidized to the desired phosphodiester using iodine in aqueous pyridine.[28]

3.2. Steroid Conjugates

Zhao's group[29] has developed H-phosphonate chemistry primarily for the preparation of nucleoside phosphodiesters, including a steroid derivative. The synthetic pathway was based on the two-step one-pot transesterification of diphenylphosphite (DPP) in pyridine by adding the steroid and then the nucleoside analogue (or in the reverse order) (Figure 13). After purification by silica gel column chromatography, the H-phosphonate diester intermediate was oxidized using iodine in aqueous pyridine.

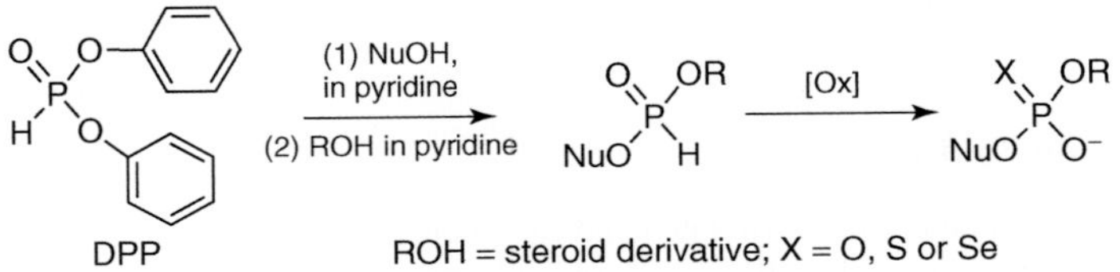

Figure 13. Synthesis of steroid conjugates.

In addition, this synthetic approach has also given access (from a single intermediate, the H-phosphonate diester) to the corresponding phosphorothioate and phosphoroselenoate derivatives, depending on the oxidation reagent (I_2, S_8, or Se), or, eventually, to the corresponding phosphoramidate (see Section 4) using the Atherton–Todd reaction.[30]

4. PHOSPHORAMIDATE PRODRUGS OF NUCLEOSIDE ANALOGUES

Analogous to the development of phosphotriesters, a large number of phosphoramidate-based prodrugs have been developed.[31] The initial study was performed by McGuigan and co-workers in 1992 and was related to arylphosphoramidate diesters incorporating amino acid residues.[32] Later, our laboratory and others have investigated the utility of phosphoramidate derivatives, including a SATE[33] or an haloalkyl/nitrofurannyl[34] group as biolabile moieties, and aminoacylphosphoramidate monoesters[35] as potential pronucleotides.

4.1. Phosphoramidate Diesters

The synthesis of nucleoside phosphoramidate diesters (Figure 14) relied mainly on the use of phosphorus oxychloride, or alkoxy- or aryloxydichlorophosphates

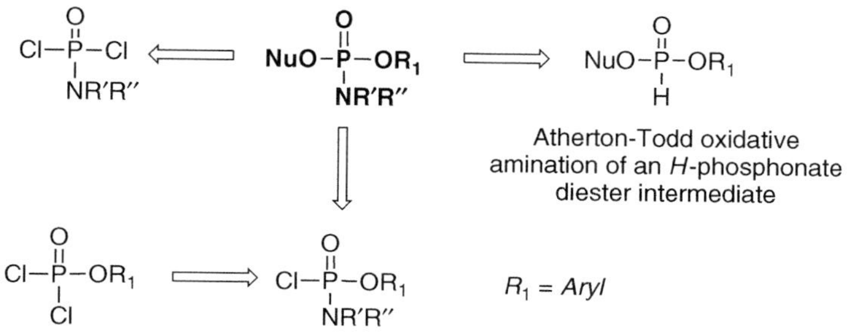

Figure 14. Main strategies for phosphoramidate diester synthesis.

as starting materials (PV chemistry) or H-phosphonate intermediates (PIII chemistry).

4.1.1. Aryloxyphosphoramidates. Most often, aryloxyphosphoramidate monoesters of nucleoside analogues were synthesized in a few steps from phosphorus oxychloride or commercially available phenyldichlorophosphate (Figure 15).[31c,g] The synthesis of the aryloxydichlorophosphate involved phosphorylation of the required phenol analogue with phosphorus oxychloride in the presence of triethylamine. The resulting intermediate was engaged directly (as a crude material) in the next step (i.e., coupling of the esterified amino acid), to afford the aryloxyaminoacylchlorophosphate, which was eventually purified by flash chromatography. The latter was reacted with the nucleoside analogue in the presence of a base such as N-MeIm,[36] triethylamine,[37] t-BuMgCl,[38] or t-BuMgBr.[39]

Another alternative using PIII intermediates (Figure 16) and a three-component reaction has also been reported.[40] Briefly, an arylphosphodichloridite was treated with the selected nucleoside analogue in the presence of *tert*-butanol to afford an aryl H-phosphonate intermediate, in situ oxidized by N-chlorosuccinimide (NCS) in the presence of the amino acid ester and triethylamine. This one-pot procedure led to isolating the desired nucleoside aryl(aminoacyl)phosphoramidate in good yields (>60%).

One should note that within the arylphosphoramidate diester series, replacement of the natural α-amino acid with nonnatural β-amino acid or alkyl amine resulted in loss of the biological activity and was associated with the mechanism of decomposition of such pronucleotides.[33,45]

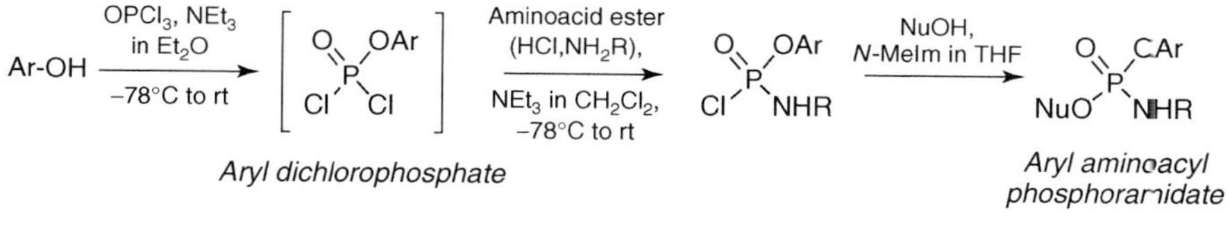

Figure 15. Synthesis of nucleoside aryl(aminoacyl)phosphoramidate diesters.

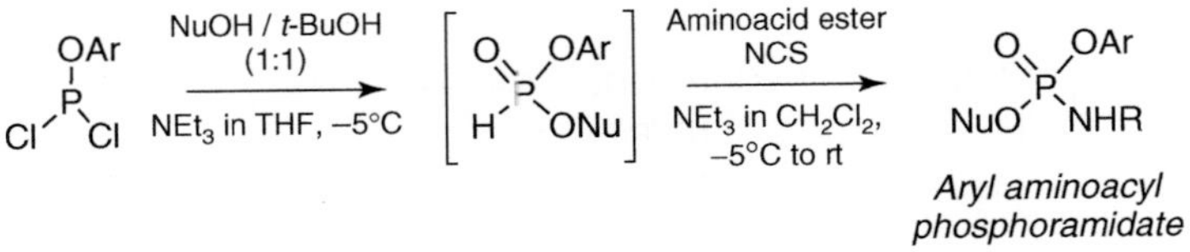

Figure 16. One-pot synthesis of nucleoside aryl(aminoacyl)phosphoramidate diesters.

4.1.2. Alkoxyphosphoramidates

Halogenoalkylamino Phosphoesters. An alternative strategy using potential alkylating phosphoramidate diesters to release nucleoside monophosphates has been explored by Borch and co-workers.[34,41] Initially, synthesis of the nucleoside haloethylaminophosphodiester[42] was accomplished using PV chemistry (Figure 17). Thus, phosphorus oxychloride was reacted with the appropriate amine in the presence of triethylamine and afforded a dichlorophosphoramidate that was subsequently activated by 1-hydroxybenzotriazole (HOBT) and coupled in situ with the nucleoside analogue in the presence of N-MeIm. In the last step, the alcohol derivative (substituted benzyl or furfuryl alcohols) was incorporated using a strong base such as n-butyl lithium (n-BuLi), hexamethyldisilazane lithium salt (LiHMDS),[41] 4-(dimethylamino)pyridine (DMAP).[31e,43]

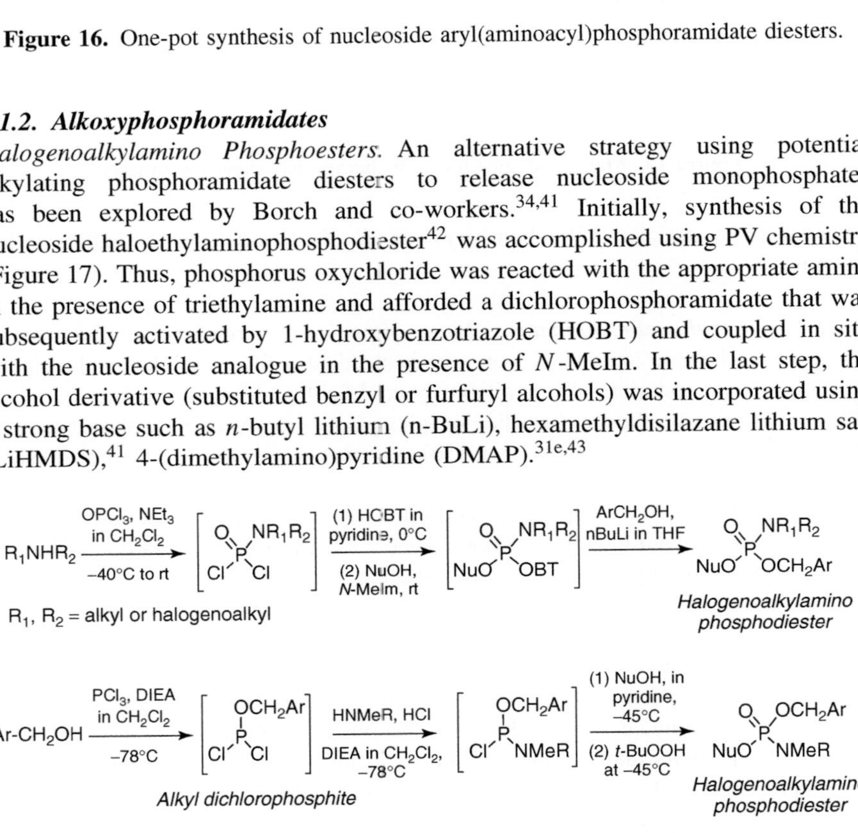

Figure 17. Synthetic pathways to haloalkylamino pronucleotides.

Later, a one-pot four-step reaction was proposed for the preparation of various halogenoethyl or tert-butyl phosphoramidate diesters using PIII intermediates.[34,44] Briefly, phosphorus chloride is reacted successively, at low temperature (Figure 17), with the benzyl alcohol in the presence of diisopropylethylamine (DIEA), the required amine, and the nucleoside analogue to generate the corresponding phosphoramidite intermediate, which was finally oxidized with t-BuOOH.

SATE Phosphoramidate Diesters. We combined the aryloxyphosphoramidate approach with our SATE strategy to evaluate the potential of new types of pronucleotides containing an S-pivaloyl-2-thioethyl (t-BuSATE) group and various amino acid esters, alkyl or aryl amines.[45] They have been synthesized

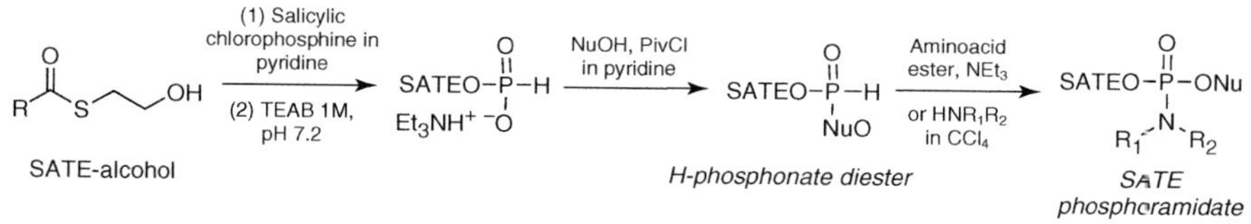

Figure 18. Synthesis of SATE phosphoramidate diesters using PIII intermediates.

using a common H-phosphonate monoester intermediate oxidized following the Atherton–Todd procedure[30] (Figure 18).

The SATE alcohol was first converted into the corresponding H-phosphonate monoester with salicyl chlorophosphite and in situ hydrolysis of the cyclic phosphite intermediate. The latter was then condensed with the required nucleoside analogue in the presence of PivCl as a condensing agent, leading to the nucleoside H-phosphonate diester. Finally, the oxidative amination was accomplished in the presence of CCl_4 and various amines.[33,46]

4.2. Phosphoramidate Monoesters

Another phosphoramidate approach is based on anionic nucleotides incorporating either amino acids or aryl groups as biolabile moieties. The synthesis of the first ones was performed mainly using nucleoside 5′-monophosphate, whereas the second relies on H-phosphonate chemistry (Figure 19).

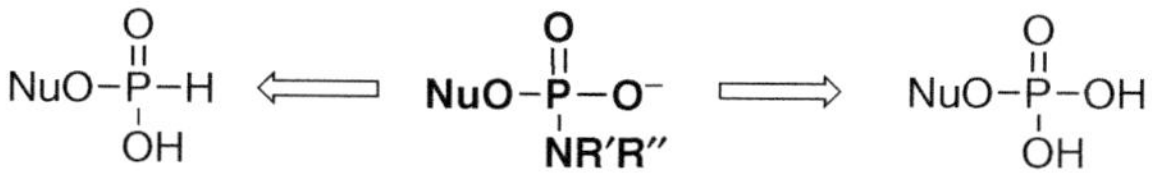

Figure 19. Two synthetic pathways to phosphoramidate monoester derivatives.

4.2.1. N-Aminoacylphosphoramidates. α-Amino acid phosphoramidate analogues were obtained according to the method initially reported by Wagner et al.[47] involving the preparation of the nucleoside 5′-monophosphate and then coupling with the aminoacyl esters in the presence of DCC[2,48] (Figure 20).

Zhao et al.[49] have also proposed a synthetic route based on the transesterification of diphenylphosphite (DPP) with fluorenylmethanol (Fm-OH, used as a phosphorus-protecting group) and then the nucleoside in pyridine, followed by the oxidative amination of the mixed H-phosphonate diester intermediate and removal of the protecting group. A similar pathway was used for the preparation of phenylphosphoramidate derivatives[50] and consisted of reacting the selected nucleoside analogue with DPP in pyridine (Figure 20); then the resulting H-phosphonate diester intermediate was sylilated to a phosphite triester in situ oxidized in the presence of the amino acid and iodine.

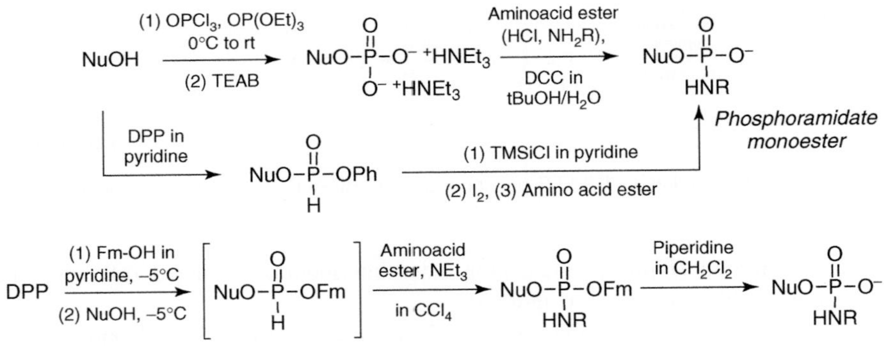

Figure 20. Synthetic pathways to phosphoramidate monoester derivatives.

4.2.2. N-Aryl Phosphoramidates.
Recently, the preparation of nucleoside (N-aryl)phosphoramidate monoesters was described by Kraszewski and co-workers.[51] Coupling of the preformed nucleoside H-phosphonate monoester with aromatic or heteroaromatic amines using DPCP as an activating reagent (Figure 21) was followed by oxidation of the resulting nucleoside H-phosphonamidate with iodine in aqueous pyridine. A similar protocol had been reported previously by the same group for the preparation of N-alkylphosphoramidate monoesters.[52]

4.3. Phosphorodiamidate Monoesters

Among other types of compounds, phosphorodiamidates have not greatly interested the community until very recently. Indeed, McGuigan et al. reported the first synthesis of a diamide pronucleotide in the early 1990s.[53] Since then, only a few groups have explored synthetic pathways for their preparation.[54] These derivatives have been obtained primarily using a one-pot reaction, including phosphorylation of the nucleoside analogue with phosphorus trichloride (PIII reagent) or oxychloride (PV-reagent) and in situ substitution of the chlorine atoms by the required amines (Figure 22).

Shipitsyn et al. reported[54a] the synthesis of diamidate by direct phosphorylation of the nucleoside analogue by phosphorus oxychloride in triethylphosphate,

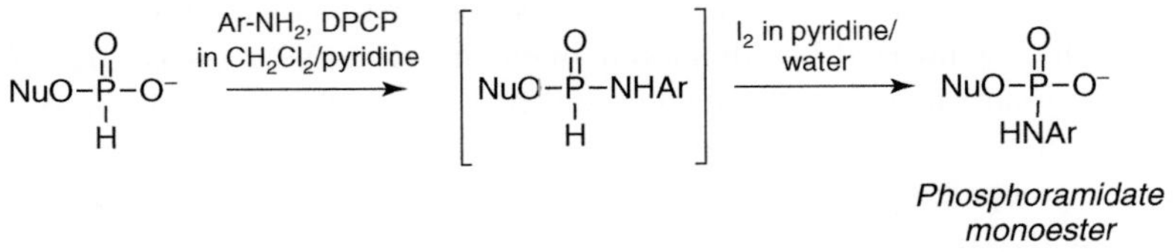

Figure 21. Synthesis of N-arylphosphoramidate monoesters using H-phosphonoamidates.

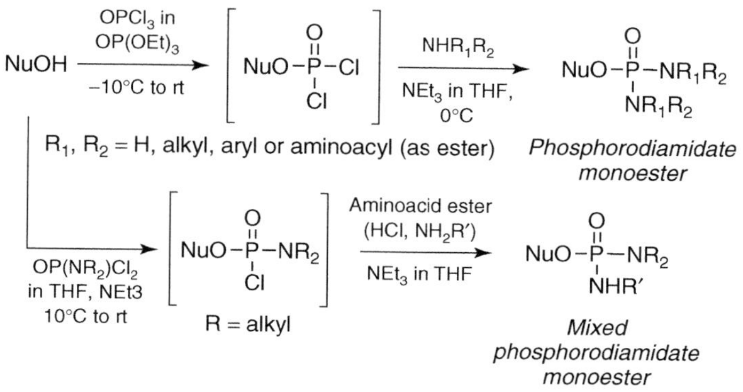

Figure 22. Synthetic pathways to nucleoside phosphorodiamidate monoesters.

followed by the addition of an excess of amines. The desired nucleoside phosphorodiamidates were obtained in variable yields, depending on the nature of the amine used.

Eventually, the nucleoside analogue was reacted with a dichlorophosphoramidate[54c] in the presence of triethylamine (Figure 22); then the required amino acid ester was added, leading to the compound desired. The advantage of this protocol is the possibility of obtaining mixed pronucleotides, incorporating two different amino residues.

Recently, a one-pot three-step synthesis of diamidate phosphoesters was reported[54b] and involved the reaction of a nucleoside analogue with phosphorus trichloride (Figure 23), then with amino acid esters or aniline in the presence of Hünig's base, leading to the corresponding phosphoramidite. The latter was finally oxidized with m-CPBA to afford the targeted prodrug.

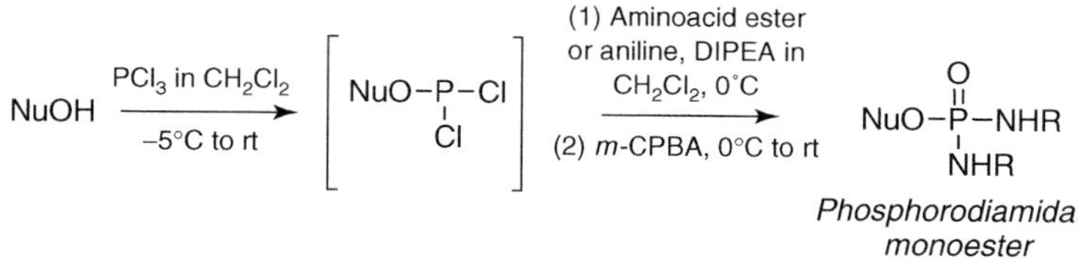

Figure 23. Synthesis of phosphorodiamidate monoesters using PIII chemistry.

5. MISCELLANEOUS SYNTHESES

With the goal of proposing new types of prodrugs, several research groups have reported efficient protocols for the synthesis of nucleoside 5′-phosphoesters containing heteroatoms such as sulfur (thiono- and phosphorothiolates), selenium (phosphoroselenoates), and boron (boranophosphates).

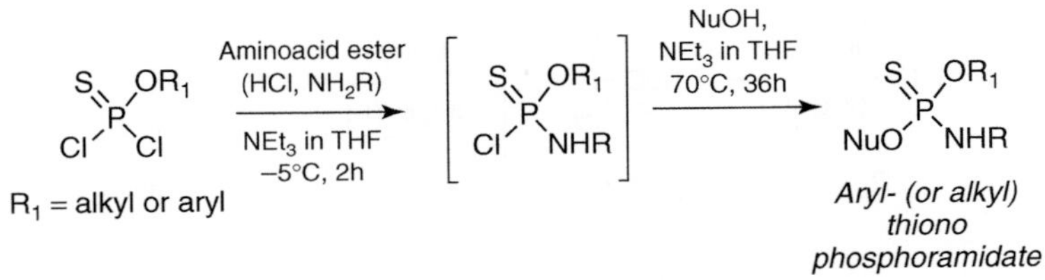

Figure 24. Synthesis of thionophosphoramidate monoester derivatives.

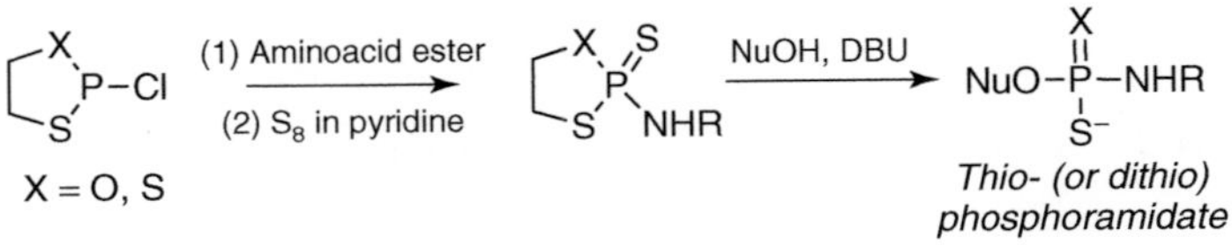

Figure 25. Synthesis of thio- or dithiophosphoramidate monoesters using 1,3,2-oxathia (or dithia)-phospholane ring-opening condensation.

5.1. Thiono- and Phosphorothiolate Derivatives

Nucleoside alkyl- and arylthiophosphoramidate derivatives have been prepared using the corresponding thionophosphodichloridates.[55] Commercially available O-aryl- and O-alkylthionophosphodichloridates were reacted with the amino acid ester selected in the presence of triethylamine at low temperature (Figure 24). The reaction process was monitored by ^{31}P-NMR spectroscopy, and when the peak corresponding to the starting material was minimal, a solution of the nucleoside analogue and triethylamine in THF was added to the reaction mixture.

Based on 1,3,2-oxathiaphospholane chemistry developed by Stec et al.,[56] nucleoside aminoacyl phosphorothioate or phosphorodithioate derivatives have been obtained in satisfactory yield (Figure 25).[57] The synthetic pathway involved first the reaction of 2-chloro-1,3,2-oxathiaphospholane with the required amino acid ester, in situ oxidation of the PIII intermediate, and then the ring opening of the 1,3,2-oxathia (or dithia)-phospholane intermediate with the nucleoside in the presence of 1,8-diazabicyclo[5.4.0]undec-7-ene (DBU).

5.2. Boranophosphoesters

Borane containing phosphate derivatives have been introduced by Shaw and co-workers.[58] Thus, nucleoside 5′-boranophosphoesters are isoelectronic to phosphates, when one of the two nonbridging oxygen atoms of phosphate is replaced by a borane group (BH_3). Nucleotides containing an aryl residue (tyrosine) and a boranephosphate or boranephosphorothioate group were prepared via the oxathiaphospholane and dithiaphospholane methodology.[59] The cyclic PIII intermediate (2-chloro-1,3,2-oxathiaphospholane) was reacted with the phenol function in the presence of bis(N,N-diisopropyl)ethylamine (Figure 26), followed

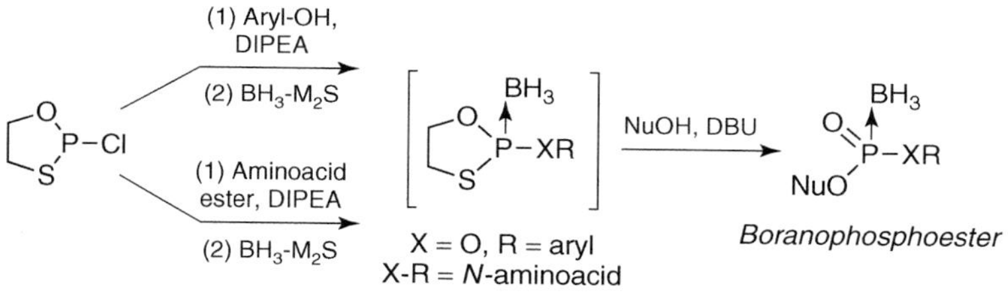

Figure 26. Synthesis of nucleoside boranophosphoesters using 1,3,2-oxathiaphospholane ring-opening condensation.

by in situ boronation using the borane–dimethyl sulfide (Me_2S–BH_3) complex, affording the oxathiaphospholane–borane complex, which was finally coupled with the nucleoside analogue in the presence of DBU. This methodology was also applied successfully to the synthesis of boranephosphoramidates.

H-phosphonate chemistry has been developed extensively for the preparation of boranophosphoesters[58b]; therefore, an attempted was also made to synthesize nucleoside boranophosphoramidate using an H-phosphonate diester derivative as the key intermediate[60] (Figure 27). Briefly, an aryl nucleoside H-phosphonate was silylated and in situ condensed with the required amino acid ester to afford a nucleoside aminoacylphosphoramidite, which was directly engaged in the boronation step. Final hydrolysis with H_2O resulted in formation of the desired nucleoside boranophosphoramidate. One should note that this procedure is no longer used by the authors, due to the difficulties in removing side products (coming from the coupling reagent) from the final products.

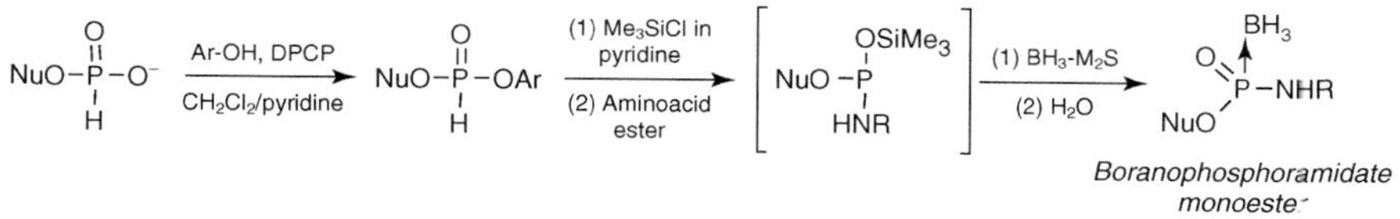

Figure 27. Synthesis of boranophosphoramidate derivatives using H-phosphonate intermediates.

6. RECENT DEVELOPMENTS IN STEREOSELECTIVE SYNTHESIS

In most of the pronucleotides described above, except for symmetrical phosphotriester and phosphorodiamidate derivatives, the phosphorus atom behaves as a stereogenic center. Therefore, the prodrug may exist as two diastereomers (in variable ratio), and in the perspective of a pharmaceutical development, it is more suitable to develop a single stereoisomer than a mixture of diastereomers. Indeed, it has been shown that the two forms are unequal substrates for the enzymes responsible for the cleavage of biolabile groups[61]; consequently, one of the two enantiomers is, by definition, "favored."[62]

Of the various classes of prodrugs discussed herein, only the HepDirect type could easily be isolated as a single diastereomeric form (>95%). During the synthetic pathway developed,[21] the *cis*-isomer (related to the dioxaphosphinane ring) is generally activated much more readily than the *trans*-isomer. Due to the cyclic nature of the phosphoester intermediate, the kinetics of formation of the two isomers are significantly different, allowing the facile and stereoselective synthesis of the preferred isomer. Concerning, the other types of pronucleotides, a mixture of the two diastereomers is generated and could hardly be separated by regular column chromatography, only enriched fractions of one of the other isomer were obtained,[36b,63] or eventually using preparative HPLC.[64]

The field of the diastereoselective synthesis of pronucleotide has merely been explored in the past. In 2010, the use of chiral auxiliaries based on (*S*)-4-isopropylthiazolidine-2-thione was explored for this purpose by Meier's group (Figure 28).[65] The diastereoselective synthesis of the (aryl)phosphoramidate derivative of thymidine containing nucleosides (in almost pure form) was performed in few steps.[65,66]

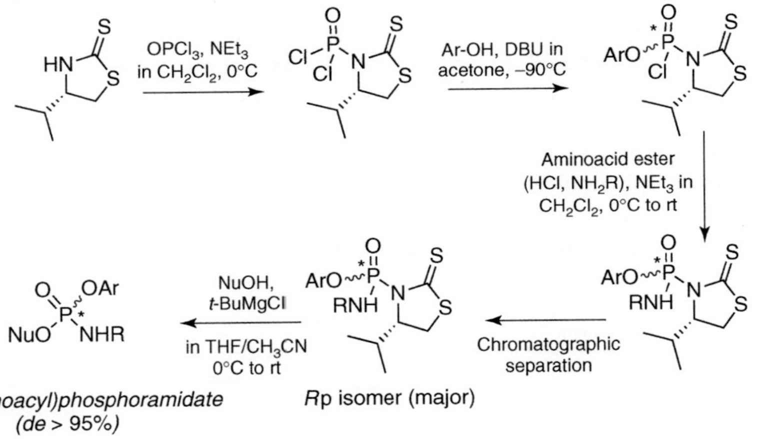

Figure 28. Synthetic pathway to diastereomerically pure arylphosphoramidate derivatives.

The (*S*)-4-isopropylthiazolidine-2-thione was reacted with an excess of phosphorus oxychloride and triethylamine and afforded the achiral phosphorodichloridate, which was used in the next step without purification. The phenol derivative was coupled at low temperature in the presence of a base and the corresponding chiral phosphorochloridate, the ratio of the two diastereomers being strongly dependent on the substitution of phenol derivatives and the reaction conditions. Conversion of phosphorochloridates to phosphorodiamidates was performed by adding the required amino acid ester in the presence of triethylamine. At this stage the stereochemically pure diastereomers of the phosphorodiamidates were stable enough to be purified and separated by silica gel column chromatography. Finally, each diastereochemically pure pronucleotide could be obtained by coupling the required phosphorodiamidate with the nucleoside analogue in the presence of an excess of *t*-BuMgCl.

This approach was adapted for the preparation of another series of pronucleotides, such as cyclo-Sal derivatives.[67] At first, the chiral thiazoline derivative reported previously was used to prepare nonsubstituted cyclo-Sal phosphotriesters in modest to good diastereomeric excess (48 < de < 95%). To obtain cyclo-Sal pronucleotides substituted on the aromatic ring, other chiral auxiliaries based on the N-cyaniminooxazolidine scaffold were developed.

In the meantime, another group[68] has reported an efficient strategy for the preparation of diastereomerically pure phenyl phosphoramidates using a bi(aryl)phosphoramidate reagent as a key intermediate (Figure 29). This reagent contained the aminoacyl residue and two phenolic groups, one unsubstituted (the aryl counterpart of the pronucleotide) and the other with electron-withdrawing substituents, and was isolated as a single diastereomer by crystallization. The phosphoramidate reagent was synthesized, as a racemic mixture, from commercially available phenylphosphorodichloridate with subsequent coupling of both the amino acid ester and then the aryl group (or in the reverse order) in the presence of triethylamine, at low temperature. Then, selective crystallization of the (S,Sp)-diastereomer was achieved in modest to good yields. Finally, through a selective nucleophilic displacement of the substituted phenol (pentafluorophenol was selected as the optimal leaving group) by the 5′-hydroxyl group of various nucleoside analogues in the presence of t-BuMgCl, the desired diastereomers of the phosphoramidate pronucleotides were obtained.

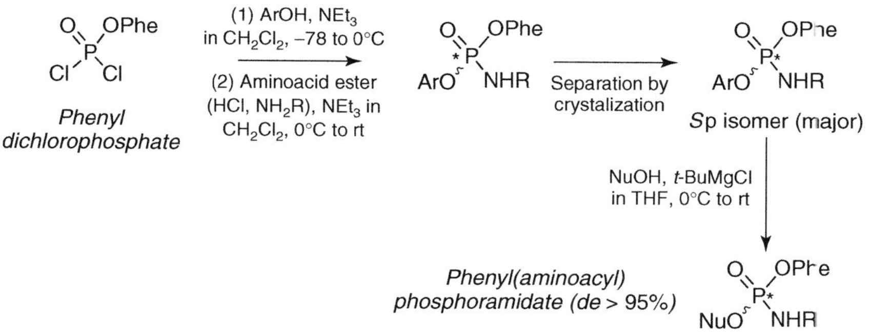

Figure 29. Synthetic pathway to diastereomerically pure phenylphosphoramidate derivatives.

7. CONCLUDING REMARKS

Pronucleotides were the subject of intensive studies due to their promising prodrug potential against various viruses and in cancer therapy. Among the compounds described herein, bis(pivaloyloxymethyl)phosphotriesters (POMs), S-acyl-2-thioethylphosphotriesters (SATEs), cyclo-Saligenyl phosphotriesters (cyclo-Sal nucleotides), cyclic 1-aryl-1,3-propanyl phosphotriesters (HepDirect), and phosphoramidate diesters have received the most attention.

Despite remarkable achievements as to the diastereoselective synthesis of aryloxyphosphoramidate and cyclo-Sal pronucleotides, these strategies still require further development, especially regarding the diversity of the pronucleotide structures. Diamide prodrugs present two advantages relative to other pronucleotide series. First, the by-products generated upon cleavage of the pro-moieties consist exclusively of nontoxic amino acids. Second, since two identical groups may be attached to the phosphorus atom, there are no issues related to the presence of diastereomers. It is worth mentioning that some of these prodrug approaches have also been applied successfully to nonnucleosidic substrates,[69] demonstrating their potentiality and broaden application to various phosphorylated effectors.

8. EXPERIMENTAL PROCEDURES

General Procedure for the Preparation of the Bis(POM)phosphotriesters. A stirred suspension of the nucleoside 5′-monophosphate [1 equiv, usually as bis(triethyl- or tributyl)ammonium salt] in anhydrous N-methylpyrrolidinone (eventually, DMF or CH_3CN) and NEt_3 (14 equiv) was treated with chloromethyl pivalate (18 equiv) and heated for a few hours at 60°C under a nitrogen atmosphere. Once cooled, the reaction mixture was diluted with AcOEt and a white precipitate (corresponding to $Et_3N \cdot HCl$ and unreacted starting material) was filtered off. The organic filtrate was washed with H_2O, and the H_2O wash was back-extracted with AcOEt. The combined organic layers were dried and evaporated using an oil pump to give viscous oil, which was purified by column chromatography. The desired phosphotriester is usually isolated in low yield (<40%).

General Procedure for the Preparation of the Bis(alkyl)phosphotriesters Using Mitsunobu Conditions. The bis(alkyl)phosphodiester intermediate (1.5 equiv), the nucleoside analogue (1 equiv), and triphenylphosphine (1.5 equiv) were dissolved in DMA and the reaction mixture was stirred for 10 min. A solution of diethylazodicarboxylate (1.5 equiv) in DMA was added, and the reaction mixture was heated at 60°C for 5 days under a nitrogen atmosphere. The solvent was evaporated under high vacuum, the remaining oil was triturated several times with a mixture of n-hexane/EtOAc (3:1 v/v), and the supernatants were decanted and then purified on silica gel chromatography.

General Procedure for the Preparation of the Bis(SATE)phosphotriesters
PV Pathway. A mixture of the nucleoside 5′-monophosphate selected (1 equiv, usually as bis(triethylammonium salt or acidic form), and the required SATE alcohol (5 equiv) was coevaporated twice with dry pyridine and then dissolved in the same anhydrous solvent. While stirring, TPSCl (5 equiv) was added and the reaction was allowed to perform overnight at room temperature. The reaction was quenched by adding aqueous $NaHCO_3$ solution and then diluted with CH_2Cl_2. The organic layer

was washed twice with water, dried over anhydrous Na_2SO_4, and evaporated to dryness. The residue was coevaporated with toluene before purification on silica gel chromatography.

PIII Pathway. To a suspension of the nucleoside analogue (1 equiv) and $1H$-tetrazole (2 equiv) in anhydrous acetonitrile was added the bis(S-acyl-2-thioethyl) N,N-diisopropylphosphoramidite (1.2 equiv), and the reaction mixture was stirred at room temperature under an inert atmosphere until TLC indicated almost complete conversion of the starting material. The reaction mixture was cooled to $-40°C$ and a solution of 3-chloroperbenzoic acid (2 equiv) in CH_2Cl_2 was added. The solution was allowed to warm to room temperature over 1 h, then quenched by adding an aqueous sodium hydrogen sulfite solution (10 wt%). The reaction mixture was diluted with CH_2Cl_2, then washed with saturated aqueous $NaHCO_3$ and water. The organic layer was dried over Na_2SO_4 and evaporated to dryness. The residue obtained was purified by silica gel column chromatography, and eventually 5'- and 3'-[bis(S-acyl-2-thioethyl)]phosphotriester was isolated as a secondary product.

General Procedure for the Preparation of Mixed (SATE)phosphotriesters
PV Pathway. To a stirred solution of the required SATE alcohol (1 equiv) and freshly distilled triethylamine (2 equiv) in dry THF at $-78°C$ was added the phenylphosphorodichloridate (1.02 equiv). The reaction mixture was stirred at room temperature for 4 h, filtered, and the filtrate concentrated on vacuum. The residue was diluted in anhydrous carbon tetrachloride and filtered again, to yield after evaporation of the volatiles to the SATE, phenylphosphorochloridate, which was used without further purification. To a stirred solution of the phosphorochloridate obtained previously (3 equiv) in dry THF were added successively the nucleoside analogue selected (1 equiv) and N-MeIm (6 equiv), and the reaction mixture was stirred 4 h at room temperature. Then dichloromethane was added and the reaction was quenched by adding diluted 0.1 N HCl solution. The resulting organic layer was separated and washed with an aqueous saturated solution of $NaHCO_3$ and then water. The organic layer was dried over Na_2SO_4, filtered, and concentrated under vacuum; finally, the residue was purified on silica gel column chromatography.

PIII Pathway. To a solution of the aryl residue selected (1 equiv) in dry CH_3CN containing 3-Å activated sieves, at $0°C$, a solution of (S-acyl-2-thioethyl)-N,N-bis(diisopropylamino)phosphine (1.5 equiv) and diisopropylamine (2.5 equiv) in dry CH_3CN was added. A solution of $1H$-tetrazole (2 equiv, 0.45 M solution in CH_3CN) was added and the reaction mixture was stirred for 2 h at room temperature and then diluted with acid-free EtOAc. The resulting organic phase was washed with brine and water, dried over $MgSO_4$, and concentrated under reduced pressure. The product was purified by silica column chromatography (cyclohexane/EtOAc containing 1% TEA), yielding the desired mixed phosphoramidite. The required nucleoside analogue (1 equiv) was dissolved in dry CH_3CN in a flame-dried bottle containing 3-Å molecular sieves, then $1H$-tetrazole solution (4 equiv,

0.45 M solution in CH_3CN) and a solution of the mixed phosphoramidite (2 equiv) in dry CH_3CN were added successively. The reaction mixture was stirred at room temperature for 2 h, cooled to 0°C, and a solution of *tert*-butyl hydroperoxide [2.4 equiv, anhydrous solution in toluene (3 M) or decane (5.7 M)] was added. The reaction mixture was stirred for 1 h and then diluted with CH_2Cl_2. The organic phase was washed with an aqueous solution of $Na_2S_2O_3$ (10% w/v) and water, dried over $MgSO_4$, and concentrated under reduced pressure. The mixed pronucleotide desired was purified by silica column chromatography.

General Procedure for the Preparation of Mixed Diarylphosphotriesters

PIII Pathway. A mixture of the nucleoside H-phosphonate monoester (1 equiv) and the required phenol (Ar_1OH, 1.5 equiv) was rendered anhydrous by coevaporation with dry pyridine, and the residue was dissolved in CH2Cl2 containing pyridine (9:1 v/v). The diphenyl chlorophosphate (1.2 equiv) was added and completion of the reaction was checked by ^{31}P-NMR. Thus, addition of the second phenol derivative (Ar_2OH, 1.5 equiv) was performed, followed by iodine (3 equiv) in pyridine. After 5 min the excess iodine was decomposed with ethanethiol, and the solvents were removed by evaporation. The oily residue was dissolved in CH2Cl2, and the organic layer was washed with aqueous 0.1 M KH_2PO_4 buffer (pH 6) dried over Na_2SO_4. The mixed di(aryl)phosphotriester expected was purified by silica gel chromatography.

General Procedure for the Preparation of the cyclo-Sal-phosphotriesters

PIII Pathway. DIEA (2 equiv) was added to a solution of the nucleoside analogue (1 equiv) in CH_3CN at room temperature. Under stirring at 0°C, the desired chlorophosphite (2 equiv), was added and the reaction was allowed to proceed until TLC indicated that formation of the cyclic phosphite triester was complete. Oxidation was carried out by addition of *tert*-butylhydroperoxide (2.1 equiv) and the reaction was allowed to proceed at room temperature for 1 h. Removal of the solvent in vacuum gave a residue that was purified by silica gel column chromatography to give a mixture of the two diastereomeric products, in approximatively a 1:1 ratio.

General Procedure for the Preparation of the HepDirect Phosphotriesters

PV Pathway: Synthesis of Trans-4-(aryl)-2-(4-nitrophenoxy)-2-oxido-1,3,2-dioxaphosphorinanes. A solution of the 1-aryl-1,3-propane diol selected (1 equiv) and TEA (3 equiv) in THF was added to a solution of 4-nitrophenylphosphorodichloridate (1.1 equiv) in THF at room temperature, and the resulting solution was heated at reflux. After 2–2.5 h, TLC indicated complete consumption of the starting diol and formation of the *cis*- and *trans*-isomers in a 50:50 to 60:40 ratio (determined by HPLC). The yellow solution was cooled to 30°C, sodium 4-nitrophenoxide (3 equiv) was added, and the reaction mixture was heated at 40°C to achieve isomerization of the *cis*-isomer. Alternatively, the isomerization may be carried out using 4-nitrophenol (3 equiv) and DBU (4 equiv). After 1.5 h, the reaction mixture was cooled to room temperature, quenched with a saturated solution of NH₄Cl, and diluted with ethyl acetate. The organic layer was washed several times with a 0.3 N solution of NaOH, then, by brine, dried over Na₂SO₄ and concentrated under reduced pressure. The residue was purified by column chromatography.

Synthesis of the cis prodrugs. A solution of *tert*-BuMgCl (1.5 equiv) in THF was added to a solution of the protected nucleoside analogue (1 equiv) in THF at room temperature. The reaction mixture was stirred at room temperature for 30 min and the *trans*-PV-reagent (1.3 equiv) was added in one portion. After stirring overnight at room temperature, the reaction mixture was quenched with a saturated solution of NH₄Cl and extracted with ethyl acetate. The combined organic layers were washed with a 1 N solution of NaOH, brine, dried over Na₂SO₄, and concentrated under reduced pressure. The residue was purified by flash column chromatography to give the protected pronucleotide.

General Procedure for the Preparation of Phospholipid Conjugates

PV Pathway. The phosphatidic acid required was prepared by reaction of the glycerol derivative (1 equiv) with POCl₃ (1.25 equiv) in the presence of pyridine, in THF at 0°C. After 5 h of stirring, the reaction was quenched by adding a saturated aqueous solution of NaHCO₃. The resulting solution was poured over ice and acidified with 6 N HCl solution, then extracted by diethyl ether. The combined organic layers were washed by water, dried over Na₂SO₄, and concentrated under reduced pressure. The crude was purified by silica gel chromatography using HCCl₃/ MeOH/NH₄OH as eluent.

The two partners (the phosphatidic acid and the nucleoside analogue) were mixed (equimolar mixture), dissolved in dry pyridine, and evaporated to complete dryness. This mixture was dissolved in dry pyridine, DCC, or TPSCl (2.3 equiv) was added, and the reaction mixture was stirred at room temperature for 3 days. After removal of the volatiles in vacuum, the residue was chromatographed on a silica gel column.

PIII Pathway. To a solution of imidazole (14 equiv) and NEt₃ (8 equiv) in dry CH₂Cl₂ at 0°C was added dropwise a solution of PCl₃ (3 equiv) in the same solvent. The reaction mixture was stirred for 15 min, cooled to −5°C, and treated dropwise with a solution of the glycerol derivative (1 equiv) in dry CH₂Cl₂, stirred at room temperature for 1 h, then quenched with a saturated aqueous NaHCO₃ solution. The organic layer was washed with water, dried over Na₂SO₄, and concentrated on vacuum. The residue was purified on silica gel to afford the desired *H*-phosphonate monoester in good yield (ca. 60%). The latter (1 equiv) and the nucleoside analogue (2.1 equiv) were dried by evaporation with anhydrous pyridine, then dissolved in the same solvent and treated with pivaloylchloride (3 equiv). After stirring at room temperature for 20 min, a solution of iodine (2 equiv) in a mixture of pyridine and water (98:2 v/v) was added. The reaction mixture was stirred a further 15 min, diluted with CH₂Cl₂, and washed with NaHSO₄ aqueous solution (5% w/v). The organic layer was dried over Na₂SO₄, concentrated, and the resulting solid was purified by silica gel chromatography.

General Procedure for the Preparation of Steroid Conjugates. A solution of the steroid (1 equiv) in dry pyridine was slowly added to a stirred solution of DPP (1 equiv) in the same solvent at 0°C, and stirring was pursued for 5 h. Then the nucleoside analogue (1 equiv) selected was added in one portion, and the reaction mixture was stirred overnight at room temperature. The solvent was removed by codistillation with toluene, and the resulting residue was purified by silica gel column chromatography to afford the H-phosphonate diester intermediate. The latter was oxidized using I_2 (2 equiv) in pyridine/H_2O (49:1), and the reaction mixture was stirred for 30 min. After removal of the solvent, the targeted phosphodiesters were obtained in good yields (>75%) as solids after purification by chromatography.

General Procedure for the Preparation of Aryl(aminoacyl)phosphoramidate Diesters

PV Pathway. To a stirred solution of phosphorus oxychloride (1 equiv) in anhydrous diethyl ether was added, dropwise, a solution of the phenol derivative (1 equiv) and anhydrous triethylamine (1 equiv) in anhydrous diethyl ether at −78°C. After 1 h of stirring, the reaction mixture was left to rise to room temperature and stand overnight. Then the triethylamine hydrochloride salt formed was filtered off and the filtrate evaporated to dryness to afford the crude product, which was engaged directly in the next step. A suspension of the aryl dichlorophosphate (1 equiv) and the appropriate amino acid ester hydrochloride salt (1 equiv) in anhydrous dichloromethane was cooled to −78°C and a solution of anhydrous triethylamine (2 equiv) in anhydrous CH_2Cl_2 was added dropwise. The reaction was then left to rise to room temperature overnight. Volatiles were removed under reduced pressure, and the resulting white solid was washed with anhydrous diethyl ether, filtered, and the filtrate was reduced to dryness to give the desired product. This intermediate was stored as a solution (0.5 to 1 M) in anhydrous THF and used without further purification. To a solution of the required nucleoside analogue (1 equiv) in THF were successively added N-MeIm (4 equiv) and aryl(aminoacyl)phosphorochloridate (1.5 equiv). After 4 h at room temperature, the reaction mixture was concentrated and the crude was purified by column chromatography.

PIII Pathway. To a solution of the arylphosphorodichloridite (1 equiv) in CH_2Cl_2, at −5°C was added dropwise the selected nucleoside analogue (1 equiv) dissolved in THF, t-BuOH (1 equiv), and NEt_3, within 40 min under a nitrogen atmosphere. The reaction mixture was stirred at 0 °C for another 30 min. Then added successively dropwise were the amino acid ester (1 equiv), NCS (1 equiv), and NEt_3 (4 equiv) in CH_2Cl_2. The resulting solution was warmed to room temperature and stirred for another 2 h. The volatiles were evaporated, the residue was suspended in EtOAc, and the triethylammonium hydrochloride was filtered off. The filtrate was concentrated and purified by silica gel column chromatography using EtOAc/MeOH (40:1) as eluent.

General Procedure for the Preparation of Halogenoalkylamino Phosphodiesters.
PV Pathway. Phosphoramidic dichloride was obtained by reaction of the appropriate haloalkylamine (1 equiv) and phosphorus oxychloride (1 equiv) in CH_2Cl_2 in the presence of triethylamine (2.2 equiv), at 0°C under a nitrogen atmosphere. The reaction mixture was allowed to stir overnight and then poured over an ice–water mixture. The layers were separated and the aqueous layer was extracted with CH_2Cl_2. The organic layers were combined, dried over $MgSO_4$, and concentrated under reduced pressure. The residue was purified by chromatography to afford the phosphoramidic dichloride as oil. The latter (1 equiv) was dissolved in dry THF and added dropwise to a stirred solution of 1-hydroxybenzotriazole (2 equiv) and pyridine (2 equiv) in dry THF at 0°C under an argon atmosphere. The reaction mixture was stirred at room temperature for 2 hours, transferred to adapted tubes, and the pyridine hydrochloride was removed by centrifugation under an argon atmosphere. The supernatant containing the reactive bis(benzotriazoyl) intermediate was added in one portion to a stirred solution of the required nucleoside analogue (0.5 equiv) in dry pyridine, at room temperature. Then N-MeIm (0.5 equiv) was added dropwise, and stirring was continued at room temperature overnight. The volatiles were removed under reduced pressure and the residue was dissolved in $CHCl_3$ and washed with an aqueous saturated solution of $NaHCO_3$, then with saturated NH_4Cl solution. The aqueous layers were combined and extracted with $CHCl_3$. The combined organic layers were dried over Na_2SO_4 and concentrated under reduced pressure. The residue was purified by silica gel chromatography.

PIII Pathway (One-Pot Procedure). Benzyl alcohol (1 equiv) was dissolved in CH_3CN/CH_2Cl_2 (5:20 v/v) and cooled to −78°C. To this solution, phosphorus trichloride (1 equiv, 2.0 M in CH_2Cl_2) and diisopropylethylamine (1.5 equiv) were added slowly. The reaction mixture was stirred at −78°C for 15 min and a solution in anhydrous CH_3CN of the N-methyl-N-(haloalkyl)amine salt (1 equiv) selected was added dropwise. After an extra addition of diisopropylethylamine (3 equiv), the reaction mixture was warmed to −60°C and stirred for 20 min. A solution of the nucleoside analogue (0.5 equiv, coevaporated with anhydrous pyridine) in anhydrous pyridine was cooled to −45°C and transferred to the previous reaction mixture. The disappearance of the nucleoside was monitored by TLC; the reaction mixture was oxidized by dropwise addition of *tert*-butyl hydroperoxide (1 equiv, 5.0–6.0 M solution in decane) at −45°C and warmed to 0°C over 30 min. A saturated aqueous NH_4Cl solution was added and the aqueous layer was extracted with CH_2Cl_2. The combined organic layers were dried over Na_2SO_4, concentrated, and coevaporated with toluene several times. Finally, the crude mixture was evaporated to dryness and the residue was passed through a short plug of silica gel to remove the remaining amine hydrochloride salts and then purified by chromatography on silica gel.

General Procedure for the Preparation of SATE Phosphoramidate Diesters. To a solution of the SATE H-phosphonate monoester as triethylammonium (1.5 equiv), and the selected nucleoside analogue (1 equiv) in anhydrous pyridine was added pivaloyl chloride (2.5 equiv). After 2 h, the reaction was quenched by the addition of ammonium acetate buffer (0.5 M solution, pH 5.6), and dichloromethane was added. The organic layer was washed successively with ammonium acetate buffer (0.01 M solution), water, dried over Na_2SO_4, and evaporated under reduced pressure. Column chromatography of the residue on silica gel using a stepwise gradient of methanol (0–2%) in dichloromethane containing 0.2% acetic acid afforded a diastereomeric mixture (1:1) of the nucleoside SATE H-phosphonate diester.

To a solution of the nucleoside SATE H-phosphonate diester (1 equiv) in dry CCl_4 was added either a solution or a suspension of the appropriate amino acid ester as hydrochloride salt (10 equiv) and freshly distilled triethylamine (9 equiv) in dry pyridine. The reaction mixture was stirred for 3 h at room temperature and diluted with CH_2Cl_2. The organic layer was washed successively with 1 M aqueous HCl solution, saturated Na_2SO_4 aqueous solution, and water, then dried over Na_2SO_4, filtered, and evaporated to dryness. Purification of the residue was carried out by column chromatography on silica gel.

General Procedure for the Preparation of Aminoacylphosphoramidate Monoesters

PV Pathway. To a solution of the required nucleoside analogue (1 equiv) in triethylphosphate was added at 0°C phosphoryl oxychloride (2.6 equiv). The reaction mixture was stirred at room temperature for 1.5 h, and eventually another portion of phosphoryl oxychloride (0.5 equiv) was added and the mixture was stirred overnight at room temperature. The reaction was quenched by adding TEAB (1 M solution, pH 7.4). The volatiles were evaporated in vacuum, and the residue was purified by silica column chromatography eluting with i-PrOH/NH_3/H_2O (7:1:1 v/v/v). The resulting nucleoside 5'-monophosphate (ammonium salt) ammonium and the amino acid ester selected (3.3 equiv) were suspended in a mixture of t-BuOH and water (eventually, only water), followed by the addition of the coupling agent (3.5 equiv DCC or EDC,HCl). The reaction mixture was stirred for several hours at room temperature, concentrated in vacuum, and further freeze-dried. The residue was purified by silica column chromatography eluting with $CHCl_3$/MeOH (5:1 v/v) containing 0.5% NH_4OH.

PIII Pathway. To a solution of Fm-OH (1 equiv) in anhydrous pyridine at −5°C was added dropwise diphenylphosphite (1.2 equiv), and the solution was stirred for 10 min. The nucleoside analogue (1.4 equiv) dissolved in anhydrous pyridine was added to the previous reaction mixture and was stirred for 1 h. The crude nucleoside H-phosphonate diester was obtained after evaporation of the volatiles under reduced pressure and then dissolved in anhydrous THF. A suspension of the amino acid ester hydrogen chloride (1.2 equiv) and NEt_3 (2 equiv) in CCl_4 was added dropwise to the previous solution maintained at −5°C, and the reaction mixture was stirred for 30 min at room temperature. The solvents were removed by evaporation, the residue was dissolved in CH_2Cl_2, and the resulting organic layer was washed with HCl solution (0.1 M) and water, dried over Na_2SO_4 and concentrated. The residue was purified by silica gel column chromatography, affording the fluorenylmethyl nucleoside phosphoramidate intermediate. The latter was dissolved in CH_2Cl_2, treated by an excess of piperidine, and the solution was stirred for 5 min at room temperature. The volatiles were removed on vacuum and the residue purified by silica gel column chromatography, eluting with i-PrOH/H_2O/NH_4OH (7:1:2 v/v/v).

General Procedure for the Preparation of N-Arylphosphoramidate Monoesters.
The nucleoside H-phosphonate monoester (1 equiv) and the aryl amine (1.5 equiv) required were dried by codistillation with anhydrous pyridine and then dissolved in a mixture (9:1 v/v) of CH_2Cl_2 and pyridine. To this solution DPCP (1.5 equiv) was added, and the reaction mixture was stirred for 5 min at room temperature. The intermediate was in situ oxidized using a solution of iodine (2 equiv) in pyridine containing water. After 8 h of stirring, the excess iodine was decomposed with the addition of ethanethiol, and the volatiles were evaporated. The resulting residue was dissolved in water, washed with CH_2Cl_2, and the aqueous layer was evaporated. Purification of the desired derivative was carried out on silica gel flash chromatography.

General Procedure for the Preparation of Phosphorodiamidate Monoesters
PIII Pathway. To a solution of PCl_3 (10 equiv) in dry dichloromethane, the required nucleoside analogue (1 equiv) was added portionwise at $-5°C$, the solution was stirred for 30 min, and then it was warmed to room temperature. After 3 h of stirring, the volatiles were removed under reduced pressure and the residue was dissolved in dry dichloromethane and cooled to $0°C$, followed by the addition of DIPEA (2.5 equiv), then the amine or the amino acid ester (2 equiv). The reaction mixture was stirred for 30 min and the oxidation step was carried out in situ using *m*-CPBA (2.5 equiv). After concentration, the residue was purified by column chromatography on silica gel.

General Procedure for the Preparation of Mixed Phosphorodiamidate Monoesters.
To a cooled (at $10°C$) solution of the nucleoside analogue (1 equiv) in dry THF and TEA (1 equiv) was added dropwise a solution of bis(alkyl)phosphoramidic dichloride (1 equiv) in dry THF over a period of 20 min. The reaction mixture was stirred at room temperature for 1 h and then a solution of the required amino acid ester, HCl (1 equiv) in dry THF was added slowly and stirred further for 1 h more. The progress of the reaction was monitored by TLC analysis (ethyl acetate/hexane 1:1), and after completion of the reaction, the salts were separated by filtration and the volatiles were evaporated. The resulting crude product was recrystallized from 2-propanol.

General Procedure for the Diastereoselective Preparation of Arylphosphoramidate Monoesters Using a Chiral Auxiliary.
To a cooled ($0°C$) solution of (4S)-4-isopropyl-1,3-thiazolidine-2-thione (1 equiv) and phosphorus oxychloride (3 equiv) in dry CH_2Cl_2 was slowly added anhydrous NEt_3 (1.1 equiv) in dry CH_2Cl_2. The reaction mixture was warmed to room temperature and stirred overnight. The solvent was removed under reduced pressure and the residue was suspended in a

minimum of CH_2Cl_2 to precipitate the salts in diethyl ether. After filtration, the filtrate was concentrated under reduced pressure to afford the phosphorodichloridate as an oil (^{31}P-NMR spectrum of the crude showed only one peak at $\delta = 3.70$ ppm).

A solution of the phosphorodichloridate (1 equiv) obtained previously and the selected phenol derivative (1 equiv) in anhydrous acetone was cooled to $-91°C$, and DBU (1.0 equiv) was added dropwise. After stirring 30 min to 1 h, depending on the aromatic moiety, the reaction was quenched (keeping the temperature at $-91°C$) by adding an aqueous saturated ammonium chloride solution, transferred, and extracted with CH_2Cl_2. The combined organic layers were combined, dried over magnesium sulfate, and concentrated under reduced pressure. The crude was purified immediately by flash column chromatography on silica gel using petroleum ether/EtOAc (2:1 v/v).

A suspension of the aryl[(4S)-4-isopropyl-2-thioxo-1,3-thiazolidin-3-yl]phosphonochloridate (1.0 equiv) and the selected amino acid ester hydrochloride (1.0 equiv) in dry CH_2Cl_2 was cooled to $0°C$, then anhydrous NEt_3 (3.0 equiv) was added dropwise and the reaction mixture was warmed to room temperature and stirred overnight. The reaction was quenched with an aqueous saturated ammonium chloride solution and the mixture extracted with CH_2Cl_2. The combined organic layers were dried over magnesium sulfate and the volatiles were evaporated. The resulting residue was purified by column chromatography on silica gel using petroleum ether/EtOAc (2:1 v/v) and afforded the Rp-isomer as a major phosphorodiamidate (20% < yield <70%) and the Sp-isomer as a minor one (yield <5%).

A solution of the thymine-containing nucleoside (1.5 equiv) in a mixture of anhydrous THF and CH_3CN (1:1 v/v) was cooled to $0°C$ and *tert*-butylmagnesium chloride (3.0 equiv, 1.7 M solution in THF) was added dropwise. The reaction mixture was warmed to room temperature and stirred for 30 min. The required diastereochemically pure phospho-reagent (1 equiv) in solution of anhydrous THF and CH_3CN (1:1 v/v) was added to the nucleoside suspension at $0°C$, and the reaction mixture was warmed to room temperature and stirred for 5 days. The reaction was quenched with an aqueous saturated ammonium chloride solution and the mixture extracted with CH_2Cl_2. The combined organic layers were dried over magnesium sulfate and concentrated under reduced pressure. The crude was purified by column chromatography on silica gel using CH_2Cl_2/MeOH (39:1 v/v).

General Procedure for the Diastereoselective Preparation of Phenylphosphoramidate Monoesters Using Aryl(phenyl)phosphoramidates as Intermediates. The amino acid ester hydrochloride (1 equiv) in dry CH_2Cl_2 was cooled to $-70°C$; anhydrous NEt_3 (2.05 equiv) and then diphenyl dichlorophosphate (1 equiv) in solution in dry CH_2Cl_2 were added dropwise. The reaction mixture was stirred at low temperature for 30 min and warmed to $0°C$ over 2 h and stirred further for 1 h, followed by the addition of pentaflurophenol (1 equiv) and NEt_3 (1.1 equiv) in solution in dry CH_2Cl_2. The reaction mixture was stirred at $0°C$ for 4 h before the filtration of the salts was carried out. The filtrate was concentrated, triturated with *t*-butylmethyl ether and filtrated. After evaporation under reduced pressure, the mixture of the two diastereomers was triturated with EtOAc/hexanes (2:8 v/v) allowing isolation of the diastereochemically pure pentafluorophenyl(phenyl)phosphoramidate. The latter (1.2 equiv) was dissolved in anhydrous THF and added slowly to a previously prepared mixture of the required nucleoside analogue (1 equiv) treated by *t*-BuMgCl (1.2 equiv, 1.7 M solution) in THF and stirred for 30 min before the addition. The reaction mixture was stirred at room temperature overnight, and the reaction was quenched by adding methanol. The volatiles were evaporated; the crude was purified by silica gel chromatography.

Acknowledgments

The authors are grateful to past and present members of the Imbach, Gosselin, and Périgaud laboratories for their contributions to some of the studies described herein. Special thanks to M.-C. Bergogne for providing bibliographic assistance.

REFERENCES

1. (a) Parang, K.; Wiebe, L. I.; Knaus, E. E. *Curr. Med. Chem.* **2000**, *7*, 995–1039; (b) Zemlicka, J. *Biochim. Biophys. Acta Mol. Basis Dis.* **2002**, *1587*, 276–286; (c) Schultz, C. *Bioorg. Med. Chem.* **2003**, *11*, 885–898; (d) Poijarvi-Virta, P.; Lonmberg, H. *Curr. Med. Chem.* **2006**, *13*, 3441–3465; (e) Hecker, S. J.; Erion, M. D. *J Med. Chem.* **2008**, *51*, 2328–2345002E

2. Wagner, C. R.; Iyer, V. V.; McIntee, E. J. *Med. Res. Rev.* **2000**, *20*, 417–451.

3. Lohrmann, R; Khorana, H. G. *J. Am. Chem. Soc.* **1966**, *88*, 829–833.

4. (a) Peyrottes, S.; Périgaud, C. In Current Protocols in Nucleic Acid Chemistry; Beaucage, S. L., Bergstrom, D. E., Herdewijn, P., Matsuda, A., Eds. Wiley, Hoboken, NJ, **2007**; Sec. 15.3; (b) Romanowska, J.; Szymanska-Michalak, A.; Boryski, J.; Stawinski, J.; Kraszewski, A.; Loddo, R.; Sanna, G.; Collu, G.; Secci, B.; La Colla, P. *Bioorg. Med. Chem.* **2009**, *17*, 3489–3498.

5. (a) Lannuzel, M.; Egron, D.; Imbach, J. L.; Gosselin, G.; Périgaud, C. *Nucleosides Nucleotides* **1999**, *18*, 1001–1002; (b) Peyrottes, S.; Périgaud, C.; Aubertin, A. M.; Gosselin, G.; Imbach, J. L. *Antivir. Chem. Chemother.* **2001**, *12*, 223–232.

6. Rose, J. D.; Parker, W. B.; Someya, H.; Shaddix, S. C.; Montgomery, J. A.; Secrist, J. A. *J. Med. Chem.* **2002**, *45*, 4505–4512.

7. Hwang, Y. S.; Cole, P. A. *Org. Lett.* **2004**, *6*, 1555–1556.

8. Khan, S. R.; Kumar, S. K.; Farquhar, D. *Pharm. Res.* **2005**, *22*, 390–396.

9. (a) Lefebvre, I.; Périgaud, C.; Pompon, A.; Aubertin, A. M.; Girardet, J. L.; Kirn, A.; Gosselin, G.; Imbach, J. L. *J. Med. Chem.* **1995**, *38*, 3941–3950; (b) Schlienger, N.; Périgaud, C.; Gosselin, G.; Imbach, J. L. *J. Org. Chem.* **1997**, *62*, 7216–7221; (c) Périgaud, C.; Gosselin, G.; Girardet, J. L.; Korba, B. E.; Imbach, J. L. *Antivir. Res.* **1999**, *40*, 167–178; (d) Egron, D.; Périgaud, C.; Gosselin, G.; Aubertin, A. M.; Gatanaga, H.; Mitsuya, H.; Zemlicka, J.; Imbach, J. L. *Bioorg. Med. Chem. Lett.* **2002**, *12*, 265–266.

10. (a) Prakash, T. P.; Prhavc, M.; Eldrup, A. B.; Cook, P. D.; Carroll, S. S.; Olsen, D. B.; Stahlhut, M. W.; Tomassini, J. E.; MacCoss, M.; Galloway, S. M.; Hilliard, C.; Bhat, B. *J. Med. Chem.* **2005**, *48*, 1199–1210; (b) Sun, Y.-W.; Chen, K.-M.; Kwon, C.-H. *Mol. Pharm.* **2006**, *3*, 161–173; (c) Gunic, E.; Chow, S.; Rong, F.; Ramasamy, K.; Raney, A.; Li, D. Y. Z.; Huang, J. F.; Hamatake, R. K.; Hong, Z.; Girardet, J. L. *Bioorg. Med. Chem. Lett.* **2007**, *17*, 2456–2458; (d) Agarwal, H. K.; Doncel, G. F.; Parang, K. *Tetrahedron Lett.* **2008**, *49*, 4905–4907.

11. Schlienger, N.; Peyrottes, S.; Aubertin, A. M.; Gosselin, G.; Imbach, J. L.; Périgaud, C. *Nucleosides Nucleotides* **1999**, *18*, 1025–1026.

12. (a) Schlienger, N.; Beltran, T.; Périgaud, C.; Lefebvre, I.; Pompon, A.; Aubertin, A. M.; Gosselin, G.; Imbach, J. L. *Bioorg. Med. Chem. Lett.* **1998**, *8*, 3003–3006; (b) Villard, A. L.; Coussot, G.; Lefebvre, I.; Augustijns, P.; Aubertin, A. M.; Gosselin, G.; Peyrottes, S.; Périgaud, C. *Bioorg. Med. Chem.* **2008**, *16*, 7321–7329.

13. (a) Schlienger, N.; Peyrottes, S.; Kassem, T.; Imbach, J. L.; Gosselin, G.; Aubertin, A. M.; Périgaud, C. *J. Med. Chem.* **2000**, *43*, 4570–4574; (b) Peyrottes, S.; Coussot, G.; Lefebvre, I.; Imbach, J. L.; Gosselin, G.; Aubertin, A. M.; Périgaud, C. *J. Med. Chem.* **2003**, *46*, 782–793.

14. Petersen, L.; Jorgensen, P. T.; Nielsen, C.; Hansen, T. H.; Nielsen, J.; Pedersen, E. B. *J. Med. Chem.* **2005**, *48*, 1211–1220.

15. (a) Meier, C. *Mini-Rev. Med. Chem.* **2002**, *2*, 219–234; (b) Meier, C.; Ruppel, M. F. H.; Vukadinovic, D.; Balzarini, J. *Mini-Rev. Med. Chem.* **2004**, *4*, 383–394; (c) Meier, C. *Eur. J. Org. Chem.* **2006**, 1081–1102.

16. (a) Ducho, C.; Balzarini, J.; Naesens, L.; De Clercq, E.; Meier, C. *Antiviral Chem. Chemother.* **2002**, *13*, 129–141; (b) Meier, C.; Ducho, C.; Jessen, H.; Vukadinović-Tenter, D.; Balzarini, J. *Eur. J. Org. Chem.* **2006**, *2006*, 197–206; (c) Gisch, N.; Balzarini, J.; Meier, C. *J. Med. Chem.* **2007**, *50*, 1658–1667.

17. Sun, W. Y.; Zhou, A. H.; Wiebe, L. I.; Knaus, E. E. *Nucleosides Nucleotides Nucleic Acids* **2003**, *22*, 2121–2132.

18. Mugnier, F.; Meier, C. *Nucleosides Nucleotides* **1999**, *18*, 941–942.

19. Meier, C.; Lorey, M.; De Clercq, E.; Balzarini, J. *Bioorg. Med. Chem. Lett.* **1997**, *7*, 99–104.

20. Ludek, O. R.; Kramer, T.; Balzarini, J.; Meier, C. *Synthesis* **2006**, 1313–1324.

21. Erion, M. D.; Reddy, K. R.; Boyer, S. H.; Matelich, M. C.; Gornez-Galeno, J.; Lemus, R. H.; Ugarkar, B. G.; Colby, T. J.; Schanzer, J.; van Poelje, P. D. *J. Am. Chem. Soc.* **2004**, *126*, 5154–5163.

22. (a) Reddy, K. R.; Boyer, S. H.; Erion, M. D. *Tetrahedron Lett.* **2005**, *46*, 4321–4324; (b) Bookser, B. C.; Raffaele, N. B. *J. Comb. Chem.* **2008**, *10*, 567–572.

23. Boyer, S. H.; Sun, Z. L.; Jiang, H. J.; Esterbrook, J.; Gomez-Galeno, J. E.; Craigo, W.; Reddy, K. R.; Ugarkar, B. G.; MacKenna, D. A.; Erion, M. D. *J. Med. Chem.* **2006**, *49*, 7711–7720.

24. (a) Hecker, S. J.; Reddy, K. R.; van Poelje, P. D.; Sun, Z. L.; Huang, W. J.; Varkhedkar, V.; Reddy, M. V.; Fujitaki, J. M.; Olsen, D. B.; Koeplinger, K. A.; Boyer, S. H.; Linemeyer, D. L.; MacCoss, M.; Erion, M. D. *J. Med. Chem.* **2007**, *50*, 3891–3896; (b) Bookser, B. C.; Raffaele, N. B.; Reddy, K. R.; Fan, K.; Huang, W. J.; Erion, M. D. *Nucleosides Nucleotides Nucleic Acids* **2009**, *28*, 969–986.

25. Alexander, R. L.; Morris-Natschke, S. L.; Ishaq, K. S.; Fleming, R. A.; Kucera, G. L. *J. Med. Chem.* **2003**, *46*, 4205–4208.

26. Kucera, G. L.; Goff, C. L.; Iyer, N.; Morris-Natschke, S.; Ishaq, K. S.; Wyrick, S. D.; Fleming, R. A.; Kucera, L. S. *Antivir. Res.* **2001**, *50*, 129–137.
27. Pickin, K. A.; Alexander, R. L.; Morrow, C. S.; Morris-Natschke, S. L.; Ishaq, K. S.; Fleming, R. A.; Kucera, G. L. *J. Drug Deliv. Sci. Technol.* **2009**, *19*, 31–36.
28. Lonshakov, D. V.; Baranova, E. O.; Lyutik, A. I.; Shastina, N. S.; Shvets, V. I. *Pharm. Chem. J.* **2011**, *44*, 557–563.
29. (a) Jin, P. Y.; Ji, S. H.; Ju, Y.; Zhao, Y. F. *Lett. Org. Chem.* **2007**, *4*, 189–192; (b) Jin, P. Y.; Liu, K.; Ji, S. H.; Ju, Y.; Zhao, Y. F. *Synthesis* **2007**, 407–411; (c) Jin, P. Y.; Liu, K.; Ju, Y.; Zhao, Y. F. *Phosphorus Sulfur Silicon Relat. Elem.* **2008**, *183*, 538–542.
30. Atherton, F. R.; Openshaw, H. T.; Todd, A. R. *J. Chem. Soc.* **1945**, 660–663.
31. (a) Freel Meyers, C. L.; Borch, R. F. *J. Med. Chem.* **2000**, *43*, 4319–4327; (b) Peyrottes, S.; Egron, D.; Lefebvre, I.; Gosselin, G.; Imbach, J. L.; Perigaud, C. *Mini-Rev. Med. Chem.* **2004**, *4*, 395–408; (c) Cahard, D.; McGuigan, C.; Balzarini, J. *Mini-Rev. Med. Chem.* **2004**, *4*, 371–381; (d) Drontle, D. P.; Wagner, C. R. *Mini-Rev. Med. Chem.* **2004**, *4*, 409–419; (e) Wu, W. D.; Borch, R. F. *Mol. Pharm.* **2006**, *3*, 451–456; (f) Mehellou, Y.; Balzarini, J.; McGuigan, C. *ChemMedChem* **2009**, *4*, 1779–1791; (g) Yang, S.; Pannecouque, C.; Lescrinier, E.; Giraut, A.; Herdewijn, P. *Org. Biomol. Chem.* **2012**, *10*, 146–153.
32. McGuigan, C.; Pathirana, R. N.; Mahmood, N.; Hay, A. J. *Bioorg. Med. Chem. Lett.* **1992**, *2*, 701–704.
33. Egron, D.; Imbach, J. L.; Gosselin, G.; Aubertin, A. M.; Périgaud, C. *J. Med. Chem.* **2003**, *46*, 4564–4571.
34. Tobias, S. C.; Borch, R. F. *J. Med. Chem.* **2001**, *44*, 4475–4480.
35. McIntee, E. J.; Remmel, R. P.; Schinazi, R. F.; Abraham, T. W.; Wagner, C. R. *J. Med. Chem.* **1997**, *40*, 3323–3331.
36. (a) Harris, S. A.; McGuigan, C.; Andrei, G.; Snoeck, R.; De Clercq, E.; Balzarini, J. *Antivir. Chem. Chemother.* **2001**, *12*, 293–300; (b) Congiatu, C.; Brancale, A.; Mason, M. D.; Jiang, W. G.; McGuigan, C. *J. Med. Chem.* **2006**, *49*, 452–455.
37. Venkatachalam, T. K.; Qazi, S.; Uckun, F. M. *Bioorg. Med. Chem.* **2006**, *14*, 5161–5177.
38. (a) Gudmundsson, K. S.; Daluge, S. M.; Johnson, L. C.; Jansen, R.; Hazen, R.; Condreay, L. D.; McGuigan, C. *Nucleosides Nucleotides Nucleic Acids* **2003**, *22*, 1953–1961; (b) Derudas, M.; Brancale, A.; Naesens, L.; Neyts, J.; Balzarini, J.; McGuigan, C. *Bioorg. Med. Chem.* **2010**, *18*, 2748–2755.
39. McGuigan, C.; Kelleher, M. R.; Perrone, P.; Mulready, S.; Luoni, G.; Daverio, F.; Rajyaguru, S.; Le Pogam, S.; Najera, I.; Martin, J. A.; Klumpp, K.; Smith, D. B. *Bioorg. Med. Chem. Lett.* **2009**, *19*, 4250–4254.
40. Jiang, P.; Guo, J.; Fu, H.; Jiang, Y. Y.; Zhao, Y. F. *Synlett* **2005**, 2537–2539.
41. Meyers, C. L. F.; Hong, L. P.; Joswig, C.; Borch, R. F. *J. Med. Chem.* **2000**, *43*, 4313–4318.
42. Fries, K. M.; Joswig, C.; Borch, R. F. *J. Med. Chem.* **1995**, *38*, 2672–2680.
43. Wu, W.; Sigmond, J.; Peters, G. J.; Borch, R. F. *J. Med. Chem.* **2007**, *50*, 3743–3746.
44. Tobias, S. C.; Borch, R. F. *Mol. Pharm.* **2004**, *1*, 112–116.
45. Jochum, A.; Schlienger, N.; Egron, D.; Peyrottes, S.; Périgaud, C. *J. Organomet. Chem.* **2005**, *690*, 2614–2625.
46. Beltran, T.; Egron, D.; Pompon, A.; Lefebvre, I.; Périgaud, C.; Gosselin, G.; Aubertin, A.; Imbach, J. *Bioorg. Med. Chem. Lett.* **2001**, *11*, 1775–1777.

47. Abraham, T. W.; Kalman, T. I.; McIntee, E. J.; Wagner, C. R. *J. Med. Chem.* **1996**, *39*, 4569–4575.
48. (a) Choy, C. J.; Drontle, D. P.; Wagner, C. R. In Current Protocols in Nucleic Acid Chemistry; Beaucage, S. L., Bergstrom, D. E., Herdewijn, P., Matsuda, A., Eds. Wiley, Hoboken, NJ, **2006**; Sec. 15.1; (b) Adelfinskaya, O.; Herdewijn, P. *Angew. Chem., Int. Ed.* **2007**, *46*(23), 4356–4358.
49. Zhu, J. G.; Fu, H.; Jiang, Y. Y.; Zhao, Y. F. *Synlett* **2005**, 1927–1929.
50. Wang, R. F.; Corbett, T. H.; Cheng, Y. C.; Drach, J. C.; Kern, E. R.; Mitsuya, H.; Zemlicka, J. *Bioorg. Med. Chem. Lett.* **2002**, *12*, 2467–2470.
51. Romanowska, J.; Sobkowski, M.; Szymanska-Michalak, A.; Kolodziej, K.; Dabrowska, A.; Lipniacki, A.; Piasek, A.; Pietrusiewicz, Z. M.; Figlerowicz, M.; Guranowski, A.; Boryski, J.; Stawinski, J.; Kraszewski, A. *J. Med. Chem.* **2011**, *54*, 6482–6491.
52. Sobkowska, A.; Sobkowski, M.; Cieślak, J.; Kraszewski, A.; Kers, I.; Stawiński, J. *J. Org. Chem.* **1997**, *62*, 4791–4794.
53. (a) Jones, B. C. N. M.; McGuigan, C.; O'Connor, T. J.; Jeffries, D. J.; Kinchington, D. *Antiviral Chem. Chemother.* **1991**, *2*, 35–39; (b) Kinchington, D.; Harvey, J. J.; O'Connor, T. J.; Jones, B.; Devine, K. G.; Taylor-robinson, D.; Jeffries, D. J.; McGuigan, C. *Antiviral Chem. Chemother.* **1992**, *3*, 107–112.
54. (a) Shipitsyn, A. V.; Zakirova, N. F.; Belanov, E. F.; Pronyaeva, T. R.; Fedyuk, N. V.; Kukhanova, M. K.; Pokrovsky, A. G. *Nucleosides Nucleotides Nucleic Acids* **2003**, *22*, 963–966; (b) Hu, A. F.; Xu, P. X.; Zhu, M. X.; Ji, T.; Tang, G.; Zhao, Y. F. *Synth. Commun.* **2009**, *39*, 1342–1354; (c) Reddy, S. S.; Rao, V. K.; Venkataramana, K.; Reddy, C. S.; Ghosh, S. K.; Raju, C. N. *Pharma Chem.* **2010**, *2*, 1–9.
55. (a) Miao, Z. W.; Fu, H.; Han, B.; Chen, Y.; Zhao, Y. F. *Synth. Commun.* **2002**, *32*, 1159–1167; (b) Miao, Z. W.; Fu, H.; Tu, G. Z.; Zhao, Y. F. *Synth. Commun.* **2002**, *32*, 3301–3309.
56. Stec, W. J.; Grajkowski, A.; Kobylanska, A.; Karwowski, B.; Koziolkiewicz, M.; Misiura, K.; Okruszek, A.; Wilk, A.; Guga, P.; Boczkowska, M. *J. Am. Chem. Soc.* **1995**, *117*, 12019–12029.
57. (a) Baraniak, J.; Kaczmarek, R.; Stec, W. J. *Tetrahedron Lett.* **2000**, *41*, 9139–9142; (b) Baraniak, J.; Kaczmarek, R.; Wasilewska, E.; Stec, W. J. *Phosphorus Sulfur Silicon Relat. Elem.* **2002**, *177*, 1667–1670.
58. (a) Summers, J. S.; Shaw, B. R. *Curr. Med. Chem.* **2001**, *8*, 1147–1155; (b) Li, P.; Sergueeva, Z. A.; Dobrikov, M.; Shaw, B. R. *Chem. Rev.* **2007**, *107*, 4746–4796.
59. Baraniak, J.; Kaczmarek, R.; Wasilewska, E. *Tetrahedron Lett.* **2004**, *45*, 671–675.
60. Li, P.; Shaw, B. R. *J. Org. Chem.* **2005**, *70*, 2171–2183.
61. (a) Mesplet, N.; Saito, Y.; Morin, P.; Agrofoglio, L. A. *Bioorg. Chem.* **2003**, *31*, 237–247; (b) Siccardi, D.; Gumbleton, M.; Omidi, Y.; McGuigan, C. *Eur. J. Pharm. Sci.* **2004**, *22*, 25–31; (c) Venkatachalam, T. K.; Samuel, P.; Uckun, F. M. *Bioorg. Med. Chem.* **2005**, *13*, 1763–1773.
62. Siccardi, D.; Kandalaft, L. E.; Gumbleton, M.; McGuigan, C. *J. Pharmacol. Exp. Ther.* **2003**, *307*, 1112–1119.
63. (a) Allender, C. J.; Brain, K. R.; Ballatore, C.; Cahard, D.; Siddiqui, A.; McGuigan, C. *Anal. Chim. Acta* **2001**, *435*, 107–113; (b) Gao, X. A.; Chen, W. Z.; Zhu, G. T.; Yi, R. Z.; Wu, Z.; Xu, P. X.; Zhao, Y. F. *J. Chromatogr. A* **2011**, *1218*, 1416–1422.

64. (a) Mesplet, N.; Saito, Y.; Morin, P.; Agrofoglio, L. A. *J. Chromatogr. A* **2003**, *983*, 115–124; (b) Goossens, J. F.; Roux, S.; Egron, D.; Périgaud, C.; Bonte, J. P.; Vaccher, C.; Foulon, C. *J. Chromatogr. B* **2008**, *875*, 288–295; (c) Foulon, C.; Tedou, J.; Peyrottes, S.; Périgaud, C.; Bonte, J. P.; Vaccher, C.; Goossens, J. F. *J. Chromatogr. B* **2009**, *877*, 3475–3481.
65. Roman, C. A.; Balzarini, J.; Meier, C. *J. Med. Chem.* **2010**, *53*, 7675–7681.
66. Roman, C. A.; Wasserthal, P.; Balzarini, J.; Meier, C. *Eur. J. Org. Chem.* **2011**, 4899–4909.
67. (a) Rios Morales, E. H.; Balzarini, J.; Meier, C. *Chem. Eur. J.* **2011**, *17*, 1649–1659; (b) Morales, E. H. R.; Roman, C. A.; Thomann, J. O.; Meier, C. *Eur. J. Org. Chem.* **2011**, 4397–4408.
68. Ross, B. S.; Reddy, P. G.; Zhang, H. R.; Rachakonda, S.; Sofia, M. J. *J. Org. Chem.* **2011**, *76*, 8311–8319.
69. (a) Ruda, G. F.; Alibu, V. P.; Mitsos, C.; Bidet, O.; Kaiser, M.; Brun, R.; Barrett, M. P.; Gilbert, I. H. *ChemMedChem* **2007**, *2*, 1169–1180; (b) Boyer, S. H.; Jiang, H. J.; Jacintho, J. D.; Reddy, M. V.; Li, H. Q.; Li, W. Y.; Godwin, J. L.; Schulz, W. G.; Cable, E. E.; Hou, J. Z.; Wu, R. R.; Fujitaki, J. M.; Hecker, S. J.; Erion, M. D. *J. Med. Chem.* **2008**, *51*, 7075–7093.

7 Synthesis of C-Nucleosides

OMAR BOUTUREIRA, M. ISABEL MATHEU, YOLANDA DÍAZ, and SERGIO CASTILLÓN

Departamento de Química Analítica i Química Orgànica, Universitat Rovira i Virgili, Tarragona, Spain

1. INTRODUCTION

C-Nucleosides are analogues of nucleosides in which the C–N anomeric link between the sugar and the heterocyclic base moieties has been replaced by a C–C bond. As a consequence of this structural modification, C-nucleosides became stable to enzymatic and acid-catalyzed hydrolysis, in addition to the change in acid–base properties of the heterocyclic base and in the potential hydrogen-bond interactions. A few nucleosides, such as pseudouridine, isolated from transfer RNA, formycins, and showdomycin have been isolated from natural sources (Figure 1).[1]

Showdomycin possesses antibiotic properties, and other C-nucleosides, such as formycins, show a variety of biological properties, such as antibiotic, cytotoxic, and enzyme inhibition. Inmunicillins, a family of C-nucleosides containing aza-sugar moieties, inhibit human purine nucleoside phosphorilase. C-Nucleosides containing a carboxamide function such as tiazofurin can inhibit some oxoreductases (Figure 2).

C-Nucleosides have also attracted the interest of researchers looking for hydrogen-bond interactions alternative to those produced in the classical Watson–Crick model. More recently, research in the field has focused on the study of base pairing by hydrophobic interaction (Figure 3), such as that produced by halogenated or nonhalogenated aromatics, particularly polyaromatic compounds, which are used as analogues of nucleobases. These nucleoside analogues have been used as chemical probes for the study of numerous biological processes.

Several reviews dealing with C-nucleosides covering biological applications[2,3] and synthetic methods[2,4] have been published recently. In this chapter we focus on chemical synthetic procedures addressing C-nucleosides, paying special attention to the more recently developed and those considered as reliable procedures.

Chemical Synthesis of Nucleoside Analogues, First Edition. Edited by Pedro Merino.
© 2013 John Wiley & Sons, Inc. Published 2013 by John Wiley & Sons, Inc.

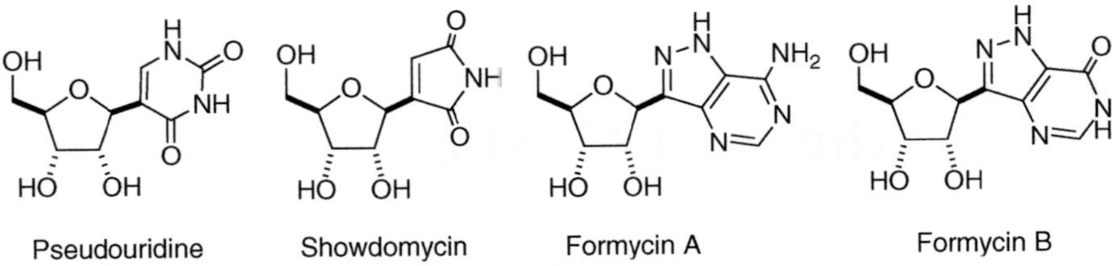

Figure 1. C-nucleosides isolated from natural sources.

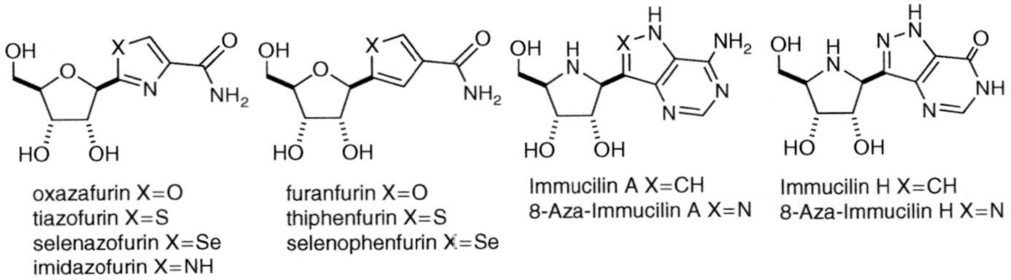

Pseudouridine   Showdomycin   Formycin A   Formycin B

oxazafurin X=O
tiazofurin X=S
selenazofurin X=Se
imidazofurin X=NH

furanfurin X=O
thiphenfurin X=S
selenophenfurin X=Se

Immucilin A X=CH
8-Aza-Immucilin A X=N

Immucilin H X=CH
8-Aza-Immucilin H X=N

Figure 2. Biologically relevant C-nucleosides.

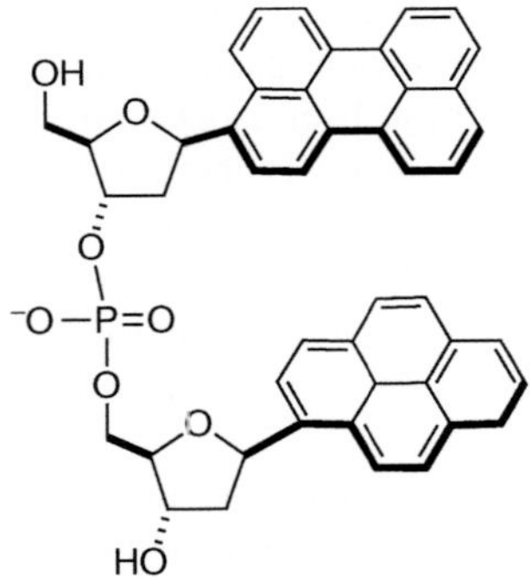

Figure 3. Example of a base-pairing base with hydrofobic interactions.

2. SYNTHESIS OF SYNTHETIC INTERMEDIATES BY C–C FORMATION PROCEDURES AT C1 AND FURTHER INTEGRATION OF THE HETEROCYCLE

2.1. Nucleophilic Substitution at C1

2.1.1. Cyanide Salts: Synthesis of Glycosyl Cyanides. One of the most general and useful method for preparing C-nucleosides is the multistep elaboration of the heterocycles desired from a C-glycosyl derivative functionalized at the anomeric

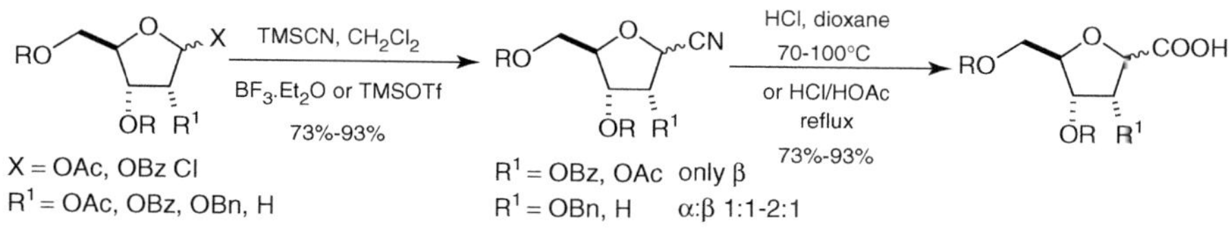

Scheme 1. Synthesis of glycosyl cyanides and carboxylic acids.

position. In this sense, one of the most important types of C-glycosyl intermediates is the glycosyl cyanide, which is usually obtained from 1-O-acyl or 1-chloro carbohydrates by reaction with trimethyl silyl cyanide in a polar aprotic solvent and in the presence of a Lewis acid as catalyst (Scheme 1).[5] The 1,2-*trans*-glycosyl cyanide is formed when the starting sugar has a participating 2-O-acyl substituent, whereas a mixture of cyanide anomers is obtained with nonparticipating groups or for 2-deoxycarbohydrates. In most cases, either the nitrile or the acid resulting from its hydrolysis, which is often also a key precursor, can be separated chromatographically.

This precursor affords an α-chloroketone derivative through a modified Arndt–Einstert procedure that involves acid activation via an acyl chloride, displacement of the chloride by diazomethane, and HCl treatment (Scheme 2).[6] Reaction of the α-chloroketone obtained with various thioamide nucleophiles leads to the formation of substituted thiazoles, whereas cyclization with semithiocarbazide gives aminothiadiazole derivatives.

Only the α-acid derivative is used to obtain the novel disubstituted pyrimidinyl C-nucleosides shown in Scheme 3. Formation of the Weinreib amide and reaction with lithium acetylides yield propargyl ketones, which by reacting with substituted amidines results in formation of the desired compounds.[7] However, both the cyclization and subsequent Lewis acid–catalyzed deprotection cause epimerization of the products. This fact makes unnecessary the use of a single anomer of the starting material.

C-glycosyl enaminoketoesters obtained from protected 1-cyano-α-ᴅ-ribofuranose result in intermediates useful in the synthesis of pyrazole and pyrimidine C-nucleosides of Scheme 4 (via a).[8] Thus, after a Reformatsky reaction of glycosyl cyanide and further hydrolysis, the key step is the copper(II) reaction of the β-ribofuranosyl ketoesters obtained with alkylcyanoformates.[9] Subsequent cyclization with amidoimidates or benzylhydrazine furnishes the protected nucleosides. When treated with various hydrazonyl chlorides in the presence of Lewis acids such as ytterbium triflate the same starting material affords 1,2,4-triazoles in good yields (Scheme 4, via b).[10]

The coupling of toluyl-protected glycosyl carboxylic acid indicated in Scheme 5 with different *ortho*-diamines using Mukaiyama's reagent (chloromethylpyridinium iodide) affords the corresponding amides in good yields.[11] These compounds are cyclized by heating in acetic acid, although epimerization takes place in some cases, depending on the nature of the aglycone.[12] After deprotection, 2′deoxy-C-nucleoside derivatives are obtained in moderate overall yields.

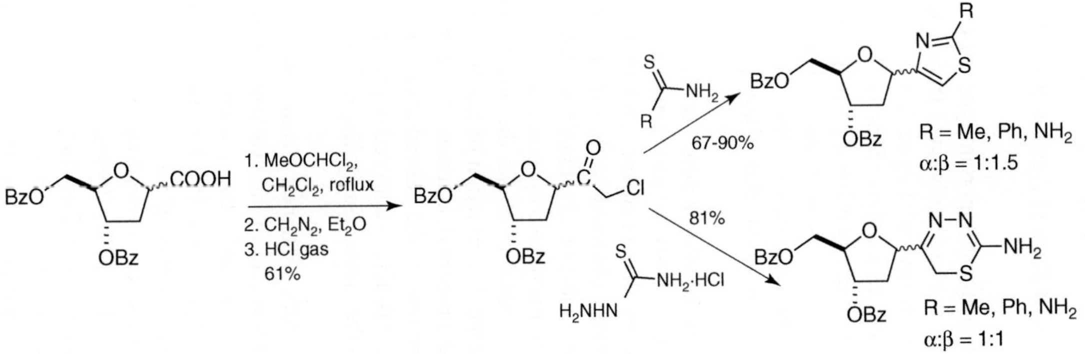

Scheme 2. Synthesis of thiazol and aminothiadiazole *C*-nucleosides from glycosyl cyanides.

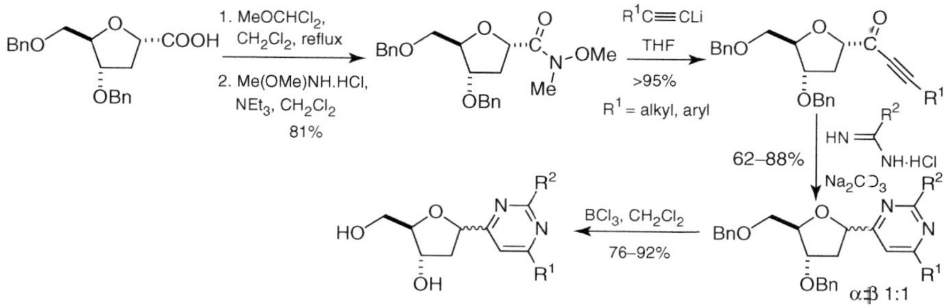

Scheme 3. Synthesis of disubstituted pyrimidinyl C-nucleosides from glycosyl cyanides.

A similar strategy is used in the synthesis of dideoxynucleosides. The pure β-acid derivative is obtained in this case from D-glucosamine in seven steps via reduction of an enenitrile (Scheme 6).[13] Cyanation and Wittig-type reaction affords different functionalized cyanoesters as common intermediates. Thus, subsequent cyclization and deprotection give dideoxyshowdomycin, whereas reaction with ethyl diazoacetate leads to dideoxyformycin B after five final steps.[14]

The purified β-anomer of a protected acid is used again in the synthesis of the benzopyrimidine 2'-deoxyriboside indicated in Scheme 7.[15] In the synthetic sequence the starting acid is converted into the corresponding acyl chloride, which undergoes Stille coupling with N-(tert-butoxycarbonyl)-2-(trimethylstannyl)aniline. After removing the protecting group, the key step is the elaboration of the 2-aminoquinazoline heterocycle through a low-temperature fusion reaction with cyanamide. Final deprotection furnishs the desired C-nucleoside.

The synthesis of 2',3'-anhydrotiazofurin (Scheme 8) also starts from a glycosyl cyanide, but requires the presence of a mesylate group in the carbohydrate moiety to mediate the final epoxide ring closure. The suitable ribofuranosyl nitrile is obtained from D-glucose and converted into the corresponding thioamide to incorporate the sulfur atom in the heterocycle (Scheme 8). The Hantzsch reaction of the amide obtained with ethyl bromopyruvate affords the corresponding thiazole. Final treatment with methanolic ammonia provides the target compound as a result of successive ester aminolysis, debenzoylation, and epoxide formation via mesylate displacement.[16]

Nitrile oxides are also used as starting materials in the synthesis of C-nucleosides. As they are prone to dimerization, they are usually generated in situ, either by base-induced dehydrohalogenation of hydroxymoyl halides or by dehydration of nitromethyl compounds. In the synthesis of benzothiazol, benzimidazol, and benzoxazol C-nucleosides, the first approach is used (Scheme 9).[17] The hydroxymoyl chloride obtained from D-glucose is reacted with 2-aminothiophenol, 1,2-diaminobenzene, and 2-aminophenol. The first equivalent of these reagents, acting as a base, releases the nitrile oxide. Nucleophilic attack by a second equivalent forms the amidoxime, and finally, intramolecular nucleophilic displacement of the hydroxyl amine yields the nucleosides desired.

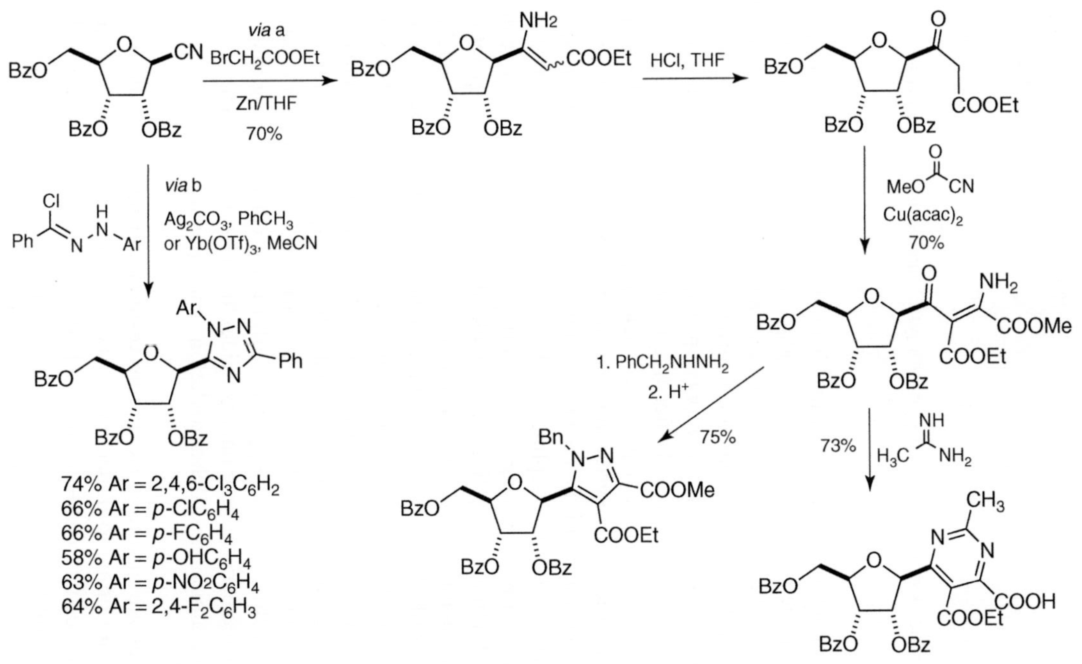

Scheme 4. Synthesis of pyrazole and pyrimidine *C*-nucleosides from glycosyl cyanides via enaminoketoesters.

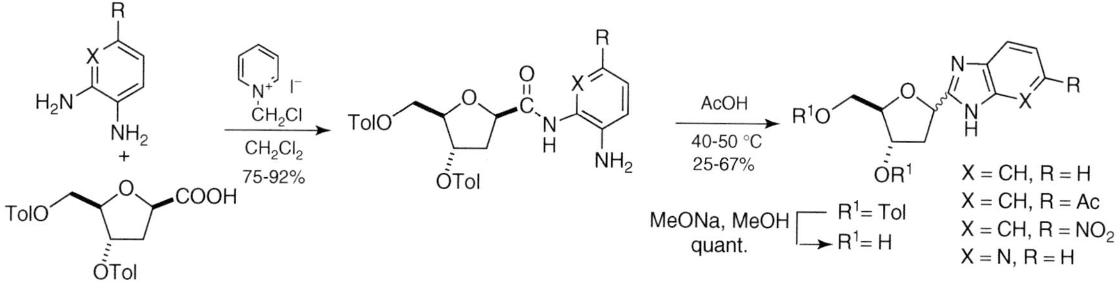

Scheme 5. Synthesis of 2′-deoxybenzo- and pyridoimidazole *C*-nucleosides.

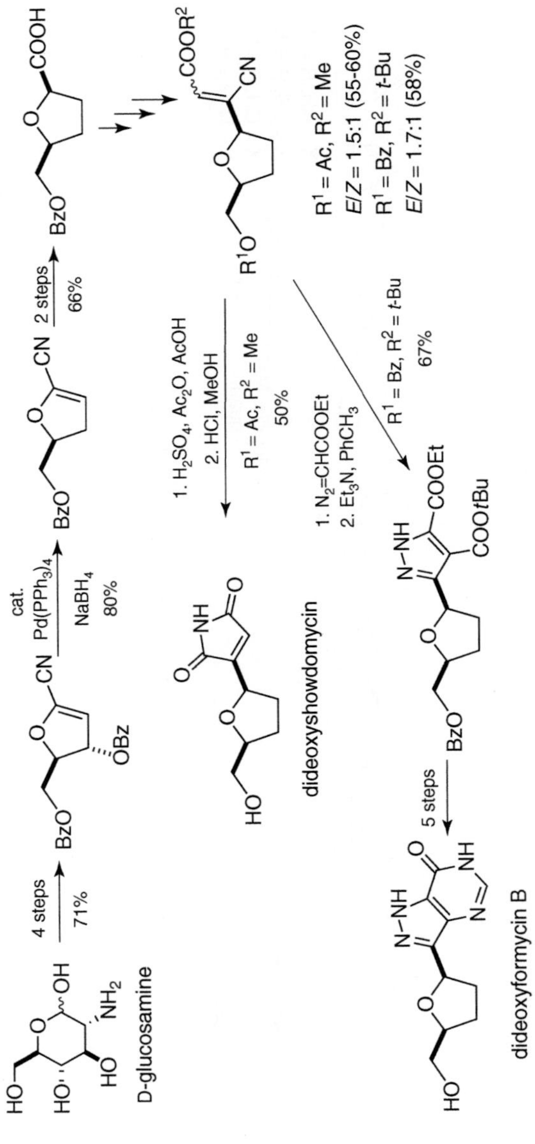

Scheme 6. Synthesis of dideoxyshowdomycin and dideoxyformycin B.

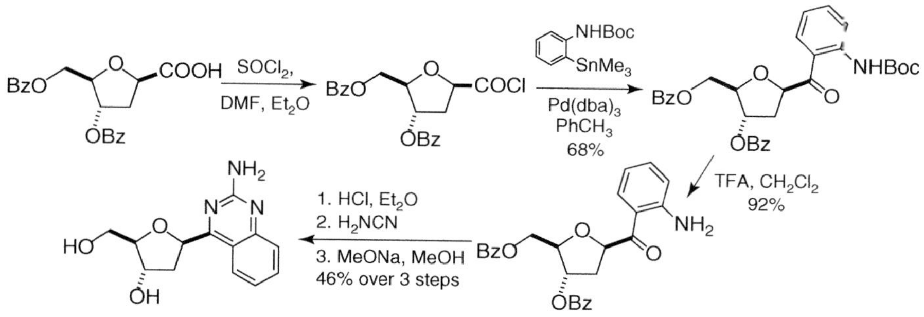

Scheme 7. Synthesis of benzopyrimidin C-nucleosides.

2.1.2. Organometallic Reagents: Alkynyl and Alkenyl Glycosides.

Alkynyl and alkenyl carbohydrates are also used efficiently in the synthesis of C-nucleosides. The alkynyl group at the anomeric position is usually introduced by the reaction of a Grignard reagent with a suitable protected glycosyl halide (Scheme 10, via a, method I) or a protected furanose (Scheme 10, via a, method II). In the latter case, the addition of alkynylmagnesium bromide is followed by treatment with p-toluenesulfonyl chloride to promote further cyclization to the furanose. Although the ribofuranosylethine derivative can be prepared with various degrees of stereoselectivity, the α-anomer is the predominant product in both procedures. The synthesis of these compounds is also described by indium-mediated alkynylation from acetyl glycosides (Scheme 10, via a, method III).[18] Method II, using vinylmagnesium bromide, is reported used to obtain alkenyl glycosides. (Scheme 10, via b).[19] At present, the best pathway to prepare a β-allyl C-glycoside of D-ribofuranose proceeds via the 1,5-di-O-acetyl-2,3-O-isopropylidene-β-D-ribofuranoside by treatment with allyltrimethylsilane and zinc bromide (Scheme 10, via c).[20]

The ethynyl C-glycosides are suitable precursors for various types of cycloaddition reactions that allow construction of the heterocycle unit. For example, [4 + 2]-cycloaddition reactions afford pyridazine C-nucleosides by reaction with dimethyl[1,2,4,5]tetrazine-3,6-dicarboxylates (Scheme 11, via a).[21] Pyridazine C-nucleosides in turn lead to pyrrole C-nucleosides by nitrogen extrusion.[22] On the other hand, [3 + 2]-cycloaddition reaction with azides sometimes affords equimolar mixtures of two possible triazole regioisomers in moderate yields,[21] or good yields of only one regioisomer in most cases[23,24] (Scheme 11, via b).

A very efficient two-reaction *one-pot* procedure based on a cross-enyne metathesis, followed by a Diels–Alder cycloaddition reaction, is developed for the synthesis of C-aryl glycosides.[25] In this methodology, protected ribofuranosylethyne derivatives are reacted with ethylene in the presence of a Grubbs second-generation catalyst to afford a diene that on reaction with dimethylethynyldicarboxylate or benzoquinone affords α-C-nucleosides (Scheme 11, via c).

Another cycloaddition reaction used from ethynyl C-glycosides is the [2 + 2 + 2]-cyclotrimerization process. Although various transition metal complexes have been used for this proposal (i.e., Rh, Ni, Ir, Co, and Ru), the most general catalysts

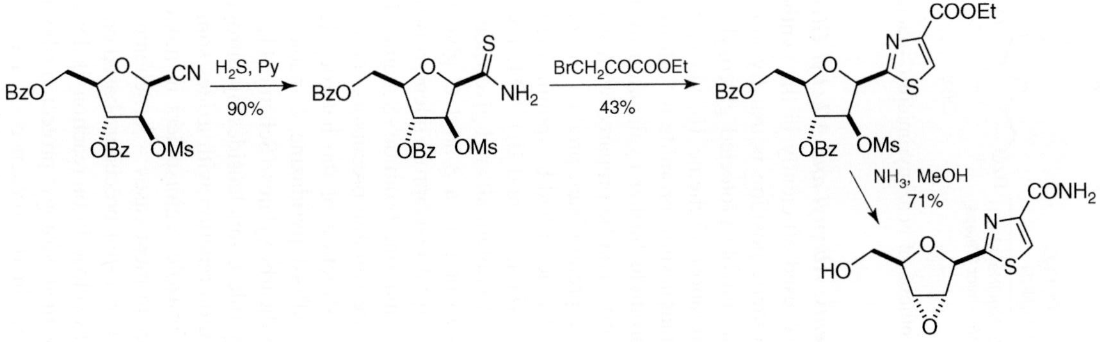

Scheme 8. Synthesis of 2′,3′-anhydrotiazofurin.

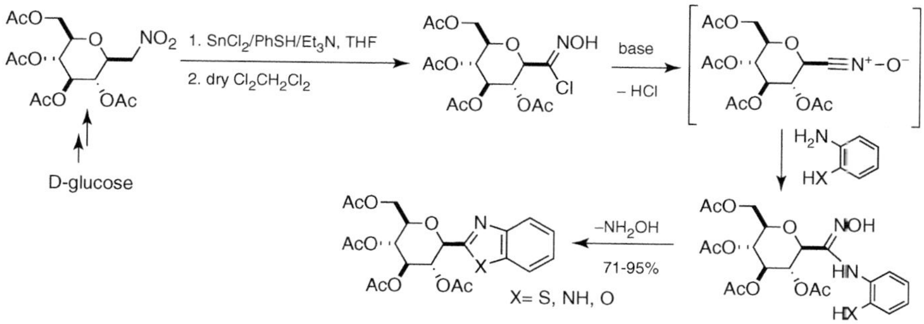

Scheme 9. Synthesis of pyranosyl benzothiazoles, benzimidazoles, and benzoxazoles, from pyranosyl nitrile oxides.

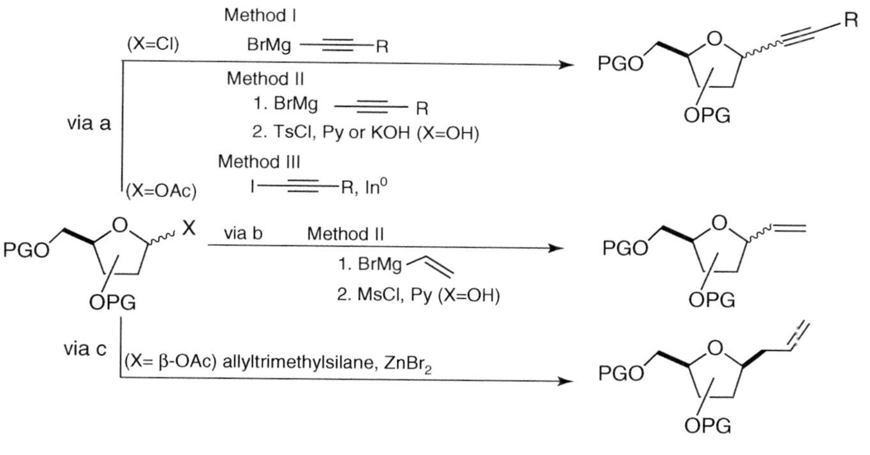

Scheme 10. Synthesis of alkynyl and alkenyl glycosides.

prove to be based on Rh complexes such as RhCl(Ph$_3$P)$_3$ or Cp*RuCl(cod) (Cp* = η-C$_5$Me$_5$, cod = 1,5-cyclooctadiene). In this sense, different 1,6-diynes are used in the construction of C-aryl glycosides (Scheme 11, via d).[23,26] This strategy is also used for the synthesis of spirocyclic C-nucleosides (Scheme 12).[27] In these cases the starting material is a gluconolactone or ribonolactone derivative, which after lithium acetylide addition followed by glycosylation is converted to silylated diynes. After desilylation, subsequent cycloaddition with alkynes or chloroacetonitrile give spirocyclic C- nucleosides.

A different approach involves the use of Sonogashira cross-coupling to derivatize the starting alkynyl glycoside. In these cases the nucleobase is connected at the anomeric center of the final product by an ethynyl linker. For this purpose, the use of 5-iodocytosine,[28] 2-bromopyridine,[23] 2-,[29,30] or 3-iodopyridines,[30] or different fluorophore[31] and chromophore[32] bromohydrocarbons is reported (Scheme 13).

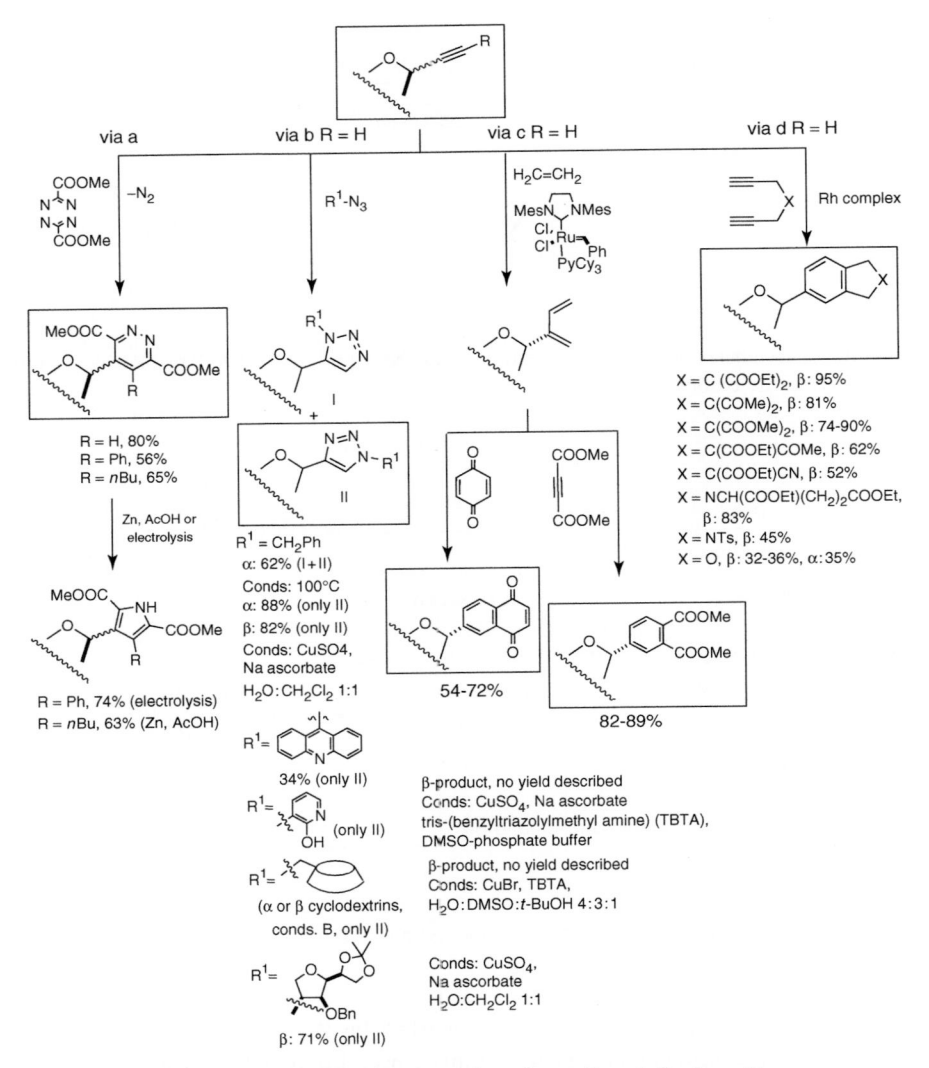

Scheme 11. Cycloaddition reactions from alkynyl C-glycosides.

With regard to vinyl glycosides, the reaction of anomerically pure C1-vinyl 2-deoxypiranoses with unprotected heterocycles via olefin metathesis makes it possible to obtain heterocyclic ethenyl C-nucleosides, as indicated in Scheme 14.[19]

β-allyl C-glycosides of D-ribofuranose allow the preparation of various intermediates of nucleoside analogues. For example, transformation of the double bond in the terminal alcohol by hydroboration–oxidation, mesylation of the alcohol obtained, and mesylate displacement using the imidazole sodium salt give the heterocyclic β-C-glycosidic moiety showed in Scheme 15 in good yields.[20]

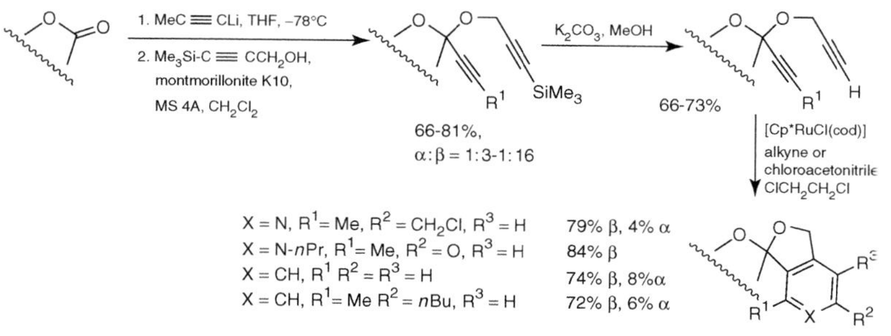

Scheme 12. Cycloaddition reactions in the synthesis of spyrocyclic C-nucleosides.

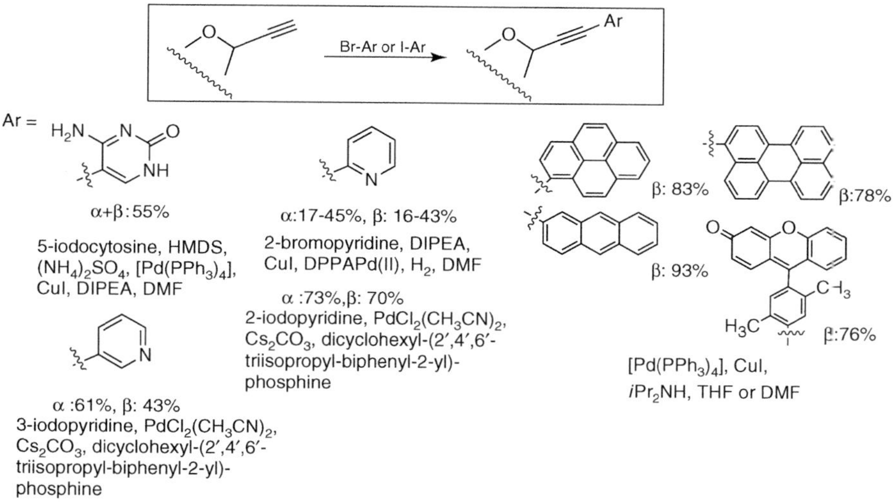

Scheme 13. Sonogashira reaction from alkynyl C-glycosides.

Finally, the synthesis of a series of acyclo-C-nucleoside analogues bearing different heterocycles is achieved through $CeCl_3 \cdot 7H_2O$-NaI-catalyzed one-pot condensation of a variant of the Bohlmann–Rahtz reaction using β-enaminones derived from L-rhamnose, acyclic and cyclic dicarbonyl compounds, and ammonium acetate (Scheme 16).[33] A plausible mechanism involves cerium-catalyzed sequential Michael addition–cyclodehydration–elimination reactions.

2.1.3. Radical Processes. Only one radical process is described for the construction of the aglycon unit in aromatic alkenyl C-glycosides.[34] The method is general and achieves the stereospecific introduction of a styryl group at the anomeric center of a particular carbohydrate by the radical-induced cyclization of a 3-phenylethynyl group tethered, via a temporary silicon connection, to a suitable hydroxyl group of the carbohydrate (Scheme 17).

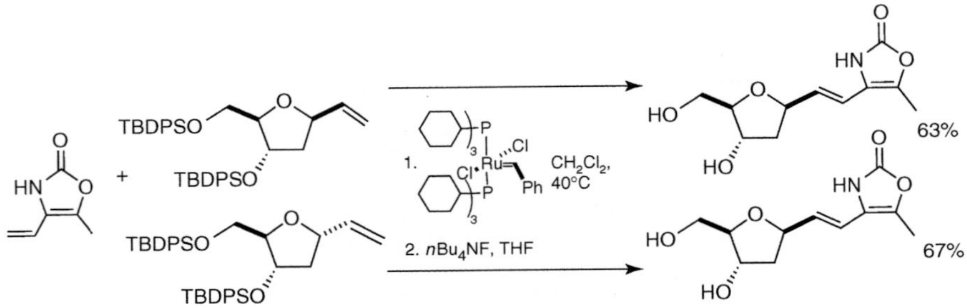

Scheme 14. Vinyl glycosides in the synthesis of C-nucleosides via olefin metathesis.

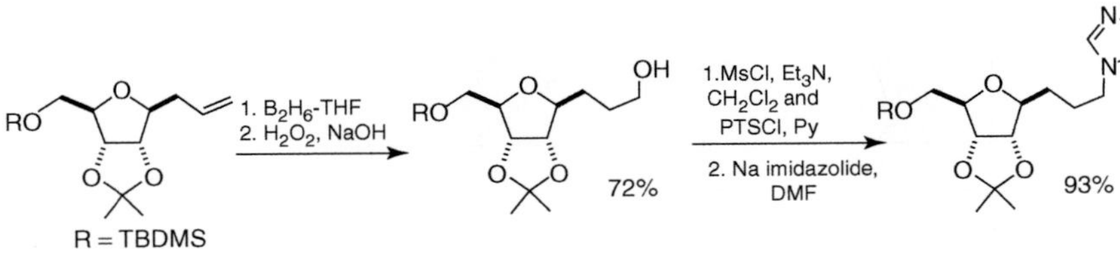

Scheme 15. Allyl glycosides in the synthesis of C-nucleosides.

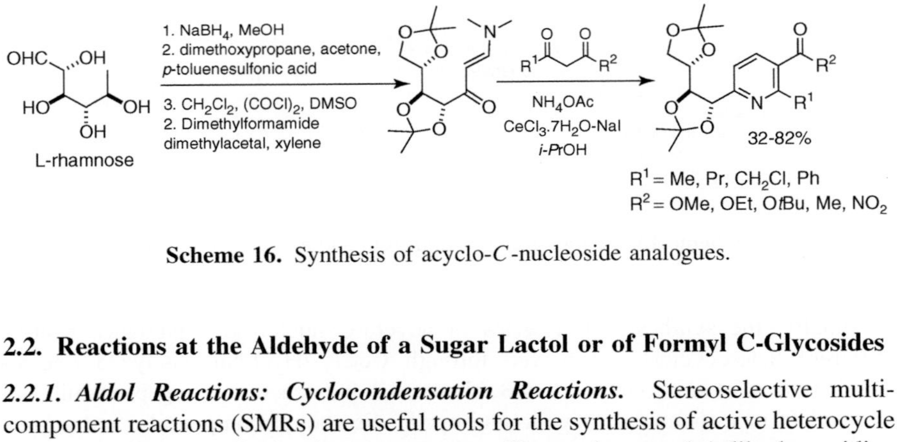

Scheme 16. Synthesis of acyclo-C-nucleoside analogues.

2.2. Reactions at the Aldehyde of a Sugar Lactol or of Formyl C-Glycosides

2.2.1. Aldol Reactions: Cyclocondensation Reactions.
Stereoselective multi-component reactions (SMRs) are useful tools for the synthesis of active heterocycle C-glycoconjugates. Both three-component Hantzsch-type 1,4-dihydropyridine (DHP) synthesis[35] or three-component Biginelli dihydropyrimidinone (DHPM) synthesis[36] have been used as cyclocondensation processes for construction of the heterocyclic moiety from readily available anomeric sugar aldehydes[37] (formyl C-glycosides) (Scheme 18).

The Hanztsch reaction, which involves reaction of a sugar aldehyde, a β-ketoester, and an enamine, gives access to a collection of structurally and stereochemically diversified 1,4-DHP C-glycoconjugates, among which the C2-glycosylated nifedipine analogue is a target molecule. Although the asymmetric

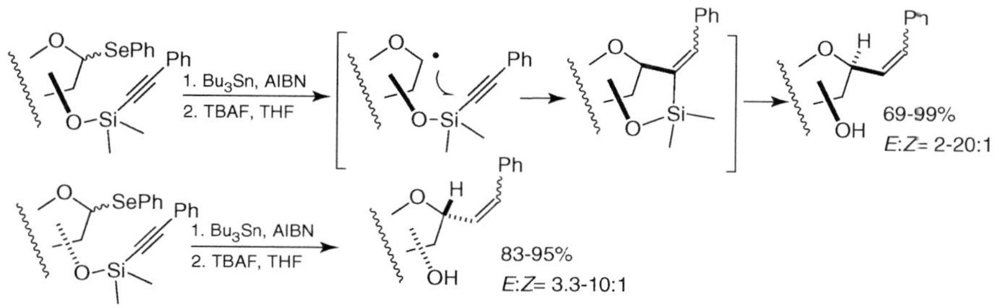

Scheme 17. Radical-mediated reaction for the stereospecific synthesis of aromatic alkenyl C-glycosides.

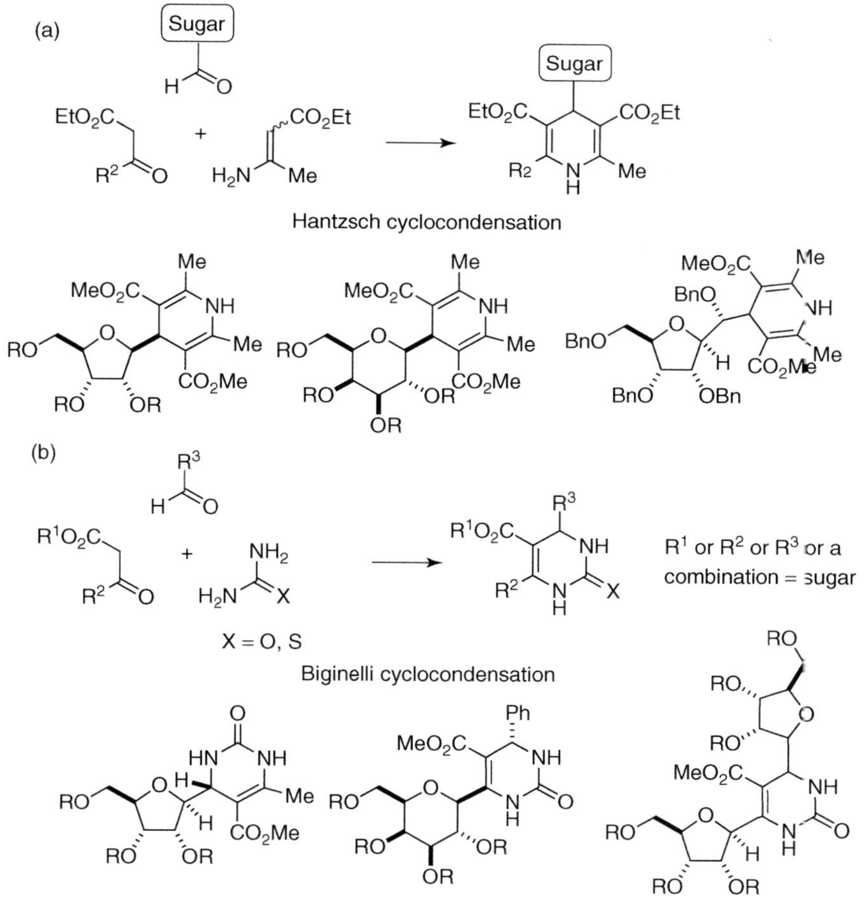

Scheme 18. Synthesis of C-nucleosides via (a) Hantzsch and (b) Bignelli cyclocondensation.

induction in this process is provided by the carbohydrate chiral auxiliary, typically appended to the aldehyde reactant, obtaining epimers at C4 with a modest degree of selectivity, the asymmetric version using proline as an organocatalyst renders dihydropyridine C-glycoconjugates with a high degree of stereoselectivity (de > 95%).

On the other hand, the three-component Biginelli reaction, which involves acid-catalyzed condensation of a sugar aldehyde, a β-ketoester, and urea, can be applied to the synthesis of different mono- and bis-C-glycosylated DHPS. The process with glycosylated ketoesters and aldehydes occurs with satisfactory asymmetric induction leading to chiral products with a given configuration at the C4 stereocenter of the DHPM ring.

In the search for multicomponent reactions for the rapid construction of chiral heterocycle C-glycoconjugates, a Staudinger [2 + 2] imine–ketene cycloaddition reaction provides rapid access to C-glycosyl-β-lactams by one-pot reaction of the corresponding acetyl chloride, the primary amine, and the formyl C-glycoside (Scheme 19).[38] Initial reaction of the sugar aldehyde and the amine leads to the C-glycosylimine, which is subsequently treated with the acetyl chloride and triethylamine to produce the corresponding acetoxy–ketene.

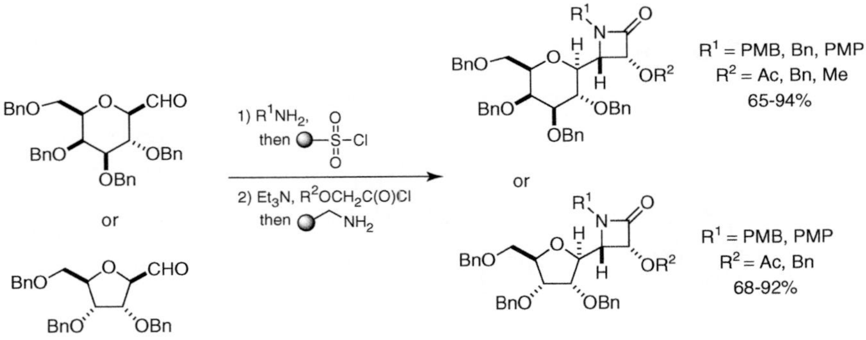

Scheme 19. Synthesis of C-glycosyl-β-lactams by Staudinger [2 + 2] imine–ketene cycloaddition reaction.

2.2.2. Wittig Reactions.

The Wittig olefination reaction can be used for the synthesis of polyhalogenated quinoline C-nucleosides.[39] In this approach, a ribofuranose derivative is coupled with a fully functionalized 2-nitrobenzylidenephosphorane to afford an alkene in good yield (85%, Z/E = 12:1) (Scheme 20). Elaboration of the alkene in order to install a functional group at the benzylic position and reduction of the nitro moiety affords a homo-C-nucleoside, from which a Wittig reaction with a keteneylidene phosphorane renders quinolin-2-one. Further elaboration finally furnishes C-nucleosides bearing trichloroquinoline and bromodichloroquinoline aglycons, although in low overall yield.

Another approach for the synthesis of C-nucleosides involves tandem Horner–Wadsworth–Emmons reaction/conjugate addition C-glycosylation starting from a furanose or pyranose derivative and sulfonylphosphonate reagent to produce

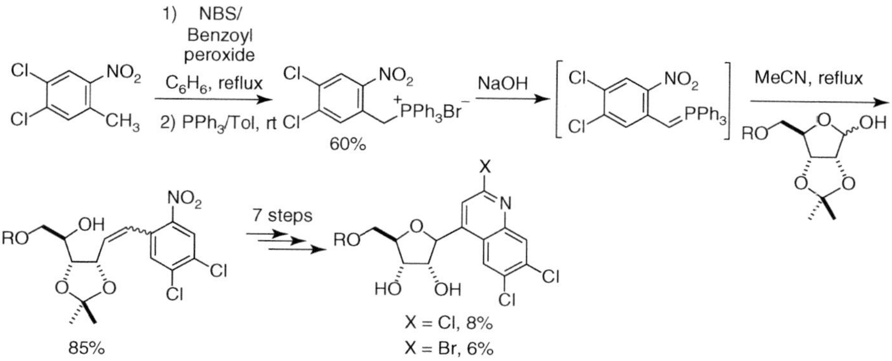

Scheme 20. Synthesis of polyhalogenated quinoline C-nucleosides using a Wittig reaction as a key step.

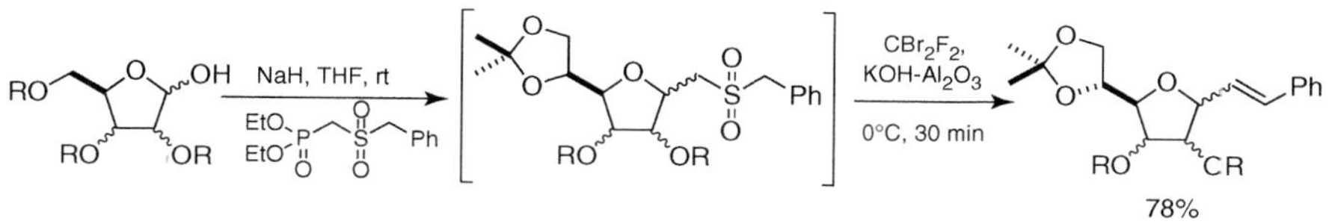

Scheme 21. Tandem Horner–Wadsworth–Emmons/ring closure/Ramberg–Bäcklund used to obtain C-nucleosides.

C-glycoside with low anomeric stereoselectivity (Scheme 21). Subsequent tandem halogenation/Ramberg–Bäcklund sequence produces the corresponding styrenyl C-glycoside.[40] The method can be also carried out in a one-pot transformation and from unprotected carbohydrate derivatives.

The synthesis of 9-deaza-2'-C-methyl- and 7,9-dideaza-7-oxa-2'-C-methyladenosine[41] is based on slight variations of a method reported for 9-deazaadenosine.[42] In this case, ribose is elaborated to render a 2'-C-methyl ribose derivative[43] that is treated with diethyl cyanomethylphosphorane in a Wittig reaction followed by intramolecular conjugate addition to produce a C-glycoside (Scheme 22). The pyrrole ring is easily constructed via a direct Claisen reaction with methyl formate and subsequent DBU-induced Kirsch[44] cyclization. Construction of the second pyrimidine ring by conventional methods renders 9-deaza-2'-C-methyladenosine. Slight modification of the procedure leads to the formation of a 7,9-dideaza-7-oxa-2'-C-methyladenosine analogue.

Using an approach similar to that used for construction of a heterocycle ring, Hong and Oh have synthesized threosyl C-nucleosides with good anti-HIV activity without significant cytotoxicity.[45] 4'-Hydroxymethylated 9-deaza-adenosine C-nucleosides are prepared easily from a deoxythreose intermediate as a starting material, via a process that involves a tandem Wittig–Michael sequence to give a C-glycoside that is elaborated in a process analogous to that mentioned earlier (Scheme 23).

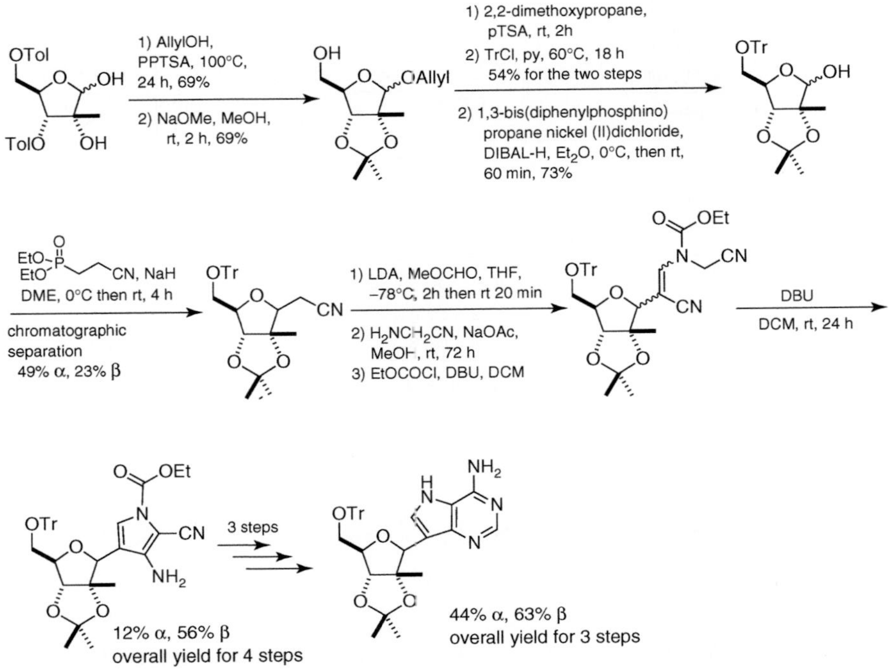

Scheme 22. Synthesis of 9-deaza-2′-C-methyladenosine involving the Horner–Wadsworth–Emmons reaction from sugar lactol.

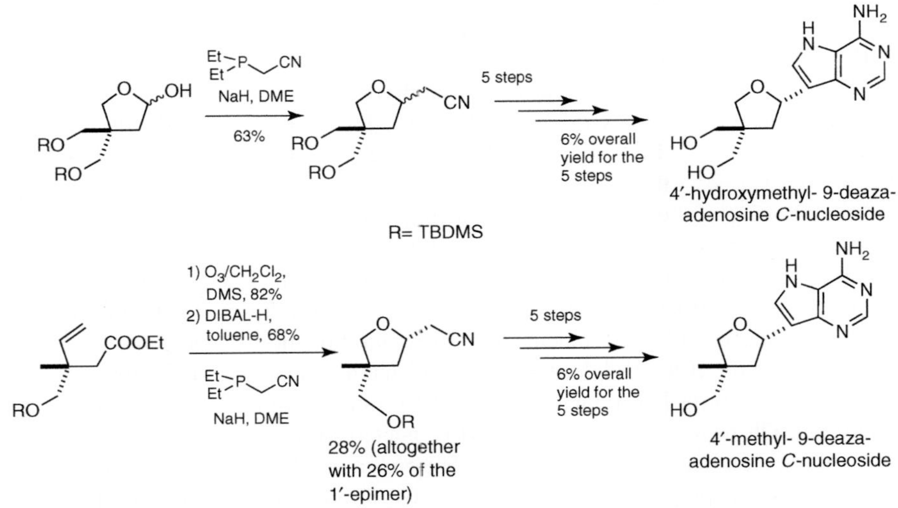

Scheme 23. Synthesis of 4′-hydroxymethylated and 4′-methylated 9-deaza-adenosine C-nucleosides tandem Wittig–Michael sequence from lactols.

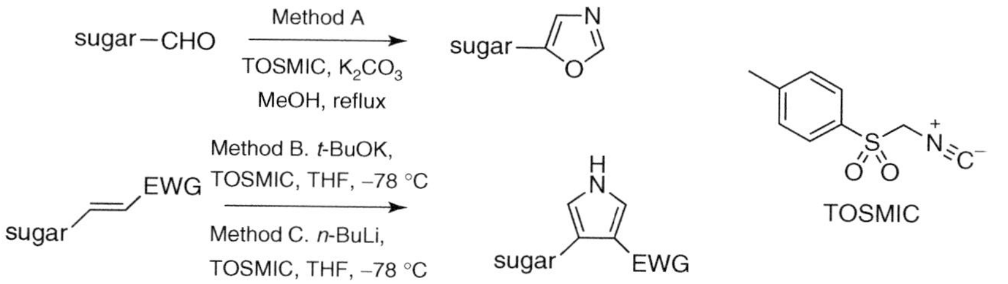

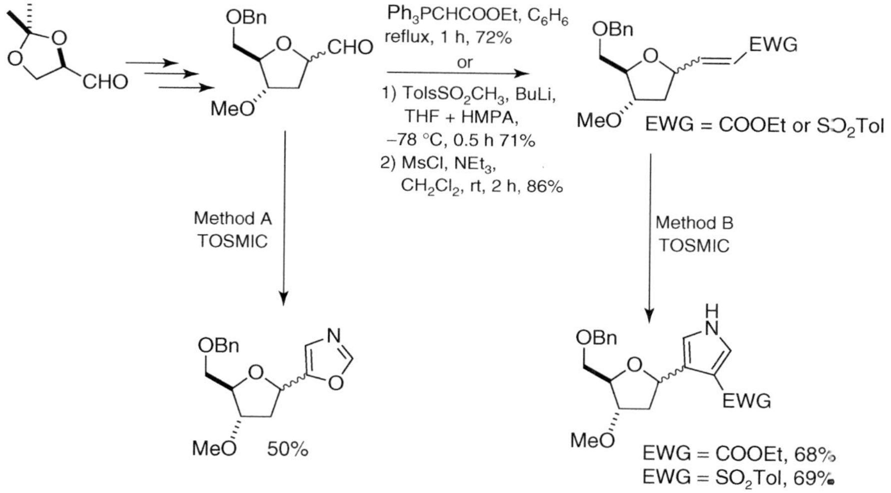

Scheme 24. Synthesis of oxazole, 3-substituted pyrrole D- and L-C-nucleosides using TOSMIC on sugar vinyl esters prepared by a Wittig reaction.

Krishna et al.[46] have developed a simple and efficient route for the synthesis of oxazole, 3-nitro-, 3-sulfonyl-, and 3-carbetoxy-substituted pyrrole D-ribo-, 2-deoxy-D-ribo-, and 2-deoxy-L-ribo-C-nucleosides involving as a key step the addition of one carbon synthon p-toluenesulfonylmethyl isocyanide (TOSMIC) anion on either sugar-derived aldehydes or sugar vinyl esters, with concomitant cyclization, resulting in the corresponding products in moderate to good yields. Sugar vinyl esters are, in turn, prepared by a Wittig reaction from C-formyl sugars (Scheme 24).[35]

Pseudo-C-nucleosides bearing a C–C bond between C4′ and C5′ of a pyranose sugar moiety and a pyrazole and pyridazinone ring have been prepared via the reaction of a β-formyl-α,β-unsaturated ester bearing a β-sugar moiety with hydrazines in neutral and acidic conditions.[47] The preparation of a β-formyl-α,β-unsaturated ester is accomplished by oxidation of the secondary hydroxyl group of 3-O-benzyl-1,2-isopropylidene-α-D-glucofuranose, followed by elongation of its carbon chain with (ethoxycarbonylmethylene)triphenylphosphorane and oxidation of the hydroxymethyl group (Scheme 25).

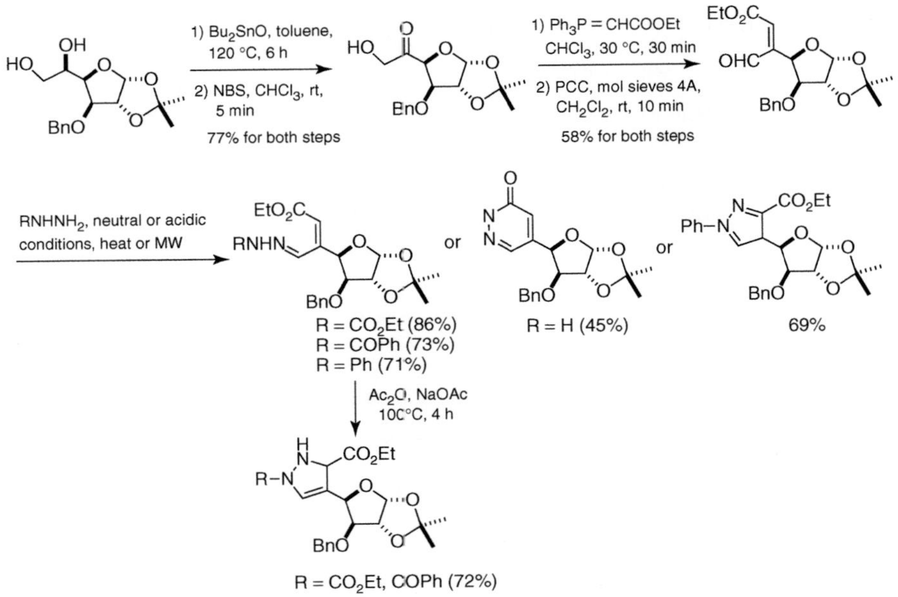

Scheme 25. Synthesis of pseudo-C-nucleosides from β-formyl-α,β-unsaturated esters bearing a β-furanosidic moiety.

3. DIRECT CONSTRUCTION OF THE C–C BOND BETWEEN THE CARBOHYDRATE AND THE HETEROCYCLE

A synthetically convergent strategy developed for the construction of C-nucleosides involves the direct coupling of a protected carbohydrate unit with a preformed nucleophilic aglycon.

3.1. Reaction of Organometallic Reagents with a Protected Carbohydrate Unit

In the present section we focus on recent advances in the preparation of C-nucleosides using the direct construction of the C–C bond between the carbohydrate and the heterocycle, in particular, using the addition of organometallic aglycones to a conveniently protected carbohydrate moiety as a key step.

3.1.1. Halofuranoses: Substitution Reaction. Most of the synthetic approaches that have traditionally been reported to access C-nucleosides start from protected ribosyl and 2-deoxyribosyl halides, mainly bromides and chlorides (Hoffer's chlorosugar **1**), using organometallic partners based on lithium,[48,49] magnesium,[49–51] cadmium,[49,52–54] zinc,[49,51,54] and mercury[49,55] (Scheme 26). Given the fact that these reactions usually afford low overall yields and employ toxic metals, alternative protocols are highly desirable. Indeed, the preference for the α-anomer instead

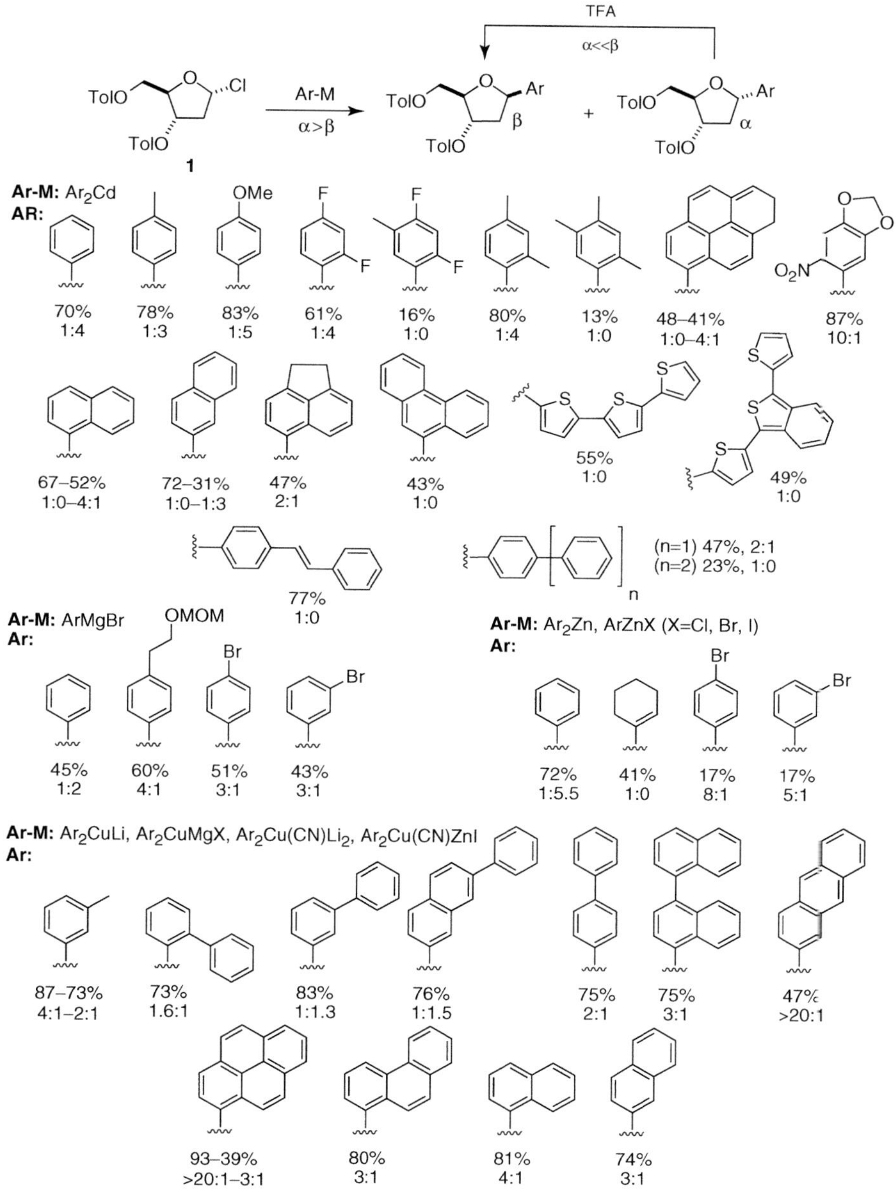

Scheme 26. Reaction of Hoffer's chlorosugar **1** with various organometallic reagents and subsequent acid-catalyzed epimerization; the yield and α/β ratio shown under the aryl moieties.

of the more desirable naturally occurring β-anomer requires an extra acid-catalyzed epimerization step that reduces the synthetic flexibility of this approach.

Successful improvement in this methodology, especially in the isolated yield, is achieved by using mixed magnesium–cuprate[53,56] organometallics in combination with the aforementioned acidic α-to-β anomerization. This optimized protocol not only allows the operationally simple preparation of a wide range of different 1-C-aryl-nucleosides in good yields but also reduces the toxic waste products otherwise resulting from previous heavy metal organometallic systems. The introduction of noncanonical base surrogates allows the study of complex biological interactions and also offers new opportunities for studying the fluorescence properties of new structures containing these moieties.[57]

Alternatively, ribosyl bromides are also employed for this transformation. For example, treatment of **2** with furanylmercury chloride in nitromethane affords a separable α/β mixture (being the α-anomer predominant), which can subsequently be epimerized to the desired β-anomer with trifluoroacetic acid[55] (Scheme 27).

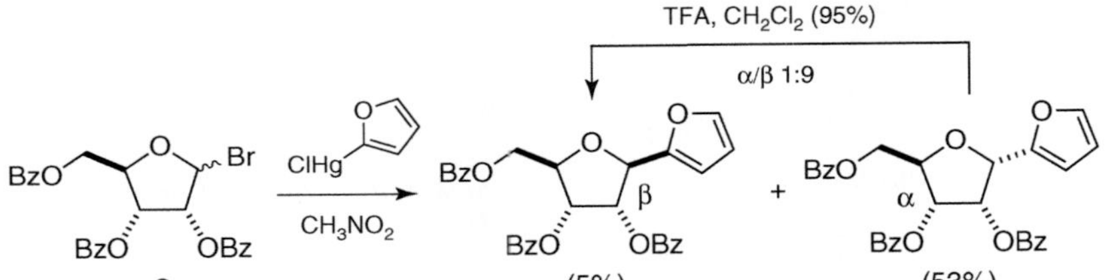

Scheme 27. Reaction of furanylmercury with benzoylated D-ribofuranosyl bromide **2** and subsequent acid-catalyzed epimerization.

3.1.2. Furanoses: Addition to the Aldehyde Followed by Cyclization.

The addition of organometallic aglycones, based on magnesium[58] and lithium,[59] to the aldehyde moiety of a protected furanose at low temperature typically affords a diastereomeric mixture of diols which are efficiently cyclized under either acidic or Mitsunobu conditions to the corresponding C-nucleoside analogue (Scheme 28). A similar sequential aryl-aldol condensation/cycloetherification approach is employed for the preparation of a novel class of fluorescent nucleosides.[60] Regioselective *ortho*-lithiation of 2-bromothiophene followed by addition of protected 2-deoxy-D-ribose leads to a inseparable mixture of diastereisomers which is subjected to an acid-catalyzed cyclization with *p*-TsOH in toluene to afford the corresponding 2-thiophene nucleoside in 64% yield over two steps and a 2:3 α/β ratio. Alternatively, diacetylated 2-deoxy-D-ribose product is also obtained in good yield (73%) and a 3:7 α/β ratio using a Friedel–Crafts dehydrative glycosylation as a key step.

Similarly, this approach can also be employed for the preparation of six-membered carbocyclic C-nucleosides after the addition of 6-chloro-2,4-dimethoxy-pyrimidin-5-yl lithium organometallic to the corresponding ketone followed by hydroxyl elimination/double-bond reduction or deprotection/intramolecular cyclization[61] (Scheme 29).

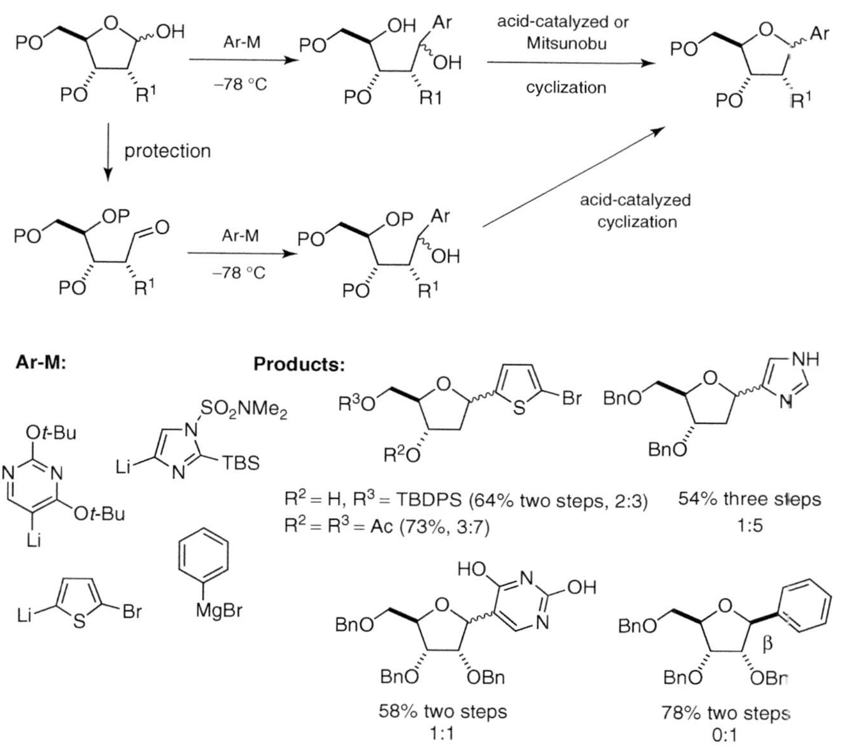

Scheme 28. Addition of organometallic aglycones to furanoses followed by acid-catalyzed or Mitsunobu cyclization. P = protecting group.

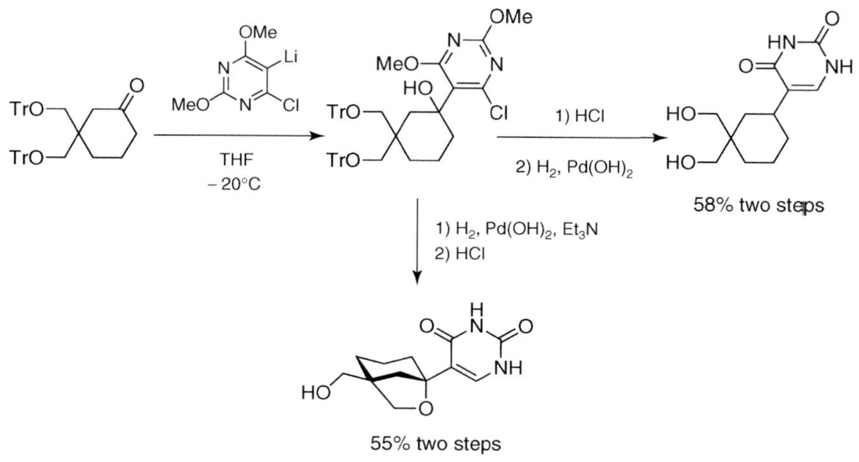

Scheme 29. Synthesis of carbocyclic C-nucleosides by the addition of organolithium reagents to ketones. Undefined solid lines denote any possible stereochemistry.

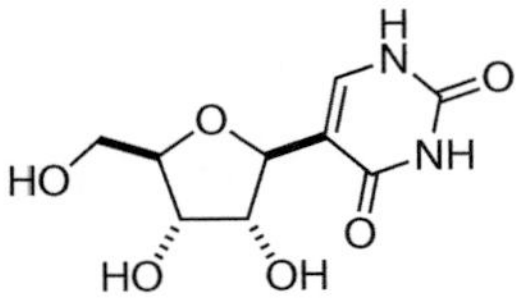

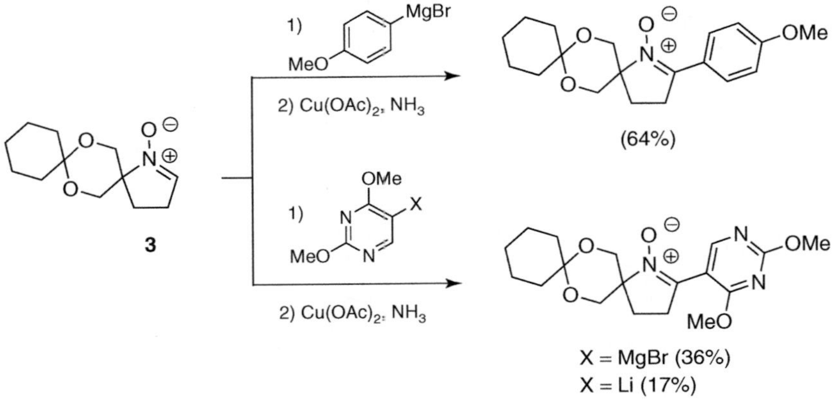

Scheme 30. Synthesis of 1-pyrroline-1-oxides as pseudouridine aza analogues.

A conceptually different approach exploits the addition of organometallic reagents to nitrones followed by oxidation that allows the preparation of aza analogues of uridine (ψ-uridine, ψ), a naturally occurring C-nucleoside involved in the preservation of RNA structure and fine-tuning of its functions during the translation process[62] (Scheme 30).

The use of Grignard reagents is favored over their lithium counterparts, yielding the corresponding coupled products in moderate yields (up to 64%). The reduced yield of the lithium reagents is attributed to the competing dimerization of **3** (up to 39%) under the conditions tested, which predominates over the desired organometallic addition. Apart from organometallics, electron-rich aromatic compounds can also be used for the direct arylation of sugar-based cyclic nitrones **4–8** by a Friedel–Crafts type of reaction affording 2-aryl polyhydroxylated N-hydroxypyrrolidines in good to excellent yields (up to 97%) and high diastereoselectivities (up to 1:99 dr). Final deprotection of N-hydroxypyrrolidines affords the corresponding 2-aryl pyrrolidines under mild conditions[63] (Scheme 31).

3.1.3. Furanones: Addition to the Lactone Followed by Hemiacetal Reduction.
The addition of organometallic aglycones typically based on lithium[50,58,64–67] and, less often, magnesium[67] to the lactone moiety of a pentofuranose is one of the most important C–C bond-forming methods for the preparation of C-ribonucleosides,

DIRECT CONSTRUCTION OF THE C–C BOND BETWEEN THE CARBOHYDRATE   287

Addition products:

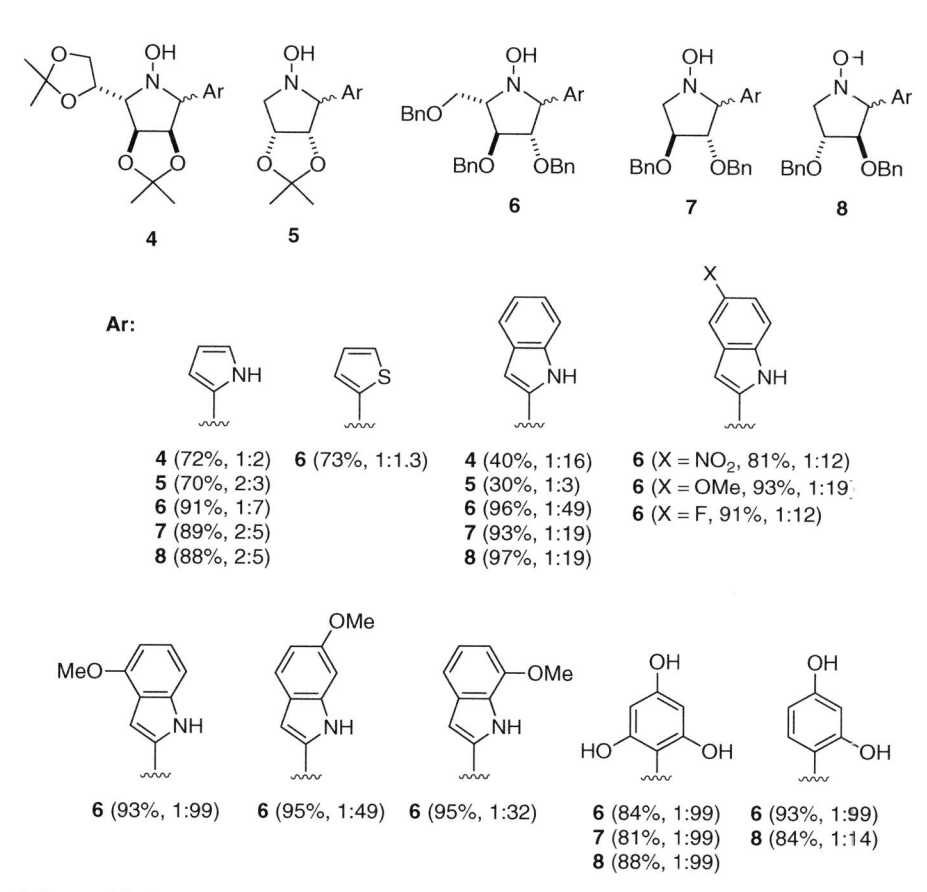

Ar:

4 (72%, 1:2)
5 (70%, 2:3)
6 (91%, 1:7)
7 (89%, 2:5)
8 (88%, 2:5)

6 (73%, 1:1.3)

4 (40%, 1:16)
5 (30%, 1:3)
6 (96%, 1:49)
7 (93%, 1:19)
8 (97%, 1:19)

6 (X = NO_2, 81%, 1:12)
6 (X = OMe, 93%, 1:19)
6 (X = F, 91%, 1:12)

6 (93%, 1:99)

6 (95%, 1:49)

6 (95%, 1:32)

6 (84%, 1:99)
7 (81%, 1:99)
8 (88%, 1:99)

6 (93%, 1:99)
8 (84%, 1:14)

Scheme 31. Stereoselective synthesis of *C*-aza-nucleosides by arylation of sugar-based cyclic nitrones; the yield and 2′,3′-*cis/trans* ratio are shown under the aryl moieties. Undefined solid lines denote any possible stereochemistry. P = protecting group.

usually with ribo or 2-deoxyribo configurations. The addition of the corresponding organometallic reagent at low temperature affords a hemiketal intermediate that is further reduced to the final *C*-nucleosides **9–16**. The first method reported for the aforementioned reduction involves treatment of the hemiketal with a Lewis acid followed by reduction of the corresponding oxonium intermediate with a

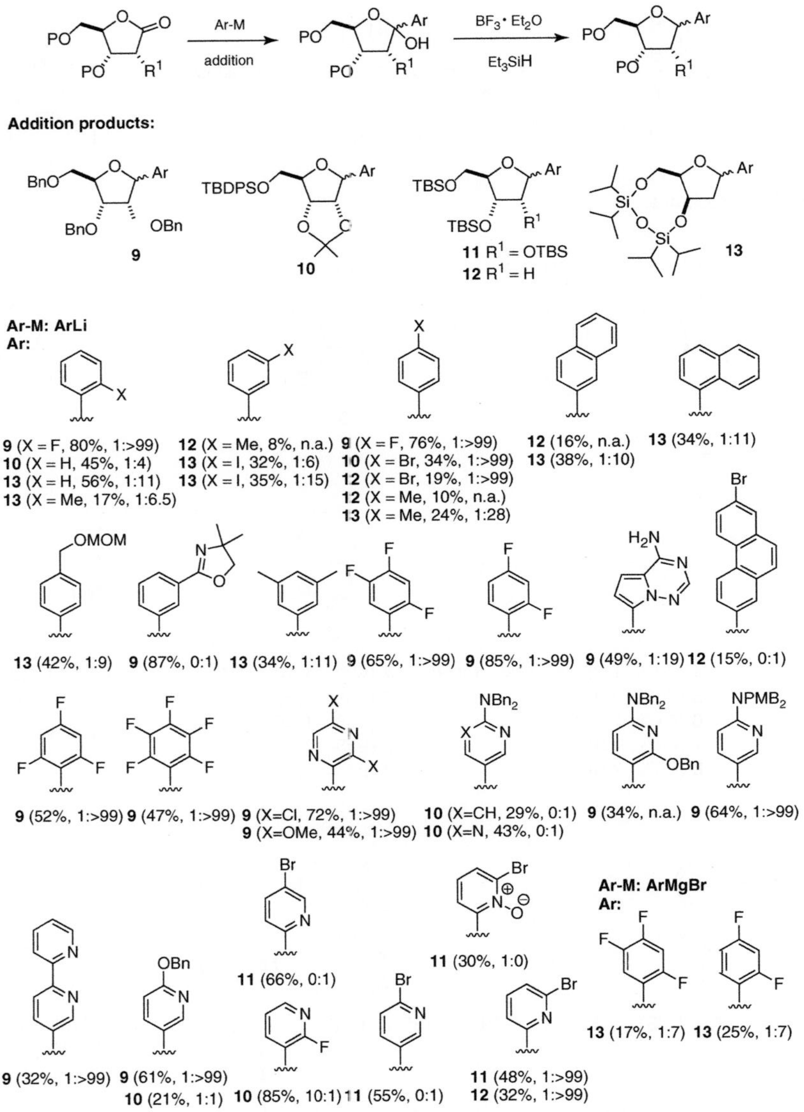

Scheme 32. Synthesis of carbocyclic C-nucleosides by the addition of organometallic reagents to furanolactones followed by silane reduction; the overall yield and α/β ratio are shown under the aryl moieties. P = protecting group.

silane.[50,58,64,65,67] The yields for this transformation are moderate to good, but the selectivities are excellent (up to a 1:> 99 α/β ratio). Interestingly, the selectivity relies primarily on the nature of the aglycone and the protecting groups present in the carbohydrate lactone (Scheme 32).

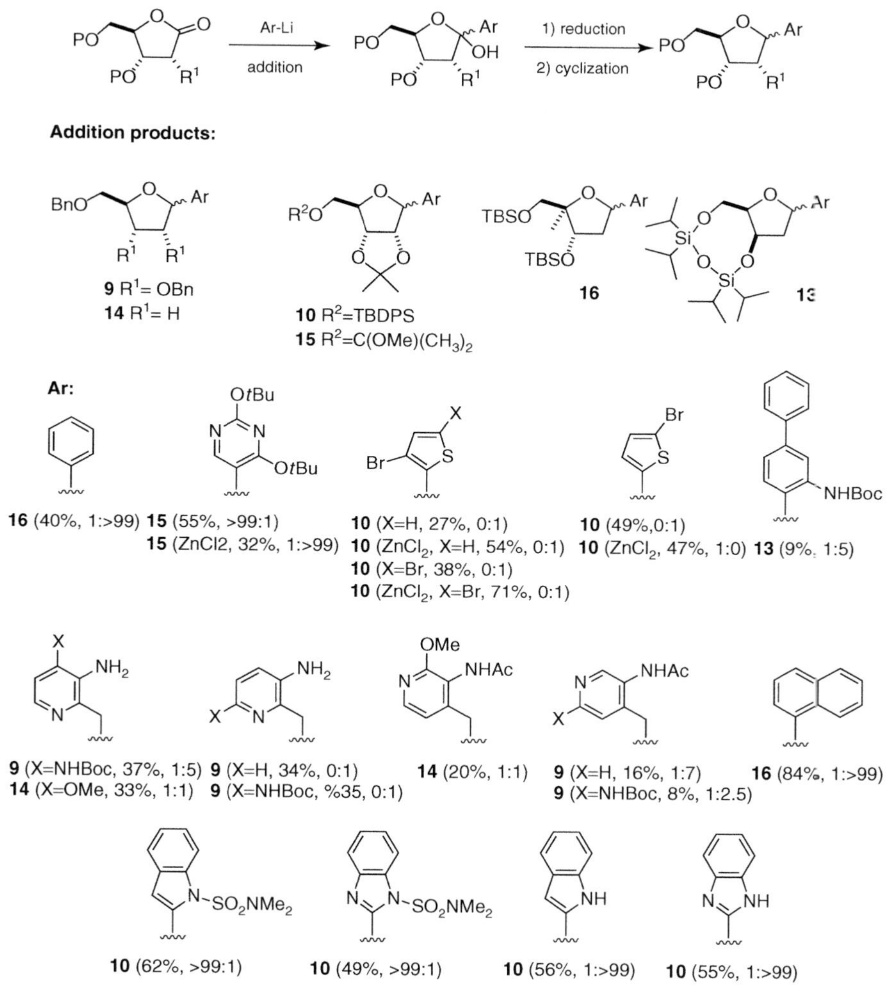

Scheme 33. Synthesis of carbocyclic *C*-nucleosides by the addition of organometallic reagents to furanolactones followed by Mitsunobu cyclization; the overall yield and α/β ratio are shown under the aryl moieties. P = protecting group.

The second general method for the hemiketal-to-ether transformation involves reduction of the hemiketal intermediate to the corresponding diol using $NaBH_4$, L-Selectride, or other complex hydrides followed by Mitsunobu cylization (Scheme 33).

Again, the addition of the corresponding organolithium alone, or, exceptionally, using $CeCl_3$ as a Lewis acid, leads to an hemiketal intermediate in moderate to good yield (44–97%). The reduction step typically affords the corresponding diol in good to excellent yield (up to 98%) and variable selectivity, ranging from

>99:1 to 1:> 99 dr, depending on the reaction conditions. For example, the addition of $ZnCl_2$ as a Lewis acid during the reduction seems to reverse the selectivity. Finally, Mitsunobu cyclization using either DIAD or DEAD with Ph_3P furnishes the expected C-nucleosides **9, 10, 13**, and **14–16** in excellent yields (up to 90%). The selectivities are excellent (up to >99:1 or 1:>99 α/β ratio) but dependent on the nature of the protecting groups on the organometallic (typically, for N-protected heterocycles). This behavior proved useful for the preparation of both anomers from a common starting material.

Alternatively, 1′-C-branched or double-substituted structures are also available after addition of an excess of nucleophilic organometallic aglycone and subsequent reduction using either Et_3SiH or Mitsunobu cyclization.[65]

3.2. Epoxide and Sulfite Opening

Glycosyl epoxides are useful glycosyl donors in glycosylation reactions and in glycosyl transfer reactions (Scheme 34). Epoxidation of ribofuranosyl glycals take place by attack *anti* to the 3-OH to afford the arabino-epoxide. The reaction with naphthylaluminum reagents in THF affords the corresponding β-arylnucleoside as a consequence of epoxide-opening *syn*. Further deoxygenation of position 2 and removal of protecting groups yields the 2-deoxy-β-arylnucleosides.[68]

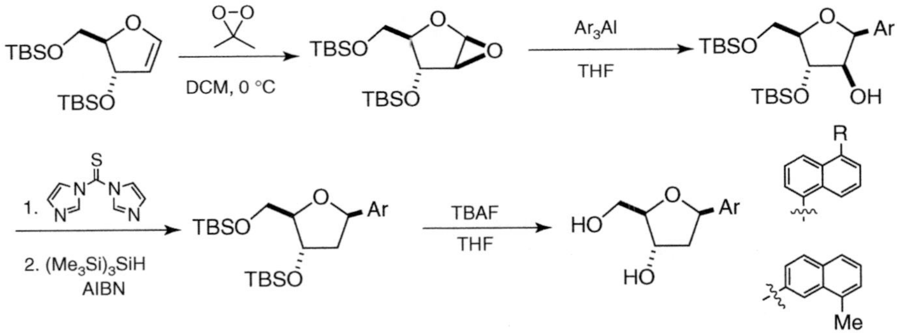

Scheme 34. Synthesis of C-nucleosides by Lewis acid–catalyzed opening of epoxides.

1,2-Cyclic sulfite xylosides were used as an alternative to 1,2-epoxides in the synthesis of different glycosyl derivatives (Scheme 35). The reaction was carried out in xylene or ionic liquids under zinc chloride catalysis to afford C-nucleosides in 60–65% yield as mixtures α/β = 1:1.[69]

3.3. Palladium-Catalyzed Allylic Alkylation

Allylic substitution reaction is a transition metal–catalyzed reaction that is a powerful tool in organic synthesis. Catalysts based in metals such as Pd, Mo, W, Ru, Rh, and Ir are actives in allylic substitution reactions. Control of the selectivity

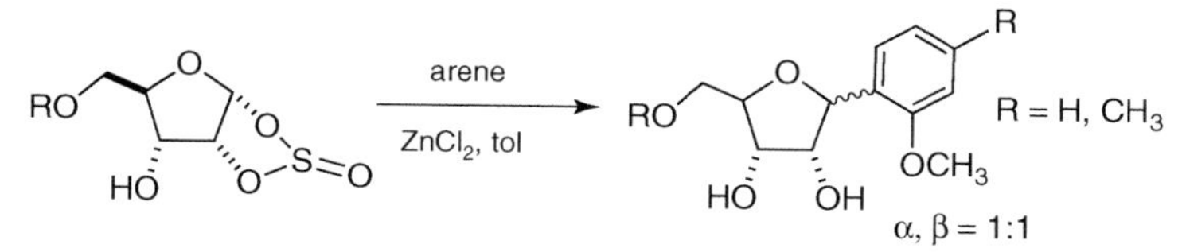

Scheme 35. Synthesis of C-nucleosides by Lewis acid–catalyzed opening of sulfites.

(regio, stereo, and enantio) is one of the challenges of this reaction. Palladium-catalyzed asymmetric allylic substitution developed by Trost is a powerful methodology for the synthesis of several types of nucleosides and also provides excellent enatioselectivity.[70] Some representative ligands used in this process are shown in Scheme 37. Esters, carbonates, and carbamates can also be present as leaving groups.

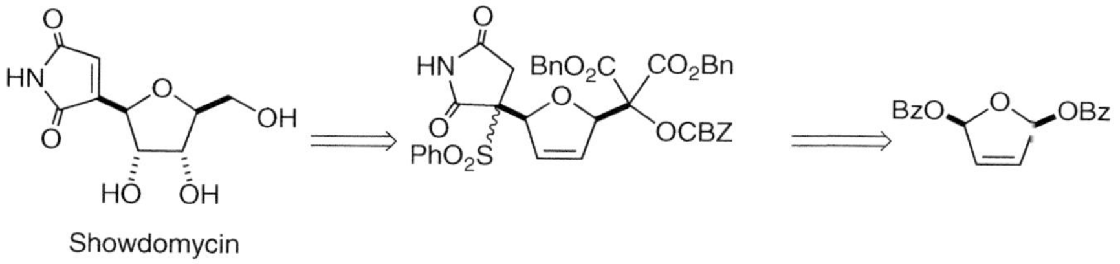

Scheme 36. Retrosynthesis of showdomycin based on asymmetric allylic substitution.

Scheme 36 shows the retrosynthetic scheme for the synthesis of L-showdomicin[71] using allylic substitution as a key step and 2,5-dibenzoyloxy-2,5-dihydrofuran[72] as the starting material. Moreover, after introduction of the aryl or heteroaryl group, the hydroxymethyl chain characteristic of nucleosides must be introduced in a further step. This chain can also be introduced by palladium-catalyzed allylic alkylation using an appropriate reagent. In this way, the key synthon is obtained by way of two allylic substitutions with a maleimide equivalent and a hydroxymethyl equivalent. In the first step the desymmetrization of the diester is carried out.

The maleimide sulfone was selected as a suitable synthetic equivalent of maleimide, and the nitrogen was protected to avoid competitive reactions in the allylic substitution step. Tetronic acid was selected as an appropriate hydroxymethyl chain equivalent. Either the base or the C1 fragment can be introduced in the first step, which provides a high degree of flexibility in the synthesis. Once the first nucleophile is introduced, the compound become chiral, and for introducing the second nucleophile a chiral catalysts is not required, although in general they provide improved results.

Thus, the *meso*-diester **17** was reacted with the maleimide sulfone sodium salt in the presence of a palladium catalyst precursor and the chiral ligand L1, affording

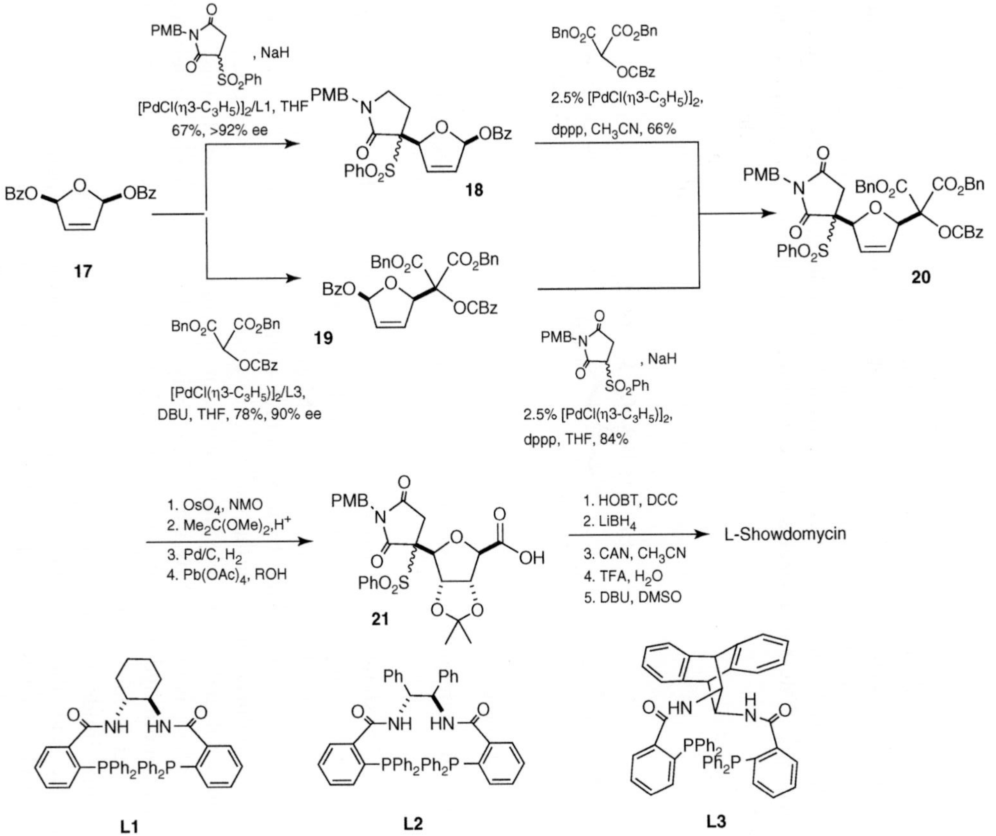

Scheme 37. Synthesis of L-showdomycin by asymmetric palladium-catalyzed allylic substitution using Trost's ligands L1–L3.

the disymmetrized product **18** in 67% yield and an ee of 92%. A second allylic substitution using tetronic acid and Pd/dppp as a catalytic system yielded the key product **20** (Scheme 37). Alternatively, the reaction with tetronic acid can be performed first using Pd/L3 as a catalytic system to afford **19** in 78% yield and a 90% ee. Further reaction with maleimide sulfone sodium salt in the presence of Pd/dppp as catalyst yielded the same key intermediate **20**.

The mechanism of allylic substitution with soft nucleophiles takes place by a double-inversion process which involves retention of the configuration. In the *meso*-diesters considered, the desymmetrization consists in determining which substituent is first replaced, but the two allylic substitution processes for installing the base and hydroxymethyl chain guarantee that they will be *cis*, as occurs in the practical totality of biologically active nucleosides.

The ribose fragment typical of showdomicin was obtained by stereoselective dihydroxylation with osmium tetroxide, which was followed by protection such as acetonide. The last steps of the synthesis consisted in transforming the tretronic acid moiety into a carboxylic acid, by removal of benzyl protecting groups and oxidation to give **21**. Further esterification and reduction with lithium borohydride allowed installation of the hydroxymethyl chain. Synthesis of the target compound was completed by removal of the *p*-methoxybenzyl group using CAN, and of the acetonide using TFA. The final step was the elimination of the sulfone in the presence of DBU to yield L-showdomicin.

Palladium is the most popular metal for allylic substitution, as shown in earlier schemes; however, the control or regioselectivity required to yield the branched product awaits solution. Iridium catalysts have attracted interest because the reactions are highly regioselective, affording the branched product and because in the presence of phosphoramidite, ligands also provide high enatioselectivities. With these antecedents, Hocek, Malkov, and Kocovský[73] carried out a broad study of the iridium-catalyzed allylic substitution reaction in order to obtain *C*-nucleosides. They envisioned the synthetic scheme presented in Scheme 38, which combines the allylic substitution reaction with the ruthenium-catalyzed ring-closing metathesis to build up the tetrahydrofuran skeleton incorporating the aryl moiety and the hydroxymethyl chain at C5 typical of nucleosides.

The articles cited analyze all possible combinations of allylcarbonates and nucleophilic alcohols presented in the scheme. Reactions were carried out using racemic alcohols and terminal or racemic carbonates in the presence of chiral ligands, enantiopure alcohols and racemic carbonates in the presence of chiral or not chiral ligands and enantiopure both alcohols and carbonates in the presence of common achiral catalysts. The following general conclusions were deduced: (1) branched carbonate is more reactive than linear carbonate; (2) the reagent combination shown in Scheme 38B was the most convenient, and the use of $[IrCl(cod)]_2$ in the absence of ligand was the most reactive catalytic system; (3) the use of chiral reagents and an achiral catalyst provides better results than the use of chiral catalyst; (4) the optimal alcohol/carbonate ratio is 1:0.8–0.9; and (5) the TBS group is the more successful protecting group.

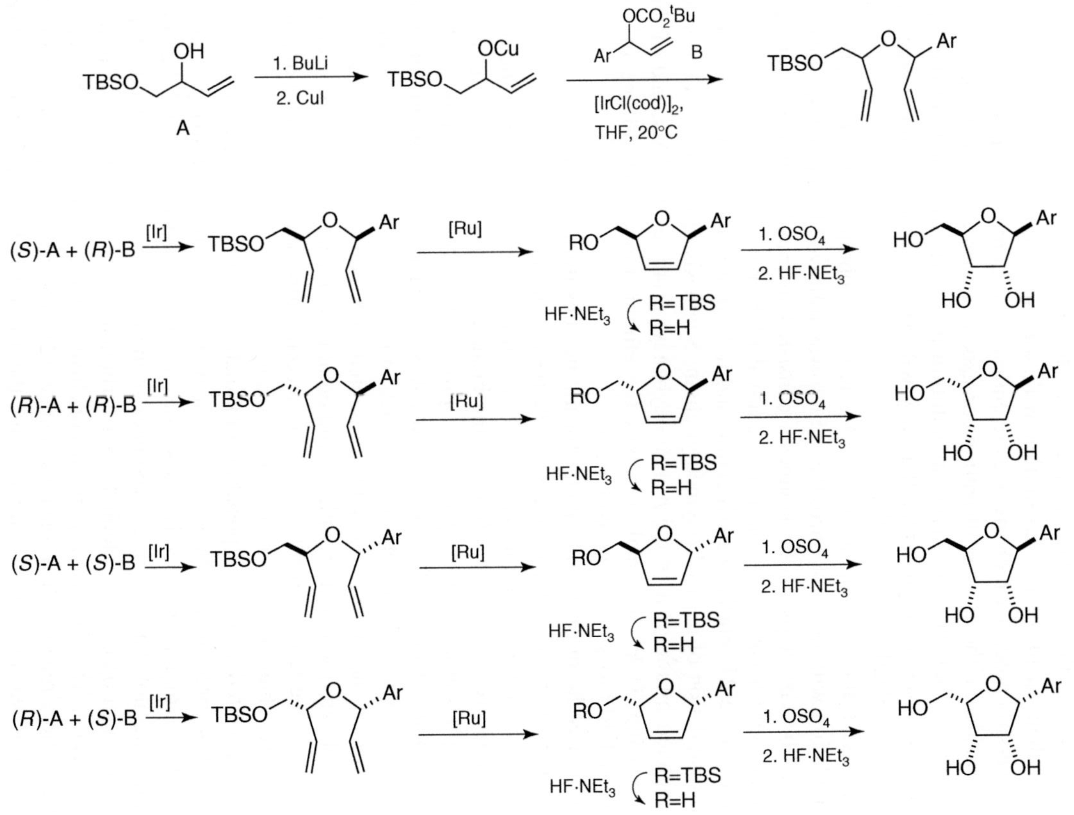

Scheme 38. Synthesis of *C*-nucleosides using iridium-catalyzed asymmetric allylic alkylation and ring-closing metathesis as key steps.

Thus, enantiomers of both A and B were reacted in the presence of [IrCl(cod)]$_2$ to afford the diallyl ether derivatives in yields ranging from 56 to 91%. These four isomers have undergone ring-closing metathesis using second-generation Grubbs' catalysts to yield the dihydrofuran derivatives. The stereochemical outcome of the allylic substitution reaction was determined in these products after deprotection, observing that the configuration at C4 reflects the absolute configuration of the starting alcohol, since this center was not involved in the reaction. The configuration at C1 is also coincident with the starting carbonate, in agreement with that expected for the iridium-catalyzed allylic substitution, which is known to proceed with retention of configuration. The dihydrofuran derivatives are already of interest since they are analogues of 2′,3′-dideoxynucleosides (R = H). The corresponding ribo derivatives were obtained by dihydroxylation of silyl-protected dihydrofuran derivatives using osmium tetraoxide followed by deprotection. The dihydroxylation reaction was also studied in detail; thus, metathesis–dihydroxylation performed in a one-pot manner, but using ruthenium tetraoxide as the dihydroxylation reaction, was tried. However, the use of osmium tetraoxide was found to be most efficient.

3.4. Palladium-Catalyzed Cross-Coupling Reactions

In the late 1970s Doyle Daves[74] employed the Heck[75,76] reaction to create a C–C bond in pyranoid glycals using organomercuric uracil derivatives as an nucleophiles. The NH of uracil was protected as anMe derivative. Doyle Daves also performed a systematic study of furanoid glycals of various configurations that made it possible to determine the regio- and stereochemical outcome.[77] From this study it was concluded that the reaction was completely regioselective, making possible the introduction of uracil at the anomeric position. As to the stereoselectivity, when both hydroxyls were protected, stereoselectivity was moderate and the major isomer resulted from the coordination of palladium *trans* to 3-OR (via b), from which the β-nucleosides were primarily obtained. When the 3-OH was unprotected, this group directed the coordination of palladium and stabilized the intermediates (via a) in such a way that reduction yielded the α-nucleoside directly (Scheme 39).

Furanoid glycals are less accessible than the corresponding pyranoid glycals, and several successful procedures for preparing pyranoid glycals are not applicable for obtaining furanoid glycals. Scheme 40 summarizes the more common strategies for synthesizing furanoid glycals. The most classical is the method of Ireland et al.,[78] which starts from protected ribonolactone and yields glycals by reduction, halogenation, and reductive elimination. The second procedure, developed by Garegg and Samuelsson,[79] obtained glycal through the reaction of riboses protected at positions 3 and 5 with iodine–Ph$_3$P–imidazol. Finally, different elimination reactions from 2-deoxyribose derivatives (1-OMs,[80] 1-SePh[81]) also yield the corresponding glycal. In this way, thymidine has been used as a starting material because of its easy and selective protection of sugar hydroxyl and because heating

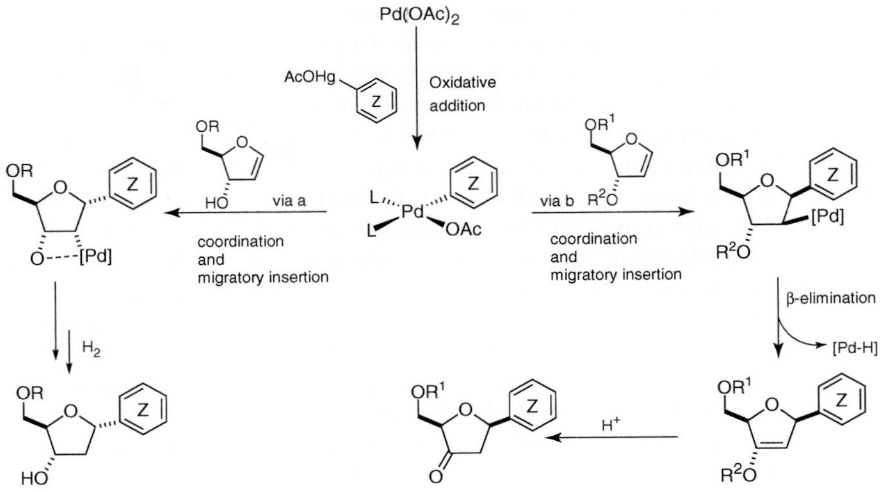

Scheme 39. Synthesis of *C*-nucleosides from glycals using the Heck reaction. Mechanistic aspects and stereochemical behavior.

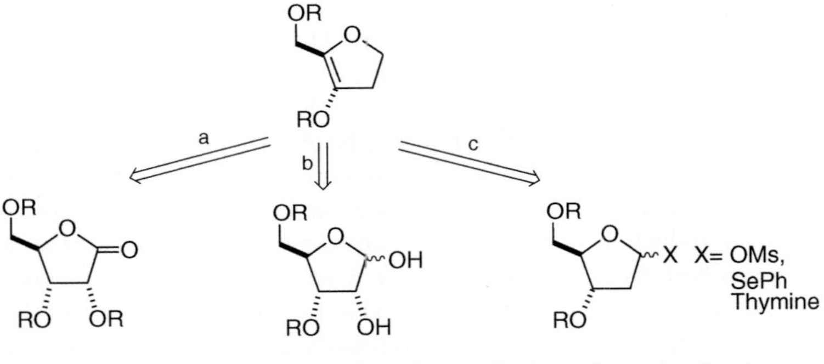

Scheme 40. General procedures for synthesizing furanoid glycals.

the protected nucleoside in hexamethyldisilazane/$(NH_4)_2SO_4$ eliminates the base, thus yielding the glycal.[82]

At this stage the procedure had several limitations: the use of mercury salts and the stereoselectivity of glycosylation and of ketone reduction. These problems were overcome by using iodo derivatives as an alternative to the mercury, using bulky protecting groups for the 3-OH, leaving the 5-OH unprotected. Moreover, the 5-OH was used advantageously to direct the ketone reduction by using $NaHB(OAc)_3$. These improvements were used in an efficient synthesis of 2′-deoxypseudouridine. Thus, when 1,4-anhydro-2-deoxy-3-*O*-(*t*-butyldiphenylsilyl)-D-erythro-pent-1-enitol, a ribofuranosyl glycal designed for stereospecific formation of β-glycosyl bonds, was treated with 5-iodouracil in the presence of $Pd(OAc)_2/AsPh_3$ followed

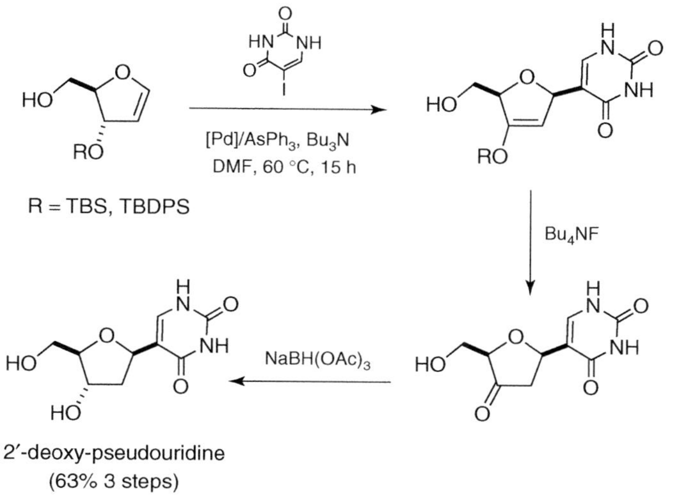

Scheme 41. Synthesis of 2′-deoxypseudouridine using the Heck reaction as a key step.

by treatment with tetrabutylammonium fluoride and sodium triacetoxyborohydride, 2′-deoxypseudouridine was obtained in 63% yield in three steps (Scheme 41).[83] Using a similar procedure, 2′-deoxy analogues of formycin B were obtained using 3-iodopyrazolo[4,3-d]pyrimidin-7-one, although in this case the NH present had to be protected as THP derivatives.

This efficient protocol for accessing 2′-deoxy-C-nucleosides is reported widely in the literature, and Scheme 42 shows some representative examples. Iodides and triflate derivatives have both been used, although iodine is preferred when heterocycles are used. Palladium acetate and Pd(dba)$_2$ are the most common palladium sources. The variety of ligands is large. Thus, Ph$_3$As,[84,85] Ph$_3$P,[86,87] or the poorly basic (C$_6$F$_5$)$_3$P[88,89] and strongly basic t-Bu$_3$P[90] were used. In one example no phosphorus ligand was present.[91] The preferred bases were tertiary amines (R$_3$N), but sodium hydrogencarbonate in the presence of phase-transfer reagents or silver triflate have also been used. Other halogens, such as chlorine, are compatible with the reaction conditions, yielding haloderivatives that can be transformed further, taking advantage of the reactivity of chloropyridines[84] or pyrimidines.[89]

In some cases diprotected, or unprotected, glycals were used. In general, the oxidative addition of electron-rich aryl halide to Pd(0) species is not easy to achieve. In fact, this was the case for pyrrol iodides that were not deactivated sufficiently. Thus, when a 3-iodopyrrol having an acyl and a nitro group was treated with the furanoid glycal (R = TBS) in the presence of Pd/P(o-MeC$_6$H$_4$)$_3$ the cross-coupling product was obtained in poor yield. The use of LiCl as an additive made it possible to achieve a 48% yield. Removal of the protecting groups resulted in a hydroxyketone similar to the one obtained in earlier schemes, which was also stereoselectively reduced to get the nucleoside (Scheme 43).[92] A furanoid glycal in which R = Bn, which was obtained from xylose, was coupled with p-iodoaniline

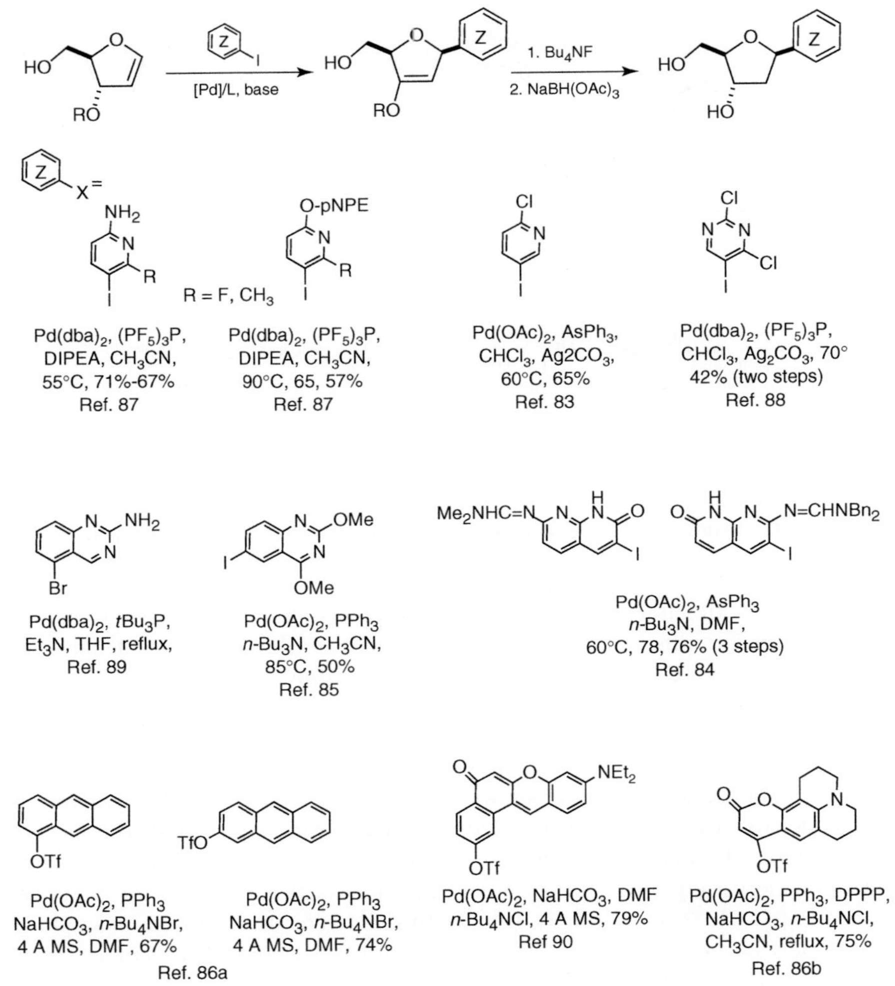

Scheme 42. *C*-nucleosides synthesized using the Heck reaction.

using Pd/AsPh$_3$ as a catalytic system. Yields were moderate to good (57%, R = H; 71%, R = F). Benzyl groups were removed under hydrogenolitic conditions, and the synthesis was completed by ketone reduction.[93] The unprotected furanoid glycal was treated with a 8-iodo[1,5-*a*]-1,3,5-triazine with catalysis of Pd/AsPh$_3$ to give β-ketonucleoside exclusively in a remarkable 75% yield.[94]

The Heck reaction has also been used to obtain 2′,3′-dideoxy-*C*-nucleosides, commonly used as chain terminators in antiviral therapies. With this objective, the ketonucleoside obtained under standard Heck conditions was deoxygenated by reaction with tosyl hydrazine and further reduction of the resulting hydrazone with sodium triacetoxyborohydride (Scheme 44).[95]

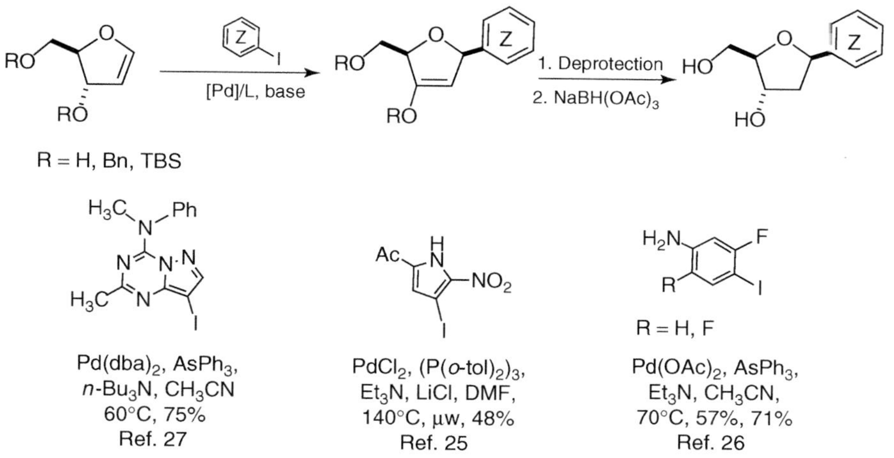

Scheme 43. *C* nucleosides synthesized using the Heck reaction and electron-rich haloaromatics.

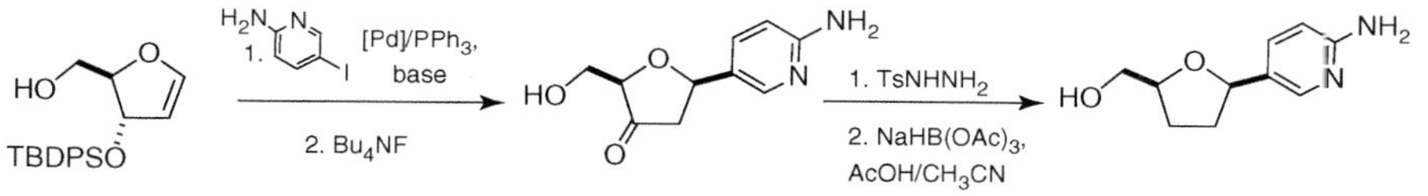

Scheme 44. 2′,3′-Dideoxynucleosides synthesized by the Heck reaction and standard glycals followed by deoxygenation.

2′,3′-Dideoxynucleosides can also been obtained by 2-hydroxymethyl-2,3-dihydrofurans appropriately protected. In the palladium-mediated coupling reaction, the group at C4 directs the coordination of palladium to the less hindered face of the glycal, so that the new bond at C2 forms stereoespecifically *trans* to the C4 substituent (sugar numbering). However, depending on the reaction conditions, the double bond formed at positions 2 and 3 isomerize to positions 3 and 4, which are thermodynamically more stable. The presence of ammonium chloride salts favorizes the decoordination of palladium and consequently the formation of a double bond at 2,3, whereas isomerization is favored in the absence of this reagent, using AsPh$_3$ as a ligand. In this way, reduction of the isomerized product yielded a 2′,3′-dideoxynucleoside with the substituents cis at positions 1 and 4 (Scheme 45).[96]

3.5. Lewis Acid–Catalyzed Reactions

The Friedel–Craft reaction is a simple approach to synthesizing *C*-nucleosides, due to the facility for the formation of oxonium cations, especially in activated sugars. However, yields are in general moderate and the stereoselectivity poor, except when participating groups are present at position 2. All leaving groups used

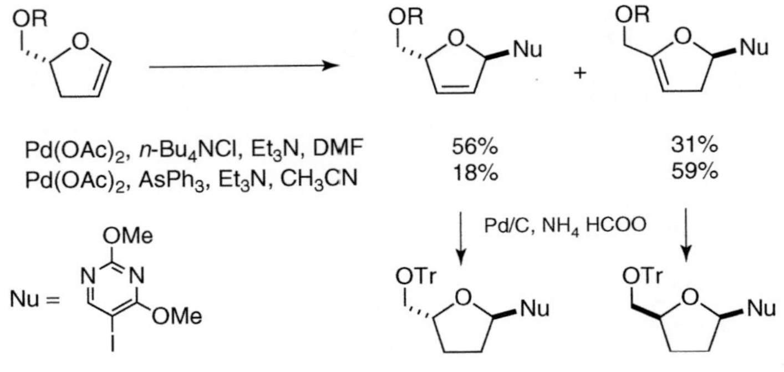

Scheme 45. 2′,3′-Dideoxynucleosides synthesized from 3-deoxyglycals by the Heck reaction.

in standard glycosylation reactions can also be used in this reaction, although the higher stability of the new bond formed makes it possible to use more drastic reaction conditions, compatible with the presence of poor leaving groups. Thus, tetra-O-acetyl-D-ribose is a readily accessible starting material for this reaction, and it reacts with 3-carboxyethylfuran catalyzed by tin tetrachloride to give the furan derivative glycosylated at positions 2 and 5 as an α/β mixture. The 5-substituted furan of β configuration was the major isomer. A similar process was carried out for the thiophene derivative.[97] The reaction of tri-O-benzoyl-1-O-acetylribose with 1-benzyl-9-deazaguanine in the presence of tin chloride also afforded the β-C-nucleoside in 73% yield. The best yields were obtained for a catalyst/nucleic base ratio of 2.8:1.[98] Other deazaguanine and pyrrolopyrimidinone derivatives were prepared using the same procedure (Scheme 46).

2-Glycosylfuran was prepared from trichloroacetamidite derived from 2,3,4,6-tetra-O-benzyl-D-glucose using zinc chloride as a catalyst, which was doubly

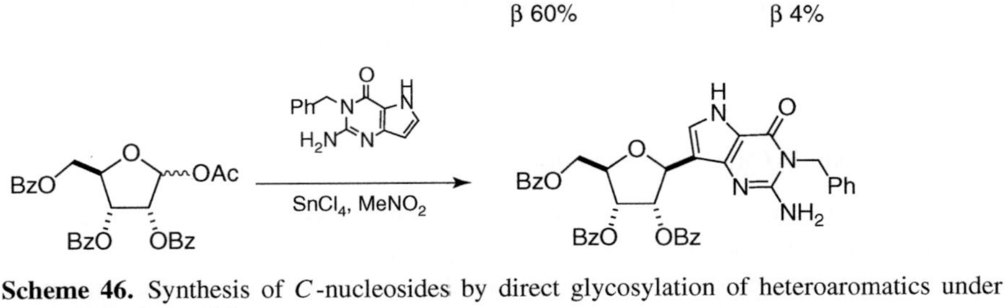

Scheme 46. Synthesis of C-nucleosides by direct glycosylation of heteroaromatics under Friedel–Craft conditions.

glycosylated by further treatment with the same thrichloroacetamidite, now with catalysis of tin tetrachloride. Only the α-anomer was formed in both cases. The [4 + 2]-cycloaddition of singlet oxygen to furan at low temperature afforded a diastereomeric mixture of endoperoxide that at room temperature underwent a migration of glucose to give a disaccharide (Scheme 47).[99]

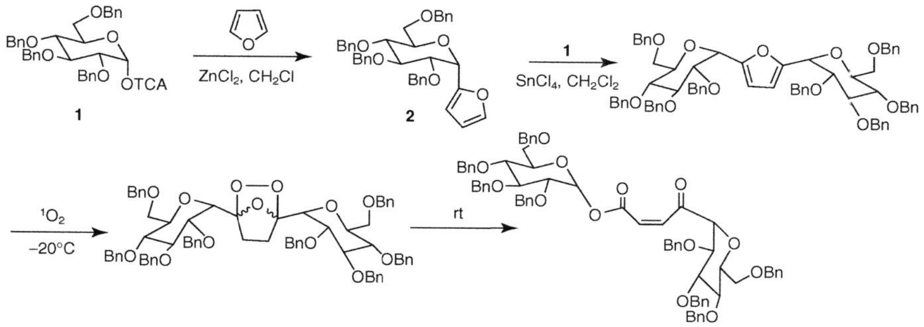

Scheme 47. Synthesis of disaccharides from glucosylfuran derivatives.

The coupling between 2,3,5-tri-O-benzyl-β-D-ribofuranosyl fluoride and 1-acetyl-2,5,6-trichloroindole catalyzed by boron trifluoride yielded glycosylated indole as a 1:4 α/β mixture that can be separated, providing the β-anomer in 45% yield (Scheme 48b). Removal of protecting groups was accomplished by hydrogenolysis using 5% Pd/BaSO$_4$ and treatment with methylamine. Alternatively, the same product was obtained through a longer reaction sequence where the coupling between the heterocycle and the sugar was carried out for 2,3-O-isopropylidene-D-ribofuranose and 5,6-dichloroindole-2-thione in basic medium in a Knoevenagel type of reaction. The sulfide sodium salt was obtained in a good yield, but as a 2:1 α/β anomeric mixture, and was benzylated in situ. Futher elaboration yielded a the thrichloro derivative (Scheme 48a).[100]

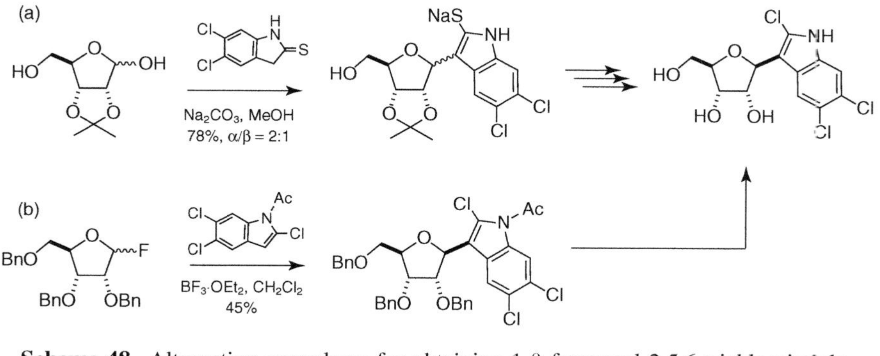

Scheme 48. Alternative procedures for obtaining 1-β-furanosyl-2,5,6-trichloroindole.

3,5-Ditoluoyl-2-deoxyribofuranosyl chloride is commonly the starting material of choice for the synthesis of 2-deoxy derivatives (see Section 2.1.1), although the more stable methyl glycoside has also been used to take advantage of the higher reactivity of the 2-deoxyderivatives. The reaction from the chloro derivative can be driven under kinetic conditions, while activation of methyl glycoside requires drastic reaction conditions and the reaction takes place under thermodynamic control. Yields and selectivities are variables and depend on the catalytic system and reaction conditions. Thus, reaction of methyl glycoside with diphenyldisulfide in dichloromethane in the presence of tin dichloride as catalyst at low temperature,[101] with 1-benzyl-9-deazaguanine in nitromethane with catalysis of tin tetrachloride,[98,102] or with 1-phenylsulfonyl indole,[103] furnished very poor yields and stereoselectivities (Scheme 49). More successful was the reaction of glycosylation of 2-bromofuran. Interestingly, in this case the reaction from the methyl glycoside using SnCl$_4$/AgOOCCF$_3$ provided a better yield than that from chloride using silver triflate as an activator, although the stereoselectivity was slightly better in this case.[104] Pyrene and other polycondensed aromatics were glycosylated using both glycosyl donors, and in this case yields and stereoselectivities were comparable. The reaction of pyrene with 2-deoxyfuranosyl chloride was activated by silver tetrafluoroborate, while the methyl glycoside required a strong Lewis acid.[105] The reaction was extended to other aromatic compounds, such as fluorine, benzothiophene, 1-methylnaphtalene, and thioanisole, and it was found that SnCl$_4$/AgOOCCF$_3$ was the most efficient catalytic system. Yields between 33 and 51%, and α/β selectivities from 1:2 to 1:5, were obtained. Phenanthrene did not react under these conditions.

Modification of the aromatic or heteroaromatic ring after glycosylation is a common strategy used to prepare compounds with related structures, thus enlarging the

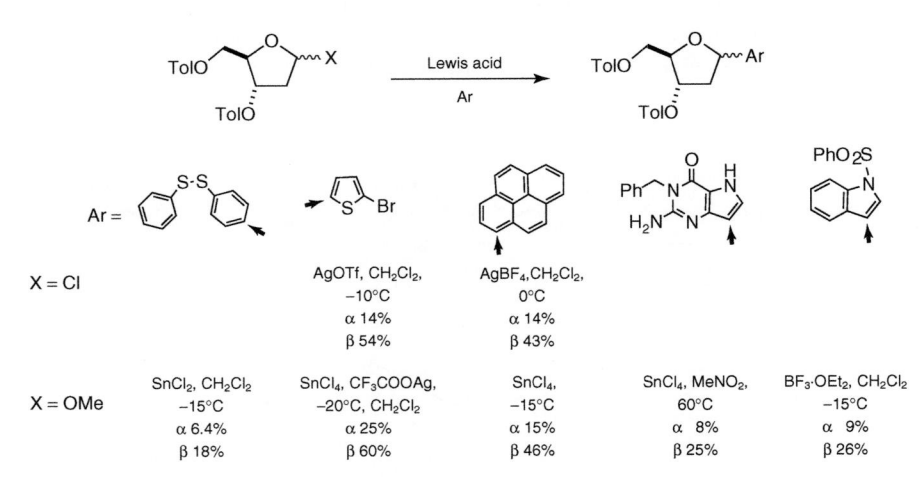

Ar =	S-S diphenyl disulfide	2-bromothiophene	pyrene	1-benzyl-9-deazaguanine	1-phenylsulfonyl indole
X = Cl		AgOTf, CH$_2$Cl$_2$, $-10°C$ α 14% β 54%	AgBF$_4$, CH$_2$Cl$_2$, 0°C α 14% β 43%		
X = OMe	SnCl$_2$, CH$_2$Cl$_2$ $-15°C$ α 6.4% β 18%	SnCl$_4$, CF$_3$COOAg, $-20°C$, CH$_2$Cl$_2$ α 25% β 60%	SnCl$_4$, $-15°C$ α 15% β 46%	SnCl$_4$, MeNO$_2$, 60°C α 8% β 25%	BF$_3$·OEt$_2$, CH$_2$Cl$_2$ $-15°C$ α 9% β 26%

Scheme 49. Glycosylation of aromatic and heteroaromatics with 2-deoxyglycosyl donors under Lewis acid activation.

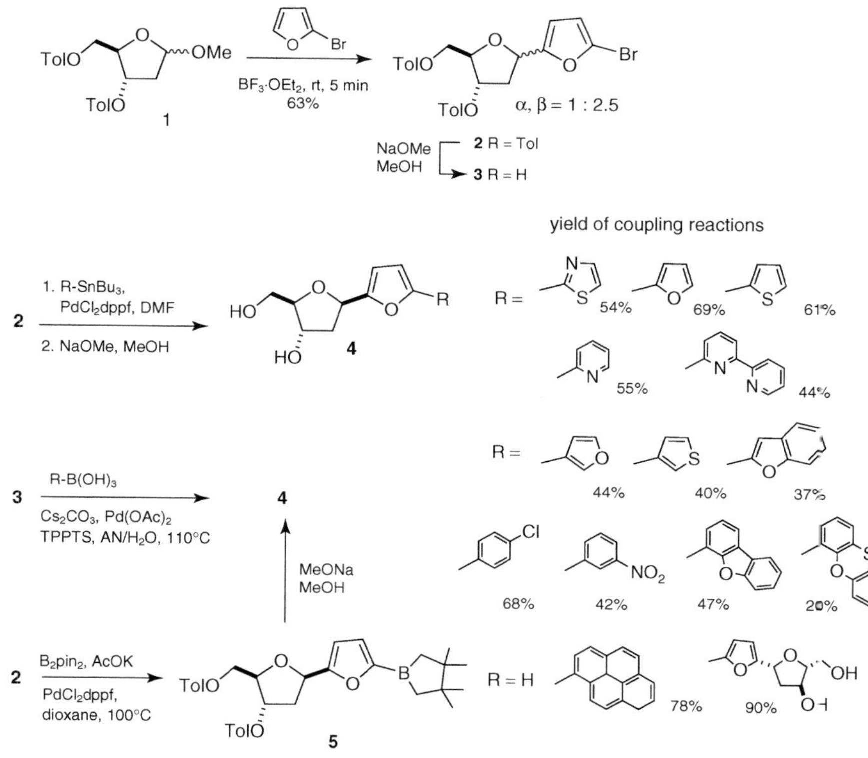

Scheme 50. Glycosylation of 2-bromofuran and further modifications by palladium-catalyzed cross-coupling reactions.

library of C-nucleosides. Palladium-catalyzed cross-coupling reactions are particularly useful for this purpose, and consequently, aromatic or heteroaromatic moieties incorporating halogen are required. A representative example is the 5-bromo-2-glycosylfuran, from which a variety of heterocycles were linked to the furan ring by a Stille reaction (R–SnBu$_3$) or a Suzuki reaction [R–B(OH)$_3$] (Scheme 50). The yields ranged from moderate to good and seem to be more dependent on the stereochemical hindrance than on the type of substituents in the aromatic ring. The Suzuki reaction was also carried out by inverting the functionality of reagents by converting the bromofuran in furylboronate and performing the reaction with an haloaromatic. In this way, pyrene was attached to the furan in good yields. It is also possible to prepare the 2,2′-bifuran bis-nucleoside in excellent yield by reaction of the nucleoside functionalized as furylboronate with furan bromide.[106]

Another interesting example of functionalization by a palladium-catalyzed cross-coupling reaction is the hydroxylation of haloarenes[107] (Scheme 51). The reaction were tested on TBS-protected 4- or 3-bromophenyl C-ribonucleosides with either CsOH or KOH as a hydroxide source. Commercially available Buchwald ligands L1 and L2 were used in combination with Pd$_2$(dba)$_3$. The reaction provides better

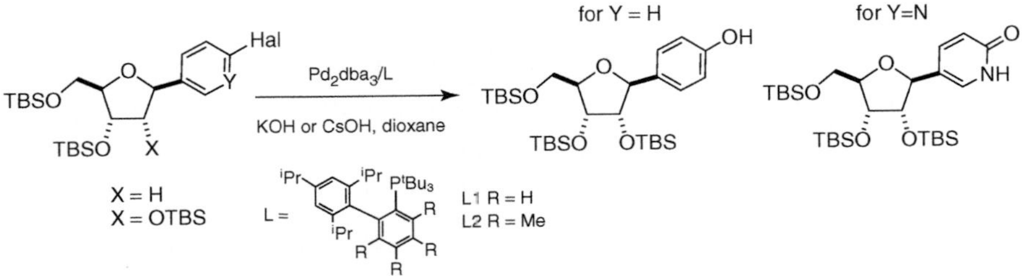

Scheme 51. Hydroxylation of haloaromatics by palladium-catalyzed cross-coupling reactions.

results when ligand L2 is used, but deprotection of the 5′-OH always competes. The deprotection is higher with CsOH than with KOH, and it can be reduced by controlling the reaction time. Thus, at short reaction times (3 h), up to 85% of the fully protected product can be obtained for Y = CH and Hal = Br. Under optimized conditions, yields of 70 to 80% were obtained even when Hal = Cl. In the 2-halopyridine derivatives, the hydroxyl compound obtained tautomerized to the pyridine.[108]

3.6. Other Reactions

Considering the exceptional reactivity of tellurides affording reaction intermediates such as radicals, carbenium ions, and carbanions, the reactivity of benzyl-protected D-ribofuranosyl tellurides and 2-deoxy-D-ribolfuranosyl tellurides in obtaining C-nucleosides was explored. The corresponding tellurides, which turned out to be moderately stable, were prepared from a 1-O-mesyl derivative by reaction with di-p-anisyl ditelluride and sodium borohydride[109] (Scheme 52).

Anomeric radicals were formed treating tellurides with Et_3B, which reacted with electron-poor heteroaromatics (EP-Ar) to afford C-nucleosides in moderate yields as α/β mixtures. Anomeric cation intermediates were generated using $BF_3·OEt_2$, which reacted with electron-rich aromatics (ER-Ar) to give the coupling products in moderate to good yields. Compared with the more radical process, stereoselectivity in D-ribofuranosyl derivatives was better, due to the epimerization induced by $BF_3·OEt_2$; however, for 2-deoxy-D-ribolfuranosyl, the stereoselectivity was moderate. Finally, anion intermediates were obtained by reaction of the thellurides with butyl lithium, and the intermediate formed was added to benzaldeyde to yield the alcohol in moderate yield (52%).

4. FUNCTIONALIZATION OF A HETEROCYCLE AND FURTHER CONSTRUCTION OF THE CARBOHYDRATE

Another approach to C-nucleosides relies on the construction of a carbohydrate moiety upon the corresponding heterocycle from a noncarbohydrate

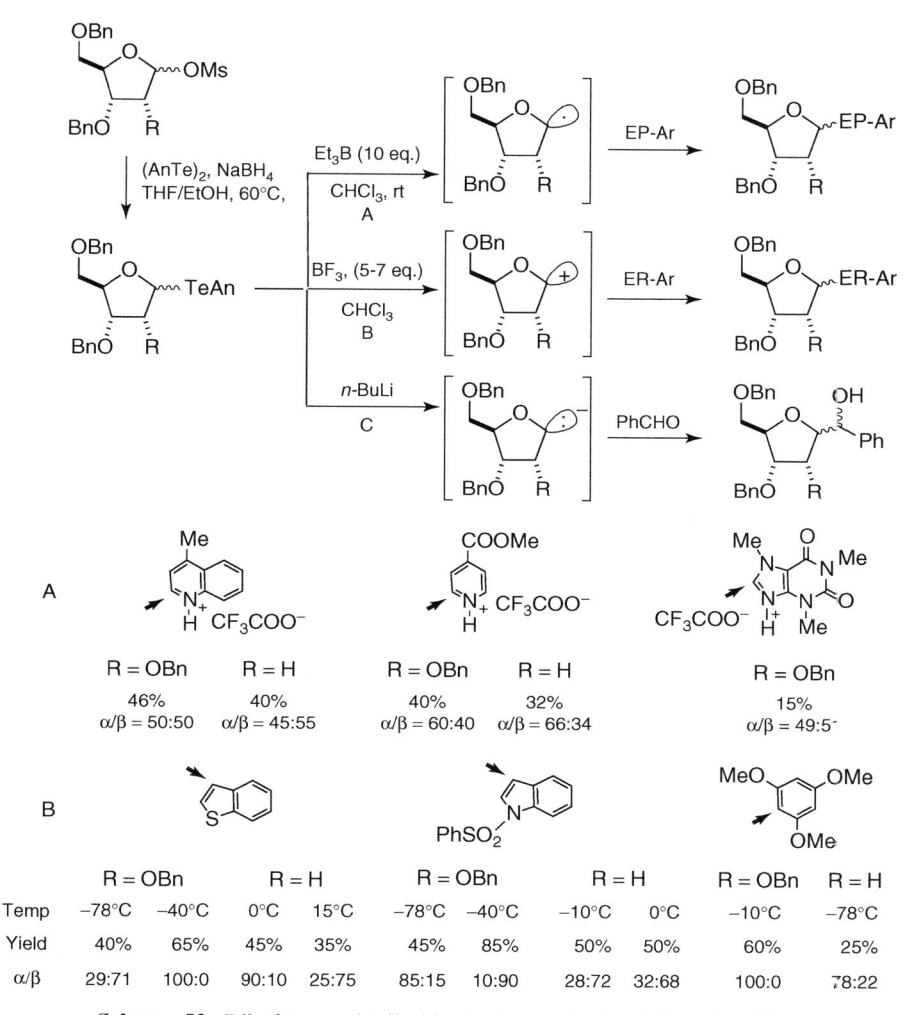

Scheme 52. Ribofuranosyl tellurides in the synthesis of *C*-nucleosides.

acyclic precursor. This represents a less conventional strategy, because it requires creating up to four stereocenters. One processes published in this context involves the reductive cyclization of enantiopure hydroxylsulfinyl esters by reduction, Weinreb's amide, and ketone formation, followed by the reductive cyclization of the resulting hydroxylsulfinyl ketones to give *cis*-2,5-disubstituted tetrahydrofuran and 2,6-disubstituted tetrahydropyran systems[110] (Scheme 53).

A cycloaddition reaction is the key step in the construction of the sugar moiety developed by Son and Fu.[111] Copper-catalyzed asymmetric [4 + 1]-cycloadditions of enones with diazo compounds can produce highly substituted 2,3-dihydrofurans

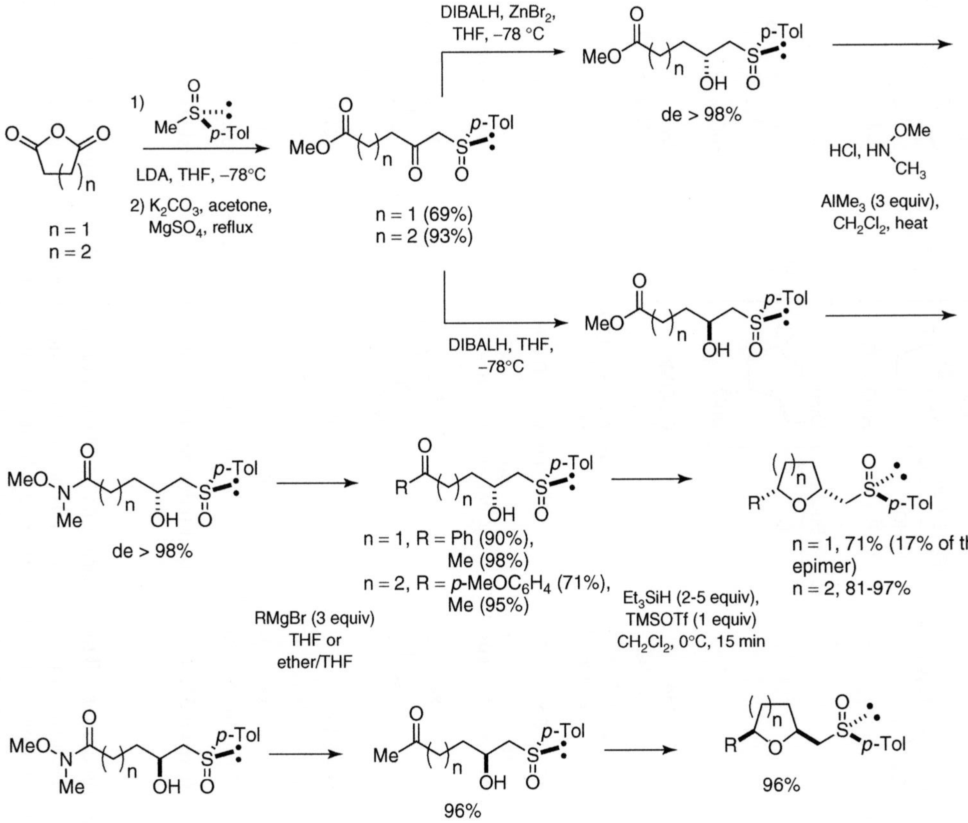

Scheme 53. Hydroxysulfinyl ketones in the stereoselective synthesis of 2,5-disubstituted tetrahydrofurans.

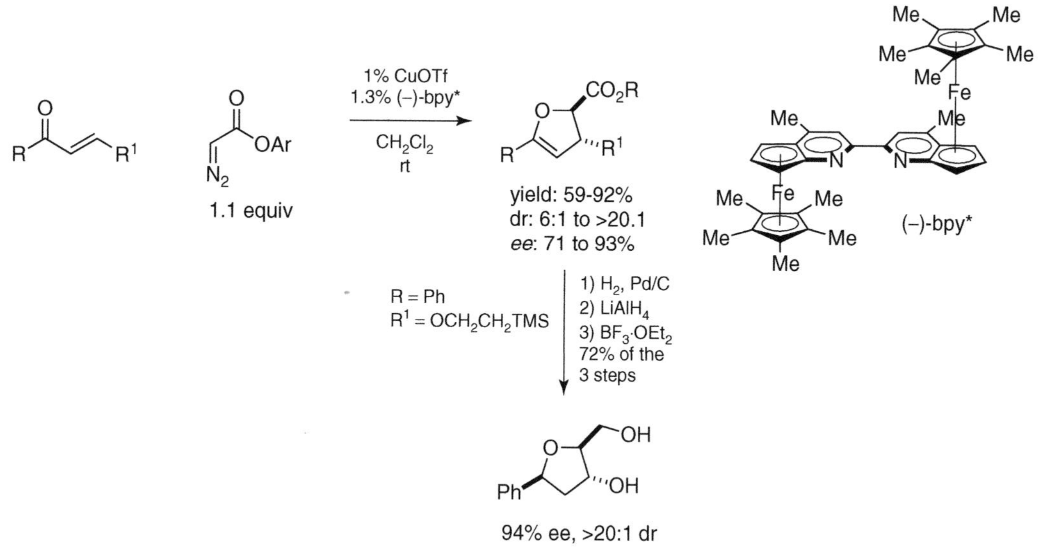

Scheme 54. Copper-catalyzed asymmetric [4 + 1]-cycloadditions of enones with diazo compounds to provide 2′-deoxy-L-C-nucleosides.

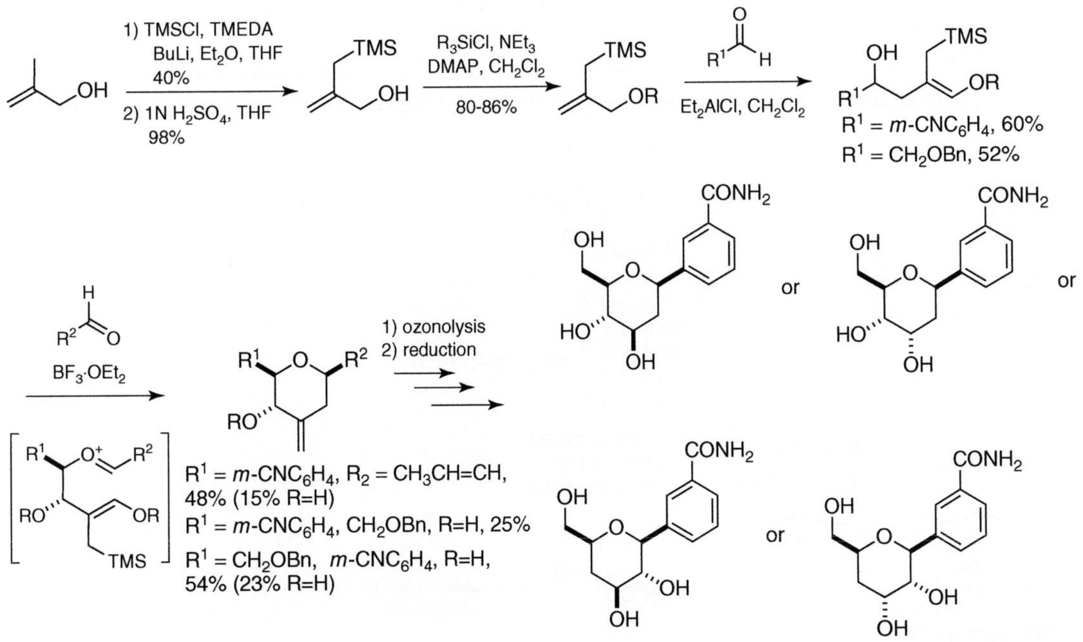

Scheme 55. Ene/intromolecular Sakurai cyclization in the synthesis of pyranosyl and furanosyl *C*-nucleosides.

FUNCTIONALIZATION OF A HETEROCYCLE AND FURTHER CONSTRUCTION 309

with good efficiency and stereoselection. Diastereoselective hydrogenation, followed by reduction of the ester group and trimethylsilylethyl group deprotection, furnishes the deoxy-C-nucleoside (Scheme 54).

Ene//intromolecular Sakurai cyclization is a simple, general regio- and stereo-controlled method for the synthesis of pyranosyl and furanosyl C-nucleosides[112] (Scheme 55). This method involves an initial ene reaction between an allylsilane derivative and an aldehyde to give a homoallylic alcohol with complete control of the geometry of the double bond. Subsequent Lewis acid–promoted condensation with a second aldehyde provides the oxacarbenium ion that undergoes Sakurai cyclization to give *exo*-methylene tetrahydropyrans, which are further elaborated to give the final C-nucleoside. Exchange in the sequence of the aldehyde addition gives access to a series of regioisomeric C-nucleosides.

An analogous Sakurai cyclization from allyl acetal gives access to *exo*-methylene tetrahydrofurans in the preparation of D/L-2-deoxy-C-nucleosides in moderate to good yields[113] (Scheme 56).

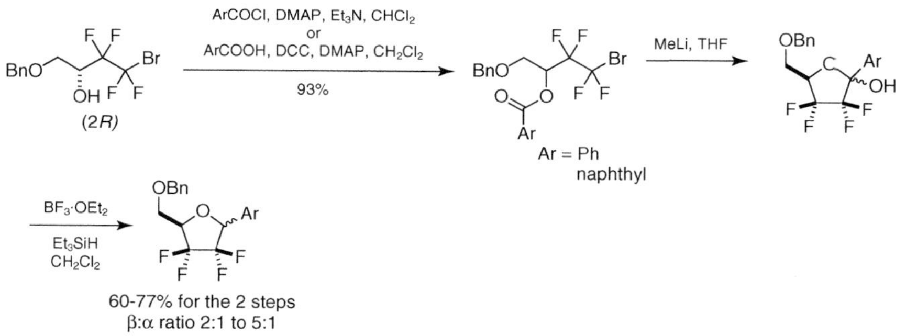

Scheme 56. Synthesis of tetrafluorinated aryl-C-nucleosides by aglycone attachment, cyclization, and reductive dehydroxylation.

Aryl-C-nucleosides incorporating a tetrafluoroethylene motif have been synthesized using a fluorinated alcohol as a starting material in a process that begins with esterification to give an ester that incorporates the aglycone unit, followed by a ring closure triggered by bromine–lithium exchange with MeLi. Subsequent Lewis acid–promotes reductive dehydroxylation with triethylsilane to give the final C-nucleoside with a stereoselectivity of up to 5:1. The failure of derivatives with electron-withdrawing groups in both the sugar and the algycon to undergo reductive dehydroxylation reveals that the presence of those groups prevents formation of the putative oxonium intermediate.[114]

The final example in this area relies on the use of a 1,3-dipolar cycloaddition as a key step for construction of the pyrrolidine ring of pyrrolidino analogues of C-nucleosides related to pseudouridine (Scheme 57). The method involves 1,3-dipolar cycloaddition of uracil-5 and 2,4-dimethoxypyrimidine-5 nitrones with allyl alcohol and methyl acrylate, and subseqüent cleavage of the corresponding isoxazolidine cycloadducts.

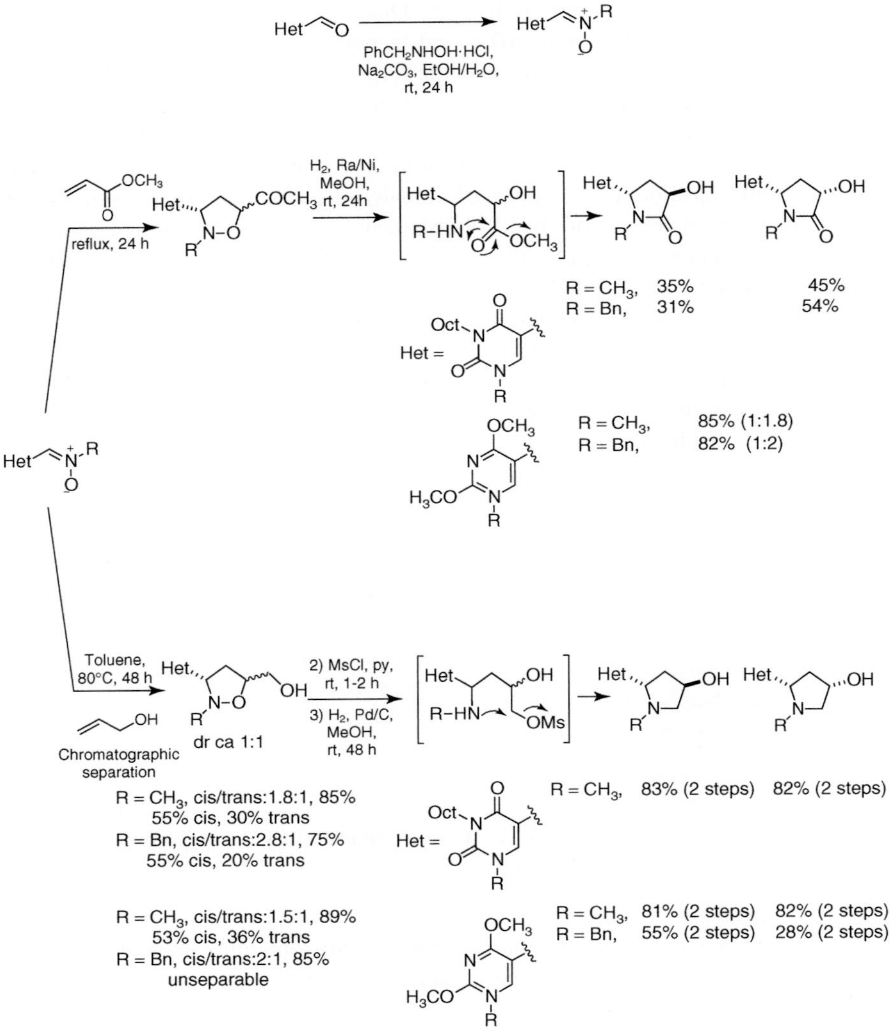

Scheme 57. 1,3-Dipolar cycloaddition as a key step for construction of the pyrrolidine ring of pyrrolidino analogues of *C*-nucleosides.

Acknowledgments

Financial support of DGI CTQ2011-01569-BQU, Ministerio de Economía y Competitividad, Spain is gratefully acknowledged. O.B. thanks the Ministerio de Ciencia e Innovación, Spain (Juan de la Cierva Fellowship) and the European Commission (Marie Curie Career Integration Grant).

REFERENCES

1. Townsend, L.B. *Chemistry of Nucleosides and Nucleotides*, Plenum Press, New York, **1994**, pp. 421–535.
2. Stambasky, J.; Hocek, M.; Kocovsky, P. *Chem. Rev.* **2009**, *109*, 6729–6764.
3. (a) Malnuit, V.; Duca, M.; Benhida, R. *Org. Biomol. Chem.* **2011**, *9*, 326–336. (b) Varghese, R.; Wagenknecht, H.-A. *Chem. Commun.* **2009**, 2615–2624.
4. Wu, Q.; Simons, C. *Synthesis* **2004**, 1533–1553.
5. De las Heras, F. G.; Fernández-Resa, P. *J. Chem. Soc. Perkin Trans. 1* **1982**, *4*, 903–907.
6. Adamo, M. F. A.; Adlington, R. M.; Baldwin, J. E.; Day, A. L. *Tetrahedron* **2004**, *60*, 841–849.
7. Adlington, R. M.; Baldwin, J. E.; Pritchard, G. J.; Spencer, K. C. *Tetrahedron Lett.* **2000**, *41*, 575–578.
8. Veronese, A. C.; Morelli, C. F. *Tetrahedron Lett.* **1998**, *39*, 3853–3856.
9. For a procedure employing iminoester and ketoester here employed in the Hantzsch cyclocondensation to *C*-nucleosides, see (a) Dondoni, A.; Minghini, E. *Helv. Chim. Acta* **2002**, *85*, 3331–3348. (b) Ducatti, D. R. B.; Massi, A.; Noseda, M. D.; Duarte, M. E. R.; Dondoni, A. *Org. Biomol. Chem.* **2009**, *7*, 1980–1986. For an analogous procedure in the Biginelli reaction leading to pyrimidine *C*-nucleosides, see (c) Dondoni, A.; Massi, A.; Sabbatini, S. *Tetrahedron Lett.* **2001**, *42*, 4495–4497.
10. Al-Masoudi, N. A.; Al-Soud, Y. A.; Ali, I. A. I. *Nucleosides Nucleotides Nucleic Acids* **2007**, *26*, 37–43.
11. Jazouli, M.; Guianvarc'h, D.; Soufiaoui, M.; Bougrin, K.; Vierling, P.; Benhida, R. *Tetrahedron Lett.* **2003**, *44*, 5807–5810.
12. Epimerization process in *C*-nucleosides is highly dependent on the nature of the aglycone. See, for example, (a) Aketani, S.; Tanaka, K.; Yamamoto, K.; Ishihama, A.; Cao, H.; Tengeiji, A.; Hiraoka, S.; Shiro, M.; Shionoya, M. *J. Med. Chem.* **2002**, *45*, 5594–5603. (b) Chen, D.-W.; Beuscher, A. E., IV; Stevens, R. C.; Wirsching, P.; Lerner, R. A.; Janda, K. D. *J. Org. Chem.* **2001**, *66*, 1725–1732. (c) Griesang, N.; Richert, C. *Tetrahedron Lett.* **2002**, *43*, 8755–8758.
13. Jung, M. E.; Trifunovich, I. D. *Tetrahedron Lett.* **1992**, *33*, 2921–2924.
14. Ohno, M.; Ito, Y.; Arita, M.; Shibata, T.; Adachi, K.; Sawai, H. *Tetrahedron* **1984**, *40*, 145–152.
15. Li, J.-S.; Gold, B. *J. Org. Chem.* **2005**, *70*, 8764–8771.
16. Popsavin, M.; Spaic, S.; Svircev, M.; Kojic, V.; Bogdanovic, G.; Pejanovic,V.; Popsavin, V. *Tetrahedron* **2009**, *65*, 7637–7645.
17. Smellie, I. A. S.; Fromm, A.; Fabbiani, F.; Oswald, I. D. H.; White, F. J.; Paton, R. M. *Tetrahedron* **2010**, *66*, 7155–7160.
18. Lubin-Garmain, N.; Baltaze, J.-P.; Coste, A.; Hallonet, A.; Lauréano, H.; Legrave, G.; Uziel, J.; Augé, J. *Org. Lett.* **2008**, *10*, 725–728.
19. Rothman, J. H. *J. Org.Chem.* **2009**, *74*, 925–928.
20. Wächtler, H.; Peña Fuentes, D.; Michalik, D.; Köckerling, M.; Villinger, A.; Kragl, U.; Arias Cedeño, Q.; Vogel, C. *Synthesis*, **2011**, 3099–3108.
21. Wamhoff, H.; Warnecke, H. *ARKIVOC* **2001**, 95–100.

22. Joshi, U.; Josse, S.; Pipelier, M.; Chevallier, F.; Pradère, J.-P.; Hazard, R.; Legoupy, S.; Huet, F.; Dubreuil, D. *Tetrahedron Lett.* **2004**, *45*, 1031–1033.
23. Adamo, M. F. A.; Pergoli, R.; Moccis, M. *Tetrahedron* **2010**, *66*, 9242–9251.
24. (a) Hari, Y.; Nakahara, M.; Pang, J.; Akabane, M.; Kuboyama, T.; Obika, S. *Biorg. Med. Chem.* **2011**, *19*, 1162–1166. (b) Fujimoto, K.; Yamada, S.; Inouye, M. *Chem. Commun.* **2009**, 7164–7166. (c) Reddy, P. V.; Bajpai, V.; Kumar, B.; Shaw, A. K. *Eur. J. Org. Chem.* **2011**, 1575–1586.
25. Kaliappan, K. P.; Subrahmanyam, A. V. *Org. Lett.* **2007**, *9*, 1121–1124.
26. Novák, P.; Pohl, R.; Kotora, M.; Hocek, M. *Org. Lett.* **2006**, *8*, 2051–2054. (b) Yamamoto, Y.; Saigoku, T.; Ohgai, T.; Nishiyama, H.; Itoh, K. *Chem. Commun.* **2004**, 2702–2073. (c) Yamamoto, Y.; Saigoku, T.; Nishiyama, H.; Ohgai, T.; Itoh, K. *Org. Biomol. Chem.* **2005**, *3*, 1768–1775. (d) Novák, P.; Číhalová, S.; Otmar, M.; Hocek, M.; Kotora, M. *Tetrahedron* **2008**, *64*, 5200–5207.
27. (a) Yamamoto, Y.; Hashimoto, T.; Hattori, K.; Kikuchi, M.; Nishiyama, H. *Org. Lett.* **2006**, *8*, 3565–3568. (b) Yamamoto, Y.; Yamashita, K.; Hotta, T.; Hashimoto, T.; Kuikuchi, M.; Nishiyama, H. *Chem. Asian J.* **2007**, *2*, 1388–1399.
28. Heinrich, D.; Wagner, T.; Diederichsen, U. *Org. Lett.* **2007**, *9*, 5311–5314.
29. Adamo, M. F. A.; Pergoli, R.; Moccis, M. *Tetrahedron* **2010**, *66*, 9242–9251.
30. Bobula, T.; Hocek, M.; Kotora, M. *Tetrahedron* **2010**, *66*, 530–536.
31. Chiba, J.; Takeshima, S.; Mishima, K.; Maeda, H.; Nanai, Y.; Mizuno, K.; Inouye, M. *Chem. Eur. J.* **2007**, *13*, 8124–8130.
32. Fujimoto, K.; Aizawa, S.; Oota, I.; Chiba, J.; Inouye, M. *Chem. Eur. J.* **2010**, *16*, 2401–2406.
33. Kantevari, S.; Putapatri, S. R. *Synlett* **2010**, 2251–2256.
34. Stork, G.; Suh, H. S.; Kim, G. *J. Am. Chem. Soc.* **1991**, *113*, 7054–7056.
35. (a) Dondoni, A.; Massi, A.; Minghini, E. *Synlett* **2002**, *28*, 2001. (b) Dondoni, A.; Massi, A.; Minghini, E. *Helv. Chim. Acta* **2002**, *85*, 3331–3347. (c) Dondoni, A.; Massi, A.; Aldhoun, M. *J. Org. Chem.* **2007**, *72*, 7677–7687. (d) Ducatti, D. R. B.; Massi, A.; Noseda, M. D.; Duarte, M. E. R.; Dondoni, A. *Org. Biomol. Chem* **2009**, *7*, 1980–1986.
36. Dondoni, A.; Massi, A.; Sabbatini, S.; Bertolasi, V. *J. Org. Chem.* **2002**, *67*, 6979–6994.
37. (a) Dondoni, A.; Scherrmann, M.-C. *J. Org. Chem.* **1994**, *59*, 6404–6412. (b) Dondoni, A. *Pure Appl. Chem.* **2000**, *72*, 1577–4588. (c) Dondoni, A.; Formaglio, P.; Marra, A.; Massi, A. *Tetrahedron* **2001**, *57*, 7719–7727.
38. Dondoni, A.; Massi, A.; Sabbatini, S.; Bertolari, V. *Adv. Synth. Catal.* **2004**, *346*, 1355–1360.
39. Chen, J. J.; Drach, J. C.; Townsend, L. B. *J. Org. Chem.* **2003**, *68*, 4170–4178.
40. (a) McAllister, G. D.; Paterson, D. E.; Taylor, R.J.K. *Angew. Chem. Int. Ed.* **2003**, *42*, 1387–1391. (b) Jeanmart, S.; Taylor, R. J. K. *Tetrahedron Lett.* **2005**, *46*, 9043–9048.
41. Butora, G.; Olsen, D. B.; Carroll, S. S.; McMasters, D. R.; Schmitt, C.; Leone, J. F.; Stahlhut, M.; Burlein, C.; MacCoss, M. *Bioorg. Med. Chem.* **2007**, *15*, 5219–5229.
42. Lim, M.-I.; Klein, R. S.; *Tetrahedron Lett.* **1981**, *22*, 25. (b) Bhattacharya, B. K.; Lim, M.-I.; Otter, B. A.; Klein, R. S. *Tetrahedron Lett.* **1986**, *27*, 815–818.

43. (a) Bio, M. M.; Xu, F.; Waters, M.; Williams, J. M.; Savary, K. A.; Cowden, C. J.; Yang, C.; Buck, E.; Song, Z. J.; Tschaen, D. M.; Volante, R. P.; Reamer, R. A.; Grabowski, E. J. J. *J. Org. Chem.* **2004**, *69*, 6257–6266. (b) Xu, F.; Simmons, B.; Savary, K.; Yang, C.; Reamer, R. A. *J. Org. Chem.* **2004**, *69*, 7783–7786.

44. Kirsch, G.; Cagniant, D.; Cagniant, P. *J. Heterocycl. Chem.* **1982**, *19*, 443–445.

45. Hong, J. H.; Oh, C. H. *Arch. Pharm. Chem. Life Sci.* **2009**, *342*, 600–604.

46. Krishna, P. R.; Reddy, V. V. R.; Sharma, G. V. M. *Synlett* **2003**, 1619–1622. (b) Krishna, P. R.; Reddy, V. V. R.; Srinivas, R. *Tetrahedron* **2007**, *63*, 9871–9880.

47. Pinheiro, J. M.; Ismael, M. I.; Figueiredo, J. A.; Silva, A. M. S. *Monatsh. Chem.* **2009** *140*, 1237–1244.

48. Shapiro, R.; Chambers, R. W. *J. Am. Chem. Soc.* **1961**, *83*, 3920–3921.

49. Chaudhuri, N. C.; Kool, E. T. *Tetrahedron Lett.* **1995**, *36*, 1795–1798.

50. (a) Hurd, C. D.; Bonner, W. A. *J. Am. Chem. Soc.* **1945**, *67*, 1972–1977. (b) Chen, D. W.; Beuscher, A. E.; Stevens, R. C.; Wirsching, P.; Lerner, R. A.; Janda, K. D. *J. Org. Chem.* **2001**, *66*, 1725–1732.

51. Hocek, M.; Pohl, R.; Klepetářová, B. *Eur. J. Org. Chem.* **2005**, 4525–4528.

52. (a) Klein, R. S.; Kotick, M. P.; Watanabe, K. A.; Fox, J. J. *J. Org. Chem.* **1971**, *36*, 4113–4116. (b) Sharma, R. A.; Bobek, M.; Bloch, A. *J. Med. Chem.* **1975**, *18*, 473–476. (c) Ren, R. X.-F.; Narayan, C.; Chaudhuri, P. L.; Rumney, S. I. V.; Kool, E. T. *J. Am. Chem. Soc.* **1996**, *118*, 7671–7678. (d) Pirrung, M. C.; Zhao, X.; Harris, S. V. *J. Org. Chem.* **2001**, *66*, 2067–2071. (e) Griesang, N.; Richert, C. *Tetrahedron Lett.* **2002**, *43*, 8755–8758. (f) Singh, I.; Hecker, W.; Prasad, A. K.; Parmar, V. S.; Seitz, O. *Chem. Commun.* **2002**, 500–501.

53. Beuck, C.; Singh, I.; Bhattacharya, A.; Hecker, W.; Parmar, V. S.; Seitz, O.; Weinhold, E. *Angew. Chem. Int. Ed.* **2003**, *42*, 3958–3960.

54. Strässler, C.; Davis, N. E.; Kool, E. T. *Helv. Chim. Acta* **1999**, *82*, 2160–2171.

55. Maeba, I.; Iwata, K.; Usami, F.; Furukawa, H. *J. Org. Chem.* **1983**, *48*, 2998–3002.

56. Hainka, S.; Singh, I.; Hemmings, J.; Seitz, O. *J. Org. Chem.* **2007**, *72*, 8811–8819.

57. Hainke, S.; Seitz, O. *Angew. Chem. Int. Ed.* **2009**, *48*, 8250–8253.

58. Krohn, K.; Heins, H.; Wielckens, K. *J. Med. Chem.* **1992**, *35*, 511–517.

59. (a) Brown, D. M.; Ogden, R. C. *J. Chem. Soc. Perkin. Trans. 1* **1981**, 4, 723–725. (b) Harusawa, S.; Murai, Y.; Moriyama, H.; Imazu, T.; Ohishi, H.; Yoneda, R.; Kurihara, T. *J. Org. Chem.* **1996**, *61*, 4405–4411. (c) Harusawa, S.; Matsuda, C.; Araki, L.; Kurihara, T. *Synthesis* **2006**, 793–798.

60. Spadafora, M.; Postupalenko, V. Y.; Shvadchak, V. V.; Klymchenko, A. S.; Mély, Y.; Burger, A.; Benhida, R. *Tetrahedron Lett.* **2009**, *65*, 7809–7816.

61. Nencka, R.; Šála, M.; Dejmek, M.; Dračínský, M.; Holý, A.; Hřebabecký, H. *Synthesis* **2010**, 4119–4130.

62. Koszytkowska-Stawińska, M.; Mironiuk-Puchalska, E.; Sas, W. *Tetrahedron Lett.* **2011**, *52*, 1866–1870.

63. (a) Su, J.-K.; Jia, Y.-M.; He, R.; Rui, P.-X.; Han, N.; He, X.; Xiang, J.; Chen, X.; Zhu, J.; Yu, C.-Y. *Synlett* **2010**, 1609–1616. (b) Li, X.; Qin, Z.; Wang, R.; Chen, H.; Zhang, P. *Tetrahedron* **2011**, *67*, 1792–1798.

64. (a) Wilcox, C. S.; Cowart, M. D. *Carbohydr. Res.* **1987**, *171*, 141–160. (b) Matulic-Adamic, J.; Beigelman, L. *Tetrahedron Lett.* **1996**, *37*, 6973–6976. (c) Matulic-Adamic, J.; Beigelman, L.; Portmann, S.; Egli, M.; Usman, N. *J. Org. Chem.* **1996**, *61*, 3909–3911. (d) Gudmundsson, K. S.; Drach, J. C.; Townsend, L. B. *Tetrahedron Lett.* **1996**, *37*, 2365–2368. (e) Gudmundsson, K. S.; Drach, J. C.; Townsend, L. B. *J. Org. Chem.* **1997**, *62*, 3453–3459. (f) Matulic-Adamic, J.; Beigelman, L. *Tetrahedron Lett.* **1997**, *38*, 1669–1672. (g) Matulic-Adamic, J.; Beigelman, L. *Tetrahedron Lett.* **1997**, *38*, 203–206. (h) Hildbrand, S.; Blaser, A.; Parel, S. P.; Leumann, C. J. *J. Am. Chem. Soc.* **1997**, *119*, 5499–5511. (i) Wichai, U.; Woski, S. A. *Org. Lett.* **1999**, *1*, 1173–1175. (j) Parsch, J.; Engels, J. W. *Helv. Chim. Acta* **2000**, *83*, 1791–1808. (k) Liu, W.; Wise, D. S.; Townsend, L. B. *J. Org. Chem.* **2001**, *66*, 4783–4786. (l) Brotschi, C.; Häberli, A.; Leumann, C. *J. Angew Chem. Int. Ed.* **2001**, *40*, 3012–3014. (m) Sollogoub, M.; Fox, K. R.; Powers, V. E. C.; Brown, T. *Tetrahedron Lett.* **2002**, *43*, 3121–3123. (n) Kourafalos, V.; Marakos, P.; Pouli, N.; Townsend, L. B. *J. Org. Chem.* **2003**, *68*, 6466–6469. (o) Živkovič, A.; Engels, J. W. *Nucleosides Nucleotides Nucleic Acids* **2003**, *22*, 1167–1170. (p) Živkovič, A.; Engels, J. W. *Nucleosides Nucleotides Nucleic Acids* **2005**, *24*, 1023–1027. (q) Brotschi, C.; Mathis, G.; Leumann, C. J. *Chem. Eur. J.* **2005**, *11*, 1911–1923. (r) Zahn, A.; Brotschi, C.; Leumann, C. J. *Chem. Eur. J.* **2005**, *11*, 2125–2129. (s) Urban, M.; Pohl, R.; Klepetářová, B.; Hocek, M. *J. Org. Chem.* **2006**, *71*, 7322–7328. (t) Zahn, A.; Leumann, C. J. *Bioorg. Med. Chem.* **2006**, *14*, 6174–6188. (u) Grigorenko, N. A.; Leumann, C. J. *Chem. Eur. J.* **2009**, *15*, 639–645. (v) Lu, J.; Li, N.-S.; Koo, S. C.; Piccirilli, J. A. *J. Org. Chem.* **2009**, *74*, 8021–8030. (w) Štefko, M.; Slavětínská, L.; Klepetářová, B.; Hocek, M. *J. Org. Chem.* **2010**, *75*, 442–449. (x) Štefko, M.; Slavětínská, L.; Klepetářová, B.; Hocek, M. *J. Org. Chem.* **2011**, *76*, 6619–6635.

65. (a) Taniguchi, Y.; Nakamura, A.; Senko, Y.; Nagatsugi, F.; Sasaki, S. *J. Org. Chem.* **2006**, *71*, 2115–2122. (b) Nasr, T.; Taniguchi, Y.; Sasaki, S. *Heterocycles* **2007**, *71*, 2659–2668. (c) Peyron, C.; Benhida, R. *Synlett,* **2009** 472–476. (d) Metobo, S. E.; Xu, J.; Saunders, O. L.; Butler, T.; Aktoudianakis, E.; Cho, A.; Kim, C. U. *Tetrahedron Lett.* **2012**, *53*, 484–486.

66. (a) Guianvarc'h, D.; Benhida, R.; Fourrey, J. L. *Tetrahedron Lett.* **2001**, *42*, 647–650. (b) Guianvarc'h, D.; Fourrey, J. L.; Huu Dau, M.; Guerineau, V.; Benhida, R. *J. Org. Chem.* **2002**, *67*, 3724–3732. (c) Hanessian, S.; Machaalani, R. *Tetrahedron Lett.* **2003**, *44*, 8321–8323. (d) Enders, D.; Hieronymi, A.; Ridder, A. *Synlett* **2005**, 2391–2393. (e) Stoop, M.; Zahn, A.; Leumann, C. J. *Tetrahedron* **2007**, *63*, 3440–3449. (f) Peyron, C.; Navarre, J. M.; Dubreuil, D.; Vierling, P.; Benhida, R. *Tetrahedron Lett.* **2008**, *49*, 6171–6174. (g) Tite, T.; Lougiakis, N.; Marakos, P.; Pouli, N. *Synlett* **2009**, 2927–2930. (h) Tite, T.; Lougiakis, N.; Skaltsounis, A.-L.; Marakos, P.; Pouli, N.; Tenta, R.; Balzarini, J. *Synlett* **2009**, 1741–1744. (i) Tite, T.; Lougiakis, N.; Myrianthopoulos, V.; Marakos, P.; Mikros, E.; Pouli, N.; Tenta, R.; Fragopoulou, E.; Nomikos, T. *Tetrahedron* **2010**, *66*, 9620–9628.

67. Tawarada, R.; Seio, K.; Sekine, M. *J. Org. Chem.* **2008**, *73*, 383–390.

68. Singh, I; Steiz, O. *Org. Lett.* **2006**, *8*, 4319–4322.

69. Batoux, N.; Hardacre, C.; Migaud, M. E.; Ness, K. A.; Norma, S. E. *Tetrahedron* **2009**, *65*, 8858–8862.

70. (a) Trost, B. M. *J. Org. Chem.* **2004**, *69*, 5813–5837. (b) Trost, B. M.; Machacek, M. R.; Aponik, A. *Acc. Chem. Res.* **2006**, *39*, 747. (c) Trost, B. M.; Fandrick, D. R. *Aldrichimica Acta* **2007**, *40*, 59–72.

71. (a) Trost, B. M.; Kallander, L. S. *J. Org. Chem.* **1999**, *64*, 5427–5435. (b) Trost, B. M.; Van Vranken, D. L.; Bingel, C. *J. Am. Chem. Soc.* **1992**, *114*, 9327–9343.
72. Elming, N.; Claason-Kass, N. *Acta Chem. Scand.* **1952**, *6*, 535.
73. Stambasky, J.; Kapra, V.; Stefko, M.; Kysilka, O.; Hocek, M.; Malkov, A. V.; Kocovsky, P. *J. Org. Chem.* **2011**, *76*, 7781–7803.
74. (a) Arai, I.; Doyle Daves, G. *J. Am. Chem. Soc.* **1978**, *100*, 287–288. (b) Doyle Daves, G. *Acc. Chem. Res.* **1990**, *23*, 201–206. (c) Doyle Daves, G.; Hallberg, A. *Chem. Rev.* **1989**, *89*, 1433–1445.
75. (a) Mizoroki, T.; Mori, K.; Ozaki, A. *Bull. Chem. Soc. Jpn.* **1971**, *44*, 581. (b) Heck, R. F.; Nolley, J. P., Jr. *J. Org. Chem.* **1972**, *37*, 2320.
76. For a review on the Heck reaction in the synthesis of C-nucleosides: Wellington, K. W.; Benner, S. A. *Nucleosides Nucleotides Nucleic Acids* **2006**, *25*, 1309–1333.
77. (a) Hacksell, U.; Doyle Daves, G. *J. Org. Chem.* **1983**, *48*, 2870–2876. (b) Chi-Y. J.; Hacksell, U.; Doyle Daves, G. *J. Org. Chem.* **1986**, *51*, 3093–3098.
78. Ireland, R. E.; Thairisvongs, S., Vanier, D.; Wilcox, C. S. *J. Org.Chem.* **1980**, *45*, 48–61.
79. Garegg, J.; Samuelsson, B. *Synthesis* **1979**, 813–814.
80. Walker, J. A.; Chen, J. J.; Wise, D. S.; Townsend, L. B. *J. Org. Chem.* **1996**, *61*, 2219–2221.
81. (a) Kassou, M.; Castillon, S. *Tetrahedron Lett.* **1994**, *35*, 5513–5516. (b) Bravo, F.; Kassou, M.; Castillon, S. *Tetrahedron Lett.* **1999**, *40*, 1187–1190. (c) Bravo, F.; Kassou, M.; Castillon, S. *Carbohydr. Res.* **2001**, *336*, 83–97.
82. (a) Cameron, M. A.; Cush, S. B.; Hammer, R. P. *J. Org. Chem.* **1997**, *62*, 9065–9069. (b) Larsen, E.; Jorgensen, P. T.; Sofan, M. A.; Pedersen, E. B. *Synthesis* **1994**, 1037–1038.
83. Zhang, H.-C.; Doyle Daves, G. *J. Org. Chem.* **1992**, *57*, 4690–4696.
84. Jouber, N.; Pohl, R.; Klepetářová, B.; Hocek, M. *J. Org. Chem.*, **2007**, *72*, 6797–6805.
85. Hikishima, S.; Minakawa, N.; Kuramoto, K.; Fujisawa, Y.; Ogawa, M.; Matsuda, A. *Angew. Chem. Int. Ed.* **2005**, *44*, 596–598.
86. Lee, A. H. F.; Kool, E. T. *J. Org. Chem.* **2005**, *70*, 132–140.
87. (a) Coleman, R. S.; Mortensen, M. A. *Tetrahedron Lett.* **2003**, *44*, 1215–1219. (b) Coleman, R. S.; Berg, M. A.; Murphy, C. J. *Tetrahedron* **2007**, *63*, 3450–3456.
88. Sun, Z.; Ahmed, S.; McLaughlin, L. W. *J. Org. Chem.* **2006**, *71*, 2922–2925.
89. Kubelka, T.; Slavětínská, L.; Klepetářová, B.; Hocek, M. *Eur. J. Org. Chem.* **2010**, 2666–2669.
90. Li, J.-S.; Fan, Y.-H.; Zhang, Y.; Marky, L. A.; Gold, B. *J. Am. Chem. Soc.* **2003**, *125*, 2084–2093.
91. Okamoto, A.; Tainaka, K.; Fujiwara, Y. *J. Org. Chem.* **2006**, *71*, 3592–3598.
92. Oda, H.; Hanami, T.; Iwashita, T.; Kojima, M.; Itoh, M.; Hayashizaki, Y. *Tetrahedron* **2007**, *63*, 12747–12753.
93. Wang, Z.-X.; Wiebe, L. I.; Balazarini, J.; De Clerq, E.; Knaus, E. E. *J. Org. Chem.* **2000**, *65*, 9214–9219.
94. Raboisson, P.; Baurand, A.; Cazenave, J.-P.; Gachet, C.; Schultz, D.; Spiess, B.; Bourguignon, J.-J. *J. Org. Chem.* **2002**, *67*, 8063–8071.
95. Fraley, A. W.; Chen, D.; Johnson, K.; McLaughlin, L. W. *J. Am. Chem. Soc.* **2003**, *125*, 616–617.

96. Zhang, H.-C.; Doyle Daves, G. *J. Org. Chem.* **1993**, *58*, 2557–2560.
97. Franchetti, P.; Cappellacci, M.; Grifantini, M.; Barzi, A.; Nocentini, G.; Yang, H.; O'Connor, A.; Jayaran, H. N.; Carrell, C.; Goldstein, B. M. *J. Org. Chem.* **1995**, *38*, 3829–3837.
98. Liu, M.-C.; Luo, M.-Z.; Mozdziesz, D. E.; Sartorelli, A. C. *Nucleosides Nucleotides Nucleic Acids* **2005**, *24*, 45–62.
99. Cermola, F.; Iesce, M. I.; Astarita, A.; Passananti, M. *Lett. Org. Chem.* **2011**, *8*, 309–314.
100. Cheng, J. J. ; Wei, Y.; Williams, J. D.; Drach, J. C.; Townsend. L. B. *Nucleosides Nucleotides Nucleic Acids* **2005**, *24*, 1417–1437.
101. Hatano, A.; Makita, S.; Kirihara, M. *Tetrahedron* **2005**, *61*, 1723–1730.
102. (a) Gibson, E. S.; Lesiak, K.; Watanabe, K. A.; Gudas, L. J.; Pankiewicz, K. W. *Nucleosides Nucleotides* **1999**, *18*, 363–376. (b) Hamm, M. L.; Parker, A. J.; Steele, T. W. E.; Carman, J. L.; Paris, C. A. *J. Org. Chem.* **2010**, *75*, 5661–5669.
103. Barbaric, J.; Wanninger-Weib, C.; Wagenknecht, H.-A. *Eur. J. Org. Chem.* **2009**, 364–370.
104. Bárta, J.; Pohl, R.; Klepetářová, Ernsting, N. P.; Hocek, M. *J. Org. Chem.* **2008**, *73*, 3798–3806.
105. Bárta, J.; Slavětínská, L.; Klepetářová, B.; Hocek, M. *Eur. J. Org. Chem.* **2010**, 5432–5443.
106. Alonso, D. A.; Nájera, C.; Pastor, I.; Yus, M. *Chem. Eur. J.* **2010**, *16*, 5274.
107. Stefko, M.; Hocek, M. *Synthesis* **2010**, 4199–4206.
108. He, W.; Togo, H.; Yokoyama, M. *Tetrahedron Lett.* **1997**, *38*, 5541–5544.
109. Carreño, M. C.; Des Mazery, R.; Urbano, A.; Colobert, F.; Solladié, G. *J. Org. Chem.* **2003**, *68*, 7779–7787.
110. Son, S.; Fu, G. C. *J. Am. Chem. Soc.* **2007**, *129*, 1046–1047.
111. Redpath, P.; Macdonald, S.; Migaud, M. E. *Org. Lett.* **2008**, *10*, 3323–3326.
112. Midtkandal, R. R.; Macdonald, S. J. F.; Migaud, M. E. *Chem. Commun.* **2010**, *46*, 4538–4540.
113. Bonnac, L.; Lee, S. E.; Giuffredi, G. T.; Elphick, L. M.; Anderson, A. A.; Child, E. S.; Mann, D. J.; Gouverneur, V. *Org. Biomol. Chem.* **2010**, *8*, 1445–1454.
114. Coutouli-Argyropoulou, E.; Trakossas, S. *Tetrahedron* **2011**, *67*, 1915–1923.

8 Methodologies for the Synthesis of Isomeric Nucleosides and Nucleotides of Antiviral Significance

MAURICE OKELLO and VASU NAIR

Center for Drug Discovery and College of Pharmacy, University of Georgia, Athens, Georgia, USA

1. INTRODUCTION

The human immunodeficiency virus (HIV), the etiological agent of AIDS, presents a variety of unique biochemical targets for the discovery and development of antiviral agents.[1–3] Nucleoside derivatives have been and continue to be discovered as inhibitors of key viral targets on HIV.[4–9] Both natural "D-related" nucleoside derivatives and, more recently, their mirror images, the "L-related" nucleoside derivatives, have been investigated for antiviral activity.[6,10] Remarkable progress has been made in advancing drugs for the treatment of HIV infection, and selected examples of nucleosides that inhibit the cytopathic effect of HIV and that are in clinical use include the following dideoxynucleosides: 2′,3′-dideoxyinosine (ddI), 2′,3′-dideoxycytidine (ddC), 2′,3′-didehydro-3′-deoxythymidine (d4T), and 3′-substituted dideoxynucleosides such as 3′-azido-3′-deoxythymidine (AZT), 2′,3′-dideoxy-3'-thiacytidine (3-TC), β-L-2′,3′-dideoxy-3′-thia-5-fluorocytidine (FTC), and abacavir.[4,6,11–16]

As precursors of intracellularly active drugs, these dideoxynucleosides must be sequentially phosphorylated in cells through three steps to produce the corresponding 5′-triphosphates, which may act at the reverse transcriptase (RT) level as competitive inhibitors and/or alternative substrates with respect to the natural deoxynucleotide substrates. The efficiency of dideoxynucleosides to be activated by phosphorylation is a key factor in determining their antiviral potency.[17]

Even though the dideoxynucleosides noted above appear to be very successful drugs, there remains the need to develop new nucleosides and other compounds that exhibit more favorable toxicity profiles, that are less susceptible to cross-resistance,

Chemical Synthesis of Nucleoside Analogues, First Edition. Edited by Pedro Merino.
© 2013 John Wiley & Sons, Inc. Published 2013 by John Wiley & Sons, Inc.

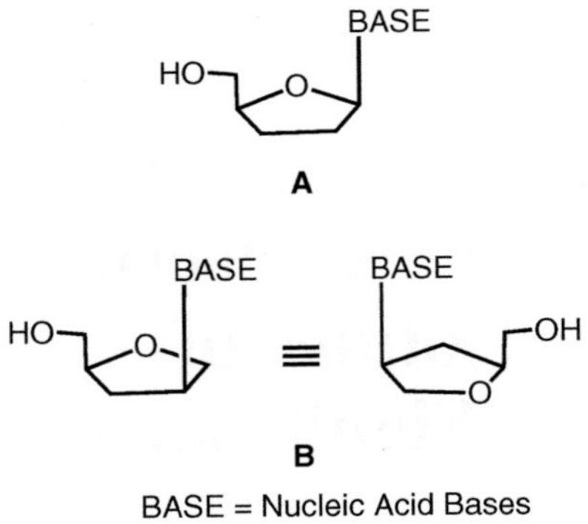

Figure 1. General structures of anti-HIV active dideoxynucleosides of natural origin (A) and L-related isomeric dideoxynucleosides of nonnatural origin (B).

and that are active against resistant strains of HIV. As noted in the examples cited above, the most common modifications of dideoxynucleosides have been strategic substitution on the carbohydrate moiety, elimination of the endocyclic oxygen of the carbohydrate moiety, or introduction of an additional endocyclic oxygen or sulfur in the carbohydrate moiety. In addition to the requirements necessary within these molecules for antiviral activity, such as appropriate structure and stereochemistry and efficiency of phosphorylation, a property that is also very important is stability under chemical and physiological conditions. Dideoxynucleosides (Figure 1A), particularly those of the purine family, are very unstable with respect to hydrolytic cleavage of the glycosidic bond under acidic conditions.[18] This inherent chemical property, which results from both the absence of the 2'- and 3'-hydroxyl groups (-I effect) and the presence of the proximal endocyclic oxygen, limits the usefulness of these compounds as antiviral agents. Stability toward catabolic enzymes is also desirable. For example, 2',3'-dideoxyadenosine is a substrate for mammalian adenosine deaminase (ADA) and is rapidly converted by enzyme-catalyzed hydrolytic deamination to 2',3'-dideoxyinosine (ddI).[19,20] Although ddI has anti-HIV activity, it has to be converted cellularly to ddATP via ddAMP, and this pathway is inefficient.[19] For these reasons, the design and synthesis of deoxynucleosides that are stable with respect to both glycosidic cleavage and enzymatic deamination was of considerable significance in this area. Our approach to achieving this was through a structural change in the general structure of dideoxynucleosides of natural origin (Figure 1A) by transposing the nucleobase from the 1'-position to the 2'-position (Figure 1B).[21] These nonnatural isomeric L-related dideoxynucleosides are remarkably stable with respect to glycosidic bond cleavage and enzymatic hydrolytic deamination involving the nucleobase.[22–29]

2. CLASSES OF ISOMERIC DIDEOXYNUCLEOSIDES AND ISOMERIC DEOXYNUCLEOSIDES

Four representative classes of isomeric dideoxynucleosides (isodideoxynucleosides) and their relationship to normal nucleosides are illustrated in Figure 2. The

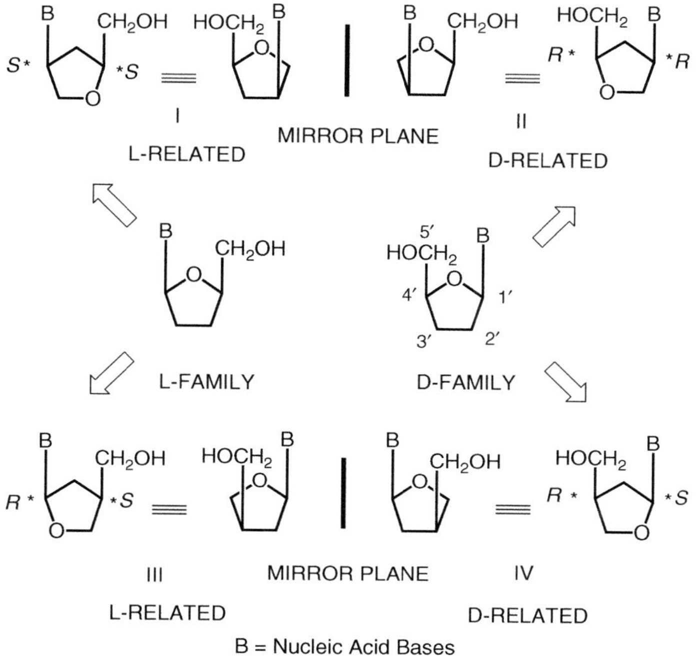

Figure 2. Four classes of isomeric dideoxynucleosides and their relationship to standard dideoxynucleosides of the D- and L-families.[21,26]

structural relationship of the enantiomeric compounds of classes I and II to the natural D- and nonnatural L-nucleosides may be explained as arising from the transposition of the nucleic acid base from C1' to C2'. These compounds can also be viewed as arising from the transposition of the endocyclic oxygen from the normal position to the 3'-position. Compounds of class I are viewed as being L-related and those of class II are D-related because of their relationship to L- and D-nucleosides, as shown in Figure 2.[21]

Compounds of classes III and IV can be viewed as emerging from the transposition of the hydroxymethylene group from C4' to C3' or the transposition of the endocyclic oxygen from the O1' to the 2'-position. Compounds of classes III and IV may also be referred to as apiodideoxynucleosides because of the relationship of the carbohydrate moiety to apio sugars. Although the *cis*-structure of the base and the 4'-CH$_2$OH and their 1,3-relationship are maintained as in D- and L-nucleosides, the corresponding *trans*-diastereoisomeric structures are also possible. It should be mentioned that compounds of classes III and IV have normal glycosidic linkages and are less stable with respect to glycosidic cleavage than compounds of classes I and II. From the antiviral perspective, the most active isomeric nucleosides were in the class I group.

In this chapter we also describe methodologies for the incorporation of new isomeric deoxynucleosides into dinucleotides. The intent of this work was to produce novel nuclease-resistant dinucleotide analogues that would exhibit greater stability under a variety of physiological conditions and that would possess critical structural features for recognition and inhibition of HIV integrase.[30–32] We designed these dinucleotides with structural features that exploited the ability of integrase to recognize the terminal sequence of the truncated viral DNA produced in the 3′-processing step of HIV-1 enzymology, prior to strand transfer. This molecular design led to the discovery of conceptually novel, nuclease-resistant, nonnatural dinucleotides with defined base sequences that are inhibitors of HIV-1 integrase.

Characterization of the structure and stereochemistry of the isomeric nucleosides and the corresponding dinucleotides was determined by ultraviolet (UV)–visible (Vis) spectroscopy, circular dichroism (CD) data, optical rotation, ^{1}H and ^{13}C nuclear magnetic resonance (NMR) spectra, high-resolution mass spectrometry (HRMS), elemental analysis, and single-crystal x-ray crystallography.

3. GENERAL INFORMATION ON EXPERIMENTAL DATA

Melting points reported are uncorrected and were determined on Electrothermal Engineering Ltd. melting-point apparatus. NMR spectra were recorded on Bruker Model AC300 and WM 360 spectrometers and a Varian Inova 500-MHz NMR instrument. Chemical shifts (δ, ppm) are relative to TMS (^{1}H and ^{13}C) or H_3PO_4 (^{31}P). HRMS data were determined at the Nebraska Center for Mass Spectrometry. UV–Vis quantitative spectral data were determined using a Varian Cary Model 3 spectrophotometer. Infrared (IR) spectra were recorded on a Mattson Cygnus 25 Fourier transform instrument. CD spectra were measured on an AVIV circular dichroism spectrometer. Purities of intermediates and final products were determined by a combination of HPLC analyses on a Waters automated 600E system or a Beckman System Gold using C_{18} columns or Partsil-10 SAX ion-exchange columns. Elemental analyses were performed at Galbraith Laboratories, Inc. or at NuMega Resonance Laboratories. Single-crystal x-ray data were obtained at the University of Georgia Crystallographic Center.

4. SYNTHESIS OF INTERMEDIATES OF L-RELATED ISOMERIC DIDEOXYNUCLEOSIDES

The key carbohydrate intermediate **6**, for the synthesis of L-related isodideoxy nucleosides, was prepared starting with D-xylose **1** and involved seven steps, with an overall yield of 17% (Scheme 1). The methodology involved initial preparation of the bisisopropylidene derivative of D-xylose **1**, followed by selective deprotection of the six-membered isopropylidene ring[33] and subsequent selective benzoylation of the primary 5-hydroxyl group to provide **2** in 39% overall yield. Compound **2** was deoxygenated at the 3-position by conversion of the 3-hydroxyl

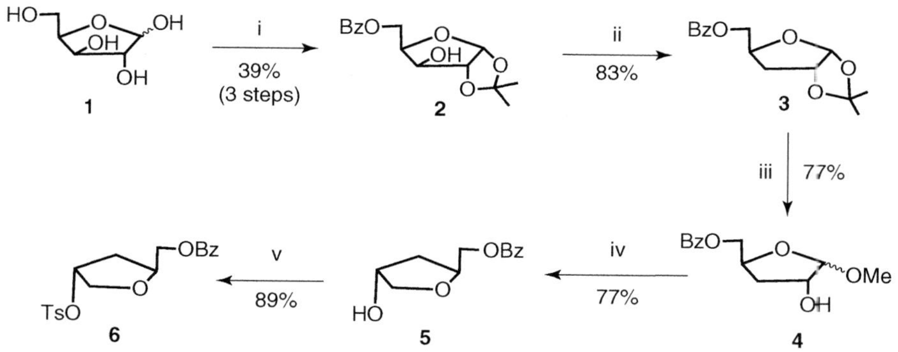

Scheme 1. *Reagents and conditions:* (i) 1, reference 33; 2, BzCl, Pyr, -10 to 0°C, 1 h; (ii) 1, Im$_2$CS, ClCH$_2$CH$_2$Cl, 83°C, 4 h; 2, *n*-Bu$_3$SnH, AIBN, toluene, 120°C, 3 h; (iii) HCl, MeOH; (iv) 1, HMDS, TMSCl, 123°C, 17 h; 2, Et$_3$SiH, TMSOTf, CH$_3$CN, rt, 17 h; (v) TsCl, Pyr, 0°C, 31 h.

group to its imidazole thiocarbonyl ester and subsequent treatment with tributyltin hydride and AIBN in refluxing toluene (83% yield in two steps). Acid-catalyzed methanolysis of the remaining acetonide group afforded the α- and β-methyl glycosides **4**. Reductive demethoxylation of the latter utilized a methodology that first involved protection of the 2-hydroxyl group by silylation (HMDS, TMSCl), followed in situ by treatment of the product with triethylsilane and TMS-triflate[22,34] in acetonitrile, which produced the tetrahydrofuran **5** in 60% yield for the two steps. Tosylation of compound **5** generated the key carbohydrate intermediate **6** (89%). Condensation of the tosylate **6** with the appropriate heterocyclic base provided the desired products in good yields and with the expected complete inversion of stereochemistry at the 2-position.

Experimental Procedures for the Synthesis of Intermediates of IsodNs and IsoddNs

5-*O*-Benzoyl-1,2-*O*-isopropylidene-α-D-xylofuranose (2).

D-xylose was converted to 1,2-*O*-isopropylidene-α-D-xylofuranose in two steps (58–80% overall yield). A solution of 1,2-*O*-isopropylidene-α-D-xylofuranose (15.95 g, 83.86 mmol) in anhydrous pyridine (54 mL) was cooled down to −20°C. At this temperature, a solution of benzoyl chloride (9.73 mL, 83.86 mmol) in 20 mL of pyridine was added dropwise into the sugar solution over a period of 30 min. The reaction was stirred continuously for an additional 30 min while allowing the mixture to warm to 0°C and was then poured into 200 mL of an ice–water mixture. The mixture was extracted with CHCl$_3$ (4 × 75 mL). The organic extracts were combined, washed, dried with Na$_2$SO$_4$, filtered, evaporated, and coevaporated with several small portions of toluene. The residue was purified by column chromatography with 0–5% MeOH/CHCl$_3$ as the eluting solvent to yield 18.75 g (63.71 mmol) of **2** as a clear light yellow viscous oil (39% yield from

D-xylose). ^{1}H-NMR (CDCl$_3$): δ 1.28 (s, 3H), 1.47 (s, 3H), 3.69 (s, 1H), 4.21 (m, 1H), 4.39 (m, 1H), 4.44 (m, 1H), 4.69 (m, 2H), 5.94 (d, 1H), 7.46 (t, 3H), 8.01 (d, 2H).

5-O-Benzoyl-3-deoxy-1,2-isopropylidene-α-D-xylofuranose (3). A solution of **2** (7.08 g, 24.06 mmol) and 1,1′-thiocarbonyldiimidazole (6.43 g, 36.08 mmol) in dry dichloroethane (75 mL) was stirred at reflux for 4 h. The solvent was evaporated, and the residue was purified by flash chromatography using 0–10% MeOH/CHCl$_3$ to provide 8.88 g (21.96 mmol, 91%) of 5-*O*-benzoyl-3-*O*-(1-imidazolylthiocarbonyl)-1,2-*O*-isopropylidene-α-D-xylofuranose as a clear, light yellow viscous oil. To a refluxing solution of the latter compound in toluene (175 mL), a nitrogen-purged solution of Bu$_3$SnH (8.86 mL, 32.93 mmol) and 2,2′-azobis(2-methylpropionitrile) (2.8 g, 17.56 mmol) in toluene (100 mL) was added dropwise over a period of 45 min. The reaction mixture was stirred at reflux for 2 h. The solvent was then evaporated under reduced pressure. The residue was partitioned between CH$_3$CN (225 mL) and hexanes (150 mL). The CH$_3$CN portion was washed several times with hexanes (4 × 50 mL). Evaporation of the acetonitrile portion gave an oily residue, which was purified by flash chromatography using hexanes followed by CHCl$_3$ to afford 5.56 g (19.98 mmol, 91%) of **3** as a viscous oil. ^{1}H-NMR (CDCl$_3$): δ 1.30 (s, 3H), 1.51 (s, 3H), 1.74 (m, 1H), 2.16 (dd, 1H), 4.53 (m, 4H), 5.84 (d, 1H), 7.47 (t, 3H), 8.03 (d, 2 H).

Methyl 5-O-benzoyl-3-deoxy-α-D-erythropentofuranose and Methyl 5-O-benzoyl-3-deoxy-β-D-erythropentofuranose (4). Hydrogen chloride (0.004 g, 0.1 mmol) was bubbled through a solution of the benzoylated sugar **3** (0.577 g, 2.07 mmol) in MeOH (10 mL). The reaction mixture was stirred at room temperature for 12 h and was then neutralized by stirring with Dowex ion-exchange resin (OH$^-$ form). The resin was filtered and the filtrate concentrated. Purification of the residue by flash chromatography using 5% MeOH/CHCl$_3$ as eluting solvent afforded 0.402 g (1.59 mmol, 77%) of **4** as a viscous, clear, light yellow oil. ^{1}H-NMR (CDCl$_3$): δ 2.07 (m, 2H), 3.3 (s, 3H), 3.68 (s, 1H), 4.28 (m, 2H), 4.44 (dd, 1H), 4.72 (m, 1H), 4.87 (s, 1H), 7.42 (t, 2H), (t, 1H), 8.08 (d, 2H).

2(S)-(O-Benzoylmethyl)tetrahydrofuran-4(R)-ol (5). A mixture of sugar **4** (5.03 g, 19.9 mmol) and chlorotrimethylsilane (1.26 mL, 9.97 mmol) in 1,1,1,3,3,3-hexamethyldisilazane (HMDS) (63 mL, 299 mmol) was refluxed for 17 h. Unreacted HMDS was evaporated under reduced pressure to afford a clear, yellow viscous oil. Dry CH$_3$CN (25 mL) was added to this residue followed by the addition of triethylsilane (9.55 mL, 59.8 mmol) and TMS-triflate (11.6 mL, 59.8 mmol). The reaction mixture was stirred at room temperature for 17 h. Excess TMS-triflate in the reaction was quenched by stirring the mixture with 8 mL of H$_2$O. The reaction mixture was neutralized by dropwise addition of 5 N NaOH while stirring the mixture vigorously. The solvent was removed under reduced pressure. The residue was purified by flash chromatography using 8% MeOH/CHCl$_3$ as the eluting solvent to give 3.44 g (15.47 mmol, 77.6%) of **5** as a clear, viscous, light yellow oil. ^{1}H-NMR (CDCl$_3$): δ 1.83 (m, 1H), 2.01 (m, 1H), 3.1 (s, 1H), 3.76 (d, 1H), 3.94 (dd, 1H), 4.23 (dd, 1H), 4.38 (dd, 1H), 4.48 (m, 2H), 7.37 (t, 2H), 7.49 (t, 1H), 7.98 (d, 2H). *Anal.* Calcd for C$_{12}$H$_{14}$O$_4$: C, 64.85; H, 6.35; found: C, 64.62; H, 6.15.

2(S)-(O-Benzoylmethyl)-4(R)-(p-O-toluenesulfonyl)tetrahydrofuran (6). A solution of the benzoylated sugar **5** (3.44 g, 15.4 mmol) in dry pyridine (20 mL) was cooled down to −10°C for 1 h. At this temperature, tosyl chloride (4.43 g, 23 mmol) was added to the sugar solution. The reaction mixture was allowed to stand at 5°C for 31 h. The solvent was evaporated under reduced pressure. The residue was taken up in dichloromethane, then washed with H$_2$O (3 × 20 mL). The organic portion was dried with Na$_2$SO$_4$, filtered, and concentrated. Purification of the residue by flash chromatography using 3% MeOH/CHCl$_3$ as the eluting solvent afforded 5.15 g (13.7 mmol, 89%) of **6** as a clear, light yellow viscous oil. ^{1}H-NMR (CDCl$_3$): δ 1.97 (m, 1H), 2.25 (m, 1H), 2.45 (s, 3H), 3.98 (m, 2H), 4.29 (m, 1H), 4.42 (m, 2H), 5.17 (m, 1H), 7.45 (m, 5H), 7.80 (d, 2H), 8.01 (d, 2H). ^{13}C-NMR (CDCl$_3$): δ 21.6, 35.2, 65.6, 72.8, 75.9, 81.3, 127.6, 128.3, 129.5, 129.6, 129.9, 133.1, 133.6, 145.1, 166.2. *Anal.* Calcd for C$_{19}$H$_{20}$O$_6$S: C, 60.62; H, 5.36; found: C, 60.54; H, 5.43.

5. SYNTHESIS OF ISODIDEOXYADENOSINE (ISODDA, 8)

Synthesis of 4(*S*)-(6-amino-9*H*-purin-9-yl)tetrahydro-2(*S*)-furanmethanol **8** was achieved by heating compound **6** and adenine in DMF in the presence of potassium carbonate and 18-crown-6 (64% purified yield), followed by deprotection with sodium methoxide (85% yield) (Scheme 2).

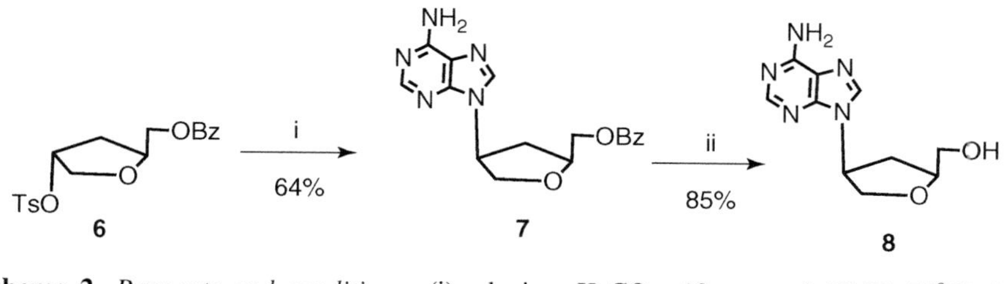

Scheme 2. *Reagents and conditions:* (i) adenine, K$_2$CO$_3$, 18-crown-6 DMF, 75°C, 9 h; (ii) NaOMe, MeOH, rt, 2 h.

Experimental Procedures for the Synthesis of IsoddA[22,35]

4(S)-(6-Amino-9 H-purin-9-yl)-2(S)-(O-benzoylmethyl)tetrahydrofuran (7). A mixture of adenine base (1.96 mmol), potassium carbonate (2.62 mmol), 18-crown-6 (2.62 mmol), and 2(*S*)-(*O*-benzoylmethyl)-4(R)-(*p-O*-toluenesulfonyl)tetrahydrofuran (1.31 mmol) (**6**) in DMF (11 mL) was stirred at 75°C for 11 h. The solvent was removed under reduced pressure and the residue was purified by flash chromatography on silica gel with 0–10% MeOH/CHCl$_3$ to afford **7** as a white solid in 64% yield. Melting point 193–195°C UV (MeOH) λ_{max} 260 nm. ^{1}H-NMR (CDCl$_3$): δ 2.17 (m, 1H), 2.79 (m, 1H), 4.12(dd, 1H), 4.29 (dd, 1H), 4.45 (m, 2H), 4.59 (dd, 1H), 5.32 (m, 1H), 6.00 (s, 2H), 7.40 (t, 2H), 7.53 (t, 1H), 7.98 (d,1H), 8.02(s, 1H), 8.31 (s, 1H).

4(S)-(6-Amino-9H-purin-9-yl)tetrahydro-2(S)-furanmethanol (8). To a solution of the 6'-O-benzoylated isonucleoside **7** (0.31 mmol) in methanol (15 mL) was added sodium methoxide (0.46 mmol). After stirring for 2 h at room temperature, the reaction mixture was neutralized by stirring with Dowex ion-exchange resins (H$^+$ form). The resin was filtered and the filtrate was then concentrated under reduced pressure. The residue was purified on silica gel plates with 10% MeOH/CHCl$_3$. Final purification of the product was done by reversed-phase HPLC on Amberlite XAD-4 resin using 10% EtOH/H$_2$O as the eluting solvent to give the title compound **8** as a white solid (85% yield). Melting point 180–182°C. Mass spectrum, m/z 235 (M$^+$). [α]$_D$ −26.6° (c = 1.0 in MeOH). UV (H$_2$O) λ_{max} 260 nm (ε = 13,788). ^{1}H-NMR (Me$_2$SO-d$_6$): δ 2.09 (m, 1H), 2.58 (m, 1H), 3.55 (m,2H), 3.99 (m, 3H), 4.95 (m, 1H), 5.17 (m, 1H), 7.25 (s, 2H), 8.15 (s, 1H), 8.26 (s, 1H). ^{13}C-NMR (Me$_2$SO-d$_6$): δ 33.9, 53.9, 62.4, 71.8, 79.6, 118.7, 138.9, 149.3, 152.3, 155.9. *Anal.* Calcd for C$_{10}$H$_{13}$N$_5$O$_2$: C, 51.06; H, 5.57; N, 29.77; found: C, 51.40; H, 5.56; N, 29.66.

6. SYNTHESIS OF ISODIDEOXYCYTIDINE (ISODDC) AND ISODIDEOXYTHYMIDINE (ISODDT)

An attempt to prepare the dideoxycytidine analog **15** via the direct displacement reaction on **6** with cytosine under conditions described previously was unsuccessful and generated only the O-alkylated product. However, isoddC **15** could be synthesized from its isoddU counterpart **13** via intermediate **12** in good yields (see Scheme 3). Treatment of compound **12** with 4-chlorophenyl phosphorodichloridate and 1,2,4-triazole in pyridine produced the 4-triazolylpyrimidinone derivative **14**.[36] Reaction of **14** with aqueous ammonia in dioxane and subsequent deprotection with sodium methoxide afforded the desired product **15**.

Experimental Procedures for the Synthesis of IsoddC and IsoddT[35]

Synthesis of IsoddC via Displacement with Uracil and Conversion of IsoddU to IsoddC

2(S)-(O-Benzoylmethyl)-4(S)-[3,4-dihydro-2,4-dioxo-1(2H)-pyrimidinyl]tetrahydrofuran *(12)*. A mixture uracil base (1.96 mmol), potassium carbonate (2.62 mmol), 18-crown-6 (2.62 mmol), and 2(S)-(O-benzoylmethyl)-4(R)-(p-O-toluenesulfonyl)tetrahydrofuran (1.31 mmol) (**6**) in DMF (11 mL) was stirred at 75°C for 11 h. The solvent was removed under reduced pressure and the residue

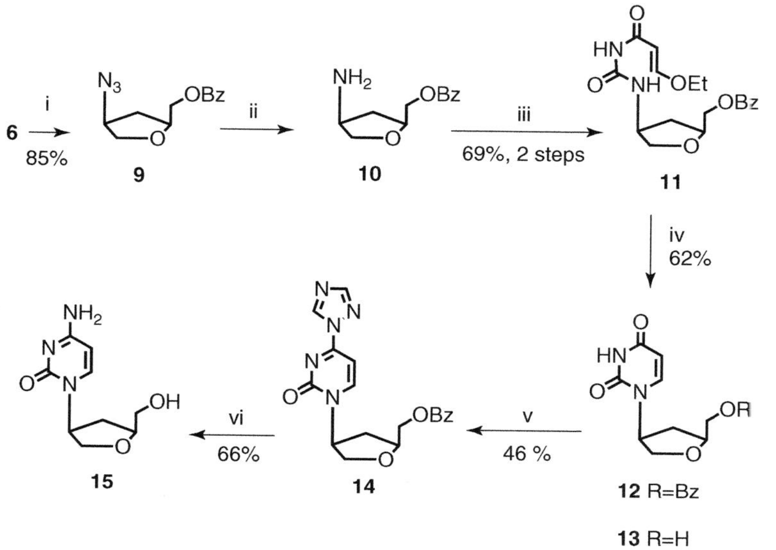

Scheme 3. *Reagents and conditions:* (i) NaN$_3$, DMF, 80°C, 3 h; (ii) H$_2$, Pd/C EtOH, 3 h; (iii) (E)-3-ethoxyacryloyl isocyanate, DMF, -15°C to rt,18 h; (iv) 2 N H$_2$SO$_4$ reflux, 3 h; (v) *p*-Cl-Ph-OPOCl$_2$, Pyr, triazole, 0°C to rt, 26 h; (vi) NH$_4$OH, dioxane, rt,14 h.

was purified by flash chromatography on silica gel with 0–10% MeOH/CHCl$_3$ to provide **12** as a white solid in 43% yield. Melting point 154–156°C. UV (MeOH) λ_{max} 266 nm. UV (0.1 N NaOH) λ_{max} 265 nm. ^{1}H-NMR (Me$_2$SO-d_6): δ 1.81 (m, 1H), 2.5 (m, 1H), 3.92 (m, 2H), 4.20 (m, 2H), 4.45 (m, 2H), 5.10 (m, 1H), 5.51 (d, 1H), 7.61 (m, 4H), 7.96 (d, 2H).

4(S)-[3,4-Dihydro-2,4-dioxo-l(2H)-pyrimidinyl]tetrahydro-2(S)-furanmethanol (13). A solution of **12** (0.44 mmol) in methanol (10 mL) was cooled down to 0°C and saturated with ammonia. The reaction mixture was allowed to stand at 0°C for 2 h and then at room temperature for 46 h. Excess ammonia was purged out with nitrogen and the solvent was then removed under reduced pressure. The residue was purified by preparative TLC with 10% MeOH/CHCl$_3$ to afford **13** as a very hygroscopic white solid in 59% yield. Mass spectrum, *m/z* 212 (M$^+$). [α]$_D$ +31.0° (c = 1.0 in MeOH). UV (H$_2$O) λ_{max} 266 nm (ε = 8786). FTIR (KBr): 1682 cm^{-1}. ^{1}H-NMR (Me$_2$SO-d_6): δ 1.73 (m, 1H), 2.42 (m, 1H), 3.52 (dd, 2H), 3.81 (m, 3H), 4.91 (m, 1H), 5.09 (m, 1H), 5.58 (d, 1H), 7.72 (d, 1H), 11.23 (s, 1H). ^{13}C-NMR (Me$_2$SO-d_6): δ 33.4, 54.6, 61.8, 71.6, 79.7, 101.7, 141.9, 150.9, 163.1. *Anal.* Calcd for C$_9$H$_{12}$N$_2$O$_4$: C, 50.94; H, 5.70; N, 13.20; found: C, 50.77; H, 5.52; N, 12.97.

2(S)-(O-Benzoylmethyl)-4(S)-[4-(l,2,4-triazolyl)-2-oxo-l(2H)-pyrimidinyl]tetrahydrofuran (14). To a stirred solution of the nucleoside **13** (0.187 g, 0.59 mmol) in pyridine (7 mL), 4-chlorophenylphosphorodichloridate (0.2 mL, 1.2 mmol) was added dropwise at 0°C. After the addition, 1,2,4-triazole (0.163 g, 2.36 mmol) was added to the mixture. The reaction mixture was stirred at room temperature for 26 h. The solvent was removed under reduced pressure. The residue was taken up in CH_2Cl_2, then washed with H_2O (3 × 25 mL). The organic portion was dried with Na_2SO_4, filtered, concentrated, and purified on a silica gel plate with 6% $MeOH/CHCl_3$ to afford 0.101 g (0.27 mmol, 46%) of **14** as a viscous yellow oil. UV (MeOH) λ_{max} 252, 316 nm. ^{1}H-NMR ($CDCl_3$): δ 1.93 (m, 1H), 2.81 (m, 1H), 3.96 (dd, 1H), 4.15 (d, 1H), 4.30 (m, 1H), 4.40 (dd, 1H), 4.68 (dd, 1H), 5.46 (m, 1H), 6.78 (d, 1H), 7.33 (t, 2H), 7.45 (t, 1H), 7.90 (d, 2H), 8.05 (s, 1H), 8.18 (d, 1H), 9.17 (s, 1H).

4(S)-[4-Amino-2-oxo-1(2H)-pyrimidinyl]tetrahydro-2(S)-furanmethanol **(15)**. A solution of nucleoside **14** (0.098 g, 0.27 mmol) in 1:6 v/v of NH_4OH/dioxane (4 mL) was stirred at room temperature for 14 h. The solvent was removed under reduced pressure and the residue was purified on a silica gel plate with 8% $MeOH/CHCl_3$ to afford 0.056 g (0.18 mmol, 66%) of 4(S)-[4-amino-2-oxo-1(2H)-pyrimidinyl]-2(S)-(O-benzoylmethyl)tetrahydrofuran as a colorless viscous oil. This nucleoside was deprotected with the same procedure as that used for compound **17** to afford 0.014 g (0.07 mmol, 39%) of the title compound as a white hygroscopic solid. Mass spectrum, m/z 211 (M^+). $[\alpha]_D$ +87.0° (c = 1.0 in MeOH). UV (H_2O) λ_{max} 274 nm (ε = 9040). FTIR (KBr): 1652 cm^{-1}. ^{1}H-NMR (Me_2SO-d_6): δ 1.67 (m, 1H), 2.38 (m, 1H), 3.50 (dd, 2H), 3.81 (m, 3H), 4.85 (m, 1H), 5.14 (m, 1H), 5.68 (d, 1H), 7.02 (s, 2H), 7.68 (d, 1H). ^{13}C-NMR (Me_2SO-d_6): δ 33.9, 55.0, 62.3, 71.9, 79.7, 94.0, 142.3, 155.7, 165.3. *Anal*. Calcd for $C_9H_{13}N_3O_3$: C, 51.18; H, 6.20; N, 19.89; found: C, 50.71; H, 5.79; N, 19.40.

Synthesis of IsoddC via Heterocyclic Ring Construction

4(S)-Azido-2(S)-(O-benzoylmethyl)tetrahydrofuran **(9)**. A mixture of the tosylate **6** (1.97 g, 5.23 mmol) and sodium azide (0.41 g, 6.27 mmol) in wet DMF (5 mL) was stirred at 80°C for 3 h. The reaction mixture was then cooled to room temperature and the solvent was evaporated under reduced pressure. The residue was partitioned between ethyl acetate and saturated aqueous sodium bicarbonate. The organic portion was dried (Na_2SO_4), and concentrated under reduced pressure. The residue was purified by preparative TLC using $CHCl_3$ as the eluting solvent to give the title compound (4.44 mmol, 85%) as a viscous oil. FTIR (KBr): 2955, 2253, 2105, 1719 cm^{-1}. ^{1}H-NMR ($CDCl_3$): δ 1.92 (m, 1H), 2.42 (m, 1H), 3.95 (m, 1H), 3.98 (m, 1H), 4.20 (m, 1H), 4.40 (m, 3H), 7.50 (dt, 3H), 8.08 (d, 2H).

N-[{[(2S,4S)-2-(O-Benzoylmethyl)tetrahydrofuranyl]amino}carbonyl]-3-ethoxy-2-propenamide **(11)**. A suspension of compound **9** (0.182 g, 0.736 mmol) and 0.052 g of 5% Pd/C in absolute EtOH (15 mL) was shaken under 40 psi of H_2 for 3 h. The reaction mixture was filtered and the filtrate evaporated. The residue was used for the next step without purification. To a solution of the crude amino sugar **10** (0.207 g, 1.472 mmol) in dry DMF (5 mL), 3-ethoxy-2-propenoyl isocyanate

(0.163 g, 0.736 mmol) was added dropwise at −15°C. The temperature was not allowed to exceed −10°C during the addition. After the addition, the reaction mixture was slowly warmed to room temperature and was stirred for an additional 18 h. The mixture was filtered and the filtrate was evaporated under reduced pressure. Ethanol was added to the residue and then evaporated. Purification of the residue by preparative TLC on silica gel with 5% MeOH/CHCl$_3$ afforded compound **11** in 69% yield (for two steps). ^{1}H-NMR (CDCl$_3$): δ 1.36 (t, 3 H), 1.77 (m, 1 H), 2.53 (m, 1 H), 3.86 (m, 1H), 3.98 (m, 1 H), 4.34 (m, 1H), 4.50 (m, 2H), 5.30 (d, 1H), 7.50 (dt, 3H), 7.79 (d, 1H), 8.09 (d, 2H), 9.12 (br s, 1H).

4(S)-[3,4-Dihydro-2,4-dioxo-l(2H)-pyrimidinyl]tetrahydro-2(S)-furanmethanol (13). Compound **11** (0.182 g, 0.502 mmol) was refluxed in 2 N H$_2$SO$_4$ (10 mL) for 3 h. The reaction mixture was then treated with 2 N NaOH. The solvent was evaporated under reduced pressure and the residue was purified on a silica gel plate with 9% MeOH/CHCl$_3$ to provide 0.100 g of the deprotected nucleoside **13** (0.31 mmol, 62%) as a white solid. The physical data obtained for this product were identical to the data collected for the directly coupled product.

Synthesis of IsoddT (17) (Scheme 4)
2-(S)-(O-Benzoylmethyl)-4(S)-[3,4-dihydro-2,4-dioxo-5-methyl-l(2H)-pyrimidinyl] tetrahydrofuran (16). Thymine was condensed with 2(S)-(O-benzoylmethyl)-4(R)- (p-O-toluenesulfonyl)tetrahydrofuran (**6**) using a procedure similar to that used in the preparation of intermediate **7** to give the 6′-benzoate of compound **16** in 30% yield: UV (MeOH) λ$_{max}$ 271 nm. ^{1}H-NMR (Me$_2$SO-d$_6$): δ 1.63 (s, 3H), 1.84 (m, 1H), 2.5 (m, 1H), 3.84 (dd, 1H), 3.97 (dd, 1H), 4.21 (m, 1H), 4.40 (m, 1H), 4.55 (dd, 1H), 5.13 (m, 1H), 7.41 (s, 1H), 7.52 (t, 2H), 7.67 (t, 1H), 7.95 (d, 2H), 11.27 (s, 1 H).

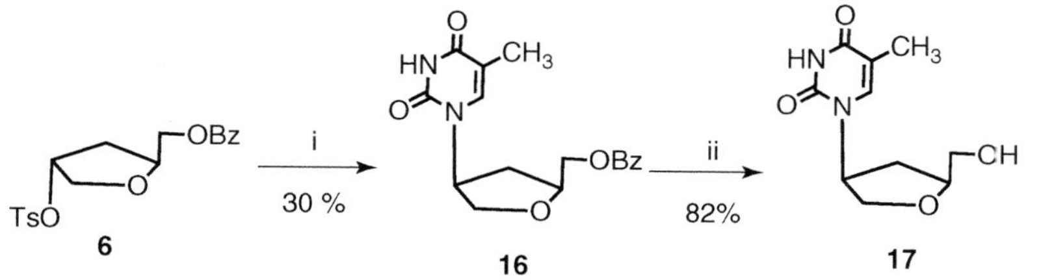

Scheme 4. *Reagents and conditions:* (i) thymine, K$_2$CO$_3$, 18-crown-6, DMF, 75°C, 9 h; (ii) NaOMe, MeOH, rt. 1 h.

4(S)-[3,4-Dihydro-2,4-dioxo-5-methyl-l(2H)-pyrimidinyl]tetrahydro-2(S)-furanmethanol (17). The protected nucleoside **16** was debenzoylated using a procedure similar to that used in the synthesis of compound **8** to afford compound **17** as a very hygroscopic white solid in 82% yield. Mass spectrum, m/z 226 [M$^+$]. [α]$_D$ +26.0° (c = 1.0 in MeOH). UV (H$_2$O) λ$_{max}$ 271 nm (ε = 9554). FTIR (KBr): 1670 cm^{-1}. ^{1}H-NMR (Me$_2$SO-d$_6$): δ 1.76 (m, 4H), 2.35 (m, 1H), 3.48 (m, 1H),

3.65 (m, 1H), 3.80 (m, 3H), 4.95 (t, 1H), 5.10 (m, 1H), 7.61 (s, 1H), 11.21 (s, 1H). ^{13}C-NMR (Me$_2$SO-d_6): δ 12.3, 35.0, 54.2, 61.8, 71.6, 79.8, 109.3, 137.8, 151.0, 163.9. *Anal.* Calcd for C$_{10}$H$_{14}$N$_2$O$_4$: C, 53.09; H, 6.24; N, 12.38; found: C, 52.85; H, 6.48; N, 12.21.

7. SYNTHESIS OF MODIFIED DINUCLEOTIDES INCORPORATING ISODDNS

The synthesis of two of the target molecules, **23** and **24**, are summarized in Scheme 5. These and other molecules of this paper were designed with a natural

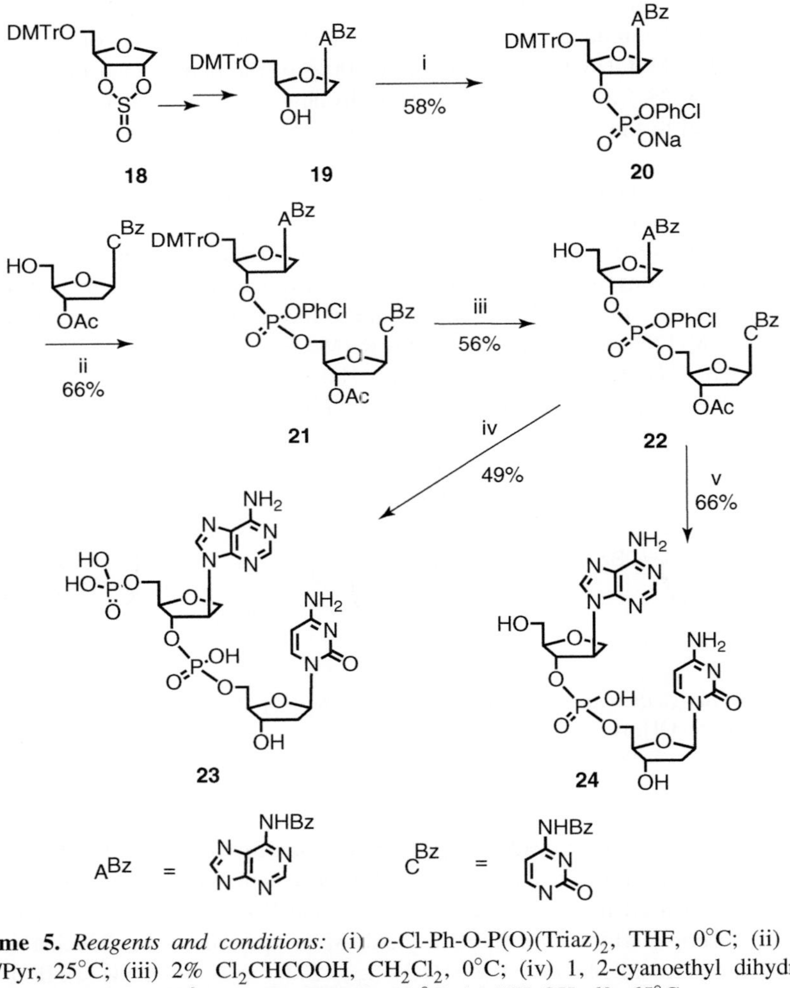

Scheme 5. *Reagents and conditions:* (i) *o*-Cl-Ph-O-P(O)(Triaz)$_2$, THF, 0°C; (ii) TPS-TAZ/Pyr, 25°C; (iii) 2% Cl$_2$CHCOOH, CH$_2$Cl$_2$, 0°C; (iv) 1, 2-cyanoethyl dihydrogen phosphate, DCC/Pyr, 20°C; 2, NH$_4$OH, 60–65 °C; (v) NH$_4$OH, 60–65°C.

D-deoxynucleoside (3'-OH terminus) bonded through a 5' → 3' internucleotide phosphate linkage to an isomeric L-related deoxynucleoside.[21,22,37-39] Dinucleotide 23 and its precursor 24 were synthesized in several steps from protected (S,S')-isodeoxyadenosine (IsodA), a structural analogue of natural 2'-deoxyadenosine. The protected IsodA, although previously synthesized by us,[22] was prepared for these studies by a more convenient route using the cyclic sulfite 18. Synthesis of the dinucleotides utilized the solution-phase phosphotriester approach.[40,41] The solid-phase synthesis using a DNA synthesizer was useful only for submilligram quantities of target molecules. The solution-phase approach was preferred because this allowed the synthesis of these compounds in sufficient quantities (>50 mg) to allow for complete structural studies as well as reproducible enzyme assays. Thus, isodeoxynucleoside 19 was phosphorylated with the bifunctional phosphorylating agent 2-chlorophenylphosphoro-bistriazolide[42,43] to give 20 in 58% yield after purification (Scheme 5).[38]

The internucleotide coupling was carried out with triisopropylbenzenesulfonyl tetrazolide (TPS-TAZ)[44-46] to give the protected dinucleotide 21 in 66% yield. The formation of 21 was confirmed by its NMR and HRMS data. Deprotection with 2% dichloroacetic acid gave 22 (56% yield), and this key intermediate was fully characterized by ^{1}H- and ^{31}P-NMR and HRMS data. Further deprotection of 22 with NH$_4$OH gave target molecule 24 (66%), after purification by reversed-phase HPLC (C$_{18}$, ethanol/water).[38] The product was initially identified by comparison of HPLC ion-exchange retention times with those of natural dinucleotides (Partisil-10 SAX ion-exchange column, phosphate buffer system, retention time = 70 min). The complete structure was established by multinuclear NMR data (^{1}H-, ^{13}C-, ^{31}P-NMR spectra and COSY, HMQC, and HMBC data), quantitative UV spectra, CD spectra, and ESI HRMS data (calculated for 24: 539.1404 [M − H]$^-$, found: 539.1402). The quantitative UV spectral data (λ_{max} = 263 nm, ε = 19,400) gave evidence of hypochromicity in these molecules. These and the CD data suggested the existence of base-stacking interactions. Target molecule 23 was synthesized by phosphorylation of intermediate 22 with 2-cyanoethylphosphate[47,48] in the presence of DCC followed by deprotection with NH$_4$OH.[38] This compound was characterized in the same way as described above for 24.

Experimental Procedures for the Synthesis of Dinucleotides Incorporating IsoddNs[38]

N-*Benzoyl-5'-O-dimethoxytrityl-isodeoxyadenosine-3'-(o-chlorophenyl) phosphate (20).*
Compound 19 (1.784 mmol, 1.176 mg) was kept overnight under vacuum over P$_2$O$_5$ to remove traces of water. 2-Chlorophenyldichlorophosphate [2-Cl-Ph-O-P(O)Cl$_2$] (2 mmol, 0.32 mL) and 1,2,4-triazole (4.0 mmol, 276 mg)[42,43,49,50] were dissolved under nitrogen in anhydrous THF (15–20 mL) and cooled on ice. TEA (4 mmol, 545 mL) was added dropwise, resulting in a copious precipitate. The reaction mixture was stirred for 1–2 h and then filtered into the flask containing 19. It was kept for 2 h on ice and then quenched with water (2 mL). After 0.5 h, ethyl acetate (200 mL) was added and the organic layer

was washed with saturated aqueous $NaHCO_3$ (10 mL) and brine (10 mL). The solvent was evaporated and pyridine was removed with toluene. The residue was purified by column chromatography on silica gel (CH_2Cl_2/10% MeOH) to give **20** (227 mg, 58%) as a white foam (R_f = 0.1, CH_2Cl_2/10% MeOH). ^{1}H-NMR ($CDCl_3$): δ 8.64 (br s, 1H, NHBz), 8.33 (s, 1H, H-8), 7.96 (s, 1H, H-2), 7.95 (s), 7.58–7.12 (m) and 6.75–6.70 (d) (22H, Arom-H), 5.44 (br s, 1H, 2′), 5.25 (br s, 1H, 3′), 4.27 (br s, 2H, 1′), 3.70 (s, 7H, CH_3O, 4′), 3.46 (br s, 2H, 5′). ^{13}C-NMR ($CDCl_3$): δ 167.2 (CO), 153.1 (C-6), 152.2 (C-2), 151.1 (C-4), 142.0 (C-8), 145.5, 138.4, 136.4, 135.8, 135.4, 134.1, 133.1, 130.1, 128.9, 127.9, 127.6 (Arom), 128.2 (C-5), 115.0, 114.8 (DMTr), 86.0 (Ph_3C), 84.9 (4′), 75.3 (2′), 70.1 (1′), 62.8 (5′), 62.4 (3′), 55.0 (OCH_3). FAB HRMS: [M + H]$^+$ calcd for $C_{44}H_{40}ClN_5NaO_9P$, 848.2243; found, 848.2250.

***N*-Benzoyl-1-deoxy-2-isoadenosine-3′-[(2-chlorophenyl)(*N*-benzoyl-3′-O-acetyl-cytidine-5′-yl)]phosphate (22).**[38] A mixture of **20** (0.770 mmol, 670 mg) and *N*-benzoyl-3′-*O*-acetyl-cytidine (0.770 mmol, 270 mg) was evaporated repeatedly with anhydrous pyridine and then dissolved in pyridine (3 mL) under a nitrogen atmosphere. 2,4,6-Triisopropylbenzensulfonyltetrazolide (TPS-TAZ) (1.54 mmol, 555 mg) was added. After 1 h at room temperature, the reaction mixture was quenched with water (0.5 mL), stirred for 1.5 h, and concentrated. The residue was dissolved in ethyl acetate (100 mL) and washed with saturated aqueous $NaHCO_3$ (2 × 20 mL). Solvents were removed by evaporation and coevaporation with toluene, and the residue was purified by silica gel column chromatography (CH_2Cl_2/5% MeOH) to give 600 mg (66%) of *N*-benzoyl-5′-*O*-dimethoxytrityl-1-deoxy-2-isoadenosine-3′-[(2-chlorophenyl)(*N*-benzoyl-3′-*O*-acetylcytidine-5′-yl)]phosphate **21** as a white foam (R_f = 0.35, CH_2Cl_2/5% MeOH). ^{1}H-NMR($CDCl_3$): δ 8.67 (s, 1H, A H-8), 8.22 (s, 1H, A H-2), 8.05–8.03 (d, 2H, C H-6, Bz), 7.85–7.83 (d, 1H, C H-5), 7.83–7.80 (m, 3H, Arom), 7.55–7.10 (m, 14H, Arom), 6.78–6.70 (m, 5H, DMTr), 6.20 (q, 1H, C 1′), 5.62–5.50 (m, 1H, A 2′), 5.48–5.38 (m, 1H, C 3′), 5.27–5.22 (d, 1H, A 3′), 4.52–4.35 (m, 2H, A 1′), 4.30 (m, 1H, C 4′), 4.25–4.13 (m, 2H, C H 5′), 3.70 (s, 6H, CH_3O), 3.53–3.35 (m, 2H, A 5′). The protected dinucleotide **21** (500 mmol, 600 mg) was dissolved in 3 mL of anhydrous methylene chloride and cooled on an ice bath. Cold 2% dichloroacetic acid (DCA) in methylene chloride (10–15 mL) was added. The crimson red reaction mixture was kept on the ice bath for 3 min and then poured onto saturated aqueous $NaHCO_3$ (20 mL), extracted with methylene chloride (2 × 50 mL), and the solvent evaporated. The residue was purified by column chromatography on silica gel (CH_2Cl_2/7% MeOH) to give **22** (225 mg, 56%) as a white foam (R_f = 0.35, CH_2Cl_2/7% MeOH). ^{1}H-NMR (D_2O): δ 8.31 (brs, 1H, A H-8), 8.11 (s, 1H, A H-2), 7.43 (d, J = 7.8 Hz, 1 H, C H-6), 5.97 (t, J = 6.0 Hz, 1H, C 1′), 5.75 (brs, 1H, C H-5), 5.24 (q, J = 5.4 Hz, 1H, A 2′), 4.76–4.75 (m, 1H, A 3′), 4.40–4.37 (m, 1H, A 1′), 4.36–04.34 (m, 1H, C 3′), 4.31–4.28 (m, 1H, A 1′), 4.22 (s, 1H, A 4′), 4.14 (dt, 1H, A 5′a), 4.10 (dt, 1H, A $5'_b$), 3.95 (s, 1H, C 4′), 3.35 (s, 1H, C $5'_a$), 2.95 (s, 1H, C $5'_b$), 2.24–2.20 (m, 1H, C $2'_a$), 2.01–1.97 (m, 1H, C $2'_b$). ^{13}C-NMR (D_2O): δ 166.6 (C C-4), 157.6 (C

C-2), 157.6 (A C-6), 154.5 (A C-2), 154.0 (A C-4, not observable), 143.0 (C C-6), 142.8 (A C-8), 120.0 (A C-5), 98.1 (C C-5), 87.5 (A 4'), 87.1 (C 1'), 87.0 (C 4'), 82.3 (A 3'), 73.3 (A 1'), 71.8 (C 3'), 66.8 (C 5'), 66.1 (A 2'), 63.4 (A 5'), 42.0 (C 2'). ^{31}P-NMR (D$_2$O) −11.13 (s), −12.22 (s), ratio 1:1. FAB HRMS: [M + H]$^+$ calcd for C$_{41}$H$_{39}$ClN$_8$O$_{12}$P, 901.2105; found, 901.2121.

2'-Deoxycytidylyl(5' → 3')-1'-deoxy-2'-isoadenosine 5'-phosphate (23).[38] Compound **22** (0.16 mmol, 150 mg) and β-cyanoethylphosphate (0.8 mmol, 0.6 mL),[38,47,51] prepared from β-cyanethylphosphate Ba-salt, were evaporated with anhydrous pyridine three times. Dicyclohexylcarbodiimide (DCC) (110 mmol, 230 mg) was added, together with 4 mL of pyridine. A copious white precipitate of dicyclohexylurea formed within 10 min. After 48 h, the reaction mixture was filtered, the precipitate was washed with pyridine, and the combined filtrate was evaporated and coevaporated with water to remove pyridine. Cold NH$_4$OH (29%, d 0.88, 40 mL) was added and the suspension was heated in a Hastelloy bomb reactor at 60–65°C for 6 h. The reactor was cooled and opened and the reaction mixture was filtered and the filtrate volume was reduced by one-fifth to remove excess NH$_3$. The pH was adjusted to 10 by the addition of 29% NH$_4$OH (a drop). The reaction mixture was extracted with diethyl ether (2 × 20 mL) and the aqueous layer was evaporated to a small volume, filtered and the filtrate, treated with LiOH (to saturation). Anhydrous ethanol (10 volumes) was added to the resulting slurry, which was centrifuged at 10,000 rpm, 0°C for 15 min. The supernatant was discarded and the precipitate was stirred with Dowex 50 H$^+$ resin (5 mL) and filtered through the same resin (20 mL). The UV-absorbing fractions (60 mL) were concentrated and filtered. The filtrate was purified by reversed-phase HPLC (flow rate 5 mL/min, H$_2$O) to give **23** (49 mg, 49%) (retention time 35 min, single peak) as a fluffy white solid after lyophilization. Ion-exchange chromatography on a Partisil 10 SAX column, buffers: A: 2 mM KH$_2$PO$_4$, 8 mM KCl, 0.05 M MgCl$_2$, 0.05% acetonitrile, pH 3.0; B: 0.2 M KH$_2$PO$_4$, 0.9 M KCl, 0.8 M MgCl$_2$, 5% acetonitrile, pH 3.0; retention time 70 min.[38,1] H-NMR (D$_2$O): δ 8.31 (br s, 1H, A H-8), 8.11 (s, 1H, A H-2), 7.43 (d, J = 7.8 Hz, 1 H, C H-6), 5.97 (t, J = 6.0 Hz, 1 H, C 1'), 5.75 (br s, 1H, C H-5), 5.24 (q, J = 5.4 Hz, 1H, A 2'), 4.76–4.75 (m, 1H, A 3'), 4.40–4.37 (m, 1H, A 1'), 4.36–4.34 (m, 1H, C 3'), 4.31–4.28 (m, 1H, A 1'), 4.22 (s, 1H, A 4'), 4.14 (dt, 1H, A 5'$_a$), 4.10 (dt, 1H, A 5'$_b$), 3.95 (s, 1H, C 4'), 3.35 (s, 1H, C 5'$_a$), 2.95 (s, 1H, C 5'$_b$), 2.24–2.20 (m, 1H, C 2'$_a$), 2.01–1.97 (m, 1H, C 2'$_b$). ^{13}C-NMR (D$_2$O): δ 166.6 (C C-4), 157.6 (C C-2), 157.6 (A C-6), 154.5 (A C-2), 154.0 (A C-4, not observable), 143.0 (C C-6), 142.8 (A C-8), 120.0 (A C-5), 98.1 (C C-5), 87.5 (A 4'), 87.1 (C 1'), 87.0 (C 4'), 82.3 (A 3'), 73.3 (A 1'), 71.8 (C 3'), 66.8 (C 5'), 66.1 (A 2'), 63.4 (A 5'), 42.0 (C 2'). ^{31}P-NMR (D$_2$O): δ 8.8 (s, phosphomonoester), −0.9 (s, phosphodiester), ratio 1:1. λ$_{max}$ 263 nm (ε = 19,500). FAB HRMS: [(M − H]$^-$ calcd for C$_{19}$H$_{25}$N$_8$O$_{12}$P$_2$, 619.1067; found, 619.1056; [M − 2H + Na]$^-$ calcd for C$_{19}$H$_{23}$N$_8$NaO$_{18}$P$_2$, 641.0887; found, 641.0906.

2'-Deoxycytidylyl(5' → 3')-1'-deoxy-2'-isoadenosine (24). This compound (21 mg, 66%) was prepared using **22** (0.06 mmol, 54 mg) and conc. NH$_4$OH (5 mL) in a procedure similar to that described for **23**. Reversed-phase HPLC purification of **24** was carried out at a flow rate 5 mL/min, using water (solvent A) and ethanol (solvent B) in the following linear binary gradient: 100% A (100 min), 0 to 60% B (160 min); retention time 150 min, single peak. Ion-exchange HPLC on Partisil SAX column: retention time 22 min. ^{1}H-NMR (D$_2$O): δ 8.25 (s, 1H, A H-8), 8.10 (s, 1H, A H-2), 7.40–7.39 (d, $J = 7.2$ Hz, 1H, C H-6), 5.94–5.92 (t, $J = 6.3$ Hz, 1H, C 1′), 5.66–5.65 (d, $J = 7.2$ Hz, 1H, C H-5), 5.26–5.24 (qn, $J = 3.6$ Hz, 1H, A 2′), 4.84–4.82 (m, 1H, A 3′), 4.39–4.36 (m, 1H, A 1′), 4.29–4.26 (q, $J = 5.6$ Hz, C 3′), 4.24–4.22 (dd, $J_1 = 6.6$ Hz, $J_2 = 4.2$ Hz, A 1′), 4.12–4.09 (m, 1H, A 4′), 4.04–4.02 (m, 2H, A $5'_a$ and C $5'_a$) and 3.97–3.924 (m, 2H, A $5'_b$ and C $5'_b$), 3.89–3.88 (m, 1H, C 4′), 2.25–2.21 (dt, $J_1 = 13.8$ Hz, $J_2 = 6.0$ Hz, 1H, C $2'_a$) and 1.98–1.93 (dt, $J_1 = 13.8$ Hz, $J_2 = 6.0$ Hz, 1H, C $2'_b$). ^{13}C-NMR (D$_2$O): δ 164,2 (C C-4), 157.5 (C C-2), 157.4 (A C-6), 154.6 (A C-2), 154.0 (A C-4), 143.5 (A C-6), 143.4 and 143.0 (A C-8), 120.8 (A C-5), 97.5 (C C-5), 87.70 and 87.48 (A 4′), 87.5 (C 1′), 87.14 and 87.02 (C 4′), 82.02 and 81.95 (A 3′), 72.99 (A 1′), 71.74 (C 3′), 66.71 and 66.64 (C 5′), 64.05 and 63.97 (A 2′), 62.68 (A 5′), 42.16 (C 2′). ^{31}P-NMR (D$_2$O): δ −0.85 (s). λ$_{max}$ 263 nm (ε = 19,400). FAB HRMS: [M − H]$^-$ calcd for C$_{19}$H$_{24}$N$_8$O$_9$P, 539.1404; found, 539.1402.

8. NEW PHOSPHORYLATION METHODOLOGY IN THE SYNTHESIS OF ISODINUCLEOTIDE 28[52]

Three other dinucleotides were synthesized by multistep procedures for these studies. Compound **26** (Scheme 6) was prepared using methodologies similar to the synthesis of **24** above. However, 5′-phosphorylation of **25a** with 2-cyanoethylphosphate as well as with other reagents (e.g., phosphoroimidazolides, phosphorotriazolides, phosphoramidites) was inefficient. We therefore developed a new reagent, **25c**, a derivative of 2,2′-sulfonyldiethanol, for this transformation. Treatment of **25a** with **25c** in pyridine in the presence of TPS-TAZ at room temperature for 0.5 h followed by deprotection of the trityl group with dichloroacetic acid (3 min) gave **27** in 70% overall yield (for two steps). The yield of the phosphorylation step was consistently between 80 and 90%. *Reagent **25c** is an excellent new phosphorylating agent!* Compound **27** was isolated, purified, and completely characterized. Deprotection with NH$_4$OH gave target molecule **28**.

Experimental Procedures Using New Phosphorylation Methodology[52]

N-Benzoyl-1-deoxy-2-isoadenosine-3′-[(2-chlorophenyl)(3′-O-acetylthymidine-5′-yl)]phosphate (25a). 5′-*O*-Dimethoxytritylderivative of **25a** (423 mg, 50%) was prepared using **20** (770 mmol, 670 mg), 3′-*O*-acetylthymidine (770 mmol, 210 mg), and TPS-TAZ (1.54 mmol, 555 mg) in a procedure similar to that described for **21**. Detritylation with methylene chloride/2% DCA furnished **25a** (261 mg, 85% yield) as a mixture of diastereomers. It was purified by silica column

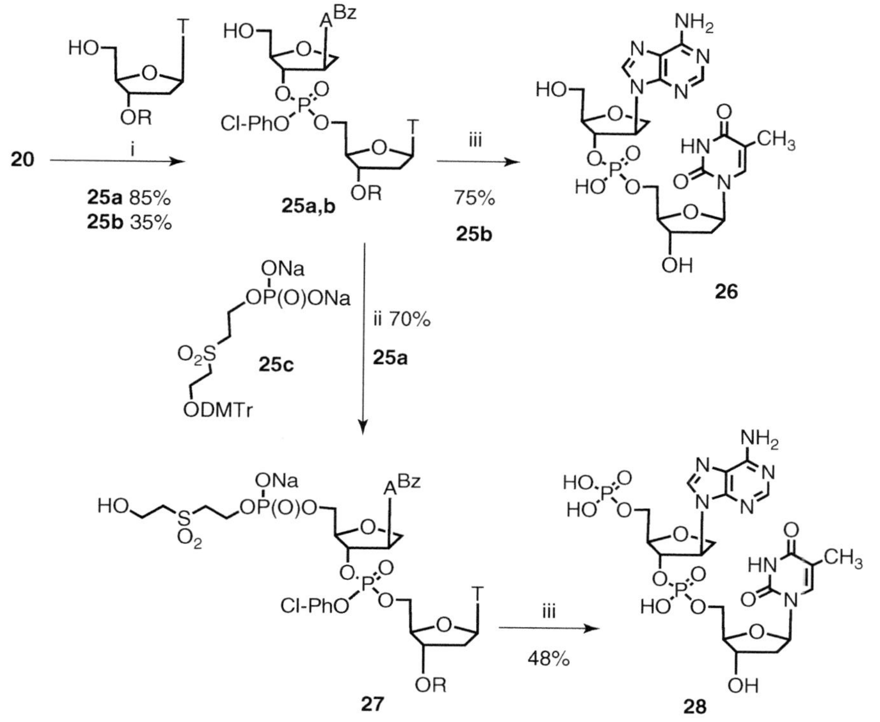

Scheme 6. *Reagents and conditions:* (a.) R = Ac; (b) R = Bz. (i) (1) TPS-TAZ/Pyr, 25°C; (2) 2% $Cl_2CHCOOH/CH_2Cl_2$, 0°C; (ii) (1) **25c**, TPS-TAZ/Pyr, 25°C; 2. Pyr/H_2O, 20–40°C, (workup); (3) 2% $Cl_2CHCOOH/CH_2Cl_2$, 0°C; (iii) NH_4OH, 60–65 °C.

chromatography (CH_2Cl_2/7% MeOH). The slower-moving diastereomer of **25a** [125 mg, 40%, R_f = 0.40 (CH_2Cl_2/MeOH)]. ^{1}H-NMR ($CDCl_3$): δ 8.37 (s, 1H, A H-2), 8.10 (d, 2H, Bz), 7.57–7.10 (m, 8H, Arom), 7.51 (s, 1H, T H-6), 6.19 (dd, J_1 = 5.8 Hz, J_2 = 2.9 Hz, T H-1'), 5.42–5.39 (m, 2H, A 2' and T 3'), 5.12–5.10 (m, 1H, A 3'), 4.70–4.64 (m, 1H, A $1'_a$), 4.59–4.52 (m, 1H, A $5'_b$), 4.41–4.37 (m, 1H, A 4'), 4.20–4.11 (m, 2H, A 5'), 4.06–4.01 (m, 1H, T 4'), 3.93–3.83 (m, 2H, T 5'), 2.24 (dd, J_1 = 9 Hz, J_2 = 5.7 Hz, T $2'_a$), 2.01–1.96 (m, 1H, T $2'_t$), 2.09 (s, 3H, Ac), 1.64 (s, 1H, T CH_3). Impurities: 0.5–1% TEA. ^{31}P-NMR ($CDCl_3$, 85% H_3PO_4 for external reference): δ −0.50. The faster-migrating diastereomer of **25a** [136 mg, 45% yield, R_f = 0.41 (CH_2Cl_2/MeOH)]. ^{1}H-NMR ($CDCl_3$): δ 10.08 (br s, 1H, NH), 8.78 (s, 1H, A H-8), 8.46 (s, 1H, A H-2), 8.16 (d, 2H Bz), 7.56 (s, 1H, T H-6), 7.54–7.17 (m, 8H, Arom), 5.89 (dd, J_1 = 5.8 Hz, J_2 = 2.9 Hz, T 1'), 5.46–5.44 (m, 1H, A 2'), 5.25–5.21 (m, 1H, T 3'), 5.09–5.07 (m, 1H, A 3'), 4.53–4.44 (m, 2H, A 1'), 4.22–4.21 (m, 1H, A 4'), 4.30–4.29 (m, 2H, A 5'), 3.95–3.88 and 3.81–3.76 (m, 3H, T 4' and 5'), 2.23–2.16 (m, 1H, T H-$2'_a$), 2.02–1.96 (m, 1H, T H-$2'_b$), 2.10 (s, 3H, Ac), 1.73 (s, 3H, T CH_3). ^{31}P-NMR

(CDCl$_3$): δ −0.72. FAB HRMS: [M + H]$^+$ calcd for C$_{35}$H$_{36}$ClN$_7$O$_{12}$P, 812.1840; found, 812.1845.

N-Benzoyl-1-deoxy-2-isoadenosine-3'-[(2-chlorophenyl)(3'-O-benzoylthymidine-5'-yl)]phosphate (25b). 5'-O-Dimethoxytritylderivative of **25b** (118 mg, 100%) was prepared using **20** (0.12 mmol, 109 mg), 3'-O-benzoylthymidine (0.1 mmol, 35 mg) and TPS-TAZ (0.3 mmol, 100 mg) in a procedure similar to that described for **21**. Detritylation (CH$_2$Cl$_2$/2% DCA, 15 mL) and purification by column chromatography on silica gel (CH$_2$Cl$_2$/5%MeOH) gave **25b** (30 mg, 35%) (mixture of diastereomers, ratio 1:1) as a white foam, $R_f = 0.35$ (CH$_2$Cl$_2$/5%MeOH). ^{1}H-NMR (CDCl$_3$): δ 11.06 (br s, 1H, NHBz), 10.17 (br s, 1H, T NH), 8.76 and 8.74 (s and s, 1H, A H-8), 8.39 and 8.37 (s and s, 1H, A H-2), 8.04–7.99 (m, 4H, Bz), 7.62 and 7.59 (s and s, 1H, T-6), 7.69–7.14 (m, 10H, Arom), 6.54–6.37 (m, 1H, T 1'), 5.88 (br s, 1H, 5'-OH), 5.58 (br s, 2H, A 2' and T 3'), 4.66–4.59 and 3.99–3.91 (9H, unresolved sugar protons signal), 2.66–2.36 (m, 2H, T 2'), 1.84 and 1.82 (d and d, 3H, CH$_3$). ^{31}P-NMR (CDCl$_3$): δ −8.95 (s), −9.93 (s). HRMS (FAB): [M + H]$^+$ calcd for C$_{40}$H$_{38}$ClN$_7$O$_{12}$P, 874.1996; found, 874.2021.

Thymidylyl(5' → 3')-1'-deoxy-2'-isoadenosine-5'-phosphate (28). This compound was prepared from **25a** in four steps. Sulfonyldiethanol (20 mmol, 3.08 g) was evaporated with anhydrous pyridine several times and dissolved in 10 mL of pyridine under nitrogen. 4,4'-Dimethoxytrityl chloride (10 mmol, 3.39 g) was added and the reaction was left overnight at room temperature. The reaction mixture was partitioned between cold methylene chloride (300 mL) and saturated aqueous NaHCO$_3$ (100 mL), and the organic phase was concentrated. The residue was purified by column chromatography on silica gel (CH$_2$Cl$_2$/3% MeOH) to give 2-O-(4,4'-dimethoxytrityl) sulfonyldiethanol (4.0 g, 87%) as a colorless oil. ^{1}H-NMR (CDCl$_3$) δ 7.43–7.24 (9H, Arom), 6.35–6.30 (m, 4H, Arom), 4.12 (m, 2H, DMTrOCH$_2$), 3.81 (s, 6H, CH$_3$O), 3.68 (t, $J = 6$ Hz, 2H, CH$_2$OH), 3.40 (t, $J = 6$ Hz, 2H, DMTrOCH$_2$CH$_2$), 3.20 (t, $J = 6$ Hz, 2H, CH$_2$CH$_2$OH). FAB HRMS: [M + Na]$^+$ calcd for C$_{25}$H$_{28}$O$_6$NaS, 874.1996; found, 874.2021.

The compound from the preceding step (5 mmol, 2.28 g) was evaporated with anhydrous pyridine several times and dissolved in 5 mL of pyridine. 1,2,4-Triazole (40 mmol, 2.76 g) and TEA (40 mmol, 545 mL) were dissolved in 100 mL of anhydrous THF under nitrogen and cooled down. POCl$_3$ (10 mmol, 0.92 mL) was added dropwise at −10°C. After stirring for 10 min, the reaction mixture was filtered under nitrogen into the solution prepared above. After 1 h at 0°C, the reaction mixture was partitioned between cold ethyl acetate (300 mL) and NaHCO$_3$ (saturated, 100 mL), washed with brine (1 × 100 mL), and concentrated. The residue was purified by column chromatography on silica gel (CH$_2$Cl$_2$/12%MeOH) to give 2-O-(4,4'-dimethoxytrityl)sulfonyldiethanol disodium phosphate as a colorless glassy material (3.0 g, 100%). ^{1}H-NMR (CDCl$_3$): δ 7.45–7.25 (m, 9H, Arom), 6.35–6.30 (m, 4H, Arom), 4.20–4.10 (m, 2H, DMTrOCH$_2$), 3.65–3.45 (m, 2H, CH$_2$OH), 3.20–3.15 (m, 2H, DMTrOCH$_2$CH$_2$), 3.10–3.05 (m, 2H, CH$_2$CH$_2$OH). ^{31}P-NMR (CDCl$_3$): δ −4.90 (br s).

{3'-[(3''-O-Acetylthymidine-5''-yl)-(2-chlorophenyl)phosphoro]-1'-deoxy-2'-isoadenosine}-5'-(2-hydroxyethylsulfonylethan-2'-yl)phosphate (27). The compound from the preceding step (0.5 mmol, 300 mg) and 25a (0.1 mmol, 80 mg) were evaporated several times with anhydrous pyridine and dissolved in 4 mL of pyridine under nitrogen. TPS-TAZ (1 mmol, 340 mg) was added and the reaction mixture was kept for 0.5 h at room temperature. A workup of the reaction mixture and the following steps were carried out as for 22 to give 27 as a colorless glassy material (0.85 mg, 70% from 25a). Compound 27 (0.7 mmol, 85 mg) was fully deprotected as described for 22 and worked up as described for 23. Reversed-phase HPLC purification (flow rate 5 mL/min, H_2O) gave 28 (retention time 30 min, major peak) as a white fluffy material after lyophilization (21 mg, 48%). ^{1}H-NMR (D_2O): δ 8.34 (s, 1H, A H-8), 8.10 (s, 1H, A H-2), 7.24 (s, 1H, T H-6), 6.01 (t, J = 6.1 Hz, 1H, T 1'), 5.27 (qn, J = 3.6 Hz, A 2'), 4.98 (qn, J = 3.6 Hz, A 3'), 4.41–4.32, 4.25–4.09. and 4.02–3.95 (m, m, m, 9H, unresolved sugar signals), 2.21–2.14 (m, 1H, T $2'_a$), 1.97–1.89 (m, 1H, T $2'_b$), 1.67 (s, 1H, CH_3-T). ^{31}P-NMR (D_2O): δ −0.19 (br s, phosphomonoester), −1.94 (s, phosphodiester), ratio 1:1. λ_{max} 263 nm (ε = 18,000). HRMS (ESI): [M − H]$^-$ calcd for $C_{20}H_{25}N_7O_{10}P$, 635.1051; found, 635.1078.

Thymidylyl(5' → 3')-1'-deoxy-2'-isoadenosine (26). Compound 26 (15 mg, 75%) was synthesized using 25b (0.035 mmol, 30 mg) and 29% NH_4OH (10 mL) and purified by reversed-phase HPLC as described for 23 (retention time 120 min, single peak). ^{1}H-NMR (D_2O): δ 8.30 (s, 1H, A H-8), 8.13 (s, 1H, A H-2), 7.21 (s, 1H, T H-6), 5.98 (t, J = 7.5 Hz, 1H, T 1'), 5.30–5.26 (m, 1H, A 2'), 4.25–4.19 and 4.10–3.92 (m, m, m, 8H, unresolved sugar proton signals), 2.21–2.15 (m, 1H, T $2'_a$), 2.08–1.99 (m, 1H, T $2'_b$), 1.67 (s, 3H, T CH_3). ^{31}P-NMR (D_2O): δ 1.93 (s). λ_{max} 263 nm (ε = 17,990). HRMS (ESI): [M − H]$^-$ calcd for $C_{20}H_{26}N_7O_{13}P_2$, 554.0995; found, 554.0980.

9. SYNTHESIS OF DINUCLEOTIDES (29,30) BEARING TWO ISODIDEOXYNUCLEOSIDE COMPONENTS

The preparation of dinucleotides utilized protected isodeoxy nucleosides as starting materials (Figure 3). Thus, 6-N-benzoylisodeoxyadenosine 29[39] was 5'-protected using dimethoxytrityl chloride in pyridine and benzoylated with benzoyl chloride to give intermediate 32 (Scheme 7). Addition of concentrated ammonium hydroxide selectively removed one 6-N-benzoyl group in the tribenzoyl intermediate 32, but the benzoate group at the 3'-position was not affected. Detritylation of the resulting compound 33 gave intermediate 34 (73% yield from 31).

Isodeoxycytidine 38 (Scheme 8), another key intermediate, cannot be synthesized from the direct reaction of the sulfate 18 and cytosine, because O-alkylation, and not N-alkylation, of cytosine is the predominant reaction. However, it was synthesized from isodeoxyuridine 35 as described previously.[35,53] Thus, the dibenzoyl derivative 36 of isodeoxyuridine was converted to its corresponding 4-O-triisopropylbenzenesulfonyl derivative 37. Ammonolysis of this intermediate and

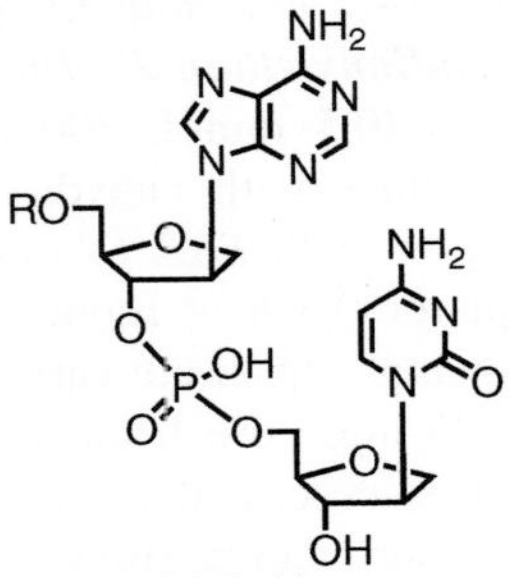

29 R=H
30 R=(HO)$_2$P(O)

Figure 3. Isodeoxydinucleotides with two IsodN components.

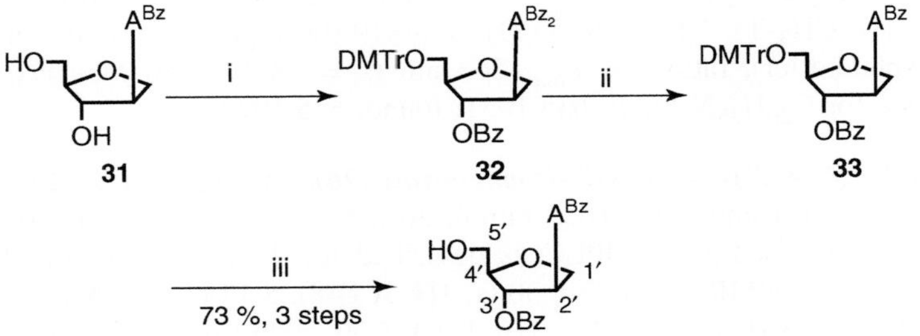

31 **32** **33** **34**

Scheme 7. *Reagents and conditions:* (i) (1) DMTrCl/Pyr; (2) BzCl; (ii) 2 M NH_4OH; (iii) H^+.

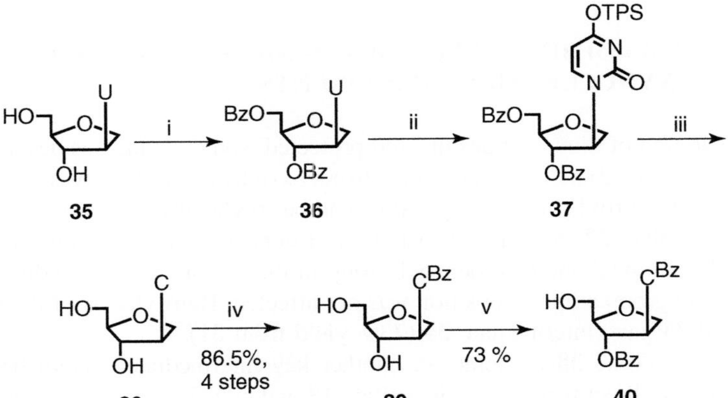

35 **36** **37** **38** **39** **40**

Scheme 8. *Reagents and conditions:* (i) BzCl/Pyr; (ii) TPSCl/Pyr; (iii) (1) NH_4OH; (2) NH_3/MeOH; (iv) 1) TMSCl/Pyr; (2) BzCl; (3) NH_4OH; (v) (1) DMTrCl/Pyr; (2) BzCl; (3) H^+.

subsequent deprotection of the benzoyl group with methanolic ammonia afforded isodeoxycytidine **38**, which was fully characterized through its 4-*N*-benzoyl derivative **39** (76% yield from **35**). Compound **39** can be tailored for coupling by its conversion to **40** in three steps (5'-*O*-tritylation, benzoylation, and detritylation, 73% yield).

The dinucleotides were synthesized by the phosphoramidite method[54] (Scheme 9). Thus, for example, the free 5'-hydroxyl group of isonucleoside **40** was condensed with the reagent, 2-cyanoethyl tetraisopropylphosphorodiamidite, in the presence of 1*H*-tetrazole to give the intermediate **41**, which was directly coupled with nucleoside **42**.[39] Subsequent oxidation with iodine and detritylation provided the phosphotriester **44** (62% yield from **40**). 5'-Phosphorylation of **44**

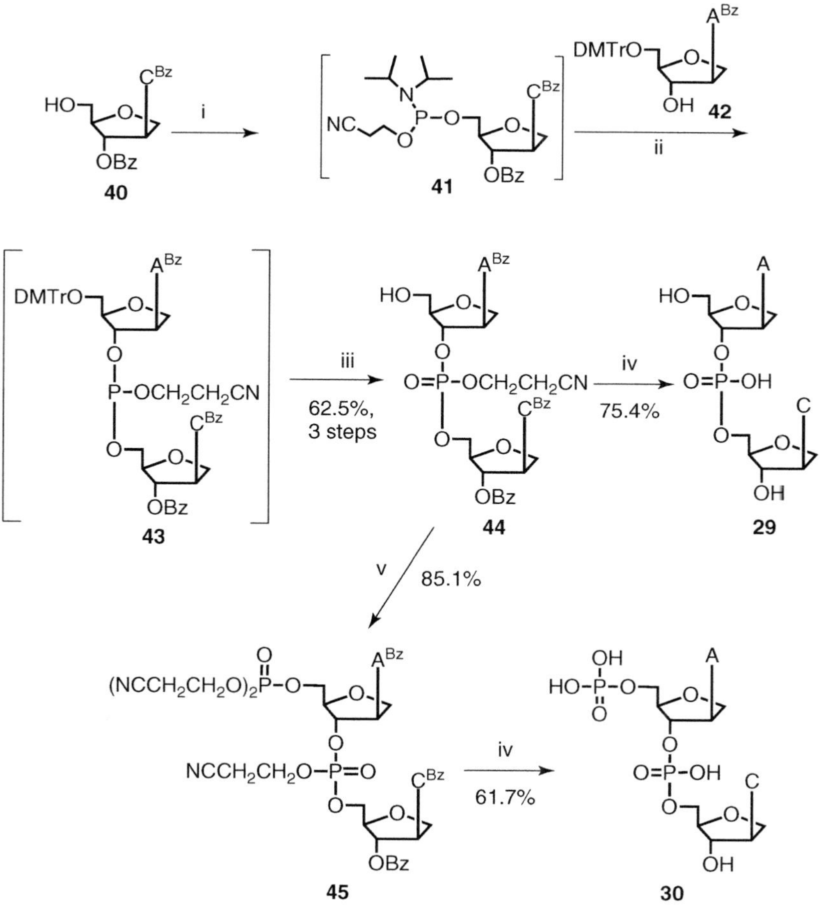

Scheme 9. *Reagents and conditions:* (i) (*i*-Pr$_2$N)$_2$POCH$_2$CH$_2$CN, 1*H*-tetrazole, CH$_2$Cl$_2$, rt; (ii) 1*H*-tetrazole; (iii) (1) I$_2$; (2) H$^+$; (iv) NH$_4$OH; (v) (1) *i*-Pr$_2$NP(OCH$_2$CH$_2$CN)$_2$, 1*H*-tetrazole, CH$_2$Cl$_2$, rt; (2) I$_2$.

was performed using bis(2-cyanoethyl) N,N-diisopropyl-phosphoramidite and 1H-tetrazole. Compound **45** was obtained in 85% yield after oxidation with iodine. The protected dinucleotides, **44** and **45** (Scheme 9), were deprotected using concentrated ammonium hydroxide at room temperature for 24 h. The benzoyl group and the 2-cyanoethyl group were removed simultaneously. Purification was performed by HPLC using a C_{18} column with elution involving MeOH and 10 mM aqueous AcOH. The residual acetic acid was removed completely by coevaporation several times with water. Lyophilization produced the target compounds as white spongy solids. The yields of the target compounds **29** and **30** (Figure 3) were in the range of 58–89% for the deprotection step. The compounds were characterized by their quantitative UV spectral data, ^{1}H-, ^{13}C-, and ^{31}P-NMR spectra and HRMS data.

Experimental Procedures for the Synthesis of Dinucleotides with Two IsodNs

1′-Deoxy-2′-isocytidine (38). This compound was prepared from the dibenzoyl derivative of its uracil counterpart 1′-deoxy-2′-isouridine (**35**) using methods developed in our laboratory.[22,39,55]

4-N-Benzoyl-1′-deoxy-2′-isocytidine (39). This compound was synthesized in 86.5% yield over four steps from 1′-deoxy-2′-isocytidine as we described previously.[35] ^{1}H-NMR (DMSO-d_6) δ 11.21 (s, 1H), 8.16 (d, 1H, J = 7.0 Hz), 7.99 (d, 2H, J = 7.5 Hz), 7.49–7.63 (m, 3H), 7.32 (d, 1H, J = 7.0 Hz), 5.71 (d, 1H, J = 5.5 Hz), 4.90 (m, 2H), 4.17 (m, 1H), 4.07 (dd, 1H, J = 10.0 and 6.5 Hz), 3.96 (dd, 1H, J = 10.0 and 3.5 Hz), 3.59–3.65 (m, 2H), 3.53 (m, 1H). ^{13}C-NMR (DMSO-d_6): δ 167.7, 163.0, 155.6, 147.7, 133.6, 133.2, 128.94, 128.9, 96.7, 86.8, 76.4, 70.2, 65.3, 60.9. ESI-HRMS: [M + H]$^+$ calcd for $C_{16}H_{18}N_3O_5$, 332.1246; found, 332.1260.

4-N, 3′-O-Dibenzoyl-1′-deoxy-2′-isocytidine (40). This compound was synthesized in 73.0% yield from **39** by a method similar to that for **34**, but without the addition of concentrated ammonia solution. ^{1}H-NMR (DMSO-d_6): δ 11.26 (s, 1H), 8.24 (d, 1H, J = 7.5 Hz), 8.00 (m, 4H), 7.49–7.71 (m, 6H), 7.36 (d, 1H, J = 7.5 Hz), 5.47 (dd, 1H, J = 5.5 and 3.5 Hz), 5.21 (m, 1H), 5.09 (t, 1H, J = 5.5 Hz), 4.22 (m, 2H), 4.05 (m, 1H), 3.73 (m, 2H). ^{13}C-NMR (DMSO-d_6): δ 167.8, 165.7, 163.4, 155.4, 148.2, 134.3, 133.6, 133.2, 129.9, 129.5, 129.3, 128.9, 96.8, 85.1, 79.4, 70.2, 63.9, 60.7. ESI-HRMS: [M + H]$^+$ calcd for $C_{23}H_{22}N_3O_6$, 436.1509; found, 436.1511.

6-N-Benzoyl-P-(2-cyanoethyl)-1′-deoxy-2′-isoadenylyl-(3′ → 5′)-4-N, 3′-O-dibenzoyl-1′-deoxy-2′-isocytidine (44). To a suspension of **40** (261 mg, 0.60 mmol) and 1H-tetrazole (63 mg, 0.90 mmol) in dry CH_2Cl_2 (5 mL), 2-cyanoethyl $N,N,N′,N′$-tetraisopropylphosphorodiamidite (271 mg, 0.90 mmol) was added. After stirring for 2 h, 1H-tetrazole (63 mg, 0.90 mmol) and **42**[39] (513 mg, 0.78 mmol) were added. The reaction mixture was stirred for 4 h and then iodine (460 mg, 1.81 mmol) in THF/H$_2$O/pyridine (66:33:1, 4.6 mL) was added. After 10 min, the mixture was poured into CH$_2$Cl$_2$ (100 mL) and washed with 0.2 M sodium sulfite (30 mL × 2). The organic layer was dried, filtered, and concentrated. The residue in 5 mL of dichloromethane was stirred with 2% dichloroacetic acid in CH$_2$Cl$_2$ (20 mL) for 20 min. The reaction mixture was poured into CH$_2$Cl$_2$ (100 mL) and washed with saturated aqueous sodium bicarbonate (50 mL). The organic layer was dried, filtered, and concentrated. The residue was purified over silica gel column (CH$_2$Cl$_2$/CH$_3$OH 25:1) to give **44** (340 mg, 62.5%) as an amorphous solid. ^{1}H-NMR (CDCl$_3$): δ 9.30 (br, 2H), 8.70 and 8.68 (s and s, 1H), 8.45 and 8.44 (s and s, 1H), 7.77–8.02 (m, 7H), 7.38–7.60 (m, 10H), 5.32–5.58 (m, 3H), 5.14 (m, 1H), 3.98–4.58 (m, 12H), 2.76 (m, 2H). ^{31}P-NMR (CDCl$_3$): δ −1.57, −2.28. ESI-HRMS: [M + H]$^+$ calcd for C$_{43}$H$_{41}$N$_9$O$_{12}$P, 906.2612; found, 906.2627.

6-N-Benzoyl-5′-O-[di(2-cyanoethoxy)phosphinyl]-P-(2-cyanoethyl)-1′-deoxy-2′-isoadenylyl-(3′ → 5′)-4-N, 3′-O-dibenzoyl-1′-deoxy-2′-isocytidine (45). To a solution of **44** (272 mg, 0.30 mmol) and 1*H*-tetrazole(42 mg, 0.60 mmol) in CH$_2$Cl$_2$ (5 mL), di(2-cyanoethyl) *N,N*-diisopropylphosphoramidite (163 mg, 0.60 mmol) was added. The resulting mixture was stirred at room temperature for 4 h. Iodine (300 mg, 1.18 mmol) in THF/H$_2$O/pyridine (66:33:1, 3.0 mL) was added. After 10 min, the mixture was poured into CH$_2$Cl$_2$ (100 mL) and washed with 0.2 M sodium sulfite (30 mL × 2). The organic layer was dried, filtered, and concentrated. The residue was purified on a silica gel column (CH$_2$Cl$_2$/CH$_3$OH, 20:1) to give **45** (279 mg, 85.1%) as an amorphous solid. ^{1}H-NMR (CDCl$_3$): δ 9.11 (br, 2H), 8.74 and 8.72 (s and s, 1H), 8.36 and 8.32 (s and s, 1H), 7.85–8.03 (m, 7H), 7.39–7.63 (m, 10H), 5.43–5.60 (m, 2H), 5.16–5.31 (m, 2H), 4.07–4.65 (m, 16H), 2.67–2.85 (m, 6H). ^{31}P-NMR (CDCl$_3$): δ −0.70, −1.75, −1.85, −2.27. ESI-HRMS: [M + H]$^+$ calcd for C$_{49}$H$_{48}$N$_{11}$O$_{15}$P$_2$, 1092.2807; found, 1092.2798.

General Procedure for a Deprotection Reaction

The protected dinucleotide (100–200 mg) in concentrated ammonium hydroxide (8 mL) was capped and stirred at room temperature for 24 h. Then the reaction solution was evaporated to dryness and the residue was dissolved in water (50 mL) and washed with ether (30 mL). The aqueous layer was concentrated and purified by HPLC to give the product as a white spongy solid after lyophilization.

1′-Deoxy-2′-isoadenylyl-(3′ → 5′)-1′-deoxy-2′-isocytidine (29). (75.4% yield). UV (H$_2$O): λ$_{max}$ 264 nm (ε = 19,300).^{1}H-NMR (D$_2$O): δ 8.13 (s, 1H), 8.02 (s, 1H), 7.22 (d, 1H, *J* = 7.5 Hz), 5.65 (d, 1H, *J* = 7.5 Hz) , 5.08 (m, 1H), 4.81 (m, 1H), 4.68 (m, 1H), 4.24 (dd, 1H, *J* = 10.5 and 7.5 Hz), 4.10 (dd, 1H, *J* = 10.5 and 5.0 Hz), 3.96 (m, 2H), 3.81–3.87 (m, 2H), 3.63–3.77 (m, 4H), 3.47 (s, br, 1H). ^{13}C-NMR (D$_2$O): δ 165.2, 157.4, 155.3, 152.3, 148.8, 142.9, 140.5, 118.5, 96.0, 84.3, 83.7, 78.5, 76.0, 70.0, 69.9, 64.1, 64.0, 61.1, 60.3. ^{31}P-NMR (D$_2$O): δ −0.10. FAB-HRMS: [M + H]$^+$ calcd for C$_{19}$H$_{26}$N$_8$O$_9$P, 541.1560; found, 541.1552.

5'-O-Phosphoryl-1'-deoxy-2'-isoadenylyl-(3' → 5')-1'-deoxy-2'-isocytidine (30).
(61.7% yield). UV (H$_2$O): λ_{max} 265 nm(ϵ = 19,700).^{1}H-NMR (D$_2$O): δ 8.34 (s, 1H), 8.24 (s, 1H), 7.46 (d, 1H, J = 7.5 Hz), 5.98 (d, 1H, J = 7.5 Hz), 5.20 (m, 1H), 4.78 (m, 1H), 4.63 (m, 1H, partially hidden in water peak), 4.25 (dd, 1H, J = 10.0 and 6.5 Hz), 4.18 (dd, 1H, J = 10.0 and 4.0 Hz), 4.14 (dd, 1H, J = 5.5 and 2.5 Hz), 4.05–4.09 (m, 2H), 4.00 (m, 1H), 3.90 (dd, 1H, J = 11.5 and 6.0 Hz), 3.78–3.83 (m, 3H), 3.58 (m, 1H). ^{13}C-NMR (D$_2$O): δ 159.2, 150.6, 149.2, 148.7, 146.0, 145.4, 143.0, 118.3, 95.1, 84.3, 83.8, 79.4, 75.9, 70.6, 69.0, 64.8, 64.0, 63.9, and 61.9. ^{31}P-NMR (D$_2$O): δ 0.96, −0.33. FAB-HRMS: [M + H]$^+$ calcd for C$_{19}$H$_{27}$N$_8$O$_{12}$P$_2$, 621.1224; found, 621.1234.

10. CONCLUDING REMARKS

In this chapter we have presented concise synthetic methodologies for the preparation of isomeric deoxy- and dideoxynucleosides. The methodologies developed will allow accessibility to many more novel isomeric nucleosides. These compounds have opened a new chapter in the field of antiviral nucleosides. We have also developed methodologies for the incorporation of isomeric deoxynucleosides into novel nuclease-stable dinucleotides of antiviral relevance.

Acknowledgments

We thank the National Institutes of Health for support of this research work on the discovery of novel anti-HIV compounds. It is a pleasure to acknowledge the contributions of my many able co-workers and collaborators whose names appear in the publications cited.

ABBREVIATIONS

A	adenine
Ac	acetyl
AIBN	azobis(isobutyronitrile), radical initiator
BzCl	benzoyl chloride
C	cytosine
CD	circular dichroism
COSY	correlation spectroscopy
D	dextrorotatory
DCA	dichloroacetic acid
DCC	dicyclohexylcarbodiimide
ddAMP	dideoxyadenosine monophosphate
ddATP	dideoxyadenosine triphosphate
DMF	dimethylformamide
ESI HRMS	electrospray ionization high-resolution mass spectrometry

EtOH	ethanol
FAB-HRMS	fast-atom bombardment high-resolution mass specrometry
FTIR	Fourier transform infrared
HMBC	heteronuclear multiple bond correlation (NMR)
HMDS	hexamethyldisilazide
HMQC	proton-detected heteronuclear multiquantum coherence (NMR)
HPLC	high-performance liquid chromatography
Im_2CS	1,1'-thiocarbonyldiimidazole
IsoddA	isodideoxyadenosine
IsoddC	isodideoxycytidine
IsoddN	isodideoxynucleoside
IsoddT	isodideoxythymidine
IsodN	isodeoxynucleoside
L	levorotatory
MeOH	methanol
Me_2SO-d_6	dimethyl sulfoxide–deuterated
NMR	nuclear magnetic resonance
o-Cl-Ph-O-P(O)(Triaz)$_2$	2-chlorophenylphosphorobistriazolide
Pyr	pyridine
rt	room temperature
T	thymidine
THF	tetrahydrofuran
TLC	thin-layer chromatography
TMSCl	trimethylsilyl chloride
TMSOTf	trimethylsilyl triflate
TPS-TAZ	triisopropylbenzenesulfonyl tetrazolide
TsCl	tosyl chloride
U	uracil
UV	ultraviolet light

REFERENCES

1. Frankel, A. D.; Young, J. A. T. *Annu. Rev. Biochem.* **1998**, *67*, 1–25.
2. De Clercq, E. *Annu. Rev. Pharmacol. Toxicol.* **2011**, *51*, 1–24.
3. Katz, R. A.; Skalka, A. M. *Annu. Rev. Biochem.* **1994**, *63*, 133–173.
4. De Clercq, E. *J. Med. Chem.* **1995**, *38*, 2491–2517.
5. De Clercq, E. *Nat. Rev. Drug Discov.* **2007**, *6*, 1001–1018.
6. Frontiers in Nucleosides and Nucleic Acids; Schinazi, R. F., Liotta, D. C., Eds.; IHL Press: Tucker, GA, **2004**; pp. 431–445.
7. Nair, V.; Chi, G. *Rev. Med. Virol.* **2007**, *17*, 277–295.
8. Nair, V.; Chi, G. In HIV-1 Integrase; Neamati, N., Ed.: Wiley: Hoboken, NJ, **2011**; pp. 379–388.

9. Summa, V.; Petrocchi, A.; Bonelli, F.; Crescenzi, B.; Donghi, M.; Ferrara, M.; Fiore, F.; Gardelli, C.; Paz, O. G.; Hazuda, D. J.; Jones, P.; Kinzel, O.; Laufer, R.; Monteagudo, E.; Muraglia, E.; Nizi, E.; Orvieto, F.; Pace, P.; Pescatore, G.; Scarpelli, R.; Stillmock, K.; Witmer, M. V.; Rowley, M. *J. Med. Chem.* **2008**, *51*, 5843–5855.

10. Mathe, C.; Gosselin, G. *Antiviral Res.* **2006**, *71*, 276–281.

11. Mitsuya, H.; and Broder, S. *Proc. Natl. Acad. Sci. USA* **1986**, *83*, 1911–1915.

12. Mitsuya, H.; Weinhold, K. J.; Furman, P. A.; St. Clair, M. H.; Lehrman, S. N.; Gallo, R. C.; Bolognesi, D.; Barry, D. W.; Broder, S. *Proc. Natl. Acad. Sci. USA* **1985**, *82*, 7096–7100.

13. Coates, J. A. V.; Cammack, N.; Jenkinson, H. J.; Mutton, I. M.; Pearson, B. A.; Storer, R.; Cameron, J. M.; Penn, C. R. *Antimicrob. Agents Chemother.* **1992**, *36*, 202–205.

14. Martin, J. C.; Hitchcock, M. J. M.; De Clercq, E.; Prusoff, W. H. *Antiviral Res.* **2009**, *85*, 34–38.

15. Daluge, S. M.; Good, S. S.; Faletto, M. B.; Miller, W. H.; St. Clair, M. H. *Antimicrob. Agents Chemother.* **1997**, *41*, 1082–1093.

16. Richman, D. D. *Antiviral Ther.* **2001**, *6*, 83–88.

17. Pal, S.; Nair, V. *Biochem. Pharmacol.* **2000**, *60*, 1505–1508.

18. Nair, V.; Buenger, G. S. *J. Org. Chem.* **1990**, *55*, 3695–3697.

19. Nair, V.; Sells, T. B. *Biochim. Biophys. Acta* **1992**, *1119*, 201–204.

20. Ahluwalia, G.; Cooney, D. A.; Mitsuya, H.; Fridland, A.; Flora, K. P.; Hao, Z.; Dalal, M.; Broder, S.; Johns, D. G. *Biochem. Pharmacol.* **1987**, *36*, 3797–3800.

21. Nair, V.; Jahnke, T. S. *Antimicrob. Agents Chemother.* **1995**, *39*, 1017–1029.

22. Nair, V.; Nuesca, Z. M. *J. Am. Chem. Soc.* **1992**, *114*, 7951–7953.

23. Zheng, X. P.; Nair, V. *Tetrahedron* **1999**, *55*, 11803–11818.

24. Nair, V.; Chun, B. K.; Vadakkan, J. J. *Tetrahedron* **2004**, *60*, 10261–10268.

25. Nair, V.; Piotrowska, D. G.; Okello, M.; Vadakkan, J. *Nucleosides Nucleotides Nucleic Acids* **2007**, *26*, 687–690.

26. Nair, V. In *Recent Advances in Nucleosides: Chemistry and Chemotherapy*; Chu, C. K., Ed.; Elsevier Science: Amsterdam, **2002**; pp. 149–166.

27. Nair, V.; St. Clair, M. H.; Reardon, J. E.; Krasny, H. C.; Hazen, R. J.; Paff, M. T.; Boone, L. R.; Tisdale, M.; Najera, I.; Dornsife, R. E.; Averett, D. R.; Borrotoesoda, K.; Yale, J. L.; Zimmerman, T. P.; Rideout, J. L. *Antimicrob. Agents Chemother.* **1995**, *39*, 1993–1999.

28. Guenther, S.; Balzarini, J.; De Clercq, E.; Nair, V. *J. Med. Chem.* **2002**, *45*, 5426–5429.

29. de Cienfuegos, L. A.; Mota, A. J.; Robles, R. *Org. Lett.* **2005**, *7*, 2161–2164.

30. Chi, G. C.; Nair, V. *Nucleosides Nucleotsides Nucleic Acids* **2005**, *24*, 1449–1468.

31. Nair, V.; Pal, S. *Bioorg. Med. Chem. Lett.* **2004**, *14*, 289–291.

32. Newton, M. G.; Campana, C. F.; Chi, G. C.; Lee, D.; Liu, Z. J.; Nair, V.; Phillips, J.; Rose, J. P.; Wang, B. C. *Acta Crystallogr* **2005**, *C61*, o518–o520.

33. Nair, V.; Emanuel, D. J. *J. Am. Chem. Soc.* **1977**, *99*, 1571–1576.

34. Barton, D. H. R.; Subramanian, R. *J. Chem. Soc. Perkin Trans. 1* **1977**, 1718–1723.

35. Bolon, P. J.; Sells, T. B.; Nuesca, Z. M.; Purdy, D. F.; Nair, V. *Tetrahedron* **1994**, *50*, 7747–7764.

36. Shealy, Y. F.; Odell, C. A. *J. Heterocycl. Chem.* **1976**, *13*, 1015–1020.

37. Jahnke, T. S.; Nair, V. *Bioorg. Med. Chem. Lett.* **1995**, *5*, 2235–2238.
38. Taktakishvili, M.; Neamati, N.; Pommier, Y.; Pal, S.; Nair, V. *J. Am. Chem. Soc.* **2000**, *122*, 5671–5677.
39. Wenzel, T.; Nair, V. *Bioconjug. Chem.* **1998**, *9*, 683–690.
40. Reese, C. B. *Tetrahedron* **1978**, *34*, 3143–3179.
41. Chattopadhyaya, J. B.; Reese, C. B. *Nucleic Acids Res.* **1980**, *8*, 2039–2053.
42. van Boom, J. H.; van der Marel, G. A.; van Boeckel, C. A. A.; Wille, G.; Hoyng, C. F. In Chemical and Enzymatic Synthesis of Gene Fragments; Gassen, H. G., Lang, A., Eds.; Verlag Chemie: Weinheim, germany, **1982**; pp. 53–70.
43. Wreesmann, C. T. J.; Fidder, A.; van der Marel, G. A.; van Boom, J. H. *Nucleic Acids Res.* **1983**, *11*, 8389–8405.
44. Muller, S.; Wolf, J.; Ivanov, S. A. *Curr. Org. Synth.* **2004**, *1*, 293–307.
45. Lesnikowski, Z. J. *Bioorgan. Chem.* **1993**, *21*, 127–155.
46. Beaucage, S. L.; Iyer, R. P. *Tetrahedron* **1992**, *48*, 2223–2311.
47. Torrence, P. A.; Witkop, B. In Nucleic Acids Chemistry; Part 2 ed.; Townsend, L. B., Tipson, R. S., Eds.; Wiley: New York, **1978**; Vol. 166, pp. 977–988.
48. Smrt, J. In Nucleic Acids Chem; Part 2 ed.; Townsend, L. B., Tipson, R. S., Eds.; Wiley, **1978**; Vol. 154, pp. 885–888.
49. van der Marel, G.; van Boeckel, C. A. A.; Wille, G.; van Boom, J. H. *Tetrahedron Lett.* **1981**, *22*, 3887–3890.
50. Marugg, J. E.; Tromp, M.; Jhurani, P.; Hoyng, C. F.; van der Marel, G. A.; van Boom, J. H. *Tetrahedron* **1984**, *40*, 73–78.
51. Moffatt, J. G.; Khorana, H. G. *J. Am. Chem. Soc.* **1961**, *83*, 649.
52. Taktakishvili, M.; Nair, V. *Tetrahedron Lett.* **2000**, *41*, 7173–7176.
53. Kakefuda, A.; Shuto, S.; Nagahata, T.; Seki, J.; Sasaki, T.; Matsuda, A. *Tetrahedron* **1994**, *50*, 10167–10182.
54. Zhu, X. F.; Scott, A. I. *Nucleoside Nucleotide Nucleic Acids* **2001**, *20*, 197–211.
55. Bera, S.; Nair, V. *Tetrahedron Lett.* **2001**, *42*, 5813–5815.

9 Synthesis of Conformationally Constrained Nucleoside Analogues

ESMA MAOUGAL
Université de Nantes, Faculté des Sciences et des Techniques, Nantes, France

JEAN-MARC ESCUDIER
Université Paul Sabatier, Laboratoire de Synthèse et Physicochimie des Molécules d'Intérêt Biologique, Toulouse, France

CHRISTOPHE LEN
Université de Technologie de Compiègne, Ecole Supérieure de Chimie Organique et Minérale, Transformaciones Intégrées de la Matière Renouvelable, Centre de Recherche Royallieu, Compiègne, France

DIDIER DUBREUIL and JACQUES LEBRETON
Université de Nantes, Faculté des Sciences et des Techniques, Nantes, France

1. INTRODUCTION

Conformationally locked nucleosides constitute an important class of molecules for antisense and antigene therapies, including more recent RNA interference, due to preorganization of the nucleoside sugar moiety[1] in an appropriate conformation leading to better binding ability.[2] Nevertheless, since the pioneering works of Stephenson and Zamecnik[3] to inhibit gene expression using antisense oligonucleotides, this attractive chemotherapy has found relatively little success to date. Only one antisense drug, marketed as Vitravene, has been approved so far by the U.S. Food and Drug Administration, in August 1998, for the treatment of cytomegalovirus retinitis (CMV) in immunocompromised patients.[4] It should also be pointed out that the development of new chemotherapeutic agents, effective against viral diseases[5] and cancer,[6] based on locked nucleosides continues to be of great importance.[7] Synthesis of modified nucleic acids has been widely investigated by organic chemists since the 1960s and the focus of more extensive effort in the last two decades. The development of new methodologies, in particular, in the field of metal- and free radical–mediated carbon–carbon bond formation and group-protecting manipulation has substantially opened up straightforward access

Chemical Synthesis of Nucleoside Analogues, First Edition. Edited by Pedro Merino.
© 2013 John Wiley & Sons, Inc. Published 2013 by John Wiley & Sons, Inc.

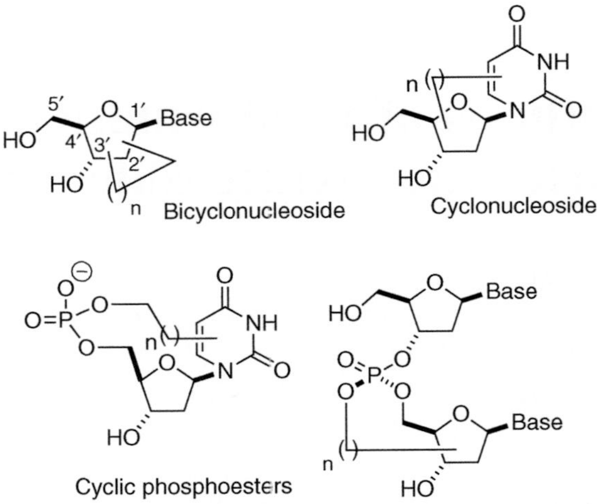

Figure 1. Different types of nucleoside analogues with a restricted conformation.

to a new generation of modified nucleosides, as well as greatly improving the preparation of those already described in the literature.

The bicyclic nucleosides prepared so far can be divided into three categories (Figure 1): (1) bicyclonucleosides,[8] obtained by linking two atoms of the furanose moiety via an alkylene unit or analogues; (2) cyclonucleosides,[9] obtained by bonding one atom of the furanose moiety and one atom of the nucleobase via an alkylene unit or analogues; and (3) cyclic phosphoesters,[10] obtained by forming an alkylene bridge or analogues between the phosphorus atom and the nucleobase or the furanose moiety.

The main purpose of this chapter is to provide the reader with pathways toward the different families of locked nucleosides, emphasizing the key methodologies used to reach them. The most significant examples of preparation will be presented, but neither biological nor biophysical aspects concerning these nucleoside analogues will be discussed in detail. For more information, the reader is referred to earlier reviews.[11] As most of the work done so far has been devoted to this first group, we also opted to present the synthesis of selected relevant bicyclic nucleoside analogues following this order.

2. PREPARATION OF BICYCLIC NUCLEOSIDES

2.1. Introduction

In this part we have opted to present the synthesis of selected relevant bicyclic nucleoside analogues classified according to the position of the bridge on the furanose ring (refer to the bicyclonucleoside structure for numbering as presented in Figure 1). For the synthesis of locked bicyclic nucleosides described here, both convergent and linear strategies have been applied. In the first approach, the locked aglycone moiety is prepared following a multistep sequence from an inexpensive natural sugar analogue (e.g., diacetone-D-glucose). At the final stage, the nucleobase is introduced applying Vorbrüggen coupling.[12]

2.2. Preparation of 1′,2′-Linked Bicyclic Nucleosides

Chattopadhyaya et al. have contributed largely to developing an efficient and practical gram-scale route to 1′,2′-oxetane locked nucleosides containing all natural nucleobases. Their first contribution in this series is outlined in Scheme 1.[13] The key starting material **1** was prepared in four steps from inexpensive D-fructose following well-established procedures.[14] Protection of the primary hydroxyl group in **1** as the 4-toluoyl derivative **2** followed by coupling with silylated thymine in the presence of TMSOTf as a Lewis acid catalyst afforded an equimolar α/β mixture of psicothymidine **3** in 60% yield. Both anomers were separated by careful chromatography on silica gel. Then, alcohol β-**3** was mesylated to give **4**, which was treated under acidic conditions, leading to removal of the 3′,4′-O-isopropylidene protecting group (refer to structure **1** for numbering used throughout the discussion in sections 2.1 to 2.3), affording the precursor **5** of the oxetane ring formation in 90% overall yield. At this stage, the ring-closing step was achieved by simple treatment of **5** with an excess of NaH to provide the protected 1′,2′-oxetane locked thymidine **6** in 60% yield. Finally, cleavage of the 4-toluoyl group with methanolic ammonia furnished the target nucleoside analogue **7** in high yield. It should be pointed out

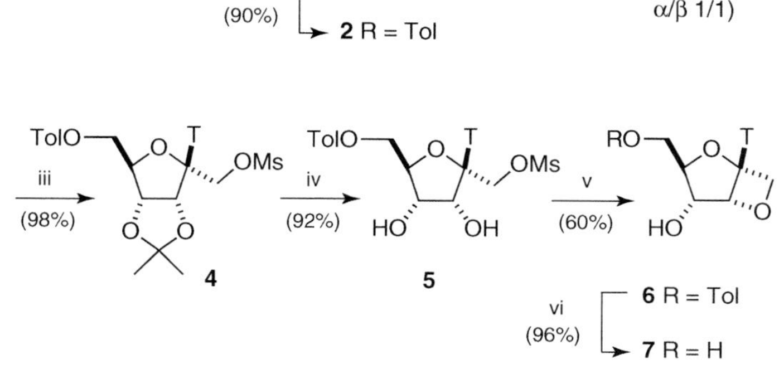

Scheme 1. *Reagents and conditions:* (i) 4-toluoyl chloride, pyridine, rt, overnight; (ii) persilylated thymine, TMSOTf, ACN, 4°C for 1 h, rt for 18 h; (iii) MsCl, pyridine, 4°C, overnight; (iv) 90% aq. TFA, rt, 20 min; (v) NaH, DMF, 4°C, 9 h; (vi) NH$_3$, MeOH, rt, 2 days.

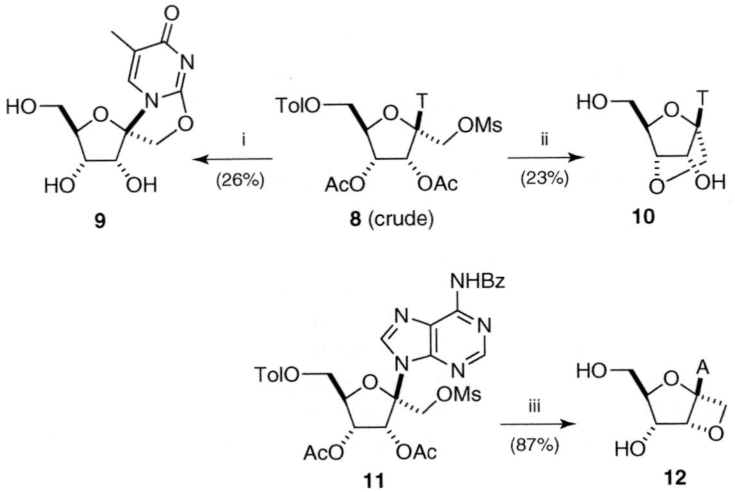

Scheme 2. *Reagents and conditions:* (i) NH$_3$, MeOH, monitored by HPLC; (ii) NaOMe, MeOH, 50°C, 18 h; (iii) NH$_3$, MeOH, 20°C, 18 h.

that due to the vulnerability of the oxetane ring in Vorbrüggen N-glycosylations, in all the convergent approaches, its formation should be planned after this key step.

Interestingly enough, the oxetane ring formation in this series seems to depend on the nature of the bases used, as pointed out by Mikhailopulo et al.[15] and outlined in Scheme 2, with isolation of the anhydro derivative **9** or the 1′,3′-linked thymidine analogue **10** from the mesylate intermediate **8**. Also in the purine series, the mesylate **11** was cleanly converted into the desired 1′,2′-oxetane locked adenosine **12**, with concomitant cleavage of the acyl protecting groups, in 87% yield, by simple treatment with NaOMe in MeOH.

As indicated by the authors, this route to oxetane nucleoside **7** suffered from poor stereoselectivity of the key glycosylation step.[16] After exploring different pathways to increase the glycosylation selectivity, Chattopadhyaya et al. proposed an elegant solution to this critical β-stereoselective introduction of purine bases, as illustrated in Scheme 3.[17] The appropriately protected glycosyl donor **13** was prepared in four steps from the previous intermediate **2** (see Scheme 1) following a set of known procedures. The bromosugar **14**, prepared by treatment of **13** with HBr in AcOH, was reacted directly, without purification, with persilylated N^6-benzoyladenine in the presence of SnCl$_4$ as a Lewis acid to give a mixture of products, including a very small amount of α-anomer with many nonnucleosidic impurities. After chromatographic purification, the pure β-adenine derivative **11** was isolated in 42% yield (6.1 g). From the latter compound, removal of the acyl protecting groups with methanolic ammonia followed by selective protection of the 4′,6′-hydroxyl groups with the bidentate protecting group TIPDSCl$_2$ led cleanly to the oxetane precursor **15**, which, upon treatment with NaHMDS, gave the protected target adenosine nucleoside **16** in 60% yield over the three steps.

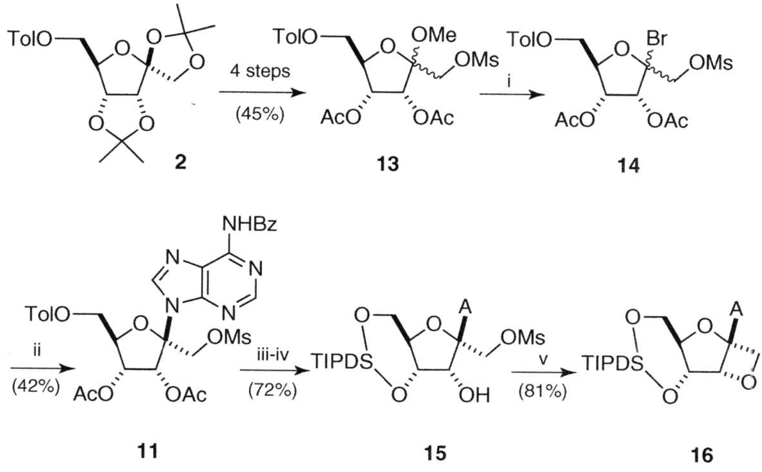

Scheme 3. *Reagents and conditions:* (i) HBr, AcOH, 0°C to rt, 12 h; (ii) persilylated N^6-benzoyladenine, SnCl$_4$, ACN-DCE, 70°C for 5 h, rt for 12 h; (iii) NH$_3$, MeOH, rt, 48 h; (iv) TIPDSCl$_2$, pyridine, 4°C, 30 min; (v) NaHMDS, THF, 4°C, 2 h.

Chattopadhyaya et al. have extended their strategy to the synthesis of 1′,2′-azetine pyrimidine nucleosides as outlined in Scheme 4.[18] The 2-O-acetyl-D-psicofuranose **17**, prepared in eight steps from D-fructose, was coupled with the persilylated uridine in a Vorbrüggen protocol to afford the desired β-anomer **18** in 65% yield with good stereoselectivity (β/α 9:1). It should be noted that the stereoselectivity observed in these conditions with purine bases is lower (β/α, 7:3).[16] Saponification of the ester protecting groups of **18**, followed by selective protection of the primary alcohol function as a 4-monomethoxytrityl ether derivative, afforded the intermediate **19** in 75% overall yield. The implementation of a leaving group with the correct stereochemistry to promote formation of the azetine ring was carried out via the 2,3′-anhydrouracil intermediate **20**. This key intermediate **20** was obtained in high yield by treatment of **19** with diphenyl carbonate, which was found to be more efficient than 1,1′-thiocarbonyldiimidazole,[19] and NaHCO$_3$ in PhMe at 110°C. Subsequent benzylation of the remaining alcohol function at C4′ in **20** followed by acidic cleavage of the trityl ether and activation of the liberated hydroxyl group as mesylate led to compound **23** in 60% overall yield. Nucleophilic displacement of the 1′-O-mesylate group of **23** with azide ion, followed by opening of the O-3′-anhydro bond upon treatment with NaOH, furnished the desired D-fructofuranosyl nucleoside **25** in 74% overall yield. To complete the synthesis of the required azetine precursor, intermediate **25** was mesylated under classical conditions. Finally, Staudinger reduction of azide **26**, utilizing trimethylphosphine with aqueous THF, afforded the corresponding amine **27**, which by simple heating in an Et$_3$N/pyridine mixture led by an intramolecular cyclization reaction to the protected azetine-locked nucleoside **28** in 60% yield over the two steps.

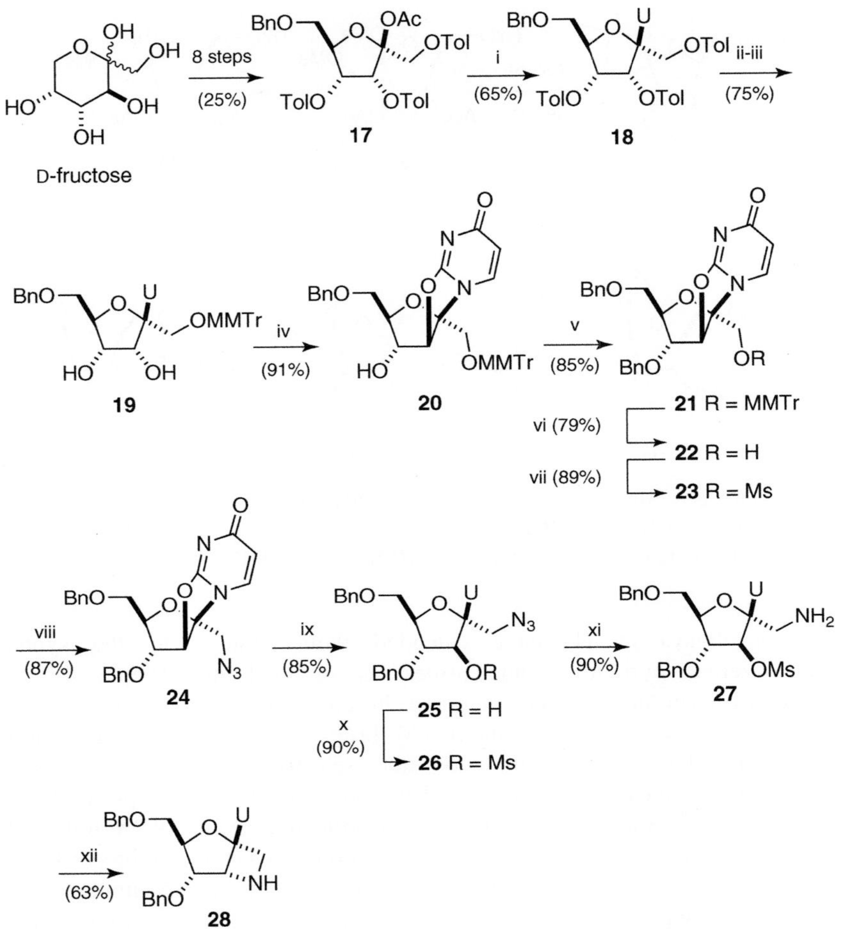

Scheme 4. *Reagents and conditions:* (i) uridine, BSA, ACN, then TMSOTf, 40°C, 2 h; (ii) NaOEt, EtOH, rt, 10 min; (iii) MMTrCl, pyridine, rt, overnight; (iv) (PhO)$_2$CO, NaHCO$_3$, PhMe, 110°C, 2 h; (v) BnBr, NaH, ACN, rt, overnight; (vi) 80% aq AcOH, rt, 78 h; (vii) MsCl, pyridine, 0°C, 1.5 h; (viii) NaN$_3$, DMF, 100°C, 60 h; (ix) 1 N NaOH, MeOH, rt, 2 h; (x) MsCl, pyridine, 0°C, 24 h; (xi) PMe$_3$, THF/H$_2$O (1:3), rt, 1 h; (xii) Et$_3$N/pyridine (1:25), 90°C, 48 h.

2.3. Preparation of 1′,3′-Linked Bicyclic Nucleosides

Wightman et al. developed the synthesis of locked pyrimidine nucleosides **41** and **42** with a C1′–O3′ link from the psicofuranosyl-uracil **35**, as shown in Scheme 5.[20] The known 1-(3′-deoxy-β-D-psicofuranosyl)uracil **35** was prepared from D-fructose as described in the literature.[21] Treatment of D-fructose with 2 equiv of cyanamide in methanolic ammonia afforded the oxazoline **29**, which was engaged in the next step without further purification. Then, construction of the uracil nucleobase was

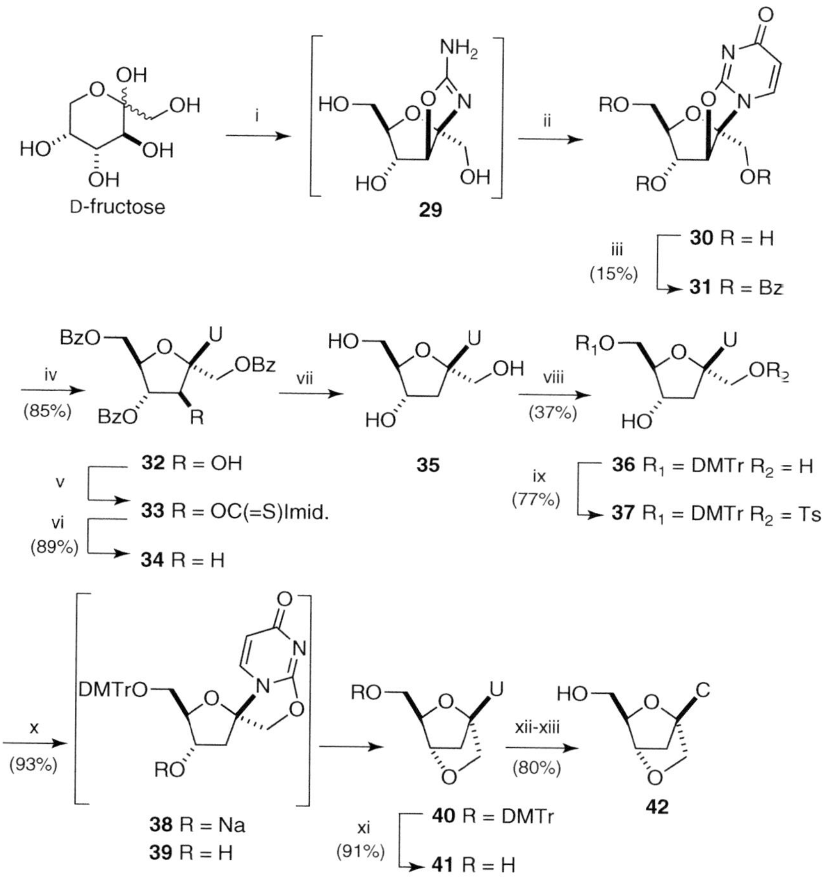

Scheme 5. *Reagents and conditions:* (i) H$_2$N-CN, conc. NH$_3$, MeOH, rt, overnight; (ii) ethyl propiolate, EtOH, reflux, 5 h; (iii) BzCN, Et$_3$N, ACN, rt, 1 h; (iv) 2 N HCl/DMF (2:5), rt, 2 days; (v) N,N'-thiocarbonyldiimidazole, DMF, rt, 4 days; (vi) Bu$_3$SnH, AIBN, PhMe, reflux, 40 min; (vii) NaOMe, MeOH, rt, 13 h; (viii) DMTrCl, pyridine, DCM, 0°C, 4 h, rt, 21 h; (ix) TsCl, pyridine, DCM, rt, 48 h; (x) NaH, DMF, rt, 3 days; (xi) 80% AcOH, rt, 40 min; (xii) TMSCl, N-methylpyrrolidine, ACN, 0°C, 1 h, then Tf$_2$O, p-nitrophenol, rt, 4.5 h; (vi) NH$_3$, dioxane, sealed tube, 50°C, 24 h.

readily achieved by heating the crude oxazoline **29** with ethyl propiolate to provide the 2,3'-anhydro intermediate **30**, which was perbenzoylated with benzoyl cyanide, leading to isolation of the expected nucleoside derivative **31** in 15% overall yield (17 g) from D-fructose, after purification on silica gel. Subsequent hydrolysis of the O,3'-anhydro linkage of **31** under acidic conditions, followed by a classical Barton–McCombie deoxygenation on the resulting secondary alcohol **32** via the (thiocarbonyl)imidazolide **33** furnished the 3'-deoxy derivative **34** in 76% yield over the three steps. Cleavage of the benzoyl esters using NaOMe in MeOH resulted in the formation of the key psicofuranosyluracil **35**. Selective protection of the 6′-hydroxyl group in **35** was an arduous task, due to its similar reactivity with the other primary alcohol function. In optimized conditions, treatment of **35** with 1.1 equiv of DMTrCl gave the desired monoprotected alcohol **36** in only 37% yield, after tedious chromatography on silica gel. Surprisingly, treatment with 2 equiv of DMTrCl led to the formation of side products, including protected secondary alcohol derivatives. This result is in sharp contrast with the tosylation step, in which the intermediate **36** was reacted with 3 equiv of TsCl to provide the monotosyl derivative **37** in 77% yield. At this stage, the latter compound **37** was converted into the cyclized nucleoside analogue **40** in 93% yield, via the 2,1′-anhydro nucleoside as its alkoxide **38**, by treatment with NaH over a period of 3 days. It should be pointed out that under the previous experimental conditions, the 2,1′-anhydro nucleoside **39** intermediate could be isolated in 41% yield after reaction overnight. Removal of the DMTr-protecting group in **40** led to the isolation of the bicyclic uracil nucleoside **41** in 91% yield. Transformation of uracil **41** to the corresponding cytidine analogue **42** was accomplished by sequential treatment with TMSCl, Tf_2O, and p-nitrophenol followed by aminolysis. Using this one-pot sequence described by Reese et al.,[22] the target cytosine **42** was isolated in 80% overall yield.

The corresponding protected 1′,3′-azamethylene **53** analogue of the previous bridged nucleoside **41** (Scheme 5) was synthesized from 1-(3′-deoxy-β-D-psicofuranosyl)uracil **35** by Wengel et al. as depicted in Scheme 6.[23] Selective protection of both primary alcohols in **35** with DMTrCl in correct yield proved to be difficult (see below). However, slow addition of DMTrCl in DCM at 0°C followed by stirring and the resulting mixture at room temperature repeated twice with 2.2 and 0.9 equiv, respectively, afforded after purification the diprotected nucleoside **43** in good yield (77% on a 10-mmol scale). The configuration at C4′ before the introduction of a nitrogen atom was inverted by standard mesylation of **43** followed by treatment of the crude material **44** with 1 N aqueous NaOH at 90°C to furnish, after purification, the desired configured nucleoside **46** in 93% overall yield. In this last step under basic conditions, the inversion of configuration resulted from the formation of the anhydro intermediate **45**, which, by alkaline hydrolysis of the $O,4'$-anhydro bond, afforded the desired nucleoside **46**. Then, when the latter compound **46** was treasted with MsCl, it afforded the expected 4′-O-mesylated intermediate **47**, which, by heating with NaN_3 in DMF at 70°C, went smoothly to give **48** in nearly quantitative yield over the two steps. It is worth mentioning that the "anhydro" approach from intermediate **45** turned out to be inefficient. The nucleophilic opening of the corresponding 2,4′-anhydro bond at C4′ of **45** (prepared in 80% yield from **44** upon treatment with NaH in anhydrous DMF at room temperature for 3 days) with azide ion gave the corresponding azide **48** in only 26% yield. At this stage, acidic cleavage of the DMTr ethers of **48** followed by mesylation of the liberated primary alcohol functions yielded the bicyclic precursor **50** in 75% overall yield. Exposure of azide **50** to a hydrogen atmosphere in the presence of Pd on carbon led to the corresponding primary amine, which was not characterized but was heated directly at 60°C in DMF to promote intramolecular cyclization, leading to the bicyclic amine nucleoside **51** in

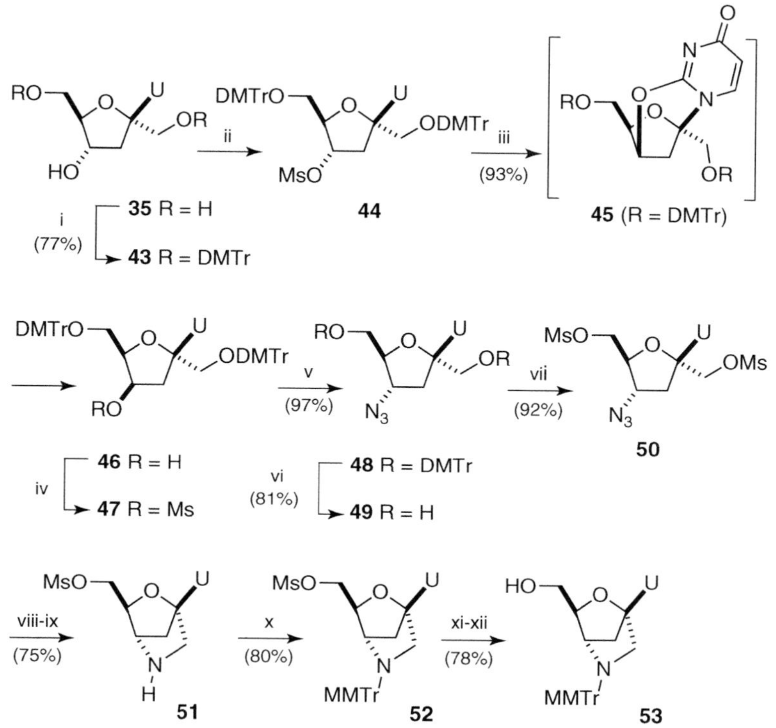

Scheme 6. *Reagents and conditions:* (i) 2.2 equiv DMTrCl, pyridine, 0°C, addition over 3 h, then rt for 16 h, 0.9 equiv DMTrCl, pyridine, 0°C, addition over 1.5 h, then rt for 10 h; (ii) MsCl, pyridine, 0°C, 2 h; (iii) 1 M NaOH, dioxane, 90°C, 10 h; (iv) MsCl, pyridine, rt, 2 h; (v) NaN$_3$, DMF, 70°C, 2 days; (vi) 80% aq. AcOH, rt, 4 h; (vii) MsCl, pyridine, 0°C, 1.5 h; (viii) H$_2$, 10% Pd/C, MeOH, rt, 25 h; (ix) DMF, 60°C, 19 h; (x) MMTrCl, pyridine, rt, 21 h; (xi) NaOBz, DMF, 100°C, 22 h; (xii) NaOMe, MeOH, rt, 13 h.

75% overall yield. To pursue the synthesis, the secondary amine **51** was first protected with an MMTr group to give **52**. Attempts to cleave the mesyl group of **52** through nucleophilic attack by hydroxide ion failed. Instead, removal of the mesyl group of **52** was achieved using a two-step sequence by subsequent treatment with NaOBz followed by saponification leading to the N-protected 1′,3′-azamethylene bridged nucleoside **53** in 78% overall yield.

2.4. Preparation of 1′,4′-Linked Bicyclic Nucleosides

Kværnø and Wengel accomplished the synthesis of a unique C1′–C4′ fused bicyclonucleoside **62** from the known 1-(3′-deoxy-β-D-psicofuranosyl)uracil **36** as depicted in Scheme 7.[24] First, from the DMTr-protected nucleoside **36**, both alcohol functions were protected as benzyl ethers to give **54**, which, under classical aqueous AcOH treatment, led to the intermediate **55** in 80% overall yield.

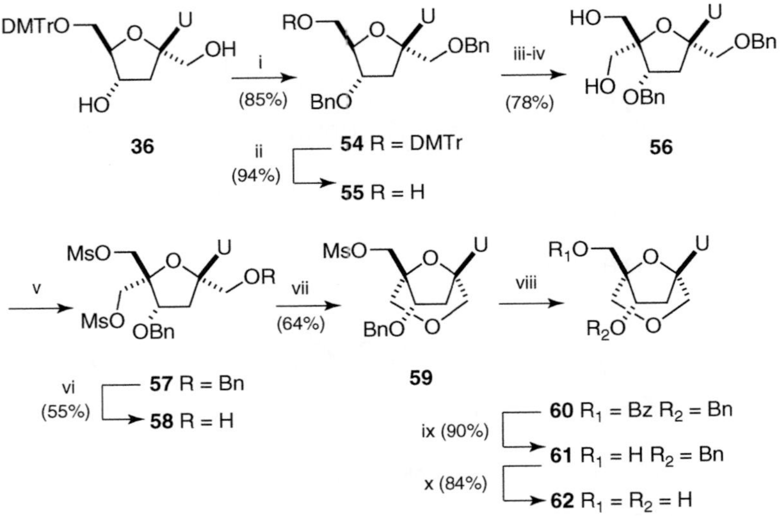

Scheme 7. *Reagents and conditions:* (i) BnBr, NaH, THF, rt, 8 h, workup, then BnBr, NaH, THF, rt, 15 h; (ii) 80% aq. AcOH, rt, 1.5 h; (iii) DMP, DCM, rt, 40 min; (iv) 37% aq. H$_2$CO, 1 N NaOH, dioxane, rt, 22 h; (v) MsCl, pyridine; 0°C, 1.3 h; (vi) H$_2$, 20% Pd(OH)$_2$/C, EtOH, rt, 7 h; (vii) 1 N NaOH, dioxane, 90°C, 22 h; (viii) NaOBz, DMF, 120°C, 36 h; (ix) NaOMe, MeOH, rt, 1.5 h; (x) BCl$_3$, DCM, −78°C, 7 h.

It is worth mentioning that to drive the benzylation reaction of **36** to completion while avoiding the N-benzylation of the uracil base, the protection was run once, and after workup, the crude material was treated again with correct amounts of NaH and BnBr. Oxidation of **55** to the corresponding aldehyde with Dess–Martin reagent, followed by an aldol condensation and Cannizzaro reaction sequence, furnished the diol **56** in 78% yield for the two steps. It should be pointed out that for the preparation of bicyclic nucleosides, in both linear and convergent strategies, standard oxidation of the C5 (or C5′) alcohol followed by a one-pot two-step aldol condensation with formaldehyde and aqueous NaOH, a Cannizzaro reduction or NaBH$_4$ reduction is one of the most popular methodologies to introduce a hydroxymethyl group at C4 (or C4′).[25] Treatment of this latter diol **56** with MsCl, followed by selective removal of benzyl ether on the primary hydroxyl group using atmospheric hydrogenolysis in the presence of Pearlman's catalyst, gave the desired intermediate **58** in 55% yield over the two steps, while only 11% of the corresponding fully debenzylated compound was isolated. Subsequent cyclization of dimesylate **58** under treatment with aqueous NaOH in refluxing dioxane for 22 h provided the target ether **59** in 64% yield. No side products resulting from direct hydrolysis of the mesyl groups were detected under these rather harsh conditions. Regeneration of the alcohol function at C5′ from the mesylate **59** was accomplished by treatment with NaOBz to give **60**, followed by saponification to give **61** in 90% overall yield. Attempts to cleave the benzyl protecting group in **61** under the previous hydrogenolysis conditions led to extensive hydrogenation of the double bond of the uracil base.[26] Some reports in the literature mention this problem. Although the use of transfer hydrogenation conditions could prevent this potential side reaction,[27] an attempt to apply this procedure to a DMTr-protected derivative of compound **61** led to complete reduction of the nucleobase! Finally, Lewis acid–mediated debenzylation of **61** with BCl_3 in DCM at low temperature provided the target nucleoside **62** in 84% yield.

2.5. Preparation of 2′,3′-Linked Bicyclic Nucleosides

An original strategy for the synthesis of cytosine methanonucleoside analogue **72** from the mesylate **69** was reported in 1983 by Okabe and Sun, as outlined in Scheme 8.[28] The synthesis of the key mesylate **69** was accomplished in six steps

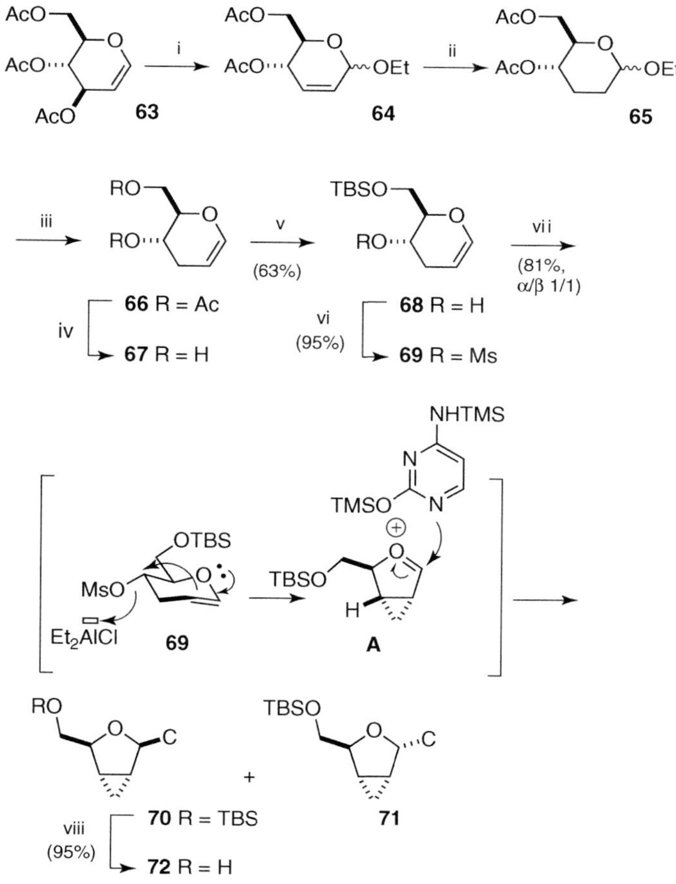

Scheme 8. *Reagents and conditions:* (i) BF_3-OEt_2, EtOH, PhMe; (ii) H_2, Pd/C, EA; (iii) Ac_2O, AcOH; (iv) cat. *t*-BuOK, MeOH; (v) TBSCl, Imid, DCM; (vi) MsCl, DMAP; (vii) persilylated cytosine, 2 equiv Et_2AlCl, ACN, reflux, 48 h; (vii) cat. PTSA, MeOH, H_2O.

from the commercially available tri-*O*-acetyl-D-glucal **63** in 60% overall yield. Thus from **63**, Lewis acid–mediated Ferrier rearrangement provided the unsaturated intermediate **64**, which was sequentially hydrogenated and then refluxed in an Ac$_2$O/AcOH mixture to give the pure deoxyglucal **66**. Subsequent classical deacetylation of **66**, followed by selective silylation of the primary alcohol and mesylation of the remaining secondary alcohol, led to the pivotal D-glucal **69**. This latter mesylate **69** was reacted with persilylated cytosine in refluxing ACN in the presence of 2 equiv of Et$_2$AlCl to afford, after purification on silica gel column, an anomeric mixture in a 1:1 ratio of the nucleosides **70** and **71** in 81% yield. In this one-pot procedure, it was assumed that activation of the leaving group of the D-glucal **69** by the Lewis acid promoted a diastereoselective S$_N$2-like intramolecular displacement, leading to the formation of an oxonium intermediate **A** which was then trapped by the silylated nucleobase to afford an anomeric mixture of nucleosides **70** and **71**. By recrystallization of the latter mixture, the pure β-anomer **70** was isolated in 39% yield from the key intermediate **69**. Subsequent cleavage of the silyl ether of **70** under acidic conditions furnished the cyclopropano derivative of ddC **72** in 95% yield.

The thymidine analogue **75** was prepared a few years later by Sard using the same strategy as depicted in Scheme 9.[29] Interestingly, the reaction of **69** with the silylated thymine was completed in 1 h, compared with 48 h for the cytosine series (see Scheme 8), and the corresponding β- and α-anomers **73** and **74** were isolated in 19% and 29% yields, respectively, after separation by chromatography on silica gel. Fluoride-induced removal of the silyl protecting group of **73** furnished the bicyclic nucleoside **75** in high yield. It is worth noting that all attempts at cyclopropanation on the silylated d4T **76** following Furukawa's protocol failed.

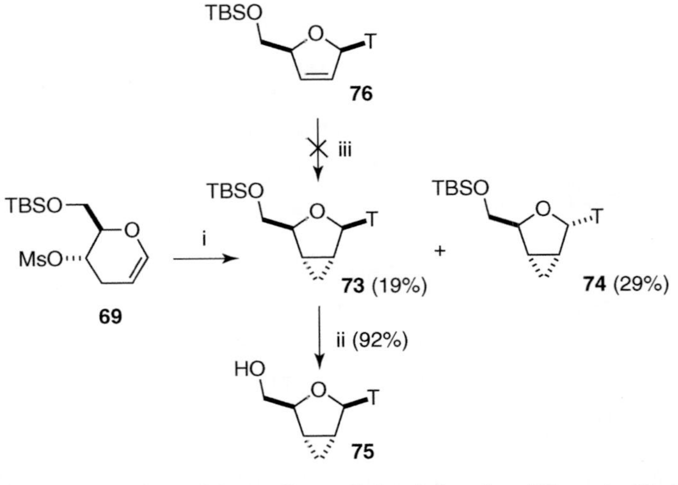

Scheme 9. *Reagents and conditions:* (i) persilylated thymine, 2.2 equiv Et$_2$AlCl, ACN, reflux, 1 h; (ii) TBAF, THF; (iii) CH$_2$I$_2$, ZnMe$_2$, THF or Et$_2$O; or CH$_2$ICl, ZnMe$_2$, DCE.

It is clear that cyclopropyl ring formation via Lewis acid–mediated intramolecular rearrangement of D-glucal **69** represents an efficient alternative to the olefin cyclopropanation reaction in this series.

Experimental Procedure

5'-O-(t-Butyldimethylsilyl)-2',3'-cyclopropano-2,3'-dideoxy-β-thymidine (73) and 5'-O-(t-Butyldimethylsilyl)-2',3'-cyclopropano-2,3'-dideoxy-β-thymidine (74).[29]
A solution of the mesylate **69** (507 mg, 1.57 mmol) was dissolved in 20 mL of anhydrous ACN under a nitrogen atmosphere at room temperature followed by persilylated thymine (470 mg, 1.74 mmol, 1.1 equiv); then a 1 M solution of Et_2AlCl in hexanes (3.46 mL, 3.46 mmol, 2.2 equiv) was added dropwise. The resulting homogeneous reaction mixture was refluxed and the reaction was monitored by TLC. After 1 h the reaction was completed and the mixture was treated at room temperature by addition of EA and water. The layers were separated and the organic phase was then dried over Na_2SO_4 and filtered. The resulting organic mixture was concentrated and then purified on an SiO_2 column (25:75 EA/hexanes to 50:50 EA/hexanes) to yield the α-anomer **74** (159 mg, 29%) as a white solid, R_f (EA/hexane, 1:1) = 0.27, mp 118–120°C. ^{1}H-NMR (300 MHz, $CDCl_3$): δ 8.68 (s, 1H), 7.32 (d, 1H, J = 2.7 Hz), 4.17 (t, 1H, J = 4.2 Hz), 3.67–3.70 (m, 2H), 2.10–2.13 (m, 1H), 1.90 (s, 3H) 1.70–1.74 (m, 1H), 0.98 (s, 9H), 0.48–0.94 (m, 2H), 0.07 and 0.06 (2s, 6H), followed by the β-anomer **73** (106 mg, 19%) as an orange oil that solidified, R_f (EA/hexane, 1:1) = 0.18, mp 50–53°C. ^{1}H-NMR (300 MHz, $CDCl_3$): δ 8.53 (s, 1H), 7.53 (s, 1H), 5.89 (s, 1H), 4.08 (t, 1H, J = 6 Hz), 3.61–3.64 (m, 2H), 1.88–1.94 (m, 1H), 1.92 (s, 3H) 1.65–1.70 (m, 1H), 1.00–1.05 (m, 1H), 0.88 (s, 9H), 0.47–0.49 (m, 1H), 0.06 (2s, 6H).

Inspired by their previous work on the preparation of bicyclonucleosides bearing a cyclobutane ring, Alibès et al.[30] extended their approach to the synthesis of cyclobutene derivatives, as well as their fluoro- and chlorocyclobutane analogues. This work is illustrated through the synthesis of the fluoro-β-adenosine analogue **86** as highlighted in Scheme 10.[31] The required fluoro-bicyclic lactone derivative **80** was obtained from the chiral fluorofuranone **79** (prepared in two steps from D-glyceraldehyde acetonide **77**)[32] using a photochemical [2 + 2]-cycloaddition reaction. Thus, light irradiation of **79** in acetone solution saturated with ethylene using a high-pressure 125-W mercury lamp at −20°C afforded the corresponding desired cycloadduct **80** in 62% yield (on a 1.02-mmol scale), after removal of the diastereoisomer **81** by column chromatography. Next, intermediate **80** was converted into an anomeric mixture of the fluoroacetate **82** in 78% yield over three steps, involving silylation of the primary alcohol, reduction of the lactone to the lactol with DIBAL-H, and acetylation of the resulting lactol. In contrast to their previous results, N-glycosylation with C2 halogenated glycone partners and 6-chloropurine was accomplished with some difficulty; in all attempts, the undesired α-anomer was formed as the major adduct. After some investigations,

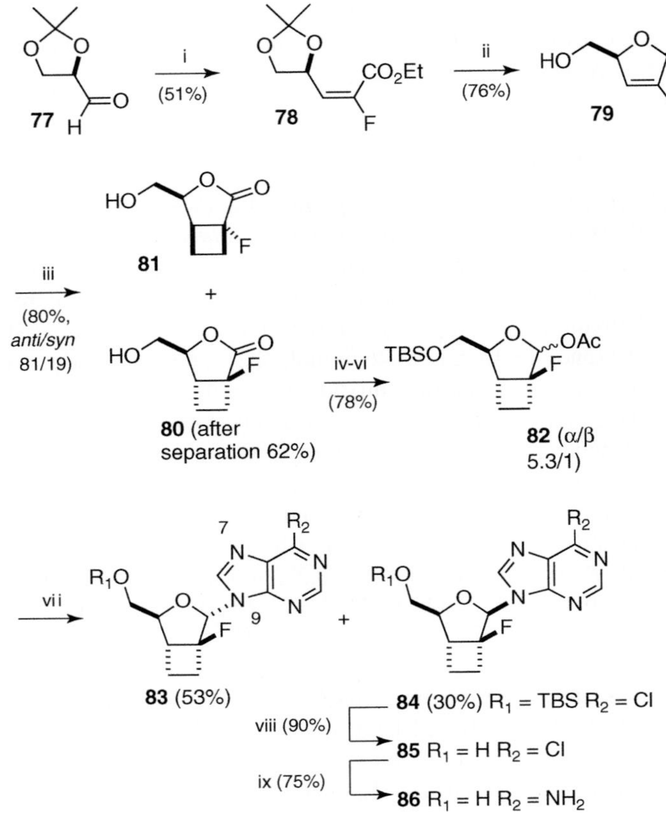

Scheme 10. *Reagents and conditions:* (i) (EtO)$_2$P(O)CHFCO$_2$Et, BuLi, THF, $-78°$C for 1 h, rt for 6 h; (ii) cat. PTSA, MeOH, rt, 2 h; (iii) $h\nu$, ethylene, Pyrex, acetone, $-20°$C, 3.5 h; (iv) TBSCl, Imid, DCM, rt, 16 h; (v) DIBAL-H, DCM, $-78°$C, 2.5 h; (vi) Ac$_2$O, pyridine, DCM, rt, overnight; (vii) 6-chloropurine, BSA, TMSOTf, PhMe, 100°C, 1 h; (viii) Et$_3$N-3HF, THF, rt, 7 h; (ix) NH$_3$, MeOH, sealed tube, 90°C, 40 h.

N-glycosylation of the anomeric acetates **82** using a modified Vorbrüggen protocol in PhMe at 100°C proved to be more efficient conditions to minimize the formation of the undesired α/β N-7 isomers (less than 10%). Both anomers **83** and **84** were obtained in pure form, after separation by column chromatography on silica gel, in 53% and 30% yields, respectively. It should be pointed out that an α/β ratio of 5.6:1 was obtained with the chloro derivative compared to 1.6:1 in the fluorinated series, suggesting that the steric effects of the halogen atom at C2 plays an important role in this N-glycosylation. Finally, on the pure minor β-anomer **84**, the cleavage of the silyl ether to give **85** followed by aminolysis afforded the desired fluoro β-adenosine analogue **86** in 68% overall yield.

Wengel et al. developed efficient access to various $2'$-O,$3'$-C linked bicyclic nucleosides as exemplified through synthesis of the analogue **95** with an oxetane

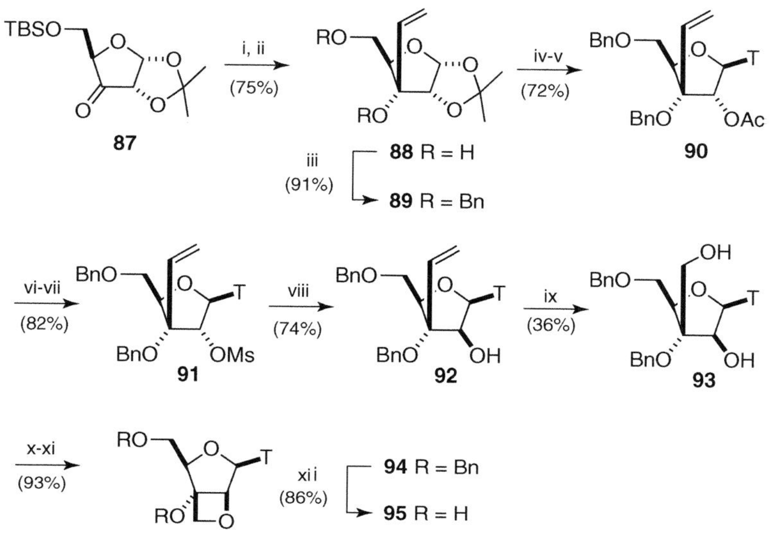

Scheme 11. *Reagents and conditions:* (i) vinyl-MgBr, Et$_2$O, THF, rt, 2 h; (ii) TBAF, THF, rt, 20 min; (iii) BnBr, NaH, DMF, rt, 20 h; (iv) 80% aq. AcOH, 90°C, 8 h, then Ac$_2$O, pyridine, rt, 48 h; (v) thymine, BSA, TMSOTf, ACN, rt, 24 h; (vi) NaOMe, MeOH, rt, 42 h; (vii) MsCl, pyridine, rt, 30 min; (viii) NaOH, EtOH, H$_2$O, reflux, 16 h; (ix) cat. OsO$_4$, NaIO$_4$, THF/H$_2$O, rt, 18 h, then NaBH$_4$, rt, 1 h; (x) MsCl, pyridine, rt, 17 h; (xi) NaH, DMF, rt, 30 min; (xii) H$_2$, 20% Pd(OH)$_2$/C, EtOH, rt, 6 h.

ring (see Scheme 11).[33] Subsequent diastereoselective addition of a vinyl Grignard reagent to the 5-protected 1,2-*O*-isopropylidene-α-D-xylofuranose derivative **87** (prepared in six steps from D-glucose)[34] followed by the cleavage of the silyl ether gave the desired furanose **88** in 75% overall yield. Next, the latter diol **88** was protected as a benzyl ether **89** and submitted to an acetolysis to provide the diacetate intermediate, which, under standard Vorbrüggen conditions, was converted exclusively into the β-nucleoside **90** in 66% yield over the three steps. The attack of the nucleobase from the favorable β-face is due to anchimeric assistance by participation of the neighboring 2′-*O*-Ac group. At this stage, deprotection of the 2′-*O*-acetate group of **90** followed by mesylation of the resulting secondary alcohol led to the intermediate **91**. Mesylate **91** was submitted to aqueous NaOH treatment to afford in situ the corresponding 2,2′-anhydro nucleoside intermediate, which was then hydrolyzed to furnish the nucleoside **92** with an arabino configuration in 60% yield from acetate **90**. At the final stage, oxidative cleavage of the vinyl double bond in **92**, by using OsO$_4$ in catalytic amounts and NaIO$_4$ as a cooxidant followed by treatment with NaBH$_4$, afforded the hydroxymethyl nucleoside derivative **93** in 36% overall yield. Selective monomesylation of the primary alcohol function of **93** followed by treatment with NaH promoted cleanly the formation of the oxetane ring of **94**, and subsequent hydrogenolysis of the benzyl protecting groups gave the target nucleoside **95** in good overall yield.

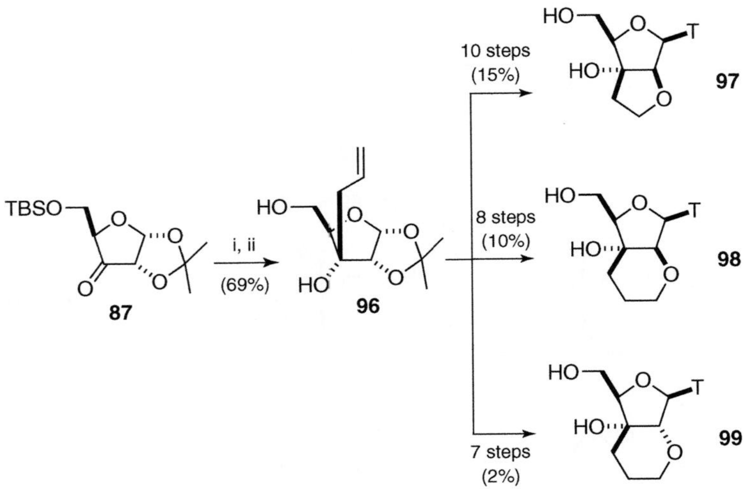

Scheme 12. *Reagents and conditions:* (i) allylMgBr, Et$_2$O, THF, 0°C, 2 h; (ii) TBAF, THF, rt, 1 h.

In the light of their previous work, the same authors[35] extended this strategy to the *cis*-furan and *cis*-pyran analogues **97** and **98**, respectively, as well as the *trans*-pyran analogue **99**, using allyl Grignard reagent instead of the corresponding vinyl, as summarized briefly in Scheme 12. In this approach, the allyl substituent should be regarded as a "hidden protecting group" in which the double bond following hydroboration/oxidation or oxidative cleavage/reduction sequences led to the formation of 3-hydroxypropyl or 2-hydroxyethyl groups. It is worth noticing that synthesis of the *trans*-pyran analogue **99** was also accomplished following a similar route from diol **96**, without the inversion sequence at C2′. However, the ring closure to form the pyran of **99** proved difficult for this *trans*-fused bicyclic skeleton, and the desired cyclized intermediate was isolated in around 20% yield.

Gotor and Theodorakis et al.[36] described the synthesis of a set of various 2′-3′-cyclohexene bicyclic nucleoside derivatives as pure β-anomers or as anomeric mixtures. The synthesis of the β-adenosine **106** is shown in Scheme 13 as an illustrative example. Construction of the required bicyclic furanosyl donor **104** was based on a Lewis acid–catalyzed Diels–Alder reaction[37] between the chiral silyl-*O*-protected butenolide **100** (prepared in 30% overall yield from D-mannitol in five steps)[38] and butadiene, which reacted exclusively by the less hindered face to provide the cycloadduct **101** in 85% yield. From the latter intermediate, subsequent reduction with DIBAL-H to the corresponding lactol **102**, followed by sequential acetylation, gave **103**. Fluoride-induced desilylation and another acetylation gave the key diacetate **104** in 90% overall yield. Then 6-chloropurine was treated sequentially with BSA, and the glycosyl donor **104** followed by addition of DBU, and finally, TMSOTf, to afford an α/β anomeric mixture in a 1:1.5 ratio of the

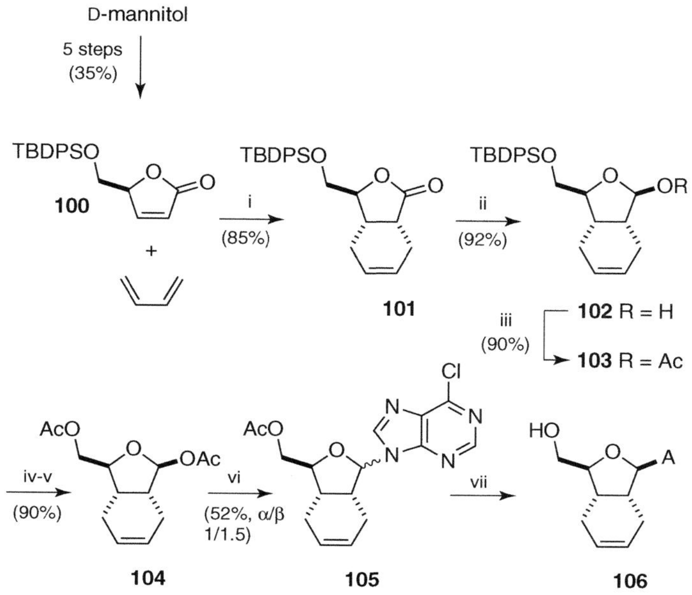

Scheme 13. *Reagents and conditions:* (i) AlCl$_3$, DCM, 60°C, 6 days; (ii) DIBAL-H, DCM, −78°C, 1 h; (iii) Ac$_2$O, pyridine, 25°C, 12 h; (iv) TBAF, THF, 25°C, 30 min; (v) Ac$_2$O, pyridine, 25°C, 12 h; (vi) 6-chloropurine, ACN, BSA, 70°C, 1.5 h, DBU, 0°C, then TMSOTf, 0°C, 2 h; (vii) NH$_3$, MeOH, Parr reactor, 110°C, 36 h.

nucleosides **105** in 52% yield. It should be pointed out that the *N*-glycosylation reaction on the 5′-*O*-silylated derivative **103** failed, presumably due to the steric hindrance of the silyl ether. Finally, subsequent treatment of **105** with methanolic ammonia at 110°C in a Parr reactor for 36 h followed by separation on silica gel chromatography furnished the desired β-anomer nucleoside **106**.

A series of 2′-*C*- and 3′-*C*-dibranched nucleosides with a benzo[*c*]furan core as an analogue of d4T has been synthesized by a convergent route starting from phthalaldehyde **107** as depicted in Scheme 14. After preparation of benzo[*c*]furan nucleoside analogue under a racemic form **115**, using epoxidation[39,40] or hydroxycyanation,[41] the enantio- or diastereoselective synthesis of the nucleoside **115** was reported.[42–46]

Starting from phtalaldehyde **107**, selective protection of one of the formyl groups followed by Wittig homologation of the remaining formyl group furnished the corresponding ethene derivative **109** in 58% overall yield.[14] The styrene **109** was converted into the corresponding dihydro derivatives **110** in 85% yield (ee > 99%) using the commercial Sharpless reagent, AD-mix α. After selective benzoylation of the primary hydroxyl group of **110**, treatment of the corresponding esters **111** in the presence of methanolic HCl afforded an anomeric mixture of 2,3-didehydro-2,3-dideoxyfuranosides **112**, which was used without purification in the next step. Using standard Vorbrüggen chemistry on **112**, both anomers of the

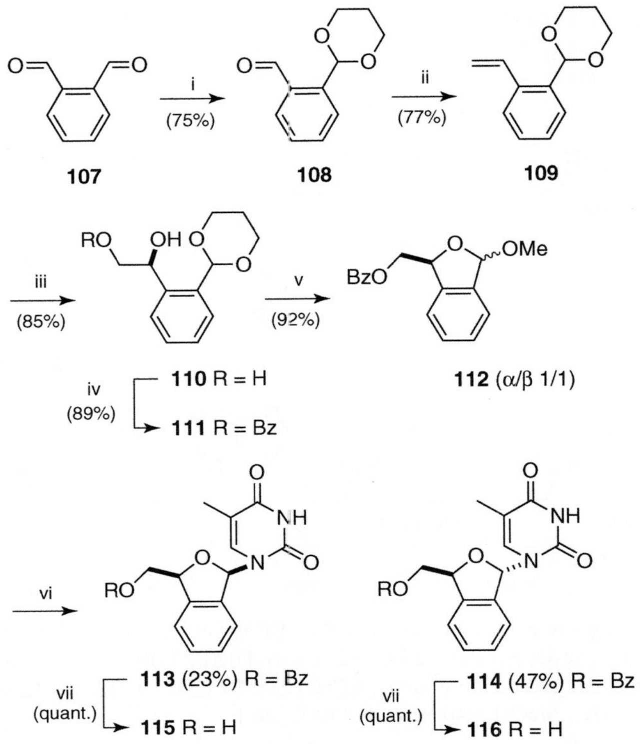

Scheme 14. *Reagents and conditions:* (i) propan-1,3-diol, cat. PTSA, PhMe, reflux, 1 h; (ii) Ph$_3$PCH$_3$Br, BuLi, THF, rt, 1 h; (iii) AD-mix α, *t*-BuOH, H$_2$O, 0°C, 1 h; (iv) BzCl, pyridine, −10°C, 2 h, then rt, 12 h; (v) HCl, MeOH, rt, 1 h; (vi) persilylated thymine, TMSOTf, DCE, −35°C, 1 h; (vii) NH$_3$, MeOH, rt, 24 h.

thymine derivatives β-**113** and α-**114** were obtained due to the lack of neighboring group participation to direct stereoselectivity. After removal of the benzoyl protection and subsequent silica gel chromatography, the target nucleosides **115** and **116** in the D-series were obtained enantiomerically pure in 23% and 47% overall yield, respectively. The related enantiomer analogues to L-nucleosides were synthesized with similar enantiomeric excess using the same strategy but employing AD-mix β.

2.6. Preparation of 2′,4′-Linked Bicyclic Nucleosides

Independently introduced by Imanishi et al.[47] and Wengel et al.,[48] locked nucleic acids (LNAs), in which the sugar adopts a perfect mimic of the N-type nucleoside conformation, present an unprecedented binding affinity toward complementary DNA and RNA sequences. This family of now commercially available locked nucleosides displays promising therapeutic properties and, consequently, is the focus of extensive research, which has been largely covered by recent reviews.[49]

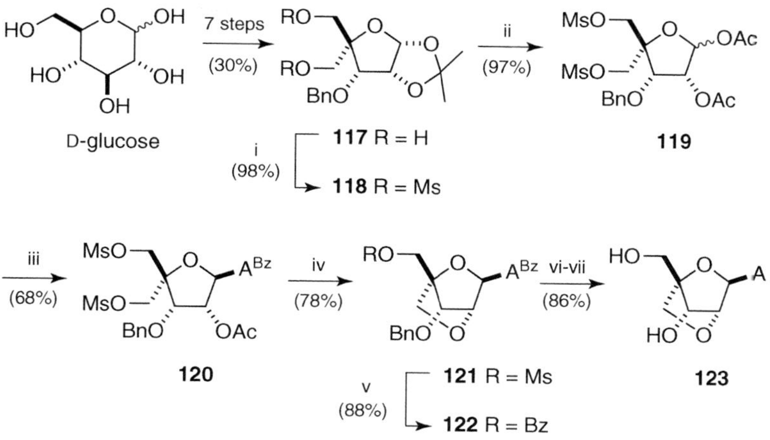

Scheme 15. *Reagents and conditions:* (i) MsCl, pyridine, rt, 1 h; (ii) Ac$_2$O, AcOH, cat. H$_2$SO$_4$, rt, 36 h; (iii) N^6-benzoyladenine, BSA, ACN, reflux, 1 h, then TMSOTf, reflux, 5 h; (iv) aq. LiOH, THF, rt, 3.5 h; (v) NaOBz, DMF, 90°C, 7 h; (vi) MeNH$_2$, NH$_3$, MeOH, rt, 2 days; (vii) H$_2$, 20% Pd(OH)$_2$/C, HCO$_2$NH$_4$, EtOH, reflux, 5 h.

The LNA nucleosides containing all the natural nucleobases have been prepared using a convergent strategy optimized by Koshkin et al.[50] from the 4-*C*-branched pentafuranose **117** derived from D-glucose, as exemplified in Scheme 15 through the synthesis of the adenosine derivative **123**. In the convergent Wengel's approach, the regioselective benzylation of the less hindered primary hydroxyl group of the ribofuranose **117** was identified as a bottleneck of the entire sequence. The maximum yield of the desired benzyl ether was 71% (18.5 g) but with a general yield of 45–50% on a larger scale. An elegant solution to this problem appeared to be the permesylation of the diol **117**. From a synthetic point of view, the mesyl on the C5' (β) hydroxyl group could be regarded as a protecting group. To this end, treatment of diol **117** with MsCl provided the corresponding di-*O*-mesylated intermediate **118**, which was engaged in the next step without purification. Then, acetolysis and acetylation in a one-pot procedure provided the glycosyl donor **119** as an anomeric mixture in nearly quantitative yield over the entire sequence, without purification. The latter universal glycosyl donor **119** was successfully engaged in Vorbrüggen coupling with 6-*N*-benzoyladenine, resulting in the isolation of the desired β-anomer **120** in 68% yield (18.1 g) after purification on a silica gel column. Treatment with aqueous LiOH promoted the intramolecular ether ring closure to furnish **121** in 78% yield. Subsequent displacement of the mesylate group in **121** by a benzoate to give **122** followed by removal of all the protecting groups, using ammonia–methylamine treatment and catalytic hydrogen transfer hydrogenolysis sequentially, afforded the target adenine–LNA parent **123** in 76% yield (2.54 g) over the three steps.

Stimulated by the impressive increased thermal affinities of LNA-modified oligonucleotides toward both DNA and RNA strands, various analogues have

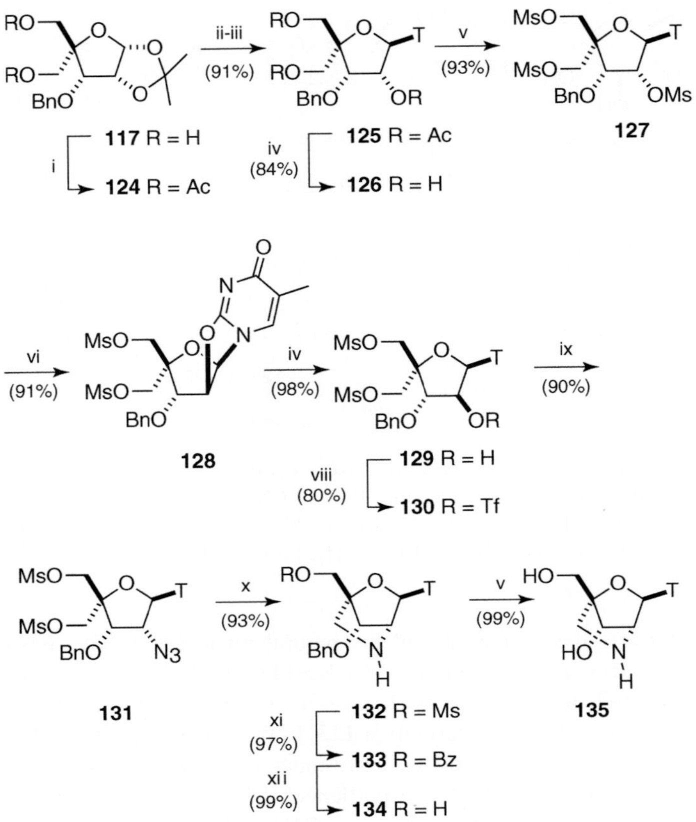

Scheme 16. *Reagents and conditions:* (i) Ac$_2$O, pyridine, rt, overnight; (ii) AcOH/Ac$_2$O/H$_2$SO$_4$ (100:10:0.1), rt, 2 h; (iii) thymine, BSA, TMSOTf, ACN, reflux, 1.5 h, then 0°C, TMSOTf, reflux, 3.5 h; (iv) sat. NH$_3$, MeOH, rt, overnight; (v) MsCl, pyridine, DCM, 15°C, 3 h; (vi) DBU, ACN, rt, 2 h; (vii) acetone, 0.1 M H$_2$SO$_4$, reflux, overnight; (viii) Tf$_2$O, DMAP, pyridine, DCM, 0°C, 2 h; (ix) NaN$_3$, DMF, rt, 4 h; (x) PMe$_3$, 2 M NaOH, THF, rt, 8 h; (xi) NaOBz, DMF, 80°C, 5.5 h; (xii) NH$_3$, MeOH, rt, 30 h; (xiii) H$_2$, 20% Pd(OH)$_2$/C, EtOH, rt, overnight.

been synthesized, including 2′-amino-, 2′-thio-, and 2′-carba-LNA nucleosides **135**, **142**, and **149**, respectively (Schemes 16 to 18).

First, Koch et al.[51] developed an efficient route for the synthesis of 2′-amino-LNA pyrimidinic nucleosides **135** as highlighted in Scheme 16. Peracetylation of the starting material **117** to afford **124**, followed by acidic hydrolysis and in situ acetylation under a standard protocol, gave the corresponding tetra-*O*-acetate intermediate. The latter was subsequently used, without further purification, as a glycosyl donor in a Vorbrüggen-type coupling to afford, after purification on a silica gel column, the β-thymine nucleoside intermediate **125** in 91% overall yield. Then, deacylation of the intermediate **125** provided the corresponding triol **126**, which was treated with MsCl to afford the permesylated nucleoside **127** in 78% yield over the two steps. At this point, the double inversion at C2′ to introduce an amino group precursor was performed via the formation of the 2,2′-anhydro nucleoside **128** which could be efficiently obtained from **127** by simple treatment with DBU. It should be pointed out that although this approach is straightforward, its applicability is restricted to pyrimidine nucleosides. Attempts to achieve a nucleophilic attack at C2′ on the 2,2′-anhydro **128** with either benzylamine or NaN_3 under various conditions have met with limited success. To circumvent this problem, this rigid anhydro-nucleoside **128** was first hydrolyzed under aqueous acidic conditions to provide the corresponding threo-configured nucleoside **129**, which was then reacted with Tf_2O to give the required triflate **130** in 80% yield after silica gel chromatography. However, the crude previous material **130** could be treated directly with NaN_3 to afford, via a regioselective nucleophilic displacement at C2′, the desired azido intermediate **131** in 90% yield after purification over the two steps. It is noteworthy that in the previous step, the mesylate groups implanted on both primary alcohol functions were not affected by treatment with an excess of NaN_3. Next, to avoid partial cleavage of the 3′-O-benzyl protecting group in **131** during the reduction of the azido function with hydrogen in the presence of a palladium catalyst, this step was carried out using a modified Staudinger reaction. Thus, the azide **131** was reacted with trimethylphosphine and aqueous NaOH in THF, leading to the corresponding amine, which cyclized spontaneously, by intramolecular nucleophilic displacement of the mesylate on the α-face, to afford the protected 2′-amino-LNA **132**. Following a sequence similar to that described in Scheme 15, the 2′-amino-LNA **135** was isolated in 95% yield from **132** over three steps.

The chemistry developed above (see Scheme 16) was next applied by the same authors to the synthesis of 2′-thio-LNA **142** from the key building block **128** (see Scheme 17).[52] In this route, attention was turned to the exchange of the 3′-O-benzyl ether protecting group by benzoyl ester. Indeed, probably due to catalyst poisoning, Pd-mediated hydrogenolysis failed when the sulfur atom was present in the substrate. After some trials it was found that hydrogenolysis of **128** under a hydrogen atmosphere (balloon) and 10% Pd/C in a 1:1 mixture of acetone and MeOH gave the desired alcohol **136** almost exclusively. This crude material was converted into its corresponding benzoate **137**, which was directly submitted to aqueous H_2SO_4 treatment to promote hydrolysis of the 2,2′-anhydro linkage, affording, after purification, the arabinothymidine nucleoside **138**. Reaction of **138** with Tf_2O under standard conditions provided the desired activated nucleoside **139**, which was transformed by simple treatment with Na_2S in DMF into the sulfur derivative **140** in 72% overall yield. Then, nucleophilic displacement of the 5′-O-mesylate group of nucleoside **140** with benzoate to give **141** followed by subsequent treatment using methanolic ammonia provided the 2′-thio-LNA parent **142** in 61% overall yield.

Using an intramolecular free-radical ring-closure reaction between a radical generated at C2′ and a chain with a terminal unsaturation as the radical trap at C4′, Chattopadhyaya et al. opened up an avenue for the preparation of five- and six-membered conformationally restricted 2′,4′-carbocyclic nucleosides as highlighted

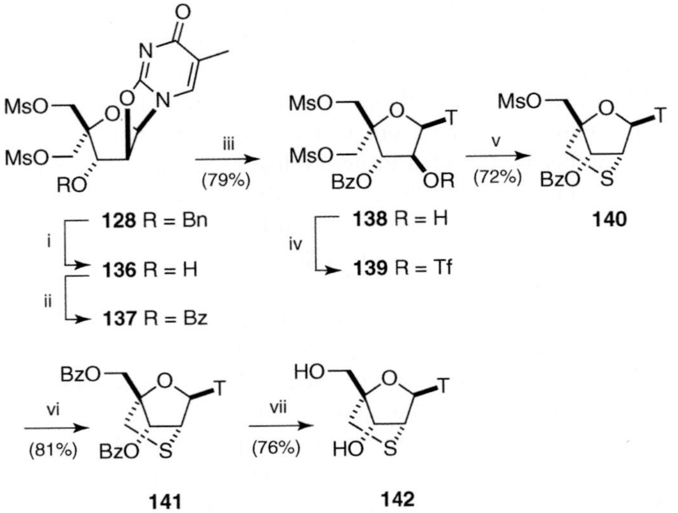

Scheme 17. *Reagents and conditions:* (i) H$_2$, 10% Pd/C, acetone/MeOH 1:1, rt, 23 h; (ii) BzCl, pyridine, DMF, rt, 18 h; (iii) 0.25 M H$_2$SO$_4$, DMF, 80°C, 22 h; (iv) Tf$_2$O, DMAP, pyridine, DCM, 0°C, 1 h; (v) Na$_2$S, DMF, rt, 3 h; (vi) NaOBz, DMF, 100°C, 24 h; (vii) NH$_3$, MeOH, rt, 20 h.

in Schemes 18 and 19. This strategy was used effectively by researchers from Isis Pharmaceuticals for the preparation of carba-LNA derivatives.[53]

The first carba-LNA **149** was prepared successfully as outlined in Scheme 18.[54] Exposure of the known modified diol **117** (Scheme 16) to NaH and BnBr at −5°C overnight afforded the expected monobenzylated adduct **143** in 66% yield (10.6 g). The vinyl chain in **144** was introduced from **143** in 87% overall yield using a classical Swern oxidation/Wittig methylenation sequence. Then, homologation of **144** to the corresponding allyl derivative **146** was performed in 67% overall yield via a hydroboration/oxidation sequence to deliver the alcohol **145**, which was oxidized under Swern conditions followed by Wittig methylenation. The *O*-phenyl thionocarbonate **147** was obtained in 58% overall yield from **146**, following a series of four straightforward reactions. Next, this key radical precursor **147** was refluxed in PhMe in the presence of AIBN and slow addition of Bu$_3$SnH, under highly diluted conditions, to avoid the formation of uncyclized side products. The expected 2′,4′-*cis*-fused bicyclic thymidine **148**, formed via a 5-hexenyl-type *exo*-mode cyclization, was isolated in 73% yield (1.6 g) after purification on silica gel as an inseparable mixture of diastereoisomers at C7′ in a 7:3 ratio in favor of the 7′-*R* stereoisomer. Cleavage of the benzyl ether protecting groups of **148** by transfer hydrogenolysis gave the targeted nucleoside analogue **149** as a mixture of diastereoisomers in 76% yield (Scheme 18).

In agreement with the chemistry presented in Scheme 18, preparation of the non-substituted parent carba-LNA **160** was described via the 5-*exo* radical cyclization of the *O*-benzyl oxime intermediate **154** as depicted in Scheme 19.[55] From the

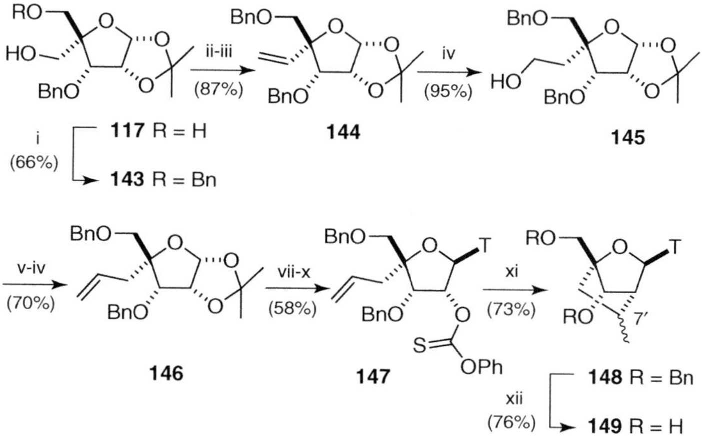

Scheme 18. *Reagents and conditions:* (i) NaH, BnBr, ACN, −5°C, overnight; (ii) (COCl)$_2$, DMSO, −78°C, DIPEA, DCM; (iii) Ph$_3$PCH$_3$Br, BuLi, THF, −78°C to rt, overnight; (iv) 9-BBN, THF, rt, overnight, then 3 N NaOH, 33% aq. H$_2$O$_2$, 50°C, 30 min; (v) (COCl)$_2$, DMSO, −78°C, DIPEA, DCM; (vi) Ph$_3$PCH$_3$Br, BuLi, THF, −78°C, overnight; (vii) AcOH, Ac$_2$O, cat. TfOH, 0°C, 30 min; (viii) thymine, BSA, ACN, 0°C, overnight; (ix) NH$_3$, MeOH, rt, overnight; (x) DMAP, PTC-Cl, pyridine, rt, overnight; (xi) Bu$_3$SnH PhMe, AIBN, reflux, 4 h; (xii) 20% Pd(OH)$_2$/C, HCO$_2$NH$_4$, MeOH, reflux, 8 h.

protected sugar **143**, the nitrile group precursor of the aldehyde function was installed by a classical sequence to furnish **150** in 56% yield over the two steps. Then, acetolysis of **150** followed by *N*-glycosylation with thymine furnished the compound **151**, which was then deacetylated to give the nucleoside intermediate **152** in 40% overall yield. Subsequent reduction of the cyano-nucleoside **152** with DIBAL-H and an acidic aqueous workup, followed by condensation of the corresponding aldehyde with *O*-benzylhydroxylamine, gave **153**, which was finally treated with PTC-Cl to yield the key radical precursor **154** in 42% yield for this sequence. Then, tin-mediated radical cyclization applied on the intermediate **154** proceeded exclusively in a 5-*exo* mode with a good level of diastereoselectivity to provide the C7′-*R* **155** in 60% yield after purification on silica gel column along with the minor C7′-*S* adduct isolated in 4% yield. At this stage, the *O*-benzyl oxime ether **155** was oxidized with MCPBA to give the corresponding oxime **156**, which was treated with Dess–Martin reagent to regenerate the carbonyl group, followed by reduction with NaBH$_4$ to furnish the alcohol **157** as the sole diastereoisomer in 60% overall yield. Deoxygenation of the secondary alcohol **157** was accomplished by the Barton–McCombie protocol: namely, formation of the C7′-*O*-(methylthio)thiocarbonate derivative **158**, which was more accessible than phenyl thionocarbonate due to steric hindrance, and reduction with Bu$_3$SnH to provide the protected carba-LNA **159** in 40% yield. Finally, hydrogenolysis of the benzyl ethers using Pearlman's catalyst in the presence of a hydrogen donor yielded the target carba-LNA **160** (Scheme 19).

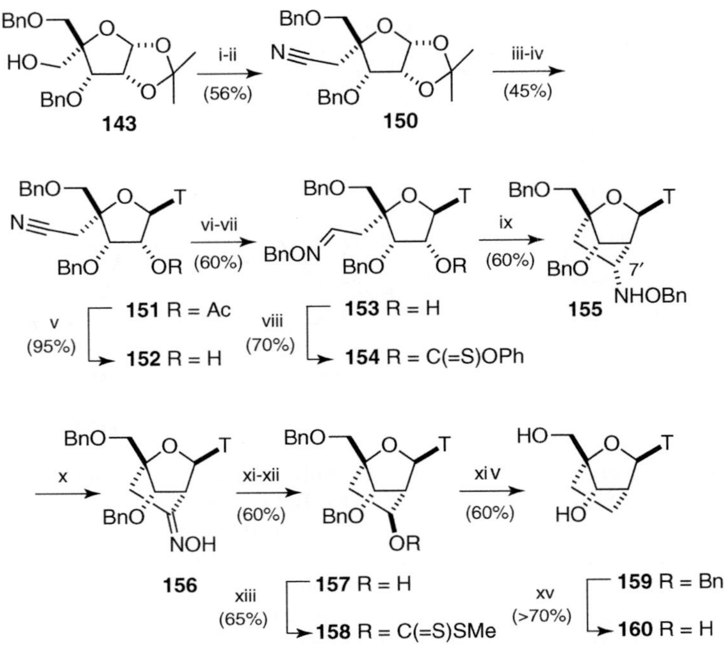

Scheme 19. *Reagents and conditions:* (i) Tf$_2$O, pyridine, DCM, 0°C, 3 h; (ii) LiCN, DMF, rt, 3 days; (iii) AcOH, Ac$_2$O, TfOH, rt, 3 h; (iv) thymine, BSA, TMSOTf, ACN, 80°C, overnight; (v) MeNH$_2$, MeOH, rt, 1 h; (vi) DIBAL-H, DCM, −78°C, 4 h; (vii) HCl·NH$_2$OBn, DCM, pyridine, reflux, 2 h; (viii) PTC-Cl, pyridine, rt, 2 h; (ix) Bu$_3$SnH, AIBN, PhMe, reflux, 8 h; (x) MCPBA, K$_2$CO$_3$, EA, rt, 1 h; (xi) DMP, NaOAC, DCM, rt, 30 min; (xii) NaBH$_4$, EtOH, rt, 3 h; (xiii) CS$_2$, MeI, NaH, THF, rt, 3 h; (xiv) Bu$_3$SnH, AIBN, PhMe, reflux, 1.5 h; (xv) 10% Pd/C, HCO$_2$NH$_4$, MeOH, reflux, 2 h.

Following this unique radical cyclization strategy, Chattopadhyaya et al. have prepared various carba-LNAs with different substituents in the fused carba-ring as presented in Figure 2.

The development of six-membered ethylene-bridged nucleic acids (ENAs) was boosted by the previous results on LNAs. Koizumi et al. reported the first preparation of this new family of bicyclic nucleosides containing all natural nucleobases via a convergent approach from the known pentofuranose **145** (Scheme 18) as exemplified in Scheme 20 with ENA-thymidine **175**.[56] The hydroxyethyl of the starting material **145** was converted into its corresponding tosylate **172**, which led to the pivotal nucleoside derivative **173** in 56% overall yield, according to the sequence shown in Scheme 19. Cleavage of the acetate in **173** by simple treatment with a mixture of 1 M aqueous NaOH and pyridine triggered an intramolecular etherification to furnish the cyclized adduct **174** in 79% yield. Finally, catalytic hydrogenolysis afforded the ENA-thymidine analogue **175** in 64% yield (Scheme 20).

The aza-ENA-thymidine analogue **182** was synthesized by Chattopadhyaya et al. as shown in Scheme 21 from the cyano-nucleoside intermediate **152** (prepared in

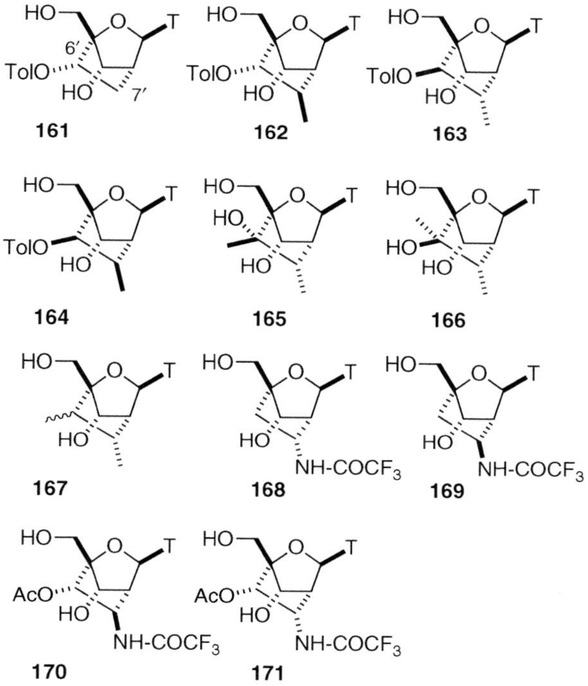

Figure 2. Carba-LNA analogues **161–167** with C6′-OH and/or C6′-Me; carba-LNA amino- and amino alcohol analogues **168–171**.

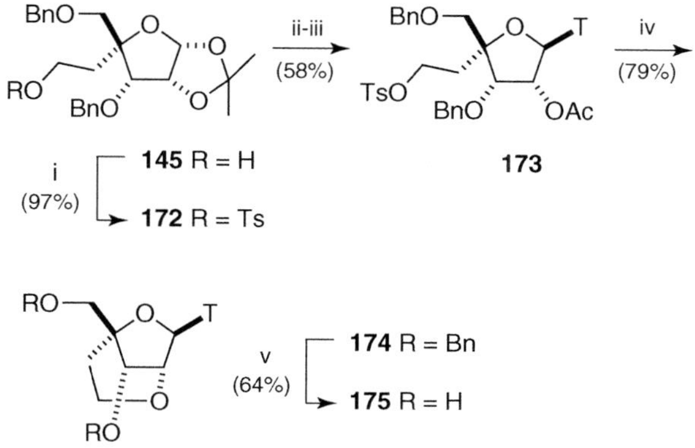

Scheme 20. *Reagents and conditions:* (i) TsCl, Et$_3$N, DMAP, DCM, rt, overnight; (ii) AcOH, Ac$_2$O, H$_2$SO$_4$, rt, 1 h; (iii) thymine, BSA, TMSOTf, DCE, 50°C, 1 h; (iv) 2 N NaOH, pyridine, H$_2$O, rt, 15 min; (v) H$_2$, 20% Pd(OH)$_2$/C, MeOH, rt, 5 h.

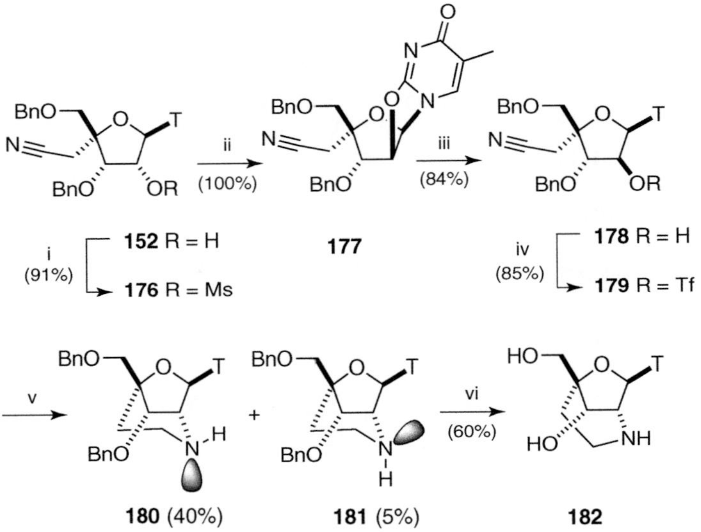

Scheme 21. *Reagents and conditions:* (i) MsCl, pyridine, 0°C, 6 h; (ii) DBU, ACN, rt, 1 h; (iii) 0.1 M H_2SO_4, acetone, reflux, overnight; (iv) Tf_2O, pyridine, DMAP, DCM, 0°C, 2.5 h; (v) $NaBH_4$, TFA, THF, rt, overnight; (vi) 20% $Pd(OH)_2/C$, HCO_2NH_4, MeOH, reflux, overnight, followed by 1 M BCl_3 in DCM, −78°C, 3 h.

Scheme 19).[57] The arabino-configured nucleoside **178** was synthesized from **152** in a way similar to that described for **129** in Scheme 16. Thus, from **152**, C2′ inversion was achieved in three steps and 76% overall yield to give **178**. Subsequent treatment of the resulting alcohol **178** with Tf_2O provided the triflate **179** in 85% yield. Cyano derivative **179** was reduced using trifluoroacetoxy borohydride formed in situ to deliver the primary amine, which underwent a spontaneous intramolecular cyclization to afford a mixture of two diastereomeric aza-ENA-T **180** and **181** isomers, isolated for characterization in 40 and 5% yield, respectively, after purification on silica gel chromatography column. Removal of the benzyl groups on this diastereomeric mixture was conducted in two steps: first by classical transfer hydrogenolysis using $Pd(OH)_2/C$ and HCO_2NH_4 in MeOH, followed by exposure of the latter crude mixture to BCl_3 to give the aza-ENA-T **182** in 60% overall yield.

Chattopadhyaya et al.[58] used their previous intramolecular free-radical ring-closure reaction on the compound **183** to synthesize the carba-ENA thymidine analogue **185** as outlined in Scheme 22. A sequence identical to that applied in the synthesis of carba-LNA **149** in Scheme 18 was utilized to install the 4′-*C*-homoallylic chain on the starting material **146**. The key 6-*exo*-heptenyl radical cyclization on **183** proceeded efficiently to provide the 8′-*R* bicyclic nucleoside **184** in 76% yield as a single detectable diastereoisomer. Removal of the benzyl ethers under catalytic transfer hydrogenation provided the target carba-ENA thymidine analogue **185**.

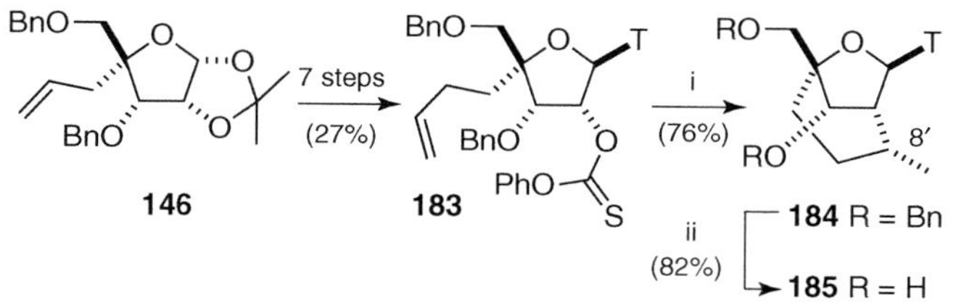

Scheme 22. *Reagents and conditions:* (i) Bu$_3$SnH, PhMe, AIBN, reflux, 4 h; (ii) 20% Pd(OH)$_2$/C, HCO$_2$NH$_4$, MeOH, reflux, 12 h.

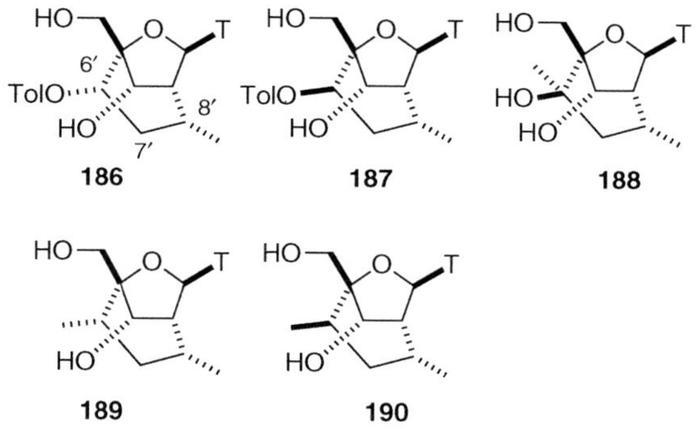

Figure 3. Carba-ENA analogues **186–190** with C6′-OH and/or C6′-Me, C8′-Me.

As already mentioned for the LNA series, this radical cyclization reaction provided an efficient access to various carba-ENA derivatives, as outlined in Figure 3.

An elegant linear synthesis of the nonsubstituted parent carba-ENA **201** from 3′,5′-di-*O*-TBS-protected uridine **191**, published by Nielsen et al., involved an RCM reaction[59] as a key step. It is depicted in Scheme 23.[60]

The secondary alcohol **191** was converted to the corresponding 2′-*O*-phenoxythiocarbonyl radical precursor **192**, which was treated with allyltributyltin and AIBN in refluxing PhMe to provide the 2′-*C*-allyl derivative **193** in 51% yield over the two steps. The high diastereoselectivity of this radical allylation is rationalized by the reaction of allyltributyltin on the less hindered α-face of the 2′-*C*-centered radical intermediate. Then, treatment of **193** with 80% acetic acid over 3 days led to smooth and selective cleavage of the 5′*O*-silyl ether and isolation of the desired primary alcohol **194** in 75% yield. Next, the key intermediate **198** was obtained in around 20% overall yield from **194** using well-established chemistry presented in this part and involving standard protecting-group manipulations. The diene **198** was submitted to an RCM reaction using Grubbs' catalyst II (for the chemical structure of Grubbs' catalysts I and II, see Figure 4) to furnish the expected bicyclic nucleoside **199** in nearly quantitative yield. Furthermore, fluorine-induced desilylation of **199** gave the unsaturated ENA-uridine **200**, which was hydrogenated using Adams' catalyst to provide the

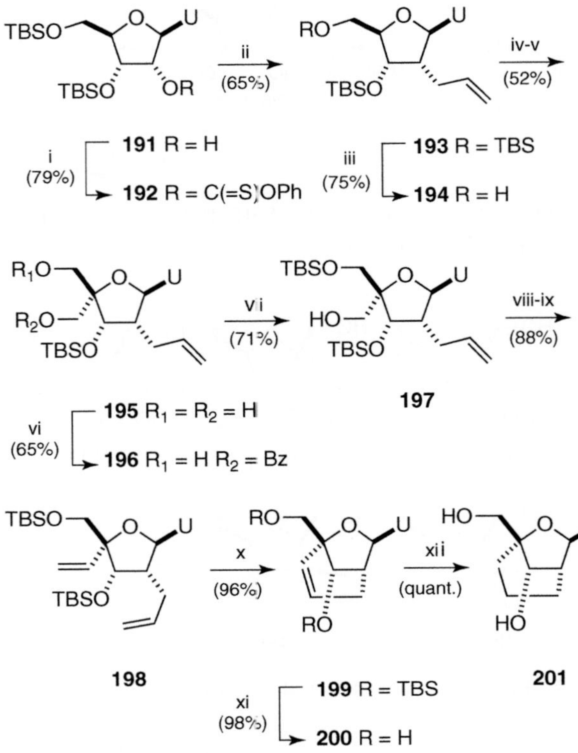

Scheme 23. *Reagents and conditions:* (i) PCT-Cl, DMAP, ACN, 19 h; (ii) allyl-SnBu$_3$, AIBN, PhMe, reflux, 3 h; (iii) 80% aq. AcOH, rt, 3 days; (iv) DMP, DCM, rt, 2 h; (v) H$_2$CO, 2 M NaOH, dioxane, rt, 17 h, then NaBH$_4$, rt, 5 h; (vi) BzCl, pyridine, rt, 1.75 h; (vii) TBSCl, Imid, DMF, rt, 15 h; (viii) DMP, DCM, rt, 2 h; (ix) BuLi, Ph$_3$PMeBr, THF, 0°C, 45 min; (x) Grubbs' catalyst II, DCM, reflux, 25 h; (xi) TBAF, THF, rt, 5 h; (xii) H$_2$, PtO$_2$, MeOH, rt, 90 min.

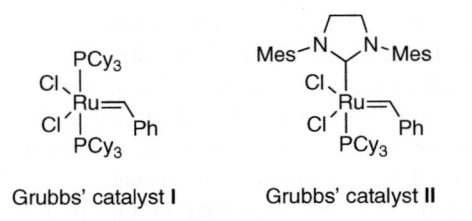

Figure 4. Chemical structure of Grubbs' catalysts I and II used in this chapter.

nonsubstituted carba-ENA **201** in quantitative yield. It should be pointed out that a carba-ENA-thymidine analogue of **201** has been synthesized by Chattopadhyaya et al.[61] from an endo-hexenyl cyclized side product isolated in low yield (less than 10%).

To end this part, starting from the nucleoside derivative **197** (Scheme 23), the same authors described the preparation of a substituted carba-ENA **205** as a mixture of diastereoisomers (see Scheme 24).[62] The alcohol **197** was subjected to the Dess–Martin reagent followed by homologation of the resulting aldehyde using the Bestmann–Ohira reagent to provide the enyne **202** in 65% overall yield. The enyne derivative **202** was then treated with Grubbs' catalyst II, using microwave heating, to afford the diene **203** in 82% yield. Deprotection of the hydroxyl groups was carried out using KF and 18-crown-ether-6 at 100°C under microwave irradiation to afford the unprotected nucleoside analogue **204** in 71% yield. Hydrogenation of the diene **204** was performed with Adams' catalyst under a hydrogen atmosphere to give a mixture of diastereoisomers **205** ($6'R/6'S$, 8:1) in 70% yield.

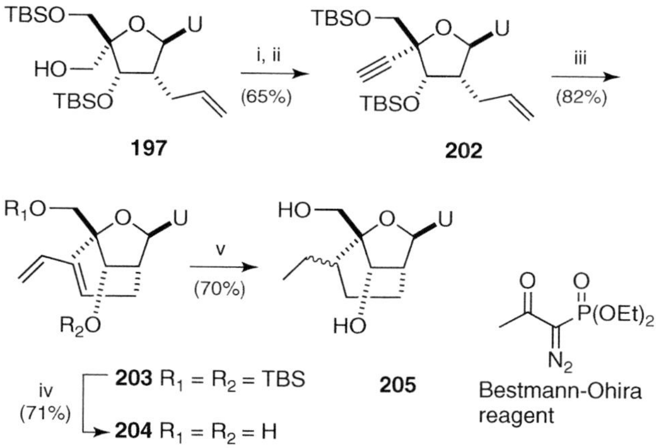

Scheme 24. *Reagents and conditions:* (i) DMP, DCM, rt, 2 h; (ii) Bestmann–Ohira reagent, K_2CO_3, MeOH, rt, 24 h; (iii) Grubbs' catalyst II, DCM, 100°C, MW, 2 h; (iv) KF, 18-crown-ether-6, ACN, 100°C, MW, 1 h; (v) H_2, PtO_2, MeOH, rt, 24 h.

2.7. Preparation of 3′,4′-Linked Bicyclic Nucleosides

A large family of bicyclic nucleosides based on a furanose moiety with linkers between C3′ and C4′ has been described as presented in this section. The preparation of many of these 3′,4′-linked bicyclic nucleosides was greatly facilitated by the easy implantation of a hydroxymethyl group at C4 (or C4′) from an aldehyde precursor using the well-known aldol condensation/reduction sequence.[25]

To prepare the four natural methano-nucleoside analogues, as illustrated through the synthesis of the adenosine derivative **214** (see Scheme 25), Mathé et al.[63]

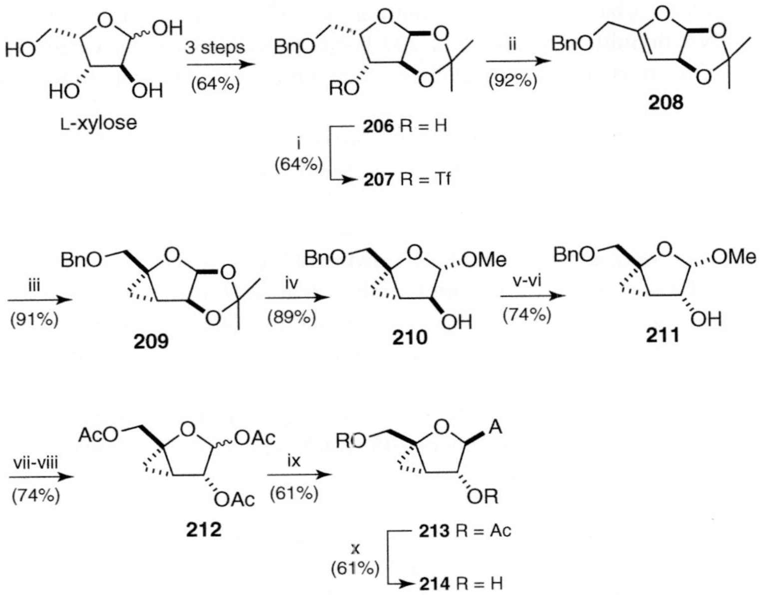

Scheme 25. *Reagents and conditions:* (i) Tf$_2$O, pyridine, DCM, −15°C, rt, 1 h; (ii) DBU, ACN, reflux, 1 h; (iii) ZnEt$_2$, CH$_2$I$_2$, PhMe, rt, 4 h; (iv) 4 N HCl/dioxane MeOH, rt, overnight; (v) CrO$_3$, Ac$_2$O, pyridine, DCM, rt, 4 h; (vi) NaBH$_4$, EtOH, rt, few minutes; (vii) Ac$_2$O, pyridine, rt, 2 h; (viii) AcOH, Ac$_2$O, cat. H$_2$SO$_4$, rt, overnight; (ix) adenine, SnCl$_4$, ACN, rt, 1 h; (x) NaOMe, MeOH, rt, overnight.

proposed an elegant and straightforward strategy based on a diastereoselective addition of carbene to the electron-rich 3,4-unsaturated 3-deoxy sugar **208** followed by a classical glycosylation step. The starting material **206** was prepared from L-xylose according to a procedure described on the D-counterpart.[64] Activation of the secondary alcohol in **206** as a triflate **207** followed by DBU-induced β-elimination led to the 4,5-dihydrofurane derivative **208** in 59% overall yield. From electron-rich enol ether **208**, a diastereoselective cyclopropanation reaction using Furukawa's procedure provided, after purification, the desired isomer **209** in 91% yield. Less than 2% of the other diastereoisomer was isolated; it was thus assumed that the cyclopropanation reaction occurred on the less hindered α-face opposite the isopropylidene protecting group. Then, removal of the acetonide group under acidic conditions with HCl in a dioxane–MeOH mixture provided the α-anomer of the methyl furanoside **210** in 89% yield as the sole isomer. At this stage, the absolute configuration at C2 was inverted using a standard oxidation–reduction sequence to yield the required alcohol **211** in 74% overall yield. The reduction of the ketone intermediate was totally diastereoselective, due to the presence of both OMe and cyclopropyl on the α-face. Since acetyl furanosides are more efficient intermediates in β-coupling reactions with nucleobases, the free alcohol at C2 in **211** was first acetylated, then the anomeric methoxy was converted to the corresponding acetate under acid-catalyzed conditions in the presence of Ac_2O to furnish **212** as an anomeric mixture (α/β, 9:91 ratio) in 74% overall yield. It is worth noting that under this H_2SO_4-mediated acetolysis in Ac_2O, the *O*-benzyl protecting group was cleaved and the resulting primary alcohol acetylated in situ without formation of the pyranose isomer. Finally, this triacetate **212** was coupled with adenine under Vorbrüggen conditions using $SnCl_4$ as a catalyst to provide the protected β-anomer **213**, which was then submitted to NaOH in MeOH to afford the desired adenosine analogue **214** in 37% yield for the two steps.

In 1999, Imanishi et al. developed a general and practical route to 3′-*O*,4′-*C*-methyleneribonucleosides containing all the natural nucleobases from 3-*O*-benzyl-4-hydroxymethyl-1,2-*O*-isopropylidene-α-D-ribofuranose **117** (Scheme 15) as outlined in Scheme 26.[65] It should be noted that Walker et al. first isolated the 3′-4′-oxetane thymidine as a side product during the preparation of 4′-substituted nucleosides.[66] A stereoselective silylation of the diastereotopic hydroxy groups in the common starting material **117** (prepared from D-glucose in six steps) to give **215**, followed by tosylation of the remaining primary alcohol function, led to the isolation of the fully protected compound **216** in 65% overall yield. Subsequent acetolysis of the 1,2-*O*-isopropylidene intermediate **216**, followed by classical hydrogenolysis of the benzyl protecting group and direct acetylation of the resulting triol, furnished the key glycosyl donor **218** in a good overall yield of 78%. Treatment of compound **218** with persilylated N^6-benzoyladenine and TMSOTf afforded exclusively the corresponding β-nucleoside, which was subjected to methanolysis to give the diol **219** in 63% overall yield. Finally, intermediate **219** upon treatment with excess NaHDMS afforded exclusively

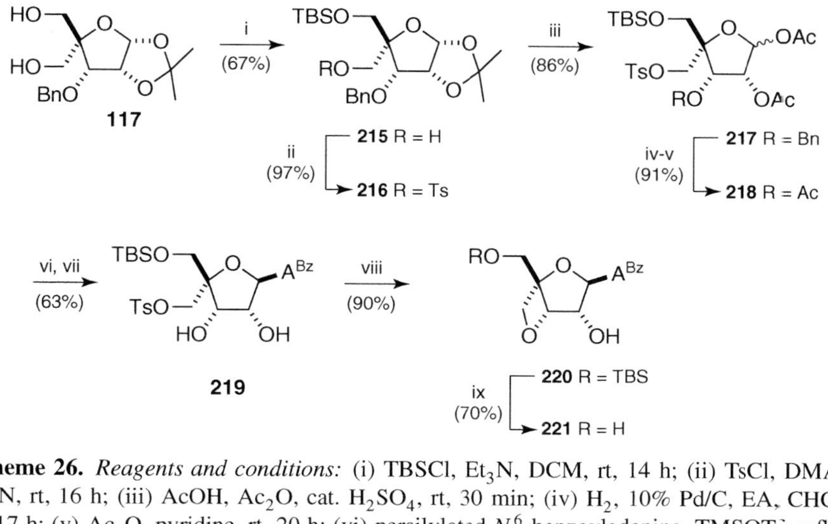

Scheme 26. *Reagents and conditions:* (i) TBSCl, Et_3N, DCM, rt, 14 h; (ii) TsCl, DMAP, Et_3N, rt, 16 h; (iii) AcOH, Ac_2O, cat. H_2SO_4, rt, 30 min; (iv) H_2, 10% Pd/C, EA, $CHCl_3$, rt, 17 h; (v) Ac_2O, pyridine, rt, 20 h; (vi) persilylated N^6-benzoyladenine, TMSOTf, reflux, 8–18 h; (vii) K_2CO_3, MeOH, rt, 15 min; (viii) NaHMDS, THF, rt, 1 h; (ix) TBAF, THF, rt, 15 min.

(see below) the desired oxetane **220** in 90% yield. Cleavage of the silyl group in **220** yielded the targeted oxetane-locked nucleoside **221**.

It should be emphasized that the exclusive oxetane ring formation in the latter example could be attributable to the predominant S-conformation of the ribonucleoside **219**, in which only the 3′-OH positioned near the 4′-methylene carbon center led to the cyclization reaction, unlike the 2′-OH, which is too far away to attack the 4′-methylene carbon. In sharp contrast, under the same conditions, the 1′-methoxy congener **222**, which should exist mainly in the N-conformation due to the anomeric effect, led to a mixture of the corresponding 2′-O,4′-C-methylene and 3′-O,4′-C-methylene derivatives **223** and **224** in almost equal amounts (see Scheme 27).[67]

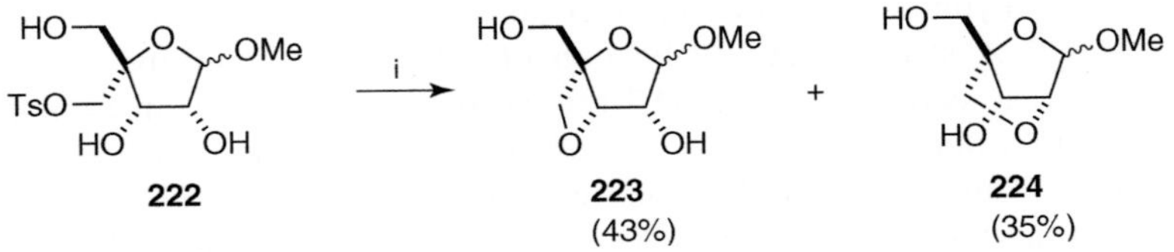

Scheme 27. *Reagents and conditions:* (i) NaHMDS, THF, 20°C.

Continuing their efforts in this series, Imanishi et al. reported the synthesis of the azetidine thymidine analogue **232** from the known 3-azido protected sugar **226**[68] prepared in four steps from **225**, as depicted in Scheme 28.[69] Diastereoselective silylation of the diol **226** led to the monosilylated derivative **227** in 62% yield. The key azetidine intermediate **228** was obtained directly from the azido alcohol **227** in nearly quantitative yield by simple treatment with Ph$_3$P in refluxing o-xylene. Very

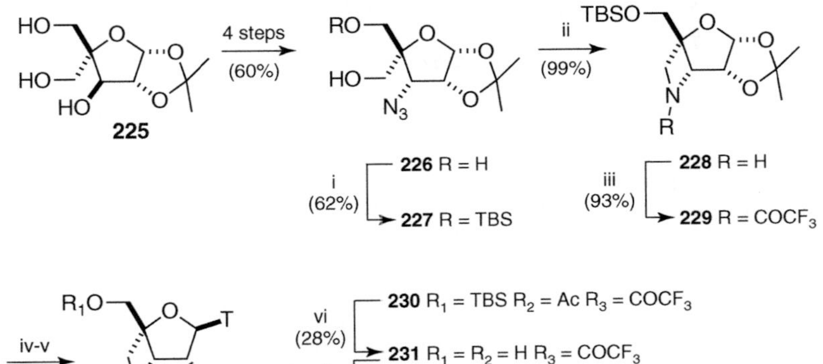

Scheme 28. *Reagents and conditions:* (i) TBSCl, Et$_3$N, DCM, rt; (ii) Ph$_3$P, o-xylene, reflux; (iii) TFAA, DMAP, DCM, 0°C; (iv) Ac$_2$O. AcOH, cat. H$_2$SO$_4$, rt; (v) persilylated thymine, TMSOTf, DCE, r; (vi) K$_2$CO$_3$, MeOH, then TBAF, THF, rt; (vii) aq. NH$_3$, dioxane, rt.

few examples have been reported concerning the direct formation of azetidines by the Staudinger reaction.[70] With azido alcohol **227** it was assumed that the azaylide intermediate reacted with the alcohol function to form the azetidine ring. Then, protection of azetidine **228** as trifluoracetate (**229**), and acetolysis and N-glycosylation of the resulting diacetate intermediate, provided the β-protected nucleoside **230** in 65% yield over the entire sequence. Finally, all the protecting groups from **230** were removed under standard protocols to give the desired azetidine **232** in 25% overall yield.

More recently, a synthesis of novel potential anti-HCV 3′,4′-oxetane cytidine **242** and adenosine **248** nucleosides were described using an original oxetane ring formation as outlined in Schemes 29 and 30.[71] From the known 2′-deoxy-2′-fluoro-2′-C-methyluridine **233**,[72] the key diol intermediate **235** was obtained following a Pfitzner–Moffatt oxidation/aldol-reduction sequence on derivative **234** in a very moderate yield of 22%. Next, the less hindered C5′ primary alcohol on the α-face was protected as its DMTr-ether **236** followed by silylation of the remaining alcohol on the β-face to provide the protected nucleoside **237** in 30% yield over the two steps. After cleavage of the DMTr-ether of **237** with CAN,[73] activation of the liberated 5′-α-hydroxyl group in **238** as a triflate was carried out under

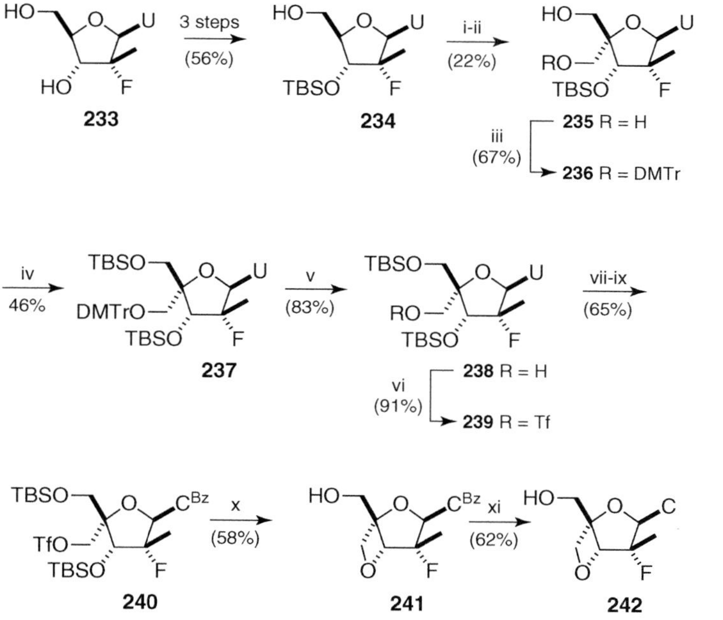

Scheme 29. *Reagents and conditions:* (i) EDC, DMSO, pyridine, TFA, rt, 30 min; (ii) CH$_2$O, aq. NaOH, dioxane, rt, 30 min, then NaBH$_4$, 0°C to rt, 1 h; (iii) DMTrCl, pyridine, 0–10°C, 8 h; (iv) TBSCl, Imid, rt, 18 h; (v) CAN, ACN, RT, 16 h; (vi) Tf$_2$O, 2,6-lutidine, DCM, 16 h; (vii) TIPBSCl, DIPEA, DMAP, ACN, rt, 1 h; (viii) aq. NH$_3$, rt, 0.5 h; (ix) BzCl, DIPEA, DMAP, rt, 1 h; (x) TBAF, THF, reflux, 3 h; (xi) 7 N NH$_3$, MeOH, rt, 6 h.

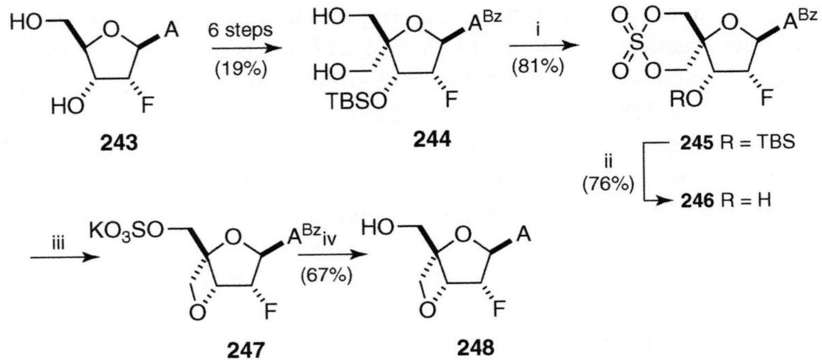

Scheme 30. *Reagents and conditions:* (i) SOCl$_2$, pyridine, DCM, −10 to 0°C, then RuCl$_3$, NaIO$_4$, DCM/MeCN/H$_2$O, rt, 1 h; (ii) TBAF, THF, rt, 1 h; (iii) *t*-BuOK, THF, 0−10°C, 0.5 h; (iv) H$_2$SO$_4$, rt, 48 h, then NH$_3$, MeOH, rt, 16 h.

classical conditions to furnish **239** in good overall yield. Subsequent treatment of uridine derivative **239** with 2,4,6-triisopropylbenzenesulfonyl chloride followed by mild aminolysis and, finally, benzoylation of the resulting amino group led to the desired compound **240**, in 65% yield for the entire process, without affecting the triflate functionality. At this point, the triflate derivative **240** was subjected to TBAF treatment and the resulting free secondary alcohol at C4′, under these basic conditions, triggered the ring closure to give oxetane **241** in 58% yield. The latter intermediate underwent standard methanolic ammonia treatment to afford the target 3′,4′-oxetane cytidine−modified nucleoside **242** in correct yield.

In the purine series, the key diol intermediate **244** was obtained from known fluoroadenosine intermediate **243**[74] (see Scheme 30) following chemistry similar to that outlined in Scheme 29. However, the selective derivatization of the 5′-α-hydroxyl group of compound **244** proved too difficult compared to the introduction of the DMTr protecting group from the diol **235** in the pyrimidine nucleoside series (Scheme 29). Tosylation of diol **244** with TsCl led to a complex mixture of the starting material, mono- and ditosylate derivatives, as well as the corresponding oxetane. This failure prompted the authors to investigate an elegant alternative route to the oxetane **248**, which is based on selective ring closure of the cyclic sulfate **246**. To this end, the diol **244** was treated with sulfuryl chloride to furnish the cyclic sulfite, which was then oxidized to the corresponding cyclic sulfate **245** with RuO$_4$ generated in situ from ruthenium(III) chloride in catalytic amounts and NaIO$_4$. Subsequent removal of the silyl ether followed by treatment of the liberated 3′-hydroxyl group in **246** with a strong base promoted the formation of the oxetane ring to afford the sulfonic salt derivate **247**. Finally, acidic hydrolysis of the sulfonic acid salt and cleavage of the *N*-benzoyl protection group on the nucleobase with methanolic ammonia gave the target 3′,4′-oxetane adenine nucleoside **248** in 67% overall yield.

Although the most popular pathway for preparing 4′-α-branched nucleosides is based on the aldol condensation with formaldehyde of the corresponding sugar or nucleoside intermediates leading, after reduction, to the corresponding 4'-α-C-branched hydroxymethyl moiety, as already shown, alternative strategies have also been reported.[75] For example, a regio- and stereoselective radical cyclization with a diphenylvinylsilyl group as a temporary connecting tether was used successfully in the synthesis of the bicyclic nucleoside **256**, as presented in Scheme 31.[76] The key 4'-phenylseleno derivative **250** was obtained as a sole diastereoisomer in 80% yield from the known thymidine 5'-aldehyde **249** through enolization in the presence of PhSeCl and Et$_3$N, followed by reduction with DIBAL-H, using a sequence described previously by Giese et al.[77] From compound **250**, after standard protecting group manipulation, the secondary alcohol **251** was treated with diphenylvinylchlorosilane to provide the corresponding vinylsilyl derivative **252** in 60% overall yield. Upon treatment under dilute conditions (1.6 × 10^{-2} M) with Bu$_3$SnH and AIBN, as initiator, the 5-*exo*-cyclization of the radical formed at the C4' position in **252** gave the kinetically favored radical **B**, which rearranged into the more stable ring-enlarged radical **C**.[78] The radical species **C** was trapped with

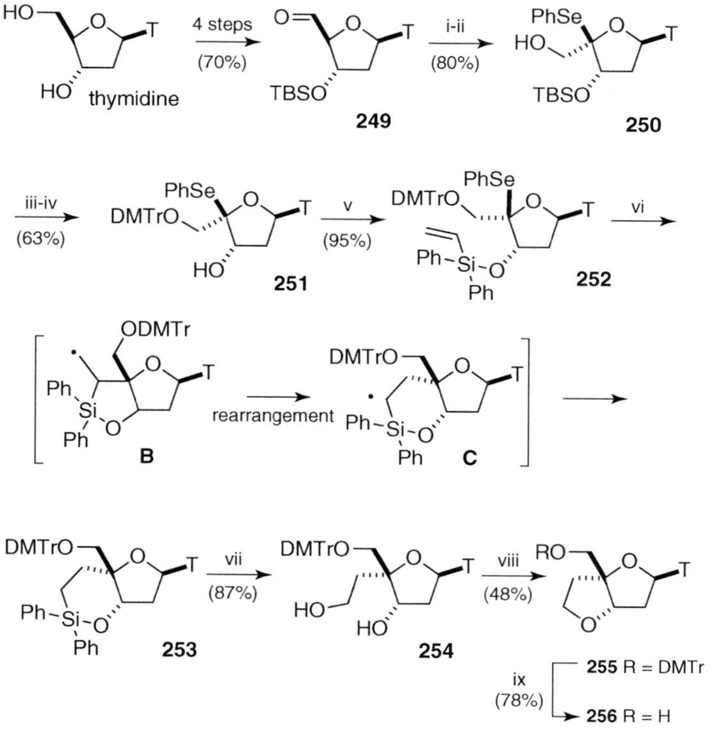

Scheme 31. *Reagents and conditions:* (i) PhSeCl, Et$_3$N, DCM, −78°C to rt; (ii) DIBAL-H, THF, −78°C; (iii) DMTrCl, pyridine, rt, 5 h; (iv) TBAF, THF, 0°C; (v) (vinyl)Ph$_2$SiCl, DMAP, Et$_3$N, PhMe, rt, 2 h; (vi) Bu$_3$SnH, AIBN (both reagents added over 4 h), PhH, 80°C; (vii) aq. H$_2$O$_2$, KF, KHCO$_3$, THF/H$_2$O 1:1, rt, 16 h; (viii) MsCl, DMAP, Et$_3$N, DCM; (ix) TFA, DCM.

Bu$_3$SnH to provide the cyclized adduct **253**, which was submitted without purification to an oxidative ring cleavage under Tamao oxidation conditions, leading to the 2-hydroxyethyl intermediate **254** in 87% yield over the entire sequence. Cyclization of diol **254** using MsCl in the presence of Et$_3$N generated the furan ring of **255**, then classical acidic cleavage of the DMTr protecting group furnished the targeted nucleoside analogue **256** in around 40% overall yield.

The thymidine analogues fused with the 3′,4′-tetrahydrofuran ring **267** were described by Chun et al.[79] starting from the ketone **257** (prepared from D-glucose according to a well-established three-step sequence)[80] as outlined in Scheme 32. The 3-keto glucoside **257** was subjected to a Wittig reaction with Ph$_3$P=CH$_2$ to yield the corresponding 3-*C-exo*-methylene intermediate **258**, which was subsequently treated with BH$_3$·SMe$_2$, followed by oxidation with alkaline H$_2$O$_2$ to provide the hydroxymethyl derivative **259** in 69% overall yield. The diastereoselective hydroboration of **258** could be explained by the delivery of the borane reagent from the less hindered β-face opposite the 1,2-*O*-isopropylidene group. From compound **259**, after benzylation to furnish **260**, regioselective acidic removal of the primary acetonide and oxidative cleavage of the resulting diol with NaIO$_4$ led to the corresponding aldehyde, which was engaged in a one-pot aldol condensation–Cannizzaro reaction sequence in the presence of an

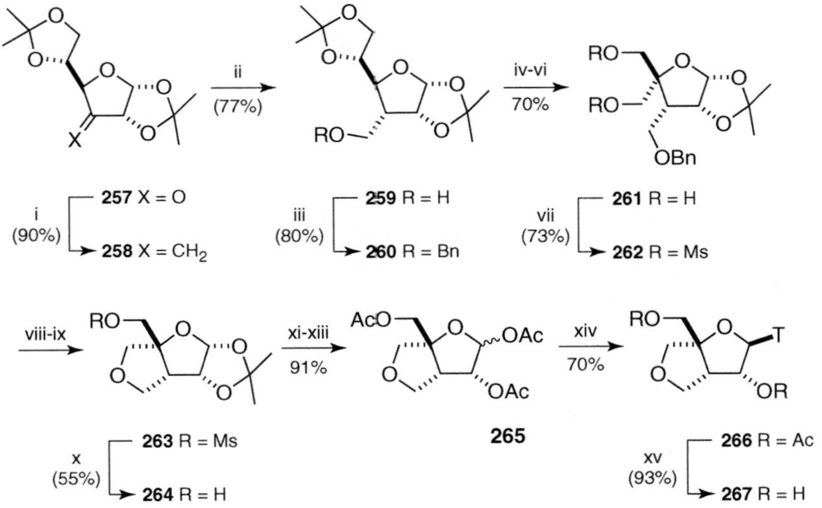

Scheme 32. *Reagents and conditions:* (i) Ph$_3$PCH$_3$Br, BuLi, THF, rt to 55°C, 3 h; (ii) BH$_3$·SMe$_2$, THF, rt, 3 h, then H$_2$O$_2$, 2 N NaOH, rt, 2 h; (iii) BnBr, NaH, TBAI, THF, rt, 16 h; (iv) 75% aq. AcOH, 55°C, 2 h; (v) NaIO$_4$, EtOH, H$_2$O, 0°C, 30 min; (vi) 37% aq. CH$_2$O, 2 N NaOH, rt, 3 days; (vii) MsCl, pyridine, rt, 14 h; (viii) H$_2$, 20% Pd(OH)$_2$/C, EtOH, rt, 14 h; (ix) NaH, THF, 55°C, 1 h; (x) 0.5 N NaOH, reflux, 12 h; (xi) Ac$_2$O, pyridine, rt, 12 h; (xii) 85% aq. HCO$_2$H, 55°C, 2 h; (xiii) Ac$_2$O, pyridine, rt, 16 h; (xiv) thymine, BSA, TMSOTf, ACN, 60°C, 4 h; (xv) NH$_3$, MeOH, rt, 14 h.

excess of formaldehyde to give the intermediate **261** in 70% overall yield. Both hydroxymethyl groups in **261** were activated as mesylates to afford **262**. Prior hydrogenolysis of the *O*-benzyl protecting group liberated the primary alcohol, which upon treatment with NaH, promoted the formation of the tetrahydrofuran ring to afford the precursor of the glycosyl donor **263**. Displacement of the mesylate group at C5 in **263** by hydroxide anion gave **264**, which by subsequent acetylation of the resulting primary alcohol, acidic removal of the acetonide group, and finally, treatment with Ac$_2$O, led to the triacetate **265** in 50% overall yield. Intermediate **265** was coupled with thymine to afford the corresponding β-isomer **266** according to Vorbrüggen's procedure; then removal of both acetyl groups provided the bicyclic nucleoside **267**.

The racemic carbanucleoside **275** analogue of the latter bicyclonucleoside **267** was obtained by Samuelsson et al. via an original approach as outlined in Scheme 33.[81] In their strategy, the bicyclic sugar skeleton was built first and the nucleobase was introduced at the final stage. Addition of *i*-PrOSiMe$_2$CH$_2$MgCl to the α-allylcyclopentanone **268** afforded a 4:1 *cis/trans* mixture of the corresponding alcohol, which was then subjected to Tamao–Flemming oxidation

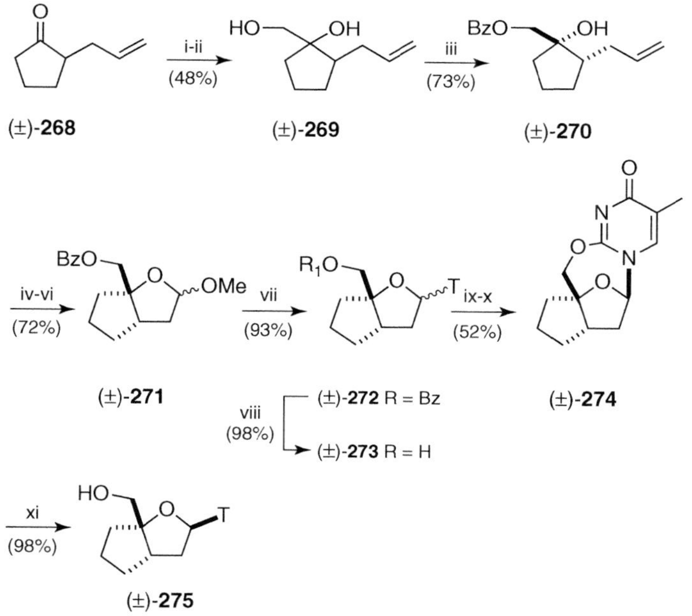

Scheme 33. *Reagents and conditions:* (i) *i*-PrOSiMe$_2$CH$_2$MgCl, THF, 0°C, 30 min; (ii) KF, KHCO$_3$, H$_2$O$_2$, MeOH/THF (1:1), rt, 2 h; (iii) BzCl, pyridine/DCM (1:4), 0°C, 45 min; (iv) cat. OsO$_4$, NMO, THF/H$_2$O 3:1, rt, overnight; (v) NaIO$_4$, THF/H$_2$O 3:1, rt, 30 min; (vi) HCl, MeOH, rt, 15 min; (vii) persilylated thymine, TBSOTf, ACN/DCM 1:4, rt, overnight; (viii) NaOMe, MeOH, rt, 20 h; (ix) TsCl, pyridine, rt, overnight; (x) DBU, ACN, reflux, 24 h; (xi) aq. NaOH, dioxane, rt, 30 min.

to give the diol **269** as a diastereomeric mixture in 48% overall yield. Then, selective benzoylation of the latter followed by removal of the *trans*-isomer by silica gel chromatography furnished the monobenzoate **270** in 73% yield. Subsequent catalytic dihydroxylation of **270** with OsO_4 and NMO as cooxidant followed by periodate cleavage of the resulting diol and, finally, treatment in acidic conditions gave the key methyl furanoside **271** as an anomeric mixture in 72% yield over the three steps. Glycosyl donor **271** was coupled with persilylated thymine according to Vorbrüggen's procedure with TBSOTf as the Lewis acid to give an inseparable mixture of β/α nucleosides **272** in a 2.8:1 ratio in 93% yield. At this stage, the desired β-anomer **275** was obtained from the previous mixture efficiently via the formation of the 2,5'-O-anhydro intermediates **274**. Toward this end, the mixture of α/β nucleosides **272** was debenzoylated and the resulting primary alcohol was activated as the tosylate and treated with DBU to afford the corresponding 2,5'-O-anhydro intermediate **274** in 52% overall yield. The latter intermediate was subjected to aqueous NaOH to furnish the expected racemic bicyclic thymidine analogue **275** in 98% yield.

As outlined in Scheme 32, following similar chemistry starting from the ketone **257**, Imanishi et al. described the synthesis of a *trans*-fused pyran-ribothymidine analogue **288** (see Scheme 34).[82] Diastereoselective addition of the allyl Grignard reagent to the 3-ulose **257** occurred from the sterically less hindered β-face to afford the corresponding *ribo*-configured furanose exclusively, which was benzylated to give **276** in 70% yield.[83] Following the sequence presented in Scheme 32 to provide the diol **277**, introduction of a hydroxymethyl moiety at C3 was achieved from **276**. At this stage, regioselective benzylation under standard conditions gave **278**, and the remaining primary alcohol on the more hindered α-face was activated as tosylate to yield **279** in 45% overall yield. It should be pointed out that on a large scale, this selective benzylation is problematic. Oxidative cleavage of the allyl group in **279** and subsequent reduction of the aldehyde were achieved with the standard $OsO_4/NaIO_4$ protocol and $NaBH_4$ sequence, respectively, to afford the compound **280** in 63% overall yield. Subsequent acetolysis and acetylation in a one-pot procedure furnished the triacetate **281** as an anomeric mixture in nearly quantitative yield, which was then engaged in a coupling reaction with in situ persilylated thymine following the method of Vorbrüggen to give the corresponding β-anomer of **282** in 87% yield. A key point in this synthetic pathway was to distinguish the two acetoxy groups in the intermediate **282**. A solution was found with the selective removal of the 2'-acetyl group upon treatment with aqueous methylamine followed by protection of the resulting secondary alcohol **283** as the 2-methoxy-2-propyl derivative **284**. Exposure of the β-nucleoside analogue **284** to aqueous NaOH to give **285** followed by treatment with NaHMDS yielded the desired cyclized intermediate **286** in 65% yield over the two steps. The targeted nucleoside analogue **288** was isolated in 60% yield after removal of the MOP protecting group under acidic conditions and Pd-mediated hydrogenolysis of the benzyl ethers using cyclohexene as the hydrogen donor.

Lebreton et al. achieved a synthesis of the pyran ribonucleoside **299** via the glycosyl donor **295** as described in Scheme 35.[84] Oxidative cleavage of the diol **289**

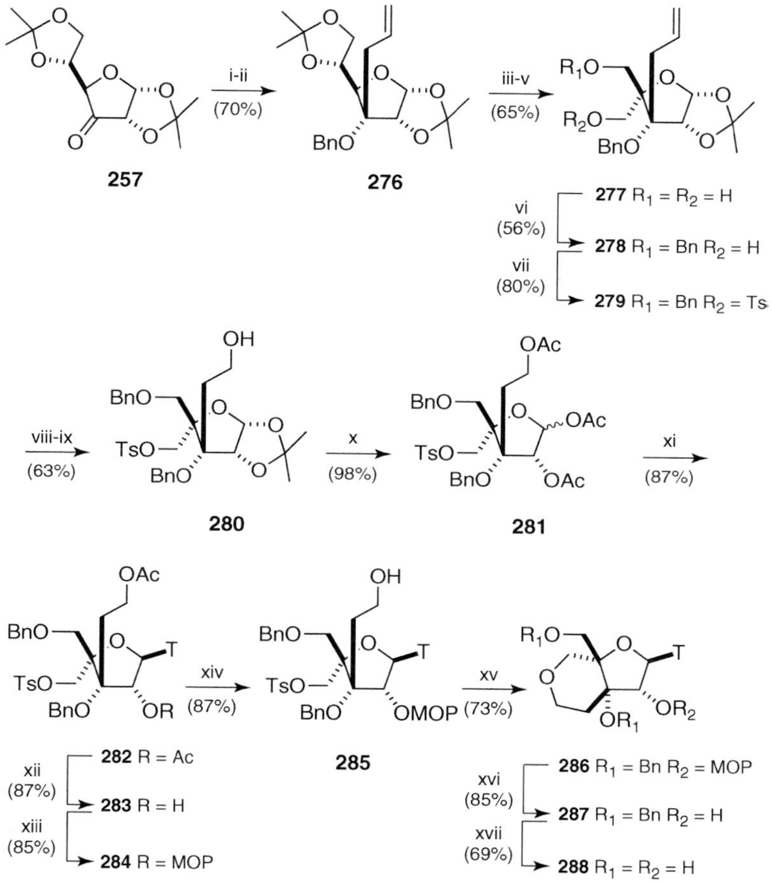

Scheme 34. *Reagents and conditions:* (i) allylMgBr, THF; (ii) BnBr, NaH, THF; (iii) 80% aq. AcOH, 60°C; (iv) NaIO$_4$, THF/H$_2$O, 0°C; (v) 37% aq. HCHO, 1 N aq. NaOH, THF/H$_2$O, rt; (vi) NaH, BnBr, DMF, 0°C; (vii) TsCl, Et$_3$N, DMAP, DCM, rt; (viii) cat. OsO$_4$, NaIO$_4$, THF/H$_2$O, rt; (ix) NaBH$_4$, THF/H$_2$O, 0°C; (x) Ac$_2$O, AcOH, cat. H$_2$SO$_4$, rt; (xi) thymine, BSA, TMSOTf, DCE, reflux; (xii) 40% aq. MeNH$_2$, THF, 0°C; (xiii) 2-methoxypropene, cat. PTSA, DCM, 0°C; (xiv) 2 N NaOH, MeOH/THF, rt; (xv) 1 M NaHMDS, THF, reflux; (xvi) cat. PTSA, THF/MeOH, 0°C; (xvii) 20% Pd(OH)$_2$/C, cyclohexene, EtOH, reflux.

(prepared in large-scale synthesis in four steps from a glucose derivative following known procedures) by standard treatment with NaIO$_4$ furnished the aldehyde **290**. This crude material was reacted with an excess of formaldehyde and NaOH to yield the corresponding aldol, which was then trapped in situ by addition of NaBH$_4$ to afford the desired diol **291**. According to examples reported previously, the less hindered 5′β-hydroxyl group of diol **291** was regioselectively protected as the silyl ether **292** in 64% yield. Then, oxidation of the resulting primary alcohol of **292** using the Swern procedure and subsequent standard Wittig methylenation afforded

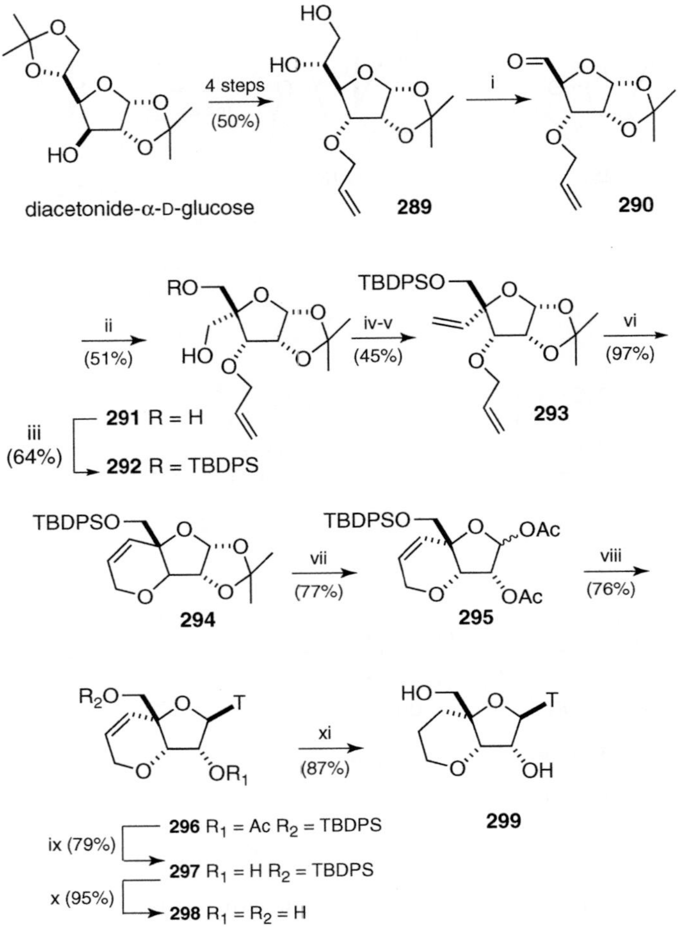

Scheme 35. *Reagents and conditions:* (i) NaIO$_4$, H$_2$O/dioxane 1:1, 0°C, 30 min; (ii) aq. HCHO, aq. NaOH, rt, dioxane, 6 h, then NaBH$_4$, rt, 30 min; (iii) TBDPSCl, Et$_3$N, DCM, 0°C, then rt, overnight; (iv) (COCl)$_2$, DMSO, DIPEA, DCM, −78°C; (v) Ph$_3$PMeBr, BuLi, THF, −78°C, rt, 3 h; (vi) Grubbs' catalyst II, DCM, rt, 16 h; (vii) AcOH, Ac$_2$O, cat. H$_2$SO$_4$, rt, 3 h; (viii) thymine, HMDS, TMSCl, TMSOTf, DCE, −35°C, then rt, 24 h; (ix) K$_2$CO$_3$, MeOH, rt, 3 h; (x) TBAF, THF, rt, 4 h; (xi) 1 atm. H$_2$, 10% Pd/C, EtOH, rt, 24 h.

the diene **293** in 45% yield. The RCM reaction was carried out on the previous intermediate **293** with Grubbs' catalyst II to afford the desired cyclized compound **294** in almost quantitative yield. From intermediate **294**, acetolysis followed by *N*-glycosylation with in situ silylated thymine in the presence of TMSOTf led exclusively to the β-anomer **296** in 60% yield over the two steps. Finally, cleavage of the protecting groups of **296** and classical hydrogenation of the double bond led to the targeted thymidine analogue **299** in 65% yield over the last three steps.

Experimental Procedures

3-O-Allyl-4-C-hydroxymethyl-1,2-O-isopropylidene-α-D-erythrofuranose (87).[84] To a stirred solution of diol **291** (14.2 g, 54.6 mmol) in 38 mL of water and dioxane (1:1) was added at 0°C NaIO$_4$ (14.0 g, 65.5 mmol, 1.2 equiv). After stirring for 30 min, 1.5 mL of ethylene glycol was added and the resulting mixture was extracted with EA (3 × 50 mL) and then with DCM. The organic layer was then dried over MgSO$_4$, filtered, and concentrated under reduced pressure to obtain the aldehyde **290**. To the crude aldehyde dissolved in 185 mL of dioxane was added a 37% aqueous formaldehyde solution (6.0 mL) and a 2 M aqueous solution of NaOH (27.3 mL). After stirring for 6 h at room temperature, the mixture was cooled to 0°C and NaBH$_4$ (3.8 g, 99.1 mmol, 1.8 equiv) was slowly added. The solution was stirred for 30 min at room temperature and 200 mL of a pyridine/AcOH mixture (4:1) was added. After stirring for a further 30 min at 0°C, the crude mixture was concentrated under reduced pressure. Purification by chromatography on silica gel (petroleum ether/EA, 3:7) afforded the diol **291** as a colorless syrupy oil (6.6 g, 51% over two steps). Spectroscopic data of this compound were consistent with those reported by Kierzek et al.[85] $[\alpha]_D^{25}$ +78.3 (c 1.0, CHCl$_3$). IR (neat) 3429, 2975, 2930, 2880, 1644, 1456, 1386, 1377, 1106, 1055, 876 cm^{-1}. ^{1}H-NMR (300 MHz, CDCl$_3$): δ 5.94–5.84 (m, 1H, H$_2$C=C*H*–CH$_2$), 5.73 (d, J = 3.6 Hz, 1H, H$_{1'}$), 5.28 (syst. AB*MXY*, J = 17.1 Hz, 1.2 Hz, 1H, H–(*H*)C=CH–CH$_2$), 5.20 (syst. AB*MX*Y, J = 10.5 and 1.2 Hz, 1H, *H*–(H)C=CH–CH$_2$), 4.63 (dd, J = 5.1 and 3.6 Hz, 1H, H$_{2'}$), 4.22 (d, J = 5.1 Hz, 1H, H$_{3'}$), 4.21 and 4.06 (syst. *AB*MXY, J = 12.7 and 5.8 Hz, 2H, H$_2$C=CH–C*H*$_2$), 3.92 and 3.60 (syst. AB, J = 12.0 Hz, 2H, H$_{5'\alpha}$), 3.83 and 3.78 (syst. AB, J = 5.8 Hz, 2H, H$_{5'\beta}$), 1.57 [s, 3H, (C(CH$_3$)–C*H*$_3$(endo)], 1.29 [s, 3H, (C(CH$_3$)–C*H*$_3$(exo)]. ^{13}C-NMR (75 MHz, CDCl$_3$): δ 134.0 (H$_2$C=*C*H–), 118.3 (H$_2$*C*=CH–), 113.5 [2 × O–*C*(O)CH$_3$], 104.3 (C$_{1'}$), 86.2 (C$_{4'}$), 78.5 (C$_{2'}$), 78.2 (C$_{3'}$), 71.9 (O–*C*H$_2$–CH=CH$_2$), 64.0, 63.1 (2 × *C*H$_2$–OH), 26.5 [C(CH$_3$)–*C*H$_3$(endo)], 25.8 [C(CH$_3$)–*C*H$_3$(exo)].

2-O-Acetyl-3-O-4-α-(3,6-dihydro-2H-pyrano)5-O-(t-butyldiphenylsilyl)ribothymidine (296).[84] To a suspension of thymine (475 mg, 3.8 mmol, 2.1 equiv) in 50 mL of dry HMDS was added TMSCl (7.5 mL, 59.0 mmol, 33.0 equiv). The reaction mixture was heated under reflux overnight and then concentrated under reduced pressure. The crude mixture was diluted with dry DCE (5 mL), and a solution of compound **295** (935 mg, 1.8 mmol) in dry DCE (10 mL) was added. The reaction mixture was cooled to −35°C and TMSOTf (720 μL, 4.0 mmol, 2.2 equiv) was added dropwise. After 24 h of stirring at room temperature, an aqueous saturated solution of NaHCO$_3$ was added and the resulting mixture was stirred for 40 min at room temperature. After extraction with DCM, the organic layers were dried over MgSO$_4$, filtered, and concentrated under reduced pressure. The crude residue was purified by chromatography on silica gel (petroleum ether/EA, 2:3) to give the thymidine derivative **296** as a white solid (790 mg, 76%); mp 82–83°C. $[\alpha]_D^{25}$ +103.3 (c 1.0, CHCl$_3$). IR (KBr) 3360, 2874, 1700, 1458, 1129, 1060 cm^{-1}. ^{1}H-NMR (300 MHz, CDCl$_3$): δ 8.90 (br s, 1H, NH),

7.72–7.66 (m, 4H, H$_{ar}$), 7.53 (s, 1H, HC=C$_{thy}$), 7.47–7.37 (m, 6H, H$_{ar}$), 6.40 (d, J = 8.6 Hz, 1H, H$_{1'}$), 6.18–6.10 (syst. ABMX, J = 10.3 and 3.8 Hz, 1H, HC=CH–CH$_2$), 5.61–5.51 (m, 2H, H$_{2'}$, HC=CH–CH$_2$), 4.40 (d, J = 4.9 Hz, 1H, H$_{3'}$), 4.23 (syst. ABMX, J = 16.3 and 4.7 Hz, 1H, HC=CH–C(H)–H), 3.91–3.82 (m, 1H, HC=CH–C(H)–H), 3.86 and 3.50 (syst. AB, J = 11.2 Hz, 2H, H$_{5'}$), 2.16 (s, 3H, O–C(=O)–CH_3), 1.68 (s, 3H, CH$_{3thy}$), 1.11 (s, 9H, tBu). ^{13}C-NMR (75 MHz, CDCl$_3$): δ 170.6 (O–C(=O)–CH$_3$), 163.6, 150.7 (C=O), 135.6 (CH$_{ar}$), 135.3 (CH$_{thy}$), 131.8, 132.6 (C$_{ar}$), 131.5 (HC=CH–CH$_2$), 128.1, 130.2, 130.3 (CH$_{ar}$), 123.4 (HC=CH–CH$_2$), 111.8 (C$_{thy}$), 83.7 (C$_{1'}$), 79.3 (C$_{4'}$), 75.0 (C$_{3'}$), 74.3 (C$_{2'}$), 62.6 (C$_{5'}$), 60.4 (HC=CH–CH$_2$), 27.1 [(CH$_3$)$_3$C–Si], 20.7 [O–C(=O)–CH$_3$], 19.4 [(CH$_3$)$_3$$C$–Si], 11.7 (CH$_{3thy}$). HRMS (CI): calcd. for C$_{31}$H$_{37}$N$_2$O$_7$Si, [M + H]$^+$ 577.2371; found, 577.2370.

2.8. Preparation of 3′,5′-Linked Bicyclic Nucleosides

Leumann[86] reported the synthesis and a complete biological study of various bicyclonucleosides with an ethylene bridge between C3′ and C5′ using a chemoenzymatic approach. This important piece of work is illustrated through preparation of the analogue **308**, as outlined in Scheme 36.[87] The commercially available racemic ketone **300** was first submitted to a Wittig–Horner reaction using 2 equiv of TBD to afford the more stable deconjugated β,γ-unsaturated ester **301** in high yield. It should be noted that later, the same authors developed a large-scale preparation of the chiral ketone **300** in 10 steps from D-mannose with 23% overall yield.[88] Then, epoxidation of the substituted cyclopentene **301** with MCPBA led to isolation of the required *exo*-**302** and the *endo*-epoxide in 77% and 10% yields, respectively, after purification on silica gel chromatography. This good diastereoselectivity in favor of the desired *exo*-epoxide **302** could be rationalized by the reaction of MCPBA from the convex side of this bicyclic skeleton. Resolution of the racemic ethyl ester **302** by hog-liver esterase on a 10-g scale led to the chiral acid **303**, in 53% yield and in 72% ee, which was treated without further purification with LAH to promote the regioselective ring opening of the epoxide and the reduction of the carboxylic acid furnishing the diol **304** in 84% yield, after purification on silica gel. Subsequent recrystallization of this latter compound provided the enantiomerically pure (97% ee) diol **304** in 61% yield from **303**. The key glycosyl donor **306** was prepared as an anomeric mixture from diol **304** in 60% overall yield, by oxidation of the primary hydroxyl group with Dess–Martin reagent followed by direct treatment of the resulting unstable aldol with a strong acidic ion-exchange resin, leading to sequential hydrolysis of the acetonide, then spontaneous intramolecular cyclization, and finally, acetylation under classical conditions. Vorbrüggen-type coupling of the latter acetylated sugar **306** with in situ persilylated thymine in the presence of SnCl$_4$ afforded an inseparable mixture of the corresponding α/β-anomers **307** in a 2:3 ratio. From this mixture, the pure β-thymidine analogue **308** was isolated in 43% overall yield after sequential deacetylation, silylation of the primary alcohol, and anomeric separation on silica gel chromatography, followed by desilylation. Following these steps from the pivotal glycosyl donor, **306**, three other natural nucleobases were introduced successfully.

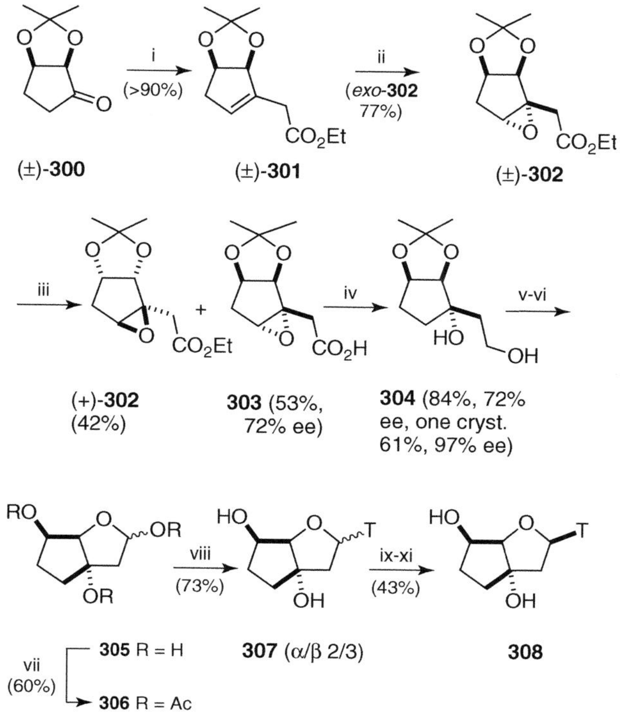

Scheme 36. *Reagents and conditions:* (i) TBD, $(EtO)_2P(O)CH_2CO_2Et$, DCM, rt, overnight; (ii) MCPBA, DCM, rt, 22 h; (iii) HLE, 0.1 M NaH_2PO_4, pH 7.75, rt; (iv) LAH, Et_2O, reflux, 6 h; (v) DMP, DCM, rt, 2 h; (vi) Amberlite IR-120, H_2O, 55°C, 2 h; (vii) Ac_2O, pyridine, DMAP, rt, 3 h; (viii) thymine, HMDS, TMSCl, $SnCl_4$, DCM, rt, 35 min, then 50°C, 35 min; (ix) 0.2 M NaOH, THF/MeOH/H_2O 5:4:3, 0°C, 75 min; (x) TBSOTf, pyridine, 0°C, 30 min, then separation on SiO_2; (xi) TBAF, THF, rt, 4 days.

An alternative route for the preparation of the previous C3′–C5′ locked bicyclic thymidine analogues **308** was described by Nielsen and Ravn from diacetonide-D-glucose using an RCM reaction as a key step (Scheme 37).[89] The aldehyde **309**, obtained according to well-established procedures,[90] was treated with vinyl magnesium bromide to give an epimeric mixture of the corresponding allylic alcohol **311** and its undesired isomer **310** in 19 and 69% yields, respectively, after separation by silica gel chromatography. It should be pointed out that the desired epimeric alcohol at C5 was preferentially obtained in a ratio of 3:1 but in lower yield (57%) by addition of the Grignard reagent to compound **309** with a free tertiary alcohol at C3. However, the latter sequence was found to be less efficient, and the major allylic alcohol **312** could easily be converted into the required allylic alcohol **313** by inversion of the configuration at C5 of the cyclopentene derivative **312** after RCM reaction of **310**. So, both allylic alcohols **310** and **311** were separately subjected to an RCM reaction. It is interesting to mention that for diene **310**, twice the amount of 5 mol% Grubbs' catalyst I was necessary to complete the reaction. On the other

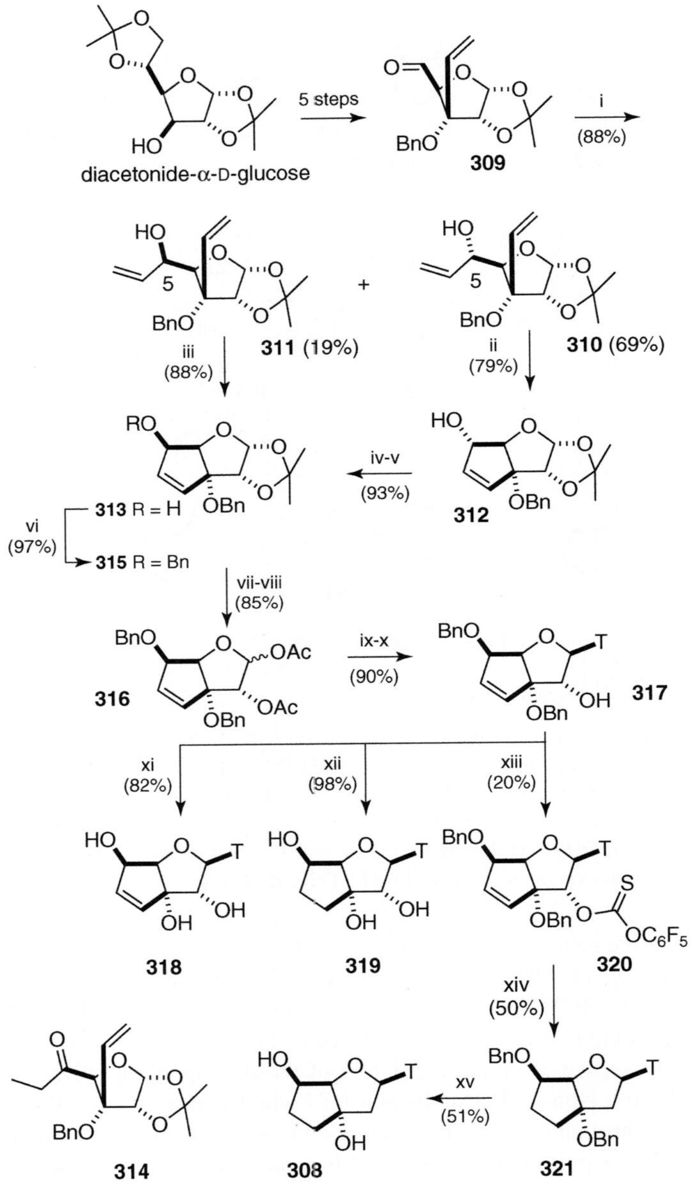

Scheme 37. *Reagents and conditions:* (i) vinyl-MgBr, THF, 0°C, 30 min, then rt, 48 h; (ii) 5 mol% Grubbs' catalyst I, DCM, rt, 48 h, then addition of 5 mol% Grubbs' catalyst I, rt, 48 h; (iii) 5 mol% Grubbs' catalyst II, DCM, reflux, 1 h; (iv) PCC, DCM, rt, 14 h; (v) CeCl$_3$·7H$_2$O, MeOH, 0°C, then NaBH$_4$, 0°C, 30 min; (vi) BnBr, NaH, DMF, rt, 16 h; (vii) 80% aq. AcOH, 90°C, 16 h; (viii) Ac$_2$O, pyridine, rt, 16 h; (ix) thymine, BSA, TMSOTf, ACN, 50°C, 16 h; (x) NaOMe, MeOH, rt, 16 h; (xi) BCl$_3$, DCM, −78°C, 5 h; (xii) H$_2$, 20% Pd(OH)$_2$/C, MeOH, rt, 2 h; (xiii) C$_6$F$_5$OC(=S)Cl, DMAP, DCM, rt, 16 h; (xiv) Bu$_3$SnH, AIBN, PhH, reflux, 16 h; (xv) H$_2$, 20% Pd(OH)$_2$/C, cyclohexa-1,4-diene, MeOH, rt, 16 h.

hand, from diene **311**, the desired cyclized compound **313** was obtained in low yield along with the saturated ketone **314** formed by a known ruthenium-catalyzed isomerization of the allylic alcohol.[91] When the RCM reaction was performed with Grubbs' catalyst II, the desired intermediate **313** was isolated in 88% yield. To invert the C5 configuration, allylic alcohol **312** was subsequently oxidized with PCC reagent to its corresponding enone, which was then reduced using Luche's protocol to afford the desired pure epimeric alcohol **313** in 93% yield for the two steps. The excellent diastereoselectivity of the reduction step could be explained by the delivery of the hydride from the less hindered convex α-face of the enone intermediate. After benzylation of **313** to give **315**, subsequent cleavage of the 1,2-isopropylidene group and acetylation of the resulting diol yielded the key bicyclic glycosyl donor **316** as an anomeric mixture in 85% overall yield. Condensation of this latter mixture with in situ persilylated thymine in the presence of TMSOTf followed by removal of the acetate furnished the corresponding β-nucleoside **317** as the sole anomer in 88% yield over the two steps. To cleave the benzyl protecting groups without affecting the double bond, the key intermediate **317** was subjected to BCl_3 to give the unsaturated analogue **318** in 82% yield. On the other hand, hydrogenolysis of the benzyl ethers and hydrogenation of the double bond of **317** under standard conditions furnished the bicyclic ribonucleoside analogue **319** in nearly quantitative yield. Finally, 2′-deoxygenation of **317** was carried out using the Barton–McCombie reaction. After some experimentation, the pentafluorophenylthiothionocarbonate ester intermediate **320** was found to be the more efficient radical precursor, even if its formation from the alcohol **317** proceeded in only 20% yield. Then, standard radical-mediated deoxygenation of compound **320** using Bu_3SnH and AIBN as a radical initiator to provide **321**, followed by transfer hydrogenation with Pearlman's catalyst and cyclohexene under a hydrogen atmosphere, gave the target molecule **308** in 25% overall yield over the two steps.

A six-membered carbocyclic analogue of the previous bicyclothymidine derivative **308** (see Schemes 36 and 37) was prepared from diacetone-D-glucose following a route similar to that developed by Nielsen et al. This work, described by Stauffiger and Leumann, is presented briefly in Scheme 38.[92] Treatment of **276** (see Scheme 34) with periodic acid promoted in situ regioselective hydrolysis of the primary acetonide while subsequent cleavage of the vicinal diol afforded the aldehyde **322**, which was reacted, without further purification, with Grignard reagent. Nevertheless, this strategy was once again hampered by diastereoselective addition of the vinyl Grignard reagent to the aldehyde **322** in favor of the undesired allylic alcohol **323**, which was isolated after purification in 78% yield along with only 8% of the targeted intermediate **324** (in the previous series, the corresponding epimeric allylic alcohols **310** and **311** were isolated in 69% and 19% yields, respectively; see Scheme 37). Moreover, inversion of the C5 configuration following an oxidation–reduction sequence was less efficient; Luche reduction of the enone intermediate was not diastereoselective compared to **312** (Scheme 37). The desired cyclic allylic alcohol **325** was isolated in 55% yield as well as its epimer in 28% yield. At the final stage, this strategy was also hampered by the low yield of the Barton–McCombie deoxygenation. Nevertheless, the thiocarbonyimidazol group was installed, as the radical precursor, on the remaining secondary alcohol **326** in high yield. After some investigation, treatment of this intermediate **327** with an excess of Bu$_3$SnH and 1 equiv of AIBN in hot PhMe provided the deoxygenated product **328** in 37% yield, along with 19% of the alcohol **326** resulting from the hydrolysis of the hemiacetal intermediate **329**, which was also isolated in 23% yield.[93] However, this two-step sequence, which was carried out in 35% overall yield, presented an improvement compared to the deoxygenation sequence (10%) reported on **317** in Scheme 37. Reductive debenzylation of **328** gave the targeted nucleoside **330** in 56% yield (Scheme 38).

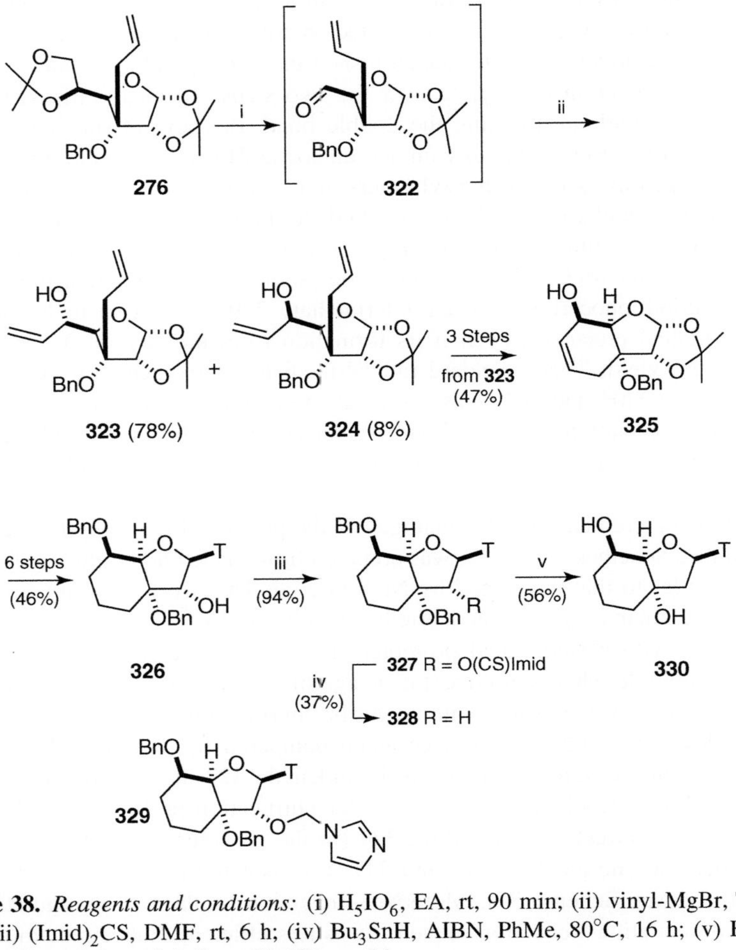

Scheme 38. *Reagents and conditions:* (i) H$_5$IO$_6$, EA, rt, 90 min; (ii) vinyl-MgBr, THF, rt, 16 h; (iii) (Imid)$_2$CS, DMF, rt, 6 h; (iv) Bu$_3$SnH, AIBN, PhMe, 80°C, 16 h; (v) H$_2$, 20% Pd(OH)$_2$/C, cyclohexa-1,4-diene, MeOH, rt, 6 h.

Experimental Procedure

2,4(1H,3H)-Pyrimidinedione, 5-Methyl-1-[(2R,3aS,7R,7aR)-octahydro-3a,7-bis(phenylmethoxy)-2-benzofuranyl] (328).[92] Thioester **327** (3.9 g, 6.63 mmol) was dissolved in PhMe (100 mL) and flooded with argon for 15 min. AIBN (544 mg, 3.31 mmol, 0.5 equiv) was then added, and the solution was again flushed with argon. Next, Bu$_3$SnH (3.50 mL, 13.25 mmol, 2 equiv) was added dropwise, and the clear solution was stirred at 80°C under an argon atmosphere for 4 h. Another portion of AIBN (544 mg, 3.31 mmol, 0.5 equiv) and of Bu$_3$SnH (1.75 mL, 6.63 mmol, 1 equiv) was added, and the solution was stirred for a further 12 h. Then, PhMe was evaporated in vacuo followed by purification on SiO$_2$ column (EA/hexane 1:1) to give compound **328** (1.14 g, 37.2%) as a white solid and compounds **329** (920.8 mg, 23.5%) and **326** (593.4 mg, 18.7%) as white foams. Data for **328**: R_f (EA/hexane, 1:1) = 0.38. ^{1}H-NMR (300 MHz, CDCl$_3$): δ 8.20 (br s, 1 H, NH), 7.72 [d, J = 1.2 Hz, 1 H, H-C(6)], 7.35 (m, 10 H, Ph), 6.21 [dd, J = 7.6 and 6.0 Hz, 1 H, H-C(1′)], 4.59 (2 × dd, 4 H, 2 × CH$_2$Ph), 4.12 [d, J = 4.4 Hz, 1 H, H-C(4′)], 3.93 [m, 1 H, H-C(5′)], 2.70 [dd, J = 13.1, 5.9 Hz, 1 H, H-C(2′eq)], 2.14 [dd, J = 13.1, 7.6 Hz, 1 H, H-C(2′ax)], 1.95–1.31 [m, 6 H, H-C(6′), H-C(7′), H-C(8′)], 1.56 [d, J = 0.8 Hz, 3 H, C(5)–CH$_3$] ppm. ^{13}C-NMR (75 MHz, CDCl$_3$): δ 163.3, 151.1 (quart. thymine), 137.9 (Ph quart), 136.2 [C(6)], 128.7–126.1 (phenyl), 110.5 [C(5)], 84.8 [C(1′)], 83.3, 75.0 [C(4′), C(5′)], 70.6 (CH$_2$Ph), 64.8 (CH$_2$Ph), 40.4 [C(2′)], 30.1, 25.2, 17.1 [C(6′), C(7′), C(8′)], 12.1 [C(5)–CH$_3$] ppm. HRMS (ESI+): calcd. for C$_{27}$H$_{30}$N$_2$O$_5$Na, [M + Na]$^+$ 485.2025; found, 485.2059.

2.9. Preparation of 5′,5′-Spiro Bicyclic Nucleosides

A unique family of conformationally restricted spirocyclic nucleosides in the D-enantiomeric series has been described by Paquette et al. as highlighted in this section with the synthesis of both diastereoisomers of ribothymine spironucleosides **342** and **354** (see Schemes 39 and 40). The preparation of the spiro aglycone moiety is based on acid- or bromonium ion–induced rearrangement of the racemic carbinol **333** intermediate.[94] In the first approach, the racemic bromo spiro ketone **334** was efficiently prepared via a highly diastereoselective oxonium ion–promoted rearrangement with NBS in the presence of propylene oxide as an acid scavenger (see Scheme 39). The bromo derivative **334** was isolated in 96% yield (21 g) from **331**. On the other hand, exposure of the pivotal carbinol **333** to an acidic ion-exchange resin in DCM afforded the spiro ketone **343** in 87% yield (24.4 g) as the sole detectable diastereoisomer (see Scheme 40). These racemic spiro ketone derivatives were resolved, in multigram quantities, via either the corresponding Johnson sulfoximine derivative for **334** or ketalization with (R)-mandelic acid for **343**. In this work, subsequent diastereoselective reduction of the carbonyl at different stages of the synthesis with suitable reducing reagents and functional group

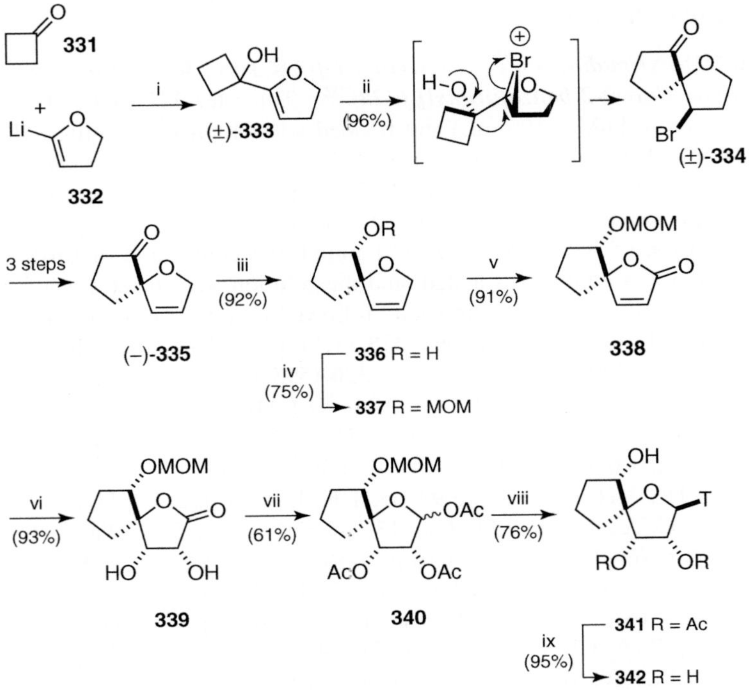

Scheme 39. *Reagents and conditions:* (i) THF, −78°C, 6 h; (ii) propylene oxide, NBS, THF, −78°C to rt, overnight; (iii) LAH, Et$_2$O, 0°C, then rt, overnight; (iv) MOMBr, DIPEA, DCM, rt, 24 h, reflux, 24 h; (v) PCC, CrO$_3$, 3,5-DMP, DCM, −20°C, 1 h; (vi) cat. OsO$_4$, NMO, pyridine, THF, H$_2$O, rt, overnight; (vii) DIBAL-H, DCM, −78°C, 1.5 h, then Ac$_2$O, AcOH, −78°C to rt, overnight; (viii) persilylated thymine, TMSOTf, THF, rt, overnight; (ix) K$_2$CO$_3$, MeOH, rt, 24 h.

manipulations provided a route toward the chiral spirocyclic aglycone moieties, which were then employed in the synthesis of various nucleoside derivatives.[95]

This important piece of work is illustrated through the preparation of both spirocyclic thymidine nucleosides **342** and **354** bearing, respectively, an α- or β-configured hydroxyl group at C5′. Stereoselective reduction of the chiral ketone **335** with LAH generated the α-carbinol **336** exclusively, which was protected as its MOM derivative **337** in 70% overall yield (see Scheme 39).[96] Then, allylic oxidation of the dihydrofuran **337** was carried out efficiently by treatment with a large excess of CrO$_3$, PCC and 3,5-DMP to provide the spirocyclic α,β-butenolide **338** in high yield. Next, the latter unsaturated lactone **338** was diastereoselectively dihydroxylated from the less hindered α-face, and the resulting diol **339** was reduced with DIBAL-H to give the corresponding lactol aluminate which, without aqueous workup, was directly trapped with Ac$_2$O in the presence of pyridine, giving rise to the triacetate **340** in 57% overall yield. Vorbrüggen-type coupling of acetylated partner **340** with persilylated thymine in the presence of TMSOTf gave the β-anomer **341** selectively in 76% yield, as the sole diastereoisomer. It is interesting to mention that in these previous N-glycosylation conditions, the removal of the MOM protecting group was promoted when an excess of Lewis acid, around 4 equiv, was used. Finally, cleavage of the acetate group by treatment with methanolic potassium carbonate afforded the C5′-α hydroxyspirocyclic ribosylthymine derivative **342**.

Starting from the chiral spiroketone (+)-**343**, Paquette et al. synthesized the other C5′-β hydroxyl spirocyclic thymidine diastereoisomer **354** as depicted in Scheme 40.[97] The reduction of the spiroketone (+)-**343** was carried out under Meerwein–Ponndorf–Verley conditions to afford the β-carbinol **344** in 70% yield and in correct diastereoselectivity (β/α 85:15). Subsequent protection of the secondary alcohol function as its corresponding silylether **345** followed by oxidation with RuO_4 generated in situ gave the lactone **346** in 90% overall yield. Next, treatment of **346** with an excess of LiHMDS and phenyl benzenethiosulfonate led to the diphenylsulfenyl derivative **347**, which was reacted with EtMgBr to furnish the monoreduced intermediate **348** in 70% yield throughout this sequence. It should be noted that direct attempts to prepare **348** from the ketone **346** afforded a mixture of mono- and diphenylsulfenyl adducts. Next, the phenylsulfenyl derivative **348** was treated with DIBAL-H to provide its corresponding lactol, which was directly acetylated, leading to the glycosyl donor **349** in 94% yield. $SnCl_4$-catalyzed condensation of **349** with persilylated thymine gave only the desired β-nucleoside **350** (β/α > 97:3) in 59% yield. At this stage, the phenylsulfenyl derivative **350** was subsequently oxidized into its corresponding sulfoxide with Davis oxaziridine reagent, followed by reflux in xylene in the presence of pyridine, as a scavenger of PhSOH, to furnish the pivotal spironucleoside derivative **351** in 88% yield. Removal of the silyl ether on the latter intermediate **351** proved to be rather difficult. The best results were obtained using KF in the presence of 18-crown-ether-6, leading to the d4T analogue **352** in only 33% yield, along with 68% of unreacted starting material. Epimerization was observed with classical TBAF desilylation. Pd-mediated hydrogenation of **351** followed by cleavage of the silyl protecting group using the previous conditions led to the dihydro compound **353**. An alternative shorter route involving desulfurization of **350** with Raney-Ni failed. Finally, compound **351** was treated with a catalytic amount of OsO_4 in the presence of NMO to provide the corresponding diol as a single diastereoisomer, as described for compound **349**, which was then subjected to fluoride-induced desilylation with TBAF to afford the spironucleoside **354** in 42% overall yield.

Alternative strategies to prepare spirocyclic nucleosides have been explored by Wendeborn et al.[98] at Syngenta Ltd. and more recently by Mandal et al.[99]

3. PREPARATION OF CYCLONUCLEOSIDES

3.1. Introduction

Among nucleosides with a restricted conformation, cyclonucleosides have been obtained by bonding one atom of the furanose moiety and one atom of the

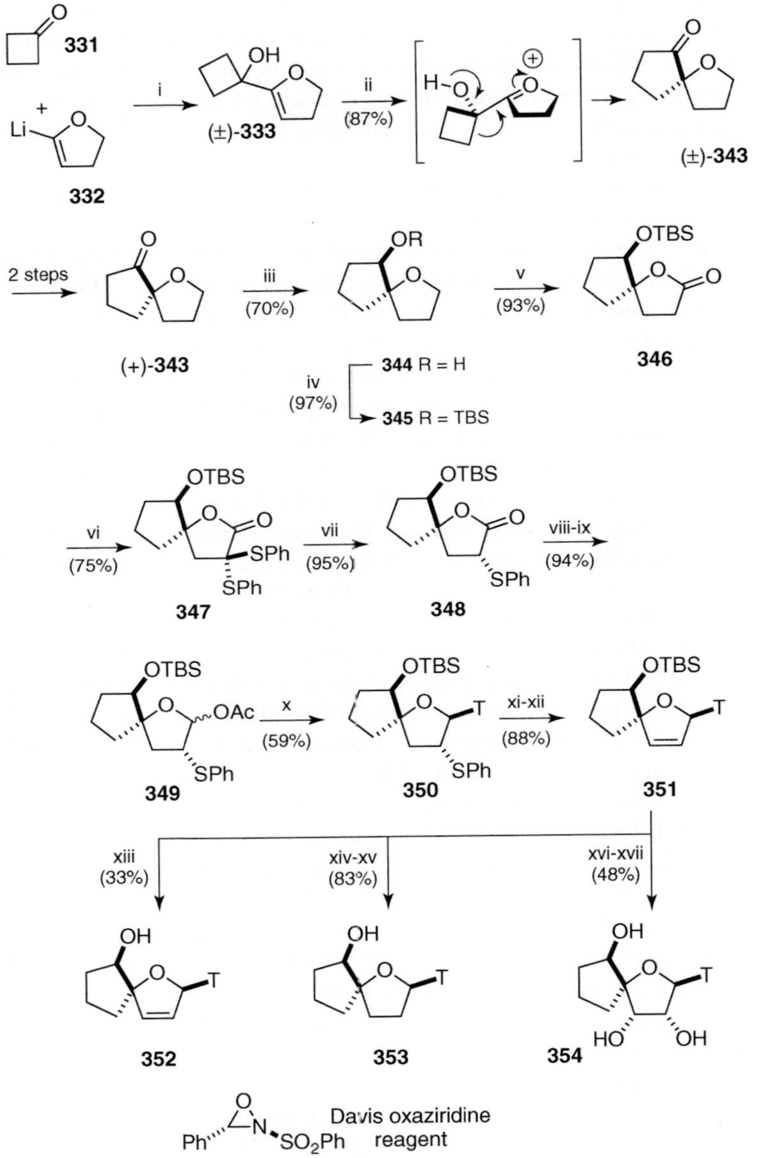

Scheme 40. *Reagents and conditions:* (i) THF, −78°C, 6 h; (ii) Amberlyst, DCM, rt, 2 h; (iii) (i-PrO)$_3$Al, ^{i}PrOH, reflux, 1 h; (iv) TBSCl, imid., DMAP, DCM, rt, 2 h; (v) cat. RuCl$_3$, NaIO$_4$, CCl$_4$/ACN 1:1, rt, 30 min; (vi) LiHMDS, PhSO$_2$-SPh, THF, −25°C, 5 h; (vii) EtMgBr, THF, −10°C, 1 h; (viii) DIBAL-H, DCM, −78°C, 30 min; (ix) Ac$_2$O, DMAP, DCM, rt, 30 min; (x) persilylated thymine, SnCl$_4$, ACN, −78°C for 30 min, rt for 30 min; (xi) Davis oxaziridine reagent, CHCl$_3$, rt, 16–24 h; (xii) xylene, pyridine, reflux, 4 h; (xiii) KF, 18-crown-6, THF, rt, 7 days; (xiv) H$_2$, 5% Pd/C, EtOH, rt, 24 h; (xv) KF, 18-crown-6, THF, rt, 7 days; (xvi) cat. OsO$_4$, NMO, acetone/H$_2$O 5:1 THF, rt, 48 h; (xvii) TBAF, THF, rt, 1 h.

nucleobase via an alkylene unit or analogue. Two different cyclonucleosides can be described: (1) those having a C–C bridge with an alkylene group between the glycone moiety and the nucleobase,[9a,b] and (2) those having a carbon atom and a heteroatom (C–O, C–N, or C–S) bridge between the sugar moiety and the heterocyclic ring.[9c,d]

3.2. Preparation of Cyclonucleosides Having a C–C Bridge

The main synthesis of cyclonucleosides having a C–C bridge has been carried out by the generation of a radical using chemical or photochemical initiation. In the first case, the formation of a radical at a position of the glycone moiety and intramolecular radical addition at a carbon atom of the nucleobase were described. In the second case, the formation of a radical at a position of either the glycone moiety or the nucleobase followed by intramolecular radical addition was developed. Other syntheses of cyclonucleosides having a C–C bridge have been described using (1) trans-N-glycosylation by condensation of either a ulose derivative or a pyrimidinyl derivative followed by intramolecular glycosylation, (2) substitution by lithiation chemistry, or (3) addition by an intramolecular aldol reaction.

3.2.1. Preparation of 5′,6-Linked Pyrimidine Cyclonucleosides. Pyrimidine cyclonucleosides having a C5′–C6 bridge have been obtained by convergent approaches either via radical reaction, substitution, and addition starting from nucleosides or via trans-N-glycosylation starting from carbohydrates. Ueda et al.[100] reported the synthesis of 5′,6-cyclo-5′-deoxyuridine **358** having a fixed anti form of uridine starting from the 5-chloro derivative **355** (see Scheme 41)

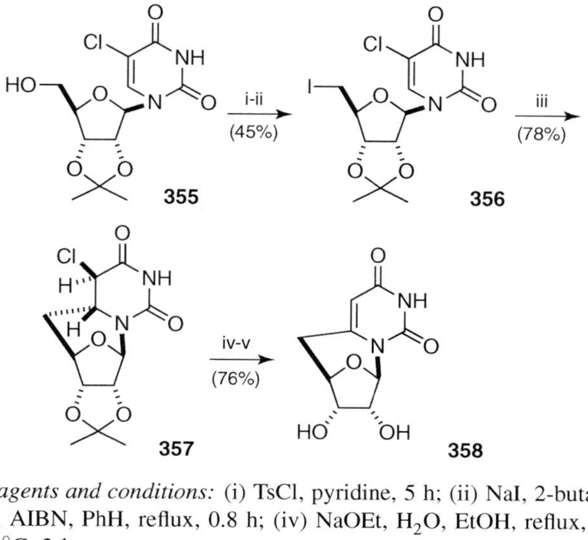

Scheme 41. *Reagents and conditions:* (i) TsCl, pyridine, 5 h; (ii) NaI, 2-butanone, reflux, 2 h; (iii) Bu₃SnH, AIBN, PhH, reflux, 0.8 h; (iv) NaOEt, H₂O, EtOH, reflux, 0.16 h; (v) aq. HCl, MeOH, 90°C, 2 h.

by a radical cyclization procedure. After tosylation and iodination of the primary hydroxyl group of compound **355**, treatment of compound **356** with Bu$_3$SnH and AIBN afforded the cyclonucleoside **357** in 78% yield. The relative configuration of the protons H$_5$ and H$_6$ of compound **357** was *trans* and the cyclization was a stereospecific *cis*-addition. Then, classical elimination of the chlorine atom of compound **357** with NaOEt in EtOH and subsequent cleavage of the acetonide protecting group gave the corresponding 6,5'-cyclonucleoside **358** in 76% yield over the two steps.

To prepare a 5',6-cyclonucleoside having a secondary 5'-OH group, Yoshimura et al.[101] developed a radical 1,6-hydrogen transfer-cyclization process as depicted in Scheme 42. After protection of the uridine analogue **359** (prepared in two steps from uridine)[102] at the *N*-3 position with a PMB protecting group, the 6-phenylseleno derivative **360** was obtained in 89% yield using a C6 lithiation method by treatment of compound **359** with LiHMDS in the presence of diphenyl diselenide. Compound **360** was then treated with TTMSS in the presence of AIBN in PhMe to afford the 5'-*S*-epimer **361** and the 5'-*R*-epimer **362** in 60% and 11% yield, respectively. The authors reported two possible reaction mechanisms. A plausible one was formation of the C6 radical intermediate **D**, which gave rise to 1,6-hydrogen transfer to produce the C5' radical intermediate **E**. The intermediate **E** preferentially adopted an *anti* conformation around the glycosidic bond with a *trans*-orientation around the C4'–C5' bond, giving the corresponding 5'-*S*-epimer **361** as the major cyclonucleoside. Treatment of compound **361** with LiHMDS at −80°C followed by PhSO$_2$Cl afforded a diastereoisomeric mixture of cyclonucleosides **363**, which was further treated with DBU in dioxane, producing the 6,5'-cyclonucleoside **364** in 69% overall yield. Classical deprotection of the *N*-3 nitrogen atom under oxidative conditions and subsequent removal of the silyl group in **364** furnished the target cyclonucleoside **365** in 58% yield over the two steps.

The lithiation chemistry of uridine derivatives was developed by Miyasaka et al.[103] Recently, Yoshimura et al.[104] have reported an efficient intramolecular alkylation of 5'-deoxy-5'-iodouridine (prepared in two steps from uridine)[105] to furnish the 5',6-*C*-cyclonucleoside via lithiation at the 6-position of uridine. Starting from the 5'-deoxy-5'-iodouridine derivative **366**, treatment with LiHMDS in the presence of Ph$_2$SiCl$_2$ gave the 6,5'-*C*-cyclonucleoside **369** in 88% yield and a trace of the 6-silyl derivative **370** (see Scheme 43). It is notable that the use of LDA, having a higher basicity, gave a complex mixture of products from which **369** was isolated in only 8% yield, while the use of simple LiHMDS gave none of the desired cyclonucleoside **369**. Yoshimura et al.[104] explained that the LiHMDS combined silylating agent led to the compound **367** by silylation of the *O*-4 of the uracil moiety. Due to the loss of the negative charge on the *N*-3 position, LiHMDS was able to generate the C6 lithio derivative **368**, which gave the cyclonucleoside **369**.

Experimental Procedure

5',6-Cyclo-1-(5-deoxy-2,3-O-isopropylidene-β-D-ribofuranosyl)uracil 369.[104] To a solution of **366** (50 mg, 0.127 mmol) and Ph$_2$SiCl$_2$ (53 μL, 0.381 mmol, 3 equiv)

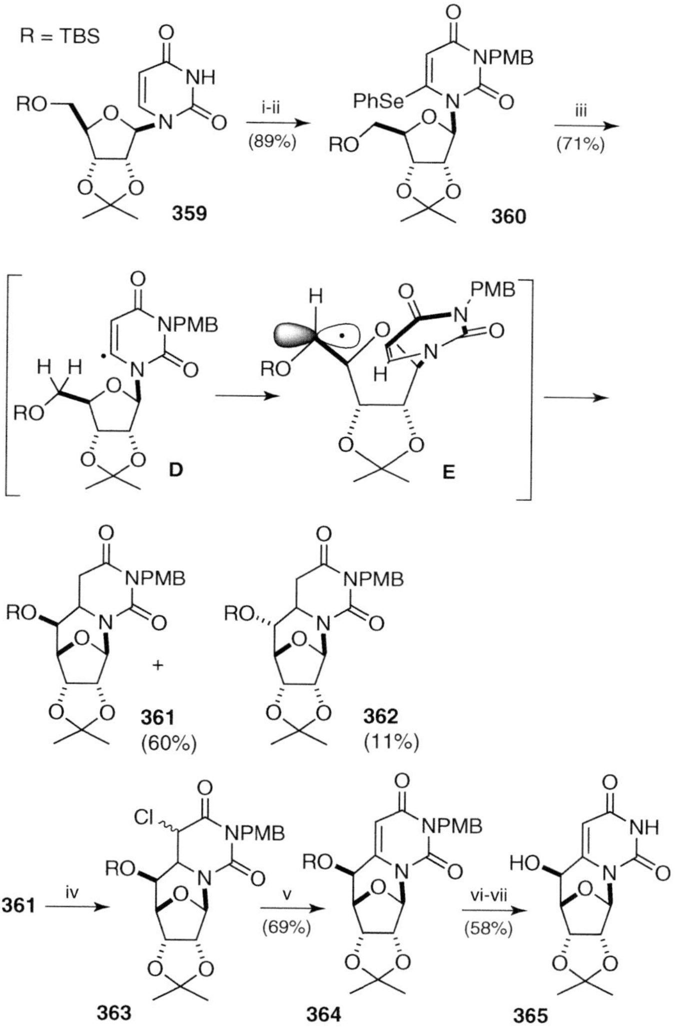

Scheme 42. *Reagents and conditions:* (i) PMBCl, DBU, DMF, rt; (ii) (PhSe)$_2$, LiHMDS, THF, −80°C; (iii) TTMSS, AIBN, PhMe, reflux, 6 h; (iv) LiHMDS, THF, −80°C, then PhSO$_2$Cl; (v) DBU, dioxane, 80°C; (vi) CAN, ACN, H$_2$O; (vii) TBAF, THF.

in 2 mL of anhydrous THF was added dropwise a THF solution of LiHMDS (0.38 mL, 0.608 mmol, 4.8 equiv) at −80°C. After the mixture was stirred at the same temperature for 1.5 h, the reaction was quenched with saturated NH$_4$Cl solution. The entire mixture was extracted with EA and dried over Na$_2$SO$_4$. Then, the filtrate was concentrated under reduced pressure, the residue was purified on an SiO$_2$ column (0–1% MeOH in CHCl$_3$) to yield **369** (30 mg, 88%) as a white solid, mp 267°C. ^{1}H-NMR (300 MHz, CDCl$_3$): δ 8.83 (br s, 1H, NH), 6.20 (s, 1H, H$_{1'}$), 5.48

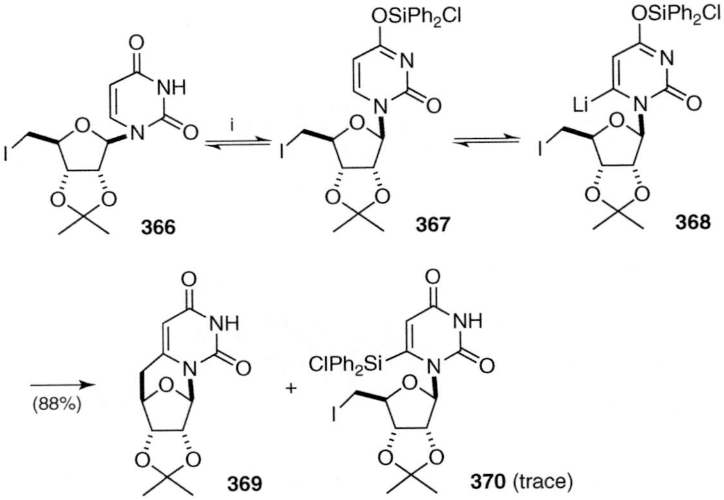

Scheme 43. *Reagents and conditions:* (i) Ph$_2$SiCl$_2$, LiHMDS, THF, −80°C; 1.5 h.

(s, 1H, H$_5$), 4.66 (d, J = 5.5 Hz, 1H, H$_{2'}$), 4.63 (d, J = 5.5 Hz, 1H, H$_{3'}$), 4.61 (d, J = 6.7 Hz, 1H, H$_{4'}$), 3.20 (ddd, J = 18.3, 6.7 and 1.8 Hz, 1H; H$_{5'\alpha}$), 2.56 (d, J = 18.3 Hz, 1H, H$_{5'\beta}$), 1.51 (s, 3H, CH$_3$), 1.32 (s, 3H, CH$_3$). *Anal.* Calcd for C$_{12}$H$_{14}$N$_2$O$_5$: C, 54.13; H, 5.30; N, 10.52; found: C, 54.17; H, 5.25; N, 10.25.

Intramolecular aldol reactions, in which cyclization is generated between the nucleobase and a formyl group in the 5′-position, have been used to a lesser extent.[106] This strategy furnished a potential secondary 5′-hydroxyl group mimicking the primary OH group of the nucleoside. The method developed by Otter et al.[106] depended on the fact that 5-hydroxyuracils were susceptible to electrophilic substitution at C6 and could undergo base-catalyzed hydroxymethylation at this position. Starting from 5-acetoxy-2′,3′-*O*-isopropylideneuridine **371**, oxidation with DMSO and DCC in the presence of pyridinium trifluoroacetate gave the 5′-aldehyde **372** (see Scheme 44). Next, treatment of compound **372** with NaOH generated compound **373** and allowed a diastereoselective hydroxyalkylation to afford the cyclonucleoside **374** with the 5′-(*S*) absolute configuration in 44% yield (two steps). Subsequent removal of the pyrimidine 5-hydroxyl group of **374** by sequential mesylation followed by classical hydrogenation in the presence of 1 equiv of Et$_3$N, and finally, acidic hydrolysis of the acetonide protecting group in acetic acid afforded the desired 6,5′-cyclouridine **375** in 37% overall yield.

Trans-*N*-glycosylation enabled the nucleobase and the glycone moiety to be condensed using carbohydrate chemistry instead of nucleoside chemistry. Starting from the aldehyde **376** (prepared in two steps from D-ribose),[107] addition of (2,4-dimethoxypyrimidin-6-yl)methyllithium yielded a diastereoisomeric mixture of alcohol **377** in 65% yield as outlined in Scheme 45.[108] To prevent dehydration during the intramolecular *N*-glycosylation, conversion of the alcohol

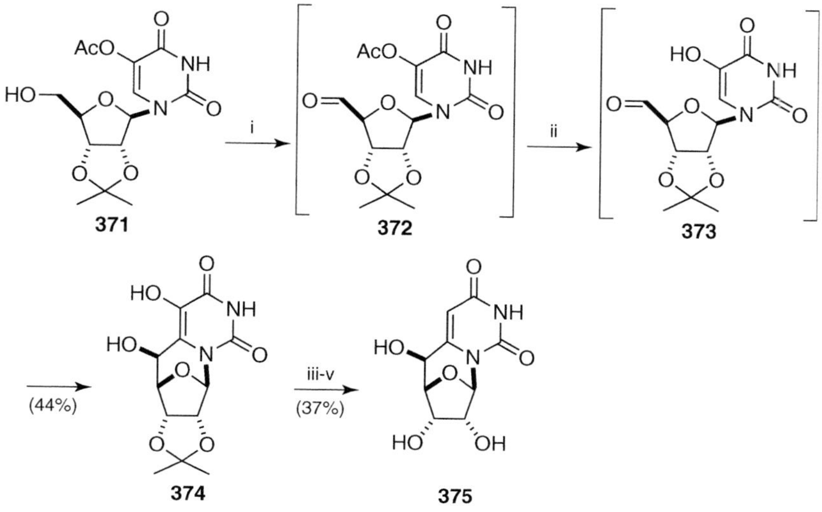

Scheme 44. *Reagents and conditions:* (i) DCC, TFA, pyridine, DMSO, rt, 15 h; (ii) 1 N NaOH, MeOH, rt, 0.5 h; (iii) MsCl, PhH, pyridine, 0°C, 1 h; (iv) H$_2$, Pd/C, H$_2$O, Et$_3$N, MeOH, rt, 3 h; (v) AcOH, H$_2$O, reflux, 5 h.

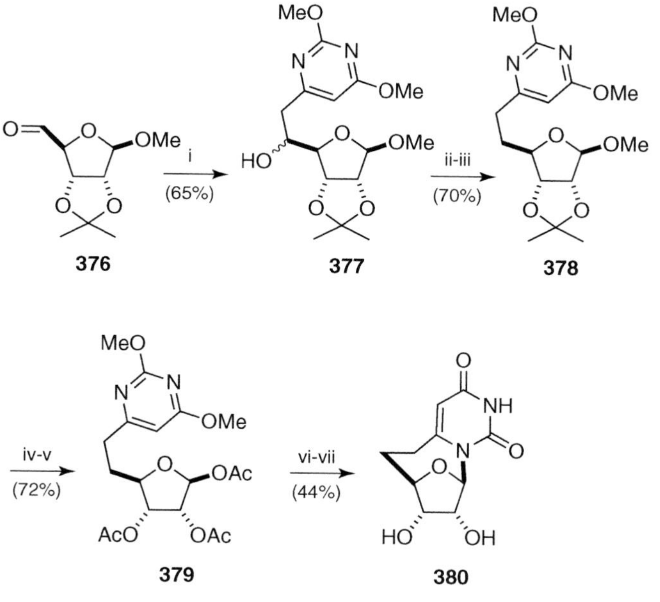

Scheme 45. *Reagents and conditions:* (i) 2,4-dimethoxy-6-methylpyrimidine, BuLi in hexane, THF, −48°C, 30 min, then −43°C, 2 h; (ii) (C=S)Imid$_2$, DMF, rt, 48 h; (iii) Bu$_3$SnH, AIBN, PhMe, 100°C, 1 h; (iv) aq. TFA, rt, 4 h; (v) Ac$_2$O, ACN, Et$_3$N, rt, 12 h; (vi) SnCl$_4$, ACN, rt, 2.5 h; (vii) 2 N NaOH, dioxane, 100°C, 20 min.

377 into the corresponding alkane **378** was carried out via the corresponding 5-imidazolylthiocarbonate, followed by treatment with Bu_3SnH and AIBN as radical initiator. Next, acidic hydrolysis of compound **378** and subsequent acetylation afforded the triacetate **379** in 72% yield (two steps). Treatment of compound **379** with $SnCl_4$ and aqueous NaOH in dioxane furnished the target cyclonucleoside **380** in 44% yield over the two steps.

3.2.2. Preparation of 5′,8-Linked Purine Cyclonucleosides.

Matsuda et al. described the synthesis of the 5′-deoxy-8,5′-cycloadenosine **383** in 39% yield, starting from the corresponding 5′-deoxy-5′-thiophenyl derivative **381**[109] as outlined in Scheme 46.[110] Compound **381** was treated with $P(OMet)_3$ in ACN and irradiated with a 100-W high-pressure mercury lamp to give, after deprotection of the diol **382**, the cyclonucleoside **383** in 50% overall yield. The authors proposed that the thiophenyl group, rather than the adenine moiety, in compound **381** was photo-activated initially to yield a 5′-methylene intermediate **F** and a thiophenyl radical. The intermediate cyclized to the next aminyl radical intermediate **G** followed by the release of a hydrogen radical to give the targeted cyclonucleoside **382**.

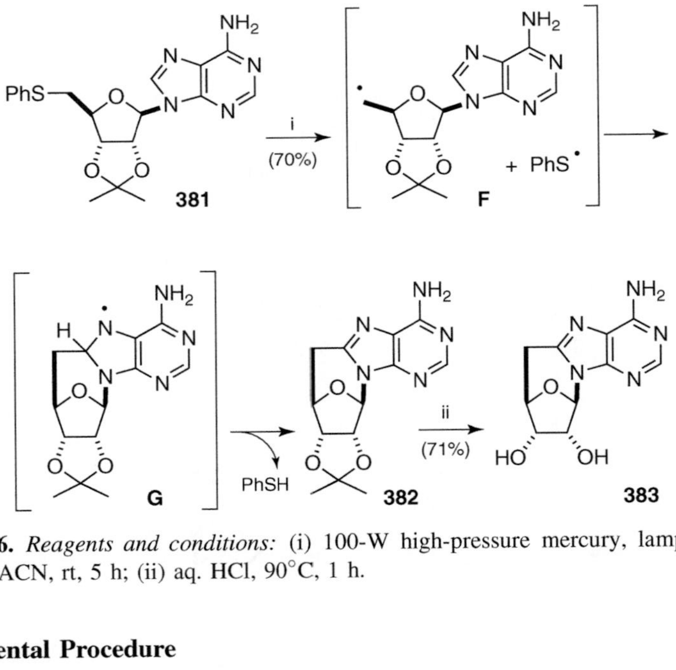

Scheme 46. *Reagents and conditions:* (i) 100-W high-pressure mercury, lamp, 254 nm, $P(OMe)_3$, ACN, rt, 5 h; (ii) aq. HCl, 90°C, 1 h.

Experimental Procedure

2′,3′-O-Isopropylidene-5′-deoxy-8,5′-cycloadenosine (382).[110] Compound **381** (416 mg, 1.041 mmol) and $P(OMe)_3$ (0.5 mL, 3.81 mol, 3.7 equiv) was dissolved in ACN (500 mL) and argon gas was bubbled through the solution for 30 min. This was irradiated with a 100-W low-pressure mercury lamp for 5 h under argon bubbling. Then, the filtrate was concentrated under reduced pressure and the residue was taken up in hot EtOH to yield **382** (205 mg, 70%) as a white solid, mp 231°C. ^{1}H-NMR (300 MHz, DMSO-d_6): δ 8.10 (s, 1H, H$_2$), 7.07 (br s, 2H, NH$_2$), 6.15 (s, 1H, H$_{1'}$), 4.79 (dd, $J = 5.5$ and 1.0 Hz, 1H, H$_{4'}$), 4.82 (d, $J = 6$ Hz, 1H, H$_{2'}$), 4.62 (d, 1H, H$_{3'}$), 3.43 (dd, $J = 18.0$ Hz, 1H, H$_{5'β}$), 2.93 (dd, $J = 18.0$ Hz, 1H, H$_{5'α}$), 1.46 (s, 3H, CH$_3$), 1.25 (s, 3H, CH$_3$). *Anal.* Calcd for C$_{13}$H$_{15}$N$_5$O$_3$ 1/4 H$_2$O: C, 53.12; H, 5.33; N, 23.84; found: C, 53.21; H, 5.25; N, 24.01.

Ueda et al. described the synthesis of 8,5'-cycloadenosines **388** and **389** using photochemical cyclization *via* the 8-phenylthioadenosine analogue **385** as presented in Scheme 47.[111] In contrast to the previous strategy described by the same group,[110] the thiophenyl was introduced onto the C8 carbon atom of the adenine instead of onto the C5' carbon atom of the glycone part. Starting from 2',3'-*O*-isopropylidene-8-bromoadenosine **384** (prepared in two steps from adenosine),[112] treatment with PhSNa in MeOH gave the 8-phenylthioadenosine analogue **385** in 85% yield. Irradiation of compound **385** with a 400-W high-pressure mercury lamp in the presence of *t*-BuOOH afforded the 8,5'-(*S*)-cycloadenosine **386** and the 8,5'-(*R*)-cycloadenosine **387** in 19% and 10% yield, respectively, along with 6% of the corresponding adenosine. Classical deprotection of compounds **386** and **387** gave the target cyclonucleosides **388** and **389** in high yields. The cyclization seen in the conversion of **385** into **386** and **387** was probably due to the formation of the radical intermediate **H** and a phenylthiyl radical and the subsequent formation of the biradical intermediate **I**, which cyclized onto the cyclonucleosides **386** and **387**, as proposed by the authors.

3.3. Preparation of Cyclonucleosides Having a C–O, C–S, or C–N Bridge

The main synthesis of cyclonucleosides having a C–O, C–S, or C–N bridge has used a substitution methodology. Two different routes have been described in the literature: (1) formation of the cyclonucleoside during a halogenation reaction; (2) formation of the cyclonucleoside starting from activated 5-halogenonucleoside analogues followed by elimination of the halogen atom. Various other approaches, such as radical addition, have also been reported.

3.3.1. Preparation of 5',6-Linked Pyrimidine Cyclonucleosides.
Pyrimidine cyclonucleosides having an O5'–C6 bridge have been obtained by convergent approaches of either halogenation-cyclization or dehalogenation-cyclization.

Lipkin and Rabi[113] reported the synthesis of 6,5'-*O*-cyclonucleoside **393** starting by the iodination of thymidine (see Scheme 48). Treatment of thymidine with NIS and TFA in anhydrous DMSO furnished, after 15 h at room temperature in the dark, a diastereoisomeric mixture of *trans*-5-iodo-5,6-dihydro-6,5'-*O*-anhydro analogues **390** and **391** in 25% and 75% yields, respectively. After photo-isomerization of the inseparable diastereoisomers **390** and **391** with an iodine quartz lamp, the major diastereoisomer **391** isomerized to the (5*S*,6*R*)-isomer **392** in 84% yield (based on

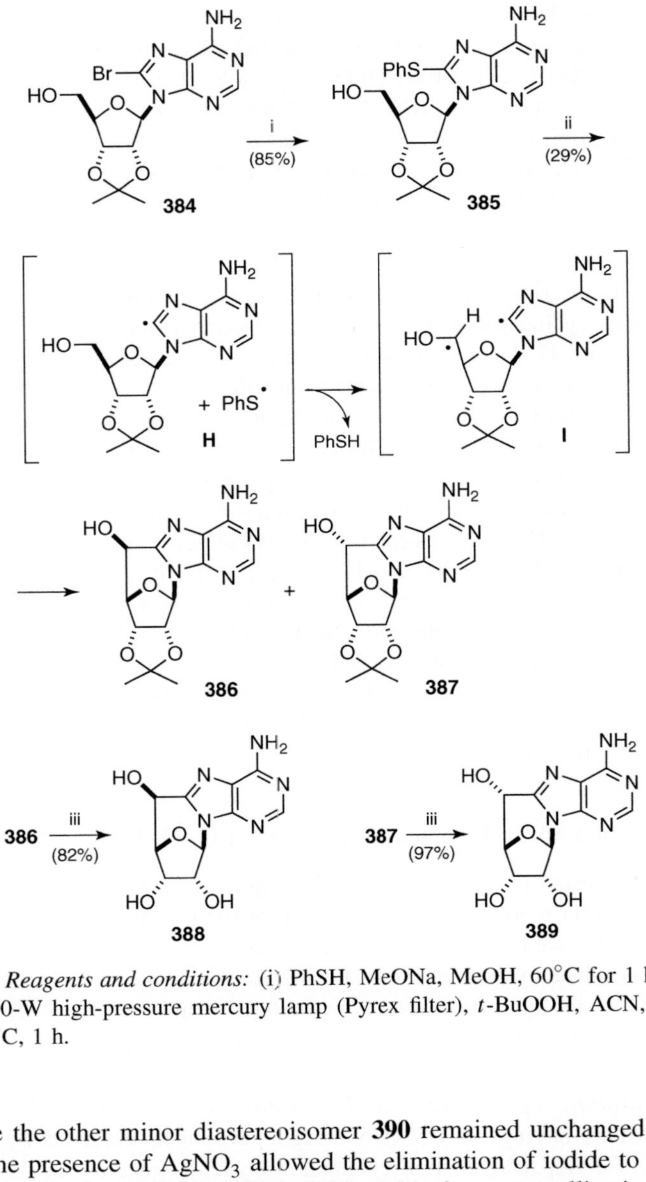

Scheme 47. *Reagents and conditions:* (i) PhSH, MeONa, MeOH, 60°C for 1 h, then rt for 12 h; (ii) 400-W high-pressure mercury lamp (Pyrex filter), *t*-BuOOH, ACN, rt, 4 h; (iii) aq. HCl, 90°C, 1 h.

391), while the other minor diastereoisomer **390** remained unchanged. Treatment of **392** in the presence of AgNO$_3$ allowed the elimination of iodide to provide the target 6,5′-*O*-anhydrothymidine **393** in 90% yield after recrystallization in water.

3.3.2. Preparation of 5′,9-Linked Purine Cyclonucleosides.
Chun et al.[114] reported the synthesis of 5′,9-anhydroxanthine derivatives **396** and **399** via the key intermediate, 6-amino-5′,6-*N*-anhydrouridine **395** as outlined in Scheme 49. Starting from the 5′-azido-5-bromo-5′-deoxy-2′,3′-*O*-isopropylidene-uridine **394** (prepared from uridine),[115] the one-pot cyclization was achieved by

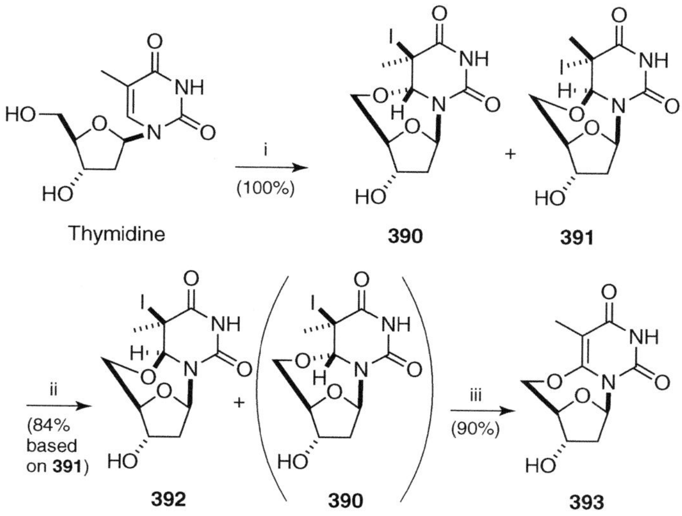

Scheme 48. *Reagents and conditions:* (i) NIS, TFA, DMSO, rt, 15 h; (ii) light from an iodine quartz lamp, 24°C, 0.4 h; (iii) AgNO$_3$.

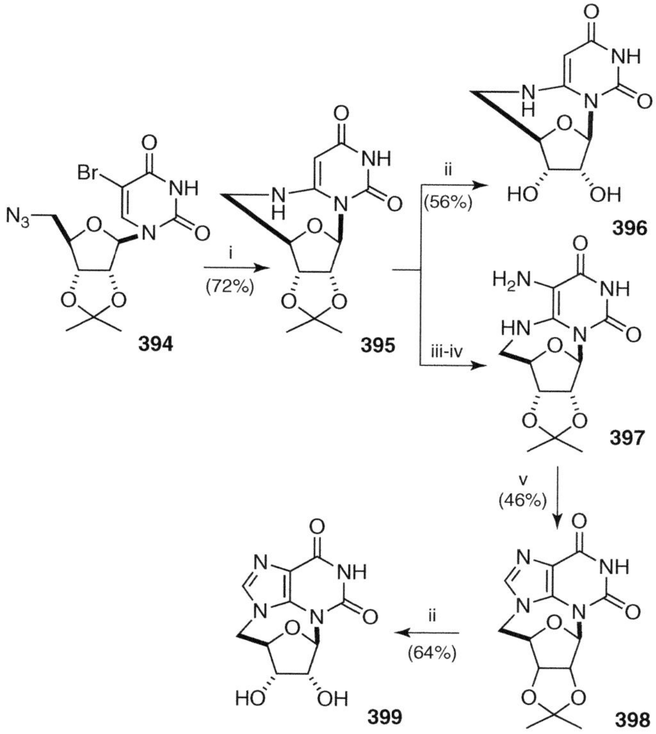

Scheme 49. *Reagents and conditions:* (i) Ph$_3$P, THF, rt, 6 h then aq. NH$_3$, rt, 6 h; (ii) aq. HCl; (iii) NaNO$_2$, HCl, aq. THF, 0°C, 15 min then NaHCO$_3$; (iv) Zn, HCl, MeOH, 90°C, 0.5 h, then NaHCO$_3$; (v) diethoxymethyl acetate, DMF, 90°C, 0.5 h.

treatment of azido **394** with Ph$_3$P followed by addition of aqueous NH$_3$ to furnish the cyclonucleoside **395** in 72% yield. At this stage, removal of the acetonide protecting group of **395** under acidic conditions gave the first target cyclonucleoside **396** in 56% yield. On the other hand, the derivative **397** was sequentially treated with NaNO$_2$ and HCl in aqueous THF to afford a 5-nitro intermediate, which was then directly reduced, without purification, with Zn/HCl in MeOH to provide the 5-amino intermediate **397**. Cyclization of the amino derivative **397** was carried out with diethoxymethylacetate in DMF at 90°C to give 5′,9-anhydro-2′,3′-*O*-isopropylidene-isoxanthosine **398** in 46% overall yield from **395**. Classical deprotection of the acetonide in compound **398** furnished the second target cyclonucleoside **399** in 64% yield.

4. SYNTHESIS OF CYCLIC PHOSPHOESTER NUCLEOTIDES

4.1. Introduction

While most of the conformationally restricted oligonucleotides to date have focused on the sugar puckering restriction,[11,86,116] the approach of restricting and structurally changing the P–O bonds is very promising because it should provide tools to help the folding and stabilization of unusual RNA and DNA motifs. In this context, in this section we describe the latest proposals and recent advances toward the introduction of conformational constraints in the sugar–phosphate backbone of nucleotides by means of cyclic phosphate structures connecting a phosphate to a base, a sugar moiety, or another phosphate of the same strand.[10]

These cyclic phosphate-modified nucleotides can be classified into two main groups: the macrocyclouridylic acid derivatives (Figure 5) and the constrained dinucleotides in which one of the phosphate oxygens is connected to a sugar carbon atom (Figure 6).

To obtain a new insight into the conformational significance of the U-turn structural element found in the anticodon loop of tRNAs and in the active site of the hammerhead ribozyme, Sekine et al. developed the synthesis of sterically fixed U-turn mimics.[117] First, macrocyclic dinucleotides were prepared by connecting the downstream nucleobase with the 2′-oxygen of the upstream nucleoside (not shown).[118] Then, with the aim of obtaining derivatives with an N-type sugar conformation, they synthesized 5′-cyclouridylic acids and derivatives (Figure 5, compounds **400** and **402**).[118b] Later, Nielsen et al. provided a new approach to these compounds based on RCM methodology and gave access to macrocyclicphosphate di- and trinucleotides in which a phosphate oxygen was connected either to a 2′-oxygen or to another phosphate oxygen (Figure 5, compounds **401**, **403**, and **404**).[119]

Following the same strategy, they prepared constrained dinucleotides with a seven- to nine-membered phosphotriester or phosphonate connecting the phosphate residue to C5′ or C2′ of the ribose moiety (Figure 6, compounds **405**, **406**, and **407**).[120] Analogous structures featuring six-membered cyclic phosphotriesters (dioxaphosphorinane) were synthesized by connecting one phosphate oxygen to

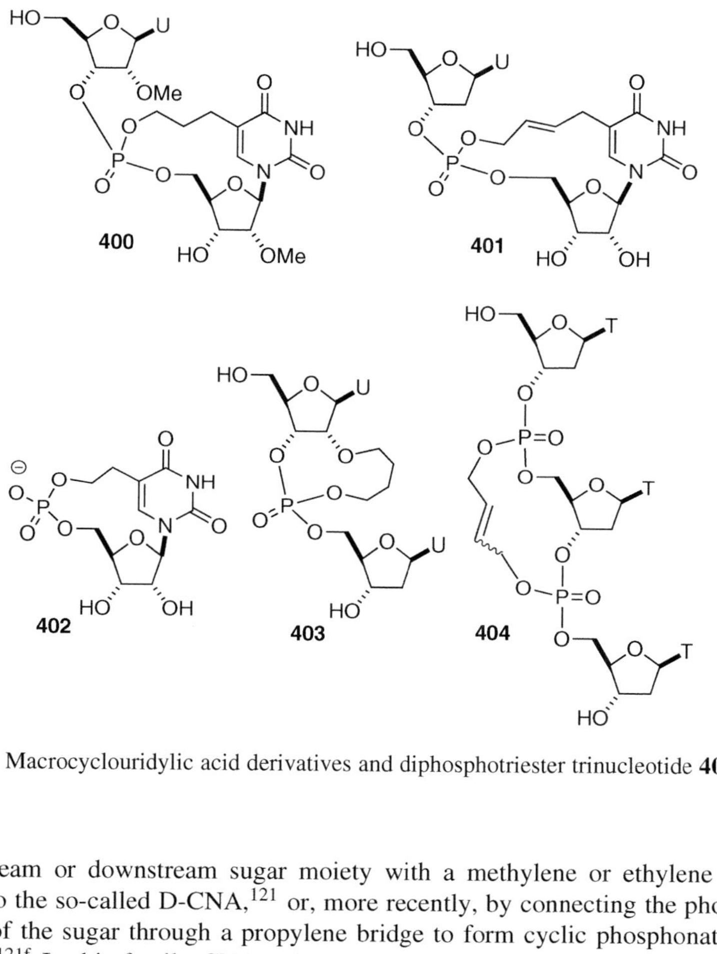

Figure 5. Macrocyclouridylic acid derivatives and diphosphotriester trinucleotide **400–404**.

the upstream or downstream sugar moiety with a methylene or ethylene bridge leading to the so-called D-CNA,[121] or, more recently, by connecting the phosphate to 5′-C of the sugar through a propylene bridge to form cyclic phosphonate (α,β-P-CNA).[121f] In this family CNA, all the torsional angles of the sugar–phosphate backbone can be controlled alternately. Taken together, these constrained nucleotide units provide an interesting tool box to fine-tune the shape of the sugar–phosphate backbone of nucleic acids and thus design transient conformations.

4.2. Preparation of Macrocyclic Diuridine Nucleotides

The "second" generation of U-turn mimics was proposed with the idea that cyclic phosphotriesters with a defined stereochemistry at the phosphorus would be efficient for obtaining a preorganized DNA single strand exhibiting a severe kink similar to that of the target structure.[122]

Starting from 5′-O-dimethoxytrityl-2′-O-methyl uridine, acetylation with acetic anhydride of the 3′-hydroxyl function and immediate acidic treatment led to the 3′-O-acetyluridine derivative, which was treated with iodine chloride in dioxane

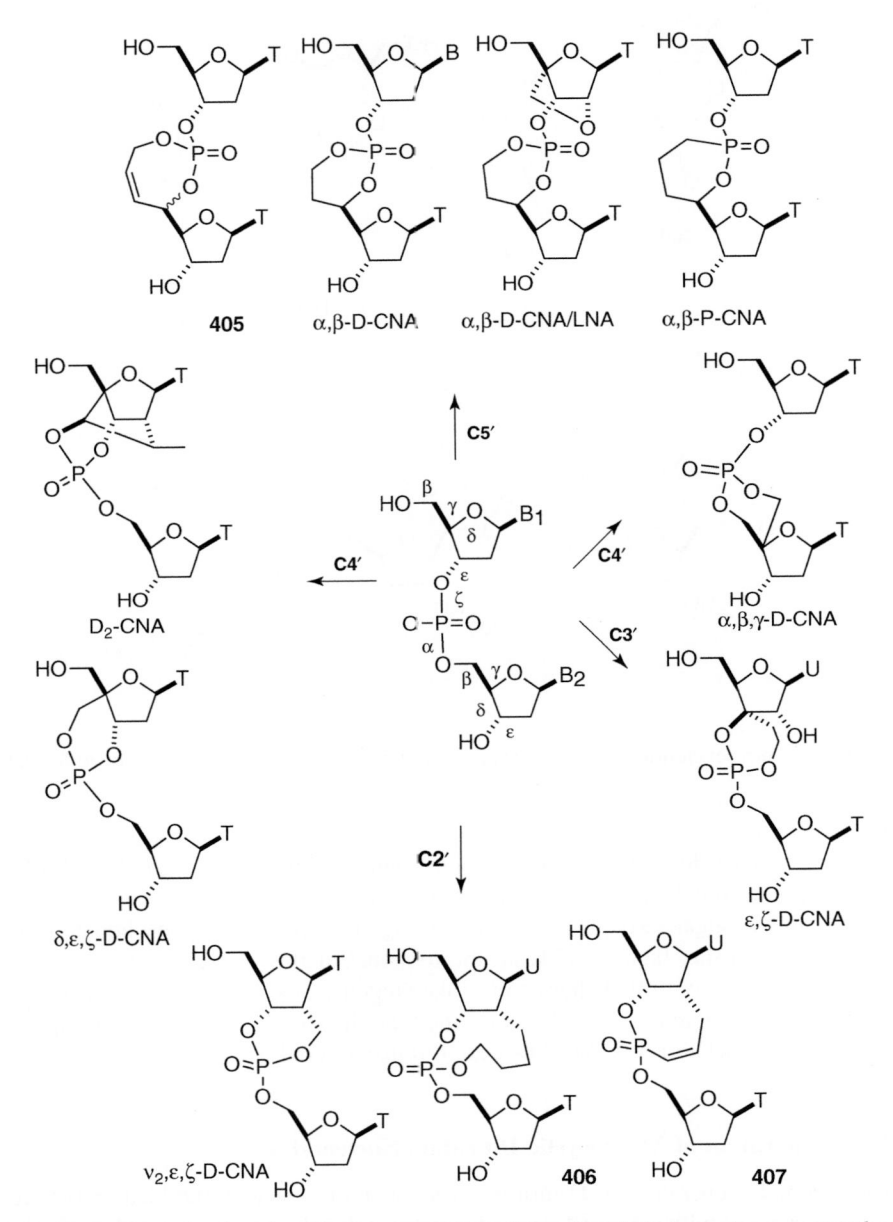

Figure 6. Constrained nucleotides by connection of the phosphate group to a sugar carbon atom.

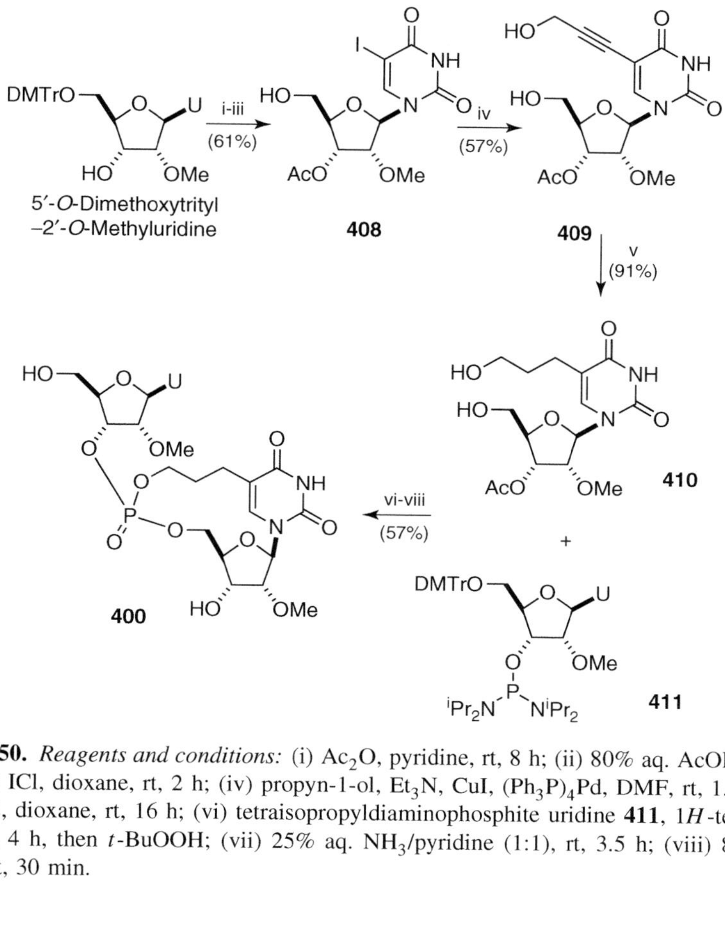

Scheme 50. *Reagents and conditions:* (i) Ac$_2$O, pyridine, rt, 8 h; (ii) 80% aq. AcOH, rt, 30 min; (iii) ICl, dioxane, rt, 2 h; (iv) propyn-1-ol, Et$_3$N, CuI, (Ph$_3$P)$_4$Pd, DMF, rt, 1.5 h; (v) H$_2$, Pd/C, dioxane, rt, 16 h; (vi) tetraisopropyldiaminophosphite uridine **411**, 1*H*-tetrazole, ACN, rt, 4 h, then *t*-BuOOH; (vii) 25% aq. NH$_3$/pyridine (1:1), rt, 3.5 h; (viii) 80% aq. AcOH, rt, 30 min.

to afford the 5-iodouridine[123] **408** in 61% overall yield (Scheme 50). The hydroxylpropinyl chain at the 5 position of **408** was installed through a (Ph$_3$P)$_4$Pd/CuI coupling in the presence of Et$_3$N to give **409** in 57% yield.[118b]

Pd-catalytic hydrogenation of the triple bond of **409** gave the diol derivative **410** in a good 91% yield. The macrocyclic phosphotriester was built by reaction of **410** with the tetraisopropyldiaminophosphite uridine derivative **411** in standard phosphoramidite coupling conditions in 80% yield as an S_P and R_P diastereoisomeric mixture. Next, 3′-*O*-acetate removal in ammonia enabled the separation of the two phosphorus epimers **400** (46 and 43% of R_P and S_P, respectively), which were treated independently with 80% AcOH to remove the 5′-*O*-DMTr group in 82% and 79% yield, respectively.

The S_P isomer of **400** induced a strong bending in the oligonucleotide that was rather unfavorable for duplex formation and was therefore identified as a good motif to mimic the biologically relevant U-turn motif.

4.3. RCM Approach Toward Constrained Dinucleotides

Intramolecular ring-closing metathesis has proved to be a powerful methodology to build cyclophosphates larger than seven-membered rings.[59] A first example of this approach was the synthesis of the four diastereoisomers of the seven-membered unsaturated cyclophosphate dithymidine derivatives **405** (Scheme 51).[120a,b] They were obtained by RCM applied to a dinucleotide precursor **414** bearing an allyl function on the phosphate and a vinyl substituent at 5'-C of the downstream nucleoside. The two fragments carrying the allyl or vinyl moiety, **412** and **413**, respectively, were prepared from thymidine. Selective protection of the 5'-hydroxyl function with TBSCl in pyridine followed by phosphitilation of the 3'-hydroxyl function with (allyloxy)bis(diisopropylamino)phosphine gave the desired thymidine phosphite **412** in an excellent 95% overall yield.[124] Thymidine aldehyde **249** (Scheme 31) was synthesized following a classical three-step protection/deprotection sequence from thymidine, ending with a Dess–Martin oxidation of the primary hydroxyl function to produce the aldehyde in a good 70% yield.[125] Addition of the Grignard reagent vinyl magnesium bromide to **249** proceeded without diastereoselectivity and in a modest 41% yield to give a

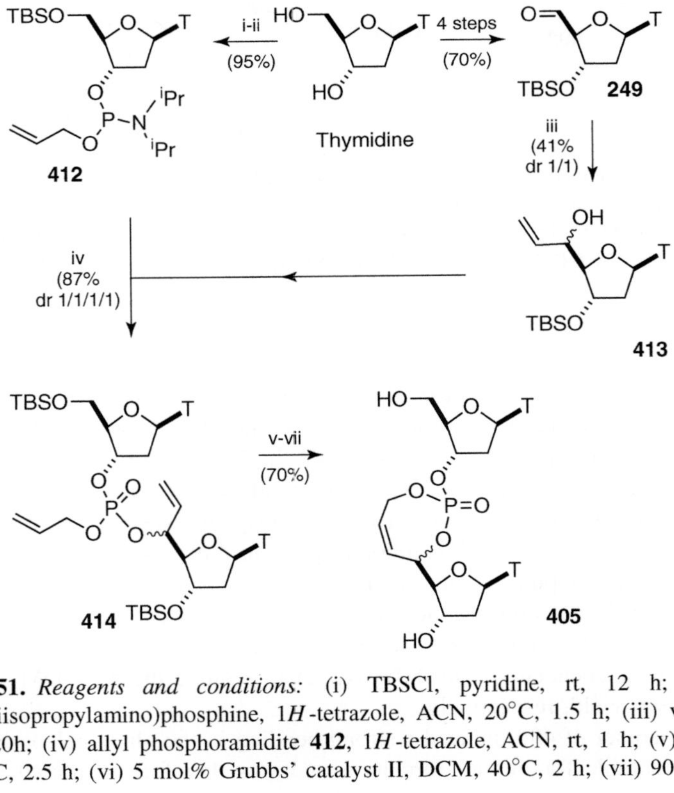

Scheme 51. *Reagents and conditions:* (i) TBSCl, pyridine, rt, 12 h; (ii) (allyloxy)bis(diisopropylamino)phosphine, 1*H*-tetrazole, ACN, 20°C, 1.5 h; (iii) vinyl-MgBr, THF, rt, 20h; (iv) allyl phosphoramidite **412**, 1*H*-tetrazole, ACN, rt, 1 h; (v) *t*-BuOOH, PhMe, 0°C, 2.5 h; (vi) 5 mol% Grubbs' catalyst II, DCM, 40°C, 2 h; (vii) 90% aq. TFA, rt, 1 h.

5′-C epimeric mixture of **413**. The dinucleotide **414** was obtained by mixing the phosphoramidite **412** and the alcohol **413** under 1H-tetrazole activation.[126] The resulting phosphotriester was isolated in 87% yield as an equimolecular mixture of the four diastereoisomers arising from the introduction of a 5′-stereocenter and an asymmetric phosphorus atom. The RCM reaction with **414** as the substrate and a Grubbs' catalyst II gave the expected cyclic phosphotriesters in 91% yield and in a short reaction time (2 h) in refluxing DCM instead of 48 h with Grubbs' catalyst I with only 65% yield.[127] The protecting groups of **414** were removed by treatment with a 90% aqueous TFA solution to furnish **405**.

This was the first example of the use of the RCM methodology for the construction of nucleotide derivatives. Although the diastereoisomers were not separated, molecular calculations indicated that one of these constrained dinucleotides presented α and β torsional angle values that could favor base stacking and therefore duplex formation.

Following a similar approach, Nielsen et al.[120c] undertook the synthesis of new uridine-thymidine dinucleotides in which the constraint was applied to ε and ζ torsional angles together with the upstream sugar pucker (Figure 6, compound **406**). To obtain the uridine-thymidine derivative **421** (Scheme 52) suitable for the RCM reaction, they constructed a 2′-deoxyuridine derivative **418** bearing an allyl substituent at the C2′ position and a 5′-O-allylphosphoramidite thymidine **420** independently.

The 2′-C-allyluridine derivative **417** was prepared following a similar route to that described in Scheme 23 for its congener **194**. Concomitant protection of the 5′- and 3′-hydroxyl functions of uridine with TIPDSCl$_2$ provided the uridine derivative **415** with a free 2′-position in 95% yield. The phenylthiocarbonate ester **416** was cleanly obtained by reaction with PTC-Cl in the presence of DMAP and Et$_3$N. The radical reaction between **416** and allyltributyltin with AIBN as initiator gave the 2′-C-allyl-uridine derivative **417** with absolute retention of configuration in 67% yield under standard conditions.[128] The yield could be increased to 84% by irradiation with a mercury medium-pressure lamp under an inert atmosphere for 9 h.[129] Removal of the TIPDS protective group with TBAF in acidic medium followed by 5′-hydroxyl function protection with TBSCl activated with AgNO$_3$ provided the uridine derivative **418** with a free 3′-hydroxyl function in 76% overall yield. The 3′-O-protected thymidine **419** reacted with allyloxybis(diisopropylamino)phosphite activated with 1H-tetrazole to provide the 5′-O-allylphosphoramidite thymidine **420** in a moderate 66% yield. Compound **419** was prepared in three steps from thymidine in 80% overall yield following well-known chemistry.[130] With the uridine derivative **418** and the thymidine phosphoramidite **420** in hand, they next ran the coupling in 85% yield with 1H-tetrazole as activating agent and the oxidation of the intermediate phosphite by t-BuOOH. The dinucleotide **421** was then submitted to a tandem RCM-hydrogenation protocol[131] using Grubbs' catalyst II which gave the cyclic phosphotriester **422** in 65% yield in a diastereoisomeric ratio R/S of 3/1. After separation by chromatography column on silica gel both phosphorus epimers were isolated in 48% and 17% yields. TBS protective groups were removed with an aqueous TFA solution to give the constrained dinucleotides **406** quantitatively.

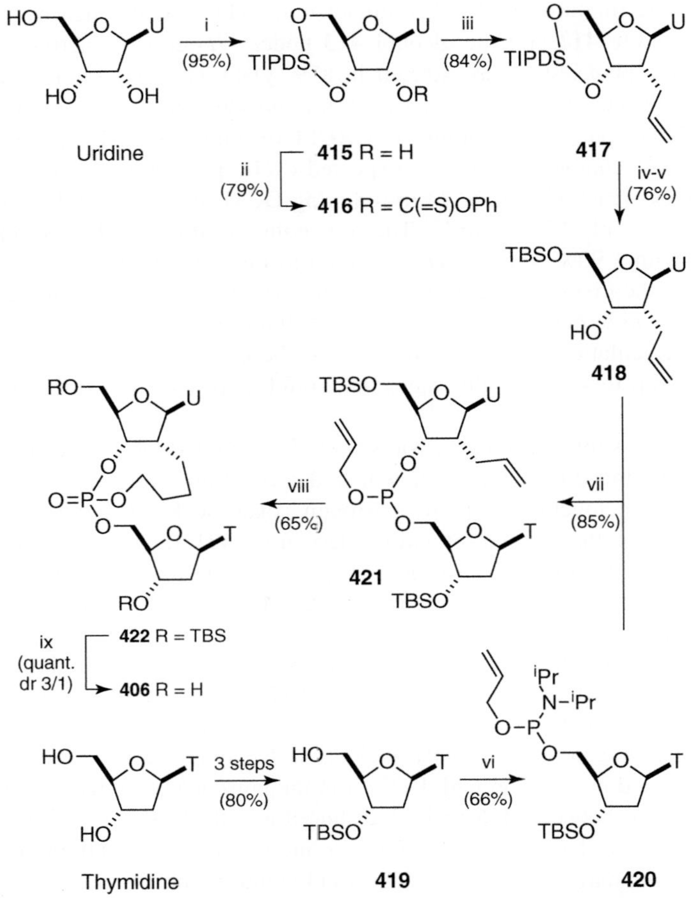

Scheme 52. *Reagents and conditions:* (i) TIPSCl$_2$, Imid, DMF, rt, 24 h; (ii) PTC-Cl, Et$_3$N, DMAP, DCM, rt, 16 h; (iii) allyl-SnBu$_3$, AIBN, PhH, $h\nu$, rt, 9 h; (iv) TBAF, AcOH, THF, rt, 20 h; (v) TBSCl, AgNO$_3$, pyridine, rt, 25 h; (vi) CH$_2$=CHCH$_2$OP(Ni-Pr$_2$)$_2$, 1H-tetrazole, ACN, rt, 1 h; (vii) 1H-tetrazole, ACN, rt, 2.5 h, then t-BuOOH, rt, 40 min; (viii) Grubbs' catalyst II, DCM, reflux, 48 h, then 1000 psi H$_2$, 50°C, 12 h; (ix) 90% aq. TFA, rt, 5 h.

When incorporated within oligonucleotides, only the (*R*)-isomer of **406** showed good stability toward ammonia treatment. This modification was able to stabilize slightly a three-way junction ($\Delta T_m \sim +2°C$).

Experimental Procedure

3 R- and 3 S-(1 S,9 R,10 R,12 R)-3-(3(S)-(t-Butyldimethylsilyl)oxy-5(R)-(thymin-1-yl)tetrahydrofuran-2(R)-yl)methoxy)-3-oxo-12-(t-butyldimethylsilyl)oxymethyl-10-(thymin-1-yl)-2,4,11-trioxa-3-phosphabicyclo[7.3.0]dodecane (422).[120c] To a solution of dinucleotide **421** (0.199 g, 0.236 mmol) in DCM (23 mL) was added

Grubbs' catalyst II (10 mg, 12 µmol). After 48 h under reflux, the reaction mixture was subjected to 1000 psi H_2, in a Parr bomb, at 50°C for 12 h. After removal of the solvent under reduced pressure, the crude mixture was submitted to silica gel chromatography with EA in petroleum ether from 60 to 100% followed with 2–5% MeOH in EA as solvent. The slower-moving compound was identified as the (S)-configured phosphorus (32 mg, 17%) and the fast-moving one as the (R)-configured phosphorus (92 mg, 48%). Data for R_P **422**: ^{1}H-NMR (300 MHz, CDCl$_3$): δ 9.31 (s, 1H, NH), 9.06 (s, 1H, NH), 7.84 (d, J = 8.1 Hz, H_{6U}), 7.41 (s, 1H, H_{6T}), 6.31 (t, 1H, J = 6.6 Hz, $H_{1'T}$), 6.07 (d, J = 8.3 Hz, 1H, $H_{1'U}$), 5.73 (d, J = 8.1 Hz, H_5), 5.21 (m, 1H, $H_{3'U}$), 4.38–4.51 (m, 2H, $H_{3'T}$ and $H_{9'bU}$), 4.18–4.32 (m, 4H, $H_{5'T}$ and $H_{9'aU}$), 4.03 (m, 1H, $H_{4'T}$), 3.89 (br s, 2H, $H_{5'U}$), 2.25–2.35 (m, 2H, $H_{2'U}$ and $H_{2'bT}$), 2.15 (m, 1H, $H_{2'aT}$), 1.95 (s, 3H, Me), 1.80–1.95 (m, 4H, $H_{6'U}$ and $H_{7'U}$), 1.60–1.75 (m, 2H, $H_{8'U}$), 0.90–0.93 (2s, 18H, tBu), 0.10–0.14 (m, 12H, MeSi). ^{31}P-NMR (121.5 MHz, CDCl$_3$): δ 0.44. HiRes MALDI FT-MS m/z [M + Na] found/calcd: 837.3323/837.3298. Data for S_P **423**: ^{1}H-NMR (300 MHz, CDCl$_3$): δ 8.51 (s, 1H, NH), 8.48 (s, 1H, NH), 7.78 (d, J = 8.1 Hz, H_{6U}), 7.28 (d, J = 1.0 Hz, 1H, H_{6T}), 6.25 (m, 1H, $H_{1'T}$), 6.04 (d, J = 9.1 Hz, 1H, $H_{1'U}$), 5.71 (dd, J = 1.8 and 8.1 Hz, H_5), 4.83 (dd, J = 4.8 and 9.6 Hz, 1H, $H_{3'U}$), 4.51 (m, 1H, $H_{9'bU}$), 4.17–4.33 (m, 5H, $H_{4'U}$, $H_{3'T}$, $H_{5'T}$ and $H_{9'aU}$), 4.04 (m, 1H, $H_{4'T}$), 3.73–3.87 (m, 2H, $H_{5'U}$), 2.00–2.35 (m, 3H, $H_{2'U}$ and $H_{2'T}$), 2.15 (m, 1H, $H_{2'aT}$), 1.96 (s, 3H, Me), 1.55–1.90 (m, 6H, $H_{6'U}$, $H_{7'U}$ and $H_{8'U}$), 0.90–0.92 (2s, 18H, tBu), 0.10–0.12 (m, 12H, MeSi). ^{31}P-NMR (121.5 MHz, CDCl$_3$): δ −1.62. HiRes Maldi FT-MS m/z [M + Na] found/calcd. 837.3262/837.3298.

4.4. Dioxaphosphorinane-Constrained Dinucleotides

All the constrained nucleotides presented previously involved large and rather flexible cyclicphosphate structures (7 < n < 13). The conformations of these molecules are therefore not well defined in terms of sugar–phosphate torsional angle values. To circumvent this problem, dimeric building units have been developed by introducing six-membered dioxaphosphorinane rings at selected positions along the sugar–phosphate backbone (Figure 6).[131] In this family, referred to as D-CNA (dioxaphosphorinane-constrained nucleic acid), the dioxaphosphorinane structure adopts a rigid conformation and, consequently, the torsion angles exhibit well-defined and fixed values.

The dioxaphosphorinane ring formation has been achieved by two methodologies depending on the target D-CNA; either by an intramolecular nucleophilic substitution of a tosylate by a phosphate oxyanion (for the α,β-, α,β,γ- and ε,ζ-D-CNA) or by a standard phosphotriester methodology using triazole derivatives as activators (for the δ,ε,ζ- and ν_2,ε,ζ-D-CNA).

The representative diastereoisomers of α,β-D-CNA: ($S_{C5'}$, R_P) **428**, ($R_{C5'}$, S_P) **432** and ($R_{C5'}$, R_P) **433** (Scheme 53) were synthesized from thymidine aldehyde **423**. The latter aldehyde was obtained in 90% yield from thymidine in a similar fashion to that for **249** (Scheme 31) but with TBDPS as the 3′-O protective group and with

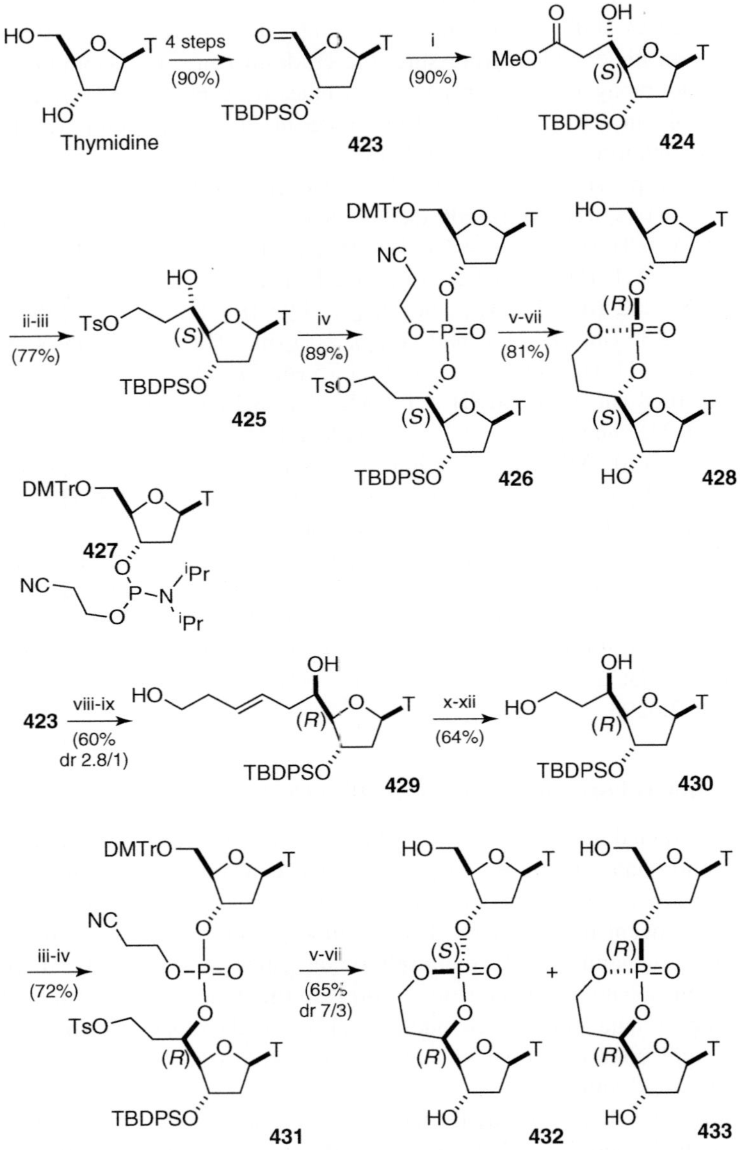

Scheme 53. *Reagents and conditions:* (i) cat. BiCl$_3$, ZnI$_2$, methyl acetate silylketene, DCM, rt, 2 h; (ii) NaBH$_4$, EtOH, rt, 12 h; (iii) TsCl, CHCl$_3$, pyridine, 0°C, 12 h; (iv) 5′-O-DMTr-3′-O-diisopropylaminocyanoethoxyphosphite thymidine **427**, 1H-tetrazole, ACN, rt, 20 min, then I$_2$/H$_2$O, rt, 10 min; (v) Et$_3$N, DMF, 90°C, 2 h; (vi) TBAF, THF, rt, 1 h; (vii) 3% TFA/DCM, rt, 15 min; (viii) ω-t-butyldimethylsilyloxyallyltrimetrisilane, Et$_2$O-BF$_3$, DCM, −78°C, 3 h; (ix) TiCl$_4$, DCM, rt, 1 min; (x) cat. OsO$_4$, NMO, dH$_2$O, rt, 16 h; (xi) NaIO$_4$, MeOH, 0°C, 1 h, then rt, 2 h; (xii) NaBH$_4$, EtOH, 0°C, 16 h.

a Pfitzner–Moffatt oxidation protocol[132] to convert the 5′-hydroxyl function into aldehyde.

A diastereoselective Mukaiyama reaction between **423** and methyl acetate silylketene catalyzed by $BiCl_3/ZnI_2$ gave the thymidine ester derivative **424** with an (S)-configuration at C5′ in 90% yield.[133] Further reduction of the ester function with $NaBH_4$ in EtOH and selective tosylation of the newly generated primary hydroxyl function with TsCl in chloroform/pyridine led to the activated 5′-C-tosyloxyethyl thymidine **425** in 77% overall yield. The dinucleotide **426** was obtained in 89% yield by standard phosphoramidite coupling technology with commercially available thymidine phosphoramidite **427**. The cyanoethyl phosphate protective group of **426** was removed under basic conditions (Et_3N/DMF/90°C or K_2CO_3/DMF/rt) leading to the formation of the phosphate, which cleanly displaced the tosylate leaving group to form the ioxaphosphorinane ring in a diastereoselective manner.[134] α,β-D-CNA ($S_{C5'}$, R_P) **428** was recovered as a single diastereoisomer in 81% yield after sequential removal of the protective groups with fluoride ion and acidic treatment. Conformational analysis showed that this particular constrained dinucleotide featured a gauche (+)-conformation for the α torsional angle which is different from that observed in regular nucleotides involved in an A- or B-type duplex.[135]

The diastereoisomers **432** and **433** were prepared starting from a Sakurai condensation involving the aldehyde **423** and an ω-t-butyldimethyl-silyloxyallyltrimetrisilane in the presence of BF_3-Et_2O as a Lewis acid.[136] The crude product of the reaction was then submitted to $TiCl_4$ in DCM and the 5′-C(R)-hydroxypentenyl thymidine **429** was separated from its C5′ (S) epimer (60% yield and dr R/S = 2.8:1). The 5′-C hydroxylethylthymidine **430** resulted from a three-step oxidative cleavage of the double bond with $OsO_4/NaIC_4$ and finally a reduction of the carbonyl with $NaBH_4$ in 64% yield. The subsequent steps were identical to those described for the preparation of **428** by a tosylation and a phosphoramidite coupling to give **431** in 72% overall yield. Here the cyclization proceeded with less diastereoselectivity, providing a separable 7/3 phosphate epimeric mixture. After removal of the protective groups, the conformations of α,β-D-CNA **432** and **433** were determined. For dimer **432**, the α and β torsional angle values were assigned as gauche (−)/trans, identical to those observed in an A- or B-type duplex, whereas for **433** α deviated to the trans conformation with no change for β.

The synthesis of $v_2,ε,ζ$-D-CNA illustrated in Scheme 54 used the sugar chiral pool as the starting material. The 3-C-hydroxymethyl-D-ribose derivative **259**, which presents the necessary extra hydroxymethyl at C3 for further connection with the phosphate moiety, was prepared from the commercially available diacetonide-α-D-glucose as described in Scheme 32.[137]

Protection of the primary hydroxyl function of **259** with benzoyl chloride in pyridine provided the corresponding benzoate **434**. A selective deprotection of the 5,6-isopropylidene of **434** in diluted HCl followed by a difficult selective protection of the primary hydroxyl function with benzoyl chloride gave **435** in a modest 44% yield. Cleavage of the remaining acetal with AcOH and periodic oxidation led,

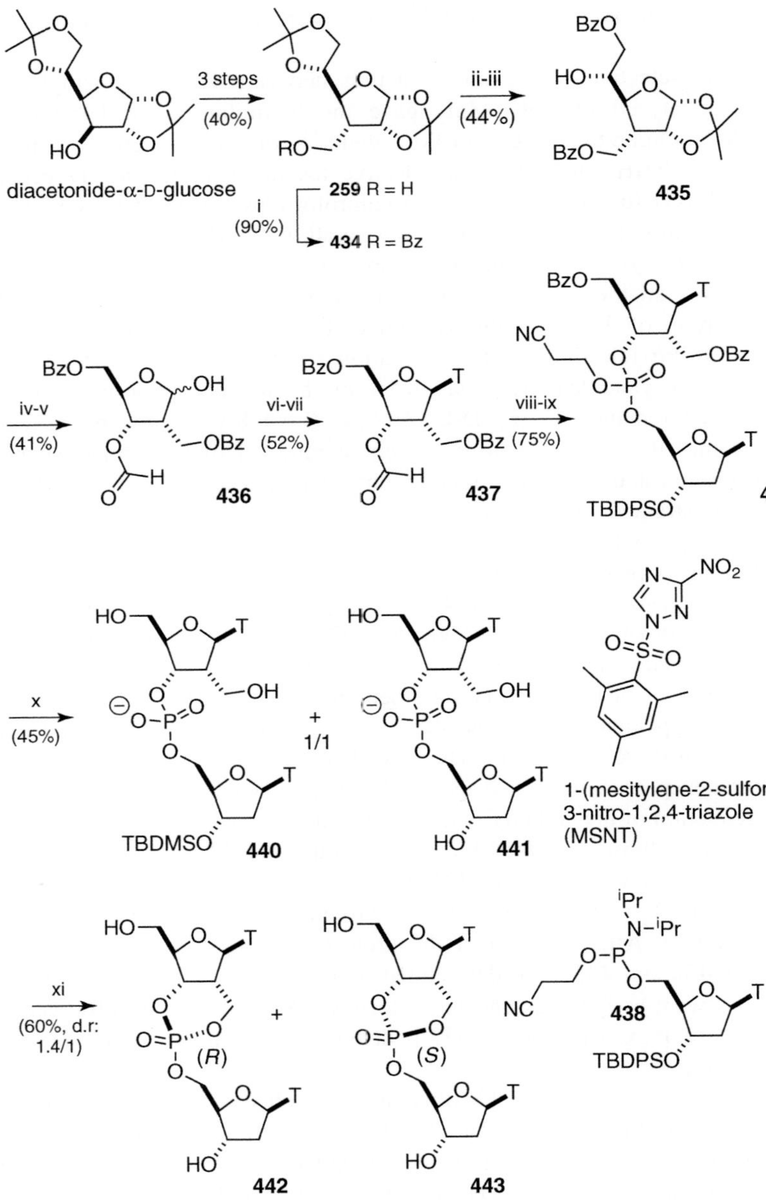

Scheme 54. *Reagents and conditions:* (i) BzCl, pyridine, DMAP, rt, 12 h; (ii) HCl 0.06 N, MeOH, 55°C, 2 h; (iii) BzCl, pyridine, 0°C, 2 h; (iv) 80% aq. AcOH, 80°C; (v) KIO$_4$, EtOH/H$_2$O 1:1, rt, 18 h; (vi) Ac$_2$O, pyridine; (vii) persilylated thymine, TMSOTf, ACN, reflux, 2 h; (viii) 28% aq. NH$_3$, rt, 2 h; (ix) 3′-*O*-silylthymidine-5′-*O*-phosphoramidite **438**, 1*H*-tetrazole, ACN, then collidine, I$_2$/H$_2$O, rt, 3 h; (x) K$_2$CO$_3$, MeOH/H$_2$O 4:1, rt, 6 h; (xi) MSNT, pyridine, rt for 1 h, then 80°C for 30 min and 90°C for 30 min.

after rearrangement of the sugar, to the 2-deoxy-2-hydroxymethyl-D-ribose derivative **436** in 41% yield over the two steps. Acetylation of the anomeric hydroxyl with acetic anhydride in pyridine enabled the thymine moiety to be introduced by Vorbrüggen's procedure[12] with good 9:1 selectivity in favor of the β-anomer **437** in 52% yield after separation. Ammonia treatment removed the residual formyl group at O3 and coupling with the 5'-O-phosphoramidite thymidine derivative **438** with 1H-tetrazole activation and iodine oxidation gave the dinucleotide **439** as a 1:1 mixture of diastereoisomers in 75% yield. The benzoyl and cyanoethyl protective groups were eliminated with K_2CO_3 in MeOH/water, inducing a partial cleavage of the 3'-O-TBS group. Consequently, dinucleotides **440** and **441** were isolated in 45% yield as a 1:1 mixture of 3'-O-TBDPS/3'-OH, respectively. After chromatographic separation, intermediate **441** was submitted to activation by MSNT in pyridine to form the dioxaphosphorinane ring.[138] The cyclization occurred in 60% yield with a dr = 1.4:1 in favor of the ($R_{C2'}$, S_P) isomer **443** that was separated from its diastereoisomer ($R_{C2'}$, R_P) **442** by silica gel chromatography.

The structural studies of these constrained dinucleotides revealed that the upstream sugar puckering adopted a C2'-*endo* conformation (south) and that the torsional angles ε and ζ of isomers ($R_{C2'}$, S_P) **443** and ($R_{C2'}$, R_P) **442** were in the range of (a$^+$/t, g$^+$) and (a$^+$/t, a$^-$), respectively. Both of these value sets deviate from those depicted in an A- or B-type duplex.

Experimental Procedure

v_2,ε,ζ-*D-CNA* ($R_{C3'}$, R_P) *(442) and* ($R_{C3'}$, S_P) *(443).*[121d] To compound **441** (40 mg, 0.049 mmol) in anhydrous pyridine (1.2 mL) was added, under an inert argon atmosphere, 1-(mesitylene-2-sulfonyl)-3-nitro-1,2,4-triazole (30 mg, 0.1 mmol) and stirred for 1 h at rt, 1 h at 80°C, then 30 min at 90°C. After addition of EA/MeOH 10% (5 mL), the solvent was removed under high vacuum and compounds **442** (10 mg) and **443** (13 mg) were isolated (60% yield) by means of HPLC (Kromasil C_{18}, 7 μm, 250 × 4.6 mm; water/ACN gradient from 95:5 to 60:40 in 30 min), **442**: T_r = 27.91 min, **443**: T_r = 27.12 min. Data for **442**: ^{1}H-NMR (500 MHz, CD$_3$OD): δ 7.89 (d, J = 1.0 Hz, 1H, 6α-H), 7.53 (d, J = 1.0 Hz, 1H, 6β-H), 6.47 (d, J = 9.0 Hz, 1H, 1'α-H), 6.31 (t, J = 7.0 Hz, 1H, 1'β-H), 5.31 (m, J = 5.0 Hz, $J_{H/P}$ = 2.5 Hz, 1H, 3'α-H), 4.76 [A part of an ABX(Y) syst.], J = 3.7 and 12.5 Hz, $J_{H/P}$ = 6.9 Hz, 1H, 6'α-H], 4.49–4.39 (m, 4H, 6'α-H, 5'β-H, and 3'β-H), 4.30 (b s, 1H, 4'α-H), 4.09 (m, J = 3.6 Hz, $J_{H/P}$ = 2.2 Hz, 1H, 4'β-H), 3.84 (d, J = 3.0 Hz, 2H, 5'α-H), 2.89 (m, 1H, 2'α-H), 2.33–2.24 (m, J = 6.5, 6.5, and 7.0 Hz, 2H, 2'β-H), 1.90 (d, J = 1.0 Hz, 3H, Me), 1.89 (d, J = 1.0 Hz, 3H, Me). ^{13}C-NMR (125 MHz, CD$_3$OD): δ 164.9, 164.8, 151.5, 150.9, 136.1, 135.9, 111.0, 110.5, 86.0, 85.9, 84.9, 84.7, 84.6, 84.4, 82.7, 70.5, 68.9, 68.8, 63.9, 63.8, 61.4, 61.3, 43.8, 43.7, 39.0, 11.1. ^{31}P-NMR (202 MHz, CD$_3$OD): δ −6.6. MS (ESI): 559.3 ([M + H]$^+$), 581.2 ([M + Na]$^+$), 597.5 ([M + K]$^+$). $C_{21}H_{27}N_4O_{12}P$ (558.43): calcd. C 45.17, H 4.87, N 10.03; found: C 45.23, H 4.78, N 9.92. Data for **443**: ^{1}H-NMR (500 MHz, CD$_3$OD): δ 7.92 (d, J = 1.0 Hz, 1H, 6a-H), 7.55 (d, J = 1.0 Hz, 1H, 6b-H), 6.51 (d, J = 9.5 Hz, 1H, 1'a-H), 6.28 (t, J = 7.0 Hz, 1H, 1'b-H), 5.27 (d, J = 4.5 Hz,

1H, 3′a-H), 4.69 [A part of an ABX(Y) syst.], $J = 3.2$ and 12.5 Hz, $J_{H/P} = 1.5$ Hz, 1H, 6′a-H), 4.46 (m, 1H, 3′b-H), 4.42 [B part of an ABX(Y) syst.], $J = 12.5$ Hz, $J_{H/P} = 24.0$ Hz, 1H, 6′a-H), 4.40 [A part of an ABX(Y) syst.], $J = 11.3$ Hz, $J_{H/P} = 3.3$ Hz, 1H, 5′b-H), 4.36 [B part of an ABX(Y) syst.], $J = 5.8$ and 11.3 Hz, $J_{H/P} = 7.7$ Hz, 1H, 5′β-H), 4.27 (b s, 1H, 4′a-H), 4.11 (m, $J = 3.3$, 3.4 and 5.8 Hz, 1H, 4′b-H), 3.81 (ABX syst., $J = 2.5$, 3.0 and 12.0 Hz, 2H, 5′a-H), 2.84 (m, 1H, 2′a-H), 2.32 (m, $J = 6.5$ and 7.0 Hz, 2H, 2′b-H), 1.92 (d, $J = 1.0$ Hz, 3H, Me), 1.90 (d, $J = 1.0$ Hz, 3H, Me). ^{13}C-NMR (125 MHz, CD$_3$OD): δ 164.9, 164.8, 151.5, 150.8, 136.6, 136.1, 110.9, 110.5, 86.2, 86.1, 85.6, 84.7, 84.6, 84.1, 83.8, 83.7, 70.3, 67.1, 67.0, 64.7, 64.6, 61.5, 43.1, 43.0, 38.9, 11.1, 11.0. ^{31}P-NMR (202 MHz, CD$_3$OD): δ −7.8. MS (ESI): 559.3 ([M + H]$^+$), 581.2 ([M + Na]$^+$), 597.5 ([M + K]$^+$). C$_{21}$H$_{27}$N$_4$O$_{12}$P (558.43): calcd: C 45.17, H 4.87, N 10.03; found: C 44.97, H 4.92, N 10.14.

To date, only two α,β-D-CNA have been evaluated in the oligonucleotide context: **432**, featuring the canonical values of B-DNA, and its diastereoisomer, **428**, with the unusual α in a gauche (+)-conformation. The former showed very good duplex formation ability with an increase in stability of $\Delta T_m = +5°C$ per modification toward DNA counterparts and $1 < \Delta T_m < +3°C$ toward RNA, making it a better B-type mimic.[139] On the other hand, it was shown that when included in the loop moiety of a hairpin structure, **428** was able to preorganize the unpaired nucleotides and induce an overall structure stabilization.[140]

4.5. Conclusions

The seminal work of Sekine has opened the way to the design and synthesis of conformationally constrained nucleotides with the aim of mimicking biologically relevant or transient conformations of nucleic acids.

Being dinucleotidic units, their main drawback is that in each case, 16 compounds need to be prepared to describe all the base combinations. Nevertheless, all these structures exhibit various backbone conformations and are therefore unique tools for distinguishing between the influence of backbone constraint and sugar puckering on the folding and stability of nucleic acid secondary structures. There is still a large field to be explored concerning their incorporation within oligonucleotides in order to stabilize the secondary structural elements of aptamers, riboswitches, or even siRNA.

5. CONCLUDING REMARKS

In this chapter we have reviewed various strategies to prepare a wide variety of conformationally restricted nucleosides, and it is clear that tremendous progress has been made in the past decade. Both convergent and linear multistep syntheses have been employed with careful control of the stereochemistry at various stages to reach the target nucleoside or nucleotide derivatives. No matter which approach is used, developing shorter and large-scale syntheses remains a continuing and significant challenge. The design and the synthesis of new original locked nucleosides will undoubtedly have an impact not only on the future development of modified oligonucleotides as therapeutic agents but also on DNA nanotechnology. In addition, these newly designed building blocks should open the way to the study of unusual nucleic acid structures.

Acknowledgments

The authors are deeply indebted to the Ministère de l'Enseignement Supérieur et de la Recherche, the Centre National de la Recherche Scientifique (CNRS), the Université de Nantes, the Université de Technologie de Compiègne, and the Université de Toulouse for their constant support. E.M. wishes to acknowledge the Algerian Government for a Ph.D. grant (Programme National Exceptionnel). We also wish to thank Carol Wrigglesworth (Scientific English, Nantes, France) for linguistic assistance. One of us (J.L.) would like to thank his colleagues, André Guingant and François-Xavier Felpin (CEISAM-UMR-CNRS 6320), for their invaluable comments and suggestions.

ABBREVIATONS

A	adenine
A^{Bz}	N^6-benzoyladenine
Ac	acetyl
ACN	acetonitrile
AD-mix	asymmetric dihydroxylation mixture
AIBN	2,2′-azobisisobutyronitrile
All	allyl
aq.	aqueous
B	nucleobase
9-BBN	9-borabicyclo[3.3.1]nonane
Bn	benzyl
BOM	benzyloxymethyl
BSA	N,O-bis(trimethylsilyl)acetamide
Bu	butyl
Bz	benzoyl
C	cytosine
CAN	ceric ammonium nitrate
carba-ENA	C2′-C4′-ethylene bridged nucleic acid
carba-LNA	C2′-C4′-methylene locked nucleic acid
cat.	catalytic
CNA	constrained nucleic acid
conc.	concentrated
Cy	cyclohexyl
DBU	1,8-diazabicyclo[5,4,0]undecen-7-ene

DCC	dicyclohexylcarbodiimide
DCE	1,2-dichloroethane
DCM	dichloromethane
D_2-CNA	double-constrained nucleic acid
ddC	2′,3′-dideoxycytidine
DDQ	2,3-dichloro-5,6-dicyano-1,4-benzoquinone
(DHQ)$_2$PHAL	1,4-bis(dihydroquininyl)phthalazine
DIBAL-H	diisobutylaluminum hydride
DIPEA	N,N-diisopropylethylamine
DMAP	4-(N,N-dimethylamino)pyridine
DMF	N,N-dimethylformamide
DMP	Dess–Martin periodinane
3,5-DMP	3,5-dimethylpyrazole
DMSO	dimethyl sulfoxide
DMTr	dimethoxytrityl
DNA	deoxyribonucleic acid
dr	diastereoisomeric ratio
ΔT_m/mod	[T_m(modified oligonucleotide) − T_m(wild-type)]/(number of modifications)
d4T	2′,3′-didehydro-2′,3′-dideoxythymidine
EC_{50}	half maximal effective concentration
ee	enantiomeric excess
ENA	O2′-C4′-ethylene bridged nucleic acid
equiv	equivalent
Et	ethyl
HIV	human immunodeficiency virus
HLE	hog liver esterase
HMDS	1,1,1,3,3,3-hexamethyldisilazane
HMPA	hexamethylphosphoramide
HPLC	high-pressure liquid chromatography
Imid	imidazol-1-yl
i-Pr	isopropyl
LAH	lithium aluminum hydride
LDA	lithium diisopropylamide
LNA	locked nucleic acid
MCPBA	m-chloroperoxybenzoic acid
Me	methyl
Mes	mesityl
MMTr	monomethoxytrityl
MOM	methoxymethyl
Ms	mesyl
MSNT	1-(mesitylene-2-sulfonyl)-3-nitro-1,2,4-triazole
MW	microwave
NBS	N-bromosuccinimide
NIS	N-iodosuccinimide

NMO	N-methylmorpholine-N-oxide
NMR	nuclear magnetic resonance
NOS	N-hydroxysuccinimide
PCC	pyridinium chlorochromate
Ph	phenyl
PMB	p-methoxybenzyl
PTC-Cl	phenoxythiocarbonylchloride
PTSA	p-toluenesulfonic acid
quant	quantitative
RCM	ring-closure metathesis
RNA	ribonucleic acid
rt	room temperature
siRNA	silencer ribonucleic acid
T	thymine
TBAF	tetrabutylammonium fluoride
TBD	1,5,7-triazabicyclo[4.4.0]dec-5-ene
TBS	*tert*-butyldimethylsilyl
TBDPS	*tert*-butyldiphenylsilyl
t-Bu	*tert*-butyl
Tf	trifluoromethanesulfonate
TFA	trifluoroacetic acid
TFAA	trifluoroacetic anhydride
THF	tetrahydrofuran
TIPBSCl	2,4,6-triisopropylbenzenesulfonyl chloride
TIPDS	tetraisopropyldisiloxan-1,3-diyl
TMS	trimethylsilyl
TMSOTf	trimethylsilyl trifluoromethanesulfonate
Tol	toluoyl
Ts	tosyl
TTMS	tris(trimethylsilyl)silane
U	uracyl

REFERENCES

1. Altona, C.; Sundaralingam, M. *J. Am. Chem. Soc.* **1972**, *94*, 8205–8212. (b) Saenger, W. Principles of Nucleic Acid Structure; Springer: Verlag: New York, **1984**.
2. For recent general reviews, see (a) Khakshoor, O.; Kool, E. T. *Chem. Commun.* **2011**, *47*, 7018–7024. (b) Yamamoto, T.; Nakatani, M., Narukawa, K.; Obika, S. *Future Med. Chem.* **2011**, *3*, 339–365. (c) Shukla, S.; Sumaria, C.; Pradeepkumar, P. I. *Chem. Med. Chem.* **2010**, *3*, 328–349. (d) Bell, N. M.; Micklefield, J. *ChemBioChem* **2009**, *10*, 2691–2703. (e) Cobb A. J. A. *Org. Biomol. Chem.* **2007**, *5*, 3260–3275.
3. Stephenson, M. L.; Zamecnik, P. C. *Proc. Natl. Acad. Sci. USA* **1978**, *75*, 285–288.
4. Eckstein, F. *Expert Opin. Biol. Ther.* **2007**, *7*, 1021–1034.
5. De Clercq, E. *J. Clin. Virol.* **2004**, *30*, 115–133.
6. Matsuda, A.; Sasaki, T. *Cancer Sci.* **2004**, *95*, 105–111.

7. Nicolaou, K. C.; Ellery, S. P.; Rivas, F.; Saye, K.; Rogers, E.; Workinger, T. J.; Schallenberger, M.; Tawatao, R.; Montero, A.; Hessell, A.; Romesberg, F.; Carson, D.; Burton, D. *Bioorg. Med. Chem.* **2011**, *19*, 5648–5669.
8. For recent reviews in this field, see (a) Mathé, C.; Périgaud, C. *Eur. J. Org. Chem.* **2008**, 1489–1505. (b) Zhou, C.; Chattopadhyaya, J. *Curr. Opin. Drug Discov. Dev.* **2009**, *12*, 876–898. (c) Lebreton, J.; Escudier, J.-M.; Arzel, L.; Len, C. *Chem. Rev.* **2010**, *110*, 3371–3418. (d) Obika, S.; Rahman, S. M. A.; Fujisaka, A.; Kawada, Y.; Baba, T.; Imanishi, T. *Heterocycles* **2010**, *81*, 1347–1392.
9. For recent reviews in this field, see (a) Len, C.; Mondon, M.; Lebreton, J. *Tetrahedron* **2008**, *64*, 7453–7475. (b) Chatgilialoglu, C.; Ferreri, C.; Terzidis, M. A. *Chem. Soc. Rev.* **2011**, *40*, 1368–1382. (c) Mieczkowski, A.; Roy, V.; Agrofoglio, L. A. *Chem. Rev.* **2010**, *110*, 1828–1856. (d) Mieczkowski, A.; Agrofoglio, L. A. *Curr. Med. Chem.* **2010**, *17*, 1527–1549.
10. For a recent review in this field, see Dupouy, C.; Payrastre, C.; Escudier, J.-M. In *Targets in Heterocyclic Systems: Chemistry and Properties; Attanasi*, O. A.; Spinelli, D., Eds.; Società Chimica Italiana: Rome, **2008**; Vol. 12, pp. 185–211.
11. (a) Kurreck, J. *Eur. J. Biochem.* **2003**, *270*, 1628–1644. (b) Meldgaard, M.; Wengel, J. *J. Chem. Soc. Perkin Trans. 1* **2000**, 1, 3539–3554. (c) Leumann, C. J. *Bioorg. Med. Chem.* **2002**, *10*, 841–854. (d) Kool, E. T. *Chem. Rev.* **1997**, *97*, 1473–1487. (e) Bell, N. M.; Micklefield, J. *ChemBioChem* **2009**, *10*, 2691–2703. See also reviews cited in references 9 and 10.
12. Vorbrüggen, H.; Krolikiewicz, K.; Bennua, B. *Chem. Ber.* **1981**, *114*, 1234–1255. (b) Vorbrüggen, H.; Höfle, G. *Chem. Ber.* **1981**, *114*, 1256–1268.
13. Pradeepkumar, P. I.; Zamaratski, E.; Földesi; A.; Chattopadhyaya J. *J. Chem. Soc. Perkin Trans.* **2001**, *2*, 402–408.
14. (a) Perali, R. S.; Mandava, S.; Bandi R. *Tetrahedron* **2011**, *67*, 4031–4035. (b) Prisbe, E. J.; Smejkal, J.; Verbeyden, J. P. H.; Moffatt J. G. *J. Org. Chem.* **1976**, *41*, 1836–1846.
15. Roivainen, J.; Vepsäläinen, J.; Azhayeva, A.; Mikhailopulo, I. A. *Tetrahedron Lett.* **2002**, *43*, 6553–6555.
16. Bogucka, M.; Naus, P.; Pathmasiri, W.; Barman J.; Chattopadhyaya J. *Org. Biomol. Chem.* **2005**, *3*, 4362–4372.
17. Pradeepkumar, P. I.; Cheruku, P.; Plashkevych, O.; Acharya, P.; Gohil, S.; Chattopadhyaya J. *J. Am. Chem. Soc.* **2004**, *126*, 11484–11499.
18. Honcharenko, D.; Varghese, O. P.; Plashkevych, O.; Barman, J.; Chattopadhyaya, J. *J. Org. Chem.* **2006**, *71*, 299–314.
19. Fox, J. J.; Miller, N.; Wempen, I. *J.Med. Chem.* **1966**, *9*, 101–105.
20. Kværnø, L.; Nielsen, C.; Wightman, R. H. *J. Chem. Soc. Perkin Trans. 1* **2000**, 2903–2906.
21. (a) Holy, A. *Nucleic Acids Res.* **1974**, *1*, 289–298. (b) Ono, A.; Dan, A.; Matsuda, A. *Bioconjugate. Chem.* **1993**, *4*, 499–508. (c) Ossipov, D.; Chattopadhyaya, J. *Tetrahedron* **1998**, *54*, 5667–5682.
22. Frieden, M.; Giraud, M.; Reese, C. B.; Song Q. *J. Chem. Soc. Perkin Trans. 1* **1998**, 2827–2832.
23. Kværnø, L.; Wightman, R. H.; Wengel, J. *J. Org. Chem.* **2001**, *66*, 5106–5112.
24. Kværnø, L.; Wengel J. *J. Org. Chem.* **2001**, *66*, 5498–5503.

25. Youssefyeh, R. D.; Verheyden, J. P. H.; Moffatt, J. G. *J. Org. Chem.* **1979**, *44*, 1301–1309.
26. For a discussion on this point, see (a) Kumar, S.; Kumar, P.; Sharma, P. K.; Erdlicka, P. J. *Tetrahedron Lett.* **2008**, *49*, 7168–7170. For selected articles on hydrogenation of uridine derivatives using H_2 and Pd/C, in alcohols as solvents without affecting the nucleobase, see (b) Chan, M. Y.; Fairhurst, R. A.; Collingwood, S. P.; Fisher, J. J.; Arnold, R. P.; Cosstick, R. I.; O'Neil. A. *J. Chem. Soc. Perkin Trans. 1* **1999**, *3*, 315–320. (c) Seth, P. P.; Allerson, C. R.; Berdeja, A.; Siwkowski, A.; Pallan, P. S.; Gaus, H.; Prakash, T. P.; Watt, A. T.; Egli, M.; Swayze, E. E. *J. Am. Chem. Soc.* **2010**, *132*, 14942–14950.
27. Johnson, D. C.; Widlanski, T. S. *Org. Lett.* **2004**, *6*, 4643–4646. (b) Cho, J. H.; Amblard, F.; Coats, S. J.; Schinazi, R. F. *Tetrahedron* **2011**, *67*, 5487–5493.
28. Okabe, M.; Sun, R.-C. *Tetrahedron Lett.* **1989**, *30*, 2203–2206.
29. Sard, H. *Nucleosides Nucleotides* **1994**, *13*, 2321–2328.
30. Alibès, R.; Alvarez-Larena, A.; De March, P.; Figueredo, M.; Font, J.; Parella, T.; Rustullet, A. *Org. Lett.* **2006**, *8*, 491–494.
31. Flores, R.; Rustullet, A.; Alibès, R.; Alvarez-Larena, A.; de March, P.; Figueredo, M.; Font, J. *J. Org. Chem.* **2011**, *76*, 5369–5383.
32. Patrick, T. B.; Lanahan, M. V.; Yang, C.; Walker, J. K.; Hutchinson, C. L.; Neal, B. E. *J. Org. Chem.* **1994**, *59*, 1210–1212.
33. Christensen, N. K.; Petersen, M.; Nielsen, P.; Jacobsen, J. P.; Olsen C. E.; Wengel, J. *J. Am. Chem. Soc.* **1998**, *120*, 5458–5463.
34. Yoshimura, Y.; Sano, T.; Matsuda, A.; Ueda, T. *Chem. Pharm. Bull.* **1988**, *36*, 162–167.
35. Nielsen, P.; Pfundheller, H. M.; Olsen, C. E.; Wengel, J. *J. Chem. Soc. Perkin Trans. 1* **1997**, 3423–3433.
36. Díaz-Rodríguez, A.; Sanghvi, Y. S.; Fernández, S.; Schinazi, R. F.; Theodorakis, E. A.; Ferrero, M.; Gotor, V. *Org. Biomol. Chem.* **2009**, *7*, 1415–1423.
37. Brady, T. P.; Kim, S. H.; Wen, K.; Theodorakis, E. A. *Angew. Chem. Int. Ed.* **2004**, *43*, 739–741.
38. (a) Schmid, C. R.; Bryant, J. D.; Dowlatzedah, M.; Phillips, J. L.; Prather. D. E.; Shantz, R. D.; Sear, N. L.; Vianco, C. S. *J. Org. Chem.* **1991**, *56*, 4056–4058. (b) Fazio, F.; Scheider, M. P. *Tetrahedron: Asymmetry* **2000**, *11*, 1869–1876. It should be noted that (*S*)-5-hydroxymethyl-2(5*H*)-furanone is also commercially available and its silylation into **100** is described in reference 36.
39. Ewing, D. F.; Fahmi, N.; Len, C.; Mackenzie, G.; Ronco, G.; Villa, P.; Shaw, G. *Collect. Czech. Chem. Commun.* **1996**, *61*, S145–S147.
40. Ewing, D. F.; Fahmi, N.; Len, C.; Mackenzie, G.; Ronco, G.; Villa, P.; Shaw, G. *Nucleosides Nucleotides* **1999**, *18*, 2613–2630.
41. Ewing, D.F.; Fahmi, N.; Len, C.; Mackenzie, G.; Pranzo, A. *J. Chem. Soc. Perkin Trans. 1* **2000**, 3561–3565.
42. Belloli, E.; Len, C.; Mackenzie, G.; Ronco, G.; Bonte, J. P.; Vaccher, C. *J. Chromatogr. A* **2001**, *943*, 91–100.
43. Ewing, D. F.; Len, C.; Mackenzie, G.; Ronco, G.; Villa, P. *Tetrahedron: Asymmetry* **2000**, *11*, 4995–5002.

44. Selouane, A.; Vaccher, C.; Villa, P.; Postel, D.; Len, C. *Tetrahedron: Asymmetry* **2002**, *13*, 407–413.
45. Pilard, S.; Riboul, D.; Glaçon, V.; Moitessier, N.; Chapleur, Y.; Postel, D.; Len, C. *Tetrahedron: Asymmetry* **2002**, *13*, 529–537.
46. Egron, D.; Périgaud, C.; Gosselin, G.; Aubertin, A. M.; Faraj, A.; Selouane, A.; Postel, D.; Len, C. *Bioorg. Med. Chem. Lett.* **2003**, *13*, 4473–4475.
47. The LNA nucleosides containing uracil and cytosine nucleobases have been synthesized by a linear approach; see Obika, S.; Nanbu, D.; Hari, Y.; Morio, K.; In, Y.; Ishida, T.; Imanishi, T. *Tetrahedron Lett.* **1997**, *38*, 8735–8738.
48. Koshkin, A. A.; Singh, S. K.; Nielsen, P.; Rajwanshi, V. K.; Kumar, R.; Meldgaard, M.; Olsen, C. E.; Wengel, J. *Tetrahedron* **1998**, *54*, 3607–3630.
49. For an excellent review on this topic, see Kaur, H.; Babu, B. R.; Maiti, S. *Chem. Rev.* **2007**, *107*, 4672–4697.
50. Koshkin, A. A.; Fensholdt, J.; Pfundheller, H. M.; Lomholt, C. *J. Org. Chem.* **2001**, *66*, 8504–8512.
51. Rosenbohm, C.; Christensen, S. M.; Sørensen, M. D.; Pedersen, D. S.; Larsen, L.-E.; Wengel, J.; Koch, T. *Org. Biomol. Chem.* **2003**, *1*, 655–663.
52. Pedersen, D. S.; Koch, T. *Synthesis* **2004**, 578–582.
53. Seth, P. P.; Allerson, C. R; Berdeja, A.; Siwkowski, A.; Pallan, P. S.; Gaus, H.; Prakash, T. P.; Watt, A. T.; Egli, M.; Swayze, E. E. *J. Am. Chem. Soc.* **2010**, *132*, 14942–14950 and cited literature.
54. Srivastava, P.; Barman, J.; Pathmasiri, W.; Plashkevych, O.; Wenska, M.; Chattopadhyaya, J. *J. Am. Chem. Soc.* **2007**, *129*, 8362–8379.
55. Xu, J.; Liu, Y.; Dupouy, C.; Chattopadhyaya J. *J. Org. Chem.* **2009**, *74*, 6534–6554.
56. (a) Morita, K.; Hasegawa, C.; Kaneko, M.; Tsutsumi, S.; Sone, J.; Ishikawa, T.; Imanishi, T.; Koizumi, M. *Bioorg. Med. Chem. Lett.* **2002**, *12*, 73–76. (b) Morita, K.; Takaji, M.; Hasegawa, C.; Kaneko, M.; Tsutsumi, S.; Sone, J.; Ishikawa, T.; Imanishi, T.; Koizumi, M. *Bioorg. Med. Chem.* **2003**, *1*, 2211–2226.
57. Varghese, C. P.; Barman, J.; Pathmasiri, W.; Plashkevych, O.; Hondcharenko, D.; Chattopadhyaya, J. *J. Am. Chem. Soc.* **2006**, *128*, 15173–15187.
58. Srivastava, P.; Barman, J.; Pathmasiri, W.; Plashkevych, O.; Wenska, M.; Chattopadhyaya, J. *J. Am. Chem. Soc.* **2007**, *129*, 8362–8379.
59. For an excellent review on metathesis strategy in nucleoside chemistry, see Amblard, F.; Nolan, S. P.; Agrofoglio L. A. *Tetrahedron* **2005**, *61*, 7067–7080.
60. Albæk, N.; Petersen, M.; Nielsen, P. *J. Org. Chem.* **2006**, *71*, 7731–7740.
61. Zhou, C.; Plashkevych, O.; Chattopadhyaya, J. *Org. Biomol. Chem.* **2008**, *6*, 4627–4633.
62. Kumar, S.; M. H. Hansen, M. H.; Albæk, N.; Steffansen, S. I.; Petersen, M.; Nielsen, P. *J. Org. Chem.* **2009**, *74*, 6756–6769.
63. Gagneron, J.; Gosselin, G.; Mathé, C. *J. Org. Chem.* **2005**, *70*, 6891–6897.
64. Tino, J. A.; Clark, J. M.; Field, A. K.; Jacobs, G. A.; Lis, K. A.; Michalik, T. L.; McGeever-Rubin, B.; Slusarchyk, W. A.; Spergel, S. H.; Sundeen, J. E.; Tuomari, A. V.; Young, M. G.; Zalher, R. *J. Med. Chem.* **1993**, *36*, 1221–1229.
65. (a) Obika, S.; Morio, K.-I.; Hari, Y.; Imanishi, T. *Chem. Commun.* **1999**, 2423–2424. (b) Obika, S.; Morio, K.-I.; Daishu, N.; Hari, Y.; Itoh, H.; Imanishi, T. *Tetrahedron* **2002**, *58*, 3039–3049.

66. O-Yang, C.; Wu, H. Y.; Fraser-Smith, E. B.; Walker, K. A. M. *Tetrahedron Lett.* **1992**, *33*, 37–40.
67. Obika, S.; Morio, K.-I.; Daishu, N.; Imanishi, T. *Chem. Commun.* **1997**, 1643–1644.
68. Obika, S.; Andoh, J.-I; Sugimoto, T.; Miyashita, K.; Imanishi, T. *Tetrahedron Lett.* **1999**, *40*, 6465–6468.
69. Obika, S.; Andoh, J.-I.; Onoda, M.; Nakagawa, O.; Hiroto, A.; Sugimoto, T.; Imanishi, T. *Tetrahedron Lett.* **2003**, *44*, 5267–5270.
70. Marples, B. A.; Toon, R. C. *Tetrahedron Lett.* **1999**, *40*, 4873–4876.
71. Chang, W.; Du, J.; Rachakonda, S.; Ross, B. S.; Convers-Reignier, S.; Yau. W. T.; Pons, J.-F.; Murakami, E.; Bao, H.; Steuer, H. M.; Furman, P. A.; Otto, M. J.; Sofia, M. J. *Bioorg. Med. Chem. Lett.* **2010**, *20*, 4539–4543.
72. Clark, J. L.; Hollecker, L.; Mason, J. C.; Stuyver, L. J.; Tharnish, P. M.; Lostia, S.; McBrayer, T. R.; Schinazi, R. F.; Watanabe, K. A.; Otto, M. J.; Furman, P. A.; Stec, W. J.; Patterson, S. E.; Pankiewicz, K. W. *J. Med. Chem.* **2005**, *48*, 5504–5508.
73. Khalafi-Nezhad, A.; Alamdari, F. R. *Tetrahedron* **2001**, *57*, 6805–6807.
74. Pankiewicz, K. W.; Krzeminski, J.; Ciszewski, L. A.; Ren, W. Y.; Watanabe. K. A. *J. Org. Chem.* **1992**, *57*, 553–559.
75. Haraguchi, K.; Takeda, S.; Tanaka, H. *Org. Lett.* **2003**, *5*, 1399–1402 and literature cited.
76. Sugimoto, I.; Shuto, S.; Mori, S.; Shigeta, S.; Matsuda, A. *Bioorg. Med. Chem. Lett.* **1999**, *9*, 385–388.
77. Giese, B.; Erdmann, P.; Giraud, L.; Göbel, T.; Petretta, M.; Schäfer, T.; von Raumer, M. *Tetrahedron Lett.* **1994**, *35*, 2683–2686.
78. (a) Ueno, Y.; Nagasawa, Y.; Sugimoto, I.; Kojima, N.; Kanazaki, M.; Shuto, S.; Matsuda, A. *J. Org. Chem.* **1998**, *63*, 1660–1667. For previous investigations using this methodology, see (b) Shuto, S.; Kanazaki, M.; Ichikawa, S.; Matsuda, A. *J. Org. Chem.* **1997**, *62*, 5676–5677. (c) Shuto, S.; Kanazaki, M.; Ichikawa, S.; Minakawa, N.; Matsuda, A. *J. Org. Chem.* **1998**, *63*, 746–754.
79. Kim, J. M.; Kim, H. O.; Kim, H.-D.; Kim, J. H.; Jeonge, L. S.; Chun, M. W. *Bioorg. Med. Chem. Lett.* **2003**, *13*, 3499–3501.
80. Tzioumaki, N.; Tsoukala, E.; Manta, S.; Kiritsis, C.; Balzarini, J.; Komiotis, D. *Carbohydr. Res.* **2011**, *346*, 328–333 and cited literature.
81. Bjorsne, M.; Szabo, T.; Samuelsson, B.; Classon, B. *Bioorg. Med. Chem.* **1995**, *3*, 397–402.
82. Obika, S.; Sekiguchi, M.; Osaki, T.; Shibata, N.; Masaki, M.; Hari, Y.; Imanishi, T. *Tetrahedron Lett.* **2002**, *43*, 4365–4368.
83. Bar, N. C.; Roy, A.; Achari, B.; Mandal, S. B. *J. Org. Chem.* **1997**, *62*, 8948–8951.
84. Hatton, W.; Hunault, J.; Egorov, M.; Len, C.; Pipelier, M.; Blot, V.; Silvestre, V.; Fargeas, V.; Ané, A.; McBrayer, T.; Detorio, M.; Cho, J.-H.; Bourgougnon, N.; Dubreuil, D.; Schinazi, R. F.; Lebreton, J. *Eur. J. Org. Chem.* **2011**, 7390–7399.
85. Carlucci, M.; Kierzek, E.; Olejnik, A.; Turner, D. H.; Kierzek, R. *Biochemistry* **2009**, *48*, 10882–10893.
86. Leumann, C. J. *Chimia* **2005**, *59*, 776–779.
87. Tarköy, M.; Bolli, M.; Schweizer, B.; Leumann, C. *Helv. Chim. Acta* **1993**, *76*, 481–510.

88. (a) Vonlanthen, D.; Leumann, C. *J. Synthesis* **2003**, 1087–1090. (b) For recent improvements concerning the preparation of the methyl glycoside analogue of **305** as a potential glycosyl donor, see Haziri, A. I.; Silhar, P.; Renneberg, D.; Leumann, C. *J. Synthesis* **2010**, 823–827.
89. Ravn, J.; Nielsen, P. *J. Chem. Soc. Perkin Trans. 1* **2001**, 985–993.
90. (a) Patra, R.; Bar, N. C.; Roy, A.; Achari, B.; Mandal S. B. *Tetrahedron* **1996**, *52*, 11265–11272. (b) Nielsen, P.; Petersen, M.; Jacobsen, P. *J. Chem. Soc. Perkin Trans. 1* **2000**, 3706–3713.
91. (a) Bäckvall, J.-E.; Andreasson, U. *Tetrahedron Lett.* **1993**, *34*, 5459–5462. (b) Nakashima, K.; Okamoto, S.; Sono, M.; Tori, M., *Molecules* **2004** *9*, 541–549. (c) Liu, P. N.; Ju, K. D.; Lau, C. P. *Adv. Synth. Catal.* **2011**, *353*, 275–280.
92. Stauffiger, A.; Leumann C. *Eur. J. Org. Chem.* **2009**, 1153–1162.
93. Chochrek, P.; Wicha, J. *Org. Lett.* **2006**, *8*, 2551–2553.
94. Paquette, L. A.; Owen, D. R.; Bibart, R. T.; Seekamp, C. K.; Kahane, A. L.; Lanter, J. C.; Corral, M. A. *J. Org. Chem.* **2001**, *66*, 2828–2834.
95. Paquette, L. A.; Owen, D. R.; Bibart, R. T.; Seekamp, C. K. *Org. Lett.* **2001**, *3*, 4043–4045. (b) Paquette, L. A.; Seekamp, C. K.; Kahane, A. L.; Hilmey, D. G.; Galluci, J. *J. Org. Chem.* **2004**, *69*, 7442–7447. (c) Paquette, L. A.; Kahane, A. L.; Seekamp, C. K. *J. Org. Chem.* **2004**, *69*, 5555–5562. (d) Dong, S.; Paquette, L. A. *J. Org. Chem.* **2006**, *71*, 1647–1653.
96. Paquette, L. A.; Bibart, R. T.; Seekamp, C. K.; Kahane, A. L. *Org. Lett.* **2001**, *3*, 4039–4041.
97. Paquette, L. A.; Seekamp, C. K.; Kahane, A. L. *J. Org. Chem.* **2003**, *68*, 8614–8624.
98. Wendeborn, S.; Binot, G.; Nina, M.; Winkler, T. *Synlett* **2002**, 1683–1687.
99. Maity, J. K.; Achari, B.; Drew, M. G. B.; Mandal, S. B. *Synthesis* **2010**, 2533–2542.
100. Ueda, T.; Usui, H.; Shuto, S.; Inoue, H. *Chem. Pharm. Bull.* **1984**, *32*, 3410–3416.
101. Yoshimura, Y.; Yamazaki, Y.; Wachi, K.; Satoh, S.; Takahata, H. *Synlett* **2007**, 111–114.
102. Bello, A. M.; Poduch, E.; Fujihashi, M.; Amani, M.; Li, Y.; Crandall, I.; Hui, R.; Lee, P. I.; Kain, K. C.; Pai, E. F.; Kotra, L. P. *J. Med. Chem.* **2007**, *50*, 915–921.
103. Tanaka, H.; Hayakawa, H.; Miyasaka, T. *Tetrahedron* **1982**, *38*, 2635–2642.
104. Yoshimura, Y.; Kumamoto, H.; Baba, A.; Takeda, S.; Tanaka, H. *Org. Lett.* **2004**, *6*, 1793–1795.
105. Verheyden, J. P. H.; Moffatt, J. G. *J. Org. Chem.* **1970**, *35*, 2319–2326.
106. Otter, B. A.; Falco, E. A.; Fox, J. J. *J. Org. Chem.* **1976**, *41*, 3133–3137.
107. Barrett, A. G. M., Lebold, S. A. *J. Org. Chem.* **1990**, *55*, 3853–3857.
108. Yoshimura, Y.; Matsuda, A.; Ueda, T. *Chem. Pharm. Bull.* **1989**, *37*, 660–664.
109. Kuhn, R.; Jahn, W. *Chem. Ber.* **1965**, *98*, 1699–1704.
110. Matsuda, A.; Muneyama, K.; Nishida, T.; Sato, T.; Ueda, T. *Nucleic Acids Res.* **1976**, *3*, 3349–3357.
111. Matsuda, A.; Tezuka, M.; Ueda, T. *Tetrahedron* **1978**, *34*, 2449–2452.
112. Ikehara, M.; Tada, H.; Kaneko, M. *Tetrahedron* **1968**, *24*, 3489–3498.
113. Lipkin, D.; Rabi, J. A. *J. Am. Chem. Soc.* **1971**, *93*, 3309–3310.
114. Chun, B. K.; Wang, P.; Hassan, A.; Du, J.; Tharnish, P. M.; Stuyver, L. J.; Otto, M. J.; Schinazi, R. F.; Watanabe, K. A. *Tetrahedron Lett.* **2005**, *46*, 2825–2827.

115. Inoue, H.; Ueda, T. *Chem. Pharm. Bull.* **1978**, *26*, 2664–2667.
116. Wengel, J. A. *Acc. Chem. Res.* **1999**, *32*, 301–310.
117. Seio, K.; Wada, T.; Sakamoto, K.; Yokoyama, S.; Sekine, M. *J. Org. Chem.* **1998**, *63*, 1429–1443.
118. (a) Seio, K.; Wada, T.; Sakamoto, K.; Yokoyama, S.; Sekine, M. *J. Org. Chem.* **1996**, *61*, 1500–1504. (b) Sekine, M.; Kurrasawa, O.; Shohda, K-I.; Seio, K.; Wada, T. *J. Org. Chem.* **2000**, *65*, 3571–3578.
119. (a) Børsting, P.; Nielsen, P. *Chem. Commun.* **2002**, 2140–2141. (b) Børsting, P.; Christensen, M. S.; Steffansen, S. I.; Nielsen, P. *Tetrahedron* **2006**, *62*, 1139–1149. (c) Børsting, P.; Sørensen, A. M.; Nielsen, P. *Synthesis* **2002**, 797–801.
120. (a) Sørensen, A. M.; Nielsen, P. *Org. Lett.* **2000**, *2*, 4217–4219. (b) Sørensen, A. M.; Nielsen, K. E.; Vogg, B.; Jacobsen, J. P.; Nielsen, P. *Tetrahedron* **2001**, *57*, 10191–10201. (c) Børsting, P.; Nielsen, K. E.; Nielsen, P. *Org. Biomol. Chem.* **2005**, *3*, 2183–2190.
121. (a) Dupouy, C.; Le Clezio, I.; Lavedan, P.; Gornitzka, H.; Escudier, J.-M.; Vigroux, A. *Eur. J. Org. Chem.* **2006**, 5515–5525. (b) Le Clezio, I.; Gornitzka, H.; Escudier, J.-M.; Vigroux, A. *J. Org. Chem.* **2005**, *70*, 1620–1629. (c) Dupouy, C.; Lavedan, P.; Escudier, J.-M. *Tetrahedron* **2007**, *63*, 11235–11243. (d) Dupouy, C.; Lavedan, P.; Escudier, J.-M. *Eur. J. Org. Chem.* **2008**, 1285–1294. (e) Zhou, C.; Plashkevych, O.; Chattopadhyaya, J. *J. Org. Chem.* **2009**, *74*, 3248–3265. (f) Catana, D. A.; Maturano, M.; Payrastre, C.; Lavedan, P.; Tarrat, N.; Escudier, J.-M. *Eur. J. Org. Chem.* **2011**, 6857–6863.
122. (a) Sekine, M.; Kurrasawa, O.; Shohda, K.-I.; Seio, K.; Wada, T. *J. Org. Chem.* **2000**, *65*, 6515–6524. (b) Seio, K.; Wada T.; Sekine, M. *Helv. Chim. Acta* **2000**, *83*, 162–180.
123. Robbins, M. J.; Barr, P. J. *J. Org. Chem.* **1983**, *48*, 1854–1862.
124. Bhat, B.; Leonard, N. J.; Robinson, H.; Wang, A. H.-J. *J. Am. Chem. Soc.* **1996**, *118*, 10744–10751.
125. Dess, D. B.; Martin., J. C. *J. Am. Chem. Soc.* **1991**, *113*, 7277–7287. For a recent preparation of this aldehyde following this route, see (b) Szilagyi, A.; Fenyvesi, F.; Majercsik, O.; Pelyvas, I.; F. Bacskay, I.; Feher, P.; Varadi, J. Vecsernyes, M.; Herczegh, P. *J. Med. Chem.* **2006**, *49*, 5626–5630.
126. Caruthers, M. H. *Acc. Chem. Res.* **1991**, *24*, 278–284.
127. (a) Sholl, M.; Ding, S.; Lee, C. W.; Grubbs, R. H. *Org. Lett.* **1999**, *1*, 953–956. (b) Morgan, J. P.; Grubbs, R. H. *Org. Lett.* **2000**, *2*, 3153–3155.
128. Cicero, D. O.; Neuner, P. J. S.; Franzece, O.; D'Onofrio, C.; Iribarren, A. M. *Bioorg. Med. Chem. Lett.* **1994**, *4*, 861–866.
129. De Mesmaeker, A.; Lebreton, J.; Hoffmann, P.; Freier, S. M. *Synlett* **1993**, 677–679.
130. For a recent reference concerning the preparation of this alcohol **419**, see Angeloff, A.; Dubey, I.; Pratviel, G.; Bernadou, J.; Meunier, B. *Chem. Res. Toxicol.* **2001**, *14*, 1413–1420.
131. Louie, J.; Bielawski, C. W.; Grubbs, R. H. *J. Am. Chem. Soc.* **2001**, *123*, 11312–11313.
132. (a) Pfitzner, K. E.; Moffatt, J. G. *J. Am. Chem. Soc.* **1965**, *87*, 5661–5678. (b) Jones, G. H.; Taniguchi, M.; Tegg, D.; Moffatt, J. G. *J. Org. Chem.* **1979**, *44*, 1309–1317.
133. (a) Le Roux, C.; Gaspard Iloughmane, H.; Dubac, J.; Jaud, J.; Vignaux, P. *J. Org. Chem.* **1993**, *58*, 1835–1839. (b) Escudier, J.-M.; Tworkowski, I.; Bouziani, I.; Gorrichon, L. *Tetrahedron Lett.* **1996**, *50*, 4689–4692.

134. Le Clezio, I.; Escudier J.-M.; Vigroux, A. *Org. Lett.* **2003**, *5*, 161–164.
135. Schneider, B.; Neidle, S.; Berman, H. M. *Biopolymers* **1997**, *42*, 113–124.
136. (a) Banuls, V.; Escudier, J.-M. *Tetrahedron* **1999**, *55*, 5831–5838. (b) Banuls, V.; Escudier, J.-M.; Zedde, C.; Claparols, C.; Donnadieu, B.; Plaisancié, H. *Eur. J. Org. Chem.* **2001**, 4693–4700.
137. Tseng, C. K. H.; Marquez, V. E.; Milne, G. W. A.; Wysocki, R. J.; Mitsuya, H.; Shirasaki, T.; Driscoll, J. S. *J. Med. Chem.* **1991**, *34*, 343–349.
138. (a) Reese, C. B.; Titmas, R. C.; Yan, L. *Tetrahedron Lett.* **1978**, *34*, 2727–2730. (b) Jones, S. S.; Rayner, B.; Reese, C. B.; Ubasawa, A.; Ubasawa, M. *Tetrahedron* **1980**, *36*, 3075–3085.
139. (a) Dupouy, C.; Iché-Tarrat, N.; Durrieu, M. P.; Rodriguez, F.; Escudier, J.-M.; Vigroux, A. *Angew. Chem. Int. Ed.* **2006**, *45*, 3623–3627. (b) Dupouy, C.; Iché–Tarrat, N.; Durrieu, M. P.; Vigroux, A.; Escudier, J.-M. *Org. Biomol. Chem.* **2008**, *6*, 2849–2851. (c) Boissonnet, A.; Dupouy, C.; Millard, P.; Durrieu, M. P.; Iché-Tarrat, N.; Escudier, J.-M. *New J. Chem.* **2011**, *35*, 1528–1533.
140. (a) Le Clezio, I.; Dupouy, C.; Lavedan, P.; Escudier, J.-M. *Eur. J. Org. Chem.* **2007**, 3894–3900. (b) Dupouy, C.; Boissonet, A.; Millard, P.; Escudier, J.-M. *Chem. Commun.* **2010**, *46*, 5142–5144.

10 Synthesis of 3′-Spiro-Substituted Nucleosides: Chemistry of TSAO Nucleoside Derivatives

MARÍA-JOSÉ CAMARASA, SONSOLES VELÁZQUEZ,
ANA SAN-FÉLIX, and MARÍA-JESÚS PÉREZ-PÉREZ
Instituto de Química Médica, CSIC, Madrid, Spain

1. INTRODUCTION

Nature has always been an important source of inspiration for the synthesis of structurally similar synthetic analogues for the development of new therapeutic agents. Many nucleoside analogues designed to mimic naturally occurring nucleosides show important antiviral, antitumor, antibacterial, and antifungal activities.[1,2] Moreover, they are also of importance in other fields, such as in regulation of gene expression or in immunomodulation. Therefore, they are considered promising leads for drug development. With this aim, nucleosides have been modified extensively either in the heterocyclic base or in the sugar part. Among the various modifications, and inspired on the presence of spiroacetals as a common structural element in many natural products isolated from different sources that display a plethora of biological activities,[3–9] several spirocyclic substituted nucleosides have been reported (Figures 1 to 3 show some examples of such nucleosides).

The isolation in 1991 of (+)-hydantocidin **1** (Figure 1), the first natural spironucleoside (with herbicidal and plant growth regulatory activities)[10,11] has stimulated a growing interest in the chemistry and biological evaluation of spironucleosides. These relatively new classes of anomeric or C1′-spironucleosides are characterized by a structurally constrained conformation, which is locked by sharing the one common atom between the nucleobase and the sugar moiety. Examples of synthetic anomeric spironucleoside analogues of (+)-hydantocidin are described. Thus, 2′-deoxyhydantocidin **2**, or the spiroanalogue in which the hydantoin ring is replaced by the barbiturate ring (**3**), has been prepared.[12] Like the hydantoin ring, the barbiturate ring possesses a thymine-like hydrogen-bonding capacity against adenine derivatives.[13] Anomeric spiroisoxazolines were described by Hyrosova et al.[14]

Chemical Synthesis of Nucleoside Analogues, First Edition. Edited by Pedro Merino.
© 2013 John Wiley & Sons, Inc. Published 2013 by John Wiley & Sons, Inc.

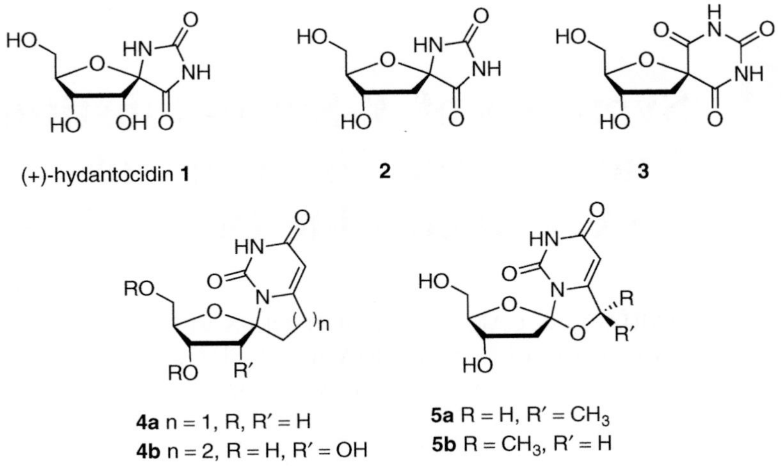

(+)-hydantocidin 1   **2**   **3**

4a n = 1, R, R' = H
4b n = 2, R = H, R' = OH

5a R = H, R' = CH$_3$
5b R = CH$_3$, R' = H

Figure 1. Selected examples of C1'-spironucleosides.

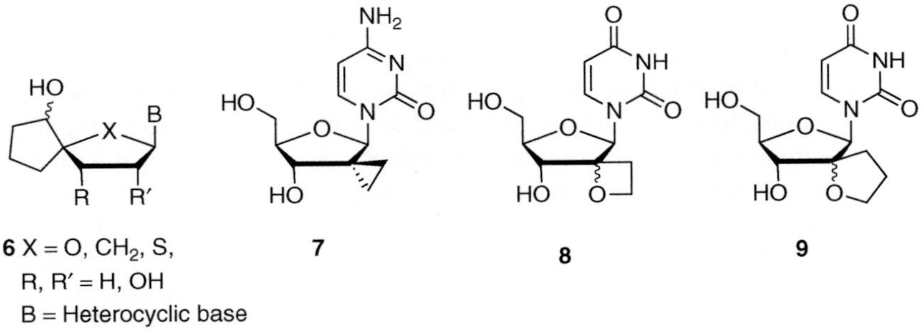

6 X = O, CH$_2$, S,
R, R' = H, OH
B = Heterocyclic base

7   **8**   **9**

Figure 2. Some examples of C2'-spiro and C4'-spironucleosides.

4 via 1,3-dipolar cycloadditions of TBDPS- substituted 7-methylenepyrrolo[1,2-c]pyrimidin-1(5H)-one with methoxycarbonyl and cyanonitrile oxide.[14] Anomeric spirouracil nucleosides with orthoester structure **5** were synthesized by Tanaka and co-workers.[15] from 2'-deoxy-6-(hydroxyalkyl)uridines by hypoiodite-initiated cyclization.

Later, the concept of spirocyclic restriction in nucleosides was introduced at other positions of the sugar ring (Figure 2). Thus, Paquette et al.[16–18] prepared C4'-spirocyclic nucleosides **6**[16] that show a restricted conformation in the sugar through insertion of a carbocyclic ring at C4' of the furanose ring (Figure 2). These nucleosides may have an enhanced lipophilicity and resistance to alkaline hydrolysis as well as stability toward nucleobase degradation.[16] Such spiro-fused nucleosides may be incorporated into oligomers (oligonucleotides), thus representing a new direction in the design of antisense molecules by improving the binding properties and the base-pairing preference. Thereafter, considerable

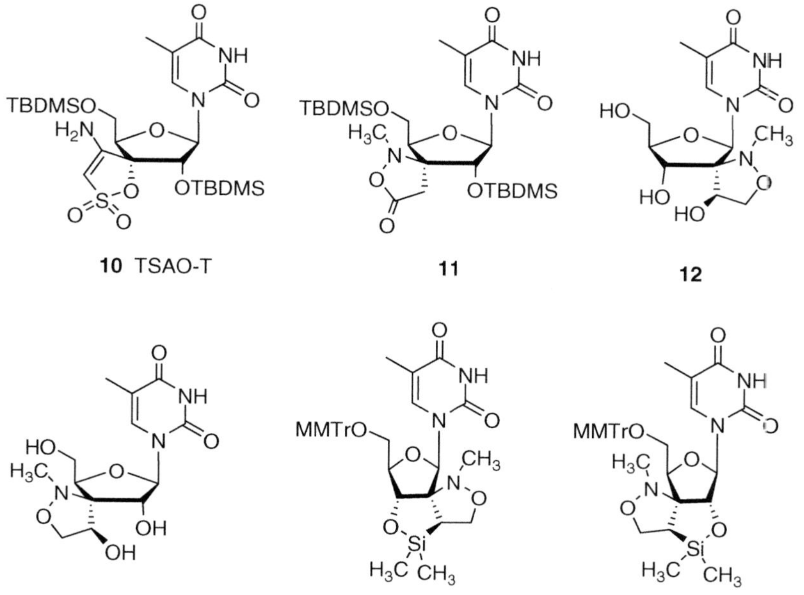

Figure 3. Chemical structures of TSAO-T and related examples of C2'- and C3'-spironucleosides.

attention was paid to the synthesis of a number of spiro derivatives, including C2'-spironucleosides (Figure 2). Ueda and co-workers first described the synthesis of spiroepoxide derivatives in nucleosides.[19] On the other hand, Robins et al.[20] prepared 2'-spiropyrazolidine-substituted nucleosides by the dipolar cycloaddition of 2'-deoxy-2'-methylene nucleosides with diazomethane. Inspired by the mechanism of action of ribonucleoside diphosphate reductase (RDPR), an attractive target for new anticancer agents, Czernecki et al.[21] reported the synthesis of 2'-deoxy-2'-spirocyclopropyl cytidine **7** as a potential inhibitor of RDPR (Figure 2). The keysynthetic step was the condensation of diazometane with properly protected 2'-methylene nucleosides, which by light-induced nitrogen extrusion afforded the cyclopropane ring.[21] Wengel et al.[22] prepared 2'-spirocyclic nucleosides **8** and **9** as novel conformationally restricted probes for ribonucleotide reductases (Figure 2).

Little attention was given to the preparation of 3'-spironucleosides. In 1992, in collaboration with a group of virologists at the Rega Institute for Medical Research, we reported on an entirely novel class of highly functionalized C3'-spironucleosides that were highly specific and potent inhibitors of human immunodeficiency virus type 1 (HIV-1) replication. The prototype compound is the [1-[2',5'-bis-*O*-(*tert*-butyldimethylsilyl)-β-D-ribofuranosyl]thymine]-3'-spiro-5''-(4''-amino-1'',2''-oxathiole-2'',2''-dioxide), designated as TSAO-T (**10**, Figure 3).[23–25] These compounds are specifically targeted at the HIV-1-encoded reverse transcriptase (RT), with which they interact at an allosteric nonsubstrate binding site, a hydrophobic pocket located near the polymerase active site. Thus, mechanistically TSAO derivatives behave as do all the other nonnucleoside HIV-1-specific RT inhibitors (NNRTIs), but structurally they are highly functionalized nucleosides. This can be seen as the first feature of their "unique" character. From a structural point of view, a unique and relevant structural feature of TSAO derivatives is the presence of an unusual 4-amino-1,2-oxathiole-2,2-dioxide (or β-amino-γ-sultone)spiro ring inserted at the C3'-position of the ribose moiety.

Following our first papers on TSAO derivatives, other 3'-spiro-substituted nucleosides (**11**, Figure 3) were described by Tronchet et al.[26,27] Also, the group of Chattopadhyaya[28] prepared 2' and 3'-spiro-substituted nucleosides (**12** and **13**, Figure 3) and 2'- and 3'-hypermodified tricyclic *cis*-fused-spiroisoxazolidine nucleosides (**14** and **15**, Figure 3) by intramolecular 1,3-dipolar cycloadditions of several *C*-alkenyl nitrones of nucleosides.[28] Others, such as Wengel and co-workers,[29] described the synthesis of 3'-spiro-γ-lactone nucleoside as structural analogues of the anti-HIV-1 TSAO compounds.

In this chapter we deal with the synthetic methods for the preparation of spiro-substituted TSAO compounds, which, as mentioned, have been tested for anti-HIV activity and turn out to be a unique family of NNRTIs with a mode of action different from that of the other NNRTIs described so far. TSAO derivatives show a dual mechanism of action against HIV-1 replication by acting as alosteric inhibitors (like the other NNRTIs) but also interfere in the dimerization process of the enzyme (HIV-1 RT), thus behaving as dimerization inhibitors (unlike the rest of the NNRTIs).[35b] We believe that the TSAO molecules represent an important tool available to study many biological aspects and implications of RT in HIV-1 replication and inhibition.

2. CHEMICAL SYNTHETIC APPROACHES TO THE PROTOTYPE TSAO-T

The synthesis of xylo- and ribo-TSAO-T (**21** and **10**) derivatives was initially accomplished by treatment of tertiary *O*-mesylcyanohydrins of furanosid-3'-ulosylthymine under non-nucleophilic basic conditions (Scheme 1).[23,24] The method involves abstraction of one proton from the mesylate group, followed by nucleophilic attack of the resulting carbanion at the nitrile carbon atom through a novel intramolecular aldol-type condensation.[30,31] Thus, reaction of the 3'-ketonucleoside **16** (Scheme 1) with sodium cyanide and sodium hydrogen carbonate followed by mesylation (mesyl chloride/pyridine) of the 3'-cyanohydrin epimers **17** and **18** gave the 3'-*C*-cyano-3'-*O*-mesyl-β-D-xylo- and ribofuranosyl thymine nucleosides **19** and **20**. The major diastereomer **19** results by the attack of the cyanide ion from the sterically less hindered α-face of the furanose ring opposite the base. Treatment of **19** and **20** with cesium carbonate in dry acetonitrile at room temperature afforded the xylo- and ribospiro nucleosides **21** and **10**, respectively.[23,24]

The ribospiro nucleoside **10** (TSAO-T), obtained as a minor product in the pathway in Scheme 1, exhibited potent and selective inhibition of HIV-1 replication in

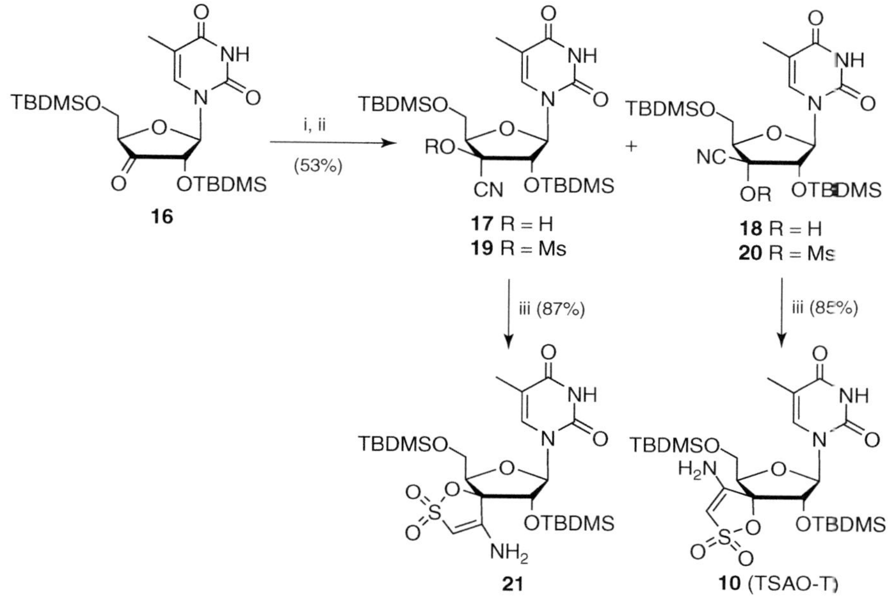

Scheme 1. *Reagents and conditions:* (i) NaCN, NaHCO$_3$, ethyl ether/water, rt, 16 h; (ii) MsCl, pyridine, 8–10°C, 48 h; (iii) Cs$_2$CO$_3$, acetonitrile, rt, 6 h.

vitro while its xylo-isomer was devoid of antiretroviral activity.[24] We next devised a new synthetic route, which provided active ribo derivatives exclusively (Scheme 2). Ribo-TSAO-T **10** was prepared stereoselectively by glycosylation of a tertiary cyanomesylate of ribose with persilylated thymine followed by basic treatment of the cyanomesyl nucleoside to give exclusively the β-D-ribospiro nucleoside.[25] The ribo-configuration of the nucleoside is determined by the configuration of the starting cyanohydrin used in the preparation of the cyanomesylate of ribose **24**, as demonstrated clearly in earlier papers.[30,31] Thus, reaction of the ribofuranosulose **22**[32] with sodium cyanide and sodium hydrogen carbonate afforded exclusively the kinetically controlled ribocyanohydrin **23**, which was transformed, without further purification, to the 3-*C*-cyano-3-*O*-mesyl derivative **24** by reaction with mesyl chloride in pyridine. This ribo-configuration is in agreement with the approach of the cyanide ion from the sterically less hindered β-face of the ulose **22** opposite the 1,2-isopropylidene group, as stated previously for other additions of nucleophiles to uloses.[33]

Hydrolysis of the 1,2-isopropylidene group of **24** with TFA, followed by acetylation with acetic anhydride/pyridine, provided a 3:2 mixture of the diacetate derivative **25**. Glycosylation of **25** with a previously silylated thymine base in refluxing acetonitrile in the presence of trimethylsilyl triflate as condensing reagent[34] afforded the 3′-cyanomesylate **26**. Due to the presence of a 2-*O*-acyl participating group, the product obtained was exclusively the β-anomer. Treatment of the cyanomesylate

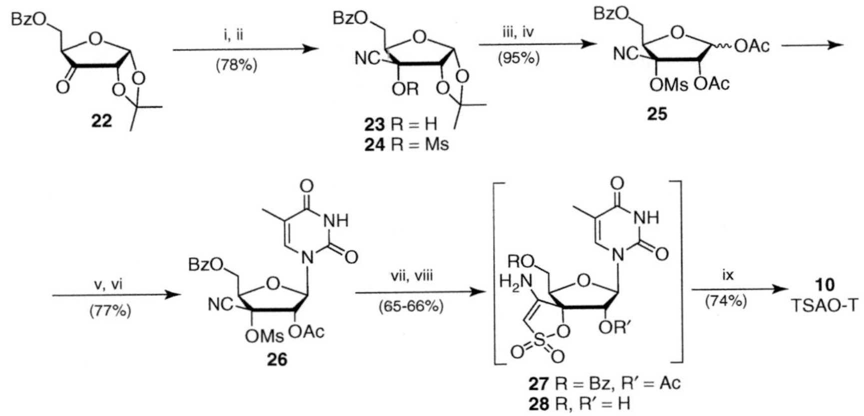

Scheme 2. *Reagents and conditions:* (i) NaCN, NaHCO$_3$, ethyl ether/water, rt, 16 h rt, 4 h; (ii) MsCl, pyridine, 8–10°C, 16 h; (iii) TFA/water, rt, 4 h; (iv) Ac$_2$O/pyridine, rt, overnight; (v) thymine, HMDS, reflux; (vi) TMSOTf, acetonitrile, 80°C, 5 h; (vii) Cs$_2$CO$_3$, acetonitrile, rt, 3 h; (viii) NH$_3$/MeOH, rt, overnight; (ix) TBDMSCl, DMAP, rt, 24 h.

26 with cesium carbonate afforded the spiroderivative **27** (in 65% yield). Deprotection of **27** with saturated methanolic ammonia, followed by reaction with an excess of *tert*-butyldimethylsilyl chloride, yielded the 2′,5′-bis-*O*-silylated nucleoside TSAO-T **10** in 74% yield (Scheme 2).

Since we first synthesized TSAO-T in 1992,[23,24] a wide range of modifications have been carried out on this family of compounds to reveal the structural requirements for their optimal interaction with HIV-1 RT, and a considerable number of TSAO derivatives (more than 800) have been prepared and studied.[35] In the next section we cover synthetic aspects of the chemistry of TSAO nucleosides that has been carried out to date on both the base and the sugar part of these unique spiro-substituted nucleosides. Most relevant aspects of structure–activity relationships are also commented on briefly.

Experimental Procedures

[1-[(2′,5′-Bis-O-tert-butyldimethylsilyl)-3′-cyano-3′-O-mesyl-β-D-xylofuranosyl] thymine and [1-[(2′,5′-Bis-O-tert-butyldimethylsilyl)-3′-cyano-3′-O-mesyl-β-D-ribofuranosyl]thymine[24] *(19 and 20, Scheme 1).* A mixture of the 3′-ketonucleoside **16** (4 mmol), water (16 mL), ethyl ether (32 mL), NaHCO$_3$ (0.64 g, 8 mmol), and sodium cyanide (0.2 g, 4 mmol) was stirred vigorously at room temperature for 16 h. The organic phase was separated and the aqueous phase was washed with ethyl ether (2 × 50 mL). The combined ethereal phases were dried over Na$_2$SO$_4$, filtered, and evaporated to dryness. The residue, a mixture of the two epimeric cyanohydrins, was dissolved in dry pyridine (8 mL). To this solution was added mesyl chloride (1.6 mL, 20 mmol). The mixture was stirred at 8–10°C for 48 h, poured into ice and water, and extracted with CH$_2$Cl$_2$

(2 × 50 mL). The combined extracts were washed with 1 N HCl (50 mL), dried over Na_2SO_4, filtered, and evaporated to dryness. The residue was purified by column chromatography with hexane/ethyl acetate, 1:4. The fastest-moving fractions afforded 1.18 g (53%) of **19** as a white foam: IR (KBr) 1375, 1185 cm^{-1} (SO_2). ^{1}H-NMR (CDCl$_3$, 90 MHz): δ 1.96 (s, 3H, CH$_3$-5), 3.33 (s, 3H, CH$_3$SO$_2$), 4.05 (m, 2H, H-5'), 4.51 (m, 1H, H-4'), 4.87 (d, 1H, H-2', $J_{1',2'}$ = 2 Hz), 6.03 (d, 1H, H-1'), 7.28 (s, 1H, H-6), 9.12 (b s, 1H, NH-3). ^{13}C-NMR (CD$_2$Cl, 50 MHz): δ 12.26 (CH$_3$-5), 40.24 (CH$_3$SO$_2$), 59.17 (C-5'), 81.51, 83.55 (C-2', C-4'), 82.51 (C-3'), 91.57 (C-1'), 111.87, 112.63 (CN, C-5), 134.68 (C-6), 150.22 (C-2), 163.55 (C-4).

The slowest-moving fractions afforded 0.22 g (10%) of **20** as a white foam: IR (KBr) 1375, 1180 cm^{-1} (SO_2). ^{1}H NMR (CDCl$_3$, 300 MHz): δ 1.94 (s, 3H, CH$_3$-5), 3.26 (s, 3H, CH$_3$SO$_2$), 4.04 (m, 2H, H-5', $J_{5'a,5'b}$ = 12.0 Hz, $J_{4',5'a}$ = 1.2 Hz, $J_{4',5'b}$ = 2.1 Hz), 4.51 (d, 1H, H-2', $J_{1',2'}$ = 8.3 Hz), 4.73 (m, 1H, H-4'), 6.24 (d, 1H, H-1'), 7.38 (s, 1H, H-6), 8.48 (b s, 1H, NH-3). ^{13}C-NMR (CD$_3$Cl, 50 MHz): δ 12.05 (CH$_3$-5), 40.33 (CH$_3$SO$_2$), 62.09 (C-5'), 80.34 (C-3'), 78.20, 84.10, 84.28 (C-2', C-4', C-1'), 112.24, 114.19 (CN, C-5), 133.68 (C-6), 150.42 (C-2), 163.18 (C-4).

*1-[(2',5'-Bis-**O**-tert-butyldimethylsilyl)-β-**D**-ribofuranosyl]thymine-3'-spiro-5''-(4''-amino-1'',2''-oxathiole-2'',2''-dioxide)*[24] *(10, Scheme 1).* A solution of the 3'-*C*-cyano-3'-*O*-mesylribofuranosyl nucleoside **20** (1 mmol) in dry acetonitrile (10 mL) was treated with Cs_2CO_3. The mixture was stirred at room temperature for 6 h and then filtered and evaporated to dryness. The residue was purified by column chromatography (hexane/ethyl acetate, 1:3) to give **10** as a white solid, which was crystallized from hexane (0.57 g, 87%): mp > 230°C. IR (KBr) 3420, 3350 cm^{-1} (NH$_2$), 1655 (C=C–N). ^{1}H-NMR [(CD$_3$)$_2$SO, 300 MHz]: δ 1.81 (s, 3H, CH$_3$-5), 3.79 (m, 2H, H-5', $J_{5'a,5'b}$ = 11.7 Hz), 4.43 (d, 1H, H-2', $J_{1',2}$ = 5.7 Hz), 4.53 (dd, 1H, H-4', $J_{4',5'a}$ = 6.4 Hz, $J_{4',5'b}$ = 3.4 Hz), 5.56 (s, 1H, H-3''), 6.01 (d, 1H, H-1'), 6.95 (b s, 2H, NH$_2$), 7.44 (s, 1H, H-6), 11.52 (b s, 1H, NH-3).^{13}C-NMR [(CD$_3$)$_2$CO, 50 MHz]: δ 12.53 (CH$_3$-5), 61.53 (C-5'), 82.54, 84.47, 90.05 (C-1', C-2', C-4', C-3''), 93.22 (C-3'), 112.31 (C-5), 135.83 (C-6), 151.42, 153.57 (C-2, C-4''), 163.95 (C-4).

*1-(2'-**O**-Acetyl-5'-**O**-benzoyl-3'-cyano-3'-**O**-mesyl-β-**D**-xylofuranosyl)thymine*[24] *(26, Scheme 2).* Thymine (0.52 g, 4.08 mmol) was silylated with hexamethyldisilazane (HMDS) (24 mL) under reflux in the presence of ammonium sulfate (20 mg), and the reaction was refluxed until the solution became clear. The excess of HMDS was removed by distillation under reduced pressure. A solution of the sugar derivative **25** (1.52 g, 3.4 mmol) in dry acetonitrile (30 mL) was added to the syrupy silylated base, followed by the addition of trimethylsilyl triflate (7.3 mL, 3.7 mmol). The resulting mixture was heated to reflux. After 2 h, an additional portion of trimethylsilyl triflate (7.3 mL, 3.7 mmol) was added and the refluxing was continued for 3 h. The reaction was allowed to cool to room temperature, then CH_2Cl_2 (50 mL) was added and poured into cold, saturated, aqueous NaHCO$_3$. The organic phase was separated, the aqueous phase was washed with CH$_2$Cl$_2$ (2 × 20 mL), dried (Na$_2$SO$_4$), and the solvent was removed. The residue was purified by column chromatography (hexane/ethyl acetate, 1:2) to yield **26** (1.34 g, 77%) as a white foam; IR (KBr) 1760 (C=O), 1375, 1180 cm^{-1} (SO$_2$). ^{1}H-NMR (CDCl$_3$, 90 MHz): δ 1.74 (s, 3H, CH$_3$-5), 2.21 (s, 3 H, OAc), 3.26 (s, 3H, CH$_3$SO$_2$), 5.13–5.43 (m, 3 H, H-4′, H-5′), 5.85, 6.30 (2 d, 2 H, H-1′, H-2′, $J_{1',2'}$ = 7.0 Hz), 7.10 (s, 1 H, H-6), 7.40–8.25 (m, 5 H, Ph), 9.44 (b s, 1 H, NH-3). ^{13}C-NMR (CDCl$_3$, 50 MHz): δ 11.65 (CH$_3$-5), 19.95 (OAc), 40.34 (CH$_3$SO$_2$), 62.31 (C-5′), 73.44, 80.82, 84.64 (C-2′, C-4′, C-1′), 76.32 (C-3′), 110.71 (C-5), 113.02 (CN), 128.72, 129.20, 133.65, 135.52 (C-6, Ph), 150.24 (C-2), 163.18, 164.95, 168.87 (C-4, C=O).

3. MODIFICATIONS OF THE PROTOTYPE TSAO-T

3.1. Base Modifications

3.1.1. Synthesis of Pyrimidine and Purine Analogues.
A large body of work has been directed toward the synthesis of base-modified analogues of the prototype compound TSAO-T **10**. Thus, the above-mentioned stereoselective synthesis of TSAO-T[25] allowed us to prepare a variety of TSAO analogues of pyrimidines and purines **30a–j** (Scheme 3).[36,37] In the pyrimidine series,[37] a similar reaction sequence with 3′-cyanomesyl nucleosides of 5-bromo or 5-iodo **29e** and **29f** gave the desired spiro derivatives **30e** and **30f** in low yields, together with the cyclospironucleoside **31**. Formation of **31** could be explained by an intramolecular addition of 4″-NH$_2$ to the C6 position of the pyrimidine base.

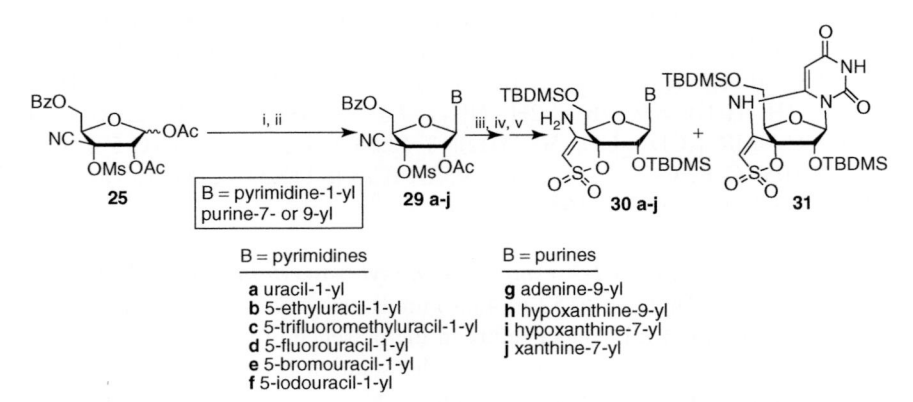

B = pyrimidine-1-yl
purine-7- or 9-yl

B = pyrimidines
a uracil-1-yl
b 5-ethyluracil-1-yl
c 5-trifluoromethyluracil-1-yl
d 5-fluorouracil-1-yl
e 5-bromouracil-1-yl
f 5-iodouracil-1-yl

B = purines
g adenine-9-yl
h hypoxanthine-9-yl
i hypoxanthine-7-yl
j xanthine-7-yl

Scheme 3. *Reagents and conditions:* (i) base, HMDS, reflux; (ii) TMSOTf, acetonitrile, 80°C, 5 h; (iii) Cs$_2$CO$_3$, acetonitrile, rt, 3 h; (iv) NH$_3$/MeOH, rt, overnight; (v) TBDMSCl, DMAP, rt, 24 h.

Spironucleosides of uracil **30a** and thymine **10** were further transformed to the corresponding derivatives of cytosine **32** and 5-methylcytosine **35**[37] (Scheme 4)

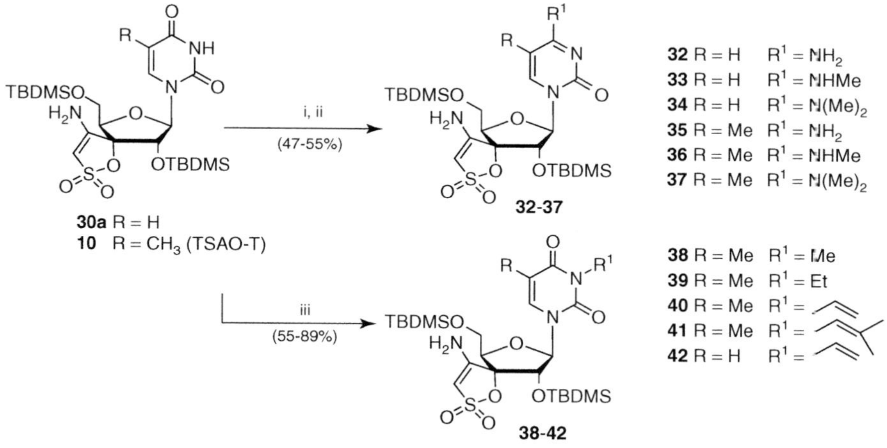

Scheme 4. *Reagents and conditions:* (i) POCl$_3$, 1,2,4-triazole, acetonitrile, 0°C; (ii) appropriate amine, rt, 4 h; (iii) R^{1}I, K$_2$CO$_3$, acetone, 60°C, 3–8 h.

following standard classical methods by treatment with phosphoryl chloride and 1,2,4-triazole followed by reaction with an excess of concentrated ammonia.[38,39] 4-*N*-alkyl nucleoside analogues of TSAO-C **33**, **34**, **36**, and **37** were prepared using a similar reaction sequence by treatment with an excess of the appropriate amines (Scheme 4).[37]

Selective N^3-alkylation[40] of TSAO-U **30a** and TSAO-T **10** afforded the N^3-substituted derivatives **38–42** (Scheme 4) by reaction with the appropriate alkyl iodide in the presence of potassium carbonate.[37] On the other hand, N^6-alkylated derivatives of adenine **44**, **45**, and hypoxanthine **46** were also synthesized by reaction of the 6-chloropurine nucleoside **43** (prepared by glycosylation of **25** with 6-chloropurine) with methylamine, dimethylamine, or sodium methoxide, respectively (Scheme 5).[36] Similar to the pyrimidine series, selective alkylation of spiro nucleosides of hypoxanthine **30h** and **30i** with methyl or ethyl iodide/K$_2$CO$_3$ gave the corresponding N1-alkyl nucleosides **47–50**.

The thymine moiety of TSAO-T can be replaced by a number of other pyrimidines and purines without a marked decrease in antiviral efficacy.[36,37] The TSAO-purine derivatives are in general three- to fivefold less effective than the most active TSAO-pyrimidine derivatives. Interestingly, introduction of an alkyl or alkenyl function at N3 of the thymine ring or an alkyl group at N1 of the purine moiety markedly decrease cytotoxicity without affecting the antiviral activity.[36,37] Particular attention has been paid to the attachment of different groups to the N3-position of the pyrimidine base.[25,41,42] Thus, novel analogues of TSAO-T bearing a variety of polar, lipophilic, or aromatic groups linked to the N3-position through flexible polymethylene linkers of different length **51–68** (Scheme 6) were prepared and evaluated for their anti-HIV activity.[42] Most of these compounds were readily obtained in good yields following the above-mentioned selective N^3-alkylation of

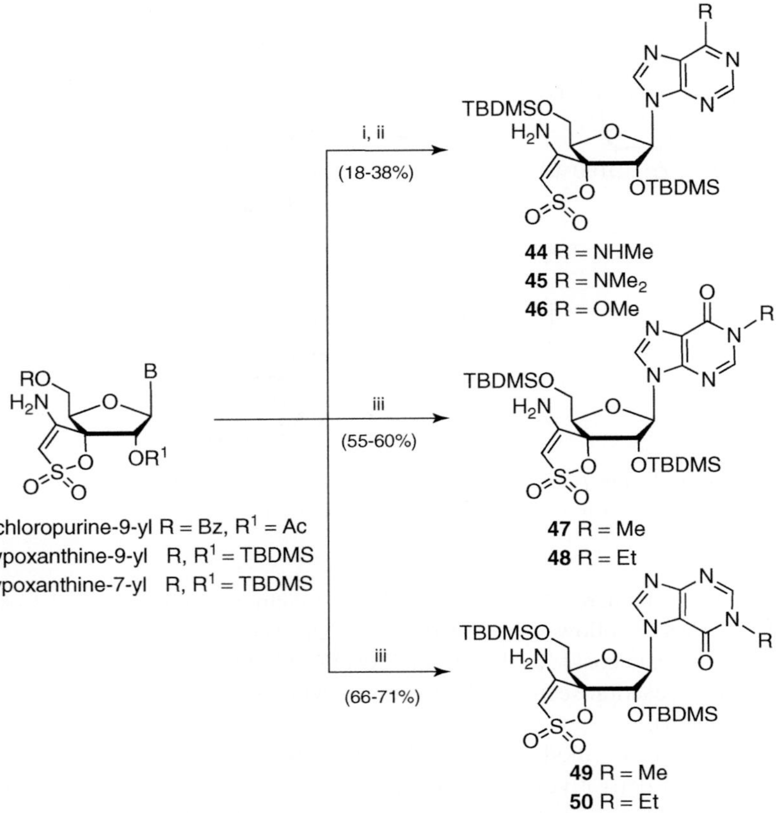

43 B = 6-chloropurine-9-yl R = Bz, R¹ = Ac
30h B = Hypoxanthine-9-yl R, R¹ = TBDMS
30i B = Hypoxanthine-7-yl R, R¹ = TBDMS

44 R = NHMe
45 R = NMe$_2$
46 R = OMe

47 R = Me
48 R = Et

49 R = Me
50 R = Et

Scheme 5. *Reagents and conditions:* (i) appropriate amine, rt, 2 h overnight or Amberlyst A-26 in MeOH, 60°C, 8 h; (ii) TBDMSCl, DMAP, rt, 24 h; (iii) R¹I, K$_2$CO$_3$, acetone, 60°C, 3–6 h.

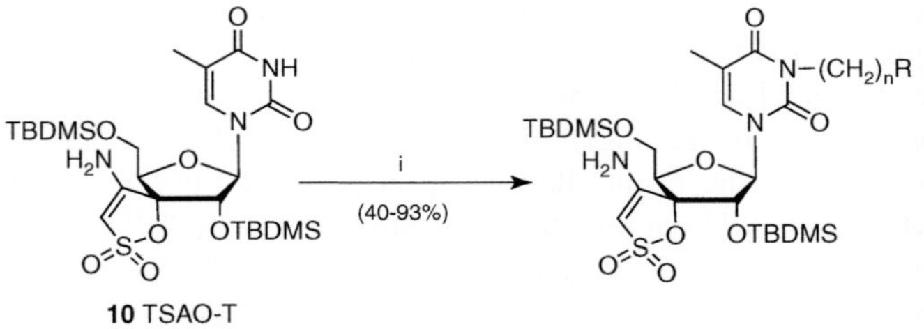

10 TSAO-T

51 R = CO$_2$Me	n = 1	
52 R = CO$_2$Bn	n = 1	
53 R = CONH$_2$	n = 1	
54 R = cyclopropyl	n = 1	
55 R = Ph	n = 1	
56 R = COOH	n = 1	
57 R = Br	n = 2	
58 R = OH	n = 2	
59 R = OMe	n = 2	
60 R = Ph	n = 2	
61 R = indolyl	n = 2	
62 R = CONHMe	n = 1	
63 R = CONMe$_2$	n = 1	
64 R = OH	n = 3	
65 R = CO$_2$Me	n = 4	
66 R = CO$_2$Et	n = 5	
67 R = CONHMe	n = 4	
68 R = CONHMe	n = 5	

Scheme 6. *Reagents and conditions:* (i) Br(CH$_2$)$_n$R, K$_2$CO$_3$, acetone, 60°C, 5–10 h.

TSAO derivatives by treatment of TSAO-T with the corresponding alkyl halides in the presence of K_2CO_3.[42]

Virtually all synthesized N^3-alkylated TSAO derivatives are endowed with a marked anti-HIV-1 activity that is in the nanomolar range. The antiviral activity of the parent prototype TSAO-T has been improved by two- to sixfold by the introduction of polar substituents such as OH (**64**), $CONH_2$ (**53**), CONHMe (**62, 67, 68**), or $CONMe_2$ (**63**). Some of the most active N^3-TSAO derivatives retain antiviral activity against the TSAO-resistant HIV-1 strain (Glu138/Lys) being, up to now, the only TSAO derivatives that showed activity against this resistant strain. Interestingly, the N3 CONHMe derivative **67** is also five- to sixfold more active against recombinant HIV-1 RT than was the lead compound TSAO-T **10**, thus becoming the most active TSAO derivative synthesized so far. Moreover, compound **62** turn out to have the highest selectivity index yet reported for this class of compounds (around ≥ 12.000).[42]

Recently, Moura et al.[43] reported an efficient methodology for direct acylation of TSAO-T at the N3-position of the base using acyl chlorides as acylating agents under phase-transfer conditions.[43] Several TSAO derivatives bearing mono- or dicarbonyl substituents (Figure 4) were prepared and evaluated against HIV-1 replication. EC_{50} values of N3-acylated TSAO-T compounds (ranging between 0.032 and 0.070 µM) were similar to that observed with TSAO-m^3T (EC_{50} = 0.037 µM). However, most of these derivatives were much more cytotoxic (CC_{50} around 8–11 µM) than N^3-alkylated TSAO derivatives such as TSAO-m^3T (CC_{50} = 115 µM), resulting in compounds with poor selectivity indexes.

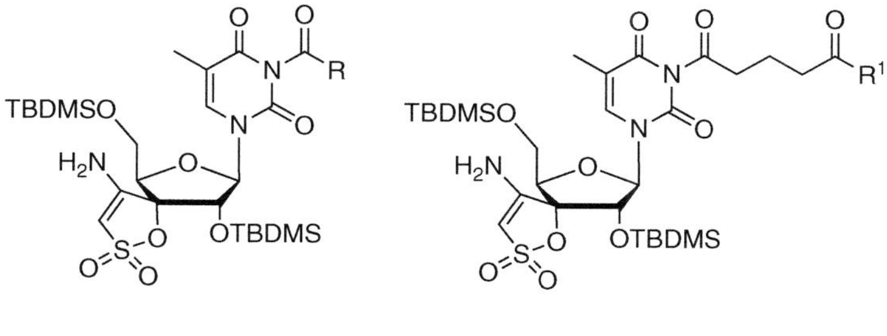

R = Ph, OMe, cyclopropyl

R^1 = OMe, NH_2, NHMe, NMe_2, SMe, O^tBu, OBn, NHBn

Figure 4. N^3-acyl-substituted TSAO-T analogues.

3.1.2. Synthesis of 1,2,3-Triazole TSAO Analogues.

A series of 1,2,3-triazole TSAO derivatives were prepared stereoselectivly by 1,3-dipolar cycloaddition of a suitable functionalized and protected ribofuranosyl azide intermediate **69** to differently substituted acetylenes to give β-D-ribospironucleosides.[44] Cycloaddition of azide **69** to unsymmetrical acetylenes gave a mixture of the corresponding 4- and 5-substituted isomers **70a–m** and **71a–m**, respectively, whereas cycloaddition of azide **69** to symmetric acetylenes afforded difunctional 1,2,3-triazole nuclecsides

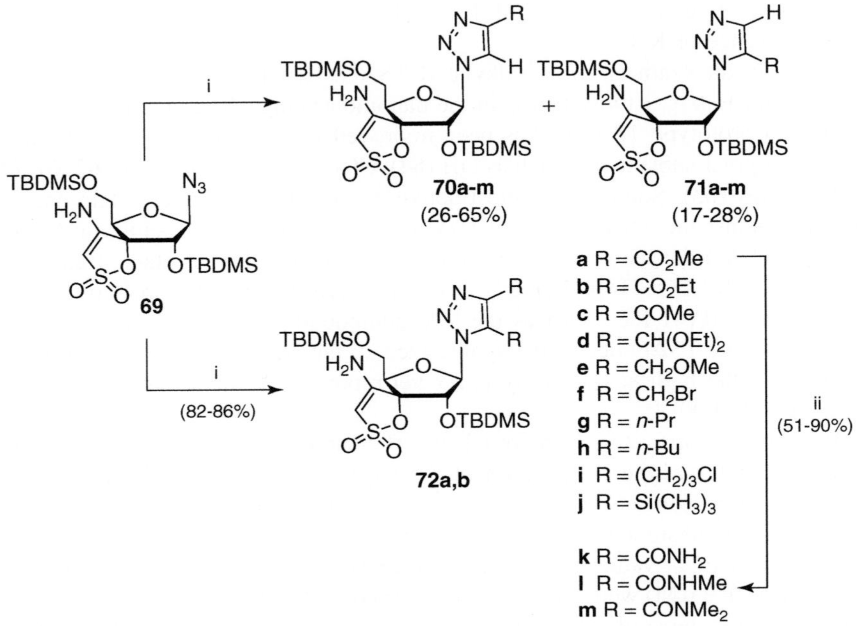

Scheme 7. *Reagents and conditions:* (i) mono- and disubstituted acetylene, dry toluene, reflux, 18 h to 3 days; (ii) amines, methanol, rt, 1–8 h.

72a,b (Scheme 7). Further aminolysis of alkyl propiolate TSAO derivatives **70a** and **71a** with amines yielded 4- and 5-carbamoyl-substituted derivatives **70k–m** and **71k–m**. Several members of this class of compounds showed potent anti-HIV-1 activity comparable to that of TSAO-T, which proves that the antiviral activity of TSAO analogues is not dependent on the presence of either a pyrimidine or purine moiety in these nucleoside analogues. In particular, the antiviral activity of the 4-carbamoyl-substituted compounds **70k–m** were decreased 10-fold in comparison with the corresponding 5-substituted triazole analogues **71k–m**.[44]

The most active 5-isomers were obtained as minor products in the aforementioned procedure, owing to steric and electronic factors in the cycloaddition reaction.[45–48] For this reason we next designed and developed a new and improved synthetic procedure that yields exclusively the desired 5-isomers **71b**,

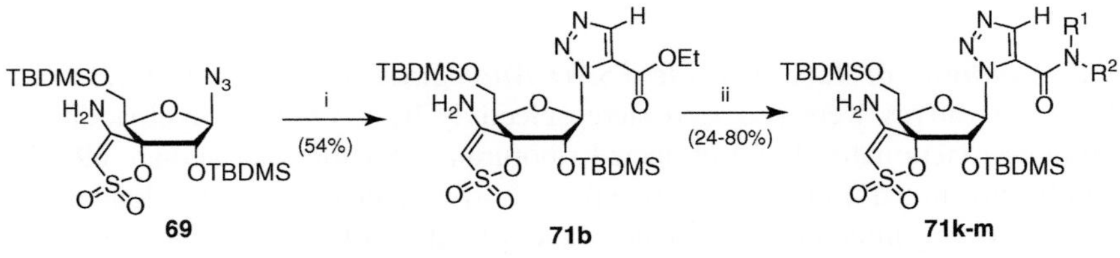

Scheme 8. *Reagents and conditions:* (i) $Ph_3P=CHCOCO_2Et$, dry xylene, reflux, 12 h; (ii) amine, ethanol, rt, 45 min to 24 h.

71k–m (Scheme 8) based on the reported cycloaddition of alkylazides[46,49,50] and glycosylazides.[51–53] Thus, reaction of the azide intermediate **69** with 2-oxoalkylidentriphenyl phosphorane[54] in refluxing xylene afforded exclusively the 5-substituted 1,2,3-triazole derivative **71b**, which was treated further with the appropriate primary or secondary amines to give the corresponding 5-carbamoyl derivatives **71k–m** in good yields.[55]

Several 5-substituted 1,2,3-triazole-TSAO derivatives proved to be potent inhibitors of HIV-1 replication with higher selectivity than that of the parent TSAO-T prototype, due to a marked decrease in cytotoxicity.[55]

Experimental Procedure

{1-[(2′,5′-Bis-O-tert-butyldimethylsilyl)-β-D-ribofuranosyl]-5-[(ethyloxy)carbonyl] 1,2,3-triazole}-3′-spiro-5″-(4″-amino-1″,2″-oxathiole-2″,2″-dioxide)[55] (**71b**, *Scheme 8*). A solution of azide **69** (0.50 g, 0.99 mmol) and Ph$_3$P=CHCOCO$_2$Et (0.45 g, 1.18 mmol) in dry xylene (7 mL) was heated to reflux under N$_2$. After complete disappearance (HPLC) of the starting material (approximately 12 h) the solvent was evaporated to dryness. The residue was purified by column chromatography (hexane:ethyl acetate, 3:1) to give 0.32 g (54%) of **71b** as a white foam. IR (KBr) 3400, 3310 (NH$_2$), 1740 (CO), 1650 cm^{-1} (C=CN). ^{1}H-NMR (CDCl$_3$, 90 MHz): δ 1.38 (t, 3H, CH$_3$CH$_2$, J = 7.2 Hz), 4.01 (m, 2H, 2H-5′), 4.43 (m, 3H, H-4′, CH$_3$*CH*$_2$), 5.74 (s, 1H, H-3″), 5.82 (d, 1H, H-2′, $J_{1′,2′}$ = 7.8 Hz), 6.62 (b s, 2H, NH$_2$), 6.68 (d, 1H, H-1′), 8.33 (s, 1H, H-4).

3.1.3. Synthesis of 3-Spiro-Sugar Abasic Analogues.

To determine which is the minimum fragment of the thymine base of TSAO-T necessary for interaction with HIV-1 RT, we designed a series of 3-spiro-sugar derivatives substituted at the anomeric position with amine, amide, urea, or thiourea moieties that mimic parts or the entire thymine base of TSAO-T.[56] The 1-urea- or thiourea-substituted 3-spiro-sugar derivatives were prepared by reaction of the corresponding isocyanates or isothiocyanates with the appropriate 1-amino sugar intermediate **73** (Scheme 9). This intermediate was synthesized by reduction of the ribosyl azide derivative **69** using the method of Staudinger.[57] In all cases we obtained mixtures of α- and β-anomers of the *N*-substituted derivatives that were separated by chromatography.

Compounds substituted at the anomeric position with an azido or amino group were devoid of marked antiviral activity (**69** or **73a,b**). However, among the substituted urea sugar derivatives, compounds **76b** and **78b**, which mimic to a large extent both the shape and the electrostatic potential of a thymine ring of the intact TSAO-T molecule, retain the highest antiviral activity.[56] The acyclic substituent at the anomeric position of the most active acyl ureas can exist in different conformations (Figure 5). The spectroscopic data were compatible with six-membered hydrogen-bonded pseudocycle conformer II but not with conformers I or III. These urea derivatives represent the first examples of sugar derivatives that interact with HIV-1 RT in a specific manner.[56]

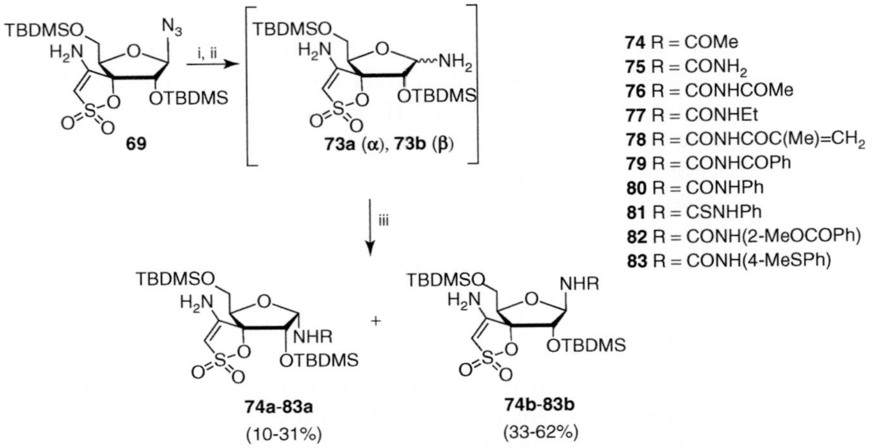

Scheme 9. *Reagents and conditions:* (i) PMe₃, dry THF, rt, 20 min; (ii) NH₃/MeOH, rt 3 h; (iii) RNCO or RNCS, dry acetonitrile, rt, overnight.

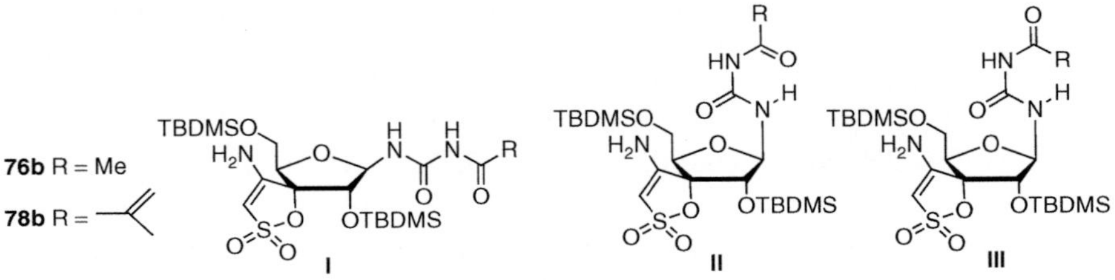

Figure 5. Conformations for the most active acylureas, **76b** and **78b**.

3.2. Sugar Modifications

When the thymine moiety of the prototype TSAO-T was held unmodified but the sugar part was modified, a series of interesting features were observed with respect to the structural requirements of TSAO molecule to keep its antiviral potency. The presence of the bulky silyl groups at both C2' and C5' of the ribose part seems to be a prerequisite for anti-HIV-1 activity.[23–25] Removal of these lipophilic groups or replacement by other lipophilic entities resulted in antivirally inactive TSAO compounds.[58] The synthesis of these compounds involves classical protection and deprotection nucleoside strategies. The specific role of the 5'-position of TSAO-T has been explored further by replacing the 5'-TBDMS group by a wide variety of alkyl, alkenyl, or aromatic groups and substituted amines, carbamoyl, or (thio)acyl moieties.[58] Thus, we prepared 5'-*O*-acyl derivatives **86–88** by reaction of 5'-*O*-deprotected TSAO derivative **84** with acetic anhydride, isobutyryl chloride, or allyl chloroformate in pyridine (Scheme 10). 5'-*O*-Carbamate derivative **89** is prepared by reaction of **84** with phenyl chloroformate, followed by treatment with a solution of dimethylamine in ethanol, while 5'-*O*-thiocarbamate **90** was synthesized

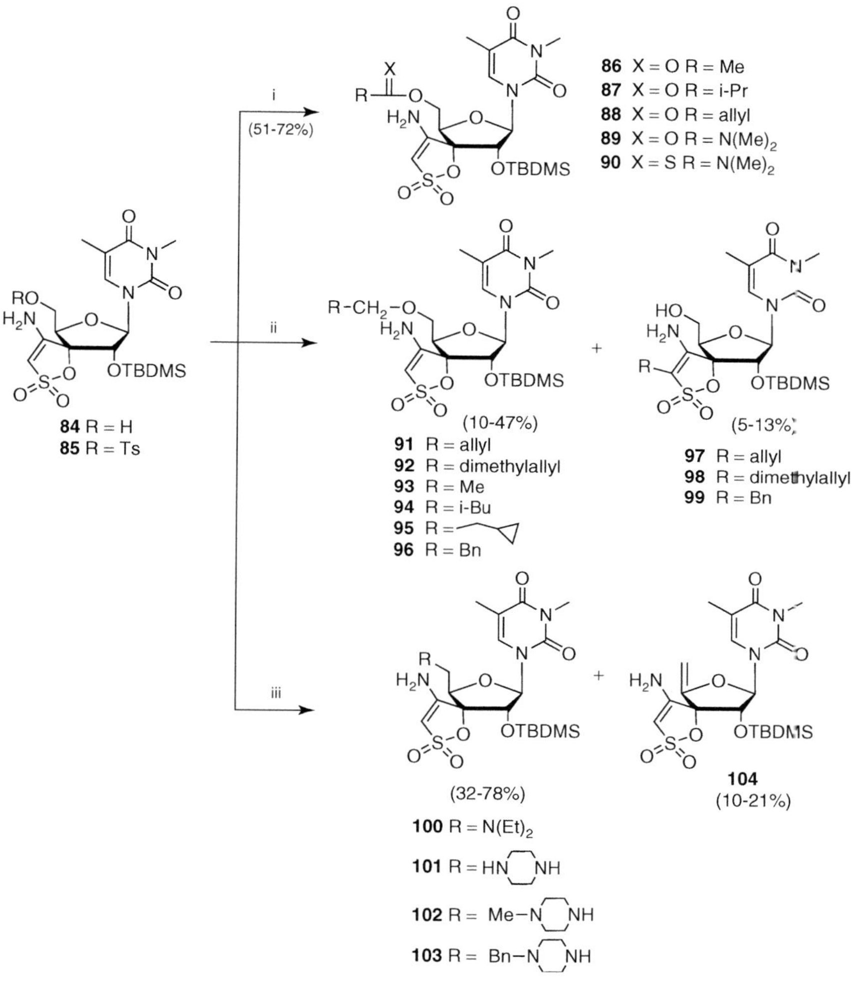

Scheme 10. *Reagents and conditions:* (i) RCOX, dry pyridine, rt or 80°C, 1–3 h; (ii) RCH$_2$I, NaH, dry DMF, rt, overnight; (iii) amines, dry acetonitrile, reflux, 6–48 h.

similarly by reaction with N,N'-thiocarbonyldiimidazole, followed by treatment with dimethylamine. The synthesis of 5'-O-ether TSAO derivatives **91–96** was performed in low-to-moderate yields by the classical Williamson procedure with the appropriate alkyl, alkenyl, or benzyl halides in the presence of sodium hydride as the base. In some cases the 5'-OH 3''-C-substituted derivatives **97–99** were also isolated (Scheme 10). 5'-N-Substituted TSAO compounds **100–103** were prepared by treatment of 5'-O-tosyl **85** with the corresponding amine. In these reactions the 4',5'-didehydronucleoside **104** was also isolated. Replacement of the 5'-TBDMS group by an acyl, aromatic, or cyclic moiety eliminates the anti-HIV activity. However, the presence of an alkyl or alkenyl chain makes possible, partial retention of the antiviral activity, although at a 20- to 100-fold lower potency than that of TSAO-T. These results suggest that the 5′-TBDMS group of the TSAO-T molecule plays a crucial role in the activity.[59]

Several other sugar-modified 3′-spiro TSAO derivatives, including the *allo*-furanosyl TSAO analogue,[60] compounds bearing L-sugars,[61] or compounds with inverted configuration at the C4′ stereocenter[62] were also prepared following the stereospecific synthesis of the prototype TSAO-T mentioned above. These modifications led to completely inactive TSAO derivatives.

Finally, change in the 3′-spiro moiety to the 2′-position of the ribose also results in annihilation of the antiviral activity.[63] The synthesis of arabino- and ribo-2′-spironucleosides involves the reaction of α-mesylcyanohydrins of furanosid-2′-ulosyl nucleosides with base following a protocol similar to that described for the preparation of 3′-spironucleoside TSAO derivatives.[24] In conclusion, modifications at the sugar part in TSAO molecules turn out to be very stringent with respect to the structural requirements necessary to retain anti-HIV-1 activity.

3.3. Synthesis of Spiro-Modified Analogues

As mentioned, one of the most characteristic features of TSAO derivatives is the presence of the 4-amino-1,2-oxathiole-2,2-dioxide spiro (or β-amino-γ-sultone) ring. As part of our medicinal chemistry program to further explore the importance of the substituent effects on the anti-HIV-1 activity and toxicity of TSAO derivatives, we next focused our attention on modifications of the 3′-spiro moiety. According to our model of interaction, this represents the closest part of the molecule to the interface between the p51 and p66 HIV-1 RT subunit domains. On the basis of this hypothesis, we introduced different functional groups at positions 3″ or 4″, which may give additional interactions with amino acids (of both subunits, p66/p51) adjacent to the Glu-B138 of the p51 subunit (key interaction point of TSAOs with HIV-1 RT).[35] Moreover, we also designed TSAO compounds bearing different carbonyl functionalities that may interact with the amino group of Lys138 in the TSAO-resistant strains.[35]

With these aims, various synthetic studies for the direct and selective functionalization of the C3″- or the C4″-amino-positions of the enamine system on TSAO compounds were explored and are described in this section. For the synthesis of spiro-modified TSAO compounds, two important issues were considered from a chemical point of view. First, the potentially rich chemical reactivity embodied in the β-amino-γ-sultone heterocyclic ring, first reported by us in 1988,[30] had scarcely been studied[64] and represented a synthetic challenge. Second, the presence of the *tert*-butyldimethylsilyl (TBDMS) groups at positions 2′ and 5′ of the sugar moiety, being crucial for antiviral activity and highly sensitive to both basic and acidic media, required the selection of smooth reaction conditions compatible with such groups.

3.3.1. Modifications at C3″. To perform modifications at the C3″-position of the β-amino-γ-sultone spiro ring of TSAO compounds, we investigated alkylation reactions of the enamine system.[65] Reaction of TSAO-m³T **38** with electrophiles such as methyl iodide, allyl bromide, or iodoacetamide in the presence of sodium hydride or potassium hydroxide as bases gave complex mixtures of 4″-*N*- and 3″-*C*-alkylated compounds **105–109** together with unreacted starting material and 5′-deprotected compounds (Scheme 11).[65]

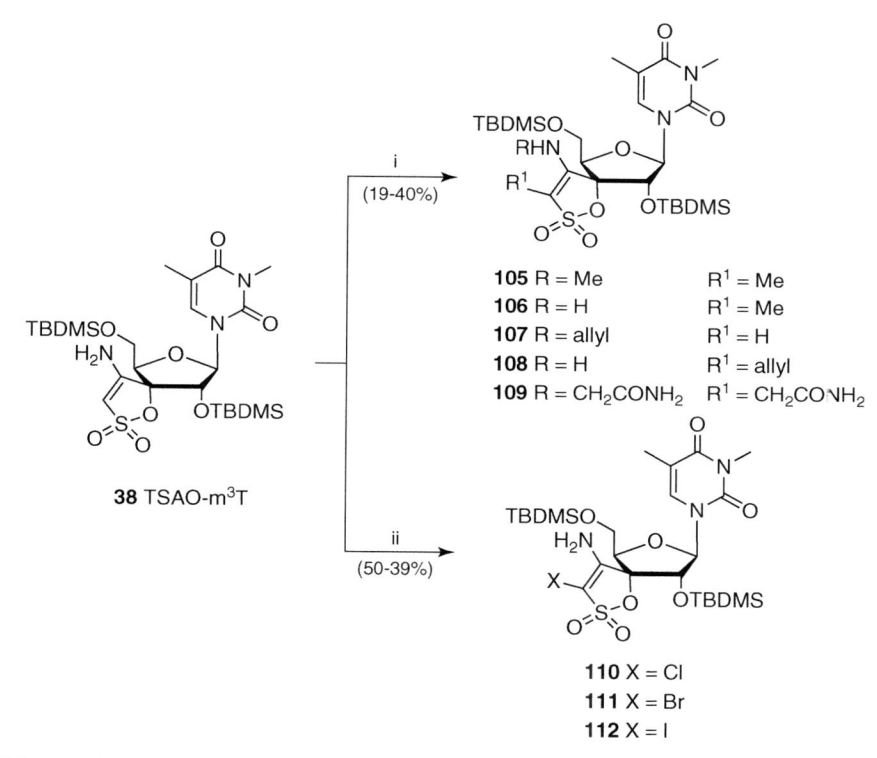

Scheme 11. *Reagents and conditions:* (i) RX, NaH, or KOH, dry THF, rt, 2–24 h; (ii) LiCl, LiBr, or I₂, CAN, Et₃N, 80°C, 1–48 h.

In contrast to the lack of selectivity of the alkylation reaction, halogenation occurs exclusively at C3″ in good-to-excellent yields. Chloro, bromo, and iodo substituents were easily introduced into the 3″-position of the spiro moiety by using the mild ceric ammonium nitrate (CAN)-mediated halogenation method,[66] which avoids the use of acidic and/or high-temperature conditions. Thus, treatment of **38** with lithium halides (LiCl or LiBr) or iodine and CAN in acetonitrile in the presence of triethylamine (to avoid deprotection of the TBDMS group at the 5′-position) afforded the 3″-chloro, bromo, and iodo derivatives **110**, **111**, and **112** in 50%, 80%, and 93% yields, respectively (Scheme 11).

The easy availability of 3″-halo-TSAO derivatives prompted us to use them as key intermediates to develop a general route for the synthesis of 3″-substituted

TSAO compounds via Pd-catalyzed cross-coupling reactions. It should be noted that palladium cross-coupling reactions in this aminosultone system had never been described. Initial attempts of the cross-coupling of the iodo derivative **112** with activated alkenes (i.e., methyl acrylate) under Heck-type coupling conditions[67,68] or with terminal alkynes (i.e., 5-chloro-1-pentyne) using the Sonosgashira modification[69] failed to give the coupled products, and only the corresponding deiodinated derivative **38** and unreacted starting material **112** were isolated. We next assayed coupling of the iodo **112** with organotin reagents under classical Stille coupling conditions (Scheme 12).

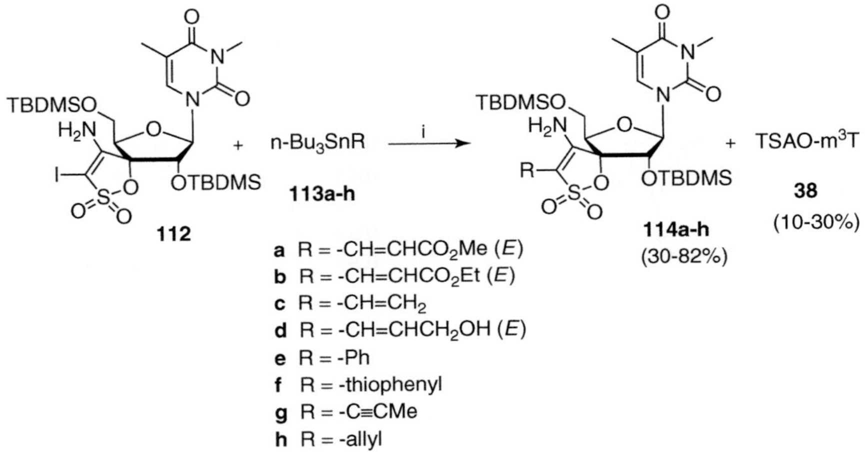

Scheme 12. *Reagents and conditions:* $Pd_2(dba)_3$, $AsPh_3$, CuI, dry NMP, 60°C, 12 h.

Treatment of iodide **112** with stannane **113a** in dimethylformamide (DMF) in the presence of $Pd(Ph_3P)_4$ as catalyst gave the desired product **114a** in low yields (<10% yield) together with the starting compound and TSAO-m³T produced by reduction of the iodo precursor. Our system appeared difficult to couple and required carefully optimized conditions.[70] To improve the yield, an exploratory coupling study of iodo **112** with stannane **113a** in *N*-methylpyrrolidinone (NMP) with different catalyst systems was performed. The yield of the reaction was improved (72% vs. <10%) by the use of the weakly coordinated $Pd_2(dba)_3$ as catalyst, the addition of copper iodide as cocatalyst, triphenylarsine as "soft ligand," and an increased number of equivalents of the stannane compared to the $Pd(Ph_3P)_4$ used initially (Scheme 12).[70]

Using these optimized reaction conditions, we studied the scope and limitations of the cross-coupling reaction of the iodide **112** with a variety of (un)substituted vinyl, alkynyl, phenyl, thienyl, and allyl stannannes **113a–h** (Scheme 12).[65] In all cases we obtained moderate-to-good yields of the coupling products **114a–h** and variable amounts of the reduction product TSAO-m³T (**38**). The ratio of coupling to reduction products depends on the nature of the stannanne. In general, the higher the transfer rate of the different groups from tin, the higher the ratio (coupling/reduction products), ranging from 10:1 in compound **114d** to 1:1 in allyl derivative **114h**. Thus, Stille cross-coupling of iodo TSAO derivative **112** with a broad range of stannanes provides an efficient and straightforward route for the direct and selective functionalization of the 3″-position of the sultone moiety via carbon–carbon bond formation.[65]

The compounds were evaluated for their inhibitory effect on HIV-1 and HIV-2 replication in cell culture. Among the 3″-halogen-substituted TSAO derivatives, the 3″-iodo derivative **112** particularly shows potent anti-HIV-1 activity, which proves comparable to that of the prototype TSAO-T. There is a clear trend toward decreased antiviral potency going from iodine to bromine and from bromine to chlorine. Surprisingly, several 3″-substituted TSAO derivatives (**114a,c** and **114d**) showed antiviral activity not only against HIV-1 but also against HIV-2 at subtoxic concentrations, an observation that is very unusual for NNRTIs and never observed before for TSAO derivatives.[65]

Experimental Procedure

{1-[(2′,5′-Bis-O-tert-butyldimethylsilyl)-β-D-ribofuranosyl]thymine}-3′-spiro-5″-[4″-amino-3″-(E)-(2-methoxyacryloil)-1″,2″-oxathiole-2″,2″-dioxide][65] *(114a, Scheme 12).* A solution of the iodonucleoside **112** (0.1 mmol) in dry NMP (5 mL) was treated with AsPh$_3$ (0.008 mmol), Pd$_2$(dba)$_3$ (0.002 mmol), CuI (0.004 mmol), and after 10 min with methyl (E)-3-(tri-n-butylstannyl)acrylate (0.2 mmol). The resulting solution was stirred under an argon atmosphere at 60°C. After 30 min and 1 h, two additional portions of the stannane (2 × 0.1 mmol) were added and the reaction was continued for 12 h at 60°C. The reaction mixture was allowed to cool to room temperature and water (15 mL) and ethyl acetate (25 mL) were added. The organic phase was separated and washed several times with water (4 × 15 mL) to remove NMP completely, dried (Na$_2$SO$_4$), filtered, and evaporated under reduced pressure. The residue was redissolved in acetonitrile (20 mL) and washed with hot hexane (4 × 15 mL) to remove the tin iodide. The final residue was purified by radial chromatography (CH$_2$Cl$_2$/methanol, 50:1). The fastest-moving fractions gave 0.014 g (20%) of the reduction product TSAO-T. The slowest-moving band gave compound **114a** (0.048 g, 72%) as a white foam. ^{1}H-NMR [(CD$_3$)$_2$CO 300 MHz]: δ 1.90 (d, 3H, CH$_3$-5, J = 1.2 Hz), 3.70 (s, 3H, OCH$_3$), 4.01 (dd, 1H, H-5′a, $J_{4′,5′a}$ = 4.2 Hz, $J_{5′a,5′b}$ = 8.6 Hz), 4.13 (dd, 1H, H-5′b, $J_{4′,5′b}$ = 4.1 Hz), 4.39 (t, 1H, H-4′), 4.78 (d, 1H, H-2′, $J_{1′,2′}$ = 8.1 Hz), 6.03 (d, 1H, H-1′), 6.08 (d, 1H, CH=CHCO$_2$CH$_3$, J_{trans} = 15.6 Hz), 7.49–7.57 (m, 4H, CH=CHCO$_2$CH$_3$, NH$_2$, H-6), 10.40 (b s, 1H, NH-3). ^{13}C-NMR [(CD$_3$)$_2$CO 75 MHz]: δ 12.93 (CH$_3$-5), 54.30 (OCH$_3$), 62.92 (C-5′), 74.80 (C-2′), 83.23, 88.90 (C-1′, C-4′), 92.42 (C-3′), 110.31 (C-3″), 111.28 (C-5), 129.22 (CH=CHCO$_2$CH$_3$), 132.12 (CH=CHCO$_2$CH$_3$), 132.40 (C-6), 151.11 (C-4″), 152.11 (C-2), 160.11 (CO$_2$CH$_3$), 163.33 (C-4).

3.3.2. Modifications at C4″. Four type of modifications of the 4″-amino group of TSAO compounds were addressed: (1) deamination reaction, (2) acid hydrolysis of the enamino-type system, (3) acylation reactions of the amino group with different electrophiles, and (4) substitution of this amino group by other conveniently functionalized amines via transamination reactions.

TSAO-Deaminated Compound. A proposed important interaction of TSAO molecules with the HIV-1 RT, as supported by molecular modeling studies, is a hydrogen bond between the 4''-amino group of the TSAO molecule and the carboxylic acid residue of Glu138 in the β7/β8 loop of the p51 subunit of HIV-1 reverse transcriptase.[35,71] Thus, we considered it of interest to remove the amino group attached to the double bond in the spiro moiety and evaluate the effect of this modification on the activity and resistance profile of the corresponding deaminated derivative.[72] This amino is a peculiar type of amino group. Experimental and theoretical studies indicated that the enamine form is the preferred tautomer both in polar and apolar solvents.[73] However, from the initial attempts of hydrogenolysis of TSAO-m^{3}T with aluminium hydride in refluxing ether (according to described protocols of hydrogenolysis of enamines derived from pyrrolidine),[74] only unreacted starting material was recovered. This result may not be surprising, due to the special primary enamine character of this nitrogen of the molecule.

Although the 4''-amino group of TSAO is not aromatic, ab initio calculations suggested that it has mainly an sp^2 character but with slight nonplanarity (certain sp^3 tetrahedral geometry).[73] We next attempted the synthesis of deaminated TSAO derivative using the anhydrous diazotation and radical deamination procedure developed by Nair and Richardson for aromatic amines.[75] These nonaqueous nonacidic reductive deamination conditions were selected to be compatible with the crucial TBDMS groups. Treatment of TSAO-m^{3}T with 40 equiv of *n*-pentyl nitrite as the nitrosating agent in dry THF as the hydrogen atom–donating solvent[75] under incandescent illumination (160-W bulb) for 10 h at 80°C afforded the desired deaminated TSAO derivative **115** in low yield (15%) together with the 3''-nitro-4''-amino spiro derivative **116** (Scheme 13). Attempts to improve the yield of **115** by using other alkyl nitrites, other hydrogen atom–donating solvents, or other reducing agents were not successful.[72]

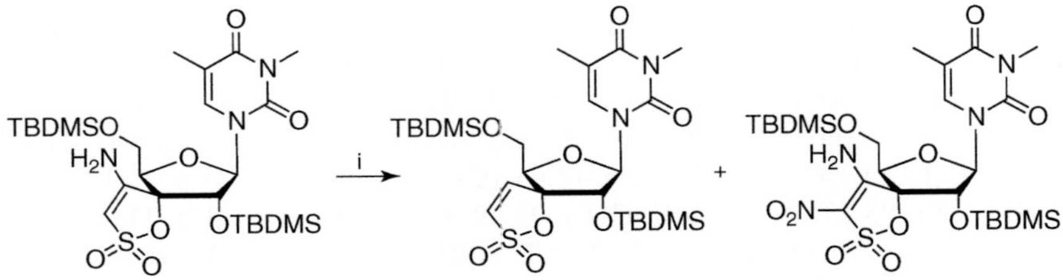

Scheme 13. *Reagents and conditions:* (i) *n*-C$_5$H$_{11}$ONO, dry THF, *hv*, 80°C, 3 h.

Unexpectedly, the deaminated compound **115** showed significant activity against HIV-1 replication. Moreover, when resistant HIV-1 strains were obtained under the pressure of compound **115**, these strains did not show the characteristic TSAO-specific mutation in the RT such as E138K. However, the compound markedly lost its antiviral potential against a variety of virus strains that contain NNRTI-characteristic mutations (including E138K). Therefore, our hypothesis is that TSAO compounds that lack the amino group at the spiro moiety must fit in the HIV-1 RT enzyme in a different way than does the prototype (TSAO-T).[72]

Acid Hydrolysis of the Enamino-Type System. To design TSAO compounds able to interact with the amino group of the Lys-B138 (HIV-1 TSAO-resistant strains), we envisaged direct replacement of the enamino group by a ketone through acid hydrolysis.[76] The suitable TSAO-deprotected nucleoside precursor **28**[25] (Scheme 14) was used as a starting material in view of the known acid lability of the 5'-TBDMS group. Treatment of **28** with 1 N HCl in methanol gave the desired keto derivative **117**, which exhibits a strong tendency toward its 5'-cyclic hemiacetal form, and therefore the isolated compound was identified as **118** (Scheme 14). Attempts at silylation of **118** to give the target compound **119** failed under all the experimental conditions investigated, and only the monosilylated 5'-cyclic hemiacetal **120** was isolated. The tricyclic 5'-cyclic hemiacetal derivatives **118** and **120** were devoid of anti-HIV activity at 100 μM in CEM and MT-4 cell cultures.

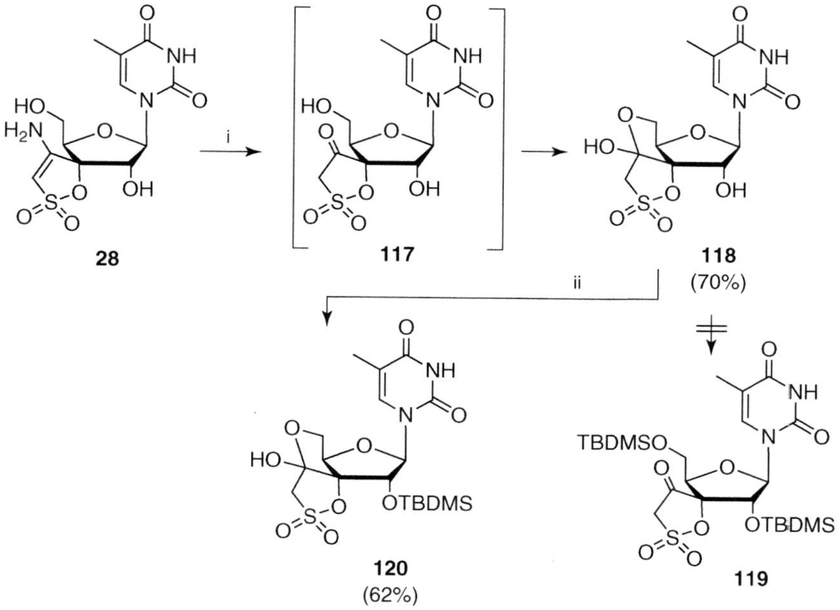

Scheme 14. *Reagents and conditions:* (i) 0.1 N HCl, MeOH, rt, 24 h; (ii) TBDMSCl, DMAP, acetonitrile, rt, 24 h.

Acylation Reactions of the Amino Group with Various Electrophiles. We next focused on the synthesis of TSAO molecules bearing at the 4″-position carbonyl functionalities directed against TSAO-resistant strains.[76] Moreover, the presence of amide or urea groups at this 4″-position may give additional interactions with highly conserved residues of the enzyme close to Glu-B138/Lys-B138 of the p51 subunit, such as the Asn-B136. As mentioned before, ab initio theoretical calculations of the aminosultone system[73] showed that the nitrogen has mainly a sp^2 character, the HOMO energy of the amine being very low (less than -9.73 eV), which corresponds to amines with very low reactivity toward electrophiles.[77] Because of this expected low reactivity, a highly reactive electrophile, methyl oxalyl chloride, was used as an acylating reagent. Thus, treatment of TSAO-m^{3}T **38** with methyloxalyl chloride in the presence of dimethylaminopyridine (DMAP) as an acid scavenger at room temperature afforded compound **121** in 35% yield together with unreacted starting material (Scheme 15). However, when this reaction is carried out in the presence of stoichiometric amounts of aluminum trichloride and 4-Å molecular sieves, the N-acylated compound **121** is obtained in higher yield (66%).

Nucleoside **121** was used as a starting material to synthesize other N-oxalyl-substituted derivatives (Scheme 15). Thus, treatment of **121** with 1 N NaOH yielded the free acid **122a** (72% yield). Reaction of **121** with the appropriate amines afforded the amides **122b–d** in 52%, 50%, and 60% yields, respectively. The methyl ester oxalyl derivative **121** and the amides **122b–d** were endowed with potent anti-HIV-1 activity similar to or even slightly higher than that of prototype TSAO-m^{3}T. To explore the role of each carbonyl group in anti-HIV-1 activity, we prepared novel N-acyl compounds **123e–h** bearing only one carbonyl group. These derivatives were synthesized by reaction of TSAO-m^{3}T **38** with the corresponding acyl chlorides in the presence of an excess of DMAP, in a pressure reaction vessel, in low-to-moderate yields (Scheme 15). (Un)substituted 4″-ureido TSAO derivatives **124i–n** bearing one or two carbonyl groups separated by different spacers at the 4″-position were prepared by reaction of **38** with differently substituted isocyanates in acetonitrile at 80°C, in good yields (Scheme 15). Reaction of ethoxycarbonylmethyl isocyanate with TSAO-m^{3}T **38** in a sealed pressure tube in the presence of a catalytic amount of triethylamine at 80°C gave the N-substituted ureido derivative **124n** in 52% yield together with the unexpected cyclic compound **125** (30%). Formation of this side product can be explained by attack of a second molecule of the isocyanate to the N-ureido derivative formed initially and subsequent cyclization to give the dioxoimidazoline ring.[76] Few reports on this type of cyclization had been described previously.[78]

Ttransamination Reactions. To study the influence of the β-carbonyl group on the anti-HIV-1 activity of N-oxalyl derivatives **122a–d**, we also assayed transamination reactions of TSAO derivatives with appropriate amines.[76] Compounds **128a,b** bearing a carboxylic ester attached to the 4″-amine group of the sultone moiety through one or two methylenes were obtained from TSAO-deprotected derivative **28**, in 20% and 28% overall yield, by the three-step protocol depicted in

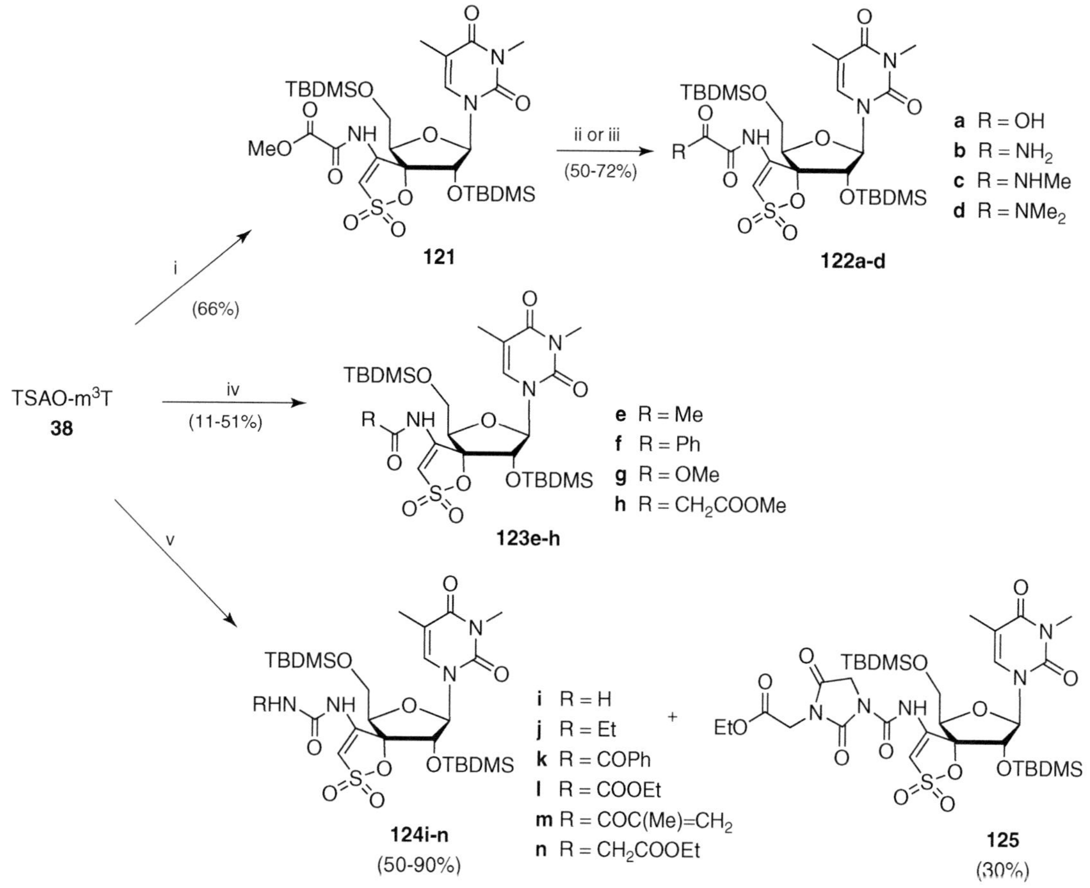

Scheme 15. *Reagents and conditions:* (i) ClCOCOOCH₃, AlCl₃, 1,2-dichloroethane, 4-Å molecular sieves, rt, 24 h; (ii) 1 N NaOH, dioxane, rt, 1 h; (iii) amine, methanol, 45 min to 4 h; (iv) RCOCl, DMAP, 1,2-dichloroethane, 80°C, 0.5–24 h; (v) RNCO, NEt₃, acetonitrile, 80°C, 3–96 h.

122a–d
- **a** R = OH
- **b** R = NH₂
- **c** R = NHMe
- **d** R = NMe₂

123e–h
- **e** R = Me
- **f** R = Ph
- **g** R = OMe
- **h** R = CH₂COOMe

124i–n
- **i** R = H
- **j** R = Et
- **k** R = COPh
- **l** R = COOEt
- **m** R = COC(Me)=CH₂
- **n** R = CH₂COOEt

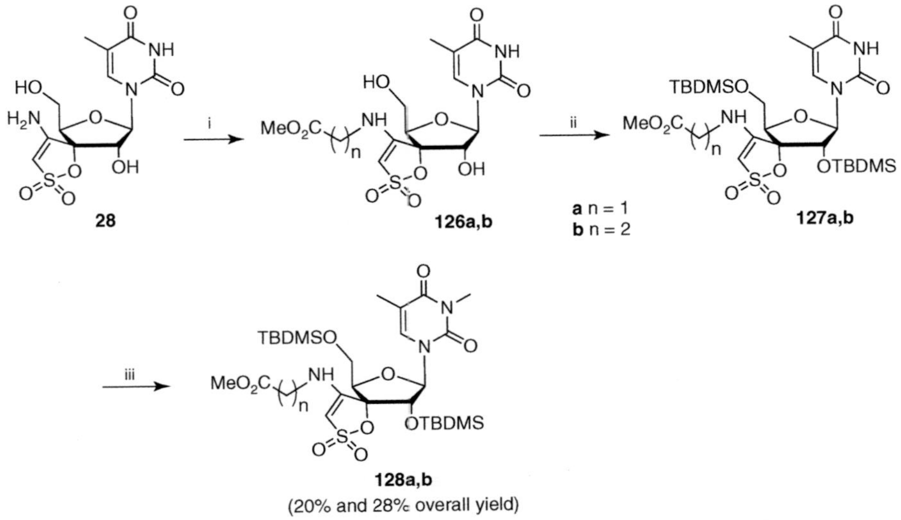

Scheme 16. *Reagents and conditions:* (i) RNH$_3$Cl, methanol, 70°C, 2–10 days; (ii) TBDMSCl, DMAP, dry acetonitrile, 80°C, 5 h; (iii) IMe, K$_2$CO$_3$, acetone, 50°C, overnight.

Scheme 16. This involves transamination reaction of **28** with the corresponding amine in refluxing methanol in a sealed tube, followed by (in situ) silylation of the 2'- and 5'-hydroxyl groups and, finally, N^3-selective methylation of the thymine ring using standard conditions.

Acid derivatives **130a,b** were obtained by transesterification of methyl ester derivatives **128a,b** to the benzyl ester **129a,b** in the presence of a catalytic amount of 1,3-disubstituted tetrabutyldistannoxane[79,80] followed by standard hydrogenation (H$_2$, Pd/C) of the benzyl moiety (Scheme 17).

On the other hand, treatment of **130a,b** with ammonia in the presence of BOP as coupling reagent and triethylamine as a base provided the amides **131a,b** in good yields (Scheme 17).

From the biological results it seems that the presence of two neighboring carbonyl groups at the 4''-position of the spiro moiety is important for anti-HIV-1 activity since monocarbonyl or nonneighboring dicarbonyl TSAO derivatives were devoid of antiviral activity (EC$_{50}$ > 50 μM). Unfortunately, no significant activity against HIV-1 TSAO-resistant viruses was observed for the novel 4''-substituted TSAO derivatives. Surprisingly, several of the 4''-ureido TSAO derivatives (inactive against HIV) gained inhibitory activity against the replication of human cytomegalovirus (HCMV) in cell culture, showing pronounced antiviral activity at concentrations in the low-μM range (i.e., 0.29–2.0 μM), that is, at a concentration well below their toxicity threshold. This had never been observed previously for TSAO derivatives.[76]

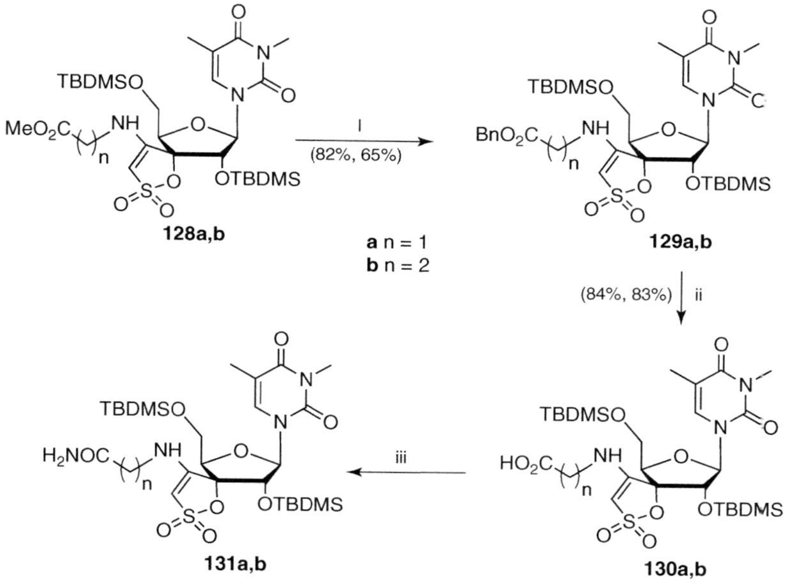

Scheme 17. *Reagents and conditions:* (i) PhCH$_2$OH, ClBu$_2$SnOSnBu$_2$OH, toluene, reflux, 24 h; (ii) H$_2$, 10% Pd/C, 30 psi, EtOAc, 30°C, 1 h; (iii) 2 M NH$_3$/methanol, NEt$_3$, BOP, rt overnight.

3.3.3. Studies of the Unprecedented Lability of the 5'-TBDMS Group on 4''-Acylamino TSAO Derivatives.

To determine whether the potent antiviral activity of the N-methoxalyl TSAO derivative **121** was due to the intact compound and not to the release of TSAO-m^{3}T **38** through hydrolysis of the amide bond at the 4''-position, the stability of **121** was studied under the conditions employed in antiviral assays. Interestingly, when a solution of **121** in DMSO was diluted by addition of 10% fetal calf serum, compound **121** was converted to a more polar compound that was identified as the 5'-deprotected N-methoxalyl TSAO analogue **132** (Scheme 18).[81]

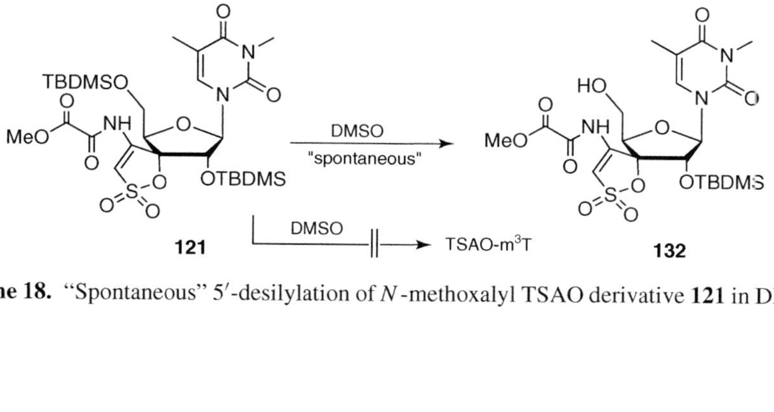

Scheme 18. "Spontaneous" 5'-desilylation of N-methoxalyl TSAO derivative **121** in DMSO.

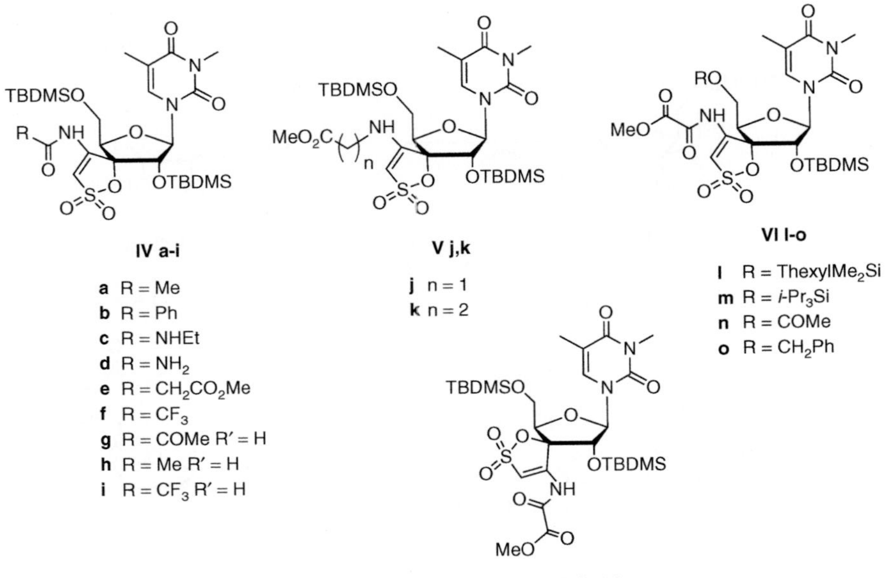

Figure 6. Structures of N-substituted TSAO derivatives **IV**, **V**, and **VI** and xylo derivative **133**.

Intriguingly, desilylation of compound **121** to its 5′-OH **132** was also observed in DMSO-d_6 solution on standing in an NMR tube. This desilylation under neutral conditions was unexpected and had never been observed in the more than 800 TSAO compounds prepared previously. This made us to consider that the desilylation could be related to the presence of the oxalyl moiety at the 4″-position. Therefore, other 4″-N-substituted TSAO derivatives with carbonyl substituents of different nature (compounds of general formulas **IV** and **V**, Figure 6) were synthesized and tested for stability studies.[81] Also, the 5′-TBDMS group of **121** was replaced by other groups (compounds **VI**).

Comparative chemical stability studies of these compounds were carried out in DMSO, and their conversion into the corresponding 5′-O-deprotected derivatives was monitored by HPLC. These studies indicate that the higher the electrophilic character of the 4″-carbonyl substituent is, faster 5′-O-TBDMS deprotection takes place. The hydrolysis also depends on the steric hindrance of the 5′-silyl group, the 5′-O-triisopropyl derivative **VIm** being the most stable silyl analogue. All these results point to direct interaction between the 4″-substituent and the 5′-silyloxy functionalities to explain the desilylation process. The polarity and basicity of the solvent also play a crucial role in the "spontaneous" release of the 5′-TBDMS group. Interestingly, the xylo derivative **133** was perfectly stable under the same conditions and no desilylation was observed. Theoretical studies using DFT (density functional theory) calculations were carried out on selected examples of 4″-acylamino-TSAO derivatives (**IVg–i**). These studies suggested a reasonable correlation between desilylation and geometric parameters such as the distance of the 4″-N–H bond (acidity of the amido group), the distance bewteen the 4″-NH and 5′-O–Si functionalities (degree of interaction between the 4″-amido hydrogen and the nucleophilic 5′-oxygen) and the distance of the 5′-O–Si bond (related to the lability of this bond). All the results prompted us to propose a silyl hydrolysis mechanism involving neighboring group participation of the acidic 4″-amide proton further assisted by certain solvents (Scheme 19). This mechanism involves the interaction between an acidic 4″-amido proton and the nucleophilic oxygen of the 5′-silyloxy functionality, and may result in a longer distance for the 5′-O–Si bond and in a higher lability of this bond toward nucleophilic attack of water assistance by molecules of solvent. Desilylation reaction should proceed faster in polar basic solvents such as DMSO or DMA, which may favor the charged transition state proposed. Moreover, solvent molecules with high dipolar moments and dielectric constant values may enclose the TBDMS moiety, further weakening the energy of the 5′-O–Si bond. Our results complement the scarce previous findings of enhanced hydrolytic cleavage of a TBDMS group by neighboring group participation of acidic hydrogen atoms of a different nature.[82,83]

3.4. Synthesis of Other 3′-Spiro-Substituted Nucleoside Analogues

In parallel, we replaced the spiro aminooxathioledioxide of TSAO-T by other spiro moieties that maintain an NH_2 group at the same position as the 4″-NH_2 group in the prototype compound (TSAO-T, **10**).[84] Figure 7 depicts the general structures of these novel spiro nucleosides bearing a spiro 4-amino-2-oxazolone or a spiro 4-amino-1,2,3-oxathiazole-2,2-dioxide moiety (compounds **VII** and **VIII**) at the 3′-position of the sugar residue.

Our strategy for the synthesis of these new series of 3′-spiro nucleosides is based on the functionalization and ring closure of the corresponding cyanohydrins (xylo and ribo) obtained from 3′-ketonucleosides. This strategy allowed us to obtain such highly functionalized nucleosides in two or three steps starting from a common precursor. Thus, reaction of 3′-ketonucleoside **16**[30] with sodium cyanide in a two-phase ethyl ether–water system in the presence of sodium bicarbonate provides a mixture of the two epimeric nucleoside 3′-cyanohydrins **17** and **18** that were used in the next step without further purification (Scheme 20). Treatment of the cyanohydrin mixture with chlorosulfonylisocyanate followed by basic treatment with saturated aqueous sodium hydrogen carbonate solution gave a 4.8:1 mixture of the xylo- and ribospiro nucleosides **134** and **135** in 43% and 9% overall yield, respectively.[84] On the other hand, treatment of the cyanohydrin mixtures **17** and **18** with sulfamoyl chloride in the presence of DMAP afforded the 3′-spiro derivatives **136** (40%) and **137** (8%) (Scheme 20).

The novel spiro nucleoside derivatives **134** and **136** in the xylo configuration were devoid of anti-HIV-1 activity, while the anti-HIV-1 activity of the corresponding ribospiro derivatives **135** and **137** was at least two orders of magnitude lower (EC_{50}: 5.3–13.3 μM) than that of the parent TSAO-m^{3}T derivative **38** (EC_{50}: 0.04–0.06 μM).[84] These data stress how subtle structural differences at the 3′-spiro

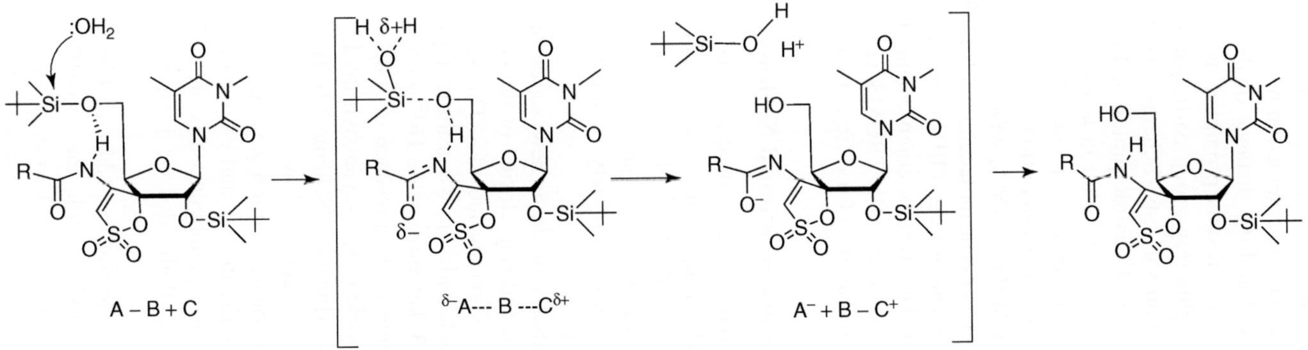

Scheme 19. Proposed mechanism of 5′-desilylation of 4″-acyl amino TSAO derivatives.

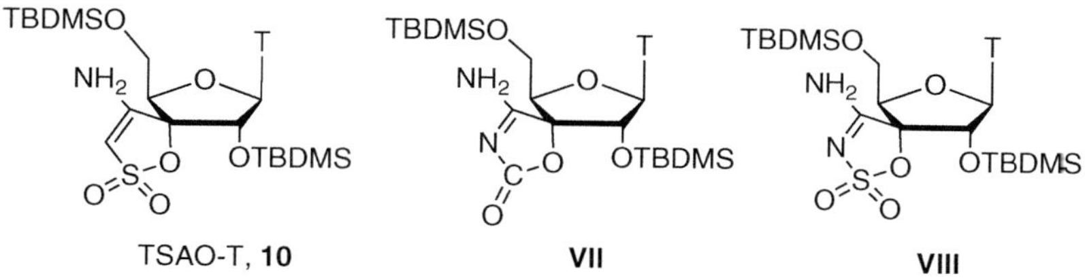

Figure 7. General structure of spiro nucleoside derivatives **VII** and **VIII**.

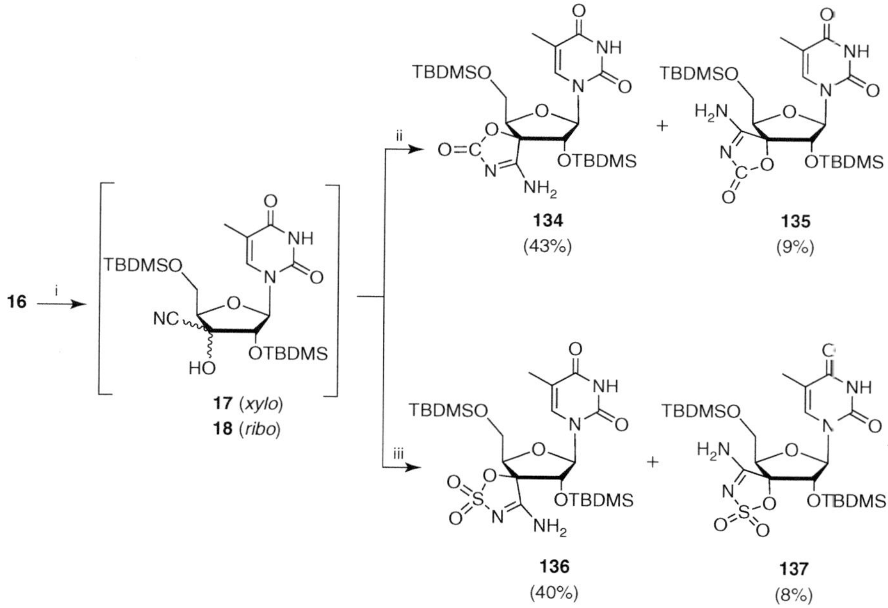

Scheme 20. *Reagents and conditions:* (i) NaCN, NaHCO$_3$, ethyl ether/water, rt, 16 h; (ii) chlorosulfonylisocyanate, NaHCO$_3$, dry CH$_2$Cl$_2$, rt, 5 h; (iii) sulfamoyl chloride, DMAP/pyridine, dry dioxane, rt, 5 days.

moiety of TSAO compounds may have a major impact on the eventual antiviral activity. A comparative study based on NMR in solution and on theoretical calculations[85] of the hydrophobicity, the solvation free energies, and the molecular electrostatic potentials (MEP) of TSAO-T, **135**, and **137** reveal differences in the MEPs calculated for the spiro systems. This suggests different electrostatic surroundings of the 4″-amino group in the novel TSAO analogues with respect to that of the prototype that may be responsible for a detrimental electrostatic interaction of the spiro rings with the Glu-B138 of the HIV-1 RT.

Following a stereoselective route similar to that described by us for the synthesis of TSAO-T, Postel and co-workers[86,87] described the synthesis of aza analogues of TSAO-T bearing a the dihydroisothiazole dioxide ring instead of the oxathiole

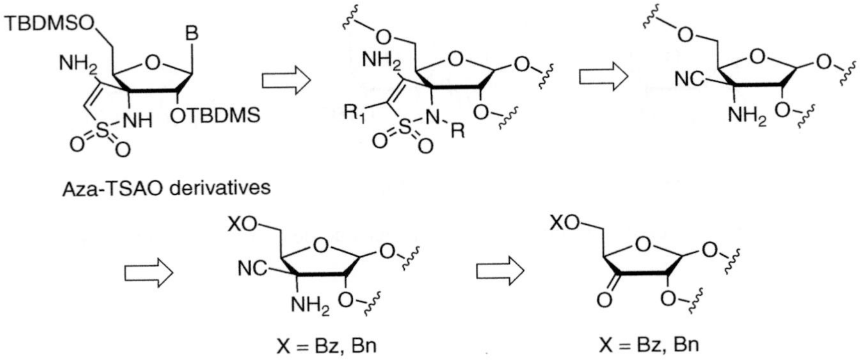

Figure 8. Key intermediates for the synthesis of aza-TSAO derivatives.

dioxide ring at the C3′-position. The synthesis of these analogues involved glyco-α-aminonitrile formation from the conveniently protected 3-keto derivative **138**, N-glycosylation with nucleic bases, and a cyclization step in the presence of Cs_2CO_3. Figure 8 shows key intermediates for the synthesis of aza-TSAO compounds.

Aza-TSAO derivatives with an unsubstituted isothiazolic ring prove less inhibitory against HIV-1 (EC_{50}: 0.22–0.43 μM) than TSAO-T (EC_{50}: 0.04–0.06 μM). Surprisingly, N^3-alkylated aza-TSAO analogues were 50-fold more cytotoxic (CC_{50}: 4.9–5.8 μM) than the corresponding TSAO-m^{3}T (CC_{50}: 240 μM).[86]

4. SYNTHESIS OF UNUSUAL POLYCYCLIC-FUSED NUCLEOSIDES FROM TSAO-SPIRO NUCLEOSIDES

In this section we describe how the "peculiar" reactivity of TSAO molecules allowed us to obtain hypermodified nucleosides [bi- and polycyclic(spiro)fused nucleosides]. Following with modifications at the 5′-position of TSAO molecules, and in an attempt to obtain 5′-amino-substituted spiro nucleosides (i.e., **140**), when 5′-O-tosyl TSAO-m^{3}T **139**[88,89] was treated with an excess of methylamine (as nucleophile), the hitherto unknown bicyclic-fused nucleosides **141** (18%) and **142** (18%) (Scheme 21), together with the 4′,5′-didehydro nucleoside **104**[90] (25%), were obtained. Formation of the expected 5′-methylamino-5′-deoxy TSAO derivative **140** was not detected.

These unexpected results, and the formation of the new bicyclic-fused nucleosides of unknown structures, prompted us to study this reaction with other primary and secondary amines. Thus, reaction of **139** with N,N-dimethylethylenediamine (Scheme 21) afforded the bicyclic nucleosides **143** (27%) and **144** (10%) together with the 4′,5′-didehydronucleoside **104** (26%).[90]

On the other hand, reaction of **139** with an excess of ethylenediamine gave the bicyclic nucleoside **145** (20%) and the tricyclic nucleoside **146** (28%) together with **104**[90] (35%). It should be pointed out that, again in the reactions so far described,

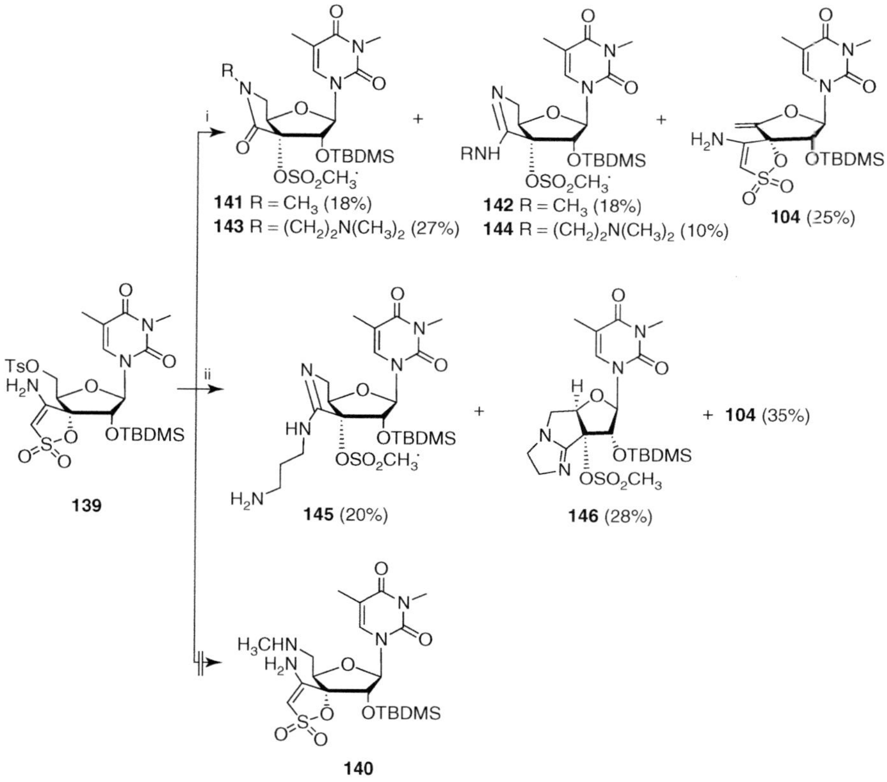

Scheme 21. *Reagents and conditions:* (i) excess of amine (MeNH$_2$ or Me$_2$NCH$_2$CH$_2$NH$_2$), acetonitrile, 70°C in a sealed tube, 20 h; (ii) NH$_2$CH$_2$CH$_2$NH$_2$, CH$_3$CN, 70°C, 20 h.

the compounds expected to result from the nucleophilic substitution of the 5'-tosyl group by the amines were not detected. However, when **139** was reacted with secondary amines such as *N,N*-dimethylamine, pyrrolidine, piperidine, or azetidine (Scheme 22), the spiro nucleosides resulting from the nucleophilic substitution of the 5'-tosyl by the corresponding amines [**147** (23%), **149** (34%), **151** (26%) and **153** (12%)] were obtained together with the bicyclic nucleosides [**148** (8%), **150** (34%), **152** (40%), and **154** (12%)]. In all these reactions the elimination 4',5'-didehydro derivative **104**[90] was also isolated.

We hypothesized that formation of various bicyclic-fused nucleosides in the reaction of **139** with alkylamines may occur through a key common precursor (compound **155**).[91] Compound **139** is the result of intramolecular attack of the amino group on the 5'-tosyl leaving group. Once formed, this intermediate experiments a Michael-type addition of the nucleophile (the corresponding alkylamine) to the conjugated double bond of the α,β-unsaturated cyclic sulfonate ester (C4″ carbon atom) and subsequent ring opening to give the novel bicyclic-fused nucleosides (Scheme 23). To demonstrate this hypothesis, **155**, obtained in 70% yield form

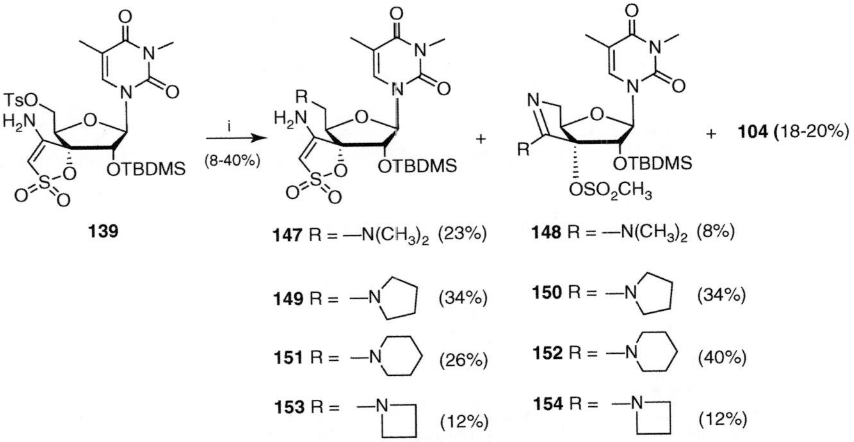

Scheme 22. *Reagents and conditions:* (i) excess of amine (Me$_2$NH, pyrrolidine, piperidine, or azetidine), CH$_3$CN, 70°C in a sealed tube, 20 h.

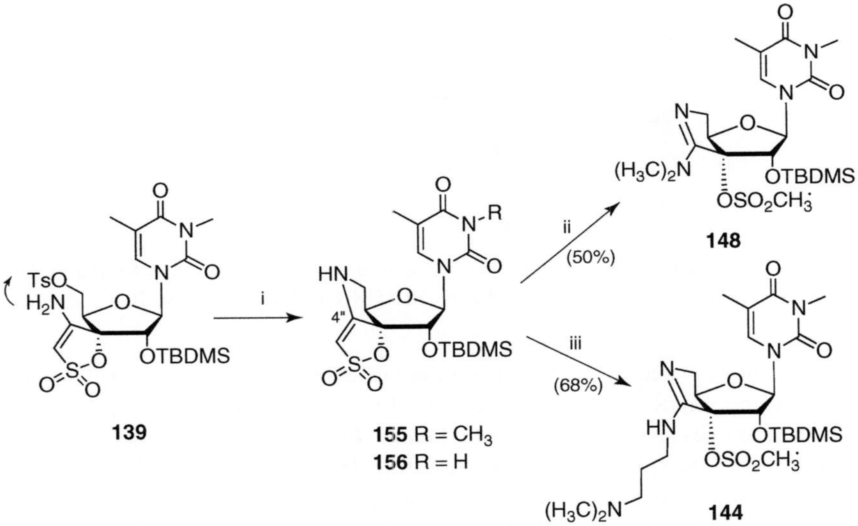

Scheme 23. *Reagents and conditions:* (i) K$_2$CO$_3$, CH$_3$CN, 80°C, 6 h; (ii) Me$_2$NH, CH$_3$CN, 70°C, sealed tube, 20 h; (iii) Me$_2$NCH$_2$CH$_2$NH$_2$, acetonitrile, 70°C in a sealed tube, 20 h.

139 under basic nonnucleophilic conditions (K$_2$CO$_3$), was reacted with an excess of dimethylamine or with N,N-dimethylethylenediamine (Scheme 23), leading to the bicyclic-fused nucleosides **148** (50%) and **144** (68%). These results not only supported the mechanism proposed but also prompted us to study the potential of such a key precursor **155** for the synthesis of different classes of policyclic-fused nucleosides.[91] It should be pointed out that under these conditions the yields of the bicyclic-fused nuecleosides improved considerably.

4.1. Reaction of Compound 156 with Nucleophiles

Next, the reactivity of the spirocyclic enaminosultone **156** toward a variety of oxygen-, sulfur-, carbon-, and nitrogen-based nucleophiles was investigated.[91]

4.1.1. Reaction of Compound 156 with Oxygen, Sulfur, and Carbon Nucleophiles.

First, reactions of **156** with (thio)alcohols were studied. Thus, treatment of **156** with ethanol gave the tricyclic nucleoside **157a** (70%) resulting from the nucleophilic attack of the alcohol to the C4″ carbon atom (Scheme 24). Similarly, reaction of **156** with different primary alcohols, such as propanol, pentanol, isobutanol, neopentanol, or ethanethiol, gave the corresponding tricyclic-fused spiro nucleosides **157b–f** in moderate to high yields (60–83%) (Scheme 24). These fused nucleosides are the result of the attack of the nucleophile at the C4″ carbon atom of the spiroaminooxathiole dioxide moiety and subsequent proton transfer in the intermediate **IX**.[91] In these reactions, opening of the spirosultone ring did not take place, and only tricyclic-fused nucleosides were obtained.

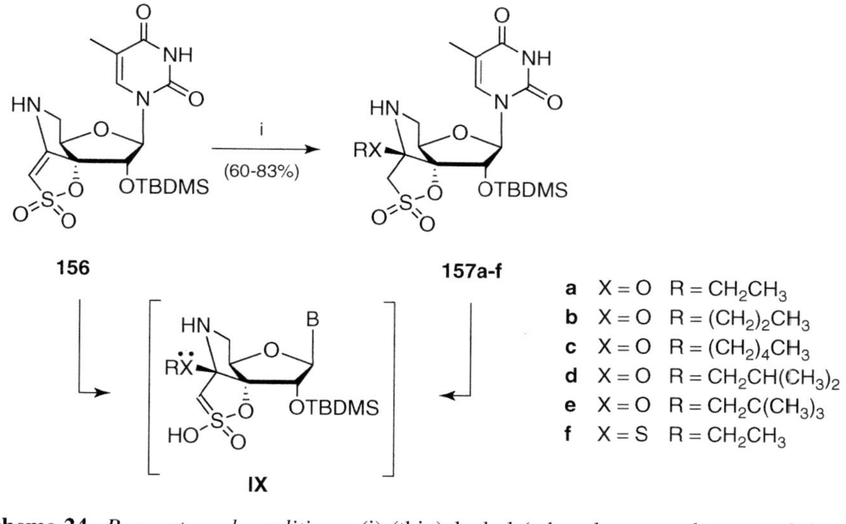

a X = O R = CH$_2$CH$_3$
b X = O R = (CH$_2$)$_2$CH$_3$
c X = O R = (CH$_2$)$_4$CH$_3$
d X = O R = CH$_2$CH(CH$_3$)$_2$
e X = O R = CH$_2$C(CH$_3$)$_3$
f X = S R = CH$_2$CH$_3$

Scheme 24. *Reagents and conditions:* (i) (thio)alcohol (ethanol, propanol, pentanol, isobutanol, neopentanol, or ethanethiol), CH$_3$CN, 80°C, 16–24 h.

On the other hand, when **156** was reacted with the more nucleophilic sodium ethoxide (Scheme 25), the bicyclic-fused nucleoside **158** was obtained in 72% yield. In this case the sodium ethoxide attacks at the sulfur atom of the SO$_2$ group instead of at the C4″ carbon atom.[91] This behavior had previously been observed in the reaction of sodium methoxide with other spiro derivatives of carbohydrates.[92]

Next, the reaction with a carbon-based nucleophile (NaCN) was investigated. Thus, reaction of **156** with sodium cyanide gave the tricyclic-fused nucleoside **159** (Scheme 26).[91]

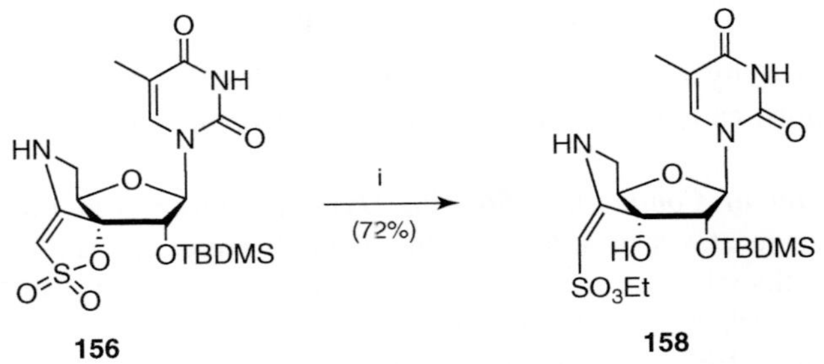

Scheme 25. *Reagents and conditions:* (i) NaOEt, CH$_3$CN, rt, 5 min, then (5%) HCl in MeOH (pH 7).

Hydrolysis of the cyano group, under phase-transfer conditions,[93] by treatment of **159** with aqueous sodium hydroxide and hydrogen peroxide, gave the carboxamide **160** (Scheme 26).[93] Alcoholysis of the cyano group of **159** by reaction with HCl in methanol gave the methoxycarbonyl derivative **161** in 60% yield (Scheme 26). Finally, catalytic hydrogenation (10% Pd/C, 40°C) of **159** gave the tetracyclic nucleoside **164** instead of the expected amino compound **162** (Scheme 27). Formation of **164** could be explained by the participation of the imine intermediate **163** formed by reductive alkylation[94] of **162** by the methanol used as a solvent in the reaction. An intramolecular attack of the NH of the pyrrolidine ring to the carbon atom of the imino moiety of **163** would give compound **164**.

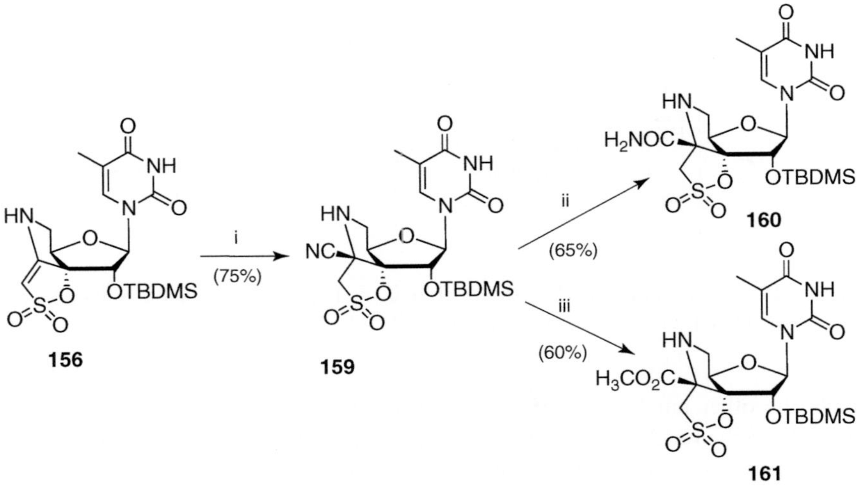

Scheme 26. *Reagents and conditions:* (i) NaCN, NaHCO$_3$, ethyl acetate/water, rt, 10 h; (ii) 30% H$_2$O$_2$, (Bu)$_4$N(HSO$_4$), 0.5 M NaOH, CH$_2$Cl$_2$, rt, 2 h; (iii) saturated HCl/MeOH, rt, 5 h.

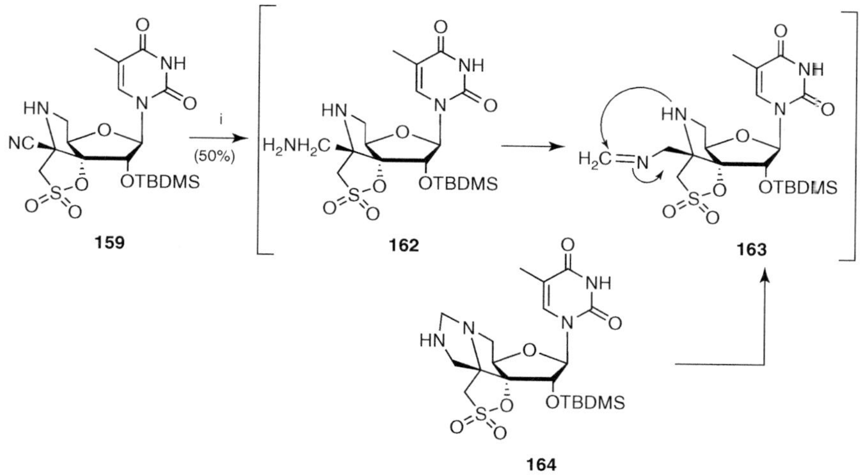
i
(50%)
159
162
163
164
OTBDMS

Scheme 27. *Reagents and conditions:* (i) H_2, 10% Pd/C, 40°C, 6 h.

4.1.2. Reaction of Compound 156 with Amines and Amino Acids. The reaction of **156** with primary and secondary amines and with amino acids was studied next. When **156** was reacted with an excess of propylamine, isopropylamine, or dimethylamine (Scheme 28), the bicyclic nucleosides **165a–c** were obtained, in good yields (63–69%), by the attack of the amine on the conjugated double bond of the α,β-unsaturated cyclic sulfonate ester followed by ring opening.[88]

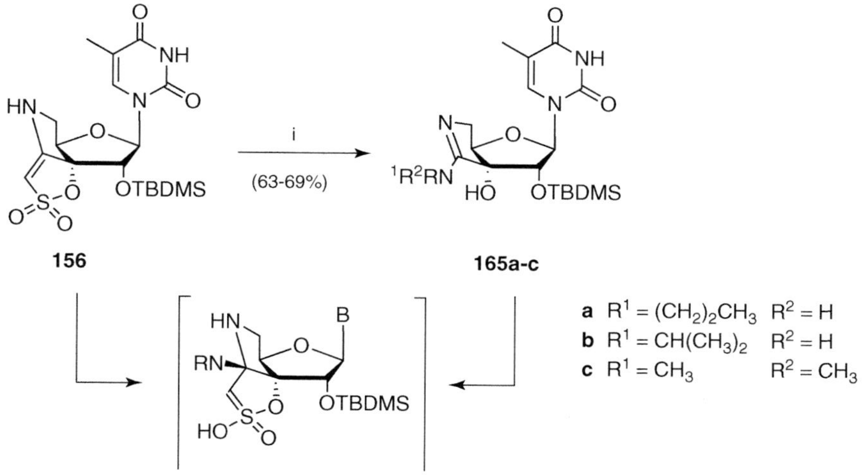
i
(63-69%)
156
165a-c
a R1 = (CH2)2CH3 R2 = H
b R1 = CH(CH3)2 R2 = H
c R1 = CH3 R2 = CH3

Scheme 28. *Reagents and conditions:* (i) amine ($PrNH_2$, $i PrNH_2$, or Me_2NH), acetonitrile, 70°C in a sealed tube, 20 h.

Compound **156** was also reacted with the benzyl or methyl ester derivatives of α-amino acids (L-valine, L-alanine, and L-glycine). The reactions were carried out in the presence of triethylamine (TEA), to give the E-vinylsulfonamide bicyclic-fused nucleosides **166a–c** in good yields (60–63%) (Scheme 29). In contrast to the results observed in the reaction of **156** with amines, in this case **166a–c** are the result of the attack of the amino group of the α-amino acid to the sulfur atom of the SO_2. However, when **156** reacted with β-L-alanine (a β-amino acid) afforded, exclusively, compound **167**, resulting from the attack of the less hindered amino group of the amino acid to the C4″ position (Scheme 29).

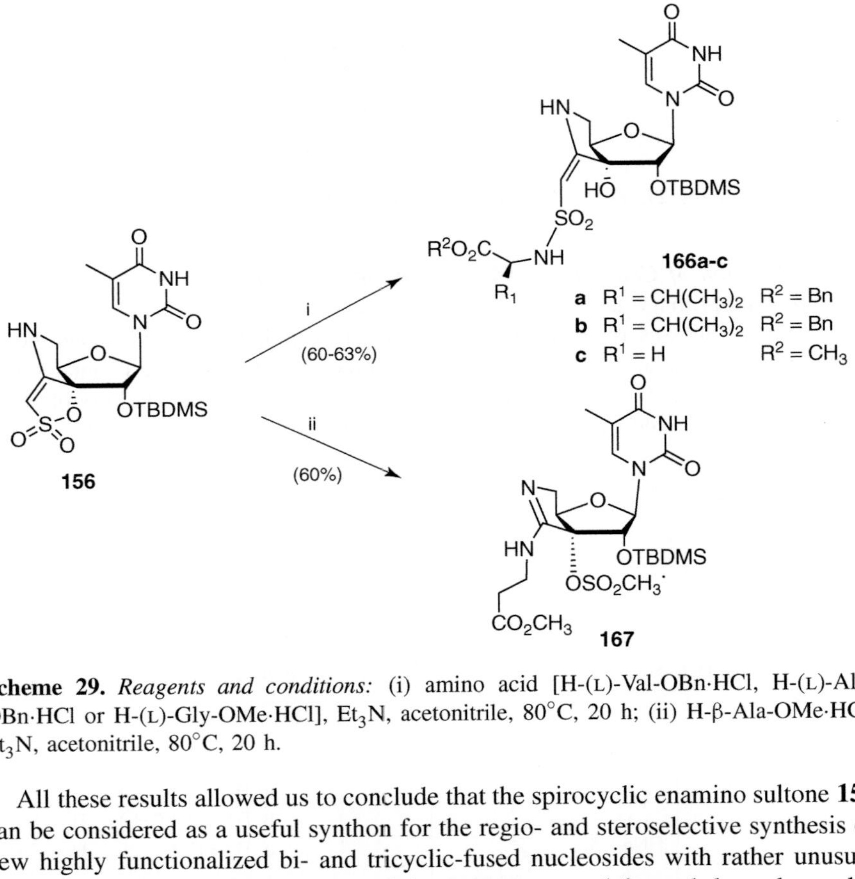

Scheme 29. *Reagents and conditions:* (i) amino acid [H-(L)-Val-OBn·HCl, H-(L)-Ala-OBn·HCl or H-(L)-Gly-OMe·HCl], Et$_3$N, acetonitrile, 80°C, 20 h; (ii) H-β-Ala-OMe·HCl, Et$_3$N, acetonitrile, 80°C, 20 h.

All these results allowed us to conclude that the spirocyclic enamino sultone **156** can be considered as a useful synthon for the regio- and steroselective synthesis of new highly functionalized bi- and tricyclic-fused nucleosides with rather unusual molecular skeletons. The structure of the new compound formed depends, exclusively, on the nature of the nucleophile present in the reaction. These compounds are very stable and are obtained efficiently, in high yields, by a one-step reaction of the cyclic enamine **156** with different nucleophiles.

4.1.3. Reaction of Compound 156 with Water. A special case studied was the reaction of **156** with water. Thus, treatment of **156** with a mixture of water/acetonitrile (1:1) in the presence of potassium carbonate afforded the bicyclic nucleoside **169** (68% yield), in which a γ-lactam ring was fused to the ribose moiety (Scheme 30).[91,95] Formation of **169** was explained by a Michael-type addition of the nucleophile (H$_2$O) to the conjugated double bond of the α,β-unsaturated cyclic sulfonate ester in **156** (position 4″) to give **169** through the ring opening of intermediate **168**.[91] On the other hand, when the hydrolysis of enamine sultone **156** (1:1, water/acetonitrile) was carried out under acidic conditions (acetic acid, pH 5–6), compound **168** was obtained in 93% yield (Scheme 30). Compound **168** rendered **169** under basic conditions (K$_2$CO$_3$, pH > 7). This experimental result suggests that compound **168** is involved in the formation of **169**.[91,95]

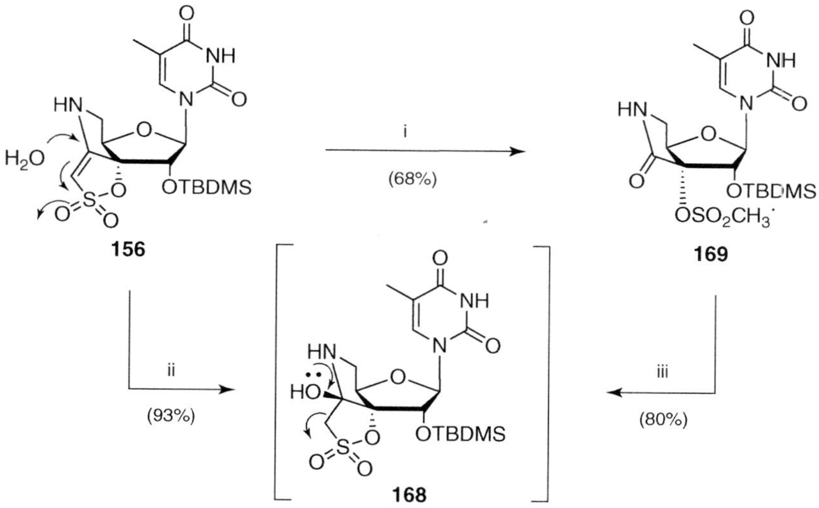

Scheme 30. *Reagents and conditions:* (i) water/acetonitrile (1:1), K$_2$CO$_3$ (pH > 7), 80°C, 10 h; (ii) water/acetonitrile (1:1), AcOH (pH 5–6), 80°C, 8 h; (iii) acetonitrile, K$_2$CO$_3$ (pH > 7), 80°C, 10 h.

Unexpectedly, **168** was transformed spontaneously into two novel tetracyclic-fused nucleosides **171** and **173** in an acetone-d_6 solution in a NMR tube. The same compounds were obtained in 35% and 30% yield, respectively, when **168** was treated with a large excess of acetone (>20 equiv). Moreover, when acetic acid (pH adjusted to 5–6) was added to the reaction of **168** with acetone, the yield of **171** increased from 35% to 80% and only traces of compound **172** were detected (Scheme 31).[95]

Formation of nucleosides **171** and **172** indicated that **168** underwent an unusual reaction with acetone in which two new carbon–carbon bonds were formed to give in a one-pot way a novel six-membered ring in which the nitrogen and carbon adjacent to the OSO$_2$ of the starting nucleoside **168** are ring atoms (Scheme 31).

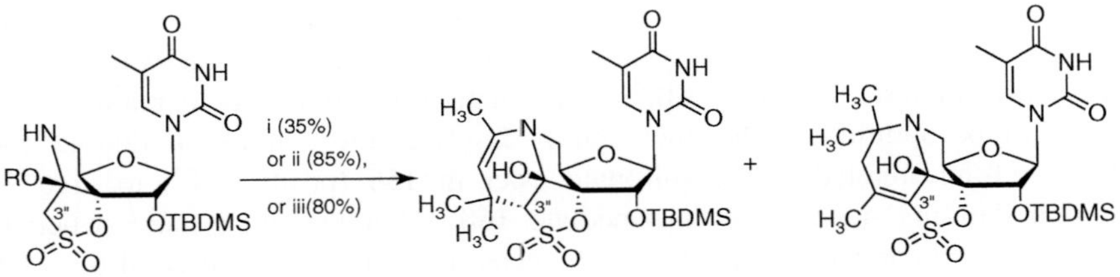

Scheme 31. *Reagents and conditions:* (i) acetone, rt, 24 h; (ii) acetone, AcOH up to pH 5–6, rt, 24 h; (iii) acetone, Et$_3$N, rt, 4 h.

A combination of NMR and HPLC–ESI–MS techniques revealed that the formation of **171** and **172** proceeded through an enamine–iminium mechanism[95] which was supported by the presence of an iminium carbon signal at δ 191.2 ppm in their ^{13}C-NMR spectra. It was postulated[95] that the reaction is initiated by the nucleophilic attack of the NH of the pyrrolidine ring of **168** to the carbonyl of the acetone. Then several proton transfers take place mediated by the hydroxyl group at the α-position of the pyrrolidine ring of **168**. This suggests that this OH is essential for the progress of the reaction. To check this hypothesis, the methoxy derivative **170** (Scheme 31) was reacted with acetone, and in that case no reaction was observed, and even traces of other compounds were not detected. This result proves the crucial role of the OH at the α-position of the pyrrolidine ring for the progress of the reaction. Finally, it should be pointed out that the presence of either acetic acid or triethylamine promotes the enamine pathway (formation of **171**) and improves the reaction rate.

In view of these results, the reaction of **168** with other aliphatic ketones (2-butanone, 3-pentanone), as well as with the α,β-unsaturated ketone (methyl vinyl ketone) was explored to study the scope of the reaction (Schemes 32 and 33). The reaction of **168** with 2-butanone required the addition of triethylamine to give tetra-cyclic-fused spironucleosides **173** (14%) and **174** (16%) together with bicyclic-fused nucleoside **169** (16%) (Scheme 32), whereas reaction of **168** with 3-pentanone required heating at 80°C overnight to yield tetracyclic-fused nucleoside **175** in 30% yield.[95]

Finally, treatment of **168** with 3-buten-2-one afforded the hydroxy tricyclic nucleoside **177** (59%) (Scheme 33). In this case, the 1,4 attack is favored over the 1,2 addition to give the enol **176**, which tautomerizes to the keto derivative **177**. In polar solvents, **177** underwent a spontaneous ring opening of the spiro sultone to give substituted α-lactam **178**.

These results suggested that the α-hydroxypyrrolidine tricyclic nucleoside **168** is a useful synthetic intermediate for the efficient creation of molecular complexity. Different types of ketones can be attacked by the NH of the pyrrolidine ring present in **168**, giving rise to different fused cyclic nucleoside derivatives with rather unusual molecular skeletons; however, this attack and the subsequent cyclization proceeds spontaneously only with acetone.

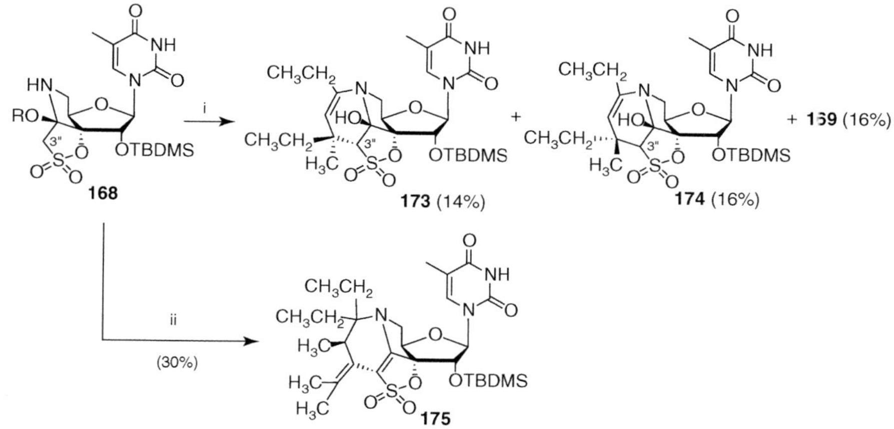

Scheme 32. *Reagents and conditions:* (i) 2-butanone, Et$_3$N, rt, 24 h; (ii) 3-pentanone, Et$_3$N, rt, 24 h, then 80°C, 24 h.

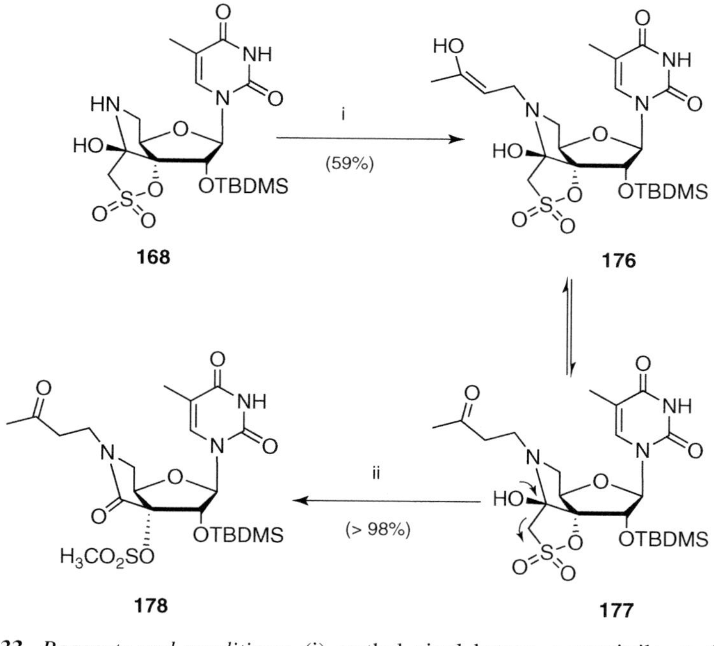

Scheme 33. *Reagents and conditions:* (i) methyl vinyl ketone, acetonitrile, rt, 24 h; (ii) MeOH, rt, 15 min.

Polycyclic-fused nucleosides were tested against a broad panel of RNA and DNA viruses; only the tricyclic nucleoside **157a** was found to specifically inhibit the HIV-1 replication.[96] Structure–activity relationship studies within this family of compounds (the prototype being **157a**) indicated that the length of the alkoxy moiety at C4″ is important for the activity and toxicity. This results in a six- to sevenfold increase in selectivity with respect to the prototype **157a**.[96] Drug resistance studies revealed the presence of characteristic nonnucleoside reverse transcriptase mutations at amino acid positions 101, 102, 108, 138, and 181. This finding, together with the lack of anti-HIV-2 activity, strongly suggested that these compounds behave as nonnucleoside reverse transcriptase inhibitors (NNRTIs) and probably bind in the NNRTI-binding pocket. It therefore was concluded that these tricyclic nucleosides represent a novel type of selective anti HIV-1 inhibitors, targeted at the HIV-1-encoded reverse transcriptase.

Experimental Procedures

5',N$^{4''}$-Cyclo-{1-[2'-O-(tert-butyldimethylsilyl)-5'-deoxy-β-D-ribofuranosyl] thymine}-3'-spiro-5''-(4''-amino-1'',2''-oxathiole-2'',2''-dioxide)^{91} (156, Scheme 23). To a solution of 5'-O-tosyl-TSAO-T **139** (0.30 g, 0.50 mmol) in dry acetonitrile (5 mL) was added potassium carbonate (0.08 g, 0.58 mmol). The solution was refluxed for 6 h and evaporated to dryness. The residue was dissolved in ethyl acetate (20 mL) and washed with water (2 × 20 mL). The organic layer was dried (Na$_2$SO$_4$), filtered, and evaporated to dryness. The residue was purified by CCTLC on a chromatotron (hexane/ethyl acetate, 1:2) to give 0.07 g (70%) of **156** as a white foam: ^{1}H-NMR [300 MHz, (CD$_3$)$_2$CO]: δ 1.87 (s, 3H), 3.92 (dd, 1H, $J_{5'a,5'b}$ = 10.8 Hz), 4.06 (m, 1H), 4.94 (dd, 1H, $J_{4',5'a}$ = 5.7 Hz, $J_{4',5'a}$ = 8.9 Hz), 5.14 (d, 1H), 5.64 (s, 1H), 5.98 (d, 1H, $J_{1',2'}$ = 8.9 Hz), 6.65 (b s, 1H), 7.69 (s, 1H), 10.51 (b s, 1H). ^{13}C-NMR [75 MHz, (CD$_3$)$_2$CO]: δ 12.1, 54.4, 74.1, 80.9, 88.6, 93.7, 111.9, 138.2, 157.2, 163.8, 168.8. MS (ES+) m/z 458.1 [M + H]$^+$.

General Procedure for the Reaction of Compound 156 with (Thio)alcohols (Scheme 24).[91] To a solution of **156** (0.2 g, 0.46 mmol) in dry acetonitrile (10 mL) was added the corresponding (thio)alcohol (2.15 mmol). Reaction was heated at 80°C for 18–24 h. After evaporation of the solvent, the residue was purified by CCTLC on the chromatotron (hexane/ethyl acetate, 1:2). Reaction with ethanol was chosen as a model.

N$^{4''}$-Cyclo-{1-[2'-O-(tert-butyldimethylsilyl)-5'-deoxy-β-D-ribofuranosyl] thymine}-3'-spiro-5''-(4''-amino-4''-S-ethoxy-1'',2''-oxathiolan-2'',2''-dioxide)^{91} (157a, Scheme 24). A solution of **156** (0.2 g, 0.46 mmol) in ethanol (6 mL) was heated in a sealed tube at 80°C for 16 h. After evaporation of the solvent, the residue was purified by CCTLC on the chromatotron (hexane/ethyl acetate, 1:2) to give 0.15 g (70%) of **157a** as a white foam. ^{1}H-NMR [300 MHz, (CD$_3$)$_2$CO]: δ 1.26 (t, 3H, J = 7.1 Hz), 1.85 (s, 3H), 3.18 (m, 1H), 3.44 (m, 1H), 3.57 (m, 1H), 3.73 (d, 1H, J = 13.9 Hz), 3.83 (t, 1H), 3.90 (m, 1H), 3.98 (d, 1H), 4.62 (dd, 1H, $J_{4',5'_a}$ = 4.3 Hz, $J_{4',5'_b}$ = 6.1 Hz), 4.94 (d, 1H), 6.02 (d, 1H, $J_{1',2'}$ = 6.8 Hz), 7.48 (s, 1H), 10.15 (b s, 1H). ^{13}C-NMR [75 MHz, (CD$_3$)$_2$CO]: δ 12.4, 15.6, 49.4, 54.7, 60.9, 72.1, 85.4, 92.2, 97.7, 99.5, 111.8, 137.5, 152.3, 166.2. MS (ES+) m/z 526.2 [M + Na]$^+$.

5. CONCLUDING REMARKS

Since Fernández-Bolaños et al.[11] first described the term *spironucleoside* in the literature in 1990 to designate a class of sugar derivative with a spiranic structure in which the anomeric carbon belonged simultaneously to the sugar ring and to a heterocyclic base, and the discovery of this class of structures in natural products as hydantocidin and related compounds (with herbicidal properties), many synthetic efforts have been pursued to get spironucleosides not only at anomeric but also at other positions (2′, 3′, or 4′) of the sugar moiety. The results of such efforts rendered compounds with different biological activities, and also locked nucleosides (conformationally constrained at the sugar moiety).

A special group of spironucleosides are the TSAO family of anti-HIV-1-specific RT inhibitors (being the prototype compound the thymine derivative named TSAO-T). The challenging chemistry carried out within these compounds afforded not only more active and less cytotoxic anti AIDS compounds but also new families of anti-HCMV inhibitors (4″-ureido derivatives), or a new family of anti-HIV-specific inhibitors (tricyclic-fused spironucleosides).

From a chemical point of view it has been possible to functionalize the 3″ and 4″-positions of the 3′-spiro-β-amino-γ-sultone ring (the "unique" structural feature of TSAO nucleosides). The chemistry has been very challenging, and special attention has been paid to develop smooth chemical procedures compatible with the TBDMS groups present at the 2′- and at 5′-positions of TSAO molecules, which are essential for the activity of these compounds. It has also been possible to obtain a plethora of hypermodified nucleosides [bi- and polycyclic(spiro)fused nucleosides] with rather unusual molecular skeletons by a one-step reaction of a cyclic enamine common precursor with different nucleophiles. This intermediate can be considered a useful synthon to generate molecular diversity.

Acknowledgments

We wish to thank Jan Balzarini (Rega Institute for Medical Research, K.U. Leuven, Belgium) for a fruitful collaboration over the years that led to the discovery of TSAO family of spironucleosides. We also thank Federico Gago (Universidad de Alcalá, Madrid, Spain) for the excellent molecular modeling and theoretical studies. Finally, we thank the Ministry of Science and Innovation (MICINN, Spain, project SAF2009-13914-C02-01) and the Comunidad de Madrid (CAM, Programa Biociencias, projects BIPEDD-CM, S-BIO-0214-2006 and BIPEDD-2-CM, S2010/bmd-2457) for financial support.

ABBREVIATIONS

BOP	(benzotriazol-1-yl-oxy)tris(dimethylamino)phosphoniumhexafluorphosphate
CAN	ceric ammonium nitrate
CCTLC	circular centrifugal thin-layer chromatography

DFT	density functional theory
DMA	dimethylacetamide
DMAP	4-(N,N-dimethylamino)pyridine
DMF	N,N-dimethylformamide
DMSO	dimethyl sulfoxide
DNA	deoxyribonucleic acid
HCMV	human cytomegalovirus
HIV	human immunodeficiency virus
HIV-1	human immunodeficiency virus type 1
HIV-2	human immunodeficiency virus type 2
HMDS	hexamethyldisilazane
HOMO	highest occupied molecular orbital
HPLC	high-performance liquid chromatography
MEP	molecular electrostatic potentials
MsCl	methanesulfonyl chloride (mesylchloride)
NMP	N-methylpyrrolidinone
NMR	nuclear magnetic resonance
NNRTIs	nonnucleoside reverse transcriptase inhibitors
RDPR	ribonucleoside diphosphate reductase
RNA	ribonucleic acid
RT	reverse transcriptase
TBDMS	*tert*-butyldimethylsilyl
TBDMSCl	*tert*-butyldimethylsilyl chloride
TBDPS	*tert*-butyldiphenylsilyl
TEA	triethylamine
TFA	trifluoroacetic acid
THF	tetrahydrofuran
TMSOTf	trimethylsilyltrifluoromethanesulfonate (trimethylsilyltriflate)
TSAO	1-[2′,5′-bis-*O*-(*tert*-butyldimethylsilyl)-β-D-ribofuranosyl]-3′-spiro-5″-(4″-amino-1″,2″-oxathiole-2″,2″-dioxide)
TSAO-C	1-[2′,5′-bis-*O*-(*tert*-butyldimethylsilyl)-β-D-ribofuranosyl)cytosine]-3′-spiro-5″-(4″-amino-1″,2″-oxathiole-2″,2″-dioxide)
TSAO-T	1-[2′,5′-bis-*O*-(*tert*-butyldimethylsilyl)-β-D-ribofuranosyl)thymine]-3′-spiro-5″-(4″-amino-1″,2″-oxathiole-2″,2″-dioxide)
TSAO-m^{3}T	1-[2′,5′-bis-*O*-(*tert*-butyldimethylsilyl)-β-D-ribofuranosyl)-3-*N*-methylthymine]-3′-spiro-5″-(4″-amino-1″,2″-oxathiole-2″,2″-dioxide)
TSAO-U	1-[2′,5′-bis-*O*-(*tert*-butyldimethylsilyl)-β-D-ribofuranosyl)uracil]-3′-spiro-5″-(4″-amino-1″,2″oxathiole-2″,2″-dioxide)

REFERENCES

1. Antibiotics and Chemotherapy: Anti-Infective Agents and Their Use in Therapy, Finch R. G.; Greenwood, D.; Norrby, S. R.; Whitley, R. J. (Eds.). Churchill Livingstone: Edinburgh, UK, **2003**.

2. *Pharmacology*. Rang, H. H.; Dale, M. M.; Ritter, J. M. (Eds.). Churchill Livingstone: Edinburgh, UK, **1999**.
3. Perron, F.; Albizati, K. F. *Chem. Rev.* **1989**, *89*, 1617–1661.
4. Francke, W.; Kitching, W. *Curr. Org. Chem.* **2001**, *5*, 233–251.
5. Tachibana, K.; Schever, P. J.; Tsukitani, Y.; Kikuchi, H.; Van Engen, D.; Clardy, J.; Gopichand, Y.; Schimitz, E. J. *J. Am. Chem. Soc.* **1981**, *103*, 2469–2471.
6. Klon Kanen, R. E.; Golden, T. *Curr. Med. Chem.* **2002**, *9*, 2055–2075.
7. Donnay, A. B.; Forsyth, C. J. *Curr. Med. Chem.* **2002**, *9*, 1939–1980.
8. Singh, S. B.; Zink, D. L.; Heimbach, B.; Genilloud, O.; Teran, A.; Silverman, K. C.; Lingham, R. B.; Felock, P.; Hazuda, D. J. *Org. Lett.* **2002**, *4*, 1123–1126.
9. Choi, K. W.; Brimble, M. A. *Org. Biomol. Chem.* **2009**, *7*, 1424–1436.
10. (a) Haruyama, H.; Takayanna, T.; Kinoshita, T.; Kondo, M.; Nakajima, M.; Haneishi, T. *J. Chem. Soc., Perkin Trans. 1* **1991**, 1637–1640. (b) Renard, A.; Kotera, M.; Lhomme, J. *Tetrahedron Lett.* **1998**, *39*, 3129–3132. (c) Renard, A.; Kotera, M.; Brochier, M. C.; Lhomme, J. *Eur. J. Org. Chem.* **2000**, 1831–1840. (d) Nakajima, M.; Itoi, K.; Takamatsu, Y.; Kinoshita, T.; Okazaki, T.; Kawakubo, K.; Shindo, M.; Honma, T.; Tohjigamori, M.; Haneishi, T. *J. Antibiot.* **1991**, *44*, 293–300.
11. (a) Fernández-Bolaños, J.; Blasco López, A.; Fuentes Mota, J. *Carbohydr. Res.* **1990**, *199*, 239–242. (b) Gimisis, T.; Castellari, C.; Chatgilialoglu, C. *Chem. Commun.* **1997**, 2089–2090. (c) Sano, H.; Mio, S.; Kitagawa, J.; Shindou, M.; Honma, T.; Sugai, S. *Tetrahedron* **1995**, *46*, 12563–12572. (d) Tronchet, J. M. J.; Kovács, I.; Barbalat-Rey, F.; Holm, M. V. *Nucleosides Nucleotides* **1998**, *17*, 1115–1123.
12. Renard, A.; Lhomme, J.; Kotera, M. *J. Org. Chem.* **2002**, *67*, 1302–1307.
13. (a) Kyogoku, Y.; Lord, R. C.; Rich, A. *Nature* **1968**, *218*, 69.(b) Kyogoku, Y.; Lord, R. C.; Rich, A. *Proc. Natl. Acad. Sci. USA*. **1967**, *57*, 251–263.
14. Hyrosova, E.; Vrábel, M.; Fisera, L.; Hametner, C. *Jordan J. Chem.* **2006**, *1*, 85–93.
15. Kittaka, A.; Tanaka, H.; Kato, H.; Nonaka, Y.; Nakamura, K. T.; Miyasaka, T. *Tetrahedron Lett.* **1997**, *38*, 6421–6424.
16. Paquette, L. A. *Aust. J. Chem.* **2004**, *57*, 7–17.
17. Dong, S.; Paquette, L. A. *J. Org. Chem.* **2005**, *70*, 1580–1596.
18. Dong, S.; Paquette, L. A. *J. Org. Chem.* **2006**, *71*, 1647–1652.
19. Sano, T.; Shuto, S.; Inoue, H.; Ueda, T. *Chem. Pharm. Bull.* **1985**, *33*, 3617–3622.
20. Samano, V.; Robins, M. J.. *J. Am. Chem. Soc.* **1992**, *144*, 4007–4008.
21. Czernecki, S.; Mulard, L.; Valery, J. M.; Commerson, A. *Can. J. Chem.* **1993**, *71*, 413–416.
22. Babu, B. R.; Keinicke, L.; Petersen, M.; Nielsen, C.; Wengel, J. *Org. Biomol. Chem.* **2003**, *1*, 3514–3526.
23. Balzarini, J.; Pérez-Pérez, M.-J.; San-Félix, A.; Schols, D.; Perno, C. F.; Vandamme, A. M.; Camarasa, M.-J.; De Clercq, E. *Proc. Natl. Acad. Sci. USA* **1992**, *89*, 4392–4396.
24. Camarasa, M. J.; Pérez-Pérez, M.-J.; San-Félix, A.; Balzarini, J.; De Clercq, E. *J. Med. Chem.* **1992**, *35*, 2721–2727.
25. Pérez-Pérez, M.-J.; San-Félix, A.; Balzarini, J.; De Clercq, E.; Camarasa, M.-J. *J. Med. Chem.* **1992**, *35*, 2988–2995.
26. Tronchet, J. M.; Benhamza, R.; Dolatshahi, N.; Geoffroy, M.; Turler, H. *Nucleosides Nucleotides* **1988**, *7*, 249–259.

27. (a) Tronchet, M. J.; Kovács, I.; Barbalat-Rey, F.; Vega Holm, M. *Nucleosides Nucleotides* **1998**, *17*, 1115–1123. (b) Tronchet, M. J.; Kovács, I.; Barbalat-Rey, F.; Dolatshahi, N. *Nucleosides Nucleotides* **1996**, *15*, 337–347.
28. Rong, J.; Roselt, P.; Plavec, J.; Chattopadhyaya, J. *Tetrahedron* **1994**, *50*, 4921–4936.
29. Nielsen, P.; Larsen, K.; Wengel, J. *Acta Chem. Scand.* **1996**, *50*, 1030–1035.
30. Calvo-Mateo, A.; Camarasa, M.-J.; Díaz-Ortiz, Ade las Heras, F. G. *J. Chem. Soc. Chem. Commun.* **1988**, 1114–1115.
31. Pérez-Pérez, M.-J.; Camarasa, M.-J.; Díaz-Ortiz, A.; San-Félix, A.; de las Heras, F. G. *Carbohydr. Res.* **1991**, *216*, 399–411.
32. Hollemberg, D. H.; Klein, R. S.; Fox, J. J. *Carbohydr. Res.* **1978**, *67*, 491–494.
33. Yoshimura, J. *Adv. Carbohydr. Chem. Biochem.* **1984**, 69–134.
34. Vorbrüggen, H.; Krolikiewicz, K.; Bennua, B. *Chem. Ber.* **1981**, *114*, 1234–1256.
35. (a) Camarasa, M.-J.; San-Félix, A.; Pérez-Pérez, M.-J.; Velázquez, S.; Alvarez, R.; Chamorro, C.; Jimeno, M. L.; Pérez, C.; Gago, F.; De Clercq, E.; Balzarini, J. *J.Carbohydr. Chem.* **2000**, *19*, 451–469. (b) Camarasa, M.-J.; San-Félix, A.; Velázquez, S.; Pérez-Pérez, M.-J.; Gago, F.; Balzarini, J. *Curr. Top. Med. Chem.* **2004**, *4*, 945–963 and references therein.
36. Velázquez, S.; San-Félix, A.; Pérez-Pérez, M.-J.; Balzarini, J.; De Clercq, E.; Camarasa, M.-J. *J. Med. Chem.* **1993**, *36*, 3230–3239.
37. San-Félix, A.; Velázquez, S.; Pérez-Pérez, M.-J.; Balzarini, J.; De Clercq, E.; Camarasa, M.-J. *J. Med. Chem.* **1994**, *37*, 453–460.
38. Van Aerschot, A.; Everaert, D.; Balzarini, J.; Augustyns, K.; Jie, L.; Janssen, G.; Peeters, O.; Blaton, N.; De Ranter, C.; De Clercq, E.; Herdewijn, P. *J. Med. Chem.* **1990**, *33*, 1833–1839.
39. Divakar, K. J.; Reese, C. B. *J. Chem. Soc. Perkin Trans. 1* **1982**, 1171–1176.
40. Sasaki, T.; Minamoto, K.; Suzuki, H. *J. Org. Chem.* **1973**, *38*, 598–607.
41. Chamorro, C.; De Clercq, E.; Balzarini, J.; Camarasa, M.-J.; San-Félix, A. *Antivir. Chem. Chemother.* **2000**, *11*, 61–69.
42. Bonache, M. C.; Chamorro, C.; Velázquez, S.; De Clercq, E.; Balzarini, J.; Rodríguez-Barrios, F.; Gago, F.; Camarasa, M.-J.; San-Félix, A. *J. Med. Chem.* **2005**, *48*, 6653–6660.
43. Moura, M.; Josse, S.; Nguyen VanNhien, A.; Foirnier, C.; Duverlie, G.; Castelain, S.; Soriano, E.; Marco-Contelles, J.; Balzarini, J.; Postel, D. *Eur. J. Med. Chem.* **2011**, 1–11.
44. Alvarez, R.; Velázquez, S.; San-Félix, A.; Aquaro, S.; De Clercq, E.; Perno, C.-F.; Karlsson, A.; Balzarini, J.; Camarasa, M.-J.; *J. Med. Chem.* **1994**, *37*, 4185–4194.
45. Bastide, J.; Hamelin, J.; Texier, F.; Quang, Y.V. *Bull. Soc. Chim. Fr.* **1973**, 2998–3003.
46. L'Abbé, G.; Hassner, A. *Annu. Rev. Biochem.* **1993**, *63*, 133–173.
47. García-López, M. T.; García-Muñoz, G.; Iglesias, J.; Madroñero, R. *J. Heterocycl. Chem.* **1969**, *6*, 639–642.
48. Alonso, R.; Camarasa, M.-J.; Alonso, G.; De las Heras, F. G. *Eur. J. Med. Chem.* **1980**, *15*, 105–109.
49. Harvey, G. R. *J. Org. Chem.* **1966**, *31*, 1587–1590.
50. Zbiral, E. *Synthesis* **1974**, 775–800.
51. Schorkuber, W.; Zbiral, E. *Liebigs Ann.* **1980**, 1455–1469.

52. Schorkuber, W.; Zbiral, E. *Chem. Ber.* **1981**, *114*, 3165–3169.
53. Hammerschmidt, F.; Polsterer, J. P.; Zbiral, E. *Synthesis* **1995**, 415–418.
54. Le Corre, M. C. R. *Acad. Sci. Ser. C* **1970**, *270*, 11312–11314.
55. Velázquez, S.; Alvarez, R.; Pérez, C.; Gago, F.; De Clercq, E.; Balzarini, J.; Camarasa, M.-J. *Antivir. Chem. Chemother.* **1998**, *9*, 481–489.
56. Velázquez, S.; Chamorro, C.; Pérez-Pérez, M.-J.; Alvarez, R.; Jimeno, M. L.; Martín-Domenech, A.; Pérez, C.; Gago, F.; De Clercq, E.; Balzarini, J.; San-Félix, A.; Camarasa, M.-J. *J. Med. Chem.* **1998**, *41*, 4636–4647.
57. For a review see: Gololobov, Y. G.; Kasukhin, L. F. *Tetrahedron* **1992**, *48*, 1353–1406 and references therein.
58. Ingate, S.; Pérez-Pérez, M.-J.; De Clercq, E.; Balzarini, J.; Camarasa, M.-J. *Antivir. Res.* **1995**, *27*, 281–299.
59. Chamorro, C.; Pérez-Pérez, M.-J.; Rodríguez-Barrios, F.; Gago, F.; De Clercq, E.; Balzarini, J.; San-Félix, A.; Camarasa, M.-J. *Antivir. Res.* **2001**, *50*, 207–222.
60. Alvarez, R.; San-Félix, A.; De Clercq, E.; Balzarini, J.; Camarasa, M.-J. *Nucleosides Nucleotides* **1996**, *15*, 349–359.
61. Ingate, S.; Camarasa, M.-J.; De Clercq, E.; Balzarini, J. *Antivir. Chem. Chemother.* **1995**, *6*, 365–370.
62. Ingate, S.; De Clercq, E.; Balzarini, J.; Camarasa, M.-J. *Antivir. Res.* **1996**, *32*, 149–164.
63. Velázquez, S.; Jimeno, M. L.; Camarasa, M.-J. *Tetrahedron* **1994**, *50*, 11013–11022.
64. (a) Ingate, S. T.; Marco, J. L.; Witvrouw, M.; Pannecouque, C.; De Clercq, E. *Tetrahedron* **1997**, *53*, 17795–17814, (b) Marco, J. L.; Ingate, S. T.; Chinchón. P. M. *Tetrahedron* **1999**, *55*, 7625–7644. (c) Marco, J. L.; Ingate, S. T.; Jaime, C.; Beá, I. *Tetrahedron* **2000**, *56*, 2523–2531.
65. Lobatón, E.; Rodríguez-Barrios, F.; Gago, F.; Pérez-Pérez, M.-J.; De Clercq, E.; Balzarini, J.; Camarasa, M.-J.; Velázquez, S. *J. Med. Chem.* **2002**, *45*, 3934–3945.
66. (a) Asakura, J.; Robins, M. J. *Tetrahedron Lett.* **1988**, *29*, 2855–2858. (b) Asakura, J.; Robins, M. J. *J. Org. Chem.* **1990**, *55*, 4928–4933.
67. Heck, R. F. *Acc. Chem. Res.* **1979**, *12*, 146–151.
68. Heck, R. F. *Org. React.* **1982**, *27*, 345–389.
69. (a) Sonogashira, K.; Tohda, J.; Hagihara, N. *Tetrahedron Lett.* **1975**, 4467–4470. (b) Robins, M. J.; Barr, P. J. *J. Org. Chem.* **1983**, *48*, 1854–1862.
70. Lobatón, E.; Camarasa, M.-J. Velázquez, S. *Synlett* **2000**, *9*, 1312–1314.
71. Rodríguez-Barrios, F.; Pérez, C.; Lobatón, E.; Velázquez, S.; Chamorro, C.; San-Félix, A.; Pérez-Pérez, M.-J.; Camarasa, M.-J.; Pelemans, H.; Balzarini, J.; Gago, F. *J. Med. Chem.* **2001**, *44*, 1853–1865.
72. De Castro, S.; García-Aparicio, C.; Van Laethem, K.; Gago, F.; Lobatón, E.; De Clercq, E.; Balzarini, J.; Camarasa, M.-J.; Velázquez, S. *Antivir. Res.* **2006**, *71*, 15–23.
73. Camarasa, M.-J.; Jimeno, M. L.; Pérez-Pérez, M.-J.; Alvarez, R.; Velázquez, S.; Lozano, A. *Tetrahedron* **1999**, *55*, 12187–12200.
74. Coulter, J. M.; Lewis, J. W.; Lynch, P. P. *Tetrahedron* **1968**, *24*, 4489–4500.
75. Nair, V.; Richardson, S. G. *J. Org. Chem.* **1980**, *45*, 3969–3974.
76. De Castro, S.; Lobatón, E.; Pérez-Pérez, M.-J.; San-Félix, A.; Cordeiro, A.; Andrei, G.; Snoeck, R.; De Clercq, E.; Balzarini, J.; Camarasa, M.-J.; Velázquez, S. *J. Med. Chem.* **2005**, *48*, 1158–1168.

77. This value is even higher than the HOMO energy value of typical low nucleophilic amines such as 4-nitroaniline ($E_{HOMO} = -9.97$ eV) according to HF/6-31G** Fopt studies.
78. Hurd, C. D.; Prapas, A. G. *J. Chem. Soc.* **1959**, *24*, 388–392.
79. Otera, J.; Dan-oh, N.; Nozaki, H. *J. Org. Chem.* **1991**, *56*, 5307–5311.
80. Bonache, M. C.; Chamorro, C.; Lobatón, E.; De Clercq, E.; Balzarini, J.; Velázquez, S.; Camarasa, M.-J.; San-Félix, A. *Antivir. Chem. Chemother.* **2003**, *14*, 249–262.
81. De Castro, S.; Lozano, A.; Jimeno, M. L.; Pérez-Pérez, M. J.; San-Félix, A.; Camarasa, M.-J.; Velázquez, S. *J. Org. Chem.* **2006**, *71*, 1407–1415.
82. Kawahara, S.; Wada, T.; Sekine, M. *J. Am. Chem. Soc.* **1996**, *118*, 9461–9468.
83. Greco, M. N.; Zhong, H. M.; Marynoff, B. E. *Tetrahedron Lett.* **1998**, *39*, 4959–4962.
84. Alvarez, R.; Jimeno, M. L.; Pérez-Pérez, M. J.; De Clercq, E.; Balzarini, J.; Camarasa, M.-J. *Antivir. Chem. Chemother.* **1997**, *8*, 507–517.
85. Alvarez, R.; Jimeno, M. L.; Gago, F.; Balzarini, J.; Pérez-Pérez, M. J.; Camarasa, M.-J. *Antivir. Chem. Chemother.* **1998**, *9*, 333–340.
86. Nguyen Van Nhien, A.; Tomassi, C.; Len, C.; Marco-Contelles, J. L.; Balzarini, J.; Pannecouque, C.; De Clercq, E.; Postel, D. *J. Med. Chem.* **2005**, *48*, 4276–4284.
87. Soriano, E.; Marco-Contelles, J. L.; Tomassi, C.; Nguyen Van Nhien, A.; Postel, D. *J. Chem. Inf. Model.* **2006**, *46*, 1666–1677.
88. Chamorro, C.; Luengo, S. M.; Bonache, M. C.; Velázquez, S.; Pérez-Pérez, M. J.; Camarasa, M. J.; Gago, F.; Jimeno M. L.; San-Félix. A. *J. Org. Chem.* **2003**, *68*, 6695–6704.
89. Ingate, S.; Pérez-Pérez, M. J.; De Clercq, E.; Balzarini, J.; Camarasa, M. J. *Antivir. Res.* **1995**, *27*, 281–299.
90. Chamorro, C.; Pérez-Pérez, M. J.; Rodriguez Barrios, F.; Gago, F.; De Clercq, E.; Balzarini, J.; San-Félix, A.; Camarasa, M. J. *Antivir. Res.* **2001**, *50*, 207–222.
91. Bonache, M. C.; Chamorro, C.; Cordeiro, A.; Camarasa, M. J.; Jimeno, M. L.; San-Félix, A. *J. Org. Chem.* **2004**, *69*, 8758–8766.
92. Pérez Pérez, M. J; Camarasa, M. J.; Díaz Ortíz, A.; San Félix, A.; de las Heras, F. G. *Carbohydr. Res.* **1991**, *216*, 399–411.
93. (a) Nair, V.; Purdy, D. F. *Tetrahedron* **1991**, *47*, 365–382. (b) Cacchi, S.; Misiti, D.; LaTorre, F. *Synthesis* **1989**, 243–251.
94. *Hydrogenation Methods*. Rylander, P. N. (Ed.). Academic Press: London, **1985**, pp. 82–93.
95. Bonache, M. C.; Cordeiro, A.; Carrero, P.; Quesada, E.; Camarasa, M. J.; Jimeno, M. L.; San-Félix, A. *J. Org. Chem.* **2009**, *74*, 9071–9081.
96. Bonache, M. C.; Cordeiro, A.; Quesada, E.; Vanstreels, E.; Daelemans, D.; Camarasa, M. J.; Balzarini, J.; San-Félix, A. *Antivir. Res.* **2011**, *92*, 37–44.

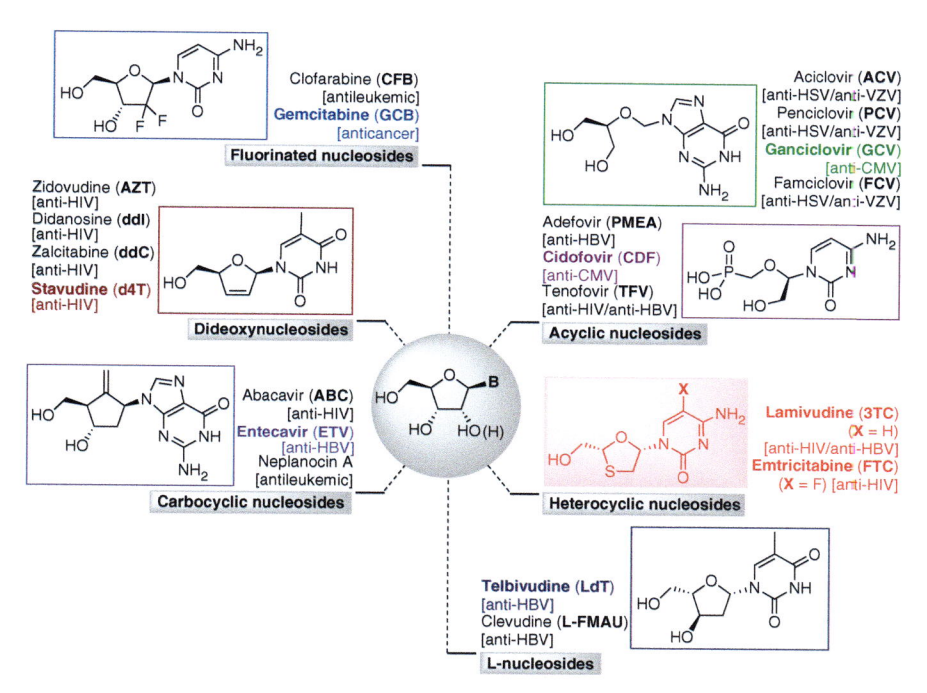

Chapter 16, Figure 1. Marketed sugar-modified nucleosides in use for antiviral and antitumor chemotherapy.

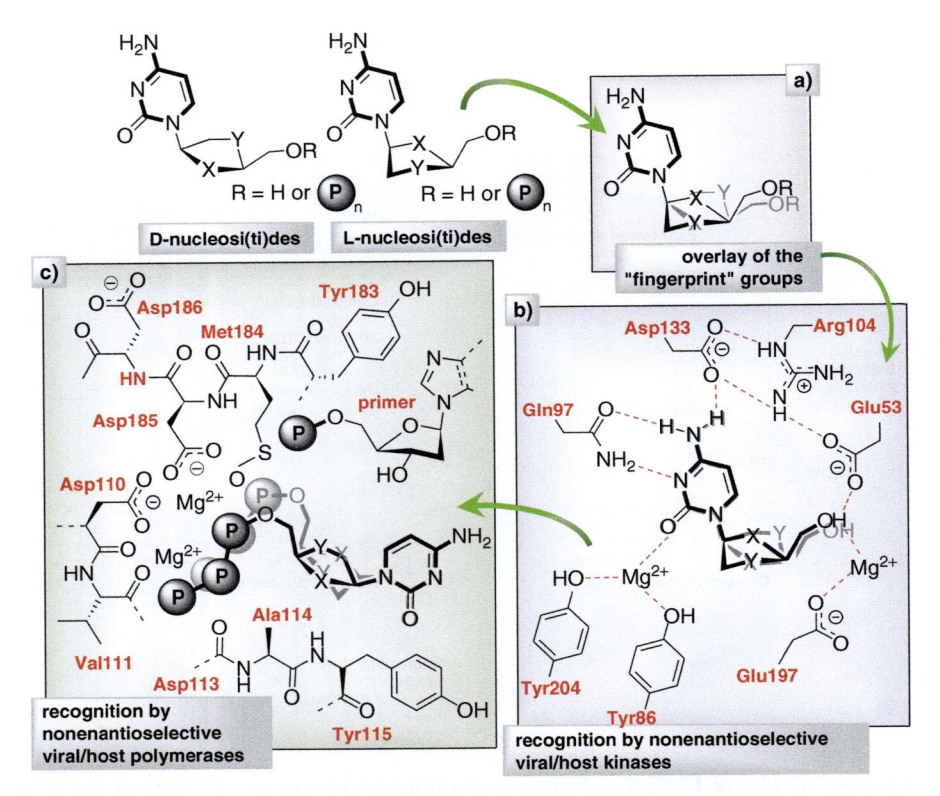

Chapter 16, Figure 5. Schematic representation of the structural factors determining the biological activity of L-heterocyclic nucleosides. (a) Conformational flexibility of D- and L-heterocyclic nucleosi(ti)des allows accurate overlay of "fingerprint" groups [i.e., nucleobase (cytosine in the example) and 5′-OR]. (b) Non-enantioselective properties of host/virally encoded kinases enable phosphorylation of either D- and L-heterocyclic nucleosides. (c) Nonenantioselective properties of polymerases (e.g., HIV-RT) enable recognition of either D- and L-heterocyclic nucleoside triphosphates.

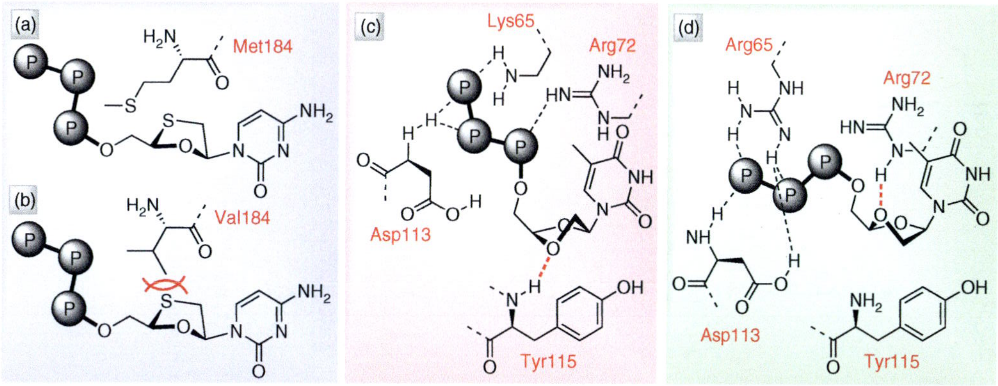

Chapter 16, Figure 6. Schematic representation of the binding modes of oxathiolane and dioxolane nucleosides in WT and drug-resistant HIV-RTs (reproduced from refs. 44, 45, and 47). (a) Nonbonding interaction between 3TC-TP and Met184 in WT HIV-RT. (b) Disfavored interaction between 3TC-TP and Val184 in M184V mutant of HIV-RT. (c) Binding mode of DOT-TP in WT HIV-RT. (d) Binding mode of DOT-TP in the K65R mutant of HIV-RT.

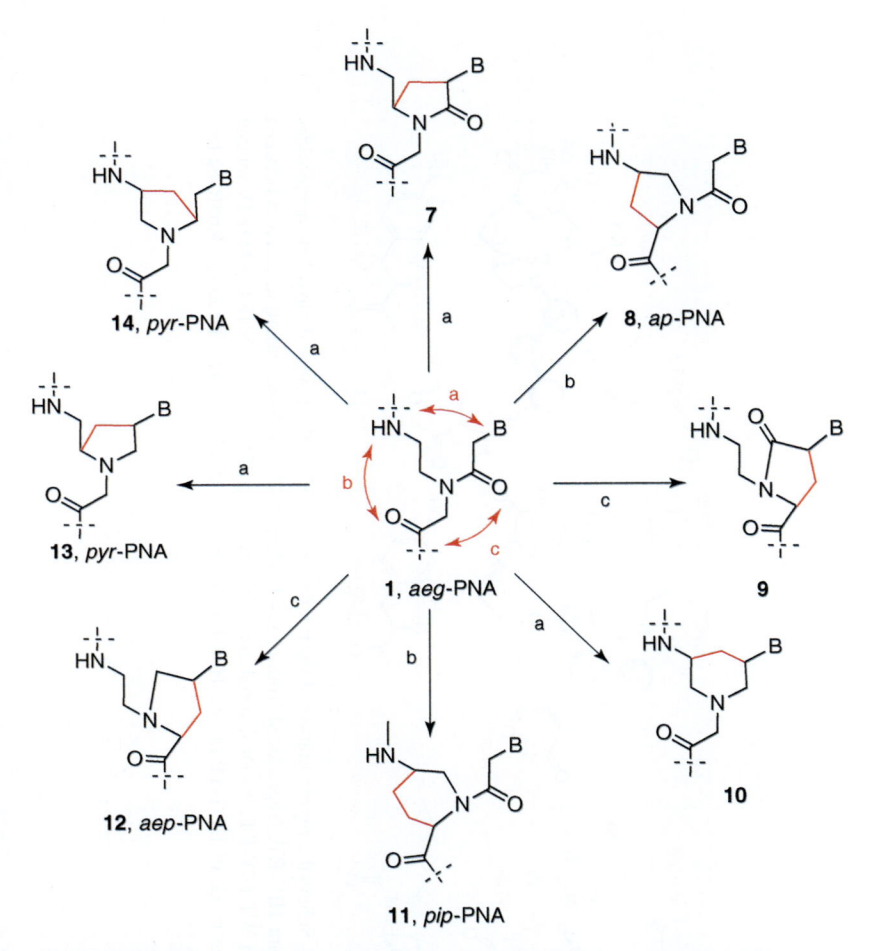

Chapter 19, Figure 3. PNA analogues containing saturated nitrogen heterocycles.

11 L-Nucleosides

DANIELA PERRONE
Dipartimento di Scienze Chimiche e Farmaceutiche, Università di Ferrara, Ferrara, Italy

MASSIMO L. CAPOBIANCO
Istituto per la Sintesi Organica e la Fotoreattività, Consiglio Nazionale delle Ricerche, Bologna, Italy

1. INTRODUCTION

L-Nucleosides are the enantiomers of the natural nucleosides: namely, the nonsuperimposable mirror image of D-nucleosides, in which all stereocenters have an inverted configuration. By analogy with the natural nucleosides found in nucleic acids, the nucleobase of an L-nucleoside is designated to be β-oriented if it is *cis* to the 4′-hydroxymethyl group of the sugar moiety in its furanosyl configuration (Figure 1).

The first syntheses of the enantiomeric form of natural DNA and RNA monomers were focused inside the mechanism of chiral "evolution": why all the natural nucleosides in the world were of the D-form[1-4] and how D- and L-oligonucletides could recognize each other.[5-10] Taking into account the fact that the efficacy of nucleoside analogues depends on their ability to mimic natural nucleosides, it was long assumed that only nucleoside analogues having the natural D-configuration could exhibit biological activity.[11,12] At the beginning of the 1990s, this paradigm had been challanged and unnatural L-nucleosides emerged as a new class of therapeutic agents. Nevertheless, little attention had been given to L-nucleosides until the discovery of lamivudine (**3TC**), the enantiomer of **BCH-189** (Figure 2), currently used for the treatment of human immunodeficiency types 1 and 2 and hepatitis B viral infections.[13,14] Since then, the discovery that several viruses could accept L-nucleosides as a substrate for their own replication, leading to stalled synthesis of their DNA, promoted the search for L-nucleoside analogues to be used as more powerful and more selective antivirals.[15-21]

The biological features and the status of preclinical and clinical studies of the chemotherapeutics of L-nucleosides have been the object of previous reviews.[22,23] Our intent in this chapter is to provide an overview concerning the synthesis of L-nucleosides and their analogues.

Chemical Synthesis of Nucleoside Analogues, First Edition. Edited by Pedro Merino.
© 2013 John Wiley & Sons, Inc. Published 2013 by John Wiley & Sons, Inc.

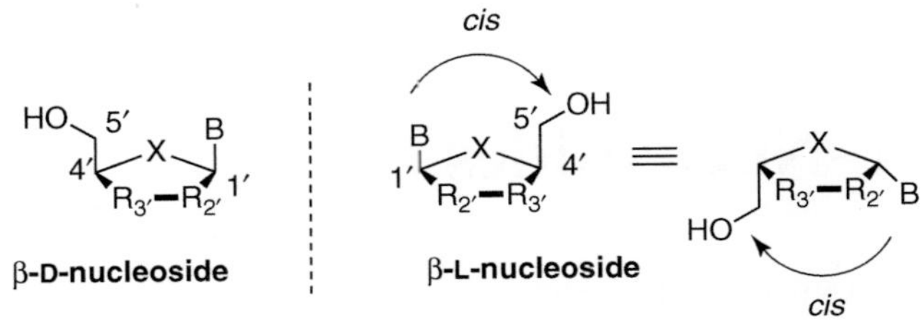

Figure 1. D,L-β-Nucleosides in their furanosic forms (in the ribo-derivatives; for example, X = O, $R_{2'} = R_{3'}$ = CHOH, with the OH groups directed toward the α-face).

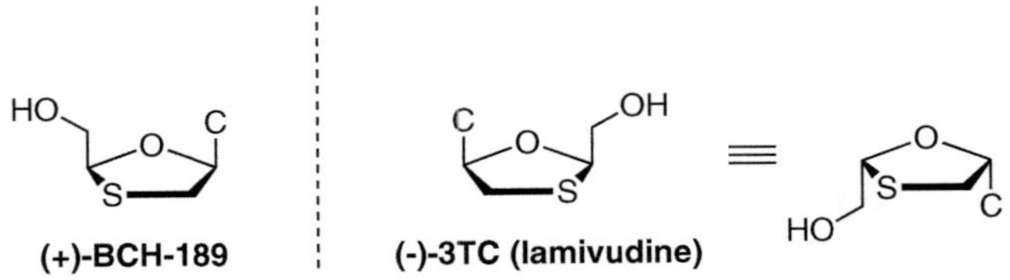

Figure 2. BCH-189 and 3TC.

2. RIBOSE AND DEOXYRIBOSE NUCLEOSIDES

This section deals with L-nucleosides, which mantain a close relationship with ribose and deoxyribose derivatives. Data have been ordered according to the starting sugar utilized to synthesize the target nucleosides. Some fluoronucleosides and C-nucleoside derivatives are also reported at the end of the section.

2.1. From L-Arabinose

L-Arabinose is the most commonly available L-sugar, being a natural constituent of hemicellulose and pectine, so it is not surprisingly that the first synthesis began with this sugar. Probably the first synthesis of L-thimidine (**2**) was reported by Smejkal and Sorm in 1964 (Scheme 1).[24] Compound **2** is obtained starting from L-arabinose by the glycosylation of thymidinyl mercury with the crystalline L-ribofuranosyl chloride **1** and successive deacylation.

The first reported synthesis of L-riboadenosine (**6**) starts from the dibenzoylated compound **3**, which is obtained with a yield of 22% from L-arabinose as well as from the tosylated **4** that can be obtained in 66% yield from L-lyxose (Scheme 2).[25] The two syntheses converge to the tribenzoyl derivative **5**, from which the L-riboadenosine **6** is obtained employing the glycosilation method of Davoll and Lowy,[26] coupling **5** with the chloromercury benzamidopurine.

Independently, one year later the synthesis of a racemic mixture of the D/L riboadenosine **7** has been reported from which the L-enantiomer **6** can be isolated after enzymatic destruction of the other isomer (Scheme 3).[27] After the enzymatic treatment, the complete deamination and hydrolysis of D-adenosine is observed,

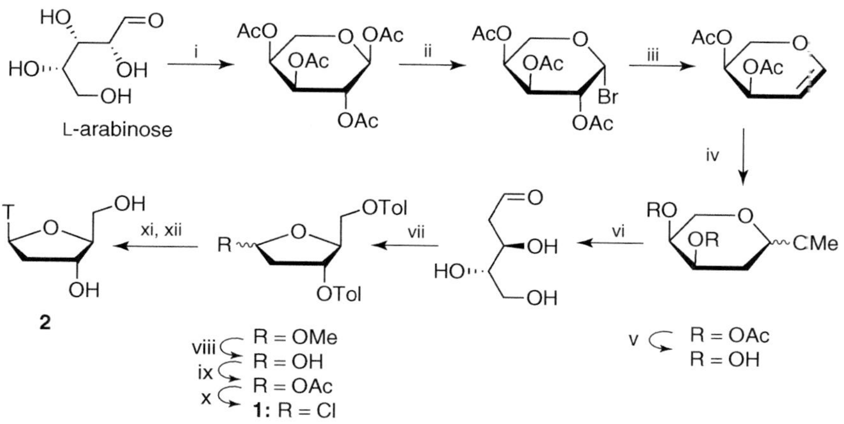

Scheme 1. *Reagents and conditions:* (i) Ac$_2$O, Py, 0°C, (12%); or activation of Ac$_2$O with AcONa at 80°C, then reaction with L-arabinose at 100°C (24%); (ii) 28% HBr, AcOH, 0°C; (iii) Zn, AcOH (85%); (iv) HCl, benzene, Ag$_2$CO$_3$; (v) 0.1 N NaOMe, MeOH, 0°C; (vi) [after step (v)]: BzOH; (vii) 1% HCl, MeOH to 1'-O-Me-2'-deoxy-α/β-L-ribofuranose; then TsCl, Py [24% from step (iv)]. (viii) HCl, dioxane, 100°C; (ix) Ac$_2$O, Py; (x) HCl, Et$_2$O, 0°C (86%); (xi) from **1**: thymidyl mercury, CdCO$_3$, toluene, reflux, 90 min (10%); (xii) 0.1 N NaOMe, MeOH.

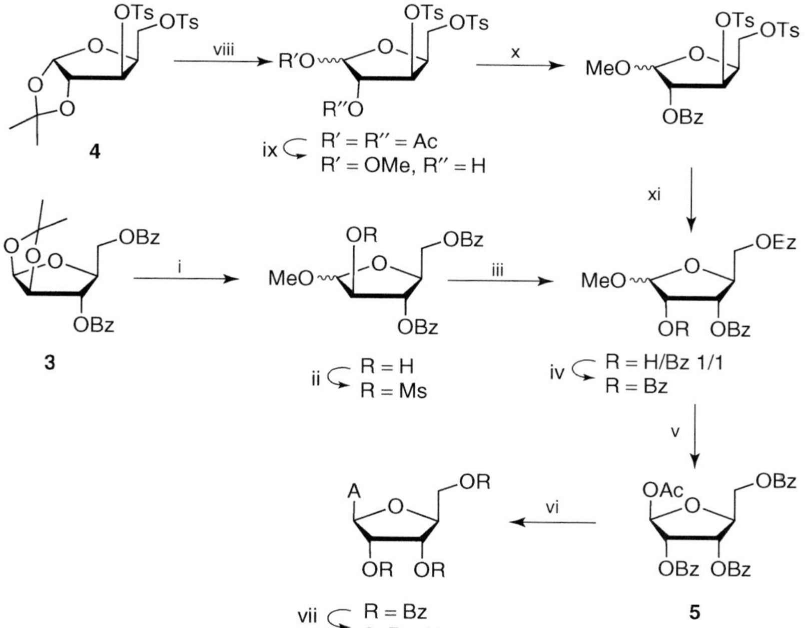

Scheme 2. *Reagents and conditions:* (i) HCl, MeOH; (ii) MsCl, Py [86% from step (:)]; (iii) after step (ii): NaOBz, DMF, reflux; (iv) BzCl, Py [10% from step (ii)]; (v) [after step (iv)]: AcOH (48%); (vi) chloromercuri-6-benzamidopurine, xylene, reflux; (vii) NaOH, MeOH (43%); (viii) AcOH; (ix) MeOH (88% from **4**); (x) [after step (ix)]: BzCl, Py (67%); (xi) NaBz, DMF, reflux.

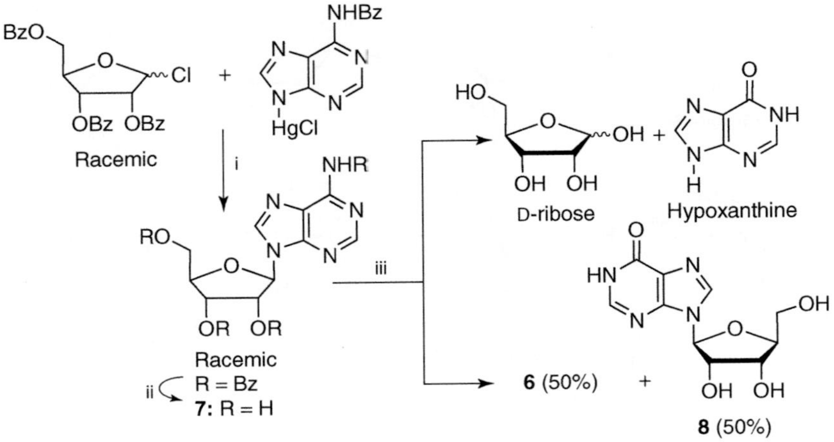

Scheme 3. *Reagents and conditions:* (i) xylene, reflux (79%); (ii) NaOH or NaOMe; (iii) [after step (ii)]: bacterial cell suspension of PS-264 strain 2–3 days at 38°C, pH 7.0.

whereas about 50% of the L-adenosine **6** remains unattacked and the rest is deaminated to the corresponding L-inosine **8**.

By using the fusion methodology, adenine and trimethylsilyl uridine are attached to the protected 1-bromo-2,3-dibenzoyl-5-fosfate of L-ribose.[28] This method has been extended by Fujimori et al.,[29] (Scheme 4) and independently by Urata et al.,[30] who report the synthesis of the L-thymidine **2** and nucleosides **9–11** containing all the natural bases, obtained by glycosylation on the α-chloro-3,5-di-p-toluyl-L-deoxyribose (**1α**).

As an expansion of applicability of this coupling methodology, the synthesis of α/β mixtures of 2′-deoxyadenosine and 2′-deoxyguanosine has been reported by the direct fusion of 2,6-dichloropurines with 1-*O*-methyl-3,5-di-*O*-p-toluyl-2-deoxy-L-erytropentofuranose.[31]

An alternative to the use of 1-Cl-2-deoxy-3,5-protected-ribose as glycosidic donor is its replacement with the epoxyde intermediate **13**, which has been exploited for the synthesis of all the "natural" L-ribonucleosides **14**, starting from the benzylarabinopyranose **12** (Scheme 5).[32]

A linear strategy based on the construction of the uracil ring directly on the sugar has been developed starting from L-arabinose (Scheme 6).[33,34] In this work, L-arabinose is treated with cyanamide to give the 2′-amino-1,2-oxazoline **15**, which is reacted with methyl propiolate to give the anhydrouridine **16**. To favor a cleaner hydrolysis of the anhydrous bond, the hydroxyl groups of **16** are protected as benzoates; then the opening of the anhydrous bond is carried out by halogen ions and the resulting chloro derivative is reduced by the use of *n*-Bu₃SnH, and the protected nucleoside is finally hydrolyzed to give the L-β-deoxyuridine **18** with an overall yield of about 19% from the L-arabinose. From the L-dibenzoyldeoxyuridine **17**, L-dibenzoylthymidine can easily be obtained,[34] and L-deoxycytidine is obtainable using the method of Sung developed for the D-anomer.[35] An improvement in

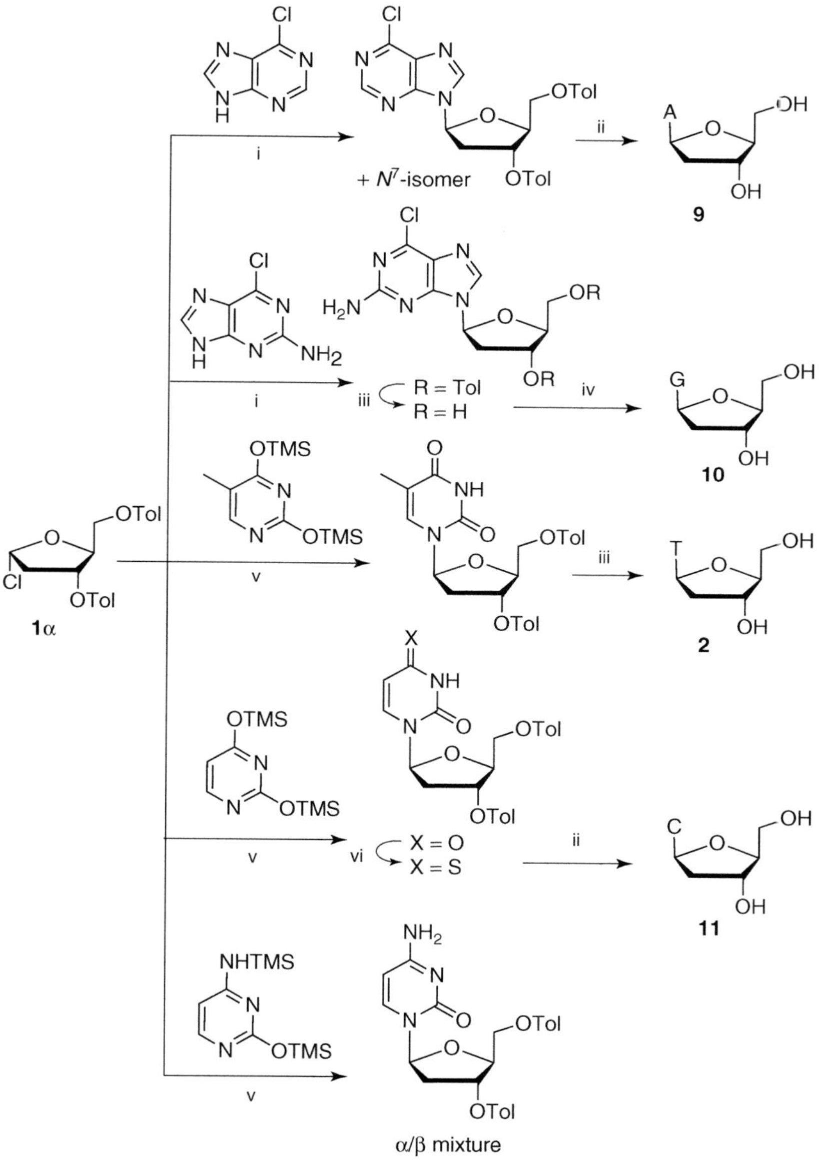

Scheme 4. *Reagents and conditions:* (i) NaH, MeCN; (ii) NH$_3$, MeOH, 100°C; (iii) NH$_3$, MeOH, rt; (iv) 2 N KOH, dioxane; (v) *p*-nitrophenol, CHCl$_3$; (vi) Lawesson's reagent, dioxane.

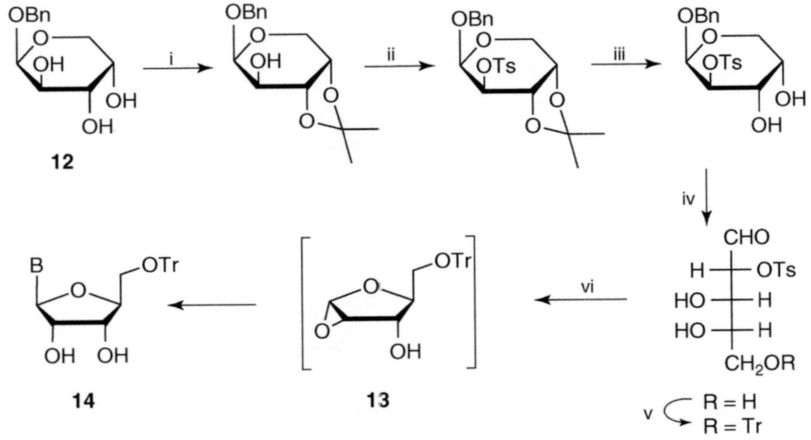

Scheme 5. *Reagents and conditions:* (i) triethylorthoformate, acetone; (ii) TsCl, Py; (iii) aq. HCl; (iv) H$_2$/Pd; (v) TrCl, Py; (vi) [after step (v)]: B$^-$Na$^+$ with B = U, C, A, G, T, 4-methoxy-2-pyrimidinone, and aza-U.

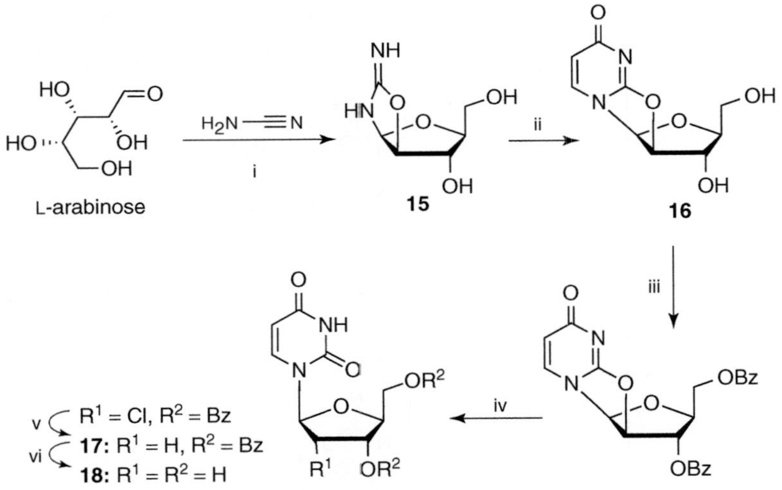

Scheme 6. *Reagents and conditions:* (i) NH$_3$, H$_2$O/MeOH (72%); (ii) methyl propiolate, EtOH/H$_2$O, reflux, (65%); (iii) BzCN, DMF, TEA (95%); (iv) 6 M HCl, DMF 100°C (87%); (v) Raney Ni, CaCO$_3$, EtOH, reflux; or n-Bu$_3$SnH, AIBN, benzene, reflux (22 or 84%, respectively); (vi) 1 M NaOMe, MeOH (93%).

Holy's synthesis was later reported by the group of Garbesi using a more efficient reducing agent (trimethylsilylsilane) and discovering that L-dibenzoyldeoxyuridine **17** can be used for transglycosylations with properly protected purines, affording preparation of all the L-deoxynucleosides.[9,36]

The availability of L-arabinose as a starting sugar allows the synthesis of L-ribonucleosides containing all four natural bases by epimerization of the C2-OH by use of molibdenum dioxide acetylacetonate (Scheme 7).[37] The reaction of L-arabinose with $MoO_2(acac)_2$ produces a mixture of L-arabinose/L-ribose 3:2 from which L-ribose (**19**) can be isolated by crystallization. This compound, once benzoylated and acetylated to **20**, is coupled with the suitably protected bases using trimethyltrifluoromethane sulfonate as catalyst, to give predominantly the β-nucleoside, in a methodology published by Visser et al.,[2] then extended and ameliorated by Garbesi et al.[10]

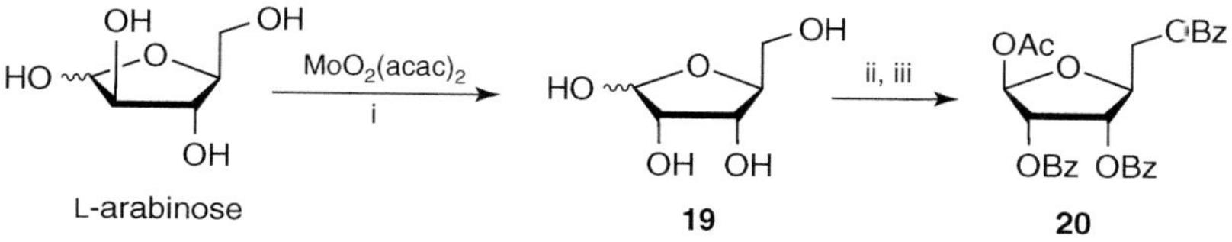

Scheme 7. *Reagents and conditions:* (i) DMF, 50°C (38%); (ii) HCl, MeOH, then BzCl, Py; (iii) AcOH, Ac_2O [28% from step (i)].

Recently, a convenient preparation of 1α-Cl-3,5-ditoluenyl-L-2-deoxyribose (**1α**) from L-arabinose has been reported according to Scheme 8.[38] The ethylthio ortho ester **21** produces the triester **22** by generation of a dialkoxyalkyl radical with tributyltin hydride in hot toluene. The chlorosugar **1α** obtained by treatment of **22** with HCl in acetic acid can be used as a glycosyl donor in the Vorbrüggen methodology to prepare various L-nucleosides.

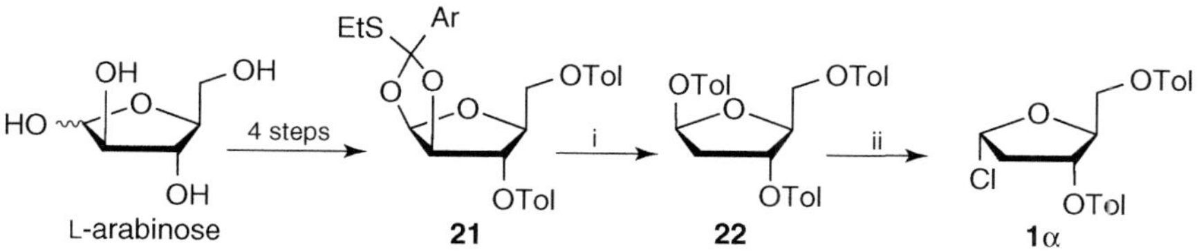

Scheme 8. *Reagents and conditions:* (i) Bu_3SnH, AIBN, toluene, reflux (85%); (ii) HCl, AcOH (70%).

Finally, it is worth noting that an enzymatic methodology can also be applied for the synthesis of (modified) nucleosides, as shown in Scheme 9.[39] The L-α-arabinofuranosylthimidine **23** can be transformed into the anhydro derivative **24**, then to the 2′-azidonucleoside **25**, hence to the 2′-azidodiester **26**, and finally, selectively deacetylated by CAL-B (*Candida antartica* lipase-B) enzyme to give the 5′-free OH nucleoside **27**.

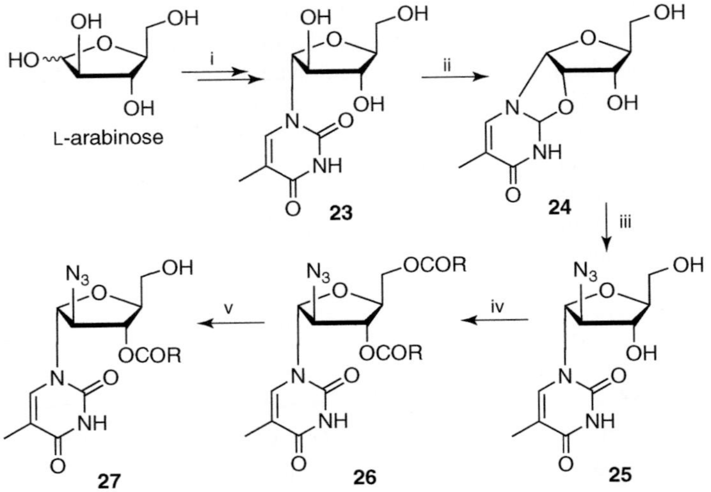

Scheme 9. *Reagents and conditions:* (i) vinyl acetate, *Candida antartica* lipase-B (CAL-B), then Ac$_2$O, Py; then T, BSA, TMSOTf, MeCN, reflux, then NH$_3$, MeOH; (ii) DPC, NaHCO$_3$, 150°C; (iii) NaN$_3$, DMF, 150°C; (iv) (RCO)$_2$O, DMA, MeCN; (v) CAL-B MeCN, *n*-BuOH, where R = Me, Et, or *n*-Pr.

2.2. From D-Glucose

An entry to L-ribose from the inexpensive D-glucose (**28**) is shown in Scheme 10.[40] The diol **29** is transformed into the corresponding di-*O*-mesylate, from which the olefin **30** is obtained by treatment with NaI. Acid-catalyzed hydrolysis of the remaining ketal group in **30**, followed by reduction of the resulting hemiacetal, yields the hexenitol **31**, which is selectively protected at the primary OH group. Finally, ozonization of the resulting intermediate, followed by reductive workup and perbenzoylation, leads to a mixture of α/β derivatives **32** in an overall yield of 48% from D-glucose. The mixture **32** is then glycosilated with the Vorbrüggen methodology to afford protected β-L-ribonucleosides. A similar approach has been reported more recently for the synthesis of L-2′-deoxynucleoside derivatives of thymine.[41]

2.3. From D-Galactose

L-ribose (**19**) was obtained starting from the D-galactose diisopropylidene **33**, following the synthesis depicted in Scheme 11.[42] Inversion of configuration at the C2-OH and hydrolysis of the diacetate **34** take place with NaOMe/MeOH for the intramolecular S$_N$2 reaction of methanesulfonyloxy group with C1 alkoxide. Then, debenzylation of **35** with 10% Pd/C followed by hydrolysis of methyl glycoside affords L-ribose. The corresponding (not shown) 1-*O*-acetyl derivative of L-ribose **19** is then used to obtain the L-ribonucleosides by use of TMSOTf, conveniently protected bases, and BSA.

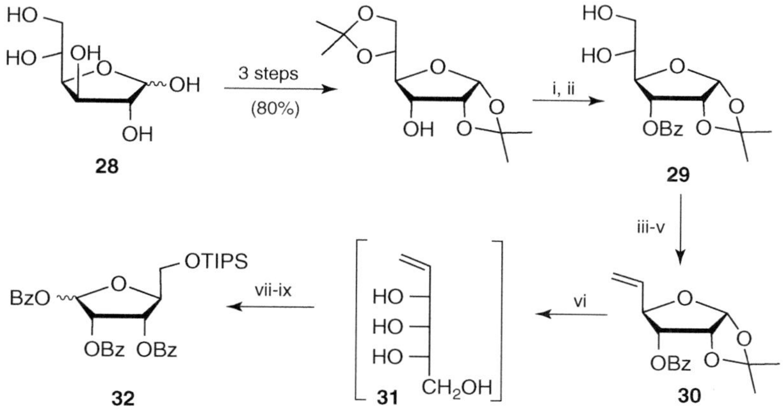

Scheme 10. *Reagents and conditions:* (i) BzCl, DMAP, Py, CH$_2$Cl$_2$; (ii) AcOH, HCOOH, H$_2$O [88% from step (i)]; (iii) MsCl, TEA, CH$_2$Cl$_2$, 4°C; (iv) NaI, 3-pentanone, 100°C; (v) NaOH, MeOH [83% from step (iii)]. (vi) resin-H$^+$, H$_2$O, THF. 80°C, then NaBH$_4$, H$_2$O; (vii) TIPS-Cl, Im, DMF, 4°C to rt; (viii) Me$_2$S, MeOH, 4°C; (ix) BzCl, Py, CH$_2$Cl$_2$ [81% from step (vi)].

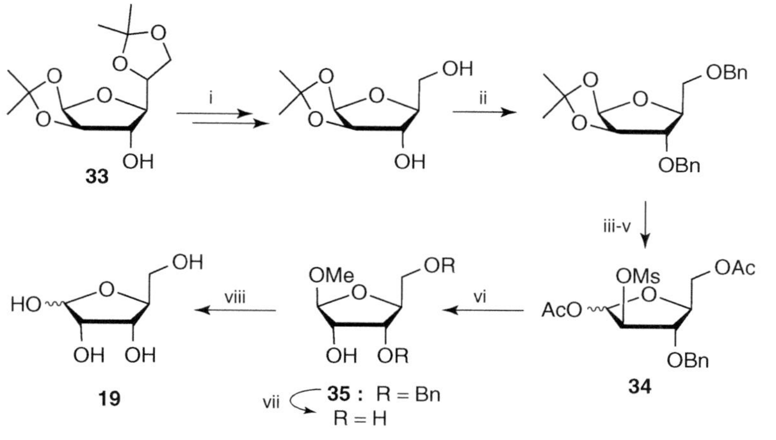

Scheme 11. *Reagents and conditions:* (i) AcOH, H$_2$O, then NaIO$_4$, MeOH H$_2$O, then NaBH$_4$, MeOH, H$_2$O (92%); (ii) KOH, BnCl, dioxane, reflux (95%); (iii)10% HCl, MeOH (96%); (iv) MsCl, TEA (98%); (v) Ac$_2$O, AcOH, H$_2$SO$_4$, 4°C (89%); (vi) NaOMe, MeOH (83%); (vii) H$_2$, Pd-C, MeOH; (viii) [after step (vii)]: Dowex H$^+$, H$_2$O, 50°C (95% from **35**).

2.4. From L-Xylose

The synthesis of several L-xylofuranosyl nucleosides can be achieved using the xylofuranose **38** synthesized by acetolysis of the xylofuranose derivative **37** as a glycosyl donor in the Vorbrüggen glycosylation (Scheme 12).[43,44]

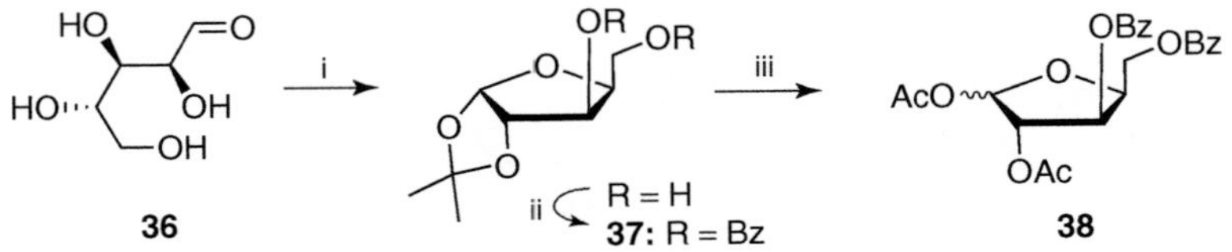

Scheme 12. *Reagents and conditions:* (i) ref. 44; (ii) BzCl, Py, CHCl$_3$; (iii) from **37**: AcOH, Ac$_2$O, H$_2$SO$_4$.

Using this strategy, several authors[45–48] made L-nucleoside derivatives despite the high price of this sugar. For example, L-xylose (**36**) was exploited as a starting sugar for synthesis of the four natural base nucleosides **40a–d** by its conversion into the L-ribose derivative **39** via an oxidoreduction procedure (Scheme 13).[47]

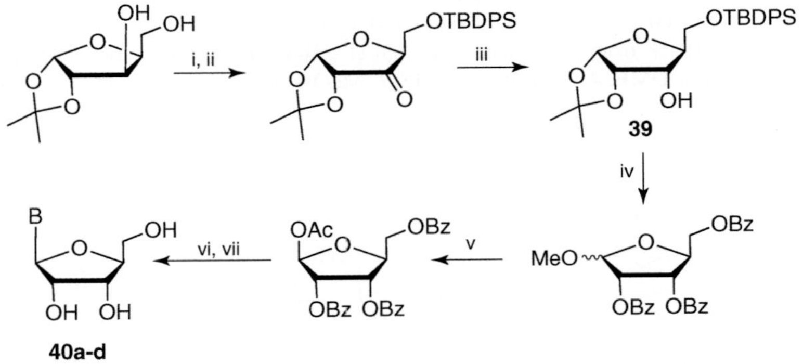

Scheme 13. *Reagents and conditions:* (i) TBDPSCl, Im, DMF (97%); (ii) CrO$_3$, Py, CH$_2$Cl$_2$ (97%); (iii) LiAlH$_4$, CH$_2$Cl$_2$ (96%); (iv) H$_2$SO$_4$, MeOH, then BzCl, Py (86%); (v) Ac$_2$O, AcOH, H$_2$SO$_4$ (77%, α/β = 1:1); (vi) protected B, TMSOTf; (vii) NaOMe, MeOH, or NH$_3$, H$_2$O [**40a**: B = U (84%); **40b**: B = C (76%); **40c**: B = A (77%); **40d**: B = G (63%)].

2.5. From L-Ascorbic Acid

A versatile strategy has been reported for the synthesis of the precursor L-2′-deoxyribose (**44**) that can be obtained starting from the commercially available unsaturated lactone **41**, which can also be synthesized from the inexpensive L-ascorbic acid (Scheme 14).[49] The key step is the 1,4-addition of (PhMe2Si)$_2$Cu(CN)Li$_2$ to unsaturated **42**, which furnishes the silyl derivative **43** with complete diastereoselection.

2.6. From D-Ribose

More recently, the precursor L-ribofuranose **47** was synthesized from D-ribose (**45**) (Scheme 15).[50] Deblocking of the 3,4-diol function of the aldehyde **46** with trifluoroacetic acid at room temperature followed by standard acetylation gives the

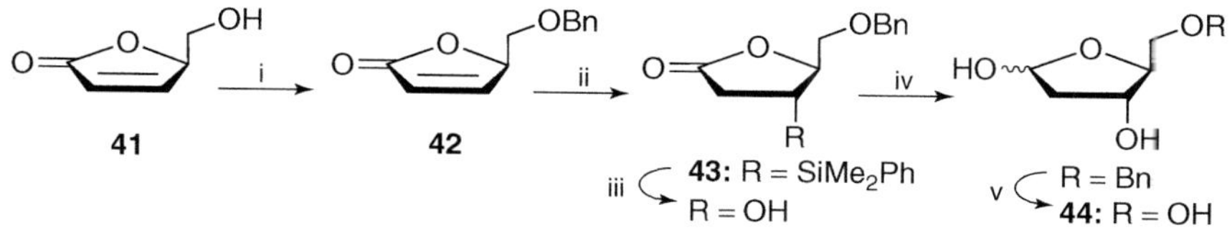

Scheme 14. *Reagents and conditions:* (i) BnOC(=NH)CCl$_3$, CF$_3$SO$_3$H, CH$_2$Cl$_2$/Cy (2:1) (84%); (ii) (PhMe$_2$Si)$_2$Cu(CN)Li$_2$, THF, 45°C (90%); (iii) Br$_2$, AcOOH, AcOH (65%); (iv) [after step (iii)]: disiamylborane, THF (60%); (v) HCOOH, Pd/C, MeOH (80%).

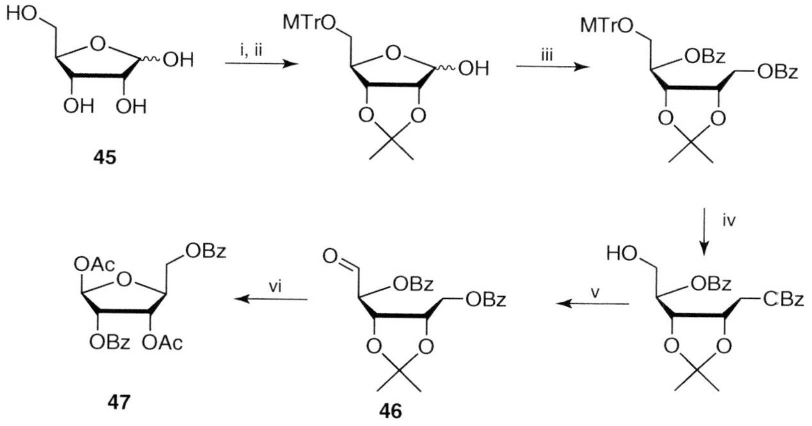

Scheme 15. *Reagents and conditions:* (i) acetone, TsOH, CaH$_2$ (80–90%); (ii) MTrCl, Py (80–90%); (iii) NaBH$_4$, EtOH, then BzCl in Py (80–95%); (iv) AcOH (76%); (v) PCC, CH$_2$Cl$_2$; 0°C (85%); (vi) TFA, then Ac$_2$O, Py (52%).

desired peracyl derivative of β-L-ribofuranose **47** as the main product. Compound **47** is employed for the preparation of 1-(β-L-ribofuranosyl)thymine and -cytosine.

2.7. From D-Xylose

The intermediate 1α-Cl-L-2′-deoxyribose **1α** can also be synthesized starting from D-xylose **48** following the synthetic pathway outlined in Scheme 16.[51] The key steps are the C2 debromination to achieve lactone **49** and the inversion of configuration at C4, producing the L-configured lactone **50**.

2.8. Fluorofuranosylnucleosides

The easiest way to prepare fluoronucleosides, with fluorine on the sugar, is through the nucleophilic displacement of a hydroxyl derivative with dietylaminosulfur trifluoride (DAST) in dichloromethane, as for the synthesis of the nucleosides **51** and **52** in Scheme 17[52] and **56** in Scheme 18,[53] respectively. The condensation

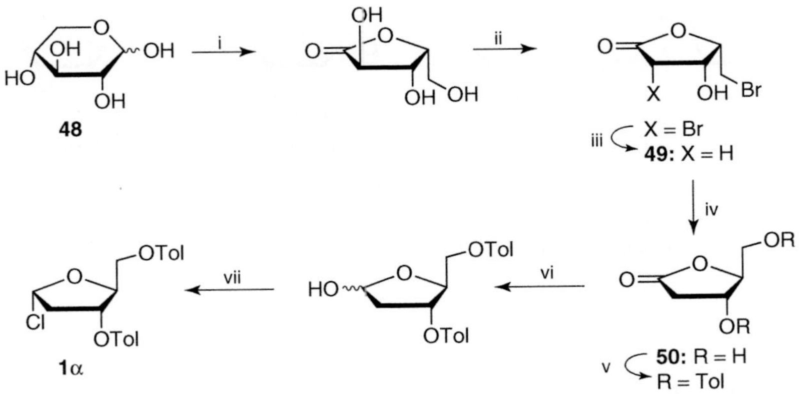

Scheme 16. *Reagents and conditions:* (i) Br$_2$, K$_2$CO$_3$, H$_2$O; then HCOOH; (ii) 30% HBr, AcOH, 45°C, then MeOH (40% from D-xylopyranose); (iii) H$_2$, Pd/C, EtOAc, TEA (96%); (iv) [after step (iii)]: KOH, H$_2$O; then H$^+$-resin; (v) TolCl, Py, DME (37% from **49**); (vi) [after step (v)]: DIBAL-H, DME, −60°C; (vii) HCl, AcOH, TBME (65% from **50**).

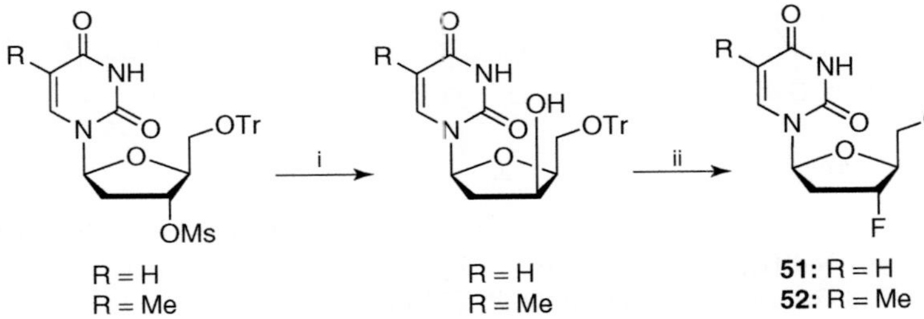

Scheme 17. *Reagents and conditions:* (i) NaOH, EtOH, reflux (47–70%); (ii) DAST, CH$_2$Cl$_2$ (48% for **51** and 51% for **52**).

product **53** is deprotected and cyclized to the anhydronucleoside **54**. Protection by the tetrahydropyranyl group followed by hydrolysis gives the arabinonucleoside **55**, and fluorination with DAST followed by deprotection affords the L-fluorothymidine derivative **56**. Various 2′-deoxy-2′-fluoro-β-L-arabinofuranosyl pyrimidine nucleosides are also prepared using KHF$_2$ for the preparation of fluoride compounds.[54]

A linear synthesis of imidazole nucleosides **61** involves construction of the heterocyclic moiety on the 1′-azidofluorosugars **57** (Scheme 19).[55] Upon hydrogenation, azide **57** is converted to the glycosylamine **58**, which is immediately treated with freshly prepared formimidate **59** to give an anomeric mixture of separable α and β isomers **60** in a 43% overall yield.

Several L-β-3′-deoxy-3′,3′-difluoronucleosides (**68–72**) are prepared starting from D-glyceraldehyde acetonide (D-**62**), from which the difluorohomoallyl alcohols **63** are achieved with good *syn/anti* selectivity (Scheme 20).[56,57] Conversion

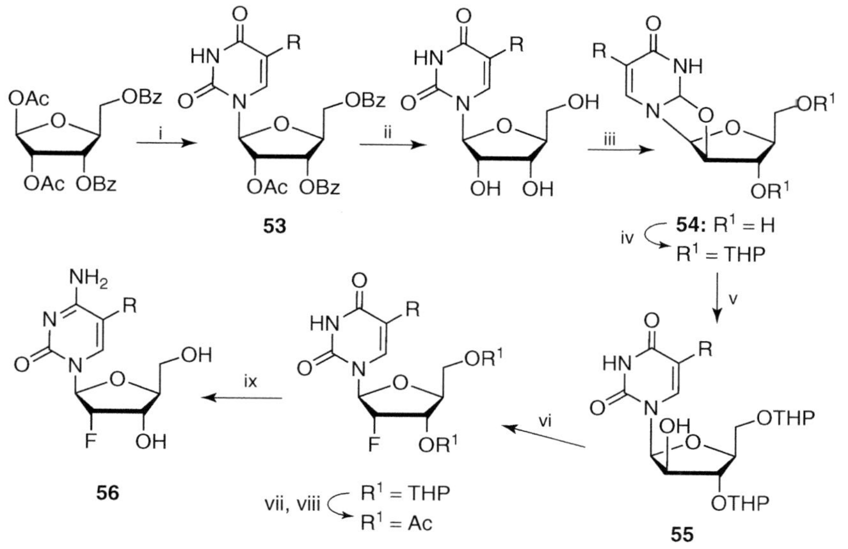

Scheme 18. *Reagents and conditions:* (i) HMDS, base, reflux, then TMSOTf, MeCN; (ii) NaOMe, MeOH; (iii) DMF, Ph$_2$CO, NaHCO$_3$, 150°C; (iv) DMF, DHP, TsOH, 0°C; (v) NaOH, MeOH, H$_2$O; (vi) DAST, CH$_2$Cl$_2$, Py, −70°C to reflux, 4 h; (vii) TsOH, MeOH; (viii) Ac$_2$O, Py (41% from **54**); (ix) [after step (viii)]: 1,2,4-triazole, POCl$_3$, Py, then NH$_3$, MeOH. In compounds **53–56**, R = Me or F.

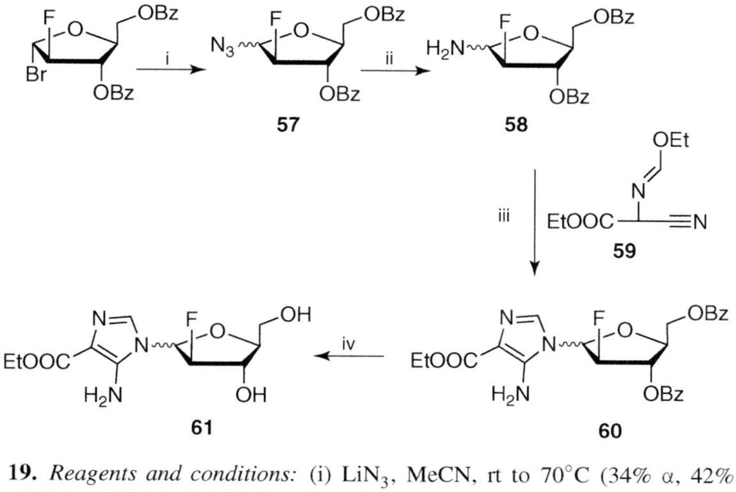

Scheme 19. *Reagents and conditions:* (i) LiN$_3$, MeCN, rt to 70°C (34% α, 42% β); (ii) PtO$_2$, H$_2$, EtOAc; (iii) MeCN (25% α, 21% β); (iv) NH$_3$, MeOH (74% α, 51% β from the corresponding anomer).

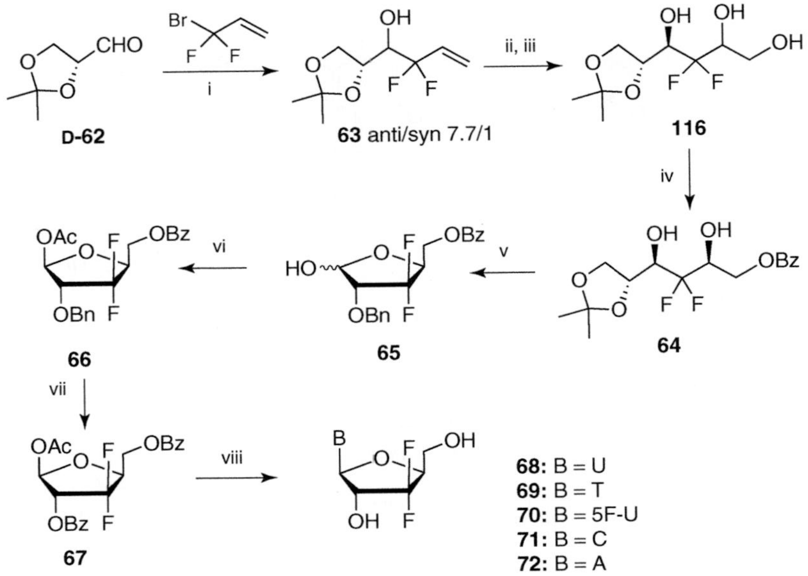

Scheme 20. *Reagents and conditions:* (i) DMF, rt (90%); (ii) NaH, BnBr, TBAI, THF; (iii) OsO$_4$, NMNO, acetone/H$_2$O (95%); (iv) BzCl, Py, CH$_2$Cl$_2$ (90%); (v) 75% AcOH, 50°C, then NaIO$_4$, acetone/H$_2$O, rt (94%); (vi) Ac$_2$O, DMAP, CH$_2$Cl$_2$, (94%); (vii) NaBrO$_3$, Na$_2$S$_2$O$_4$, EtOAc/H$_2$O, then Bz$_2$O, DMAP, Et$_3$N, CH$_2$Cl$_2$ (86%); (viii) BSA, pyrimidine, or 6-chloropurine, then NH$_3$, MeOH.

of **64** to furanose **65** is achieved by acidic hydrolysis with 75% acetic acid and oxidation with sodium periodate, followed by cyclization. The *O*-acetylation of furanose **65** affords the corresponding acetate **66** nearly as a single β-anomer, due to the assistance of neighboring large-group participation. Coupling of **67** with various persilylated pyrimidines and 6-chloropurine is carried out using Vorbrüggen conditions. A series of 2′-deoxy-2′,2′-difluoro-L-erythropentofuranosyl nucleosides have also been reported, starting from L-gulono-γ-lactone.[58]

2.9. *C*-Nucleosides

The first syntheses of L-*C*-nucleoside analogues were accomplished as a mixture of D- and L-enantiomers starting from achiral materials.[59–66] A multistep stereoselective approach to L-*C*-nucleosides (Scheme 21)[67] has been reported from L-xylose (**36**) following a synthetic pathway developed previously.[54] Protected L-ribose **73** is transformed to the key intermediate **74** by treatment with diethyl cyanomethylenephosphonate, followed by formylation with bis(dimethylamino)-*t*-butyloxymethane. The reaction of compound **74** with hydrazine and the following cyclization yields the aminopyrazole **75** as a mixture of anomers. Finally, the reaction of **75** with methyl *N*-cyanoformimidate gives an anomeric mixture of

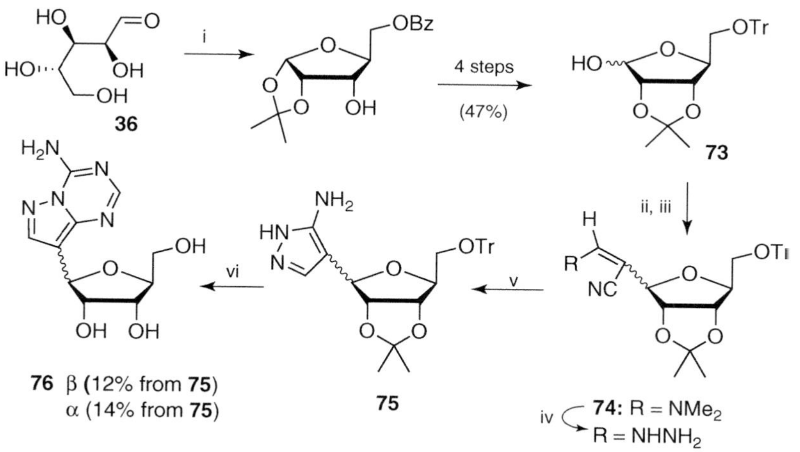

Scheme 21. *Reagents and conditions:* (i) ref. 54; (ii) (EtO)$_2$OPCH$_2$CN, DME (58%); (iii) tBuOCH(NMe$_2$)$_2$, Me$_2$NCHO, CH$_2$Cl$_2$ (93%); (iv) NH$_2$NH$_2$·NH$_2$NH$_2$·HCl MeOH, H$_2$O; (v) [after step (iv)]: MeCN, reflux (yield 57% from **74**); (vi) NCNHCHOCH$_3$, then 10% HCl, MeOH.

C-nucleosides, which is deprotected to obtain the free nucleosides **76** as an α/β anomeric mixture.

A strategy toward L-C-nucleoside analogues has been explored in the context of the synthesis of L-showdomycin using an enantioselective palladium-catalyzed allylic alkylation reaction (Scheme 22).[68] The heterocyclic dibenzoate **77** is reacted with the anion of the succinimide derivative **80**, using the allylpalladium chloride complex **78** as a catalyst and the (R,R)- configuration of chiral ligand **79**, to achieve the desired carbon adduct **81** almost exclusively when THF is used as a solvent. A second regio- and diastereoselective allylic alkylation reaction of **81** with the protected tartronic acid **82** affords the 2,5-dihydrofuran **83**. To make the synthesis more general for all C-nucleosides, the authors also explored the use of compound **82** in the desymmetrization step. *cis*-Dihydroxylation of **83** with osmium tetraoxide and N-methylmorpholine N-oxide (NMO) as the cooxidant affords the diol **84**. To remove the benzyl and benzyloxycarbonyl protecting groups from **85**, catalytic hydrogenolysis furnishes the desired hydroxy diacid **86** in essentially a quantitative yield. To convert the hydroxy diacid of **86** to the hydroxymethyl moiety of **88**, a decarboxylation with pure ground lead tetraacetate followed by direct reduction of the resulting acid **87** can be exploited. Completion of the synthesis of showdomycin **89** from **88** requires removal of the protecting groups.

2.10. Other Ribose and Deoxyribose Nucleosides

Other L-nucleosides are prepared by using the Vorbrüggen methodology with suitable glycosyl donors and natural bases, such as 3′-deoxyribo-L-pentofuranosylnucleosides,[69,70] L-lyxofuranosyl-nucleosides,[71] arabino- and

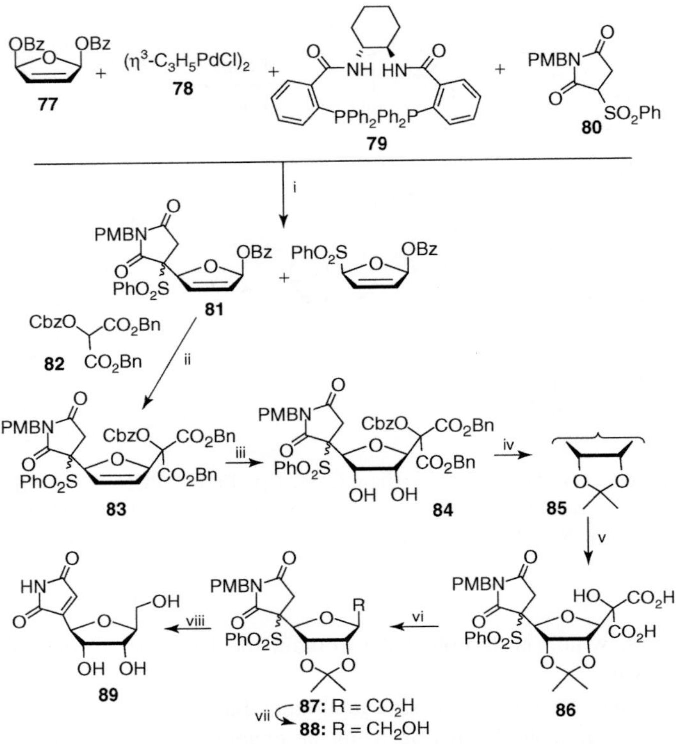

Scheme 22. *Reagents and conditions:* (i) NaH, THF, 0°C, 4 h (67% of **81**, 97% ee); (ii) Cs$_2$CO$_3$, **78**, MeCN (66%); (iii) OsO$_4$, NMO, aq. KH$_2$PO$_4$, CH$_2$Cl$_2$ (89%); (iv) Me$_2$C(OMe)$_2$, PPTS, CH$_2$Cl$_2$ (92%); (v) Pd/C, H$_2$, EtOAc (97%); (vi) Pb(OAc)$_4$, H$_2$O, acetone; (vii) HOBT, DCC, THF, then LiBH$_4$ (63% from **86**); (viii) from **88**: CAN, MeCN, H$_2$O (72%), then TFA, H$_2$O (84%), then DBU, DMSO (32%).

xylonucleosides,[72] and modified bases such as 5-azacytidines,[73] 1,2,4-triazolines,[74] 5-fluorouracil, and 5-fluorocytidine,[75] trifluoromethyluridine,[76,77] 3′-*C*-cyano-3′-deoxyribonucleosides,[78] 7-iodo-5-aza-7 deazaguanine,[79] pyrrolo[2,3-*d*]pyrimidines, pyrazolo[3,4-*d*]pyrimidines, benzimidazoles, imidazo[1,2-*a*]-s-triazines,[80] 7-functionalyzed-7-deazapurines,[79,81] and substituted benzimidazoles.[82]

3. 2′,3′-DIDEOXYNUCLEOSIDES

This section covers furanose nucleosides without the 2′ and 3′ hydroxyl groups, which can be replaced by hydrogen (dideoxy; also known as dd-nucleosides) by a C=C double bond (dideoxy-didehydro or D4-nucleosides) or by other heteroatoms, such as thiol, amino, fluorine, and azide groups. Some of the compounds that also have other structural characteristics (such as carbocyclic D4-nucleosides) are described elsewhere (section 4). Many of these compounds, if not all, have been prepared and studied for their antiviral properties after discovery of the activity of AZT in the treatment of AIDS.

The most "trivial" method of dideoxynucleoside preparation is transforming the 2' or 3'- deoxyribonucleoside analogues into the corresponding thionoester, then reducing it (e.g., by use of trimethylsilylsilane) to the dideoxynucleoside, as reported by several authors.[17,83] Other very different approaches have been made from a variety of organic compounds, often ending with racemic mixtures of D- and L- derivatives. Some examples are reported in this section.

A variety of pyrimidinic and purinic dd-nucleosides can be prepared using Vorbrüggen condensation on 1-O-acetyl-2,3-dideoxy-L-ribofuranoses **90** and **91**, as in Scheme 23,[84,85] even if other protecting groups can be used.

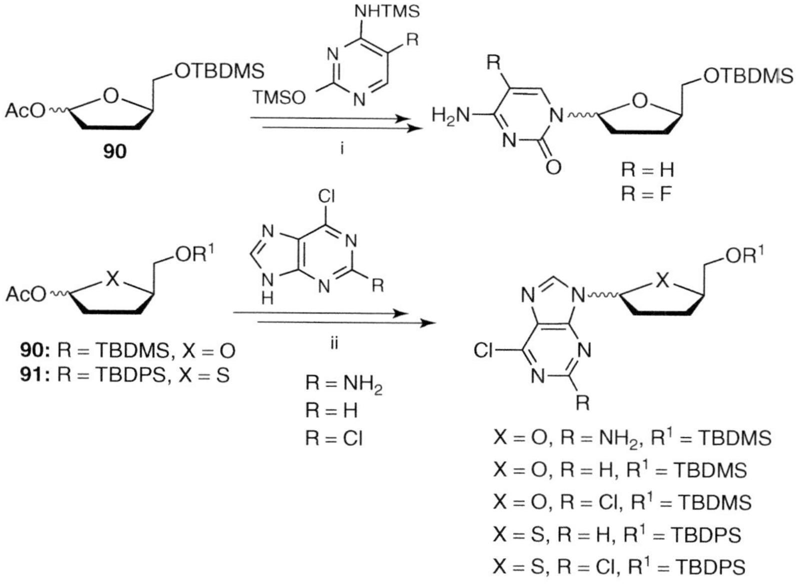

Scheme 23. *Reagents and conditions:* (i) EtAlCl$_2$, CH$_2$Cl$_2$, N$_2$, then n-Bu$_4$F; (ii) NaH, MeCN, then Et$_2$AlCl, MeCN.

The examples above reflect the use of disconnective analysis (Figure 3) showing that the simplest way to obtain 2',3'-β-L-dideoxynucleosides **92** is a condensation between a base and the acetate **93** derived from the lactol **94** obtained from the reduction of the γ-lactone **95**, whose preparation has been the target of several papers.[86–89]

Many chemists have also demonstrated that the presence of a bulky, removable substituent in the C2α position is a useful expedient to orient glycosylation of the base on the β-face of the cycle,[90–97] as shown in Scheme 24.[98] Here, the syntheses

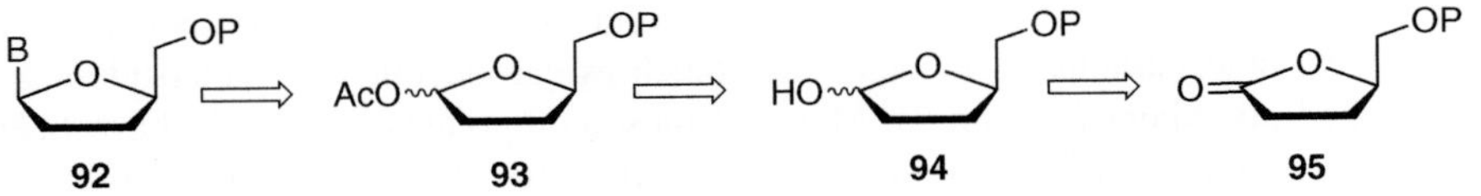

Figure 3. Disconnection analysis of β-L-dd-nucleosides. P = protecting group.

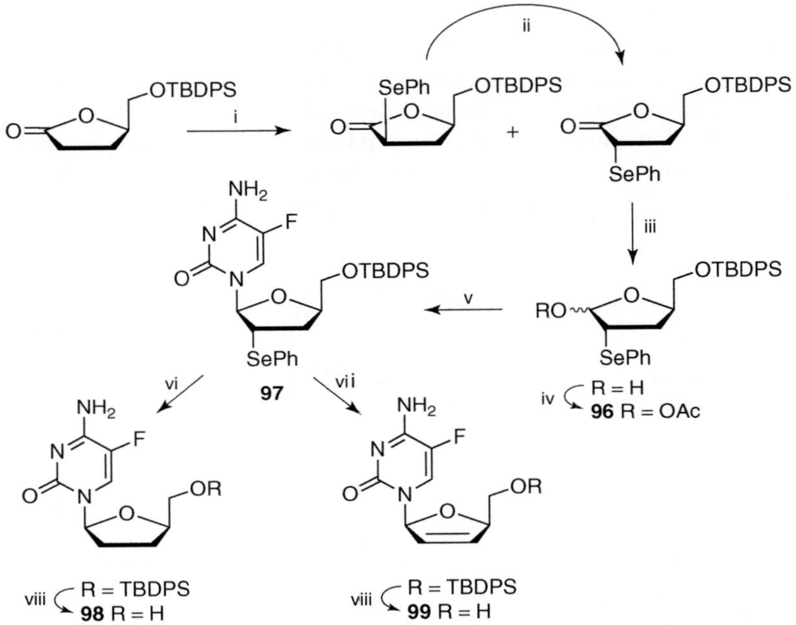

Scheme 24. *Reagents and conditions:* (i) LiHMDS, TMSCl, THF, then PhSePht, −78°C (70% yield; α/β 40:1); (ii) DBU, THF; (iii) DIBAL-H, toluene −78°C (87–98 %); (iv) Ac$_2$O, TEA (98%); (v) from **96**: bistrimethylsilyl(5-fluoro)cytosine, TMSOTf (95–100 %); (vi) Bu$_3$SnH, AIBN (95–100%); (vii) H$_2$O$_2$, Py (86–95%); (viii) TEA-HF (yields: 78% and 87–95% for **98** and **99**, respectively).

of both β-L-FD4C (**99**) and β-L-FddC (**98**) are completed in three steps from the acetate **96**. Treatment of nucleoside **97** with hydrogen peroxide in the presence of pyridine affords upon triethylamine trihydrofluoride-mediated desilylation the D4 nucleoside **99**. Reductive removal of the phenylseleno moiety in **97** using tributyltin hydride as the reductant affords the dd-nucleoside **98** in almost quantitative yield.

In a similar way, 2′-fluoro-2′,3′-dideoxynucleosides,[99,100] 3′-fluoro-2′,3′-dideoxynucleosides,[101] and 2′-fluoro-2′,3′-unsaturated analogues[102,103] can be prepared from the corresponding 1-*O*-acetylated or benzoylated sugar with a suitable protecting group (TBDMS or Bz) on the 5-hydroxyl group.

The 2′,3′-dideoxy-3′,3′-difluoropyrimidine nucleosides **103–106** and 3′-fluoro-D4 nucleosides **107–110** have been prepared according to Scheme 25.[104]

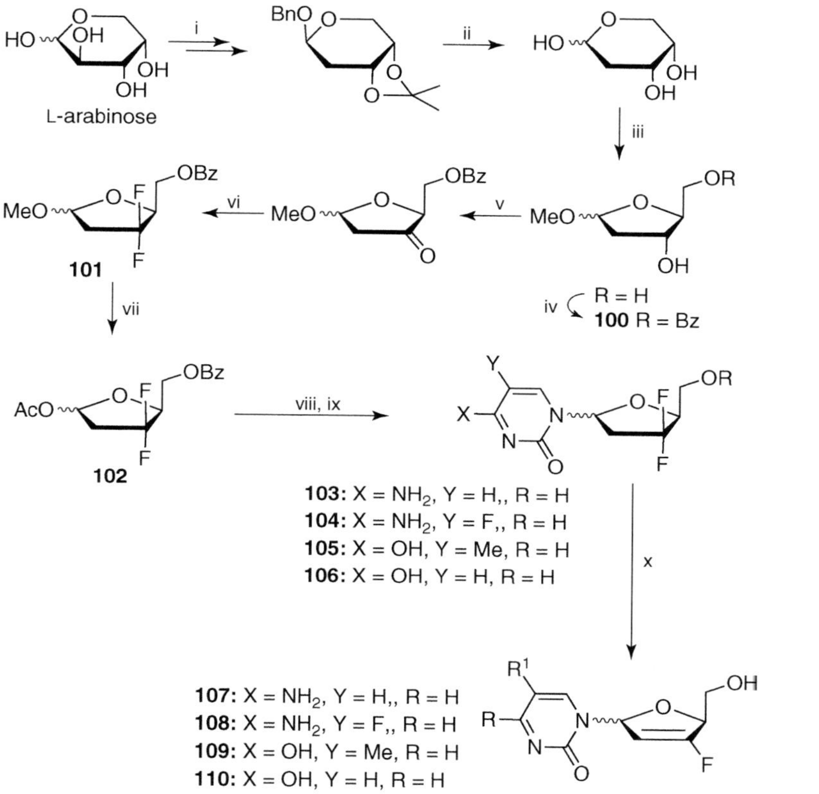

Scheme 25. *Reagents and conditions:* (i) BnOH, HCl, then DMP, TsCl, acetone, then CSCl$_2$, PhOH, CH$_2$Cl$_2$, then n-Bu$_3$SnH, AIBN, toluene, reflux; (ii) 4% TFA, 40°C; (iii) HCl·MeOH; (iv) BxCl, Py; (v) [after step (iv)]: PDC, Ac$_2$O, CH$_2$Cl$_2$; (vi) DAST, CH$_2$Cl$_2$, reflux; (vii) Ac$_2$O, H$_2$SO$_4$, AcOH; (viii) HMDS, MeCN, pyrimidines; (ix) NH$_3$, MeOH, rt; (x) NaOMe, DMF (or 2:1 DMF/dioxane), rt.

Pyridinium dichromate–mediated oxidation of compound **100** followed by difluorination with DAST provides the difluororibofuranoside **101**, which is transglycosylated to the key intermediate 1-*O*-acetylribofuranoside **102**, in high yield, by treatment with acetic anhydride and acetic acid in the presence of sulfuric acid. For the preparation of purine nucleosides, the key intermediate **102** is condensed with 6-chloropurine and 2-fluoro-6-chloropurine. This method is also applicable to 2′,2′-difluoro-2,3-dideoxynucleosides[105,106] and to carbocyclic nucleosides[107] starting from the corresponding analogues of compound **102**.

An interesting synthetic strategy based on the use of furan, pyrrole, and thiophene derivatives arises from the retrosynthetic analysis shown in Figure 4.[108] The analysis shows that D- and L-2′,3′-dideoxynucleosides **111** can be synthesized straightforwardly from intermediates deriving from the reaction of

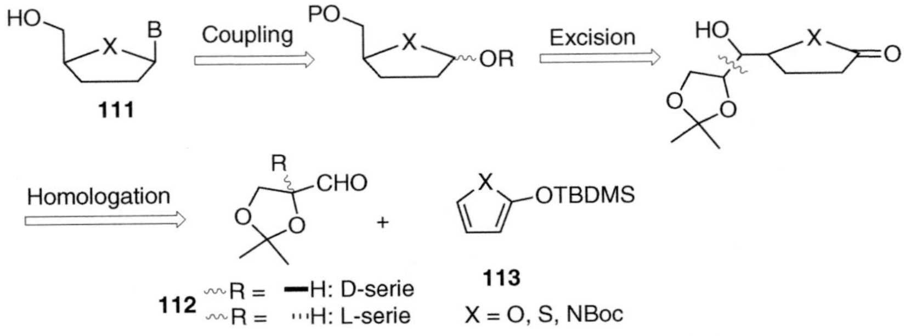

Figure 4. Retrosynthetic analysis of dideoxynucleosides.

D- and L-glyceraldehyde diacetonide (**112**) with a derivative of five-membered eterocycles (**113**). This method allows, for example, the preparation of 2′,3′-dideoxy-4′-thionucleosides or Boc-protected 4′-aza-nucleosides, and "normal" 2′,3′-dideoxynucleosides.[109]

An interesting approach produces both D- and L-(diethoxyphosphoryl)difluoromethyl nucleoside analogues,[110] as shown in Scheme 26 for the synthesis of the thymidine analogue **121**. R_S-2-Methyl-5-(4-methylphenylsulfinyl)pent-2-ene (**114**) is condensed with diethyl(ethoxycarbonyl)(difluoromethyl)phosphonate (**115**) to give compound **116**, which is reduced by $NaBH_4$ to a mixture of diastereoisomers **117** that can be separated (to give L- and D-derivatives) into mixtures of 3-epimers. After the sulfinyl function of the (2S)-epimer **117** is reduced and removed to give the (2R)-alcohol **118**, the duplicity on C3 is removed and the compound can be cyclized into lactol **120** via the formation of an intermediate aldehyde through oxidative cleavage of the C=C bond of compound **118**. The γ-lactone **119** arising from excessive oxidation of the aldehyde is formed in low yield (6%) and can be reconverted to the lactol **120** by reduction with DIBAL-H.

The L-2′,3′-dideoxyisonucleosides (oxa and thio) derivatives **124–127** have been prepared according to Scheme 27,[111,112] starting from the monobromobenzyl ester **122** or from the diepoxypentane **123**.

The anomeric mixture of C-nucleosides **129** is prepared from lactone **128** obtained in six steps from L-gulonic-γ-lactone, building the modified pyrimidine system step by step as shown in Scheme 28.[113] In the same paper, compounds **130** and **131** are also prepared by modification of the synthetic pathway (Figure 5).

The synthesis of the 2′-fluoro-4′-thio-D4-nucleoside **141** via the fluorobutenolide **132**, in turn derived from the diacetonide of L-glyceraldehyde L-**62**, combines several interesting chemical steps[114,115] (Scheme 29). (R)-2-Fluorobutenolide **132** is hydrogenated by treatment with 5% Pd-C under a hydrogen atmosphere to allow complete conversion to β-2-fluorolactone **133** in a quantitative yield. The lactone **133** is converted to the iodoester **134** in three consecutive steps, then a thioacetate group is introduced by nucleophilic displacement of the iodide group with potassium thioacetate to give the corresponding thioacetate **135**. DIBAL-H-induced cyclization of the thioacetate **135**, followed by Moffatt-type

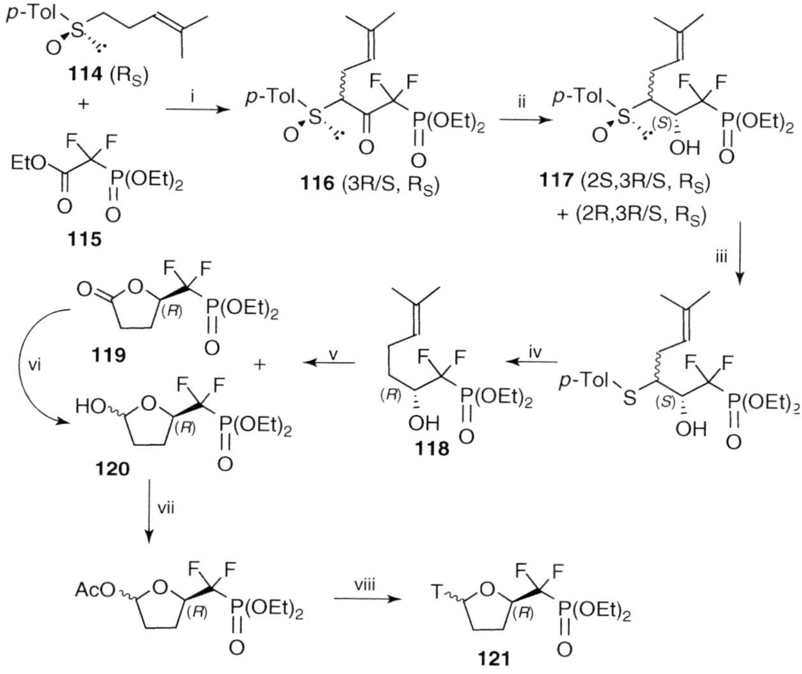

Scheme 26. *Reagents and conditions:* (i) LDA, THF, −70°C; (ii) NaBH$_4$, MeOH, NH$_3$ (85% from **114**); (iii) NaI, TFAA, acetone, −20°C; (iv) Raney-Ni, EtOH, 80°C; (v) RuCl$_3$, CCl$_4$/MeCN/H$_2$O (1:1:2), NaIO$_4$, 0°C; (vi) DIBAL-H, toluene, −60°C; (vii) Ac$_2$O, Py, 0°C; (viii) from **120**: T, HMDS, (NH$_4$)$_2$SO$_4$, TMSOTf, (CH$_2$Cl)$_2$, 45°C (overall yield 45% from **114**).

oxidation of the resulting thiolactol, gives the thiolactone **136**. Compound **136** is deprotonated by LiHMDS and trapped as a TMS enol ether, which is phenylselenylated using PhSeBr to introduce the 2-phenylselenyl group exclusively at the R-position of lactone **137**. Finally, the thiolactone **137** is reduced by DIBAL-H to give the corresponding lactol, which is acetylated to afford the acetate **138**. Condensation of the acetate **138** with N^4-benzoylcytosine under Vorbrüggen conditions gives the corresponding cytidine analogue **139**, which undergoes MCPBA oxidation to give the unsaturated 2′-fluoro-4′-thiocytidine **140**. After deprotection of TBDPS and benzoyl groups, the target compound **141** is at last obtained.

2′,3′-Dideoxy-3′-tris(methylthio)methyl-β-L (and D-) pentofuranosyl nucleosides can be prepared in a similar way, by Vorbrüggen glycosilation of compound **143**, also starting from D- and L-glyceraldehyde **62**, where the bulky tris(methyltio)methyl group is introduced by a Michael addition to the vinylic lactone **142** (Scheme 30).[116]

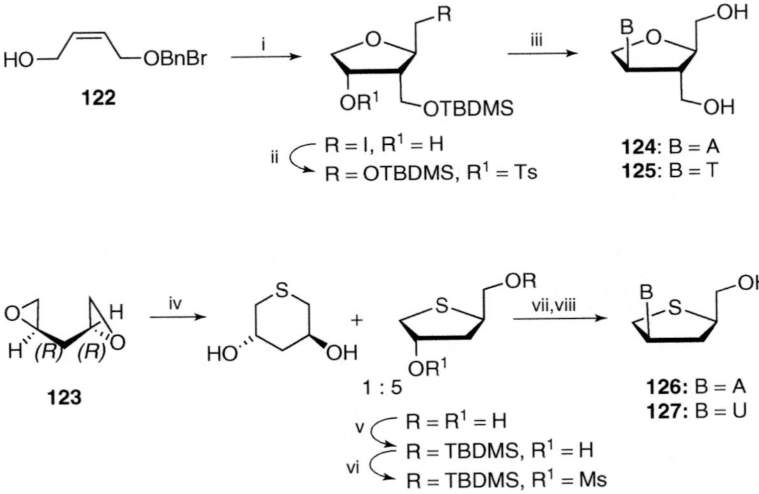

Scheme 27. *Reagents and conditions:* (i) *L*-(+)-DIPT, Ti(O-*i*-Pr)$_4$, *t*-BuOOH, CH$_2$Cl$_2$, molecular sieves, −25°C (81%, 96% ee); then CH$_2$=CHMgBr, CuI, THF, −78°C to rt (73%), then TBDMSCl, TEA, DMAP, CH$_2$Cl$_2$ (79%), then Ag(coll)$_2$ClO$_4$, I$_2$, CH$_2$Cl$_2$ (52%); (ii) AcCl, Py, 0°C (81%), then KOAc (10 equiv.) DMSO 87°C, (41%), then NH$_3$, MeOH, 0°C to rt (93%), then TBDMSCl, TEA, DMAP, CH$_2$Cl$_2$ (62%), then TsCl, Py, 0°C to rt, 18 h (68%); (iii) [after step (ii)]: A or T, 18-crown-6, K$_2$CO$_3$, DMF, 90°C, 18 h, then TBAF, THF (35% for **124**, 22% for **125**); (iv) Na$_2$S, aq. EtOH, 0°C, 7 h; (v) TBDMSCl, DMAP, TEA [57% from step (iv)]; (vi) MsCl, DMAP, TEA (89%); (vii) [after step (vi)]: A or U, 18-crown-6, K$_2$CO$_3$, DMF, 105°C, 18 h (45–36%); (viii) TBAF, THF (99–100%).

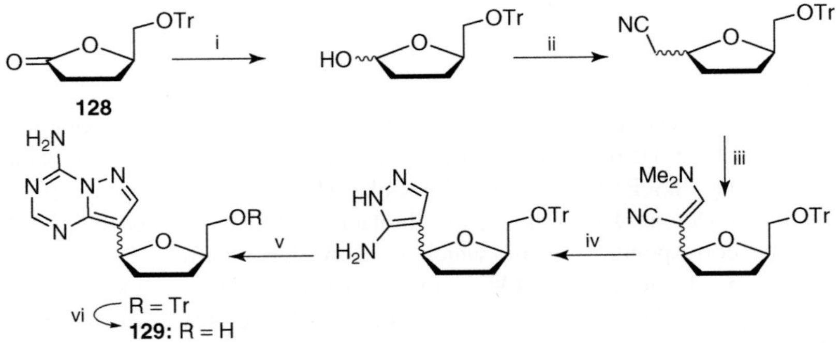

Scheme 28. *Reagents and conditions:* (i) DIBAL-H, CH$_2$Cl$_2$, −78°C (97%); (ii) NaH, NCCH$_2$PO(OEt)$_2$, 1,2-dimethoxyethane, 0°C (80%); (iii) HC[(NMe$_2$)$_2$]O-*t*-Bu, DMF rt to 70°C (71%); (iv) NH$_2$NH$_2$, NH$_2$NH·HCl, MeOH, H$_2$O, 70°C, 16 h, then MeCN, 80°C, 21 h (71%); (v) NCN=CHOCH$_3$, benzene, 74°C, 17 h (57%); (vi) 12% aq. HCl in MeOH (1:1) (43% **129β** + 31% **129α**).

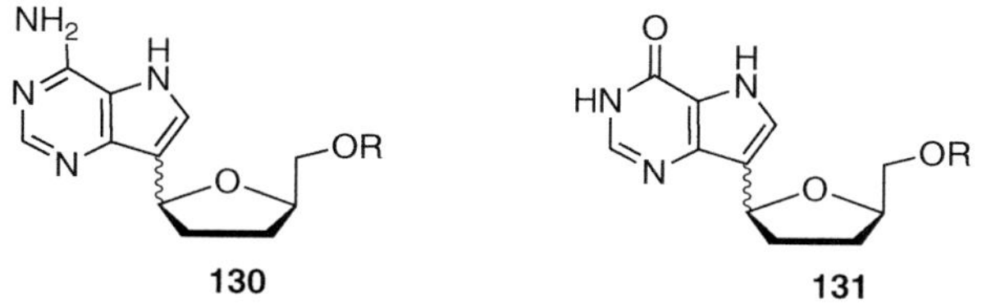

Figure 5. Structure of C-nucleosides **130** and **131**.

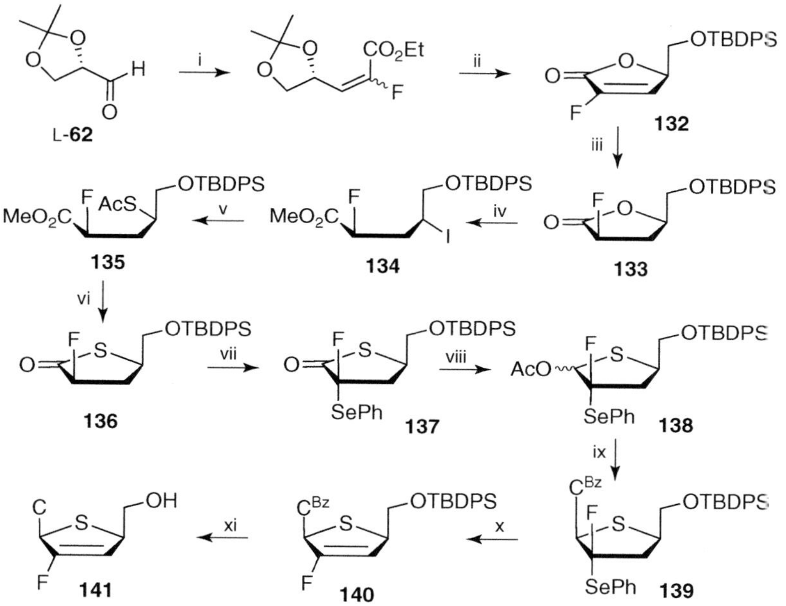

Scheme 29. *Reagents and conditions:* (i) (EtO)$_2$P(O)CHFCO$_2$Et, NaHMDS (80%); (ii) HCl, EtOH, then TBDPSCl (70%); (iii) H$_2$, Pd(0), EtOAc (100%); (iv) NaOH aq. EtOH, then Me$_2$SO$_4$, DMSO, then I$_2$, Ph$_3$P, Im, toluene, 60°C (83%); (v) KSAc, DMF (91%); (vi) DIBAL-H, toluene, −78°C, then Ac$_2$O, DMSO (54%); (vii) LiHMDS, TMSCl, PhSeBr, THF, −78°C (74%); (viii) DIBAL-H, toluene, −78°C; then Ac$_2$O, TEA, CH$_2$Cl$_2$ (90%); (ix) silylated N^4-benzoylcytosine, TMSOTf, MeCN (51%); (x) MCPBA, CH$_2$Cl$_2$, −78°C; Py, rt (80%); (xi) TBAF, THF, then NH$_3$, MeOH.

Also, compounds such as L-3′-C-cyano-2′,3′-unsaturatednucleosides **144** and L-3′-C-cyano-3′-deoxyribonucleosides **145**[78] and 3′-fluoro-2′,3′-didehydro-4′-ethynyl-L (and D)- nucleosides **146**,[117] depicted in Figure 6, have been synthesized and studied.

4. CARBOCYCLIC NUCLEOSIDES

Carbovir (**147**, CBV, Figure 7) has been reported as the first carbocyclic nucleoside analogue, with potent anti-HIV activity in vitro; its discovery provided a base

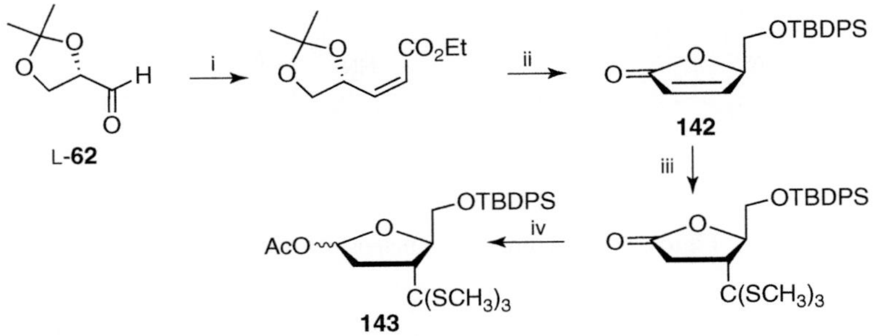

Scheme 30. *Reagents and conditions:* (i) (EtO)$_2$P(O)CH$_2$CO$_2$Et, NaHMDS; (ii) HCl, EtOH, then TBDPSCl; (iii) (MeS$_3$)CH, *n*-BuLi, THF, −78°C (81%); (iv) DIBAL-H, toluene, −70°C, then AcCl, Py, CH$_2$Cl$_2$ 0°C to rt (62%).

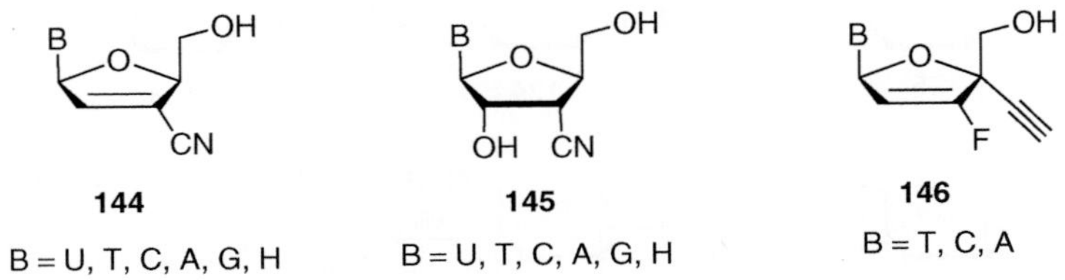

Figure 6. Structure of 3′-*C*-cyano-2′,3′-unsaturated, L-3′-*C*-cyano-3′-deoxyribo, and 3′-fluoro-2′,3′ didehydro-4′-ethynyl-L-nucleosides.

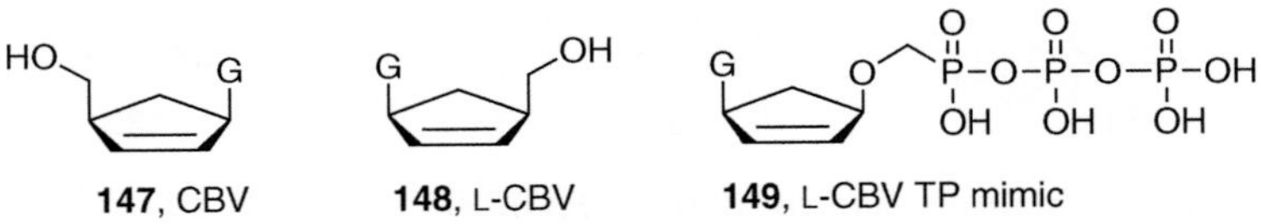

Figure 7. Structures of carbovir and its mimic.

for the synthesis of other carbocyclic analogues. The first synthesis of carbovir had been accomplished in 1990 by Vince et al. as a racemic mixture of the two enantiomers.[118,119] Afterward, a chemoenzymathic synthesis of both enantiomers of carbovir has been reported.[120] The natural (−)-enantiomer of CBV (**147**) is primarily responsible for the antiviral activity,[121] whereas the L-CBV (**148**) is weakly active against HBV.[122] Moreover, the carbovir triphosphate mimic and its L-enantiomer **149** have also been synthesized, and the L-enantiomer is surprisingly the more active compound.[123,124]

A stereoselective synthesis of L-carbovir (**148**) can be achieved through asymmetric [2 + 3]-cycloaddition of chiral oxazoline-*N*-oxide **150** with cyclopentadiene (Scheme 31).[125] The exo-adduct **151**, the only product isolated, is transformed into the dimethoxyacetal **152** after a sequential two-step treatment: namely, oxidation and acidic methanolysis. Transformation of acetal **152** into the diol

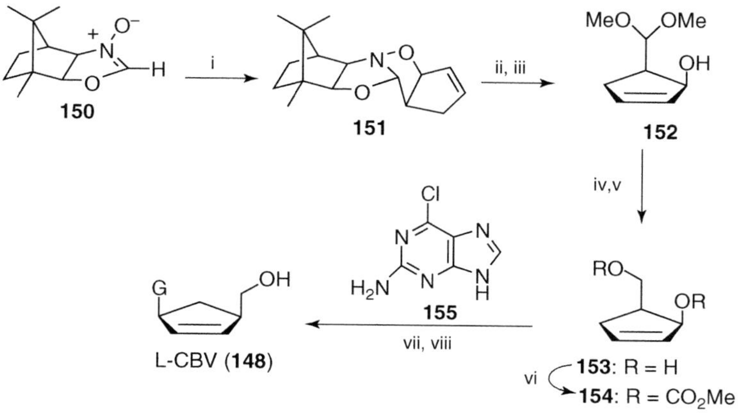

Scheme 31. *Reagents and conditions:* (i) cyclopentadiene, CH$_2$Cl$_2$, 40°C, 24 h (75%); (ii) MCPBA, 0°C, 3 h; (iii) CSA, MeOH, 20°C (65% from **151**); (iv) CSA, MeCN–6% H$_2$O, 20°C; (v) NaBH$_4$, 0°C; (vi) ClCO$_2$Me, Py, 0–20°C, 15 min (60% from **152**); (vii) 5% Pd(Ph$_3$P)$_4$, THF/DMSO 1:1; (viii) 0.5 N NaOH, reflux, 4 h.

153 is accomplished by treatment with CSA, followed by reduction with NaBH$_4$. By using a Pd(0)-catalyzed Trost condensation between compound **154** and 2-amino-6-chloropurine, L-carbovir **148** is at last obtained.

The synthesis of various carbocyclic analogues of L-ribofuranosides (i.e., L-aristeromycin **157** and L-neplanocin A **158**) has been reported, exploiting both L2-cyclopentenone L-**155** and its enantiomer D-**156** as the key compound (Figure 8).[126]

For the synthesis of L-aristeromycin (**157**) and other carbocyclic L-ribofuranoside analogues, the cyclopentenone L-**155** can be employed (Scheme 32).[127,128] Reaction of L-**155** with lithium bis(*t*-butoxymethyl)cuprate gives the optically pure cyclopentanone **159**, which is selectively reduced to alcohol **160** with DIBAL-H. The triflate derivative **162** reacts with the sodium salt of adenine to provide the protected L-aristeromycin **163** in low yield. Due to this drawback, a linear approach is adapted for the synthesis of the carbocyclic nucleosides **152–154**. Treatment of mesyl derivative **161** with lithium azide gives, after hydrogenation in the presence of 5% Pd/C, the amino derivative **164**. The reaction of **164** with β-methoxy-α-methacryloyl isocyanate and β-methoxyacryloyl isocyanate, followed by basic treatment and removal of the protective groups, affords the thymine **166** and uracil **167**, respectively. The uracil analog **165** may also be converted into the cytosine derivative **168**.

The L-2-cyclopentenone L-**155** is also employed for the synthesis of various β-L-2′,3′-idehydro-2′,3′-dideoxy and β-L-2′,3′-dideoxy pyrimidine and purine nucleoside analogues.[129] The thymidine analogue **174** (Scheme 33) is obtained from cyclopentene **172**, derived by diol **157** using both thermal elimination (via cyclic ortho ester **170**) or deoxygenation (via cyclic thionocarbonate **171**) reactions. After removal of the benzoyl group, the alcohol **173** is finally reacted

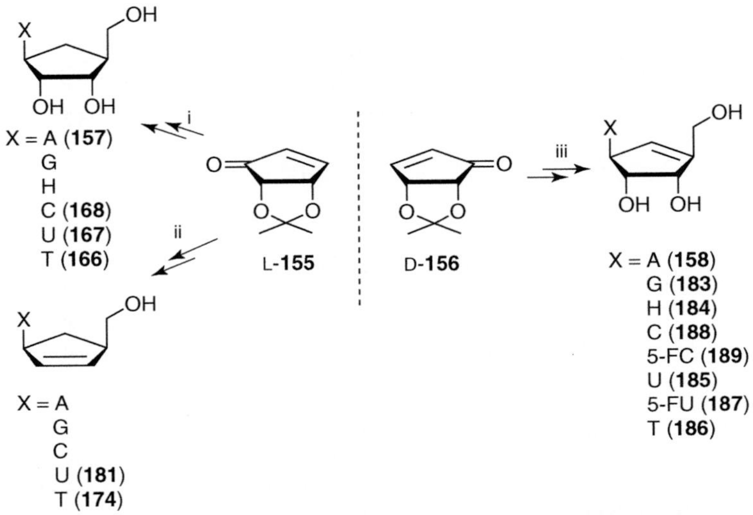

Figure 8. Cyclopentenones L-155 and D-156 as intermediates for various carbocyclic nucleosides: (i) refs. 127 and 128; (ii) ref. 129; (iii) ref. 130.

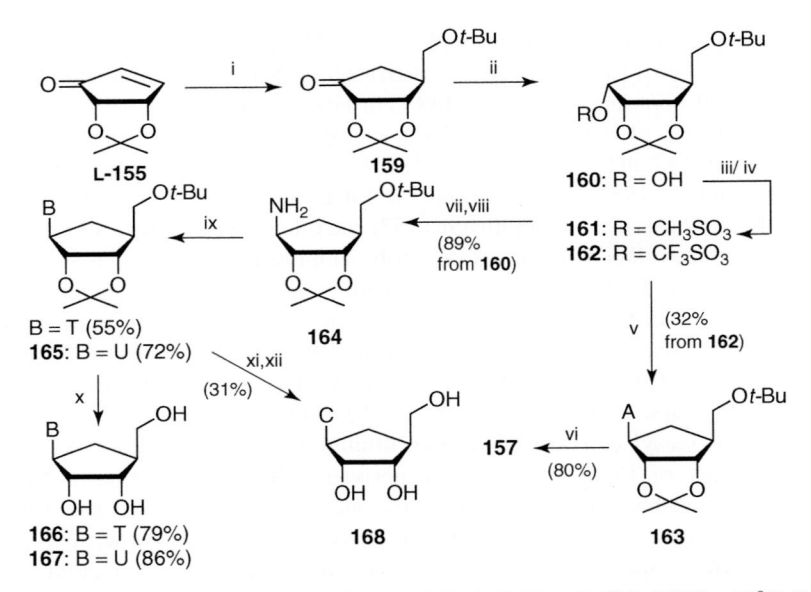

Scheme 32. *Reagents and conditions:* (i) (t-BuOCH$_2$)$_2$CuLi, t-BuOMe/THF, −30°C (87%); (ii) DIBAL–H, CH$_2$Cl$_2$, −78°C (82%); (iii) MsCl, Et$_3$N, CH$_2$Cl$_2$, 0°C, (for **161**); (iv) Tf$_2$O Py, 0°C, (for **162**); (v) A/NaH, 18-crown-6, DMF, 0–20°C;(vi) TFA/H$_2$O (2:1), 50°C; (vii) LiN$_3$, DMF, 140°C; (viii) 5%Pd/C, EtOH, rt, 20 psi; (ix) β-methoxy-α-methacryloyl isocyanate, DMF, −20 to 20°C, (for B = T) or β-methoxyacryloyl isocyanate, DMF, −20 to 20°C, (for **165**); (x) 30% NH$_4$OH, EtOH/dioxane 1:1, 80–100°C; (xi) *p*- chlorophenylphosphorodichloridate, 1,2,4-triazole, rt, 2 days; (xii) 30% NH$_4$OH, dioxane.

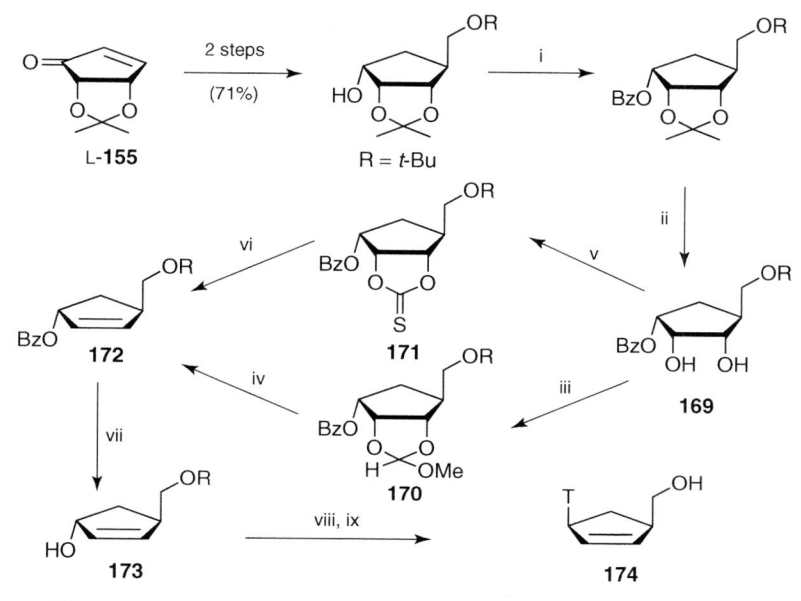

Scheme 33. *Reagents and conditions:* (i) BzCl, Py, (93%); (ii) HCl/MeOH 1:70 v/v (93%); (iii) CH(OMe)$_3$, PPTS; (iv) Ac$_2$O, 120–130°C (68% from **169**); (v) 1-1′-thiocarbonyldiimidazole, MeCN (78%); (vi) 1,3-dimethyl-2-phenyl-1,3,2-diazaphospholidine, THF (17%); (vii) 2 N NaOH/MeOH (93%); (viii) N^3-benzoylthymine, Ph$_3$P, DEAD, dioxane; (ix) TFA/H$_2$O 2:1, 50°C (49% from **172**).

with N-benzoylthymine under Mitsunobu conditions to achieve, after deprotection, the desired compound **174**. Thymidine and uridine analogues **174** and **181** are also prepared by a thermal elimination reaction via the cyclic ortho esters **179** and **180**, respectively (Scheme 34). The reaction of acrylurea **175** and **176** with 30% NH$_4$OH at 80–100°C gives the corresponding thymine **177** and uracil **178** derivatives, which after removal of the isopropylidene group and reaction with trimethyl orthoformate afford the cyclic ortho esters **179** and **180**. Finally, heating those compounds with acetic anhydride at 120–130°C and deprotecting the resulting intermediates with trifluoroacetic acid provides the target nucleosides **174** and **181** in a good yield. In a similar manner, several carbocyclic β-L-2′,3′-didehydro-2′,3′-dideoxypurine nucleoside analogues were prepared.

Starting from cyclopentenone D-**156**, (+)-neplanocin A (**158**) and various L-(+) cyclopentenyl nucleosides have been synthesized (Figure 8).[130] Coupling of **182** with appropriately blocked purine (Scheme 35) and pyrimidine (Scheme 36) bases via a Mitsunobu reaction followed by removal of the protecting groups affords the target L-(+)-cyclopentenyl nucleosides **158**, **183**, **184**, and **185–189**. The synthesized compounds have been evaluated for their antiviral activity against two RNA viruses: HIV and West Nile. Among L-(+)-nucleosides, only the cytosine analogue **188** exhibits weak anti-HIV activity.[130]

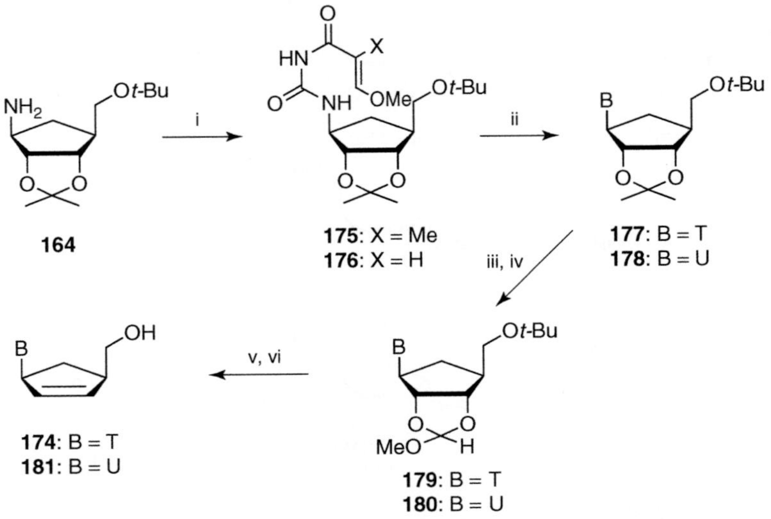

Scheme 34. *Reagents and conditions:* (i) β-methoxy-α-methacryloyl isocyanate, DMF, −20 to 20°C (for **175**, 88%) or β-methoxyacryloyl isocyanate, DMF, −20 to 20°C (for **176**, 82%); (ii) 30% NH$_4$OH, EtOH, or/and dioxane, 80–100°C, (62–90%); (iii) concentrated HCl/MeOH (81–100%); (iv) CH(OMe)$_3$, PPTS; (v) Ac$_2$O, 120–130°C (69–79%); (vi) TFAH/H$_2$O 2:1, 50°C (93–87%).

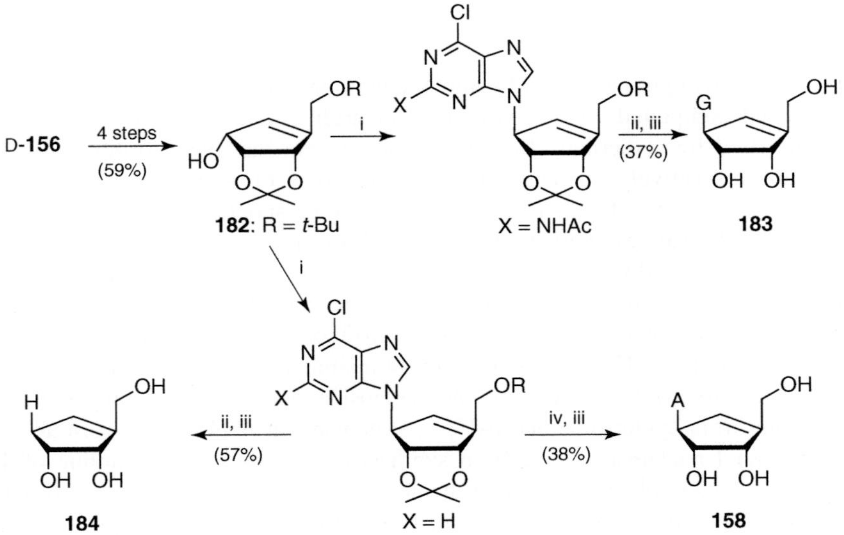

Scheme 35. *Reagents and conditions:* (i) Ph$_3$P, DEAD, N^2-acetylamino-6-chloropurine (75%) or 6-chloropurine (70%); (ii) HSCH$_2$CH$_2$OH, NaOCH$_3$, reflux, MeOH; (iii) TFA/H$_2$O 2:1, 50°C h; (iv) sat. NH$_3$/MeOH, steel bomb, 80°C.

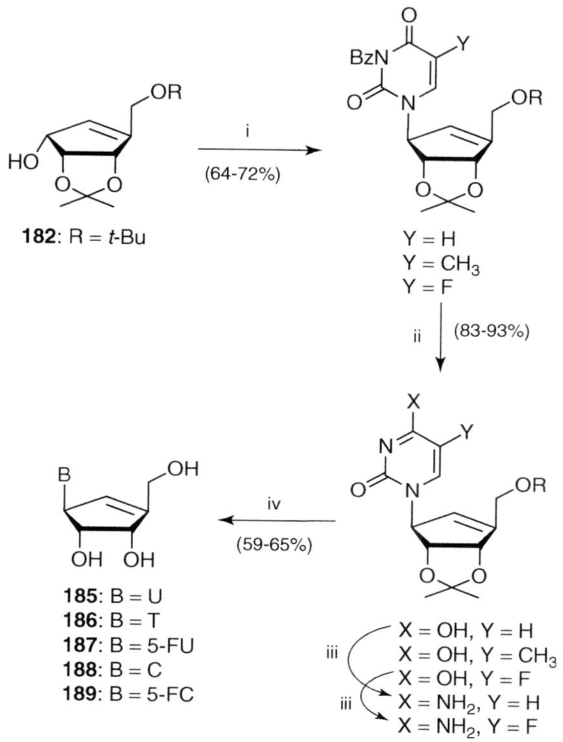

Scheme 36. *Reagents and conditions:* (i) Ph$_3$P, DEAD, N^3-benzoyluracil, N^3-benzoylthymine, or N^3-benzoyl-5-fluorouracil; (ii) sat. NH$_3$/MeOH, 0°C; (iii) 2,4,6-triisopropylbenzenesulfonyl chloride, DMAP, Et$_3$N, MeCN, 0°C to rt, then 30% NH$_4$OH (84–76%); (iv) [after step (iii)]: TFA/H$_2$O 2:1, 50°C.

Another synthetic approach to (+)-neplanocin A (**158**) involves the coupling of adenine with the mesylate **192** derived from the reduction of the protected cyclopentenone **190** with NaBH$_4$/CeCl$_3$ (Scheme 37).[131] The conversion of the alcohol **191** into mesylate **192** allows the insertion of adenine by nucleophilic displacement. Removal of the two protecting groups of **193** in a single step with BCl$_3$ completes the synthesis of (+)-neplanocin A (**158**).

A total enantioselective synthetic method using a combination of ring-closing metathesis and Pd(0)-catalyzed reactions has been reported for synthesis of the L-cyclopentenyl nucleosides **186** and **198–199**, closely related to neplanocin (Scheme 38).[132] The ring-closing metathesis, first key step in this synthesis, is accomplished by exposure of diene **194** to 10 mol% of a second-generation ruthenium catalyst, to yield the chiral cyclopentenyl analogue **195**. Removal of the benzyl group and subsequent reaction with dimethoxypropane gives the optically active allylic alcohol **196**, which is converted to the corresponding allylic carbonate **197**. Coupling of the appropriate purine and pyrimidine bases is carried

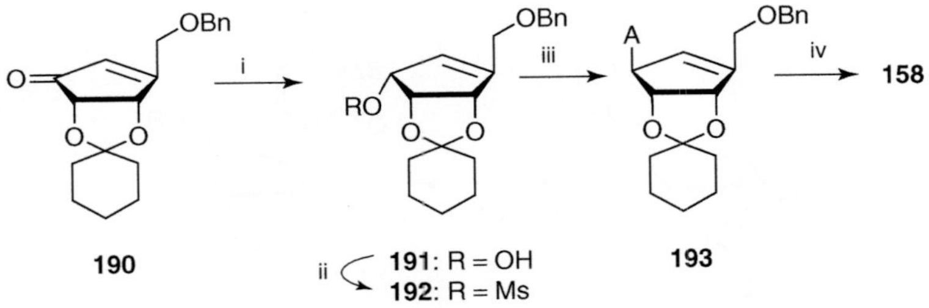

Scheme 37. *Reagents and conditions:* (i) NaBH$_4$/CeCl$_3$, MeOH, 0°C (96%); (ii) MsCl, Et$_3$N, CH$_2$Cl$_2$, 0°C; (iii) [after step (ii)]: Adenine/K$_2$CO$_3$/18-crown-6, DMF, 70°C (54%); (iv) BCl$_3$, CH$_2$Cl$_2$, −78°C (41%).

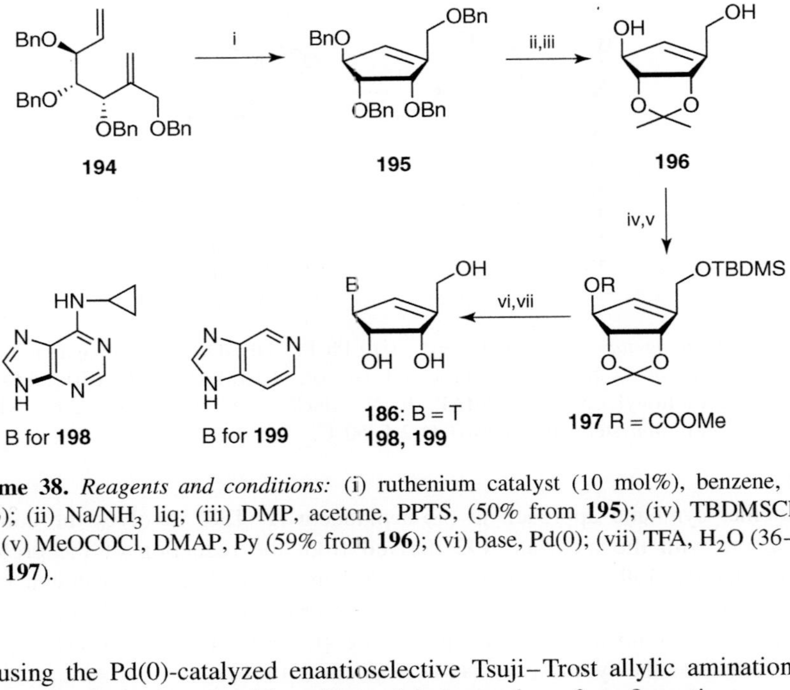

Scheme 38. *Reagents and conditions:* (i) ruthenium catalyst (10 mol%), benzene, 80°C (90%); (ii) Na/NH$_3$ liq; (iii) DMP, acetone, PPTS, (50% from **195**); (iv) TBDMSCl, Py, 0°C; (v) MeOCOCl, DMAP, Py (59% from **196**); (vi) base, Pd(0); (vii) TFA, H$_2$O (36–69% from **197**).

out using the Pd(0)-catalyzed enantioselective Tsuji–Trost allylic amination and leads to the desired nucleosides with a global retention of configuration.

The synthesis of several carbocyclic L-2′,3′-dideoxy purines[133] and pyrimidines[134] has been reported using a Mitsunobu condensation. The Mitsunobu protocol has also been reported as a convergent method for the synthesis of several L-6-azapyrimidine nucleosides[135] and for the synthesis of carbocyclic L-2-deoxynucleosides containing all of the natural basis, through a solid-phase synthetic approach.[136] The solid-phase approach is reported in comparison with a solution-phase synthesis, and the regioselectivity of the condensation has also been evaluated; the L-2-deoxynucleosides **202–206** are synthesized in four steps from

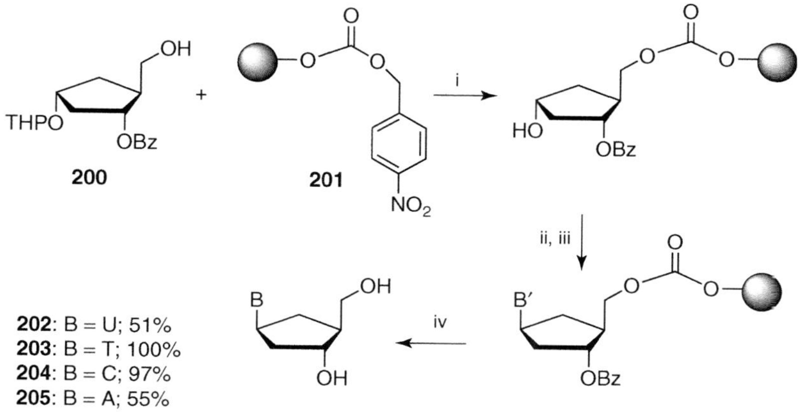
THPO
OBz
OH
+
O
O
O
i
HO
OBz
O
O
O
200
201
NO2
ii, iii
B
OH
OH
iv
B′
OBz
O
O
O
202: B = U; 51%
203: B = T; 100%
204: B = C; 97%
205: B = A; 55%
206: B = G; 62%

Scheme 39. *Reagents and conditions:* (i) DMAP, DIPEA, CH_2Cl_2, 40°C; (ii) PPTS, *n*-Bu/DCE, 60°C; (iii) blocked purine or pyrimidine, DEAD, Ph_3P, DMF/dioxane; (iv) K_2CO_3, THF/MeOH.

the appropriately protected intermediate **200**, which is loaded onto *p*-nitrophenyl carbonate resin **201** (Scheme 39).

The isosteric relationship between oxygen and a fluoromethylene group has prompted various syntheses of fluoro-substituted carbocyclic nucleosides. Several 4′-fluorocarbocyclic nucleosides,[114,137] and 2′,3′-unsaturated-2′-fluoro- and 3′-fluorocarbocyclic analogues[138,139] have been reported by the group of Chu. For the synthesis of the 2′-fluorocarbocyclic nucleosides **215–220**, the Mitsunobu reaction is used to condense the key intermediate **207** with the appropriately protected purine and pyrimidine bases (Scheme 40).[139] Crude thymidine and uridine derivatives are treated directly with methanolic ammonia to give the debenzoylated compounds **208** and **209**, respectively. To synthesize the cytidine analogue, the uridine derivative **209** is further subjected to the ammonolysis to give **210**. Finally, the thymidine, uridine, and cytidine analogues **215–217** are obtained by deprotection of the silyl group. The synthesis of purine analogues follows a similar procedure. The key intermediate **207** is condensed with 6-chloropurine to give the corresponding nucleoside **211**, which is further treated with methanolic ammonia (**212**) to obtain, after deprotection of silyl group, the adenine derivative **218**. Compound **211** is also treated with 2-mercaptoethanol and sodium methoxide in refluxing methanol (**213**), to give the inosine analogue **219** after removal of the silyl group. The guanosine analogue **220** is obtained by treatment of **214** with formic acid followed by ammonium hydroxide solution.

The L-carbocyclic *C*-nucleosides **223–225** bearing the benzoxazolo, benzimidazolo, and benzothiazolo bases are reported through the reaction of *C*-chlorooxime **222**, obtained via C-chlorination with *N*-chlorosuccinimide of *C*-oxime **221**, with the proper α-amino aromatic compounds (Scheme 41).[140]

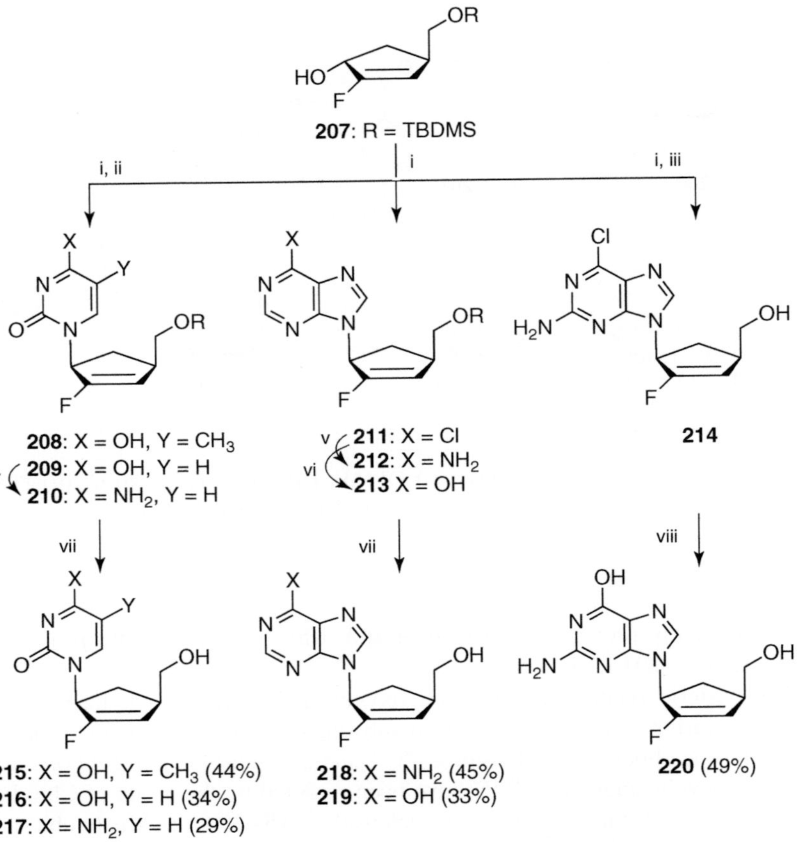

Scheme 40. *Reagents and conditions:* (i) DIAD, Ph$_3$P, pyrimidines or 6-chloropurine, THF; (ii) NH$_3$/MeOH; (iii) TBAF, THF; (iv) 2,4,6-triisopropylbenzenesulfonyl chloride, DMAP, Et$_3$N, MeCN, then NH$_4$OH, then 3 N HCl, MeOH; (v) NH$_3$, MeOH, steel bomb, 110°C; (vi) HOCH$_2$CH$_2$SH, NaOMe, 70°C; (vii) 3 N HCl, MeOH; (viii) HCOOH, 80°C, then NH$_4$OH, MeOH.

5. PYRROLIDINYL NUCLEOSIDES

Among all types of aza-nucleosides, five-membered pyrrolidine-type nucleosides have been investigated particularly because of their high similarity with naturally occurring nucleosides. Aza-sugars have been considered a convenient entry to pyrrolidinyl nucleosides. Accordingly, the pyranose **226** derived from D-lyxose can be transformed to the peracetylated L-ribofuranose **227** in a two-step process: namely, the cleavage of the acetonide group and the acetylation acid catalyzed. N-Glycosilation of intermediate **227** with the appropriate base using Vorbrüggen's method affords the L-4'-aza-β-ribofuranosyl pyrimidine derivatives **228–232** (Scheme 42).[141]

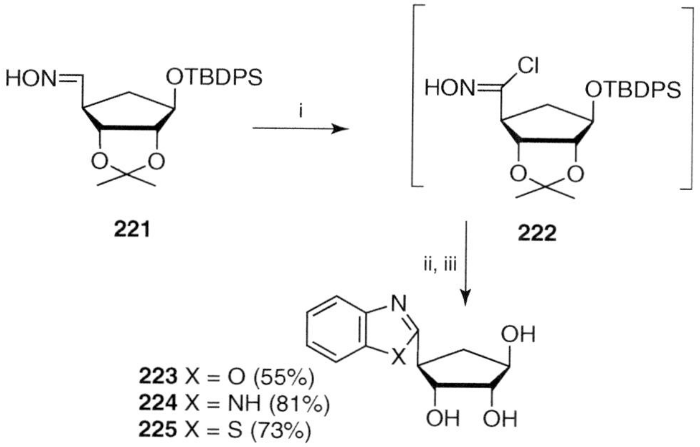

Scheme 41. *Reagents and conditions:* (i) NCS, Py, CHCl$_3$, 40°C; (ii) 2-aminophenol, EtOH, 60°C, 2 h (for **223**) or benzene-1,2-diamine, EtOH, 60°C, 2 h (for **224**) or 2-aminobenzenethiol, EtOH, 60°C, 2 h (for **225**); (iii) 2 M HCl, dioxane.

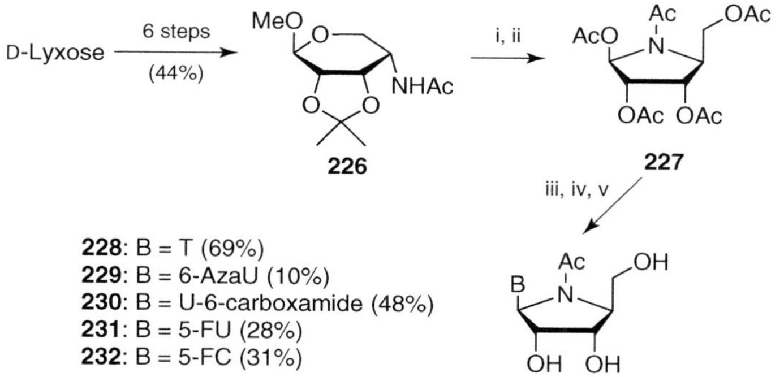

Scheme 42. *Reagents and conditions:* (i) AcOH, H$_2$O; (ii) H$_2$SO$_4$, Ac$_2$O, AcOH (52% from **226**); (iii) base, HMDS, (NH$_4$)$_2$SO$_4$; (iv) SnCl$_4$, DCE; (v) MeOH, NH$_3$.

In addition, compound **228** is converted to the corresponding 2′-deoxy and 2′,3′-dideoxy nucleosides **235** and **236** through the radical-mediated deoxygenation of **233** and **234** with tri-*n*-butyltin hydride in the presence of AIBN (Scheme 43). Finally, the 5′-hydroxyl group of compound **228** is selectively silylated, and the dimesylate **237** is treated with a telluride dianion to give the target compound **238** after desilylation and hydrogenation using Pd/C.

A similar N-glycosylation process using the mixed catalyst system SnCl$_4$/TMSOTf has been employed to transform aza-sugar **240**, derived from reduction with the superhydride of lactam **239**, into the series of *N*-Boc-protected-2′,3′-dideoxy-4′-L-aza-nucleosides **241–243** (Scheme 44).[108]

All the pyrrolidine analogues synthesized so far, having a protected nitrogen in position 4′, show negligible or nonexistent antiviral activity. This is probably due to the presence of the nitrogen-protecting group, which on the other hand, cannot be

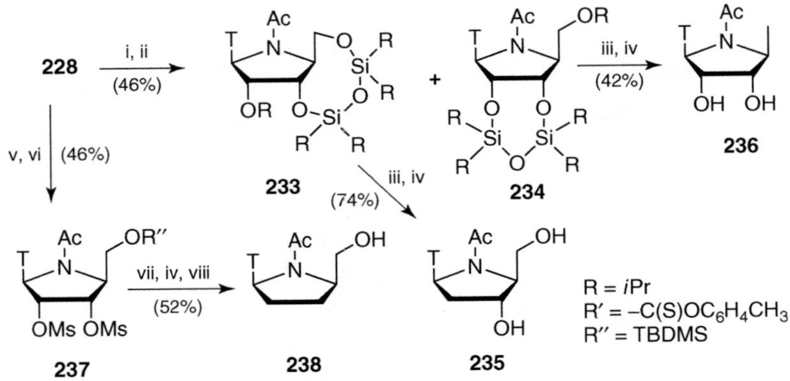

Scheme 43. *Reagents and conditions:* (i) TIPSCl, Py; (ii) $CH_3C_6H_4OC(S)Cl$, Py; (iii) Bu_3SnH, AIBN, toluene, reflux; (iv) TEA·3HF, CH_2Cl_2; (v) TBDMSCl, Py; (vi) MsCl; (vii) Li_2Te, THF; (viii) 10% Pd/C, MeOH.

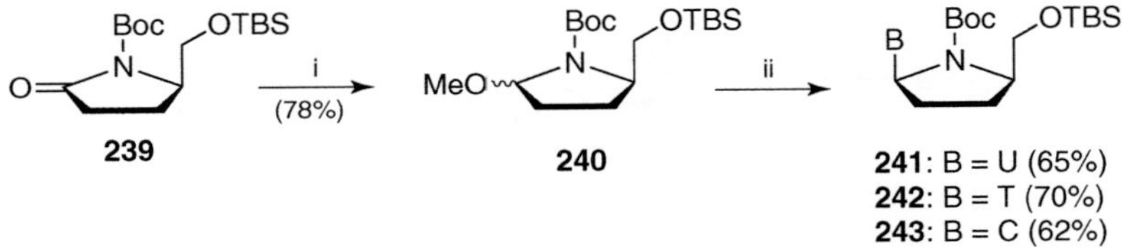

Scheme 44. *Reagents and conditions:* (i) $LiEt_3BH$, THF, $-78°C$, then $CH(OMe)_3$, $BF_3·OEt_2$, Et_2O, 4-Å molecular sieves; (ii) silylated B, $SnCl_4$/TMSOTf 1:1, DCE, then TBAF, THF.

removed without affecting the stability of the resulting nucleosides, which would have a geminal diamine system at C1′. For this reason, chemists moved toward a search for derivatives in which the inter-sugar amino atom was shifted in a different position: for example, in the 3′-position, as in some uracil-pyrrolidine-L-nucleoside analogues.[142] Scheme 45 shows the synthesis of one such derivative: the bromovinyluracil-nucleoside analogue **249**. The condensation of aminopyrrolidine **244** with butyramide **245** requires heating a solution of the reagents in dioxane at 100°C for 10 h, providing compound **246** after acidic treatment. Removal of the *p*-toluenesulfonamide protecting group from **246** gives the free nucleoside derivative **247**. Finally, the synthesis of **249** is accomplished by a radical bromination with bromine/AIBN of the carbocyclic ethyl group of acetate **248**.

Another interesting class of compounds are *C*-aza-nucleosides, in which the furanosyl moiety is linked to the aromatic heterocycles through a carbon–carbon bond, resulting in a pseudoglycosidic bond more stable toward hydrolysis and enzymic reaction. A stereoselective synthesis of *C*-aza-L-lyxo-nucleosides has been reported starting from D-pentofuranose **250** (Scheme 46).[143] Compound **250** is allowed to react with organolithium reagents of thiophene,

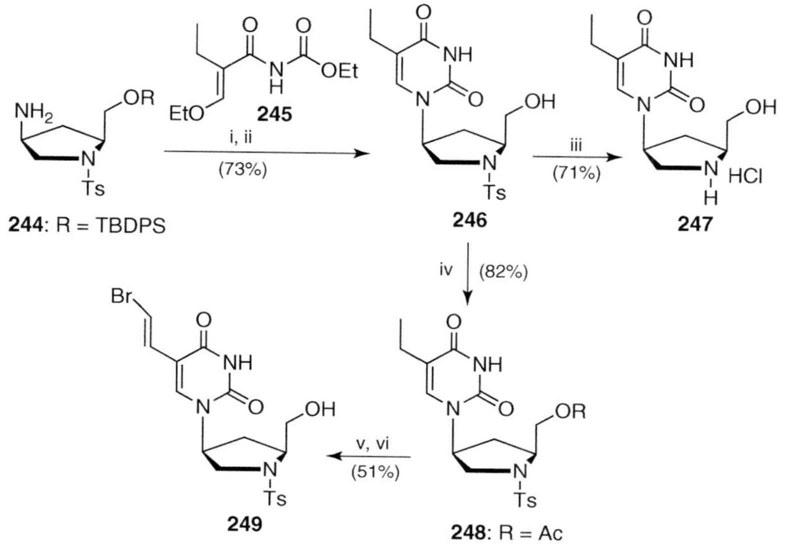

Scheme 45. *Reagents and conditions:* (i) **20**, TEA, dioxane, 100°C; (ii) 2 M HCl, dioxane, 90°C; (iii) PhOH, AcOH, HBr, 90°C; (iv) TEA, DMAP, CH$_2$Cl$_2$, Ac$_2$O; (v) Br$_2$, AIBN, CHCl$_3$, reflux, then TEA; (vi) NH$_3$, MeOH, CH$_2$Cl$_2$, NH$_4$OH.

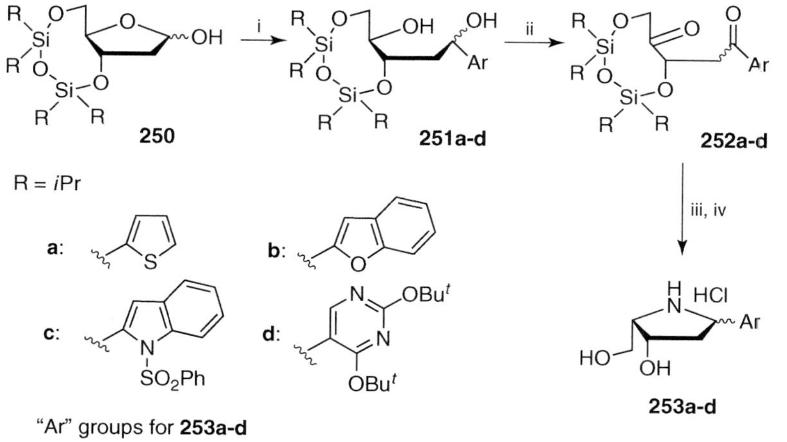

Scheme 46. *Reagents and conditions:* (i) aryllithium, THF, rt (44–97%); (ii) DMSO, TFAA, Et$_3$N, CH$_2$Cl$_2$, −78°C to rt (64–88%); (iii) HCO$_2$NH$_4$, NaBH$_3$CN, NaBH$_4$, MeOH (27–42%); (iv) 6 M HCl, MeOH, 60°C.

N-(phenylsulfonyl)indole, benzofuran, and 2,4-di(*t*-butoxy)pyrimidine to give the corresponding diol derivatives **251a–d**. After Swern oxidation, the resulting diones **252a–d** are subjected to a reductive aminocyclization. The process of Swern oxidation is required for a simple synthesis of *C*-aza-2-deoxynucleosides, but the loss of chirality in the hydroxy groups upon aminocyclization results in formation of the two epimers α/β **253a–d** in a 1:1 ratio.

A stereoselective approach to protected L-*C*-aza-nucleoside involves the addition of an organolithium reagent to the *N*-benzylnitrones **254** derived from tri-*O*-benzylarabinose (Scheme 47).[144] After reduction of the *N*-benzylhydroxylamine **255**, a ring-closure reaction with triflic anhydride furnishes the fully protected L-*C*-aza-nucleoside **256** as a major epimer.

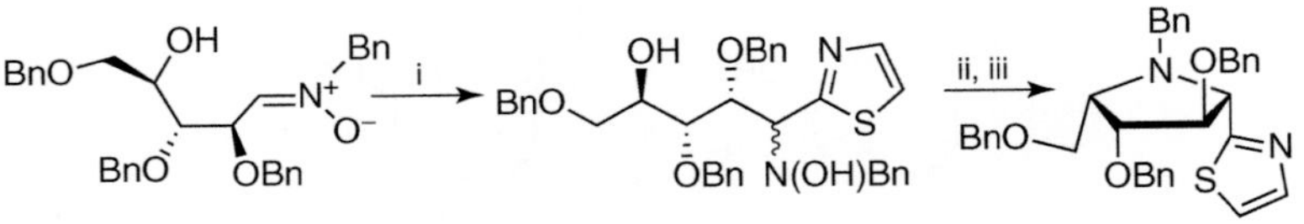

Scheme 47. *Reagents and conditions:* (i) 2-lithiothiazole, Et$_2$O, −78°C, 6 h (75%); (ii) (AcO)$_2$Cu, Zn, AcOH, H$_2$O, 70°C 1 h (78%); (iii) Tf$_2$O, Py, 40°C, 30 min (65%).

The synthesis and bioactivities of the L-enantiomers (+)-**257** and (−)-**258** (Figure 9) of two of the most biologically potent members of transition-state analogue inhibitors of purine nucleoside phosphorylases D-ImmH and D-DADMe-ImmH, respectively, have been reported.[145]

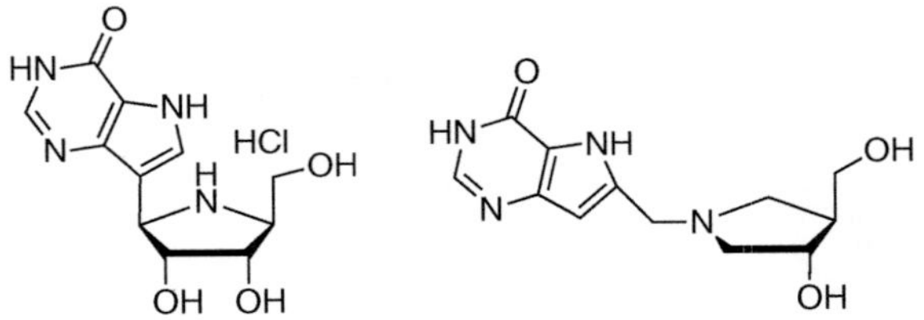

Figure 9. L-Enantiomers of D-ImmH and D-DADMe-ImmH.

The synthesis of L-ImmH **257** (Scheme 48) is achieved through the coupling of lithiated deazapurine derivative **260** to imine **259** following a procedure reported previously for making D-ImmH.[146]

The L-DADMe-immucillin-H **258**, derived by Mannich coupling of amine **261** with deazahypoxanthine **262** in the presence of formaldehyde, is illustrated in Scheme 49.

Kinetic studies of the interactions between compounds (+)-**257** and (−)-**258** and human, plasmodial, and bovine PNPases reveals that the L-enantiomers bind to the

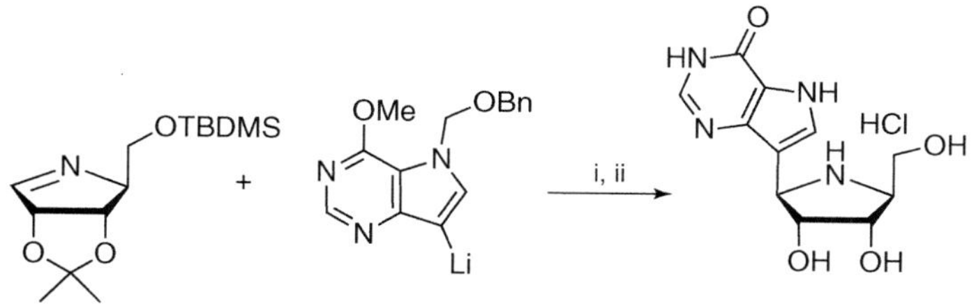

Scheme 48. *Reagents and conditions:* (i) *n*-BuLi, Et$_2$O, anisole, −70 to 0°C (41%); (ii) H$_2$, 10% Pd/C, EtOH, then HCl.

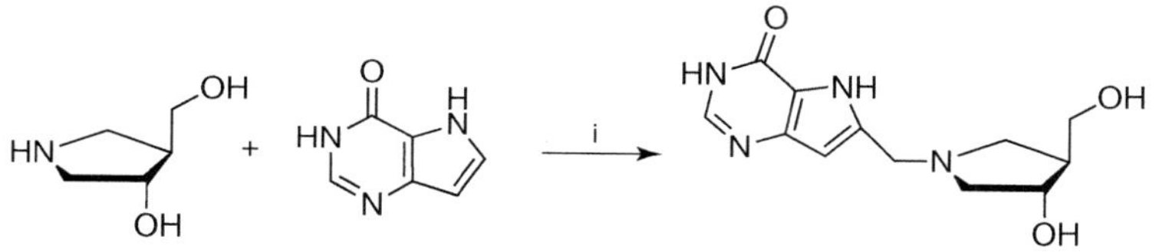

Scheme 49. *Reagents and conditions:* (i) HCHO (30%, aq.), H$_2$O, 85.°C, 15 h (≈8%).

PNPases approximately 5 to 600 times less well than do the D-compounds, but nevertheless remain powerful inhibitors with nanomolar dissociation constants.

Isoxazolidinyl nucleosides have been depicted as another emerging important class of nucleoside analogues. A synthetic approach to isoxazolidinyl nucleosides involving the glycosylation of isoxazolidines **265** with silylated thymine under Vorbrüggen conditions has been reported (Scheme 50).[147,148] The reaction affords the isoxazolidinyl nucleoside analogues *trans*-**266** and *cis*-**267**, which after deprotection of isopropylidene group followed by sequential treatment with sodium periodate and sodium borohydride are converted into the isoxazolidinyl nucleosides **268** and **269**, respectively. The key intermediate isoxazolidine **265** is obtained as the major compound from vinyl acetate addition to nitrone **263**, together with the minor epimer **264**. Compounds **266** and **267** are also the major products resulting from the 1,3-dipolar cycloaddition between vinyl thymine **270** and nitrone **263** (Scheme 50).

6. THIONUCLEOSIDES

The first synthesis of L-4′-thionucleosides was reported in 1964 by Reist et al.[149] starting from D-lyxose. The approach involves a large number of steps and correspondingly low yields. The tetra-acetoxy **271** is obtained as a mixture of the corresponding α- and β-anomers in 13.6% overall yield from six synthetic steps (Scheme 51). The intermediate **271** is then converted into the bromide derivative

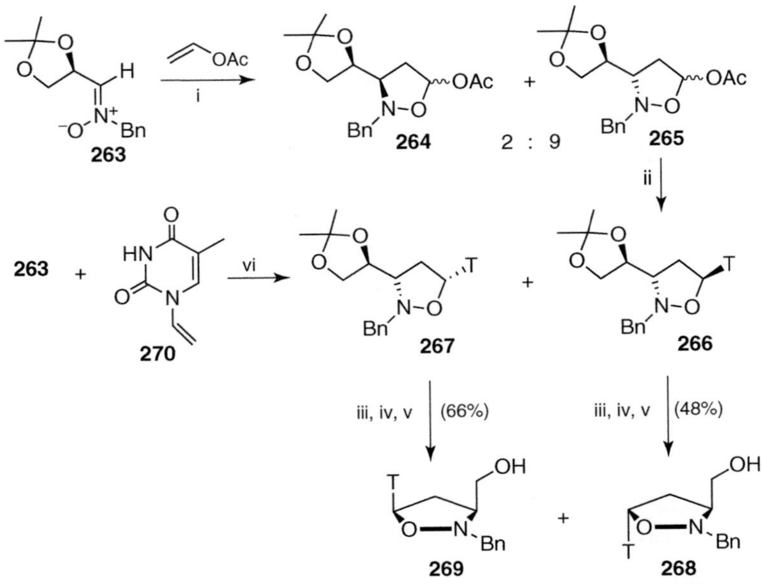

Scheme 50. *Reagents and conditions:* (i) neat, reflux, 12 h (94% combined yield); (ii) TMS-T, TMSOTf, CH$_2$Cl$_2$, rt, 4 h (**267/266** 3:1; 80% combined yield); (iii) TsOH, MeOH, reflux; (iv) NaIO$_4$, MeOH·H$_2$O, 0°C; (v) NaBH$_4$, MeOH, 0°C; (vi) toluene, reflux (**267/266** 4:1).

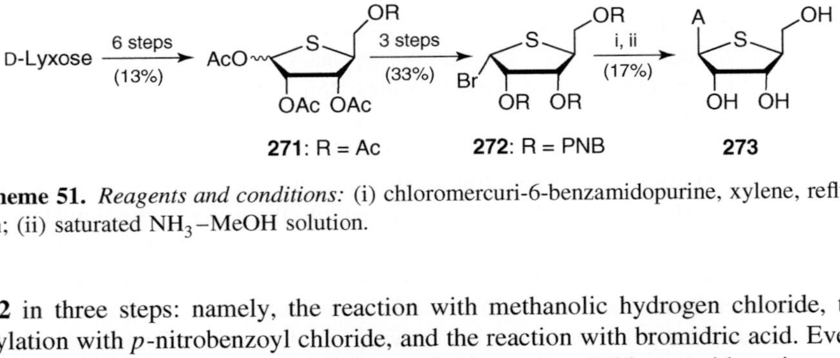

Scheme 51. *Reagents and conditions:* (i) chloromercuri-6-benzamidopurine, xylene, reflux, 6 h; (ii) saturated NH$_3$–MeOH solution.

272 in three steps: namely, the reaction with methanolic hydrogen chloride, the acylation with *p*-nitrobenzoyl chloride, and the reaction with bromidric acid. Eventually, after ammonolysis, condensation of chloromercuri-6-benzamidopurine with L-thioribofuranosyl bromide **272** gives the thioadenosine **273**.

By modifying this reported procedure,[149] pure anomer **271β** can be prepared starting from methyl-α-D-lyxopyranoside **274**, readily available from D-lyxose or D-xylose (Scheme 52).[150] Therefore, the 4-*S*-acetyl derivative **277** is synthesized from alcohol **275** via the triflic ester **276**, followed by treatment with potassium thioacetate. Hydrolytic removal of the isopropylidene protective group in **277** gives a 2:1 mixture of the expected diol **278** and the 4-*O*-acetyl derivative **279**. The mixture is subjected to acetolysis under the Whistler reaction conditions, to afford the crystalline anomer **271β**. Coupling of **271β** with 5-fluorouracil, 5-fluorocytosine,

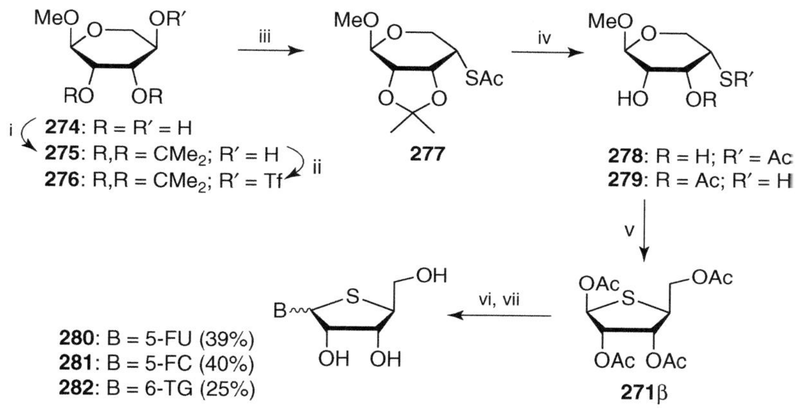

Scheme 52. *Reagents and conditions:* Me$_2$C(OMe)$_2$, TsOH, Me$_2$CO, (78%); (ii) Tf$_2$O, DMAP, Py, CH$_2$Cl$_2$, −15°C; (iii) after step (ii): KSAc, DMF (47% from **275**) (vi) TFA/H$_2$O 9:1; (v) AcOH/Ac$_2$O/H$_2$SO$_4$ 15:15:1, 0°C (58% from **277**); (vi) 5-fluorouracil, 5-fluorocytosine, or 6-mercaptoguanine, TMSCl, (NH$_4$)$_2$SO$_4$, HMDS, then TMSOTf, MeCN; (vii) NaOMe, MeOH, rt, for **280** and **281**, NH$_3$/MeOH, 0°C.

and 6-mercaptoguanine gives rise to the L-4'-thioribonucleosides **280–282** as a mixture of α- and β-anomers.

To achieve the cleavage of the exocyclic carbon–sulfur bond and the subsequent connection of the pyrimidine base, dithiofuranoside **283** is coupled directly with several trimethylsilyluracil derivatives in the presence of *N*-iodosuccinimide affording the target thionucleosides **284–290** as a mixture of anomers (Scheme 53).[151,152]

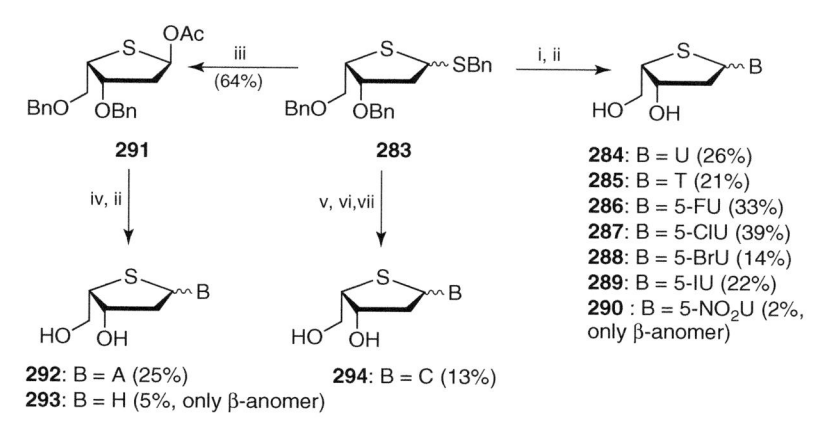

Scheme 53. *Reagents and conditions:* (i) 2,4-bis-*N,O*-trimethylsilylpyrimidine derivatives, NIS, MeCN, rt, 2 h; (ii) BBr, CH$_2$Cl$_2$, −90°C, 1 h; (iii) Hg(OAc)$_2$, AcOH, rt, 2,5 h (iv) A, TMSOTf, MeCN, −18°C to rt, 2 h; (v) 2,4-bis-*N,O*-trimethylsilyluracil, NIS, MeCN, rt, 2 h; (vi) 1,2,4-triazole, POCl$_3$, Et$_3$N, MeCN, rt, 2 h; (vii) NH$_3$ aq., dioxane, rt, 24 h, then BBr, CH$_2$Cl$_2$, −90°C, 1 h.

The corresponding cytidine analogue **294** is prepared via the triazolo derivative of protected **284**, since attempts to synthesize it by coupling the thiosugar **283** with 2,4-bis-N,O-trimethylsilylcytosine in the presence of NIS or TMSOTf fail. Coupling of glycosyl donor **291** with adenine and hypoxanthine in the presence of TMSOTf yields the corresponding purine derivatives **292** and **293**. Analogously, the same authors reported the synthesis of several L-4′-thioarabino nucleosides[153,154] and L-4′-thioarabino-5-halopyrimidine nucleosides.[155]

A synthesis of L-thioarabinopyrimidine nucleosides involves the oxidation of thioarabitol **295** with MCPBA to afford a diastereo mixture of sulfoxides, which after Pummerer rearrangement gives rise to the thioarabinofuranose **296** (Scheme 54).[156] The glycosylation reaction is allowed by using persilylated thymine and N^4-acetylcytosine derivatives and provides the L-thionucleosides **297** and **298**, respectively, as separable anomeric mixtures.

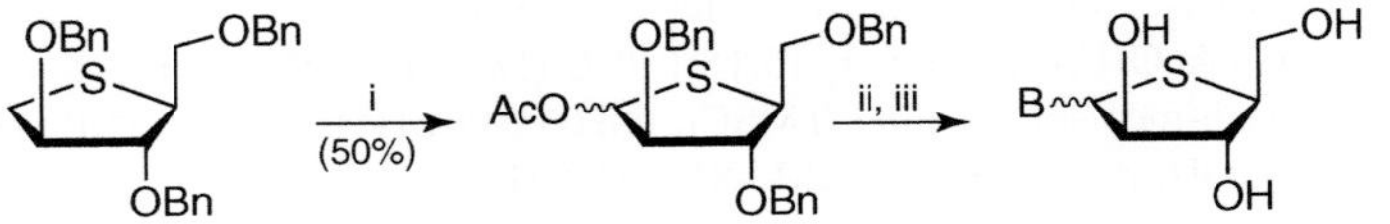

Scheme 54. *Reagents and conditions:* (i) MCPBA, CH_2Cl_2, $-78°C$, then Ac_2O; (ii) silylated thymine (for **297**) or silylated N^4-acetylcytosine (for **298**), TMSOTf; (iii) BCl_3, CH_2Cl_2, $-20°C$, then NH_4OH, MeOH (only for **298**).

Direct coupling of nucleobases with cyclic sulfoxides has been reported for the synthesis of the L-2′-deoxy-2′,2′-disubstituted thionucleosides **304** and **305** (Scheme 55).[157] L-Thiosugar **299** is converted to acetate after treatment with $Hg(OAc)_2$/AcOH, and the removal of the acetate is surprisingly performed with Et_3SiH and TMSOTf to give the L-thioarabitol derivative **300**. Oxidation of **300** with DMSO/Ac_2O affords the corresponding ketone, which is smoothly converted either to the difluoro derivative **301** after DAST treatment or into the methylidene **302** by Wittig reaction. Derivative **301** is oxidized to the sulfoxide before condensing it with persilylated N^4-benzoylcytosine, to give the thionucleoside **304** as a separable α/β mixture. Under the same conditions, the reaction with **303** produces the thionucleoside **305** as a separable anomeric mixture.

An efficient synthesis of L-β-2′-deoxy-4′-thiopurine nucleosides has been accomplished utilizing the anchimeric effect of the C2′-benzoyl group of L-4′-thiosugar.[158] To obtain the desired nucleosides **310** and **311** (Scheme 56), the thiosugar **306** is condensed with silylated 6-chloropurine, affording the desired β-anomer **307** as the predominant product. Treatment of **307** with sodium methoxide affords alcohol **308**, which is deoxygenated using modified Barton's conditions to give the 2′-deoxy nucleoside **309**, which is easily converted to the adenine derivatives **310** and **311**.

The 2′-azido-2′,3′-dideoxy-4′-thiopyrimidines **313–316** have been prepared using glycosyl donor **312** and silylated pyrimidine in the presence

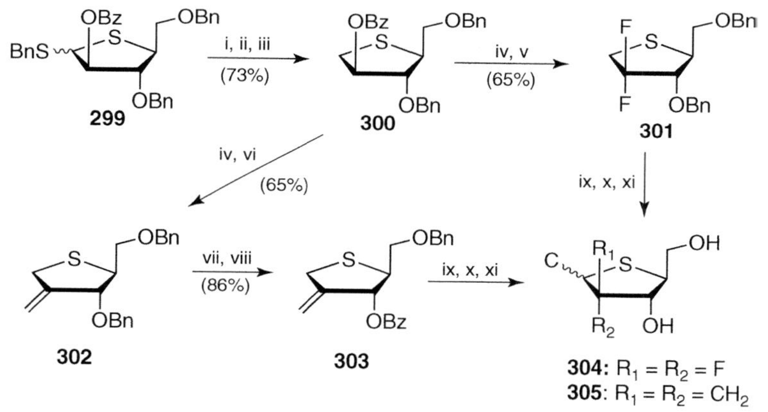

THIONUCLEOSIDES 513

Scheme 55. *Reagents and conditions:* (i) Hg(OAc)$_2$, AcOH, rt; (ii) Et$_3$SiH, TMSOTf, rt; (iii) NaOMe, MeOH, CH$_2$Cl$_2$, rt; (iv) DMSO, Ac$_2$O, rt; (v) DAST, CH$_2$Cl$_2$, rt; (vi) Ph$_3$PCH$_3$Br, NaH, *t*-amyl alcohol, rt; (vii) BBr$_3$, CH$_2$Cl$_2$, −40°C; (viii) BzCl, Py, 50°C; (ix) MCPBA, CH$_2$Cl$_2$, −40°C; (x) silylated *N*-benzoylcytosine, TMSOTf, DCE, 0°C to rt; (xi) BBr$_3$, CH$_2$Cl$_2$, −40°C, then NaOMe, MeOH for **304**; NaOMe, MeOH for **305**.

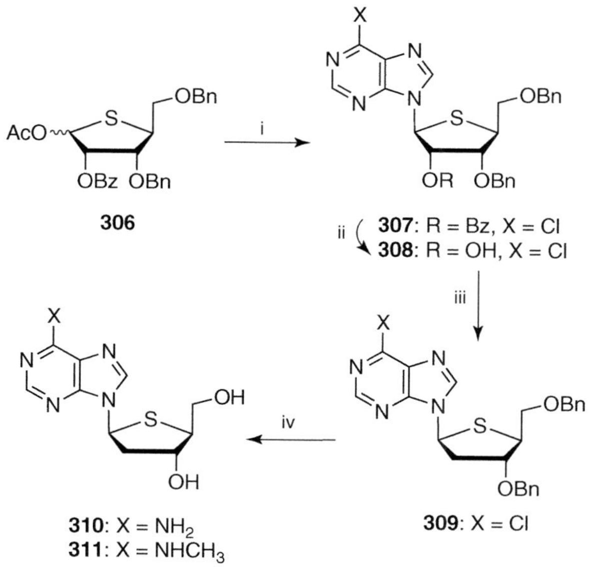

Scheme 56. *Reagents and conditions:* (i) silylated 6-chloropurine, TMSOTf, rt to 80°C (60%); (ii) NaOMe, MeOH (86%); (iii) [after step (ii)]: PhOC(S)Cl, DMAP, then Et$_3$B, *n*-Bu$_3$SnH (63%); (iv) BBr$_3$, then NH$_3$, MeOH (for **310**, 76%), CH$_3$NH$_2$, MeOH (for **311**, 77%).

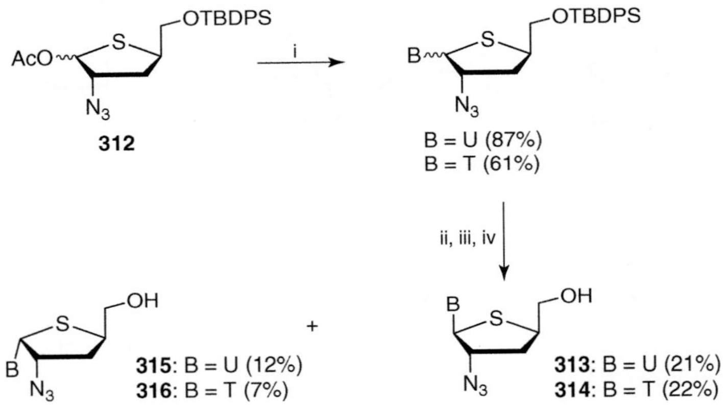

Scheme 57. *Reagents and conditions:* (i) silylated uracil or thymine, TMSOTf, DCE, (60%); (ii) TBAF, THF; (iii) BzCl, Py, 70°C, 24 h; (iv) NH$_3$, MeOH.

of TMSOTf as a Lewis catalyst (Scheme 57).[159] In a similar way, several 2′-azido-2′,3′-dideoxy-4′-cytosine and purine derivatives have been prepared.

By using the acetate **317**, a large number of L-2′,3′-unsaturated 2′-fluoro-4′-thionucleosides have been synthesized and evaluated for their antiviral activity (Scheme 58).[160] The acetate **317** is condensed with various pyrimidine bases under Vorbrüggen conditions to give the corresponding pyrimidine derivatives **318–321**. The β-anomer is obtained exclusively by virtue of the bulky (*R*)-phenylselenyl group. Oxidation of the phenylselenyl group by MCPBA followed by removal of the protecting groups gives the corresponding unsaturated compounds **322–325** in moderate-to-good yields. In a similar manner, various purine analogues are prepared.

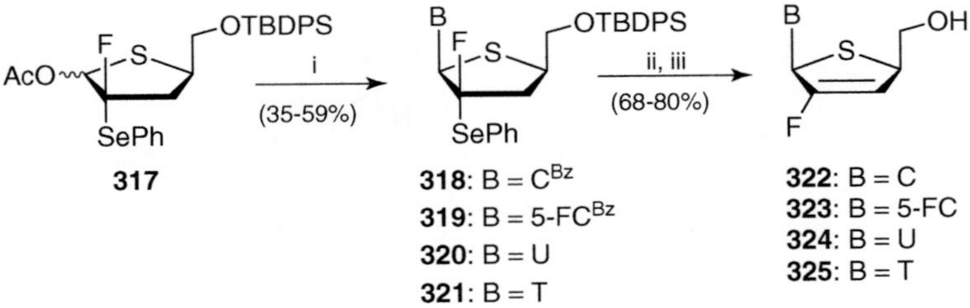

Scheme 58. *Reagents and conditions:* (i) HMDS, CH$_3$CN, pyrimidines, TMSOTf, rt; (ii) MCPBA, −78°C; then Py, rt; (iii) TBAF, THF, then NH$_3$, MeOH (only for **322** and **323**).

The authors of the synthesis above also reported the preparation of a large series of L-2′,3′-unsaturated 3′-fluoro-4′-thionucleosides by coupling the difluoro-4′-thiofuranose **326** with various silyl-protected bases in the presence of TMSOTf (Scheme 59).[161] The β-difluoronucleosides **327–330** are converted to the target unsaturated nucleosides **331–334** by treatment with potassium *t*-butoxide in THF.

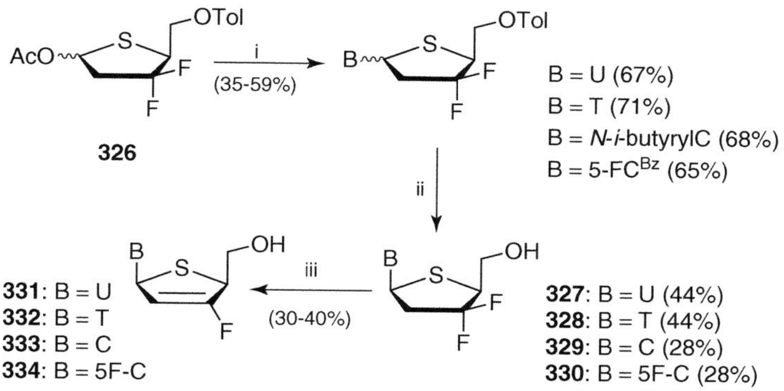

Scheme 59. *Reagents and conditions:* (i) BSA, pyrimidines, MeCN or dioxane, TMSOTf, heat; (ii) NH$_3$/MeOH; (iii) *t*-BuOK/THF.

The L-β-3′-deoxy-3′,3′-difluoro-4′-thionucleosides **336–338** analogues of highly bioactive gemcitabine have been described through the coupling of difluoro-4′-thiofuranose **335** with various persilylated pyrimidines under Vorbrüggen conditions (Scheme 60).[162] All the reactions furnish the β-anomers as main products and only a trace of α-anomers.

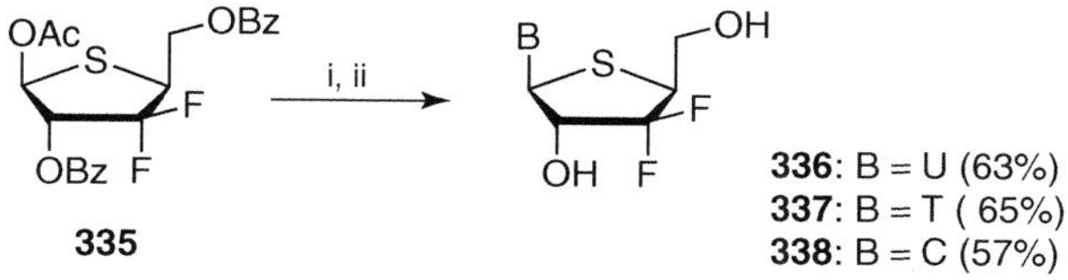

Scheme 60. *Reagents and conditions:* (i) BSA, pyrimidines, MeCN, TMSOTf, heat; (ii) NH$_3$/MeOH.

A novel approach to synthesis of the L-4′-thionucleosides **343–350** from the chiral acyclic thioaminals **339–342**, bearing the nucleobase, provides "S$_N$2-like" displacement of the mesyloxy group at C4′ by the nucleophile sulfur at C1′ (C1′–C4′ cyclization) (Scheme 61).[163] The acyclic 1,2-*syn* isomers lead to the C1′-C2′ *cis*-nucleosides, and the 1,2-*anti*-isomers to the C1′–C2′ *trans*-geometry.

To achieve the synthesis of several L-5′-deoxy-4′-thio-α- and β-nucleosides, the gylcosyl donor **351** is condensed with 2,6-dichloropurine in the presence of EtAlCl$_2$, giving rise to the nucleoside derivative **352** as a mixture of α/β-isomers with the approximate ratio 3:1 (Scheme 62).[164] The mixture **352** is exploited for the substitution of the N^6-position with the appropriate amines, furnishing after deprotection the purines **353–363** as separable anomeric mixtures. Moreover, condensation of acetate **351** with persilylated pyrimidine derivatives under Vorbrüggen conditions with TMSOTf as a catalyst furnishes after deprotection the pyrimidines

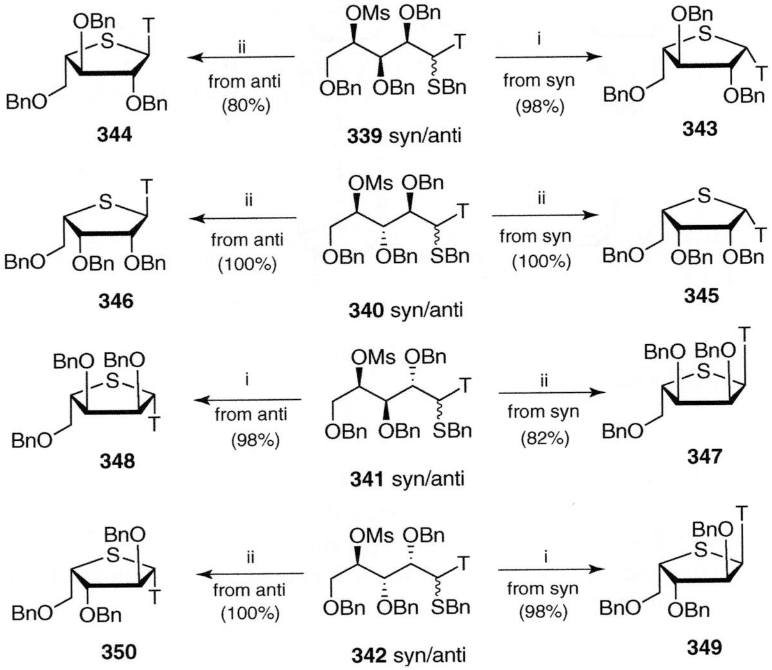
344
339 syn/anti
343
346
340 syn/anti
345
348
341 syn/anti
347
350
342 syn/anti
349
from anti (80%)
from syn (98%)
from anti (100%)
from syn (100%)
from anti (98%)
from syn (82%)
from anti (100%)
from syn (98%)

Scheme 61. *Reagents and conditions:* (i) NaI, pinacolone, 106°C; (ii) NaI, 2,6-lutidine, 145°C.

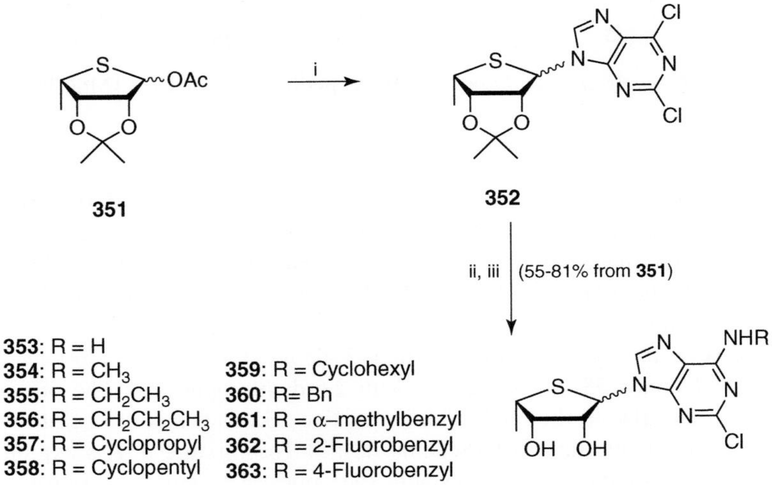
351
352
ii, iii (55-81% from 351)
353: R = H
354: R = CH3
355: R = CH2CH3
356: R = CH2CH2CH3
357: R = Cyclopropyl
358: R = Cyclopentyl
359: R = Cyclohexyl
360: R= Bn
361: R = α–methylbenzyl
362: R = 2-Fluorobenzyl
363: R = 4-Fluorobenzyl

Scheme 62. *Reagents and conditions:* (i) 2,6-dichloropurine, $EtAlCl_2$, MeCN, rt; (ii) appropriate amine, rt, THF; (iii) 85% aq. HCOOH, rt.

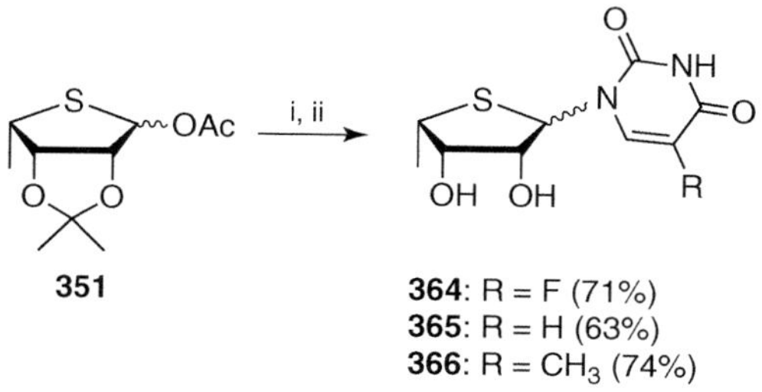

Scheme 63. *Reagents and conditions:* (i) silylated pyrimidine, TMSOTf, MeCN, rt; (ii) 85% aq. HCOOH, rt.

364–366 as separable α/β anomeric mixtures in which the α-isomers are predominant (Scheme 63). Purine derivatives **358** and **359**, in both α- and β-form, exhibit strong inhibition against the human leukemia cell line.

7. DIOXOLANYL NUCLEOSIDES

Several purine and pyrimidine-L-dioxolane nucleosides have been reported using the acetoxy dioxolane **370** as a glycosyl donor for condensation under classical Vorbrüggen conditions.[165–167] The synthesis of acetate **370** is achieved by oxidative decarboxylation of carboxylic acid **368**, which is derived from both L-gulofuranose diacetonide (**367**)[165,166] and L-ascorbic acid (**369**),[167] as shown in Scheme 64. Condensation of **370** with properly protected pyrimidines using TMSOTf as the Lewis acid catalyst, affords the corresponding L-dioxolane nucleosides **371–377** as separable α/β mixtures (Scheme 65). Among the compounds synthesized, the 5-fluorocytosine derivative **373** has been found to exhibit the most potent anti-HIV activity. In a similar way, several purine L-dioxolane nucleosides are prepared.[166,167] Finally, dioxolane **370** was employed for the synthesis of the 5-azacytosine **378** and 6-azathymine **379** nucleosides, as reported in Scheme 66.[158,169]

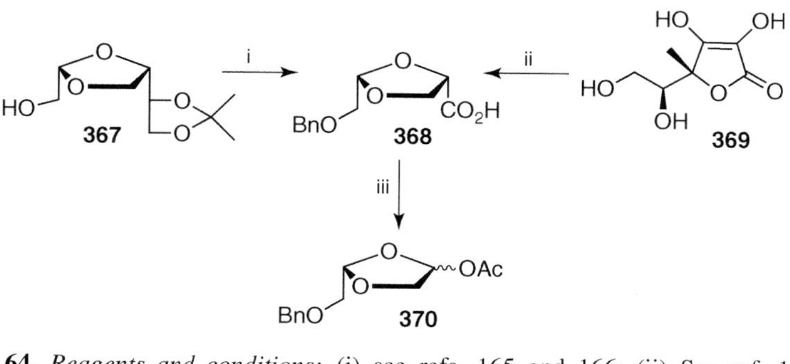

Scheme 64. *Reagents and conditions:* (i) see refs. 165 and 166; (ii) See ref. 167; (iii) Pb(OAc)$_4$, THF (70–80%).

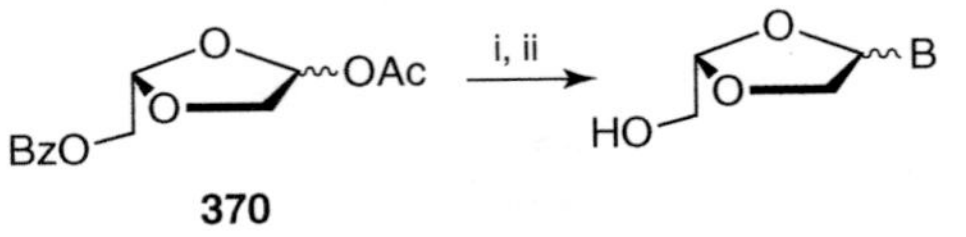

Scheme 65. *Reagents and conditions:* (i) silylated N^4-benzoylcytosine, N^4-benzoyl-5-substituted cytosine, or silylated thymine TMSOTf, DCE; (ii) NH₃/MeOH.

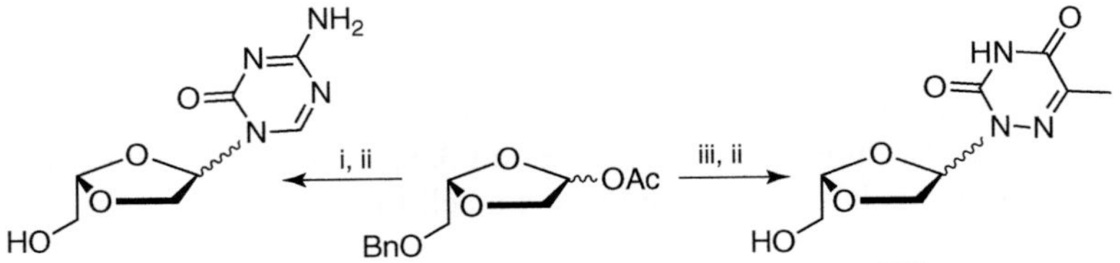

371: B = C
372: B = 5-MeC
373: B = 5-FC
374: B = 5-ClC
375: B = 5-BrC
376: B = 5-IC
377: B = T

Scheme 66. *Reagents and conditions:* (i) 5-azacytosine, HMDS, reflux, then TMSOTf, DCE; (ii) PdO, PdO hydrate, Cy, EtOH, reflux, 6 h; (iii) 6-azathymine, HMDS, reflux, then TMSOTf, DCE.

The coupling of acetate **381** with silylated N^4-benzoylcytosine or N^4-benzoyl-5-fluorocytosine with TMSOTf as a catalyst affords the respective 4′-C–methyloxacytidine **382**, and its 5-fluoro analog **383**, together with the corresponding α-isomers **384** and **385** (Scheme 67).[170] Conversion of the carboxyl group of **380** to the acetoxy group of **381** is obtained by oxidative decarboxylation with lead tetraacetate.

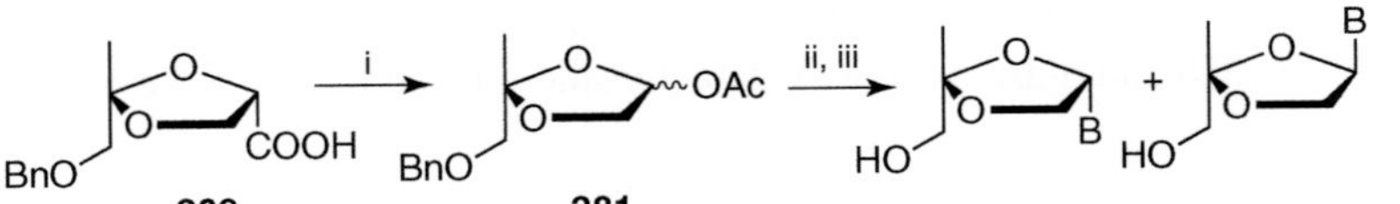

382 B = C
383: B = 5-FC
384: B = C
385: B = 5-FC

Scheme 67. *Reagents and conditions:* (i) Pb(OAc)₄, Py, MeCN; (ii) cytosine or 5-fluorocytosine, HMDS, reflux 6 h, then TMSOTf, DCE; (iii) NH₃/MeOH, then PdO hydrate, cyclohexene, EtOH.

The guanine and adenine L-dioxolane nucleosides **391** and **392** containing a 2′S-methyl substituent have been synthesized as reported in Scheme 68.[171] To synthesize the nucleoside **391**, acetate **386** is treated with the silylated diphenylcarbamoyl purine derivative **387** in the presence of TMSOTf to furnish a 1:1 anomeric mixture of **389**. Chromatographic separation of α- and β-anomers followed by removal of the protective groups leads to the separated α- and β-guanine

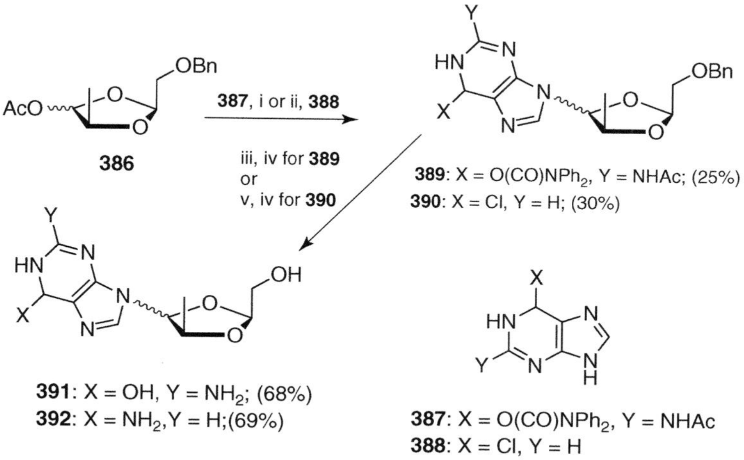

Scheme 68. *Reagents and conditions:* (i) BSA, DCE, then TMSOTf, reflux (for **389**); (ii) bromotrimethylsilane, CH_2Cl_2, 0°C to rt, then **388**, TDA-1, KOH, MeCN; (iii) $NH_2NH_2 \cdot H_2O$, THF, reflux; (iv) Pd(OH)/C, EtOH, cyclohexene; (v) NH_3, 80°C.

derivatives **391**. For the synthesis of adenine nucleoside **392**, treatment of **386** with bromotrimethylsilane affords the bromo derivative, which upon reaction with 6-chloropurine **388**, under phase-transfer conditions, gives an inseparable anomeric mixture of dioxolane derivatives **390** in a 1:2 ratio. Removal of protective groups affords the adenine derivatives **392**, which are separated by preparative HPLC.

The synthesis of the dioxolanyl analogue of tiazofurin **396** has been reported from L-gulonic-γ-lactone (Scheme 69).[172] Treatment of amide **393** with P_2S_5 affords the thioamide **394**, which after refluxing with ethyl bromopyruvate gives

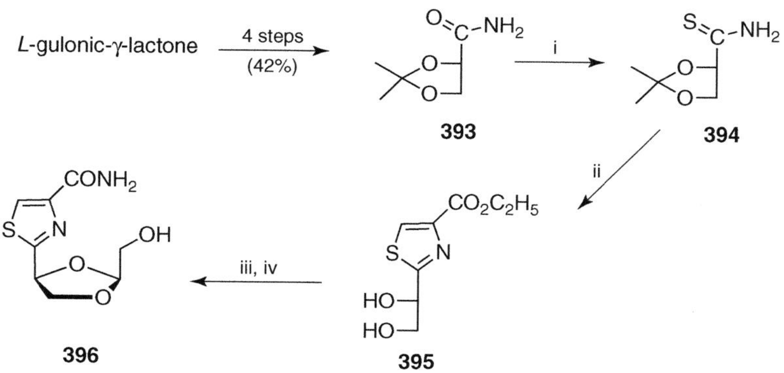

Scheme 69. *Reagents and conditions:* (i) P_2S_5, dioxane, rt; (ii) $BrCH_2COCO_2Et$, EtOH, reflux, 2 h (31% from **393**); (iii) $BzOCH_2CH(OMe)_2$, TsOH, benzene; (iv) NH_3/MeOH (71% from **395**).

the thiazole derivative **395**. Condensation of **395** with 2-benzoyloxy acetaldehyde dimethyl acetal furnishes, after treatment with methanolic ammonia, the target *C*-nucleoside **396** as a predominant anomer.

In a similar manner, the same authors, starting from the key intermediate **397**, reported the synthesis of 1,3-dioxolanyl triazole *C*-nucleosides **398** as a separable anomeric mixture (Scheme 70).[173]

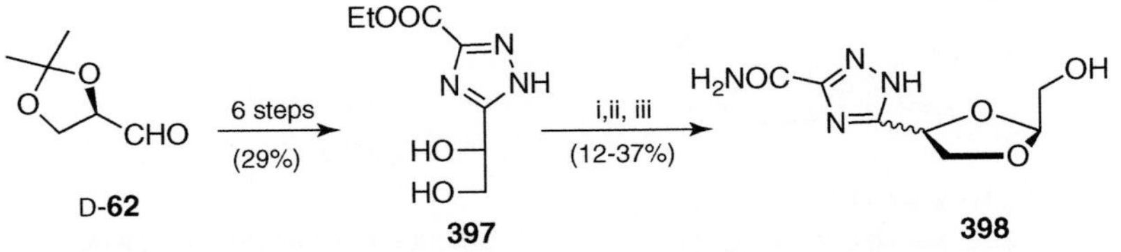

Scheme 70. *Reagents and conditions:* (i) BzOCH$_2$H(OMe)$_2$, TsOH, benzene, reflux; (ii) H$_2$, PdCl$_2$, EtOH, 50 psi, 6 h; (iii) NH$_3$/MeOH, steel bomb, 110°C, 24 h.

The synthesis of L-dioxolanyl-5-fluorouracil **401**, with the heterocyclic base substituted at the 2′-position in the dioxolane ring, has been reported from the dioxolane **400** obtained by cyclization with trimethyl orthoformate of diol **399**, derived from (*R*)-2,3-*O*-isopropylideneglyceraldehyde (D-**62**) (Scheme 71).[174] It was found that this type of nucleoside rearranges to the ring-opened product **402** during chromatography.

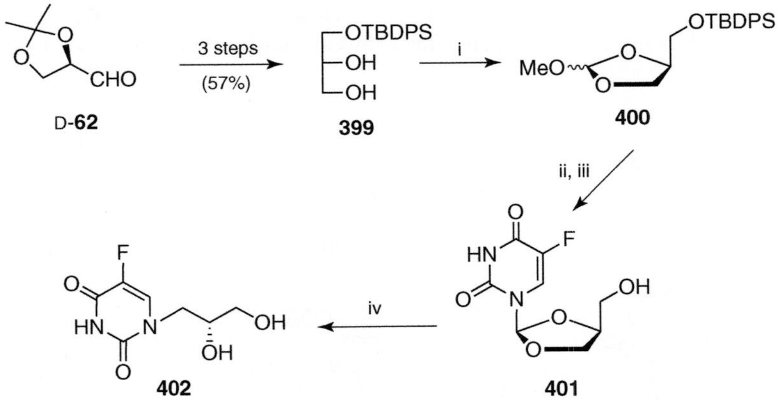

Scheme 71. *Reagents and conditions:* (i) (MeO)CH, BzOH, 130°C, 1,5 h (75%); (ii) silylated 5-FU, TMSOTf, MeCN, rt, 36 h; (iii) TBAF, AcOH, THF, rt, 30 min (61% from **400**); (iv) silica gel, MeOH (70%).

8. OXATHIOLANYL NUCLEOSIDES

Lamivudine, also known as (−)-3TC, the minus form of BCH-189 (Figure 2), is currently used for the treatment of human immunodeficiency virus types 1 and 2 and hepatitis B viral infections.[13,14] Several syntheses of (±)-lamivudine have been reported, starting from racemic starting material.[175,176] Stereospecific synthesis of the enantiomerically pure lamivudine was first reported by Chu and co-workers.[177,178] Condensation of **404** synthesized via 1,6-thioanhydro-L-gulopyranose (**403**) from L-gulose, with silylated N^4-acetylcytosine using Vorbrüggen conditions, leads to an α,β-mixture (1:2) of anomers which after separation by silica gel chromatography followed by deacetylation and desilylation affords the final compounds: (2′R,5′R)-(+) isomer **405** and lamivudine (**406**), respectively (Scheme 72). The use of stannic chloride in place of TMSOTf as the Lewis acid gives the β-isomer exclusively, but unfortunately, it was found to be optically inactive.[179] Using the same strategy, acetate **404** is condensed with thymine, 5-substituted uracils and cytosines, 6-chloropurine, and 6-chloro-2-fluoropurine to give a series of enantiomerically pure L-oxathiolanyl pyrimidine and purine nucleosides.[178]

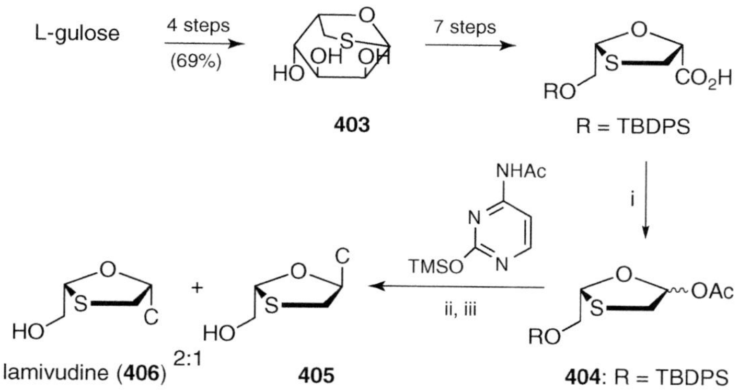

Scheme 72. *Reagents and conditions:* (i) Pb(OAc)$_4$, THF (25% from **403**); (ii) TMSOTf, ClCH$_2$CH$_2$Cl, 0°C to rt (64% from **404**); (iii) NH$_3$/MeOH, then TBAF, THF (58% for **405**; 55% for **406**).

A stereoselective synthesis of lamivudine (**406**) using inexpensive starting materials and without recourse to tedious chromatographic separation of isomers involves the coupling of persilylated cytosine with acetoxy derivative **418** as described in Scheme 73.[180] Reaction of glyoxylic acid monohydrate and dithiane-1,4-diol (**407**) affords the *trans*-hydroxy acid **408**, which after treatment with acetic anhydride provides a 1:2 mixture of *cis*-**409** and *trans*-acetoxy acid **410**. Esterification of acid **410** with L-menthol affords a 1:1 diastereomeric

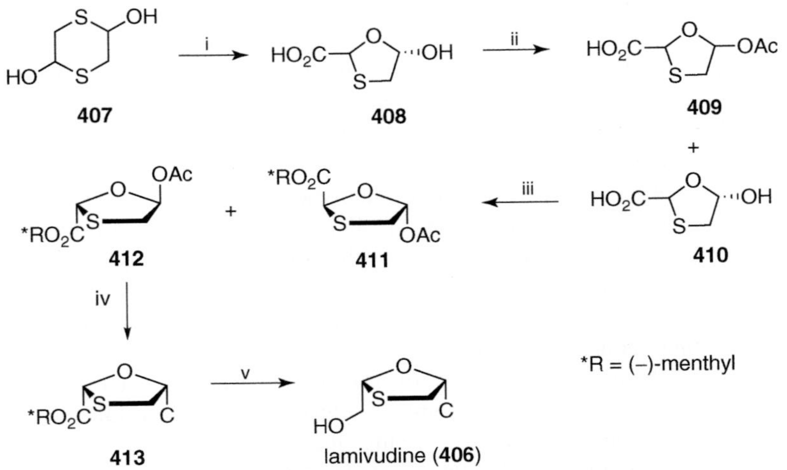

Scheme 73. *Reagents and conditions:* (i) glyoxylic acid hydrate, *t*-BuOMe, reflux (58%); (ii) Ac$_2$O, MeSO$_3$H, then recrystallization (23% for **409**; 22% for **410**); (iii) (−)-menthol, DCC, DMAP, CH$_2$Cl$_2$, (75%), then recrystallization; (iv) (TBDMS)$_2$-cytosine, TMSI, CH$_2$Cl$_2$; (v) LiAlH$_4$, THF (94%).

mixture of the corresponding esters **411** and **412** in quantitative yield. Low-temperature (−78°C) recrystallization of this mixture provides compound **412**. Finally, after reduction with LiAlH$_4$, recrystallization of ester **413** (β/α, 23:1) furnishes the desired lamivudine (**406**) (>98% ee based on chiral HPLC). An N^4-aminobutyllamivudine derivative has also been reported[181] using acetate **412**.

A chemoenzymatic approach to lamivudine involves the resolution of racemic α-acetoxysulfide (±)-**414** with *Pseudomonas fluorescens* lipase (Scheme 74).[182] The enantiomeric pure (−)-**414** is converted into the oxathiolane **415**, and then the synthesis of lamivudine **406** follows conventional nucleoside protocol. A later-stage resolution has been accomplished using *Mucor michel* lipase for the bioconversion of racemic benzoate **416** into its enantiomericallyenriched (−)-2*R* enantiomer.[183]

An enantioselective approach suitable for a large-scale synthesis of lamivudine (**406**) via a crystallization-induced dynamic kinetic resolution is reported (Scheme 75).[184] By heating a mixture of menthyl glyoxylate hydrate **417** with dithiane diol in toluene, the desired diastereomer **418** is obtained in an 80% yield after crystallization. Following chlorination and coupling with persilylated cytosine, **418** is converted into derivative **419**, which is transformed into pure **406** after reduction of menthyl ester with sodium borohydride.

Lamivudine of very high purity (with an enantiomeric excess of more than 99.9%) is obtained through resolution of racemic lamivudine by cocrystal formation with (*S*)- BINOL (Scheme 76).[185] *Cis*- and *trans*-isomers **420** are reacted with (*S*)-(+) mandelic acid. The mandelate salt of *cis*-(±)-**421** is precipitated as a solid, leaving behind *trans*-(±)-**422**-(*S*)-(+)-mandelate salt in mother liquor. The salt **421** is treated with methanolic ammonia to give (±)-lamivudine (**423**), which is heated

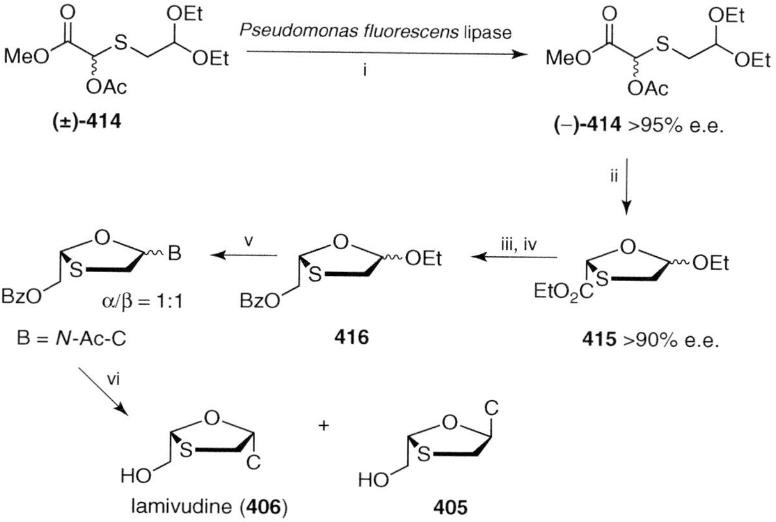

Scheme 74. *Reagents and conditions:* (i) pH 7 phosphate buffer, 30°C (49%); (ii) HCl/EtOH (95%); (iii) LiBH$_4$, *i*-PrOH (cat.), THF, (96%); (iv) BzCl, Py, CH$_2$Cl$_2$, (88%); (v) silylated *N*-acetyl cytosine, TMSOTf, MeCN (43%); (vi) NH$_3$, MeOH (98% both anomers).

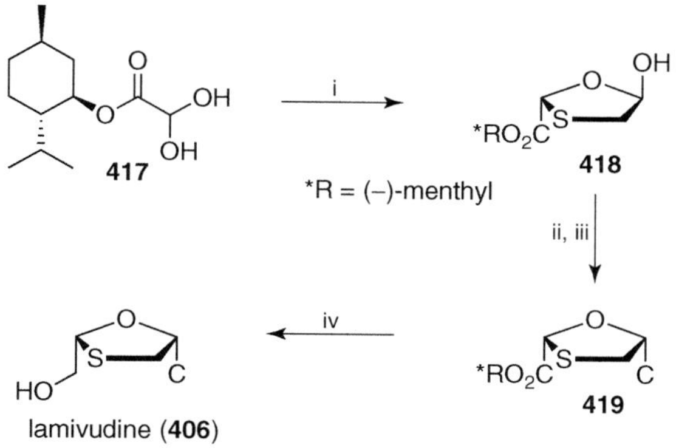

Scheme 75. *Reagents and conditions:* (i) toluene, reflux, then dithiane diol (80%); (ii) SOCl$_2$, DMF, CH$_2$Cl$_2$; (iii) (TMS)$_2$-C, Et$_3$N, toluene (66%); (iv) NaBH$_4$, EtOH (83%).

at 55–60°C with *S*-(−)-BINOL to get the cocrystal of (*S*)-BINOL with the (−)-enantiomer as a crystalline solid **425**. Further treatment with dilute HCl affords lamivudine (**406**). Interestingly, the enantiomer (+)-**424**, which does not form the cocrystal with (*S*)-BINOL, is isolated from the mother liquor.

L-Homolamivudine **433** and its 5-fluoro congener **434** are prepared via alcohol **428** from (*R*)-(−)-3-mercapto-1,2-propanediol (**426**) (Scheme 77).[186] After

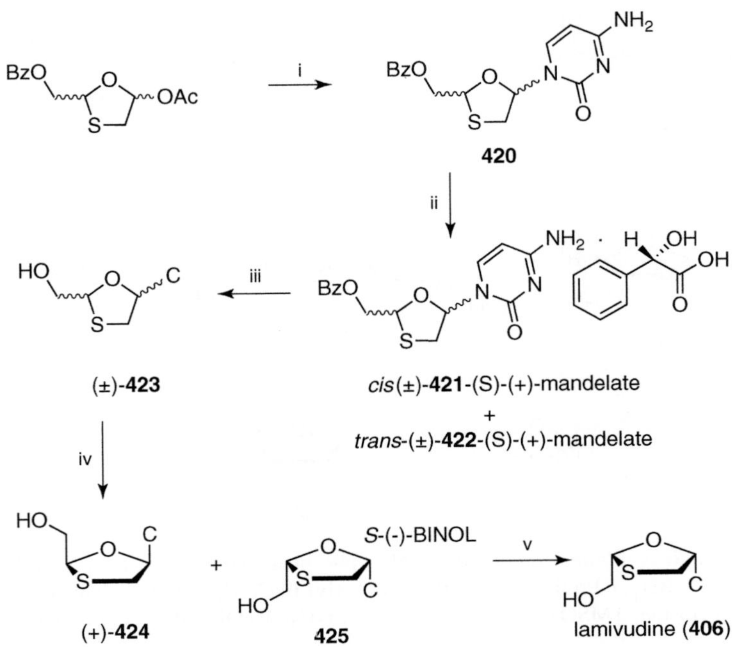

Scheme 76. *Reagents and conditions:* (i) cytosine, HMDS, TMSCl (47%); (ii) *S*-(+)-mandelic acid, EtOAc, 70°C to rt (47% for **421**); (iii) 25% NH$_3$/MeOH, EtOH (72%); (iv) (*S*)-BINOL, MeOH, 65°C (67%); (v) EtOAc, H$_2$O, HCl, NaOH, 25°C (70%).

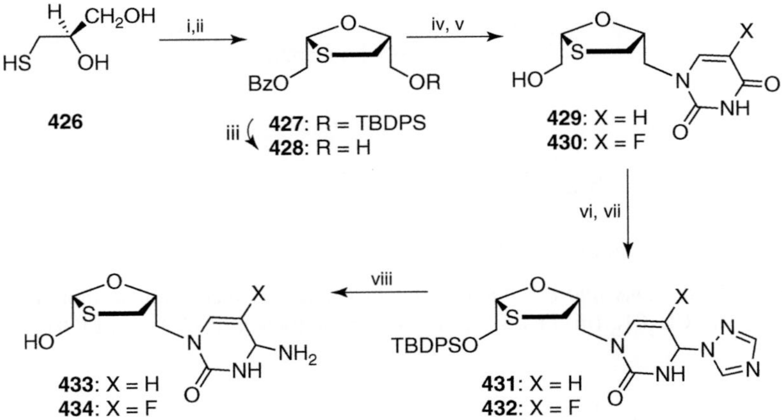

Scheme 77. *Reagents and conditions:* (i) TBDPSCl, DMAP, THF; (ii) BzOCH$_2$CHO, TsOH, 80°C, 1–2 mmHg; (iii) TBAF, THF (69% from **426**); (iv) after step (iii): Ph$_3$P, DEAD, 3-benzoyluracil or 3-benzoyl-5-fluorouracil; (v) NH$_3$, MeOH (61% for **429**; 45% for **430**); (vi) *p*-ClC$_6$H$_4$OPO(Cl)$_2$, 1,2,4-triazole, Py; (viii) NH$_4$OH, dioxane (28% for **433**; 36% for **434**).

selective protection of the primary hydroxyl group, the acid-catalyzed cyclothioacetalization with benzoyloxyacetaldehyde affords the *cis*-(2*R*,5*R*)-oxathiolane **427** along with small amounts of *trans*-(2*S*,5*R*)-diastereomer. Mitsunobu condensation of **428** with 3-benzolyluracil or 3-benzoyl-5-fluorouracil gives after complete debenzoylation the corresponding L-uracil homonucleosides **429** and **430**, which, following protection of the hydroxyl group, are converted to cytosines **433** and **434**, respectively, via their 1,2,4-triazole derivatives **431** and **432**.

A synthesis of oxathiolane nucleosides in which the nucleobase is vicinal to the sulfur atom has been developed by condensing. silylated *N*-acetylcytosine with the oxathiolane **436** derived from the chiral oxathiolane **435** (Scheme 78).[137] The coupling products are a mixture of *cis*- and *trans*-nucleosides which after separation by preparative TLC and removal of the protective group furnish the oxathiolane nucleosides **437–440**.

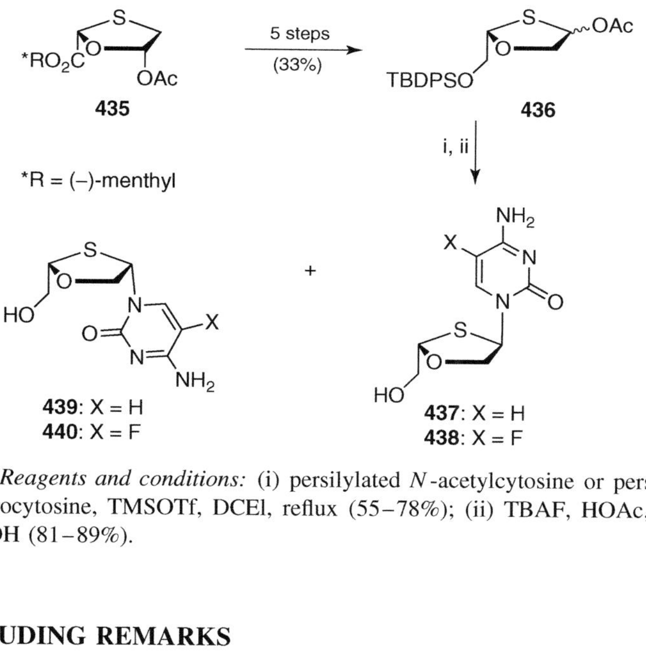

Scheme 78. *Reagents and conditions:* (i) persilylated *N*-acetylcytosine or persilylated *N*-acetyl-5-fluorocytosine, TMSOTf, DCEl, reflux (55–78%); (ii) TBAF, HOAc, THF, then K_2CO_3, MeOH (81–89%).

9. CONCLUDING REMARKS

This chapter is the result of a year's work. It started with the collection of more than 500 initial bibliographic references from our private collections and from bibliographic Internet database searches using instruments such as Web of Science and Scifinder. Despite our efforts, we cannot guarantee that all pertinent articles have been found. Our next step was to decide how to organize the material. Our choice was to refer to the nature of the sugar or ring to which the base is attached. As synthetic chemists we thought it was easier to find analogies in the preparation of the compounds following this criterion, hence easier to organize the data. A drawback, was that, at times, we ended up with some derivatives that could be placed into several groups. Due to space limitations, we decided not to review the antiviral properties of L-nucleosides and the hybridizing properties of L-oligodeoxyribonucleotides. We apologize to many authors for having overlooked this important motivation, which was beyond the synthesis of so many derivatives. At the end we have grouped our material in to about 80 reaction schemes incorporating more than 600 reactions, leading to as many different intermediates and final compounds, despite having chosen not to deal with, for example, bicyclic L-nucleosides and other hypermodified nucleosides. We arranged the material ronghly chronologically, to credit the pioneers in the field and to show how some syntheses were ameliorated in the course of years. In synthesizing nucleosides, as in the synthesis of all natural compounds, the main scope is to obtain the desired product, so there is still space for a talented chemist to optimize many reactions, and perhaps to find new compounds or new ways to synthesize known compounds. We tried to report the most varied strategies, and we were delighted to see how chemists have found elegant chemical pathways to use to synthesize their targets. We do hope that we have transmitted, at least a part of this sense of wonder to our readers.

L-Nucleosides were initially thought of as tools used to investigate why nature chose D-sugars as components of genetic information, and some of the oldest papers reviewed indeed referred to this type of quest. However, the fact that viral enzymes were found to be more tolerant of stereospecificty than human analogues paved the way for the use of L-nucleosides as antiviral compounds, a line of research that is still actively explored. Finally, thanks are due to the systematic work of Eschenmoser's group, which in a series of papers on the pairing of hexose nucleosides dissected the energy components of oligonucleotide hybridization, allowing a deeper comprehension of the chemistry of life.

ABBREVIATIONS

A	adenine
AIBN	azobis(isobutyronitrile)
BINOL	1,1′-bi(2-naphthol)
BSA	O,N-bistrimethylsilyl acetamide
C	cytosine
CSA	camphorsulfonic acid
Cy	cyclohexane
DAST	(diethylamino)sulfur trifluoride
DCC	dicyclohexyl carbodiimide
DCE	1,2-dichloroethane
DEAD	diethyl azodicarboxylate
DHP	dihydropyran
DIBAL–H	diisobutylaluminum hydride
DIPEA	N,N-diisopropylethylamine
DMAP	4-(N,N-dimethylamino)pyridine
DMF	N,N-dimethylformamide

DMP	2,2-dimethoxypropane
DMT	4,4'-dimethoxytrityl
G	guanine
H	hypoxanthine
HMDS	hexamethyldisilazane
Im	imidazole
MCPBA	m-chloroperoxybenzoic acid
Ms	methanesulfonyl
NIS	N-iodosuccinimide
PDC	pyridinium dichromate
PNB	p-nitrobenzoate
PPTS	pyridinium p-toluenesulfonate
Py	pyridine
T	thymine
TBAF	tetra-n-butylammonium fluoride
TBDMS	t-butyldimethylsilyl
TBDPS	t-butyldiphenylsilyl
TDA-1	tris-[2-(2-methoxyethoxy)ethyl]amine
TEA	triethylamine
Tf	triflate
TFA	trifluoroacetic acid
TFAA	trifluoroacetic anhydride
THF	tetrahydrofurane
THP	tetrahydropyranyl
TIPSCL	1,3-dichloro-1,1,3,3-tetraisopropyldisiloxane
TMS	trimethylsilyl
Ts	tosyl
U	uracil

REFERENCES

1. Joyce, G.F.; Visser, G.M.; van Boeckel, C.A.A.; van Boom, J.H.; Orgel, L.E.; van Westrenen, J. *Nature* **1984**, *310*, 602–604.
2. Visser, G.M.; Westrenen, J.V.; van Boeckel, C.A.A.; van Boom, J.H. *Recl. Trav. Chim. Pays-Bas* **1986**, *105*, 528–537.
3. Boeckel, van C.A.A.; Visser, G.M.; Hegstrom, R.A.; van Boom, J.H. *J. Mol. Evol.* **1987**, *25*, 100–105.
4. Kozlov, I.A.; Pitsch, S.; Orgel, L.E. *Proc. Natl. Acad. Sci. USA* **1998**, *95*, 13448–13452.
5. Fric, I.; Guschlbauer, W.; Holý, A. *FEBS Lett.* **1970**, *9*, 261–264.
6. Fujimori, S.; Shudo, K.; Hashimoto, Y. *J. Am. Chem. Soc.* **1990**, 7436–7438.
7. Urata, H.; Shinohara, K.; Ogura, E.; Ueda, Y.; Akagi, M. *J. Am. Chem. Soc.* **1991**, *113*, 8174–8175.
8. Blommers, M.J.J.; Tondelli, L.; Garbesi, A. *Biochemistry* **1994**, *33*, 7886–7896.
9. Garbesi, A.; Capobianco, M.L.; Colonna, F.P.; Tondelli, L.; Arcamone, F.; Manzini, G.; Hilbers, C.W.; Aelen, J.M.E.; Blommers, M.J.J. *Nucleic Acids Res.* **1993**, 4159–4165.

10. Garbesi, A.; Capobianco, M.L.; Coloma, F.P.; Maffini, M.; Niccolai, D.; Tondelli, L. *Nucleosides Nucleotides* **1998**, *1275*–1287.
11. Maury, G. *Antivir. Chem. Chemother.* **2000**, *11*, 165–190.
12. Focher, F.; Spadari, S.; Maga, G. *Infect. Disord.: Drug Targets* **2003**, *3*, 41–53.
13. Doong, S.-L.; Tsai, C.-H.; Schinazi, R.F.; Liotta, D.C.; Chen, Y.-C. *Proc. Natl. Acad. Sci. USA* **1991**, *88*, 8495–8499.
14. Coates, J.A.V.; Cammack, N.; Jenkinson, H.J.; Mutton, I.M.; Peaerson, B.A.; Storer, R.; Cameron, J.M.; Penn, C.R. *Antimicrob. Agents Chemother.* **1992**, *36*, 202–220.
15. Genu-Dellac, C.; Gosselin, G.; Puech, F.; Henry, J.C.; Aubertin, A.M.; Obert, G.; Kirn, A.; Imbach, J.L. *Nucleosides Nucleotides* **1991**, *10*, 1345–1376.
16. Schinazi, R.F.; Gosselin, G.; Faraj, A.; Korba, B.E.; Liotta, D.C.; Chu, C.K.; Mathe, C.; Imbach, J.L.; Sommadossi, J.P. *Antimicrob. Agents Chemother.* **1994**, *38*, 2172–2174.
17. Gosselin, G.; Boudou, V.; Griffon, J.F.; Pavia, G.; Pierra, C.; Imbach, J.L.; Aubertin, A.M.; Schinazi, R.F.; Faraj, A.; Sommadossi, J.P. *Nucleosides Nucleotides* **1997**, *16*, 1389–1398.
18. Shafiee, M.; Boudou, V.; Griffon, J.F.; Pompon, A.; Gosselin, G.; Eriksson, S.; Imbach, J.L.; Maury, G. *Nucleosides Nucleotides* **1997**, *16*, 1767–1770.
19. Verri, A.; Focher, F.; Priori, G.; Gosselin, G.; Imbach, J.L.; Capobianco, M.; Garbesi, A.; Spadari, S. *Mol. Pharmacol.* **1997**, 132–138.
20. Gosselin, G.; Boudou, V.; Griffon, J.F.; Pavia, G.; Pierra, C.; Imbach, J.L.; Faraj, A.; Sommadossi, J.P. *Nucleosides Nucleotides* **1998**, *17*, 1731–1738.
21. Maga, G.; Amacker, M.; Hubscher, U.; Gosselin, G.; Imbach, J.L.; Mathé, C.; Faraj, A.; Sommadossi, J.P.; Spadari, S. *Nucleic Acids Res.* **1999**, *27*, 972–978.
22. Wang, P.Y.; Hong, J.H.; Cooperwood, J.S.; Chu, C.K. *Antivir. Res.* **1998**, *40*, 19–44.
23. Mathe, C.; Gosselin, G. *Antivir. Res.* **2006**, *71*, 276–281.
24. Smejkal, J.; Sorm, F. *Coll. Czech. Chem. Commun.* **1964**, *29*, 2809–2813.
25. Acton, E.M.; Ryan, K.J.; Goodman, L. *J. Am. Chem. Soc.* **1964**, *86*, 5352–5354.
26. Davoll, J.; Lowy, B.A. *J. Am. Chem. Soc.* **1951**, *73*, 1650–1655.
27. Shimizu, B.; Asai, M.; Hieda, H.; Miyaki, M.; Okazaki, H. *Chem. Pharm. Bull.* **1965**, *13*, 616–618.
28. Shimizu, B.; Saito, A.; Nishimura, T.; Miyaki, M. *Chem. Pharm. Bull.* **1967**, *15*, 2011–2014.
29. Fujimori, S.; Iwanami, N.; Hashimoto, Y.; Shudo, K. *Nucleosides Nucleotides* **1992**, 341–349.
30. Urata, H.; Ogura, E.; Shinohara, K.; Ueda, Y.; Akagi, M. *Nucleic Acids Res.* **1992**, 3325–3332.
31. Robins, M.J.; Khwaja, T.A.; Robins, R.K. *J. Org. Chem.* **1970**, *35*, 636–639.
32. Holý, A.; Sorm, F. *Coll. Czech. Chem. Commun.* **1969**, *34*, 3383–3401.
33. Holý, A. *Tetrahedron Lett.* **1971**, 189–192.
34. Holý, A. *Coll. Czech. Chem. Commun.* **1972**, *37*, 4072–4087.
35. Sung, W.L. *J. Chem. Soc. Chem. Commun.*, **1981**, 1089.
36. Spadari, S.; Maga, G.; Focher, F.; Ciarrocchi, G.; Manservigi, R.; Arcamone, F.; Capobianco, M.; Carcuro, A.; Colonna, F.; Iotti, S.; Garbesi, A. *J. Med. Chem.* **1992**, *35*, 4214–4220.
37. Abe, Y.; Takizawa, T.; Kunieda, T. *Chem. Pharm. Bull.* **1980**, *28*, 1324–1326.

38. Jung, M.E.; Xu, Y. *Org.Lett.* **1999**, *1*, 1517–1519.
39. Sharma, D.; Khandelwal, A.; Sharma, R.K.; Bhatia, S.; Reddy, L.C.; Olsen, C.E.; Wengel, J.; Parmar, V.S.; Prasad, A.K. *Indian J. Chem., Sect. B: Org. Chem. Incl. Med. Chem.* **2009**, *48B*, 1712–1720.
40. Pitsch, S. *Helv. Chim. Acta* **1997**, *80*, 2286–2314.
41. Sivets, G.G. *Nucleosides Nucleotides Nucleic Acids* **2007**, *26*, 1241–1244.
42. Shi, Z.D.; Yang, B.H.; Wu, Y.L. *Tetrahedron Lett.* **2001**, *42*, 7651–7653.
43. Gosselin, G.; Bergogne, M.C.; Imbach, J.L. *J. Heterocycl. Chem.* **1993**, *30*, 1229–1233.
44. Gosselin, G.; Imbach, J.L. *J. Heterocycl. Chem.* **1982**, *19*, 597–602.
45. Chelain, E.; Floch, O.; Czernecki, S. *J. Carbohydr. Chem.* **1995**, *14*, 1251–1256.
46. Botta, O.; Moyroud, E.; Lobato, C.; Strazewski, P. *Tetrahedron* **1998** *54*, 13529–13546.
47. Moyroud, E.; Strazewski, P. *Tetrahedron* **1999**, *55*, 1277–1284.
48. Boudou, V.; Gosselin, G.; Imbach, J.L. *Nucleosides Nucleotides* **1999**, *18*, 607–609.
49. Fazio, F.; Schneider, M.P. *Tetrahedron: Asymmetry* **2000**, *11*, 1869–1876.
50. Sivets, G.G.; Klennitskaya, T.V.; Zhernosek, E.V.; Mikhailopulo, I.A. *Synthesis* **2002**, 253–259.
51. Chaudhuri, N.C.; Moussa, A.; Stewart, A.; Wang, J.; Storer, R. *Org. Process Res. Dev.* **2005**, *9*, 457–465.
52. Vonjantalipinski, M.; Costisella, B.; Ochs, H.; Hubscher, U.; Hafkemeyer, P.; Matthes, E. *J. Med. Chem.* **1998**, *41*, 2040–2046.
53. Shi, J.; Du, J.; Ma, T.; Pankiewicz, K.W.; Patterson, S.E.; Tharnish, P.M.; McBrayer, T.R.; Stuyver, L.J.; Otto, M.J.; Chu, C.K.; Schinazi, R. F.; Watanabe, K. A. *Bioorg. Med. Chem.* **2005**, *13*, 1641–1652.
54. Ma, T.; Pai, S.B.; Zhu, Y.L.; Lin, J.S.; Shanmuganathan, K.; Du, J.; Wang, C.; Kim, H.; Newton, M.G.; Cheng, Y.C.; Chu, C.K. *J. Med. Chem.* **1996**, *39*, 2835–2843.
55. Olgen, S. *Turk. J. Chem.* **2002**, *26*, 255–261.
56. Zhang, X.; Xia, H.; Dong, X.; Jin, J.; Meng, W.-D.; Qing, F.-L. *J. Org. Chem.* **2003**, *68*, 9026–9033.
57. Xu, X.-H.; Qiu, X.-L.; Zhang, X.; Qing, F.-L. *J. Org. Chem.* **2006**, *71*, 2820–2824.
58. Kotra, L.P.; Xiang, Y.J.; Newton, M.G.; Schinazi, R.F.; Cheng, Y.C.; Chu, C.K. *J. Med. Chem.* **1997**, *40*, 3635–3644.
59. Just, G.; Reader, G. *Tetrahedron Lett.* **1973**, 1525–1528.
60. Just, G.; Ramjeesingh, M. *Tetrahedron Lett.* **1975**, 985–988.
61. Just, G.; Martel, A.; Grozinger, K.; Ramjeesingh, M. *Can. J. Chem.* **1975**, *53*, 131–137.
62. Just, G.; Kim, S. *Tetrahedron Lett.* **1976**, 1063–1066.
63. Just, G.; Kim, S.G. *Can. J. Chem.* **1976**, *54*, 2935–2939.
64. Just, G.; Ramjeesingh, M.; Liak, T.J. *Can. J. Chem.* **1976**, *54*, 2940–2947.
65. Just, G.; Kim, S. *Can. J. Chem.* **1977**, *55*, 427–434.
66. Just, G.; Lim, M.-I. *Can. J. Chem.* **1977**, *55*, 2993–2997.
67. Liang, C.Y.; Ma, T.W.; Cooperwood, J.S.; Du, J.F.; Chu, C.K. *Carbohydr. Res.* **1997**, *303*, 33–38.
68. Trost, B.M.; Kallander, L.S. *J. Org. Chem.* **1999**, *64*, 5427–5435.

69. Mathé, C.; Gosselin, G.; Bergogne, M.C.; Aubertin, A.M.; Obert, G.; Kirn, A.; Imbach, J.L. *Nucleosides Nucleotides* **1995**, *14*, 549–550.
70. Mathé, C.; Gosselin, G. *Nucleosides Nucleotides Nucleic Acids* **2000**, *19*, 1517–1530.
71. Migawa, M.T.; Girardet, J.-L.; Walker, J.A., II; Koszalka, G.W.; Chamberlain, S.D.; Drach, J.C.; Townsend, L.B. *J. Med. Chem.* **1998**, *41*, 1242–1251.
72. Baddi, L.; Smietana, M.; Sebti, S.; Vasseur, J.-J.; Lazrek, H.B. *Lett. Org. Chem.* **2010**, *7*, 196–199.
73. Gaubert, G.; Mathé, C.; Imbach, J.L.; Eriksson, S.; Vincenzetti, S.; Salvatori, D.; Vita, A.; Maury, G. *Eur. J. Med. Chem.* **2000**, *35*, 1011–1019.
74. Ramasamy, K.S.; Tam, R.C.; Bard, J.; Averett, D.R. *J. Med. Chem.* **2000**, *43*, 1019–1028.
75. Griffon, J.F.; Mathé, C.; Faraj, A.; Aubertin, A.M.; De Clercq, E.; Balzarini, J.; Sommadossi, J.P.; Gosselin, G. *Eur. J. Med. Chem.* **2001**, *36*, 447–460.
76. Salvetti, R.; Marchand, A.; Pregnolato, M.; Verri, A.; Spadari, S.; Focher, F.; Briant, M.; Sommadossi, J.P.; Mathe, C.; Gosselin, G. *Bioorg. Med. Chem.* **2001**, *9*, 1731–1738.
77. Salvetti, R.; Pregnolato, M.; Verri, A.; Focher, F.; Spadari, S.; Marchand, A.; Mathé, C.; Gosselin, G. *Nucleosides Nucleotides Nucleic Acids* **2001**, *20*, 1123–1125.
78. Zhu, W.; Gumina, G.; Schinazi, R.F.; Chu, C.K. *Tetrahedron* **2003**, *59*, 6423–6431.
79. Lin, W.; Zhang, X.; Seela, F. *Helv. Chim. Acta* **2004**, *87*, 2235–2244.
80. Seela, F.; Lin, W.; Kazimierczuk, Z.; Rosemeyer, H.; Glacon, V.; Peng, X.; He, Y.; Ming, X.; Andrzejewska, M.; Gorska, A.; et al. *Nucleosides Nucleotides Nucleic Acids* **2005**, *24*, 859–863.
81. Seela, F.; Ming, X. *Tetrahedron* **2007**, *63*, 9850–9861.
82. Budow, S.; Kozlowska, M.; Gorska, A.; Kazimierczuk, Z.; Eickmeier, H.; La Colla, P.; Gosselin, G.; Seela, F. *ARKIVOC (Gainesville, FL)* **2008**, 225–250.
83. Marchand, A.; Lioux, T.; Mathé, C.; Imbach, J.L.; Gosselin, G. *J. Chem. Soc. Perkin Trans. 1* **1999**, 2249–2254.
84. Lin, T.S.; Luo, M.Z.; Liu, M.C.; Pai, S.B.; Dutschman, G.E.; Cheng, Y.C. *J. Med. Chem.* **1994**, *37*, 798–803.
85. Lin, T.S.; Luo, M.Z.; Zhu, J.L.; Liu, M.C.; Zhu, Y.L.; Dutschman, G.E.; Cheng, Y.C. *Nucleosides Nucleotides* **1995**, *14*, 1759–1783.
86. Taniguchi, M.; Koga, K.; Yamada, S. *Tetrahedron* **1974**, *30*, 3547–3552.
87. Ravid, U.; Smith, L.R.; Silverstein, R.M. *Tetrahedron* **1978**, *34*, 1449–1452.
88. Takano, S.; Goto, E.; Hirama, M.; Ogasawara, K. *Heterocycles* **1981**, *16*, 951–954.
89. Lundt, I.; Pedersen, C. *Synthesis* **1986**, 1052–1054.
90. Wang, P.Y.; Bolon, P.J.; Newton, M.G.; Chua, C.K. *Nucleosides Nucleotides* **1999**, *18*, 2819–2835.
91. Chu, C.K.; Babu, J.R.; Beach, J.W.; Ahn, S.K.; Huang, H.Q.; Jeong, L.S.; Lee, S.J. *J. Org. Chem.* **1990**, *55*, 1418–1420.
92. Beach, J.W.; Kim, H.O.; Jeong, L.S.; Nampalli, S.; Islam, Q.; Ahn, S.K.; Babu, J.R.; Chu, C.K. *J. Org. Chem.* **1992**, *57*, 3887–3894.
93. Niedballa, U.; Vorbrüggen, H. *J. Org. Chem.* **1974**, *39*, 3654–3660.
94. Vorbrüggen, H.; Krolikiewicz, K. *Angew. Chem., Int. Ed.* **1975**, *14*, 421–422.

95. Bolon, P.J.; Wang, P.Y.; Chu, C.K.; Gosselin, G.; Boudou, V.; Pierra, C.; Mathé, C.; Imbach, J.L.; Faraj, A.; Alaoui, M.A.; et al. *Bioorg. Med. Chem. Lett.* **1996**, *6*, 1657–1662.
96. Wilson, L.J.; Liotta, D. *Tetrahedron Lett.* **1990**, *31*, 1815–1818.
97. Young, R.J.; Shaw-Ponter, S.; Thomson, J.B.; Miller, J.A.; Cumming, J.G. Pugh, A.W.; Rider, P. *Bioorg. Med. Chem. Lett.* **1995**, *5*, 2599–2604.
98. Chen, S.-H.; Li, X.; Li, J.; Niu, C.; Carmichael, E.; Doyle, T.W. *J. Org. Chem.* **1997**, *62*, 3449–3452.
99. Chen, S.H.; Wang, Q.; Mao, J.; King, I.; Dutschman, G.E.; Gullen, E.A.; Cheng, Y.C.; Doyle, T.W. *Bioorg. Med. Chem. Lett.* **1998**, *8*, 1589–1594.
100. Cavalcanti, S.C.H.; Xiang, X.J.; Newton, M.G.; Schinazi, R.F.; Cheng, Y.C.; Chu, C.K. *Nucleosides Nucleotides* **1999**, *18*, 2233–2252.
101. Chun, B.K.; Schinazi, R.F.; Cheng, Y.C.; Chu, C.K. *Carbohydr. Res.* **2000**, *328*, 49–59.
102. Choi, Y.S.; Lee, K.O.; Hong, J.H.; Schinazi, R.F.; Chu, C.K. *Tetrahedron Lett.* **1998**, *39*, 4437–4440.
103. Lee, K.Y.; Choi, Y.S.; Gullen, E.; Schlueter-Wirtz, S.; Schinazi, R.F.; Cheng, Y.C.; Chu, C.K. *J. Med. Chem.* **1999**, *42*, 1320–1328.
104. Chong, Y.; Gumina, G.; Mathew, J.S.; Schinazi, R.F.; Chu, C.K. *J. Med. Chem.* **2003**, *46*, 3245–3256.
105. Kotra, L.P.; Newton, M.G.; Chu, C.K. *Carbohydr. Res.* **1998**, *306*, 69–80.
106. Xiang, Y.; Kotra, L.P.; Chu, C.K.; Schinazi, R.F. *Bioorg. Med. Chem. Lett.* **1995**, *5*, 743–748.
107. Wang, J.; Jin, Y.; Rapp, K.L.; Schinazi, R.F.; Chu, C.K. *Abstracts of Papers, 233rd ACS National Meeting, Chicago, IL, United States, March 25–29*, **2007**, 2007, MEDI-073.
108. Rassu, G.; Zanardi, F.; Battistini, L.; Gaetani, E.; Casiraghi, G. *J. Med. Chem.* **1997**, *40*, 168–180.
109. Albert, M.; De Souza, D.; Feiertag, P.; Hoenig, H. *Org. Lett.* **2002**, *4*, 3251–3254.
110. Arnone, A.; Bravo, P.; Frigerio, M.; Mele, A.; Vergani, B.; Viani, F. *Eur. J. Org. Chem.* **1999**, 2149–2157.
111. Jung, M.E.; Nichols, C.J.; Kretschik, O.; Xu, Y. *Nucleosides Nucleotides* **1999**, *18*, 541–546.
112. Jung, M.E.; Nichols, C.J. *J. Org. Chem.* **1998**, *63*, 347–355.
113. Lee, C.S.; Du, J.; Chu, C.K. *Nucleosides Nucleotides* **1996**, *15*, 1223–1236.
114. Gumina, G.; Chong, Y.; Choi, Y.; Chu, C.K. *Org. Lett.* **2000**, *2*, 1229–1231.
115. Choi, Y.; Choo, H.; Chong, Y.; Lee, S.; Olgen, S.; Schinazi, R.F.; Chu, C.K. *Org. Lett.* **2002**, *4*, 305–307.
116. Mugnaini, C.; Botta, M.; Coletta, M.; Corelli, F.; Focher, F.; Marini, S.; Renzulli, M.L.; Verri, A. *Bioorg. Med. Chem.* **2003**, *11*, 357–366.
117. Chen, X.; Zhou, W.; Schinazi, R.F.; Chu, C.K. *J. Org. Chem.* **2004**, *69*, 6034–6041.
118. Vince, R.; Hua, M. *J. Med. Chem.* **1990**, *33*, 17–21.
119. Vince, R.; Brownell, J. *Biochem. Biophys. Res. Commun.* **1990**, *168*, 912–916.
120. Evans, C.T.; Roberts, S.M.; Shoberu, K.A.; Sutherland, A.G. *J. Chem. Soc., Perkin Trans. 1* **1992**, 589–592.
121. Coates, J.A.V.; Inggall, H.J.; Pearson, B.A.; Penn, C.R.; Storer, R.; Williamson, C.; Cameron, J.M. *Antivir. Res.* **1991**, *15*, 161–168.

122. Furman, P.A.; Wilson, J.E.; Reardon, J.E.; Painter, G.R. *Antivir. Chem. Chemother.* **1995**, *6*, 345–355.
123. Merlo, V.; Roberts, S.M.; Storer, R.; Bethell, R.C. *J. Chem. Soc., Perkin Trans. 1* **1994**, 1477–1481.
124. Davis, M.G.; Wilson, J.E.; VanDraanen, N.A.; Miller, W.H.; Freeman, G.A.; Daluge, S.M.; Boyd, F.L.; Aulabaugh, A.E.; Painter, G.R.; Boone, L.R. *Antivir. Res.* **1996**, *30*, 133–145.
125. Berranger, T.; Langlois, Y. *Tetrahedron Lett.* **1995**, *36*, 5523–5526.
126. Jin, Y.H.; Liu, P.; Wang, J.; Baker, R.; Huggins, J.; Chu, C.K. *J. Org. Chem.* **2003**, *68*, 9012–9018.
127. Wang, P.Y.; Agrofoglio, L.A.; Newton, M.G.; Chu, C.K. *J. Org. Chem.* **1999**, *64*, 4173–4178.
128. Wang, P.; Agrofoglio, L.A.; Newton, M.G.; Chu, C.K. *Tetrahedron Lett.* **1997**, *38*, 4207–4210.
129. Wang, P.Y.; Gullen, B.; Newton, M.G.; Cheng, Y.C.; Schinazi, R.F.; Chu, C.K. *J. Med. Chem.* **1999**, *42*, 3390–3399.
130. Song, G.Y.; Paul, V.; Choo, H.; Morrey, J.; Sidwell, R.W.; Schinazi, R.F.; Chu, C.K. *J. Med. Chem.* **2001**, *44*, 3985–3993.
131. Hegedus, L.S.; Geisler, L. *J. Org. Chem.* **2000**, *65*, 4200–4203.
132. Agrofoglio, L.A.; Amblard, F.; Nolan, S.P.; Charamon, S.; Gillaizeau, I.; Zevaco, T.A.; Guenot, P. *Tetrahedron* **2004**, *60*, 8397–8404.
133. Marce, P.; Diaz, Y.; Matheu, M.I.; Castillon, S. *Org. Lett.* **2008**, *10*, 4735–4738.
134. Jessel, S.; Meier, C. *Eur. J. Org. Chem.* **2011**, 1702–1713.
135. Liu, P.; Chu, C.K. *Can. J. Chem.* **2006**, *84*, 748–754.
136. Choo, H.; Chong, Y.; Chu, C.K. *Org. Lett.* **2001**, *3*, 1471–1473.
137. Chong, Y.; Gumina, G.; Chu, C.K. *Tetrahedron: Asymmetry* **2000**, *11*, 4853–4875.
138. Wang, J.; Jin, Y.; Rapp, K.L.; Schinazi, R.F.; Chu, C.K. *J. Med. Chem.* **2007**, *50*, 1828–1839.
139. Wang, J.; Jin, Y.; Rapp, K.L.; Bennett, M.; Schinazi, R.F.; Chu, C.K. *J. Med. Chem.* **2005**, *48*, 3736–3748.
140. Pradere, U.; Kumamoto, H.; Roy, V.; Agrofoglio, L.A. *Eur. J. Org. Chem.* **2010**, 749–754, S749/741–S749/713.
141. Varaprasad, C.V.; Averett, D.; Ramasamy, K.S.; Wu, J.J. *Tetrahedron* **1999**, *55*, 13345–13368.
142. Westwood, N.B.; Walker, R.T. *Tetrahedron* **1998**, *54*, 13391–13404.
143. Yokoyama, M.; Ikeue, T.; Ochiai, Y.; Momotake, A.; Yamaguchi, K.; Togo, H. *J. Chem. Soc., Perkin Trans. 1* **1998**, 2185–2191.
144. Dondoni, A.; Perrone, D. *Tetrahedron Lett.* **1999**, *40*, 9375–9378.
145. Clinch, K.; Evans, G.B.; Fleet, G.W.J.; Furneaux, R.H.; Johnson, S.W.; Lenz, D.H.; Mee, S.P.H.; Rands, P.R.; Schramm, V.L.; Ringia, E.A.T.; Tyler, P.C. *Org. Biomol. Chem.* **2006**, *4*, 1131–1139.
146. Evans, G.A.; Furneaux, R.H.; Hutchison, T.L.; Kezar, H.S.; Morris, P.E.J.; Schramm, V.L.; Tyler, P.C. *J. Org. Chem.* **2001**, *66*, 5723–5730.
147. Merino, P.; Del Alamo, E.M.; Franco, S.; Merchan, F.L.; Simon, A.; Tejero, T. *Tetrahedron: Asymmetry* **2000**, *11*, 1543–1554.

148. Merino, P.; del Alamo, E.M.; Bona, M.; Franco, S.; Merchan, F.L.; Tejero, T.; Vieceli, O. *Tetrahedron Lett.* **2000**, *41*, 9239–9243.
149. Reist, E.J.; Gueffroy, D.E.; Goodman, L. *J. Am. Chem. Soc.* **1964**, *86*, 5658–5663.
150. Pejanovic, V.; Stokic, Z.; Stojanovic, B.; Piperski, V.; Popsavin, M.; Popsavin, V. *Bioorg. Med. Chem. Lett.* **2003**, *13*, 1849–1852.
151. Birk, C.; Voss, J.; Wirsching, J. *Carbohydr. Res.* **1997**, *304*, 239–247.
152. Wirsching, J.; Voss, J.; Adiwidjaja, G.; Giesler, A.; Kopf, J. *Eur. J. Org. Chem.* **2001**, 1077–1087.
153. Wirsching, J.; Voss, J. *Eur. J. Org. Chem.* **1999**, 691–696.
154. Wirsching, J.; Voss, J.; Adiwidjaja, G.; Balzarini, J.; De Clercq, E. *Bioorg. Med. Chem. Lett.* **2001**, *11*, 1049–1051.
155. Wirsching, J.; Voss, J.; Balzarini, J.; De Clercq, E. *Bioorg. Med. Chem. Lett.* **2000**, *10*, 1339–1341.
156. Satoh, H.; Yoshimura, Y.; Sakata, S.; Miura, S.; Machida, H.; Matsuda, A. *Bioorg. Med. Chem. Lett.* **1998**, *8*, 989–992.
157. Jeong, L.S.; Moon, H.R.; Choi, Y.J.; Chun, M.W.; Kim, H.O. *J. Org. Chem.* **1998**, *63*, 4821–4825.
158. Kim, H.O.; Jeong, L.S.; Lee, S.N.; Yoo, S.J.; Moon, H.R.; Kim, K.S.; Chun, M.W. *J. Chem. Soc. Perkin Trans. 1* **2000**, *9*, 1327–1329.
159. Kim, H.O.; Kim, Y.H.; Suh, H.; Jeong, L.S. *Bioorg. Med. Chem. Lett.* **2001**, *11*, 599–603.
160. Choo, H.; Chong, Y.H.; Choi, Y.S.; Mathew, J.; Schinazi, R.F.; Chu, C.K. *J. Med. Chem.* **2003**, *46*, 389–398.
161. Zhu, W.; Chong, Y.; Choo, H.; Mathews, J.; Schinazi, R.F.; Chu, C.K. *J. Med. Chem.* **2004**, *47*, 1631–1640.
162. Zheng, F.; Zhang, X.-H.; Qiu, X.-L.; Zhang, X.; Qing, F.-L. *Org. Lett.* **2006**, *8*, 6083–6086.
163. Chapdelaine, D.; Cardinal-David, B.; Prevost, M.; Gagnon, M.; Thumin, I.; Guindon, Y. *J. Am. Chem. Soc.* **2009**, *131*, 17242–17245.
164. Cong, L.; Zhou, W.; Jin, D.; Wang, J.; Chen, X. *Chem. Biol. Drug Des.* **2010**, *75*, 619–627.
165. Kim, H.O.; Shanmuganathan, K.; Alves, A.J.; Jeong, L.S.; Beach, J.W.; Schinazi, R.F.; Chang, C.N.; Cheng, Y.C.; Chu, C.K. *Tetrahedron Lett.* **1992**, *33*, 6899–6902.
166. Kim, H.O.; Schinazi, R.F.; Shanmuganathan, K.; Jeong, L.S.; Beach, J.W.; Nampalli, S.; Cannon, D.L.; Chu, C.K. *J. Med. Chem.* **1993**, *36*, 519–528.
167. Evans, C.A.; Dixit, D.M.; Siddiqui, M.A.; Jin, H.; Tse, H.L.A.; Cimpoia, A.; Bednarski, K.; Breining, T.; Mansour, T.S. *Tetrahedron: Asymmetry* **1993**, *4*, 2319–2322.
168. Liu, M.C; Luo, M.Z; Mozdziesz, D.E; Lin, T.S; Dutschman, G.E; Gullen, E.A.; Cheng, Y.C.; Sartorelli, A.C. *Nucleosides Nucleotides Nucleic Acids* **2000**, *19*, 603–618.
169. Luo, M.Z.; Liu, M.C.; Mozdziesz, D.E.; Lin, T.S.; Dutschman, G.E.; Guller, E.A.; Cheng, Y.C.; Sartorelli, A.C. *Bioorg. Med. Chem. Lett.* **2000**, *10*, 2145–2148.
170. Liu, M.C.; Luo, M.Z.; Mozdziesz, D.E.; Lin, T.S.; Dutschman, G.E.; Guller, E.A.; Cheng, Y.C.; Sartorelli, A.C. *Bioorg. Med. Chem. Lett.* **2001**, *11*, 2301–2304.
171. Bera, S.; Malik, L.; Bhat, B.; Carroll, S.S.; Hrin, R.; MacCoss, M.; McMasters, D.R.; Miller, M.D.; Moyer, G.; Olsen, D.B.; et al. *Bioorg. Med. Chem.* **2004**, *12*, 6237–6247.

172. Du, J.; Qu, F.; Lee, D.; Newton, M.G.; Chu, C.K. *Tetrahedron Lett.* **1995**, *36*, 8167–8170.
173. Qu, F.; Hong, J.H.; Du, J.; Newton, M.G.; Chu, C.K. *Tetrahedron* **1999**, *55*, 9073–9088.
174. Liang, C.; Lee, D.W.; Newton, M.G.; Chu, C.K. *J. Org. Chem.* **1995**, *60*, 1546–1553.
175. Dwyer, O. *Synlett* **1995**, *11*, 1163–1164.
176. Huang, J.J.; Rideout, J.L.; Martin, G.E. *Nucleosides Nucleotides* **1995**, *14*, 195–207.
177. Beach, J.W.; Jeong, L.S.; Alves, A.J.; Pohl, D.; Kim, H.O.; Chang, C.N.; Doong, S.L.; Schinazi, R.F.; Cheng, Y.C.; Chu, C.K. *J. Org. Chem.* **1992**, *57*, 2217–2219.
178. Jeong, L.S.; Schinazi, R.F.; Beach, J.W.; Kim, H.O.; Nampalli, S.; Shanmuganathan, K.; Alves, A.J.; McMillan, A.; Chu, C.K.; Mathis, R. *J. Med. Chem.* **1993**, *36*, 181–195.
179. Choi, W.B.; Wilson, L.J.; Yeola, S.; Liotta, D.C.; Schinazi, R.F. *J. Am. Chem. Soc.* **1991**, *113*, 9377–9379.
180. Jin, H.; Siddiqui, M.A.; Evans, C.A.; Tse, H.L.A.; Mansour, T.S.; Goodyear, M.D.; Ravenscroft, P.; Beels, C.D. *J. Org. Chem.* **1995**, *60*, 2621–2623.
181. Brossette, T.; Klein, E.; Créminon, C.; Grassi, J.; Mioskowski, C.; Lebeau, L. *Tetrahedron* **2001**, *57*, 8129–8143.
182. Milton, J.; Brand, S.; Jones, M.F.; Rayner, C.M. *Tetrahedron Lett.* **1995**, 6961–6964.
183. Cousins, R.P.C.; Mahmoudian, M.; Youds, P.M. *Tetrahedron: Asymmetry* **1995**, *6*, 393–396.
184. Goodyear, M.D.; Hill, M.L.; West, J.P.; Whitehead, A.J. *Tetrahedron Lett.* **2005**, *46*, 8535–8538.
185. Roy, B.N.; Singh, G.P.; Srivastava, D.; Jadhav, H.S.; Saini, M.B.; Aher, U.P. *Org. Process Res. Dev.* **2009**, *13*, 450–455.
186. Khan, N.; Bastola, S.R.; Witter, K.G.; Scheiner, P. *Tetrahedron Lett.* **1999**, *40*, 8989–8992.
187. Wang, W.; Jin, H.; Mansour, T.S. *Tetrahedron Lett.* **1994**, *35*, 4739–4742.

12 Chemical Synthesis of Carbocyclic Analogues of Nucleosides

E. LECLERC

Institut Charles Gerhardt, Ecole Nationale Supérieure de Chimie de Montpellier, Montpellier, France

1. INTRODUCTION

Although a long-standing approach, the development of nucleoside analogues is still an intense research field aiming at the development of safer and more efficient antitumoral or antiviral agents. Indeed, these agents may act as nontoxic and selective inhibitors of various enzymes involved in the cell or viral replication processes. Depending on their degree of phosphorylation, these molecules may act as inhibitors of thymidilate synthetase, ribonucleotide reductase, or DNA polymerases. The mechanism of action is either based on the competitive inhibition of the targeted enzyme or on the use of the agent as an alternative substrate.[1] A wide range of analogues based on a standard sugar backbone but featuring modifications either on the base or in the backbone substitutions has been prepared. This approach gave rise to several powerful drugs (e.g., 5-fluorouracil, gemcitabine, azidothymidine, zalcitabine) that are used widely in the treatment of various cancers and viral diseases (Figure 1).

Analogues based on an all-carbon backbone, the carbocyclic nucleosides, have acquired a growing importance in recent years. Disclosure of the highly active compounds aristeromycin and neplanocin A indeed initiated considerable efforts toward the synthesis of related structures featuring a carbocyclic surrogate to the usual carbohydrate moiety.[2] Although the mechanism of action of such compounds was not always thoroughly elucidated, their higher enzymatic stability, due to the suppression of the labile ketal function, certainly appeared as a key feature for their activity. Several lead compounds, such as carbovir, abacavir and entecavir, have been discovered since then (Figure 2).[3–5] As a consequence, the synthesis of

---

Chemical Synthesis of Nucleoside Analogues, First Edition. Edited by Pedro Merino.
© 2013 John Wiley & Sons, Inc. Published 2013 by John Wiley & Sons, Inc.

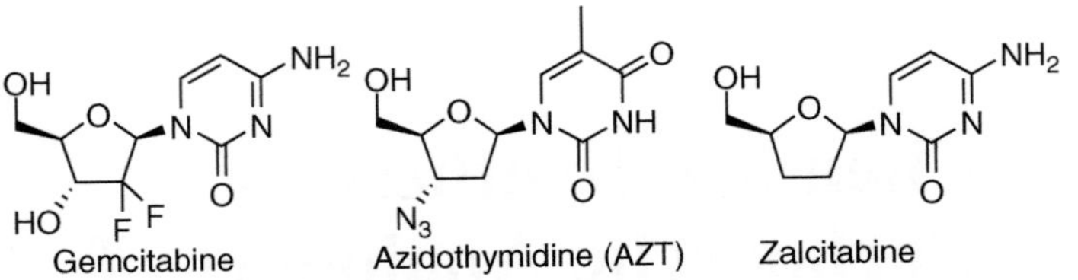

Figure 1. Nucleosidic drugs.

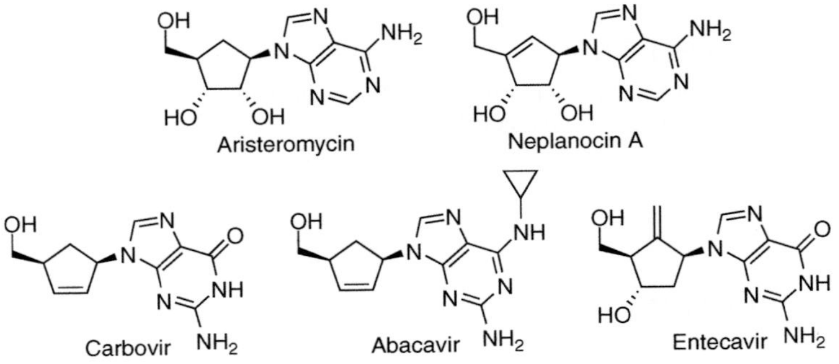

Figure 2. Carbocyclic nucleosides with antiviral or anticancer activity.

new structures and the development of new methodologies to widen the structural diversity of this family of molecules are still very active areas.

This chapter focuses on the asymmetric chemical synthesis of carbocyclic analogues of nucleosides (Figure 3), especially on five-membered ring derivatives, even if the preparation of analogues based on other ring sizes is summarized. Although the key synthetic methods developed before 1998 are described, the work reported since the last review on the topic is highlighted.[6,7] Current methods used for the introduction of the heterocyclic base on the carbasugar, including unusual or C-linked bases, are reported first. The asymmetric synthesis of cyclopentyl carbocyclic analogues of nucleosides is reviewed second. From methods using a resolved cyclopentene building block as the starting material to catalytic enantioselective syntheses, various leading strategies are reported. Finally, the synthesis of fluorinated, cyclohexylic, cyclobutylic, and cyclopropylic carbanucleosides is summarized. The highly abundant literature available on carbocyclic nucleosides added to the high structural diversity that has been reached throughout the years does not allow a comprehensive and exhaustive report. A choice thus had to be made, so only the most representative, efficient, and/or innovative methods are presented.

2. INTRODUCTION OF THE HETEROCYCLIC BASE TO THE CARBOCYCLIC CARBOHYDRATE SURROGATE

Various methods that allow the direct introduction of a heterocyclic base on an adequately functionalized carbocycle or its construction from a 1′-aminocarba-sugar

INTRODUCTION OF THE HETEROCYCLIC BASE TO THE CARBASUGAR 537

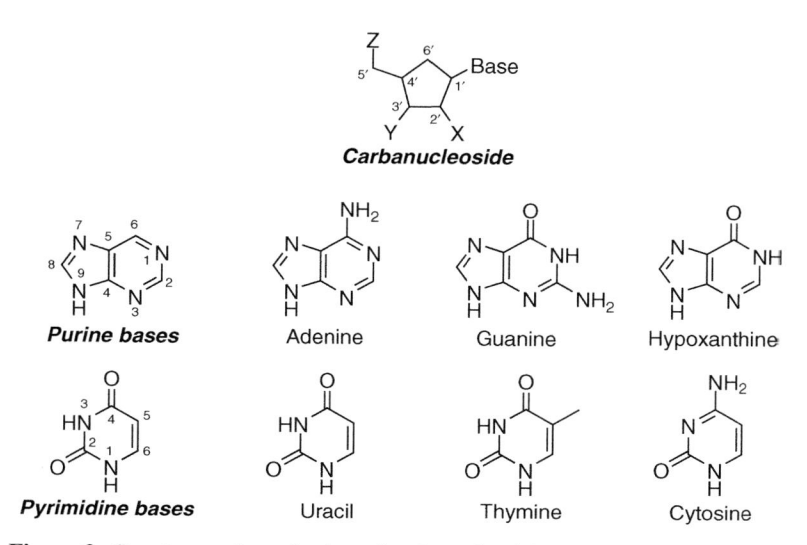

Figure 3. Structure and numbering of carbanucleoside and natural nucleobases.

are summarized. Each strategy is illustrated only by one or two examples, as most methods appear in the carbanucleoside syntheses described throughout the chapter.

2.1. Direct Introduction by Substitution Reactions

2.1.1. Substitution of an Activated Alcohol. The substitution of a mesylate or a tosylate by a purine or pyrimidine base was first described by Marquez et al.[8] The synthesis of neplanocin A and analogues was indeed performed using either the introduction of an azido group followed by base construction or the direct substitution by 6-chloropurine or uracil on tosylate **1** (Scheme 1). The yields are moderate (protected carba-nucleosides **2** and **3** are obtained in 31% and 26% yield, respectively), which is a usual drawback for this method.

One exception is the direct introduction of adenine reported by Jeong for the synthesis of an *apio* analogue of neplanocin A, for which intermediate **5** is obtained from mesylate **4** in 61% yield (Scheme 2).[9] More generally, the use of crown ethers as an additive is often beneficial to the reaction outcome. Yields of such reactions can be improved by the use of an ionic liquid as a solvent under microwave activation, as reported by Sega et al.[10]

Experimental Procedure

(1′R,2′S,3′S)-6-Amino-9-[2′,3′-O-isopropylidene-2′,3′-dihydroxy-3′-(benzyloxymethyl)cyclopent-4-en-1-yl]purine (5).[9] A stirred suspension of adenine (0.127 g, 0.94 mmol), 18-crown-6 (0.248 g, 0.94 mmol), and K_2CO_3 (0.195 mg, 1.41 mmol) in DMF (7 mL) was heated at 80°C for 30 min. A solution of mesylate **4** (0.238 g, 0.47 mmol) in DMF (3 mL) was added at 80°C and the reaction

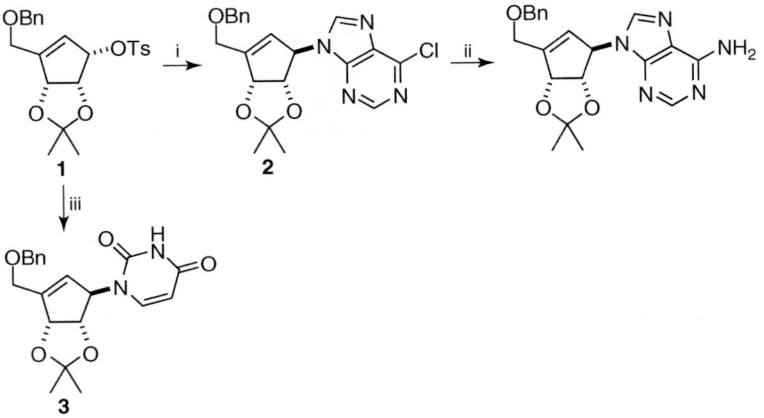

Scheme 1. *Reagents and conditions:* (i) sodium salt of 6-chloropurine, MeCN, 31%; (ii) NH₃, MeOH; (iii) uracil, K₂CO₃, DMSO, 26%.

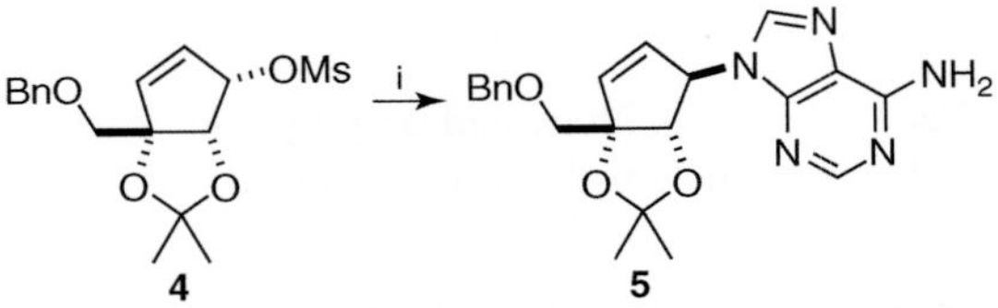

Scheme 2. *Reagents and conditions:* (i) adenine, K₂CO₃, 18-crown-6, DMF, 61%.

mixture was heated at 80°C for 12 h, cooled to room temperature, and partitioned between water (10 mL) and CH$_2$Cl$_2$ (80 mL). The organic layer was dried over MgSO$_4$, filtered, and evaporated in vacuo. The residue was purified by column chromatography (CH$_2$Cl$_2$/MeOH, 30:1) to give **5** (156 mg, 61%) as a colorless sticky oil, along with a small amount of α-isomer (22 mg, 9%). UV (CHCl$_3$): λ_{max} 260 nm. ^{1}H-NMR (400 MHz, CDCl$_3$): δ 8.19 (s, 1H), 7.70 (s, 1H), 7.35–7.23 (m, 15H), 6.82 (br s, 2H), 6.33 (dd, J = 1.2 and 6.0 Hz, 1H), 5.93 (dd, J = 2.4 and 5.6 Hz, 1H), 5.60 (br s, 1H), 4.31 (s, 1H), 3.45 (d, J = 10.4 Hz, 1H), 3.41 (d, J = 10.0 Hz, 1H), 1.46 (s, 3H), 1.39 (s, 3H); LRMS(FAB$^+$) m/z 546 [MH$^+$]. *Anal.* Calcd for C$_{33}$H$_{31}$N$_5$O$_3$: C, 72.64; H, 5.73; N, 12.84; found: C, 72.91; H, 5.80; N, 12.86.

2.1.2. Ring Opening of an Epoxide or a Cyclic Sulfate/Sulfite.

The addition of purine and pyrimidine nucleobases to functionalized cyclopentene oxides was also investigated.[11] The method often suffers from a lack of regioselectivity unless a steric or electronic bias is able to direct the addition (Scheme 3). Epoxides such as **6**, featuring no substituent in the α-position, are indeed usually opened regioselectively to furnish 2′-deoxycarba-nucleosides such as **7**.[12,13] On the other hand, the presence of a substituent in the 2′-position leads to a mixture of regioisomers **9** and **10**

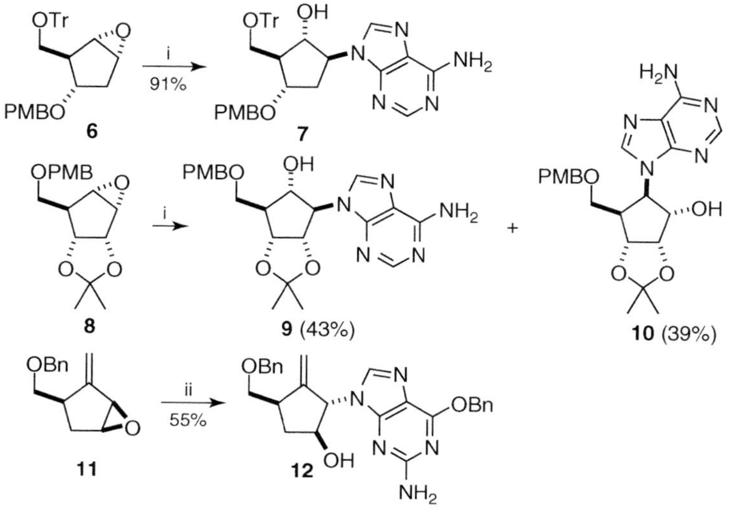

Scheme 3. *Reagents and conditions:* (i) adenine, NaH, 15-crown-5, DMF; (ii) O-benzylguanine, LiH, DMF.

from **8**.[12] The steric hindrance can be overcome, thanks to electronic effects. In the case of **11**, the more reactive allylic position is indeed regioselectively attacked by a protected guanine to furnish the entecavir analogue **12**.[14]

Despite their high sensitivity, cyclic sulfates and sulfites are reactive surrogates to epoxides and were used as such by Marquez and Jeong for the synthesis of carbanucleosides.[15,16] A recent example is the synthesis of 3′-aminocarbanucleosides by the addition of 2,6-dichloropurine to sulfite **13** to furnish **14** in 61% yield (Scheme 4).[16c]

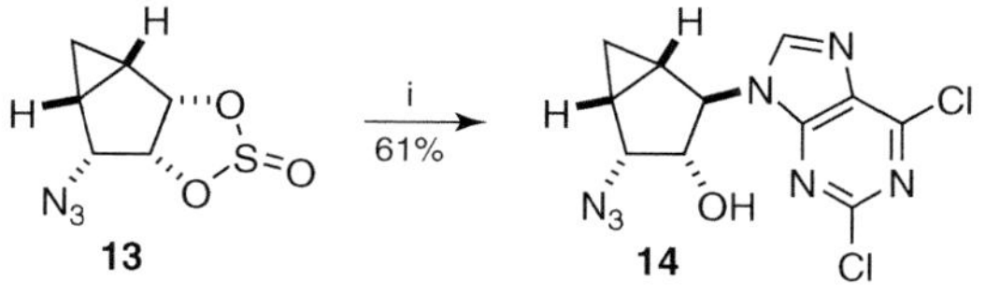

Scheme 4. *Reagents and conditions:* (i) 2,6-dichloropurine, NaH, 18-crown-6, THF, then H_2SO_4.

2.1.3. Mitsunobu Reaction.

The Mitsunobu reaction is one of the most reliable methods to introduce a nuclobase directly on an appropriately functionalized carbasugar and, as such, one of the most widely used approaches. Numerous examples of the addition of purine or pyrimidine bases, protected or not, are described throughout the chapter. The addition of 6-chloropurine **15** under Mitsunobu conditions, for example, is used extensively to prepare adenine- or hypoxanthine-derived

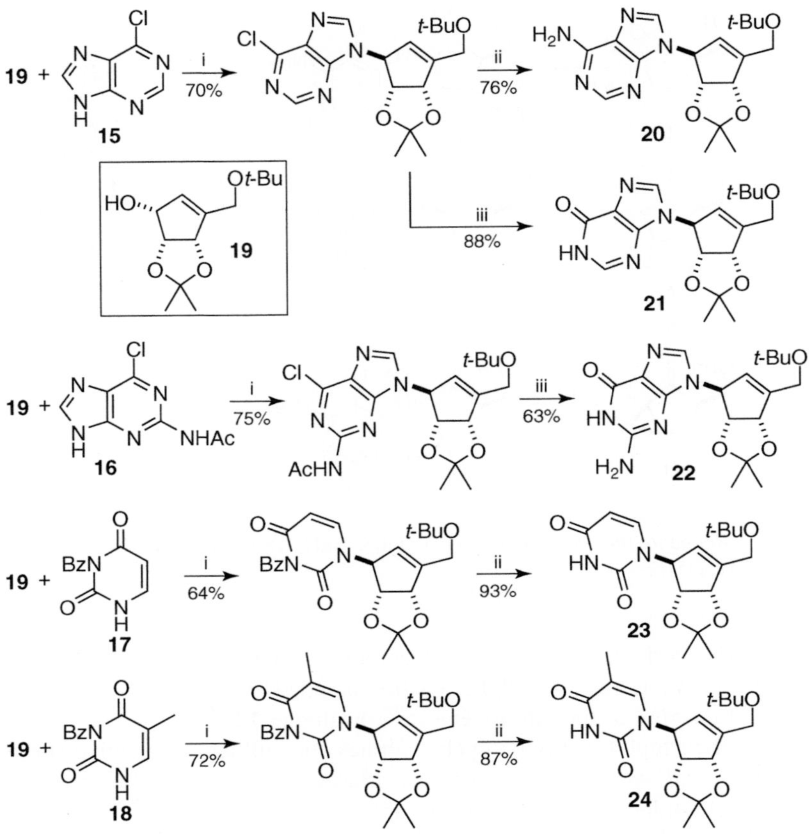

Scheme 5. *Reagents and conditions:* (i) Ph$_3$P, DEAD, THF; (ii) NH$_3$, MeOH; (iii) HSCH$_2$CH$_2$OH, MeONa, MeOH.

carbanucleosides, although subsequent aminolysis or hydroxylation is required. Schneller recently investigated the use of N^6,N^6-bis(*tert*-butoxycarbonyl)adenine as a surrogate to 6-chloropurine.[17] Similarly, 2-amino-6-chloropurine or its *N*-acetylated derivative **16** are used as a guanine precursor, whereas N^3-benzoylthymine **17** or N^3-benzoyluracil **18** can be used directly (Scheme 5).

A regioselectivity issue is often encountered with these last two pyrimidine bases, since the amount of O^2-alkylation product can be substantial compared to the desired N^1-alkylation product. Attempts to improve these results have been made by Meier but with only limited success.[18] The Mitsunobu strategy is well illustrated by the large study of Chu regarding the synthesis of various D- and L-analogues of neplanocin A, for which **15–18**, among others, were added to alcohol **19** to provide, respectively, adenine, hypoxanthine, guanine, thymine, and uracil carbanucleosides **20–24** (Scheme 5).[19]

Experimental Procedure

(1′2 S,2′2 R,3′2 S)-1-(1′S,2′R,3′S)-1-[2′,3′-O-Isopropylidene-2′,3′-dihydroxy-4′-(tert-butoxymethyl)cyclopent-4-en-1-yl]uracil (23).[19] A solution of DEAD (1.04 g, 5.98 mmol) in dry THF was added dropwise to a solution of **19** (580 mg, 2.39 mmol), triphenylphosphine (1.57 g, 5.98 mmol), and N^3-benzoyluracil (1.03 g, 4.78 mmol) in dry THF at 0°C under a nitrogen atmosphere. The mixture was stirred at room temperature for 16 h and the solvent was removed under vacuum. The residue was purified by column chromatography (50% EtOAc in n-hexane) to afford the expected compound (670 mg, 64%) as a white solid. The latter (380 mg, 0.85 mmol) was dissolved in a saturated solution of ammonia in MeOH (30 mL) and stirred for 4 h at room temperature. After removal of the solvent under vacuum, the residue was purified over silica gel (5% MeOH in $CHCl_3$) to give **23** (267 mg, 93%) as a white solid; mp 147–149°C; $[\alpha]_D^{25} = +39.28°$ (c 0.70, MeOH); UV (MeOH): λ_{max} 266 nm. ^{1}H-NMR (400 MHz, $CDCl_3$): δ 7.04 (d, J = 8.0 Hz, 1H), 5.68 (d, J = 8.0 Hz, 1H), 5.59 (s, 1H), 5.41 (s, 1H), 5.17 (d, J = 5.7 Hz, 1H), 4.55 (d, J = 5.8 Hz, 1H), 4.15 (d, J = 15.1 Hz, 1H), 4.08 (d, J = 14.9 Hz, 1H), 1.44 (s, 3H), 1.35 (s, 3H), 1.24 (s, 9H); HRMS (FAB^+) calcd for $C_{17}H_{25}N_2O_5$: m/z 337.1763 $[M + H]^+$; found: m/z 337.1762.

2.1.4. Tsuji–Trost Reaction.

Since the seminal publication of Trost, the Tsuji–Trost reaction has become, with Mitsunobu coupling, the most popular method for the direct introduction of a heterocyclic base.[20] The use of such nucleophiles revealed few drawbacks compared to the classical reaction using carbon nucleophiles but remains very effective for most pyrimidine and purine bases. Many examples of palladium-catalyzed introduction of nucleobases to functionalized cyclopentenes are decribed in this chapter, and the enantioselective version of this reaction is discussed in detail in Section 3.6.1. Only one example will therefore be described here to illustrate this strategy, especially to point out the stereochemical and regiochemical features of such a reaction. Hong reported the synthesis of several carbanucleosides based on a palladium-catalyzed addition of adenine and cytosine to carbonate **25** (Scheme 6). Intermediates **26** and **27** are obtained in satisfactory yields, as a single diasteromer and with negligible amounts of the 3′-regioisomer.[21]

The reaction thus follows the usual selectivity rules of the Tsuji–Trost process. First, the substitution proceeds with overall retention of configuration, which is complementary to the previous S_N2 and Mitsunobu reactions. Second, the least

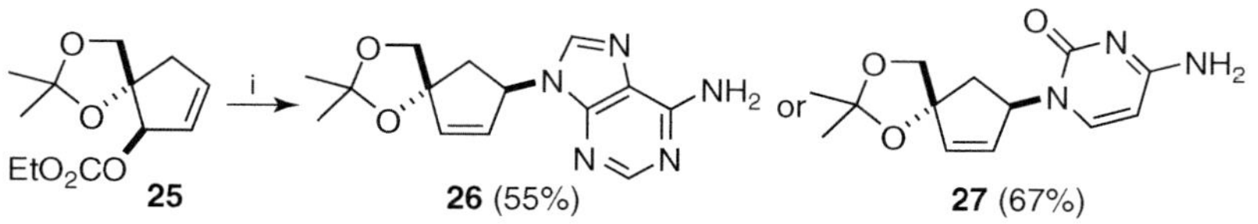

Scheme 6. *Reagents and conditions:* (i) adenine or cytosine, $Pd_2(dba)_3 \cdot CHCl_3$, P(O-i-Pr)$_3$, NaH, THF/DMSO.

hindered position will be attacked by the nucleophile with no influence of the initial position of the leaving group; the same regioisomer would have been obtained from a substrate featuring a carbonate at C1′.

2.2. Construction of the Nucleobase from a Cyclopentylamine

The linear construction of purine or pyrimidine bases from an aminocarbasugar is a long-standing and proven approach and often a secure alternative to the previous direct introduction methods. It should be mentioned that the cyclopentylamine scaffold is usually obtained by substitution reactions similar to those decribed in Sections 2.1.1 and 2.1.2 using an azide as the nucleophile and, less frequently, by a Curtius rearrangement.

2.2.1. Construction of Purine Bases.
The purine bases are prepared using the historical Traube synthesis.[22] Addition of an aminocarbasugar such as **28** to 5-amino-4,6-dichloropyrimidine yields the intermediate **29** and condensation with triethylorthoformate affords 6-chloropurine derivative **30**.[23] As described previously, aminolysis, hydrolysis, or addition of various primary amines leads to adenine, hypoxanthine, or other purine derivatives (Scheme 7).

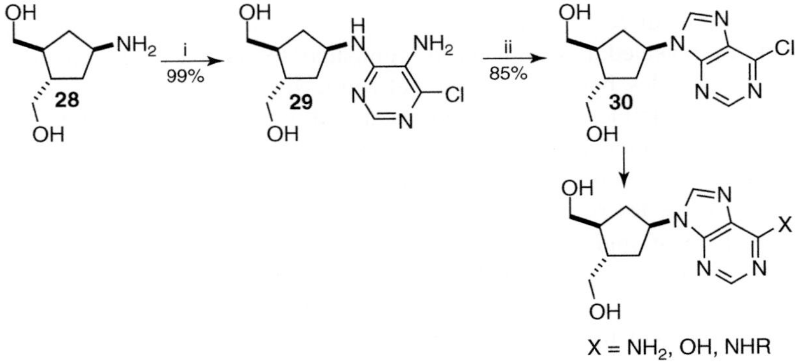

Scheme 7. *Reagents and conditions:* (i) 5-amino-4,6-dichloropyrimidine, Et$_3$N, n-BuOH; (ii) HC(OEt)$_3$, 12 N HCl.

Guanine derivatives can be prepared in a similar way by addition to 2-amino-4,6-dichloropyrimidine followed by a diazotation–reduction sequence performed on intermediate **31**.[24] The 2-amino-6-chloropurine derivative **32** obtained after action of triethylorthoformate can be converted to the corresponding guanine **33**. as mentioned earlier (Scheme 8).

Direct condensation of an amine to 2-amino-5-nitro-4,6-dichloropyrimidine followed by reduction of the nitro group was also used as an alternative to the diazotation–reduction sequence.[25]

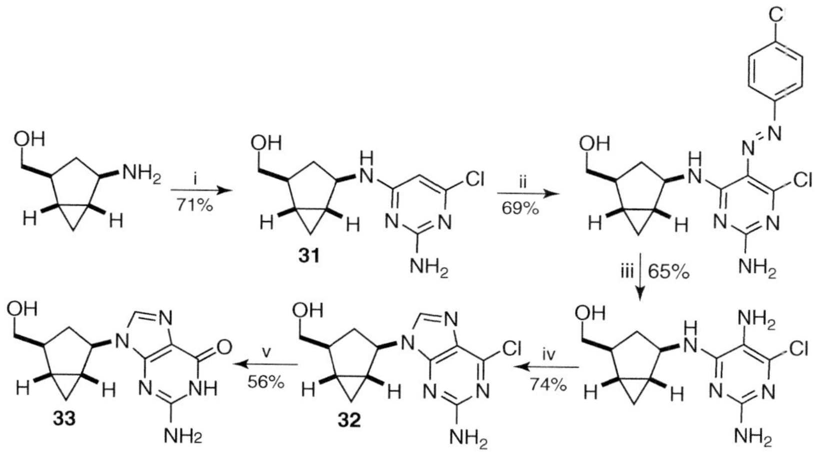

Scheme 8. *Reagents and conditions:* (i) 2-amino-4,6-dichloropyrimidine, Et$_3$N, *n*-BuOH; (ii) *p*-chloroaniline, 3 N HCl, NaNO$_2$, AcOH, NaOAc; (iii) Zn, AcOH, EtOH; (iv) HC(OEt)$_3$, 12 N HCl; (v) 1 N HCl.

Experimental Procedure

(±)-2-Amino-6-chloro-9-[4-(hydroxymethyl)bicyclo[3.1.0]hexyl]purine (*32*).[24]
A cold diazonium salt solution was prepared from *p*-chloroaniline (0.345 g, 2.7 mmol) in 3 N HCl (6 mL) and sodium nitrite (0.204 g, 2.94 mmol) in water (1.2 mL) and added to a solution of **31** (0.600 g, 2.36 mmol), acetic acid (12 mL), water (12 mL), and sodium acetate trihydrate (4.68 g). The resulting mixture was stirred overnight at room température and the yellow precipitate was filtered, washed with cold water until neutral, and air-dried under a fume hood to yield the expected diazo compound (0.638 g, 69%). To a suspension of the latter (1.227 g, 3.1 mmol) in EtOH (92 mL) and water (41 mL) was added zinc dust (2.05 g, 31.2 mmol) and acetic acid (1 mL). The reaction mixture was heated to reflux under a nitrogen atmosphere for 3 h. Zinc was filtered off and the filtrate was concentrated to give a brown residue, which was purified by column chromatography (CHCl$_3$/MeOH, 20:1) to give the expected aniline as a yellow oil (0.243 g, 65%). To a suspension of the latter (0.543 g, 2.0 mmol) in triethyl orthoformate (11 mL) was added 12 N HCl (0.5 mL), and the mixture was stirred at room temperature overnight. The suspension was then evaporated under vacuum, redissolved in 0.5 N HCl, and stirred at room temperature for 1 h. The solution pH was then adjusted to 8 with a 1 N NaOH solution and the solution was concentrated. Purification over column chromatography (CHCl$_3$/MeOH, 20:1 to 15:1) afforded **32** (0.413 g, 73.4%) as a white solid; mp 228–230°C. ^{1}H-NMR (300 MHz, DMSO-d_6): δ 8.20 (s, 1H), 6.89 (br, s, 2H), 4.65–4.62 (m, 1H), 4.57–4.52 (m, 1H), 3.29–3.14 (m, 1H), 2.99–2.93 (m, 1H), 2.09–2.06 (m, 1H), 1.84–1.79 (m, 2H), 1.77–1.72 (m, 1H), 1.63–1.57 (m, 1H), 0.71–0.67

(m, 1H), 0.33–0.28 (m, 1H). *Anal*. Calcd for $C_{12}H_{14}ClN_5O \cdot 0.6H_2O$: C, 49.61; H, 5.27; N, 24.09; found: C, 49.89; H, 5.01; N, 23.75.

2.2.2. Construction of Pyrimidine Bases. The method of Shaw and Warrener for the construction of uracil and thymine bases is still widely used.[26] The addition of amines such as **34** to 3-methoxyacryloyl isocyanates **35** and **36** leads to ureas **37** and **38**. which after ammonia-mediated cyclization yield uracil and thymine derivatives **39** and **40**.[25] The cytosine derivative **41** can be obtained from **39** through sulfonylation and aminolysis (Scheme 9).

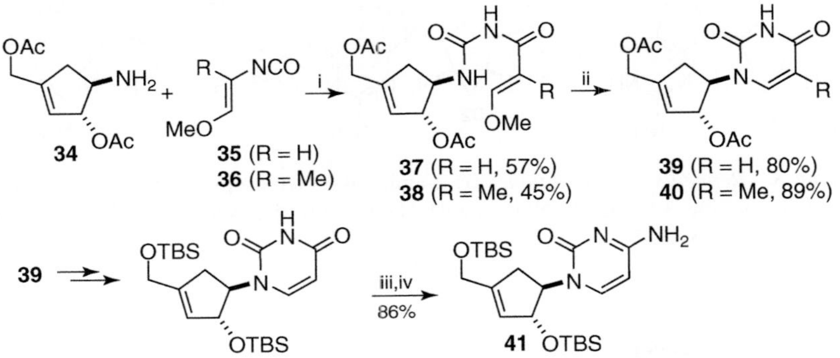

Scheme 9. *Reagents and conditions:* (i) Et_3N, toluene; (ii) NH_4OH, MeOH; (iii) 2,4,6-tri(isopropylsilyl)benzenesulfonyl chloride, Et_3N, MeCN; (iv) NH_4OH, MeCN.

Experimental Procedures

(±)-1-[3-(3-Methoxyacryloyl)ureido]-2'-acetoxy-4'-acetoxymethylcyclopent-3'-ene (37).[25] Thionyl chloride (2 mL) was added to a solution of 3-methoxyacryloyl chloride (1.6 g, 15.6 mmol) in dry dichloromethane (13 mL) and the mixture was refluxed for 3 h and concentrated. The residue was dissolved in dry toluene (6 mL) and silver cyanate (4 g, 26.6 mmol) was added. The suspension was refluxed for 30 min, then cooled to 0°C. The supernatant liquor containing **35** was transferred to a solution of **34** in dichloromethane (20 mL) and triethylamine (2 mL). The mixture was stirred at 0°C and warmed slowly to room temperature overnight. The solution was then washed with saturated $NaHCO_3$ (2 × 10 mL), dried, and purified by silica gel chromatography (ethyl acetate/hexane 1:2) to afford **37** (120 mg, 57%), mp 160–162°C; ^{1}H-NMR (500 MHz, $CDCl_3$): δ 9.68 (d, J = 12.5 Hz, 1H), 9.03 (d, J = 7.0 Hz, 1H), 7.66 (d, J = 12.5 Hz, 1H), 5.69 (br s, 2H), 5.33 (d, J = 12.5 Hz, 1H), 4.59 (dd, J = 14.0 Hz, 2H), 4.37 (m, 1H), 3.73 (s, 3H), 2.88 (dd, 1H), 2.35 (m, 1H), 2.09 (s, 3H), 2.05 (s, 3H). ^{13}C-NMR (125 MHz, $CDCl_3$): δ 170.7, 170.5, 167.9, 163.7, 155.0, 143.2, 124.6, 97.6, 83.9, 62.2, 57.9, 55.73, 38.3, 21.0, 20.7. *Anal*. Calcd for $C_{15}H_{20}N_2O_7$: C, 52.94; H, 5.92; N, 8.23; found: C, 52.88; H, 5.88; N, 8.19; MS: $[M + Na]^+$: m/z 363.11; found: m/z 363.08.

(±)-1-(2′-Hydroxy-4′-hydroxymethylcyclo-pent-3′-en-1-yl)uracil 39.[25] A suspension of **37** (0.1 g, 0.29 mmol), ethanol (2 mL), and NH_4OH (28%, 5 mL) was heated at 80–100°C in a steel bomb for 16 h. The solution was evaporated and the crude product was purified by silica gel chromatography (CH_2Cl_2/MeOH, 4:1) to afford **39** as a white solid (52 mg, 80%): mp 188–189°C; ^{1}H-NMR (300 MHz, DMSO-d_6): δ 11.24 (s, 1H), 7.55 (d, J = 7.6 Hz, 1H), 5.53 (dd, J = 2.0, 8.0 Hz, 1H), 5.50 (dd, J = 1.6, 3.2 Hz, 1H), 5.24 (d, J = 6.0 Hz, 1H), 4.86 (t, J = 5.6 Hz, 1H), 4.79 (m, 1H), 4.53 (m, 1H), 3.93 (m, 2H), 2.55 (dd, J = 8.8 and 16.2 Hz, 1H), 2.32 (dd, J = 7.2 and 16.0 Hz, 1H). ^{13}C-NMR (125 MHz, DMSO-d_6): δ 163.8, 151.5, 145.5, 144.2, 126.4, 101.8, 79.2, 66.2, 60.3, 35.8. *Anal.* Calcd for $C_{10}H_{12}N_2O_4$: C, 53.57; H, 5.39; N, 12.49; found: C, 53.62; H, 5.27; N, 12.50; UV (MeOH) λ_{max} 208, 267 nm; HRMS: $[M + H]^+$: m/z 247.0674; found: m/z 247.0687.

2.3. Introduction of C-Linked or Other Unusual Bases

The quest for more active and selective compounds prompted investigations on the use of unconventional bases in nucleosidic drugs, and this was the case for carbanucleosides as well. The chemical introduction of C-linked or unusual bases on carbocyclic structures was explored, and results in this area are summarized below.

2.3.1. Construction of C-Linked Bases. The first carbocyclic *C*-nucleosides appeared with the synthesis of carbocyclic analogues of nucleosidic drugs such as oxazinomycin and showdomycin.[27,28] Afterward, Chu initiated the preparation of carbanucleosides featuring a C-linked 9-deazapurine base, beginning with the synthesis of 9-deazaaristeromycin.[29] The preparation of analogues featuring a 9-deazaguanine or a 9-deazahypoxanthine then followed.[30] The 9-deazapurine base is constructed stepwise from a carbasugar such as **42** featuring a cyanoacetate group in the 1′-position. Reduction of the ester and condensation with aminoacetonitrile or ethyl aminoacetate afforded cyanoenaminonitrile **43** or cyanoenaminoester **44**. respectively (Scheme 10). Protection and base-induced cyclization furnished pyrroles **45** and **46**. which, in turn, gave adenine derivative **47** and hypoxanthine derivative **48**. respectively, upon reaction with formamidine. A three-step procedure involving *N*-benzoylisothiocyanate was necessary to obtain guanine derivative **49**.

Similar C-linked neplanocin A or 5′-noraristeromycin analogues were obtained using the same synthetic sequence.[31] It should be mentioned that the starting cyanoacetates are obtained either by addition of ethyl cyanoacetate to the corresponding tosylate or by olefination of the corresponding cyclopentanone with ethyl cyanoacetate and hydrogenation.

2.3.2. Construction of Triazole Bases. The development of click chemistry based on the Huisgen reaction inspired the synthesis of new carba-nucleosides bearing unusual triazole-derived bases. The first example of this type was reported by

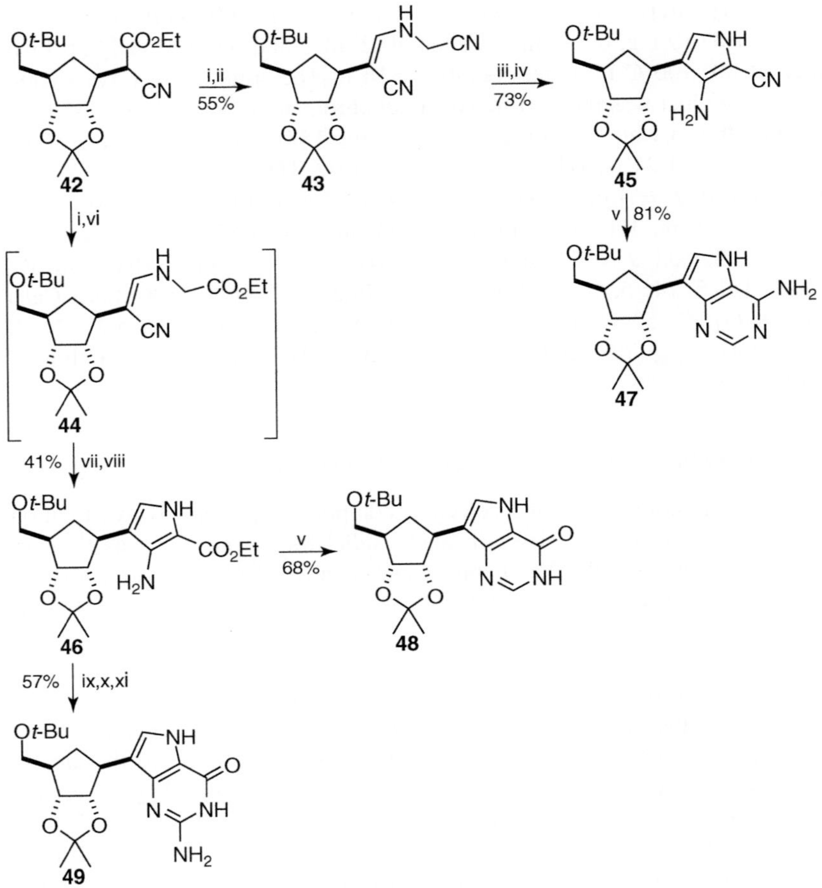

Scheme 10. *Reagents and conditions:* (i) (*i*-Bu)$_2$AlH, Et$_2$O; (ii) H$_2$NCH$_2$CN·H$_2$SO$_4$, MeOH; (iii) ClCO$_2$Et, DBN, CH$_2$Cl$_2$; (iv) Na$_2$CO$_3$, MeOH; (v) HC(=NH)NH$_2$·AcOH, EtOH; (vi) H$_2$NCH$_2$CO$_2$Et·HCl, MeOH; (vii) ClCO$_2$Et, DBU, CH$_2$Cl$_2$; (viii) NaOMe, MeOH; (ix) BzNCS, CH$_2$Cl$_2$; (x) MeI, DBN; (xi) NH$_3$, MeOH.

Kuang et al. for the synthesis of a regioisomer **50** of carbocyclic ribavirin.[32] As carba-sugar-derived azides are easily obtained by substitution from the corresponding alcohols, this approach was developed further by Agrofoglio et al. and Chu et al., and several compounds (e.g., **51** and **52**) were prepared using the same 1,3-dipolar cycloaddition (Scheme 11).[33]

2.3.3. Introduction of Other Modified Purine Bases. Among the miscellaneous modifications brought to purine bases, the use of 7-deazaadenine derivatives has been reported several times. This moiety can easily be introduced by substitution reaction, the absence of a second nitrogen atom avoiding the regioselectivity

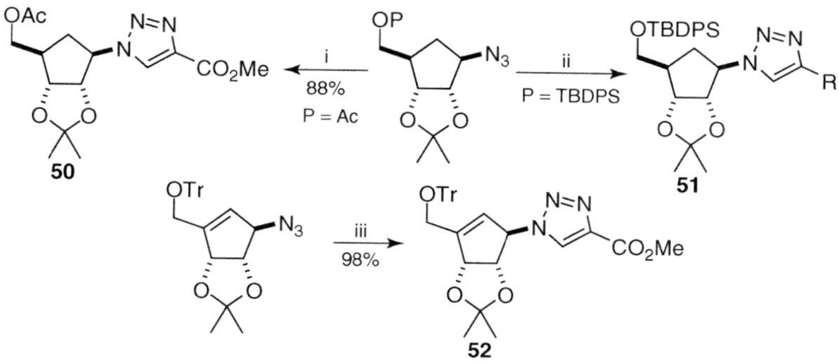

Scheme 11. *Reagents and conditions:* (i) methyl propynoate, neat; (ii) RCCH, Cu/CuSO$_4$, t-BuOH/H$_2$O; (iii) methyl propynoate, CuI, Et$_3$N, THF.

issue.[34] Several 7-deaza analogues of neplanocin A (e.g., **54**) were prepared by Chu from **53** using a highly efficient Mitsunobu reaction (Scheme 12).[34b] MLN4924, an anticancer agent under clinical trial, has recently been synthesized by Jeong using a nucleophilic opening of a cyclic sulfate with a 7-deazapurine (see Scheme 33, Section 3.2.2).[34c] Other modified purines, such as the triazole analogue of adenine, 8-azaadenine, or the 1-oxygenated analogue of guanine, have also been introduced on carba-sugars.[35]

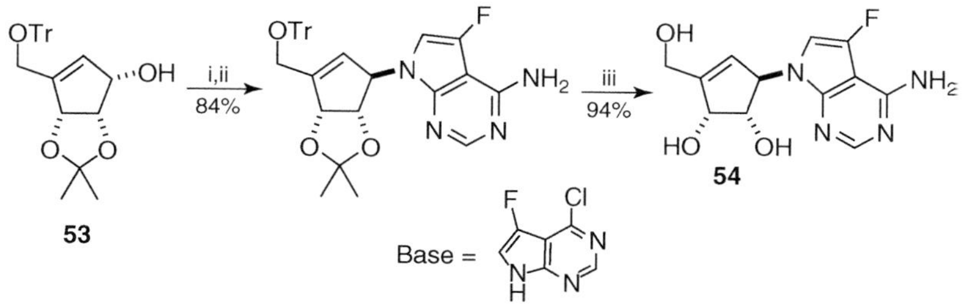

Scheme 12. *Reagents and conditions:* (i) Base, Ph$_3$P, DIAD, THF; (ii) NH$_3$, MeOH; (iii) HCl, MeOH, THF.

3. SYNTHESIS OF CYCLOPENTYL CARBOCYCLIC ANALOGUES OF NUCLEOSIDES

3.1. Synthesis from a Resolved Cyclopentene Building Block

Enantio-enriched functionalized cyclopentanes obtained by chemical or enzymatic resolution of racemic materials appeared to be attractive building blocks for the preparation of carbanucleosides. Surprisingly, the chemical resolution is not the most widespread method, although it has been used on several occasions.[36] In contrast, enzymatic resolution provided several very useful enantio-enriched building blocks. Examples of chemical synthesis from these intermediates are described below. For more precise information regarding their enzymatic preparation and the biocatalytic synthesis of nucleoside analogues in general, the reader is invited to consult a thorough review devoted to this topic.[37]

3.1.1. Cyclopent-2-ene-1,4-diols and Derivatives.

Enantio-enriched cyclopentene (+)-**55**, obtained from the corresponding meso diol or diacetate by action of lipases or esterases, has been used widely in carbanucleoside synthesis. A very efficient acetylcholinesterase-catalyzed hydrolysis of the corresponding meso-diacetate **56** was first reported by Deardorff et al.[38] They applied this methodology to the synthesis of a carba-nucleoside precursor **57** through a double Tsuji–Trost reaction performed on **55** followed by a dihydroxylation reaction (Scheme 13).[39]

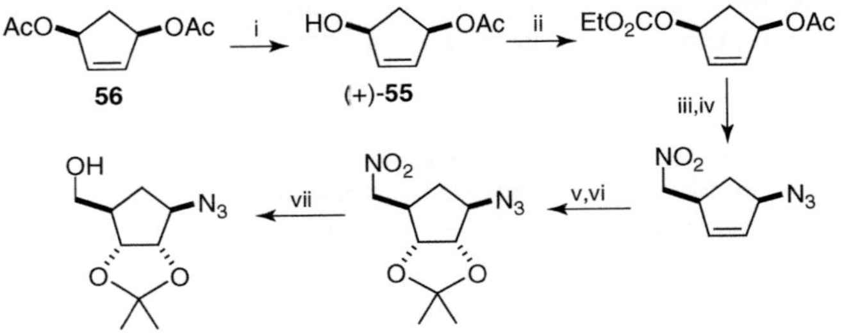

Scheme 13. *Reagents and conditions:* (i) electric eel AchE, 86%, 96% ee; (ii) ClCO$_2$Et, pyridine, 97%; (iii) CH$_3$NO$_2$, Ph$_3$P, Pd$_2$dba$_3$, CH$_2$Cl$_2$, 63%; (iv) NaN$_3$, Pd(Ph$_3$P)$_4$, THF/H$_2$O, 68%; (v) OsO$_4$, NMO; (vi) TsOH, (CH$_3$)$_2$C(OMe)$_2$, acetone, 57% (two steps); (vii) KMnO$_4$, KOH, MeOH, then NaBH$_4$, i-PrOH, 46%.

A synthesis of carbovir and analogues was performed by Nokami starting from the same building block, **55**, prepared using their own enzymatic route.[40] The hydroxymethyl group in the 5′-position was introduced though a chloromethylation–epoxidation–reductive opening sequence performed on **58**. A Tsuji–Trost reaction allowed the introduction 2-amino-6-chloropurine, and a final deprotection sequence afforded carbovir (Scheme 14).

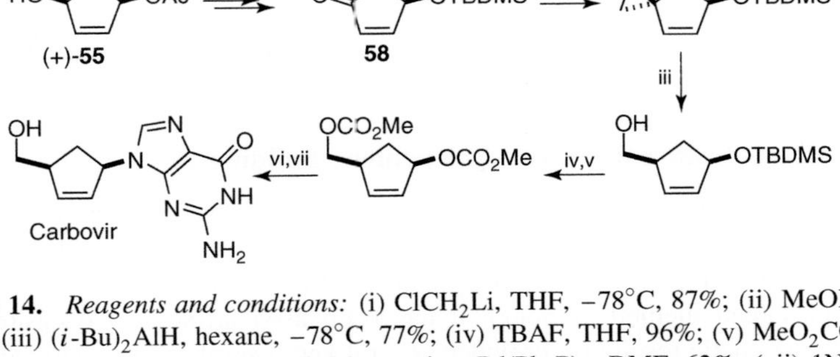

Scheme 14. *Reagents and conditions:* (i) ClCH$_2$Li, THF, −78°C, 87%; (ii) MeOK, THF, −78°C; (iii) (i-Bu)$_2$AlH, hexane, −78°C, 77%; (iv) TBAF, THF, 96%; (v) MeO$_2$CCl, pyridine, CH$_2$Cl$_2$, 91%; (vi) 2-amino-6-chloropurine, Pd(Ph$_3$P)$_4$, DMF, 62%; (vii) 1N NaOH, reflux, 83%.

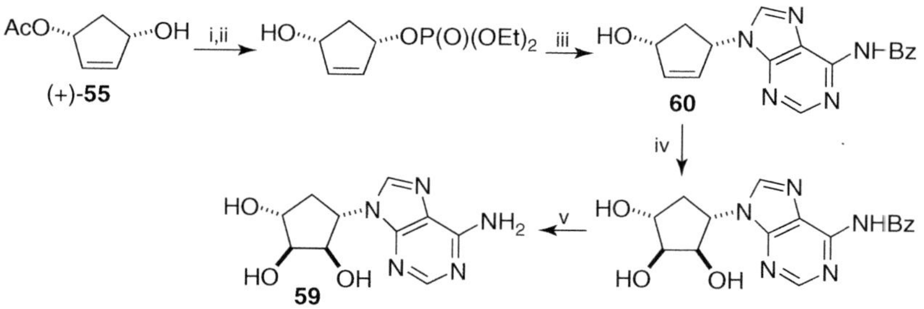

Scheme 15. *Reagents and conditions:* (i) (EtO)$_2$P(O)Cl, pyridine, 100%; (ii) NH$_3$, MeOH, 99%; (iii) N^6-benzoyladenine, NaH, Pd(Ph$_3$P)$_4$, THF, 52%; (iv) OsO$_4$, NMO, THF/H$_2$O, 86%; (v) NH$_4$OH, MeOH, 83%.

An expedient synthesis of 5′-noraristeromycin **59** (L-series) was reported by Schneller et al. using (+)-**55** as the starting material.[41] A Tsuji–Trost reaction of its phosphate derivative using N^6-benzoyladenine as the nucleophile afforded intermediate **60**. The target compound **59** was obtained in four steps from **55** after an osmylation and a deprotection reaction (Scheme 15).

This popular and commercially available building block is still used in numerous syntheses of 5′-noraristeromycin analogues (L-series).[42] Among them, a synthesis of conformationally locked analogues has been performed by Marquez using a Simmons–Smith cyclopropanation reaction of (+)-**55**. Introduction of 6-chloropurine via a Mitsunobu reaction followed by chlorine substitution and deprotection afforded compound **61** (Scheme 16).

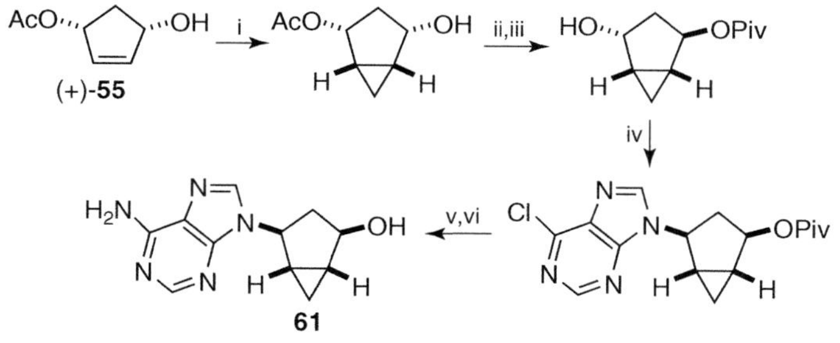

Scheme 16. *Reagents and conditions:* (i) Et$_2$Zn, CH$_2$I$_2$, CH$_2$Cl$_2$, 97%; (ii) PivOH, DIAD, Ph$_3$P, THF, 95%; (iii) K$_2$CO$_3$, MeOH, 92%; (iv) 6-chloropurine, DIAD, Ph$_3$P, THF, 55%; (v) NH$_4$OH; (vi) (*i*-Bu)$_2$AlH, 85% (two steps).

3.1.2. 4-Aminocyclopent-2-ene-1-ols and Derivatives.
4-Aminocyclopent-2-ene-1-ol **62**, or its cBz-protected analogue, is easily obtained by a nitroso-Diels–Alder reaction between the corresponding *N*-hydroxycarbamate and cyclopentadiene, producing cycloadducts such as **63**, followed by a reductive ring-opening reaction (for asymmetric versions of the cycloaddition, see Section 3.4.1). They are useful building blocks for carbanucleoside synthesis since the amino group is already in place

and allows the construction of heterocyclic bases. Their enzymatic resolution was performed by Miller et al. through a *Candida antartica* B–mediated acetylation reaction.[43] The carbocyclic analogue of uracil polyoxin C **64** was synthesized using a Tsuji–Trost reaction with a nitroacetate to settle the amino acid function.[44] The uracil base was introduced by addition of the free amine to acyl isocyanate **65** and subsequent acid-mediated cyclization (Scheme 17). The dihydroxylation and deprotection of **66** eventually furnished the targeted polyoxin C analogue **64**. It should be noted that the dihydroxylation reaction is not diastereoselective, the undesired isomer undergoing an immediate lactonization reaction to afford **67**, which could easily be separated from the desired compound **68**.

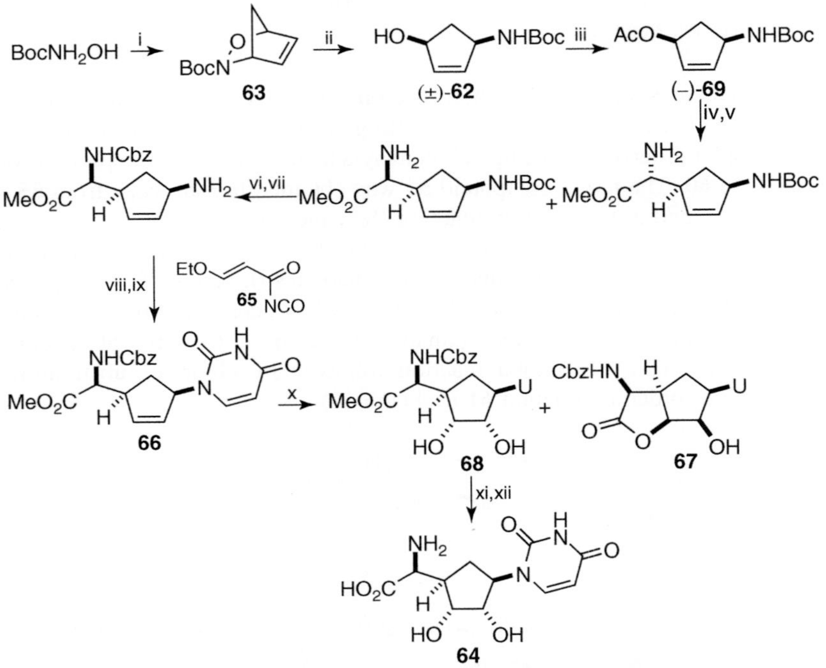

Scheme 17. *Reagents and conditions:* (i) cyclopentadiene, NaIO$_4$, MeOH/H$_2$O; (ii) NaBH$_4$, Mo(CO)$_6$, MeCN/H$_2$O, 60% (two steps); (iii) *C. antartica* B, vinyl acetate, 43%, 98% ee; (iv) O$_2$NCH$_2$CO$_2$Me, Pd(OAc)$_2$, Ph$_3$P, 95%, 1:1 mixture of epimers; (v) NaBH$_4$, TiCl$_3$, tartaric acid, 71%; (vi) CbzCl, NaHCO$_3$, 95%; (vii) TFA; (viii) **65**. (ix) 1 M H$_2$SO$_4$, MeOH, 53% (three steps); (x) OsO$_4$, NMO, THF/H$_2$O, 90%, **67**/**68** 1:1; (xi) LiOH; (xii) H$_2$, Pd/C.

Building block **69** (Scheme 17) and its bicyclic precursor have served as a starting material in many carbanucleoside syntheses by Miller et al.[45] A recent example is the synthesis of 5′-homocarbovir and analogues using a Tsuji–Trost type of decarboxylative rearrangement of allylic malonate **70** (Scheme 18). Reduction of the ester followed by step-by-step construction of the guanine base of carbovir or of the modified heterocyclic base of abacavir allowed the preparation of their 5′-homologated analogues **71** and **72**.[46]

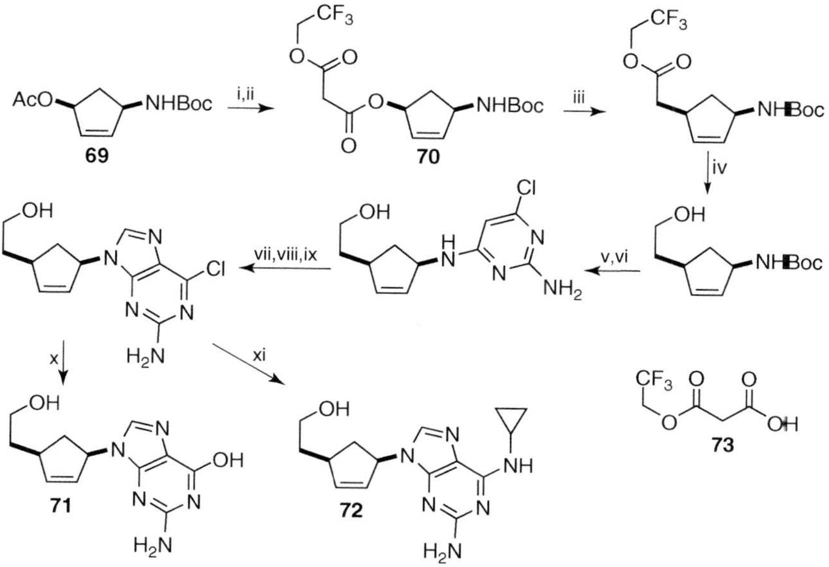

Scheme 18. *Reagents and conditions:* (i) K_2CO_3, MeOH, 99%; (ii) **73**, EDC·HCl, DMAP, CH_2Cl_2, 96%; (iii) $Pd(dba)_2$, dppe, THF, 82%; (iv) $(i\text{-}Bu)_2AlH$, THF, 85%; (v) 12 M HCl, EtOH; (vi) 2-amino-4,6-dichloropyrimidine, NEt_3, 50% (two steps); (vii) 4-chlorobenzenediazonium chloride, $H_2O/AcOH$, 80%; (viii) Zn, AcOH, 75%; (ix) $HC(OEt)_3$, 12 M HCl, 72%; (x) 0.33 M NaOH, 60%; (xi) cyclopropylamine, 80%.

3.1.3. 5-Hydroxymethylcyclopent-2-en-1-ol and Its Parent Lactone.

The kinetic resolution of lactone **74** by *Pseudomonas fluorescens* lipase (Pfl) provided an entry to a versatile enantio-enriched building block for carba-nucleoside synthesis.[47] The racemate was easily obtained by a Prins reaction between cyclopentadiene and glyoxalic acid and resolved through an acetylation reaction. A concise synthesis of carbovir from **74** was reported by these authors (Scheme 19). Reduction, oxidative cleavage, and another reduction afforded 5-hydroxymethylcyclopent-2-en-1-ol **75**. Tritylation of the primary alcohol, acetylation of the remaining OH group, and a Tsuji–Trost reaction using 2-amino-6-chloropurine as the nucleophile provided the intermediate **76**. The target molecule was finally obtained after complete deprotection.

Other carba-nucleoside syntheses were based on this convenient starting material, including 5′-homologated analogues.[48] A synthesis of a surrogate to cyclic ADP-ribose was accomplished by Shuto et al.[49] An allylic acetate transposition on intermediate **77** followed by a dihydroxylation reaction and diol protection afforded carbasugar **78** (Scheme 20). After inversion of the pseudoanomeric center by reduction of the corresponding ketone, triflate **79** was coupled to a protected inosine, and further elaboration furnished the target molecule **80**.

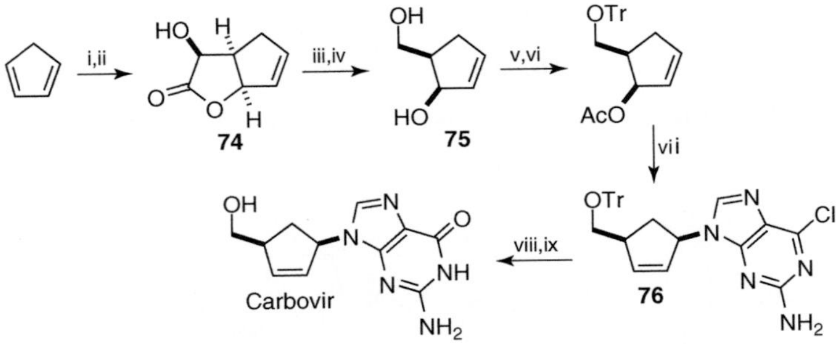

Scheme 19. *Reagents and conditions:* (i) glyoxalic acid, 65%; (ii) Pfl, vinyl acetate, 47%, 100% ee; (iii) LiAlH₄, THF; (iv) NaIO₄, Et₂O/H₂O, then NaBH₄, EtOH, 73% (two steps); (v) TrCl, NEt₃, 89%; (vi) Ac₂O, pyridine, 100%; (vii) 2-amino-6-chloropurine, NaH, Pd(Ph₃P)₄, THF, 49%; (viii) AcOH, H₂O, 96%; (ix) NaOH, H₂O, 66%.

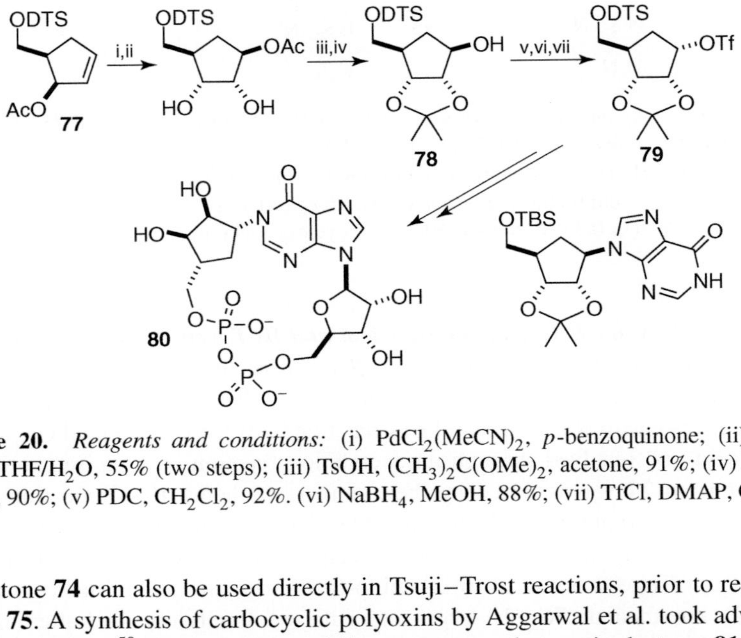

Scheme 20. *Reagents and conditions:* (i) PdCl₂(MeCN)₂, *p*-benzoquinone; (ii) OsO₄, NMO, THF/H₂O, 55% (two steps); (iii) TsOH, (CH₃)₂C(OMe)₂, acetone, 91%; (iv) K₂CO₃, MeOH, 90%; (v) PDC, CH₂Cl₂, 92%. (vi) NaBH₄, MeOH, 88%; (vii) TfCl, DMAP, CH₂Cl₂.

Lactone **74** can also be used directly in Tsuji–Trost reactions, prior to reduction to diol **75**. A synthesis of carbocyclic polyoxins by Aggarwal et al. took advantage of this reactivity.[50] Hydroxylactone **74** was converted to aminolactone **81**, which underwent a Tsuji–Trost reaction using uracil as the nucleophile. Dihydroxylation of **82** followed by complete deprotection afforded carbocyclic uracil polyoxin C **83** (Scheme 21).

3.1.4. 2-Azabicyclo[2.2.1]hept-5-en-3-one and Derivatives. Racemic lactam **84** is easily obtained by a cycloaddition between cyclopentadiene and tosyl cyanide. The enzymatic resolution of this building block through hydrolytic cleavage of the amide bond was accomplished by several groups. The use of *Pseudomonas*

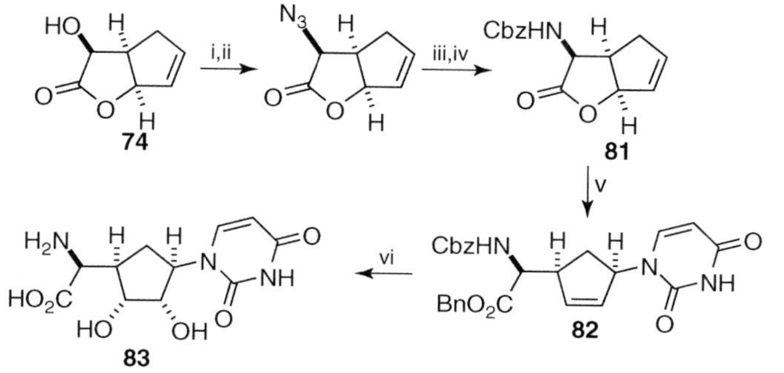

Scheme 21. *Reagents and conditions:* (i) ZnBr$_2$, PPh$_3$, DEAD, THF, 60%; (ii) NaN$_3$, DMSO, 87%; (iii) Ph$_3$P, THF/H$_2$O, 92%; (iv) CbzCl, NaHCO$_3$, THF/H$_2$O, 94%; (v) uracil, HMDS, TMSCl, MeCN, then BnBr, NaHCO$_3$, DMF, 50%; (vi) H$_2$, Pd/C, EtOH/H$_2$O, 100%.

fluorescens for **84** or of savinase (a classical and widely available enzyme) for its NAc or NBoc derivative proved to be extremely efficient.[51,52] (−)-**84** or its *N*-acetyl derivative (−)-**85** are therefore easily prepared, and (−)-**84** is also commercially available. A synthesis of carbovir was reported by Roberts et al. as an illustration of their enzymatic resolution process.[53] Carba-sugar **86** was easily obtained from (−)-**84** and converted to carbovir using a stepwise construction of the guanine base (Scheme 22).

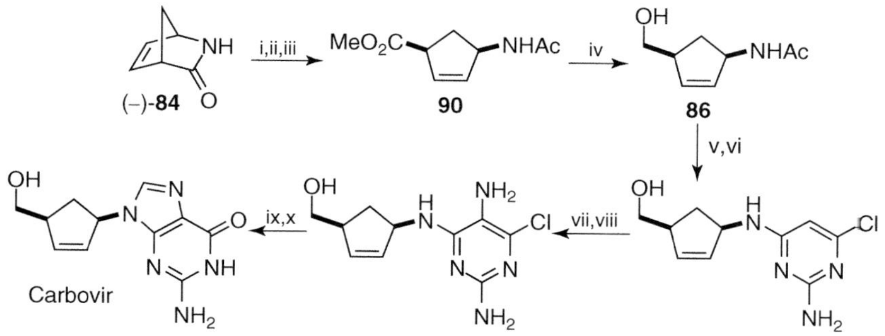

Scheme 22. *Reagents and conditions:* (i) HCl, H$_2$O; (ii) (CH$_3$)$_2$C(OMe)$_2$, HCl, MeOH; (iii) Ac$_2$O, pyridine, 96% (three steps); (iv) Ca(BH$_4$)$_2$, THF, 73%; (v) HCl, H$_2$O, EtOH; (vi) 2-amino-4,6-dichloropyrimidine, (*i*-Pr)$_2$NEt, *n*-BuOH, 80% (two steps); (vii) 4-chlorobenzenediazonium chloride, H$_2$O/AcOH, 69%; (viii) Zn, AcOH, 50%; (ix) HC(OEt)$_3$, 12 M HCl; (x) 0.33 M NaOH, 74% (two steps).

Another synthesis of carbovir was performed by Katagiri et al. using a Tsuji–Trost reaction directly on an *N*-protected derivative of (−)-**84** to introduce the heterocyclic base.[54] The use of electron-withdrawing protection on the nitrogen atom indeed allowed cleavage of the C–N bond to form the π-allylpalladium species. The same group reported the synthesis of the antiviral agent (+)-cyclaradine **87** through an epoxidation of the *N*-acetyl derivative (−)-**85**, followed by a NAc-assisted reductive opening to afford the carbocyclic L-aminosugar analog **88** (Scheme 23). Stepwise construction of the required adenine eventually delivered the desired carbanucleoside.

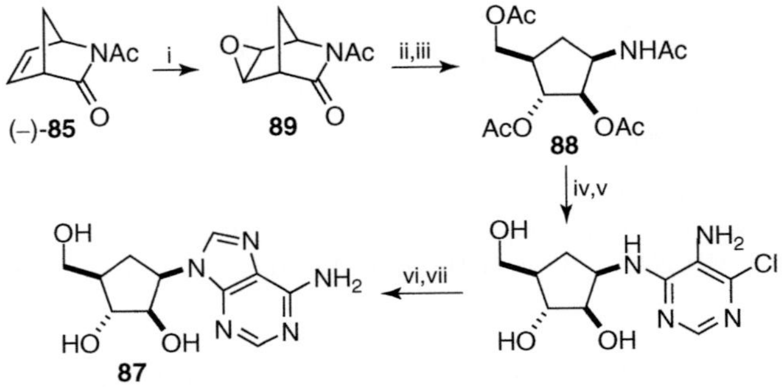

Scheme 23. *Reagents and conditions:* (i) *m*-CPBA, CHCl$_3$, 68%; (ii) NaBH$_4$, MeOH; (iii) Ac$_2$O, pyridine, 63% (two steps); (iv) 2 M HCl; (v) 5-amino-4,6-dichloropyrimidine, NEt$_3$, *t*-BuOH, 78% (two steps); (vi) HC(OEt)$_3$, 12 M HCl, 70% yield; (vii) NH$_3$, dioxane, 100%.

Epoxide **89** was also used by Cullis et al. in 2-deoxycarbanucleoside syntheses.[55] It should also be mentioned that the racemic aminoester **90** (obtained from bicyclic lactam **84**) was enzymatically resolved by Csuk and Dörr, using *Candida rugosa* lipase.[56] A synthesis of (−)-aristeromycin from (±)-**84** was performed to illustrate their strategy (Scheme 24). The dihydroxylation of (−)-**91**, obtained by reduction and acetylation of (−)-**90**, was efficient but unfortunately not diastereoselective. A stepwise construction of the adenine base, identical to the sequence described in Scheme 23 from **88** to **87**, allowed separation of the undesired diastereomer and the isolation of (−)-aristeromycin.

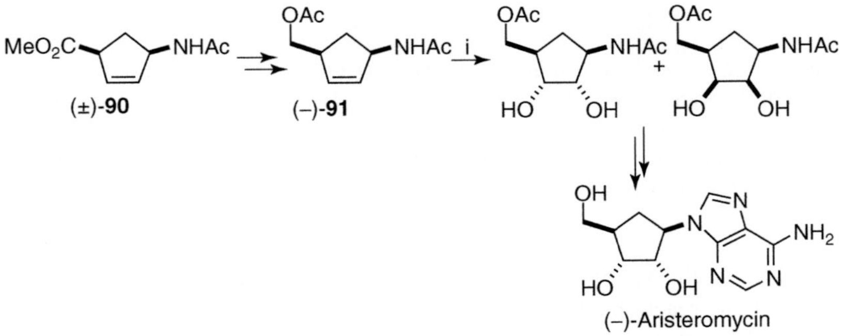

Scheme 24. *Reagents and conditions:* (i) OsO$_4$, NMO, acetone/H$_2$O, 98%, dr = 1:1.

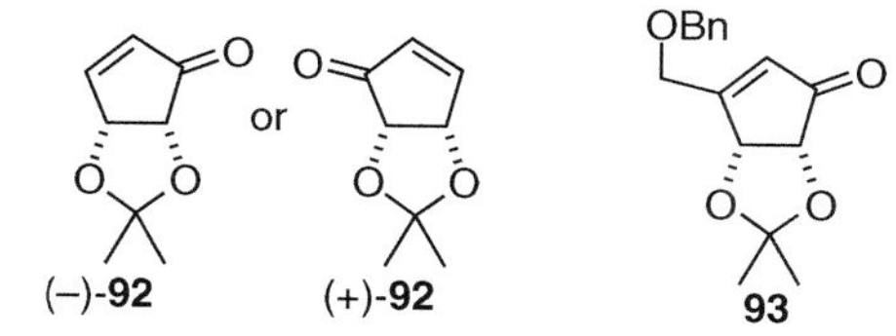

Figure 4. Cyclopentenone building blocks.

3.2. Synthesis from Cyclopentenones Derived from the Chiral Pool

Carbohydrate-derived cyclopentenones **92** and **93** are widely used chiral building blocks for carbanucleoside synthesis (Figure 4). Although several methods were reported, they are essentially prepared from D-ribose according to three great procedures reported by Borchardt, Jeong, and Chu for **92** and by Jeong for **93**. The preparation of these building blocks and their use in the synthesis of carbocyclic nucleosides is decribed below.

3.2.1. Preparation and Use of Cyclopentenone 92.

The synthesis of carbanucleosides from cyclopentenones such as **92** was first explored by Borchardt. The preparation of this starting material was adapted from known synthetic methods and starts from D-ribose or D-ribonolactone.[57] Oxidative degradation of the C4–C5 bond yields lactone **94**, which is subjected to lithiated trimethylphosphonate, resulting in a lactone opening followed by an intramolecular Horner–Wadsworth–Emmons reaction to yield (+)-**92** (Scheme 25).[58] Enantiomeric enone (−)-**96** was prepared either using exactly the same sequence from D-lyxose or from D-ribonolactone through another oxidative cleavage of the C4–C5, this time leading to lactone **95**.[58,59]

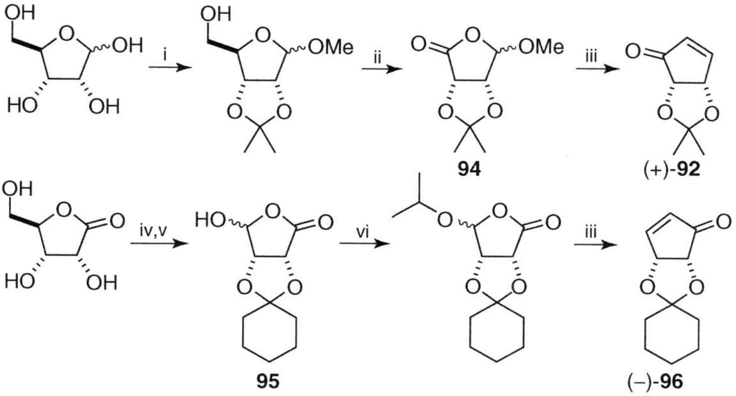

Scheme 25. *Reagents and conditions:* (i) $HClO_4$, $(CH_3)_2C(OMe)_2$, MeOH, 91%; (ii) PCC, benzene, 56%; (iii) $(MeO)_2P(O)CH_2Li$, 80% for **92** and **96**; (iv) cyclohexanone, $FeCl_3$; (v) $NaIO_4$, NaOH, H_2O, 85% (two steps); (vi) PPTS, *i*-PrOH, 95%.

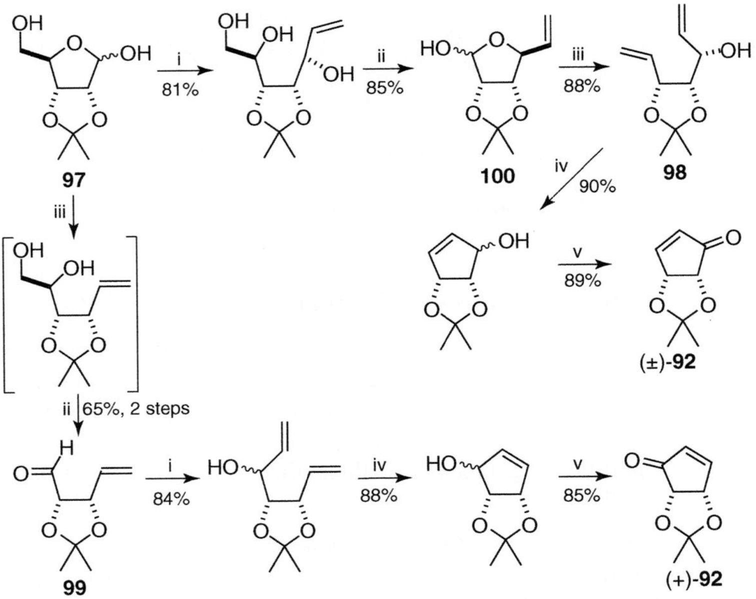

Scheme 26. *Reagents and conditions:* (i) vinylmagnesium bromide, THF; (ii) NaIO$_4$, CH$_2$Cl$_2$, H$_2$O; (iii) CH$_2$Ph$_3$P, THF; (iv) first-generation Grubbs' catalyst, CHCl$_3$; (v) MnO$_2$, CH$_2$Cl$_2$.

The development of ring-closing metathesis (RCM) revolutionized the synthesis of small rings and offered many solutions to the synthesis of carbanucleosides (see Section 3.3). Chu and Jeong took advantage of this methodology to offer a new route to the two enantiomers of **92** from D-ribose (Scheme 26). According to Jeong's procedure, addition of a vinyl Grignard reagent to hemiketal **97** followed by oxidative cleavage of the C4–C5 bond and Wittig methylenation afforded 1,6-diene **98**. Cyclopentenone (−)-**92** was obtained in 45% overall yield from D-ribose after RCM mediated by Grubbs' first-generation catalyst and oxidation.[60] A simple switch in the order of introduction of the two double bonds (i.e., methylenation of **97**), followed by the addition of vinylmagnesium bromide to aldehyde **99**, allowed the synthesis of (+)-**92** in 38% yield from D-ribose. Chu's synthesis is almost identical, with the exception of protection of the primary alcohol of **97**, adding two steps to the synthesis but with a slightly better overall yield.[61] A synthesis of the cyclohexylidene-protected enone **96** following the same route was also described by Liu and Chu.[62]

An alternative synthesis of **98** from D-ribose was reported by Schneller, and a preparation of **100** from D-mannose was described by Kumamoto et al.[63,64] The use of D-isoascorbic acid as a chiral source for a synthesis of both enantiomers of **92**, via D-erythronolactone, was also published by Jeong et al.[65] Finally, a chemoenzymatic synthesis of (+)- and (−)-**92** was decribed by Schneller et al. using a lipase-mediated resolution of a cyclopent-2-ene-1,4-diol derivative.[66]

The synthesis of (−)-aristeromycin by Borchardt was one of the first applications of **92** for carbanucleoside synthesis.[67] Addition of *tert*-butoxymethylcuprate to (−)-**96** followed by reduction of the carbonyl group afforded intermediate **101** with the appropriate relative configuration. Conversion to the triflate, substitution by adenine, and deprotection yielded aristeromycin in only nine steps from D-ribonolactone (Scheme 27).

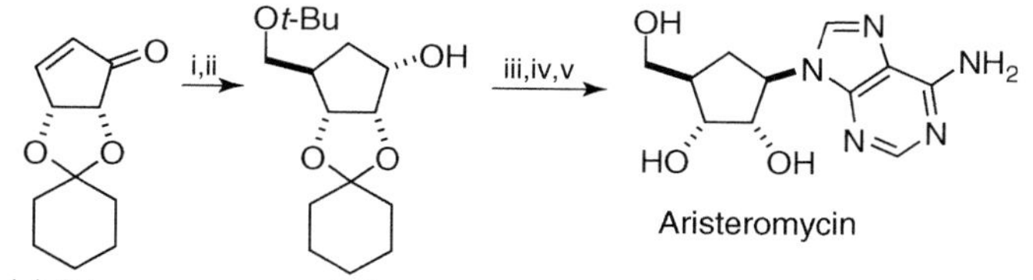

Scheme 27. *Reagents and conditions:* (i) (*t*-BuOCH$_2$)$_2$CuLi, 81%; (ii) (*i*-Bu)$_2$AlH, CH$_2$Cl$_2$, 96%; (iii) Tf$_2$O, pyridine, CH$_2$Cl$_2$, 95%; (iv) adenine, NaH, 18-crown-6 DMF, 30%; (v) TFA/H$_2$O, 79%.

This cuprate addition–reduction strategy was applied extensively to the synthesis of various D-carbanucleosides from (−)-**92**, or from (+)-**92** for the preparation of analogues from the L-series.[68] A similar approach involving a conjugate addition ethyl (trimethylsilyl)acetate to (−)-**92** provided an entry to 5′-homoaristeromycin.[69]

A conformationally locked bicyclo[3.1.0]hexyl nucleoside with an unconventional substitution pattern was also prepared from (−)-**92**.[70] The synthesis began with a Baylis-Hillman reaction to introduce the hydroxymethyl group on **92** followed by protection, reduction of the carbonyl group, and Simmons–Smith cyclopropanation to afford **102** (Scheme 28). Dehydroxylation by reduction of the corresponding tosylate and transprotection with a benzoate group yielded diol **103**. Chlorination followed by a S$_N$2 reaction with sodium azide led to a 5:1 mixture of regioisomers, from which the major compound **104** was isolated. Reduction of the azido group and thymine construction afforded the target compound **105**.

Access to the neplanocin family as well as to analogues is also possible from **92**. Inspired by the pioneering work of Johnson and his (−)-neplanocin A synthesis, 71 Chu reported the preparation of a large family of D- and L-carbocyclic nucleosides featuring an endocyclic double bond.[19,68c] The key step in those syntheses is an allylic rearrangement of acetate **107**, obtained from the 1,2-addition of *tert*-butoxymethyllithium on **92** followed by acetylation of the resulting alcohol (Scheme 29). A Mitsunobu or S$_N$2 type of reaction on **108** allowed introduction of the desired heterocyclic base with the appropriate relative configuration. The same approach was used to prepare (−)-neplanocin F, the allylic isomer of (−)-neplanocin A.[72]

Intermediates such as **108** are, of course, really structurally close to cyclopentenone **93**. More straightforward routes to this important building block have, however, been devised.

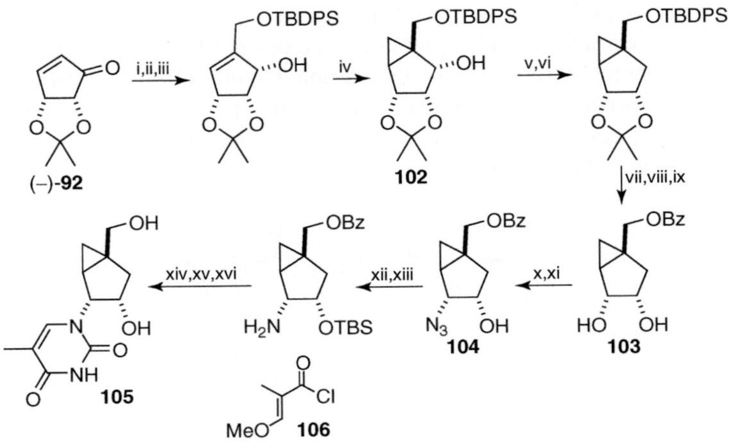

Scheme 28. *Reagents and conditions:* (i) HCOH, imidazole, THF/H$_2$O, 79%; (ii) TBDPSCl, imidazole, CH$_2$Cl$_2$, 91%; (iii) NaBH$_4$, CeCl$_3$, MeOH, 90%; (iv) Et$_2$Zn, CH$_2$I$_2$, CH$_2$Cl$_2$, 74%; (v) TsCl, pyridine, DMAP, CH$_2$Cl$_2$, 98%; (vi) LiEt$_3$BH, THF, 93%; (vii) TBAF, THF; (viii) BzCl, pyridine, CH$_2$Cl$_2$; (ix) AcOH, 71% (three steps); (x) SOCl$_2$, Et$_3$N, 96%; (xi) NaN$_3$, DMF, 82%, 5:1 mixture of regioisomers; (xii) TBSOTf, pyridine, CH$_2$Cl$_2$, 95%; (xiii) H$_2$, Lindlar's catalyst, CH$_2$Cl$_2$/MeOH, 100%; (xiv) **106**, AgNCO, C$_6$H$_6$, 94%; (xv) 2 N HCl, EtOH, 83%; (xvi) NH$_3$, MeOH, 89%.

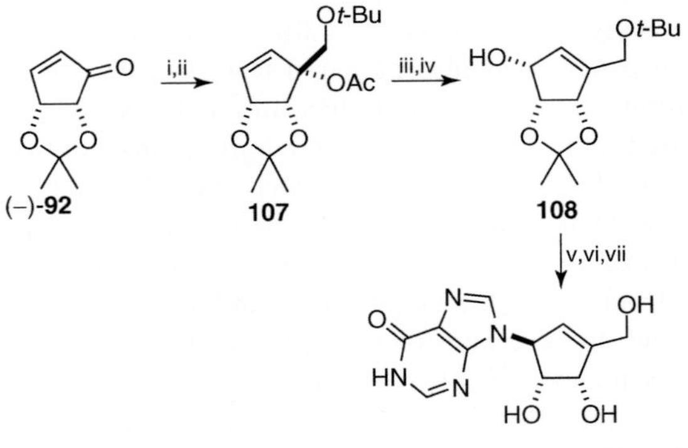

Scheme 29. *Reagents and conditions:* (i) *t*-BuOCH$_2$Li, THF, 78%; (ii) Ac$_2$O, Et$_3$N, DMAP, CH$_2$Cl$_2$, 94%; (iii) PdCl$_2$(MeCN)$_2$, *p*-benzoquinone, THF, 91%; (iv) K$_2$CO$_3$, MeOH, 87%; (v) 6-chloropurine, Ph$_3$P, DEAD, 70%; (vi) HSCH$_2$CH$_2$OH, MeONa, MeOH, 88%; (vii) TFA/H$_2$O, 65%.

3.2.2. Preparation and Use of Cyclopentenone 93.

The preparation of cyclopentenone **93** provided direct entry to neplanocin A and analogues as well as to conformationally locked carbanucleosides based on a bicyclo[3.1.0]hexane backbone. As for cyclopentenone **92**, to the "historical" synthesis using an intramolecular Horner–Wadsworth–Emmons reaction succeeded an RCM-based approach. The original synthesis reported by Marquez started from a protected D-ribonolactone **109**, to which was added lithiated trimethylphosphonate.[8b] After ring opening of the resulting lactol and oxidation of the alcohol, the intramolecular Horner–Wadsworth–Emmons reaction of diketone **110** could proceed to afford **93** in only four steps from **109**. However, diketone **110** partially racemized during this last step, due to a base-promoted epimerization of both C3 and C4 centers. Despite this major drawback, a fortunate selective crystallization of the racemate allowed the isolation of enantiopure **93** in 42% yield. This intermediate was converted in a few steps to (−)-neplanocin A, achieving a straightforward synthesis of this key carbanucleoside. A synthesis of (−)-neplanocin C, very much inspired from this work, was reported by Comin and Rodriguez.[73] Another synthesis of (−)-neplanocin A from **93** was reported by Michel and Strazewski, using a more efficient Mitsunobu reaction to introduce the adenine base (Scheme 30).[74] In that case the cyclopentenone was prepared using the second-generation synthesis described below.

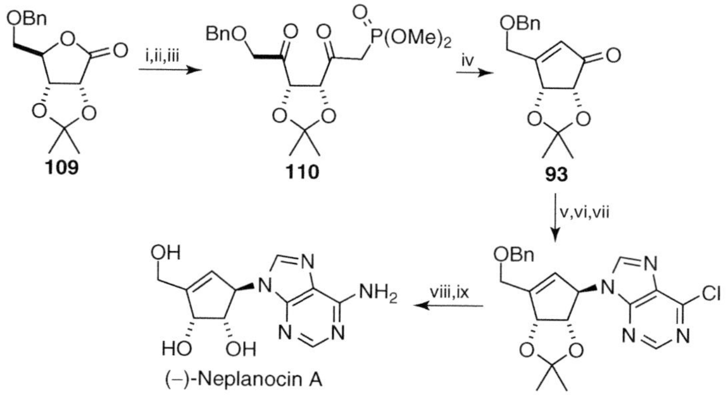

Scheme 30. *Reagents and conditions:* (i) $(MeO)_2P(O)CH_2Li$, THF, 99%; (ii) MeONa, MeOH, 95%; (iii) CrO_3, pyridine, CH_2Cl_2, 80%; (iv) K_2CO_3, 18-crown-6, 42%; (v) $NaBH_4$, $CeCl_3$, MeOH, 97%; (vi) TsCl, NEt_3, CH_2Cl_2; (vii) 6-chloropurine, NaH, MeCN, 31%; (viii) NH_3, MeOH, 70%; (ix) BCl_3, CH_2Cl_2, 50%.

A longer but secure route to the TBS-protected analogue of **93** has been reported by Jeong et al.[75] After protection and methylenation of D-ribose-derived acetonide **97**, oxidation and addition of vinylmagnesium bromide provided precursor **111** as an inseparable mixture of diastereomers (Scheme 31). The major diastereomer **112** could be isolated from the RCM reaction products and converted to **113** after an oxidative rearrangement. TBDPS and trityl-protected cyclopentenones were also obtained through this method, whereas their benzyl-protected homologue **93** could not be. The opposite diastereomer was obtained from vinyl Grignard addition, and the resulting RCM product was unable to undergo the PDC-mediated rearrangement due to an unfavorable relative configuration.

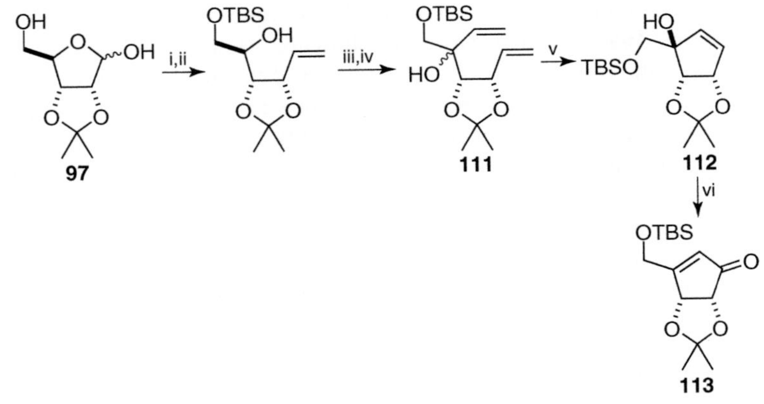

Scheme 31. *Reagents and conditions:* (i) TBSCl, 90%; (ii) CH$_2$Ph$_3$P, THF, 85%; (iii) (COCl)$_2$, DMSO, CH$_2$Cl$_2$, 95%; (iv) vinylmagnesium bromide, THF, 87%; (v) first-generation Grubbs' catalyst, CH$_2$Cl$_2$, 75%; (vi) PDC, DMF, 84%.

Beside the preparation of neplanocin A analogues, **93** is a useful starting material for the synthesis of conformationally locked bicyclo[3.1.0]hexyl carba-nucleosides.[16b,76] A synthesis of a cyclopropyl-fused carbocyclic ring-expanded analogue of oxetanocin A was based on this starting material.[16b] Reduction and cyclopropanation performed on **93**,[76a] followed by acetonide migration and oxidation, afforded ketone **114** (Scheme 32). The hydroxymethyl group was introduced at C3′ through methylenation and hydroboration. Protection and replacement of the acetonide group by a cyclic sulfite led to **115** on which several bases, such as adenine, were introduced. A Barton reduction of the C2′ alcohol and deprotection eventually afforded the target compound **116**.

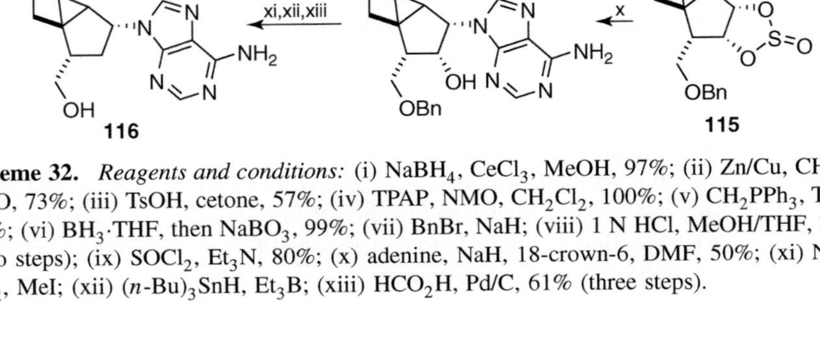

Scheme 32. *Reagents and conditions:* (i) NaBH$_4$, CeCl$_3$, MeOH, 97%; (ii) Zn/Cu, CH$_2$I$_2$, Et$_2$O, 73%; (iii) TsOH, cetone, 57%; (iv) TPAP, NMO, CH$_2$Cl$_2$, 100%; (v) CH$_2$PPh$_3$, THF, 90%; (vi) BH$_3$·THF, then NaBO$_3$, 99%; (vii) BnBr, NaH; (viii) 1 N HCl, MeOH/THF, 88% (two steps); (ix) SOCl$_2$, Et$_3$N, 80%; (x) adenine, NaH, 18-crown-6, DMF, 50%; (xi) NaH, CS$_2$, MeI; (xii) (*n*-Bu)$_3$SnH, Et$_3$B; (xiii) HCO$_2$H, Pd/C, 61% (three steps).

SYNTHESIS OF CYCLOPENTYL CARBOCYCLIC ANALOGUES 561

A synthesis of MLN4924, a potent anticancer agent, was recently performed.[34c] TBDPS-protected cyclopentenone **117** was subjected to diastereoselective hydrogenation and reduction reactions, followed by a regioselective opening of the acetonide and conversion to the corresponding cyclic sulfate **118** (Scheme 33). Introduction of the modified adenine to afford **119** was followed by Barton reduction of the C2′ alcohol. Sulfonylation of the C5′ alcohol and deprotection yielded the target compound **120**.

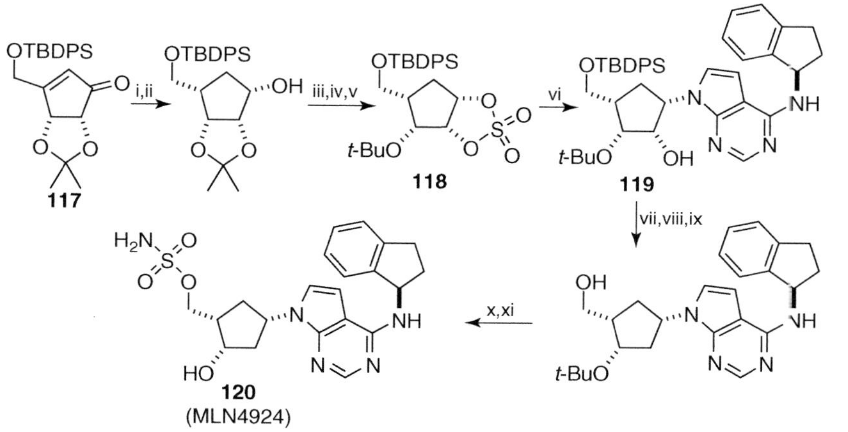

Scheme 33. *Reagents and conditions:* (i) H$_2$, Pd/C, MeOH, 100%; (ii) NaBH$_4$, CeCl$_3$, MeOH, 98%; (iii) Me$_3$Al, CH$_2$Cl$_2$, 62%; (iv) SOCl$_2$, Et$_3$N, CH$_2$Cl$_2$, 97%; (v) RuCl$_3$·3H$_2$O, NaIO$_4$, CCl$_4$/MeCN/H$_2$O, 90%; (vi) purine base, NaH, 18-crown-6, THF, 65%; (vii) PhOC(S)Cl, DMAP, CH$_2$Cl$_2$, 99%; (viii) (n-Bu)$_3$SnH, AIBN, toluene, 82%; (ix) HF·pyridine, THF/pyridine, 99%; (x) NH$_4$SO$_4$Cl, Et$_4$N, MeCN, 92%; (xi) TFA, 90%.

A third synthesis of **113**, based on a zirconium-mediated ring contraction of a D-allose derivative, was also reported by Paquette et al.[77]

Experimental Procedure

(1′S,2′S,3′R,4′S)-7-Deaza-6-(indan-1-ylamino)-[3′-tert-butoxy-4′-(tert-butyldiphenylsilanyloxymethyl)-2′-hydroxycyclopentan-1′-yl]purine (119).[34c] A suspension of N^6-indanyl-7-deazaadenine (8.80 g, 35.2 mmol), NaH (1.38 g, 45.7 mmol), and 18-crown-6 (9.11 g, 45.7 mmol) in THF (200 mL) was stirred at 80°C. To this reaction mixture was added a solution of **118** (13.36 g, 26.5 mmol) in THF (150 mL), and the mixture was stirred at 80°C overnight. The reaction mixture was cooled to 0°C, concentrated HCl was added slowly until the pH reached 1–2, and the reaction mixture was stirred at 80°C for another 2 h. The mixture was neutralized with saturated aqueous NaHCO$_3$ and partitioned between EtOAc and water. The organic layer was then washed with brine, dried over MgSO$_4$, filtered, and evaporated. The residue was purified by silica gel column chromatography (hexane/ethyl acetate, 2:1) to give **119** (11.62 g, 65%) as a white foam. UV (CH$_2$Cl$_2$) λ$_{max}$ 272.5 nm; [α]$_D^{20}$ = −8.89° (c 0.45, MeOH); HRMS (ESI) calcd for C$_{41}$H$_{51}$N$_4$O$_3$Si [M + H]$^+$ m/z 675.3730; found, m/z 675.3717. ^{1}H-NMR (400 MHz, CDCl$_3$): δ 8.38 (s, 1H), 7.70 (m, 4H), 7.41 (m, 6H), 6.92 (d, J = 3.6 Hz, 1H), 6.29 (d, J = 3.2 Hz, 1H), 5.91 (dd, J = 7.6, 14.8 Hz, 1H), 5.14 (br d, J = 6.8 Hz, 1H), 4.77 (m, 1H), 4.36 (t, J = 6.0 Hz, 1H), 4.22 (dd, J = 5.2 and 10.8 Hz, 1H), 3.84 (dd, J = 5.6 and J = 10.4 Hz, 1H), 3.73 (dd, J = 8.4 and 10.4 Hz, 1H), 3.37 (d, J = 5.6 Hz, 1H), 3.06 (m, 1H), 2.95 (m, 1H), 2.75 (m, 1H), 2.75 (m, 1H), 2.58 (m, 1H), 2.38 (m, 1H), 2.15 (m, 1H), 1.98 (m, 1H), 1.65 (s, 1H), 1.55 (s, 1H), 1.16 (s, 9H), 1.07 (s, 9H). ^{13}C-NMR (100 MHz, CDCl$_3$): δ 156.4, 151.8, 150.3, 144.1, 143.8, 135.9, 134.0, 129.9, 128.2, 127.9, 127.9, 127.0, 125.1, 124.4, 123.3, 103.8, 97.4, 77.8, 77.6, 77.2, 76.9, 74.9, 72.4, 63.5, 62.1, 56.3, 43.9, 34.9, 30.5, 30.5, 28.5, 27.2, 19.5. *Anal*. Calcd for C$_{41}$H$_{50}$N$_4$O$_3$Si: C, 72.96; H, 7.47; N, 8.30; found: C, 73.01; H, 7.45; N, 8.36.

3.3. Synthesis by Ring-Closing Metathesis

The finding of efficient and easy-to-handle catalysts has made ring-closing metathesis (RCM) the method of choice in preparing small and medium-sized rings.[78] As illustrated by the preparation of cyclopentenones **92** and **113** described above, the synthesis of carbanucleosides has evolved considerably, thanks to this method. From the transformation of carbohydrate precursors into RCM substrates to their asymmetric synthesis using chiral auxiliaries, various methods involving metathesis as the key cyclization step are discussed below.

3.3.1. Preparation of the RCM Precursor from a Carbohydrate.
Variations around the preparation of **92** and **113** developed by Jeong and Chu (see Sections 3.2.1 and 3.2.2) have led to many carbanucleoside syntheses starting from D-ribose. The synthetic elaboration of this single starting material has been modulated to provide RCM precursors that are structurally close to intermediates **98** (Scheme 26) or **111** (Scheme 31) but lead to great structural diversity for the final carbanucleoside. The synthesis of various apiocarbanucleosides was accomplished by Jeong through the metathesis reaction of intermediate **121** (Scheme 34).[79] The hydroxymethyl group, which in this particular subclass of nucleosides has moved from the 4′-position to C3′, was introduced by an aldol-type reaction performed on **100** (Scheme 34). Methylenation of the resulting lactol afforded **121** that underwent a quantitative RCM reaction. Precursor **122** was then converted, after protection, to apio-neplanocin A and analogues through Mitsunobu reactions. The internal double bond was also reduced or subjected to a cyclopropanation reaction to produce apio-aristeromycin or cyclopropyl-fused derivatives.

The introduction of an ethynyl instead of a vinyl group in the 1′-position provided an enyne metathesis precursor that, after cyclization, ultimately led to 6′-isoneplanocin A.[80] Addition of ethynylmagnesium bromide to aldehyde **123**, which has already been used by Schneller et al. for an alternative synthesis of compound **98** (Scheme 26) and cyclopentenone **92**,[63] and subsequent enyne metathesis

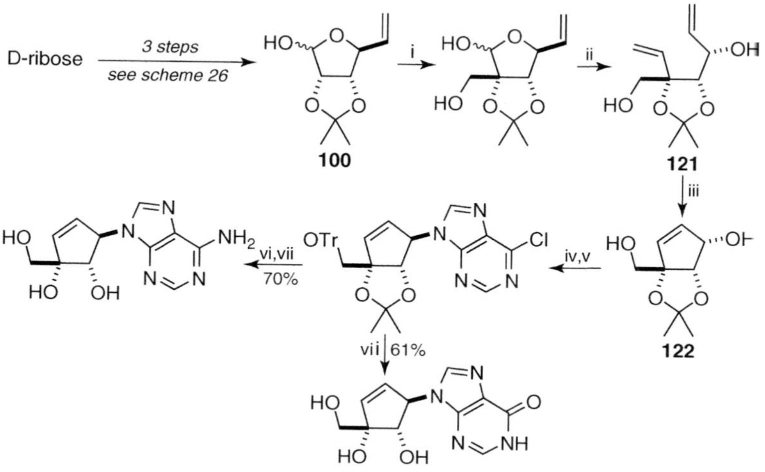

Scheme 34. *Reagents and conditions:* (i) HCHO, K_2CO_3, MeOH, 95%; (ii) CH_3Ph_3PBr, *t*-BuOK, THF, 81%; (iii) second-generation Grubbs' catalyst, CH_2Cl_2, 99%; (iv) TrCl, DMAP, pyridine, 78%; (v) 6-chloropurine, Ph_3P, DEAD, THF, 72%; (vi) NH_3, MeOH; (vii) 3 N HCl, THF.

afforded diene **124** (Scheme 35). Oxidative cleavage of the double bond and reduction of the resulting aldehyde yielded cyclopentene **125**. Protecting group modification, Mitsunobu coupling with adenine and final deprotection eventually provided the targeted 6′-isoneplanocin A **126**.

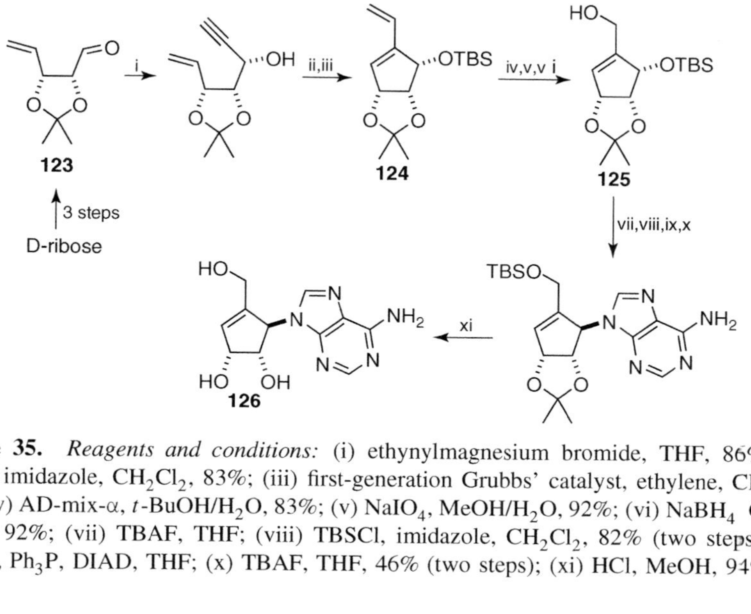

Scheme 35. *Reagents and conditions:* (i) ethynylmagnesium bromide, THF, 86%; (ii) TBSCl, imidazole, CH_2Cl_2, 83%; (iii) first-generation Grubbs' catalyst, ethylene, CH_2Cl_2, 86%; (iv) AD-mix-α, *t*-BuOH/H_2O, 83%; (v) $NaIO_4$, MeOH/H_2O, 92%; (vi) $NaBH_4$ $CeCl_3$, MeOH, 92%; (vii) TBAF, THF; (viii) TBSCl, imidazole, CH_2Cl_2, 82% (two steps); (ix) adenine, Ph_3P, DIAD, THF; (x) TBAF, THF, 46% (two steps); (xi) HCl, MeOH, 94%.

The use of 2-deoxy-D-ribose as a starting material in a sequence similar to Jeong's synthesis of **113** (Scheme 31) allowed Marquez to report an optimized synthesis of conformationally locked antiviral agent north-methanocarbathymidine.[81] Starting from D-galactose provided a straightforward entry to L-cyclopentenyl nucleosides and analogues of (+)-neplanocin, as reported by Agrofoglio et al.[82] The classical conversion of methyl-α-D-galactose to tetra-*O*-benzylgalactopyranoside was followed by a Wittig reaction to afford alkene **127** (Scheme 36). An oxidation–methylenation sequence swiftly provided diene **128**, which was converted after RCM using Nolan's catalyst to **129**. Protecting group modifications led to L-cyclopentenylribose **130**.[83] This intermediate was afterward coupled to various heterocyclic bases through a Tsuji–Trost reaction to yield compounds such as **131** and, after deprotection, L-cyclopentenyl nucleosides such as **132**.

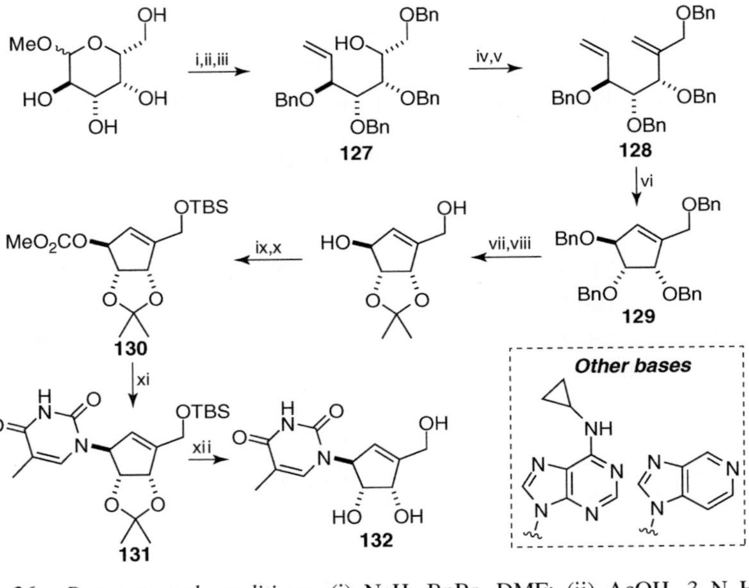

Scheme 36. *Reagents and conditions:* (i) NaH, BnBr, DMF; (ii) AcOH, 3 N H_2SO_4; (iii) CH_3Ph_3PBr, *n*-BuLi, THF, 51% (three steps); (iv) PCC, AcONa, CH_2Cl_2, 87%; (v) CH_3Ph_3PBr, *n*-BuLi, THF, 77%; (vi) Nolan's catalyst, benzene, 90%; (vii) Na/NH_3; (viii) $(CH_3)_2C(OMe)_2$, TsOH, acetone, 50% (two steps); (ix) TBSCl, pyridine, 62%; (x) MeO-COCl, DMAP, pyridine, 99%; (xi) thymine, Et_3Al, $Pd(Ph_3P)_4$, dppf, DMF/THF, 88%; (xii) TFA/H_2O, 79%.

Experimental Procedure

(1′S,2′R,3′S)-1-[2,3-(Isopropylenedioxy)-4-(tert-butyldimethylsilyloxymethyl)-4-cyclopenten-1-yl]thymine (131).[83] To a solution of thymine (0.11 mmol) in DMF (2 mL) was added Et_3Al (111 mL, 1 N solution in hexane). After stirring at 60°C for 45 min, a solution of **130** (0.05 mmol) in THF (2 mL), $Pd(Ph_3P)_4$ (10 mol%), and dppf (5 mol%) were added and the reaction was stirred for

6 h at 60°C. After evaporation of the volatiles, the residue was purified by flash chromatography on silica gel to afford **130** (88%). $[\alpha]_D^{20} = +105.2°$ (c 0.9, MeOH); UV (MeOH) λ_{max} 265 nm. ^{1}H-NMR (CDCl$_3$): δ 8.46 (s, 1H), 6.80 (s, 1H), 5.56 (s, 1H), 5.41 (s, 1H), 5.16 (d, J = 6 Hz, 1H), 4.57 (d, J = 6 Hz, 1H), 4.38 (s, 2H), 1.89 (s, 3H), 1.43 (s, 3H), 1.34 (s, 3H), 0.93 (s, 9H), 0.10 (s, 6H). ^{13}C-NMR (CDCl$_3$): δ 163.9, 153.1, 150.8, 137.1, 121.5, 112.7, 110.9, 84.7, 83.4, 67.3, 60.5, 27.4, 26.0, 25.9, 18.5, 12.6, −5.2; HRMS: calcd for C$_{20}$H$_{32}$N$_2$O$_5$NaSi m/z 431.5642; found, m/z 431.5638.

3.3.2. Preparation of the RCM Precursor from Other Chiral Sources.

Among other chiral substrates that have been used to prepare the RCM precursor, diisopropyl L-tartrate is a noteworthy example.[84] The C_2-symmetry of this substrate is exploited judiciously since both ester functions are elaborated to provide cyclization precursor **133** (Scheme 37). Indeed, a Horner–Wadsworth–Emmons elongation followed by a reduction and a diastereospecific [3,3]-sigmatropic rearrangement

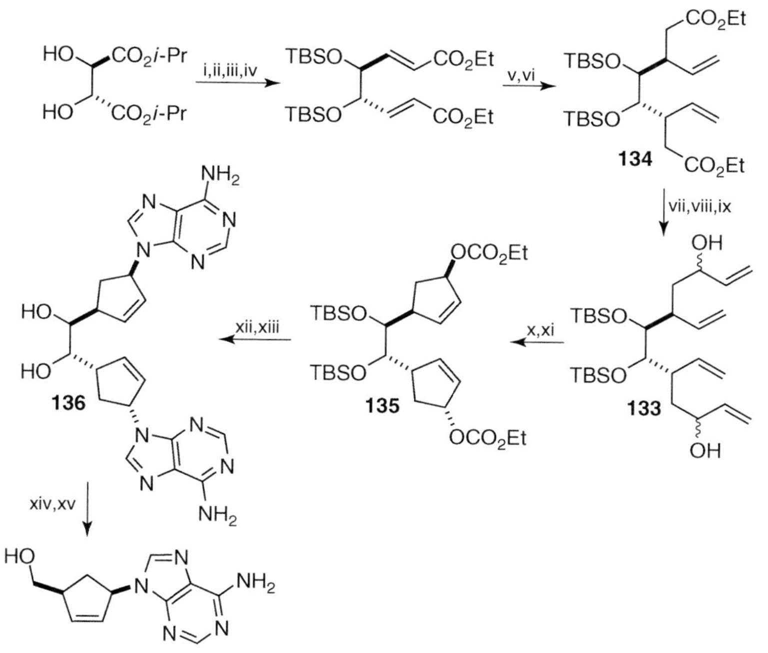

Scheme 37. *Reagents and conditions:* (i) (CH$_3$)$_2$C(OMe)$_2$, TsOH, benzene, 97%; (ii) (*i*-Bu)$_2$AlH, toluene, then (*i*-PrO)$_2$P(O)CH$_2$CO$_2$Et, THF, 98%; (iii) 2 N HCl, EtOH, 84%; (iv) TBSOTf, Et$_3$N, CH$_2$Cl$_2$, 99%; (v) (*i*-Bu)$_2$AlH, CH$_2$Cl$_2$, 99%. (vi) (EtO)$_3$CCH$_3$, propionic acid, 86%; (vii) (*i*-Bu)$_2$AlH, CH$_2$Cl$_2$, 97%; (viii) PCC, CH$_2$Cl$_2$, 90%; (ix) vinylmagnesium bromide, THF, 72%; (x) second-generation Grubbs' catalyst, CH$_2$Cl$_2$, 43%; (xi) ClCO$_2$Et, DMAP, DMAP, pyridine, 84%; (xii) adenine, Pd$_2$(dba)$_3$·CHCl$_3$, P(O-*i*-Pr)$_3$, NaH, THF/DMSO, 42%; (xiii) TBAF, THF, 80%; (xiv) NaIO$_4$, MeOH/H$_2$O; (xv) NaBH$_4$, EtOH, 76% (two steps).

afforded diester **134**. Conversion of **134** to the corresponding dialdehyde and a nonselective addition of vinyl Grignard reagent afforded **133** as an unseparable mixture of diastereomers. The diastereomer desired was isolated after RCM using a second generation Grubbs' catalyst and converted to carbonate **135**. Adenine was introduced via a Tsuji–Trost reaction and deprotection afforded intermediate **136**. An oxidative cleavage of the C_2-symmetric diol followed by a reduction of the resulting aldehyde eventually provided the desired carbanucleoside.

A similar synthesis, starting from D-glyceraldehyde and also involving a Claisen [3,3]-sigmatropic rearrangement, was performed by Ghosh to produce the cyclopentanyl core of a carbanucleoside.[85] Another D-glyceraldehyde derivative was recently used as a starting material for the synthesis of the carbocyclic core of D- and L-carbovir.[86] Several racemic syntheses of carbanucleosides from propionaldehyde or ethyl glycolate, which involved a key RCM reaction, were also reported by Jong et al.[87]

A synthesis of conformationally locked L-carbanucleosides from (R)-epichlorohydrin was reported by Moon et al.[88] Cyclopropyllactone **137** was obtained through a double-substitution reaction of diethylmalonate to epichlorohydrin followed by a selective reduction of the ester group (Scheme 38). Compound **137** was protected and reduced to the corresponding lactol, which was, in turn, subjected to a methylenation reaction and to a Swern oxidation to afford aldehyde **138**. The addition of vinylmagnesium bromide to **138** was not diastereoselective

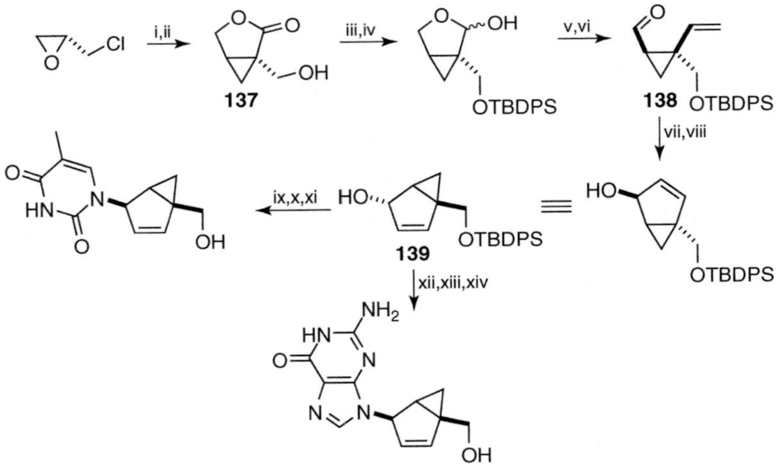

Scheme 38. *Reagents and conditions:* (i) $(EtO_2C)_2CH_2$, EtONa, EtOH, 67%. (ii) NaOH, EtOH, then $NaBH_4$, then 2 N HCl, 62%; (iii) TBDPSCl, imidazole, CH_2Cl_2, 84%; (iv) (i-Bu)$_2$AlH, CH_2Cl_2, 97%; (v) CH_3Ph_3PBr, t-BuOK, THF, 88%; (vi) $(COCl)_2$, DMSO, Et_3N, CH_2Cl_2, 90%; (vii) vinylmagnesium bromide, THF, 39% of the (2S)-epimer; (viii) second-generation Grubbs' catalyst, CH_2Cl_2, 85%; (ix) N-benzoylthymine, Ph_3P, DEAD, THF, 28%; (x) NH_4OH, MeOH; (xi) TBAF, THF, 76% (two steps); (xii) 2-amino-6-chloropurine, Ph_3P, DEAD, THF, 17%; (xiii) TBAF, THF; (xiv) 2-mercaptoethanol, 1 N NaOMe, 97% (two steps).

and afforded the (2R)-epimer in 48% yield and the (2S)-epimer in 39% yield. The RCM reaction of the (2S)-epimer afforded **139**, which was converted to various carbanucleosides by a Mitsunobu coupling and a complete deprotection.

3.3.3. Preparation of the RCM Precursors Using an Asymmetric Aldol Reaction or an Addition to Chiral Sulfinylimine.

Crimmins et al. developed a general and flexible route to the cyclopentanyl core of carbanucleosides.[89] A phenylalanine-derived oxazolidinethione was N-acylated with pivaloyl 4-pentenoate to provide **140** that was subjected to the Evans-type asymmetric aldol reaction developed by Crimmins et al.[90] Addition to crotonaldehyde delivered the "non-Evans" *syn*-aldol **141** in good yield and with excellent diastereoselectivity (Scheme 39). Ring closure and removal of the chiral auxiliary by a LiBH$_4$-mediated reduction provided 5-hydroxymethylcyclopent-2-en-1-ol **75** (Scheme 19, Section 3.1.3), a well-known intermediate that previously was prepared by chemoenzymatic synthesis (see Section 3.1.3). A synthesis of carbovir, similar to the one described by Roberts et al.,[47b] and of abacavir was completed from **75**. The 2′-methyl analogues of these drugs were also prepared by this method. Indeed, Crimmins anticipated that this substitution would force the system to adopt the bioactive northern conformation.

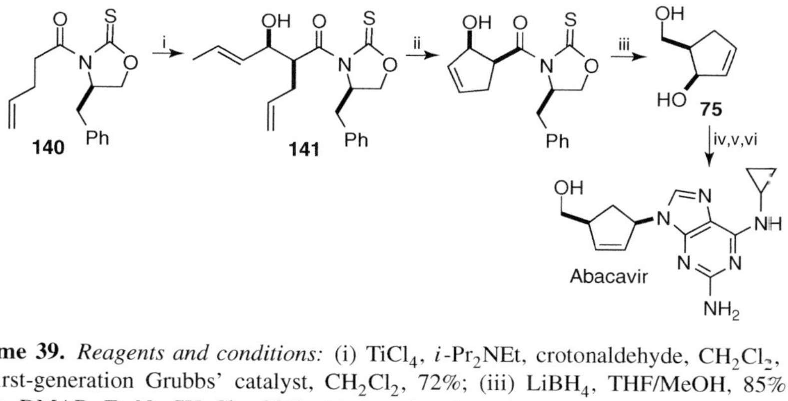

Scheme 39. *Reagents and conditions:* (i) TiCl$_4$, *i*-Pr$_2$NEt, crotonaldehyde, CH$_2$Cl$_2$, 79%; (ii) first-generation Grubbs' catalyst, CH$_2$Cl$_2$, 72%; (iii) LiBH$_4$, THF/MeOH, 85%; (iv) Ac$_2$O, DMAP, Et$_3$N, CH$_2$Cl$_2$, 90%; (v) 2-amino-6-(cyclopropylamino)purine, Pd(Ph$_3$P)$_4$, NaH, THF/DMSO, 62%; (vi) NaOH, H$_2$O.

A solid-phase version of Tsuji–Trost coupling from **75** was also reported.[91] A similar strategy was used by Kuang et al. for synthesis of the carbocyclic analogue of ribavirin.[92] The conditions used for the aldol condensation were those described by Heathcock et al. and provided the *anti*-aldol **142** in moderate yield (Scheme 40).[93] Epoxide **143** was obtained after RCM, removal of the chiral auxiliary, epoxidation of the resulting double bond, and protection. Epoxide opening with the required 1,2,4-triazole and complete deprotection afforded the desired carbanucleoside **144**. A synthesis of an analogue of formycin A, based on a very similar approach, was recently reported by Schneller et al.[94]

The asymmetric synthesis of a 4-aminocyclopent-2-ene-1-ol derivative through a Mannich reaction on a chiral sulfinylimine and an RCM reaction was reported

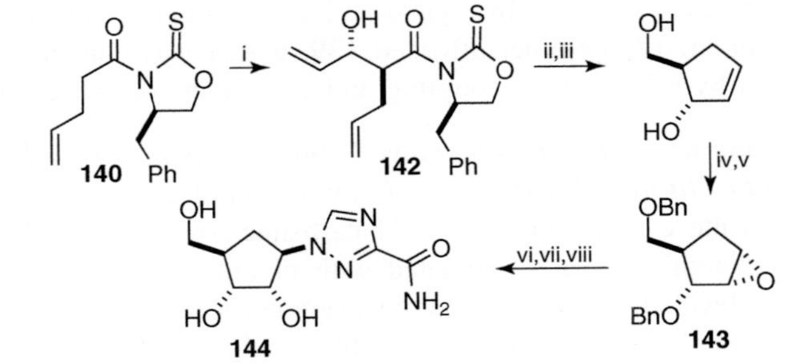

Scheme 40. *Reagents and conditions:* (i) (*n*-Bu)$_2$BOTf, *i*-Pr$_2$NEt, acrolein, 53%; (ii) first-generation Grubbs' catalyst, 96%; (iii) LiBH$_4$, THF/MeOH, 83%; (iv) *m*-CPBA, 97%; (v) NaH, BnBr, THF, 74%; (vi) 1,2,4-triazol-3-carboxylic acid methyl ester, NaH, DMF, 40%; (vii) NH$_3$, MeOH, 100%; (viii) H$_2$, Pd/C.

by Davis and Wu.[95] Intermediates such as **145** were usually prepared by enzymatic resolution of the racemate obtained from a nitroso-Diels–Alder reaction or by an asymmetric version of the latter (see Sections 3.1.2 and 3.4.1). The addition of methyl acetate to *p*-tolunesulfinylimine **146** was efficient and highly diastereoselective (Scheme 41). The resulting β-aminoester was converted to enone **147** by addition of phosphonate followed by an HWE reaction and transprotection of the amino group. RCM cyclization and diastereoselective reduction eventually afforded **145**.

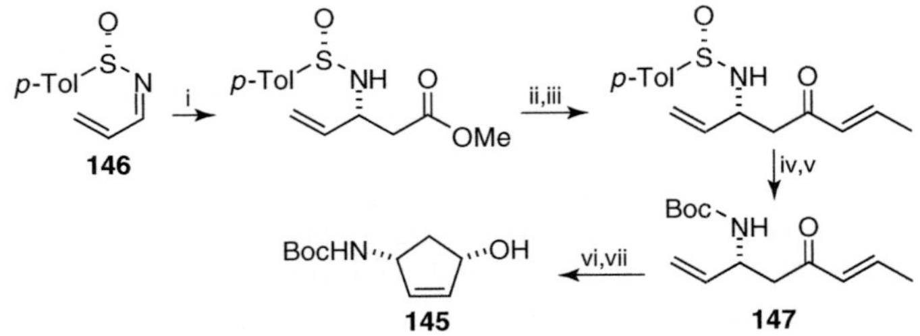

Scheme 41. *Reagents and conditions:* (i) CH$_3$CO$_2$Me, NaHMDS, THF, 87%; (ii) CH$_3$P(O)(OMe)$_2$, *n*-BuLi, THF, 83%; (iii) CH$_3$CHO, DBU, THF, 94%; (iv) TFA, MeOH; (v) (Boc)$_2$O, Et$_3$N, DMAP, THF, 85% (two steps); (vi) first-generation Grubbs' catalyst, CH$_2$Cl$_2$, 97%; (vii) NaBH$_4$, CeCl$_3$, MeOH, 76%.

3.4. Synthesis Using Cycloaddition and Intramolecular Carbene Insertion Reactions

Beside the intramolecular HWE and RCM reactions described in Sections 3.2 and 3.3, a great variety of cyclization reactions were employed for the preparation of carbanucleosides. From nitroso-Diels–Alder reactions to dipolar cycloadditions, from anionic to radical cyclizations, the most efficient methods leading to enantio-enriched carba-nucleosides or precursors are discussed in the two next sections.

3.4.1. Asymmetric Nitroso-Diels–Alder Reaction.

As mentioned in Section 3.1.2, 4-aminocyclopent-2-ene-1-ol derivatives are easily prepared through a nitroso-Diels–Alder reaction and can be obtained in enantio-enriched form by enzymatic resolution. However, asymmetric versions of this [4 + 2]-cycloaddition have been disclosed and included in synthetic routes to carbanucleosides devoid of chemoenzymatic steps. This approach has been explored by Miller and was based on the reaction between a mandelic acid–derived acylnitroso compound and cyclopentadiene, developed previously by Kirby et al. and Procter et al.[96] The method developed by Miller has evolved and been optimized through the years and can be summed up as follows. Swern oxidation of the alanine-derived hydroxamic acid **148** delivered the corresponding nitroso compound that was trapped in situ by cyclopentadiene to afford cycloadduct **149** in a practically useful diastereomeric ratio of 5.9:1.[97] This level of diastereoselectivity is higher than the one obtained with the ammonium periodate oxidation method used in the original publication.[98] Compound **149** was either subjected directly to a sequence of osmylation, protection, and reductive removal of the chiral auxiliary to provide **150**, or reduced to afford cyclopentene **151**.[97,99] A rapid synthesis of nucleoside precursor **152** was performed from **150**, whereas 4-aminocyclopent-2-ene-1-ol derivatives such as **69** were obtained from **151** through an Edman degradation.[100] A Tsuji–Trost coupling to introduce a heterocyclic base to ent-**151** also allowed the preparation of various 5′-azacarba-nucleosides via compound **153**.[99] Compounds such as **151** were also engaged in a Tsuji–Trost reaction with nitromethane followed by a Nef reaction to introduce a hydroxymethyl group in the 5′-position[101] (Scheme 42).

Attempts were made to improve the diastereoselectivity of the [4 + 2]-cycloaddition reaction and to facilitate removal of the chiral auxiliary. The α-chloronitroso compound **154**, derived from D-mannose, showed very high levels of stereoselectivity in its reaction with cyclohexadiene (Scheme 43).[102] A slightly lower selectivity was observed by Miller in the reaction with cyclopentadiene, but the spontaneous cleavage of the chiral auxiliary after cyclization gives a lot of credit to this approach.[103] The Boc-protected 4-aminocyclopent-2-ene-1-ol ent-**145** was indeed obtained in two steps and with an enantiomeric excess up to 84%, depending on the Lewis acid used for the cycloaddition.

A very efficient chiral auxiliary was reported by Yan et al.[104] Camphor-derived hydroxamic acid **155** underwent a periodate-promoted oxidation to the corresponding nitroso compound, which was trapped immediately by cyclopentadiene (Scheme 44). Cycloadduct **156** was obtained in 94% yield and with an almost complete diastereoselectivity. Osmylation, protection, and removal of the chiral auxiliary eventually afforded carbanucleoside precursor **150**. A conversion of **150** to both enantiomers of cyclopentenone **92** was also reported in this study.

The development of a catalytic asymmetric nitroso-Diels–Alder reaction is still a very challenging task and, to date, no methodology applicable to the preparation of carbanucleosides has been reported.[105]

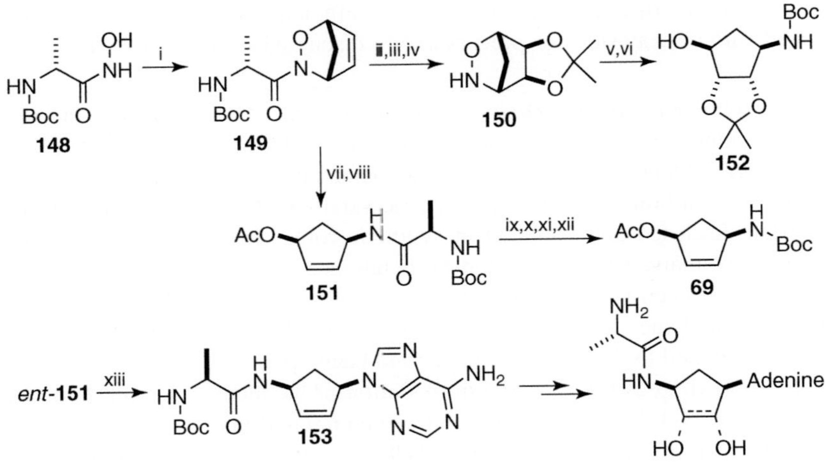

Scheme 42. *Reagents and conditions:* (i) (COCl)$_2$, DMSO, cyclopentadiene, pyridine, CH$_2$Cl$_2$, 78%, dr = 5.9:1; (ii) OsO$_4$, NMO, THF/H$_2$O; (iii) (CH$_3$)$_2$C(OMe)$_2$, TsOH, 95% (two steps); (iv) NaBH$_4$, MeOH, 76%; (v) H$_2$, Pd/C; (vi) (Boc)$_2$O, Na$_2$CO$_3$, THF/H$_2$O, 83% (two steps); (vii) Ac$_2$O, pyridine, CH$_2$Cl$_2$, 87%; (viii) Mo(CO)$_6$, MeCN/H$_2$O, 95%; (ix) TFA; (x) PhNCS, Et$_3$N; (xi) TFA; (xii) (Boc)$_2$O, Na$_2$CO$_3$, 71% (four steps); (xiii) adenine, NaH, Pd(Ph$_3$P)$_4$, DMF, 62%.

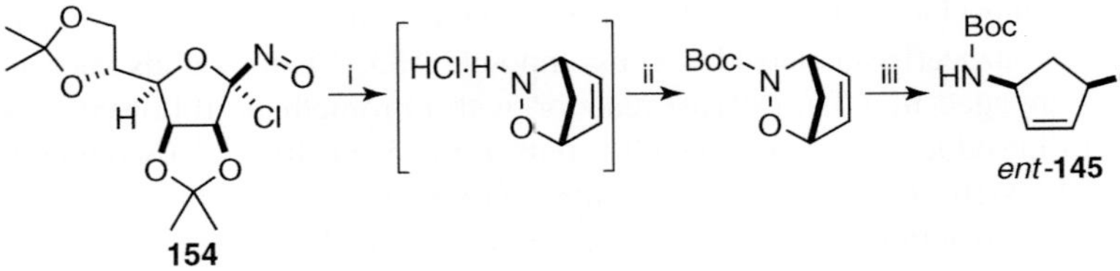

Scheme 43. *Reagents and conditions:* (i) cyclopentadiene, Et$_2$AlCl, toluene, EtOH; (ii) (Boc)$_2$O, Na$_2$CO$_3$, THF/H$_2$O, 61% (two steps), 84% ee; (iii) Mo(CO)$_6$, NaBH$_4$, MeCN/H$_2$O, 80%.

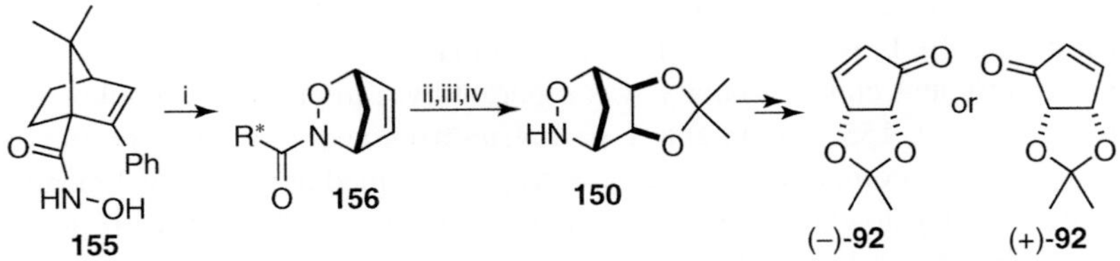

Scheme 44. *Reagents and conditions:* (i) Et$_4$NIO$_4$, cyclopentadiene, CH$_2$Cl$_2$, 95%, dr = 99:1; (ii) OsO$_4$, NMO, THF/acetone/H$_2$O; (iii) (CH$_3$)$_2$C(OMe)$_2$, TsOH, 96% (two steps); (iv) LiAlH$_4$, THF, 85%.

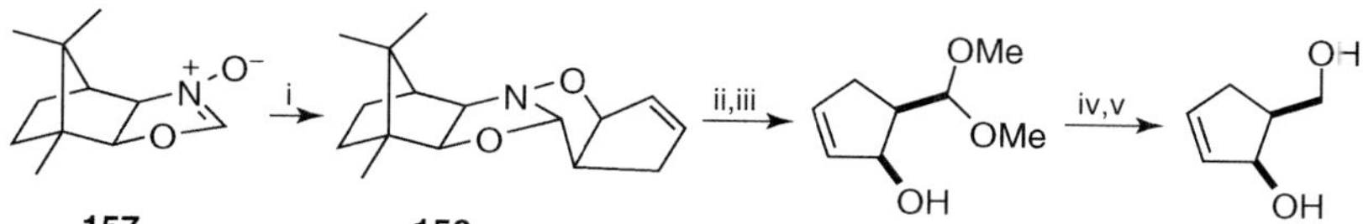

Scheme 45. *Reagents and conditions:* (i) cyclopentadiene, CH$_2$Cl$_2$, 75%; (ii) *m*-CPBA; (iii) CSA, MeOH, 65% (two steps); (iv) CSA, MeCN/H$_2$O; (v) NaBH$_4$, 60% (two steps).

3.4.2. Miscellaneous Cycloaddition Reactions.

Although less studied than the nitroso-Diels–Alder reaction, several other cycloaddition reactions were employed for carbanucleoside synthesis. Few studies demonstrated, for example, the synthetic utility of nitrone or nitrile oxide 1,3-dipolar cycloadditions. Gallos reported the synthesis of an aminocyclopentitol building block via an intramolecular cycloaddition of the nitrone derived from enal **123** (Section 3.3.1).[106] Shortly after, Berranger and Langlois reported a very efficient preparation of (+)-carbovir, broadly applicable to the synthesis of various L-carbanucleosides.[107] A stereoselective 1,3-dipolar cycloaddition between a chiral oxazoline *N*-oxide **157** and cyclopentadiene afforded cycloadduct **158** in good yield and as a single diastereomer (Scheme 45). This reagent-controlled approach allowed the preparation, after removal and recovery of the chiral auxiliary followed by reduction, of diol *ent*-**75**. The chemoenzymatic synthesis of this useful building block was been described in Section 3.1.3, and *ent*-**75** was converted to (+)-carbovir using a sequence similar to that used by Roberts (Scheme 19, Section 3.1.3).

Mandal et al. prepared many unusual spirocyclic carbanucleosides using an intramolecular nitrone 1,3-dipolar cycloaddition on a glucose-derived substrate as the key step.[108] Very recently, Zhou and Li reported an efficient preparation of entecavir from epoxide **159**[109] which was prepared in five steps from 1,3-propanediol using a Sharpless epoxidation to introduce the stereogenic centers, and converted to isoxazoline **160** though an intramolecular nitrile oxide cycloaddition (Scheme 46). Reductive ring opening and elimination provided cyclopentenone **161** which after reduction and Mitsunobu coupling was converted to entecavir.

Asymmetric Diels–Alder reactions were also investigated and Ortuño et al. reported, for example, the diastereoselective reaction between cyclopentadiene and a chiral enoate.[110] A related catalytic asymmetric Diels–Alder reaction between cyclopentadiene and ethyl bromopropenoate **162** was the first step in an efficient, formal synthesis of aristeromycin described by Boyer and Leahy.[111] Cycloadduct **163** was obtained in excellent yield and enantiomeric excess, and was converted to bicyclic compound **164** after osmylation of the double bond, elimination of bromide, and protection with benzyl groups (Scheme 47). Reductive ozonolysis followed by oxidative cleavage of the resulting 1,2-diol, oxidation, and protection provided **165**. The ester group of **165** was converted to the corresponding acyl azide, which underwent a Curtius rearrangement to yield the desired isocyanate and **166** after reaction with benzyl alcohol. Deprotection afforded aminocyclopentitol **167**, a precursor of aristeromycin that was obtained by Ohno et al. from an enzymatically resolved Diels–Alder adduct.[112]

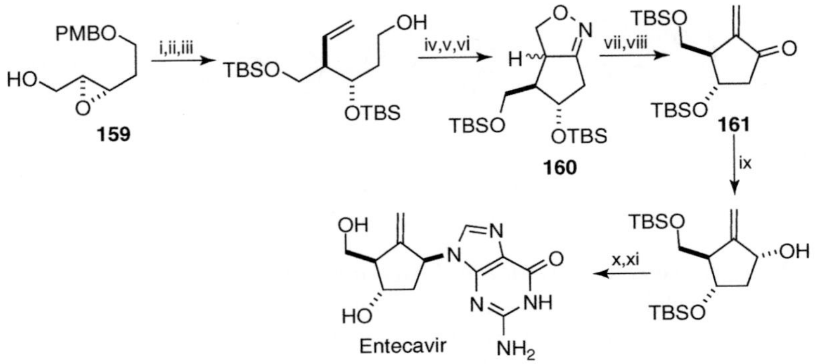

Scheme 46. *Reagents and conditions:* (i) vinylmagnesium bromide, CuI, THF, 84%; (ii) TBSCl, imidazole, DMF; (iii) DDQ, CH_2Cl_2, 90% (two steps); (iv) $(COCl)_2$, DMSO, Et_3N, CH_2Cl_2, 78%; (v) $NH_2OH·HCl$, AcONa, MeOH; (vi) NaOCl, CH_2Cl_2, 82% (three steps); (vii) H_2, $B(OH)_3$, Pd/C, THF/H_2O; (viii) MsCl, Et_3N, CH_2Cl_2, 76% (two steps); (ix) $LiBHEt_3$, THF, 90%; (x) 2-amino-6-benzyloxypurine, Ph_3P, DEAD, THF, 80%; (xi) conc. HCl, THF, MeOH, 96%.

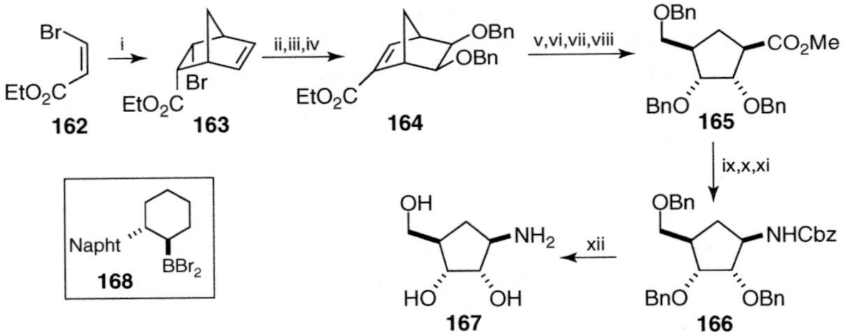

Scheme 47. *Reagents and conditions:* (i) cyclopentadiene, **168**, CH_2Cl_2, 94%, 95.4% ee; (ii) OsO_4, NMO, acetone/H_2O, 74%; (iii) DBU, Et_2O, 97%; (iv) BnBr, Ag_2O, 3-Å molecular sieves, benzene, 80%; (v) O_3 then $LiBH_4$, CH_2Cl_2/MeOH; (vi) $NaIO_4$, THF/H_2O; (vii) Br_2, $NaHCO_3$, MeOH/H_2O, 66% (three steps); (viii) BnBr, Ag_2O, 3-Å molecular sieves, benzene, 81%; (ix) H_2NNH_2, EtOH; (x) N_2O_4, CCl_4; (xi) BnOH, benzene, 67% (three steps); (xii) Na, NH_3, THF/MeOH, 61%.

A catalytic, asymmetric synthesis of bicyclic β-lactone reported by Romo et al., based on the intramolecular reaction between an in situ-generated ketene and an aldehyde function, offered a new entry to carbanucleoside precursors.[113] A formal intramolecular [2 + 2]-cycloaddition performed on **169** and catalyzed by *O*-acetylquinidine provided lactone **170** in 37% yield and 92% ee (Scheme 48). Lactone **170** was easily converted to cyclopentenone **171** and to building block **172**, a known precursor of carbanucleosides (see Section 3.2.3).

Brown and Hegedus reported the synthesis of various carbanucleosides from chiral cyclopentenone **174**[114] which was derived from cyclobutanone **175**, a versatile intermediate prepared from chromium carbene **176** using a photolysis reaction

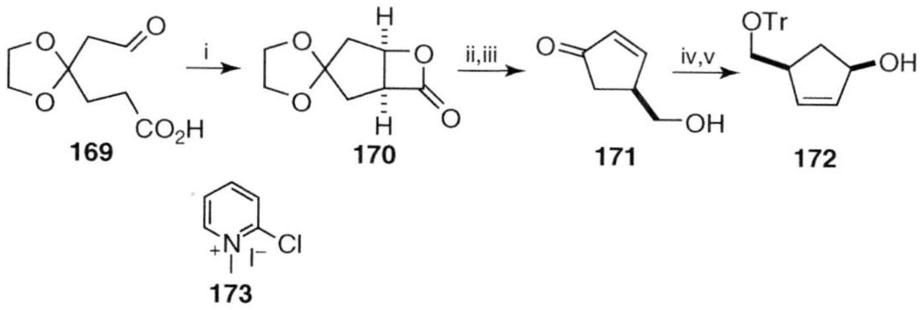

Scheme 48. *Reagents and conditions:* (i) *O*-acetylquinidine, **173**, *i*-Pr$_2$NEt, MeCN, 37%, 92% ee; (ii) BH$_3$·DMS, THF; (iii) 1 N HCl, THF, 51% (two steps); (iv) TrCl, DMAP, pyridine, CH$_2$Cl$_2$, 90%; (v) (*i*-Bu)$_2$AlH, THF, 87%.

in the presence of the chiral enecarbamate **177** (Scheme 49).[115] **175** arose from a [2 + 2]-cycloaddition reaction between the in situ–generated ketene and **177** and was ring-expanded to provide **174**. Hydrogenation, elimination of the chiral auxiliary, and reduction of the carbonyl group led to enantio-enriched cyclopentene **178**, a useful intermediate that was easily converted to some L-carbanucleosides, such as (+)-carbovir and (+)-aristeromycin.

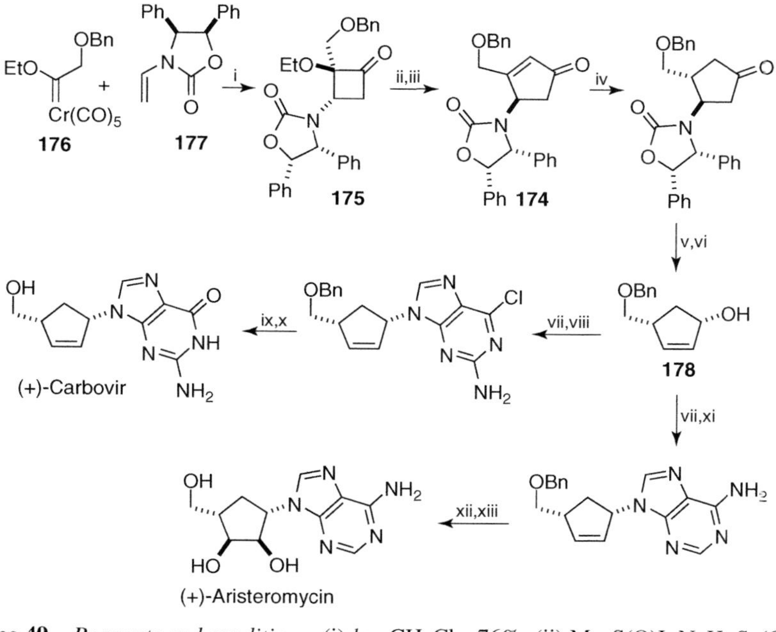

Scheme 49. *Reagents and conditions:* (i) *hv*, CH$_2$Cl$_2$, 76%; (ii) Me$_3$S(O)I, NaH, Sc(OTf)$_3$, DMF; (iii) Li$_2$CO$_3$, THF/MeOH, 74% (two steps); (iv) H$_2$, [Rh(COD)dppb]BF$_4$, DMF, 77%; (v) LDA, THF; (vi) (*i*-Bu)$_2$AlH, THF, 50% (two steps); (vii) ClCO$_2$Et, pyridine, CH$_2$Cl$_2$, 87%; (viii) 2-amino-6-chloropurine, Pd(Ph$_3$P)$_4$, DMF, 63%; (ix) BCl$_3$, CH$_2$Cl$_2$, 83%; (x) 0.5 N NaOH, 72%; (xi) adenine, Pd(Ph$_3$P)$_4$, DMF, 65%; (xii) BCl$_3$, CH$_2$Cl$_2$, 64%; (xiii) OsO$_4$, DMF, 43%.

It should also be mentioned that a Bayer–Williger oxidation of **175** led to an enantio-enriched butenolide that was converted to the enantiomer of cyclopentenone **93** and to (+)-neplanocin A.[116]

3.4.3. Intramolecular C–H or C=C Carbene Insertion.

A straightforward synthesis of neplanocin A based on an intramolecular C–H insertion of a vinylidenecarbene was reported by Ohira et al.[117] The addition of lithiated trimethylsilyldiazomethane to ribose-derived ketone **179** resulted in the in situ generation of the carbene, which immediately underwent the insertion (Scheme 50). The process is efficient since **180** was obtained in 55–65% yield but, unfortunately, in favor of the undesired epimer. Despite this drawback, the diastereomer required was obtained through an oxidation–selective reduction sequence performed on the mixture of epimers and easily converted to neplanocin A. A similar selectivity but with a lower yield for the carbene insertion was obtained by Nguyen Van Nhien et al. using an α-cyanomesylate as the carbene precursor.[118]

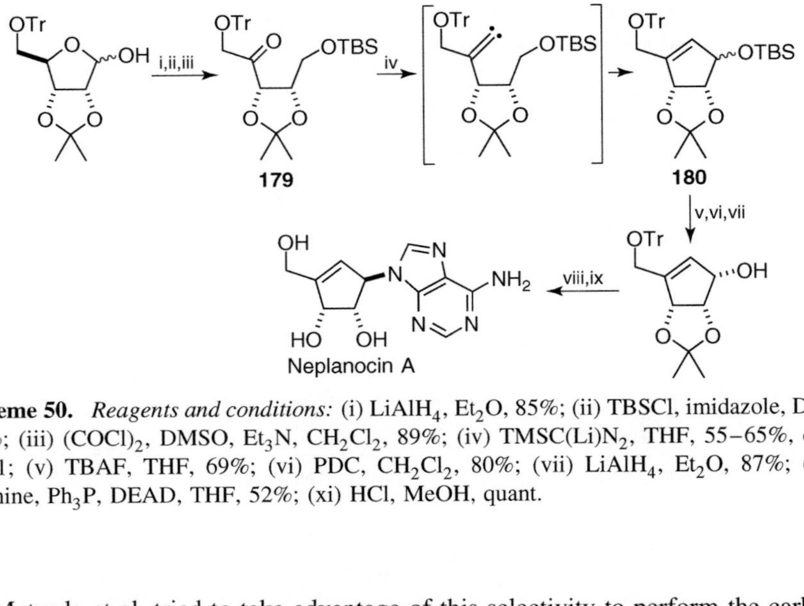

Scheme 50. *Reagents and conditions:* (i) LiAlH$_4$, Et$_2$O, 85%; (ii) TBSCl, imidazole, DMF, 97%; (iii) (COCl)$_2$, DMSO, Et$_3$N, CH$_2$Cl$_2$, 89%; (iv) TMSC(Li)N$_2$, THF, 55–65%, dr = 2.7:1; (v) TBAF, THF, 69%; (vi) PDC, CH$_2$Cl$_2$, 80%; (vii) LiAlH$_4$, Et$_2$O, 87%; (viii) adenine, Ph$_3$P, DEAD, THF, 52%; (xi) HCl, MeOH, quant.

Matsuda et al. tried to take advantage of this selectivity to perform the carbene insertion after introduction of the base, which was supposed to lead this time to the rightly configured compound.[119] The β-epimer **181** was indeed obtained selectively in a 4:1 ratio but, unfortunately, in low yield (Scheme 51).

Marquez et al. reported the preparation of south- and north-locked carbanucleosides by intramolecular cyclopropanation of diazo compounds.[120] Racemic bicyclo[3.1.0]hexyl carbanucleosides such as **182** were indeed prepared from ethyl acetoacetate and were ultimately resolved, at least for north-locked nucleosides, by an enzymatic route (Scheme 52).

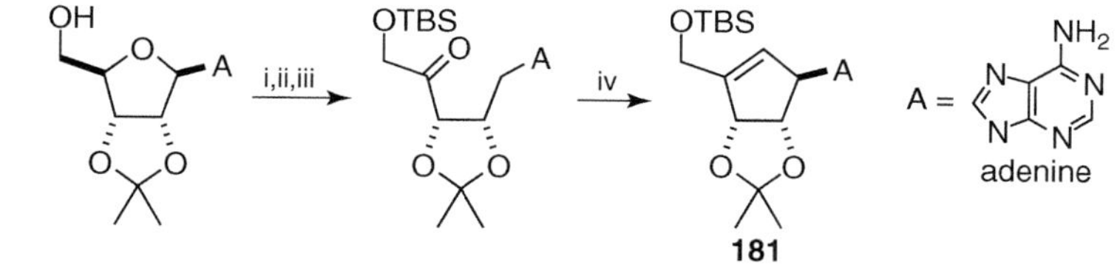

Scheme 51. *Reagents and conditions:* (i) (i-Bu)$_2$AlH, THF, 85%; (ii) TBSCl, imidazole, DMF, 78%; (iii) Dess–Martin periodinane, CH$_2$Cl$_2$, 71%; (iv) TMSC(Li)N$_2$, THF, 20%, dr = 4:1.

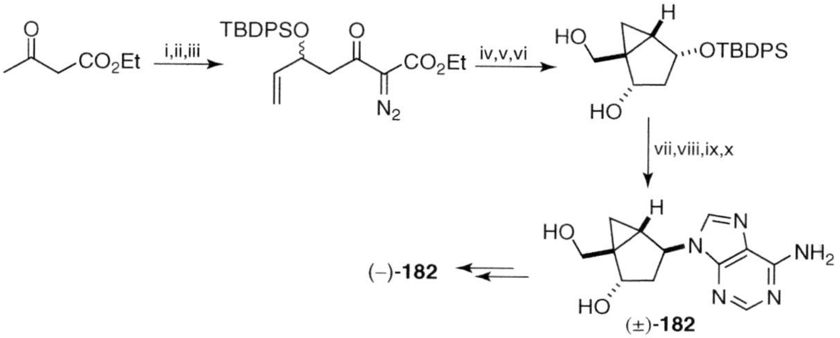

Scheme 52. *Reagents and conditions:* (i) LDA, acrolein, THF; (ii) TBDPSCl, imidazole, CH$_2$Cl$_2$, 53% (two steps); (iii) TsN$_3$, Et$_3$N, MeCN, 99%; (iv) CuSO$_4$, cyclohexane, 61%; (v) NaBH$_4$, MeOH; (vi) LiAlH$_4$, Et$_2$O, 79% (two steps); (vii) PhCOCl, 98%; (viii) TBAF, THF, 95%; (ix) 6-chloropurine, Ph$_3$P, DEAD, THF, 80%; (x) NH$_3$, MeOH.

3.4.4. Pauson–Khand Reaction.

Although it is an efficient method for the preparation of cyclopentenones, the Pauson–Khand (PK) reaction was used for the first time in a carbanucleoside synthesis in 2002. Schmalz et al. indeed reported the preparation of unusual unsaturated carbanucleosides **183** using a PK reaction of an allylpropargyl ether **184**.[121] Classical conditions requiring stoechiometric amounts of cobalt complex efficiently provided racemic cyclopentenone **185** as a single diastereomer (Scheme 53). Attempts to perform a catalytic asymmetric PK reaction unfortunately failed and could not provide enantio-enriched **185**. This drawback was, however, judiciously overcome by performing a kinetic resolution of **185** through a Corey reduction. The oxazaborolidine-catalyzed borane reduction of **185** indeed provided alcohol (+)-**186** in 58% yield and 83% ee and left (−)-**185** in 34% yield and > 99% ee. Introduction of nucleobases by a Tsuji–Trost reaction followed by the acetal hydrolytic cleavage and the reduction of the resulting aldehyde provided carbanucleosides **183**, which exhibited interesting cytotoxicities.

An asymmetric version of the Pauson–Khand reaction was eventually applied to the synthesis of carbovir and abacavir.[122] Verdaguer and Riera have indeed developed a stoichiometric enantioselective PK reaction based on the chelation of an alkyne–dicobaltcarbonyl complex by a chiral bidentate ligand followed by a thermal reaction with norbornadiene of the resulting diastereomerically pure complex

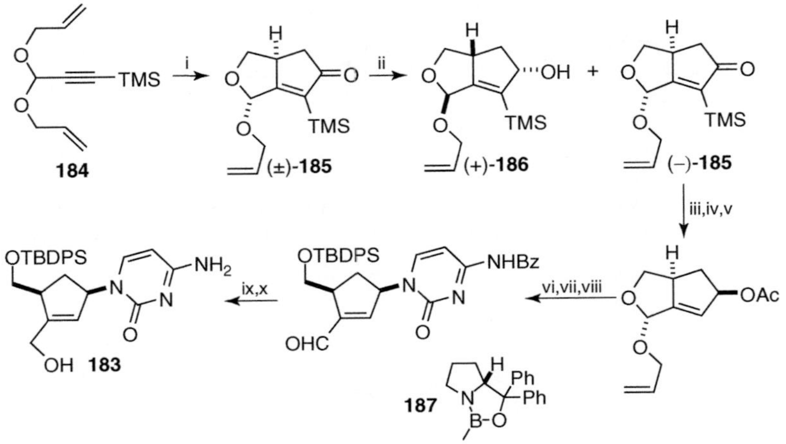

Scheme 53. *Reagents and conditions:* (i) $Co_2(CO)_8$, trimethylamine N-oxide, CH_2Cl_2, 76%, dr > 98:2; (ii) **187**, catecholborane, THF/toluene; (−)-**185**: 34%, > 99% ee; (+)-**186**: 58%, 83% ee; (iii) $NaBH_4$, $CeCl_3$, MeOH, 100%; (iv) t-BuOK, DMSO/H_2O, 87%; (v) Ac_2O, DMAP, Et_3N, CH_2Cl_2, 99%; (vi) N^4-benzoylcytosine, NaH, Pd(dba)$_2$, P(O-i-Pr)$_3$, DMSO, 84%; (vii) PPTS, acetone; (viii) TBDPSCl, pyridine, 57% (two steps); (ix) $NaBH_4$, MeOH/CH_2Cl_2, 53%; (x) NH_3, 76% (two steps).

(Scheme 54).[123] The conjugate addition to the resulting cyclopentenone **188** of a hydroxymethyl radical under photochemical conditions followed by desilylation yielded **189** as a single diastereomer. Protection of the primary alcohol followed by a retro-Diels−Alder delivered a cyclopentenone that was subsequently reduced and converted to carbonate **190**. A Tsuji−Trost coupling to introduce 2-chloro-6-aminopurine afforded compound **191**. which was converted, after reaction with cyclopropylamine or sodium hydroxide and deprotection, to abacavir and carbovir.

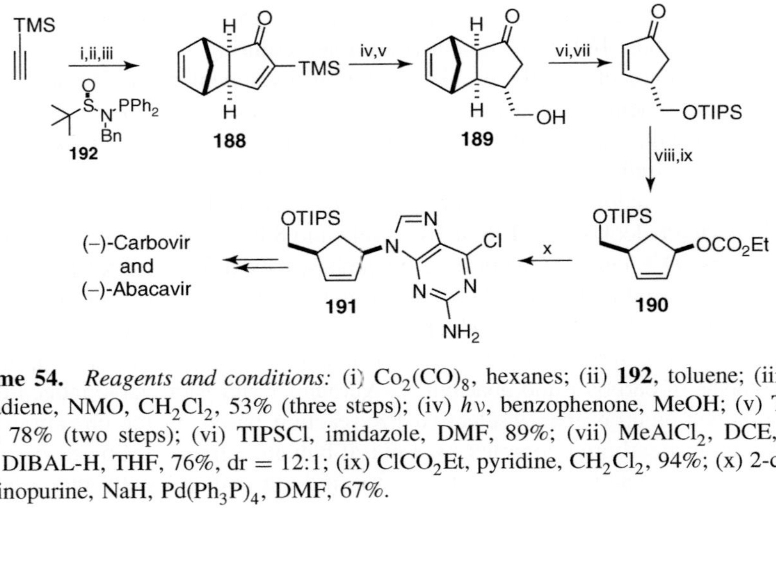

Scheme 54. *Reagents and conditions:* (i) $Co_2(CO)_8$, hexanes; (ii) **192**, toluene; (iii) norbornadiene, NMO, CH_2Cl_2, 53% (three steps); (iv) $h\nu$, benzophenone, MeOH; (v) TBAF, THF, 78% (two steps); (vi) TIPSCl, imidazole, DMF, 89%; (vii) MeAlCl$_2$, DCE, 86%; (viii) DIBAL-H, THF, 76%, dr = 12:1; (ix) ClCO$_2$Et, pyridine, CH_2Cl_2, 94%; (x) 2-chloro-6-aminopurine, NaH, Pd(Ph$_3$P)$_4$, DMF, 67%.

3.5. Synthesis Using Radical or Anionic Cyclization Reactions

3.5.1. Radical Cyclizations.
Although some works regarding the synthesis of carbocyclic pentose analogues by radical cyclization have been reported, such an approach was only rarely applied to the preparation of carba-nucleosides. Indeed, Wilcox et al., and, more recently, Gallos et al., demonstrated that the 5-exo-trig radical cyclization of carbohydrate-derived unsaturated esters **193** and **194** could efficiently provide carba-sugar **195**, but no synthesis of 5′-homocarba-nucleosides was reported through this method (Scheme 55).[124,125]

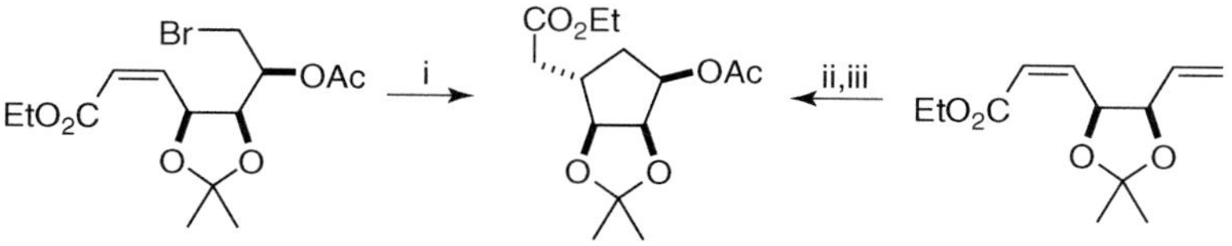

Scheme 55. *Reagents and conditions:* (i) $(n\text{-Bu})_3$SnH, AIBN, benzene, 80%, dr = 5:1; (ii) Hg(OAc)$_2$, AcOH; (iii) NaBH(OMe)$_3$, CH$_2$Cl$_2$, 53% (two steps), dr > 98:2.

Radical cyclization of carbohydrate-derived aldoximes also allowed Martinez-Grau and Marco-Contelles to prepare various aminocarba-sugars.[126] Desire and Prandi devised a very nice route to carbapentofuranoses, which involved the oxidative radical cyclization of hexose-derived iodides such as **196** (Scheme 56).[127] The diastereoselectivity of the cyclization depends strongly on the relative configurations of the starting material, which might explain why no carba-nucleosides were prepared from this method. The carbocyclic analogue of D-arabinose **197** was, however, prepared in good yield and high selectivity.

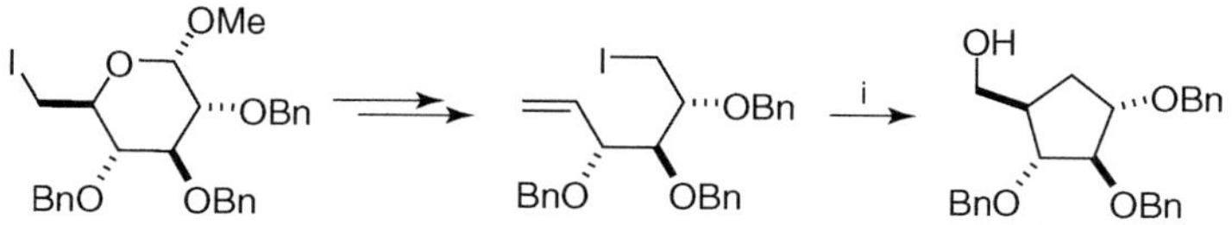

Scheme 56. *Reagents and conditions:* (i) NaOH, NaBH$_4$, Co(salen), air, EtOH, 69%, dr = 12:1.

The carbocyclic core of entecavir was prepared by a Ti(III)-mediated 5-exo-dig radical cyclization from D-glucose-derived epoxide **198**, as reported by Ziegler and Sarpong (Scheme 57).[128] A similar strategy was used by Shuto et al. to synthesize methylenecarba-nucleosides such as **199** from xanthate **200** (Scheme 57).[129]

3.5.2. Anionic Cyclizations.
Several strategies to build up the cyclopentane backbone through an intramolecular addition of a nucleophilic site to an internal electrophile were devised. Rapoport et al. and Ho et al. reported various preparations of functionalized cyclopentylamines using either Dieckmann condensations of

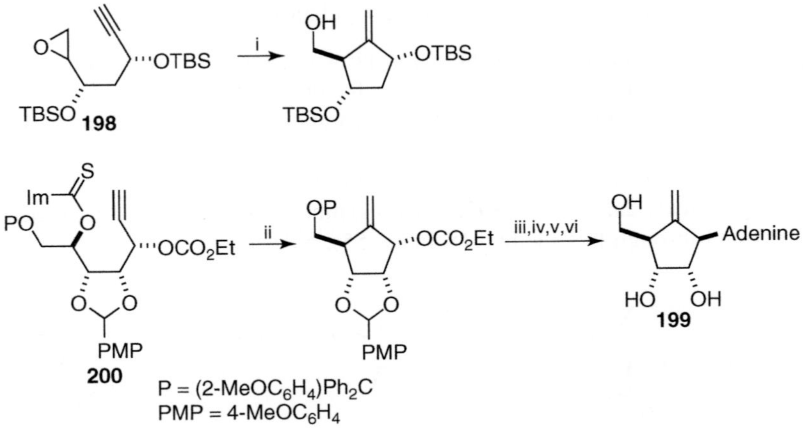

Scheme 57. *Reagents and conditions:* (i) Cp$_2$TiCl, THF, 82%, dr > 98:2. (ii) (n-Bu)$_3$SnH, AIBN, benzene, 63%, dr = 3:1; (iii) NaOMe, MeOH, THF, 93%; (iv) 6-chloropurine, Ph$_3$P, DIAD, THF, 79%; (v) NH$_3$, MeOH, 95%; (vi) AcOH, 100%.

D-gluconolactone–derived or L-aspartic acid–derived amides or an intramolecular α-alkylation of a L-serine–derived amide.[130,131] This last approach provided an entry to enantio-enriched 2-azabicyclo[2.2.1]hept-5-en-3-one derivatives such as **201** that were previously obtained by enzymatic resolution (see Section 3.1.4). Configurationally stable *N*-(9-phenyl-9-fluorenyl)-L-serinal **202** reacted with titanium homoenolate **203** to provide lactone **204** in excellent yield and good diastereoselectivity (Scheme 58). Lactamization and bromination afforded intermediate **205**, which underwent base-mediated cyclization. Subsequent deprotection and elimination eventually afforded **201**, which could be transprotected and opened reductively to the corresponding cyclopentenylamine.

A synthesis of the antiviral agent (+)-cyclaradine **87** was performed by Yoshikawa et al. using an intramolecular Henry reaction as a key step in assembling the carba-sugar backbone.[132] A D-arabinose–derived ketone was submitted to a first Henry reaction with nitromethane to provide, after elimination, conjugate reduction and hydrolysis of the ketal, intermediate **206** (Scheme 59). A fluoride-promoted intramolecular Henry reaction afforded a ring-contracted product that was subsequently subjected to an acetylation–elimination sequence to yield **207**. Adenine was introduced through a previously reported methodology involving a conjugate addition to afford **208**.[133] A radical denitrohydrogenation and full deprotection afforded the carbanucleoside targeted.

Experimental Procedure

(1′R,2′R,3′R,4′S,5′S)-N^6-Benzoyl-6-amino-9-[2′,3′-(benzyloxy)-4′-(hydroxymethyl)-5′-nitrocyclopentan-1′-yl]purine **(208)**.[133] A solution of **207** (0.500 g, 1.26 mmol) in DMF (12.5 mL) was treated with N^6-benzoyladenine (0.361 g,

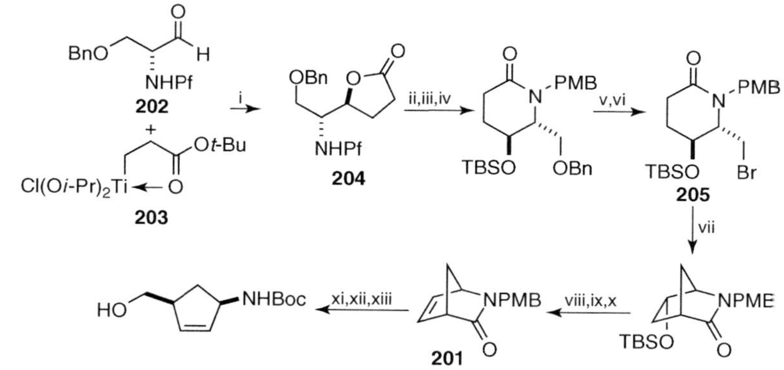

Scheme 58. *Reagents and conditions:* (i) DMA, CH$_2$Cl$_2$, 93%, dr = 6.5:1; (ii) H$_2$, Pd/C, THF/MeOH; (iii) TBSCl, imidazole, DMF, 92%; (iv) NaH, PMBBr, THF, 97%; (v) H$_2$, Pd/C, EtOH, 100%; (vi) CBr$_4$, Ph$_3$P, CH$_2$Cl$_2$, 89%; (vii) KHMDS, THF, 90%; (viii) TBAF, THF, 97%; (ix) I$_2$, Ph$_3$P, toluene, 98%; (x) DBU, xylenes, 98%; (xi) CAN, MeN/H$_2$O; (xii) Boc$_2$O, DMAP, pyridine, 59% (two steps); (xii) NaBH$_4$, MeOH, 93%.

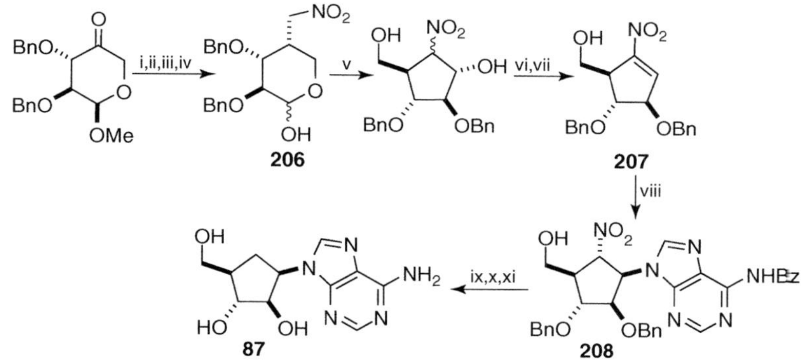

Scheme 59. *Reagents and conditions:* (i) CH$_3$NO$_2$, KF, 18-crown-6, DMF, 72%; (ii) TsOH, Ac$_2$O; (iii) NaBH$_4$, EtOH, 93% (two steps); (iv) conc. HCl, AcOH, 57%; (v) CsF, DMF, 86%, 1:1 mixture of epimers; (vi) TsOH, Ac$_2$O; (vii) pyridine, 82% (two steps); (viii) N^6-benzoyladenine, CsF, DMF, 85%; (ix) (n-Bu)$_3$SnH, AIBN, toluene, 21%; (x) NaOMe, MeOH, 88%; (xi) H$_2$, Pd/C, AcOH/MeOH, 94%.

1.51 mmol) and CsF (0.230 g, 1.51 mmol) and stirred at 0°C for 1 h. The mixture was then poured into ice–water and NaCl was added. Extraction with AcOEt, followed by a standard workup, delivered the crude product, which was purified by column chromatography (*n*-hexane/acetone 2:1) to afford pure **208** (0.679 g, 85%) as a white powder. [α]$_D^{20}$ = +65.7° (c 1.0, CHCl$_3$); HRMS (FAB): Calcd for C$_{34}$H$_{32}$N$_6$O$_7$ [M + H]$^+$ *m/z* 637.2411, found: *m/z* 637.2412. IR (KBr): 1740, 1700, 1610, 1590, 1560, 1370, 700 cm^{-1}. ^{1}H-NMR (270 MHz, CDCl$_3$): δ 9.10 (br s, 1H), 8.77 (s, 1H), 8.24 (s, 1H), 8.08–6.91 (m, 15H), 5.93 (dd, *J* = 4.9 and 10.2 Hz, 1H), 5.46 (dd, *J* = 7.9 and 10.2 Hz, 1H), 4.59 (br s, 2H), 4.39–4.25 (m,

2H), 4.33 (d, $J = 6.9$ Hz, 2H), 4.15 (d, $J = 4.9$ Hz, 1H), 3.89 (dd, $J = 1.3$ and 3.3 Hz, 1H), 3.20 (m, 1H), 2.02 (s, 3H). ^{13}C-NMR (67.5 MHz, CDCl$_3$): δ 170.4, 164.9, 152.4, 152.1, 149.5, 141.9, 136.5, 127.6, 122.4, 87.1, 80.5, 79.4, 72.3, 71.7, 63.0, 58.6, 47.6; MS (FAB$^+$) m/z 637 [M + H]$^+$.

The strategy of addition to a dielectrophile followed by cyclization on the remaining electrophilic site was also performed using tandem processes. The thiolate-initiated Michael addition/aldol cyclization developed by Tomioka et al. provided a rapid assembly of the cyclopentane backbone from tartrate-derived ester **209** (Scheme 60).[134] The sulfide moiety was used to introduce the double bond of neplanocin A, the synthesis of which was completed from **210** using proven transformations. A similar approach, involving a tandem rhodium-catalyzed hydrosilylation aldol cyclization on an aldehyde structurally close to **209**, was used by Freiría et al.[135]

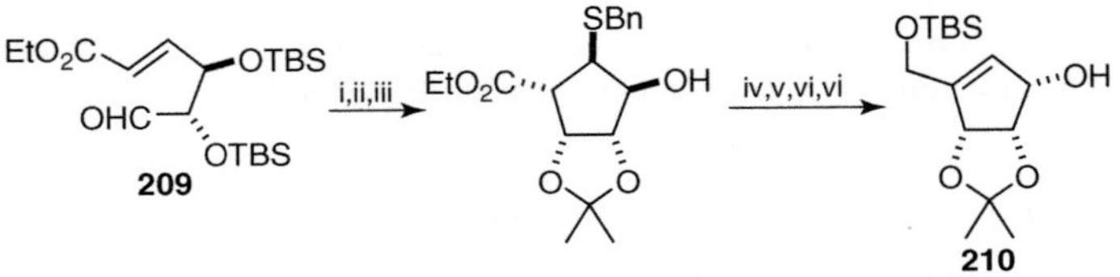

Scheme 60. *Reagents and conditions:* (i) BnSLi, THF, 62%; (ii) HF, MeCN, 99%; (iii) TsOH, acetone, 89%; (iv) NaIO$_4$, EtOH then decalin, 180°C, 68%; (v) PDC, CH$_2$Cl$_2$, 89%; (vi) DIBAL-H, toluene; (vii) TBSCl, Et$_3$N, DMAP, CH$_2$Cl$_2$, 59% (two steps).

Linclau et al. developed an intramolecular version of the linchpin coupling strategy disclosed by Smith et al., using readily available chiral bisepoxides such as **211**.[136,137] This strategy relies on the addition of the lithiated anion of silylated dithiane **212** onto one of the epoxide functions of **211** (Scheme 61). The resulting alkoxide instantly undergoes a 1,4-Brook rearrangement, regenerating the dithianyl anion and allowing the cyclization to occur via addition to the second epoxide. Cyclopentane **213** was obtained in 76% yield and was readily converted to L-carbanucleoside **214** through simple transformations. Other analogues were obtained through this method.

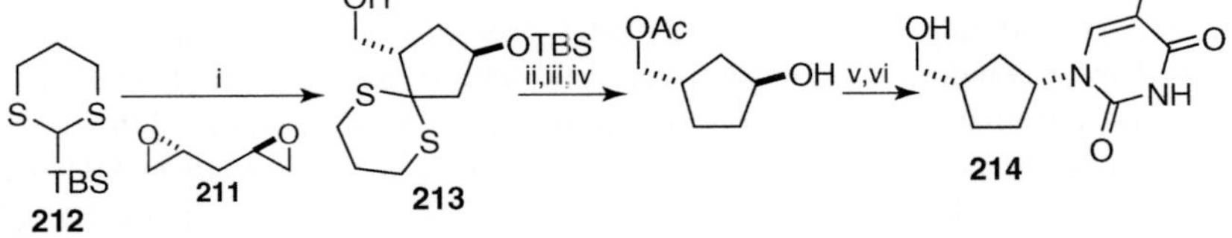

Scheme 61. *Reagents and conditions:* (i) t-BuLi, HMPA/THF, then **211**, 76%; (ii) Raney-Ni, EtOH, 72%; (iii) Ac$_2$O, pyridine; (iv) TBAF, THF, 89%; (v) N^3-benzoylthymine, DEAD, Ph$_3$P, DMF, 64%; (vi) NH$_3$, MeOH, 89%.

A very efficient and elegant approach disclosed by Castillón et al. involved an asymmetric rhodium-catalyzed intramolecular hydroacylation of aldehyde **215**.[138] Cyclopentanone **216** was obtained in 85% yield and > 95% ee and was diastereoselectively reduced, provided that the primary alcohol was deprotected prior to reduction (Scheme 62). Protection, Mitsunobu introduction of the base, and deprotection delivered carbanucleoside **217**, which through this method could also be obtained as the other enantiomer.

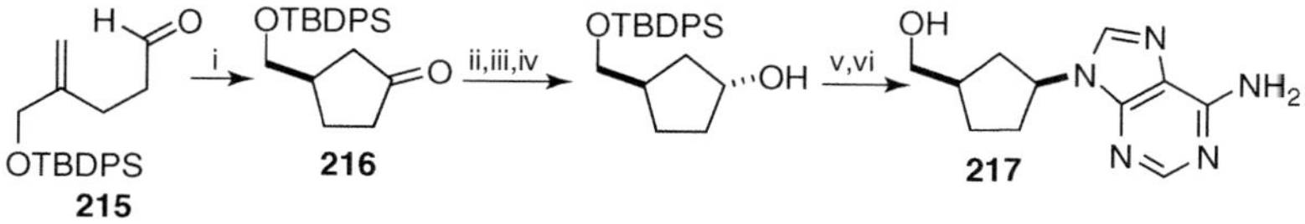

Scheme 62. *Reagents and conditions:* (i) [Rh(NBD)(*R*,*R*)-Me-Duphos], acetone, 85%, >95% ee; (ii) TBAF, THF, 80%; (iii) NaBH(OAc)$_3$, EtOAc/MeCN, 70%; (iv) TBDPSCl, Et$_3$N, DMAP, CH$_2$Cl$_2$, 65%; (v) adenine, DEAD, Ph$_3$P, dioxane, 70%; (vi) TBAF, THF, 64%.

3.6. Synthesis Using Desymmetrization Reactions

The use of asymmetric transformations to desymmetrize *meso* or achiral functionalized cyclopentanes has generated a great deal of effort for the preparation of carba-nucleosides. From the asymmetric version of the Tsuji–Trost reaction to enantioselective opening of mesoepoxides via asymmetric hydroboration of 3-substituted cyclopentadienes, these methods often allow rapid construction of the carbocyclic core of the nucleoside.

3.6.1. Asymmetric Allylic Alkylation or Amination (AAA, Tsuji–Trost) Reactions.
Numerous examples of the palladium-catalyzed introduction of heterocyclic bases at C1′ or, to a lesser extent, carbon nucleophiles at C4′ (see Schemes 13, 17, and 18) on enantio-enriched cyclopentene derivatives have been mentioned throughout this chapter. Another major contribution of Trost to the synthesis of carba-nucleosides was the application of his powerful asymmetric allylic alkylation method to *meso*-cyclopent-2-ene-1,4-diol derivatives. The desymmetrization of such substrates by adding either a nucleic base or a carbon nucleophile indeed provides enantio-enriched cyclopentenes that can be converted in a few steps to carbanucleosides.

The first approach developed by Trost et al. consisted in the asymmetric addition of an acyl anion equivalent (umpolung approach) to bisbenzoate **218** that will ultimately allow installation of the hydroxymethyl group at C4′ (Scheme 63) Lithiated phenylsulfonylnitromethane was selected and its palladium-catalyzed addition to **218** using diphosphine **219** as the chiral ligand provided isoxazoline *N*-oxide **220** in 94% yield and 96% ee.[139] Its conversion to ester **221** required three steps, and the palladium-catalyzed introduction of the heterocyclic base was performed on biscarbonate **222**. Addition of 2-amino-6-chloropurine allowed the synthesis

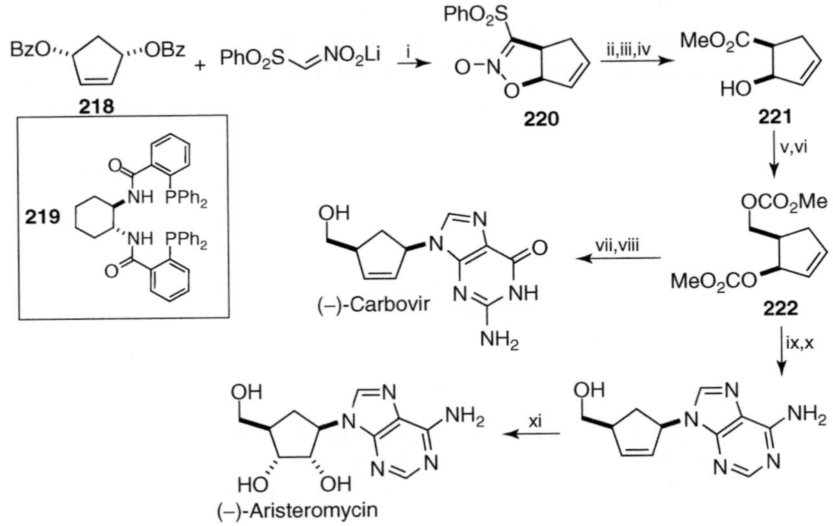

Scheme 63. *Reagents and conditions:* (i) Pd$_2$(dba)$_3$·CHCl$_3$, **219**, THF, 94%, 96% ee; (ii) SnCl$_2$·2H$_2$O, MeCN, 94%; (iii) K$_2$CO$_3$, MeOH, 91%; (iv) Mo(CO)$_6$, H$_3$BO$_3$, MeCN, H$_2$O, 84%; (v) LiAlH$_4$, Et$_2$O, 95%; (vi) *n*-BuLi, ClCO$_2$CH$_3$, THF, 98%; (vii) (allylPdCl)$_2$, Ph$_3$P, 2-amino-6-chloropurine, THF, 77%; (viii) NaOH, H$_2$O, 50%; (ix) *n*-BuLi, Pd(OAc)$_2$, P(O-*i*-Pr)$_3$, adenine, THF/DMSO, 96%; (x) NaOH, EtOH; (xi) OsO$_4$, NMO, THF/H$_2$O, 88%, dr = 2.4:1.

of (−)-carbovir after basic treatment, whereas (−)-aristeromycin was obtained by substitution with adenine followed by deprotection and a moderately diastereoselective osmylation reaction. It should be mentioned that the diastereoselectivity outcome in the dihydroxylation of such cyclopentene derivatives is often difficultly predictable.[140] Indeed, a strong dependence on the nature of the substituents of the cyclopentene and on the coordinating ability of the metal oxide involved in the reaction has been noted.[139–144]

This first strategy was recently improved by using acetoxy Meldrum's acid as an umpolung nucleophile.[141] Indeed, its addition to biscarbonate **223** was still very efficient and enantioselective, and this moiety could be converted to an ester group by simple treatment with CAN, followed by esterification with TMSCHN$_2$ (Scheme 64). A second palladium-catalyzed allylic substitution performed with trimethylsilyl azide prior to Meldrum's acid degradation allowed the preparation of ester **224**. Reduction of both functional groups and protection yielded the carbanucleoside precursor **225**.

The use of a nucleobase as a nucleophile in the desymmetrization step was explored in a second approach.[142] This strategy proved to be more than just a trivial switch in the sequence order since it required a thorough methodological study to optimize the results of the AAA reaction. Indeed, if most nucleophiles get along very well with the standard reaction conditions, the ability of nucleobase to

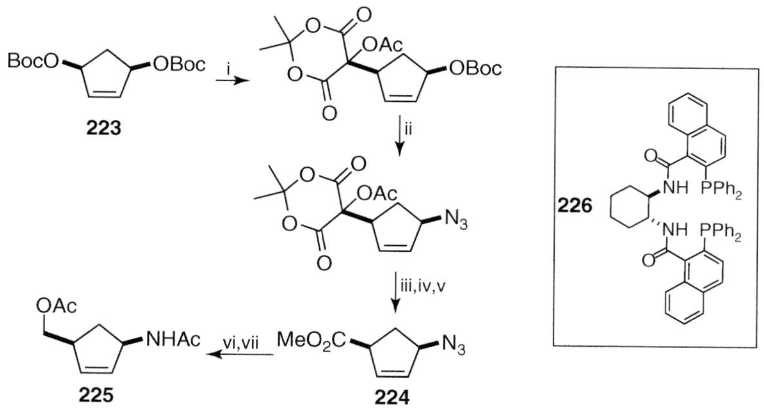

Scheme 64. *Reagents and conditions:* (i) Pd$_2$(dba)$_3$·CHCl$_3$, **226**, Cs$_2$CO$_3$, DCE, 97%, 98% ee; (ii) TMSN$_3$, Pd$_2$(dba)$_3$·CHCl$_3$, (±)-**219**, THF, 99%; (iii) LiOH, THF/H$_2$O; (iv) CAN, MeCN/H$_2$O; (v) TMSCHN$_2$, MeOH, benzene, 67% (three steps); (vi) LiAlH$_4$, THF; (vii) Ac$_2$O, DMAP, Et$_3$N, CH$_2$Cl$_2$, 91% (two steps).

coordinate palladium had a profound impact on the reaction course. The reaction conditions have been optimized for each base, and an efficient and enantioselective introduction of a guanine equivalent, of 6-chloropurine, and of an uracil precursor has finally been achieved on bisbenzoate **218** (Scheme 65). All compounds were obtained in 54–62% yield and more than 96% ee and a second-generation synthesis of (−)-carbovir was completed in four steps via compound **227**. Compound **228**, which resulted from the addition of 6-chloropurine to **218**, was afterward engaged in a second allylic substitution with phenylsulfonylnitromethane to deliver an intermediate which had previously been converted to (−)-aristeromycin and (−)-neplanocin A.[20,143] Alternatively, the desymmetrization of bisbenzoate **218** using trimethylsilyl azide as the nucleophile and the subsequent preparation of carba-nucleoside precursor **224** was also reported.[144]

Experimental Procedure

(1'R,4'S)-6-Chloro-9-[4'-(benzoyloxy)-cyclopenten-2'-yl]purine (228).[142]
Pd$_2$dba$_3$·CHCl$_3$ (650 mg, 0.63 mmol) was added to a degassed solution of ligand **230** (1.5 g, 1.90 mmol) in THF (50 mL). This solution was stirred for 15 min and then added to a degassed mixture of bisbenzoate **218** (17.5 g, 56.8 mmol), 6-chloropurine (11 g, 71.2 mmol), and triethylamine (32 mL, 230 mmol) in THF (170 mL). The reaction was stirred for 5 h at room temperature to give a clear brown solution. The mixture was concentrated and purified by flash chromatography (EtOAc/n-hexane 60:40, then 100% EtOAc, then 100% acetone) to afford 6.02 g of recovered bisbenzoate **218** together with dibenzylideneacetone, and 10.3 g (53%, 76% based on recovered **218**, 94% ee) of the desired product **228** as a solid; mp 90–92°C; $[\alpha]_D^{20}$ −106.8° (c 1.71, CH$_2$Cl$_2$); IR (CDCl$_3$): 1717,

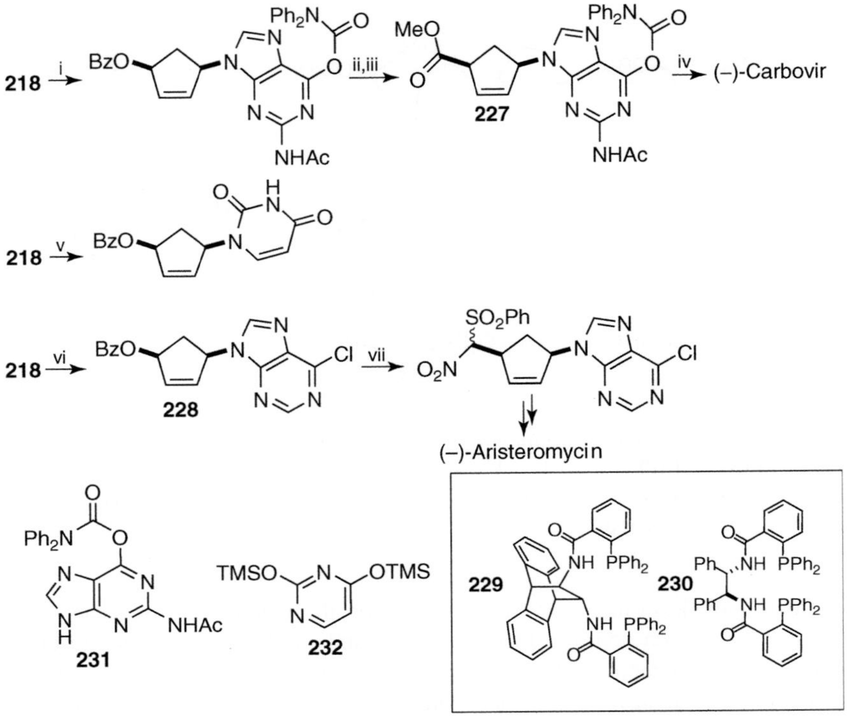

Scheme 65. *Reagents and conditions:* (i) (allylPdCl)$_2$, **229**, **231**, pempidine, THF/DMSO, 54%, 96% ee; (ii) Pd$_2$(dba)$_3$·CHCl$_3$, Ph$_3$P, PhSO$_2$CH$_2$NO$_2$, Et$_3$N, THF, 97%; (iii) *N,N,N′,N′*-tetramethylguanidine, tetrabutylammonium oxone, Na$_2$CO$_3$, MeOH/CH$_2$Cl$_2$, 71%; (iv) Ca(BH$_4$)$_2$, THF, then NH$_4$OH, 61%; (v) Pd$_2$(dba)$_3$·CHCl$_3$, ent-**219**, **232**, Et$_3$N, DMF, 55%, 96% ee; (vi) Pd$_2$(dba)$_3$·CHCl$_3$, **230**, 6-chloropurine, Et$_3$N, THF, 53%, 94% ee; (vii) Pd$_2$(dba)$_3$·CHCl$_3$, PhP$_3$, PhSO$_2$CH$_2$NO$_2$, Et$_3$N, THF, 95%.

1590, 1561, 1334 cm^{-1}. ^{1}H-NMR (300 MHz, CDCl$_3$): δ 8.78 (s, 1H), 8.28 (s, 1H), 8.02 (m, 2H), 7.60 (dt, *J* = 1.3 and 7.4 Hz, 1H), 7.46 (m, 2H), 6.57 (dt, *J* = 2.0 and 5.6 Hz, 1H), 6.30 (dd, *J* = 2.4 and 5.6 Hz, 1H), 6.04 (dt, *J* = 2.5 and 7.4 Hz, 1H), 5.87 (m, 1H), 3.25 (ddd, *J* = 7.7, 8.0 and 15.3 Hz, 1H), 2.16 (dt, *J* = 3.0, 15.3 Hz, 1H). ^{13}C-NMR (75 MHz, CDCl$_3$): δ 165.7, 151.8, 151.2, 150.9, 143.3, 136.4, 133.6, 133.3, 131.6, 129.4, 129.3, 128.5, 77.3, 57.4, 38.5; *Anal*. Calcd for C$_{17}$H$_{13}$ClN$_4$O$_2$: C, 59.92; H, 3.85; N, 16.44; found: C, 60.16; H, 4.00; N, 16.67.

3.6.2. Enantioselective Ring-Opening Reactions of meso-Cyclopentene Oxides.

The enantioselective β-deprotonation/ring-opening sequence of a *meso*-cyclopentene oxide was pioneered by Asami and applied to 4-(hydroxymethyl)cyclopent-1-ene oxide derivatives by Hodgson et al. and Asami et al.[145,146] A synthesis of (−)-carbovir was performed by Asami et al. using a proline-derived lithium amide to perform the enantioselective rearrangement

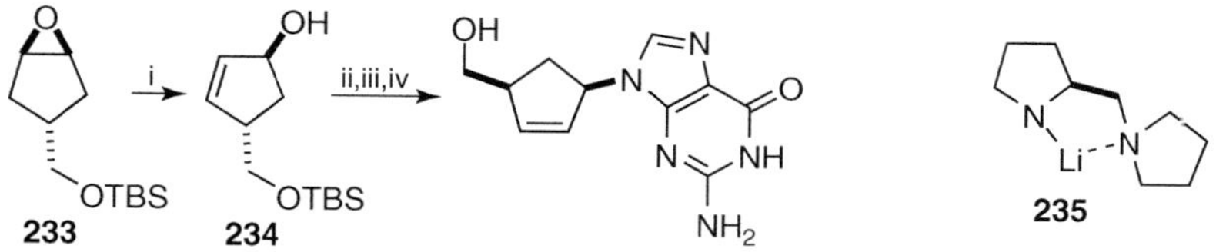

Scheme 66. *Reagents and conditions:* (i) **235**, DBU, THF, 74%, 83% ee; (ii) 2-amino-6-chloropurine, DEAD, Ph$_3$P, 1,4-dioxane, 35%; (iii) TBAF, THF, 92%; (iv) NaOH, 89%.

of epoxide **233** (Scheme 66). Allylic alcohol **234** was obtained in 74% yield and 83% ee, and the synthesis of (−)-carbovir was completed using a classical Mitsunobu approach. It should be mentioned that the elimination reaction was performed in 95% ee by Hodgson et al. on the deprotected derivative, using dilithiated norephedrine as the chiral base.[146a]

The asymmetric β-deprotonation of epoxide **236** reported by Hodgson et al. was used by Kim and Jacobson to perform the synthesis of conformationally locked carbanucleoside **237**.[147,148] After osmylation and appropriate protection, intermediate **238** was cyclized under basic conditions to provide bicyclic compound **239**, along with 30% of the oxetane derivative resulting from cyclization on the C3' hydroxy group (Scheme 67). Inversion of the C1' stereogenic center followed by nucleophilic displacement of the corresponding triflate by adenine afforded **237**. The enantioselective rearrangement of 4-aminocyclopent-1-ene oxide derivatives was also reported by O'Brien et al. and Seki and Asami.[149]

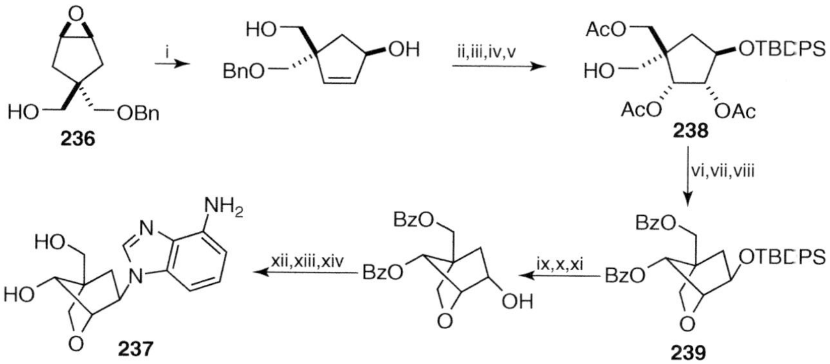

Scheme 67. *Reagents and conditions:* (i) (1*S*,2*R*)-norephedrine, *n*-BuLi, benzene, THF, 76%, 89% ee; (ii) TBDPSCl, Et$_3$N, DMAP, CH$_2$Cl$_2$, 76%; (iii) OsO$_4$, NMO, acetone/H$_2$O, 94%; (iv) Ac$_2$O, Et$_3$N, DMAP, CH$_2$Cl$_2$, 95%; (v) Pd/C, HCO$_2$H, MeOH, 87%; (vi) MsCl, Et$_3$N, CH$_2$Cl$_2$; (vii) K$_2$CO$_3$, MeOH, 89% (for two steps and two regioisomers); (viii) BzCl, Et$_3$N, CH$_2$Cl$_2$, 53% (+30% of regioisomer); (ix) TBAF, THF, 83%; (x) PDC, CH$_2$Cl$_2$; (xi) NaBH$_4$, EtOH, 87% (two steps); (xii) Tf$_2$O, DMAP, CH$_2$Cl$_2$, 83%; (xiii) adenine, K$_2$CO$_3$, 18-crown-6, DMF, 76%; (xiv) K$_2$CO$_3$, MeOH, 91%.

Jacobsen et al. applied his asymmetric ring-opening of *meso*-epoxides using TMSN$_3$ and a chiral Cr(III)complex to the synthesis of the carbocyclic core **240**.[150] The azide derivative was obtained in 95% yield and 96% ee and converted to cyclopentene **241** in four steps (Scheme 68). The latter was either osmylated to provide an aristeromycin precursor or reduced to afford **240**.

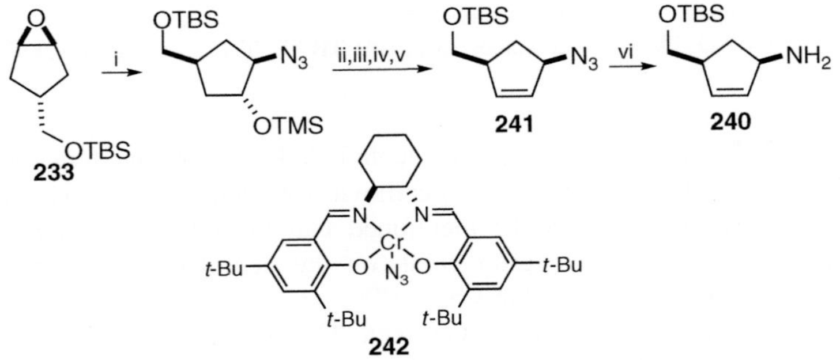

Scheme 68. *Reagents and conditions:* (i) TMSN$_3$, **242**, Et$_2$O, 95%, 96% ee; (ii) CSA, MeOH; (iii) TsCl, pyridine, 81% (two steps); (iv) NaSePh, THF, HMPA; (v) H$_2$O$_2$, pyridine, 81% (two steps); (vi) LiAlH$_4$, Et$_2$O, 88%.

3.6.3. Asymmetric Hydroboration of 3-Benzyloxycyclopenta-1,4-diene.

This reaction was first developed by Biggadike et al., inspired by the work of Partridge et al. on 3-methylcyclopenta-1,4-diene, who applied it to the synthesis of 5'-fluorocarba-nucleosides.[151,152] Several syntheses of 2'-deoxycarba-nucleosides were based on this approach, and a preparation of entecavir was performed, for example, by Bisacchi et al.[153] The action of (−)-diisopinocampheylborane on 3-benzyloxycyclopenta-1,4-diene, generated in situ from sodium cyclopentadienide and benzyl(chloromethyl)ether, yielded cyclopentene **243** in 75% yield and 96.6 to 98.8% ee. Diastereoselective epoxidation followed by ring-opening reaction with 6-benzyloxy-2-aminopurine and N-protection afforded compound **244**. Oxidation and methylenation gave intermediate **245**. which was fully deprotected to produce entecavir (Scheme 69).

Ludek and Meier performed the synthesis of various 2'-deoxycarbanucleosides through ring-opening reaction of epoxide **246** with a heterocyclic base followed by deoxygenation of the 5'-hydroxy group (Scheme 70).[13b] The same group also reported the preparation L-carbanucleosides from *ent*-**243**, obtained by a similar hydroboration with (+)-Ipc$_2$BH.[154] A second hydroboration of the corresponding mesylate followed by elimination afforded cyclopentene **248** (Scheme 70). Introduction of the heterocyclic base by a double Mitsunobu reaction followed by dihydroxylation afforded the desired L-carbanucleoside.

As mentioned earlier, many syntheses were based on this reaction and will not be described here.[155] The desymetrization of a *meso*-norbornene by asymmetric hydroboration was also reported by Mohar and Kobe for the synthesis of carbocyclic 4-deoxypyrazofurin.[156]

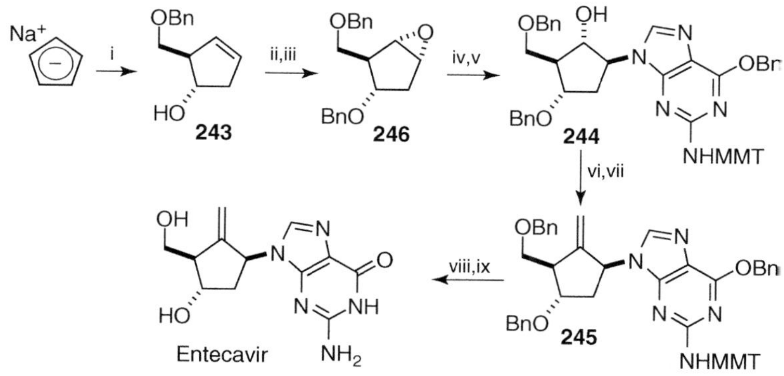

Scheme 69. *Reagents and conditions:* (i) BnOCH$_2$Cl, THF, then (−)-Ipc$_2$BH, THF, then NaOH, H$_2$O$_2$, 75%, 96.6–98.8% ee; (ii) VO(acac)$_2$, *t*-BuOOH, CH$_2$Cl$_2$; (iii) BnBr, NaH, DMF, 83% (two steps); (iv) 6-benzyloxy-2-aminopurine, LiH, DMF, 60%; (v) MMTCl, Et$_3$N, DMAP, CH$_2$Cl$_2$, 82%; (vi) Dess–Martin periodinane, *t*-BuOH, CH$_2$Cl$_2$; (vii) CH$_2$Br$_2$, Zn, TiCl$_4$, THF, 75% (two steps); (viii) HCl, THF, MeOH, 92%; (ix) BCl$_3$, CH$_2$Cl$_2$, 89%.

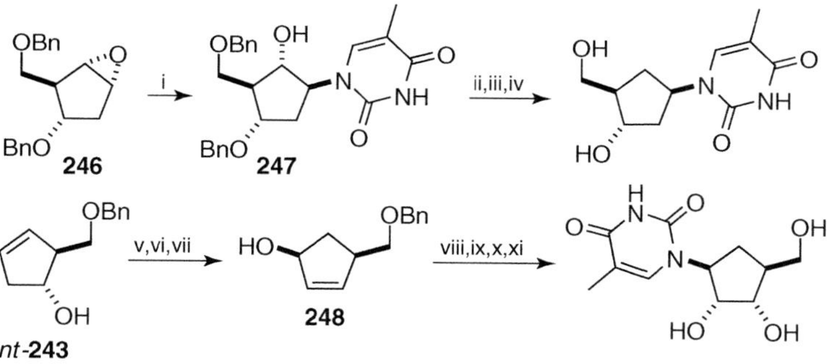

Scheme 70. *Reagents and conditions:* (i) thymine, Et$_3$Al, ultrasound, THF, 46%; (ii) (PhO)C(S)Cl, pyridine, DMAP; (iii) (*n*-Bu)$_3$SnH, AIBN, toluene, 55% (two steps); (iv) FeCl$_3$, CH$_2$Cl$_2$, 92%; (v) MsCl, Et$_3$N, THF, 100%; (vi) 9-BBN, THF, then oxone, 77%; (vii) *t*-BuOK, DMF, 82%; (viii) benzoic acid, DIAD, Ph$_3$P, Et$_2$O, 99%; (ix) N^3-benzoylthymine, DIAD, Ph$_3$P, MeCN, 62%; (x) AD-mix α, *t*-BuOH/H$_2$O, 53%; (xi) H$_2$, Pd/C, EtOH, 82%.

Experimental Procedure

(1′S,3′S,4′S,5′S)-1-[3′-(Benzyloxy)-4′-(benzyloxymethyl)-5′-hydroxycyclopent-1′-anyl]thymine (247).[13b] Triethylaluminum (1 M solution in *n*-hexane, 12.8 mL, 12.8 mmol) was added slowly to a suspension of thymidine (1.61 g, 12.8 mmol) in THF (60 mL) at room temperature and the mixture was stirred at room temperature for 1 h. Epoxide **246** (2.00 g, 6.40 mmol) was then added and the mixture was sonicated at room temperature for 48 h. The mixture was quenched with AcOH (2 mL), diluted with water (200 mL), and extracted with EtOAc (3 × 75 mL). The combined organic extracts were washed with saturated aqueous

NaHCO$_3$ (100 mL) and concentrated under reduced pressure. Purification by column chromatography on silica gel (CH$_2$Cl$_2$/MeOH, 20:1) afforded **247** (1.28 g, 46%) as a light yellow oil. ^{1}H-NMR (400 MHz, DMSO-d_6): δ 11.22 (br s, 1H), 7.60 (q, J = 1.0 Hz, 1H), 7.36–7.26 (m, 10H), 5.28 (d, J = 4.5 Hz, 1H), 4.86–4.79 (m, 1H), 4.58–4.48 (m, 4H), 3.95–3.90 (m, 2H), 3.53 (dd, J = 6.9 and 9.4 Hz, 1H), 3.46 (dd, J = 6.1 and 9.4 Hz, 1H), 2.11–2.01 (m, 3H), 1.80 (d, J = 1.0 Hz, 3H). ^{13}C-NMR (101 MHz, DMSO-d_6): δ = 164.1, 150.5, 138.8, 138.6, 138.0, 128.7, 128.6, 128.5, 128.3, 128.0, 127.8, 109.6, 77.3, 73.3, 72.4, 70.2, 53.9, 44.3, 35.9, 32.5, 12.4; HRMS (FAB): calcd for C$_{25}$H$_{28}$N$_2$O$_5$ [M + H]: m/z 437.2076; found: m/z 437.2094.

4. SYNTHESIS OF FLUORINATED OR NONCYCLOPENTYLIC CARBA-NUCLEOSIDES

The quest for more active and more selective nucleosidic drugs has prompted chemists to explore carbocyclic structures with extended modifications compared to the sole replacement of the intracyclic oxygen by a methylene group. Fluorination of various positions of the carba-sugar backbone and modification of the ring size are the most common. The synthesis of these unusual carbanucleosides is summarized below.

4.1. Fluorinated Carba-nucleosides

The use of fluorine atoms and fluoromethyl groups as surrogates to hydroxy groups or oxygen atoms has led to the development of various fluorinated carbocyclic structures. Only a few examples of such compounds are reported here; for more details, the reader is invited to consult reviews devoted to fluorinated nucleosides.[157]

4.1.1. 2′-, 3′-, 4′, or 5′-Monofluorinated Carba-nucleosides. The replacement of hydroxy groups by fluorine atoms is a long-standing strategy either to probe the interaction between an active compound and its biological target or to increase the stability of glycosidic compounds.[157b,158] Biggadike et al. were the first to prepare a fluorinated carbocyclic nucleoside, and many 2′-deoxy-2′-fluorocarba-nucleoside syntheses have been published since then.[159] Most of these compounds are prepared by nucleophilic fluorination of the corresponding alcohol, and the synthetic challenge is therefore to prepare an adequately protected precursor. Compound **249** was prepared by RCM reaction from D-ribonolactone using a route similar to those described in Section 3.3.1. Regioselective reductive opening of the benzylidene group and selective protection of the 1′-hydroxy group afforded intermediate **250** (Scheme 71). Fluorination and deprotection yielded fluorinated carba-sugar **251**. to which protected adenine was added. Neplanocin analogue **252** was then obtained after complete deprotection.

Few examples of 3′-deoxy-3′-fluorocarba-nucleosides were reported using similar synthetic approaches.[160] Fluorinated 2′,3′-didehydrocarba-nucleosides were also

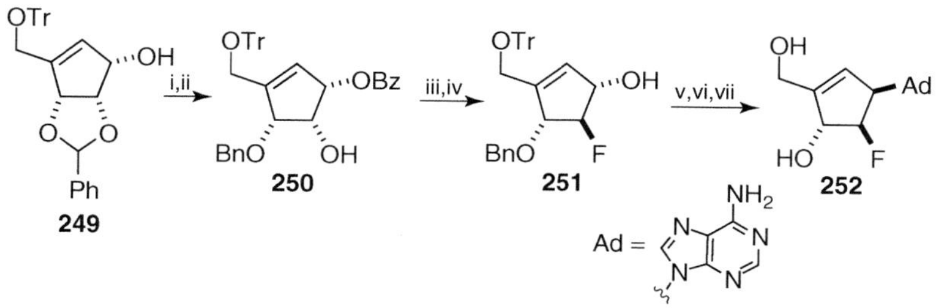

Scheme 71. *Reagents and conditions:* (i) (*i*-Bu)$_2$AlH, CH$_2$Cl$_2$, 61%; (ii) BzCl, pyridine, 78%; (iii) DAST, CH$_2$Cl$_2$, 74%; (iv) NaOH, MeOH, 77%; (v) N^6-(Boc)$_2$AdH, DIAD, Ph$_3$P, THF, 77%; (vi) BCl$_3$, CH$_2$Cl$_2$; (vii) HCl/MeOH, 77%.

prepared by difluorination of a cyclopentenone followed by a base-promoted elimination of a fluorine atom.[161] Finally, the synthesis of 4'-fluorocarba-nucleosides and 5'-deoxy-5'-fluorocarba-nucleosides was reported by Chu et al. and Schneller et al., respectively.[162,163]

4.1.2. Synthesis of Fluoroneplanocin A and Analogues.
Jeong et al. reported the preparation of fluoroneplanocin A **253**, a neplanocin A analogue with improved *S*-adenosylhomocysteine hydrolase inhibition.[164] Other analogues were prepared, but all those syntheses were based on an electrophilic fluorination reaction either from cyclopentenone **93** or from its saturated homologue.[165] Original Jeong's synthesis proceeded from **93** by iodination followed by reduction and protection to afford iodocyclopentene **254** (Scheme 72). Iodine/lithium exchange and reaction with an F$^+$ source yielded, after deprotection, intermediate **255**. Adenine was introduced by substitution of the corresponding mesylate and complete deprotection afforded **253**.

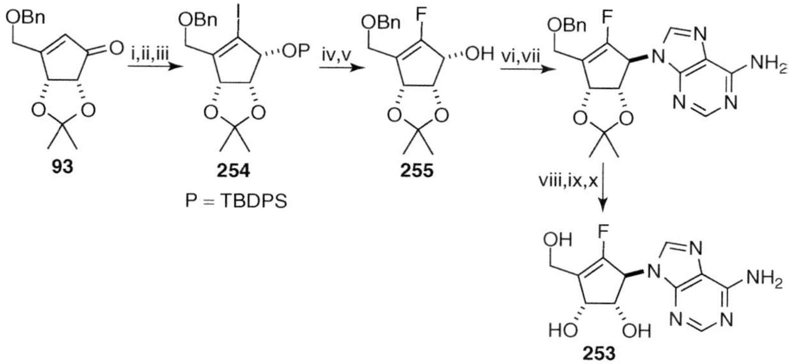

Scheme 72. *Reagents and conditions:* (i) I$_2$, CCl$_4$, pyridine, 93%; (ii) NaBH$_4$, CeCl$_3$, MeOH, 55%; (iii) TBDPSCl, imidazole, DMF, 97%; (iv) *n*-BuLi, *N*-fluorobenzenesulfonimide, THF; (v) TBAF, THF, 63% (two steps); (vi) MsCl, Et$_3$N, CH$_2$Cl$_2$, 86%; (vii) adenine, K$_2$CO$_3$, 18-crown-6, DMF, 69%; (viii) BBr$_3$, CH$_2$Cl$_2$; (ix) Ac$_2$O, pyridine; (x) NH$_3$, MeOH, 77% (three steps).

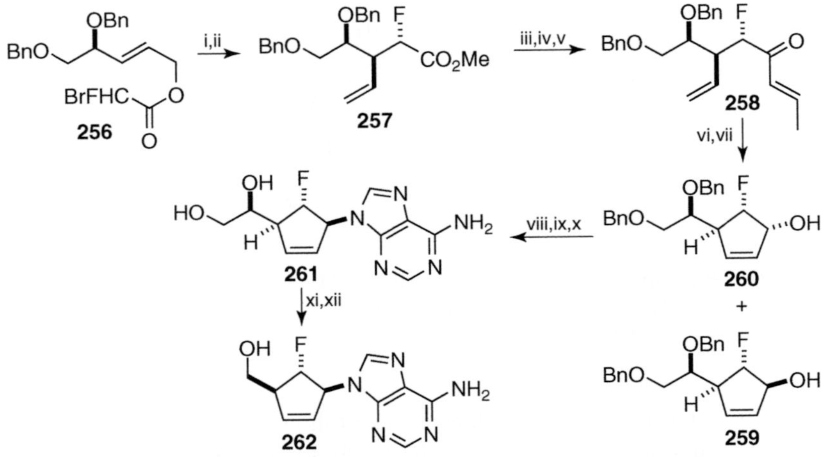

Scheme 73. *Reagents and conditions:* (i) Zn, TMSCl, MeCN; (ii) SOCl$_2$, MeOH, 45% (two steps); (iii) MeNHOMe·HCl, Me$_3$Al, CH$_2$Cl$_2$, 87%; (iv) allylMgCl, THF; (v) Et$_3$N, THF, 94%; (vi) second-generation Grubbs' catalyst, toluene, 95%; (vii) NaBH$_4$, CeCl$_3$·7H$_2$O, MeOH, 22% (**260**) + 70% (**259**); (viii) 6-chloropurine, DEAD, Ph$_3$P, THF, 55%; (ix) NH$_3$, MeOH, 80%; (x) BCl$_3$, CH$_2$Cl$_2$, 94%; (xi) NaIO$_4$, MeOH, H$_2$O; (xii) NaBH$_4$, MeOH, 79%.

4.1.3. Synthesis of Nucleosides Based on a 6'-Fluorinated Carba-sugar Scaffold.

Replacement of the intracyclic oxygen atom of carbohydrates by a fluoromethyl or difluoromethyl group has been the subject of many studies, since such moieties were expected to be better surrogates to oxygen than a methylene group would be. This strategy was applied to carbanucleosides by Biggadike and Qing for monofluorinated derivatives and by Qing and Kumamoto for difluorinated derivatives.[151,166,167] If a classical nucleophilic reaction was used by Biggadike, Qing's approach relied on a Reformatsky–Claisen rearrangement of ester **256** that afforded **257** (Scheme 73). Formation of the corresponding Weinreb amide followed by addition of an allyl-Grignard reagent and isomerization yielded RCM precursor **258**. Ring closure followed by Luche reduction of the resulting cyclopentenone afforded a mixture of epimers from which **259** was isolated as the major epimer. Since attempts to introduce nucleobases on **259** by a Tsuji–Trost reaction failed, the minor α-epimer **260** was subjected to a Mitsunobu reaction using 6-chloropurine as the nucleophile. Aminolysis and deprotection yielded **261** which was converted to adenine-derived fluorocarbanucleoside **262** by an oxidative cleavage–reduction sequence.

Using a related Claisen rearrangement, Qing reported a racemic synthesis of a difluorinated carbanucleoside, whereas two successive electrophilic fluorination reactions were involved in Kumamoto's synthesis of a difluoromethylene analogue of an anti-HIV agent.[167a,c] Several other methods of preparing difluorinated carbocyclic analogues of pentoses were disclosed, but conversion to carbanucleosides was never reported.[168]

4.2. Cyclohexenyl and Cyclohexanyl Carba-nucleosides

Following the disclosure of antiviral 1,5-anhydrohexitol nucleosides, Herdewijn extended his six-membered ring strategy to carbocyclic analogues and reported many syntheses of cyclohexanyl or cyclohexenyl carba-nucleosides.[169,170] One example is the asymmetric synthesis of 2-deoxyribo-type analogues from (R)-carvone.[170b,c] Epoxide **263**, obtained in eight steps from carvone, was converted to cyclitol **264** by elimination and hydroboration (Scheme 74). Protection and deprotection steps and mesylation led to alcohol **265** which was oxidized to the corresponding ketone and yielded cyclohexenone **266** after spontaneous elimination of the mesylate. Diastereoselective Luche reduction, Mitsunobu reaction with adenine, and deprotection afforded cyclohexenyl carbanucleoside **267**. Its saturated counterpart was also obtained by simple hydrogenation of **267**.

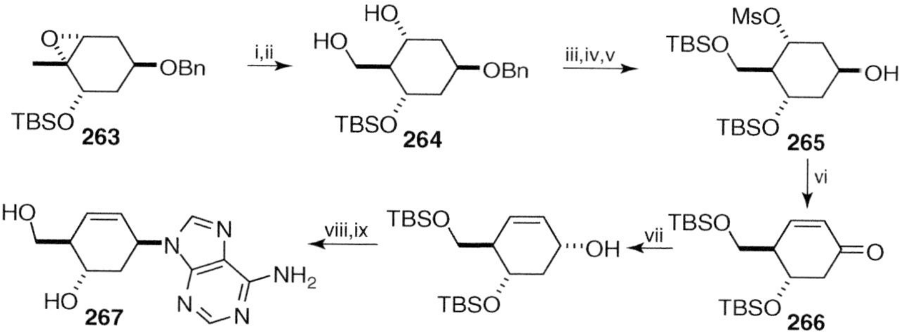

Scheme 74. *Reagents and conditions:* (i) LiTMP, Et$_2$AlCl, toluene, 71%; (ii) 9-BBN, THF, then H$_2$O$_2$, NaOH, 74%; (iii) TBSCl, imidazole, DMF, 70%; (iv) MsCl, Et$_3$N, CH$_2$Cl$_2$, 92%; (v) HCO$_2$NH$_4$, Pd/C, MeOH, 76%; (vi) MnO$_2$, CH$_2$Cl$_2$, 48%; (vii) NaBH$_4$, CeCl$_3$·7H$_2$O, MeOH, 91%; (viii) adenine, DEAD, PPh$_3$, dioxane; (ix) TFA/H$_2$O, 54% (two steps).

The synthesis of many structures has been reported since this pioneering work, including racemic syntheses, asymmetric synthesis through substrate-controlled reactions, or enantioselective chemoenzymatic preparations.[171–173] The latter approach is illustrated by the straightforward synthesis of cyclohexenyl carbanucleoside **268** from enzymatically resolved diol **269** (Scheme 75). Protection, epoxidation, and rearrangement of **269** provided allylic alcohol **270** which was readily converted to **268** by a Pd-catalyzed introduction of adenine and deprotection.

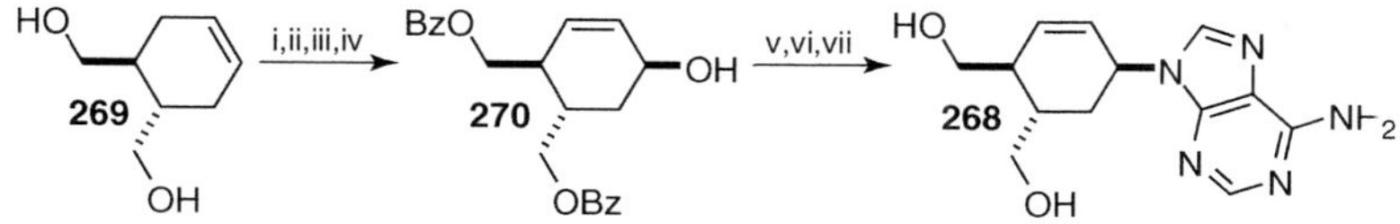

Scheme 75. *Reagents and conditions:* (i) BzCl, pyridine, 98%; (ii) *m*-CPBA, CH$_2$Cl$_2$, 87%; (iii) TMSOTf, DBU, toluene; (iv) 2 M HCl, MeOH, 86% (two steps); (v) Ac$_2$O, pyridine, 96%; (vi) adenine, NaH, Pd(Ph$_3$P)$_4$, DMF; (vii) NH$_3$, MeOH, 69% (two steps).

4.3. Cyclobutyl and Cyclopropyl Carba-nucleosides

The potent anti-HIV activity of oxetanocin, a four-membered ring nucleoside, generated considerable effort toward the synthesis of analogues, especially its carbocyclic analogue, cyclobut-A. Ichikawa et al. reported the first preparation of cyclobut-A, which is still the most efficient and elegant synthesis of this analogue.[174] An enantioselective [2 + 2]-cycloaddition between acrylate derivative **271** and 1,1-bis(methylthio)ethylene mediated by a chiral titanium catalyst afforded the cyclobutane scaffold **272** in excellent yield and enantioselectivity (Scheme 76). Cleavage of the oxazolidinone moiety followed by reduction of both esters and release of the ketone function led to **273**. Reduction of the ketone unfortunately led to the undesired epimer and required a Mitsunobu inversion to obtain the desired alcohol **274**. Mesylation followed by a substitution reaction with adenine or protected guanine afforded cyclobut-A or cyclobut-G, respectively. Another asymmetric synthesis based on a related diastereoselective [2 + 2]-cycloaddition reaction was also reported.[175]

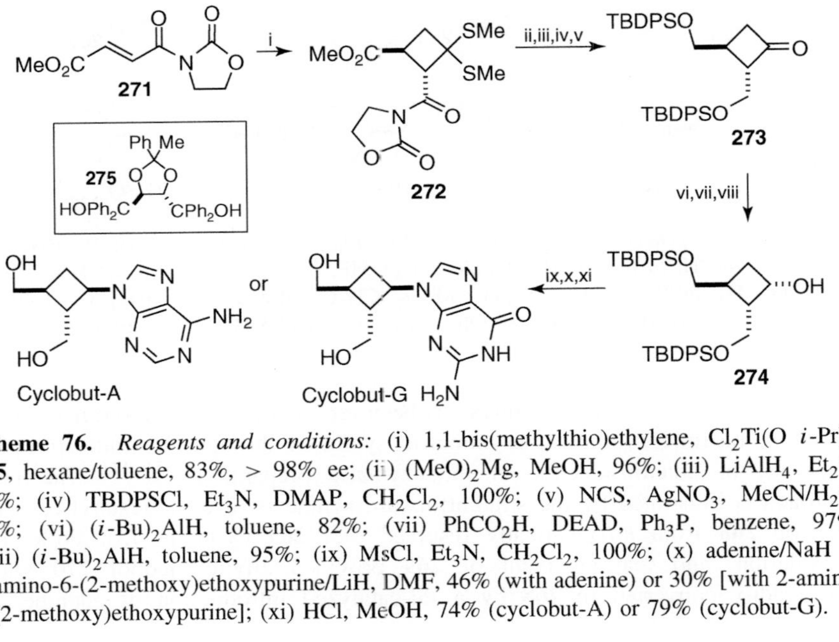

Scheme 76. *Reagents and conditions:* (i) 1,1-bis(methylthio)ethylene, $Cl_2Ti(O\ i\text{-Pr})_2$, **275**, hexane/toluene, 83%, > 98% ee; (ii) $(MeO)_2Mg$, MeOH, 96%; (iii) $LiAlH_4$, Et_2O, 99%; (iv) TBDPSCl, Et_3N, DMAP, CH_2Cl_2, 100%; (v) NCS, $AgNO_3$, $MeCN/H_2O$, 93%; (vi) $(i\text{-Bu})_2AlH$, toluene, 82%; (vii) $PhCO_2H$, DEAD, Ph_3P, benzene, 97%; (viii) $(i\text{-Bu})_2AlH$, toluene, 95%; (ix) MsCl, Et_3N, CH_2Cl_2, 100%; (x) adenine/NaH or 2-amino-6-(2-methoxy)ethoxypurine/LiH, DMF, 46% (with adenine) or 30% [with 2-amino-6-(2-methoxy)ethoxypurine]; (xi) HCl, MeOH, 74% (cyclobut-A) or 79% (cyclobut-G).

An enantioselective synthesis of cyclobut-A, based on the aforementioned methodology, was also reported by Brown and Hegedus (see Section 3.4.2). In that case, cyclobutanone **175** (Scheme 49, Section 3.4.2) was, of course, not ring-expanded but converted in 10 steps to the target.[176] Many racemic syntheses of this compound or of related analogues were also published.[177] Finally, enantioselective chemoenzymatic approaches were reported, such as the Chu synthesis of spiro[2.3]hexane compounds **276**.[178] Compound **277** was indeed obtained in enantio-enriched form, thanks to enzymatic resolution of a precursor, and converted in a few classical steps to **276** (Scheme 77).

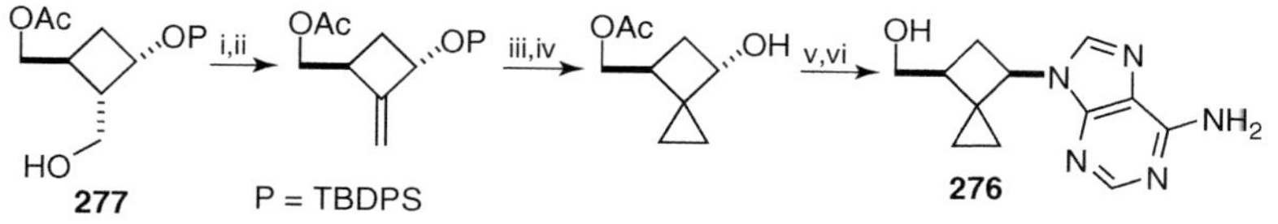

Scheme 77. *Reagents and conditions:* (i) MeP(OPh)$_3$I, DMF, 80%; (ii) DBU, THF, 68%; (iii) TBAF, THF, 95%; (iv) Et$_2$Zn, CH$_2$I$_2$, Et$_2$O, 53%; (v) 6-chloropurine, DIAD, Ph$_3$P, THF, 63%; (vi) NH$_3$, MeOH, 92%.

The first racemic syntheses of cyclopropyl carba-nucleosides, such as the one related to oxetanocin reported by Katagiri et al.,[179] were soon followed by an asymmetric approach to these structures. One example is the synthesis of uracil-, thymine-, and cytosine-bearing nucleoside analogues **278** and **279** from D-glyceraldehyde achieved by Chu et al.[180] Cyclopropane **280** was obtained in three steps via a Wittig olefination, a reduction of the ester group, and a diastereoselective Simmons–Smith cyclopropanation (Scheme 78). Oxidation of the primary alcohol to the acid and conversion to the corresponding acylazide was followed by a Curtius rearrangement, affording an isocyanate that was immediately trapped by ammonia to yield cyclopropylurea **281**. The uracil or thymine base was constructed step by step from **281** by addition to the appropriate 3-methoxyacryloyl chloride followed by a NH$_4$OH-mediated cyclization and removal of the isopropylidene protecting group by HCl. The resulting compounds **282** and **283** were converted to **278** and **279**, respectively, by an oxidative cleavage–reduction sequence. The cytosine derivative was obtained in two steps from **278** and the adenine and guanine analogues were prepared using a similar route.

The same approach was used in the preparation of related compounds, either homologated of one carbon on the lateral chain or belonging to the L-series.[181] Other asymmetric syntheses relied on the enzymatic resolution of a cyclopropane carboxylate or on other approaches from the chiral pool.[182]

Experimental Procedure

(1'S,2'R)-1-{2'-[(1 S)-1,2-Dihydroxyethyl]cyclopropyl}thymine (282).[180]
3-Methoxyacryloyl chloride (1.35 g, 10 mmol) was added to a solution of urea **281** (1.0 g, 5 mmol) in CH$_2$Cl$_2$ (24 mL) and pyridine (12 mL) and the mixture was stirred at room temperature for 24 h, poured into ice–water (40 mL), and extracted with CHCl$_3$ (3 × 40 mL). The combined organic layers were dried over MgSO$_4$, evaporated, and purified by silica gel chromatography (toluene/EtOH, 10:1), affording the intermediate *N*-acyl urea (2.02 g). A mixture of the latter, 30% aqueous NH$_4$OH (5 mL), and EtOH (50 mL) was heated at 85–90°C for 5 h and the solvents were then evaporated. The residue was dissolved in MeOH (30 mL),

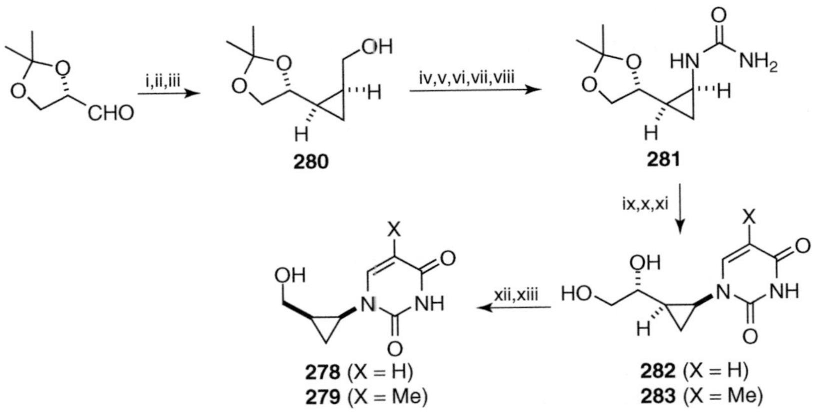

Scheme 78. *Reagents and conditions:* (i) Ph$_3$P=CHCO$_2$Me, MeOH, 81%; (ii) (*i*-Bu)$_2$AlH, CH$_2$Cl$_2$, 84%; (iii) ICH$_2$Cl, Et$_2$Zn, DCE, 70%; (iv) NaIO$_4$, RuO$_2$, KHCO$_3$, MeCN/CHCl$_3$/H$_2$O; (v) ClCO$_2$Et, Et$_3$N, acetone; (vi) NaN$_3$, acetone/H$_2$O; (vii) toluene, 100°C; (viii) NH$_3$, Et$_2$O, 38% (five steps); (ix) 3-methoxyacryloyl chloride or 3-methoxy-2-methylacryloyl chloride, pyridine, CH$_2$Cl$_2$; (x) NH$_4$OH, EtOH; (xi) 12 N HCl, MeOH, 41% (X = H, three steps) or 60% (X = Me, three steps); (xii) NaIO$_4$, MeOH/H$_2$O; (xiii) NaBH$_4$, MeOH/H$_2$O, 95% for **278**, 85% for **279**.

12 N HCl (0.5 mL) was added, and the mixture was stirred at room temperature overnight. Neutralization with Et$_3$N, evaporation, and chromatography on a silica gel column (CHCl$_3$/MeOH, 100:8) afforded **282** as a white solid (0.68 g, 60%); mp 161–162.5°C; $[\alpha]_D^{24}$ −106.4° (c 0.43, MeOH). ^{1}H-NMR (DMSO-d_6): δ 11.18 (s, 1H), 7.43 (s, 1H), 4.48 (t, J = 5.7 Hz, 1H), 4.43 (d, J = 5.2 Hz, 1H), 3.36 (ddd, J = 4.8, 5.7 and 11.4 Hz, 1H), 3.27 (ddd, J = 5.7, 8.4 and 11.4 Hz, 1H), 3.08 (m, 2H), 1.73 (s, 3H), 1.19 (m, 2H), 0.95 (ddd, J = 5.4, 6.4 and 8.4 Hz, 1H). *Anal.* Calcd for C$_{10}$H$_{14}$O$_4$N$_2$: C, 53.09; H, 6.24; N, 12.38; found: C, 53.10; H, 6.23; N, 12.37.

5. CONCLUDING REMARKS

Disclosure of the first carbocyclic analogues of nucleosides and of their antibiotic and antitumoral activities has thus attracted a great deal of interest toward such structures. This overview of the literature concerning this topic exhibited only a small part of the myriads of methods and structures that have been investigated. Their asymmetric synthesis itself has been the subject of an incredible number of studies, without mentioning the numerous structural variations introduced on the carbocyclic backbone or the nucleobase. But the most striking feature of this overwhelming literature is that despite the number of molecules prepared and tested and despite the number of SAR parameters that have already been investigated, the need for new, more active and more selective compounds is still very high. The potency of such structures toward viral diseases or cancers will certainly continue to generate research in this area.

ABBREVIATIONS

AIBN	aza-bis(isobutyronitrile)
9-BBN	9-borabicyclo[3.3.1]nonane
Bn	benzyl
Boc	*tert*-butoxycarbonyl
BOM	benzyloxymethyl
Bz	benzoyl
CAN	cerium(IV) ammonium nitrate
Cbz	benzyloxycarbonyl
m-CPBA	metachloroperbenzoic acid
CSA	camphorsulfonic acid
DAST	diethylaminosulfur trifluoride
DBU	1,8-diazabicycloundec-7-ene
DCC	dicyclohexylcarbodiimide
DCE	1,2-dichloroethane
DDQ	2,3-dichloro-5,6-dicyano-1,4-benzoquinone
DEAD	diethylazodicarboxylate
DIAD	diisopropylazodicarboxylate
DMA	*N,N*-dimethylacetamide
DMAP	4-(*N,N*-dimethylamino)pyridine
DMF	*N,N*-dimethylformamide
DMPU	*N,N'*-dimethyl-*N,N'*-propylene urea
DMSO	dimethyl sulfoxide
EDC	1-ethyl-3-(3-dimethylaminopropyl)carbodiimide
HMDS	hexamethyldisilazane
LDA	lithium di(isopropyl)amide
MMT	monomethoxytrityl
Ms	mesyl
NCS	*N*-chlorosuccinimide
NIS	*N*-iodosuccinimide
NMO	*N*-methylmorpholine *N*-oxide
PCC	pyridinium chlorochromate
PDC	pyridinium dichromate
Pht	phtalimido
PMB	*p*-methoxybenzyl
PPTS	pyridine *p*-toluenesulfonate
Py	pyridine
TBAF	tetrabutylammonium fluoride
TBDPS	*tert*-butyldiphenylsilyl
TBS	*tert*-butyldimethylsilyl

TEMPO	2,2,6,6-tetramethylpiperidinyl-1-oxy
Tf	trifluoromethanesulfonyl (triflate)
TFA	trifluoroacetic acid
TFAA	trifluoroacetic anhydride
THF	tetrahydrofurane
THP	tetrahydropyranyl
TIP	tri(isopropyl)silyl
TMP	2,2,6,6-tetramethylpiperidine
TMS	trimethylsilyl
TPAP	tetrapropylammonium perruthenate
Tr	triphenylmethyl (trityl)
Ts	tosyl

REFERENCES

1. (a) Mathé, C.; Gosselin, G. *Antivir. Res.* **2006**, *71*, 276–281. (b) Matsuda, A.; Sasaki, T. *Cancer Sci.* **2004**, *95*, 105–111.
2. (a) Kusaka, T.; Yamamoto, H.; Shibata, M.; Muroi, M.; Kishi, T.; Mizuno, K. *J. Antibiot.* **1968**, *21*, 255–271. (b) Yaginuma, S.; Muto, N.; Tsujino, M.; Sudate, Y.; Hayashi, M.; Otani, M. *J.Antibiot.* **1981**, *34*, 359–366.
3. Vince, R.; Hua, M. *J.Med. Chem.* **1990**, *33*, 17–21.
4. (a) Daluge, S. M. U.S. Patent 5,034,394. **1991** (b) Crimmins, M. T.; King, B. W. *J. Org. Chem.* **1996**, *61*, 4192–4193.
5. Bisacchi, G. S.; Chao, S. T.; Bachard, C.; Daris, J. P.; Innaimo, S.; Jacobs, G. A.; Kocy, O.; Lapointe, P.; Martel, A.; Merchant, Z.; Slusarchyk, W. A.; Sundeen, J. E.; Young, M. G.; Colonno, R.; Zahler, R. *Bioorg. Med. Chem. Lett.* **1997**, *7*, 127–132.
6. Crimmins, M. T. *Tetrahedron* **1998**, *54*, 9229–9272.
7. For previous reviews on this topic, see (a) Borthwick, A. D.; Biggadike, K. *Tetrahedron* **1992**, *48*, 571–623. (b) Agrofoglio, L.; Suhas, E.; Farese, A.; Condom, R.; Challand, S. R.; Earl, R. A.; Guedj, R. *Tetrahedron* **1994**, *50*, 10611–10670.
8. (a) Tseng, C. K. H.; Marquez, V. E. *Tetrahedron Lett.* **1985**, *26*, 3669–3672. (b) Marquez, V. E.; Lim, M.-I.; Tseng, C. K.-H.; Markovac, A.; Priest, M. A.; Khan, M. S.; Kaskar, B. *J.Org. Chem.* **1988**, *53*, 5709–5714.
9. Moon, H. R.; Kim, H. O.; Lee, K. M.; Chun, M. W.; Kim, J. H.; Jeong, L. S. *Org. Lett.* **2002**, *4*, 3501–3503.
10. Paoli, M. L.; Piccini, S.; Rodriquez, M.; Sega, A. *J.Org. Chem.* **2004**, *69*, 2881–2883.
11. (a) Madhavan, G. V. B.; Martin, J. C. *J.Org. Chem.* **1986**, *51*, 1287–1293. (b) Biggadike, K.; Borthwick, A. D.; Exall, A. M.; Kirk, B. E.; Roberts, S. M.; Youds, P. *J. Chem. Soc. Chem. Commun.* **1987**, 1083–1084. (c) Biggadike, K.; Borthwick, A. D.; Exall, A. M. *J. Chem. Soc. Chem. Commun.* **1990**, 458–459. (d) Baumgartner, H.; Marschner, C.; Pucher, R.; Griengl, H. *Tetrahedron Lett.* **1991**, *32*, 611–614.
12. Yin, X.-Q.; Schneller, S. W. *Tetrahedron Lett.* **2005**, *46*, 1927–1929.
13. For other examples, see (a) Urata, H.; Miyagoshi, H.; Yumoto, T.; Akagi, M. *J. Chem. Soc. Perkin Trans. 1* **1999**, 1833–1838. (b) Ludek, O. R.; Meier, C. *Synthesis* **2003**, 2101–2109.

14. Ruediger, E.; Martel, A.; Meanwell, N.; Solomon, C.; Turmel, B. *Tetrahedron Lett.* **2004**, *45*, 739–742.
15. Byun, H.-S.; He, L.; Bittman, R. *Tetrahedron* **2000**, *56*, 7051–7091.
16. (a) Jeong, L. S.; Marquez, V. E. *Tetrahedron Lett.* **1996**, *37*, 2353–2356. (b) Moon, H. R.; Kim, H. O.; Chun, M. W.; Jeong, L. S. *J.Org. Chem.* **1999**, *64*, 4733–4741. (c) Choi, M. J.; Chandra, G.; Lee, H. W.; Hou, X.; Choi, W. J.; Phan, K.; Jacobson, K. A.; Jeong, L. S. *Org. Biomol. Chem.* **2011**, *9*, 6955–6962.
17. Yin, X.-Q.; Li, W.-K.; Schneller, S. W. *Tetrahedron Lett.* **2006**, *47*, 9187–9189.
18. Ludek, O. R.; Meier, C. *Eur. J. Org. Chem.* **2006**, 941–946.
19. Song, G. Y.; Paul, V.; Choo, H.; Morrey, J.; Sidwell, R. W.; Schinazi, R. F.; Chu, C. K. *J.Med. Chem.* **2001**, *44*, 3985–3993.
20. Trost, B. M.; Kuo, G.-H.; Benneche, T. *J.Am. Chem. Soc.* **1988**, *110*, 621–622.
21. Hong, J. H.; Shim, M. J.; Ro, B. O.; Ko, O. H. *J.Org. Chem.* **2002**, *67*, 6837–6840.
22. Traube, W.; Schottländer, F.; Goslich, C.; Peter, R.; Meyer, F. A.; Schlüter, H.; Steinbach, W.; Bredow, K. *Liebigs Ann. Chem.* **1923**, *432*, 266–296.
23. Blanco, J. M.; Caamaño, O.; Fernández, F.; Rodríguez-Borges, J. E.; Balzarini, J.; De Clercq, E. *Chem. Pharm. Bull.* **2003**, *51*, 1060–1063.
24. Bhushan, R. G.; Vince, R. *Bioorg. Med. Chem.* **2002**, *10*, 2325–2333.
25. Zhang, H.; Schinazi, R. F.; Chu, C. K. *Bioorg. Med. Chem.* **2006**, *14*, 8314–8322.
26. Shaw, G.; Warrener, R. N. *J. Chem. Soc.* **1958**, 157–161.
27. Cookson, R. C.; Dudfield, P. J.; Scopes, D. I. C. *J. Chem. Soc. Perkin Trans. 1* **1986**, 393–398.
28. Takahashi, T.; Kotsubo, H.; Koizumi, T. *Tetrahedron: Asymmetry* **1991**, *2*, 1035–1040.
29. Chun, B. K.; Chu, C. K. *Tetrahedron Lett.* **1999**, *40*, 3309–3312.
30. Chun, B. K.; Song, G. Y.; Chu, C. K. *J.Org. Chem.* **2001**, *66*, 4852–4858.
31. (a) Tuncbilek, M.; Schneller, S. W. *Bioorg. Med. Chem.* **2003**, *11*, 3331–3334. (b) Rao, J. R.; Schinazi, R. F.; Chu, C. K. *Bioorg. Med. Chem.* **2007**, *15*, 839–846.
32. Kuang, R.; Ganguly, A. K.; Chan, T.-Z.; Pramanik, B. N.; Blythin, D. J.; McPhail, A. T.; Saksena, A. K. *Tetrahedron Lett.* **2000**, *41*, 9575–9579.
33. (a) Broggi, J.; Kumamoto, H.; Berteina-Raboin, S.; Nolan, S. P.; Agrofoglio, L. A. *Eur. J. Org. Chem.* **2009**, 1880–1888. (b) Cho, J. H.; Bernard, D. L.; Sidwell. R. W.; Kern, E. R.; Chu, C. K. *J.Med. Chem.* **2006**, *49*, 1140–1148.
34. (a) Arumugham, B.; Kim, H.-J.; Prichard, M. N.; Kern, E. R.; Chu, C. K. *Bioorg. Med. Chem. Lett.* **2006**, *16*, 285–287. (b) Kim, H.-J.; Sharon, A.; Bal, C.; Wang, J.; Allu, M.; Huang, Z.; Murray, M. G.; Bassit, L.; Schinazi, R. F.; Korba, B.; Chu, C. K. *J.Med. Chem.* **2009**, *52*, 206–213. (c) Lee, H. W.; Nam, S. K.; Choi, W. J.; Kim, H. O.; Jeong, L. S. *J.Org. Chem.* **2011**, *76*, 3557–3561.
35. (a) Cowart, M.; Bennet, M. J.; Kerwin, J. F., Jr., *J. Org. Chem.* **1999**, *64*, 2240–2249. (b) Saito, Y.; Nakamura, M.; Ohno, T.; Chaicharoenpong, C.; Ichikawa, E.; Yamamura, S.; Kato, K.; Umezawa, K. *J. Chem. Soc. Perkin Trans. 1* **2001**, 298–304.
36. (a) Madhavan, G. V. B.; Martin, J. C. *J.Org. Chem.* **1986**, *51*, 1287–1293. (b) Chen, J.; Grim, M.; Rock, C.; Chan, K. *Tetrahedron Lett.* **1989**, *30*, 5543–5546. (c) Hildbrand, S.; Troxler, T.; Scheffold, R. *Helv. Chim. Acta* **1994**, *77*, 1236–1240. (d) Burlina, F.; Favre, A.; Fourrey, J.-L.; Thomas, M. *Bioorg. Med. Chem. Lett.* **1997**, *7*, 247–250.
37. Ferrero, M.; Gotor, V. *Chem. Rev.* **2000**, *100*, 4319–4347.

38. (a) Deardorff, D. R.; Windham, C. Q.; Craney, C. L. *Org. Synth.* **1996**, *73*, 25–30. (b) Deardorff, D. R.; Mathews, A. J.; McMeekin, D. S.; Craney, C. L. *Tetrahedron Lett.* **1986**, *27*, 1255–1256.
39. Deardorff, D. R.; Savin, K. A.; Justman, C. J.; Karanjawala, Z. E.; Sheppeck, J. E.; Hager, D. C.; Aydin, N. *J.Org. Chem.* **1996**, *61*, 3616–3622.
40. Nokami, J.; Matsuura, H.; Nakasima, K.; Shibata, S. *Chem. Lett.* **1994**, 1071–1074.
41. Siddiqi, S. M.; Chen, X.; Schneller, S. W.; Ikeda, S.; Snoeck, R.; Andrei, G.; Balzarini, J.; De Clercq, E. *J.Med. Chem.* **1994**, *37*, 551–554.
42. (a) Watson, T. J. N.; Curran, T.; Hay, D. A.; Shah, R. S.; Wenstrup, D. L.; Webster, M. E. *Org. Process Res. Dev.* **1998**, *2*, 357–365. (b) Roy, A.; Schneller, S. W.; Keith, K. A.; Hartline, C. B.; Kern, E. R. *Bioorg. Med. Chem.* **2005**, *13*, 4443–4449. (c) Boojamra, C. J.; Parrish, J. P.; Sperandio, D.; Gao, Y.; Petrakovsky, O. V.; Lee, S. K.; Markevitch, D. Y.; Vela, J. E.; Laflamme, G.; Chen, J. M.; Ray, A. S.; Barron, A. C.; Sparacino, M. L.; Desai, M. C.; Kim, C. U.; Cihlar, T.; Mackman, R. L. *Bioorg. Med. Chem.* **2009**, *17*, 1739–1746.
43. Mulvihill, M. J.; Gage, J. L.; Miller, M. J. *J. Org. Chem.* **1998**, *63*, 3357–3363.
44. Li, F.; Brogan, J. B.; Gage, J. L.; Zhang, D.; Miller, M. J. *J. Org. Chem.* **2004**, *69*, 4538–4540.
45. (a) Jiang, M. X.-W.; Jin, B.; Gage, J. L.; Priour, A.; Savela, G.; Miller, M. J. *J. Org. Chem.* **2006**, *71*, 4164–4169. (b) Cesario, C.; Tardibono, L. P., Jr.; Miller, M. J. *Tetrahedron Lett.* **2010**, *51*, 3050–3052.
46. (a) Tardibono, L. P., Jr.; Patzner, J.; Cesario, C.; Miller, M. J. *Org. Lett.* **2009**, *11*, 4076–4079. (b) Tardibono, L. P., Jr.; Miller, M. J.; Balzarini, J. *Tetrahedron* **2011**, *67*, 825–829.
47. (a) McKeith, R. A.; McCague, R.; Olivo, H. F.; Palmer, C. F.; Roberts, S. M. *J. Chem. Soc. Perkin Trans. 1* **1993**, 313–314. (b) McKeith, R. A.; McCague, R.; Olivo, H. F.; Roberts, S. M.; Taylor, S. J. C.; Xong, H. *Bioorg. Med. Chem.* **1994**, *2*, 387–394.
48. (a) Olivo, H. F.; Yu, J. *Tetrahedron: Asymmetry* **1997**, *8*, 3785–3788. (b) Yu, J.; Olivo, H. F. *J. Chem. Soc. Perkin Trans. 1* **1998**, 391–392.
49. Shuto, S.; Shirato, M.; Sumita, Y.; Ueno, Y.; Matsuda, A. *J. Org. Chem.* **1998**, *63*, 1986–1994.
50. (a) Aggarwal, V. K.; Monteiro, N.; Tarver, G. J.; Lindell, S. D. *J. Org. Chem.* **1996**, *61*, 1192–1193. (b) Aggarwal, V. K.; Monteiro, N. *J. Chem. Soc. Perkin Trans. 1* **1997**, 2531–2537.
51. For *P. fluorescens*: (a) Evans, C.; McCague, R.; Roberts, S. M.; Sutherland, A. G. *J. Chem. Soc. Perkin Trans. 1* **1991**, 656–657. (b) Taylor, S. J. C.; McCague, R.; Wisdom, R.; Lee, C.; Dickson, K.; Ruecroft, G.; O'Brien, F.; Littlechild, J.; Bevan, J.; Roberts, S. M.; Evans, C. T. *Tetrahedron: Asymmetry* **1993**, *4*, 1117–1128.
52. For savinase: Mahmoudian, M.; Lowdon, A.; Jones, M.; Dawson, M.; Wallis, C. *Tetrahedron: Asymmetry* **1999**, *10*, 1201–1206.
53. (a) Taylor, S. J. C.; Sutherland, A. G.; Lee, C.; Wisdom, R.; Thomas, S.; Roberts, S. M.; Evans, C. *J. Chem. Soc. Chem. Commun.* **1990**, 1120–1121. (b) Evans, C. T.; Roberts, S. M.; Shoberu, K. A.; Sutherland, A. G. *J. Chem. Soc. Perkin Trans. 1* **1992**, 589–592.
54. Katagiri, N.; Takebayashi, M.; Kokufuda, H.; Kaneko, C.; Kanehira, K.; Torihara, M. *J. Org. Chem.* **1997**, *62*, 1580–1581.

REFERENCES

55. Dominguez, B. M.; Cullis, P. M. *Tetrahedron Lett.* **1999**, *40*, 5783–5786.
56. Csuk, R.; Dörr, P. *Tetrahedron* **1995**, *51*, 5789–5798.
57. (a) Beer, D.; Meuwly, R.; Vasella, A. *Helv. Chim. Acta* **1982**, *65*, 2570–2582. (b) Bélanger, P.; Prasit, P. *Tetrahedron Lett.* **1988**, *29*, 5521–5524.
58. Ali, S. M.; Ramesh, K; Borchardt, R. T. *Tetrahedron Lett.* **1990**, *31*, 1509–1512.
59. Borcherding, D. R.; Scholtz, S. A.; Borchardt, R. T. *J. Org. Chem.* **1987**, *52*, 5457–5461.
60. Moon, H. R.; Choi, W. J.; Kim, H. O.; Jeong, L. S. *Tetrahedron: Asymmetry* **2002**, *13*, 1189–1193.
61. (a) Jin, Y. H.; Chu, C. K. *Tetrahedron Lett.* **2002**, *43*, 4141–4143. (b) Jin, Y. H.; Liu, P.; Wang, J.; Baker, R.; Huggins, J.; Chu, C. K. *J. Org. Chem.* **2003**, *68*, 9012–9018.
62. Liu, P.; Chu, C. K. *Can. J. Chem.* **2006**, *84*, 748–754.
63. Yang, M., Ye, W.; Schneller, S. W. *J. Org. Chem.* **2004**, *69*, 3993–3996.
64. Kumamoto, H.; Deguchi, K.; Wagata, T.; Furuya, Y.; Odanaka, Y.; Kitade, Y.; Tanaka, H. *Tetrahedron* **2002**, *65*, 8007–8013.
65. Choi, W. J.; Park, J. G.; Yoo, S. J.; Kim, H. O.; Moon, H. R.; Chun, M. W.; Jung, Y. H.; Jeong, L. S. *J. Org. Chem.* **2001**, *66*, 6490–6494.
66. Siddiqi, S. M.; Schneller, S. W.; Ikeda, S.; Snoeck, R.; Andrei, G.; Balzarini, J.; De Clercq, E. *Nucleosides Nucleotides* **1993**, *12*, 185–198.
67. Wolfe, M. S.; Borcherding, D. R.; Borchardt, R. T. *Tetrahedron Lett.* **1989**, *30*, 1453–1456.
68. (a) Wang, P.; Agrofoglio, L. A.; Newton, M. G.; Chu, C. K. *Tetrahedron Lett.* **1997**, *38*, 4207–4210. (b) Wang, P.; Agrofoglio, L. A.; Newton, M. G.; Chu, C. K. *J. Org. Chem.* **2001**, *66*, 6490–6494. (c) Jin, Y. H.; Liu, P.; Wang, J.; Baker, R.; Huggins, J.; Chu, C. K. *J. Org. Chem.* **2003**, *68*, 9012–9018. (d) Gadthula, S.; Rawal, R. K.; Sharon, A.; Wu, D.; Korba, B.; Chu, C. K. *Bioorg. Med. Chem. Lett.* **2011**, *21*, 3982–3985.
69. Yang, M.; Schneller, S. W. *Bioorg. Med. Chem. Lett.* **2005**, *15*, 149–151.
70. (a) Comin, M. J.; Agbaria, R.; Ben-Kasus, T.; Huleihel, M.; Liao, C.; Sun, G.; Nicklaus, M. C.; Deschamps, J. R.; Parrish, D. A.; Marquez, V. E. *J.Am. Chem. Soc.* **2007**, *129*, 6216–6222. (b) Comin, M. J.; Vu, B. C.; Boyer, P. L.; Liao, C.; Hughes, S. H.; Marquez, V. E. *ChemMedChem* **2008**, *3*, 1129–1134.
71. Medich, J. R.; Kunnen, K. B.; Johnson, C. R. *Tetrahedron Lett.* **1987**, *28*, 4131–4134.
72. Rodriguez, S.; Edmont, D.; Mathé, C.; Périgaud, C. *Tetrahedron* **2007**, *63*, 7165–7171.
73. Comin, M. J.; Rodriguez, J. B. *Tetrahedron* **2000**, *56*, 4639–4649.
74. Michel, B. Y.; Strazewski, P. *Tetrahedron* **2007**, *63*, 9836–9841.
75. Choi, W. J.; Moon, H. R.; Kim, H. O.; Yoo, B. N.; Lee, J. A.; Shin, D. H.; Jeong, L. S. *J. Org. Chem.* **2004**, *69*, 2634–2636.
76. (a) Altmann, K.-H.; Kesselring, R.; Francotte, E.; Rihs, G. *Tetrahedron Lett.* **1994**, *35*, 2331–2334. (b) Lee, J. A.; Kim, H. O.; Tosh, D. K.; Moon, H. R.; Kim, S.; Jeong, L. S. *Org. Lett.* **2006**, *8*, 5081–5083. (c) Michel, B. Y.; Strazewski, P. *Chem. Eur. J.* **2009**, *15*, 6244–6257.
77. Paquette, L. A.; Tian, Z.; Seekamp, C. K.; Wang, T. *Helv. Chim. Acta* **2005**, *88*, 1185–1198.
78. For a special issue dedicated to olefin metathesis, including Grubbs, Schrock, and Chauvin Nobel lectures, see *Adv. Synth. Catal.* **2007**, 349, issues 1–2.

79. (a) Moon, H. R.; Kim, H. O.; Lee, K. M.; Chun, M. W.; Kim, J. H.; Jeong, L .S. *Org. Lett.* **2002**, *70*, 3501–3503. (b) Lee, J. A.; Moon, H. R.; Kim, H. O.; Kim, K. R.; Lee, K. M.; Kim, B. T.; Hwang, K. J.; Chun, M. W.; Jacobson, K. A.; Jeong, L .S. *J. Org. Chem.* **2005**, *70*, 5006–5013. (c) Kim, J.-H.; Kim, H. O.; Lee, K. M.; Chun, M. W.; Moon, H. R.; Jeong, L .S. *Tetrahedron* **2006**, *62*, 6339–6342.
80. Ye, W.; He, M.; Schneller, S.W. *Tetrahedron Lett.* **2009**, *50*, 7156–7158.
81. Ludek, O. R.; Marquez, V. E. *Tetrahedron* **2009**, *65*, 8461–8467.
82. (a) Gillaizeau, I.; Charamon, S.; Agrofoglio, L. A. *Tetrahedron Lett.* **2001**, *42*, 8817–8819. (b) Agrofoglio, L. A.; Amblard, F.; Nolan, S. P.; Charamon, S.; Gillaizeau, I.; Zevaco, T.A.; Guenot, P. *Tetrahedron* **2004**, *60*, 8397–8404.
83. Huang, J.; Stevens, E. D.; Nolan, S. P.; Petersen, J. L. *J.Am. Chem. Soc.* **1999**, *121*, 2674–2678.
84. Fang, Z.; Hong, J. H. *Org. Lett.* **2004**, *6*, 993–995.
85. Banerjee, S.; Ghosh, S.; Sinha, S.; Ghosh, S. *J. Org. Chem.* **2005**, *70*, 4199–4202.
86. Chattopadhyay, A.; Tripathy, S. *J. Org. Chem.* **2011**, *76*, 5856–5861.
87. (a) Liu, L. J.; Ko, O. H.; Jong, J. H. *Bull. Korean Chem. Soc.* **2008**, *29*, 1723–1728. (b) Oh, C. H.; Yoo, K. H.; Jong, J. H. *Bull. Korean Chem. Soc.* **2010**, *31*, 2473–2478. (c) Yoo, J. C.; Li, H.; Lee, W.; Jong, J. H. *Bull. Korean Chem. Soc.* **2010**, *31*, 3348–3352.
88. (a) Park, A.-Y.; Moon, H. R.; Kim, K. R.; Chun, M. W.; Jeong, L. S. *Org. Biomol. Chem.* **2006**, *4*, 4065–4067. (b) Park, A.-Y.; Kim, W. H.; Kang, J.-A.; Lee, H. J.; Lee, C.-K.; Moon, H. R. *Bioorg. Med. Chem.* **2011**, *19*, 3945–3955.
89. Crimmins, M. T.; King, B. W; Zuercher, W. J.; Choy, A. L. *J. Org. Chem.* **2000**, *65*, 8499–8509.
90. Crimmins, M. T.; King, B. W.; Tabet, E. A. *J.Am. Chem. Soc.* **1997**, *119*, 7883–7884.
91. Crimmins, M. T.; Zuercher, W. J. *Org. Lett.* **2000**, *2*, 1065–1067.
92. Kuang, R.; Ganguly, A. K.; Chan, T.-M.; Pramanik, B. N.; Blythin, D. J.; McPhail, A. T.; Saksena, A. K. *Tetrahedron Lett.* **2000**, *41*, 9575–9579.
93. (a) Walker, M. A.; Heathcock, C. H. *J. Org. Chem.* **1991**, *56*, 5747–5750. (b) Raimundo, B. C.; Heathcock, C. H. *Synlett* **1995**, 1213–1214.
94. Zhou, J.; Yang, M.; Akdag, A.; Wang, H.; Schneller, S. W. *Tetrahedron* **2008**, *64*, 433–438.
95. Davis, F. A.; Wu, Y. *Org. Lett.* **2004**, *6*, 1269–1272.
96. (a) Kirby, G. W.; Nazeer, M. *Tetrahedron Lett.* **1988**, *29*, 6173–6174. (b) Miller, A.; McC. Paterson, T.; Procter, G. *Synlett* **1989**, 32–34.
97. Miller, M. J.; Shireman, B. T. *Tetrahedron Lett.* **2000**, *41*, 9537–9540.
98. Ritter, A. R.; Miller, M. J. *J. Org. Chem.* **1994**, *59*, 4602–4611.
99. Ghosh, A.; Ritter, A. R.; Miller, M. J. *J. Org. Chem.* **1995**, *60*, 5808–5813.
100. Vogt, P. F.; Hansel, J.-G.; Miller, M. J. *Tetrahedron Lett.* **1997**, *38*, 9537–9540.
101. Zhang, D.; Ghosh, A.; Süling, C.; Miller, M. J. *Tetrahedron Lett.* **1996**, *22*, 3799–3802.
102. Felber, H.; Kresze, G.; Braun, H.; Vasella, A. *Tetrahedron Lett.* **1984**, *25*, 5381–5382.
103. Zhang, D.; Süling, C.; Miller, M. J. *J. Org. Chem.* **1998**, *63*, 885–888.
104. Lin, C.-H.; Wang, Y.-C.; Hsu, J.-L.; Chiang, C.-C.; Su, D.-W.; Yan, T.-H. *J. Org. Chem.* **1997**, *62*, 3806–3807.
105. Bodnar, B. S.; Miller, M. J. *Angew. Chem. Int. Ed.* **2011**, *50*, 5630–5647.

106. Gallos, J. K.; Goga, E. G.; Koumbis, A. E. *J. Chem. Soc. Perkin Trans. 1* **1994**, 613–614.
107. Berranger, T.; Langlois, Y. *Tetrahedron Lett.* **1995**, *36*, 5523–5526.
108. (a) Singha, K.; Roy, A.; Dutta, P. K.; Tripathi, S.; Sahabuddin, S.; Achari, B.; Mandal, S. B. *J. Org. Chem.* **2004**, *69*, 6507–6510. (b) Sahabuddin, S.; Roy, A.; Drew, M. G. B.; Roy, B. G.; Achari, B.; Mandal, S. B. *J. Org. Chem.* **2006**, *71*, 5980–5992.
109. Zhou, B.; Li, Y. *Tetrahedron Lett.* **2012**, *53*, 502–504.
110. Díaz, M.; Ibarzo, J.; Jiménez, J. M.; Ortuño, R. M. *Tetrahedron: Asymmetry* **1994**, *5*, 129–140.
111. Boyer, S. J.; Leahy, J. W. *J. Org. Chem.* **1997**, *62*, 3976–3980.
112. Arita, M.; Adachi, K.; Ito, Y.; Sawai, H.; Ohno, M. *J.Am. Chem. Soc.* **1983**, *105*, 4049–4055.
113. (a) Cortez, G. S.; Tennyson, R. L.; Romo, D. *J.Am. Chem. Soc.* **2001**, *123*, 7945–7946. (b) Yokota, Y.; Cortez, G. S.; Romo, D. *Tetrahedron* **2002**, *58*, 7075–7080.
114. Brown, B.; Hegedus, L. S. *J. Org. Chem.* **2000**, *65*, 1865–1872.
115. Hegedus, L. S.; Bates, R. W.; Söderberg, B. C. *J.Am. Chem. Soc.* **1991**, *113*, 923–927.
116. Hegedus, L. S.; Geisler, L. *J. Org. Chem.* **2000**, *65*, 4200–4203.
117. Ohira, S.; Sawamoto, T.; Yamato, M. *Tetrahedron Lett.* **1995**, *36*, 1537–1538.
118. Nguyen Van Nhien, A.; Soriano, E.; Marco-Contelles, J.; Postel, D. *Carbohydr. Res.* **2009**, *344*, 1605–1611.
119. Niizuma, S.; Shuto, S.; Matsuda, A. *Tetrahedron* **1997**, *53*, 13621–13632.
120. (a) Shin, K. J.; Moon, H. R.; George, C.; Marquez, V. E. *J. Org. Chem.* **2000**, *65*, 2172–2178. (b) Moon, H. R.; Ford, H. Jr., Marquez, V. E. *Org. Lett.* **2000**, *2*, 3793–3796. (c) Comin, M. J.; Parrish, D. A.; Deschamps, J. R.; Marquez, V. E. *Org. Lett.* **2006**, *8*, 705–708.
121. (a) Velcicky, J.; Lex, J.; Schmalz, H.-G. *Org. Lett.* **2002**, *4*, 565–568. (b) Velcicky, J.; Lanver, A.; Lex, J.; Prokop, A.; Wieder, T.; Schmalz, H.-G. *Chem. Eur. J.* **2004**, *10*, 5087–5110.
122. Vázquez-Romero, A.; Rodrìguez, J.; Lledó, A.; Verdaguer, X.; Riera, A. *Org. Lett.* **2008**, *10*, 4509–4512.
123. Solà, J.; Revés, M.; Riera, A.; Verdaguer, X. *Angew. Chem. Int. Ed.* **2007**, *46*, 5020–5023.
124. (a) Wilcox, C. S.; Thomasco, L. M. *J. Org. Chem.* **1985**, *50*, 546–547. (b) Gallos, J. K.; Dellios, C. C.; Spata, E. E. *Eur. J. Org. Chem.* **2001**, 79–82.
125. For related reactions, see (a) Rajanbabu, T. V. *Acc. Chem. Res.* **1991**, *24*, 139–145. (b) Horneman, A.-M.; Lundt, I. *Tetrahedron* **1997**, *53*, 6879–6892. (c) Hersant, G.; Ferjani, M. B. S.; Bennet, S. M. *Tetrahedron Lett.* **2004**, *45*, 8123–8126.
126. Martinez-Grau, A.; Marco-Contelles, J. *Chem. Soc. Rev.* **1998**, *27*, 155–162.
127. Desire, J.; Prandi, J. *Eur. J. Org. Chem.* **2000**, 3075–3084.
128. Ziegler, F. E.; Sarpong, M. A. *Tetrahedron* **2003**, *59*, 9013–9018.
129. Takagi, C.; Sudeka, M.; Kim, H.-S.; Wataya, Y.; Yabe, S.; Kitade, Y.; Matsuda, A.; Shuto, S. *Org. Biomol. Chem.* **2005**, *3*, 1245–1251.
130. (a) Park, K. H.; Rapoport, H. *J. Org. Chem.* **1994**, *59*, 394–399. (b) Ho, J. Z.; Mohareb, T. M.; Ahn, J. H.; Sim, T. B.; Rapoport, H. *J. Org. Chem.* **2003**, *68*, 109–114.
131. Campbell, J. A. Lee, W. K.; Rapoport, H. *J. Org. Chem.* **1995**, *60*, 4602–4616.

132. Yoshikawa, M.; Yokokawa, Y.; Inoue, Y.; Yamaguchi, S.; Murakami, N.; Kitagawa, I. *Tetrahedron* **1994**, *50*, 9961–9974.
133. Yoshikawa, M.; Okaichi, Y.; Cha, B. C.; Kitagawa, I. *Tetrahedron* **1990**, *46*, 7459–7470.
134. Ono, M.; Nishimura, K.; Tsubouchi, H.; Nagaoka, Y.; Tomioka, K. *J. Org. Chem.* **2001**, *66*, 8199–8203.
135. Freiría, M.; Whitehead, A. J.; Motherwell, W. B. *Synthesis* **2005**, 3079–3084.
136. (a) Leung, L. M. H.; Boydell, A. J.; Gibson, V.; Light, M. E.; Linclau, B. *Org. Lett.* **2005**, *7*, 5183–5186. (b) Leung, L. M. H.; Gibson, V.; Linclau, B. *J. Org. Chem.* **2008**, *73*, 9197–9206. (c) Leung, L. M. H.; Gibson, V.; Light, M. E.; Linclau, B. *Tetrahedron: Asymmetry* **2009**, *20*, 821–831.
137. Smith, A. B., III; Pitram, S. M.; Boldi, A. M.; Gaunt, M. J.; Sfouggatakis, C.; Moser, W. H. *J.Am. Chem. Soc.* **2003**, *125*, 14435–14445.
138. Marcé, P.; Díaz, Y.; Matheu, M. I.; Castillón, S. *Org. Lett.* **2008**, *10*, 4735–4738.
139. Trost, B. M.; Li, L.; Guile, S. D. *J.Am. Chem. Soc.* **1992**, *114*, 8745–8747.
140. (a) Katagiri, N.; Ito, Y.; Kitano, K.; Toyota, A.; Kaneko, C. *Chem. Pharm. Bull.* **1994**, *42*, 2653–2655. (b) Palmer, C. F.; McCague, R.; Ruecroft, G.; Savage, S.; Taylor, S. J. C.; Ries, C. *Tetrahedron Lett.* **1996**, *37*, 4601–4604.
141. Trost, B. M.; Osipov, M.; Kaib, P. S. J.; Sorum, M. T. *Org. Lett.* **2011**, *13*, 3222–3225.
142. (a) Trost, B. M.; Madsen, R.; Guile, S. G.; Elia, A. E. H. *Angew. Chem. Int. Ed.* **1996**, *35*, 1569–1572. (b) Trost, B. M.; Madsen, R.; Guile, S. G.; Brown, B. *J.Am. Chem. Soc.* **2000**, *122*, 5947–5956.
143. Trost, B. M.; Madsen, R.; Guile, S. G. *Tetrahedron Lett.* **1997**, *38*, 1707–1710.
144. Trost, B. M.; Stenkamp, D.; Pulley, S. R. *Chem. Eur. J.* **1995**, *1*, 568–572.
145. Asami, M. *Bull. Chem. Soc. Jpn.* **1990**, *63*, 721–727.
146. (a) Hodgson, D. M.; Witherington, J.; Moloney, B. A. *Tetrahedron: Asymmetry* **1994**, *5*, 337–338. (b) Asami, M.; Takahashi, J.; Inoue, S. *Tetrahedron: Asymmetry* **1994**, *5*, 1649–1652.
147. Hodgson, D. M.; Gibbs, A. R.; Drew, M. G. B. *J.Chem. Soc. Perkin. Trans. 1* **1999**, 3579–3590.
148. Kim, H. S.; Jacobson, K. A. *Org. Lett.* **2003**, *5*, 1665–1668.
149. (a) Barret, S.; O'Brien, P.; Steffens, H. C.; Towers, T. D.; Voith, M. *Tetrahedron* **2000**, *56*, 9633–9640. (b) Seki, A.; Asami, M. *Tetrahedron* **2002**, *58*, 4655–4663.
150. Martinez, L. E.; Nugent, W. A.; Jacobsen, E. N. *J. Org. Chem.* **1996**, *61*, 7963–7966.
151. Biggadike, K.; Borthwick, A. D.; Exall, A. M.; Kirk, B. E.; Roberts, S. M.; Youds, P.; Slawin, A. M. Z.; Williams, D. J. *J. Chem. Soc. Chem. Commun.* **1987**, 255–256.
152. Partridge, J. J.; Chadha, N. K.; Uskokovic, M. R. *J.Am. Chem. Soc.* **1973**, *95*, 532–540.
153. Bisacchi, G. S.; Chao, S. T.; Bachard, C.; Daris, J. P.; Innaimo, S.; Jacobs, G. A.; Kocy, O.; Lapointe, P.; Martel, A.; Merchant, Z.; Slusarchyk, W. A.; Sundeen, J. E.; Young, M. G.; Colonno, R.; Zahler, R. *Bioorg. Med. Chem. Lett.* **1997**, *7*, 127–132.
154. Jessel, S.; Meier, C. *Eur. J. Org. Chem.* **2011**, 1702–1713.
155. (a) Ezzitouni, A.; Russ, P.; Marquez, V. E. *J. Org. Chem.* **1997**, *62*, 4870–4873. (b) Marquez, V. E.; Russ, P.; Alonso, R.; Siddiqui, M. A.; Hernandez, S.; George, C.; Nicklaus, M. C.; Dai, F.; Ford, H., Jr. *Helv. Chim. Acta* **1999**, *82*, 2119–2129. (c) Gathergood, N.; Knudsen, K. R.; Jørgensen, K. A. *J. Org. Chem.* **2001**, *66*, 1014–1017. (d) Elhalem, E.; Pujol, C. A.; Damonte, E. B.; Rodriguez, J. B. *Tetrahedron* **2010**, *66*, 3332–3340.

156. Mohar, B.; Kobe, J. *Synlett* **1997**, 1467–1468.
157. (a) Pankiewicz, K. W. *Carbohydr. Res.* **2000**, *327*, 87–105. (b) Qiu, X.-L.; Xu, X.-H.; Qing, F.-L. *Tetrahedron* **2010**, *66*, 789–843.
158. Dax, K.; Albert, M.; Ortner, J.; Paul, B. J. *Carbohydr. Res.* **2000**, *327*, 47–86
159. (a) Biggadike, K.; Borthwick, A. D.; Evans, D.; Exall, A. M.; Kirk, B. E.; Roberts, S. M.; Stephenson, L.; Youds, P.; Slawin, A. M. Z.; Williams, D. J. *J. Chem. Soc. Chem. Commun.* **1987**, 1083–1084. (b) Borthwick, A. D.; Evans, D.; Kirk, B. E.; Biggadike, K.; Exall, A. M.; Youds, P.; Roberts, S. M.; Knight, D. J.; Coates, J. A. V. *J.Med. Chem.* **1990**, *33*, 179–186. (c) Wachtmeister, J.; Classon, B.; Samuelsson, B.; Kvarnström, I. *Tetrahedron* **1997**, *53*, 1861–1872. (d) Liu, C.; Chen, Q.; Schneller, S. W. *Tetrahedron Lett.* **2011**, *52*, 4931–4933.
160. (a) Baumgartner, H.; Bodenteich, M.; Griengl, H. *Tetrahedron Lett.* **1988**, *29*, 5745–5746. (b) Roy, A.; Serbessa, T.; Schneller, S.W. *Bioorg. Med. Chem.* **2006**, *14*, 4980–4986.
161. (a) Wang, J.; Jin, Y.; Rapp, K. L.; Bennett, M.; Schinazi, R. F.; Chu, C. K. *J.Med. Chem.* **2005**, *48*, 3736–3748. (b) Wang, J.; Jin, Y.; Rapp, K. L.; Schinazi, R. F.; Chu, C. K. *J.Med. Chem.* **2007**, *50*, 1828–1839.
162. Gumina, G.; Chong, Y.; Choi, Y.; Chu, C. K. *Org. Lett.* **2000**, *2*, 1229–1231.
163. Li, W.; Yin, X.; Schneller, S. W. *Bioorg. Med. Chem. Lett.* **2008**, *18*, 220–222.
164. Jeong, L. S.; Yoo, S. J.; Lee, K. M.; Koo, M. J.; Choi, W. J.; Kim, H. O.; Moon, H. R.; Lee, M. Y.; Park, J. G.; Lee, S. K.; Chun, M. W. *J.Med. Chem.* **2003**, *46*, 201–203.
165. (a) Moon, H. R.; Lee, H. J.; Kim, K. R.; Lee, K. M.; Lee, S. K.; Kim, H. O.; Chun, M. W.; Jeong, L. S. *Bioorg. Med. Chem. Lett.* **2004**, *14*, 5641–5644. (b) Kumamoto, H.; Kobayashi, M.; Kato, N.; Balzarini, J.; Tanaka, H. *Eur. J. Org. Chem.* **2011**, 2685–2691.
166. Yang, Y.; Zheng, F.; Qing, F.-L. *Tetrahedron* **2011**, *67*, 3388–3394.
167. (a) Yang, Y.-Y.; Meng, W.-D.; Qing, F.-L. *Org. Lett.* **2004**, *6*, 4257–4259. (b) Yang, Y.-Y.; Xu, J.; You, Z.-W.; Xu, X.-H.; Qiu, X.-L.; Qing, F.-L. *Org. Lett.* **2007**, *9*, 5437–5440. (c) Kumamoto, H.; Haraguchi, K.; Ida, M.; Nakamura, K. T.; Kitagawa, Y.; Hamasaki, T.; Baba, M.; Matsubayashi, S. S.; Tanaka, H. *Tetrahedron* **2009**, *65*, 7630–7636.
168. (a) Barth, F.; O-Yang, C. *Tetrahedron Lett.* **1991**, *32*, 5873–5876. (b) Arnone, A.; Bravo, P.; Cavicchio, G.; Frigerio, M.; Viani, F. *Tetrahedron* **1992**, *48*, 8523–8540. (c) Fourrière, G.; Van Hijfte, N.; Dutech, G.; Fragnet, B.; Coadou, G.; Lalot, J.; Quirion, J.-C.; Leclerc, E. *Tetrahedron* **2010**, *66*, 3963–3972.
169. (a) Verheggen, I.; Van Aerschot, A.; Toppet, S.; Snoeck, R.; Janssen, G.; Balzarini, J.; De Clercq, E.; Herdewijn, P. *J.Med. Chem.* **1993**, *36*, 2033–2040. (b) Verheggen, I.; Van Aerschot, A.; Van Meervelt, L.; Rozenski, J.; Wiebe, L.; Snoeck, R.; Andrei, G.; Balzarini, J.; Claes, P.; De Clercq, E.; Herdewijn, P. *J.Med. Chem.* **1995**, *38*, 826–835.
170. (a) Pérez-Pérez, M.-J.; Rozenski, J.; Busson, R.; Herdewijn, P. *J. Org. Chem.* **1995**, *60*, 1531–1537. (b) Wang, J.; Busson, R.; Blaton, N.; Rozenski, J.; Herdewijn, P. *J. Org. Chem.* **1998**, *63*, 3051–3058. (c) Wang, J.; Herdewijn, P. *J. Org. Chem.* **1999**, *64*, 7820–7827. (d) Wang, J.; Viña, D.; Busson, R.; Herdewijn, P. *J. Org. Chem.* **2003**, *68*, 4499–4505.
171. (a) Calvani, F.; Macchia, M.; Rossello, A.; Gismondo, M. R.; Drago, L.; Fassina, M. C.; Cisternino, M.; Domiano, P. *Bioorg. Med. Chem. Lett.* **1995**, *5*, 2567–2572. (b) Viña, D.; Santana, L.; Uriarte, E.; Terán, C. *Tetrahedron* **2005**, *61*, 473–478. (c)

Barral, K.; Courcambeck, J.; Pèpe, G.; Balzarini, J.; Neyts, J.; De Clercq, E.; Camplo, M. *J.Med. Chem.* **2005**, *48*, 450–456.
172. Dalençon, S.; Youcef, R. A.; Pipelier, M.; Maisonneuve, V.; Dubreuil, D.; Huet, F.; Legoupy, S. *J. Org. Chem.* **2011**, *76*, 8059–8063.
173. (a) Rosenquist, Å.; Kvarnström, I.; Classon, B.; Samuelsson, B. *J. Org. Chem.* **1996**, *61*, 6282–6288. (b) Varada, M.; Kotikam, V.; Kumar, V. A. *Tetrahedron* **2011**, *67*, 5744–5749.
174. Ichikawa, Y.-I.; Narita, A.; Shiozawa, A.; Hayashi, Y.; Narasaka, K. *J. Chem.. Soc. Chem. Commun.* **1989**, 1919–1921.
175. Rustullet, A.; Alibés, R.; de March, P.; Figueredo, M.; Font, J. *Org. Lett.* **2007**, *9*, 2827–2830.
176. Brown, B.; Hegedus, L. S. *J. Org. Chem.* **1998**, *63*, 8012.
177. (a) Maruyama, T.; Hanai, Y.; Sato, Y.; Snoeck, R.; Andrei, G.; Hosoya, M.; Balzarini, J.; De Clercq, E. *Chem. Pharm. Bull.* **1993**, *41*, 516–521. (b) Gourdel-Martin, M.-E.; Huet, F. *J. Org. Chem.* **1997**, *62*, 2166–2172.
178. (a) Jung, M. E.; Sledeski, A. W. *J. Chem. Soc. Chem. Commun.* **1993**, 589–591. (b) Bondada, L.; Gumina, G.; Nair, R.; Ning, X. H.; Schinazi, R. F.; Chu, C. K. *Org. Lett.* **2004**, *6*, 2531–2534.
179. (a) Katagiri, N.; Sato, H.; Kaneko, C. *Chem. Pharm. Bull.* **1990**, *38*, 3184–3186. (b) Nishiyama, S.; Ueki, S.; Watanabe, T.; Yamamura, S.; Kato, K.; Takita, T. *Tetrahedron Lett.* **1991**, *32*, 2141–2142.
180. Zhao, Y.; Yang, T.; Lee, M.; Lee, D.; Newton, M. G.; Chu, C. K. *J. Org. Chem.* **1995**, *60*, 5236–5242.
181. (a) Yang, T.-F.; Kim, H.; Kotra, L. P.; Chu, C. K. *Tetrahedron Lett.* **1996**, *37*, 8849–8852. (b) Lee, M. G.; Du, J. F.; Chun, M. W.; Chu, C. K. *J. Org. Chem.* **1997**, *62*, 1991–1995.
182. (a) Csuk, R.; Von Scholz, Y. *Tetrahedron* **1996**, *52*, 6383–6396. (b) Muray, E.; Rifé, J.; Branchadell, V.; Ortuño, R. M. *J. Org. Chem.* **2002**, *67*, 4520–4525.

13 Uncommon Three-, Four-, and Six-Membered Nucleosides

E. GROAZ and P. HERDEWIJN

Laboratory for Medicinal Chemistry, Rega Institute for Medical Research, Leuven, Belgium

1. INTRODUCTION

Over the past two decades, nucleoside analogues exhibiting uncommon sugar moieties (Figure 1) rather than natural ribo- and 2′-deoxyribofuranose rings have received growing attention because of the increased recognition of their various biological roles. In particular, countless reports have been published that describe the value of ring-contracted analogues based on a three- or four-membered pseudosugar unit in the treatment of viral infections. Within this latest context, the pursuit of carbocyclic nucleosides structurally related to natural furanose units has been driven by their distinguishing stability toward chemical and metabolic degradation, due to the absence of an anomeric center.

Despite a reduced conformational flexibility inherent in six-membered ring nucleosides and thus a diminished freedom of interaction with activating enzymes, recent work in this field has demonstrated the great forthcoming potential that hexose-like nucleosides hold for antiviral therapy.[1] All the same, the fundamental application of synthetic nucleic acids derived from such monomers, particularly HNA, ANA, and FHNA, has been a most important development in the field of antisense or antigene therapy. It is well recognized that antisense oligonucleotides (ASOs) bind to their target mRNA according to Watson–Crick base-pairing rules modulating mRNA function. Multiple mechanisms of gene expression downregulation are operative, including steric blockage by simple binding and cleavage of the RNA strand either by introducing a catalytic core into the oligomer (ribozyme) or by induction of RNase H enzymes. In this respect, the ongoing rational drug design is devoted widely to attaining high-affinity structural modifications through the pursuit of new enzymatically stable conformationally restricted oligonucleotides able to form stable A-like duplexes with the cognate RNA.[2] The possibility of examining the role played by a specific gene in a complex biological environment, together

---

Chemical Synthesis of Nucleoside Analogues, First Edition. Edited by Pedro Merino.
© 2013 John Wiley & Sons, Inc. Published 2013 by John Wiley & Sons, Inc.

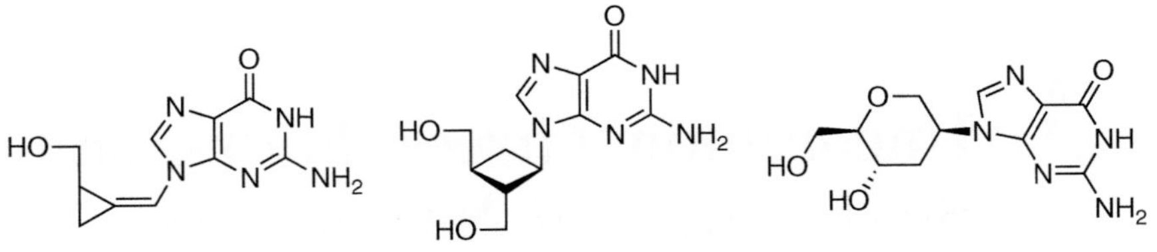

Figure 1. Examples of uncommon antiviral nucleosides.

with the therapeutic potential against diseases evolving from abnormal expression of genes (such as infectious diseases or cancer induction), has established nucleic acid–based therapeutics as a challenging but fundamental drug discovery platform.

2. THREE-MEMBERED NUCLEOSIDE DERIVATIVES

From a structural viewpoint, cyclopropane nucleosides can be classified by the nature of the association between pseudosugar and a heterocyclic base moiety, which, as shown in Figure 2, may be directly linked, for example, in (a) or separated as (b) and (c) cores, and then further elaborated in the type of spacer, whether a methylene (b) or an unsaturated group (c). While compounds of type **4–6** were found to be devoid of activity, the biological significance of the **7**, **8**, and **10–12** types has been demonstrated against a plethora of viral infections (e.g., HCMV, EBV, HSV, VZV) as well as exhibiting strong antitumoral activity.[3]

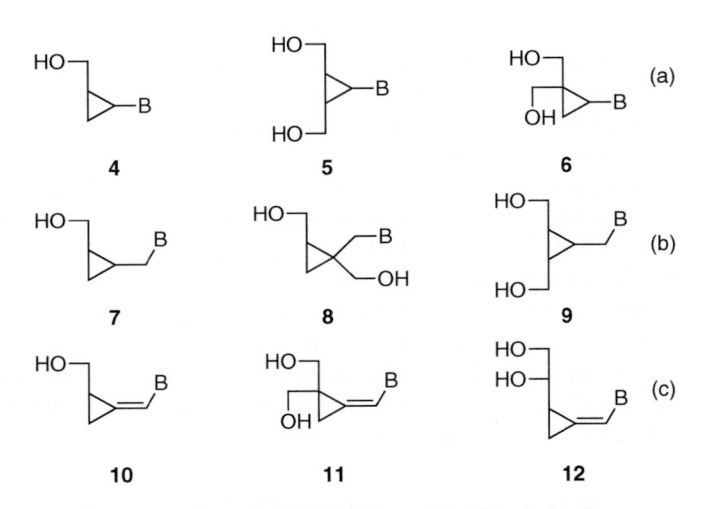

Figure 2. Three-membered ring nucleoside derivatives.

Early contributions in this area focused on the synthesis of 2-hydroxy methylcyclopropyl nucleosides **4**. Two groups reported synthetic efforts directed toward the enantioselective synthesis of those targets.[4] A chemoenzymatic resolution was the key step in the strategy elaborated by Csuk et al.[4b] A complementary route based on a chiral pool approach was developed by Chu et al.,[4a] as shown in Scheme 1. D-Glyceraldehyde acetonide **13** was reacted with a suitable phosphorane to form α,β-unsaturated ester **14**. To perform a stereoselective cyclopropanation, the product from the Wittig olefination had first to be reduced with DIBALH to cyclopropylmethyl alcohol **15**. Ring closure could be achieved using diethyl zinc and chloroiodomethane in dichloroethane at 0°C and gave the desired cyclopropane in 70% yield. Cyclopropyl alcohol **16** was then converted to acyl azide **17** via sequential $NaIO_4/RuO_2$ oxidation, amination, and aziridation without isolation of the intermediates. Curtius rearrangement of **17** and treatment of the resulting crude isocyanate with ammonia gas formed urea derivative **18**, from which optically pure D-(or 1'S,2'R)-pyrimidine nucleosides could be assembled

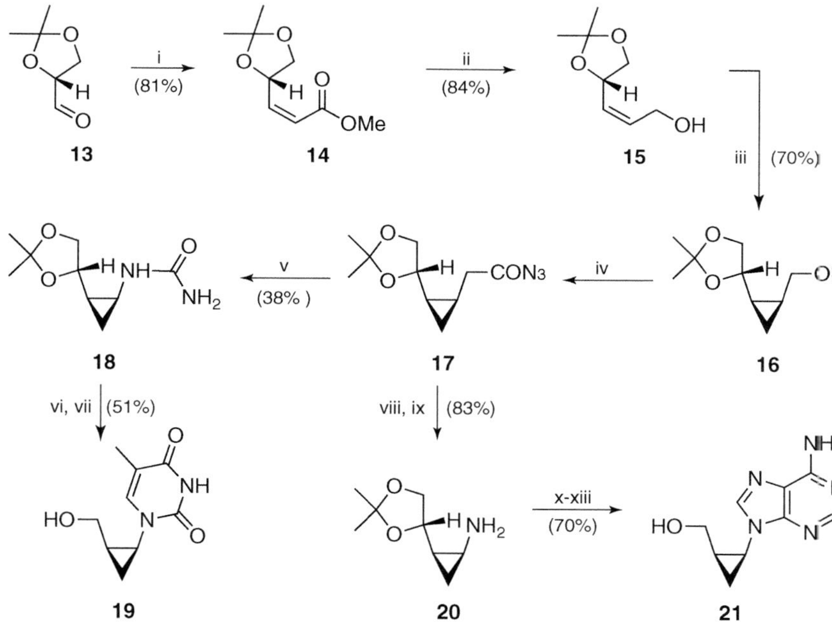

Scheme 1. *Reagents and conditions:* (i) $Ph_3P=CHCO_2Me$, MeOH, rt; (ii) DIBALH, CH_2Cl_2, −78°C; (iii) $ZnEt_2$, ICH_2Cl, DCE, 0°C; (iv) RuO_2, $NaIO_4$, K_2CO_3, $MeCN/CHCl_3/H_2O$, then $ClCO_2Et$, Et_3N, acetone, 0°C, then NaN_3, H_2O; (v) toluene, 90–100°C, then NH_3, Et_2O; (vi) $MeOCH=C(Me)COCl$, CH_2Cl_2, pyridine, rt, then NH_4OH, EtOH, 85–90°C, then HCl, MeOH; (vii) $NaIO_4$, MeOH, H_2O, then $NaBH_4$; (viii) BnOH, toluene, 100°C; (ix) H_2, Pd/C, MeOH; (x) 4,6-dichloroformamidopyrimidine, Et_3N, dioxane, heat; (xi) $(EtO)_2CO_2Me$, 120°C, then NH_4OH, MeOH; (xii) NH_3, MeOH, 90°C; (xiii) 80% AcOH, rt, then $NaIO_4$, MeOH, H_2O, 0°C, then $NaBH_4$.

stepwise. Additionally, **17** could be transformed into a benzyl carbamate before being transformed by catalytic hydrogenation into suitable purine precursor **20**. The asymmetric synthesis of L-(or $1'R,2'S$)-isomers was also described following an equivalent route from L-gulonic γ-lactone to form a cyclopropyl alcohol intermediate and standard base construction by linear approach.[5]

Cyclopropyl nucleosides can virtually be considered as rigid analogues of potent antiherpetic acyclic nucleoside agents. The steps in the systematic upgrading of the biological activity of this class of compounds were conceived and carried out in line with the principles ruling the structure–activity relationship of acyclovir and congeners, by which increased potency is likely to be related to enhanced ability to mimic the interaction of natural substrates with viral thymidine kinases, necessary for nucleoside activation. This was supported by early evidence of anti HSV-1 and HSV-2 activity shared by racemic 1-(2-hydroxymethyl)cyclopropylmethyl guanine **7**.[6] The extension of the distance between the cyclopropyl ring and base by one carbon and the concomitant introduction of an additional hydroxyl group mimicking the 3'- and 5'-OH groups of the 2'-deoxyribose moiety of nucleosides resulted in the discovery of ($1'S,2'R$)-9-[[1',2'-bis(hydroxymethyl)cycloprop-1'-yl]methyl]guanine or A-5021 **29**, a superior and more selective antiherpetic inhibitor than acyclovir (IC_{50} = 0.020 μg/mL against HSV-1 and 0.20 against VZV).[3a,7] Interesting anti-VZV activity was also found in a series of 5-substituted uracil[8] and tricyclic[9] analogues of **29**, while phosphonates derivative were completely inactive.[10]

The enantioselective synthesis of **29** was developed by Tsuji et al. and based on the reactivity of chiral cyclopropane-lactone ester **23**.[11] This compound can be assembled enantioselectively from (R)-epichlorohydrin **22** and diethyl malonate with 97% ee.[3a] Selective reduction of the ethyl ester group of **23** to give **25** was achieved via formation of a carboxylic acid intermediate **24**, leaving the lactone ring unaffected. Alcohol **25** was then converted to chloride **26**, which was reacted with guanine precursor 2-amino-6-chloropurine, giving a regioisomeric 7:1 ratio in favor of the desired N^9-alkylated product **27**. After chromatographic separation, conversion to guanine and deprotection were straightforward to yield **29** (Scheme 2).

Experimental Procedure

($1'S,2'R$)-9-{[$1',2'$-Bis(hydroxymethyl)cyclopropan-$1'$-yl]methyl}guanine (29).[3a,11] To a solution of sodium (2.42 g, 105 mmol) in EtOH (195 mL) was added diethyl malonate (16.7 mL, 110 mmol) at 0°C over 5 min. (R)-(−)-Epichlorohydrin **22** (7.8 mL, 100 mmol) was dissolved in EtOH (5 mL) and added dropwise to the solution at room temperature over 1 h. After stirring at 75°C for 20 h, the mixture was filtered, and the filtrate was concentrated in vacuo. The residue was dissolved in CH_2Cl_2 and washed with H_2O. The organic layer was concentrated in vacuo, and the residue was purified by column chromatography using 15–50% EtOAc in hexane to yield **23** as a colorless oil (12.0 g, 70%), $[\alpha]_D^{25}$ = −146.58° (c = 1.22, EtOH).[3a] Compound **23** was treated with 1 N NaOH (2.5 equiv.) for 4 h at room temperature, and concentrated HCl was added until pH < 1 to give crude lactone–carboxylic acid **24**, which was purified by absorption

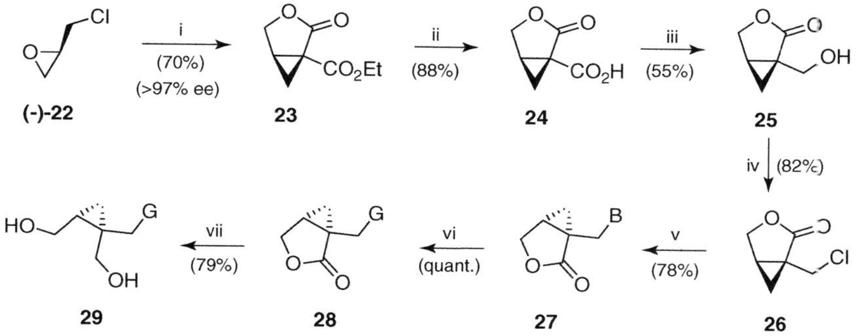

Scheme 2. *Reagents and conditions:* (i) Na, EtOH, DEM; (ii) NaOH, rt, then HCl, rt; (iii) ClCO$_2$Et, Et$_3$N, THF, $-18°$C, then NaBH$_4$, THF/H$_2$O, $-18°$C; (iv) SOCl$_2$, Et$_3$N, CH$_2$Cl$_2$, rt; (v) B = 2-amino-6-chloropurine, K$_2$CO$_3$, DMF, rt; (vi) 80% HCO$_2$H, 100°C, then 29% aq. NH$_3$, rt; (vii) NaBH$_4$, EtOH, 70°C.

to Sepabead SP-207 equilibrated with 0.1 N HCl, followed by elution with 20% MeOH to give a white solid in 88% yield. Treatment of **24** with Et$_3$N (1.2 equiv.) and ethyl chloroformate at $-18°$ C for 30 min in THF was followed by addition of a solution of NaBH$_4$ (3 equiv.) in water. After stirring for 20 min at $-18°$C, excess of 2 N HCl was added. The product was extracted and purified by silica gel chromatography to give **25** as a colorless oil in 55% yield. Treatment of **25** with SOCl$_2$ (1.5 equiv.) and Et$_3$N in CH$_2$Cl$_2$ at room temperature for 90 min gave chloride **26**. The product was extracted at pH 7 and purified by column chromatography. Compound **26** was then reacted with 2-amino-6-chloropurine (1 equiv.) and K$_2$CO$_3$ in DMF at 55°C for 20 h. The resulting 7- and 9-alkylated products were separated by column chromatography to give **27** as a white solid. Compound **27** was then heated in 80% formic acid at 100°C for 2 h and treated with 29% aqueous ammonia for 1 h at room temperature to give **28** in quantitative yield. After evaporation, **28** was treated with NaBH$_4$ (3 equiv.) in EtOH at 70°C for 3 h, and purified by reversed-phase chromatography to give **29**. Mp 297–298.5°C; $[\alpha]_D^{20} = -11.18°$ (c = 1%, DMSO). ^{1}H-NMR (DMSO-d_6): δ 0.40 (t, J = 5.1 Hz, 1H), 0.88 (dd, J = 4.8 and 8.7 Hz, 1H), 1.23 (m, 1H), 3.24–3.37 (m, 2H), 3.41 (dd, J = 6.0 and 12.0 Hz, 1H), 3.58 (dt, J = 12.0 and 6.0 Hz, 1H), 3.81 (d, J = 14.1 Hz, 1H), 4.00 (d, J = 14.1 Hz, 1H), 4.49 (m, 1H), 4.64 (m, 1H), 6.38 (b s, 2H), 7.71 (s, 1H), 10.49 (b s, 1H).

Along with the structural diversity created by introducing an additional geminal group at C1′, C3′ geminal substitution was also explored. In particular, (−)-(Z)-2,3-methanohomoserine was used as a versatile unit to prepare stereoselectively unusual nucleosides bearing an amino or a methyl group at the C3′-position.[12] An example of this chemistry is illustrated in Scheme 3. A simple, yet powerful route gave novel *gem*-amino alcohol **34** modified nucleoside in four steps from conveniently protected derivative **30**. Adenine base was installed by means of a nucleophilic displacement of mesylate **31**. Methyl ester reduction and subsequent

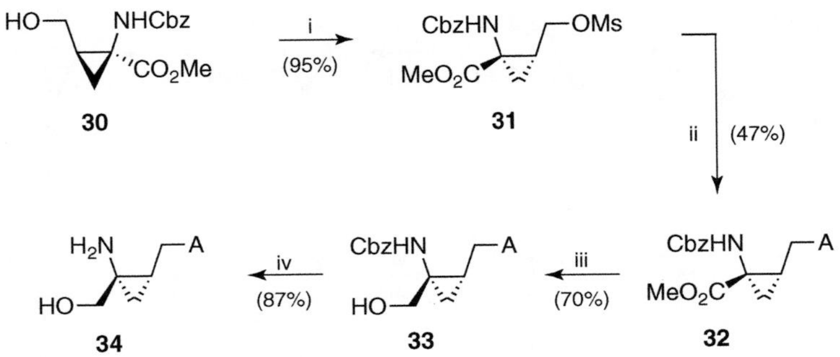

Scheme 3. *Reagents and conditions:* (i) MsCl, Et$_3$N; (ii) adenine, K$_2$CO$_3$, 18-crown-6; (iii) LiBH$_4$, THF, $-78°$C; (iv) H$_2$, Pd(OH)$_2$/C.

removal of the carboxybenzyl moiety via catalytic hydrogenation led to **34** in 27% overall yield. Other cyclopropyl homo-nucleosides appeared in the literature.[13]

Experimental Procedure

(1S,2R)-2-(6-Amino-9H-purinylmethyl)-1-amino-2-hydroxymethylcyclopropane (34).[12b] To a stirred and ice-cooled solution of alcohol **30** (415 mg, 1.5 mmol) in dry CH$_2$Cl$_2$ (10 mL) were added successively freshly distilled TEA (415 µL, 3.0 mmol) and mesyl chloride (275 µL, 3.0 mmol) under a nitrogen atmosphere. After stirring at 0°C for 10 min, water (6 mL) was added to the mixture, and the layers were separated. The aqueous phase was extracted with CH$_2$Cl$_2$ (3 × 6 mL), the combined organic layers were then dried over MgSO$_4$ and evaporated under reduced pressure. The resulting residue was chromatographed (hexane/EtOAc, 2:1) to afford crystalline (1S,2R)-mesylate **31** (502 mg, 95% yield). A solution of mesylate **31** (286 mg, 0.8 mmol) in dry DMF (5 mL) was added to a mixture of adenine (113 mg, 0.8 mmol), 18-crown-6 (211 mg, 0.8 mmol), and K$_2$CO$_3$ (121 mg, 0.9 mmol) in dry DMF (8 mL) under a nitrogen atmosphere. After stirring at 80°C for 15 h, the reaction mixture was cooled to room temperature, brine (15 mL) was added, then extracted with EtOAc (5 × 10 mL). The combined organic layers were dried over MgSO$_4$ and evaporated under reduced pressure. The residue was chromatographed (CH$_2$Cl$_2$/MeOH) to afford 149 mg (47% yield) of compound **32** as a colorless oil. A 1 M solution of LiBH$_4$ (1.4 mL, 1.4 mmol) was added to a stirred solution of **32** (139 mg, 0.4 mmol) in dry THF (10 mL) cooled at $-78°$C under a nitrogen atmosphere. The resulting mixture was stirred at room temperature for 20 h. MeOH was then slowly added to destroy excess hydride, and solvents were removed under reduced pressure. The residue was poured into water (6 mL) and ethyl acetate (6 mL), and the layers were separated. The aqueous phase was extracted with EtOAc (5 × 6 mL), the combined organic phases were dried over MgSO$_4$ and solvent was evaporated. The residue was eluted (water) through a C$_{18}$ reversed-phase cartridge to afford pure carbamate alcohol **33** (116 mg, 90% yield) as a solid. Carbamate **33** was hydrogenated under 4 atm of pressure for 17 h in the presence of 20% Pd(OH)$_2$/C. The mixture was filtered through Celite,

and the solvent was removed under reduced pressure. The residue was purified eluting (water) through a C_{18} reversed-phase cartridge to furnish pure amine **34** as a crystalline solid in 87% yield. Mp 197–199°C (MeOH·H$_2$O). $[\alpha]_D$ +22.5° (c 0.80, MeOH). ^{1}H-NMR (MeOH, d_4): δ 0.63 (t, J = 5.1 Hz, 1H), 0.84 (dd, J = 9.5 and 5.1, 1H), 1.39 (m, 1H), 3.34 (s, 2H), 4.27 (dd, J = 8.0 and 7.3 Hz, 1H), 4.45 (dd, J = 14.6 and 5.8 Hz, 1H), 8.18 (s, 1H), 8.49 (s, 1H).

Unsaturated nucleosides comprising a methylenecyclopropane unit as an alternative element of constraint (Figure 2, **10–12**) represent a widely studied class of three-membered ring analogues, currently under evaluation as clinical candidates. Since the area has been reviewed extensively in recent years,[3b,14] only selected methodologies describing optimal reaction conditions are covered here. A pioneering contribution to the development of this field of antiviral research was the first demonstration by the group of Zemlicka that the 2′-hydroxymethyl cyclopropylidene purines (Figure 2, **10**) synadenol (B = A) and synguanol (B = G) were broad-spectrum antiviral agents.[15] An in-depth structure–activity relationship analysis revealed that purine-based nucleosides with a Z-configured double bond possess broader spectrum and higher activities than their pyrimidine counterparts with an E-configuration, effective only in isolated cases. Furthermore, enantioselectivity represented an additional factor affecting antiviral potencies.[16] This finding motivated the formulation of an entire family of related 6- and 2-Z-modified purine nucleosides, together with pronucleotidic analogues, as a means for bioavailability.[3b] Monofluoro and geminal difluoro derivatives were also studied, an in-depth treatise of which is provided in a recent review.[17]

An optimized approach to first-generation methylenecyclopropanes is illustrated in Scheme 4.[18] Racemic epichlorohydrin (±)-**22** underwent alkylation in the presence of methylenetriphenylphosphorane generated in situ by the deprotonation of methylenetriphenylphosphonium bromide. Resolution of racemic oxaphospholane **35** by treatment with L-(+)-tartaric acid led to optically pure diastereoisomeric salts **36** and **37** that could readily be separated by crystallization. Access to multigram quantities of (S)-(+)-methylenecyclopropylcarbinol **39** was achieved through basic treatment of the diastereoisomeric tartarate **37**, followed by Wittig olefination between the resulting (R)-(−)-oxaphospholane **38** and paraformaldehyde, according to a modified literature procedure.[19] Following acetyl protection, treatment with pyridinium tribromide gave diastereoisomeric dibromocyclopropanes **40**, which were reacted as a mixture in the next step. An alkylation–elimination sequence allowed base insertion affording Z,E-isomeric acetates **41**. Deacetylation and the only chromatographic separation required in the entire protocol completed the synthesis of S,Z-**42** and S,E-**43** synadenol (Scheme 4). From the opposite, (S)-(−)-oxaphospholane, identical synthetic steps led to R-isomers.

Experimental Procedure

(Z,S)-(+)-2-Amino-6-chloro-9-[(2-hydroxymethyl)cyclopropylidene]methylpurine (42).[16,18b] A mixture of (R,R)-(−)-oxaphospholane **38** (308.5 g, 0.929 mol)

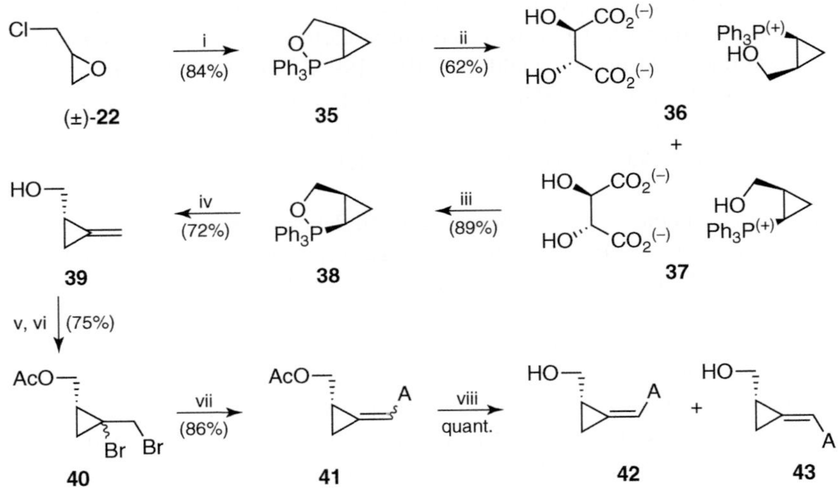

Scheme 4. *Reagents and conditions:* (i) $(Ph_3)_3PMe^+Br^-$, *t*-BuOK, toluene, heat; (ii) L-(+)-tartaric acid, acetone, then recrystallization; (iii) 2 M NaOH, CH_2Cl_2; (iv) $(CH_2O)_n$, sulfolane, heat; (v) Ac_2O, pyridine; (vi) pyridine·HBr_3, CH_2Cl_2; (vii) adenine, K_2CO_3, DMF, heat; (viii) NH_3, MeOH.

and paraformaldehyde (55.3 g, 1.84 mol) was heated in sulfolane (215 mL) at 100–110°C for 1 h. After cooling, the crude carbinol was distilled in vacuo (bp 50–85°C/20 torr, 66.3 g). Redistillation gave pure (*S*)-(+)-(methylenecyclopropyl)carbinol **39** (bp 59–66°C/20 torr, 54.3 g, 69.5%). Acetic anhydride (13.2 g, 130 mmol) was added dropwise over 30 min to a solution of (*S*)-(+)-(methylenecyclopropyl)carbinol **39** (10.35 g, 123 mmol) and pyridine (11.7 g, 148 mmol) in CH_2Cl_2 (40 mL) with stirring at room temperature. The stirring was continued for 3 h, the mixture was then quenched with water, and the product was extracted with pentane (4 × 70 mL). The combined organic layers were washed successively with saturated $CuSO_4$, 5% HCl, aqueous $NaHCO_3$, and brine and dried over Na_2SO_4. Pentane was removed at atmospheric pressure using a Vigreux column, and the acetylated product was distilled (bp 60–65°C, 5 torr, 21.0 g, 87.5%). Distillation afforded (*S*)-(+)-(methylenecyclopropyl)carbinol acetate (bp 68–70°C/20 torr, 13.2 g, 85%). Pyridine·HBr_3 (2.42 g, 7.56 mmol) was added to a solution of (*S*)-(+)-carbinol acetate (0.635 g, 5.04 mmol) in CH_2Cl_2 (20 mL) at −10°C with stirring. The stirring was continued at 0°C for 3 h. Ether (60 mL) was added, the solids were filtered off, and the organic phase was washed with saturated aqueous $Na_2S_2O_3$ followed by $NaHCO_3$. After drying over Na_2SO_4, the solvent was evaporated to give the product as a mixture of diastereoisomers (1*R*,2*R*)- + (1*R*,2*S*)-**40** as a syrup (1.3 g, 88%). The (1*R*,2*R*)- + (1*R*,2*S*)-**40** mixture (1 equiv.), K_2CO_3 (10 equiv.), and adenine (2 equiv.) in DMF were stirred at 100°C under nitrogen for 24 h. The solids were filtered off, and the solvent was evaporated to give an isomeric mixture of acetates (*Z/E*) **40** after chromatography on a silica gel column. Isomeric mixture **41** was stirred in 20%

methanolic ammonia at room temperature overnight. After removal of the volatiles and separation of the isomers by column chromatography using CH_2Cl_2/MeOH, (S)-(+)-synadenol **42** was obtained. Mp. 233–235°C; $[\alpha]_D^{25}$ 123.0° (c 0.073, MeOH). ^{1}H-NMR: δ 1.21(t, 1H, $^2J = {}^3J_{trans}$ = 6.8 Hz) and 1.49 (t, 1H, $^2J = {}^3J_{cis}$ = 8.7 Hz, H$_{3'}$), 2.14 (dq, 1H, $^3J_{cis}$ = 7.5 Hz, $^3J_{trans}$ = 5.7 Hz, H$_{4'}$), 3.10 and 3.17 (2s, 6H, N(CH$_3$)$_2$), 3.29 (t, overlapped with H$_2$O, 1H, ^2J = 10.8 Hz) and 3.74 (dt, 1H, ^2J = 10.8 Hz, ^3J = 5.3 Hz, H$_{5'}$), 5.11 (t, 1H, ^3J = 5.1 Hz, OH), 7.43 (s, 1H, H$_{1'}$), 8.42 (s, 1H, H$_2$), 8.86 (s, 1H, H$_8$), 8.90 [s, 1H, N=CHN(CH$_3$)$_2$].

The development of a second generation of geminal bishydroxymethylcyclopropanes[20] (Figure 2, **11**) led to the discovery of cyclopropavir **48**, which is currently undergoing preclinical studies as a potential therapeutic agent for the treatment of human cytomegalovirus infections.[21] The sequence of steps elected for generating methylenecyclopropane diol as key intermediate could be performed on a multigram scale starting from benzyl 2-bromoacrylate (Scheme 5).[22] A double Michael addition of benzyl alcohol provided cyclopropane dicarboxylate **44**, which was then reduced to **45** with DIBALH. Acetylation, followed by acetoxybromo elimination and debenzoylation proceeded smoothly to give **46**, which was then converted into suitably protected dibromo fragment **47**. The key alkylation–elimination of **47** with 2-amino-6-chloropurine and subsequent

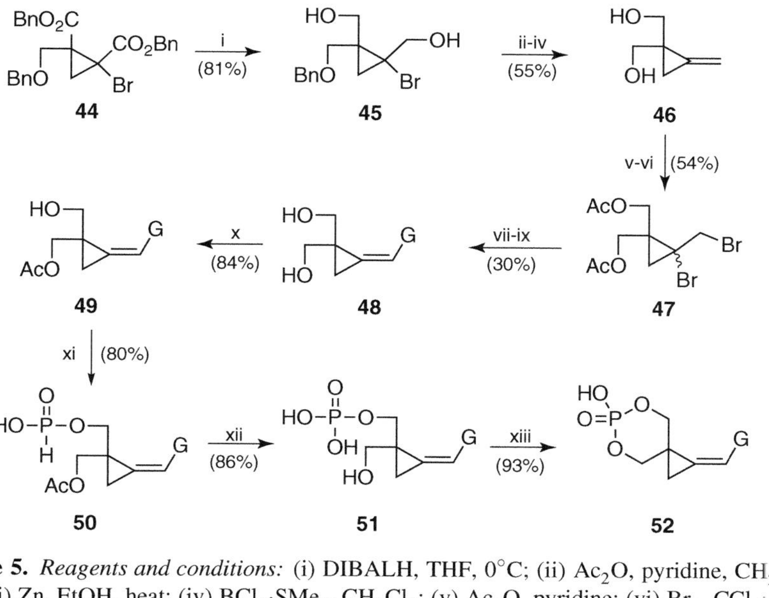

Scheme 5. *Reagents and conditions:* (i) DIBALH, THF, 0°C; (ii) Ac$_2$O, pyridine, CH$_2$Cl$_2$, 0°C; (iii) Zn, EtOH, heat; (iv) BCl$_3$·SMe$_2$, CH$_2$Cl$_2$; (v) Ac$_2$O, pyridine; (vi) Br$_2$, CCl$_4$; (vii) 2-amino-6-chloropurine, K$_2$CO$_3$, DMF, heat; (viii) K$_2$CO$_3$, MeOH/H$_2$O, then separation; (ix) H$_2$CO$_2$H, heat, then NH$_3$, MeOH; (x) MeC(OMe)$_3$, MeSO$_3$H, DMF, then 80% AcOH, rt; (xi) (PhO)$_2$P(O)H, pyridine, heat, then Et$_3$N, H$_2$O; (xii) TMSCl, Im, pyridine, then I$_2$, pyridine, then H$_2$O, then NH$_4$OH, then Dowex 50 H$^+$ form; (xiii) N,N'-dicyclohexyl-4-morpholinecarboxamide, DCC, pyridine, heat.

conversion into guanine base gave cyclopropavir **48**, which could be further reacted to produce active pronucleotides **51** and **52**.[23] Compound **48** was selectively monoacetylated to **49** via a cyclic ortho ester. Phosphitylation was carried out using diphenyl phosphite in pyridine, leading to **50**. Silylation, iodine-induced oxidation, and deacetylation furnished phosphate **51**, which underwent cyclization to **52**.

Higher functionalized species (Figure 2, **12**) represent the most recent addition to the growing family of methylenecyclopropane nucleosides.[24] Examination of the biological data collected for this series of compounds revealed trends similar to previous classes in terms of activity, although lower efficiencies were observed than of **10** and **11**.

3. FOUR-MEMBERED NUCLEOSIDE DERIVATIVES

Following the isolation from a culture filtrate of *Bacillus megaterium* of the first oxetane nucleoside, oxetanocin A,[25] and with a growing understanding of its structure–activity relationship, highly active analogues of this unique natural product have become increasingly available.[26] The sites of variations included the adenine base, which was modified with related purines as well as thymine, and the four-membered ring. The oxetanocins family was shown to exhibit a broad range of biological activities, including potent inhibitory action against HSV-1, HSV-2, HCMV, and HIV, together with antitumor and antibacterial activity.[25,27] A number of papers described synthetic studies directed toward the total synthesis of naturally occurring D-oxetanocin A.[28] A relatively efficient preparative method (12 steps, 7% overall yield) consists of a Wolff rearrangement of a ribonucleoside precursor bearing a pendant diazoketone moiety.[28d] This approach overcome a major inconvenience occurring upon Lewis acid catalyzed base–sugar couplings, since oxetanes are poorly tolerant of acidic conditions, due primarily to the sensitivity of the oxetane ring. Inspired by new evidences that appeared recently in the literature suggesting the possibility of gaining biological activity through the preparation of unnatural L-nucleosides, Gumina and Chu developed a novel route to the L-isomer of oxetanocin A, which may have notable potential as preparative method.[29]

Ribofuranose derivative **53** was synthesized on a multigram scale from L-xylose in seven steps (Scheme 6).[30] Benzylation of primary alcohol **53** using catalytic TBAI followed by an acid-induced deprotection of the isopropylidene ketal afforded lactol **54**. Subsequent oxidation with the formation of a lactone moiety required controlled pH (4.5–4.8) to be efficient in terms of isolated yield. Then mesylated lactone **55** underwent ring contraction, affording a 3:1 epimeric ratio of acids **56**, which were used as such in a one-pot domino reaction to provide an unseparable mixture of anomeric thiopyridyl oxetanes **57**. Condensation of benzoylated adenine with **57** proceeded with moderate α/β selectivity (2:3). The desired β-nucleoside **59** was easily isolated by chromatography, affording, after deprotection, **60**. On preliminary screening, no anti-HIV-1 activity was observed up to 100 mM.

The incorporation of a nucleobase into a suitably protected methyleneoxetane ring can be mediated by an electrophilic fluorine source via formation of an

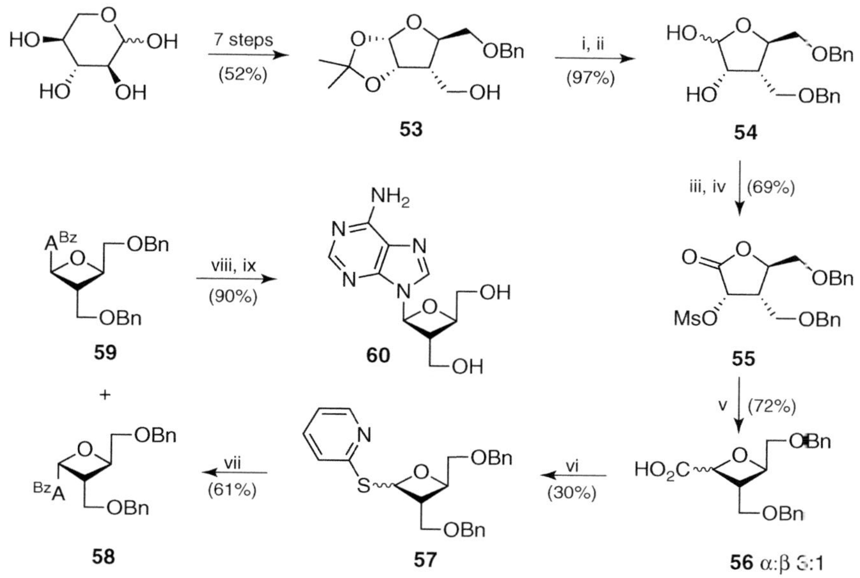

Scheme 6. *Reagents and conditions:* (i) NaH, THF, then TBAI, BnBr; (ii) 10% HCl, 1,4-dioxane; (iii) Br$_2$, BaCO$_3$, H$_2$O, 1,4-dioxane; (iv) MsCl, pyridine, CH$_2$Cl$_2$; (v) 1 N NaOH, MeOH; (vi) *i*-BuOC(O)Cl, *N*-methylmorpholine, THF, then 2-mercaptopyridine-*N*-oxide, Et$_3$N, then *hv*; (vii) N^6-benzoyladenine, Br$_2$, 4-Å molecular sieves, DMF; (viii) MeONa, MeOH; (ix) H$_2$, Pd black, EtOH.

oxonium ion intermediate.[31] Thus, starting with commercially available *cis*-butene-1,4-diol **61** (Scheme 7), protection as a bis-silyl ether and subsequent treatment with dimethyloxirane furnished epoxide **62**. Exposure of **62** to high-pressure carbonylation in the presence of an Al-based catalyst gave β-lactone **63**, whose further reaction in the presence of Petasis reagent furnished the required methyleneoxetane **64**. The efficiency of the next fluoro-mediated base insertion step in terms of overall yield was hampered by low diastereoselectivity. Purine regioisomers **65** and **66** were each obtained as mixtures of their facial diastereoisomers. Attempts to improve the stereochemical outcome of the reaction by exploitation of neighboring group anchimeric assistance were not fruitful. Nevertheless, separation of isomeric **65**, followed by conversion to adenine base and desilylation, led to the first *psico*-four-membered ring nucleoside analogue of oxetanocin A **67**.

3.1. Four-Membered Nucleoside Carbacycles

Several groups developed closely related carbocyclic analogues of oxetanocins (COXTs), some of which exhibited higher inhibitory ability as well as increased bioavailability as a result of their enhanced stability toward hydrolysis compared to their oxetanyl counterparts, while retaining the favorable broad-spectrum activity distinctive of this class of nucleosides. In particular, cyclobut-G showed activity

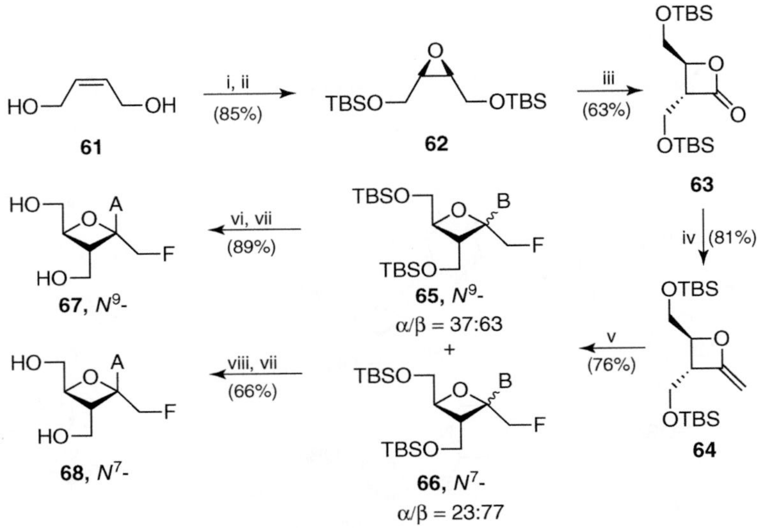

Scheme 7. *Reagents and conditions:* (i) TBDMSCl, Im, DMAP, CH$_2$Cl$_2$; (ii) dimethyldioxirane, CH$_2$Cl$_2$, rt; (iii) CO (800 psi), [(ClTPP)Al-(THF)$_2$]$^+$[Co(CO)$_4$]$^-$ 1 mol%, 60°C; (iv) Cp$_2$TiMe$_2$, toluene, 80°C; (v) 6-chloro-9-(trimethylsilyl)-9H-purine, Selectfluor, MeNO$_2$, rt; (vi) NH$_3$, i-PrOH, 100°C; (vii) TBAF, THF; (viii) NH$_3$, MeOH, 80°C.

comparable to that of acyclovir against HSV-1 and HSV-2, but was more active than acyclovir against VZV and EBV. Cyclobut-A was more potent than acyclovir against VZV, less potent for treating HSV-1 and HSV-2 infections. Excellent activity against both MCMV and HCMV was displayed by both compounds, with the adenine congener surpassing gancyclovir.[32]

Early synthetic methodologies[32b,33] for the preparation of racemic cyclobutyl purine nucleosides were soon superseded by the development of more appealing enantioselective protocols, especially in view of data pointing to the drastic divergence in potencies displayed by enantiomeric carbocyclic oxetanocins, with the natural isomer accounting solely for the antiviral activity.[32d] This suggested a role played by the absolute configuration of the 2′-hydroxymethyl group as a mimic of the configuration of the 3′-OH in natural nucleosides.

Besides chemical[32d] and enzymatic[34] resolution methods, a chiral titanium complex was employed to catalyze the asymmetric [2 + 2] cyclization of a highly functionalized cyclobutane ring as chiral precursor to optically pure cyclobut-A and cyclobut-G.[32a] Independently, Jung and Sledeski devised an expedient route to cyclobut-A based on an enzymatic desymmetrization strategy of a *meso*-cyclobutene intermediate.[35] The utility of chiral substituted cyclobutanones as synthetic intermediates for the synthesis of (−)-cyclobut-A and (±)-3′-*epi*-cyclobut-A has been exploited by Brown and Hegedus, as illustrated in Scheme 8.[36] As a result of photolysis between chromium carbene complex **69** and optically active ene-carbamate **70**, a α-heteroatom-substituted cyclobutanone

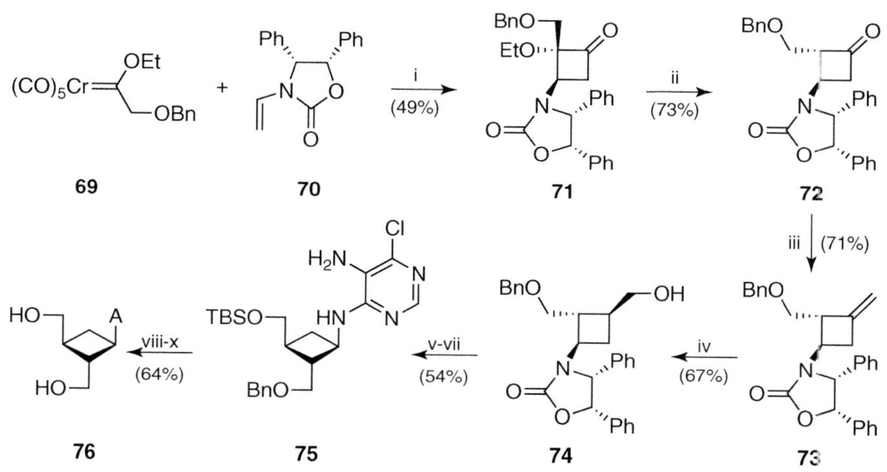

Scheme 8. *Reagents and conditions:* (i) hv, CH$_2$Cl$_2$, −30°C; (ii) SmI$_2$, MeOH, THF; (iii) Zn/CH$_2$I$_2$/TiCl$_4$; (iv) BH$_3$·THF, THF, 0°C, then (COCl)$_2$, DMSO, CH$_2$Cl$_2$, −78°C, then NaOMe, MeOH, NaBH$_4$, EtOH; (v) TBDMSCl, Im, DMF; (vi) H$_2$, Pd(OH)$_2$, Et$_3$N, EtOH; (vii) 5-amino-4,6-dichloropyrimidine, Et$_3$N, BuOH; (viii) HCl, HC(OEt)$_3$, DMF, 70°C; (ix) NH$_3$, MeOH, 70°C; (x) BCl$_3$, −78 to 25°C.

intermediate **71** is generated, which readily undergoes selective reductive cleavage to deoxygenated compound **72** by reacting with SmI$_2$. Organotitanium-induced carbonyl methylenation of **72** was achieved preferentially with Takai reagent. With the methylenecyclobutane **73** in hand, several conditions were screened to selectively access the desired *trans*-dihydroxymethylated product. Eventually, a three-component sequence with minimal intermediate stages purification, comprising hydroboration/ Swern oxidation, NaOMe epimerization, and NaBH$_4$ reduction, proved to be most efficient in terms of isolated overall yields (67%) and proportions of *trans*- and *cis*-dialkyl compounds (>17:1 ratio). Cyclobut-A **76** was then obtained from this optically pure precursor in 64% overall yield and >98% ee. The process required TBS protection of primary OH, palladium-catalyzed hydrogenolysis of the oxazolidinone ring to generate a free amine, and coupling with 5-amino-4,6-dichloropyrimidine. Further pyrimidine elaboration by standard methodology, followed by removal of the protecting groups, provided optically pure **76**.

Another remarkable example of cyclobutanone chemistry led to the enantioselective synthesis of cyclobut-G **2**.[37] The [2 + 2]-cycloaddition reaction of readily available chiral *cis*-enol ether **77**, prepared from commercially available (*S*)-stericol, to dichloroketene could be combined successfully with a dechlorination process in one-pot fashion. Such a transformation is very powerful for the construction of functionalized four-membered ring systems in a stereoselective manner. Thus, cyclobutanone **79** could be isolated efficiently in the form of a stable dehalogenated adduct and with high stereoselectivity (98:2), as shown in Scheme 9.

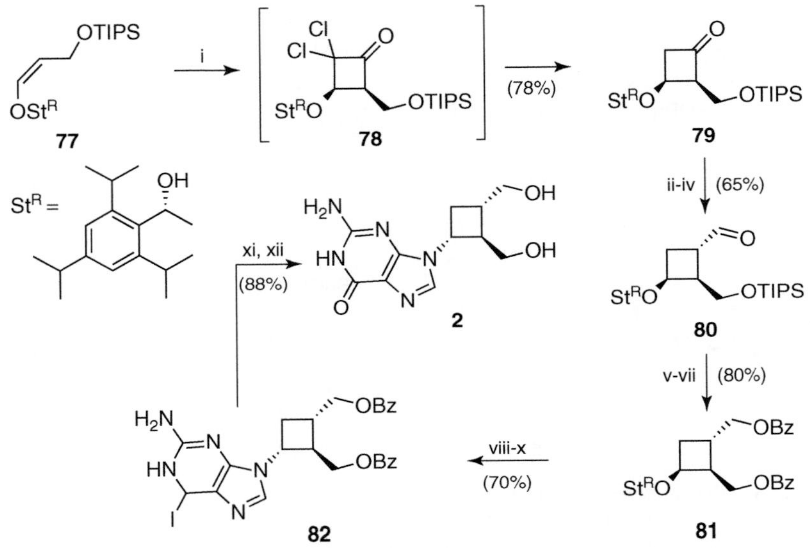

Scheme 9. *Reagents and conditions:* (i) Cl₃CCOCl, Zn/Cu, Et₂O, 20°C, then MeOH, NH₄Cl; (ii) Ph₃PCH₂OMe⁺Cl⁻, KHMDS; (iii) Cl₃CCO₂H; (iv) NaOMe; (v) NaBH₄; (vi) TBAF; (vii) BzCl, pyridine; (viii) TFA; (ix) Tf₂O, pyridine; (x) 2-amino-6-iodopurine tetrabutylammonium salt; (xi) NaOMe; (xii) HCl.

Further elaboration of **79** required Wittig olefination, hydrolysis of the so-formed enol ether, and isomerization of the resulting formyl group. Aldehyde **80** underwent sequential reduction, TIPS cleavage, and dibenzoylation to provide **81**, which upon treatment with TFA generated a cyclobutanol after stericol cleavage. Nucleophilic displacement of a convenient triflate derivative by a purine salt precursor of guanine gave **82**, which was hydrolyzed to **2** in a straightforward way.

Fused [3.2.0]bicyclic lactones are an attractive platform for the regio- and diastereoselective synthesis of cyclobutanes, as they are readily accessed by [2 + 2] photocycloaddition reactions, during which a substituted four-membered carbocycle with the desired relative stereochemistry of contiguous centers is formed. An intramolecular version of this reaction has proven to be instrumental in allowing the preparation of cyclobut-A in a virtually stereochemically pure form.[38] Similar thinking is to be found behind the photochemical preparation of chiral bicyclic lactone precursor **85** from (S)-5-substituted 2(5H)-furanone **83**, devised by Alibes et al. (Scheme 10).[39] The irradiation of a solution of cyclic enone **83** and a ketene dialkyl acetal in ether with a high-pressure mercury lamp through a quartz filter proceeded with good facial selectivity to give a 2:1 mixture of HT-*anti* and HT-*syn* configured photocycloadducts, after direct treatment with *p*-TsOH in acetone. Chromatographic purification afforded the isomerically pure *anti*-cyclobutanone **85** in 46% yield over two steps. Stereoselective reduction of **85** was possible from the less hindered face using L-Selectride. The resulting alcohol was then silylated and reduced with LiAlH₄ to **87**. The vicinal diol moiety of the resulting triol was

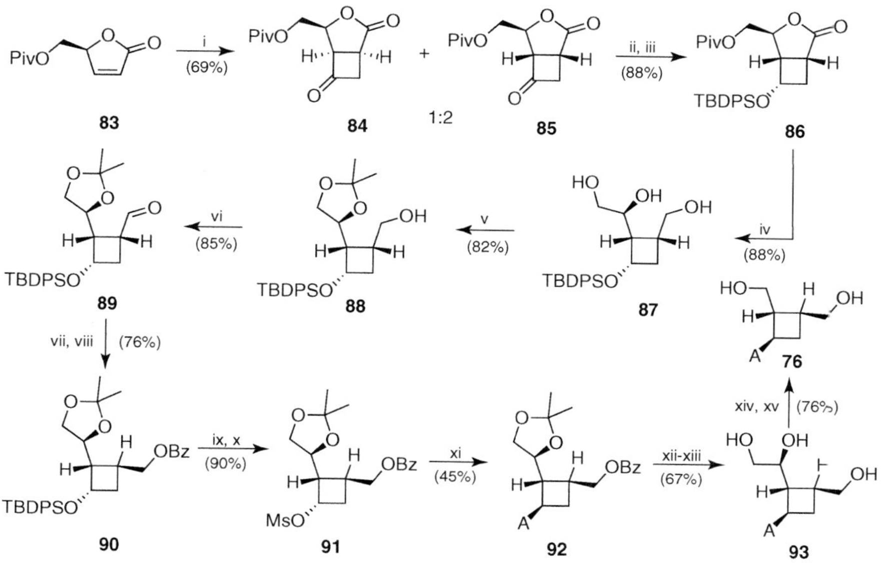

Scheme 10. *Reagents and conditions:* (i) 1,1-diethoxyethylene, $h\nu$, Et$_2$O, $-20°$C, then TsOH, acetone, 56°C; (ii) L-Selectride, THF, $-78°$C; (iii) TBDPSCl, Im, THF; (iv) LiAlH$_4$, THF, 0°C; (v) acetone, CuSO$_4$, HCl (cat.); (vi) DMP, CH$_2$Cl$_2$, rt; (vii) Na$_2$CO$_3$, MeOH, rt, then NaBH$_4$, MeOH, rt; (viii) BzCl, pyridine, CH$_2$Cl$_2$, rt; (ix) TBAF, THF; (x) MsCl, Et$_3$N, CH$_2$Cl$_2$; (xi) adenine, K$_2$CO$_3$, 18-crown-6, DMF, 120°C; (xii) Na$_2$CO$_3$, MeOH, rt; (xiii) TFA, H$_2$O, rt; (xiv) NaIO$_4$, THF, H$_2$O, rt; (xv) NaBH$_4$, MeOH, 0°C.

protected as isopropylidene acetal **88** in the presence of anhydrous copper sulfate and catalytic HCl. Epimerization at C1' occurred via an intermediate *trans*-aldehyde **89**, which was reduced to **90**. Subsequent functional group manipulation led to **91**, followed by standard base incorporation by alkylation of mesylate **91** to afford **92** with high N^9-isomeric regioselectivity. Global deprotection using the standard conditions for Bz and acetal removal, followed by an oxidative cleavage–reduction step, furnished (−)-cyclobut-A **76** in six steps and 28% overall yield.

Experimental Procedure

9-[(1 R,2 R,3 S)-2,3-Dihydroxymethylcyclobutyl]adenine (76).[39] A mixture of adenine (60 mg, 0.44 mmol), anhydrous K$_2$CO$_3$ (92 mg, 0.67 mmol), and 18-crown-6 (118 mg, 0.45 mmol) in dry DMF (3 mL) was stirred for 15 min at 80°C under an argon atmosphere. A solution of mesylate **91** (85 mg, 0.22 mmol) in dry DMF (2.4 mL) was then added dropwise and the mixture was heated at 120°C for 7 h. The mixture was cooled to room temperature and concentrated to dryness. The residue was purified by column chromatography (CH$_2$Cl$_2$/MeOH, 15:1) to give the N^9-alkylated isomer **92** (42 mg, 0.10 mmol, 45% yield) as a white solid, together with traces of the N^7-alkylated isomer contaminated with 18-crown-6 ether. To a solution of **92** (34 mg, 0.08 mmol) in MeOH (7 mL),

Na$_2$CO$_3$ (89 mg, 0.64 mmol) was added. After stirring vigorously for 2 h at room temperature, Et$_2$O (14 mL) was added and the mixture was filtered through Celite and then concentrated to dryness, affording crude product which was used in the next step without further purification. A solution of the crude material in a mixture (8:1) of trifluoroacetic acid–water (2.5 mL) was stirred overnight at room temperature. It was then diluted with EtOH and concentrated in vacuo. The residue was purified by column chromatography (CH$_2$Cl$_2$/MeOH, 9:1) to give triol **93** (18 mg, 0.06 mmol, 80% yield). To an ice-cooled solution of **93** (12 mg, 0.04 mmol) in a 1:1 mixture of THF/H$_2$O, NaIO$_4$ (10 mg, 0.05 mmol) was added. The reaction mixture was stirred for 2 h at room temperature. Then THF was added (2 mL) and the mixture was cooled to 0°C and filtered. The filtrate was maintained at 0°C and treated with NaBH$_4$ (7 mg, 0.19 mmol). After stirring for 1.5 h, the reaction mixture was treated with saturated aqueous NH$_4$Cl and 30% NH$_3$ and evaporated to dryness. Purification by column chromatography (CH$_2$Cl$_2$/MeOH, 9:1) afforded **76** (9 mg, 0.036 mmol, 84% yield). [α]$_D$ −14.3° (c 0.7, H$_2$O). ^{1}H-NMR (D$_2$O): δ 8.20 (s, 1H, H-8), 8.07 (s, 1H, H-2), 4.59 (q, $J_{1',4'}$ = 8.7 Hz, $J_{1',2'}$ = 8.7 Hz, 1H, H-1'), 3.73–3.67 (m, 4H, H-1'', H-1'''), 2.74 (dddd, $J_{2',1'}$ = 8.8 Hz, $J_{2',1''}$ = 6.0 Hz, $J_{2',1'''}$ = 6.0 Hz, $J_{2',3'}$ = 6.0 Hz, 1H, H-2'), 2.64 (ddd, J_{gem}=10.6 Hz, $J_{4',1'}$ = 7.7 Hz, $J_{4',3'}$ = 7.7 Hz, 1H, H-4'), 2.24 (m, 1H, H-3'), 2.15 (m, 1H, H-4').

Owing to the fact that the presence of a hydroxymethyl group in carbocycles mimicking the terminal carbon atom of the sugar in the natural nucleoside series is not a precondition for excellent activity, lower homologues of cyclobutane nucleosides were also prepared, mostly in a nonstereoselective manner by different groups.[40] Synthetic methods for sugar–base coupling were alternatively based on ring opening of a convenient cyclobutyl epoxide,[40a] Michael addition to cyclobutene carboxylates,[40c] and nucleophilic displacement of a cyclobutyl tosylate[40b] and more recently of bromocyclobutanones.[40d] No significant biological activity was reported.

Fernandez and co-workers studied in detail the use of natural terpenes as versatile scaffolds for the synthesis of enantiomerically pure *gem*-dimethyl cyclobutane carbonucleosides. This method relied on the initial oxidative cleavage of (−)-1S-α-pinene **94** to (−)-*cis*-pinonic acid **95** and was proved to be relevant to attain a wide range of chiral amino alcohols **96**,[41] and **98**, **99**,[42] and corresponding purine nucleosides (Scheme 11).[43] For example, **95** can be reduced to pinolic acid and then acetylated to give intermediate **97**, which upon treatment with ethyl chloroformate and reduction afforded amino alcohol **99**. Similar strategies were also developed from R-(−) pinene,[44] nopinone,[45] and S-(−)-verbenone.[46]

3.2. Four-Membered Unsaturated Carbocycles and Other Nucleoside Derivatives

Unsaturated four-membered ring nucleosides have been prepared which possess either an exocyclic[47] or an endocyclic double bond,[48] and have been found to be, in general, poorly active as antivirals.

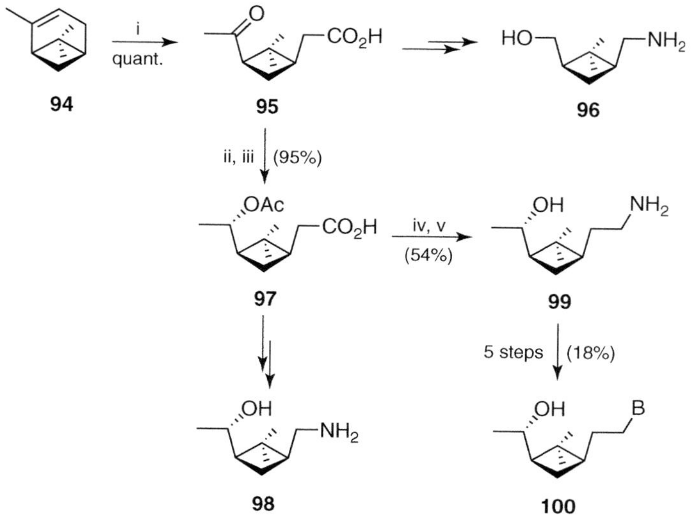

Scheme 11. *Reagents and conditions:* (i) NaIO$_4$, RuCl$_3$, CCl$_4$/MeCN/H$_2$O, rt; (ii) NaBH$_4$, 0.5 N NaHCO$_3$, EtOH, heat; (iii) Ac$_2$O, pyridine, rt; (iv) EtOCOCl, THF, Et$_3$N, −10 to 0°C, then NH$_3$; (v) LiAlH$_4$, THF, heat.

The synthetic challenge posed by the inherent instability of strained cyclobutene intermediates to thermal electrocyclic ring opening was dealt with in the route proposed by Gourdel-Martin and Huet (Scheme 12).[49] This approach allowed for the synthesis of four-membered analogues of norcarbovir unsaturated at the vinylic position. The required *cis*-substituted cyclobutene **103** was generated from the photocycloaddition product of acetylene to maleic anhydride, compound **101**, in two stages. First, cyclobutene anhydride **101** was reduced to lactone **102** by treatment with NaBH$_4$, and then exposure to ammonia led to hydroxyamide **103**. Compound **103** underwent Hofmann rearrangement in the presence of bis(acetoxy)iodobenzene with formation of a cyclic carbamate, whose protection as *tert*-butoxycarbonate gave **104**. Hydrolysis of **104** was attained under mild basic treatment, followed by formation of the corresponding hydrochloride salt, which after benzylation served as a substrate for the subsequent coupling step. By using the highly reactive adenine precursor 4,6-dichloro-5-nitropyrimidine, the substitution reaction could be carried out at room temperature. The final cyclobutene adenine **108** was then formed via linear approach in good yield.

Methylenecyclobutane nucleosides were prepared by Zemlicka using the procedure depicted in Scheme 13.[47e] Lewis acid–catalyzed [2 + 2] cycloaddition between ethyl acrylate **109** and ketene dimethyl thioacetal **110** afforded cyclobutyl disulfide **111**. Reduction of carboxylate to the hydroxymethyl derivative **112** was followed by benzyl protection and thioacetal cleavage. The resulting cyclobutanone **113** underwent Wittig olefination and subsequent epoxidation to oxirane **115**. Base insertion occurred by ring opening of intermediate **115**, giving a *Z,E*-mixture of

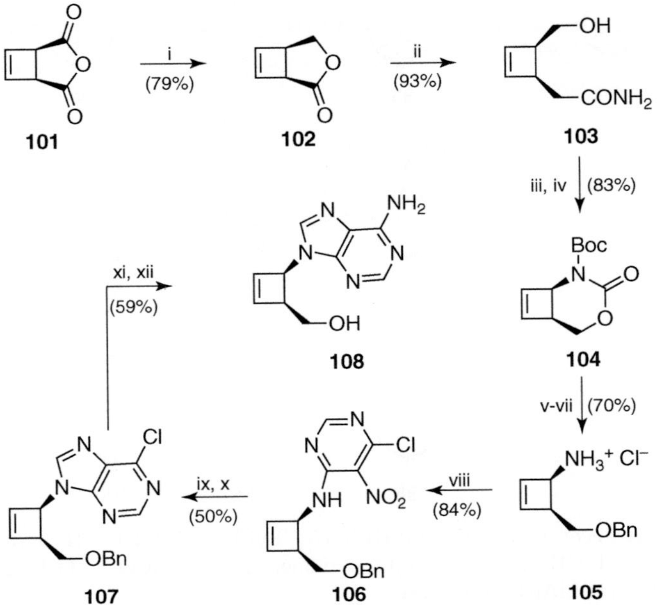

Scheme 12. *Reagents and conditions:* (i) NaBH$_4$, THF, MeOH, −78°C; (ii) NH$_3$, MeOH, rt; (iii) PhI(OAc)$_2$, KOH, MeOH, 0°C to rt; (iv) Boc$_2$O, Et$_3$N, DMAP, THF, 0°C to rt; (v) LiOH, MeOH/H$_2$O 1:1, −10°C; (vi) NaH, Bu$_4$NI, BnBr, THF, −5 to 15°C; (vii) 3 M HCl, MeOH, −5 to 15°C; (viii) 4,6-dichloro-5-nitropyrimidine, Et$_3$N, CH$_2$Cl$_2$, rt; (ix) SnCl$_2$·2H$_2$O, EtOH, 60°C; (x) HC(OEt)$_3$, HCl, rt; (xi) NH$_3$, MeOH, 10–11 bar, 40°C; (xii) BCl$_3$, CH$_2$Cl$_2$, −78°C.

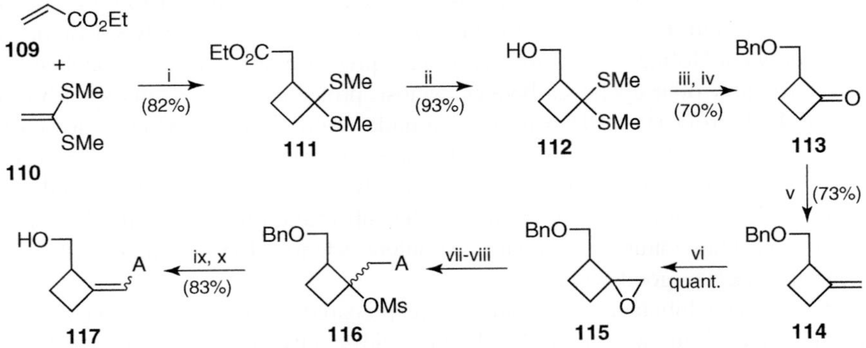

Scheme 13. *Reagents and conditions:* (i) Et$_2$AlCl, CH$_2$Cl$_2$; (ii) LiAlH$_4$, THF; (iii) NaH, THF, then BnBr, Bu$_4$NI; (iv) NCS, AgNO$_3$, MeCN; (v) Ph$_3$PMe$^+$ Br$^-$, BuLi, THF; (vi) *m*-CPBA, NaHCO$_3$, CH$_2$Cl$_2$; (vii) A, NaH, DMF, heat; (viii) MsCl, DMAP, CH$_2$Cl$_2$, pyridine; (ix) *t*-BuOK, THF; (x) BCl$_3$, CH$_2$Cl$_2$, −78°C.

hydroxy compounds which were readily protected as mesyl ethers **116**. Mesyloxy elimination upon treatment with *t*-BuOK and *O*-debenzylation proceeded smoothly to give *E*- and *Z*-isomers **117**, which could be separated only partially on alumina.

Alongside the multitude of synthetic efforts concerning four-membered carbocycles, attention has also been dedicated to replacement of the oxygen of oxetanocins with a sulfur atom, aiming at increasing nucleoside stability toward chemical and enzymatic degradation.[50] So far, the Pummerer rearrangement has been employed as a common preparative tool for four-membered thietane nucleoside synthesis, combining a cyclic sulfoxide intermediate with a silylated base in the presence of a Lewis acid.[50a-c,e] For example, Chu et al. described a procedure for the preparation of a variety of modified D- and L-thietanose purines and pyrimidines, derived, respectively, from D- (Scheme 14) and L-xylose.[50c] By using known conditions, D-xylose was transformed in five steps into ribofuranoside **118**. Mesylation of the free hydroxyl groups and selective nucleophilic substitution with potassium thioacetate furnished monothioacetate **119**. Cyclization and isopropylidene deprotection led smoothly to bicycle **120**, whose oxidative cleavage furnished key intermediate **121**. Monosilylation at the primary OH function using 2,4,6-collidine as a catalyst, followed by a sequence of Moffatt oxidation and methylenation, provided methylenethietane **122**, which then underwent hydroboration to give preferentially 2′-β-configured derivative **123**. After inversion of configuration and an additional

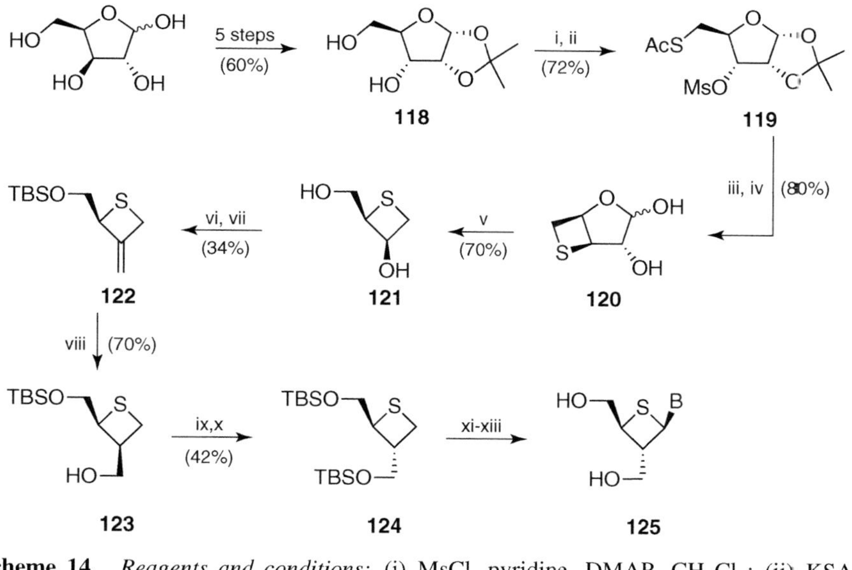

Scheme 14. *Reagents and conditions:* (i) MsCl, pyridine, DMAP, CH$_2$Cl$_2$; (ii) KSAc, DMF; (iii) NaHCO$_3$, EtOH/H$_2$O, heat; (iv) 4% TFA, OH$^-$ resin; (v) NaIO$_4$, NaBH$_4$, MeOH; (vi) TBDPSCl, TEA, DMAP, CH$_2$Cl$_2$; (vii) Ac$_2$O, DMSO, then Petasis reagent, THF/Et$_2$O 1:1, heat; (viii) BH$_3$SMe$_2$, THF, H$_2$O$_2$, 1 N NaOH; (ix) Swern oxidation, NaOMe, MeOH, NaBH$_4$; (x) TBDPSCl, Im, CH$_2$Cl$_2$; (xi) *m*-CPMA, CH$_2$Cl$_2$; (xii) pyrimidine or purine, HMDS, MeCN, TMSOTf, TEA, ZnI$_2$, toluene; (xiii) TBAF, THF.

silylation step, fully protected derivative **124** was oxidized and subjected immediately to Pummerer reaction with canonical and modified silylated heterocyclic bases, furnishing thietane nucleosides **125** with different degrees of α/β epimerization, depending on the base inserted. A moderate anti-HIV activity was observed, together with cytotoxicity.

Only recently have Nishizono et al. addressed the possibility of performing direct glycosylation between glycosyl fluoride **129** and silylated bases to obtain thietane derivatives (Scheme 15).[50d] Isopropylidene protection of 2,2-bis(bromomethyl)-1,3-propanediol **126** followed by treatment with sodium sulfide afforded intermediate **127**. Following functional group manipulation, dibenzoyl thietane **128** was fluorinated to **129**. The four-membered ring was found to be stable under the Lewis acidic conditions required for base addition.

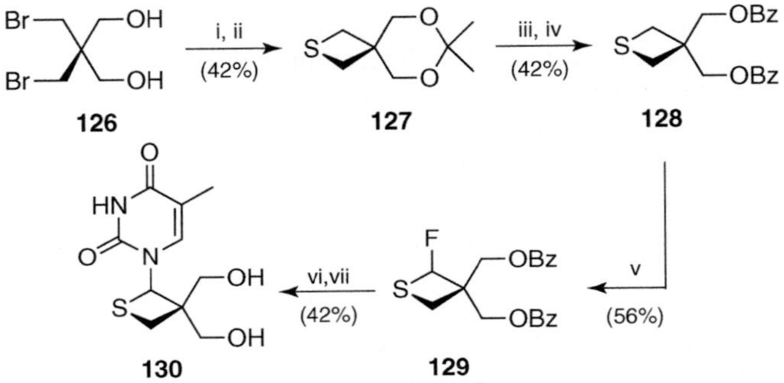

Scheme 15. *Reagents and conditions:* (i) 2,2-dimethoxypropane, *p*-TsOH, acetone; (ii) $Na_2S \cdot 9H_2O$, DMF, 100°C; (iii) *p*-TsOH, MeOH; (iv) BzCl, DMAP, Et_3N, CH_2Cl_2; (v) DAST, $SbCl_3$, CH_2Cl_2; (vi) thymine, HMDS, $SnCl_2$, $AgClO_4$; (vii) $MeNH_2$, MeOH.

Experimental Procedure

1-[3,3-Bis(hydroxymethyl)thietan-2-yl]thymine (130).[50d] To a solution of **128** (100 mg, 0.29 mmol) in dry CH_2Cl_2, we added DAST (77 mL, 0.58 mmol) and $SbCl_3$ (1 mg) at room temperature under a nitrogen atmosphere. The reaction mixture was stirred at room temperature for 20 h and was then diluted with EtOAc and washed with saturated $NaHCO_3$, water, and brine. The organic layer was dried over Na_2SO_4 and concentrated in vacuo to leave a residue that was purified by column chromatography using hexane/EtOAc (6:1) as eluent, to give **129** (101 mg, 92%) as an yellow oil. A mixture of thymine (87 mg, 0.68 mmol) and HMDS (4 mL) was refluxed for 3 h until the solution became clear. After evaporation, the residue was dissolved in dry CH_2Cl_2 (3 mL). A solution of **129** (122 mg, 0.34 mmol) in dry CH_2Cl_2 (2 mL), $AgClO_4$ (71 mg, 0.34 mmol), and $SnCl_4$ (65 mg, 0.34 mmol), was then added to the mixture at 0°C, which was stirred at room temperature for 18 h. The reaction mixture was diluted with EtOAc and washed with saturated $NaHCO_3$, water, and brine. The organic layer was dried over Na_2SO_4 and concentrated in vacuo. The residue was purified by column chromatography using hexane/EtOAc (1:1) as eluent to give protected thymine derivative (88 mg, 56%) as a white foam.

Methylamine (40% in MeOH, 50 mL) was then added (572 mg, 1.2 mmol), and the resulting mixture was stirred at room temperature for 2 h. The mixture was concentrated in vacuo, and the residue was purified by column chromatography, using 10% MeOH in CHCl$_3$ as eluent, to give **130** (269 mg, 87%) as a colorless solid. Mp 204–209°C (MeOH). ^{1}H-NMR (DMSO-d_6): δ 9.50 (br s,1H), 8.18 (s, 1H), 5.85 (s, 1H), 4.98 (dd, 1H, J = 5.4 and 5.7 Hz), 4.64 (dd, 1H, J = 4.5 and 4.6 Hz), 3.59 (dd, 1H, J = 5.7 and 10.8 Hz), 3.48–3.32 (m, 3H), 2.87 (d, 1H, J = 8.9 Hz), 2.79 (d, 1H, J = 8.9 Hz), 1.85 (s, 3H).

Kato et al. achieved the first enantioselective synthesis of azetidinyl nucleosides via an N-aminoazetidine intermediate and stepwise buildup of the base moiety.[51] In contrast to the previously studied systems, the four-membered ring is actually linked to the base via the heteroatom, giving rise to a N,N-glycosidic bond type.

4. SIX-MEMBERED HEXOPYRANOSYL NUCLEOSIDE DERIVATIVES

The development of alternative nucleic acids, especially comprising pento- and hexopyranose sugar mimics in their backbone (instead of the natural furanosyl units), for the formation of new oligomeric pairing systems represents an important avenue of research in synthetic biology and in the search for the origin of life.[52] Within this context, it is relevant to mention the pioneering demonstrations of the prebiotic relevance and therapeutic applicability of hexopyranosyl oligomers. In 1992, Eschenmoser and co-workers described the first example of an autonomous pairing system **131** in the framework of a project aiming at providing insight into the natural selection of pentose over hexose sugars, which led to the assembling of DNA and RNA.[53] Critical to the improved understanding in this regard was the parallel study of related synthetic derivatives of homo-DNA itself, aiming at defining an etiology of nucleic acids.[52a,54] Concurrently, the group of Herdewijn directed their research efforts across a broad range of six-membered carbohydrate-like oligonucleotides as potentially useful therapeutical agents able to cross-communicate, which culminated with the discovery of a new family of hexitol nucleic acids (Figure 3).[52e] The rationale behind this research was based on enthalpic and entropic factors ruling duplex formation, which is facilitated by stacking interactions and hydrogen bonding while being hindered by a reduced conformational freedom. The small loss of entropy provided by conformationally rigid pyranose oligonucleotides upon binding would be energetically advantageous over furanose oligomers upon hybridization with DNA and/or RNA targets.

4.1. β-D- and β-L-Homo-DNA Monomers and Related Dideoxy D-Erythrohexopyranosyl Nucleosides

The β-D-homo-DNA building block 1-(2,3-dideoxy-β-D-erythrohexopyranosyl) thymine **143**, as well as its congeners, can be prepared conveniently through two different routes, both starting from commercially available 3,4,6-tri-O-acetyl-D-glucal **140** (Scheme 16). Ferrier rearrangement of **140** into 2,3-unsaturated

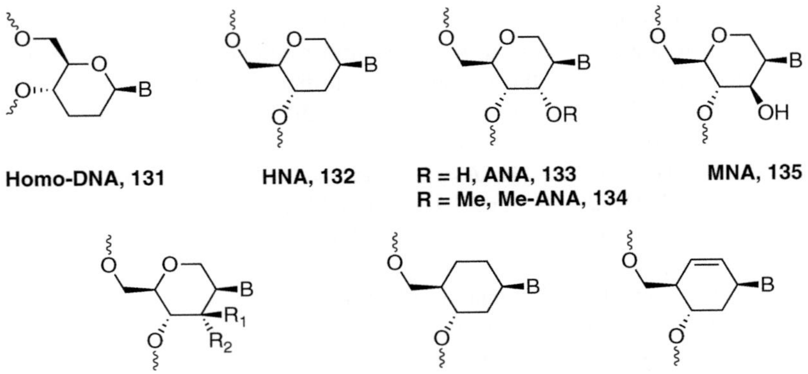

Homo-DNA, 131 HNA, 132 R = H, ANA, 133; R = Me, Me-ANA, 134 MNA, 135

R_1 = H, R_2 = F, FHNA, 136; R_1 = F, R_2 = H, Ara-FHNA, 137 CNA, 138 CeNA, 139

Figure 3. 4′→6′-Linked hexopyranosyl–NA and structural relationship among hexitol nucleic acids.

glycoside **141**, followed by catalytic hydrogenation, provided pyranose **142**, which underwent Vorbrüggen glycosylation with silylated thymine, to afford, after deacetylation, the desired β-anomer **143** as major compound of a 4:1 separable mixture with the related α-anomer.[53] Alternatively, the base can be introduced in the first step of the synthesis by direct condensation of **140** with bis(trimethylsilyl)thymine in the presence of trimethylsilyltrifluoromethanesulfonate. Compound **143** is then obtained in good yield after removal of the 3′-acetyl function and final deprotection.[55] While β-homo-DNA does not cross-pair with DNA or any other of the known artificial nucleic acid systems,[56] isomeric α-homo-DNA is a self-pairing system that can also hybridize with RNA, forming a parallel-oriented non-A-, non-B type of duplex.[57]

In recent years, there has been a further effort toward the development of the stereoselective synthesis of diastereoisomeric β-L-homo-DNA monomers.[58] In their synthesis of homoadenosine, O'Doherty et al. utilized palladium-catalyzed cross-coupling en route to their target (Scheme 17).[58a] Diastereoselective and enantioselective glycosylation of β-Boc-protected pyranone **146** was used to form the key enone **147**. This enone was then the substrate for $NaBH_4$ reduction to give allylic alcohol **148** with complete retention of stereochemical integrity. Exposure, in turn, to methanolic ammonia and diimide reduction conditions (NBSH, Et_3N) afforded **150**. Unmasking the hydroxyl functionality with TBAF eventually provided 2,3-dideoxy-β-L-ribohexopyranose **151**.

Experimental Procedure

6-Amino-9-(2,3-dideoxy-β-L-ribohexopyanosyl)purine (151).[58a] A THF (4 mL) solution of pyranone **146** (200 mg, 0.56 mmol) and 6-chloropurine (173 mg, 1.12 mmol) was cooled to 0°C. A CH_2Cl_2 (0.4 mL) solution of $Pd_2(dba)_3{\cdot}CHCl_3$

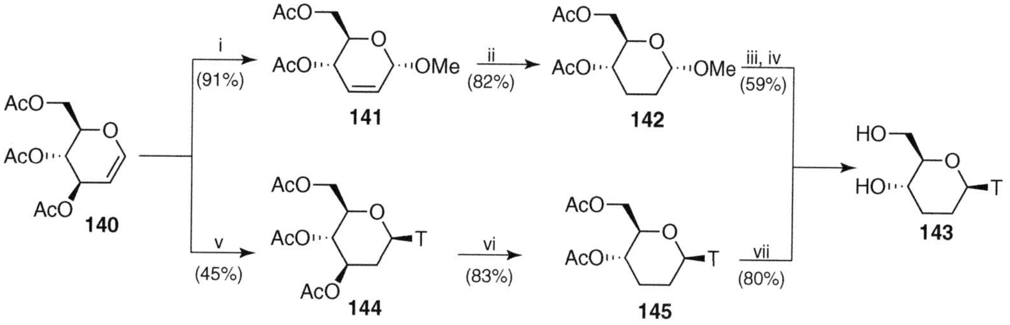

Scheme 16. *Reagents and conditions:* (i) MeOH, toluol, BF$_3$·Et$_2$O, 5°C to rt; (ii) H$_2$, Pd/C, MeOH, AcOH; (iii) thymine, HMDS, TMSCl, SnCl$_4$, MeCN, 45°C; 65% of **143** together with 18% of the α-anomer; (iv) NH$_3$, MeOH, rt; (v) thymine, HMDS, TMSOTfl, MeCN; (vi) H$_2$, Pd/C, MeOH; (viii) NH$_3$, MeOH.

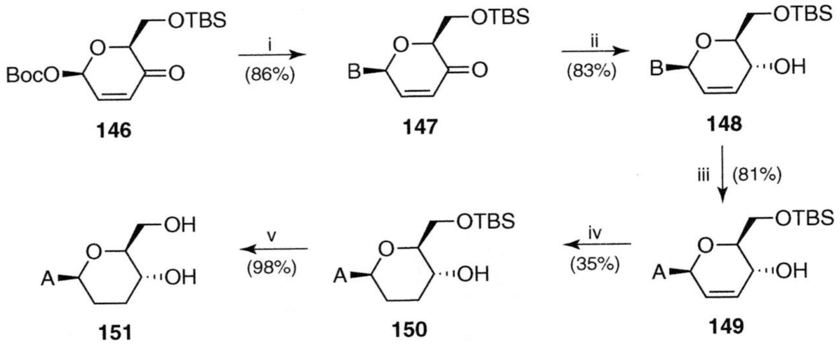

Scheme 17. *Reagents and conditions:* (i) B = 6-chloropurine, 0.5% Pd$_2$(dba)$_3$·CHCl$_3$, 2.5% Ph$_3$P, THF, 0°C; (ii) NaBH$_4$, MeOH/CH$_2$Cl$_2$, −78°C; (iii) NH$_3$, MeOH; (iv) NBSH, Et$_3$N; (v) TBAF, THF, 0°C.

(14.5 mg, 2.5 mol%) and triphenylphosphine (14.5 mg, 10 mol%) was added to the reaction mixture at 0°C. The reaction mixture was stirred at 0°C for 2 h and then quenched with 5 mL of saturated aqueous NaHCO$_3$, extracted (3 × 5 mL) with Et$_2$O, dried over Na$_2$SO$_4$, and concentrated under reduced pressure. The crude product was purified by flash chromatography, eluting with 40% EtOAc/hexane to give **147** (190 mg, 0.48 mmol, 86%) as a viscous oil. A CH$_2$Cl$_2$ (1 mL) solution of enone **147** (162 mg, 0.41 mmol) and MeOH (1 mL) was cooled to −78°C. NaBH$_4$ (16.3 mg, 0.43 mmol) was added and the reaction mixture was stirred at −78°C for 1 h. The reaction mixture was diluted with ether (10 mL) and was quenched with 5 mL of saturated aqueous NaHCO$_3$, extracted (3 × 5 mL) with EtOAc, dried over Na$_2$SO$_4$, and concentrated under reduced pressure. The crude product was purified by flash chromatography, eluting with 50% EtOAc/hexane to give **148** (136 mg, 0.34 mmol, 84%) as a colorless oil. Methanolic ammonia (3 ml) was added to chloride **148** (31 mg, 78 µmol), which was kept at room temperature for 48 h. The solvent was evaporated and the product was purified by column chromatography using 4% MeOH/EtOAc to give compound **149** as a viscous oil (24 mg, 64 µmol, 81%). Allylic alcohol **149** (35 mg, 93 µmol) and *o*-NO$_2$ArSO$_2$NHNH$_2$ (186 mg, 0.92 mmol) were dissolved in 0.3 mL of NMM in a round-bottomed flask under a nitrogen atmosphere. Triethylamine (156 µL, 112 µmol) was added and the reaction mixture was stirred at room temperature for 12 h. The reaction mixture was diluted with EtOAc (5 mL) and quenched with 3 mL of saturated NaHCO$_3$. The aqueous layer was extracted (3 × 5 mL) with EtOAc, dried over Na$_2$SO$_4$, and concentrated under reduced pressure. The crude product was purified by flash chromatography, eluting with 5% MeOH/EtOAc to give **150** (12 mg, 32 µmol, 35 %) as a viscous oil as well as recovered allylic alcohol **149** (19 mg, 50 µmol, 55%). To a THF (0.1 mL) solution of compound **150** (3 mg, 7.9 µmol) was added TBAF (8 µL, 1 M in THF). The solution was kept at room temperature for 2 h. The reaction mixture was diluted with EtOAc (5 mL) and quenched with 2 mL of H$_2$O, extracted (3 × 5 mL) with EtOAc, dried over Na$_2$SO$_4$, and concentrated under reduced pressure. The crude product was purified by flash chromatography, eluting with 15% MeOH/EtOAc to afford 6-amino-9-(2,3-deoxy-β-L-ribohexopyanosyl) purine **151** as a white solid (2 mg, 7.5 μmol, 95%). Mp: 225–226°C; $[\alpha]_D^{26}$ −30.0° (c = 0.20, MeOH). ^{1}H-NMR (CD$_3$OD/CDCl$_3$): δ 8.24 (s, 1H), 8.00 (s, 1H), 5.72 (dd, J = 10.2 and 4.2 Hz, 1H), 3.81 (dd, J = 12.6 and 3.6 Hz, 1H), 3.78 (ddd, J = 12.6 and 4.2 Hz, 1H), 3.72 (ddd, J = 11.4, 9.6 and 4.8 Hz, 1H), 3.52 (ddd, J = 9.0, 4.2 and 3.6 Hz, 1H), 2.27 (dddd, J = 12.6, 4.2 and 4.2, 4.2 Hz, 1H), 2.20–2.10 (m, 2H), 1.74 (dddd, J = 12.6, 12.6, 11.4 and 5.4 Hz, 1H).

Herdewijn et al. compared the hybridization properties of oligomeric sequences based on 2,3-dideoxy monomers with other oligodeoxynucleotides containing trisubstituted hexopyranosyl monomers with a different substitution pattern of the hydroxyl function, aiming at the generation of systems with a high cross-communication capacity.[59]

The same commercially available glucal **140** featuring in Scheme 16 was also employed as a starting material for the synthesis of 2,4-dideoxy-β-D-erythrohexopyranosyl building blocks **159**,[60] selected as a model due to their close structural resemblance to natural deoxyribose nucleosides. The key intermediate in the synthetic route illustrated in Scheme 18 is methyl 2,4-dideoxy-6-O-trityl-β-D-erythrohexopyranoside **154**, which was generated in five steps from **140**

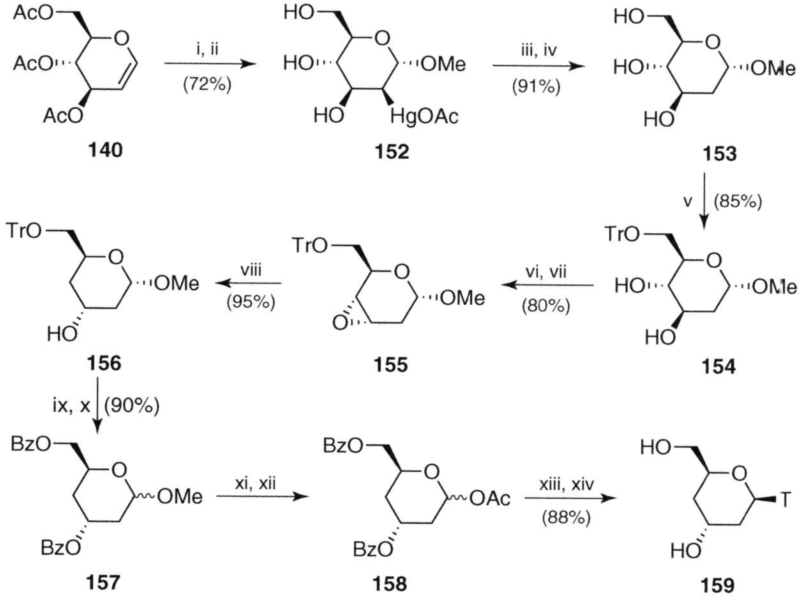

Scheme 18. *Reagents and conditions:* (i) MeONa, MeOH; (ii) Hg(OAc)$_2$; (iii) NaCl, MeOH; (iv) NaBH$_4$, i-PrOH, HCl; (v) tritylchloride (= triphenylmethyl; TrCl), pyridine; (vi) NaH, hexamethylphosphoric triamide (HMPA); (vii) i-Pr$_3$C$_6$H$_2$SO$_2$-1H-imidazole, THF; (viii) LiAlH$_4$, Et$_2$O; (ix) TsOH 0.05%, MeOH; (x) BzCl, pyridine; (xi) 80% AcOH; (xii) Ac$_2$O, pyridine; 62% of **158** together with 7% of the α-anomer; (xiii) thymine, BSA, TMSOTf, C$_2$H$_4$Cl$_2$; 71% together with 9% of the α-anomer; (xiv) NH$_3$, MeOH, rt.

according to a previously described procedure.[61] Thus, acetyl groups methanolysis and subsequent methoxymercuration and $NaBH_4$ reduction of the chloromercurial compound afforded triol **153**. Monotritylation of **153** occurred at the primary OH. The resulting trityl ether diol **154** underwent selective epoxidation to **155**, which after regioselective ring opening, afforded **156**. A multistep protecting group sequence produced an anomeric mixture of pyranoside **157**. Sugar–base condensation proceeded with yields and α/β selectivity depending on the various reaction conditions employed for each base.

The synthesis of 3,4-dideoxy-β-D-erythrohexopyranosyl nucleosides **165** was straightforward, beginning with the selective benzoylation of methyl-α-D-glucopyranoside **160** at the 2′- and 6′-positions (Scheme 19).[59] Conversion of the *trans*-vicinal diol to a double bond gave **162**, which after hydrogenation was transformed into 1-acetoxy glycoside **164** and condensed with a silylated base.

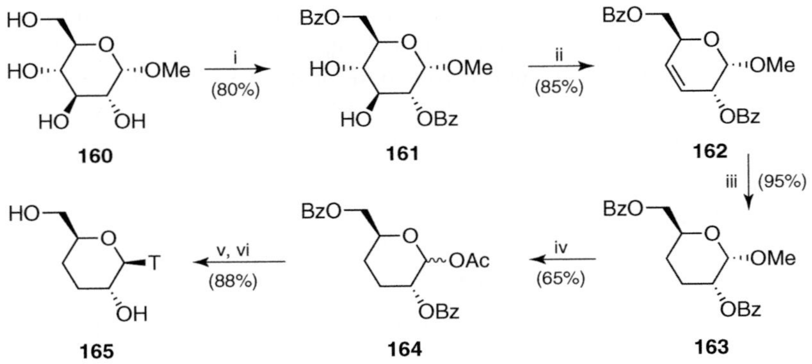

Scheme 19. *Reagents and conditions:* (i) $(Bu_3Sn)_2O$, toluene, BzCl; (ii) Ph_3P, Im, 2,4,5-triiodo-1*H*-imidazole, toluene; (iii) H_2, Pd/C, Et_3N, EtOH; (iv) AcOH, Ac_2O, H_2SO_4, AcONa; (v) thymine, $SnCl_4$, MeCN; (vi) NH_3, MeOH.

4.2. Hexitol Nucleic Acid Monomers

In the course of studies meant to explore the relationship between the geometry of six-membered carbohydrate-like rings and antiviral activity of the corresponding nucleosides, Herdewijn and co-workers prepared new analogues by shifting the base moiety from the anomeric to the 2′-position.[1a] It was anticipated that the insertion of a methylene unit between the ring oxygen and the carbon atom bearing the base moiety could be accountable for higher chemical and enzymatic stability, due to abolition of the anomeric effect, as well as leading to an overall conformational change of the molecule.

Accordingly, 1,5-anhydro-2,3-dideoxy-β-D-arabinohexitol nucleosides proved to hold marked and highly selective antiherpes (HSV-1 and HSV-2) activity.[1a] In particular, cytosine and guanine **3** analogues represent broad-range antivirals, at concentrations below the cytotoxicity threshold, and can also inhibit the growth of human T-cells. 1,5-Anhydro-5-iodouracil emerged as a potent anti-herpesvirus agent among the new derivatives at a concentration of 0.07 μg/mL.[1a,b] In the first report from the Herdewijn's group, 3-deoxy-1,5-anhydro-D-hexitol derivative **170** was envisaged as the key partner in the reaction with several heterocyclic bases. However, the initial route to prepare this intermediate involved a number of shortcomings that limited throughput on multigram scales.[1a] A more convenient preparative method (Scheme 20) employed commercially available diacetone-D-glucose as a starting material, which, once subjected to Barton deoxygenation, gave 3-deoxydiacetone-D-glucose **166**.[62] Upon sequential acid deprotection and peracetylation, **166** afforded a 2:1 α/β anomeric mixture of **167**, along with 15–20% of α/β furanoses **168**.[62a] Bromination of the mixture using HBr/AcOH, followed by reduction with Bu_3SnH in Et_2O furnished pyranose **169**. The benzylidene-protected hexitol synthon **170** was finally obtained in 30–40% overall yield after a quantitative deacetylation and treatment with benzaldehyde dimethyl acetal. This approach proved to be reproducible and gave access to multigram quantities of **170** without the need to purify intermediate stages and ready for coupling with the four standard base moieties as well as modified nucleobases.[1a,b,62] While the synthesis of nucleoside analogues bearing a purine moiety (**173**) was accomplished efficiently via

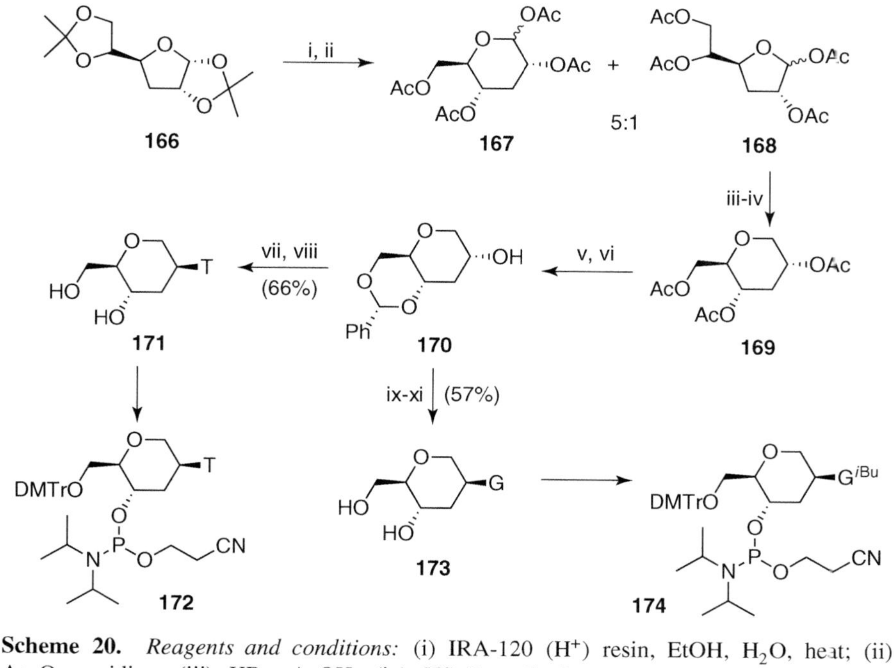

Scheme 20. *Reagents and conditions:* (i) IRA-120 (H^+) resin, EtOH, H_2O, heat; (ii) Ac_2O, pyridine; (iii) HBr, AcOH; (iv) $HSnBu_3$, Et_2O, rt; (v) NaOMe, MeOH; (vi) $PhCH(OMe)_2$, dioxane, rt; (vii) N^3-benzoylthymine, Ph_3P, DEAD, THF; (viii) NH_3, MeOH, then 80% AcOH, heat; (ix) Tf_2O, pyridine; (x) 2-amino-6-iodopurine tetrabutylammonium salt, CH_2Cl_2, rt; (xi) 10% HCl.

nucleophilic substitution reactions on the triflate derivative of **170**, higher yields were obtained in the corresponding reaction of pyrimidine bases with **170** under Mitsunobu-type conditions (**171**).[62b] HNAs were then built up by means of the phosphoramidite method and standard solid-support synthesis, once the monomers were suitably protected.[63]

In contrast to most pyranosyl nucleosides, anhydrohexitol derivatives are characterized by a energetically preferential slightly distorted chair conformation with an atypical axial orientation of the nucleobase, avoiding sterically unfavorable 1,3-diaxial repulsions,[1b,62] thus mimicking a furanose ring in an N-type conformation (A form). This structural feature seems to play a crucial role equally in the recognition by activating enzymes and in the properties demonstrated by oligonucleotides synthesized from these monomers. Despite differing from the parent homo-DNA **131** only by the position of the base moiety, HNA **132** shows a considerable change in base-pairing ability. In virtue of its preorganized structure, fitting the A-form of dsRNA,[65] HNA strongly hybridizes with natural complements in a sequence-specific fashion, forming HNA·DNA (ΔT_m/mod = +1.3°C) and HNA·RNA (ΔT_m/mod = +3°C) duplexes with increased thermal stability.[63,66] There is a high mismatch discrimination by HNA sequences. The preference for binding RNA,[63,65,66b] combined with the stability of HNA·RNA hybrids toward ribonuclease degradation,[63,66] has made **132** an attractive antisense construct, as established in cell-free translational experiments.[67] A detailed understanding of the factors that determine the cross-pairing ability of this family of hexopyranoses has recently been reviewed.[52c]

Experimental Procedure

1′,5′-Anhydro-2′,3′-dideoxy-2′-(guanin-9-yl)-D-arabinohexitol (173).[62b] A solution of triflic anhydride (9.24 mL, 36.6 mmol) in CH_2Cl_2 was added dropwise, over 5 min, to a stirred solution of 1,5-anhydro-4,6-O-benzylidene-3-deoxy-D-glucitol **170** (6 g, 25.42 mmol) and pyridine (3.7 mL, 45.76 mmol) in 36 mL of dry CH_2Cl_2 at −5°C. After 10 min of additional stirring, the reaction mixture was worked up below 20°C. The mixture was quenched with ice and diluted to 200 mL with CH_2Cl_2. The organic layer was washed with ice-cold water, cold 1 M KH_2PO_4 (200 mL), and finally, ice-cold water. The aqueous layers were extracted with CH_2Cl_2. The combined organic layers were dried over $MgSO_4$, filtered, and concentrated at low temperature to yield the corresponding triflate (24.8 mmol, 98%) as a white–yellow solid. A solution of triflate (9.13 g, 24.8 mmol) in dry CH_2Cl_2 (24 mL) was added to a stirred, ice-cold solution of the tetrabutylammonium salt of 6-iodo-9H-purin-2-amine (15.33 g, 30.5 mmol) in dry CH_2Cl_2 (36 mL). After 30 min, the ice bath was removed and the reaction mixture was stirred overnight at room temperature. A precipitate was filtered off, washed with CH_2Cl_2, and the filtrate was absorbed on silica. Purification by column chromatography with gradient elution (hexane/EtOAc, 55:45 to 45:55) afforded 8.49 g (17.67 mmol) of the coupled product. A suspension of iodide (8.18 g, 17.06 mmol) in 160 mL of 10% aqueous HCl was heated to 100°C for 2 h. After cooling to room temperature, the yellow–brown solution was washed with 60 mL of CH_2Cl_2 to remove benzaldehyde. The acidic yellow water phase was neutralized with 120 mL of 4 N NaOH using phenolphthaleine as indicator. At pH 7, the product started to precipitate. This suspension was concentrated and the white product was dissolved in boiling water (745 mL) and filtered while hot. The solution, after standing for 1 h, was cooled in the refrigerator at 4°C overnight. The crystals obtained were filtered off and washed with cold water, yielding 3.98 g of **173** (14.16 mmol, 83%). Mp > 300°C. ^{1}H-NMR (DMSO d_6): δ 1.8 (m, 1H, 3′ ax-H), 2.17 (br s, 1H, 3′ eq-H), 3.2–3.7 (m, 3H, 5′-H, 4′-H, 6A-H), 3.79 (dd, 1H, 1′ax-H, J = 12.5 and 2.2 Hz), 4.05–4.15 (m, 2H, 1′eq-H, 6B-H), 4.52 (br s, 1H, 2′-H), 4.63 (t, 1H, 6′-OH, J = 6 Hz), 4.91 (d, 1H, 4′-OH, J = 5.3 Hz), 6.46 (br s, 2H, NH$_2$), 7.87 (s, 1H, 8-H).

The corresponding β-L-1,5-anhydro hexitol nucleosides were originally prepared through a protocol identical to the one depicted in Scheme 20, but starting from L-glucose.[62a] A novel synthetic strategy featuring a powerful multicomponent process has made access to those enantiopure adenine and thymine monomers much more practical (Scheme 21).[68]

Treatment of starting acetate **175** with DDQ in a CH$_2$Cl$_2$/H$_2$O emulsion proved to affect MPM cleavage and in situ oxidation of the resulting primary alcohol simultaneously to generate an aldehyde, which, after isopropylidene group removal, engaged in an intramolecular double-cyclization reaction. The required β-L-1,6-anhydrohexose **177** was then obtained after removal of the dithioethylene bridge

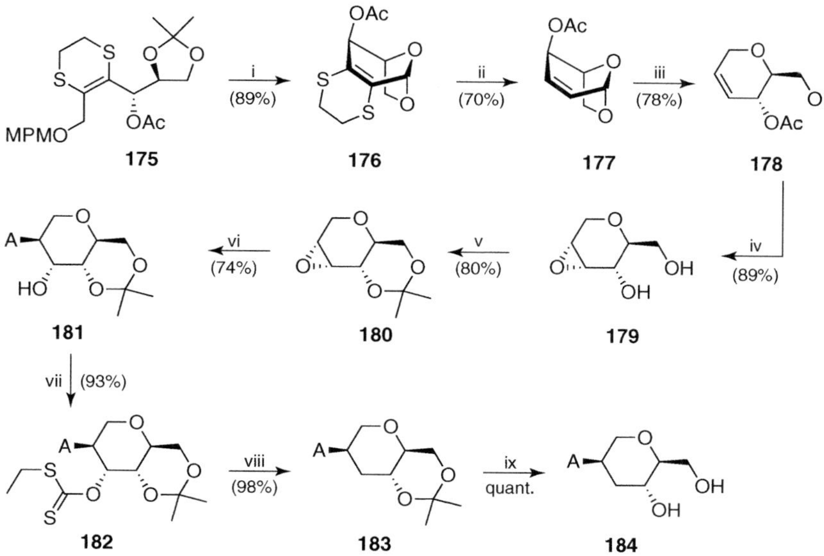

Scheme 21. *Reagents and conditions:* (i) DDQ, CH$_2$Cl$_2$/H$_2$O 18:1, heat; (ii) Ra-Ni, acetone, rt; (iii) TMSOTf, Et$_3$SiH, CH$_2$Cl$_2$, 0°C to rt; (iv) MeONa, MeOH, rt, then MCPBA, CH$_2$Cl$_2$, 0°C to rt; (v) DMP, PPTS, acetone, rt; (vi) adenine, DBU, DMF, 90°C; (vii) aq. NaOH, CS$_2$, BrCH$_2$CH$_3$, DMF, 0°C; (viii) Bu$_3$SnH, AIBN, toluene, heat; (ix) 80% AcOH, 60°C.

from intermediate **176**. Next, reductive cleavage was promoted by triethylsilane under acidic conditions to provide the key pseudoglucal **178**. Deacetylation and oxidation to *allo*-epoxide **179** was necessary to generate an electrophilic site at the C2-position as required by successive insertion of an axially oriented base. Following acetal formation, oxirane **180** underwent nucleophilic ring opening under mild conditions. To accelerate the synthesis, L-altritol nucleoside **181** was deoxygenated at the C3-position using an adapted version of the Barton reaction. Pleasingly, compound **184** was obtained stereoselectively in good yield after acetonide deprotection.

4.3. Altritol, Mannitol, FHNA, and Ara-FHNA Nucleic Acid Monomers

In view of the potential advantages associated with the use of conformationally restricted hexitol nucleic acids in antisense drug design, a variety of modified analogues were synthesized and their hybridization properties tested with natural and unnatural complementary strands. Possible avenues for further advances hinge on increasing the hydrophilicity of the monomers while retaining the favorable $^4C_1$-axial conformation distinctive of HNAs. Hypothetically, a supplementary OH group would facilitate groove solvatation and increase duplex stability, as suggested by molecular-dynamic simulations carried out on HNA hybrids.[64b] For the purpose of this study, nucleoside derivatives containing a D-allopyranose sugar (ribo-HNA or ANA) were selected as test monomers.[69] According to conditions described in the literature and involving reductive dehalogenation of tetra-*O*-acetyl-α-D-glucopyranosyl bromide **185**, followed by deacetylation and 4,6-*O*-benzylidene protection, the desired D-allopyranose precursor **186** was readily accessible in five steps and good overall yield (54%) (Scheme 22). Formation of 1,5-anhydro-2-deoxy-D-altrohexitol nucleosides proceeded smoothly and with comparable efficiency upon regioselective nucleophilic ring opening of epoxide **186** under various conditions, depending on the type of heterocyclic base envisioned. The cytosine congener was obtained from the uracil nucleoside via 1,2,4-triazolyl intermediate. Among the different attempts made to protect the 3′-OH for oligonucleotide synthesis, 3′-*O*-Bz was the most suitable in terms of deprotection and purification steps.[69b] The synthesis of **190** was readily scalable through easily handled crystalline intermediates. Migration from the 3′-axial to the 4′-equatorial position during benzylidene deprotection was avoided by carrying out the reaction with TFA at 0°C and precipitating the benzoyl-protected nucleosides from the reaction mixture. Similarly, monomethoxytrityl-protected intermediates were readily crystallized. The 3′-*O*-methylated analogue of ANA was obtained in a similar manner.[70]

Experimental Procedure

1,5-Anhydro-2-deoxy-2-(adenin-9-yl)-D-altrohexitol (188).[69b,71] Adenine (9.0 g, 67 mmol) and epoxide **186** (10 g) were dissolved in 80 mL of dry DMF under a nitrogen atmosphere. A solution of DBU (12 mL, 79 mmol) in 20 mL of dry

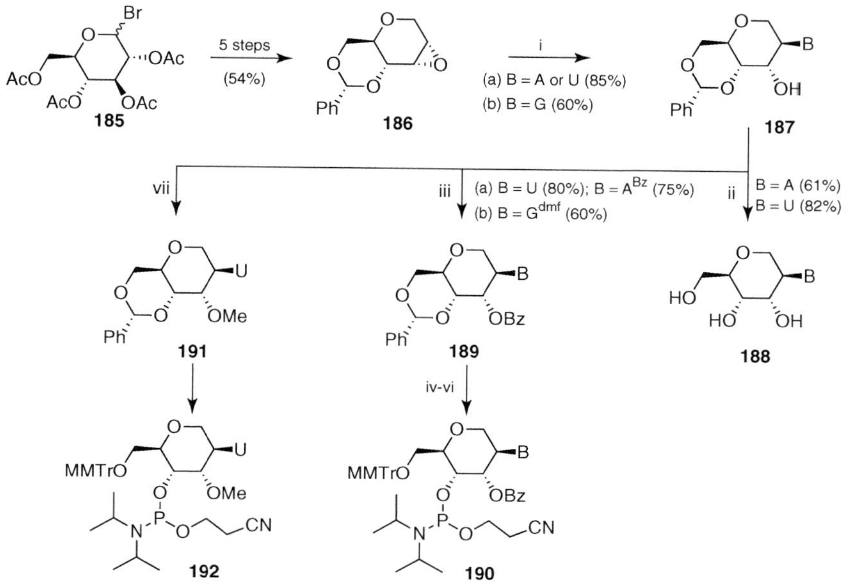

Scheme 22. *Reagents and conditions:* (i) (a) uracil or adenine, DBU, DMF; (b) 2-amino-6-chloropurine, 18-crown-6, K$_2$CO$_3$, HMPA, then 1 N NaOH, DABCO; (ii) 80% AcOH, 80°C; (iii) (a) BzCl, pyridine; (b) dmf(OEt)$_2$, DMF, then BzCN, DMF; (iv) TFA, CH$_2$Cl$_2$, 0°C; (v) MMTrCl, pyridine, CH$_2$Cl$_2$; (vi) *i*-Pr$_2$N(CN)PCl, 2,4,6-collidine, *N*-MeIm, dioxane; (vii) NaH, CH$_3$I, THF, 0°C.

DMF was then added via a dropping funnel over a period of 30 min. The reaction mixture was then stirred for 4 h at 95°C. The light-brown solution was cooled to room temperature and the volatiles were removed in vacuo. Water (300 mL) was added to suspend the crude product. The suspension was cooled using an ice–water bath and neutralized with 5 mL of AcOH. The white crystalline solid was then filtered off using a glass filter and washed with cold water (3 × 50 mL). The white solid was dried overnight in an oven at 70–80°C to give 1,5-anhydro-4,6-*O*-benzylidene-2-deoxy-2-(adenine-9-yl)-D-altrohexitol in 85% yield (13.4 g). This compound (120 mg, 0.32 mmol) was treated with 80% aqueous acetic acid (6.5 mL) at 80°C for 2 h. After evaporation and coevaporation with toluene, the residue was suspended in water. After neutralization with 4 N NaOH, evaporation and coevaporation with toluene, the resulting residue was purified by flash chromatography (CH$_2$Cl$_2$/MeOH, 80:20) to afford 68 mg (0.24 mmol, 75% yield) of **188** (B = A) as a white foam. The product was further purified by reversed-phase HPLC (H$_2$O/MeOH, 100:0 to 77:23, 4 mL/min). The pure fractions were combined, concentrated under reduced pressure, and lyophilized, yielding 55 mg (0.22 mmol, 61% yield) of **188** (B = A). ^{1}H-NMR (D$_2$O): δ 3.59 (m, 1H, 6'-Ha), 3.65–3.83 (m, 3H, 4'-H, 5'-H, 6'-He); 4.17 (br s, 3H, 1'-Ha, 1'-He, 3'-H); 4.56 (m, 1H, 2'-H); 8.06 (s, 1H, 2H); 8.30 (s, 1H, 8-H).

Aiming at further variation, diastereomeric mannitol monomers (ara-HNA or MNA) were also prepared.[72] Two routes can generally be taken into account for the synthesis of mannohexitol nucleosides, which differ in the nature of the base moiety to be incorporated into the sugar. The preparation of pyrimidine nucleosides took advantage of an anchimeric assistance of the base moiety (Scheme 23, route a). As depicted previously in Scheme 22, compound **186** is an excellent substrate for regioselective base insertion. Mesylation of the resulting altritol uracil nucleoside preceded inversion of configuration at the 3′-position upon opening of the O^2,3′-anhydro intermediate **194** under basic aqueous conditions.[72a] The cytosine nucleoside was accessible from the uracil analogue under standard conditions. As this procedure is clearly unsuitable for the synthesis of purine derivatives, another methodology was proposed, as shown in Scheme 23 (route b).[72b,c] Silyl protection of the starting 1,5-anhydro-4,6-O-benzylidene-D-glucitol **197**, prepared from commercially available bromoacetyl-α-D-glucose in three steps, gave 3′-O-TBDMS-protected intermediate **198**, which was then functionalized as a triflate prior to reaction with adenine in the form of its tetrabutylammonium salt. Guanine was more conveniently introduced by reaction of 2-amino-6-iodopurine precursor. Functional group elaboration and phosphitylation under standard conditions gave the MNA building block **201**.

Thermodynamical and molecular modeling data collected for dsANA, ANA·RNA, ANA·DNA, and MNA·RNA complexes highlighted a remarkable

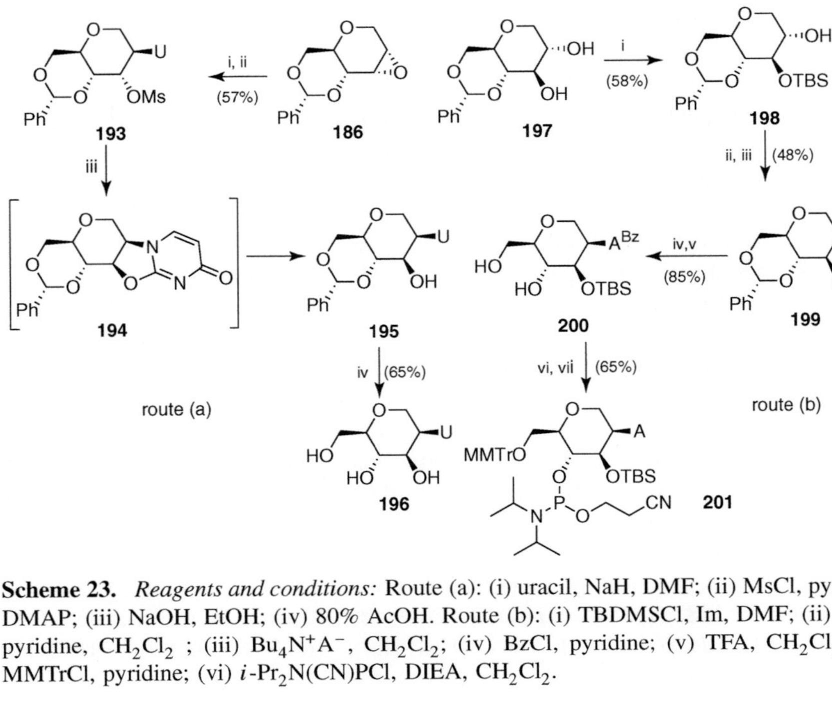

Scheme 23. *Reagents and conditions:* Route (a): (i) uracil, NaH, DMF; (ii) MsCl, pyridine, DMAP; (iii) NaOH, EtOH; (iv) 80% AcOH. Route (b): (i) TBDMSCl, Im, DMF; (ii) Tf$_2$O, pyridine, CH$_2$Cl$_2$; (iii) Bu$_4$N$^+$A$^-$, CH$_2$Cl$_2$; (iv) BzCl, pyridine; (v) TFA, CH$_2$Cl$_2$; (vi) MMTrCl, pyridine; (vi) i-Pr$_2$N(CN)PCl, DIEA, CH$_2$Cl$_2$.

difference in RNA binding affinity between D-altritol and D-mannitol oligomers due to their different levels of preorganization. The thermal stability of ANA duplexes with complementary sequences was found not only to be superior to related HNA complexes, but also to have a preference for DNA.[73] Conversely, MNA displays considerably weaker hybridization affinity for cognate RNA than does ANA or HNA.[72b] The explanation given for this evidence is that the presence of an equatorial 3'-β-OH group in MNA may lead to greater conformational changes in the corresponding duplex, letting it drift away from the A-type RNA form in a partially unwinded state, through intrastrand hydrogen-bonds formation between the 3'-OH groups and the O-atoms in the phosphate backbone of the following nucleotide. For diastereoisomeric ANA, not only is this effect suppressed, but the axial 3'-OH inherent to ANA points to the minor groove, facilitating hybridization by increasing hydratation and/or restricting conformational freedom. However, the beneficial stabilizing effect of the 3'-OH group on RNA affinity was diminished upon methylation.

These findings were recently confirmed by preliminary results concerning 3'-fluoro-modified hexitol nucleic acids **136** and **137**, published by Egli and coworkers.[74] The structural diversity created by the presence of a fluorine atom in equatorial orientation in ara-FHNA appeared detrimental to duplex stabilization in comparison to its axial substituted counterpart FHNA. This outcome has been ascribed to an increased proximity, and thus steric repulsion of the 3'-flanking nucleotide, which might disrupt stacking of nucleobases. FHNA qualifies as a better system in term of pairing affinities than its parent HNA, and most notably displays a unique antisense activity, showing higher in vivo potency than HNA and rivaling LNA's, but without producing hepatotoxicity. Parallel in vitro tests showing comparable biological activity between HNA and FHNA nicely address the valuable character of fluorination from a delivery point of view. The fluorinated monomers were synthetized according to Scheme 24, starting from the familiar synthon **186**. Fluorination at the 3'-position of mannohexitol thymine **203** using nonafluorobutanesulfonyl fluoride as fluorine source occurs with inversion of configuration to produce the desired nucleoside **204** together with 5% of an unseparable elimination species generated by the nonaflate intermediate. Pleasingly, hydrogenation of the mixture in the presence of palladium hydroxide on carbon resulted in conversion to and isolation of pure nucleoside **205**, affording a reliable 50% yield over the two steps. Standard tritylation and phosphitylation conditions led to FHNA thymine phosphoramidite building block **207**. The 5-methylcytosine analogue resulted from 4'-silylation of **206**, conversion of the nucleobase moiety via a triazolide intermediate, and benzoyl protection of the exocyclic amino group to give **208**. TBS cleavage followed by phosphitylation led to **209**.

Experimental Procedure

1,5-Anhydro-2,3-dideoxy-3-fluoro-2-(thymin-1-yl)-D-altritol (205).[74] To a suspension of thymine (16.1 g, 128.1 mmol) and epoxide **186** (10.0 g, 42.7 mmol) in acetonitrile (512 mL) was added DBU (10.0 mL, 67.1 mmol). The reaction was

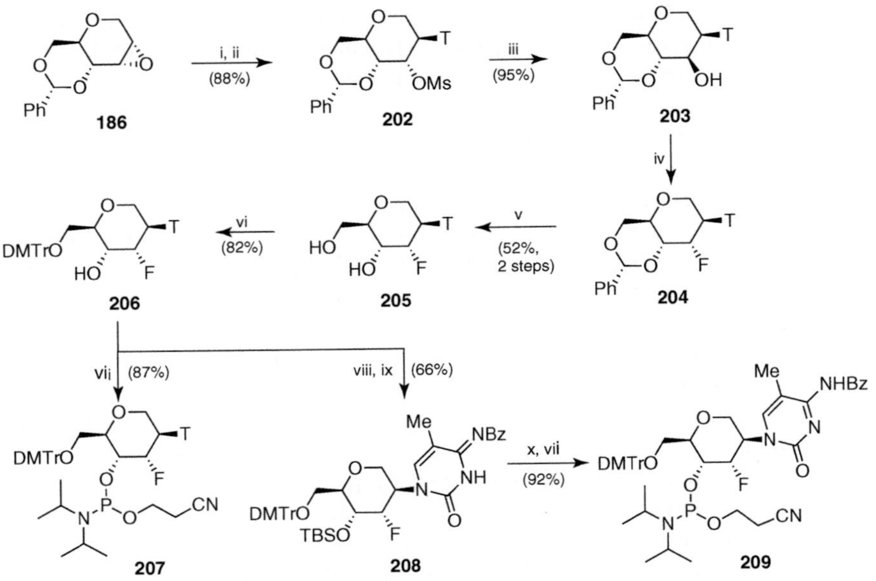

Scheme 24. *Reagents and conditions:* (i) thymine, DBU, MeCN; (ii) MsCl, pyridine; (iii) aq. NaOH, dioxane; (iv) CF$_3$(CF$_2$)$_3$SO$_3$F, DBU, THF, 35°C; (v) H$_2$, Pd(OH)$_2$, MeOH; (vi) DMTrCl, pyridine; (vii) [(i-Pr)$_2$N]$_2$P(OCH$_2$CH$_2$CN), NMI, 1H-tetrazole, DMF; (viii) TBSCl, DMF, Im; (ix) 1,2,4-triazole, POCl$_3$, Et$_3$N, then NH$_4$OH, then Bz$_2$O, DMF; (x) TBAF, THF.

heated at 85°C for 36 h after which it was cooled to room temperature and concentrated under reduced pressure. The residue was suspended in CH$_2$Cl$_2$ and the organic layer was washed with half-saturated aqueous NaHCO$_3$, brine, dried over Na$_2$SO$_4$, and concentrated. The residue was purified by column chromatography (2% MeOH in CH$_2$Cl$_2$) to afford 1,5,-anhydro-4,6-O-benzylidene-2-deoxy-2-(thymin-1-yl)-D-altritol. Methanesulfonyl chloride (1.7 mL, 21.6 mmol) was added dropwise to a cold (0°C) solution of 1,5-anhydro-4,6-O-benzylidene-2-deoxy-2-(thymin-1-yl)-D-altritol (5.0 g, 14.4 mmol) in a mixture of CH$_2$Cl$_2$/pyridine (1:1, 30 mL). The reaction was warmed to room temperature and stirred for an additional 24 h. The reaction was quenched with water and evaporated under reduced pressure. The residual oil was redissolved in EtOAC and the organic layer was washed with 2% AcOH, half-saturated NaHCO$_3$, brine, dried over Na$_2$SO$_4$, and concentrated. Purification by column chromatography (1% MeOH in CH$_2$Cl$_2$) provided mesylate **202** (5.43 g, 89%). Sodium hydroxide (18.2 mL of a 2 M solution, 36.5 mmol) was added to a solution of mesylate **202** (5.0 g, 11.4 mmol) in 1,4-dioxane (18 mL). After stirring at 65°C for 3 h, the reaction was cooled to room temperature and neutralized with glacial AcOH. The reaction was poured into half-saturated NaHCO$_3$ and the aqueous layer was extracted twice with CH$_2$Cl$_2$. The organic layers were combined, dried over Na$_2$SO$_4$, and concentrated to provide the crude product, which was suspended in hexane and collected by filtration. The crude product was washed further with hexane and then dried under reduced pressure for 16 h to provide **203** (3.92 g, 95%), which was used without further purification. Nonafluorobutanesulfonyl fluoride (2.3 mL, 13.4 mmol) was added dropwise to a solution of nucleoside **203** (3.20 g, 8.9 mmol) and DBU (2.0 mL, 13.4 mmol). The reaction was stirred at 35°C for 42 h, after which it was quenched with MeOH. The reaction was diluted with EtOAc, and the organic layer was washed with water, 5% HCl, saturated $NaHCO_3$, brine, dried over Na_2SO_4, and concentrated. Purification by column chromatography (66% EtOAc in hexane) provided compound **204** (3.2 g, contaminated with about 5% of an elimination by-product). Compound **204** (1.75 g, ca. 95% purity) was hydrogenated using a hydrogen balloon and 20% palladium hydroxide on charcoal (0.44 g) in MeOH (50 mL) for 4 h. The reaction was filtered through Celite and concentrated under reduced pressure. Purification by column chromatography (7% MeOH in CH_2Cl_2) provided **205** (1.26 g, 52% from **203**). ^{1}H-NMR (DMSO-d_6): δ 11.38 (s, 1H, NH), 7.85 (s, 1H, pyr-H_6), 5.33 (d, J = 5.8 Hz, 1H, 4'-OH), 4.99–4.73 (m, 2H, 1'H, 6'-OH), 4.68–4.59 (m, 1H, 2'H), 4.03–3.72 (m, 3H, 1'H, 1'H, 4'H), 3.72–3.51 (m, 3H, 6'H, 6'H, 5'H), 1.79 (s, 3H, pyr-CH_3).

4.4. Cyclohexanyl and Cyclohexenyl Nucleoside Derivatives

The generation of enantiomerically pure six-membered carbocycles in nucleoside chemistry is an important challenge, owing to their importance as structural motifs when incorporated into oligonucleotidic sequences. This research has contributed to an understanding of structure–conformational relationships in the nucleic acid field.[52c] Many methods have been devised successfully that allow access to cyclohexanyl analogues, including stepwise base construction from amino alcohols,[75] nucleophilic ring opening of an epoxide,[76] Michael-type base addition to conjugated cyclohexenes,[77] and Mitsunobu-type base condensation with conveniently functionalized cyclohexyl or cyclohexenyl alcohols.[78] However, the corresponding asymmetric variations have not been so forthcoming. In a few reported examples, resolution using enzymes[79] and derivatization with chiral compounds[80] was found to provide six-membered cyclohexyl nucleosides in optically pure form. This last methodology was pursued in the synthesis of D- and L-3-hydroxy-4-hydroxymethyl-1-cyclohexanyl adenine and thymine, the monomeric forms of the carba analogue of HNAs, as illustrated in Scheme 25.[80a,81] Conjugate addition of thymine to 1,3-cyclohexadienecarboxylate **210** proceeded regioselectively to give **211**, which was reduced, protected as trityl ether at the primary position, and finally, subjected to hydroboration. This last step produced a separable mixture of the desired 3',4'-*trans* **212** and 3',4'-*cis* isomers. Separation of racemic **212** was possible via acylation with (*R*)-methylmandelic acid and subsequent chromatography of the resulting diastereoisomeric esters.[80a] Deacylation of **214** followed by acid treatment gave optically pure **215**, while the opposite enantiomeric form was obtained similarly from **213**. While oligomers derived from both enantiomeric monomers form Watson–Crick-type self-duplexes, given that the backbones on both strands are homochiral, only D-CNA cross-pairs with natural nucleic acids, but with lower

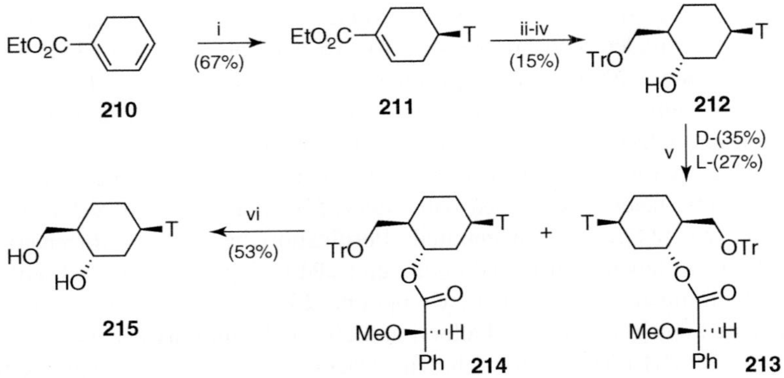

Scheme 25. *Reagents and conditions:* (i) thymine, DBU, DMF, heat; (ii) DIBALH, CH$_2$Cl$_2$; (iii) TrCl, pyridine; (iv) BH$_3$·THF, THF, 0°C to rt; (v) DCC, (R)-(−)-O-methylmandelic acid, DMAP, CH$_2$Cl$_2$, 0°C to rt, then chromatography; (vi) 0.1 M KOH, MeOH, then 80% AcOH, 60°C.

affiniy than HNA and ANA. It is suggested that the $^4C_1$ conformation adopted by the monomer, with an equatorial orientation of the base, is flipped to the $_4C^1$ conformation when the monomer is incorporated in an oligonucleotide.[80a]

Experimental Procedure

1-[(1 R,3 S,4 R)-4-Hydroxymethyl-3-hydroxycyclohexanyl]thymine (215).[80a,81]
A mixture of thymine (4 equiv.), ethyl 1,3-cyclohexadiene-1-carboxylate (12 equiv.), DBU (1 equiv.), and DMF was stirred at 75°C for 48 h. After addition of AcOH and removal of all the volatiles in vacuo, the residue was treated with MeOH, and the resulting solid was filtered off, affording **211** in 67%. To a solution of **211** (1 equiv.) in CH$_2$Cl$_2$ under a nitrogen atmosphere at 0°C was added DIBALH (1.0 M solution in hexane, 4 equiv.) over 20 min, and the reaction mixture was then stirred for 20 min. Excess of DIBALH was destroyed by slow addition of MeOH at 0°C. The resulting suspension was adsorbed on silica gel and eluted to give (±)-1-[4-(hydroxymethyl)-3-cyclohexenyl]thymine in 79% yield. A mixture of (±)-1-[4-(hydroxymethyl)-3-cyclohexenyl]thymine (1 equiv.) and trityl chloride (1.5 equiv.) in pyridine was stirred at 70°C for 3 h. The reaction mixture was quenched with MeOH and evaporated. The residue was dissolved in CH$_2$Cl$_2$ and treated with saturated NaHCO$_3$. The organic layer was then dried over MgSO$_4$, evaporated, and coevaporated with toluene. The residue was purified by column chromatography to afford (±)-1-[4-(trityloxymethyl)-3-cyclohexenyl]thymine (56%) as a foam. To a stirred solution of this compound (1 equiv.) in THF under a nitrogen atmosphere was added BH$_3$-THF (1 M solution in THF) at 0°C. After being stirred at room temperature for 1 h, the solution was diluted with H$_2$O and EtOH, made basic with 3 M NaOH, and 35% H$_2$O$_2$ was slowly added. The mixture was stirred at 45°C for 20 h, and then saturated aqueous Na$_2$SO$_3$ was

added. The mixture was extracted with CH_2Cl_2 (100 mL × 3), dried over $MgSO_4$, and evaporated. The residue was chromatographed (CH_2Cl_2/MeOH, 99:1) to obtain **212** (34%) and 3′,4′-*cis*-isomers (21%) as foams. DCC (1.10 g, 5.3 mmol) was added to a solution of **212** (2.4 g, 4.83 mmol), (*R*)-(−)-*O*-methylmandelic acid (0.88 g, 5.31 mmol), and DMAP (65 mg, 0.53 mmol) in CH_2Cl_2 (50 mL) at 0°C. The mixture was allowed to warm to room temperature over 2 h and filtered. The filtrate was washed with aqueous H_3PO_4 (1 m, 30 mL), water (30 mL), and saturated aqueous $NaHCO_3$ (15 mL), dried over $MgSO_4$, and filtered. Column chromatography (CH_2Cl_2/MeOH, 99.5:0.5 to 99:1) afforded 1.09 g (35%) of **214** and 0.84 g (27%) of **213** as foams, as well as 0.25 g (8%) of the mixture of **214** and **213**. A solution of KOH (0.1 M) in MeOH (15 mL) was added to a solution of **214** (1.24 g, 1.92 mmol) in MeOH (15 mL). The mixture was evaporated after 30 min and the residue was dissolved in CH_2Cl_2 (100 mL), washed with water (20 mL), dried over $MgSO_4$, and evaporated. The residue was dissolved in 80% AcOH and then heated at 60°C for 4 h. After removing AcOH by evaporation and coevaporation with toluene, the residue was dissolved in MeOH (10 mL), adsorbed on silica gel, and eluted with CH_2Cl_2/MeOH (95:5 to 9:1) to afford, after recrystallization (MeOH/Et_2O), 0.26 g (53%) of **215**. Mp. 201–202°C. ^1H-NMR ($CDCl_3$): δ 1.0–2.0 (10H), 3.18 (2H), 3.92 (1H), 4.60 (1H), 4.78 (1H), 6.98 (1H), 7.15–7.50 (15H), 11.17 (1H).

(*R*)-Carvone **216** featured as an enantioselective precursor in the synthesis of 2-hydroxymethylcyclohexane-1,3-diol nucleosides (Scheme 26).[82] The synthetic plan relied on the epoxidation of the *endo* double bond and reduction of the carbonyl group, both occuring stereoselectively to afford **217** after 2′-OH silylation. Oxidative cleavage of the isopropenyl side chain with retention of configuration at C4 was possible employing OsO_4/$NaIO_4$, followed by peroxide oxidation to give an acetate intermediate, which was first hydrolyzed and then protected, giving **218**. Regioselective opening of epoxide **218** by treatment with LiTMP/Et_2AlCl set up the desired allylic alcohol intermediate, featuring an *exo* double bond, which was protected as benzyl ether **219**. Hydroboration of 2-methylene cyclohexane diol **219** proceeded with good *anti*-selectivity in the presence of 9-BBN. Inversion of configuration was then achieved under Mitsunobu conditions to give suitably protected α-alcohol **222**, which then underwent a second Mitsunobu reaction in order to insert the base.

Six-membered carbocyclic nucleosides are not restricted to cyclohexanyl derivatives but include a variety of related unsaturated analogues, which have proven to be good candidates as antivirals.[1c] The attractive structural facet of cyclohexenyl nucleosides is their ability to combine a high level of chemical and enzymatic stability derived from the absence of the anomeric center with the flexibility induced by the presence of the double bond in the place of the oxygen atom of related furanoses. Most interestingly, the two unsaturated C-atoms can be regarded as a single pseudoatom, pointing to a close resemblance of the six-membered ring to a natural furanose system with two preferred conformational N- and S-type states.[83]

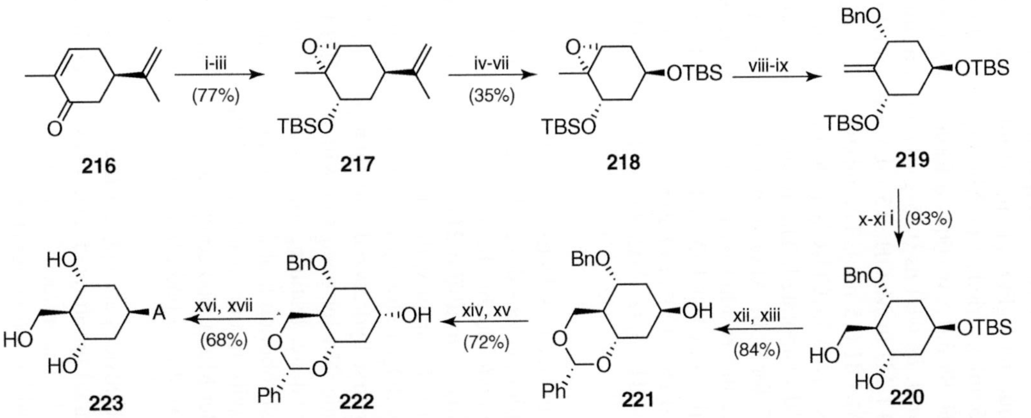

Scheme 26. *Reagents and conditions*: (i) H$_2$O$_2$/NaOH, MeOH; (ii) L-Selectride, THF, −65°C; (iii) TBDMSCl, Im, DMF; (iv) 1% OsO$_4$/KIO$_4$, THF, H$_2$O, rt; (v) *m*-CPBA, CHCl$_3$, pH 8 buffer solution, rt; (vi) K$_2$CO$_3$, MeOH; (vii) TBDMSCl, Im, DMF; (viii) LiTMP/Et$_2$AlCl, benzene, 0°C; (ix) NaH, BnBr, TBAI, THF, rt; (x) 9-BBN, THF, rt; (xi) TBAF (1 equiv.), THF, rt; (xii) PhCH(OMe)$_2$, PTSA, dioxane; (xiii) TBAF, THF; (xiv) benzoic acid, DEAD, Ph$_3$P, dioxane; (xv) K$_2$CO$_3$, MeOH; (xvi) adenine, DEAD, Ph$_3$P, dioxane, rt; (xvii) Pd(OH)$_2$/C, cyclohexene, MeOH, reflux.

Both D- and L-enantiomers of 5-hydroxy-4-hydroxymethyl-2-cyclohexenylguanine were screened against a variety of viruses and showed comparable biological activity and selectivity.[1c] The original enantioselective syntheses of those chiral cyclohex-2-enyl nucleosides were reported by Herdewijn et al., both starting from (R)-carvone **216**.[1c,84] However, the method suffered from low efficiency, being unable to provide synthetically useful quantities of material, especially in view of oligonucleotide synthesis. As a consequence, a modified route was developed, which relied on the cycloaddition between ethyl (2E)-3-acetyloxy-2-propenoate **224** and Danishefsky's diene **225** to generate a six-membered ring skeleton with the desired 4,5-*trans*-relationship, predominantly as *endo*-adduct **226** (*endo:exo* 4:1)[85] (Scheme 27). Reduction of the diastereomeric mixture using LiAlH$_4$ not only converted the two esters to alcohols, but also resolved the stereochemical ambiguity via rearrangement of the silyloxyenol ether to enone and subsequent reduction. Selective protection of triol **227** with benzaldehyde dimethyl acetal was a straightforward way to provide racemic intermediate **228**. At this stage, to make the process stereoselective, a rewarding resolution strategy was needed. After examination of several methods, enantiopure cyclohexene precursor **229** was obtained through an enzymatic resolution approach featuring 14% w/w Novozyme 435 as catalyst.[86] Once isolated, the enantiopure cyclohexyl products **232** and its isomer derived from **231** were reacted with canonical purines and pyrimidines to afford nucleosides. Further conversion to the corresponding phosphoramidate building

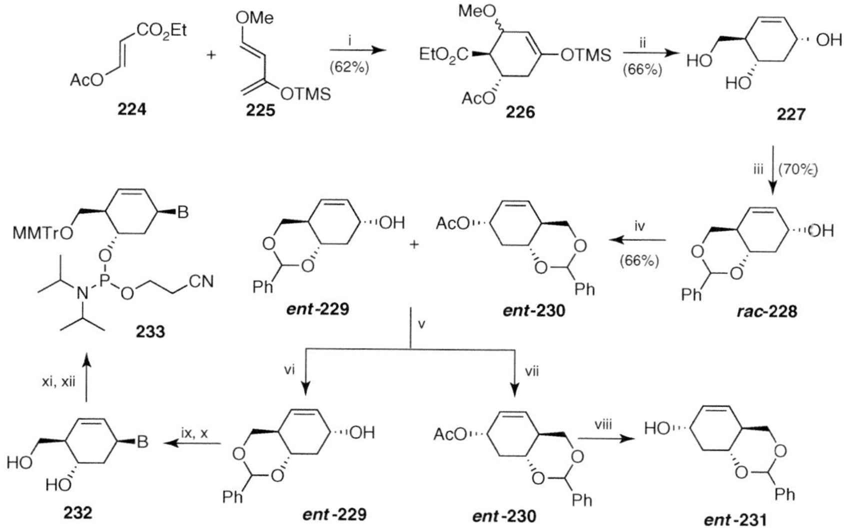

Scheme 27. *Reagents and conditions:* (i) hydroquinone, 180°C; (ii) LiAlH$_4$, THF, 0°C; (iii) PhCH(OMe)$_2$, PTSA, dioxane, rt; (iv) isopropenyl acetate, Novozyme 435 (14% w/w), CH$_2$Cl$_2$, rt; (v) column chromatography; (vi) recrystallization from 50% EtOAc/hexane; (vii) recrystallization from 20% EtOAc/hexane; (viii) NH$_3$, MeOH, then recrystallization from 50% EtOAc/hexane; (ix) B (A, G, T, or C), Ph$_3$P, DEAD, dioxane, rt; (x) 80% TFA, rt; (xi) MMTrCl, pyridine, rt; (xii) (*i*-Pr)$_2$N(CE)PCl, (*i*-Pr)$_2$NEt, CH$_2$Cl$_2$.

blocks as before occurred without incident, as well as oligonucleotide synthesis. The resulting CeNA showed higher complementary recognition of RNA than of DNA, but the hybridization strength of a CeNA·RNA duplex is lower than that of an HNA·RNA duplex.[87] Nevertheless, its preorganization was superior to CNA. Most notably, CeNA may sustain siRNA activity in mixed CeNA/RNA duplexes.[88]

Experimental Procedure

(1S,4R,5S)-9-(5-Hydroxy-4-hydroxymethyl-2-cyclohexen-1-yl)guanine.[85] To a mixture of **229** (1 equiv.), 2-amino-6-chloropurine (2 equiv.), and triphenylphosphine (2 equiv.) in dry 1,4-dioxane was added slowly a solution of DEAD (2 equiv.) in dry 1,4-dioxane. The reaction was stirred at room temperature overnight and concentrated. The residue was absorbed on silica gel and chromatographed (CH_2Cl_2/MeOH, 100:1 and 50:1) to afford crude chloride as a white solid. The crude compound was treated with TFA/H_2O (3:1) at room temperature for 2 days. The reaction mixture was concentrated and coevaporated with toluene. The residue was chromatographed on silica gel (CH_2Cl_2/MeOH, 50:1 and 10:1) to give (1S,4R,5S)-9-(5-hydroxy-4-hydroxymethyl-2-cyclohexen-1-yl)guanine after crystallization with a mixture of diisopropyl ether/MeOH (8:2). Mp 275–280°C dec. ^{1}H-NMR (CD$_3$OD): δ 1.94–2.27 (m, 3H), 3.77 (d, 2H, J = 4.7 Hz), 3.85 (m, 1H), 5.17 (m, 1H), 5.88 (dm, 1H, J = 10.2 Hz), 6.09 (dm, 1H, J = 10.2 Hz), 7.73 (s, 1H).

The $_2H^3$ (N-type) energetically preferred conformation typical of 2′-deoxycyclohexenyl nucleosides **232** was compared with those of their ribo and arabino analogues in the cyclohexenyl series. In the first case shifting toward the $^2H_3$ S-pucker was observed, while the introduction of a 2′-OH in arabino position compared to **232** for conformational preference.[83,89]

The synthetic strategy toward ribocyclohexenyl adenine[89] (Scheme 28) utilized as a key step the inverse-electron-demand Diels–Alder cycloaddition of 2,2-dimethyl-1,3-dioxole **235** to 3-bromo-2H-pyran-2-one **234**. A 4:1 mixture of regioisomers was formed, favoring the desired *endo*-adduct **236**. Radical dehalogenation at the bridgehead position was affected by treatment of **236** with Bu$_3$SnH and AIBN. Lactone **237** then underwent reduction and ring opening using LiAlH$_4$ to give diol **238**. Following TBS protection of the primary OH group, inversion of configuration of the allylic OH group by an oxidation–reduction sequence yielded **240**, the desired precursor for coupling with adenine under Mitsunobu conditions. On treatment with trifluoroacetic acid, the fully deprotected nucleoside was isolated as a racemic mixture in 50% yield over the two steps. At this stage, enzymatic desymmetrisation was achieved by using adenosine deaminase, which converted the D-like enantiomer into an inosine derivative **242**.

In Scheme 29 the chemical synthesis of racemic aracyclohexenyladenine is described.[90] A sequence involving lactonization of (−)-endo-7-oxabicyclo[2.2.1]hept-5-ene-2-carboxylic acid **244**, reduction using LiAlH$_4$ and acetylation afforded triacetate **245** on a synthetically useful multigram

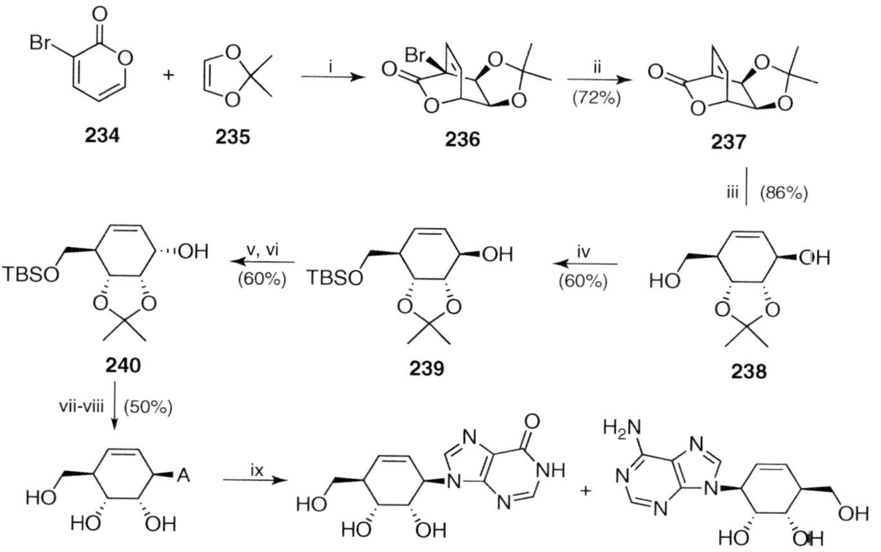

Scheme 28. *Reagents and conditions:* (i) $(i\text{-Pr})_2\text{NEt}$, CH_2Cl_2, 90°C, 4 days; (ii) Bu_3SnH, AIBN, toluene, 130°C; (iii) LiAlH_4, THF, 0°C; (iv) TBDMSCl, imidazole, DMF, 0°C; (v) MnO_2, CH_2Cl_2, rt; (vi) NaBH_4, $\text{CeCl}_3\cdot 7\text{H}_2\text{O}$, MeOH; (vii) A, Ph_3P, DIAD, dioxane; (viii) TFA, H_2O; (ix) adenosine deaminase, H_2O.

scale. Ring opening was promoted by HBr, followed by an elimination and simultaneous replacement of the primary bromo atom by an *O*-benzyl group. A deprotection–reprotection sequence gave cyclohexenyl diol **248**. Selective protection of a single OH group proved difficult, and thus the allylic OH group was oxidized selectively, and the remaining OH group was then protected as a TBS derivative. Mitsunobu reaction of base moieties with **249** led to racemic **250**.

As can be seen from previous examples in this section, the Diels–Alder reaction represents a common and efficient strategy for the construction of cyclohexenyl rings, accommodating the use of various functional motifs. One of the latest adaptations of this methodology, suggested by Legoupy et al., exploits the chirality of *endo*-cycloadduct **253** (Scheme 30).[91]

The well-defined stereochemical relationship across the cycloadduct arises from the control of both π-facial and *endo* selectivity, as a result of the approach of the dienophile being directed by the pendant bulky cyclohexyl. Three further steps count reduction of the anhydro **253**, acetylation of the triol **254** and Boc cleavage. Precursor **256** is then reacted with β-methoxy-α-methylacrylolyl isocyanate; final cyclization and complete deprotection are affected simultaneously by exposure to metabolic ammonia under pressure.

An interesting and considerably powerful approach in this field consists of the transformation of a single chiral intermediate to produce both absolute configurations of 4′-hydroxycyclohexenyl nucleosides via an enantiodivergent synthesis.[92]

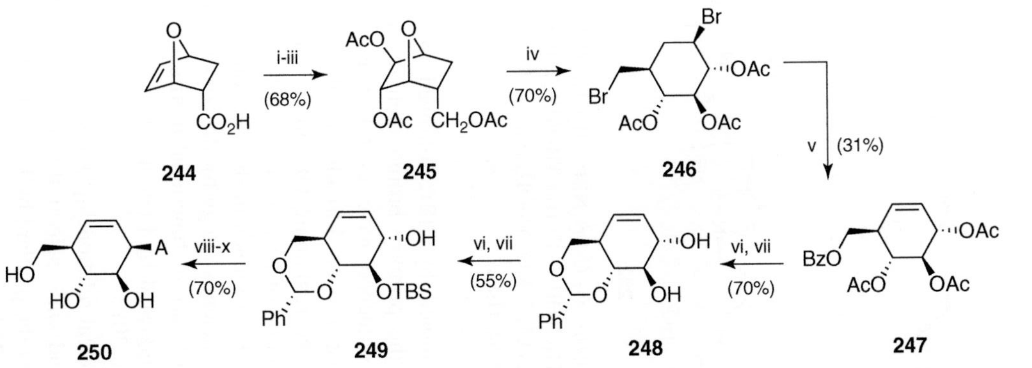

Scheme 29. *Reagents and conditions*: (i) H₂O, HCOOH; (ii) LiAlH₄, THF; (iii) Ac₂O, pyridine; (iv) 15% HBr, AcOH; (v) BzONa, LiBr, DMF; (vi) MeONa, MeOH; (vii) PhCH(OMe)₂, PTSA, DMF; (viii) MnO₂, CH₂Cl₂; (ix) TBSCl, Im, DMF; (x) NaBH₄, CeCl₃·7H₂O; (xi) adenine, DEAD, Ph₃P, dioxane; (xii) TFA, H₂O.

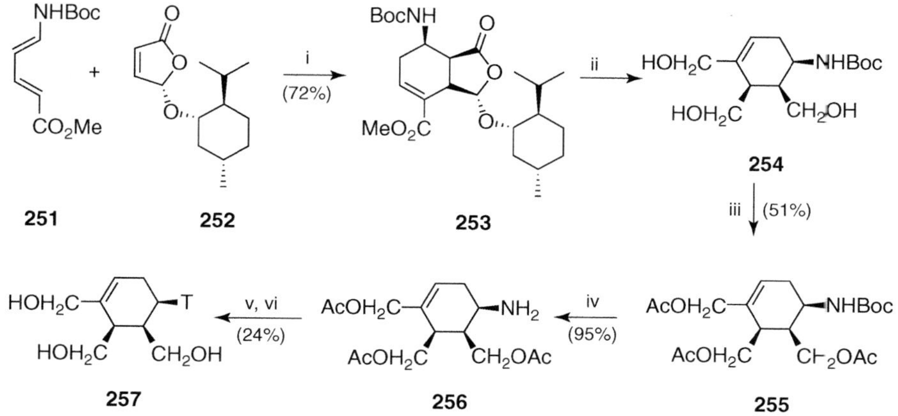

Scheme 30. *Reagents and conditions:* (i) cat. LiOH, toluene, sealed tube, 120°C; (ii) DIBALH, toluene, −78°C to rt; (iii) Ac$_2$O, pyridine, DMAP; (iv) TMSOTf, CH$_2$Cl$_2$, 0°C; (v) β-methoxy-α-methylacrylolyl isocyanate, benzene, DMF, −15°C to rt; (vi) 30% NH$_4$OH, MeOH, sealed tube, 100°C.

As illustrated in Scheme 31, enantiopure allyl alcohol **261** bearing a hydrobenzoin moiety served as a precursor for the construction of D- and L-purine and uracil derivatives. For the preparation of **261**, the authors followed a straightforward route from 1,4-cyclohexandione requiring derivatization with (R,R)-hydrobenzoin and a bromination–dehydrobromination step to obtain enone **260**. Compound **260** was then reacted with cathecolborane to afford (2R,3R,8R)-ketal **261** with good diastereoselectively (7:1). A well-established Mitsunobu protocol was applied to achieve D-series of cyclohexenyl nucleosides, while Pd-(0)-catalyzed coupling led to the L-isomers. Racemic monosubstituted *cis*- and *trans*-cyclohexenyl thymines bearing an hydroxy group at the 1′- or 4′-position have previously been prepared by Arango et al.[93]

5. CONCLUDING REMARKS

In an effort to generate more effective therapeutics through modifications of the carbohydrate moiety of nucleosides, many powerful and innovative transformations have been accomplished, which led to the creation of an important structural variety, including three-, four-, and six-membered ring analogues. In many cases, the compounds obtained exhibit exceptional biological activity. Special mention of the promising application of hexose nucleosides and their related oligomers in antisense technology is necessary. From the early development of an autonomous base-pairing system built on hexopyranosyl monomers, many duplexes have been described with a completely different geometry and shape beyond the classical Watson–Crick helical models. The ability to communicate with the natural furanose-type nucleic acids manifested by these oligomers is important from

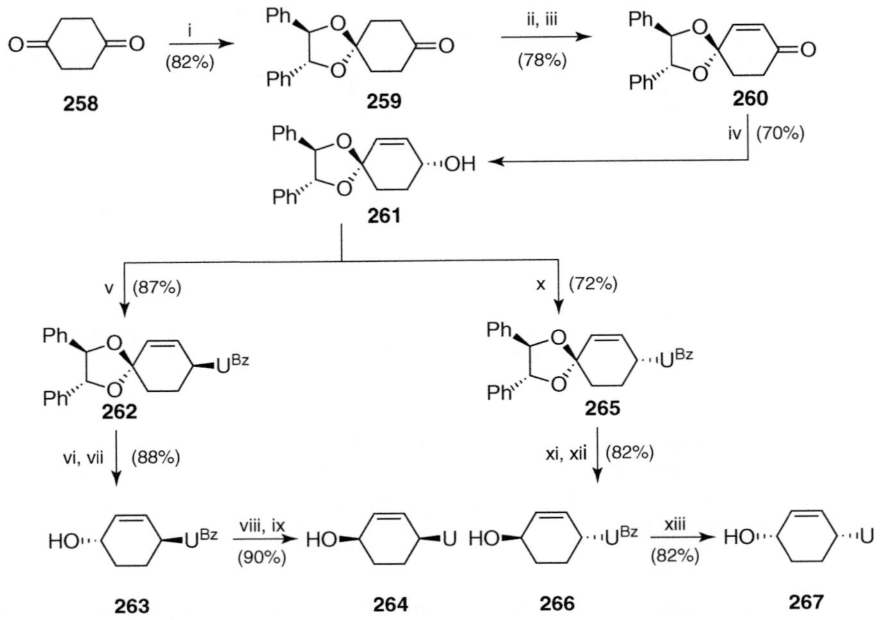

Scheme 31. *Reagents and conditions:* (i) (*R*,*R*)-hydrobenzoin, TsOH, benzene, heat; (ii) Br_2, Et_2O, −10 to 0°C; (iii) DBU, dioxane, 100°C; (iv) (*S*)-2-Me-CBS, CB, CH_2Cl_2, −78°C to rt; (v) N^3-benzoyluracil, DBAD, Ph_3P, THF, −10°C to rt; (vi) TFA/H_2O (14:1), 0°C; (vii) (*S*)-2-Me-CBS, CB, CH_2Cl_2, −78°C to rt; (viii) *p*-nitrobenzoic acid, DBAD, Ph_3P, THF, −10°C to r; (ix) Me_2NH, EtOH; (x) $ClCO_2Et$, pyridine, DMAP, CH_2Cl_2, rt, then N^3-benzoyluracil, $(\eta\text{-}C_3H_5PdCl)_2$, dppe, DMF, 80°C; (xi) TFA/H_2O (14:1), 0°C; (xii) (*R*)-2-Me-CBS, CB, CH_2Cl_2, −78°C to rt; (xiii) *p*-nitrobenzoic acid, DBAD, Ph_3P, THF, −10°C to rt, then Me_2NH, EtOH.

the biological standpoint and upgraded the understanding of the factors involved in controlling duplex stability and geometry.

ABBREVIATIONS

A	adenine
ANA	altritol nucleic acid
CeNA	cyclohexenyl nucleic acid
CNA	cyclohexanyl nucleic acid
DAST	diethylaminosulfur trifluoride
DBU	diazabicyclo[5.4.0]undec-7-ene
DEAD	diethylazodicarboxylate
DIAD	diisopropylazodicarboxylate
DIBALH	diisobutylaluminum hydride

DMAP	4-(*N*,*N*-dimethylamino)pyridine
DMF	*N*,*N*-dimethylformamide
EBV	Epstein–Barr virus
FHNA	3'-fluorohexitol nucleic acid
G	guanine
HCMV	human cytomegalovirus
HIV	human immunodeficiency virus
HMDS	hexamethyldisilazane
HNA	hexitol nucleic acid
HSV	herpes simplex virus
MCMV	murine cytomegalovirus
MNA	mannitol nucleic acid
NBSH	2-nitrobenzenesulfonylhydrazide
T	thymine
TBAF	tetrabutylammonium fluoride
TBAI	tetrabutylammonium iodide
TBS and TBDMS	*tert*-butyldimethylsilyl
TFA	trifluoroacetic acid
U	uracil

REFERENCES

1. (a) Verheggen, I.; Vanaerschot, A.; Toppet, S.; Snoeck, R.; Janssen, G.; Balzarini, J.; Declercq, E.; Herdewijn, P. *J. Med. Chem.* **1993**, *36*, 2033–2040; (b) Verheggen, I.; Vanaerschot, A.; Vanmeervelt, L.; Rozenski, J.; Wiebe, L.; Snoeck, R.; Andrei, G.; Balzarini, J.; Claes, P.; Declercq, E.; Herdewijn, P. *J. Med. Chem.* **1995**, *38*, 826–835; (c) Wang, J.; Froeyen, M.; Hendrix, C.; Andrei, G.; Snoeck, R.; De Clercq, E.; Herdewijn, P. *J. Med. Chem.* **2000**, *43*, 736–745.
2. Herdewijn, P. *Liebigs Ann.* **1996**, 1337–1348.
3. (a) Sekiyama, T.; Hatsuya, S.; Tanaka, Y.; Uchiyama, M.; Ono, N.; Iwayama, S.; Oikawa, M.; Suzuki, K.; Okunishi, M.; Tsuji, T. *J. Med. Chem.* **1998**, *41*, 1284–1298; (b) Zemlicka, J., Chen, X. In Frontiers in Nucleosides and Nucleic Acids; Schinazi, R. F., Liotta, D. C., Eds.; IHL Press: Tucker, GA, **2004**; pp. 267–307; (c) Komiotis, D.; Manta, S.; Tsoukala, E.; Tzioumaki, N. *Anti-Infect. Agents Med. Chem.* **2008**, *7*, 219–244.
4. (a) Zhao, Y. F.; Yang, T. F.; Lee, M.; Lee, D.; Newton, M. G.; Chu, C. K. *J. Org. Chem.* **1995**, *60*, 5236–5242; (b) Csuk, R.; vonScholz, Y. *Tetrahedron* **1996**, *52*, 6383–5396.
5. Lee, M. G.; Du, J. F.; Chun, M. W.; Chu, C. K. *J. Org. Chem.* **1997**, *62*, 1991–1995.
6. Ashton, W. T.; Meurer, L. C.; Cantone, C. L.; Field, A. K.; Hannah, J.; Karkas, J. D.; Liou, R.; Patel, G. F.; Perry, H. C.; Wagner, A. F.; Walton, E.; Tolman, R. L. *J Med. Chem.* **1988**, *31*, 2304–2315.
7. Iwayama, S.; Ono, N.; Ohmura, Y.; Suzuki, K.; Aoki, M.; Nakazawa, H.; Oikawa, M.; Kato, T.; Okunishi, M.; Nishiyama, Y.; Yamanishi, K. *Antimicrob. Agents Chemother.* **1998**, *42*, 1666–1670.
8. Onishi, T.; Mukai, C.; Nakagawa, R.; Sekiyama, T.; Aoki, M.; Suzuki, K.; Nakazawa, H.; Ono, N.; Ohmura, Y.; Iwayama, S.; Okunishi, M.; Tsuji, T. *J. Med. Chem.* **2000**, *43*, 278–282.

9. Ostrowski, T.; Golankiewicz, B.; De Clercq, E.; Balzarini, J. *Bioorg. Med. Chem.* **2006**, *14*, 3535–3542.
10. Onishi, T.; Sekiyama, T.; Tsuji, T. *Nucleosides Nucleotides Nucleic Acids* **2005**, *24*, 1187–1197.
11. Onishi, T.; Matsuzawa, T.; Nishi, S.; Tsuji, T. *Tetrahedron Lett.* **1999**, *40*, 8845–8847.
12. (a) Rife, J.; Ortuno, R. M. *Org. Lett.* **1999**, *1*, 1221–1223; (b) Muray, E.; Rife, J.; Branchadell, V.; Ortuno, R. M. *J. Org. Chem.* **2002**, *67*, 4520–4525.
13. (a) Csuk, R.; Eversmann, L. *Tetrahedron* **1998**, *54*, 6445–6456; (b) Csuk, R.; Kern, A. *Tetrahedron* **1999**, *55*, 8409–8422.
14. Zemlicka, J. In *Recent Advances in Nucleosides: Chemistry and Chemotherapy*; Chu, C. K., Ed.; Elsevier: Amsterdam, **2002**; pp. 327–357.
15. Qiu, Y. L.; Ksebati, M. B.; Ptak, R. G.; Fan, B. Y.; Breitenbach, J. M.; Lin, J. S.; Cheng, Y. C.; Kern, E. R.; Drach, J. C.; Zemlicka, J. *J. Med. Chem.* **1998**, *41*, 10–23.
16. Qiu, Y. L.; Hempel, A.; Camerman, N.; Camerman, A.; Geiser, F.; Ptak, R. G.; Breitenbach, J. M.; Kira, T.; Li, L.; Gullen, E.; Cheng, Y. C.; Drach, J. C.; Zemlicka, J. *J. Med. Chem.* **1998**, *41*, 5257–5264.
17. Qiu, X.-L.; Xu, X.-H.; Qing, F.-L. *Tetrahedron* **2010**, *66*, 789–843.
18. (a) Qiu, Y. L.; Geiser, F.; Kira, T.; Gullen, E.; Cheng, Y. C.; Ptak, R. G.; Breitenbach, J. M.; Drach, J. C.; Hartline, C. B.; Kern, E. R.; Zemlicka, J. *Antivir. Chem. Chemother.* **2000**, *11*, 191–202; (b) Chen, X. C.; Zemlicka, J. *J. Org. Chem.* **2002**, *67*, 286–289.
19. Lecorre, M.; Hercouet, A.; Bessieres, B. *J. Org. Chem.* **1994**, *59*, 5483–5484.
20. Zhou, S. M.; Breitenbach, J. M.; Borysko, K. Z.; Drach, J. C.; Kern, E. R.; Gullen, E.; Cheng, Y. C.; Zemlicka, J. *J. Med. Chem.* **2004**, *47*, 566–575.
21. Kern, E. R.; Bidanset, D. J.; Hartline, C. B.; Yan, Z. H.; Zemlicka, J.; Quenelle, D. C. *Antimicrob. Agents Chemother.* **2004**, *48*, 4745–4753.
22. Tiruchinapally, G.; Zemlicka, J. *Synth. Commun.* **2008**, *38*, 697–702.
23. Yan, Z. H.; Kern, E. R.; Gullen, E.; Cheng, Y. C.; Drach, J. C.; Zemlicka, J. *J. Med. Chem.* **2005**, *48*, 91–99.
24. Zhou, S.; Drach, J. C.; Prichard, M. N.; Zemlicka, J. *J. Med. Chem.* **2009**, *52*, 3397–3407.
25. Shimada, N.; Hasegawa, S.; Harada, T.; Tomisawa, T.; Fujii, A.; Takita, T. *J. Antibiot.* **1986**, *39*, 1623–1625.
26. Ortuno, R. M.; Moglioni, A. G.; Moltrasio, G. Y. *Curr. Org. Chem.* **2005**, *9*, 237–259.
27. (a) Hoshino, H.; Shimizu, N.; Shimada, N.; Takita, T.; Takeuchi, T. *J. Antibiot.* **1987**, *40*, 1077–1078; (b) Alder, J.; Mitten, M.; Norbeck, D.; Marsh, K.; Kern, E. R.; Clement, J. *Antivir. Res.* **1994**, *23*, 93–105.
28. (a) Niitsuma, S.; Ichikawa, Y.; Kato, K.; Takita, T. *Tetrahedron Lett.* **1987**, *28*, 4713–4714; (b) Niitsuma, S.; Ichikawa, Y.; Kato, K.; Takita, T. *Tetrahedron Lett.* **1987**, *28*, 3967–3970; (c) Nishiyama, S.; Yamamura, S.; Kato, K.; Takita, T. *Tetrahedron Lett.* **1988**, *29*, 4743–4746; (d) Norbeck, D. W.; Kramer, J. B. *J. Am. Chem. Soc.* **1988**, *110*, 7217–7218.
29. Gumina, G.; Chu, C. K. *Org. Lett.* **2002**, *4*, 1147–1149.
30. Cooperwood, J. S.; Boyd, V.; Gumina, G.; Chu, C. K. *Nucleosides Nucleotides Nucleic Acids* **2000**, *19*, 219–236.
31. Liang, Y.; Hnatiuk, N.; Rowley, J. M.; Whiting, B. T.; Coates, G. W.; Rablen, P. R.; Morton, M.; Howell, A. R. *J. Org. Chem.* **2011**, *76*, 9962–9974.

32. (a) Ichikawa, Y.; Narita, A.; Shiozawa, A.; Hayashi, Y.; Narasaka, K. *J. Chem. Soc, Chem. Commun.* **1989**, 1919–1921; (b) Norbeck, D. W.; Kern, E.; Hayashi, S.; Rosenbrook, W.; Sham, H.; Herrin, T.; Plattner, J. J.; Erickson, J.; Clement, J.; Swanson, R.; Shipkowitz, N.; Hardy, D.; Marsh, K.; Arnett, G.; Shannon, W.; Broder, S.; Mitsuya, H. *J. Med. Chem.* **1990**, *33*, 1281–1285; (c) Field, A. K.; Tuomari, A. V.; McGeeverrubin, B.; Terry, B. J.; Mazina, K. E.; Haffey, M. L.; Hagen, M. E.; Clark, J. M.; Braitman, A.; Slusarchyk, W. A.; Young, M. G.; Zahler, R. *Antivir. Res.* **1990**, *13*, 41–52; (d) Bisacchi, G. S.; Braitman, A.; Cianci, C. W.; Clark, J. M.; Field, A. K.; Hagen, M. E.; Hockstein, D. R.; Malley, M. F.; Mitt, T.; Slusarchyk, W. A.; Sundeen, J. E.; Terry, B. J.; Tuomari, A. V.; Weaver, E. R.; Young, M. G.; Zahler, R. *J. Med. Chem.* **1991**, *34*, 1415–1421.

33. (a) Honjo, M.; Maruyama, T.; Sato, Y.; Horii, T. *Chem. Pharm. Bull.* **1989**, *37*, 1413–1415; (b) Slusarchyk, W. A.; Young, M. G.; Bisacchi, G. S.; Hockstein, D. R.; Zahler, R. *Tetrahedron Lett.* **1989**, *30*, 6453–6456; (c) Katagiri, N.; Sato, H.; Kaneko, C. *Chem. Pharm. Bull.* **1990**, *38*, 288–290.

34. (a) Cotterill, I. C.; Roberts, S. M. *J. Chem. Soc. Perkin Trans. 1* **1992**, 2585–2586; (b) Katagiri, N.; Morishita, Y.; Yamaguchi, M. *Tetrahedron Lett.* **1998**, *39*, 2613–2616.

35. Jung, M. E.; Sledeski, A. W. *J. Chem. Soc. Chem. Commun.* **1993**, 589–591.

36. Brown, B.; Hegedus, L. S. *J. Org. Chem.* **1998**, *63*, 8012–8018.

37. Darses, B.; Greene, A. E.; Coote, S. C.; Poisson, J.-F. *Org. Lett.* **2008**, *10*, 821–824.

38. Panda, J.; Ghosh, S. *J. Chem. Soc. Perkin Trans. 1* **2001**, 3013–3016.

39. Rustullet, A.; Alibes, R.; de March, P.; Figueredo, M.; Font, J. *Org. Lett.* **2007**, *9*, 2827–2830.

40. (a) Jacobs, G. A.; Tino, J. A.; Zahler, R. *Tetrahedron Lett.* **1989**, *30*, 6955–6958; (b) Slusarchyk, W. A.; Bisacchi, G. S.; Field, A. K.; Hockstein, D. R.; Jacobs, G. A.; McGeeverrubin, B.; Tino, J. A.; Tuomari, A. V.; Yamanaka, G. A.; Young, M. G.; Zahler, R. *J. Med. Chem.* **1992**, *35*, 1799–1806; (c) Wu, J. Y.; Schneller, S. W.; Seley, K. L.; Snoeck, R.; Andrei, G.; Balzarini, J.; DeClercq, E. *J. Med. Chem.* **1997**, *40*, 1401–1406; (d) Ebead, A.; Fournier, R.; Lee-Ruff, E. *Nucleosides, Nucleotides Nucleic Acids* **2011**, *30*, 391–404.

41. Fernandez, F.; Lopez, C.; Hergueta, A. R. *Tetrahedron* **1995**, *51*, 10317–10322.

42. Hergueta, A. R.; Lopez, C.; Fernandez, F.; Caamano, O.; Blanco, J. M. *Tetrahedron: Asymmetry* **2003**, *14*, 3773–3778.

43. (a) Borges, J. E. R.; Fernandez, F.; Garcia, X.; Hergueta, A. R.; Lopez, C.; Andrei, G.; Snoeck, R.; Witvrounw, M.; Balzarini, J.; De Clercq, E. *Nucleosides Nucleotides Nucleic Acids* **1998**, *17*, 1237–1253; (b) Blanco, J. M.; Caamano, O.; Fernandez, F.; Garcia-Mera, X.; Hergueta, A. R.; Lopez, C.; Rodriguez-Borges, J. E.; Balzarini, J.; De Clercq, E. *Chem. Pharm. Bull.* **1999**, *47*, 1314–1317; (c) Fernandez, F.; Hergueta, A. R.; Lopez, C.; De Clercq, E.; Balzarini, J. *Nucleosides, Nucleotides Nucleic Acids* **2001**, *20*, 1129–1131; (d) Hergueta, A. R.; Fernandez, F.; Lopez, C.; Balzarini, J.; De Clercq, E. *Chem. Pharm. Bull.* **2001**, *49*, 1174–1177.

44. Lopez, C.; Balo, C.; Blanco, J. M.; Fernandez, F.; De Clercq, E.; Balzarini, J. *Nucleosides Nucleotides Nucleic Acids* **2001**, *20*, 1133–1135.

45. Figueira, M. J.; Blanco, J. M.; Caamano, O.; Fernandez, F.; Garcia-Mera, X.; Lopez, C. *Arch. Pharm.* **1999**, *332*, 348–352.

46. Rouge, P. D.; Moglioni, A. G.; Moltrasio, G. Y.; Ortuno, R. M. *Tetrahedron: Asymmetry* **2003**, *14*, 193–195.

47. (a) Boumchita, H.; Legraverend, M.; Guilhem, J.; Bisagni, E. *Heterocycles* **1991**, *32*, 867–871; (b) Maruyama, T.; Hanai, Y.; Sato, Y. *Nucleosides Nucleotides* **1992**, *11*, 855–864; (c) Maruyama, T.; Hanai, Y.; Sato, Y.; Snoeck, R.; Andrei, G.; Hosoya, M.; Balzarini, J.; Declercq, E. *Chem. Pharm. Bull.* **1993**, *41*, 516–521; (d) Gauvry, N.; Bhat, L.; Mevellec, L.; Zucco, M.; Huet, F. *Eur. J. Org. Chem.* **2000**, 2717–2722; (e) Guan, H. P.; Ksebati, M. B.; Kern, E. R.; Zemlicka, J. *J. Org. Chem.* **2000**, *65*, 5177–5184; (f) Danappe, S.; Pal, A.; Alexandre, C.; Aubertin, A. M.; Bourgougnon, N.; Huet, F. *Tetrahedron* **2005**, *61*, 5782–5787.
48. (a) Gharbaoui, T.; Legraverend, M.; Bisagni, E. *Tetrahedron Lett.* **1992**, *33*, 7141–7144; (b) Hubert, C.; Alexandre, C.; Aubertin, A. M.; Huet, F. *Tetrahedron* **2002**, *58*, 3775–3778.
49. GourdelMartin, M. E.; Huet, F. *J. Org. Chem.* **1997**, *62*, 2166–2172.
50. (a) Nishizono, N.; Koike, N.; Yamagata, Y.; Fujii, S.; Matsuda, A. *Tetrahedron Lett.* **1996**, *37*, 7569–7572; (b) Ichikawa, E.; Yamamura, S.; Kato, K. *Tetrahedron Lett.* **1999**, *40*, 7385–7388; (c) Choo, H.; Chen, X.; Yadav, V.; Wang, J. N.; Schinazi, R. F.; Chu, C. K. *J. Med. Chem.* **2006**, *49*, 1635–1647; (d) Nishizono, N.; Sugo, M.; Machida, M.; Oda, K. *Tetrahedron* **2007**, *63*, 11622–11625; (e) Nishizono, N.; Akama, Y.; Agata, M.; Sugo, M.; Yamaguchi, Y.; Oda, K. *Tetrahedron* **2011**, *67*, 358–363.
51. Hosono, F.; Nishiyama, S.; Yamamura, Y.; Izawa, T.; Kato, K.; Terada, Y. *Tetrahedron* **1994**, *50*, 13335–13346.
52. (a) Eschenmoser, A. *Science* **1999**, *284*, 2118–2124; (b) Leumann, C. J. *Bioorg. Med. Chem.* **2002**, *10*, 841–854; (c) Herdewijn, P. *Chem. Biodivers.* **2010**, *7*, 1–59.
53. Boehringer, M.; Roth, H. J.; Hunziker, J.; Goebel, M.; Krishnan, R.; Giger, A.; Schweizer, B.; Schreiber, J.; Leumann, C.; Eschenmoser, A. *Helv. Chim. Acta* **1992**, *75*, 1416–1477.
54. Ebert, M.-O.; Jaun, B. *Chem. Biodivers.* **2010**, *7*, 2103–2128.
55. Augustyns, K.; Vanaerschot, A.; Urbanke, C.; Herdewijn, P. *Bull. Soc. Chim. Belg.* **1992**, *101*, 119–130.
56. (a) Hunziker, J.; Roth, H. J.; Bohringer, M.; Giger, A.; Diederichsen, U.; Gobel, M.; Krishnan, R.; Jaun, B.; Leumann, C.; Eschenmoser, A. *Helv. Chim. Acta* **1993**, *76*, 259–352; (b) Egli, M.; Lubini, P.; Pallan, P. S. *Chem. Soc. Rev.* **2007**, *36*, 31–45.
57. Froeyen, M.; Lescrinier, E.; Kerremans, L.; Rosemeyer, H.; Seela, F.; Verbeure, B.; Lagoja, I.; Rozenski, J.; Van Aerschot, A.; Busson, R.; Herdewijn, P. *Chem. Eur. J.* **2001**, *7*, 5183–5194.
58. (a) Guppi, S. R.; Zhou, M. Q.; O'Doherty, G. A. *Org. Lett.* **2006**, *8*, 293–296; (b) D'Alonzo, D.; Guaragna, A.; Van Aerschot, A.; Herdewijn, P.; Palumbo, G. *J. Org. Chem.* **2010**, *75*, 6402–6410.
59. Augustyns, K.; Vandendriessche, F.; Vanaerschot, A.; Busson, R.; Urbanke, C.; Herdewijn, P. *Nucleic Acids Res.* **1992**, *20*, 4711–4716.
60. (a) Augustyns, K.; Vanaerschot, A.; Herdewijn, P. *Bioorg. Med. Chem. Lett.* **1992**, *2*, 945–948; (b) Augustyns, K.; Rozenski, J.; Vanaerschot, A.; Janssen, G.; Herdewijn, P. *J. Org. Chem.* **1993**, *58*, 2977–2982.
61. Corey, E. J.; Weigel, L. O.; Chamberlin, A. R.; Lipshutz, B. *J. Am. Chem. Soc.* **1980**, *102*, 1439–1441.
62. (a) Andersen, M. W.; Daluge, S. M.; Kerremans, L.; Herdewijn, P. *Tetrahedron Lett.* **1996**, *37*, 8147–8150; (b) De Bouvere, B.; Kerremans, L.; Rozenski, J.; Janssen, G.; Van Aerschot, A.; Claes, P.; Busson, R.; Herdewijn, P. *Liebigs Ann.Recl.* **1997**, *0*, 1453–1461.

63. Hendrix, C.; Rosemeyer, H.; Verheggen, I.; Seela, F.; VanAerschot, A.; Herdewijn, P. *Chem. Eur. J.* **1997**, *3*, 110–120.
64. (a) Declercq, R.; Herdewijn, P.; VanMeervelt, L. *Acta Crystallogr., Sect. C: Cryst. Struct. Commun.* **1996**, *52*, 1213–1215; (b) De Winter, H.; Lescrinier, E.; Van Aerschot, A.; Herdewijn, P. *J. Am. Chem. Soc.* **1998**, *120*, 5381–5394.
65. Lescrinier, E.; Esnouf, R.; Schraml, J.; Busson, R.; Heus, H. A.; Hilbers, C. W.; Herdewijn, P. *Chem. Biol.* **2000**, *7*, 719–731.
66. (a) Vanaerschot, A.; Verheggen, I.; Hendrix, C.; Herdewijn, P. *Angew. Chem. Int. Ed.* **1995**, *34*, 1338–1339; (b) Hendrix, C.; Rosemeyer, H.; DeBouvere, B.; VanAerschot, A.; Seela, F.; Herdewijn, P. *Chem. Eur. J.* **1997**, *3*, 1513–1520.
67. Vandermeeren, M.; Preveral, S.; Janssens, S.; Geysen, J.; Saison-Behmoaras, E.; Van Aerschot, A.; Herdewijn, P. *Biochem. Pharmacol.* **2000**, *59*, 655–663.
68. (a) D'Alonzo, D.; Guaragna, A.; Van Aerschot, A.; Herdewijn, P.; Palumbo, G. *Tetrahedron Lett.* **2008**, *49*, 6068–6070; (b) D'Alonzo, D.; Van Aerschot, A.; Guaragna, A.; Palumbo, G.; Schepers, G.; Capone, S.; Rozenski, J.; Herdewijn, P. *Chem. Eur. J.* **2009**, *15*, 10121–10131.
69. (a) Allart, B.; Van Aerschot, A.; Herdewijn, P. *Nucleosides, Nucleotides Nucleic Acids* **1998**, *17*, 1523–1526; (b) Allart, B.; Busson, R.; Rozenski, J.; Van Aerschot, A.; Herdewijn, P. *Tetrahedron* **1999**, *55*, 6527–6546.
70. Van Aerschot, A.; Meldgaard, M.; Schepers, G.; Volders, F.; Rozenski, J.; Busson, R.; Herdewijn, P. *Nucleic Acids Res.* **2001**, *29*, 4187–4194.
71. Abramov, M., Herdewijn, P. *Curr. Protoc. Nucleic Acid Chem.* **2007**, *1*.18.1.
72. (a) PerezPerez, M. J.; DeClercq, E.; Herdewijn, P. *Bioorg. Med. Chem. Lett.* **1996**, *6*, 1457–1460; (b) Hossain, N.; Wroblowski, B.; Van Aerschot, A.; Rozenski, J.; De Bruyn, A.; Herdewijn, P. *J. Org. Chem.* **1998**, *63*, 1574–1582; (c) Hossain, N.; Herdewijn, P. *Nucleosides Nucleotides Nucleic Acids* **1998**, *17*, 1775–1779.
73. Allart, B.; Khan, K.; Rosemeyer, H.; Schepers, G.; Hendrix, C.; Rothenbacher, K.; Seela, F.; Van Aerschot, A.; Herdewijn, P. *Chem. Eur. J.* **1999**, *5*, 2424–2431.
74. Egli, M.; Pallan, P. S.; Allerson, C. R.; Prakash, T. P.; Berdeja, A.; Yu, J.; Lee, S.; Watt, A.; Gaus, H.; Bhat, B.; Swayze, E. E.; Seth, P. P. *J. Am. Chem. Soc.* **2011**, *133*, 16642–16649.
75. (a) Schaeffer, H. J.; Liu, G.; Godse, D. D. *J. Pharm. Sci.* **1964**, *53*, 1510; (b) Schaeffer, H. J.; Kaistha, K. K.; Chakrabo, S. *J. Pharm. Sci.* **1964**, *53*, 1371; (c) Teran, C.; Santana, L.; Uriarte, E.; Vina, D.; De Clereq, E. *Nucleosides Nucleotides Nucleic Acids* **2003**, *22*, 787–789.
76. (a) Ramesh, K.; Wolfe, M. S.; Lee, Y.; Velde, D. V.; Borchardt, R. T. *J. Org. Chem.* **1992**, *57*, 5861–5868; (b) Calvani, F.; Macchia, M.; Rossello, A.; Gismondo, M. R.; Drago, L.; Fassina, M. C.; Cisternino, M.; Domiano, P. *Bioorg. Med. Chem. Lett.* **1995**, *5*, 2567–2572; (c) Mikhailov, S. N.; Blaton, N.; Rozenski, J.; Balzarini, J.; DeClercq, E.; Herdewijn, P. *Nucleosides Nucleotides* **1996**, *15*, 867–878.
77. (a) Kitagawa, I.; Cha, B. C.; Nakae, T.; Okaichi, Y.; Takinami, Y.; Yoshikawa, M. *Chem. Pharm. Bull.* **1989**, *37*, 542–544; (b) Halazy, S.; Kenny, M.; Dulworth, J.; Eggenspiller, A. *Nucleosides Nucleotides* **1992**, *11*, 1595–1606.
78. (a) Vina, D.; Santana, L.; Uriarte, E. *Nucleosides Nucleotides Nucleic Acids* **2001**, *20*, 1363–1365; (b) Barral, K.; Courcambeck, J.; Pepe, G.; Balzarini, J.; Neyts, J.; De Clercq, E.; Camplo, M. *J. Med. Chem.* **2005**, *48*, 450–456; (c) Quezada, E.; Vina, D.; Delogu, G.; Borges, F.; Santana, L.; Uriarte, E. *Helv. Chim. Acta* **2010**, *93*, 309–313.

79. Rosenquist, A.; Kvarnstrom, I.; Classon, B.; Samuelsson, B. *J. Org. Chem.* **1996**, *61*, 6282–6288.
80. (a) Maurinsh, Y.; Rosemeyer, H.; Esnouf, R.; Medvedovici, A.; Wang, J.; Ceulemans, G.; Lescrinier, E.; Hendrix, C.; Busson, R.; Sandra, P.; Seela, F.; Van Aerschot, A.; Herdewijn, P. *Chem. Eur. J.* **1999**, *5*, 2139–2150; (b) Bardiot, D.; Rosemeyer, H.; Lescrinier, E.; Rozenski, J.; Van Aerschot, A.; Herdewijn, P. *Helv. Chim. Acta* **2005**, *88*, 3210–3224.
81. Maurinsh, Y.; Schraml, J.; DeWinter, H.; Blaton, N.; Peeters, O.; Lescrinier, E.; Rozenski, J.; VanAerschot, A.; DeClercq, E.; Busson, R.; Herdewijn, P. *J. Org. Chem.* **1997**, *62*, 2861–2871.
82. Wang, J.; Busson, R.; Blaton, N.; Rozenski, J.; Herdewijn, P. *J. Org. Chem.* **1998**, *63*, 3051–3058.
83. Nauwelaerts, K.; Lescrinier, E.; Sclep, G.; Herdewijn, P. *Nucleic Acids Res.* **2005**, *33*, 2452–2463.
84. Wang, J.; Herdewijn, P. *J. Org. Chem.* **1999**, *64*, 7820–7827.
85. Wang, J.; Morral, J.; Hendrix, C.; Herdewijn, P. *J. Org. Chem.* **2001**, *66*, 8478–8482.
86. Gu, P.; Griebel, C.; Van Aerschot, A.; Rozenski, J.; Busson, R.; Gais, H. J.; Herdewijn, P. *Tetrahedron* **2004**, *60*, 2111–2123.
87. Wang, J.; Verbeure, B.; Luyten, I.; Lescrinier, E.; Froeyen, M.; Hendrix, C.; Rosemeyer, H.; Seela, F.; Van Aerschot, A.; Herdewijn, P. *J. Am. Chem. Soc.* **2000**, *122*, 8595–8602.
88. Nauwelaerts, K.; Fisher, M.; Froeyen, M.; Lescrinier, E.; Van Aerschot, A.; Xu, D.; DeLong, R.; Kang, H.; Juliano, R. L.; Herdewijn, P. *J. Am. Chem. Soc.* **2007**, *129*, 9340–9348.
89. Vijgen, S.; Nauwelaerts, K.; Wang, J.; Van Aerschot, A.; Lagoja, I.; Herdewijn, P. *J. Org. Chem.* **2005**, *70*, 4591–4597.
90. Wang, J.; Vina, D.; Busson, R.; Herdewijn, P. *J. Org. Chem.* **2003**, *68*, 4499–4505.
91. Dalencon, S.; Youcef, R. A.; Pipelier, M.; Maisonneuve, V.; Dubreuil, D.; Huet, F.; Legoupy, S. *J. Org. Chem.* **2011**, *76*, 8059–8063.
92. Ferrer, E.; Alibes, R.; Busque, F.; Figueredo, M.; Font, J.; de March, P. *J. Org. Chem.* **2009**, *74*, 2425–2432.
93. Arango, J. H.; Geer, A.; Rodriguez, J.; Young, P. E.; Scheiner, P. *Nucleosides Nucleotides* **1993**, *12*, 773–784.

14 Recent Advances in Synthesis and Biological Activity of 4′-Thionucleosides

VARUGHESE A. MULAMOOTTIL, MAHESH S. MAJIK, GIRISH CHANDRA, and LAK SHIN JEONG

Department of Bioinspired Science and Laboratory of Medicinal Chemistry, College of Pharmacy, Ewha Womans University, Seoul, Korea

1. INTRODUCTION

Nucleosides and nucleotides are the components responsible for constructing DNA and RNA. They also act as enzyme substrates or receptor ligands, or are utilized in genome analysis or gene therapy after being converted to the oligonucleotides. It is magnificent that nucleosides have made significant progress in the recent past, and currently, its greatest impact has been on the drug discovery and development process. Nucleosides are organic molecules comprising a sugar, usually ribose or deoxyribose, linked to a heterocyclic nitrogenous base, particularly a purine or a pyrimidine, especially a compound obtained by hydrolysis of a nucleic acid. Nucleoside analogues are extremely useful for the development of therapeutic agents to control viral diseases and cancer.[1] Targeting DNA and RNA has resulted in substantial success in the development of nucleoside antimetabolites as drugs. Thus, nucleosides and nucleotides can be considered to have a privileged role in interaction with target molecules as well as utilized as cofactors, second messengers, and energy donors in most important metabolic pathways. The past decades witnessed extensive efforts from both chemists and biologists to develop novel nucleoside motifs to unravel the complications involved with the therapeutic range of these biological puzzles.[2]

D- and L-Nucleoside analogues are one of the most important biological and pharmaceutically active classes of compounds. However, nucleoside analogues with modified substituents in the furanose ring had several disadvantages, including chemical instability, appearance of resistance, and clinical toxicity.[3] With the aim

---

Chemical Synthesis of Nucleoside Analogues, First Edition. Edited by Pedro Merino.
© 2013 John Wiley & Sons, Inc. Published 2013 by John Wiley & Sons, Inc.

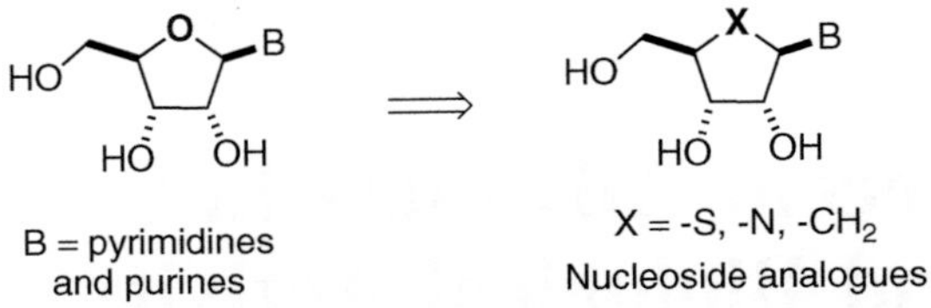

Figure 1. Development of second-generation nucleosides.

of overcoming these drawbacks, several studies were undertaken to synthesize nucleoside analogues, in which the oxygen atom of the furanose ring was replaced by different heteroatoms (e.g., S or N) or carbon itself (Figure 1). This simple rationale, which emerged from the realm of heterocyclic curiosities, led to the development of second-generation nucleosides.

Among these, 4'-thionucleosides have attracted much attention due to their potent biological activity[4] and unique metabolic stability.[5] Furthermore, when incorporated into RNA strands, certain 4'-thionucleosides led to enhanced thermal stability of the resulting modified RNA duplex.[6] The intense efforts toward the successful synthesis of these nucleoside analogues have been the subject of review on several occasions in the past.[7] In this chapter we review the chemistry that provides these privileged scaffolds and the relevant biological results. We hope the discussions contained herein and the descriptions of molecules and their synthesis will be both as enjoyable and inspiring to you as they were for us and for generations in the past.

2. DEVELOPMENT OF 4'-THIONUCLEOSIDES

4'-Thionucleoside analogues are among the most active class of heterocyclic nucleosides. In 1964, after the successful synthesis of thio analogues of D-xylose, D-glucose, etc., the synthesis leading to the first example of 4'-thionucleosides was achieved from L-lyxose.[8] Although further examples emerged in due course,[9] significant advances in this area declined due to unfavorable results of biological evaluation and difficulty in devising an efficient and large-scale preparation of the requisite 4-thiosugars.

In the early 1980s, two adenine-containing nucleosides (Figure 2), now known as fludarabine (Fludara; Berlex Oncology) and cladribine (Leustatin; Ortho Biotech),

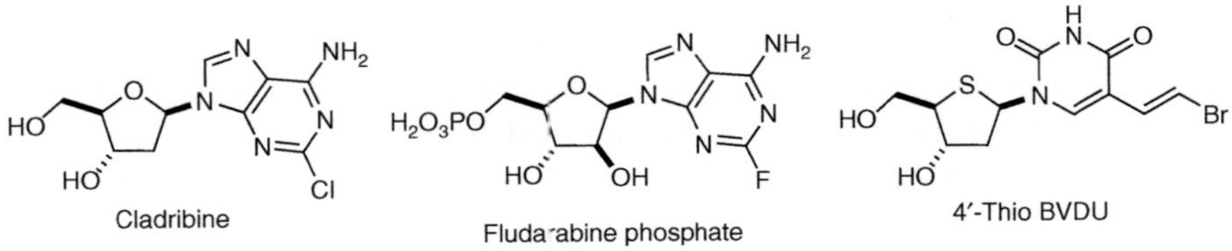

Figure 2. Modified nucleoside analogues.

were in clinical trials.[10] Both drugs were susceptible to glycosidic bond cleavage, with fludarabine subject to some phosphorylase cleavage and cladribine subject to both hydrolytic and enzymatic cleavage.[11] From the data and the structures of these two compounds, it became evident that some structural modifications could potentially enhance activity or reduce toxicity. This iterative drug discovery process involved examination of new compounds in a series of cancer cell lines. One such modification in restricting the cleavage of glycosyl bond was replacement of oxygen with sulfur in furanose ring. Corresponding to these facts, significant attempts to synthesize 4′-thionucleosides were initiated.[12] (E)-5-(2-bromovinyl)-2′-deoxyuridine (BVDU)[13] is known to be a potent and selective inhibitor of herpes simplex virus (HSV) type 1 and varicella zoster virus (VZV). However, BVDU is rapidly metabolized to the inactive E-5-(2-bromovinyl)uracil and 2-deoxyribose-1-phosphate by pyrimidine nucleoside phosphorylase.[14] In contrast, 4′-thio BVDU (Figure 2) is resistant to pyrimidine nucleoside phosphorylase and showed a higher chemotherapeutic index than BVDU.[15] Analogues, such as the 4′-thio-β-D-arabinofuranosyl (4′-thio-ara-C), have shown improved activity against solid tumors relative to their "4′-oxo" counterparts, due to an increase in stability toward enzyme-mediated hydrolysis,[16] which results in improved pharmacokinetic properties. Similarly, 4′-thio-oligonucleotides form duplexes with similar affinity to that of their 4′-oxo counterparts with the additional benefit of possessing nuclease-promoted strand scission resistance.[17] While most of these analogues exert their effects through termination of DNA elongation, alternative mechanisms include inhibition of ribonucleotide reductase and DNA methylation. Also, the significant cytotoxicity arising from replacement of the furanose ring oxygen in the naturally occurring pyrimidine 2′-deoxyribonucleosides with a sulfur atom proved to be an eye-opener for medicinal chemists. These intriguing reports revived interest in the synthesis and evaluation of 4′-thionucleosides and investigation of their L-isomers.[18] Based on the synthesis of different types of thionucleosides, Yokoyama made a systematic classification[7a] into four groups: (1) 4′-thionucleosides, (2) isothionucleosides, (3) L-thionucleosides, and (4) thioxonucleosides.

3. LOGIC FOR RATIONALE DESIGN

The inspiration to build 4′-thionucleosides is stimulated from nature (Figure 3). The limitations of modified nucleosides and the gradual evolution of thionucleosides was discussed in Section 2. The observation that replacement of oxygen with a sulfur atom, or the introduction of fluorine atoms at the 2′-position, confers resistance to cleavage of the glycosidic bond and thus results in metabolically stable nucleoside analogues broadens the scope for development of this class of nucleosides.[19]

Adenosine (1) is metabolically and structurally related to the bioactive nucleotides adenosine 5′-monophosphate (AMP), ADP, ATP, and cyclic adenosine monophosphate (cAMP) and to the biochemical methylating agent S-adenosyl-L-methionine (SAM). Adenosine receptors (ARs) consist of four subtypes, classified

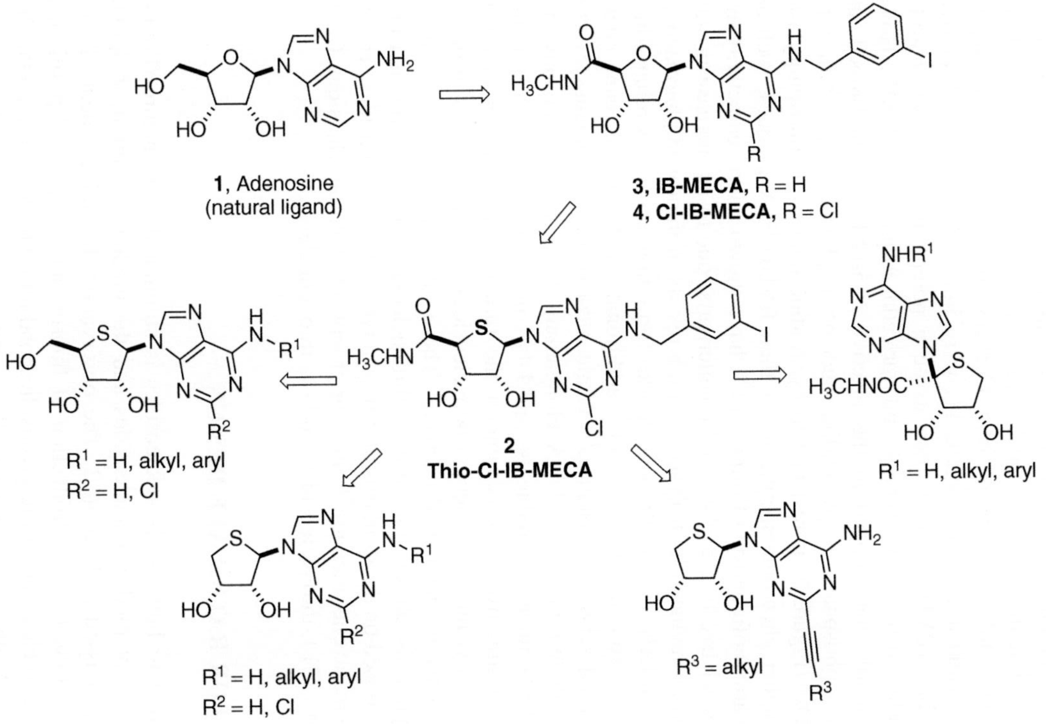

Figure 3. Rational design for the synthesis of 4′-thionucleosides.

as A_1, A_{2A}, A_{2B}, and A_3 each of which has a unique pharmacological profile, tissue distribution, and effector coupling.[20] Receptors from each of these four distinct subtypes have been cloned from a variety of species and characterized following functional expression in mammalian cells or *Xenopus* oocytes.[21] The receptor was first isolated from a rat testis cDNA library but remained without ligand identification.[22] However, the A_3 AR was first cloned, expressed, and functionally characterized from a rat striatum in 1992.[23]

Adenosine itself is a useful therapeutic agent when a short-acting response is sufficient to achieve the tissue state desired. Therefore, the discovery and development of potent and selective synthetic agonists and antagonists of ARs have been the subject of medicinal chemistry research for more than three decades. Adenine and its modified analogues (N^6- or C2-position of the adenine moiety and in the 3'-, 4'-, or 5'-position of the ribose moiety) have been explored extensively. 4'-Thionucleosides show different susceptibility to enzymes of nucleoside metabolism, and thus these exhibit better biological activity than that of the parent 4'-oxonucleosides. Although there is a bioisosteric rationale for replacement of the ring oxygen with sulfur, the potency, selectivity, and efficacy may differ with variations in the side chains.

The strategy for rational design in case of 4'-thionucleosides evolved as depicted in Figure 3. This is a modest attempt to summarize the reasoning behind the development of these novel analogues. Based on the known SARs for 4'-oxonucleosides, modifications of 4'-thionucleosides have been attempted to enhance AR affinity and selectivity by introducing selected substitutions on a nucleobase part, in combination with 5'-N-methylcarbamoyl or 5'-hydroxymethyl side chain.

3.1. Ribose Modifications

The most potent adenosine agonists with modified ribose moiety are as in Figure 3. Related studies revealed that 5'-methyl- and 5'-ethylcarbamoyl modifications usually led to an increase in A_3 AR affinity. For example, thio-Cl-IB-MECA **2**, a selective AR agonist that was the first high-affinity ligand for the A_3 AR ($K_i = 0.33$ nM for rat A_3), showed 2500- and 1400-fold rat A_3 receptor selectivity versus A_1 and A_{2A} receptors, respectively. In addition to the beneficial effect on A_3 AR affinity, 5'-modification of the ribose moiety also increased the metabolic stability against adenosine deaminase, as the 5'-hydroxyl group was essential for binding to the enzyme. Furthermore, the bioisosteric replacement of the alkylcarbamoyl moiety by any other substituent did not show any improvement in terms of affinity or selectivity.

3.2. Ribose and Adenine Modifications

The most potent adenosine agonists with modified adenine moiety are depicted as in Figure 4. Large N^6-substituents, such as substituted benzyl groups, was found to enhance the A_3 AR selectivity, but in turn reduced the maximal efficacy, which resulted in partial agonists. This problem was later overcome by the combination

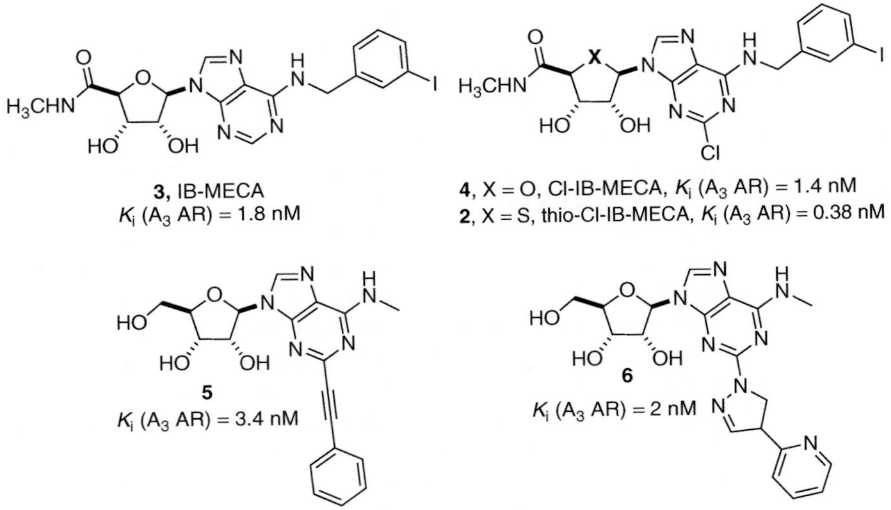

Figure 4. Examples of potent adenine-modified adenosine agonists.

of the large N^6-substitution along with 5′-alkylcarbamoyl ribose modification. The preferred N^6-benzyl substituent turned out to be a *meta*-iodine. Only smaller substituents at C2, such as 2-Cl and 2-I, were tolerated along with the N^6-iodobenzyl group. IB-MECA **3** and the more A_3 AR-selective Cl-IB-MECA **4** are very potent A_3 AR agonists containing a N^6-iodobenzyl modification. These compounds have been widely used as pharmacological probes in the elucidation of the physiological role of the A_3 AR.[24] The introduction of a small N^6-substituent, such as a methyl group, and a bulky 2-substituent resulted in very potent and selective A_3 AR ligands. Literature reveals that large 2-substituents such as 2-phenylethynyl **5**[25] and few 4-substituted-2-pyrazol-1-yl derivatives **6**[26] showed excellent A_3 AR affinity and selectivity. The combination of both large N^6- and 2-substitution failed to improve the A_3 AR affinity. The substitution at the 8-position of the ring was not well tolerated by any AR subtype.[27] Evaluation of the appropriate deaza analogues indicated that the nitrogen atoms at positions 3 and 7 are required for high affinity of adenosine at all subtypes. Thus, with all possible lessons on possible substitutions at relevant positions and the excellent selectivity exhibited by **2**, a detailed investigation for exploiting the therapeutic potential of thionucleosides became inevitable.

4. CHEMICAL SYNTHESIS AND BIOLOGICAL ACTIVITY

4.1. Synthesis of 4′-Thionucleoside-5′-uronamides

Adenosine (**1**), a natural ligand, inspired the synthesis of several analogues in the nucleoside series. Modifications at the N^6- and/or 4′-position of adenosine are a current topic of research worldwide. The outgrowth from this modification resulted in the discovery of Cl-IB-MECA **4** as a potent and selective agonist at human A_3

adenosine receptor (AR). The impressive biological properties of **4** led to the design and synthesis of 4'-thioadenosine derivatives. The desired 4'-thionucleosides were generally prepared by classical thioglycosidation of a thiosugar and an appropriate nucleobase. Matsuda and co-workers developed an alternative strategy, which involves the Pummerer condensation of a nucleobase and a sulfoxide.[28] This was a unique achievement, as this development paved the way for the large-scale synthesis of thiosugars, which otherwise was thought to be practically difficult.[29]

The synthesis of various substituted thio-Cl-IB-MECAs was reported by Jeong and co-workers.[30] In addition, valuable structure–activity relationships (SARs) of 4'-thionucleosides analogues as A_3 adenosine (AR) agonists were also established (Scheme 1). The synthesis started from D-gulonic-γ-lactone, which was converted to dimesylate via diol protection, reduction, and mesylation reactions. The cyclization of dimesylate to 4-thiofuranose was accomplished by heating with sodium sulfide in DMF. Selective hydrolysis of 5,6-acetonide without affecting the 2,3-acetonide was achieved using 30% aqueous acetic acid to give the diol. This was followed by oxidative cleavage of diol with lead tetraacetate to the corresponding aldehyde, which on reduction and benzoylation led to the formation of 4-thiosugar

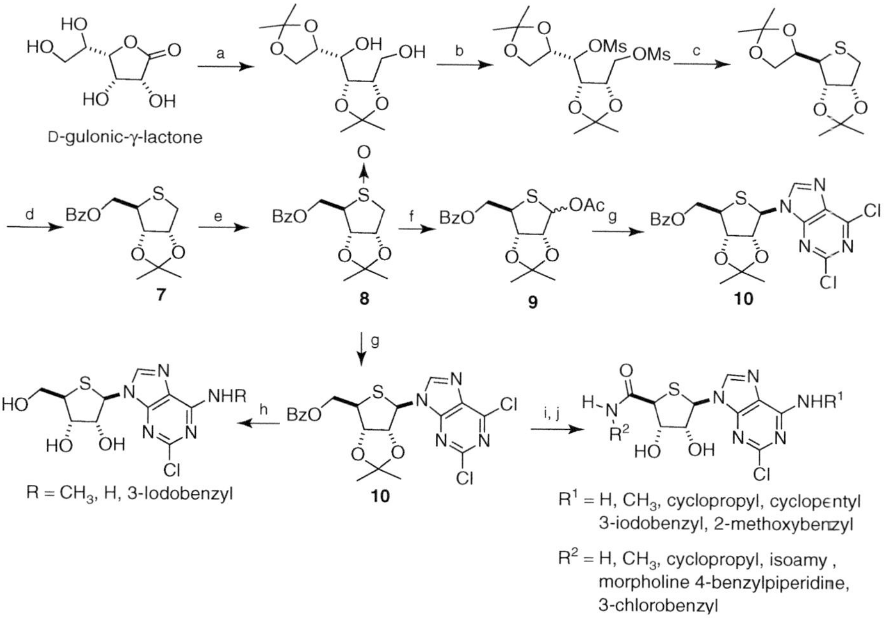

Scheme 1. Synthesis of N^6-substituted thio-Cl-IB-MECA derivatives. *Reagents and conditions*: (a) (i) acetone, H_2SO_4, $CuSO_4$, rt; (ii) $LiAlH_4$; (b) MsCl, Et_3N; (c) Na_2S, DMF, 80°C; (d) (i) 30% CH_3COOH; (ii) $Pb(OAc)_4$, EtOAc; (iii) $NaBH_4$; (iv) BzCl; (e) *m*-CPBA, CH_2Cl_2; (f) Ac_2O; (g) silylated 2,6-dichloropurine, TMSOTf, $ClCH_2CH_2Cl$, rt to 80°C; (h) (i) RNH_2, Et_3N, EtOH, rt; (ii) 80% CH_3COOH; (iii) NaOMe; (i) (i) R^1NH_2, Et_3N, EtOH, rt; (ii) 80% CH_3COOH, 70°C; (iii) TBSOTf, pyridine, 50°C; (iv) NaOMe, MeOH; (j) (i) PDC, DMF; (ii) R^2NH_2, EDC, HOBt, DIPEA, CH_2Cl_2; (iii) TBAF, THF, rt.

7. This 4-thiosugar on oxidation gave sulfoxide **8**, which on treatment with acetic anhydride led to the acetate **9**.[31] This acetate **9** was then condensed with silylated 2,6-dichloropurine in TMSOTf to the desired β–anomer **10** along with the α-anomer (9:1 ratio).

Another alternative approach involved oxidation of the 4-thiosugar to the corresponding sulfoxide **8**, which was condensed directly with 2,6-dichloropurine in the presence of TMSOTf and triethylamine to give β-anomer **10** (57%) as a major product along with a trace amount of the α-anomer. This is a better method than the former one in terms of yield (54% vs. 37% for two steps) and an additional step.

The NOE experiments between 1'-H and 4'-H led to the assignment of exact anomeric configuration. Necessary modifications were done on the condensed product in order to obtain the libraries of compounds and to examine the effect of N^6-substituent and/or 5'-hydroxymethyl or uronamide side chain, at the binding site of the A_3 AR. The required uronamide moiety at the 5'-position was introduced via an acetonide deprotection step, TBS protection, benzoyl hydrolysis, oxidation of 5'-hydroxy, followed by EDC coupling and TBS deprotection. Most synthesized 4'-thionucleosides exhibited higher binding affinity to human A_3 AR than Cl-IB-MECA, **4**. The 2-chloro-N^6-(3-iodobenzyl)adenosine-5'-methyluronamide showed the highest potent binding affinity ($K_i = 0.38$ nM).

Further modification at the 5'-ribofuran-uronamide moiety by removal of the amide hydrogen (Scheme 2) led to the successful conversion of A_3 AR agonists into selective antagonists.[32]

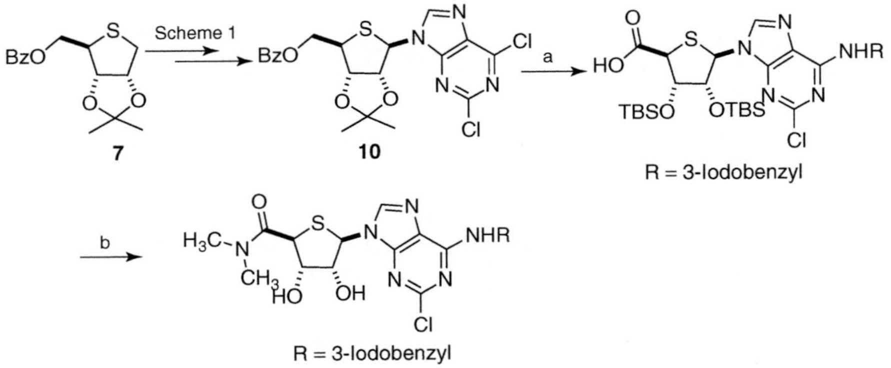

Scheme 2. Conversion of A_3 AR agonists into selective antagonists. *Reagents and conditions:* (a) (i) RNH_2, Et_3N, EtOH, rt; (ii) 80% CH_3COOH; (iii) TBSOTf, pyridine, 50°C; (iv) NaOMe, MeOH; (v) PDC, DMF; (b) (i) $NH(CH_3)_2$, EDC, HOBt, DIPEA, CH_2Cl_2; (ii) TBAF, THF.

After the successful conversion of A_3 AR full agonists into the potent and selective A_3 AR full antagonists, a systematic structure–activity relationship (SAR) study of 2-chloro-N^6-substituted-4'-thioadenosine-5'-N,N-dialkylamides[33] was inevitable. The stable sulfoxide **8** was condensed with 2,6-dichloropurine under Pummerer-type condensation to afford the desired nucleoside **10**. This on treatment with various alkyl or arylalkyl amines gave the N^6-substituted analogues **11a–j** (Scheme 3). The removal of isopropylidene group at a later stage proved to be difficult, and hence intermediates **11a–j** were first deprotected with 80% acetic acid followed by protection with TBSOTf. The resulting compounds were then treated with sodium methoxide and the key intermediates **12a–j** were obtained. These intermediates were then utilized for the synthesis of various N,N-disubstituted uronamides.

Hence, using a strategy similar to that in Scheme 1, the primary hydroxyl groups of **12a–j** were oxidized with pyridinium dichromate to the corresponding carboxylic acids, which were then immediately coupled with various dialkyl or cycloalkyl amines, in the presence of EDC and HOBt. Finally, the TBS protecting groups were removed under mild conditions using tetra-n-butylammonium fluoride (TBAF) at room temperature (Scheme 3). The synthesized $5'$-N,N-dialkyl uronamide derivatives **14a–s** were screened for their biological activity. From this study it became evident that removal of the hydrogen bond–donating ability of the $5'$-uronamide was essential for the pure A_3 AR antagonism. $5'$-N,N-Dimethyluronamide derivatives exhibited higher binding affinity than that of larger $5'$-N,N-dialkyl or $5'$-N,N-cycloalkylamide derivatives. Thus, it was clear that steric factors are crucial in binding to the human A_3 AR. The N^6-(3-bromobenzyl) derivative **14d** ($K_i = 9.32$ nM) exhibited the highest binding affinity at the human A_3 AR, with very low binding affinities to other AR subtypes. The A_3 AR antagonists synthesized during this study can be evaluated in models of several disorders related to the A_3 AR, such as glaucoma, inflammation, and asthma.

After the systematic investigation of N^6-substituted thio-Cl-IB-MECA derivatives, it was felt that a SAR study with respect to N^6-substituted thio-IB-MECA would also prove beneficial. This was due to the fact that IB-MECA **3** is a potent and selective A_3 AR agonist and currently is in clinical trials as an anticancer agent. Thus, under conditions identical to Scheme 3, a strategy for the synthesis of these derivatives was designed (Scheme 4).[34] The 6-chloropurine derivative **15** was obtained after a Pummerer-type condensation of sulfoxide **8** with 6-chloropurine. The desired N^6-methyladenine and N^6-(3-iodobenzyl)adenine derivatives were obtained after treatment with methylamine and 3-iodobenzylamine, respectively. Finally, uronamides **16a–w** were obtained after oxidation, amination, and subsequent deprotection reactions.

Most of the synthesized compounds showed very high binding affinity at the human A_3 AR, with high selectivity in comparison to other subtypes. In comparison to the $4'$-oxonucleosides, the $4'$-thionucleoside thio-IB-MECA exhibited higher binding affinity at the human A_3 AR ($K_i = 0.25$ nM) as well as higher selectivity over other subtypes, indicating that the compound has potential to be developed as a drug. This compound also showed very high binding affinity at the rat A_3 AR ($K_i = 1.86 \pm 0.36$ nM). Thus, this study established SARs of bioisosteric $4'$-thio analogues of potent and selective A_3 AR agonist IB-MECA.

Recently, an alternative method for the synthesis of thio-Cl-IB-MECA **2** starting from a cheap and commercially available D-ribose as a starting material was also established (Scheme 5).[35] Here, D-ribose was converted to

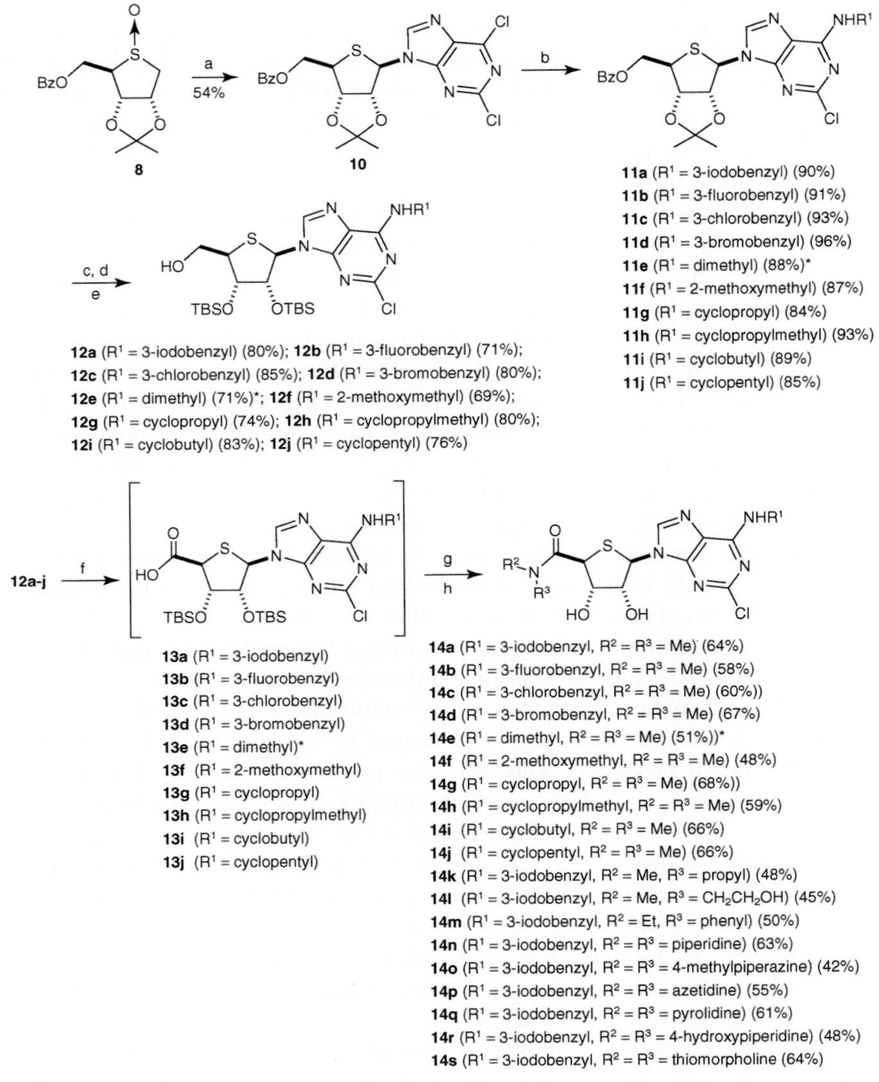

11a (R^1 = 3-iodobenzyl) (90%)
11b (R^1 = 3-fluorobenzyl) (91%)
11c (R^1 = 3-chlorobenzyl) (93%)
11d (R^1 = 3-bromobenzyl) (96%)
11e (R^1 = dimethyl) (88%)*
11f (R^1 = 2-methoxymethyl) (87%)
11g (R^1 = cyclopropyl) (84%)
11h (R^1 = cyclopropylmethyl) (93%)
11i (R^1 = cyclobutyl) (89%)
11j (R^1 = cyclopentyl) (85%)

12a (R^1 = 3-iodobenzyl) (80%); **12b** (R^1 = 3-fluorobenzyl) (71%);
12c (R^1 = 3-chlorobenzyl) (85%); **12d** (R^1 = 3-bromobenzyl) (80%);
12e (R^1 = dimethyl) (71%)*; **12f** (R^1 = 2-methoxymethyl) (69%);
12g (R^1 = cyclopropyl) (74%); **12h** (R^1 = cyclopropylmethyl) (80%);
12i (R^1 = cyclobutyl) (83%); **12j** (R^1 = cyclopentyl) (76%)

13a (R^1 = 3-iodobenzyl)
13b (R^1 = 3-fluorobenzyl)
13c (R^1 = 3-chlorobenzyl)
13d (R^1 = 3-bromobenzyl)
13e (R^1 = dimethyl)*
13f (R^1 = 2-methoxymethyl)
13g (R^1 = cyclopropyl)
13h (R^1 = cyclopropylmethyl)
13i (R^1 = cyclobutyl)
13j (R^1 = cyclopentyl)

14a (R^1 = 3-iodobenzyl, $R^2 = R^3$ = Me) (64%)
14b (R^1 = 3-fluorobenzyl, $R^2 = R^3$ = Me) (58%)
14c (R^1 = 3-chlorobenzyl, $R^2 = R^3$ = Me) (60%))
14d (R^1 = 3-bromobenzyl, $R^2 = R^3$ = Me) (67%)
14e (R^1 = dimethyl, $R^2 = R^3$ = Me) (51%))*
14f (R^1 = 2-methoxymethyl, $R^2 = R^3$ = Me) (48%)
14g (R^1 = cyclopropyl, $R^2 = R^3$ = Me) (68%))
14h (R^1 = cyclopropylmethyl, $R^2 = R^3$ = Me) (59%)
14i (R^1 = cyclobutyl, $R^2 = R^3$ = Me) (66%)
14j (R^1 = cyclopentyl, $R^2 = R^3$ = Me) (66%)
14k (R^1 = 3-iodobenzyl, R^2 = Me, R^3 = propyl) (48%)
14l (R^1 = 3-iodobenzyl, R^2 = Me, R^3 = CH_2CH_2OH) (45%)
14m (R^1 = 3-iodobenzyl, R^2 = Et, R^3 = phenyl) (50%)
14n (R^1 = 3-iodobenzyl, $R^2 = R^3$ = piperidine) (63%)
14o (R^1 = 3-iodobenzyl, $R^2 = R^3$ = 4-methylpiperazine) (42%)
14p (R^1 = 3-iodobenzyl, $R^2 = R^3$ = azetidine) (55%)
14q (R^1 = 3-iodobenzyl, $R^2 = R^3$ = pyrolidine) (61%)
14r (R^1 = 3-iodobenzyl, $R^2 = R^3$ = 4-hydroxypiperidine) (48%)
14s (R^1 = 3-iodobenzyl, $R^2 = R^3$ = thiomorpholine) (64%)

Scheme 3. Synthesis of 2-chloro-N^6-substituted-4'-thioadenosine-5'-*N,N*-dialkyl uronamides. *Reagents and conditions:* (a) 2,6-dichloropurine, TMSOTf, Et_3N, CH_3CN-$ClCH_2CH_2Cl$, rt to 83°C; (b) R^1NH_2, Et_3N, EtOH, rt; (c) 80% AcOH, 70°C; (d) TBSOTf, pyridine, 50°C; (e) NaOMe, MeOH; *N^6 $(CH_3)_2$; (f) PDC, DMF, rt, overnight; (g) R^2R^3NH, EDC, HOBt, DIPEA, CH_2Cl_2, rt, overnight; (h) TBAF, THF, rt; *$N^6(CH_3)_2$.

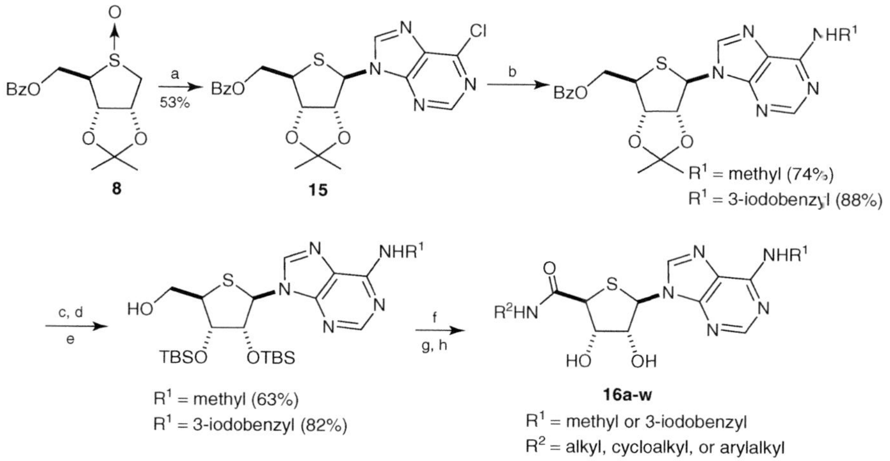

Scheme 4. Synthesis of N^6-substituted-4′-thioadenosine-5′-uronamides. *Reagents and conditions:* (a) 6-chloropurine, TMSOTf, Et$_3$N, ClCH$_2$CH$_2$Cl, rt to 80°C; (b) R^{1}NH$_2$, Et$_3$N, EtOH, rt; (c) 80% AcOH, 70°C, 12 h; (d) TBSOTf, pyridine, 50°C, 5 h; (e) NaOMe, MeOH, rt, 4 h; (f) PDC, DMF, rt, 20 h; (g) R^{2}NH$_2$, EDC, HOBt, DIPEA, CH$_2$Cl$_2$, rt, 15 h; (h) TBAF, THF, rt, 1 h.

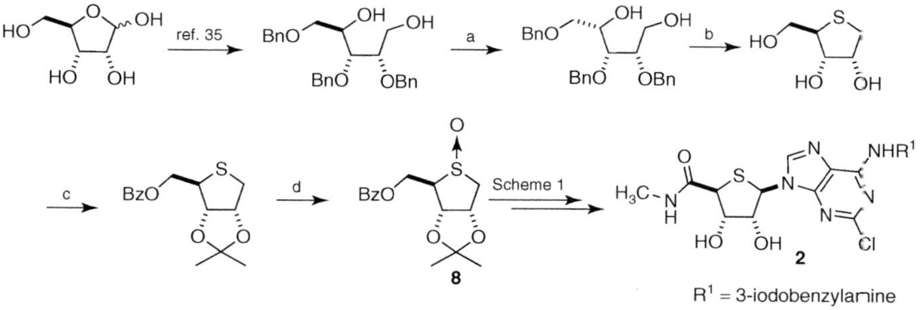

Scheme 5. Alternative strategy for the synthesis of thio-Cl-IB-MECA (**2**). *Reagents and conditions:* (a) (i) Bu$_2$SnO, toluene, reflux; (ii) TBAI, BzCl, rt; (iii) DIAD, Ph$_3$P, EzOH, THF, rt; (iv) NaOMe, rt; (b) (i) MsCl, pyridine, rt; (ii) Na$_2$S, 100°C; (iii) BCl$_3$, −90°C; (c) (i) acetone, p-TsOH, rt; (ii) BzCl, pyridine, rt; (d) m-CBPA, CH$_2$Cl$_2$, −78°C, 40 min.

2,3,5-tri-*O*-benzyl-D-ribofuranoside using the Barker and Fletcher procedure,[36] which involves benzyl protection, hydrolysis, and reduction. The selective benzoyl protection of primary alcohol followed by Mitsunobu reaction at a secondary hydroxyl group furnished dibenzoate, which on hydrolysis produced the diol. This was further subjected to mesylation and subsequent cyclization (Na$_2$S), followed by benzyl deprotection to give the trihydroxy thiosugar. Selective 2,3-hydroxy protection as acetonide followed by protection of primary hydroxy and *m*-CPBA oxidation gave the sulfoxide **8**. This was then successfully converted to thio-Cl-IB-MECA **2**.

The geometry and flexibility of hydrogen-bonding groups of the ribose moiety are necessary for activation of the A_3 AR. Based on this fact, the search for novel compounds related to thio-Cl-IB-MECA continued. Thus, two major structural modifications were reported wherein the transformation of natural D-type to L-type nucleoside and movement of 5′-uronamide from the β- to the α-position were attempted.[37] The stereoselective synthesis of 1-α-substituted-4′-thionucleosides was achieved via a stereoselective nucleophilic substitution (Scheme 6). D-Gulonic-γ-lactone was converted to 5,6-dihydroxy thiosugar using the strategy described in Scheme 1. Oxidative cleavage of diol with lead tetraacetate at 0°C gave aldehyde, which was transformed to methylamide via oxidation, methylation, and amination using methylamine. *m*-CPBA oxidation gave sulfoxide, which on condensation with 2,6-dichloropurine failed to give the product desired. The reason attributed for this observation was the destabilization of carbocation by carbonyl group.

Alternatively, it was realized that the generation of carbanion directly at the 1′-position of 4′-thionucleoside was a better choice. This was because the 1′-position is highly acidic due to sulfur atom and electron-withdrawing purine base, and thus a stereoselective electrophilic functionalization at the 1′-position was possible. The thiosugar acetate **17** was prepared from diol using an excess of lead tetraacetate and a longer reaction time. It involves oxidative cleavage of diol to aldehyde and then to acid followed by simultaneous oxidative decarboxylation. Condensation of **17** with 2,6-dichloropurine in the presence of TMSOTf and triethylamine gave exclusively β-anomer in good yield. Lithiation with LiHMDS at the 1′-position followed by treatment with alkyl chloroformate resulted in formation of 1′-α-carbonate. However, removal of an acetonide group under acidic conditions resulted in deglycosylation, possibly due to the carbonyl group and purine moiety. Further modification in strategy involved the acetonide deprotection of condensed product **18** in the presence of HCl prior to functionalization at the 1′-position. The idea was to convert the free hydroxyl groups to the corresponding 2,3-diacetate, which could be deprotected under mild conditions. However, subsequent stereoselective functionalization resulted in a substantial decrease in yield. Hence, 2,3-diol was subjected to tetrahydropyran (THP) protection followed by 1′-α-functionalization in a stereoselective manner to furnish THP-protected derivative **19**. Selective N^6-amination (3-iodobenzylamine and ethylamine) at room temperature followed by conversion of the methyl ester to amide (methyl, ethyl, benzyl amide) and THP deprotection gave the final diol. Measurement of the binding affinity of synthesized diols at adenosine receptors using radioligand binding assay revealed that the compounds were devoid of binding affinities to all subtype receptors, indicating that the 1′-uronamide group might form strong intramolecular hydrogen bonding with the 2′-hydroxyl group.

After the systematic investigation of the 5′-uronamide series of thio-Cl-IB-MECA and thio-IB-MECA, the next possible modification was the replacement of a uronamide side chain with that of a hydroxymethyl group. Thus, structure–activity relationships (SARs) were developed for 5′-hydroxyl derivatives. A series of N^6-substituted-5′-hydromethyl-4′-thioadenosines (Scheme 7) were synthesized from D-gulonic-γ-lactone.[38] The desired sulfoxide **8** was prepared using a strategy

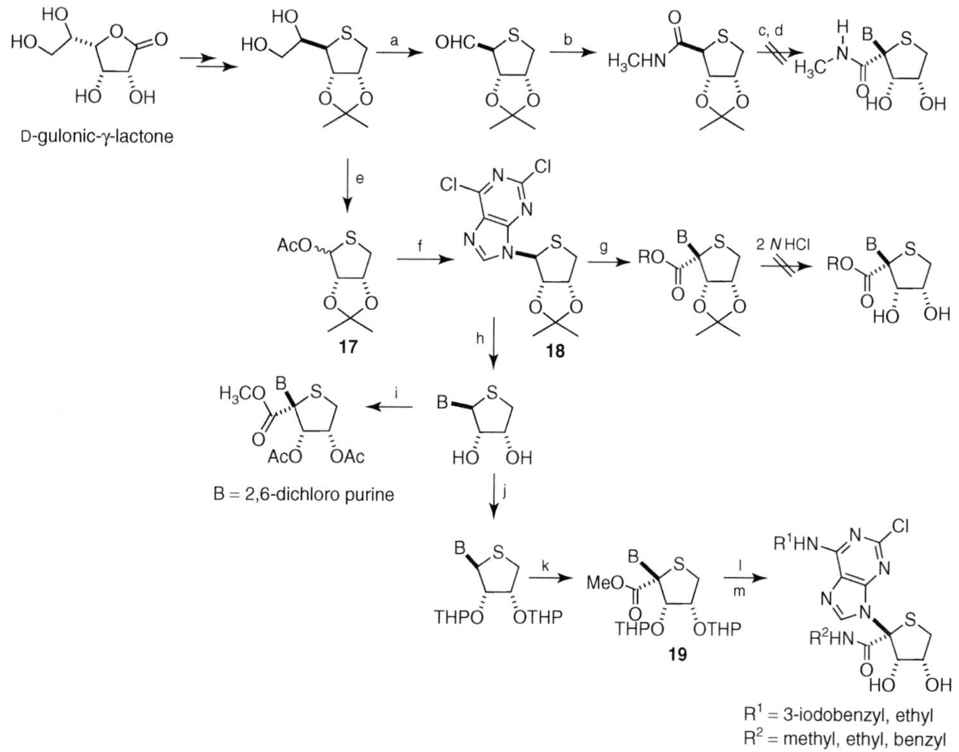

Scheme 6. Synthesis of 1′-α-substituted-4′-thionucleosides. *Reagents and conditions:* (a) Pb(OAc)₄, EtOAc, 0°C, 10 min; (b) (i) PDC; (ii) (CH₃)₂SO₄, K₂CO₃, rt; (iii) CH₃NH₂, THF; (c) *m*-CPBA; (d) 2,6-dichloropurine, TMSOTf, Et₃N, 80°C; (e) excess Pb(OAc)₄, EtOAc, rt; (f) silylated 2,6-dichloropurine, TMSOTf, 0–70°C; (g) LiHMDS, alkylchloroformate, −78°C; (h) 2 *N* HCl, THF; (i) Ac₂O, pyridine; (ii) LiHMDS, ClCOOMe, −78°C, 10%; (j) 3,4-dihydro-2*H*-pyran, PPTS; (k) LiHMDS, ClCOOMe, −78°C; (l) (i) R¹NH₂, EtOH; (ii) *p*-TsOH; (m) (i) R²NH₂, EtOH or MeNH₂, THF; (ii) *p*-TsOH.

R¹ = 3-iodobenzyl, ethyl
R² = methyl, ethyl, benzyl

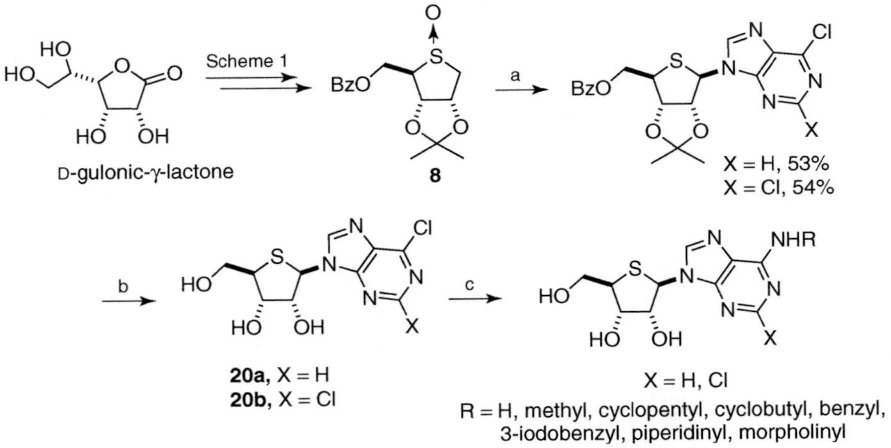

Scheme 7. Synthesis of N^6-substituted-5′-hydroxymethyl-4′-thioadenosine derivatives. *Reagents and conditions:* (a) 6-chloropurines or 2,6-dichloropurine, TMSOTf, Et$_3$N, rt to 80°C; (b) (i) 3.0 M MeMgI in ether, THF, rt; (ii) 80% CH$_3$COOH, 70°C; (c) RNH$_2$, Et$_3$N, EtOH, rt.

analogous to Scheme 1. The direct condensation of sulfoxide **8** with 6-chloropurine or 2,6-dichloropurine in the presence of TMSOTf and triethyl amine afforded β-anomer (major product) along with α-anomer (minor). At higher temperatures (i.e., on heating to 80°C), the N^3-isomer formed initially was smoothly converted to N^9-isomer. This is in agreement with observations first reported by Chu and co-workers.[39] The deprotection of benzoyl group using sodium methoxide or methanolic ammonia resulted in the formation of 6-methoxy or 6-amino derivative from the condensed product. Hence, Grignard reagent (methylmagnesium iodide in THF) was utilized for hydrolysis of the benzoyl group. Further, acetonide deprotection under acidic condition 80% acetic acid gave the triol. The amine substituent required at the N^6-position was introduced using various primary and secondary amines, such as alkyl, cycloalkyl, arylalkyl, and cyclic amines. Among the series of N^6-substituted-5′-hydromethyl-4′-thioadenosines, 2-chloro-N^6-methyl-4′-thioadenosines were found to be the most potent and selective full agonist ($K_i = 0.8 \pm 0.1$ nM) at human A$_3$ AR. In general, this class of compounds exhibited less potency than the corresponding N^6-substituted 4′-thioadenosine-5′-uronamides.

The high binding affinities of 2-hexynyl-N^6-methyladenosine[40] and N^6-substituted 4′-thioadenosine derivatives prompted the investigation of another series of novel 2-alkynyl-substituted N^6-methyl-4′-thioadenosine derivatives. It was believed that the combination of the characteristics of these two classes of nucleosides might prove to be beneficial. With this rationale, the nucleosides desired were synthesized from D-gulonic-γ-lactone, via the key intermediate **9**,[30] followed by a palladium-catalyzed cross-coupling reaction as a key step (Scheme 8).[41] Among the compounds tested in this series, only compound **21b**

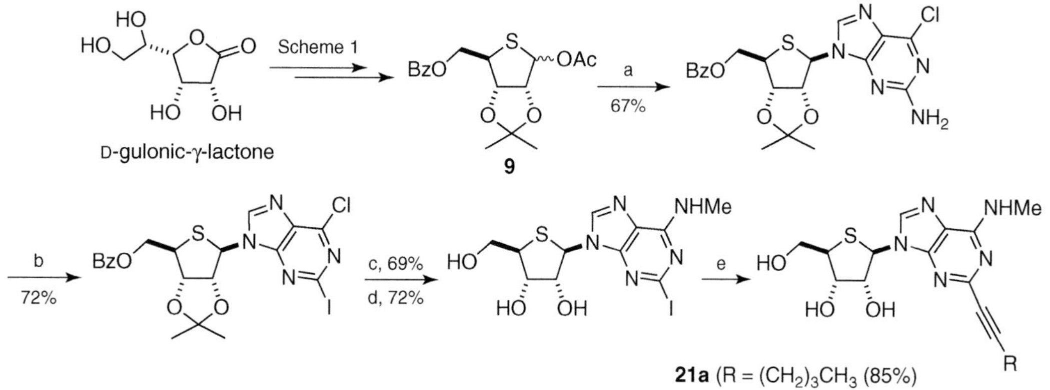

Scheme 8. Synthesis of 2-alkynyl-substituted-N^6-methyl-4′-thioadenosine derivatives. *Reagents and conditions*: (a) Silylated 2-amino-6-chloropurine, TMSOTf, ClCH$_2$CH$_2$Cl, rt to reflux, 8 h; (b) CuI, I$_2$, CH$_2$I$_2$, isoamyl nitrite, THF, 80°C, 45 min; (c) 40% MeNH$_2$, 1,4-dioxane, rt, 3 h; (d) 85% HCOOH, rt, 3 h; (e) alkyne, (Ph$_3$P)$_2$PdCl$_2$, CuI, Et$_3$N, DMF, 80°C.

21a (R = (CH$_2$)$_3$CH$_3$ (85%)
21b (R = *p*-C$_6$H$_4$ CH$_2$CH$_2$COO^{t}Bu (69%)
21c (R = *p*-C$_6$H$_4$ CH$_2$CH$_2$COOMe (78%)

showed moderate binding affinity at the human A_3 AR, without binding affinities at other subtypes.

4.2. Synthesis of Truncated 4′-Thionucleosides

The structural similarity of most adenosine analogues to that of adenosine forces them to be A_3 AR agonists. Although few nucleoside derivatives were reported to be A_3 AR antagonists, these generally exhibit weaker and less selective human A_3 AR antagonism than that of their nonpurine heterocyclic counterparts.[42] These nonpurine heterocyclic A_3 AR antagonists,[43] showed high binding affinity at the human A_3 AR, but it was found that they were weak or ineffective at the rat A_3 AR indicating that they were not ideal for evaluation in small-animal models and thus as drug candidates. Thus, the need to develop novel A_3 AR antagonists that are independent of species is highly desirable. The observations that nucleoside analogues showed minimal species dependence at the A_3 AR were the prime basis for investigations leading to the design of novel nucleoside templates.[44]

A recent molecular modeling study of the A_3 AR indicated that amide hydrogen of the 5′-uronamides of 4′-oxo and 4′-thio analogues serves as a hydrogen-bonding donor in the binding site of the A_3 AR. This is essential for the induced fit required for activation of the A_3 AR.[45] On the basis of these findings, extra alkyl groups on the 5′-uronamides of such compounds were appended to remove the hydrogen-bonding ability at this site. The expectation was that such a modification would not lead to the conformational change required for activation at the A_3 AR. As expected, the 5′-N,N-dialkylamide derivatives[31,33] displayed potent and selective A_3 AR antagonism in which steric factors were crucial for affinity in binding to the A_3 AR. After this discovery of a full A_3 AR antagonist, it was felt that the total removal of a 5′-uronamide group may minimize the steric repulsion at the binding region and that this might completely abolish the hydrogen-bonding ability. This rationale led to the development of novel truncated nucleoside analogues, discussed in this section.

A synthetic strategy to obtain the desired nucleoside analogues from D-mannose was envisaged as in Scheme 9.[46] D-Mannose was converted to the diacetonide, using 2,2-dimethoxypropane and camphorsulfonic acid in acetone. This was followed by reduction with sodium borohydride, mesylation to the dimesylate, and then cyclization with anhydrous sodium sulfide in DMF to give the desired thiosugar. Selective hydrolysis of the 5,6-acetonide with 60% aqueous acetic acid gave the protected diol, which was then converted to acetate **22** with an excess of lead tetraacetate at room temperature.[36]

The versatility of **22** permitted access to all desired final nucleosides **23a–j** under Lewis acid–catalyzed condensation reactions. The condensation reaction with 2,6-dichloropurine in the presence of trimethysilyl trifluoromethanesufonate (TMSOTf), gave the desired β-anomer exclusively, in good yield. Removal of the acetonide protection led to the formation of the diol, which on further treatment with various alkyl or arylalkylamines afforded a series of different N^6-substitued D-4′-thionucleosides **23a–j**.

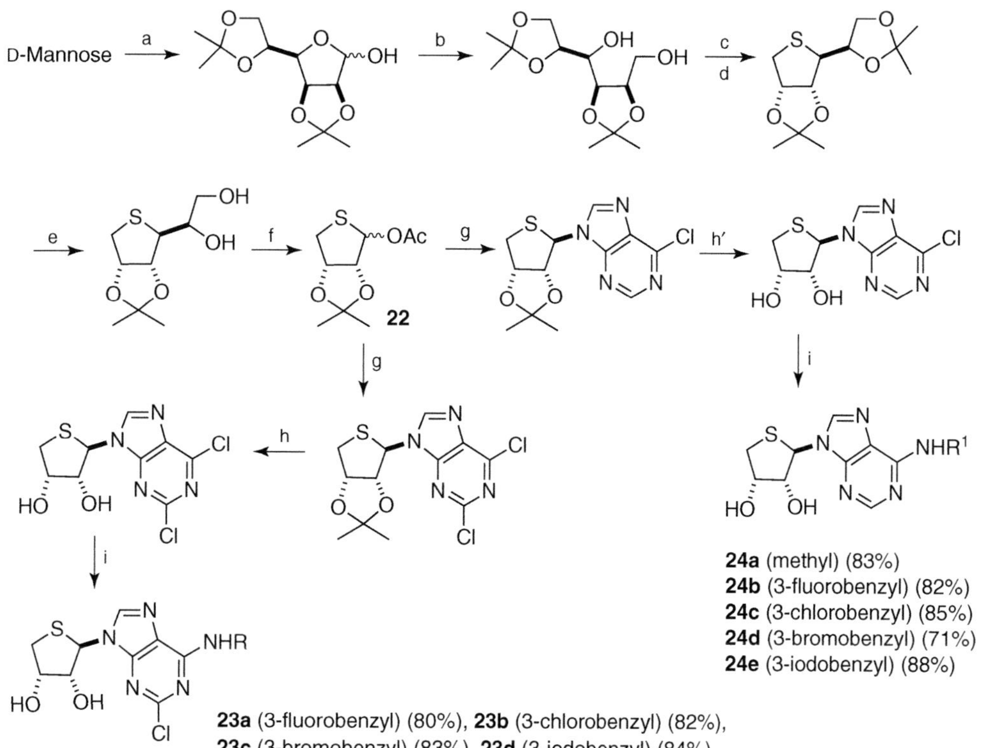

Scheme 9. Synthesis of the truncated 4′-thioadenosine nucleosides. *Reagents and conditions*: (a) 2,2-dimethoxypropane, camphorsulfonic acid, acetone, rt, 15 h; (b) NaBH₄, EtOH, rt, 2 h; (c) MsCl, Et₃N, CH₂Cl₂, rt, 1 h; (d) Na₂S, DMF, 80°C, 15 h; (e) 60% AcOH, rt, 2 h; (f) Pb(OAc)₄, EtOAc, rt, overnight. (g) 2,6-dichloropurine or 6-chloropurine, ammonium sulfate, HMDS, 170°C, 15 h, then TMSOTf, DCE, rt to 80°C, 3 h; (h) 2 N HCl, THF, rt, 15 h; (h′) 2 N HCl, THF, 45°C, 15 h; (i) RNH₂ or R¹NH₂, Et₃N, EtOH, rt, 1–3 days.

23a (3-fluorobenzyl) (80%), **23b** (3-chlorobenzyl) (82%), **23c** (3-bromobenzyl) (83%), **23d** (3-iodobenzyl) (84%), **23e** (2-chlorobenzyl) (81%), **23f** (2-methoxy-5-chlorobenzyl) (78%), **23g** (2-methoxybenzyl) (88%), **23h** (1-naphthylmethyl) (90%), **23i** (4-toluic acid) (84%), **23j** (methyl) (85%)

24a (methyl) (83%)
24b (3-fluorobenzyl) (82%)
24c (3-chlorobenzyl) (85%)
24d (3-bromobenzyl) (71%)
24e (3-iodobenzyl) (88%)

To establish SAR relationships, a detailed investigation for modifications at the C2- and N^6-positions in the purine moiety was undertaken.[47] The acetate derivative **22** was condensed with 6-chloropurine and the condensed derivative was deprotected to give the desired diol. The 2H derivative was successfully converted to the novel N^6-methyl derivative **24a** and N^6-3-halobenzyl derivatives **24b–e** by treating with methylamine and 3-halobenzylamines, respectively. Furthermore, to establish if any stereochemical preference exists for the binding affinities of D-4′-thionucleosides over those of the corresponding L-4′-thionucleosides, a series of truncated L-enantiomers were synthesized as in Scheme 10. D-Gulonic-γ-lactone was converted to the diol using a procedure published previously.[36] A one-step conversion of the diol to the L-glycosyl donor was achieved using an excess of Pb(OAc)$_4$, indicating that oxidative diol cleavage (i.e., oxidation of the resulting aldehyde to the acid) and oxidative decarboxylation occurred simultaneously. Thus, L-4′-thioadenosine derivatives **25a** and **b** were synthesized in a manner similar to Scheme 9.

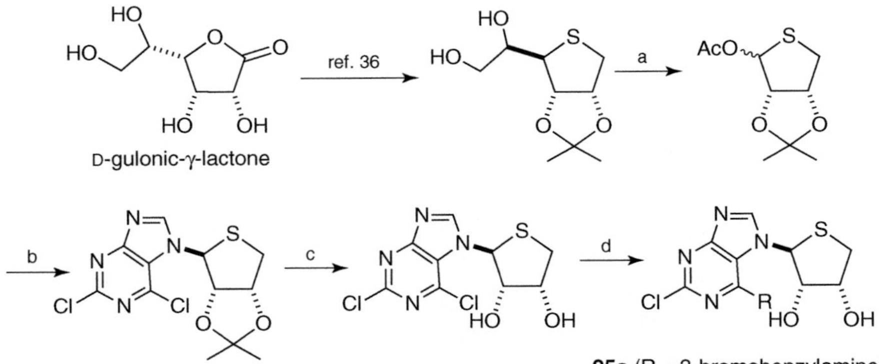

Scheme 10. Synthesis of truncated L-4′-thioadenosine derivatives. *Reagents and conditions:* (a) Pb(OAc)$_4$, AcOH, 80°C, 3 h; (b) 2,6-dichloropurine, ammonium sulfate, HMDS, 170°C, 15 h, then TMSOTf, DCE, rt to 80°C, 3 h; (c) 2 N HCl, THF, 45°C, 15 h; (d) RNH$_2$, Et$_3$N, EtOH, rt, 1–3 days.

The biological results indicated that at the human A$_3$ AR, N^6-substituted purine analogues bound potently and selectively and acted as antagonists in a cyclic AMP functional assay. Among the synthesized compounds, **23b** (K_i = 1.66 nM) and **24c** (K_i = 1.5 nM) were found to be the most potent at the human A$_3$ AR. However, the chloroderivative **24c** showed the optimal species-independent binding affinity. Also, another interesting study published recently led to an important observation that the truncated analogue **23d** was able to lower intraocular pressure in mouse and thus can be promising for further study in multiple animal models.[48] All these results led to the discovery of truncated D-4′ thioadenosine as an excellent template for the design of novel A$_3$ AR antagonists to act in both human and murine species.

The observation that truncation resulted in the 4′-thioadenosine antagonist derivative with preserved AR affinity and selectivity, the quest for design of more active analogues in this series continued. As discussed earlier, there are few reports that C2 or C8 substitution sometimes leads to substantial enhancement in the binding affinity or selectivity at the A_{2A} AR or other subtype.[49] Recently, the synthesis of novel C2- and C8- substituted 4′-thioadenosine derivatives as potent ligands for the A_{2A} AR was reported.[50] The synthesis was achieved from D-mannose as depicted in Scheme 11.

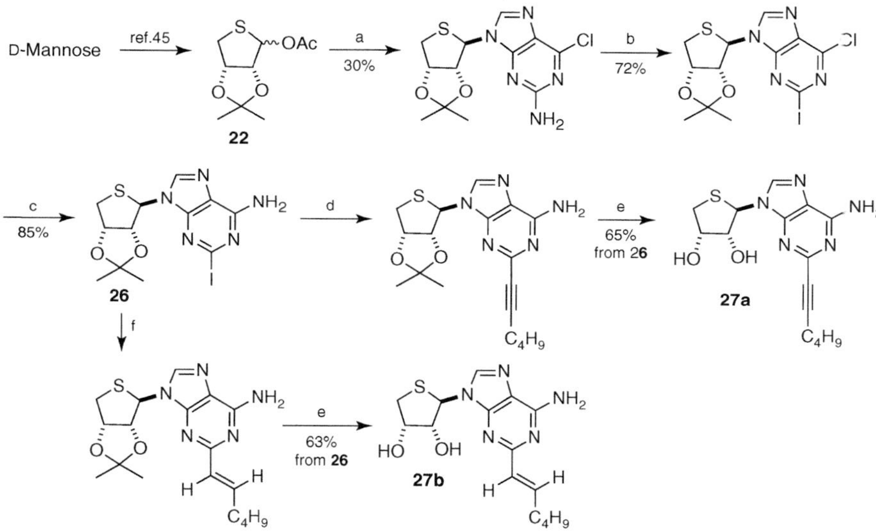

Scheme 11. Synthesis of truncated C2-substituted adenosine derivatives. *Reagents and conditions:* (a) Silylated 2-amino-6-chloropurine, TMSOTf, DCE, rt to 80°C, 3 h; (b) CuI, isoamyl nitrite, I_2, CH_2I_2, THF, 110°C, 45 min; (c) NH_3/MeOH, 80°C, 2 h; (d) 1-hexyne, CuI, TEA, DMF, bis(triphenylphosphine)palladium dichloride, rt, 3 h; (e) 1 N HCl, THF, rt, 15 h; (f) (*E*)-1-catecholboranylhexene, then tetrakis(triphenylphosphine)palladium (0), Na_2CO_3, DMF, H_2O, 90°C, 15 h.

The 2-amino-6-chloro derivative, on treatment with isoamyl nitrite, iodine, and diiodomethane in the presence of CuI, afforded the 2-iodo-6-chloro derivative, which was treated further with methanolic ammonia to give the 2-iodo-6-amino derivative, **26**. A Sonogashira coupling reaction of the iodo derivative **26** with 1-hexyne in the presence of bis(triphenylphosphine)palladium dichloride yielded the 2-hexynyl derivative. Finally, the isopropylidene group was removed with 1 N HCl to produce the final 2-hexynyl-4′-thioadenosine derivative **27a**. Similarly, a Suzuki coupling reaction of the 2-iodo derivative **26** with (*E*)-1-catecholboranylhexene in the presence of tetrakis(triphenylphosphine)palladium(0) afforded the 2-hexenyl derivative. The final deprotection of the acetonide protecting group with 1 N HCl gave the 2-hexynyl-4′-thioadenosine derivative **27b**. Using a similar strategy, 8-substituted adenosine derivatives **29a** and **b** were synthesized (Scheme 12).

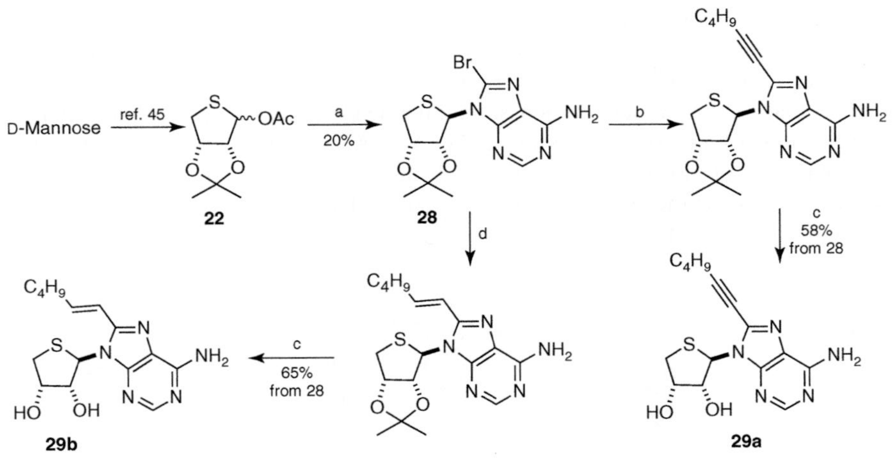

Scheme 12. Synthesis of truncated C8-substituted adenosine derivatives. *Reagents and conditions:* (a) Silylated 8-bromoadenine, TMSOTf, DCE, rt to 90°C, 2 h; (b) 1-hexyne, CuI, TEA, DMF, bis(triphenylphosphine)palladium dichloride, rt, 3 h; (c) 1 N HCl, THF, rt, 15 h; (d) (*E*)-1-catecholboranylhexene, then tetrakis(triphenylphosphine)palladium(0), Na_2CO_3, DMF, H_2O, 90°C, 15 h.

Condensation of the acetate **22** with 8-bromoadenine under Lewis acid conditions afforded the 8-bromo derivative **28**. On coupling with 1-hexyne under Sonogashira conditions, followed by deprotection of the acetonide group, this led to the 8-hexynyl derivative **29a**. The 8-hexenyl-4′-thioadenosine derivative **29b** was obtained after a Suzuki coupling reaction with (*E*)-1-catecholboranylhexene followed by deprotection of the acetonide under acidic conditions.

In this study an A_3 AR antagonist, truncated 4′-thioadenosine derivative **23b**, was converted successfully into the potent A_{2A} AR agonist **27a** ($K_i = 7.19 \pm 0.6$ nM) after appending a 2-hexynyl group at the C2-position. However, C8-substitution greatly reduced the binding affinity at the human A_{2A} AR. All synthesized compounds maintained their affinity at the human A_3 AR, except **27a**, wherein a competitive A_3 AR antagonist/A_{2A} AR agonist behavior was observed in cyclic AMP assays. From this study it was evident that the truncated C2-substituted 4′-thioadenosine derivatives **27a** and **b** can be a novel template for development of new A_{2A} AR ligands.

The introduction of a C–C single bond in a structure increases the free rotation within the molecule. The increase in free rotation allows the molecule to adopt many conformations, making it possible to induce maximum binding interaction in the binding site of the receptor. With this hypothesis, a homologated structure, in which a methylene group (CH_2) was inserted in place of the glycosidic bond of a potent and selective A_3 AR antagonist **23b**, was designed and synthesized.[51] The expectation was that this modification would perhaps show a significant enhancement in the binding interaction at the binding site of the receptor and thus serve as selective A_3 AR antagonists.

The synthetic strategy toward this goal began with the preparation of the homologated glycosyl donor **30** (Scheme 13). This started from D-mannose via a procedure established previously.[45,46] The diol obtained under mild conditions [1.2 equiv of Pb(OAc)$_4$] produced the aldehyde, as the use of excess lead tetraacetate always leads to the anomeric acetate.[36] Reduction of the aldehyde with NaBH$_4$ afforded the primary alcohol, which on further treatment with phosphorus oxychloride yielded the chloro derivative **30**.

The nucleoside analogues were synthesized from direct S$_N$2 displacement of the chloride with 6-chloropurine and 2,6-dichloropurine anions, as shown in Scheme 13. The condensation reaction, followed by deprotection under mild acidic conditions and subsequent treatment with 3-halobenzylamines, yielded the N^6-(3-halobenzyl)amine derivatives **31a–d** and the corresponding 2-Cl analogues **31e–h** in good yields. The homologated truncated nucleosides were synthesized successfully and evaluated for their biological activity. The synthesized compounds were devoid of binding affinity at all subtypes of adenosine receptors, indicating that free rotation through the single bond allowed the compound to adopt an indefinite number of conformations, disrupting all favorable binding interactions, which are responsible for receptor recognition.

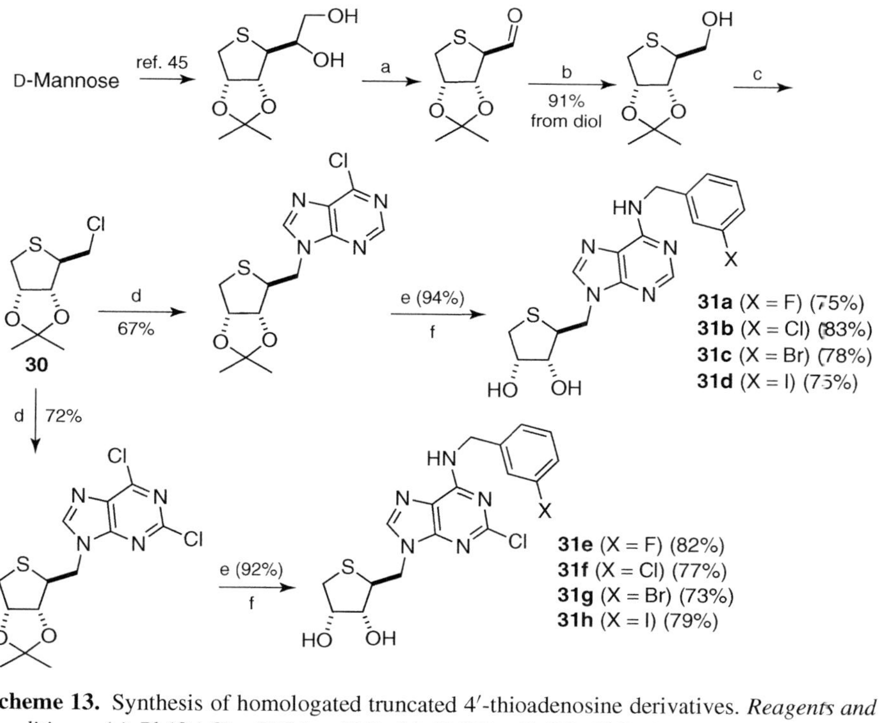

Scheme 13. Synthesis of homologated truncated 4′-thioadenosine derivatives. *Reagents and conditions:* (a) Pb(OAc)$_4$, EtOAc, 0°C; (b) NaBH$_4$, EtOH, 0°C; (c) POCl$_3$, CH$_3$CN, rt; (d) NaH, 6-chloropurine or 2,6-dichloropurine, DMF, rt; (e) 2 N HCl, THF, rt; (f) 3-halobenzylamine, Et$_3$N, EtOH.

4.3. Synthesis of Miscellaneous 4′-Thionucleosides

In previous sections we described in detail the recent developments in adenosine receptor analogues with due significance to modifications at the ribose moiety and the base. Over the years, several advancements in synthetic strategies, with emphasis on 4′-thionucleosides, also evolved. The aim of this section is to highlight a few significant syntheses of 4′-thionucleosides which may have evolved out of curiosity or due to some unusual observations during the synthetic strategy desired. The syntheses of a few fluorinated analogues are also discussed, as the fluorinated nucleoside analogues are known to exhibit enhanced biological activity. Also, the details of relevant biological data are included for better understanding of how these molecules behave in biological systems.

After the discovery of promising 1,3-dioxolanyl and 1,3-oxathiolanyl nucleosides it became clear that not only conventional but also unusual sugar structures can lead to active compounds. Entecavir (BMS-200475) was one such example and it was found to be 100 times more potent than lamivudine against hepatitis B virus (HBV).[52] It possesses an exocyclic methylene in place of the oxygen of the furanose ring. The potent anti-HBV activity of this class of compound led to the design of isodideoxynucleosides[53] with an exocyclic methylene side chain, as shown in Scheme 14.

D-Xylose was converted to the methylene derivative via 1,2-acetonide protection, selective primary alcohol protection, oxidation of secondary alcohol, Wittig reaction using methylidine phosphorane, and debenzoylation. Acid-catalyzed methanolysis followed by benzyl protection gave dibenzylated methylene derivative **32**, which on hydrolysis gave lactol. This lactol was subjected to reduction; dimesylation followed by Na_2S cyclization furnished thiosugar **33a** along with the mixture of alkene **33b** (*E* and *Z* isomers) as side products. Formation of **33a** and **33b** is dependent on the temperature and solvent used for the reaction. A higher reaction temperature gave alkene (thermodynamically stable) exclusively as a major product via S_N2 cyclization due to the presence of an allylic functional group. However, the compound desired, **33a**, can be obtained at lower temperatures (i.e., 0°C). The key intermediate **34** was prepared from **33a** via debenzylation followed by TBDPS protection. The *cis*-nucleoside derivatives were prepared using a Mitsunobu reaction with 6-chloropurine, N^3-Bz-uracil, and N^3-Bz-thymine.

Furthermore, desilylated 6-chloropurine condensed product was converted to adenine derivatives, N^6-methyladenine derivative, and hypoxanthine derivative by heating with methanolic ammonia, methanolic methylamine, and 2-mercaptoethanol/NaOMe, respectively. The *trans*-nucleoside derivative was prepared from **34**. Another strategy for the preparation of *cis*-nucleoside from acetate of glycosyl donor **35** using $NaH/Pd(Ph_3P)_4$ in DMF did not give the product expected and led to the formation of **36**. The reason for the formation of this product was attributed to the glycosylation reaction, which occurred at the least-hindered site. In terms of biological activity, few compounds exhibited weak anti-hepatitis C virus (HCV) activity among the series of synthesized derivatives.

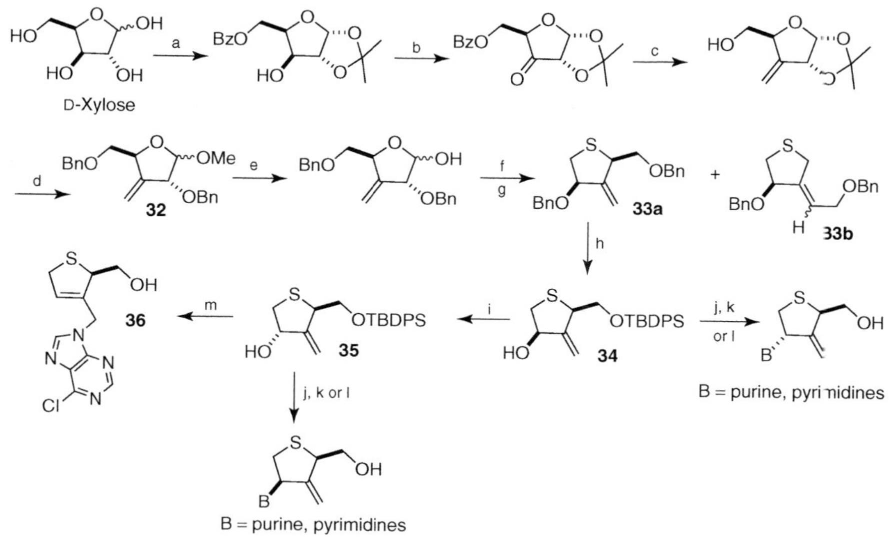

Scheme 14. Synthesis of isodideoxynucleosides with an exocyclic methylene side chain. *Reagents and conditions:* (a) (i) CuSO$_4$, conc. H$_2$SO$_4$, acetone; (ii) 0.2% HCl/H$_2$O; (iii) BzCl, pyridine; (b) PDC; (c) (i) Ph$_3$PCH$_3$Br, NaH, amyl alcohol, 0°C; (ii) BCl$_3$, CH$_2$Cl$_2$, −78°C; (d) (i) MeOH, AcCl, rt; (ii) BnBr, NaH, TBAI, THF; (e) HCl/dioxane; (f) (i) LiBH$_4$, THF; (ii) MsCl, pyridine, CH$_2$Cl$_2$; (g) Na$_2$S, DMF, 0°C; (h) (i) BCl$_3$, CH$_2$Cl$_2$; (ii) TBDPSCl, imidazole, DMF; (i) BzOH, PPh$_3$, DEAD, THF, then MeOH/NH$_3$; (j) (i) 6-chloropurine, Ph$_3$P, DEAD, THF, 0°C, rt; (ii) TBAF, THF, 0°C; (k) NH$_3$/MeOH 80°C or CH$_3$NH$_2$, MeOH, 80°C, or 2-mercaptoethanol, NaOMe, MeOH; (l) (i) N^3-Bz-uracil, or N^3-Bz-thymine, Ph$_3$P, DEAD, THF, 0°C; (ii) TBAF, THF, 0°C; (iii) NH$_3$/MeOH, rt; (m) (i) Ac$_2$O, pyridine, rt; (ii) NaH, Pd(Ph$_3$P)$_4$, DMF, 40°C.

The specific properties of 4′-thionucleosides and potent cytotoxicity of 2′-substituted cytidine prompted the synthesis of various 4′-thionucleosides containing cytidine as a nucleobase.[54] The 2′-branched nucleosides, such as 2′-C-methylcytidine and 2′-deoxy-2′(S)-methylcytidine, possess potent antitumor activity. 2′-Keto-4′-thionucleoside was used as a key intermediate in the synthesis of 2′-deoxynucleoside. Studies on the synthesis of 2′-C-methyl-4′-thiocytidine resulted in unexpected observations, which have been described in Scheme 15.

Yoshimura and co-workers have reported the synthesis of 2′-keto-4′-thionucleoside derivatives.[54a] This methodolgy involved the treatment of TIPDS-protected sugar **37** with DMSO–acetic anhydride to furnish two inseparable compounds after column chromatography (7:1, the ratio depending on the oxidation reaction time). Further, in an efforts to determine the structure of the oxidation product, the mixture (7:1 ratio) was treated with sodium borohydride, ammonium fluoride, and methylamine to give 4′-α-thiocytidine **39**. An unexpected anomerization of the 2′-keto-4′-α-thiocytidine derivative (from **38a** to **38b**, during oxidation as well as compound purification) was observed.

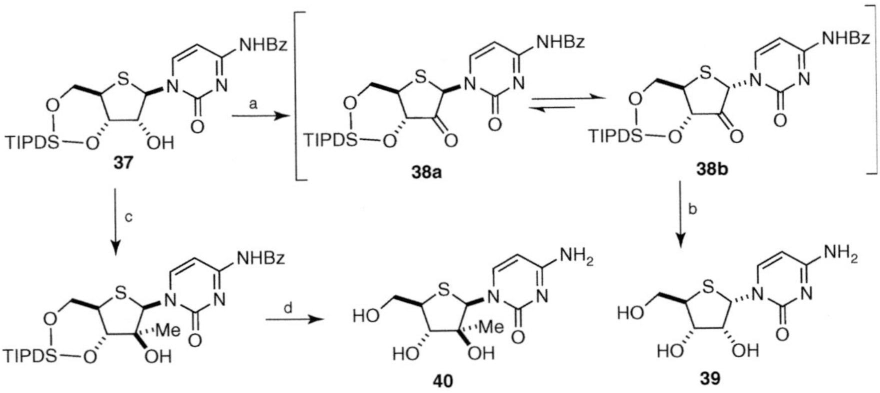

Scheme 15. Synthesis of 2′-substituted-4′-thiocytidine derivatives. *Reagents and conditions:* (a) DMSO, Ac$_2$O; (b) (i) NaBH$_4$, MeOH; (ii) NH$_4$F, MeOH, reflux; (iii) CH$_3$NH$_2$, MeOH; (c) (i) DMSO, Ac$_2$O; (ii) CH$_3$MgBr Et$_2$O, −78°C; (d) (i) NH$_4$F, MeOH, reflux; (ii) CH$_3$NH$_2$, MeOH.

The reason for the anomerization was attributed to the higher acidity of the α-hydrogen (H-1′ of **38b**) of thio ether and also the higher stability of the resulting carbanion versus those of the corresponding ether. The carbonyl methylation of the 2′-keto-4′-thionucleoside derivatives were tried using a different alkylating reagent, such as methyl lithium or trimethyl aluminum, at −78°C, which resulted in the formation of a mixture of products. The Grignard reagent was a method of choice, resulting in the formation of β-hydroxy product (α/β, 1:4), which is due to the expected chelation of Grignard reagent between 2-carbonyl oxygen of the pyrimidine ring and the 2′-carbonyl oxygen of sugar moiety. Treatment with ammonium fluoride in methanol, followed by methylamine, gave nucleoside **40** with α-2′-methyl and β-2′-hydroxy. An alternative method for the synthesis of 2′-C-methyl-4′-thiocytidine containing 2′-β-methyl and 2′-α-hydroxy was described in which a methyl group was introduced prior to the Pummerer reaction. The synthesized derivatives of 2′-substituted-4′-thiocytidine derivatives did not show any activity, which indicated that 4′-thioribocytidine derivatives are less susceptible to phosphorylation by cellular uridine–cytidine kinase.

In the design of the synthesis of novel nucleosides, one of the important modifications in the nucleobase includes its aza modification.[55] A synthesis of benzyl 3,5-di-*O*-benzyl-2-deoxy-1,4-dithio-D-erythropentofuranoside **42**, from 2-deoxy-D-ribose was reported along with some 4′-thio-2′-deoxynucleosides which have potentially useful biological activity.[56] Initially, 2-deoxy-D-ribose was converted to the corresponding methyl riboside, which was then protected with benzyl groups, followed by treatment with benzylmercaptan and HCl to the corresponding mercaptan derivative **41**. This mercaptan derivative was subjected to a Mitsunobu reaction, followed by hydrolysis of the benzoyl group, mesylation, and subsequent treatment with NaI and BaCO$_3$, which afforded the desired thiosugar **42**, which was used for further manipulation (Scheme 16).

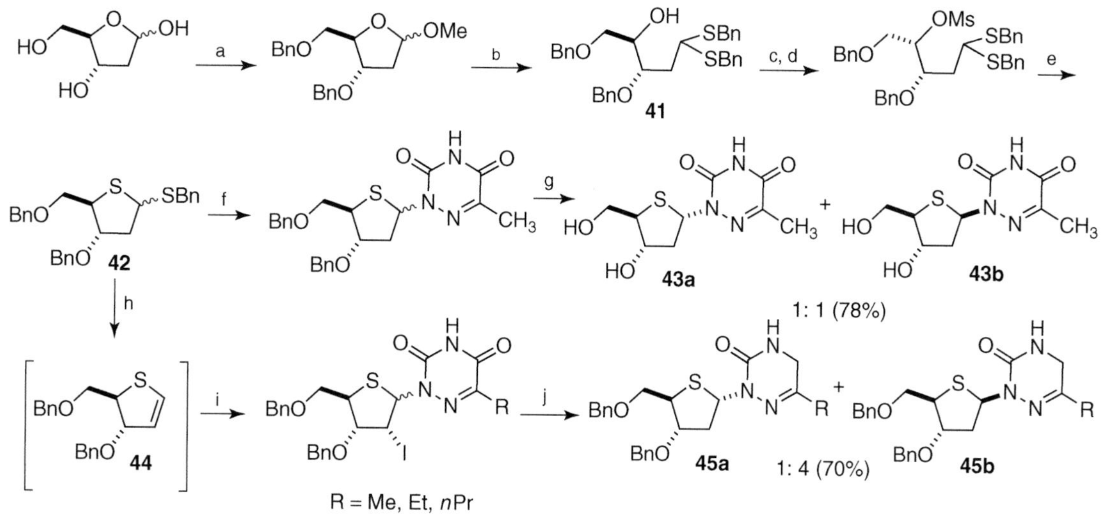

Scheme 16. Synthesis of 6-azapyrimidine-2′-deoxy-4′-thionucleosides. *Reagents and conditions:* (a) (i) MeOH, HCl; (ii) NaH, Bu$_4$NI, BnBr, THF; (b) BnSH, HCl; (c) Ph$_3$P, PhCOOH, DEAD, THF; (d) (i) NaOMe, MeOH; (ii) MsCl, pyridine; (e) NaI, CH$_3$CN, BaCO$_3$, acetone; (f) (i) silylated 5-methyl-6-azapyrimidine; (ii) ICl, CH$_3$CN, 4-Å molecular sieves, 24 h, or NIS, CH$_3$CN, 4-Å molecular sieves, 24 h; (g) BCl$_3$, CH$_2$Cl$_2$, −80°C, 4 h; (h) ICl, CH$_2$Cl$_2$, 2,6-di-*tert*-butyl-4-methylpyridine, 0°C, 5 h; (i) (6-azapyrimidine)TMS$_2$, CH$_2$Cl$_2$; (ii) ICl, CH$_2$Cl$_2$, 16 h; (j) AIBN, Bu$_3$SnH, toluene, 45–60°C.

At first, the bis-silylated 6-azapyrimidine base was prepared by reaction of the corresponding 6-azapyrimidine base (5-methyl or 5-ethyl) with N,O-bis(trimethylsilyl)acetamide (BSA) in anhydrous acetonitrile for 1 h in the presence of either iodine monochloride (ICl) or N-iodosuccinimide (NIS). The silylated base was condensed with thiosugar **42** to give benzyl-protected-2′-deoxy-4′-thionucleoside, which was later separated after debenzylation reaction to give **43a** and **43b** (1:1). The anomeric ratio of coupled nucleosides was improved further by employing an alternative method (i.e., glycal methodology). The glycosyl donor **42** was treated with iodine monochloride in the presence of 2,6-di-*tert*-butyl-4-methylpyridine at 0°C for 5 h to form the corresponding glycal. Introduction of silylated azapyrimidine base and excess of ICl led to the formation of a 2′-iodo nucleoside in a 1:4 ratio (α/β). Deiodination using the Sugiyama method provided corresponding thionucleosides **45a** and **b**. The antiviral effects of this class of compounds evaluated against a broad range of viruses indicated that only compound **43b** displayed activity that was comparable with acyclovir ($EC_{50} = 1$ μM).

With the intention of increasing the activity profile of this class of nucleosides, modification of the sugar part was attempted. Thus, the 2′,3′-dideoxy-6-azathymidine-4′-thionucleosides were synthesized and evaluated for their antiviral activity.[57] The synthesized 1-(2′-deoxy-4′-thio-β-D-erythropentofuranosyl)-(6-azathymidine) **43b** from Scheme 16 was selectively silylated at the 5′-position using TBDMSCl and imidazole. The 5′-protected nucleoside was converted to 3′-thiocarbonyl derivative using thiocarbonyldiimidazole in CH_2Cl_2 at 40°C, followed by radical deoxygenation with tributyltin hydride to the corresponding 2′,3′-dideoxy nucleoside. Removal of the 5′-protecting group using TBAF, followed by purification with Dowex 50W (H^+), gave the corresponding 2′,3′-dideoxy-4′-thionucleoside (Scheme 17). The dideoxy thionucleoside **46** was moderately active against vaccinia virus (MIC 12 μM) and herpes simplex virus strains, HSV-1 and HSV-2 (strain G).

Another possible modification in the sugar moiety resulted in the development of 5′-deoxy nucleosides. A series of 5′-deoxy-4′-thio-L-nucleosides were synthesized and their antitumor activities were tested against various cell lines (Scheme 18).[58] The intermediate **47** was first synthesized using D-ribose. Deprotection of TBDPS group of **47** followed by iodination and reduction of iodide by TBTH/AIBN provided the desired glycosyl donor **48**. This was condensed with 2,6-dichloropurine to afford a mixture of α and β anomeric products. The N^6-position of the purine base was further manipulated by treatment with different amines. Deprotection of the acetonide group provided a mixture of α and β derivatives of corresponding nucleosides **49a–k** and **49a′–k′**, respectively. Similarly, the acetate **48** was condensed with persilylated pyrimidine bases under Vorbrüggen conditions, and deprotected to a mixture of α and β nucleosides **50a–c** and **50a′–c′** in moderate yields.

All the synthesized compounds were tested for inhibition of the growth of LOVO human carcinoma of the colon, CEM human leukemia, and MDA-MB-435 human breast adenocarcinoma cells in culture with 5-FU as the reference. It was found that 6-cyclopentylamino- and 6-cyclohexylaminopurine compounds, in both α- and β-configurations, exhibited a strong inhibition to CEM.

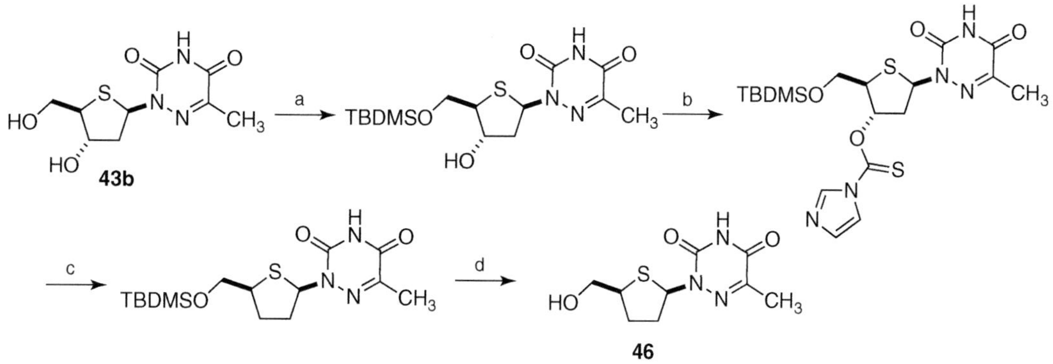

Scheme 17. Synthesis of 6-azapyrimidine-2′, 3′-dideoxy-4′-thionucleosides. *Reagents and conditions*: (a) TBDMSCl, imidazole, DMF, 3.5 h; (b) thiocarbonyldiimidazole, CH$_2$Cl$_2$, 40°C, 8 h, then rt, 12 h; (c) Bu$_3$SnH, AIBN, toluene, 100°C, 30 min; (d) (i) TBAF; (ii) Dowex 50W (H$^+$), MeOH, 24 h.

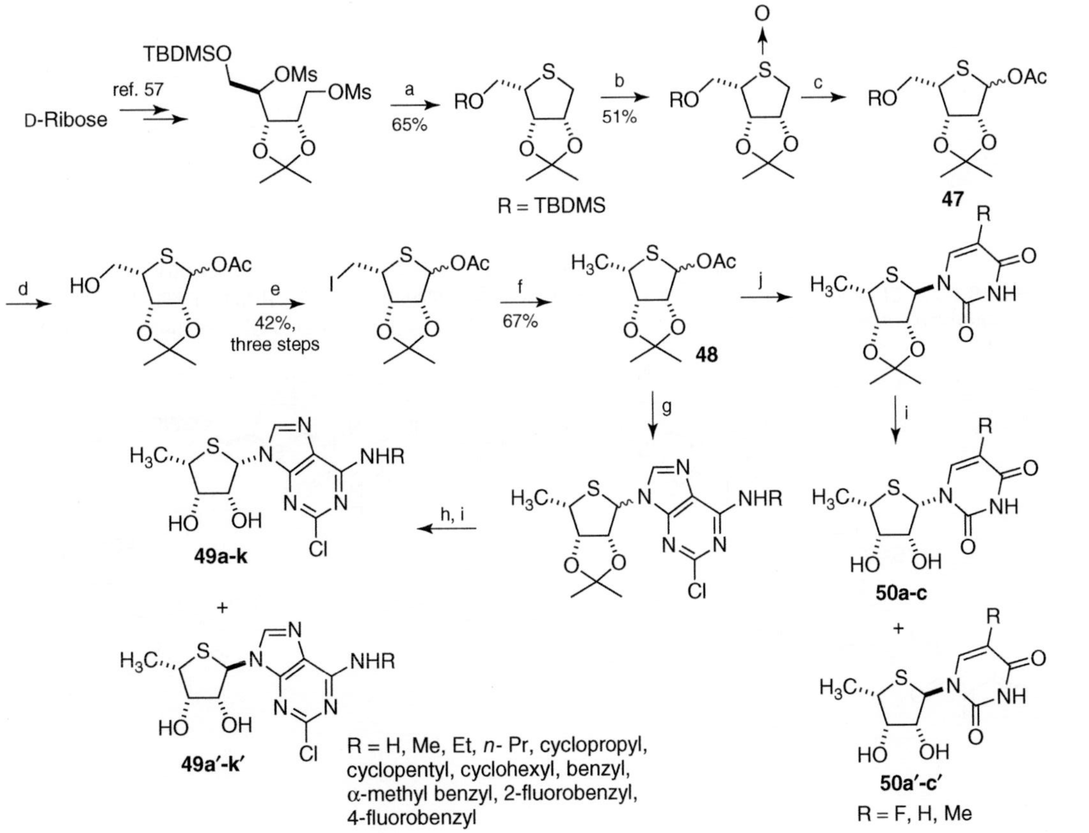
D-Ribose
ref. 57
TBDMSO
OMs
OMs
a
65%
RO
S
R = TBDMS
b
51%
O
c
OAc
47
d
HO
e
42%,
three steps
I
f
67%
H3C
48
j
R
NH
g
i
N
NHR
Cl
h, i
HO
OH
49a-k
+
49a′-k′
R = H, Me, Et, n- Pr, cyclopropyl,
cyclopentyl, cyclohexyl, benzyl,
α-methyl benzyl, 2-fluorobenzyl,
4-fluorobenzyl
50a-c
50a′-c′
R = F, H, Me

Scheme 18. Synthesis of 5′-deoxy-4′-thio-L-nucleosides. *Reagents and conditions:* (a) Na_2S, DMF, heat; (b) *m*-CPBA, CH_2Cl_2, −78°C; (c) AC_2O, heat; (d) TBAF, THF, rt; (e) (i) Ph_3P, imidazole, I_2, toluene, heat; (f) Bu_3SnH, AIBN, toluene, heat; (g) 2,6-dichloropurine, $EtAlCl_2$, CH_3CN, rt; (h) appropriate amines, rt, THF; (i) 85% aq. HCO_2H, rt; (j) silylated pyrimidines, TMSOTf, CH_3CN, rt.

Fluorinated nucleoside analogues have gained significant importance in therapeutic applications. A novel series of fluorinated iso-D-2′,3′-dideoxy-4′-thionucleosides was reported recently (Scheme 19).[59] In this case, the diacetone-D-glucose was obtained from a known one-pot synthesis,[60] which was further converted to the 3-O-benzyl-D-xylose. The reaction involved a benzylation at 3′-OH, selective acetonide deprotection, oxidation of diol, and $NaBH_4$ reduction. 3-O-Benzyl-D-xylose was subjected to acidic methanolysis to give an anomeric mixture of methyl 3-O-benzyl-D-xylose. The separated α- and β-methoxy anomer (silica gel column chromatography) was converted to dimesylate and then treated with Na_2S in DMF to give bicyclic α- and β-bicyclic thiosugar, respectively. The α-isomer is less stable and is converted to the β-form under slightly acidic conditions. The reason for this conversion was the steric repulsion between 1-OMe and H3, which makes α-isomer more unstable and accelerates the conversion of the α- to β-anomer.

Hydrolysis of the α,β-anomers followed by sodium borohydride reduction and TBDPS protection gave TBDPS-protected-4-thioarabitol as a common glycosyl donor. 4-Thioarabitol was then subjected to benzoylation at 2′-OH, followed by debenzylation (3′-position), and a stereoselective fluorination furnished the desired fluorinated compound. Removal of the benzoyl group followed by inversion of stereochemistry under Mitsunobu conditions and then hydrolysis gave an alcohol as a requisite glycosyl donor **51**. Pyrimidine nucleosides **52** and **53** were prepared using a Mitsunobu reaction of glycosyl donor with N^3-benzoyluracil or N^3-benzoylthymine, followed by benzoyl and TBDPS deprotection. Finally, **52** was converted into lamivudine analogue **54**, a cytosine nucleoside via acetylation, chlorination, and amination reactions. Similarly, for the preparation of adenine nucleoside analogue **55**, the glycosyl donor **51** was subjected to the Mitsunobu reaction with 6-chloropurine, followed by one-pot concomitant disilylation of 5′-OH and amination. These synthesized iso-D-2′,3′-dideoxy-3′-thionucleosides derivatives are bioisostere of lamivudine and hence their antiviral activities were measured against viruses such as HIV, EMCV, VSV, and HSV. Only cytosine analogues showed potent anti-VSV activity. Thus, it was thought that iso-D-2′,3′-dideoxy sugar template might serve as a sugar surrogate of nucleosides for the development of an anti-RNA virus agent.

gem-Difluoromethene (CF_2) is considered to be an isoploar and isosteric substituent for oxygen.[61] Also, the introduction of a CF_2 group usually enhances the metabolic stability and increases the lipophilicity of the compounds synthesized. In addition, introduction of the CF_2 group can profoundly change the conformational preference of the nucleosides, due to size and stereoelectronic effects. Thus, a new method for the synthesis of 2′,3′-dideoxy-6′,6′-difluoro-2′-thionucleosides (Scheme 20),[62] analogues of highly bioactive (+)-2′-deoxy-3′-oxacytidine (L-OddC)[63] and (−)-2′-deoxy-3′-thiacytidine (3TC)[64] molecules, was developed using a TMSCl/pyridine-induced stereoselective Reformatskii–Claisen rearrangement of secondary allyl chlorodifluoroacetate. The intermediate **56** for the key step was synthesized in a few steps which was further manipulated to thiosugars **57** and **58** under standard conditions. Oxidation of thiosugar followed

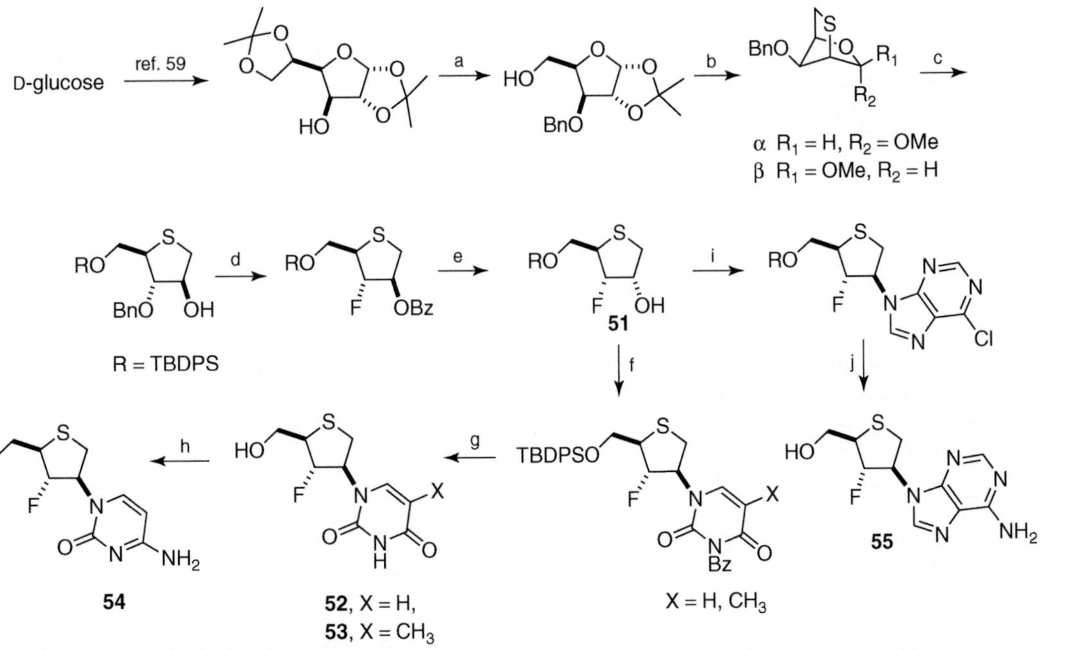

Scheme 19. Synthesis of fluorinated iso-D-2′,3′-dideoxy thionucleosides. *Reagents and conditions*: (a) (i) BnBr, NaH, DMF; (ii) 2 N HCl, THF; (iii) NaIO$_4$; (iv) NaBH$_4$; (b) (i) 5% HCl; (ii) MsCl, pyridine; (iii) Na$_2$S, DMF, 100°C; (c) (i) 4 N HCl; (ii) NaBH$_4$; (iii) TBDPSCl, imidazole; (d) (i) BzCl, DMAP, pyridine, 88°C; (ii) BCl$_3$, −78°C, 40 min; (iii) DAST, −10°C; (e) (i) 1 M NaOMe, rt, 4 h; (ii) BzOH, Ph$_3$P, DEAD; 60°C, 5 h; (iii) 1 M NaOMe, rt, 7 h; (f) N^3-benzoyl uracil or N^3-benzoyl thymine, Ph$_3$P, DEAD, rt, overnight; (g) (i) 1 M NaOMe, rt; (ii) 1 M TBAF, rt, 2 h; (h) (i) Ac$_2$O, pyridine, rt, overnight; (ii) 1,2,4-triazole, POCl$_3$, pyridine, rt, overnight; (iii) 1,4-dioxane, 28% NH$_4$OH, rt, overnight; (iv) NH$_3$, rt, overnight; (i) 6-chloropurine, Ph$_3$P, DEAD, rt, overnight; (j) NH$_3$, 80°C, overnight.

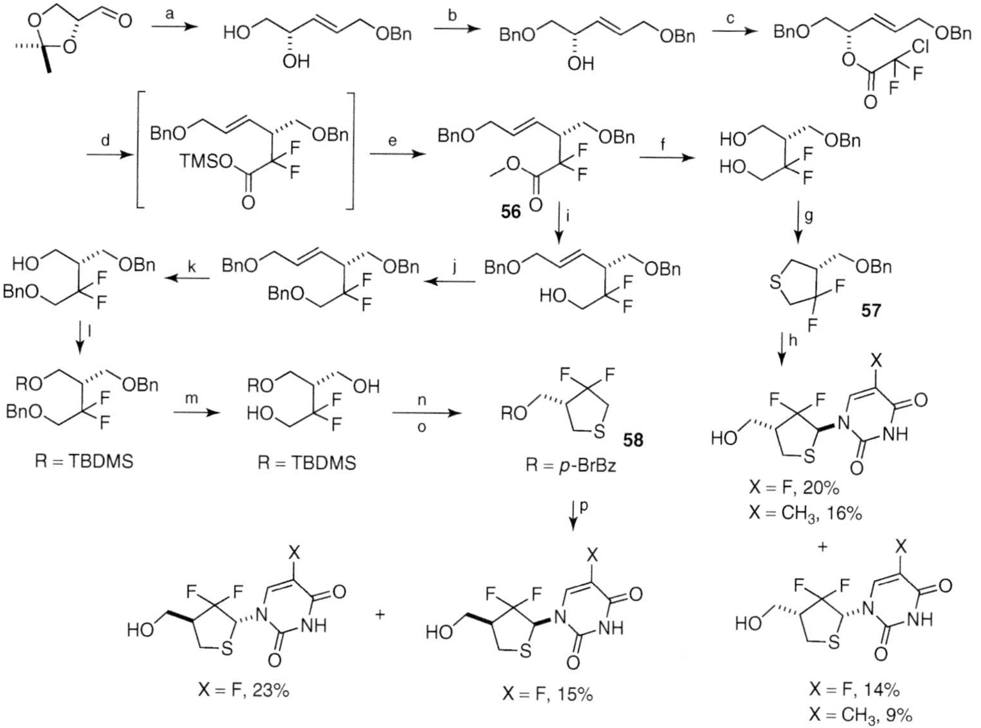

Scheme 20. Synthesis of 2′-3′-dideoxy-6′,6′-difluoro-2′-thionucleosides. *Reagents and conditions:* (a) (i) (EtO)$_2$POCH$_2$CO$_2$Et, NaH, 91%; (ii) LiAlH$_4$, AlCl$_3$, 92%; (iii) NaH, BnBr, 91%; (iv) TFA, THF, H$_2$O, 96%; (b) NaH, BnBr, 82%; (c) (COCl)$_2$, cat. DMF, ClCF$_2$CO$_2$H, 91%; (d) Zn, TMSCl, pyridine, 120°C, sealed tube; (e) SOCl$_2$, MeOH; (f) (i) O$_3$; (ii) NaBH$_4$, 92%; (g) (i) MsCl, pyridine; (ii) Na$_2$S·9H$_2$O, 81%; (h) (i) *m*-CPBA; (ii) silylated base, TMSOTf, DCE; (iii) BCl$_3$; (i) NaBH$_4$, 94%; (j) NaH, BnBr, 91%; (k) (i) O$_3$; (ii) NaBH$_4$, 92%; (l) (i) TBDMSCl, imidazole, 88%; (iii) Pd/C, H$_2$, 84%; (n) (i) MsCl, pyridine; (ii) Na$_2$S·9H$_2$O, 81%; (o) (i) TBAF, (ii) *p*-BrBzCl, 83%; (p) (i) *m*-CPBA; (ii) silylated base, TMSOTf, DCE; (iii) NH$_3$, MeOH.

by condensation with silylated bases afforded the desired nucleosides as a mixture of α/β anomers. Subsequent removal of the protecting group provided a separable mixture of α and β target thionucleosides.

Another methodology for the synthesis of L-β-3′-Deoxy-3′,3′-difluoro-4′ thionucleoside, which is an analogue of bioactive gemcitabine (Scheme 21), was developed by the same group.[65] The installation of the thioacetyl group in **59** with high efficiency and the construction of a 3-deoxy-3,3-difluorothiofuranose skeleton were the highlights of this development.

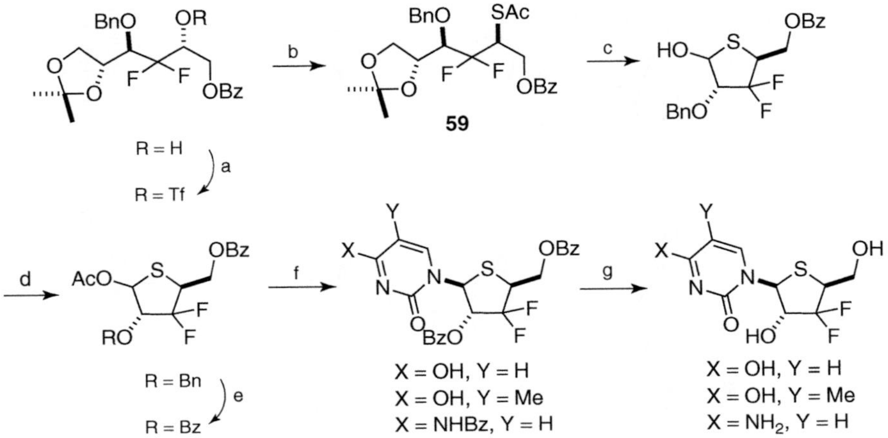

Scheme 21. Synthesis of L-β-3′-Deoxy-3′,3′-difluoro-4′-thionucleosides. *Reagents and conditions:* (a) Tf$_2$O, pyridine; (b) AcSH, CsF, DMF, rt; (c) (i) CF$_3$COOH; (ii) NaIO$_4$, MeOH; (iii) 1 M HCl, MeOH; (d) Ac$_2$O, DMAP; (e) (i) NaBrO$_3$, Na$_2$S$_2$O$_4$; (ii) Bz$_2$O, DMAP; (f) N,O-BSA, pyrimidine; (g) NH$_3$, MeOH.

Another development was the synthesis of D-2′-deoxy-2′,2′-difluoro-4′-dihydro-4′-thionucleosides (Scheme 22).[66] The conformation of this compound was studied using x-ray crystallography, NMR spectroscopy, and molecular modeling, in an attempt to explore the roles of the two *gem*-difluoro atoms in the puckering preferences of the thiosugar ring. Thiofuranose **60** was prepared in a few steps using a *gem*-difluorinated synthon. Oxidation of sulfide, followed by condensation with silylated nucleobases using Pummerer conditions, afforded the protected nucleosides, which were finally deprotected to the target thionucleosides.

A method for the synthesis of thiobutyrolactone as a precursor of modified nucleosides was reported recently from a fluoroxanthate and a protected allylic alcohol (Scheme 23).[67] This approach emphasized a new and straightforward route for the synthesis of 2′, 3′-dideoxy-2′-fluoro-4′-thionucleosides in a few steps, including the formation of a fluorothiolactone **61** and a Vorbrüggen base condensation reaction.

The synthesis of 2′-deoxy-2′-fluoro-4′-thioribonucleosides as substrates for the synthesis of novel modified RNAs was recently disclosed (Scheme 24).[68] 2′-Deoxy-2′-fluoro-4′-thiouridine, -thiocytidine, and -thioadenosine derivatives as substrates

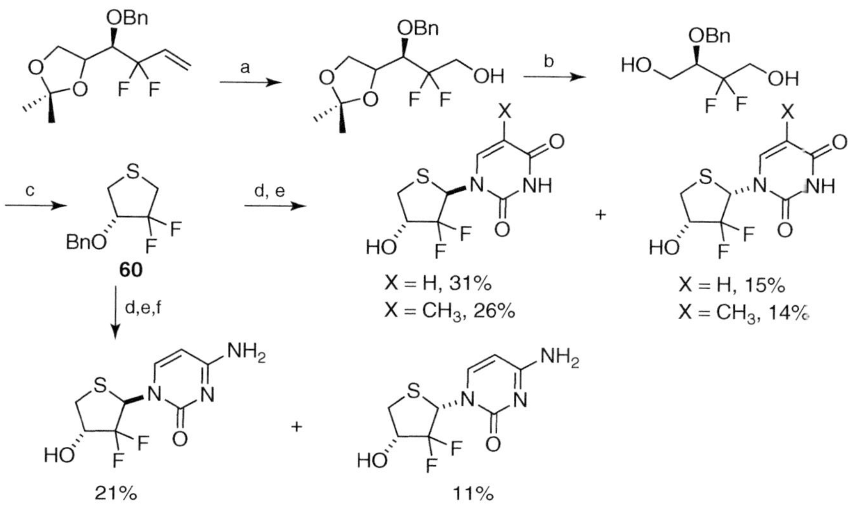

Scheme 22. Synthesis of D-2′-deoxy-2′,2′-difluoro-4′-dihydro-4′-thionucleosides. *Reagents and conditions:* (a) (i) O_3, CH_2Cl_2, −78°C; (ii) $NaBH_4$; (b) (i) TFA/H_2O/THF (1:1:1); (ii) $NaIO_4$, acetone; (iii) $NaBH_4$; (c) (i) MsCl, pyridine; (ii) $Na_2S \cdot 9H_2O$, DMF, 90°C; (d) (i) *m*-CPBA, CH_2Cl_2, −78°C; (ii) silylated base, TMSOTf, DCE; (e) BCl_3, CH_2Cl_2, −70°C; (f) NH_3, CH_3OH.

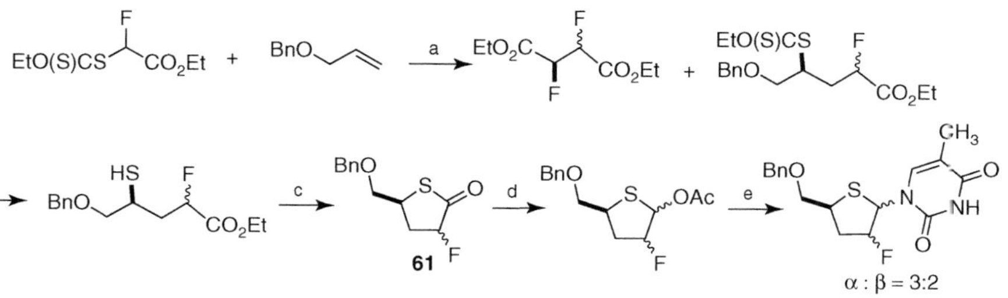

Scheme 23. Synthesis of 2′, 3′-dideoxy-2′-fluoro-4′-thionucleosides from fluoroxanthate. *Reactions and conditions:* (a) Et_3B, CH_2Cl_2, 20°C or dilauroyl peroxide, DCE, reflux, 4 h; (b) piperidine, CH_2Cl_2, 20°C, 2 h; (c) TFA, CH_2Cl_2, 20°C; (d) (i) $NaBH_4$, EtOH, −17°C; (ii) Ac_2O, Et_3N, DMAP, CH_2Cl_2, 20°C; (e) silylated thymine, TMSOTf, CH_2Cl_2, 20°C, 18 h.

for modified RNAs were developed. The x-ray crystal structural analysis revealed that 2′-deoxy-2′-fluoro-4′-thiocytidine **63** adopted predominately the same C3′-*endo* conformation as 2′-deoxy-2′-fluorocytidine. Nucleosides were synthesized by treatment of 2,2′-*O*-anhydro-4′-thiouridine **62** with HF/pyridine, in a manner similar to that of its 4′-*O* congener.

However, as the synthesis of purine analogues seemed to be trivial, a modified approach to their synthesis was employed. A glycosyl donor having a fluorine atom at the 2α-position was first synthesized (Scheme 25). Thus, the hydroxyl group of **64**

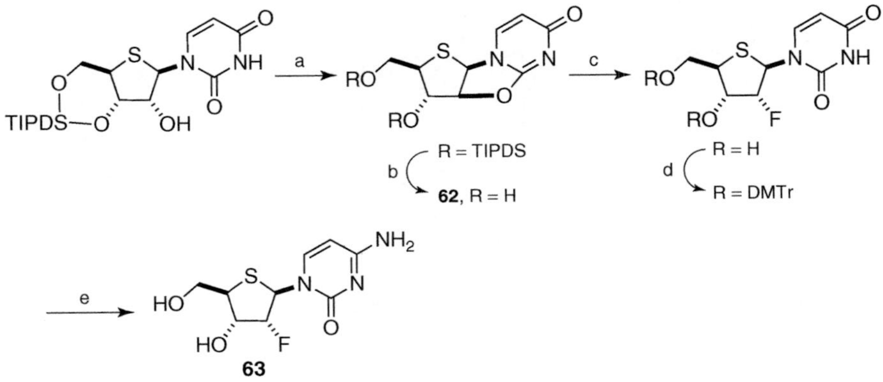

Scheme 24. Synthesis of 2′-deoxy-2′-fluoro-4′-thiocytidine. *Reagents and conditions:* (a) Tf$_2$O, DMAP, CH$_2$Cl$_2$; (b) NH$_4$F, MeOH, reflux; (c) HF/pyridine, dioxane, 125°C; (d) DMTrCl, pyridine; (e) (i) AC$_2$O, Et$_3$N, DMAP, CH$_3$CN; (ii) TPSCl, Et$_3$N, DMAP, CH$_3$CN, then NH$_4$OH; (iii) TESCl, pyridine, then BzCl; (iv) TBAF, THF; (v) 2% TFA in CH$_2$Cl$_2$.

was first inverted by standard Mitsunobu conditions. Then, TIPDS protection was replaced by a 3,5-*O*-*p*-methoxybenzylidene acetal (PMP) derivative, and deprotection of the *p*-nitrobenzoyl group gave compound **65**, which was the substrate for fluorination. Compound **65** was then converted to the desired fluorinated compound **66** by treatment with perfluoro-1-butanesulfonyl fluoride (PBSF) in the presence of 1,8-diaza-bicyclo[5.4.0]undec-7-ene (DBU) in dioxane under reflux conditions. The fluornianted analogue **66** was then oxidized to the corresponding sulfoxide by ozone and then treatment with acetic anhydride to the corresponding acetate **67**. The glycosyl donor **67** was then treated with persilylated 6-chloropurine in the presence of TMSOTf at room temperature followed by heating at reflux to give a mixture of *N*7 and *N*9 compounds as α/β anomers. The *N*9 β-derivative was manipulated further to get the desired precursor of the phorphoramidite unit for oligonucleotide (ON) synthesis.

Recently, a novel synthesis of both a D- and L-series of nucleosides from chiral acyclic thioaminal bearing the nucleobases by using stereoselective intramolecular cyclization (S$_N$2 type) was developed.[69] Acyclic thioaminals such as **A** (1,2-*syn*) and **B** (1,2-*anti*) were utilized for cyclization in the synthesis of D- and L-nucleosides simply by changing the leaving groups in the molecules (Scheme 26). The cyclization involves the intramolecular displacement by activating the thioalkyl group present at C1′ by the secondary hydroxyl group at C4′ (C4′–C1′ cyclization). This process gave the desired oxonucleosides stereoselectively with inversion of configuration at C1′. Thus, a 1,2-*syn* compound gave 1′,2′-*trans*-oxonucleosides, and 1,2-*anti* compounds gave 1′,2′-*cis*-oxonucleosides. Alternatively, the sulfur of the thioaminal was made to serve as a nucleophile when the C4′ hydroxyl was converted into a leaving group. Using this method, the authors synthesized 1′,2′-*cis*- and 1′,2′-*trans*-thionucleoside stereoselectively by using 1,2-*syn* and 1,2-*anti*

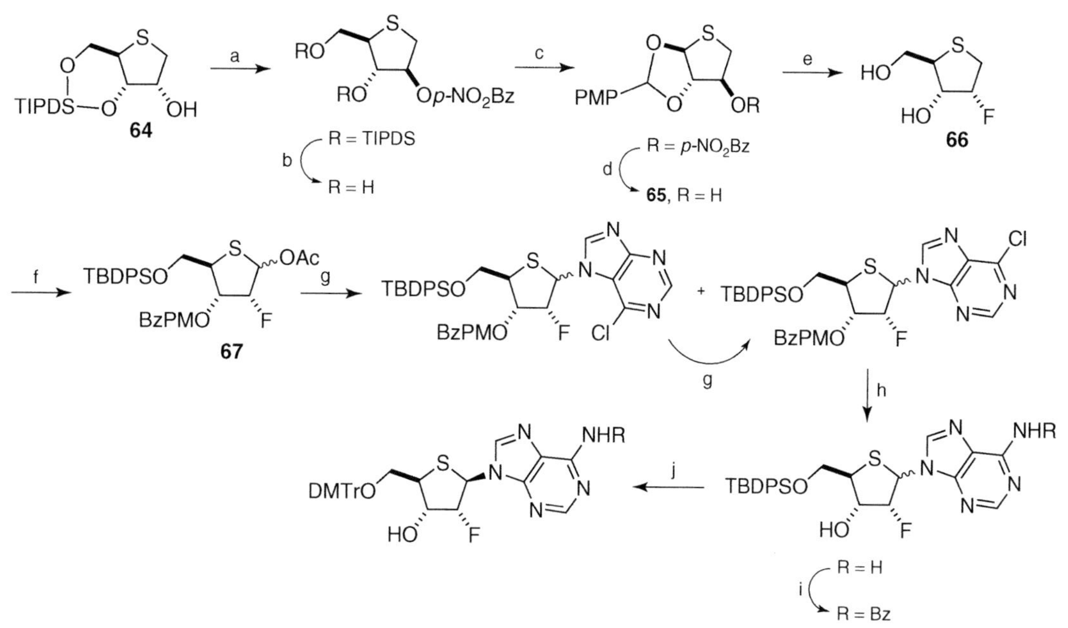

Scheme 25. Synthesis of a 2′-deoxy-2′-fluoro-4′-thioadenosine derivative as a precursor for ON synthesis. *Reagents and conditions*: (a) *p*-nitrobenzoic acid, Ph$_3$P, DIAD, THF; (b) TBAF, AcOH, THF; (c) *p*-methoxybenzaldehyde dimethyl acetal, CSA, DMF; (d) MeNH$_2$, MeOH; (e) (i) PBSF, DBU, reflux; (ii) 80% aq. AcOH; (f) (i) TIPDSCl, imidazole, DMF; (ii) PMBzCl, pyridine; (iii) O$_3$, CH$_2$Cl$_2$, −78°C; (iv) Ac$_2$O, reflux; (g) silylated 6-chloropurine, TMSOTf, CH$_3$CN, rt, then reflux; (h) (i) NH$_3$, EtOH, 80°C; (ii) MeNH$_2$, MeOH; (i) BzCl, pyridine; (j) (i) TBAF, THF; (ii) DMTrCl, pyridine.

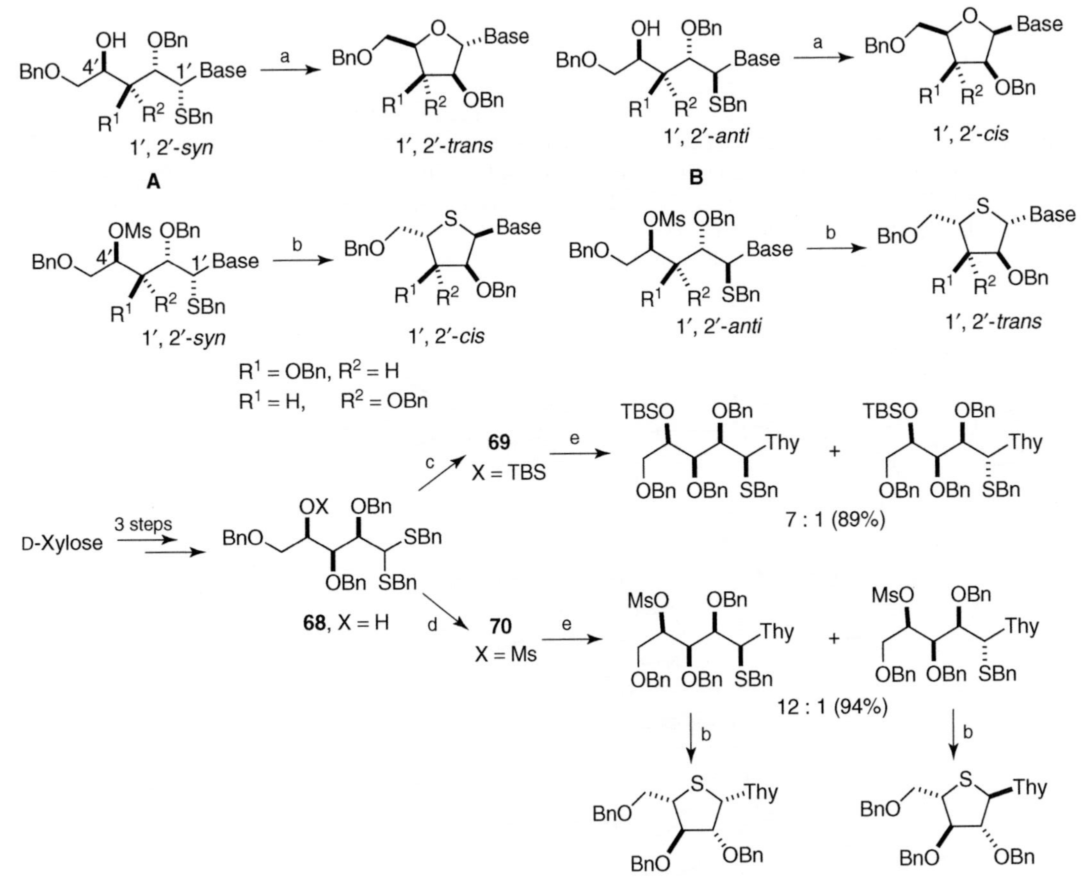

Scheme 26. Stereoselective approach to 4′-thioanalogues from acyclic precursors. *Reagents and conditions:* (a) Me₃S(SMe)BF₄, THF, rt; (b) NaI, pinacolone, 106°C, or NaI, 2,6-lutidine, 145°C; (c) TBSOTf, 2,6-lutidine; (d) MsCl, Et₃N; (e) silylated thymine, I₂.

compounds, respectively. The starting materials (i.e., acyclic thioaminals) were synthesized using D-xylose. Dithioacetal **68** was obtained by D-xylose using a standard literature method, which was then converted to the corresponding TBS-protected **69** and mesylated compound **70**. Compound **69** was then treated with a silylated base in the presence of I_2 to give the desired compounds in a 7:1 ratio. Similarly, mesylated dithioacetal **70** gave another mixture of compounds in a 12:1 ratio. Furthermore, these acyclic thioaminals with varying stereochemistry were synthesized and cyclized successfully to a series of 4′-thionucleoside derivatives in good to excellent yields.

Another promising anti-human immunodeficiency virus type 1 (HIV-1) agent was 4′-ethylstavudine,[70] which motivated several research groups, which resulted in the identification of its 4′-thio-counterpart. The synthesis and anti-HIV activity of 4′-branched (±)-4′-thiostavudines is also reported (Scheme 27).[71] Methyl 3-oxo-tetrahydrothiophene-2-carboxylate (**71**) was used as a starting material for the preparation of 4-thiofuranoid glycal **72**. A thymine base was introduced to thiofuranoid glycal **72** by electrophilic addition using N-iodosuccinimide (NIS) and bis(trimethylsilyl)thymine. The desired β-anomer **73a** obtained as a major

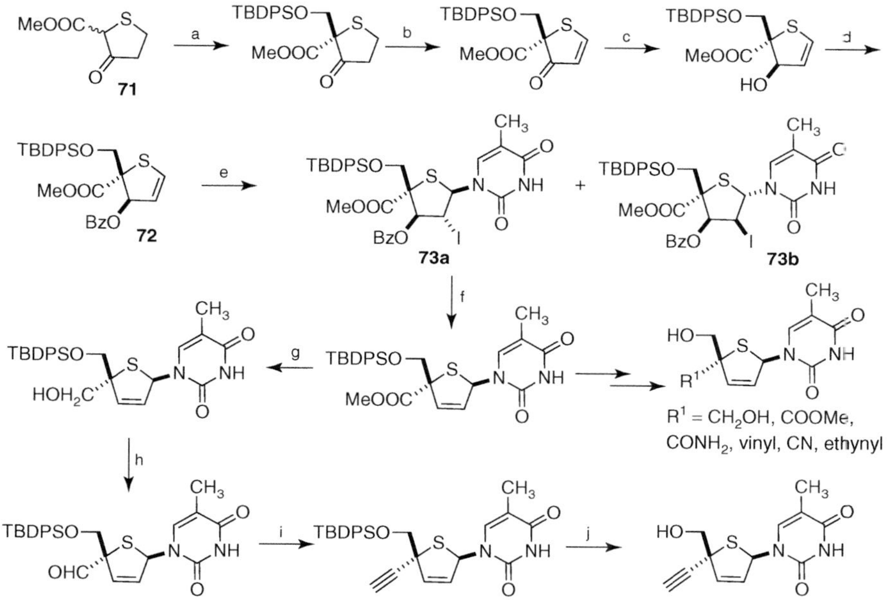

Scheme 27. Synthesis of 4′-branched (±)-4′-thiostavudines. *Reagents and conditions.* (a) (i) aq. HCHO, AcOH, rt; (ii) TBDPSCl, imidazole, DMF; (b) N-chlorosuccinimide, 0°C to rt, 2.5 h; (c) $NaBH_4/CeCl_3$, MeOH/THF, −50°C, 1 h; (d) DMAP, i-Pr_2NEt, BzCl, rt; (e) N-iodosuccinimide, bis(trimethylsilyl)thymine, CH_3CN, rt; (f) activated Zn, THF/AcOH; (g) $NaBH_4$, MeOH/THF; (h) $(CF_3CO)_2O/DMSO$, CH_2Cl_2; (i) dimethyl-1-diazo-(2-oxopropyl)phosphonate, K_2CO_3 in MeOH; (j) 1 M Bu_4NF, THF.

product was converted to the 4′-carbomethoxy derivative, which was used as a key intermediate for the synthesis of various 4′-carbon-substituted (CH_2OH, CO_2Me, $CONH_2$, $CH=CH_2$, CN, and CCH)4′-thiostavudines. Among the series of these synthesized compounds, 4′-cyano and 4′-ethynyl analogues were found to show inhibitory activity against HIV-1, with ED_{50} values of 7.6 and 0.74 μM, respectively.

The stereoselective syntheses of a group of 4′-thiaspirocyclic ribonucleosides featuring both pyrimidine and purine classes with both possible configurations at C5′ were described (Scheme 28).[72] The route involved Pummerer rearrangement as the key step. The approach comprised an acidic deprotection of acetonides of **74** and **75**, followed by subsequent regioselective introduction of a 2,4-dimethoxybenzoyl at the C2 position, followed by TIPDS protection to arrive at the sulfide **76**. After oxidation of **76** to the corresponding sulfoxide **77** with the Davis reagent, reaction with several nucleobases was attempeted with trimethylsilyltriflate and triethylamine. It was observed that under these reaction conditions, neighboring group participation was effective in delivering the β-configured ribonucleosides **79** along with the corresponding unsaturated sulfoxides **78** (major). Although there were differences in reactivity between the two stereoisomeric series, the common route led to the desirable β-anomeric sulfur-containing spiroribonucleosides with minimum formation of α-anomers. Also, another enantioselective approach to 2′-deoxy-4′-thiaspirocyclic nucleosides featuring an α- or β-hydroxyl substituent was well studied.[73]

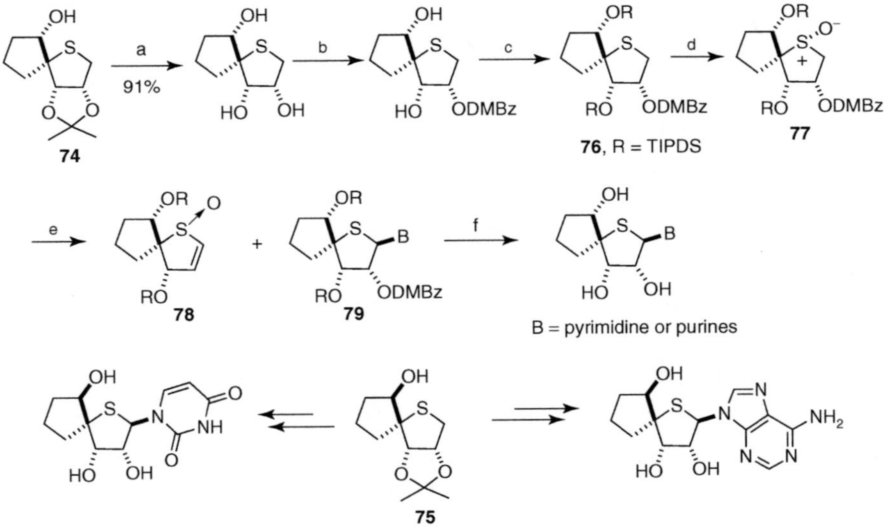

Scheme 28. Synthesis of β-anomeric 4′-thiaspirocyclic ribonucleosides. *Reaction and conditions:* (a) 6% HCl, THF, rt; (b) DMBzCl, DABCO, DMAP, CH_2Cl_2; (c) $TIPDSCl_2$, Py, $AgNO_3$, THF; (d) Davis reagent, $CHCl_3$; (e) pyrimidine or purine base, Et_3N, TMSOTf, $ClCH_2CH_2Cl$, CH_3CN, rt to 83°C; (f) (i) TBAF, AcOH, THF; (ii) NH_3, CH_3OH, rt.

Thus, it is evident that much effort has been directed toward the synthesis of these biologically active nucleosides. Haraguchi et al. developed the synthesis for 4′-substituted-4′-thionuclesoides by nucleophilic substitution at the 4′-acetoxy group.[74] Few of the synthesized derivatives exhibited potent inhibitory activity against HIV-1 and HIV-2. Matsuda and co-workers recently reported the synthesis and characterization details of 2′-modified-4′-thio-RNA.[75] All these efforts imply the growing interest of chemists and biologists in exploring in greater detail the possible therapeutic potential of 4′-thionucleosoides.

5. CONCLUSIONS

The discussions contained herein provide a comprehensive overview of important advances in the synthesis of 4′-thionucleosides in the past decade. The journey began with the development of 4′-thionucleoside analogues as AR agonists followed by successful conversion to antagonists. The focus moved from discovery, to successful rational design and synthesis, and finally to their functions in biological systems. Although the medicinal chemistry of ARs is well developed, the quest for selective agonists and antagonists for different receptor types continues. The development of novel ligands for better selectivity, affinity, and solubility holds great promise for researchers. With the current status of knowledge, serious efforts to explore and exploit the potentials of 4′-thioribonucleoside modifications in DNA and RNA chemistry are under way in several laboratories worldwide. The design of novel oligomeric compounds comprising 4′-thionucleosides for use in gene modulation, or the successful incorporation of 4′-thionucleosides into DNA or RNA as a tool for better understanding of biological functions, needs to be developed. With better understanding of the behavior of these nucleoside(tide) units in conjugation with recent developments in synthetic methodologies, there is every reason to think that new and important therapeutic applications of 4′-thionucleosides are just waiting to be discovered.

REFERENCES

1. (a) Simons, C. Nucleoside mimetics. In *Advanced Chemistry Texts*, Vol. 3; Gordon and Breach Science Publishers: Amsterdam, **2001**. (b) Isanbor, C.; O'Hagan, D. *J.Fluorine Chem.* **2006**, *127*, 303–319. (c) Chu, C. K. *Antiviral Nucleosides: Chiral Synthesis and Chemotherapy*; Elsevier Science Publishers: New York, **2003**. (d) Jeannot, F.; Gosselin, G.; Mathé, C. *Org. Biomol. Chem.* **2003**, *1*, 2096–2102.
2. (a) Herdewijn, P. *Modified Nucleosides: In Biochemistry, Biotechnology and Medicine*; Wiley-VCH, Weinheim, Germany, **2008**. (b) Romeo, G.; Chiacchio, U.; Corsaro, A.; Merino, P. *Chem. Rev.* **2010**, *110*, 3337–3370. (c) Tosh, D. K.; Jacobson, K. A.; Jeong, L. S. *Drugs Future* **2009**, *34*, 43–52.
3. Marquez, V. E.; Tseng, C. K.-H.; Mitsuya, H.; Aoki, S.; Kelley, J. A.; Ford, E., Jr.; Roth, J. S.; Broder, S.; Johns, D. G.; Driscoll, J. S. *J. Med. Chem.* **1990**, *33*, 978–985.

4. (a) Dyson, M. R.; Coe, P. L.; Walker, R. T. *J. Med. Chem.* **1991**, *34*, 2782–2786. (b) Secrist, J. A., III; Tiwari, K. N.; Riordan, J. M.; Montgomery, J. A. *J. Med. Chem.* **1991**, *34*, 2361–2366. (c) Bellon, L.; Barascut, J. L.; Imbach, J. L. *Nucleosides Nucleotides* **1992**, *11*, 1467–1479. (d) Yoshimura, Y.; Kitano, K.; Satoh, H.; Watanabe, M.; Miura, S.; Sakata, S.; Sasaki, T.; Matsuda, A. *J. Org. Chem.* **1996**, *61*, 822–823. (e) Yoshimura, Y.; Kitano, K.; Yamada, K.; Satoh, H.; Watanabe, M.; Miura, S.; Sakata, S.; Sasaki, T.; Matsuda, A. *J. Org. Chem.* **1997**, *62*, 3140–3152.

5. Parks, R. E., Jr.; Stoeckler, J. D.; Cambor, C.; Savarese, T. M.; Crabtree, G. W.; Chu, S.-H. In *Molecular Actions and Targets for Cancer Chemotherapeutic Agents*; Sartorelli, A. C., Lazo, J. S., Bertino, J. R., Eds.; Academic Press: New York, **1981**; p. 229.

6. (a) Leydier, C.; Bellon, L.; Barascut, J. L.; Morvan, F.; Rayner, B.; Imbach, J. L. *Antisense Res. Dev.* **1995**, *5*, 167–174. (b) Leydier, C.; Bellon, L.; Barascut, J. L.; Imbach, J. L. *Nucleosides Nucleotides* **1995**, *14*, 1027–1030. (c) Boggon, T. J.; Hancox, E. L.; McAuley-Hecht, K. E.; Connolly, B. A.; Hunter, W. N.; Brown, T.; Walker, R. T.; Leonard, G. A. *Nucleic Acids Res.* **1996**, *24*, 951–961.

7. (a) Yokoyama, M. *Synthesis* **2000**, *12*, 1637–1655. (b) Gunaga, P. G., Moon, H. R., Choi, W.J., Shin, D. H., Park, J. G., Jeong, L. S. *Curr. Med. Chem.* **2004**, *11*, 2585–2637.

8. Reist, E. J.; Gueffroy, D. E.; Goodman, L. *J. Am. Chem. Soc.* **1964**, *86*, 5658–5663.

9. (a) Reist, E. J.; Fisher, L. V.; Goodman, L. *J. Org. Chem.* **1968**, *33*, 189–192. (b) Bobek, M.; Whistler, R. L.; Bloch, A. *J. Med. Chem.* **1970**, *13*, 411–413. (c) Ritchie, R. G. S.; Szarek, W. A. *J. Chem. Soc. Chem. Commun.* **1973**, 686–687. (d) Bobek, M.; Bloch, A.; Parthasarathy, R.; Whistler, R. L. *J. Med. Chem.* **1975**, *18*, 784–787. (e) Pickering, M. V.; Witkowski, J. T.; Robins, R. K. *J. Med. Chem.* **1976**, *19*, 841–842.

10. Bonate, P. L.; Arthaud, L.; Cantrell, W. R.; Stephenson, K., Jr.; Secrist, J. A., III; Weitman, S. *Nat. Rev. Drug Discov.* **2006**, *5*, 855–865.

11. Lindemalm, S.; Liliemark, J.; Juliusson, J.; Larsson, R.; Albertioni, F. *Cancer Lett.* **2004**, *210*, 171–177.

12. (a) Dyson, M. R.; Coe, P. L.; Walker, R. T. *J. Med. Chem.* **1991**, *34*, 2782–2786. (b) Secrist, J. A., III; Tiwari, K. N.; Riordan, J. M.; Montgomery, J. A. *J. Med. Chem.* **1991**, *34*, 2361–2366.

13. Jones, A. S.; Verhelst, G.; Walker, R. T. *Tetrahedron Lett.* **1979**, 4415–4418.

14. Desgranges, C.; Razaka, G.; Rabaud, M.; Bricaud, H.; Balzarini, J.; De Clercq, E. *Biochem. Pharmacol.* **1983**, *32*, 3583–3590.

15. Rahim, S. G.; Littler, E.; Powell, K. L.; Collins, P.; Coe, P. L.; Dyson, M. R.; Walker, R. T. *Proc. 31st ICAAC (Chicago)* **1991**, 1232.

16. Dyson, M. R.; Coe, P. L.; Walker, R. T. *J. Med. Chem.* **1991**, *34*, 2782–2786.

17. Hancox, E. L.; Connolly, B. A.; Walker, R. T. *Nucleic Acids Res.* **1993**, *21*, 3485–3491.

18. (a) Secrist, J. A., III; Riggs, R. M.; Tiwari, K. N.; Montgomery, J. A. *J. Med. Chem.* **1992**, *35*, 533–538. (b) Tiwari, K. N.; Secrist, J. A., III; Montgomery, J. A. *Nucleosides Nucleotides* **1994**, *13*, 1819–1828. (c) Branalt, J.; Kvarnstrom, I.; Svensson, S. C. T.; Classon, B.; Samuelsson, B. *J. Org. Chem.* **1994**, *59*, 4430–4432. (d) Uenishi, J.; Takahashi, K.; Motoyama, M.; Akashi, H.; Sasaki, T. *Nucleosides Nucleotides* **1994**, *13*, 1347–1361. (e) Jeong, L. S.; Moon, H. R.; Choi, Y. J.; Chun, M. W.; Kim, H. O. *J. Org. Chem.* **1998**, *63*, 4821–4825.

19. (a) Hertel, L. W.; Kroin, J. S.; Misner, J. W.; Tustin, J. M. *J. Org. Chem.* **1988**, *53*, 2406–2409. (b) Tuttle, J. V.; Tisdale, M.; Krenitsky, T. A. *J. Med. Chem.* **1993**, *36*, 119–125.

20. Jacobson, K. A.; Gao, Z.-G. *Nat. Rev. Drug Discov.* **2006**, *5*, 247–264.
21. Ralevic V.; Burnstock, G. *Pharmacol. Rev.* **1998**, *50*(3), 413–492.
22. Meyerhof, W.; Müller-Brechlin, R.; Richter D. *FEBS Lett.* **1991**, *284*, 155–160.
23. Zhou, Q. Y.; Li, C.; Olah, M. E.; Johnson, R. A.; Stiles, G. L.; Givelli, O. *Proc. Natl. Acad. Sci. USA* **1992**, *89*, 7432–7436.
24. Jacobson, K. A. *Trends Pharmacol. Sci.* **1998**, *19*, 184–191.
25. Volpini, R.; Constanzi, S.; Lambertucci, C.; Taffi, S.; Vittori, S.; Klotz, K. N., Cristalli, G. *J. Med. Chem.* **2002**, *45*, 3271–3279.
26. Elzein, E.; Palle, V.; Wu, Y.; Maa, T.; Zeng, D.; Zablocki, J. *J. Med. Chem.* **2004**, *47*, 4766–4773.
27. (a) De Zwaart, M.; Link, R.; Von Frijtag Drabbe Künzel, J. K.; Cristalli, G.; Jacobson, K. A.; Townsend-Nichoson, A.; Ijzerman, A. P. *Nucleosides Nucleotides* **1998**, *17*, 969–985. (b) Bruns, R. T. *Can. J. Physiol. Pharmacol.* **1980**, *58*, 637–691.
28. Naka, T.; Minakawa, N.; Abe, H.; Kaga, D.; Matsuda, A. *J. Am. Chem. Soc.* **2000**, *122*, 7233–7243.
29. Dande, P.; Prakash, T. P.; Sioufi, N.; Gaus, H.; Jarres, R.; Berdeja, A.; Swayze, E. E.; Griffey, R. H.; Bhat, B. *J. Med. Chem.* **2006**, *49*, 1624–1634.
30. Jeong, L. S.; Lee, H. W.; Jacobson, K. A.; Kim, O. K.; Shin, D. H.; Lee, J. A.; Gao, Z.-G.; Lu, C.; Duong, H. T.; Gunaga, P.; Lee, S. K.; Jin, D. Z.; Chun, M. W.; Moon, H. R. *J. Med. Chem.* **2006**, *49*, 273–281.
31. Jeong, L. S.; Jin, D. Z.; Kim, H. O.; Shin, D. H.; Moon, H. R.; Gunaga, P.; Chun, M. W.; Kim, Y.-C.; Melman, N.; Gao, Z.-G.;Jacobson. K. A. *J. Med. Chem.* **2003**, *46*, 3775–3777.
32. Gao, Z.-G.; Joshi, B. V.; Klutz, A. M.; Kim, S.-K.; Lee, H. W.; Kim, H. O.; Jeong, L. S.; Jacobson, K. A. *Bioorg. Med. Chem. Let.* **2006**, *16*, 596–601.
33. Jeong, L. S.; Lee, H. W.; Kim, H. O.; Tosh, D. K.; Pal, S.; Choi, W. J.; Gao, Z.-G.; Patel, A. R.; Williams, W.; Jacobson, K. A.; Kim, H.-D. *Bioorg. Med. Chem. Lett* **2008**, *18*, 1612–1616.
34. Choi, W. J.; Lee. H. W.; Kim, H. O.; Chinn, M.; Gao, Z.-G; Patel, A.; Jacobson, K. A.; Moon, H. R.; Jung, Y. H.; Jeong, L. S. *Bioorg. Med. Chem.* **2009**, *17*, 8003–8011.
35. Hou, X.; Lee, H. W.; Tosh, D. K.; Zhao, L. X.; Jeong, L. S. *Arch. Pharm. Res.* **2007**, *30*, 1205–1209.
36. Barker, R.; Fletcher, H. G., Jr. *J. Org. Chem.* **1961**, *26*, 4605–4609.
37. Gunaga, P.; Kim, H. O.; Lee, H. W.; Tosh, D. K.; Ryu, J.-S.; Choi, S.; Jeong. L. S. *Org. Lett.* **2006**, *8*, 4267–4270.
38. Jeong, L. S.; Lee, H. W.; Kim, H. O.; Jung, J. Y.; Gao, Z.-G.; Duong, H. T.; Rao, S.; Jacobson, K. A.; Shin, D. H.; Lee, J. A.; Gunaga, P.; Lee, S. K.; Jin, D. Z.; Chun, M. W.; Moon, H. R. *Bioorg. Med. Chem.* **2006**, *14*, 4718–4730.
39. Kim, H. O.; Schinazi, R. F.; Nampalli, S.; Shanmuganathan, K.; Cannon, D. L.; Alves, A. J.; Jeong, L. S.; Beach, J. W.; Chu, C. K. *J. Med. Chem.* **1993**, *36*, 30–37.
40. Volpini, R.; Costanzi, S.; Lambertucci, C.; Taffi, S.; Vittori, S.; Klotz, K.-N.; Cristalli, G. *J. Med. Chem.* **2002**, *45*, 3271–3279.
41. Liang, C.-W.; Choi, W. J.; Jeong, L. S. *Arch. Pharm. Res.* **2008**, *31*, 973–977.

42. (a) Jacobson, K. A.; Siddiqi, S. M.; Olah, M. E.; Ji, X. D.; Melman, N.; Bellamkonda, K.; Meshulam, Y.; Stiles, G. L.; Kim, H. O. *J. Med. Chem.* **1995**, *38*, 1720–1735. (b) Volpini, R.; Costanzi, S.; Lambertucci, C.; Vittori, S.; Kliotz, K.-N.; Lorenzen, A.; Cristalli, G. *Bioorg. Med. Chem. Lett.* **2001**, *11*, 1931–1934. (c) Gao, Z.-G.; Kim, S.-K.; Biadatti, T.; Chen, W.; Lee, K.; Barak, D.; Kim, S. G.; Johnson, C. R.; Jacobson, K. A. *J. Med. Chem.* **2002**, *45*, 4471–4484.

43. Baraldi, P. G.; Tabrizi, M. A.; Romagnoli, R.; Fruttarolo, F.; Merighi, S.; Varani, K.; Gessi, S.; Borea, P. A. *Curr. Med. Chem.* **2005**, *12*, 1319–1329.

44. Gao, Z.-G.; Blaustein, J.; Gross, A. S.; Melman, N.; Jacobson, K. A. *Biochem. Pharmacol.* **2003**, *65*, 1675–1684.

45. Kim, S.-K.; Gao, Z.-G.; Jeong, L. S.; Jacobson, K. A. *J. Mol. Graphics Modell.* **2006**, *25*, 562–567.

46. Jeong, L. S.; Choe, S. A.; Gunaga, P.; Kim, H. O.; Lee, H. W.; Lee, S. K.; Tosh, D. K.; Patel, A.; Palaniappan, K. K.; Gao, Z.-G.; Jacobson, K. A.; Moon, H. R. *J. Med. Chem.* **2007**, *50*, 3159–3162.

47. Jeong, L. S.; Pal, S.; Choe, S. A.; Choi, W. J.; Jacobson, K. A.; Gao, Z.-G.; Klutz, A. M.; Hou, X.; Kim, H. O.; Lee, H. W.; Lee, S. K.; Tosh, D. K.; Moon, H. R. *J. Med. Chem.* **2008**, *51*, 6609–6613.

48. Wang, Z.; Do, C. W.; Avila, M. Y.; Yantorno, K.-P.; Stone, R. A.; Gao, Z.-G.; Joshi, B.; Besada, P.; Jeong, L. S.; Jacobson, K. A.; Civan, M. M. *Exp. Eye Res.* **2010**, *90*, 146–154.

49. (a) Matsuda, A.; Shinozaki, M.; Yamaguchi, T.; Homma, H.; Nomoto, R.; Miyasaka, T.; Watanabe, Y.; Abiru, T. *J. Med. Chem.* **1992**, *35*, 241–252. (b) Cristalli, G.; Volpini, R.; Vittori, S.; Camaioni, E.; Monopoli, A.; Conti, A.; Dionisotti, S.; Zocchi, C.; Ongini, E. *J. Med. Chem.* **1994**, *37*, 1720–1726. (c) Lambertucci, C.; Costanzi, S.; Vittori, S.; Volpini, R.; Cristalli, G. *Nucleosides Nucleotides Nucleic Acids* **2001**, *20*, 1153–1157.

50. Hou, X.; Kim, H. O.; Alexander, V.; Kim, K. L.; Choi, S.; Park, S.; Lee, J. H.; Yoo, L. S.; Gao, Z.-G.; Jacobson, K. A.; Jeong, L. S. *ACS Med. Chem. Lett.* **2010**, *1*, 516–520.

51. Lee, H. W.; Kim, H. O.; Choi, W. J.; Choi, S.; Lee, J. H.; Park, S.; Yoo, L.; Jacobson, K. A.; Jeong, L. S. *Bioorg. Med. Chem. Lett.* **2010**, *18*, 7015–7021.

52. (a) Yoo, S. J.; Kim, O. H.; Lim, Y.; Jeong, L. S. *Bioorg Med. Chem.* **2002**, *10*, 215–226. (b) Lai, C. L.; Rosmawati, M.; Lao, J.; Van Vlierberghe, H.; Anderson, F. H.; Thomas, N.; Dettertogh, D. *Gastroenterology* **2002**, *123*, 1831–1838.

53. Gunaga, P.; Baba, M.; Jeong, L. S. *J. Org. Chem.* **2004**, *69*, 3208–3211.

54. (a) Yoshimura, Y.; Kitano, K.; Satoh, H.; Watanabe, M.; Miura, S.; Sakata, S.; Sasaki, T.; Matsuda, A. *J. Org. Chem.* **1996**, *61*, 822–823. (b) Kaga, D.; Minakawa, N.; Matsuda, A. *Nucleosides Nucleotides Nucleic Acids* **2005**, *24*, 1789–1800.

55. Maslen, H. L.; Hughes, D.; Hursthouse, M.; Clercq, E. D.; Balzarini, J.; Simons, C. *J. Med. Chem.* **2004**, *47*, 5482–5491.

56. Dyson, M. R.; Coe, P. L.; Walker, R. T. *J. Chem. Soc. Chem.Commun.* **1991**, 741–742.

57. Jasamai, M.; Balzarini, J.; Simons, C. *J. Enzyme Inhib. Med. Chem.* **2008**, *23*, 56–61.

58. Cong, L.; Zhou, W.; Jin, D.; Wang, J.; Chen, X. *Chem. Biol. Drug. Des.* **2010**, *10*, 619–627.

59. (a) Kim, K. R.; Moon, H. R.; Park, A.-Y.; Chun, M. W.; Jeong, L. S. *Bioorg. Med. Chem.* **2007**, *15*, 227–234. (b) Kim, K. R.; Park, A.-Y.; Moon, H. R.; Chun, W. M.; Jeong, L. S. *Nucleosides Nucleotides Nucleic Acids* **2007**, *26*, 911–915.

60. Moravcova, J.; Capkova, J.; Stanek, J. *Carbohydr. Res.* **1994**, *263*, 61–66.
61. Blackburn, G. M.; Eckstein, F.; Kent, D. E.; Perree, T. D. *Nucleosides Nucleotides*, **1985**, *4*, 165–167.
62. Zheng, F.; Zhang, X.; Qing, F.-L. *Chem. Commun.* **2009**, 1505–1507.
63. Groove, C. L.; Guo, X.; Liu, S.-H.; Gao, Z.; Chu, C. K.; Cheng, Y. C. *Cancer Res.* **1995**, *55*, 3008–3011.
64. (a) Chang, C. N.; Doong, S. L.; Zhou, J. H.; Beach, J. W.; Jeong, L. S.; Chu, C. K.; Tasi, C. H.; Cheng, Y. C.; Liotta, D. C.; Schinazi. R. F. *J. Biol. Chem.* **1992**, *267*, 13938–13942. (b) Doong, S. L.; Tasi, C. H.; Schinazi. R. F.; Liotta, D. C.; Cheng, Y. C. *Proc. Natl. Acad. Sci. USA.* **1991**, *88*, 8495–8499.
65. Zheng, F.; Zhang, X.-H.; Qiu, X.-L.; Zhang, X.; Qing, F.-L. *Org. Lett.* **2006**, *8*, 6083–6086.
66. Zheng, F.; Fu, L.; Wang, R.; Qing, F.-L. *Org. Biomol. Chem.* **2010**, *8*, 163–170.
67. Jean-Baptiste, L.; Lefebvre, P.; Pfund, E.; Lequeux, T. *Synlett* **2008**, 817–820.
68. Takahashi, M.; Daidouji, S.; Shiro, M.; Minakawa, N.; Matsuda, A. *Tetrahedron* **2008**, *64*, 4313–4324.
69. Chapdelaine, D.; Cardinal-David, B.; Prévost, M.; Gagnon, M.; Thumin, I.; Guindon, Y. *J. Am. Chem. Soc.* **2009**, *131*, 17242–17245.
70. Haraguchi, K.; Takeda, S.; Tanaka, H.; Nitanda, T.; Baba, M.; Dutschman, G. E.; Cheng, Y.-C. *Bioorg. Med. Chem. Lett.* **2003**, *13*, 3775–3777.
71. Kumamoto, H.; Nakai, T.; Haraguchi, K.; Nakamura, K. T.; Tanaka, H.; Baba M.; Cheng, Y.-C. *J. Med. Chem.* **2006**, *49*, 7861–7867.
72. Paquette, L. A.; Dong, S. *J. Org. Chem.* **2005**, *70*, 5655–5664.
73. Paquette, L. A.; Dong, S. *J. Org. Chem.* **2005**, *70*, 1580–1596.
74. Haraguchi, K.; Shimada, H.; Tanaka, H.; Hamasaki, T.; Baba, M.; Gullen, E A.; Dutschman, G. E.; Cheng, Y.-C. *J. Med. Chem.* **2008**, *51*, 1885–1893.
75. Takahashi, M.; Minakawa, N.; Matsuda, A. *Nucleic Acids Res.* **2009**, *37*, 1353–1362.

15 Recent Advances in the Chemical Synthesis of Aza-Nucleosides

TOMÁS TEJERO

Departamento de Síntesis y Estructura de Biomoléculas, y Departamento de Química Orgànica, Instituto de Sintesis y Catalisis Homogenea, Universidad de Zaragoza, CSIC, Zaragoza, Aragón, Spain

IGNACIO DELSO

Servicio de Resonancia Magnética Nuclear, Centro de Química y Materiales de Aragón, Universidad de Zaragoza, CSIC, Zaragoza, Aragón, Spain

PEDRO MERINO

Departamento de Síntesis y Estructura de Biomoléculas, y Departamento de Química Orgànica, Instituto de Sintesis y Catalisis Homogenea, Universidad de Zaragoza, CSIC, Zaragoza, Aragón, Spain

1. INTRODUCTION

Nucleoside analogues in which the endocyclic oxygen atom of the furanose ring has been replaced by a nitrogen atom are referred to as aza-nucleosides (Figure 1). Despite the huge body of literature reported around nucleoside analogues, little attention have been devoted to those analogues, in which the furanose ring has been replaced by a different heterocycle,[1] and aza-nucleosides are not an exception. In 1999, Yokoyama and Momotake reviewed the synthesis and biological activity of aza-nucleosides.[2] Since then several new examples have been reported; in particular, recent findings on the biological properties of a type of aza-C-nucleosides (called *immucillins*) have led to a reassessment of those analogues for pharmaceutical purposes.[3] Indeed, immucillin-H is a powerful transition-state analogue inhibitor of purine nucleoside phosphorylase that also selectively inhibits human

---

Chemical Synthesis of Nucleoside Analogues, First Edition. Edited by Pedro Merino.
© 2013 John Wiley & Sons, Inc. Published 2013 by John Wiley & Sons, Inc.

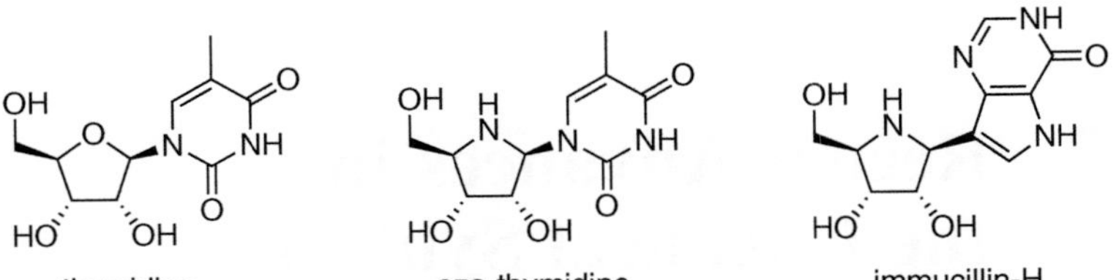

Figure 1. Aza-nucleosides.

T-lymphocytes.[4] Due to their importance, the chemistry and biology of immucillins have been reviewed elsewhere.[5]

In this chapter we focus on the most significant advances during this century in the chemical synthesis of aza-nucleosides and related compounds, including immucillins as well as other aza-C-nucleosides. Representative examples have been organized according to the type of compound.

2. AZA-NUCLEOSIDES

Correia and co-workers reported a stereoselective synthesis of the aza-analogues of the well-known antiviral agents d4T and ddU in a straightforward way. Starting from cyclic enecarbamate **1**, two and three steps were necessary to obtain N-Boc derivatives of aza-d4T **3** and aza-ddU **4**, respectively (Scheme 1).[6] The key step of the reaction was the incorporation of the base moiety promoted by phenylselenenyl bromide. Use of a promoter makes it easier to obtain a single isomer.

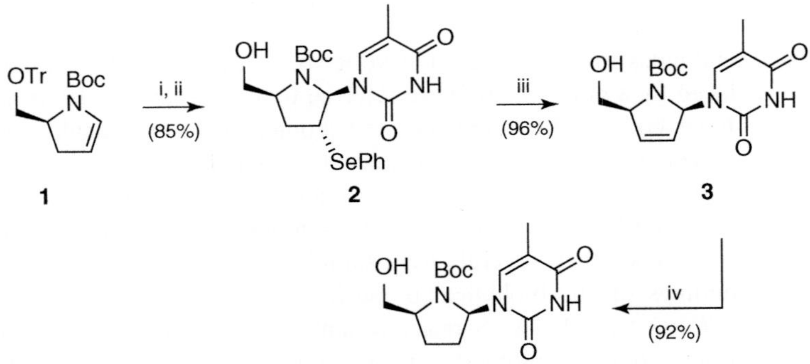

Scheme 1. *Reagents and conditions:* (i) (TMS)$_2$-thymine, PhSeBr, CH$_3$CN, $-23°$C; (ii) ZnBr$_2$, MeOH, CH$_2$Cl$_2$; (iii) H$_2$O$_2$, dioxane, NaHCO$_3$ (96%); (iv) H$_2$, Pd-C, EtOAc.

The synthesis of 2′-deoxy-aza-thymidine **10** was accomplished in 11 steps (25.2% overall yield), starting from protected maleimide **5**.[7] This compound was easily obtained from malic acid.[8] The hydroxymethyl group was introduced via chemoselective addition of furyllithium to **5**, further oxidation to carboxylic acid, and reduction of the corresponding ester **7**. The resulting compound **8** was converted into **9**, and introduction of the base moiety was achieved through the typical Hilbert–Johnson protocol.[9] Under these conditions, a 1:1.8 mixture of *cis/trans*-isomers **10** was obtained (Scheme 2). Compounds **10** can be separated by chromatographic techniques and their identity determined unambiguously using NMR techniques.

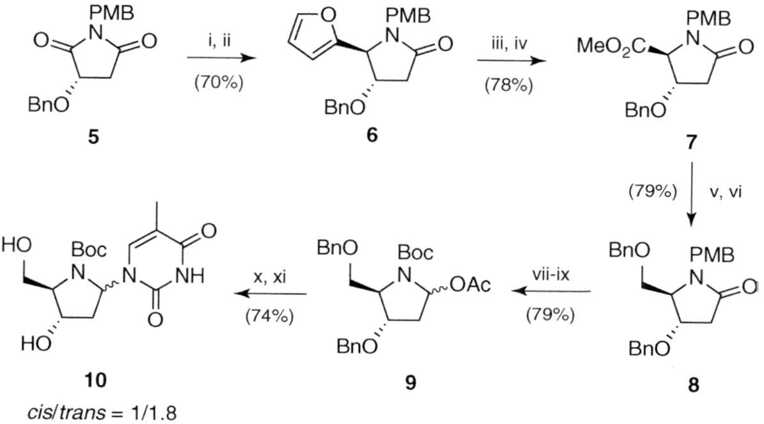

Scheme 2. *Reagents and conditions:* (i) furan, BuLi, THF, −78°C; (ii) Et$_3$SiH, BF$_3$·Et$_2$O, CH$_2$Cl$_2$, −78°C; (iii) NaIO$_4$, RuCl$_3$, MeCN–CCl$_4$–H$_2$O; (iv) CH$_2$N$_2$, Et$_2$O; (v) NaBH$_4$, CaCl$_2$, THF–EtOH; (vi) NaH, THF, BnBr, rt; (vii) CAN, MeCN–H$_2$O; (viii) Boc$_2$O, MeCN; (ix) DIBAH, CH$_2$Cl$_2$, −78°C, then Ac$_2$O, DMAP, CH$_2$Cl$_2$, Et$_3$N; (x) (TMS)$_2$-thymine, SnCl$_4$, MeCN, −25°C to rt, 1 h; (xi) H$_2$, 10% Pd-C, EtOH, rt, 9 h.

Treatment of protected 4-hydroxyproline **11** with (diacetoxy)iodobenzene (DAIB) and then methanol provided intermediate α-methoxypyrrolidine **12**, which is a typical precursor of an iminium salt. Indeed, upon the addition of a Lewis acid in the presence of a heterocyclic base, aza-nucleosides are formed (Scheme 3).[10] The procedure can be carried out without purification of any intermediate by

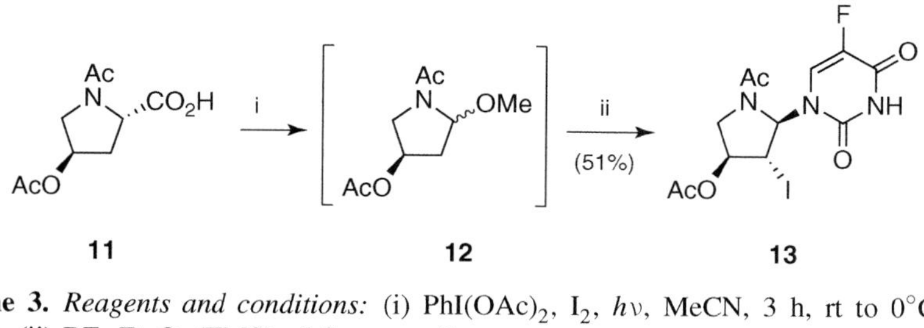

Scheme 3. *Reagents and conditions:* (i) PhI(OAc)$_2$, I$_2$, hν, MeCN, 3 h, rt to 0°C, then MeOH; (ii) BF$_3$·Et$_2$O, (TMS)$_2$-5-fluorouracil.

sequential addition of the reagents, thus achieving a very efficient protocol. Both purines and pyrimidines can be used with this methodology.[11]

Starting from lactam **14**, readily available from pyroglutamic acid, cyclobutane-fused aza-nucleosides **17** have been prepared by Alibes and co-workers.[12] The common intermediate **16** was prepared through photochemical cycloaddition between **14** and ethylene to afford **15**, which was reduced, and the emerging hemiaminal was acetylated. Condensation of **16** with pyrimidine and purine bases under modified Vorbrüggen conditions provided the target compounds in good chemical yield and selectivity (Scheme 4). The N-glycosylation reaction needed to be optimized, the ratio between compound **16**, N,O-bis(trimethylsilyl)acetamide (BSA), and the base (1:3:2), the solvent (MeCN), and the Lewis acid (SnCl$_4$) being crucial for avoiding the presence of undesired by-products.

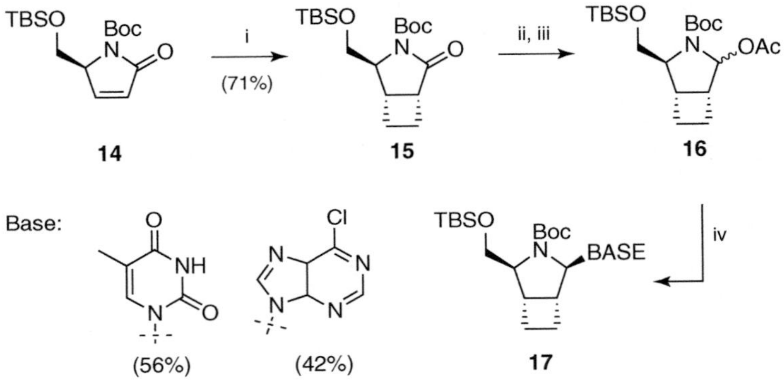

Scheme 4. *Reagents and conditions:* (i) ethylene, $h\nu$, acetone, $-20°C$; (ii) LiEt$_3$BH·THF, $-78°C$; (iii) Ac$_2$O, Et$_3$N, DMAP; (iv) BSA, thymine (or 6-chloropurine), MeCN, SnCl$_4$.

Fluorinated derivatives were prepared from *trans*-4-hydroxy-L-proline **18** in 12 steps.[13] After introduction of the fluorine-containing group through difluoromethylenation followed by catalytic hydrogenation, oxidation to generate the lactam moiety provided intermediate **20**. This compound was easily transformed into common precursor **21**. Introduction of the heterocyclic base under typical Vorbrüggen conditions afforded mixtures of *cis*- and *tran*-aza-nucleosides **22** and **23** (Scheme 5). Notably, while in the case of uracil the *cis*-aza-nucleoside was predominant, with thymine the *tran*-isomer was obtained preferentially.[14]

The same authors reported the synthesis of fluorinated aza-nucleosides in which the difluoromethyl group was incorporated into the 3′-position instead of the 2′-position as in the case of compounds **22** and **23**. The synthesis also started from *trans*-4-hydroxy-L-proline **18**, which served to prepare intermediate **24** in 11 steps and almost 20% overall yield (Scheme 6). A strategy similar to that depicted in Scheme 5 was followed with **24** to obtain aza-nucleosides **25** and **26**.[15]

Racemic branched aza-isonucleosides were prepared starting from α-bromo-γ-nitroester **29**, which was obtained through Michael reaction of susbtituted

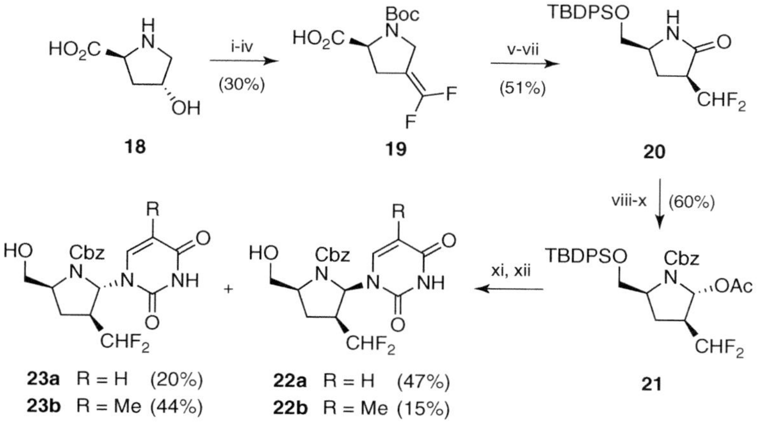

Scheme 5. *Reagents and conditions:* (i) $SOCl_2$, MeOH; (ii) Boc_2O, Et_3N, DMAP, CH_2Cl_2; (iii) Swern oxidation; (iv) CF_2Br_2, HMPT, Zn; (v) H_2, Pd-C, EtOH, rt, 1 atm. (vi) $RuO_2 \cdot xH_2O$, $NaIO_4$, EtOAc, H_2O; (vii) TFA, CH_2Cl_2, rt; (viii) LHMDS, CBzCl, THF, 10 min, $-78°C$; (ix) $LiEt_3BH \cdot THF$, $-78°C$; (x) Ac_2O, Py, DMAP; (xi) TMSOTf, MeCN, $(TMS)_2$-uracil [or $(TMS)_2$-thymine]; (xii) TBAF, THF.

1,3-dioxane **27** with methyl 2-bromoacrylate **28**. After reductive cyclization of **29** to give **30**, the base moiety was introduced through a Mitsunobu reaction (Scheme 7).[16] N^3-Benzoylated uracil and thymine were used, and further hydrolysis was required. With purine bases such as adenine, it is possible to employ the unprotected base, and the corresponding aza-isonucleoside is obtained in good chemical yield.

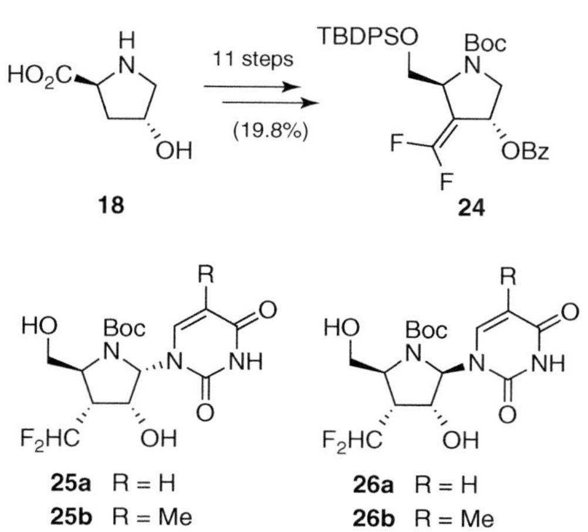

Scheme 6.

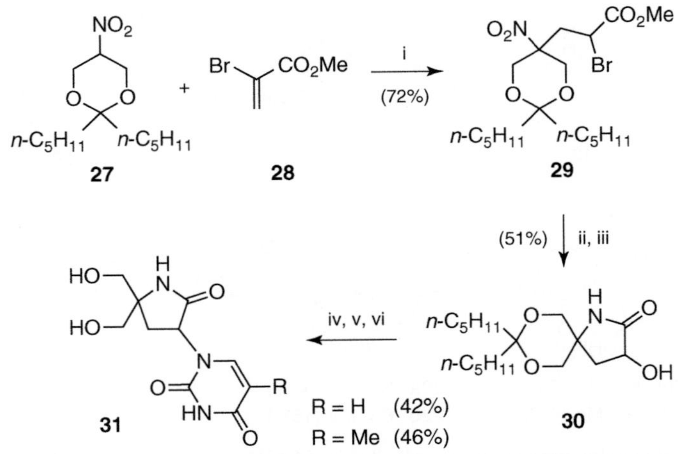

Scheme 7. *Reagents and conditions:* (i) Et$_3$N, MeOH, reflux, 4 h; (ii) CF$_3$CO$_2$K, MeCN, reflux, aq. workup; (iii) H$_2$, Pd-C. 8 bar, MeOH, 50°C, 6 h; (iv) N^3-benzoyluracil (or N^3-benzoylthymine), DEAD, Ph$_3$P, THF, 0°C, 20 h; (v) 35% MeNH$_2$, MeOH, reflux, 3.5 h; (vi) 90% TFA, rt, 1 h.

Six-membered aza-nucleosides derived from piperidines, instead of pyrrolidines, have been described by Ye and co-workers.[17] Key steps of the synthesis are the rearrangement of lactones (readily available from natural monosaccharides) and Lewisacid–catalyzed condensation of nucleobases with the corresponding glycosyl donor **34**. In Scheme 8, the synthesis of D-glucose-derived aza-nucleoside **35** is illustrated. Isomers were also prepared starting from D-mannose and D-galactose.

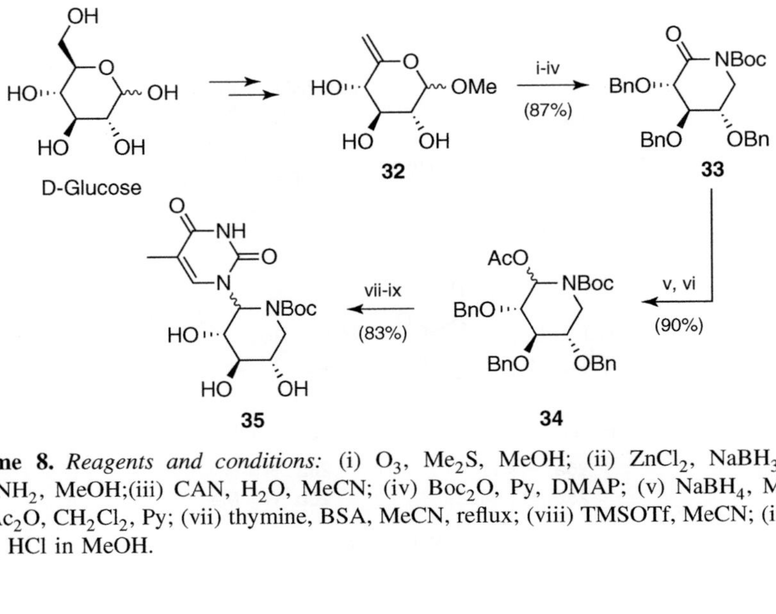

Scheme 8. *Reagents and conditions:* (i) O$_3$, Me$_2$S, MeOH; (ii) ZnCl$_2$, NaBH$_3$(CN), PMBNH$_2$, MeOH;(iii) CAN, H$_2$O, MeCN; (iv) Boc$_2$O, Py, DMAP; (v) NaBH$_4$, MeOH; (vi) Ac$_2$O, CH$_2$Cl$_2$, Py; (vii) thymine, BSA, MeCN, reflux; (viii) TMSOTf, MeCN; (ix) H$_2$, Pd-C, HCl in MeOH.

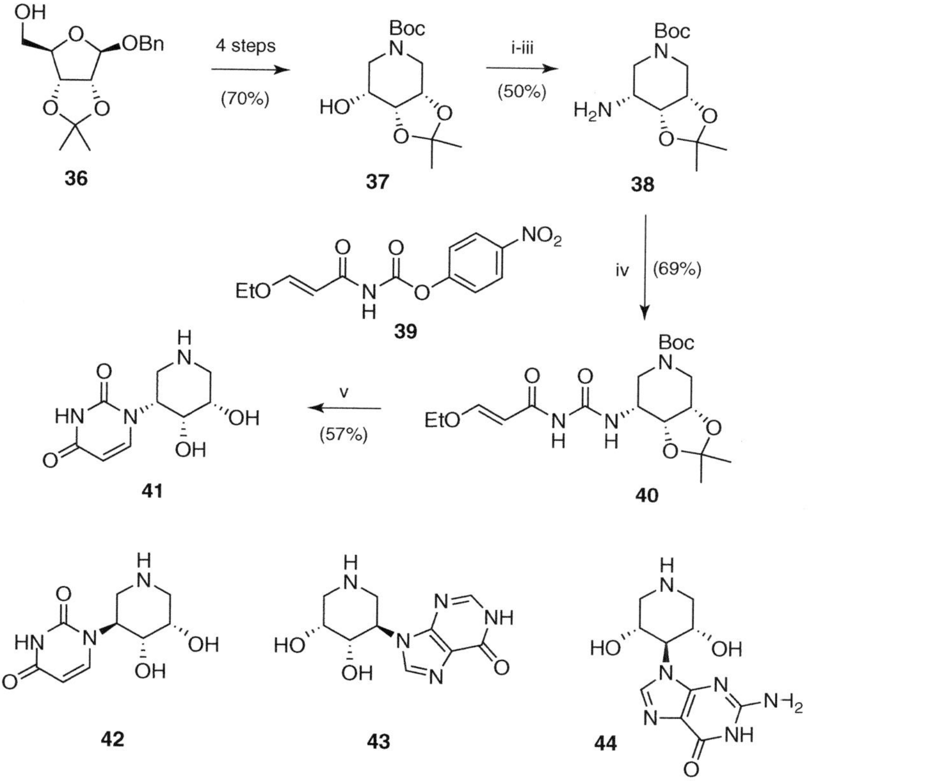

Scheme 9. *Reagents and conditions:* (i) Dess–Martin periodinane; (ii) BnNH$_2$, NaBH(OAc)$_3$; (iii) H$_2$, Pd-C; (iv) **39**, dioxane; (v) Dowex 50 H$^+$.

Another class of this sort of analogues has been reported recently by Rejman and co-workers.[18] Several aza-*C*-nucleosides were synthesized by utilizing a trihydroxypiperidine **37** as a common intermediate. Compound **37** was prepared from protected D-ribose **36** in four steps and 74% overall yield. As an example, the synthesis of the six-membered aza-*C*-nucleoside **41** is illustrated in Scheme 9. The amino group was introduced by reductive amination, and after introduction of reagent **39**, the nucleobase was constructed. Following the same strategy (i.e., introduction of an amino group in different positions and construction of the heterocyclic bases), aza-nucleosides **42–44** were also prepared (Scheme 9).

Novel analogues **46** of the antiviral agent ganciclovir in which the base moiety is linked to the nitrogen atom of the pyrrolidine ring through a methylene bridge were obtained from lactam **45** through a one-pot procedure (Scheme 10).[19] A strategy similar to that depicted in Scheme 7 was employed for preparing key intermediate **45**.

A longer bridge between the two nitrogen atoms is also possible by reacting functionalized nucleobases **48** with (*S*)-prolinol **49**. The epoxide-containing

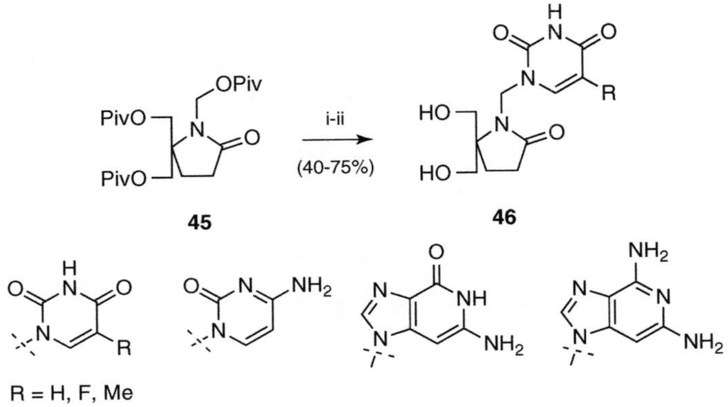

Scheme 10. *Reagents and conditions:* (i) pyrimidine nucleobase, BSA, MeCN, rt, 1 h, then TMSOTf, MeCN, rt, 2 days; (ii) NH$_3$ (aq), MeOH, sealed tube, 70°C, 1 day.

pyrimidines **48** were prepared from the corresponding heterocyclic base (thymine or uracil) in two steps (Scheme 11). The epoxide opening by **49** was promoted by sulfated zirconia under microwave irradiation.[20]

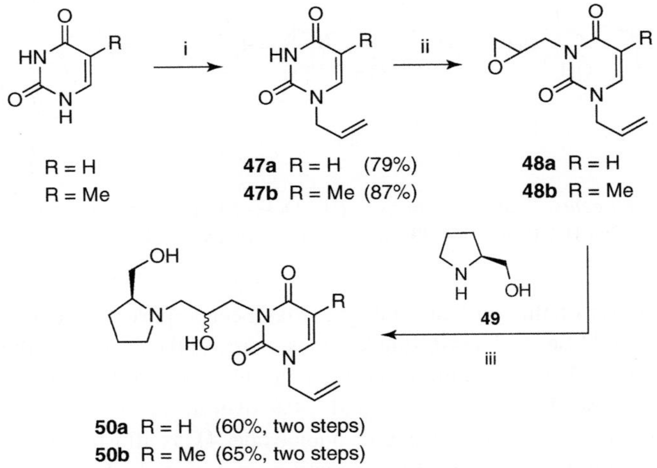

Scheme 11. *Reagents and conditions:* (i) BSA, MeCN, rt, 5 min, then allyl bromide, NaI, TMSCl, rt, 48 h; (ii) *t*-BuOK, DMF, 0°C, 5 min, then rt, 25 min, then 3-bromopropylene oxide, rt, 24 h; (iii) sulfated zirconia, microwave, 60°C, 150 min.

Experimental Procedures

(1′S,3′S,4′R)-4′-tert-Butoxyamido-2′-deoxythymidine **cis-10**. A mixture of thymine (44 mg, 0.35 mmol), (NH$_4$)$_2$SO$_4$ (15 mg, 0.12 mmol), and hexamethydisilazane (2 mL) was refluxed at 125°C for 5 h. The excess HMDS was removed by codistillation with xylene (3 × 1 mL) under reduced pressure. To the residue

was added MeCN (2 mL) and the resulting solution was transferred to compound **9** (45 mg, 0.1 mmol). To the resulting mixture was added dropwise a solution of $SnCl_4$ (0.1 mL, 0.15 mmol) in MeCN (0.6 mL) at $-25°C$ under an argon atmosphere. The resulting mixture was stirred for 1 h at the same temperature. Saturated aqueous $NaHCO_3$ was added, the resulting mixture warmed to room temperature and filtered through Celite. The aqueous layer was extracted with CH_2Cl_2 (3 × 3 mL). The combined organic layers were dried over Na_2SO_4, filtered, and evaporated in vacuo. Flash chromatography ($Et_2O/PE = 2:1$) of the residue gave β-anomer (colorless oil, 16 mg, 31%) and α-anomer (colorless oil, 29 mg, yield 56%).

To 18 mg of 20% Pd/C (18 mg) was added a solution of the pure β-anomer (26 mg, 0.05 mmol) in 3 mL of 95% ethanol. The mixture was hydrogenated under 1 atm hydrogen pressure and stirred at room temperature for 9 h. The mixture was filtered through Celite and the filtrate evaporated in vacuo. Flash chromatography of the residue using ethyl acetate as an eluent afforded *cis*-**10** (14 mg, 85%) as a white foam. $[\alpha]_D^{20}$ $-91°$ (c 1.1, MeOH).

Benzyl(2 R,3 S,5 S)-5-Hydroxymethyl-2-[2,4-dioxo-3,4-dihydropyrimidin-1(2 H)-yl]-3-(difluoromethyl)pyrrolidine-1-carboxylate (22a) and Benzyl (2 S,3 S,5 S)-5-Hydroxymethyl-2-[2,4-dioxo-3,4-dihydropyrimidin-1(2 H)-yl]-3-(difluoromethyl)pyrrolidine-1-carboxylate (23a); Typical Procedure. To a stirred solution of **21** (306 mg, 0.669 mmol) and uracil (220 mg, 1.96 mmol) in anhyd. MeCN (30 mL) was added *N*,*O*-bis(trimethylsilyl)acetamide (1.0 mL, 3.03 mmol). The reaction mixture was stirred under reflux for 30 min. After cooling to 0°C, TMSOTf (0.33 mL, 1.58 mmol) was added dropwise and the solution was stirred at room temperature for a further 30 min. The reaction was quenched with cold saturated aqueous $NaHCO_3$ and the resulting mixture was extracted with CH_2Cl_2 (3 × 50 mL). The combined organic phases were washed with brine and dried (Na_2SO_4). After removal of the solvent, the resulting residue was purified by flash chromatography (hexane/EtOAc, 4:1, 3:1, then 2:1) to give two compounds, the less polar compound (90 mg, white foam) and the more polar compound (150 mg, white foam). A stirred solution of the less polar compound (90 mg) in THF (10 mL) was treated with a 1 M solution of TBAF (0.24 mL, 0.24 mmol) at 0°C. After stirring at room temperature for 8.5 h, the reaction was quenched with H_2O and the mixture was extracted with CH_2Cl_2 (3 × 30 mL). The combined organic phases were washed with brine and dried (Na_2SO_4). After removal of the solvent, the resulting residue was purified by flash chromatography (hexane/EtOAc, 1:1) to give **22a**. White foam; yield 64 mg (34% from **21**); $[\alpha]_D^{20}$ $+15°$ (c 0.59, $CHCl_3$). The more polar compound was also treated with TBAF under the same conditions to give **22b**. White foam; yield 38 mg (14% from **18**); $[\alpha]_D^{20}$ $-75°$ (c 0.67, $CHCl_3$).

3. AZA-*C*-NUCLEOSIDES

By analogy with *C*-nucleosides, aza-*C*-nucleosides are those in which the nucleobase is linked directly to the pyrrolidine unit through a C–C bond instead the typical C–N bond. As a consequence, these derivatives are hydrolytically stable with respect to the aza-nucleosides described in Section 2.

By using an approach similar to that depicted in Scheme 7, three aza-analogues of tiazofurin were prepared.[21] Starting from intermediate **51**, the pyrrolidine ring was formed by treating that compound with aluminum amalgam. Incorporation of the cyanide functionality was achieved through the cyclic imine **53**. The thiazole ring was constructed by conventional reaction with cysteine and further oxidation of the resulting thiazoline. Three different acid-derived nitrogen-containing functionalities were prepared, and after deprotection of the hydroxyl groups the corresponding aza-*C*-nucleosides were obtained (Scheme 12).

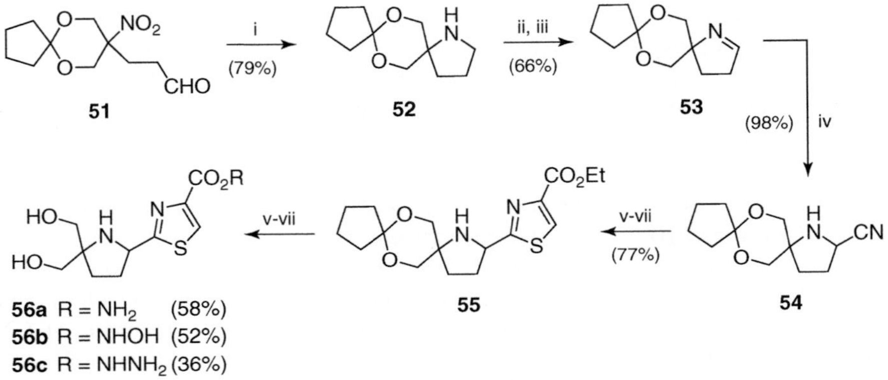

Scheme 12. *Reagents and conditions:* (i) Al-HgCl$_2$, THF, H$_2$O, rt, 3 h; (ii) NCS, THF, rt, 2 h; (iii) DBU, CH$_2$Cl$_2$, rt, 3 h; (iv) 2 M HCN in DIPE, rt, 24 h; (v) TFAA, Py, DMAP, 0°C, 2 h; (vi) L-cysteine, Et$_3$N, MeOH, rt, 1.5 h; (vii) DBU, CBrCl$_3$, CH$_2$Cl$_2$, 0°C, 24 h; (viii) for **47a**: aq. NH$_3$, EtOH, rt, 72 h; for **47b**: H$_2$NOH·HCl, EtONa, EtOH, rt, 24 h; for **47c**: N$_2$H$_4$·H$_2$O, EtOH, rt, 24 h; (ix) HCl(c), MeOH, H$_2$O, rt, 1–3 days.

Different aza-C-nucleosides were prepared from desymmetrized pyrrolidine-2,5-dicarboxamides. Desymmetrization was carried out by amidase-catalyzed hydrolysis that provided enantiomerically pure **57** and **58**. Starting from **57**, aza-nucleoside **60** was synthesized through construction of the tetrazole moiety from the 2-cyanopyrrolidine **59**. In the same paper, transformation of compound **58** into **62** and its enantiomer ent-**62** was described (Scheme 13).[22] While for synthesis of **62** the nucleobase was constructed de novo, in the preparation of ent-**62** incorporation of the base moiety was achieved by treatment with 1,2,3,4-tetrahydroisoquinoline (THIQ) in the presence of chlorodimethoxytriazine (CDMT) and *N*-methylmorpholine (NMM).

Six-membered aza-*C*-nucleosides (i.e., analogues with a piperidine ring instead a pyrrolidine ring) have been prepared by Pedersen and co-workers in a quite straightforward way.[23] The reaction of piperidines **64** with the bromo derivative of the nucleobase in pyridine at reflux furnished the corresponding aza-C-nucleosides **65** in good yield (Scheme 14).

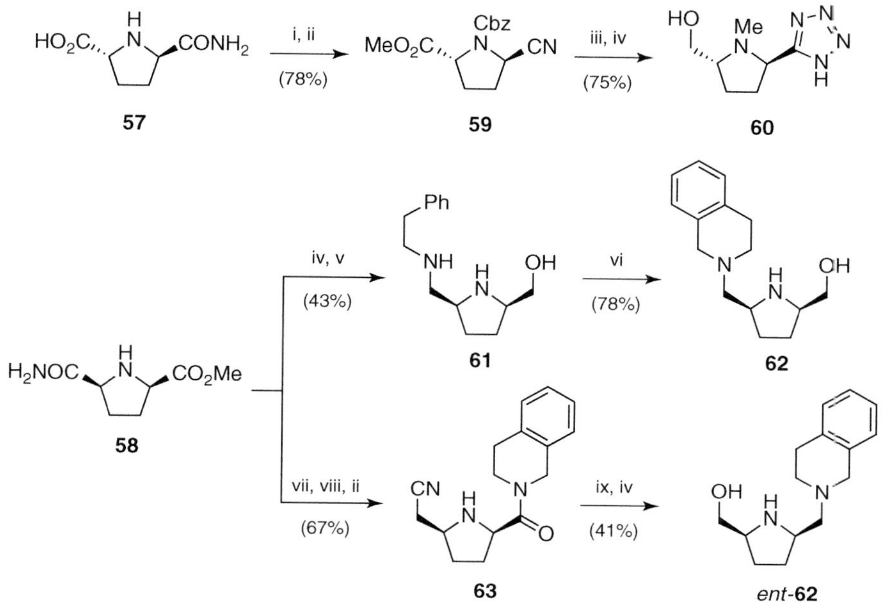

Scheme 13. *Reagents and conditions:* (i) CbzCl, then SOCl$_2$, MeOH; (ii) SOCl$_2$, DMF, 0°C; (iii) NaN$_3$, ZnBr$_2$, *i*-PrOH, H$_2$O, reflux; (iv) LiAlH$_4$, THF, reflux; (v) PhCH$_2$CHO, NaBH$_3$CN; (vi) CH$_2$O, HCl, CHCl$_3$, reflux; (vii) LiOH, THF, H$_2$O; (viii) THQ, CDMT, NMM, CH$_2$Cl$_2$; (ix) HCl(g), MeOH, CH$_2$Cl$_2$, −20°C.

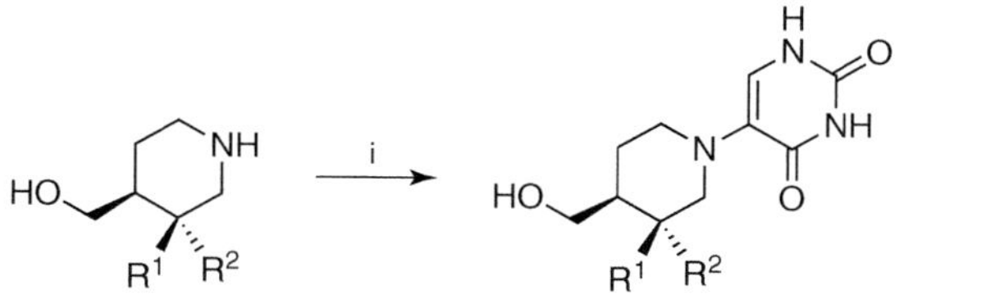

64a R^1 = OH R^2 = H
64b R^1 = H   R^2 = OH

65a R^1 = OH R^2 = H   (66%)
65b R^1 = H   R^2 = OH (54%)

Scheme 14. *Reagents and conditions:* (i) 5-bromouracil, Py, reflux, 24 h.

Aza-*C*-nucleosides related to pseudouridine have been prepared starting from nitrone **66**, derived from 3-octil-5-formyl uracil.[24] A series of isoxazolidines were prepared as key intermediates through cycloaddition reactions with allylic alcohol and methyl acrylate. Transformation of the isoxazolidine ring into the pyrrolidine ring was achieved by a known protocol[25] consisting of reduction of the N,O-bond and spontaneous intramolecular cyclization. As an example, the synthesis of aza-*C*-nucleoside **59** is illustrated in Scheme 15.

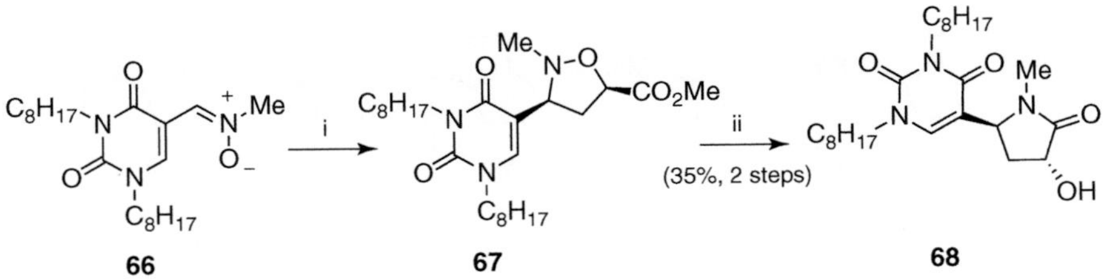

Scheme 15. *Reagents and conditions:* (i) methyl acrylate, toluene, 80°C, 48 h; (ii) H$_2$, Ni Raney, MeOH, rt, 24 h.

3,4-Dihydroxypyrrolidine derivatives **69** and **71**, prepared from D-mannose and D-ribose, respectively, served as starting materials for the synthesis of aza-*C*-nucleosides **70** and **72**, which showed inhibitory properties against several fucosidases (Scheme 16).[26] The methodology allowed the preparation of epimeric structures as well.

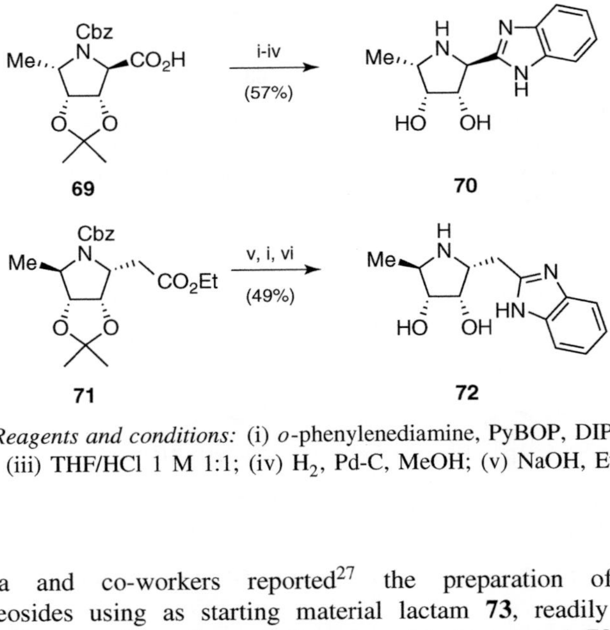

Scheme 16. *Reagents and conditions:* (i) *o*-phenylenediamine, PyBOP, DIPEA, DMF; (ii) AcOH, 50°C; (iii) THF/HCl 1 M 1:1; (iv) H$_2$, Pd-C, MeOH; (v) NaOH, EtOH; (vi) HCl, then NH$_4$OH.

Yokoyama and co-workers reported[27] the preparation of aza-*C*-2,3-dideoxynucleosides using as starting material lactam **73**, readily available in enantiomerically pure form from L-glutamic acid. Compound **73** was treated with hetaryl organometallic reagents (Grignard and organolithium) to give the open-chain derivatives **74** that were cyclized again under acidic conditions. Further reduction with sodium cyanoborohydride afforded mixtures of *cis*- and *trans*-adducts **76**, which upon additional reduction of the ester moiety with lithium aluminum hydride furnished the target compounds **77** (Scheme 17).

Direct incorporation of heterocycles to the pyrrolidine ring has been described by Li and co-workers, who reported a microwave-assisted nucleophilic addition

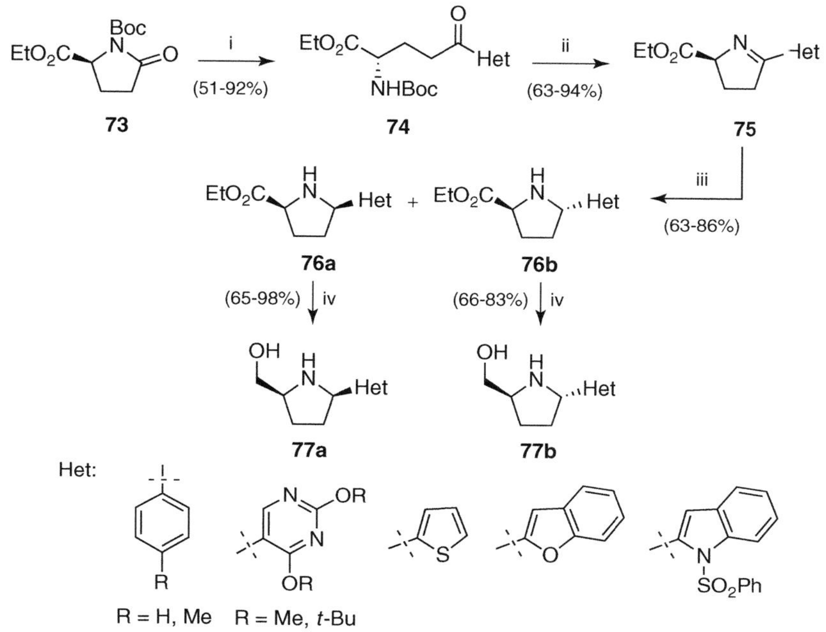

Scheme 17. *Reagents and conditions:* (i) hetaryl-metal, THF, −78 to −40°C, 1 h; (ii) TFA, CH_2Cl_2, rt, 2 h; (iii) $NaBH_3CN$, HCl, i-PrOH, rt, 2 h; (iv) $LiAlH_4$, Et_2O, 0°C, 3 h.

to sugar-derived five-membered cyclic nitrones.[28] The reaction took place with moderate chemical yields and low diastereoselectivity. By using this methodology a variety of aza-C-nucleosides analogues bearing pyrrol and indol units as nucleobases were prepared. On the other hand, good results were obtained by Yu and co-workers in the Friedel–Crafts type of reaction of several heterocycles with cyclic nitrones catalyzed by a Brönsted acid (Scheme 18).[29]

With the discovery of immucillins as transition-state analogue inhibitors for purine nucleoside phosphorylase and N-riboside hydrolases, a lot of synthetic activity on nucleoside chemistry was focused toward their synthesis and that of a large number of analogues[30] (Figure 2).

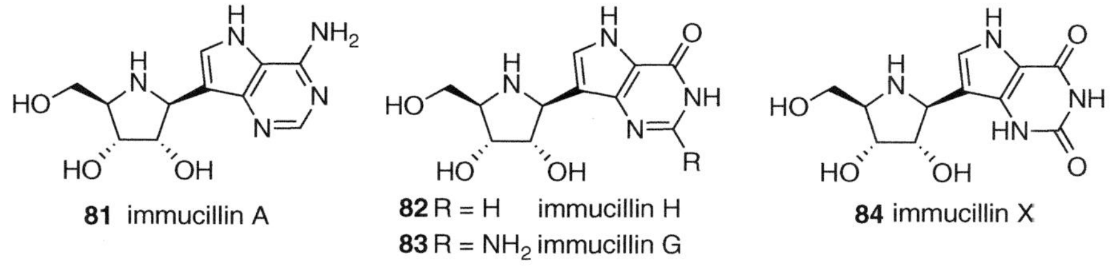

Figure 2. Immucillins.

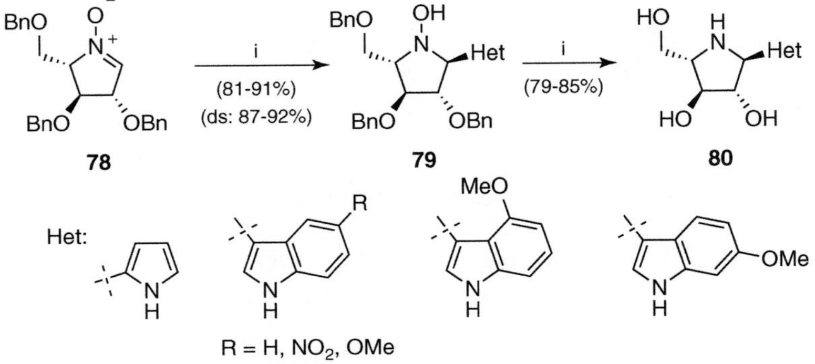

Scheme 18. *Reagents and conditions:* (i) MeCOCl, MeOH, 0°C to rt; (ii) H$_2$, Pd-C.

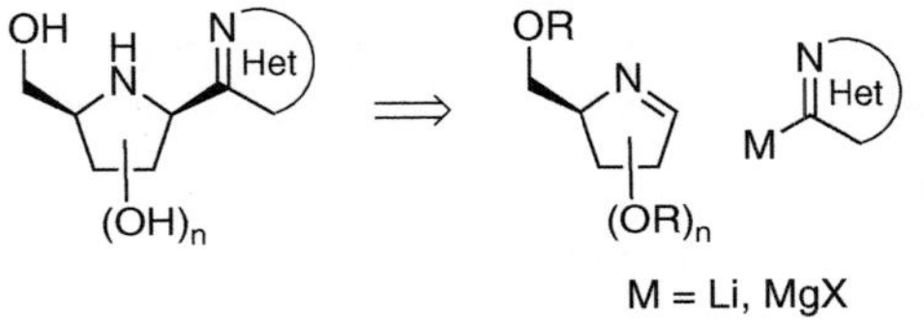

Scheme 19. Synthetic approaches to immucillins and related compounds.

Since immucillins and related compounds are aza-*C*-nucleosides, a typical approach is, indeed, direct addition of the heterocyclic base to preformed pyrrolidine bearing an electrophilic functionality, typically an imine (Scheme 19). Alternatively, and depending on the heterocycle, it is also possible to use the same substrate to incorporate an additional functionality (such as an ester, a cyanide, etc.) that allows construction of the desired nucleobase. The chemistry of immucillins has been reviewed recently[30]; consequently, in this chapter we refer only to very recent approaches (since 2008) not collected in previous reviews.

Kamath and co-workers reported the synthesis of an isomer of immucillin H **82** (also called forodesine) through the addition of a lithiated heterocycle (prepared by direct lithiation using *n*-BuLi) to the protected cyclic imine **85**.[31] The reaction took place with lower yield (22%) and after deprotection of functional groups the hydrochloride of compound **88** was obtained in good yield (Scheme 20). The synthesis of another analogue of immucillin H incorporating an additonal carboxylic acid has also been reported. The reaction of cyclic imine **85** with **89** (obtained by bromo-lithium exchange) afforded **90**, which was readily transformed into the target compound **91** by reaction of the corresponding lithiated derivative with carbon dioxide and deprotection of hydroxyl groups. The same authors prepared compound **72** by an alternative route consisting of the addition of **86** to a lactam prepared in five steps from L-pyroglutamic acid.[32]

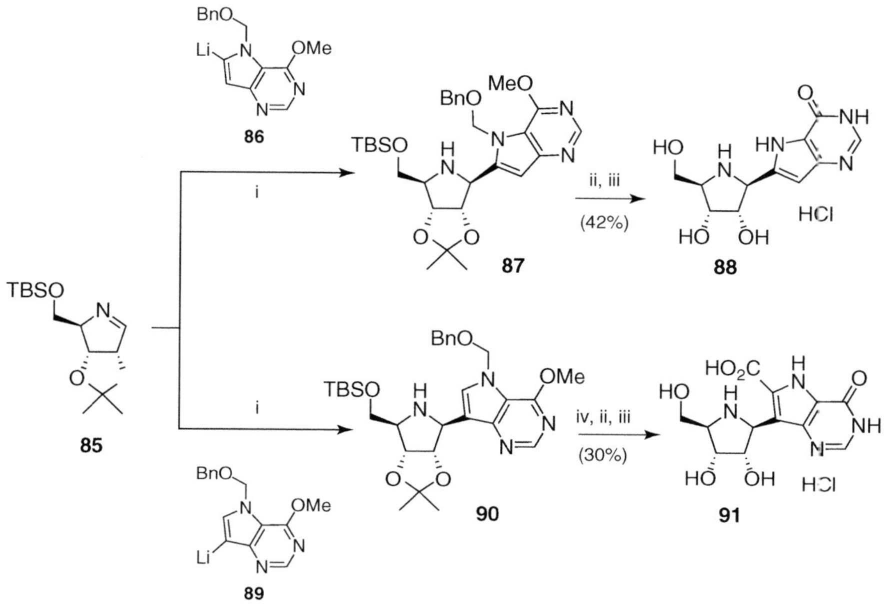

Scheme 20. *Reagents and conditions:* (i) THF, −35 to 0°C; (ii) H$_2$, Pd(OH)$_2$-C, NH$_3$, MeOH; (iii) HCl, MeOH; (iv) *n*-BuLi, −30°C, then CO$_2$.

A second-generation immucillin derivative,[33] F-DADMe-ImmH **88**, has been synthesized in seven steps starting with a 1,3-dipolar cycloaddition reaction between alkene **92** and azomethine ylide **94** generated in situ from compound **93**. After deacetylation of the resulting pyrrolidine, racemic **95** was obtained. This compound was resolved enzymatically with *Candida antarctica* (CAL-B) to give (3*R*,4*S*)-**95** in 90% ee (HPLC). Reduction of the ester moiety and exchange of the *N*-benzylk group by *N*-Boc furnished **96**. Introduction of the nucleobase was carried out though a three-component Mannich reaction, which afforded the target compound **97** (Scheme 21).[34]

Insertion of a methylene between the nitrogen atom of the pyrrolidine unit and the nucleobase has also been described recently by Gotor and co-workers.[35] These authors reported introduction of the nucleobase through a reductive amination of formyl 9-deazapurine **98** and polyhydroxylated pyrrolidine **99**. Deprotection of the resulting adduct **100** provided analogue **101** in good chemical yield (Scheme 22). In the same paper, synthesis of fluorinated analogues following the same strategy was also reported.

Acyclic analogues are considered third-generation immucillins, and they have been prepared pursuing more active compounds. Schramm and co-workers prepared acyclic analogue **105** from L-serinol **102** in six steps and 15.5% overall yield.[36] The key and last step consisted of a three-component Mannich reaction, as in the case of compound **97**. In this case, however, the chemical yield of that step was somewhat lower (Scheme 23).

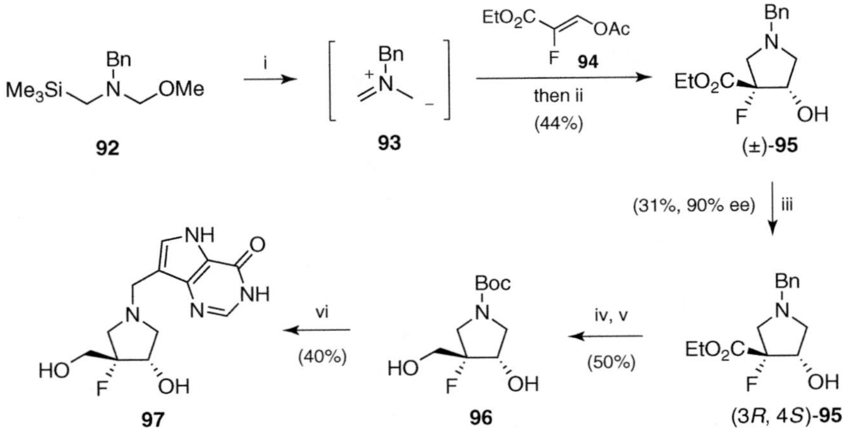

92 93 (±)-95 97 96 (3R, 4S)-95

Scheme 21. *Reagents and conditions:* (i) TFA, CH_2Cl_2, 0°C; (ii) NaOEt, EtOH, rt; (iii) CAL-B, vinyl acetate, *t*-BuOMe, 50°C; (iv) $LiBH_4$, MeOH, Et_2O, 0°C, then HCl, 50°C; (v) H_2, Pd-C, $(Boc)_2O$, MeOH; (vi) HCl, then 9-deazahypoxanthine, CH_2O, NaOAc, water–dioxane, 100°C.

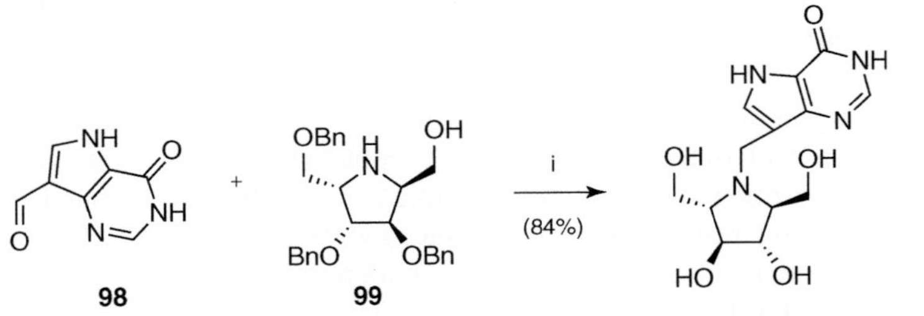

98 99 100 R = Bn 101 R = H

Scheme 22. *Reagents and conditions:* (i) THF, MeOH, 40°C, $NaBH_3CN$; (ii) HCl (conc.), reflux.

A great variety of acyclic analogues have been prepared by following the strategy depicted in Scheme 19: that, is a Mannich reaction between a free amine-bearing hydroxymethyl group, formaldehyde, and the nucleobase, typically 9-deazahypoxanthine.[37] In the same paper, the authors prepared acyclic aza-*C*-nucleoside **107** from the parent immucillin-H **82** through diol cleavage with sodium metaperiodate (Scheme 24). The procedure was quite straightforward and took place in good yield (69% for two steps).

The synthesis of Schramm's aza-*C*-nucleoside **112**, closely related to immucillins and described as a potent inhibitor of trypanosomal nucleoside hydrolase,[38] was achieved in six steps by Carretero and co-workers.[39] The key step of the

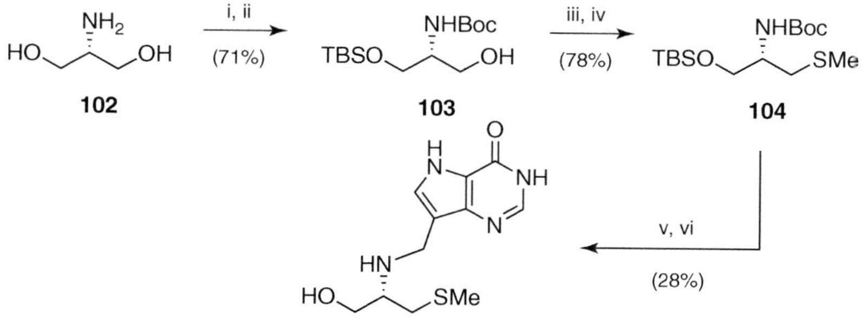

Scheme 23. *Reagents and conditions:* (i) Boc$_2$O, MeOH; (ii) NaH (1 equiv), TBSCl, THF; (iii) MsCl, Et$_3$N, CH$_2$Cl$_2$; (iv) NaSMe, DMF; (v) HCl, MeOH, H$_2$O; (vi) 9-deazahypoxanthine, CH$_2$O, NaOAc, H$_2$O, 80°C.

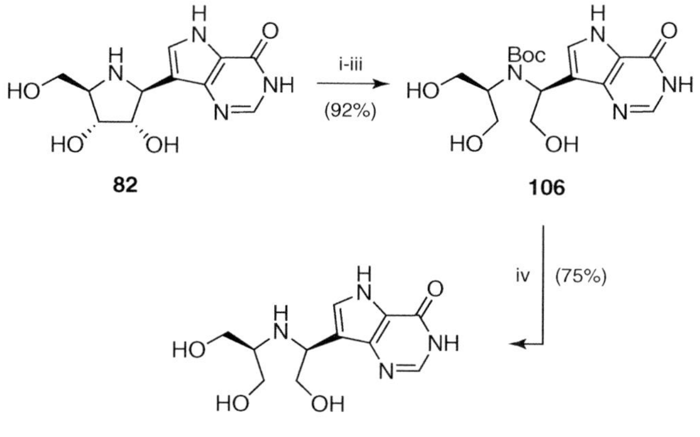

Scheme 24. *Reagents and conditions:* (i) Boc$_2$O, Et$_3$N, MeOH, H$_2$O, rt; (ii) NaIO$_4$, rt; (iii) NaBH$_4$, MeOH, rt; (iv) HCl (conc.), MeOH, rt.

synthesis was an enantioselective formal [3 + 2]-cycloaddition (actually, a stepwise process) catalyzed by a Fesulphos–copper complex. Further synthetic transformations involving elimination of the sulfonyl groups and dihydroxylation of the resulting double bond provided compound **112** after deprotection (Scheme 25).

Radicamines and related compounds, such as codonopsine and codonopsinine (Figure 3), are also considered close to aza-*C*-nucleosides because of their biological activity as inhibitors of glucosidases. A lot of synthetic activity has been reported during recent years and most of it was reviewed in 2008.[30] Very recently,

113 R¹ = OH R² = Me Radicamine A
114 R¹ = H   R² = H   Radicamine B
115 R¹ = H   Codonopsinine
116 R¹ = OMe   Codonopsinine

Figure 3. Radicamines and related compounds.

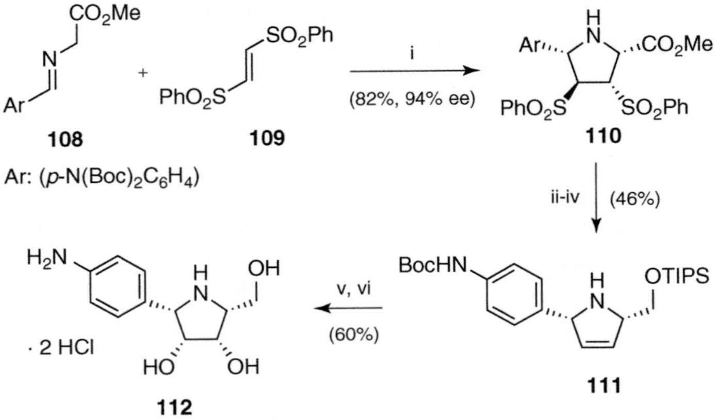

Ar: (*p*-N(Boc)$_2$C$_6$H$_4$)

Scheme 25. *Reagents and conditions:* (i) Cu(MeCN)$_4$PF$_6$ (3 mol%), (*R*)-Fesulphos (3 mol%), Et$_3$N (20 mol %), CH$_2$Cl$_2$, 48 h, −78°C; (ii) LiAlH$_4$, THF, 0°C; (iii) TIPSOTf, 2,6-lutidine, CH$_2$Cl$_2$, 0°C; (iv) Na(Hg), Na$_2$HPO$_4$, MeOH/THF; (v) OsO$_4$, TMEDA, CH$_2$Cl$_2$, −78°C; (vi) HCl/MeOH, rt.

novel approaches for some of those compounds have been communicated. Many of them make use of nitrone chemistry, taking advantage of the special characteristic of this sort of compounds (cyclic nitrones), which in general are most stable and give rise to more stereoselective reactions than imines.

Radicamine B **114** was prepared by Merino and co-workers[40] starting from cyclic nitrone *ent*-**78**, which is the correct enantiomer required for synthesizing the naturally occurring derivative. Addition of the corresponding Grignard reagent took place with very good yield and complete diastereroselectivy to give adduct **117**. Deprotection and purification afforded radicamine B **114** in 76% overall yield from the starting nitrone (Scheme 26).

Falomir and co-workers reported the synthesis of radicamine B **114** using Garner's aldehyde **118** as the starting chiral nonracemic material from which key intermediate **119** was prepared. After addition of the required Grignard

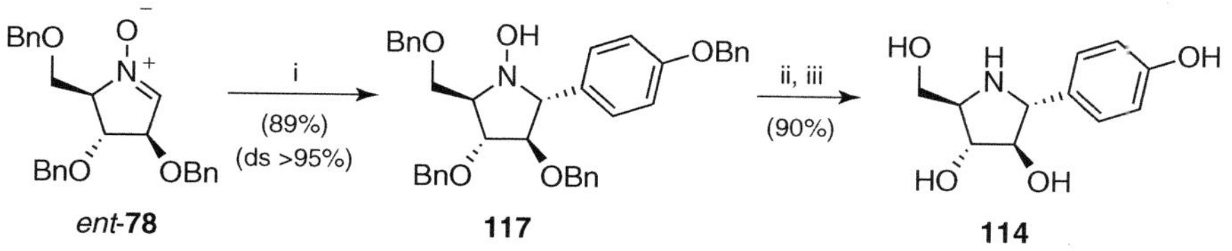

Scheme 26. *Reagents and conditions:* (i) 4-BnOC$_6$H$_4$MgBr, THF, 0°C; (ii) H$_2$, 5 atm, 5% Pd(OH)$_2$-C, HCl-MeOH; (iii) Dowex 5W×8-200, 1 N NH$_4$OH.

reagent to the Weinreb amide **120**, radicamine B **114** was obtained in three steps through stereoselective reduction of the resulting ketone and deprotection of hydroxyl and amino groups (Scheme 27).[41a]

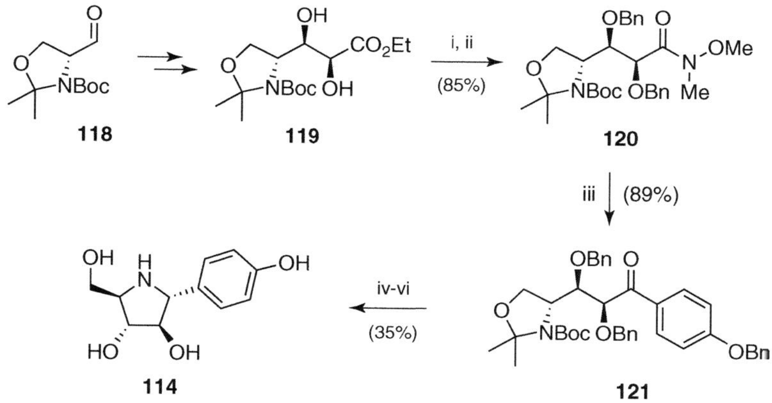

Scheme 27. *Reagents and conditions:* (i) AgO, BnBr, Et$_2$O, 24, rt; (ii) MeNHOMe·HCl, *i*-PrMgCl, THF; (iii) 4-BnOC$_6$H$_4$MgBr, THF, 0°C; (iv) L-Selectride, THF, −80°C; (v) MsCl, Et$_3$N, CH$_2$Cl$_2$, 0°C; (vi) TFA, CH$_2$Cl$_2$, 0°C.

Chandrasekhar and co-workers also employed Garner's aldehyde **118** as starting material for their synthesis of radicamine B, which started from intermediate **122** and employed Stille coupling and one-pot domino epoxidation–pyrrolidine formation as key steps (Scheme 28).[41b]

The Sharpless dihydroxylation was used in the de novo synthesis of radicamine B reported by Jagadeesh and Rao.[42] Hydroxylation of **125**, easily prepared from *p*-hydroxybenzaldehyde in two steps and 80% yield, afforded enantiomerically pure **126**, which was transformed into the advanced intermediate **127**. This compound was subjected to a Sharpless hydroxylation again, and after introduction of the azido group, **129** was obtained. Conversion into amino group, protection as Z-carbamate, and reduction afforded **130**, which, after acetylation, cyclized intramolecularly to yield the immediate precursor **131**. Deprotection of amino and hydroxyl groups

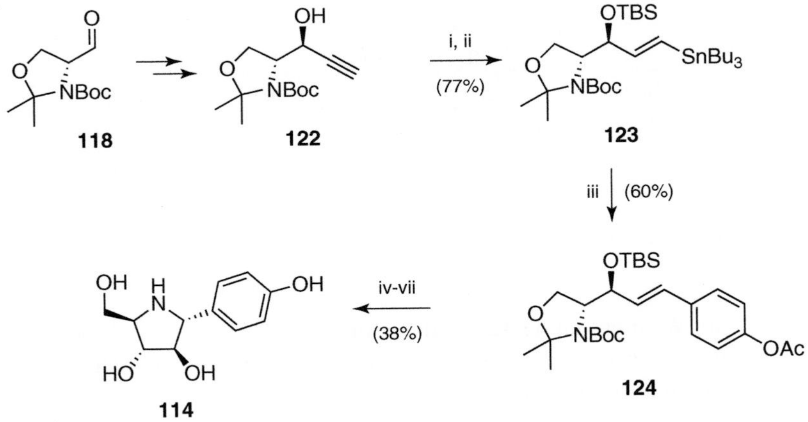

Scheme 28. *Reagents and conditions:* (i) Bu$_3$SnH, AIBN, benzene, reflux, 30 h; (ii) TBSCl, imidazole, CH$_2$Cl$_2$, rt, 5 h; (iii) 4-acetoxibromobenzene, Pd(Ph$_3$P)$_4$, toluene, reflux, 6 h; (iv) TFA, CH$_2$Cl$_2$, 4 h; (v) *m*-CPBA, CH$_2$Cl$_2$; (vi) NaHCO$_3$, MeOH, 4 h; (vii) TFA, rt, 12 h.

finally afforded radicamine B **114** in 18 steps and 10.7% overall yield from **125** (Scheme 29).

A synthesis of *ent*-**114** has been reported employing Sharpless epoxidation as a key step of the approach to obtain the required starting material **133**. After eight steps, intermediate **134** was obtained in 28% yield. Cyclization of compound **134** induced under acidic conditions provided the enantiomer of natural radicamine B (Scheme 30).[43]

The asymmetric synthesis of (−)-codonopsinine **115** was carried starting from commercially available *tert*-butylcrotonate **135**.[44] The key steps of the synthesis were the diastereoselective addition of enantiopure lithium amide **136** to **135**, followed by oxidation in situ of the resulting enolate to give **137**, and a ring-closing iodoamination of homoallylic amine **138**, which took place with concomitant debenzylation to afford **139**. Compound **139** was further elaborated to (−)-codonopsinine **115** by conventional transformations (Scheme 31). Thus, compound **115** was prepared from **135** in seven steps and 4.6% overall yield.

The unnatural 4-*epi*-(+)-codonopsinine **143** was prepared in 11 steps and 4.3% overall yield starting from D-ribose as a chiral building block.[45] Transformation of the sugar into **140** and further addition of *p*-methoxybenzylamine followed by addition of the required Grignard reagent afforded **141** with almost complete selectivity. Cyclization of this compound through mesylation of the free hydroxyl group provided **142**, which was transformed into the target compound **143** by sequential *N*-deprotection, *N*-methylation, and acetonide hydrolysis (Scheme 32).

Starting from 1,5-D-gluconolactone **144**, (−)-codonopsinol **148** and its C2 epimer **149** were prepared through a procedure that involved an acid-mediated amino cyclization as a key step.[46] Compound **145** was obtained from **14** in four steps. Introduction of the amino group in the form of an azide and further

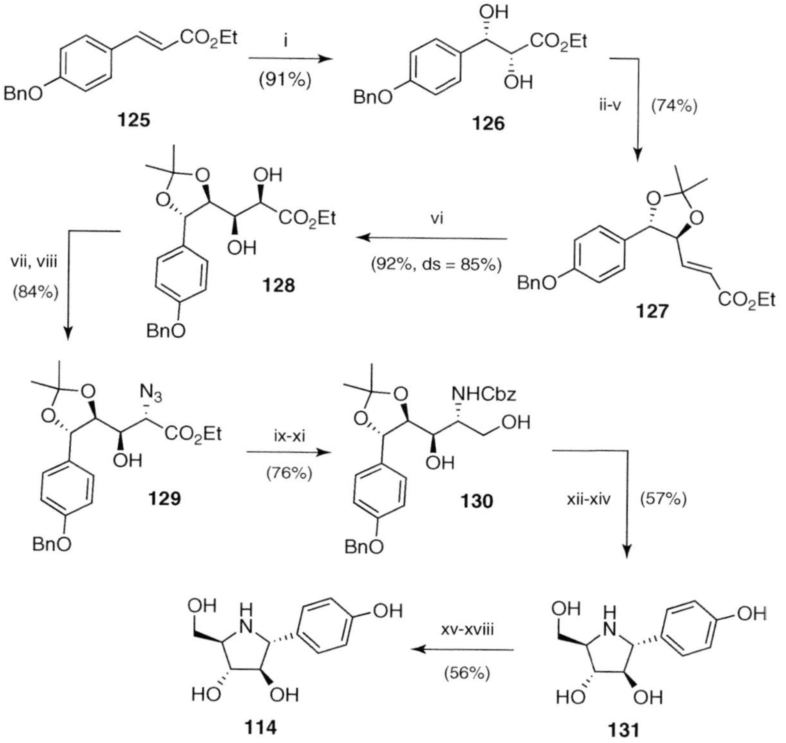

Scheme 29. *Reagents and conditions:* (i) AD-mix-α, CH$_3$SO$_2$NH$_2$, *t*-BuOH, H$_2$O, 24 h; (ii) 2,2-DMP, *p*-TSA, CH$_2$Cl$_2$, 12 h; (iii) LiAlH4, THF, 0°C to rt, 3 h; (iv) (COCl)$_2$, DMSO, CH$_2$Cl$_2$, −78°C; (v) Ph$_3$PCHCO$_2$Et, toluene, reflux, 6 h; (vi) (DHQ)$_2$PHAL, OsO$_4$, K$_2$CO$_3$, K$_3$Fe(CN)$_6$, CH$_3$SO$_2$NH$_2$, *t*-BuOH, H$_2$O; (vii) SOCl$_2$, Et$_3$N, CH$_2$Cl$_2$, 0°C to rt, 30 min; (viii) NaN$_3$, DMF, 80°C, 2 h; (ix) TPP, ethanol, 0°C to rt, 6 h; (x) CbzCl, Na$_2$CO$_3$, ethanol, 0°C to rt, 8 h; (xi) LiCl, NaBH$_4$, ethanol, THF, 0°C to rt, 3 h, 78%; (xii) 80% aq. AcOH, rt, 8 h; (xiii) Ac$_2$O, Et$_3$N, CH$_2$Cl$_2$, DMAP, 0°C to rt, 4 h; (xiv) TFA/CH$_2$Cl$_2$ 1:3, 0°C to rt, 4 h; (xv) K$_2$CO$_3$, MeOH, 0°C to rt, 1 h; (xvi) PdCl$_2$, H$_2$ MeOH, 12 h; (xvii) 6 N HCl soln, ethanol, reflux, 3 h; (xviii) Dowex 5W×8-200, 30% ammonia solution.

transformation into an *N*-Cbz group afforded **146**, which was deprotected selectively and subjected to a diol cleavage mediate by sodium periodate in situ; the resulting aldehyde was treated with the corresponding Grignard reagent to afford a mixture of diastereomers that were cyclized to compounds **147**. Both isomers were transformed, after chromatographic separation into epimers **148** and **149** (Scheme 33).

Experimental Procedures

(±)-5-[cis-3-Hydroxy-4-(hydroxymethyl)piperidin-1-yl]uracil [(±)-56a]. A mixture of **5a** (1.37 g, 10.42 mmol) and 5-bromouracil (0.66 g, 3.47 mmol) in anhydrous pyridine (10 mL) was refluxed for 24 h, cooled to room temperature and

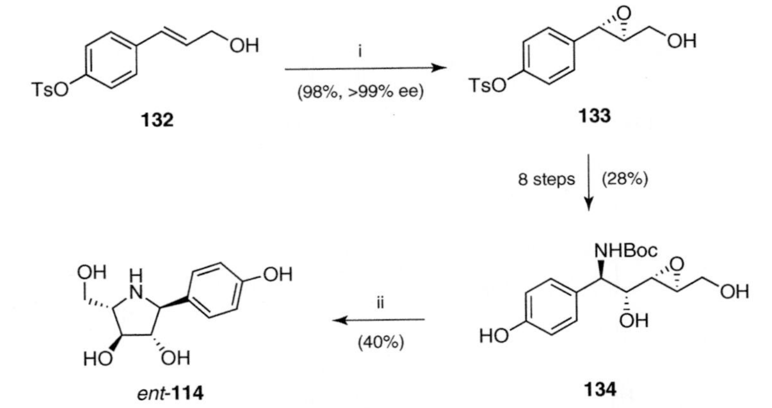

Scheme 30. *Reagents and conditions:* (i) (+)-DET, Ti(*i*-PrO)$_4$, TBHP, 4-Å molecular sieves, CH$_2$Cl$_2$, −20°C, 3 h; (ii) TFA, CH$_2$Cl$_2$, 0°C, 30 min, then solid NaHCO$_3$, 30 h.

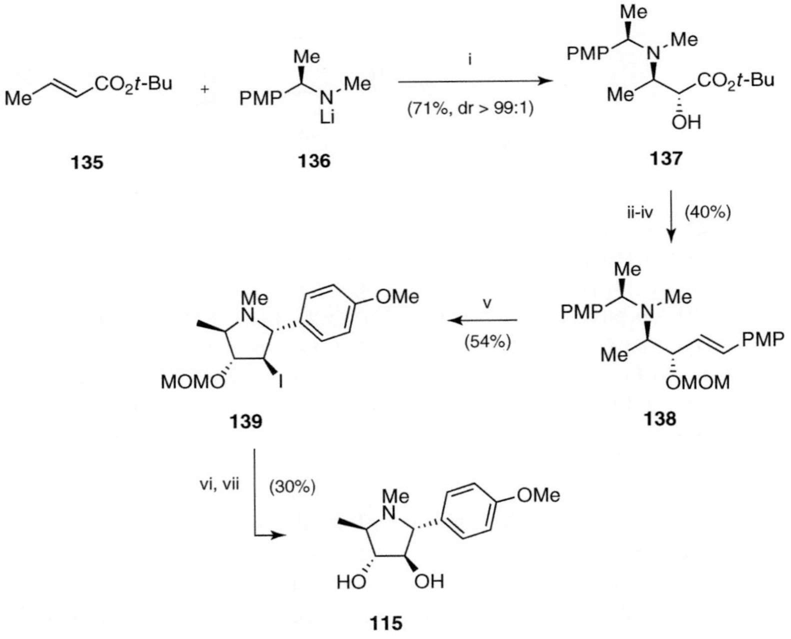

Scheme 31. *Reagents and conditions:* (i) (i) THF, −78°C, 2 h, then (−)-CSO, −78°C to rt, 12 h; (ii) NaH, DMF, 0°C, 30 min, then MOMCl, rt, 12 h; (iii) DIBAL-H, CH$_2$Cl$_2$, −78°C, 30 min; (iv) BuLi, [4-MeOC$_6$H$_4$CH$_2$PPh$_3$]$^+$ [Cl]$^-$, THF, −78°C, 30 min, then rt, 12 h; (v) I2; (vi) AgOAc, AcOH, 40°C, 24 h; (vii) HCl, MeOH, 50°C, 48 h.

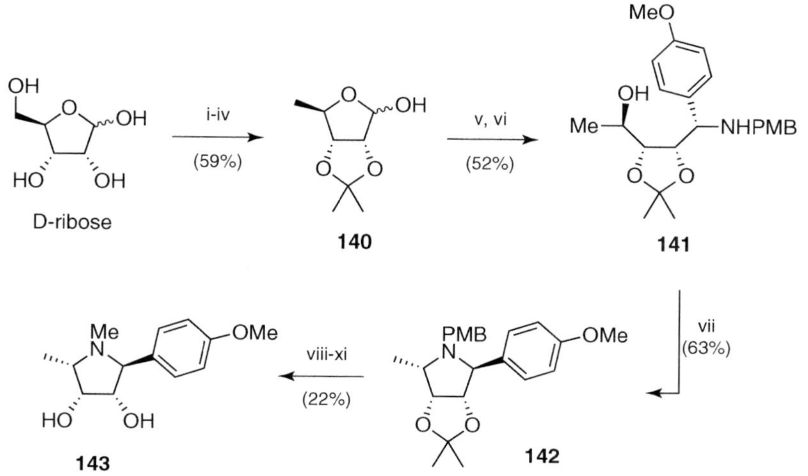

Scheme 32. *Reagents and conditions:* (i) acetone, H_2SO_4; (ii) TsCl, Py; (iii) NaI, dioxane, DMF, 80°C; (iv) H_2, Pd-C, Et_3N, EtOH; (v) $PMBNH_2$, 4Å mol. sieve, CH_2Cl_2; (vi) 4-$MeOC_6H_4MgCl$, THF; (vii) MsCl, Py; (viii) H_2, Pd-C; (ix) HCHO, HCO_2H, 80°C; (x) 1 M HCl; (xi) Amberlyst A-26 (OH^-).

evaporated in vacuo. Traces of pyridine were coevaporated with toluene. The residue was dissolved in MeOH (50 mL) under heating, and the hot solution was filtered. After cooling to 0°C the precipitate was isolated by filtration and purified by recrystallization from MeOH (40 mL) to give the title compound **56a** as a white powder; yield 556 mg (66%); mp 260–264°C.

(2S,3R,4R,5S)-3,4-Bis(benzyloxy)-5-(benzyloxymethyl)-2-(hydroxymethyl)-N-(7-(benzyloxymethyl)-6-methoxy-9-deazapurin-9-yl)methylpyrrolidine **(84)**. Imino sugar **83** (90 mg, 0.21 mmol), sodium cyanoborohydride (0.36 mmol, 22 mg), and compound **82** (56 mg, 0.19 mmol) were dissolved in a 4:1 mixture of MeOH and THF (4 mL). Three drops of acetic acid were added, and the solution was stirred at 40°C overnight. Solvents were evaporated in vacuo, and the residue was purified by flash chromatography (CH_2Cl_2/MeOH, 97:3) to give the coupled product **84** (113 mg, 84%). R_f (5% MeOH/CH_2Cl_2) = 0.44. $[\alpha]_D^{20}$ −30° (c 0.50, CH_2Cl_2).

(−)-Radicamine **(114)**. A solution of hydroxylamine **117** (0.181 g, 0.3 mmol) in methanol (3 mL) was treated with Pd(OH)$_2$-C (5 mg) and a solution of HCl in methanol (1 M). The resulting mixture was stirred under a hydrogen atmosphere (20 atm) for 6 h. The catalyst was eliminated by filtration through a pad of Celite, the filtrate was treated with HCl (3 M) in methanol, and the resulting solution was stirred at room temperature for an additional 10 min. The solvent was eliminated under reduced pressure to afford pure **114**·HCl (79 mg, 100%) as a white solid; mp > 150°C. $[\alpha]_D^{20}$ +81° (c 0.25, H_2O). An analytical sample of the free amine was

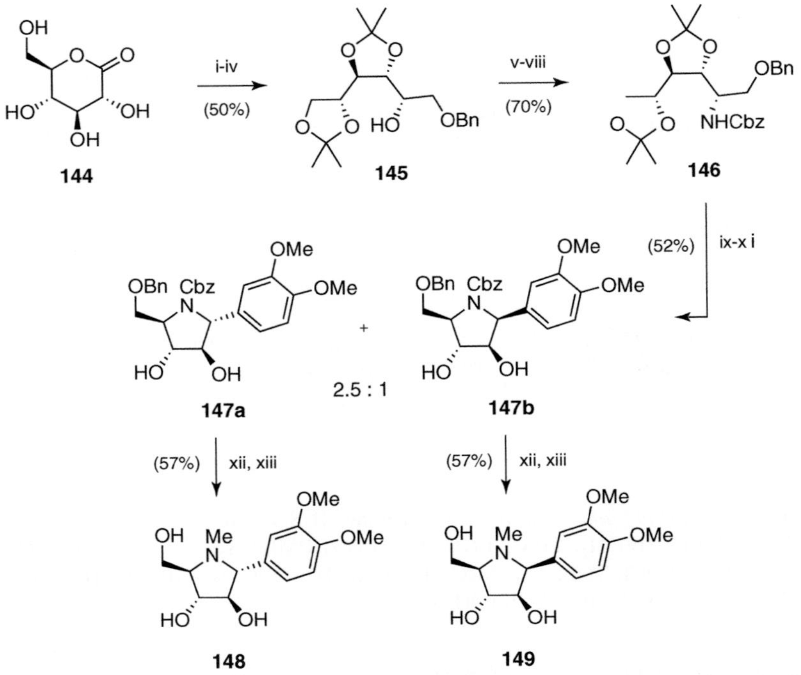

Scheme 33. *Reagents and conditions:* (i) 2,2-DMP, PTSA, acetone, MeOH, 0°C to rt, 50 h; (ii) LiAlH$_4$, THF, 0°C to rt, 4 h; (iii) Bu$_2$SnO, toluene, reflux, 8 h; (iv) BnBr, TBAI, reflux, 16 h; (v) MsCl, Et$_3$N, CH$_2$Cl$_2$, 0°C to rt, 3 h; (vi) NaN$_3$, DMF, 80°C, 24 h; (vii) LiAlH$_4$, THF, 0°C to rt, 5 h; (viii) CbzCl, Na$_2$CO$_3$, CH$_2$Cl$_2$, 0°C to rt, 8 h; (ix) H$_5$IO$_6$, Et$_2$O, 0°C to rt; (x) 3,4-dimethoxyphenyl magnesium bromide, THF, 0°C to rt, 16 h; (xi) TFA,CH$_2$Cl$_2$, 0°C to rt, 4 h; (xii) LiAlH$_4$, THF, 0–60°C, 5 h; (xiii) PdCl$_2$, H$_2$, MeOH, 12 h.

obtained by passing the hydrochloride through a Dowex 50W × 8 ion-exchange resin. Elution with ammonia in methanol (3 M) afforded the free base of **114** (61 mg, 90%) as a syrup after evaporation: $[\alpha]_D^{20}$ +74(c 0.10, H$_2$O).

Acknowledgments

The authors wish to express their gratitude to the Ministry of Science and Innovation (MICINN, Madrid, Spain, project CTQ2010-19606), FEDER Program (UE), and Government of Aragon (Zaragoza, Aragon, Spain, Research Group E-10).

ABBREVIATIONS

AIBN	azo-isobisbutyronitrile
Bn	benzyl
BSA	*N,O*-bis(trimethylsilyl)acetamide

Bz	benzoyl
CAN	ceric ammonium nitrate
CDMT	chlorodimethoxytriazine
mCPBA	*meta*-chloroperbenzoic acid
CSO	(−)-camphorsulfonyl oxaziridine
DAIB	(diacetoxy)iodobenzene
DBU	1,8-diazabicycloundec-7-ene
DEAD	diethylazodicarboxylate
DIBAH	diisobutyl aluminum hydride
DIPEA	diisopropylethylamine
DMAP	4-(N,N-dimethylamino)pyridine
DMF	N,N-dimethylformamide
HMPT	hexamethylphosphorous triamide
MOM	methoxymethyl
Ms	mesyl
NCS	N-chlorosuccinimide
NMM	N-methylmorpholine
PMB	*p*-methoxybenzyl
Py	pyridine
PyBOP	benzotriazol-1-yl-oxytripyrrolidinophosphonium hexafluorophosphate
TBAF	tetrabutylammonium fluoride
TBDPS	*tert*-butyldiphenylsilyl
TBHP	*tert*-butylhydroperoxide
TBS	*tert*-butyldimethylsilyl
TFA	trifluoroacetic acid
TFAA	trifluoroacetic anhydride
THIQ	1,2,3,4-tetrahydroisoquinoline
TIPS	triisopropylsilyl
TMEDA	tetramethylenediamine
TMS	trimethylsilyl

REFERENCES

1. For recent reviews on this topic, see (a) Romeo, G.; Chiacchio, U.; Corsaro, A.; Merino, P. *Chem. Rev.* **2010**, *110*, 3337–3370. (b) Merino, P. *Curr. Med. Chem.* **2006**, *13*, 539–545. (c) Merino, P. *Curr. Med. Chem.* **2002**, *1*, 389–411.
2. Yokoyama, M.; Momotake, A. *Synthesis* **1999**, 1541–1554.
3. Schramm, V. L. *Acc. Chem. Res.* **2003**, *36*, 588–596.
4. For a review, see Schramm, V. L. *Biochim. Biophys. Acta* **2002**, *1587*, 107–117.
5. (a) Schramm, V. L.; Tyler, P. C. *Curr. Top. Med. Chem.* **2003**, *3*, 525–540. (b) Ringia, E. A. T.; Schramm, V. L. *Curr. Top. Med. Chem.* **2005**, *5*, 1237–1258. (c) Schramm, V. L. *Nucleosides. Nucleotides.* **2004**, *23*, 1305–1311.
6. Costenaro, E. R.; Fontoura, L. A. M.; Oliveira, D. F.; Correia, C. R. D. *Tetrahedron Lett.* **2001**, *42*, 1599–1602.

7. Meng, W. H.; Wu, T.-J.; Zhang, H.-K.; Huang, P.-Q. *Tetrahedron: Asymmetry* **2004**, *15*, 3899–3910.
8. Huang, P.-Q.; Wang, S. L.; Zheng, H.; Fei, X. S. *Tetrahedron Lett.* **1997**, *38*, 271–274.
9. (a) Altmann, K. H.; Freier, S. M.; Pieles, U.; Winkler, T. *Angew. Chem., Int. Ed.* **1994**, *33*, 1654–1656. (b) Lenard, N. J.; Laursen, R. A. *Biochemistry* **1965**, *4*, 354–365. (c) Niedballa, U.; Vorbrüggen, H. *J. Org. Chem.* **1974**, *39*, 3654–3660.
10. Boto, A.; Hernandez, D.; Hernandez, F. *Tetrahedron Lett.* **2008**, *49*, 455–488.
11. Boto, A.; Hernandez, D.; Hernandez, F. *Eur. J. Org. Chem.* **2010**, 3847–3857.
12. Flores, R.; Alibes, R.; Figueredo, M.; Font, M. *Tetrahedron* **2009**, *65*, 6912–6917.
13. Qiu, X. L.; Qing, F.-L. *Synthesis* **2004**, 334–340.
14. Qiu, X. L.; Qing, F.-L.. *Bioorg. Med. Chem.* **2005**, *13*, 277–283.
15. Qiu, X. L.; Qing, F.-L. *J. Org. Chem.* **2005**, *70*, 3826–3837.
16. Mironiuk-Puchalska, E.; Kołaczkowska, E.; Sas, W. *Tetrahedron Lett.* **2002**, *43*, 8351–8354.
17. Wang, D.; Li, Y.-H.; Wang, Y.-P.; Gao, R.-M.; Zhang, L.-H.; Ye, X.-S. *Bioorg. Med. Chem.* **2011**, *19*, 41–51.
18. Rejman, D.; Pohl, R.; Dracinsky, M. *Eur. J. Org. Chem.* **2011**, 2172–2187.
19. Koszytkowska-Stawińska, M.; Kolaczkowska, E.; Adamkiewicza, E.; De Clercq, E. *Tetrahedron Lett.* **2007**, *63*, 10587–10595.
20. Hernández-Reyes, C. X.; Angeles-Beltrán, D.; Lomas-Romero, L.; González-Zamora, E.; Gaviño, R.; Cárdenas, J.; Morales-Serna, J. A.; Negrón-Silva, G. E. *Molecules* **2012**, *17*, 3359–3369.
21. Mironiuk-Puchalska, E.; Koszytkowska-Stawińska, M.; Sas, W.; De Clercq, E.; Naesens, L. *Nucleosides. Nucleotides.* **2012**, *31*, 72–84.
22. Chen, P.; Gao, M.; Wang, D.-X.; Zhao, L.; Wang, M.-X. *J. Org. Chem.* **2012**, *77*, 4063–4072.
23. Sorensen, M. D.; Khalifa, N. M.; Pedersen, E. B. *Synthesis* **1999**, 1937–1943.
24. Coutouli-Argyropoulou, E.; Trakossas, S. *Tetrahedron* **2011**, *67*, 1915–1923.
25. For the conversion of isoxazolidines into 2-pyrrolidinones see (a) Merino, P.; Mates, J. A.; Revuelta, J.; Tejero, T.; Chiacchio, U.; Romeo, G.; Iannazzo, D.; Romeo, R. *Tetrahedron: Asymmetry* **2002**, *13*, 173–190. (b) Merino, P.; Anoro, S.; Franco, S.; Merchan, F. L.; Tejero, T.; Tuñon, V. *J. Org. Chem.* **2000**, *65*, 1590–1596.
26. Moreno-Clavijo, E.; Carmona, A. T.; Vera-Ayoso, Y.; Moreno-Vargas, A. J.; Bello, C.; Vogel, P.; Robina, I. *Org. Biomol. Chem.* **2009**, *7*, 1192–1202.
27. Momotake, A.; Togo, H.; Yokoyama, M. *J. Chem. Soc. Perkin Trans. 1*, **1999**, 1193–1200.
28. Li, X.; Qin, Z.; Wang, R.; Chen, H.; Zhang, P. *Tetrahedron* **2011**, *67*, 1792–1798.
29. Su, J.-K.; Jia, Y.-M.; He, R.; Rui, P.-X.; Han, N.; He, X.; Xiang, J.; Chen, X.; Zhu, J.; Yu, C.-Y. *Synlett* **2010**, 1609–1616.
30. Merino, P.; Tejero, T.; Delso, I. *Curr. Med. Chem.* **2008**, *15*, 954–967.
31. Kamath, V. P.; Xue, J.; Juarez-Brambila, J. J.; Morris, C. B.; Ganorkar, R.; Morris, P. E. *Bioorg. Med. Chem. Lett.* **2009**, *19*, 2624–2626.
32. Kamath, V. P.; Xue, J.; Juarez-Brambila, J. J.; Morris P. E. *Tetrahedron Lett.* **2009**, *50*, 5198–5200.

33. Second-generation immucillins are those analogues in which the nucleobase is linked to the nitrogen atom through a methylene bridge. They were designed to mimic the PNP transition state by increasing the leaving-group distance and moving the base to the 1′-position.
34. Mason, J. M.; Murkin, A. S.; Li, L.; Schramm, V. L.; Gainsford, G. J.; Skeltor, B. W. *J. Med. Chem.* **2008**, *51*, 5880–5884.
35. Martinez-Montero, S.; Fernandez, S.; Sanghvi, Y. S.; Chattopadhyaya, J.; Ganesan, M.; Ramesh, N. G.; Gotor, V.; Ferrero, M. *J. Org. Chem.* **2012**, DOI:10.1021/jo3004452.
36. Murkin, A. S.; Clinch, K.; Mason, J. M.; Tyler, P. C.; Schramm, V. L. *Bioorg. Med. Chem. Lett.* **2008**, *18*, 5900–5903.
37. Clinch, K.; Evans, G. B.; Frohlich, R. F. G.; Furneaux, R. H.; Kelly, P. M.; Legentil, L.; Murkin, A. S.; Li, L.; Schramm, V. L.; Tyler, P. C.; Woolhouse, A. D. *J. Med. Chem.* **2009**, *52*, 1126–1143.
38. Miles, R.; Tyler, P. C.; Evams, G. B.; Furneaux, R. H.; Parkin, D. W.; Schramm, V. L. *Biochemistry* **1999**, *38*, 13147–13154.
39. Lopez-Perez, A.; Adrio, J.; Carretero, J. C. *J.Am. Chem. Soc.* **2008**, *130*, 10084–10085.
40. (a) Merino, P.; Delso, I.; Tejero, T.; Cardona, F.; Goti, A. *Synlett* **2007**, 2651–2654. (b) Merino, P.; Delso, I.; Tejero, T.; Cardona, F.; Marradi, M.; Faggi, E.; Parmeggiani, C.; Goti, A. *Eur. J. Chem.* **2008**, 2929–2947.
41. (a) Ribes, C.; Falomir, E.; Carda, M.; MArco, J. A. *J. Org. Chem.* **2008**, *73*, 7779–7982. (b) Mallesham, P.; Vijakumar, B. V. D.; Shin, D.-S.; Chandrasekhar, S. *Tetrahedron Lett.* **2011**, *52*, 6145–6147.
42. Jagadeesh, Y.; Rao, B. V. *Tetrahedron Lett.* **2011**, *52*, 6366–6369.
43. Shankaraiah, G.; Kumar, R. S. C.; Poornima, B.; Babu, K. S. *Tetrahedron Lett.* **2011**, *52*, 4885–4887.
44. Davies, S. G.; Lee, J. A.; Roberts, P. M.; Thomson, J. E.; West, C. J. *Tetrahedron Lett.* **2011**, *52*, 6477–6480.
45. Kotland, A.; Accadbled, R. K.; Behr, J.-B. *J. Org. Chem.* **2011**, *76*, 4094–4098.
46. Jagadeesh, Y.; Reddy, J. S.; Rao, B. V.; Swarnalatha, J. L. *Tetrahedron* **2010**, *66*, 1202–1207.

16 Stereoselective Methods in the Synthesis of Bioactive Oxathiolane and Dioxolane Nucleosides

D. D'ALONZO and A. GUARAGNA

Dipartimento di Scienze Chimiche, Università degli Studi di Napoli Federico II, Napoli, Italy

1. INTRODUCTION

One of the most challenging, extensively explored, and successful research areas in the landscape of modern antiviral and antitumor drug strategies deals with the development of sugar-modified nucleosides.[1–4] This involves replacement of the (deoxy)ribofuranose ring of natural nucleosides and nucleotides with suitable (a)cyclic bioisosteres (Figure 1), able to retain the potential for recognition by host/virally encoded kinases and polymerases while blocking the informational flow enclosed in the target genome.[5] For more than 30 years, an extremely large number of modifications of the carbohydrate core—from subtle variations of ribose backbone up to the most daring structural changes—have been devised, leading to identification of about 20 lead compounds (Figure 1) for therapeutic intervention of a wide variety of viral infections and malignant diseases.[2,4]

Among sugar-modified nucleosides, those composed of "unnatural" heterocyclic units (*heterocyclic nucleosides*), and especially those having a 1,3-oxathiolane or 1,3-dioxolane moiety as a "sugar" ring (*oxathiolane and dioxolane nucleosides*) have attracted well-deserved attention over the past two decades.[1,6,7] Indeed, oxathiolane and dioxolane nucleosides have long demonstrated exciting therapeutic potential, as antitumor agents,[8] as inhibitors of hepatitis B virus (HBV),[9] and as nucleoside reverse transcriptase inhibitors (NRTIs) in the chemotherapeutic treatment of human immunodeficiency virus (HIV).[10] As a result of their pharmacological properties, relying on a combination of excellent biological activities, good levels of bioavailabilities, and favorable toxicological profiles,[1] two oxathiolane nucleosides, lamivudine (3TC) and emtricitabine (FTC), have been licensed

---

Chemical Synthesis of Nucleoside Analogues, First Edition. Edited by Pedro Merino.
© 2013 John Wiley & Sons, Inc. Published 2013 by John Wiley & Sons, Inc.

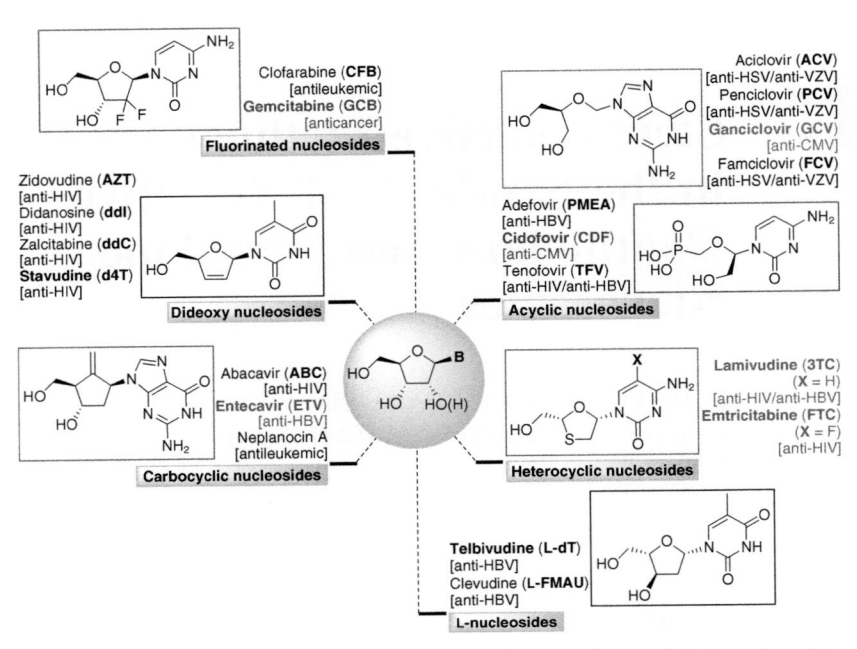

Figure 1. Marketed sugar-modified nucleosides in use for antiviral and antitumor chemotherapy. (*See insert for color representation of the figure.*)

so far for clinical use as antiretrovirals[10] (Figure 1). But a long list of other drug candidates bearing oxathiolane and dioxolane units are still awaiting approval as novel and safer therapeutic agents.[7,11]

The large clinical demand for oxathiolane and dioxolane nucleosides has long driven the development of efficient synthetic methodologies[1,12] for their preparation. The purpose of this chapter is to provide an overview of the synthetic strategies leading to such bioactive nucleosides, briefly highlighting their pharmacological importance among existing chemotherapeutic drugs. According to nomenclature of heterocyclic nucleosides and depending on the position of heteroatoms within the five-membered ring, a classification of the synthetic methods among those concerning 3′-oxanucleosides (also referred to as 2,4-disubstituted 1,3-dioxolanes), 3′-thianucleosides[13] (also referred to as 2,5-disubstituted 1,3-oxathiolanes), and 3′-oxa-4′-thionucleosides[13] (also referred to as 2,4-disubstituted 1,3-oxathiolanes), will be considered Figure 2).

2. MEDICINAL CHEMISTRY OF OXATHIOLANE AND DIOXOLANE NUCLEOSIDES

According to the World Health Organization (WHO), more than 33 million people currently need antiretroviral treatment owing to infection with HIV-1 and HIV-2, the causative agents of the acquired immunodeficiency syndrome (AIDS).[14] First-line therapies typically include use of N(t)RTIs and especially that of the

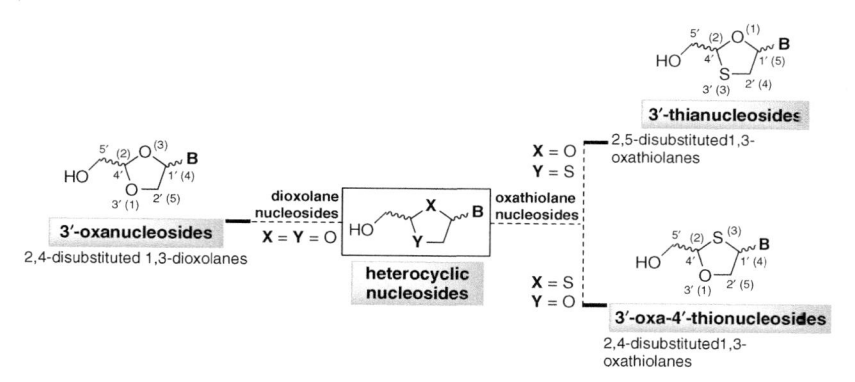

Figure 2. Nomenclature of oxathiolane and dioxolane nucleosides.

oxathiolane nucleosides 3TC and FTC (Figure 3), whether administered individually (Epivir and Emtriva) or in combination with other inhibitors [Combivir: AZT + 3TC, Epzicom/Kivexa: 3TC + ABC, Trizivir: AZT + 3TC + ABC, Truvada: FTC + TDF, Atripla: TDF + FTC + EFV (Efivarenz)].[15,16] In addition, a number of other candidates (particularly those displaying activity against 3TC/FTC-resistant HIV-RT mutants) are under evaluation (Figure 3): dioxolane fluorocytosine (FDOC),[17] dioxolane T (DOT),[18] and dioxolane aminopurine (APD)[19] are in a preclinical stage of development; amdoxovir (AMDX, DADP)[20] and racivir (RCV)[21] are undergoing phase II clinical trials, apricitabine (ATC) successfully ended phase III clinical trials in 2009, then receiving fast-track approval status from the U.S. Food and Drug Administration (FDA).[22]

Besides their main application in anti-HIV chemotherapy, oxathiolane and dioxolane nucleosides have displayed even broader therapeutic potential. 3TC has been licensed since 1998 (Epivir-HBV, Heptovir or Zeffix) for treatment of chronic HBV (CHB).[9] FTC is currently undergoing phase III clinical trials (Coviracil) to the same end;[9,23] FDOC,[24] RCV,[25] ADP, and AMDX have exhibited high anti-HBV activity in vitro and in vivo.[7] On the other hand, encouraging antitumor activity of troxacitabine (TRO), especially against solid tumors (e.g., prostatic, renal, hepatic, and colon tumors), acute myeloid leukemia (AML), and chronic myelogenous leukemia (CML),[26] has been reported. Use of TRO (Troxatyl) is currently being studied in patients (phase II/III clinical trials) with refractory lymphoproliferative diseases.[27] Eventually, from moderate to strong in vitro inhibition of the life cycles of many other viruses by a wealth of oxathiolane and dioxolane nucleosides has been documented, including simian immunodeficiency virus (SIV), feline immunodeficiency virus (FIV), duck HBV (DHBV),[1,28] SARS coronavirus (SARS-CoV),[29] Epstein–Barr virus (EBV),[30] and other human herpes viruses (HHV).[31]

The therapeutic versatility of oxathiolane and dioxolane nucleosides refers to a common mode of action dealing with the inhibition of a number of viral infections and malignancies. As widely known,[1–5] this is based on: (1) serial phosphorylation of nucleoside analogues (NAs) to the corresponding triphosphorylated

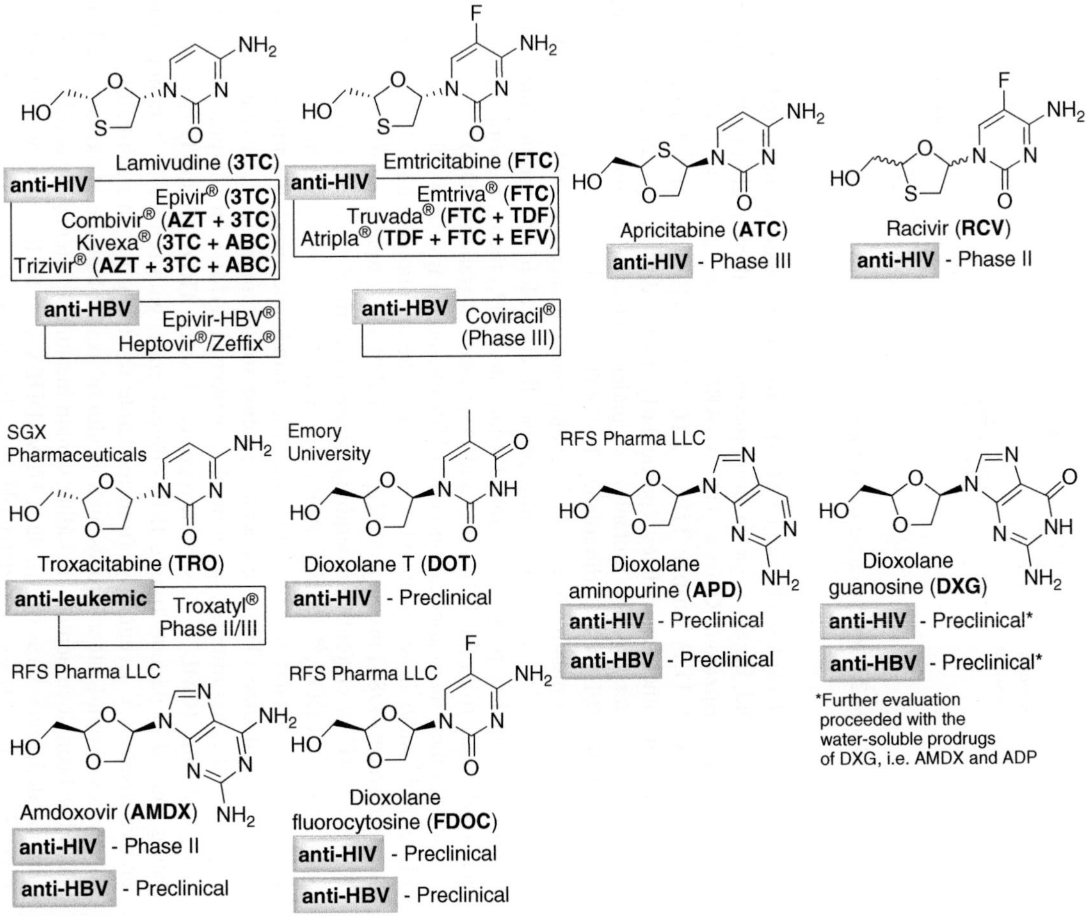

Lamivudine (3TC)
anti-HIV
Epivir® (3TC)
Combivir® (AZT + 3TC)
Kivexa® (3TC + ABC)
Trizivir® (AZT + 3TC + ABC)
anti-HBV
Epivir-HBV®
Heptovir®/Zeffix®
Emtricitabine (FTC)
anti-HIV
Emtriva® (FTC)
Truvada® (FTC + TDF)
Atripla® (TDF + FTC + EFV)
anti-HBV
Coviracil®
(Phase III)
Apricitabine (ATC)
anti-HIV - Phase III
Racivir (RCV)
anti-HIV - Phase II
SGX Pharmaceuticals
Troxacitabine (TRO)
anti-leukemic
Troxatyl®
Phase II/III
Emory University
Dioxolane T (DOT)
anti-HIV - Preclinical
RFS Pharma LLC
Dioxolane aminopurine (APD)
anti-HIV - Preclinical
anti-HBV - Preclinical
Dioxolane guanosine (DXG)
anti-HIV - Preclinical*
anti-HBV - Preclinical*
*Further evaluation proceeded with the water-soluble prodrugs of DXG, i.e. AMDX and ADP
RFS Pharma LLC
Amdoxovir (AMDX)
anti-HIV - Phase II
anti-HBV - Preclinical
RFS Pharma LLC
Dioxolane fluorocytosine (FDOC)
anti-HIV - Preclinical
anti-HBV - Preclinical

Figure 3. Clinically relevant oxathiolane and dioxolane nucleosides. *Abbreviations*: 3TC [lamivudine; (−)-BCH-189, L-(−)-SddC, BCH-790, GR-109714X, NGPB-21, (−)-β-L-2′,3′-dideoxy-3′-thiacytidine, (−)-(2*S*,5*R*)-1-[2-(hydroxymethyl)-1,3-oxathiolan-5-yl]cytosine, 4-amino-1-[(2*R*,5*S*)-2-(hydroxymethyl)-1,3-oxathiolan-5-yl]-1,2-dihydropyrimidin-2-one]. FTC [emtricitabine; L-(−)-5FSddC, (−)-β-L-2′,3′-dideoxy-5-fluoro-3′-thiacytidine, (−)-5-fluoro-1-(2*R*,5*S*)-[2-(hydroxymethyl)-1,3-oxathiolan-5-yl]cytosine, 4-amino-5-fluoro-1-[(2*S*,5*R*)-2-(hydroxymethyl)-1,3-oxathiolan-5-yl]-1,2-dihydropyrimidin-2-one]. ATC[apricitabine; AVX754, SPD754, (−)-dOTC, BCH10618, (−)-β-D-2′,3′-deoxy-3′-oxa-4′-thiocytidine, (−)-(2*R*,4*R*)-1-[2-(hydroxymethyl)-1,3-oxathiolan-4-yl]-cytosine, 4-amino-1-[(2*R*,4*R*)-2-(hydroxymethyl)-1,3-oxathiolan-4-yl]pyrimidin-2(1*H*)-one]. TRO [troxacitabine; SGX-145, L-OddC, BCH-4556, (−)-BCH-204, SPD 758, (−)-β-L-dioxolane-C, (−)-β-L-2′,3′-dideoxy-3′-oxacytidine, (−)-(2*S*,4*S*)-1-[2-(hydroxymethyl)-1,3-dioxolan-4-yl]-cytosine, 4-amino-1-[(2*S*)-2-(hydroxymethyl)-1,3-dioxolan-4-yl]pyrimidin-2-one]. RCV [racivir; (±)-FSddC, (±)-β-2′,3′-dideoxy-5-fluoro-3′-thiacytidine, (±)-5-fluoro-1-[2-(hydroxymethyl)-1,3-oxathiolan-5-yl]-cytosine, 4-amino-5-fluoro-1-[2-(hydroxymethyl)-1,3-oxathiolan-5-yl]-1,2-dihydropyrimidin-2-one]. AMDX [amdoxovir; DADP, TP-0020/96, TP-0020, diaminopurine dioxolane, (−)-(2*R*,4*R*)-(2-amino-9-[2-(hydroxymethyl)-1,3-dioxolan-4-yl]-adenine, (−)-β-D-2′,3′-dideoxy-3′-oxa-2,6-diaminopurine, [(2*R*,4*R*)-4-(2,6-diaminopurin-9-yl)-1,3-dioxolan-2-yl]methanol]. DOT {dioxolane T; D-OddT, (−)-β-D-3′-deoxy-3′-oxathymidine, (−)-(2*R*,4*R*)-1-[2-(hydroxymethyl)-1,3-dioxolan-4-yl]-thymine, 1-[(2*R*,4*R*)-2-(hydroxymethyl)-1,3-dioxolan-4-yl]-5-methyl-2,4(1*H*,3*H*)-pyrimidinedione}. DXG [dioxolane G, dioxolane guanosine, (−)-β-D-2′,3′-dideoxy-3′-oxaguanosine, (−)-(2*R*,4*R*)-9-[2-(hydroxymethyl)-1,3-dioxolan-4-yl]guanine, 2-amino-9-[(2*R*,4*R*)-2-(hydroxymethyl)-1,3-dioxolan-4-yl]-3,9-dihydro-6*H*-purin-6-one]. FDOC {dioxolane fluorocytosine; D-(+)-FOddC, (+)-5F-Dioxolane-C, (+)-5-fluoro-1-(2*R*,4*R*)-[2-(hydroxymethyl)-1,3-dioxolan-4-yl]-cytosine, 4-amino-5-fluoro-1-[(2*R*,4*R*)-2-(hydroxymethyl)-1,3-dioxolan-4-yl]-2(1*H*)-pyrimidinone}. APD {aminopurine dioxolane, (−)-(2*R*,4*R*)-(2-amino-9-[2-(hydroxymethyl)-1,3-dioxolan-4-yl]-purine, (−)-β-D-2′,3′-dideoxy-3′-oxa-2-diaminopurine, [(2*R*,4*R*)-4-(2-aminopurin-9-yl)-1,3-dioxolan-2-yl]methanol}.

form (NA-TPs) by viral/host kinases; (2) incorporation of NA-TPs into viral (acting as antiviral agents) or host (acting as antitumor drugs) genomes by successful competition with physiological deoxynucleoside triphosphates (dNTPs); and (3) inhibition of viral/host polymerases or viral/host DNA chain termination. Over the years, in-depth analysis of the molecular details determining the efficiency of such recognition processes has led to identification of some major factors coming into play, including the structure of metabolic enzymes, the conformational properties of modified nucleosides, and most notably, their stereochemical features. Particularly, study of the antiviral/antitumor properties of oxathiolane and dioxolane nucleosides has been at the core of one of the most exciting breakthrough discoveries in the arena of sugar-modified nucleosides, i.e., the assessment of the therapeutic potential of L-nucleosides[32,33] (Figure 4). Indeed, in the chemical structure of 3TC, FTC, and TRO, chirality of the stereogenic centres is opposite to that of physiological D-nucleosides. On the other hand, ATC, AMDX, FDOC, and DOT resemble the structure of D-nucleosides, while RCV is even a racemic mixture (Figure 3).

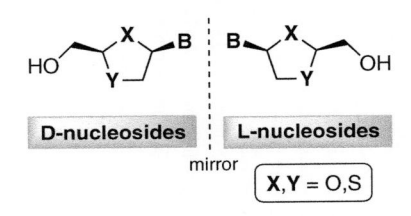

Figure 4. Oxathiolane and dioxolane nucleosides belonging to D- and L-series.

The finding that oxathiolane and dioxolane nucleosides (and most generally, 2′-deoxy- and 2′,3′-dideoxynucleosides) were able, in both enantiomeric forms, to exhibit potent biological activity represented one of the earliest notable exceptions among stereospecific biomolecular recognition processes, clearly demonstrating that the relationship between molecular stereochemistry and biological function was not always predictable. Since then, careful reevaluation of the importance of enantiospecificity in the antiviral activity of sugar-modified nucleosides was assessed,[34] and even the overall role of chirality at the chemistry–biology interface was revisited systematically, with numerous remarkable results.[35,36] The comparable therapeutic potential of D- and L-heterocyclic nucleosides was ascribed to two major factors. On the one hand, the conformational flexibility of the sugar moiety of D- and L-deoxynucleosides allowed accurate overlay of the "fingerprint" groups (i.e., nucleobase and 5′-OH) (Figure 5a).[37] On the other hand, the lack of enantiomeric selectivity of either phosphorylating enzymes,[37,38] viral polymerases,[39] and, in some cases, host polymerases,[40] resulted into unexpected catalytic activities for substrates with unnatural chirality (Figure 5b and c), even though the modes of binding of D- and L-nucleosides were slightly different.[37a]

It is worth noting that, contrarily to the relaxed stereoselectivity of the foregoing enzymes, the more rigorously enantioselective properties of other hydrolytic enzymes (mainly deoxycytidine deaminase)[41] and DNA polymerases[40] allowed

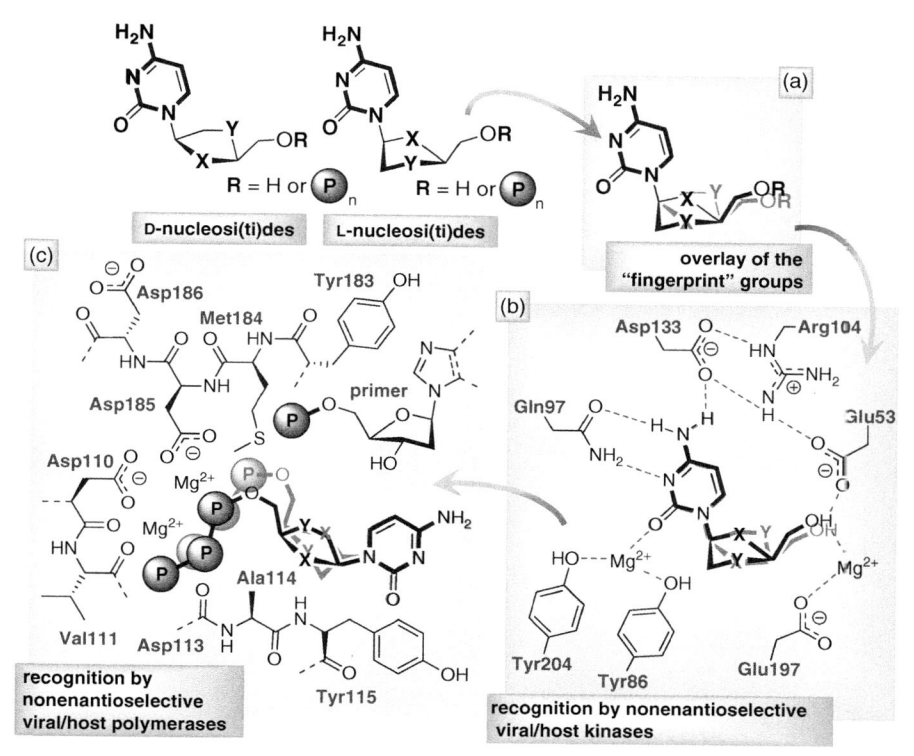

Figure 5. Schematic representation of the structural factors determining the biological activity of L-heterocyclic nucleosides. (a) Conformational flexibility of D- and L-heterocyclic nucleosi(ti)des allows accurate overlay of "fingerprint" groups [i.e., nucleobase (cytosine in the example) and 5′-OR]. (b) Nonenantioselective properties of host/virally encoded kinases enable phosphorylation of either D- and L-heterocyclic nucleosides. (c) Nonenantioselective properties of polymerases (e.g., HIV-RT) enable recognition of either D- and L-heterocyclic nucleoside triphosphates. (*See insert for color representation of the figure.*)

most L-nucleosides to display minimal cellular/mitochondrial toxicity and a superior half-life compared with their D-antipodes.[32] In addition, the presence of a 3′-heteroatom in D- and L-nucleosides was judged to play an important role in enhancing the enzyme-binding affinity, especially for viral polymerases. It was suggested that the sulfur atom of 3TC-TP produced nonbonding interaction[42,43] with the sulfur atom at the side chain of Met184 of wild-type (WT) HIV-1 RT (Figure 6a). However, replacement of Met184 with branched amino acid residues (Val, Ile, Thr) occurring in a number of HIV-RT mutations limited remarkably the accommodation of 3TC (and most generally that of L-nucleosides), owing to steric hindrance[44] on the oxathiolane unit (Figure 6b), resulting in a drop in antiretroviral activity. On the contrary, most heterocyclic D-nucleosides were significantly active against several 3TC-resistant HIV RT mutants.[45,46] As demonstrated especially for

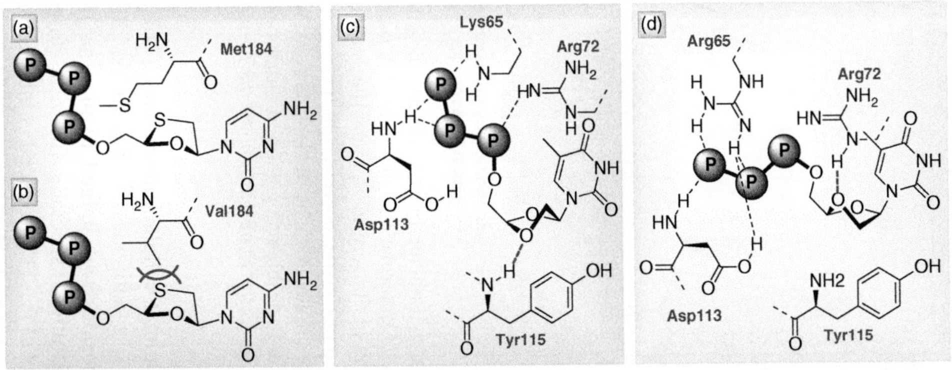

Figure 6. Schematic representation of the binding modes of oxathiolane and dioxolane nucleosides in WT and drug-resistant HIV-RTs (reproduced from refs. 44, 45, and 47). (a) Nonbonding interaction between 3TC-TP and Met184 in WT HIV-RT. (b) Disfavored interaction between 3TC-TP and Val184 in M184V mutant of HIV-RT. (c) Binding mode of DOT-TP in WT HIV-RT. (d) Binding mode of DOT-TP in the K65R mutant of HIV-RT. (*See insert for color representation of the figure.*)

D-dioxolanes,[45] versatility in antiretroviral activity was closely related to the wide variety of interactions by 3′-oxygen with the nearby enzyme residues of both WT and drug-resistant RT strains, acting in some cases as a 3′-OH mimic of natural dNTPs. In an illustrative example, 3′-oxygen interactions of DOT-TP with Arg72 and Tyr115 were found to be primarily responsible for the favorable binding affinity of the nucleoside to WT HIV-RT (Figure 6c) and the K65R mutant (Figure 6d) as well as to all most common NRTI-resistant HIV RT mutants.[47]

The data above substantiate the great synthetic interest[1,12] dealing with the preparation of oxathiolane and dioxolane nucleosides in considerable amounts and, most important, in high optical purity. As a matter of fact, the search for stereoselective methods leading to enantiomerically enriched/enantiopure material has been at the core of intense worldwide efforts for over than two decades, since the earliest seminal report given by Belleau et al.[48] at the 5th International AIDS Conference in Montreal in 1989; nevertheless, this topic continues to draw attention.[12] Indeed, despite their apparent structural simplicity, stereoselective access to oxathiolane and dioxolane nucleosides has represented a considerable challenge, as it has required stereochemistry control of two potentially epimerizable (thio)acetal stereogenic centers.[49] Accordingly, most attention is herein focused on diastereo- and enantioselective strategies toward biologically important oxathiolane and dioxolane nucleosides, whether they belong to purely chemical or chemoenzymatic methods. Synthetic approaches using chiral synthons are also considered. It is important to emphasize that a wide variety of the routes discussed below have been identified, over the years, as the methods of choice for the industrial production of these nucleosides. Therefore, the purpose of this chapter is to provide a survey of the most important scientific contributions in the field and to demonstrate how, although the pursuit of synthetic routes by now includes a large number of protocols, the area of investigation is still active, as new methodologies, new ideas, and the development of innovative and general strategies in heterocyclic chemistry have come into play.

3. OPTICALLY ACTIVE 3′-THIANUCLEOSIDES (2,5-DISUBSTITUTED 1,3-OXATHIOLANES)

Driven by the established success of the antiretroviral drugs 3TC and FTC, as well as by the therapeutic promises of RCV, 2′,3′-dideoxy-3′-thianucleosides currently represent the most prominent examples among all existing nucleoside analogues equipped with a heterocycle as the sugar ring. Since early reports, 3TC and FTC demonstrated excellent pharmacological profiles, especially compared with their D-counterparts,[32] regarding antiviral activity, cellular or mitochondrial toxicity, and cellular metabolism.[41] However, by virtue of the presence of a fluorine atom,[50] FTC exhibited even better properties in vitro and in vivo.[51,52] There is currently a driving force to switch 3TC with FTC in some combination therapies; however, full proof of the clinical equivalence of the two drugs has not yet been attained.[53] In addition, contrarily to 3TC, both enantiomeric forms of FTC have drawn attention: neither

(+)- nor (−)-FTC showed significant cytotoxicity or any detectable hepatotoxic effect,[54] and although FTC displayed a 10 to 20-fold more potent antiretroviral activity than its (+)-counterpart, both triphosphate forms displayed comparable K_i values.[55] (+)- and (−)-FTC also demonstrated the ability to select for different mutations on the HIV-1 RT gene in vitro [i.e., M184V for the (−)-enantiomer and T215Y for the (+)-enantiomer].[56] Not completely unexpected, the presence of (+)-FTC was thus found to potentiate anti-HIV activity of (−)-FTC.[57]

Given the impact that (+)- and (−)-3′-thia(fluoro)cytidines have brought to modern antiviral chemotherapy, it is not surprising that most synthetic approaches to 3′-thianucleosides have focused on their preparation, whether they deal with laboratory-scale synthesis of enantiomerically enriched mixtures or with industrial production of optically pure material. Giving a look at stereoselective strategies toward D-(+)- and L-(−)-3′-thianucleosides (Scheme 1), it must be preliminarily recognized how all approaches share preparation of 1,3-oxathiolanyl intermediates **1**, bearing a methyleneoxy or a carboxy group at C2, both acting (either in racemic or enantiomerically pure form) as electrophilic "sugars" in (stereoselective) Vorbrüggen *N*-glycosidation reactions. On one hand, synthesis of heterocyclic scaffolds has been envisaged from smaller starting materials (de novo synthesis), such as 2-mercaptoacetaldehyde [occurring in its dimeric form, i.e., as 1,4-dithiane-2,5-diol (**2**)] or thioglycolic acid (**3**), and their derivatives (Scheme 1a). Under these conditions, stereochemical control at any synthetic stage has been explored by chemical and enzymatic methods. In other cases, use of chiral starting materials (e.g., carbohydrates) to fix chirality of target 3′-thianucleosides (chiron approaches) has been devised (Scheme 1b).

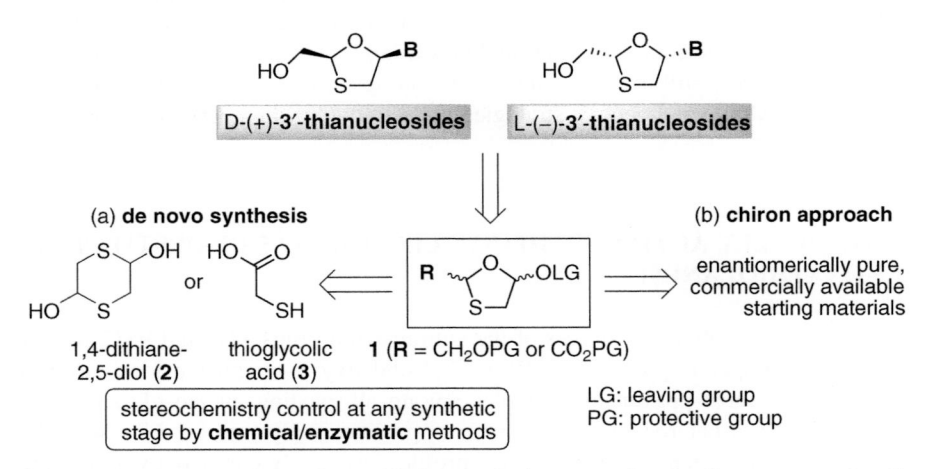

Scheme 1. Representative scheme of the synthetic approaches leading to D- and L-3′-thianucleosides.

Earliest synthetic endeavours to 3′-thianucleosides involved a worldwide rush (culminating in long and heated legal fights) among academic and industrial contenders. Belleau et al.[48] first synthesized large amounts of (±)-BCH-189 [three steps from benzoyl glycolaldehyde (**4**)] (Scheme 2) through condensation of **4** with 2-mercaptoacetaldehyde dimethylacetal, N-glycosidation of the resulting methyl glycoside (±)-**5** (TiCl$_4$), and protective group removal of nucleosides cis-(±)-**6** (NH$_3$/MeOH). Belleau also took part to early in vitro studies[48,58] reporting potent anti-HIV-1 activity of (±)-BCH-189 (Scheme 2) before the two enantiomers [(+)-BCH-189 and 3TC] were separately evaluated[59,60] after milligram-scale HPLC. Starting from Belleau's approach (with minor synthetic variations), a gram-scale resolution of (±)-BCH-189 was carried out later by Storer et al.[61] via a highly enantioselective hydrolysis of the corresponding 5'-monophosphate derivatives [POCl$_3$/PO(OMe)$_3$, 94%][62] with 5'-nucleotidase from *Crotalus atrox* venom. The enzyme preferentially recognized D-(+)-BCH-189-MP (33%), leaving the L-(−)-enantiomer unaltered (Scheme 2). The phosphate group of the latter was removed by treatment with alkaline phosphatase from *Escherichia coli* (35%), leading to pure 3TC (>99% ee).

To produce larger amounts of highly pure 3TC for further evaluation, the 5'-nucleotidase route was judged as inefficient; thus, alternative, more scalable

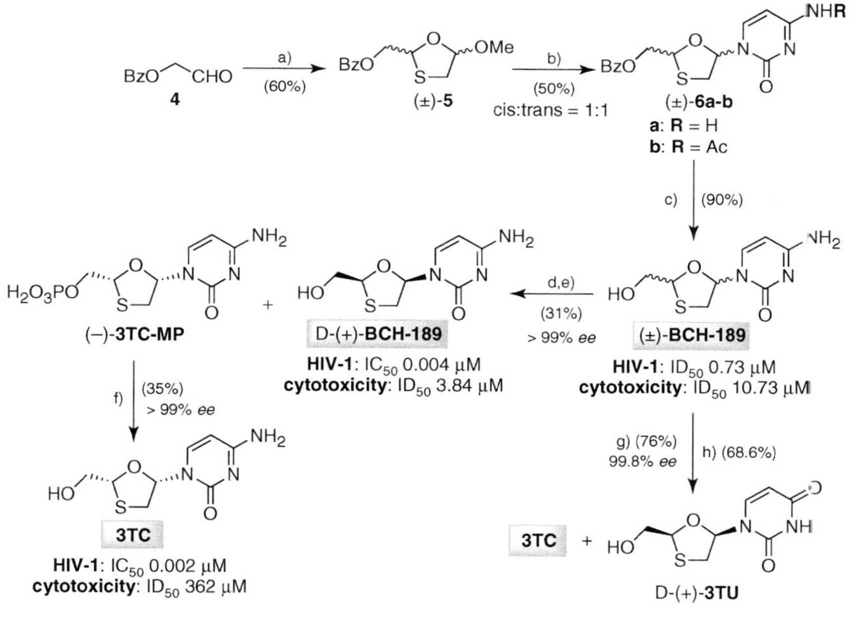

Scheme 2. *Reagents and conditions:* (a) HSCH$_2$CH(OMe)$_2$, ZnCl$_2$, toluene, reflux; (b) silylated N^4-acetylcytosine, TiCl$_4$, DCE, or silylated cytosine, TMSOTf, CH$_3$CN; (c) Amberlite (OH$^-$), MeOH, reflux; (d) P(O)Cl$_3$, PO(OMe)$_3$, 0°C, 35 min; (e) 5'-ribonucleotide phosphohydrolase (5'-nucleotidase) from *Crotalus atrox* venom, glycine, MgCl$_2$, H$_2$O, 37°C, 6 h; (f) alkaline phosphatase from *E. coli*, 37°C, H$_2$O, 1 h; (g) CDA from *E. coli*; (h) CDA from *Bacillus caldolyticus*, 65°C, 30 min. Anti-HIV-1 activity of (+)-BCH-189, 3TC, and (±)-BCH-189 was evaluated in acutely infected MT-4 cells, while cytotoxicity was evaluated in CEM cells.

strategies were devised. In an intriguing example, 3TC was prepared by treatment of (±)-BCH-189 with cytidine deaminases (CDAs). The strategy was inspired by the differential cellular metabolism of D-(+)-BCH 189 and 3TC, the former stereospecifically undergoing deamination by host deoxy-CDA[41] to corresponding D-(+)-3′-thiauridine [D-(+)-3TU]. Driven by this finding, Mahmoudian et al.[63] developed a first chemoenzymatic route, leading to large-scale production of optically pure 3TC by enantioselective deamination of the unwanted isomer from starting (±)-BCH-189 (Scheme 2). Preliminary experiments with free and immobilized CDA from *E. coli* demonstrated that the enzyme effectively recognized and readily deaminated the D-(+)-enantiomer only (>99% ee). CDA immobilized on Eupergit C was used for kilogram-scale production of 3TC (76%, 99.8% ee). Moreover, in this form, reuse[64] of the enzyme for at least 15 cycles was allowed. Further scale-up eventually provided multiton quantities of the antiviral agent for clinical trials (before chemical asymmetric methods were put in place).[65] Subsequently, a thermostable CDA from *Bacillus caldolyticus* was used by Song et al.[66] Efficacy of D-(+)-nucleoside deamination was found very similar to that of mesophilic enzyme from *E. coli* (although the authors did not investigate further the synthetic potential of the enzyme for large-scale applications). Indeed, starting from (±)-BCH-189, approximately 68.6% of D-(+)-isomer was selectively converted into the corresponding D-(+)-3TU under optimized conditions (Scheme 2).

Despite their synthetic efficiency, chemoenzymatic methodologies of Scheme 2 had two major weaknesses. On the one hand, bioresolution processes of racemic 3′-thianucleosides caused significant loss of the valuable starting materials; on the other hand, *N*-glycosidation step was characterized by a whole absence of anomeric selectivity (*cis/trans* = 1:1), using either $TiCl_4$[48] or TMSOTf[61] as Lewis acids. By contrast, Liotta et al.[67] noted that use of $SnCl_4$ in the reaction of (±)-**8** (available from protected glycolaldehydes **7**) dramatically improved diastereoselectivity toward *cis*-(±)-**9/10** (if R = TBDPS, *cis/trans* >300:1). It was proposed[67] that the stereoselective outcome of the reaction was largely influenced by an in situ complexation between stannic chloride and the sulfur atom of (±)-**8** (Scheme 3). Formation of complex **11** (*anti*-oriented to the protected hydroxymethyl group) bearing a chlorine atom at the α-anomeric position (as a result of a presumable ligand delivery from the metal to the proximal upcoming oxocarbenium ion) was proposed to drive the subsequent β-face attack of silylated nucleobase under S_N2 conditions (Scheme 3). Unfortunately, as found using enantiomerically pure 1,3-oxathiolanes, acid-labile thioacetal function was also concurrently affected.[68] By means of this strategy, synthesis of (±)-BCH-189[67a] and, inter alia, an earliest entry to large amounts of (±)-FTC[67b] (later known as RCV) were at first carried out from silyl ethers *cis*-(±)-**9a** and **10a**. Moreover, Liotta, Koszalka (Burroughs Wellcome), et al.[69] made use of this procedure to achieve a highly enantioselective, scalable production of (+)- and (−)-3′-thianucleosides via a late-stage enzymatic hydrolysis of butyrate esters (±)-**9b** and **10b** (Scheme 3). Of the 14 commercially available lipases screened, pig liver esterase (PLE) most effectively hydrolyzed D-(+)-butyrate isomers [(±)-**9b**: 54%; (±)-**10b**: 58%], leaving

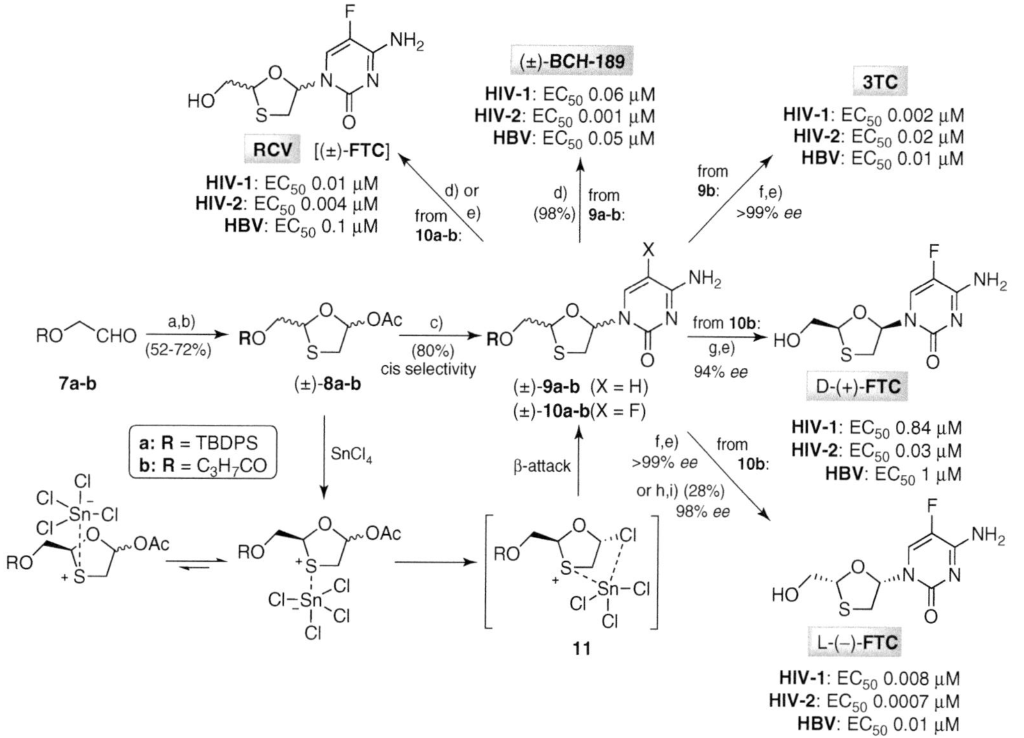

Scheme 3. *Reagents and conditions:* (a) **3**, toluene, reflux; (b) (i) DIBAL-H, toluene, −78°C or LiAlH(O-t-Bu)$_3$, THF, 0°C; (ii) Ac$_2$O, rt; (c) silylated cytosine or 5-fluorocytosine, SnCl$_4$, CH$_2$Cl$_2$, rt, 1 h; (d) TBAF, THF, rt; (e) MeONa, MeOH, rt; (f) PLE, KH$_2$PO$_4$ buffer (pH 7), CH$_3$CN, 40°C; (g) lipase PS-800, KH$_2$PO$_4$ buffer (pH 7), CH$_3$CN, 40°C; (h) cholesterol esterase (from *Candida cylindracea*) immobilized on Accurel PP, 1-pentanol, KH$_2$PO$_4$ buffer (pH 7), 30°C, 48 h; (i) Amberlite LA2, rt. Anti-HIV-1 activity of (+), (−), and (±)-3′-thianucleosides was evaluated in acutely infected PBM-LAV cells, anti-HIV-2 activity in PBM-ROD cells, and anti-HBV activity in 2.2.15 cells. For all compounds, cytotoxicity (CEM cells) was meant to be >100 μM. Stereochemistry at C2 of Sn(IV)-containing complexes (including **11**) is assigned in order to display the stereochemical outcome of the reaction.

corresponding L-(−)-enantiomers as unreacted esters (>99% ee). Then, chemical hydrolysis of the last ones (NaOMe) conveniently afforded enantiomerically pure 3TC or L-(−)-FTC (Scheme 3). Conversely, Lipase PS-800 preferentially hydrolyzed the butyrate group of protected L-(−)-FTC (58%), retrieving unreacted ester of D-(+)-FTC in 94% ee. Over the years, enantiomerically pure 3'-thia(fluoro)cytidines (along with the corresponding racemic material) obtained under these conditions were employed for a number of in vitro and in vivo studies, including extensive evaluation of their anti-HIV (HIV-1 and HIV-2)[51b,70] and HBV[25,71] potential, identification of 3TC- and FTC-resistant RT strains,[72] assessment of the different cellular metabolism by 3'-thianucleoside stereoisomers,[41] and their comparative incorporation efficiencies,[42,51a,54,73] host toxicity,[41,70] and pharmacokinetics.[55]

While performing a scale-up protocol of the PLE-mediated strategy to develop a validated manufacturing process for the production of emtricitabine, Gaede et al. pointed out that use of an animal-derived enzyme raised concerns about the risk of introducing viruses into the process stream (potentially contaminating both product and equipment), thereby identifying the essential operations to be fulfilled for adequate viral clearance.[74] In the same context, Taylor et al.[75] provided an alternative, safer, and scalable bioresolution process to resolve optical isomers of FTC butyrate (±)-**10b** based on the use of cholesterol esterase from *Candida cylindracea* immobilized on Accurel PP (Scheme 3). Kilogram-scale deacylation of (±)-**10b** resulted in preferred deprotection of required enantiomer to corresponding L-(−)-FTC (31%), which was recovered as hydrochloride salt (98% ee). Importantly, immobilized cholesterol esterase displayed excellent recycling properties, as it could be used successfully for at least 15 successive resolution processes.[75]

Experimental Procedures

cis-2-(t-Butyl-diphenylsilyloxy)-methyl-5-(5'-fluorocytosin-1'-yl)-1,3-oxathiolane [(±)-10a] (SnCl₄-based N-glycosidation).[76] To a magnetically stirred solution of racemic 2-(*t*-butyldiphenylsilyloxy)methyl-5-acetoxy-1,3-oxathiolane [(±)-**8a**, 1.70 g, 4.08 mmol] in CH$_2$Cl$_2$ (100 mL), a premixed mixture of previously silylated 5-fluorocytosine (1.22 g, 4.5 mmol) and SnCl$_4$ (8.6 mL, 1.0 M solution in CH$_2$Cl$_2$, 8.6 mmol) in CH$_2$Cl$_2$ (20 mL) was added dropwise over 20 min at room temperature. The resulting mixture was stirred at the same temperature for 3 h. Afterward, pyridine (3 mL) was added, and the solvent was removed in vacuo. Chromatography of the crude residue gave (±)-**10a** as a tan solid (1.80 g, 91% yield), which was further recrystallized from ethanol (1.75 g). Data for (±)-**10a**:[76] mp 214–215°C (EtOH). ^{1}H-NMR (300 MHz, DMSO-d_6): δ 1.09 (s, 9H, t-Bu), 3.16 (dd, $J_{2'a,1}$ = 5.4, $J_{2'a,2'b}$ = 11.5 Hz, 1H, H-2'a), 3.45 (dd, $J_{2'b,1'}$ = 5.4, $J_{2'b,2'a}$ = 11.5 Hz, 1H, H-2'b), 3.90 (dd, $J_{5'a,4'}$ = 4.3, $J_{5'a,5'b}$ = 11.5 Hz, 1H, H-5'a), 4.01 (dd, $J_{5'b,4'}$ = 3.6, $J_{5'a,5'b}$ = 11.5 Hz, 1H, H-5'b), 5.28 (t, $J_{4',5'}$ = 4.0 Hz, 1H, H-4'), 6.19 (t, $J_{1',2'}$ = 5.4 Hz, 1H, H-1'), 7.43–7.64 (m, 10H, H-arom), 7.61 (b s, 1H, NH$_a$), 7.87 (b s, 1H, NH$_b$), 7.96 (d, $J_{6,F}$ = 6.8 Hz, 1H, H-6). *Anal.* Calcd. for C$_{24}$H$_{28}$O$_3$N$_3$FSSi: C, 59.36; H, 5.81; N, 8.65; S, 6.60; found: C, 59.44; H, 5.81; 8.60; S, 6.64.

2-Hydroxymethyl-5-(5'-fluorocytosin-1'-yl)-1,3-oxathiolane [(±)-FTC, RCV] (TBDPS group removal).[76] To a magnetically stirred solution of silyl ether (±)-**10a** (1.12 g, 2.31 mmol) in THF (80 mL), a 1.0 M solution of TBAF in THF (2.50 mL, 2.50 mmol) was slowly added at room temperature. After 30 min the solution was concentrated under reduced pressure; the resulting residue was diluted with a 3:1 EtOH/pyridine mixture (4 mL) and loaded on a flash chromatography column, to give (±)-FTC (0.75 g) as a single *cis* isomer. The white solid was further recrystallized from EtOH (1.56 g, 98% yield). Data for (±)-FTC: mp 195–196°C (EtOH). ^{1}H-NMR (300 MHz, DMSO-d_6): δ 3.11 (dd, $J_{2'a,1'}$ = 4.2 Hz, $J_{2'a,2'b}$ = 11.7 Hz, 1H, H-2'a), 3.41 (dd, $J_{2'b,1'}$ = 5.7 Hz, $J_{2'b,2'a}$ = 11.7 Hz, 1H, H-2'b), 3.74 (m, 2H, H-5'), 5.17 (t, $J_{4',5'}$ = 3.6 Hz, 1H, H-4'), 5.40 (t, $J_{5',OH}$ = 5.7 Hz, 1H, OH), 6.12 (dd, $J_{1',2'a}$ = 4.2 Hz, $J_{1',2'b}$ = 5.7 Hz, 1H, H-1'), 7.57 (b s, 1H, NH$_a$) 7.81 (b s, 1H, NH$_b$), 8.18 (d, $J_{6,F}$ = 8.4 Hz, 1H, H-6). ^{13}C-NMR (75 MHz, DMSO-d_6): ppm 37.07, 62.48, 86.84, 86.90, 126.01 (d, J = 32.6 Hz), 136.12 (d, J = 241 Hz), 153.28, 157.85 (d, J = 13.4 Hz). *Anal.* Calcd. for C$_8$H$_{10}$O$_3$N$_3$SF: C, 38.86; H, 4.08; N, 17.00; S, 12.97; found: C, 38.97; H, 4.07; 16.93; S, 12.89.

As an alternative to end-stage resolution processes, Cousins et al.[77] performed a chemoenzymatic separation of racemic oxathiolane propionate *trans*-(±)-**12** mediated by *Mucor miehei* lipase (Scheme 4). Enantiomerically enriched (−)-**13** (76–80% *ee*) was then converted into 3TC under usual conditions, reasonably keeping the stereochemical integrity of the oxathiolane unit (70% ee). On the other hand, Rayner et al.[78] (University of Leeds) used *Pseudomonas fluorescens* lipase (PFL) to hydrolyze the (+)-enantiomer of acetoxysulfide (±)-**14** (available

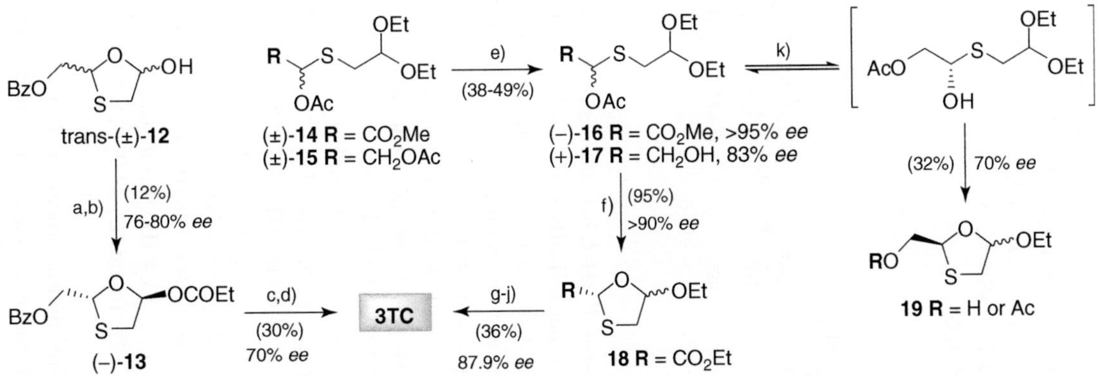

Scheme 4. *Reagents and conditions*: (a) EtCOCl, pyridine, CH$_2$Cl$_2$, 0°C; (b) *Mucor miehei*, Tris–HCl buffer (pH 8), CaCl$_2$·2H$_2$O, CH$_3$CN, 28°C; (c) silylated cytosine, TMSI, (CH$_2$Cl$_2$); (d) Amberlite IRA400 (OH$^-$), IMS, reflux; (e) *Pseudomonas fluorescens* lipase, phosphate buffer (pH 7), *t*-BuOMe, 30°C; (f) HCl, EtOH; (g) LiBH$_4$, *i*-PrOH, THF; (h) BzCl, pyridine, CH$_2$Cl$_2$; (i) *N,O*-bis(trimethylsilyl)-*N*^{4}-acetylcytosine, TMSOTf, CH$_3$CN; (j) NH$_3$/MeOH; (k) PTSA, CH$_2$Cl$_2$.

from methyl thioglycolate or thioglycolaldehyde diethylacetal),[78] retrieving (−)-**16** in a highly enantioselective manner (>95% ee). The synthesis then proceeded through de-*O*-acetylation of optically active thioacetal (−)-**16** and subsequent in situ cyclization (HCl/EtOH, 95%), obtaining 3TC from oxathiolane **18** in four steps and overall 87.9% ee (Scheme 4). In a complementary approach, PFL catalyzed the regio- and enantioselective hydrolysis of α,β-diacetoxysulfide[79] (±)-**15** (83% ee). The resulting primary alcohol (+)-**17** was used to obtain oxathiolane **19** (32%, 70% ee), functional to the preparation of D-(+)-3′-thianucleosides.

Early fully chemical methodologies to enantiomerically pure D- and L-3′-thianucleosides involved commercially available chiral starting materials. In a number of classical approaches by Chu et al.[1] use of carbohydrate templates was devised (Scheme 5a and b). Accordingly, D-(+)-BCH-189 and D-(+)-FTC were obtained earlier from D-mannose[80] (21 steps, not shown) and later from D-galactose[68] (18 steps; Scheme 5a). On the other hand, synthesis of 3TC and L-(−)-FTC was performed starting from L-gulose[81] (16 steps; Scheme 5b). Both routes broadly shared a synthetic scheme relying on iterated chemical degradations of the starting D-**20** and L-**21** [NaIO$_4$/NaBH$_4$ and Pb(OAc)$_4$], enabling their conversion into oxathioacetals **23** and *ent*-**23** via carboxylic acids **22** and *ent*-**22**. In an almost concurrent, faster approach to 3TC (four steps; Scheme 5c) developed by Jones, Mansour et al.[82] access to **26** was conceived from (+)-3-mercaptolactic acid (**24**) after condensation of the latter with aldehyde **4** (BF$_3$·OEt$_2$, 75%) and chemical degradation of the *trans*-(+)-isomer of **25** [Pb(OAc)$_4$, 64%]. As mentioned previously, in both strategies *N*-glycosidation reactions were carried out at first using the SnCl$_4$-based method. Even though from high to complete *cis*-selectivity was observed (from **23**, only *cis*-isomers detected;[68] from **26**: *cis/trans* = 10:1[82]), racemization of the oxathiolane units made this technology unsuited in these cases. On the other hand, no loss of thioacetal integrity was found with TfOTMS[68] or TMSI[82]; however, their use resulted at best in very weak stereoselectivity toward the desired *cis*-anomer (TMSOTf: *cis/trans* = 2:1; TMSI: *cis/trans* = 1.3:1). D-(+) and L-(−)-3′-thia(fluoro)cytidines in high optical purity were thus gained (e.g., synthesis of 3TC was accomplished in 96% ee from **24**)[82] after eventual protective group removal under common conditions (Scheme 5). Moreover, Chu et al. also performed the synthesis of a wide number of other 3′-thianucleosides after *N*-glycosidation of **23** and *ent*-**23** with other purines and pyrimidines; some of them were endowed with very promising antiretroviral properties.[80,81] Incidentally, the antiretroviral activity of purine and pyrimidine (+)- and (−)-3′-thianucleosides having C4′ and C1′ substituents in a *trans*-configuration [including L-(+)-**27** and L-(+)-**28**] was also examined[81] in view of SAR studies, with some noteworthy results (Scheme 5). High optical purity of D- and L-isomers allowed their use in several in vitro studies, especially those aimed at exploring the existing relationship between configuration (and conformation) of oxathiolane nucleosides and their recognition by the metabolic enzymes involved in HIV and HBV replication.[41,60]

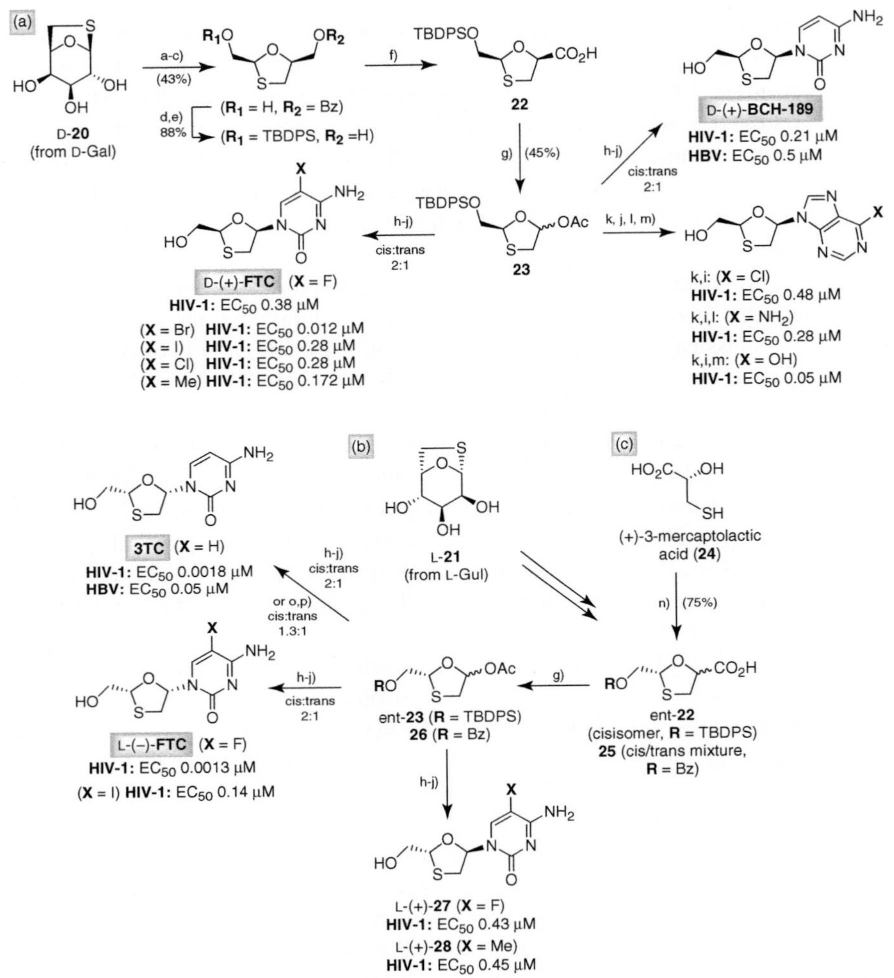

Scheme 5. *Reagents and conditions:* (a) (i) NaIO$_4$, MeOH/H$_2$O, 0°C, 10 min; (ii) NaBH$_4$, 10 min; (iii) PTSA, acetone, MeOH, rt, 1 h; (b) BzCl, pyridine, rt, 1 h; (c) (i) PTSA, MeOH, rt, 7 h; (ii) NaIO$_4$, MeOH/H$_2$O, 0°C, 10 min; (iii) NaBH$_4$, 5 min; (d) TBDPSCl, imidazole, DMF, rt, 1h; (e) MeONa, MeOH, rt, 2 h; (f) PDC, DMF, rt, 15 h; (g) Pb(OAc)$_4$, THF, pyridine, rt, 30 min; (h) (i) (5-substituted) cytosine, HMDS, (NH$_4$)$_2$SO$_4$, reflux; (ii) TMSOTf, DCE, rt, 1–3h; (i) NH$_3$, MeOH, rt; (j) TBAF, THF, rt, 0.5–1h; (k) (i) 6-chloropurine, HMDS, (NH$_4$)$_2$SO$_4$, DCE, rt; (ii) TMSOTf, DCE, −20°C to rt, 14 h; (l) NH$_3$, MeOH, 80–90°C, 15 h; (m) NaOMe, 2-mercaptoethanol, MeOH, reflux; (n) **4**, BF$_3$·OEt$_2$, CH$_3$CN; (o) silylated cytosine, TMSI, CH$_2$Cl$_2$; (p) Amberlite IRA400(OH), heat. Anti-HIV-1 activity of all the above (+)- and (−)-3′-thianucleosides was evaluated in acutely infected PBM cells, while HBV was evaluated in 2.2.15 cells. For all the compounds above, the cytotoxicity (PBM cells) was >100 μM.

Subsequent demand for more practical, cost-effective routes addressed further efforts to the use of chiral auxiliaries or other chemical resolving agents. In an early example by Tse et al.[83] later revisited by Whitehead et al.[84] resolution of acid *trans*-(±)-**30** [available, in turn, from glyoxylic acid monohydrate (**29**)] was obtained via formation of L-menthyl ester *trans*-**31b** [(−)-L-menthol/DCC/DMAP, 46% yield,[83] or (COCl)$_2$, then (−)-L-menthol, 16% yield[84]] (Scheme 6). Alternatively, (2R,4R)-**30** could be obtained through a classical resolution process based on the formation of norephedrine[84,85] salt **32** (35%). Interestingly, subsequent N-glycosidation of L-menthyl ester *trans*-**31b** using presilylated cytosine and stoichiometric TMSI in CH$_2$Cl$_2$ proceeded with high *cis*-selectivity (*cis/trans* = 23:1, 75% yield[83a]; *cis/trans* = 10:1, 70% yield[84]). Under the same conditions, N-glycosidation of the less hindered diethyl ester (±)-**31b** still provided significant anomeric selectivity (*cis/trans* >15:1). Although earlier[83a] the authors did not investigate the reaction mechanism in detail, it was later[84] conjectured that a 5α-iodooxathiolane intermediate **34** was presumably formed by reaction of I$^-$ with oxonium ion **33**, the latter undergoing stabilization through the anchimeric assistance of a C2 ester group (**33b**). Subsequent S$_N$2 attack of silylated cytosine to iodide **34** thus explained the predominant formation of *cis*-**35** (Scheme 6). Most important, this N-glycosidation method was compatible with the configurational arrangement of L- (or D-) menthyl ester *trans*-**31b**, as no enantiomeric impurities[83] were detected (>98% *ee*) after final reduction of **35b** (LiAlH$_4$,[83] 94% or NaBH$_4$,[84] 83%). Following Tse's approach (with minor synthetic variations), Chu et al.[86] later performed a synthesis of D- and L-3'-thianucleosides equipped with a (*E*)-5-(2-bromovinyl)uracil, acting as oxathiolane analogues of the anti-HHV agent brivudine (BVDU). In this case, an even more selective stereochemical outcome during the N-glycosidation step (*cis/trans* = 30:1) was found. Interestingly, the L-optical isomer was endowed with moderate anti-VZV activity.[86]

Whitehead et al.[84] provided an improved version of the procedure of Scheme 6. First, an efficient "sugar" unit assembly was achieved. Among the four diastereomeric hemiacetals **37** (obtained from L-menthyl glyoxylate hydrate **36**) rapidly interconverting in the reaction medium, equilibrium was driven to the desired *trans*-isomer (2R,5R)-**37** (80%) by a crystallization-induced dynamic kinetic resolution process (N-hexane/Et$_3$N) (Scheme 7). Then, chlorination of **37** (SOCl$_2$/DMF) followed by subsequent N-glycosidation of 5-chlorooxathiolane **38** gave nucleoside **35b** in fairly good yield (66%) and selectivity (*cis/trans* = 10:1). Eventual reduction of L-menthyl ester group of **35b** (NaBH$_4$) provided enantiomerically pure 3TC (83%). Among all synthetic procedures leading to bioactive 3'-thianucleosides, this approach so far represents the only stereoselective and scalable methodology not involving use of any resolution process. It is not surprising that this synthetic protocol (with minor variations) was chosen as the manufacturing process for both lamivudine and emtricitabine.[53,87]

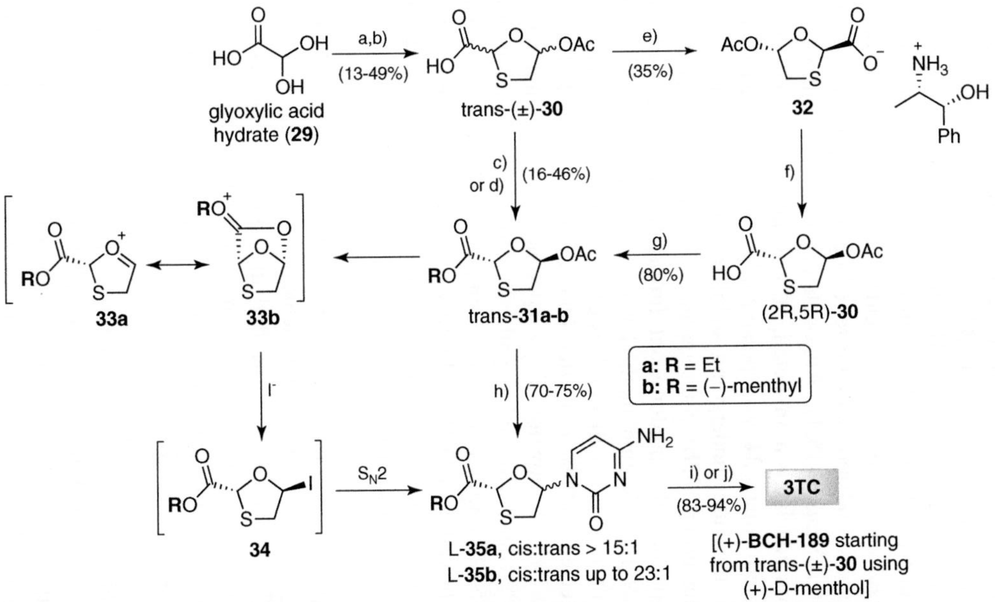

Scheme 6. *Reagents and conditions*: (a) **2**, *t*-BuOMe, reflux; (b) (i) Ac₂O, MeSO₃H; (ii) recrystallization; (c) (i) (COCl)₂, DMF, CH₂Cl₂; (−)-L-menthol, pyridine, 2,2,4-trimethylpentane, 16%; (d) (i) (−)-L-menthol, DCC, DMAP, CH₂Cl₂; (ii) recrystallization, 46%; (e) norephedrine, *i*-PrOAc; (f) 5 M HCl; (g) (i) (COCl)₂, DMF; (ii) (−)-L-menthol, pyridine, 80% from **32**; (h) (i) TBDMSOTf, 2,4,6-collidine, cytosine, CH₂Cl₂, rt, 15 min; (ii) TMSI, CH₂Cl₂, rt, 18 h; (i) LiAlH₄, THF, rt, 30 min; (j) NaBH₄, EtOH, rt.

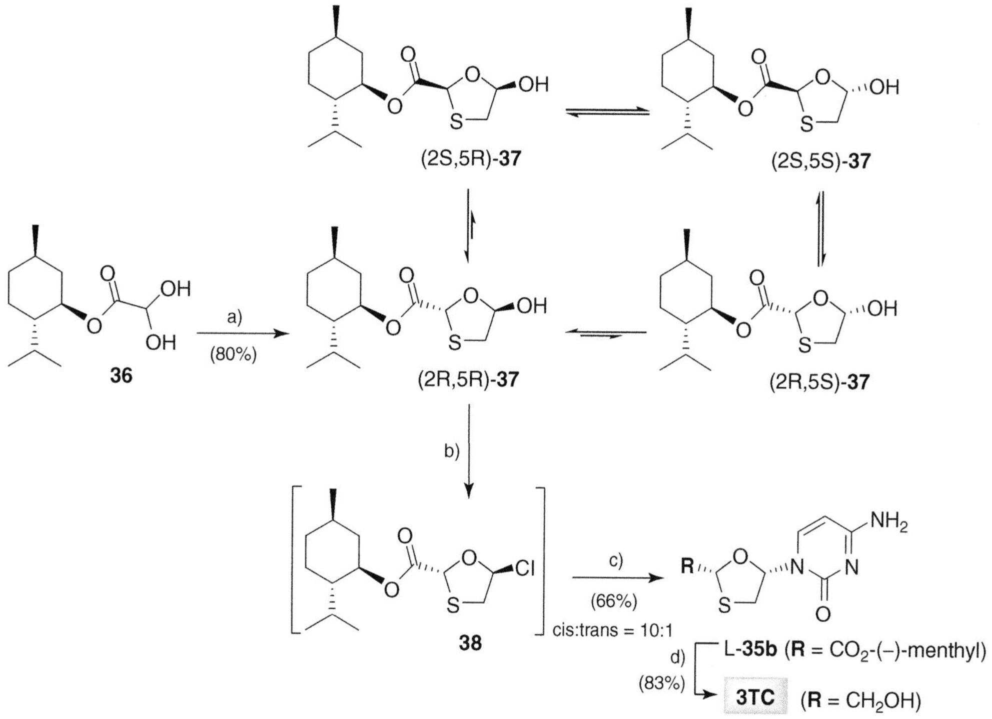

Scheme 7. *Reagents and conditions:* (a) (i) **2**, toluene, reflux; (ii) *n*-hexane, Et$_3$N; (b) SOCl$_2$, DMF, CH$_2$Cl$_2$; (c) (i) (TMS)$_2$-cytosine, Et$_3$N, toluene; (ii) Et$_3$N, H$_2$O, *n*-hexane; (d) NaBH$_4$, EtOH, rt.

Experimental Procedure

(1'R,2'S,5'R)-Menthyl 5-(S)-cytosin-1''-yl-1,3-oxathiolane-2(R)-carboxylate (L-35b) (Stereoselective N-Glycosidation).[88] Step a: Preparation of (1R,2S,5R)-menthyl 5(S)-chloro-1,3-oxathiolane-2(R)-carboxylate (**38**). To a solution of (2R,5R)-**37** (30.0 g, 110.2 mmol) and MSA (0.07 mL, 1.1 mmol) in CH$_2$Cl$_2$ (300 mL), DMF (8.5 mL) was added. To the mixture, cooled to about 8°C, SOCl$_2$ (8.0 mL, 110.2 mmol) was added over about 10 min, proceeding with the stirring for an additional 1.5 h at 10–15°C. Then the solution was concentrated by distillation under atmospheric pressure (over a period of about 1.5 h, 210 mL distillate was collected). *Step b: Preparation of silylated cytosine and coupling reaction.* To a suspension of cytosine (11.5 g, 103.6 mmol) in toluene (30 mL), MSA (0.07 mL, 1.10 mmol) and HMDS (24.2 mL, 115.0 mmol) were added. The mixture was heated at reflux for about 1.5 h until a clear solution was obtained, then TEA (14.5 mL, 103.6 mmol) and a solution of chlorooxathiolane **38** in CH$_2$Cl$_2$ was added, maintaining a gentle reflux. After 4 h at the same temperature, TEA (7.3 mL, 51.8 mmol) and H$_2$O (120 mL) were added, the temperature being kept at 30–35°C for 1.5 hours. To the suspension obtained, stirred for about 45 min, hexane (120 mL) was added over a period of 10 min at the same temperature. Stirring was continued overnight; then the suspension was filtered, the solid washed (H$_2$O, 2 × 60 mL/isopropyl acetate, 2 × 60 mL), and dried in vacuo at 40–45°C, to give pure L-**35b** (27.7 g, 66% overall yield) from a *cis/trans* = 10:1 mixture. Data for L-**35b**[83a]: mp 219°C dec. (EtOAc/hexane and MeOH in traces). $[a]_D^{25}$ −144° (c 1.02, CHCl$_3$). ^{1}H-NMR (300 MHz, CDCl$_3$): δ 0.76 (d, 3H, J = 7.0 Hz), 0.85–0.94 (m, 6H), 1.02–1.10 (m, 2H), 1.42–2.06 (m, 7H), 3.14 (d of d, 1H, J = 6.6 and J = 12.1 Hz), 3.54 (d of d, 1H, J = 4.7 and 12.1 Hz), 4.72–4.78 (m, 1H), 5.46 (s, 1H), 5.99 (d, 1H, J = 7.5 Hz), 8.43 (d, 1H, J = 7.6 Hz). ^{13}C-NMR (75 MHz, CDCl$_3$): ppm 16.1, 20.7, 21.9, 23.2, 26.4, 31.4, 34.0, 36.3, 40.7, 47.1,

76.7, 78.4, 90.3, 94.6, 141.8, 155.4, 165.6, 169.8. *Anal.* Calcd for $C_{18}H_{27}N_3O_4S$: C 56.67, H 7.13, N 11.01, S 8.41; found: C 56.84, H 7.24, N 11.22, S 8.70.

2(R)-(Hydroxymethyl)-5(S)-cytosin-1'-yl-1,3-oxathiolane (3TC)(Menthyl Ester Group Reduction).[83a] To a magnetically stirred suspension of LiAlH$_4$ (0.19 g, 5.0 mmol) in THF (20 mL), a solution of L-**35b** (0.68 g, 1.8 mmol) in THF (10 mL) was added at room temperature and under an inert atmosphere. After 30 min the reaction was stopped by adding MeOH (30 mL) and then SiO$_2$ (50 g); the residue was loaded on a short chromatographic column filled with Celite and SiO$_2$ and eluted with a EtOAc/hexane/MeOH (1:1:1, 500 mL) mixture. Solvents were concentrated and the residue purified by silica gel column chromatography (EtOAc/hexane/MeOH, 1:1:1) to furnish a sticky solid which was coevaporated with toluene to afford 3TC (0.38 g, 94% yield). Further purification by recrystallization from EtOAc-MeOH provided the product as white crystals. Data for 3TC: mp 176–177°C (EtOH/H$_2$O),[89] 156°C dec.,[90] 158–160°C (EtOAc/MeOH),[83a] 143–145°C (Et$_2$O/MeOH),[81b] 160–162°C (EtOAc/MeOH),[61] 160–162°C (EtOH).[82] $[a]_D^{25}$ −121.6° (c 1.1 MeOH),[81b] $[a]_D^{22}$ −135.0° (c 0.1 MeOH),[83a] $[a]_D^{20}$ −137.0° (c 0.21 MeOH),[90] $[a]_D^{21}$ −137.0° (c 1.0 MeOH)[61]. ^{1}H-NMR[81b,91] (300 MHz, DMSO-d_6): δ 3.03 (dd, $J_{2'a,1'}$ = 4.4 Hz, $J_{2'a,2'b}$ = 11.9 Hz, 1H, H-2'a), 3.43 (dd, $J_{2'b,1'}$ = 5.3, $J_{2'b,2'a}$ = 11.9 Hz, 1H, H-2'b), 3.80 (t, on D$_2$O exchange goes to d, $J_{5',4'}$ = 4.2 Hz, 2H, H-5'), 5.22 (t, $J_{4',5'}$ = 4.2 Hz, 1H, H-4'), 5.27 (t, $J_{OH',5'}$ = 4.6 Hz, 1H, OH), 5.88 (d, $J_{5,6}$ = 7.5 Hz, 1H, H-5), 6.21 (d, $J_{1',2'a}$ = 4.4, $J_{1',2'b}$ = 5.3 Hz, 1H, H-1'), 7.19 (b s, 2H, NH$_2$), 7.89 (d, $J_{6,5}$ = 7.5 Hz, 1H, H-6). ^{13}C-NMR (300 MHz, CD$_3$OD)[83a]: ppm 38.5, 64.1, 88.0, 88.9, 95.7, 142.8, 157.9, 167.7. IR spectra[89] [Nujol Mull] (cm^{-1}): 722, 788, 844, 927, 976, 1044, 1106, 1135, 1155, 1193, 1226, 1296, 1676, 14610, 1522, 1600, 1640, 2854, 2923, 3160, 3330. UV[81b] (H$_2$O) λ$_{max}$ 270.1 nm (ε = 9300) (pH 7). 270.0 (ε = 12150) (pH 2), 270.0 (ε = 9950) (pH 11). *Anal.* Calcd for $C_8H_{11}N_3O_3S$[83a]: C 41.91, H 4.84, N 18.33, S 13.99; found: C 41.72, H 4.90, N 18.35, S 13.91.

Among late-stage chemical resolution processes of racemic 3'-thianucleosides, Li et al.[90] reported a simple gram-scale strategy from (±)-BCH-189 (Scheme 8). The method was based on esterification of amide *cis*-(±)-**39** with (+)-menthyl chloroformate (73%), from whose diastereomeric mixture carbonate (−)-**40** could be isolated by recrystallization (MeOH). Deprotection (K$_2$CO$_3$) of (−)-**40** furnished 3TC (95%) with very high enantiomeric purity (>97% ee). Recovery of (+)-**41** from mother liquor along with its deprotection (K$_2$CO$_3$) also brought on isolation of (+)-BCH-189 (95%), albeit in very low optical purity (29% ee).[92]

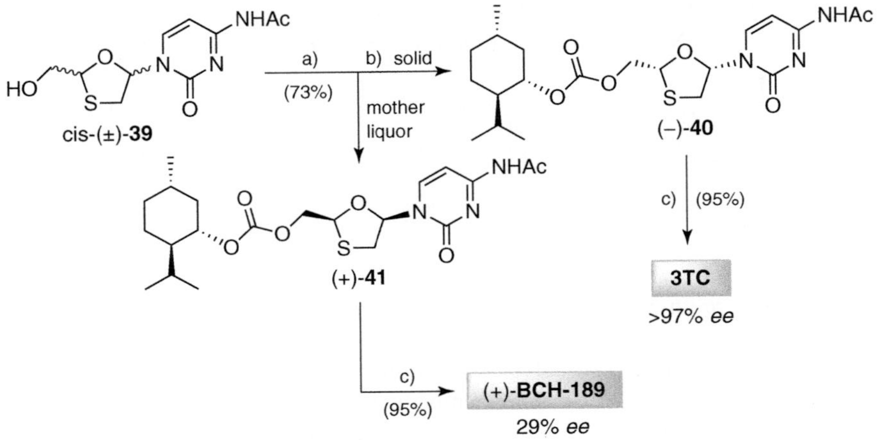

Scheme 8. *Reagents and conditions:* (a) (+)-menthyl chloroformate, CH_3CN, 0°C, 48 h; (b) recrystallization; (c) K_2CO_3, MeOH, 0°C, 10 h.

More recently, an efficient separation process of all four *cis-* and *trans-*(±)-2′,3′-dideoxy-3′-thiacytidine stereoisomers to individual stereoisomers was described by Roy et al.[89] (Scheme 9). The kilogram-scale method, relying on multiple preparations of diastereomeric salts and complexes having strikingly different solubility properties, represented a successful alternative to the wide variety of resolution techniques based on the formation of diastereomeric salts with chiral acids, including malic acid, mandelic acid, dibenzoyl tartaric acid, 3-bromocamphor-8-sulfonic acid, 10-camphorsulfonic acid, and di-*p*-toluoyltartaric acid.[89] Accordingly, after preparation of *cis/trans-*(±)-**6a** under common conditions, *cis-*isomers were first separated by formation of diastereomeric (*S*)-(+)-mandelate salts *cis-*(±)-**42** and *trans-*(±)-**43**. Indeed, mandelate salt *cis-*(±)-**42** was precipitated as solid (47%), leaving corresponding *trans-*(±)-**43** in mother liquor.

Protective group and acid counterion removal of *cis-*(±)-**42** was achieved under common conditions (NH_3/MeOH), releasing (±)-BCH-189 (72%). Then, heating of (±)-BCH-189 with a methanolic solution of (*S*)-BINOL enabled cocrystallization of the latter (1:1 stoichiometry) only with the *cis*-L-(−)-enantiomer (67%). The resulting co-crystals were finally treated with aqueous HCl, to retrieve 3TC (70%) in very high optical purity (99.9% *ee*). Interestingly, D-(+)-BCH-189 not cocrystallizing with (*S*)-BINOL could also be recovered from mother liquor and purified by recrystallization using H_2O or MeOH (98.2% *ee*). On the other hand, (*S*)-(+)-mandelate salt *trans-*(±)-**43** was deprotected (NH_3/MeOH, 72%), and treatment of resulting *trans-*(±)-**44** with (*S*)-BINOL led to cocrystal formation only with the *trans*-L-(−)-isomer (71%). Acidic treatment of the latter eventually gave (80%) pure L-(−)-**44** (98.8% *ee*); on the other hand, D-(+)-**44** was isolated from mother liquor and purified by further crystallization (98.1% *ee*; Scheme 9).

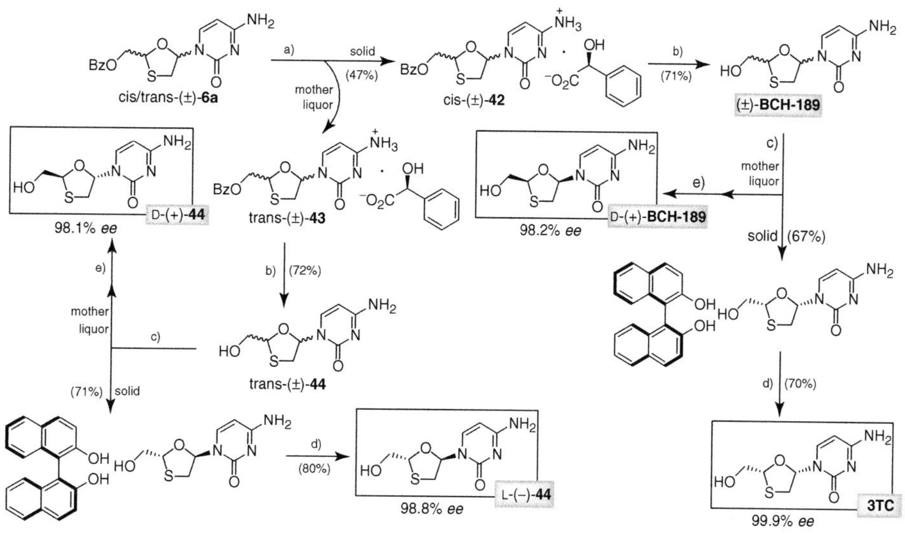

Scheme 9. *Reagents and conditions*: (a) (*S*)-(+)-mandelic acid, ethyl acetate, 70°C, then rt, 2 h; (b) (i) 15–25% NH₃/MeOH, rt, 10 h; (ii) recrystallization; (c) (*S*)-(−)-BINOL, MeOH, 65°C, 2 h; (d) ethyl acetate, H₂O, HCl$_{aq}$, rt; (e) ethyl acetate, H₂O, MeOH.

4. OPTICALLY ACTIVE 3′-OXA-4′-THIONUCLEOSIDES (2,4-DISUBSTITUTED 1,3-OXATHIOLANES)

Along with AMDX, RVT (elvucitabine), and RCV, the 2′,3′-dideoxy-3′-oxa-4′-thionucleoside apricitabine (ATC) is currently considered as one of the most attractive NRTIs among those undergoing clinical trials[7,46,93] Especially interesting is the excellent long-term safety and tolerability profile (similar to that of 3TC) in clinical trials of both treatment-naive and treatment-experienced HIV-1-infected patients. In addition, ATC displayed strong activity against WT and 3TC-resistant HIV-1 RT strains (including the M184V mutation and up to five TAMs), demonstrating a low propensity to cause resistance, even after long periods of treatment.[94] Along with the very few interactions with other antiretroviral drugs, ATC is overall a convenient candidate to be incorporated in novel HAART regimens and even to be used potentially as a replacement of 3TC and FTC in treatment-experienced patients with virus harboring the M184 mutation.[95]

Compared with the large clinical demand for ATC, the number of synthetic methods devoted to its preparation (as well as to that of its congeners) is surprisingly limited. All strategies to D- and L-3′-oxa-4′-thionucleosides have focused on de novo synthesis of key sulfoxide intermediate **45**, in either racemic or optically active form, then developing the suitable conditions for nucleobase insertion (Scheme 10).

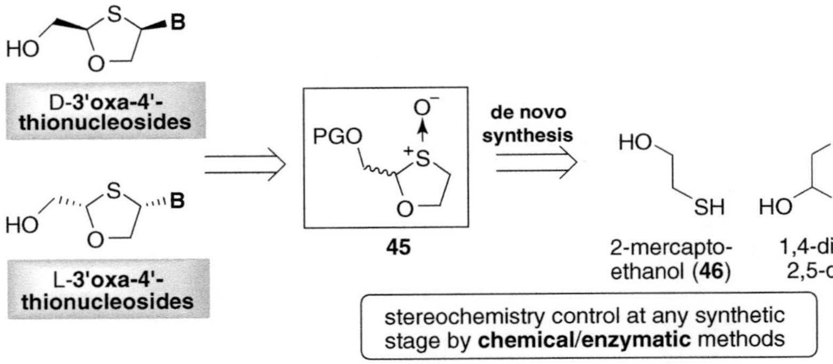

Scheme 10.

Optically active 3′-oxa-4′-thionucleosides were first obtained by Belleau et al.[96] (Scheme 11). Accordingly, racemic sulfoxide (±)-**47** was prepared starting from **4** and **46** (PTSA, then MMPP, 77% o.y.). Pummerer rearrangement (Ac$_2$O/N-Bu$_4$NOAc) of (±)-**47** then gave (±)-2-benzoyloxy-4-acetoxy-1,3-oxathiolane (**48**) as a 1:1 cis/trans mixture (66%). N-Glycosidation of **48** with persilylated purines/pyrimidines (SnCl$_4$ or TMSOTf) provided corresponding nucleosides (±)-**49** in variable yields (35–65%) without anomeric selectivity (cis/trans = 1:1). Deprotection of **49** under common conditions eventually furnished a number of racemic 3′-oxa-4′-thionucleosides, including 5-fluorocytosine nucleoside BCH-1081 [later referred to as (±)-dOFTC] and adenine nucleoside BCH-371 (Scheme 11). Notably, racemic cytosine nucleoside BCH-270 [later known as

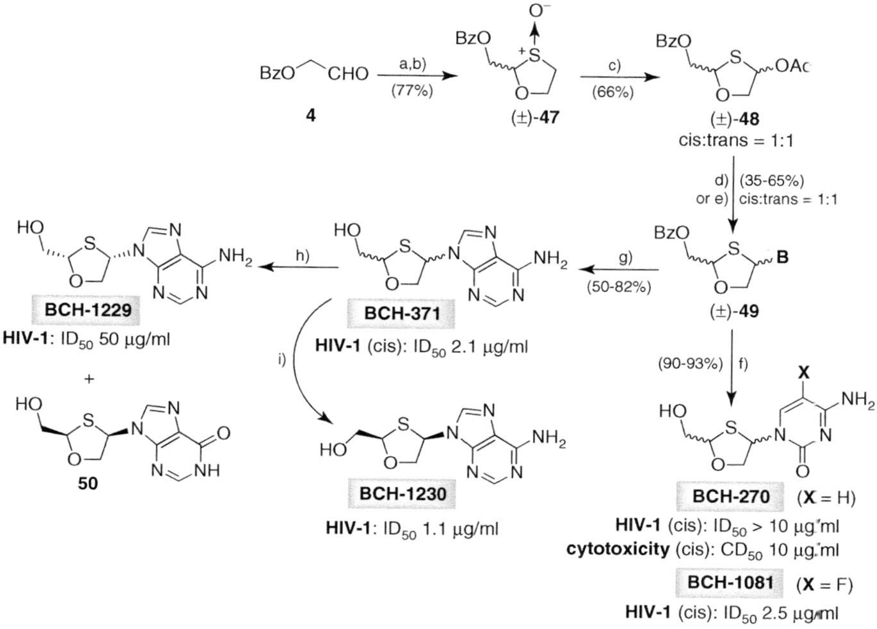

Scheme 11. *Reagents and conditions:* (a) **46**, PTSA, toluene, reflux; (b) MMPP, CH_2Cl_2-H_2O, rt; (c) Ac_2O, n-Bu_4NOAc, reflux; (d) persilylated pyrimidine (cytosine, 5-fluorocytosine, thymine, or uracil), $SnCl_4$, or TMSOTf, CH_2Cl_2, rt, 16 h; (e) persilylated purine (6-chloropurine or 2-amino-6-chloropurine), TMSOTf, DCE, reflux, 2 h; (f) NH_3/MeOH, rt; (g) NH_3/EtOH, 110°C, 6 h; (h) ADA, pH 7.2; (i) chiral HPLC. Anti-HIV-1 activity of the above (+), (−), and (±)-3′-oxa-4′-thionucleosides was evaluated in MT-4 cells; in all cases, cytotoxicity (MT-4 cells) was meant to be >100 μg/mL (unless otherwise specified).

BCH-10652 or (±)-dOTC] was not judged an interesting candidate, because of its cytotoxicity and low efficacy. However, the therapeutic potential of (±)-dOTC was reconsidered[97] in subsequent studies (see Scheme 12), even encouraging an early clinical evaluation of this compound.[98]

In the same paper, chiral 3′-oxa-4′-thionucleosides were obtained by enzymatic resolution of BCH-371, exploiting stereoselectivity properties of adenosine deaminase (ADA). Only deamination of (−)-optical isomer BCH-1230, having the hydroxymethyl group and the nucleobase in the same orientation of natural D-adenosine, was observed, leading to inosine analogue **50**. On the contrary, L-(+)-enantiomer BCH-1229 was left untouched (Scheme 11).

A short time later, Mansour et al.[99] performed the first synthetic route to enantiomerically pure (−)-dOTC (ATC), (+)-dOTC (BCH-10619), and the corresponding 5-fluorocytosine congeners (+)- and (−)-dOFTC (Scheme 12). The same method was also applied to determine unambiguously the configuration of (−)-adenine nucleoside BCH-1230[96] reported in Scheme 11 as

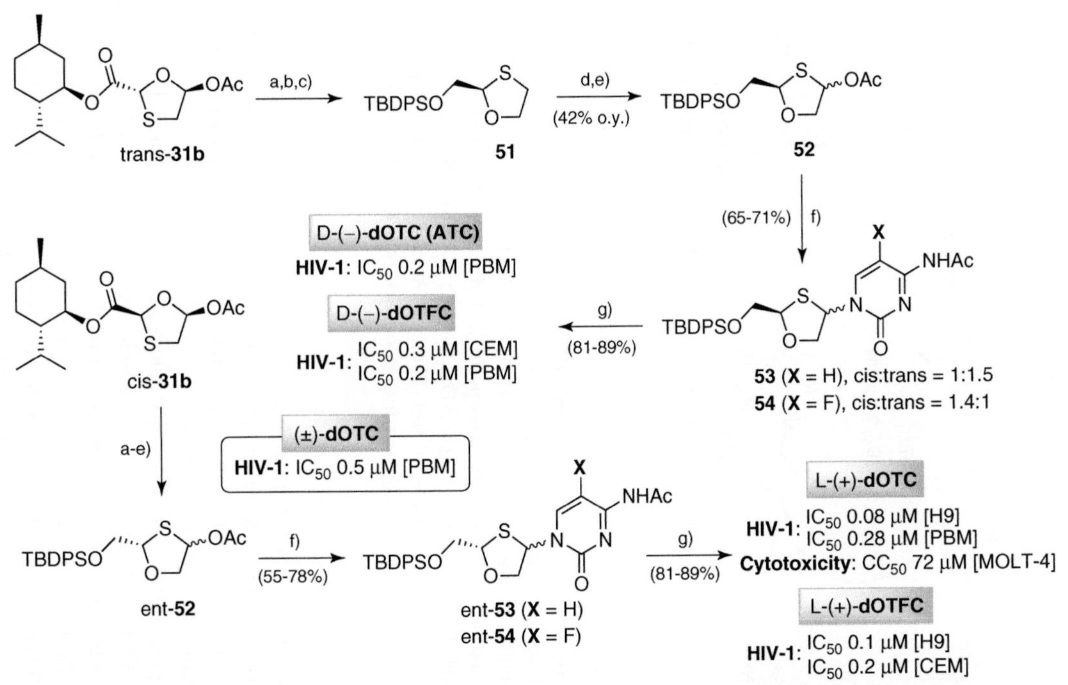

Scheme 12. *Reagents and conditions:* (a) Et$_3$SiH (neat), TMSOTf, 82–86%; (b) NaBH$_4$, EtOH, MeOH, 85–91%; (c) TBDPSCl, imidazole, THF, 90–94%; (d) *m*-CPBA, CH$_2$Cl$_2$, 89–94%; (e) *n*-Bu$_4$NOAc, Ac$_2$O, 120°C, 51–61%; (f) persilylated *N*-acetylcytosine or *N*-acetyl-5-fluorocytosine, TMSOTf, DCE, reflux; (g) TBAF, AcOH, then K$_2$CO$_3$, MeOH. Anti-HIV-1 activity of (±)-dOTC was obtained mixing equimolar amounts of (+)- and (−)-dOTC. Cytotoxicity was meant to be >100 μM (unless otherwise specified).

well as to identify oxathiolanyl BVDU analogues endowed with moderate anti-HSV-1 activity.[100] Synthesis was based on use of enantiopure 3'-thianucleoside precursors *trans*- and *cis*-**31b** to obtain two enantiomeric sugar donors **52** and *ent*-**52**. Formal transposition of acetoxy group was accomplished by a five-step sequence, involving reductive removal of acetoxy group of **31b** (Et$_3$SiH/TMSOTf), reduction of menthyl ester function (NaBH$_4$/EtOH), and protection of the resulting primary alcohol (TBDPSCl). Thioether oxidation of **51** (*m*-CPBA) and subsequent Pummerer rearrangement (*N*-Bu$_4$NOAc/Ac$_2$O) furnished oxathiolane **52** (42% over five steps) (Scheme 12). Nucleosides **53**/*ent*-**53** (*cis/trans* = 1:1.5) and **54**/*ent*-**54** (*cis/trans* = 1.4:1) were eventually synthesized by *N*-glycosylation of persilylated N^4-acetylcytosine or N^4-acetyl-5-fluorocytosine with **52** and *ent*-**52** under common conditions (TfOTMS, 55–78%). However, compared with the high optical purity of the starting **31b** (see Scheme 6), it was found that stereochemical integrity of oxathiolanyl units in the final compounds was not always entirely maintained.[99a] Not only (±)-dOTC,[97] ATC, and L-(+)-dOTC, but even D-(−)- and L-(+)-dOTFC, exhibited potent anti-HIV-1 activities, even though L-(+)-dOTC was cytotoxic to some extent.[97,101] Contrarily, anti-HBV activity of all the 3'-oxa-4'-thionucleosides noted above was much weaker.[101] Mansour's approach to optically active dOTC and congeners has been the main synthetic methodology enabling recruitment of 3'-oxa-4'-thionucleosides for pharmacology studies (before ATC underwent clinical trials), including the assessment of in vitro and in vivo activity against drug-resistant RT strains,[97,102] pharmacokinetics and metabolism of racemic and enantiomerically pure 3'-oxa-4'-thio(fluoro)cytidines,[97] drug resistance, safety and tolerability, drug–drug interactions,[103] and so on.

To the best of our knowledge, the only chemical approach of enantioselective de novo synthesis of (+)- and (−)-3'-oxa-4'-thionucleosides was that reported by Palumbo and co-workers[104] (Scheme 13). It exploited the well-established Modena's sulfoxidation protocol[105] [*t*-BuOOH, DET, Ti(*O-i*-Pr)$_4$] for the asymmetric oxidation of racemic thioether (±)-**55**. The reaction, performed using L-(+)-DET as a chiral catalyst, yielded sulfoxide **47** (90%) as a mixture of (*E*)- and (*Z*)-diastereoisomers (82:18 dr). The most abundant (*E*)-**47** was further found as a mixture of (−)- and (+)-enantiomers [60% *ee* in favor of (−)-**47**].[106] The same reaction, carried out using D-(−)-DET, gave mirrored results (Scheme 13), thus enabling synthesis of 3'-oxa-4'-thionucleosides belonging to both enantiomeric series. As an example, optically active oxathiolane-*S*-oxide (*E*)-**47** was used for the subsequent direct insertion of nucleobases (N^4-acetylcytosine or thymine) under modified Pummerer conditions (TMSOTf/Et$_3$N), with no racemization detected in the process. This yielded enantiomerically enriched 3'-oxathionucleosides D-(−)-**56** (70%) and D-(−)-**57** (76%) (*cis/trans* = 1:1 in both cases), whose *cis* anomers were eventually deprotected (MeONa, 90%).

Synthetic investigations into stereoselective conditions for nucleobase insertion were also carried out; however, contrarly to 3'-thianucleosides, *N*-glycosidation strategies leading to 3'-oxa-4'-thionucleosides did not yield remarkable results.

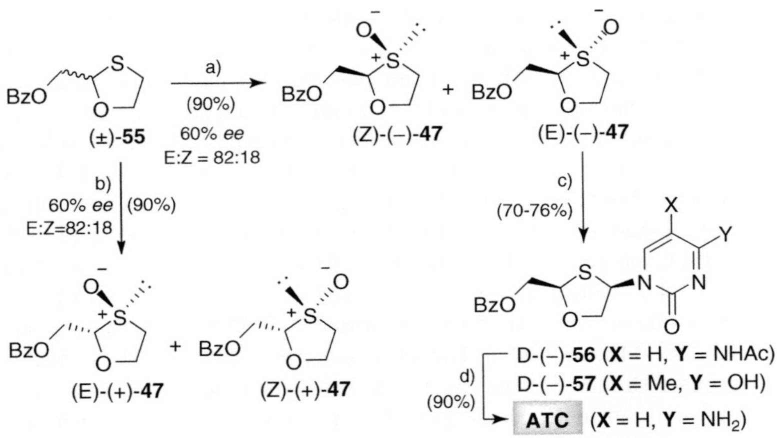

Scheme 13. *Reagents and conditions:* (a) *t*-BuOOH, L-(+)-DET, Ti(O-*i*-Pr)$_4$, CH$_2$Cl$_2$, −20°C, 14 h; (b) *t*-BuOOH, D-(−)-DET, Ti(O-*i*-Pr)$_4$, CH$_2$Cl$_2$, −20°C, 14 h; (c) N^4-acetylcytosine or thymine, TMSOTf, Et$_3$N, toluene, 0°C, 16 h; (d) MeONa, MeOH, rt, 6 h.

Compared with the lack of selectivity described previously with SnCl$_4$[96] and TfOTMS,[99] a slightly more pronounced *cis*-selectivity (*cis/trans* = 5:2) was found by Zacharie et al.[107] in the reaction of *bis*benzoyl acetal **58** with cytosine or 5-fluorocytosine in the presence of TMSI, leading to nucleosides **59** (55%) (Scheme 14). Acetal **58** was obtained directly from thioether **55** by treatment of the latter with BzOOH (50%). Removal of Bz ester at C5′ (NH$_3$/MeOH) of *cis*-**59** then gave (±)-dOTC and (±)-dOFTC (98%).

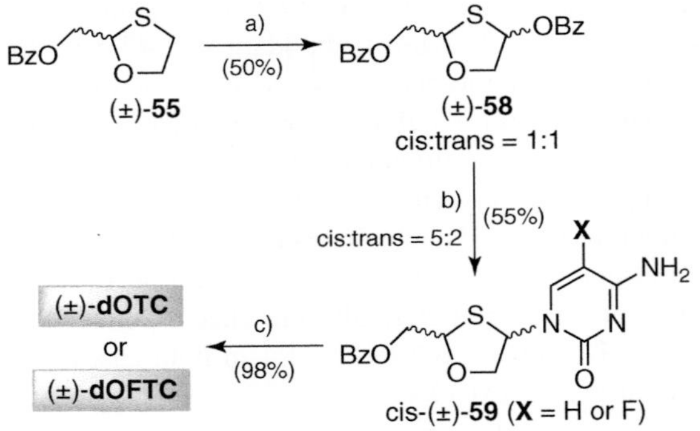

Scheme 14. *Reagents and conditions:* (a) BzOOH, C$_6$H$_6$, reflux, 4 h; (b) (i) cytosine or 5-fluorocytosine, HMDS, (NH$_4$)$_2$SO$_4$, rt, 3 h, then reflux, 2 h; (ii) TMSI, CH$_2$Cl$_2$, rt, 24 h; (c) NH$_3$/MeOH, rt, 16 h.

Within TMSI-based methodologies, Cu(II)-based catalysts were believed to play a role in inducing a β-attack by silylated nucleobases, resulting in an increase in *cis* selectivity. Yu et al.[94] tested a number of copper salts (including $CuCl_2$) in Pummerer-type *N*-glycosidations of oxathiolane sulfoxides, reaching *cis/trans* ratios of up to 3.9:1. Copper-catalyzed *N*-glycosidation of sulfoxide (*E/Z*)-(−)-**47** represented the crucial step in the manufacturing process[94] of ATC (100-kg plant scale), starting from optically active (>95% *ee*) 2-(*R*)-benzoyloxymethyl-1,3-oxathiolane [(2*R*)-**55**] (Scheme 15). Synthesis relied on thioether oxidation (H_2O_2/AcOH) of (2*R*)-**55** (90%), followed by Pummerer-type *N*-glycosidation of the resulting oxathiolane-*S*-oxides (*E*)- and (*Z*)-**47** with N^4-acetylcytosine (80%, *cis/trans* ~2:1) and eventually by deprotection (NaOMe, >99%) of nucleoside (1'*R*,4'*R*)-**56** (Scheme 15). Although undesired stereoisomeric impurities could be separated by formation of diastereomeric salts with chiral acids,[94] achiral and inexpensive PTSA was preferred, as it also allowed separation of the mixture by preferential crystallization, owing to fruitful formation of conglomerate salts. Acid counterions were then removed by treatment of PTSA salt **61** with Dowex 550A-OH (>80%), to give ATC in an overall >99% chiral purity (Scheme 15).

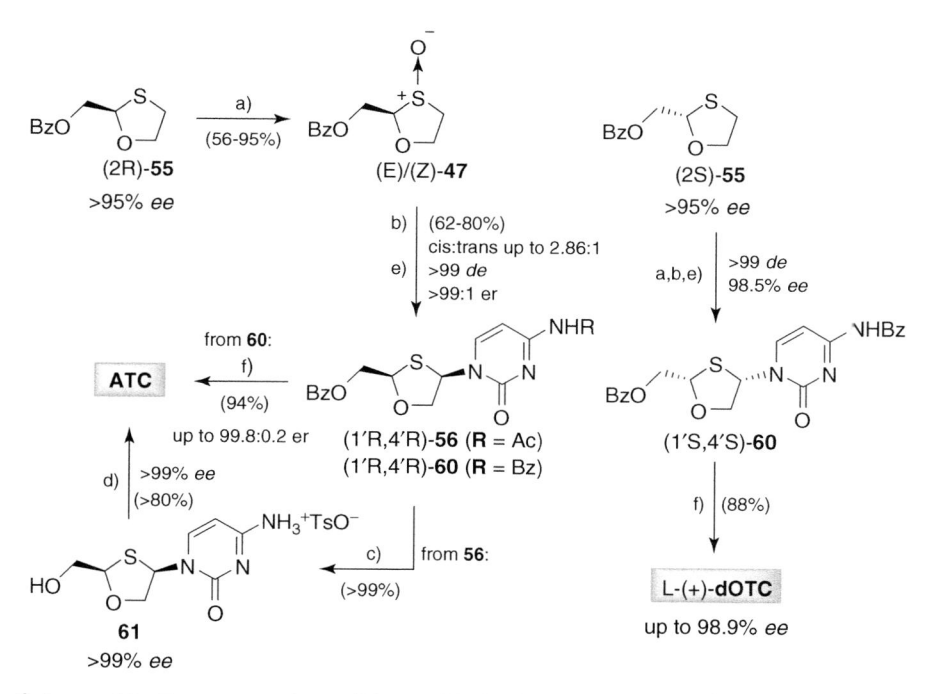

Scheme 15. *Reagents and conditions:* (a) H_2O_2, AcOH, 40–55°C, 1 h; (b) N^4-benzoylcytosine, TMSI, Et_3N, $CuCl_2$, CH_2Cl_2, −50 to 0°C, 16 h; (c) (i) MeONa, MeOH, 45°C; (ii) PTSA, 7 h; (d) Dowex 550A-OH, *i*-PrOH, 5 h; (e) recrystallization (MeOH or toluene, with or without entrainment); (f) NH_3/MeOH, rt, 16 h.

More recently, Deadman et al.[108] performed an accurate analysis of the single steps of Scheme 15, identifying some key elements to improve the synthetic protocol. Regarding the *N*-glycosidation step (initially performed using N^4-benzoylcytosine), only a minor role as a *cis*-selectivity-increasing agent was recognized to CuCl$_2$ (*cis/trans* = 2.86:1) since, even without this catalyst, a certain preferential formation of desired *cis*-isomer (1′*R*,4′*R*)-**60** was found.[108] As expected, stereoselective outcome of the reaction was not significantly influenced by protective groups at N4 of cytosine or at the CH$_2$O of oxathiolane-*S*-oxide **49**.[108] However, anomeric and (most important) optical purity only of *N,O*-bis-benzoyl nucleoside **60** could be efficiently enhanced by recrystallization, owing to the formation of conglomerates. Thus, performing further rounds of recrystallizations either exploiting the entrainment method (preferential crystallization using pure crystals as seeds) or without it, reasonably pure (1′*R*,4′*R*)-**60** was conveniently obtained (>99% *de*, up to >99:1 *ee*), and hence the desired ATC by eventual protective group removal (NH$_3$/MeOH, 94%). It was noteworthy that, to confirm the formation of true conglomerates, enantiomeric oxathiolane (2*S*)-**55** was also synthesized and subjected to coupling reaction with N^4-acetylcytosine, with mirrored results during recrystallization with entrainment of the major (1′*S*,4′*S*)-**60** (>99% *de*, up to 98:5 *ee*). The latter eventually led to formation of optically pure (+)-dOTC (98.9% *ee*, Scheme 15).

Experimental Procedures

2-(R)-Benzoyloxymethyl-4-(R)-(N-benzoylcytosin-1-yl)-1,3-oxathiolane [(1′R,4′R)-60] (N-Glycosidation by Sila–Pummerer Rearrangement).[108] To a magnetically stirred solution of 2-(*R*)-benzoyloxymethyl-1,3-oxathiolane-*S*-oxide (**47**, 95% *ee*) (12.0 g, 50.0 mmol) in CH$_2$Cl$_2$ (300 mL), at −50°C, TEA (15.3 mL, 110.0 mmol) and then TMSI (21.4 mL, 150.0 mmol) were added dropwise via a dropping funnel. Rate addition allowed keeping the internal temperature between −30 and −50°C. After 45 min, CuCl$_2$ (1.3 g, 10.0 mmol) was added to the light yellow solution; after an additional 5 min, N^4-BzC (10.1 g, 47.0 mmol) was added. Stirring was continued for a further 15 min at −50°C, then the mixture was warmed to 0°C over 1 h and left at the same temperature overnight. Finally, the reaction was left under stirring for an additional hour at room temperature. The reaction mixture was then again cooled down to 0°C and quenched (H$_2$O, 100 mL and then 5% NH$_4$OH, 100 mL). After 5 min, the mixture was diluted with CH$_2$Cl$_2$ (50 mL) and loaded on a Celite plug that was washed with additional CH$_2$Cl$_2$ (2 × 50 mL). The filtrate was transferred to a separating funnel and the organic layer was washed with 2% H₃PO₄ (2 × 60 mL) and 2.5% NH₄OH (2 × 100 mL). Reextraction of the aqueous layers was performed with CH₂Cl₂ (100 mL). The combined organic layers were dried (MgSO₄) and evaporated under reduced pressure, to give a yellowish/brownish thick oil (17.8 g, 86% recovery) consisting of a *cis/trans* mixture (2.86:1 dr, NMR: 62% purity). Recrystallization of the crude product in 250 mL MeOH (14 to 15 volumes of methanol compared with the mass of the crude residue) was performed warming the mixture to reflux until the solution was clear; then, slow mixture cooling (16 h) to room temperature led to the formation of slightly colored crystals that were filtered, washed (MeOH, 2 × 100 mL), and dried under reduced pressure. Crystals (7.2 g, 35% yield for the isomer) contained 99.1% of (1′R,4′R)-**60** and 0.9% of (1′S,4′S)-**60**. Enhancement of optical purity could be obtained by crystallization using an entrainment procedure (the stirred mixture was seeded with pure (1′R,4′R)-**60**, approximately at 55°C, then allowed to come to room temperature and stirring overnight). Data for (1′R,4′R)-**60**: mp 163–165°C. ¹H-NMR (400 MHz, CDCl₃): δ 4.05 (dd, 1H), 4.51 (d, 1H), 4.82 (m, 2H), 5.50 (t, 1H), 6.61 (d, 1H), 7.30 (poorly resolved d, 1H), 7.45 (m, 4H), 7.61 (m, 2H), 7.80 (d, 2H), 8.01 (d, 2H), 8.25 (d, 1H), 8.50 (br s, 1H).

2-(*R*)-Hydroxymethyl-4-(*R*)-(cytosin-1′-yl)-1,3-oxathiolane (ATC) (Bz Group Removal).

2-(*R*)-Benzoyloxymethyl-4-(*R*)-(*N*-benzoylcytosin-1-yl)-1,3-oxathiolane [(1′R,4′R)-**60**] (15.0 g, 34.0 mmol) was added to a 2 M NH₃/MeOH solution (250 mL). After overnight stirring, the starting muddy turned into a clear solution and was filtered on a Celite pad. Solvent was removed under reduced pressure, then acetone (100 mL) was added, yielding an off-white powdery solid. The suspension was filtered, washed (acetone, 2 × 25 mL), and dried to yield the desired ATC (6.5 g, 94% yield). Data for ATC: mp 213-215°C,[109] 210–211°C.[108] ¹H-NMR (250 MHz, DMSO-d_6)[104b]: δ 3.65–3.85 (m, 2H, H-2′), 4.08 (d, J = 11.3 Hz, 1H, H-5′a), 4.20 (d, J = 11.2 Hz, 1 H, H-5′b), 5.10 (t, J = 4.9 Hz, 1H, OH), 5.48 (t, J = 7.0 Hz, 1 H, H-4′), 5.80 (d, J = 7.4 Hz, 1H, H-5), 6.33 (d, J = 3.7 Hz, 1H, H-1′), 7.07–7.28 (m, 2H, NH₂), 7.75 (d, J = 7.4 Hz, 1H, H-6). ¹³C-NMR (300 MHz, DMSO-d_6)[109]: ppm 62.5, 62.9, 77.3, 88.9, 94.9, 142.4, 155.7, 167.9. UV[109] (CH₃OH) λ 270.0 nm.

5. OPTICALLY ACTIVE 3′-OXANUCLEOSIDES (2,4-DISUBSTITUTED 1,3-DIOXOLANES)

Although endowed with identical structural and conformational features,[37a] in many cases dioxolane and oxathiolane nucleosides differ significantly in biological properties. In an illustrative example, 3TC and troxacitabine (TRO) display comparable anti-HIV and anti-HBV potencies, but only TRO holds remarkable cytotoxicity (owing to undesired recognition by human DNA polymerases α, β, and γ).[1] This has moved attention for TRO toward antitumor rather than antiviral objectives.[110] As reported briefly earlier, the current arsenal of bioactive 2′,3′-dideoxy-3′-oxanucleosides involves a large number of interesting candidates (Figure 3); their therapeutic potential is at least as promising as that of 3′-thianucleosides and 3′-oxa-4′-thionucleosides. All 3′-oxanucleosides undergoing (pre)clinical development as antiretrovirals (DOT, FDOC, AMDX/DXG, ADP/DXG) exhibited potent activity against WT and 3TC- and AZT-resistant variants of HIV-RT.[7] DOT also drew attention because of its excellent bioavailability in rats and monkeys (ca.100%).[47] While DXG possessed limited aqueous solubility, its prodrugs AMDX and ADP were absorbed rapidly orally. AMDX demonstrated significant viral load reduction in clinical trials administered either alone and in combination with AZT.[15]

From a chemical standpoint, 2′,3′-dideoxy-3′-oxanucleosides share with 3′-thianucleosides reactivity and numerous synthetic methods; however, replacement of sulfur with oxygen expands the number of synthetic precursors available. As for 3′-thianucleosides, formation of dioxolane intermediate **62**, available in either racemic or enantiomerically enriched/pure form, depending on its preparation from starting materials **63–66** or other chiral synthons, has been of primary importance in all reported synthetic approaches leading to optically active D- and L-3′-oxanucleosides (Scheme 16).

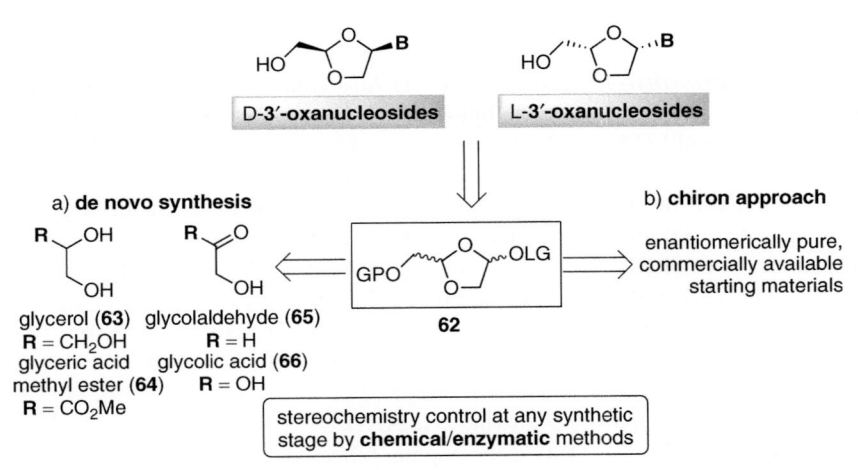

Scheme 16. Representative scheme of the synthetic approaches leading to D- and L-3′-oxanucleosides.

One of the earliest synthetic strategies to 3′-oxanucleosides is depicted in Scheme 17. Analogously to 3′-thianucleosides, Belleau et al.[48,111] performed a route to racemic (±)-**70** (yields not given) starting from **67** (available, in turn, from **63**), after chlorine displacement with a benzoyloxy group (BzOK), oxidation of primary alcohol function of **68** (PDC), and Bayer–Villiger rearrangement (*m*-CPBA) of the

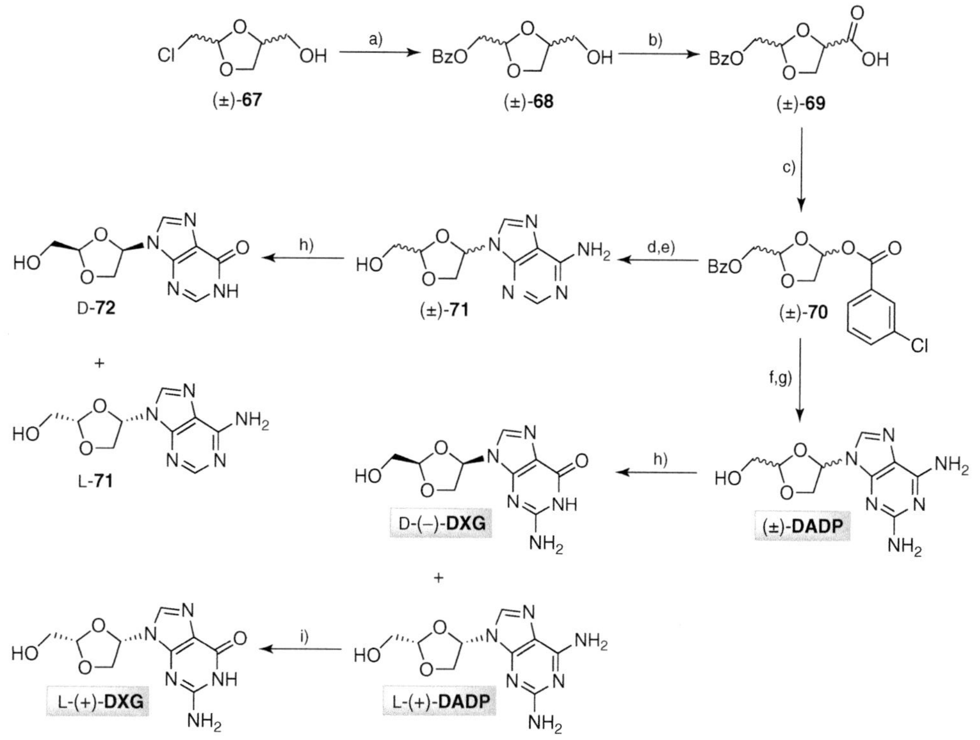

Scheme 17. *Reagents and conditions:* (a) BzOK, DMF, 18-crown-6, reflux, 24 h; (b) PDC, DMF, 0 °C; (c) EtOC(O)Cl, Et$_3$N, *m*-CPBA, CH$_2$Cl$_2$, −20 °C to rt; (d) adenine, TMSOTf, DMF, 120 °C; (e) K$_2$CO$_3$, MeOH, rt; (f) (i) 2-amino-6-chloropurine, HMDS, TMSCl, reflux, 3 h; (ii) TMSOTf, DCE, reflux, 4 h; (g) NH$_3$, MeOH, 110 °C, 16 h; (h) ADA, 0.05 M phosphate buffer, pH 7.2, rt, 24 h; (i) ADA, 0.05 M phosphate buffer, pH 7.2, rt, 20 days.

resulting carboxylic acid **69**. Coupling of acetal **70** with 2-amino-6-chloropurine (TMSOTf) and subsequent Bz group removal (NH$_3$/MeOH) finally provided, inter alia, adenine nucleoside *cis*-($\pm$)-**71** as well as bioactive diaminopurine congener [known later as ($\pm$)-DADP]. Afterward, Mansour et al.[112] conferred asymmetry to this procedure by performing an enzymatic resolution (enantiomeric excesses not given) of racemic purine nucleosides using ADA (Scheme 17). Particularly, complete deamination of *cis*-($\pm$)-**71** provided hypoxanthine D-**72**, leaving L-**71** unaltered. On the other hand, treatment of ($\pm$)-DADP with ADA furnished bioactive D-($-$)-guanine nucleoside [known later as D-($-$)-DXG] and L-($+$)-diaminopurine [L-($+$)-DADP]. The latter was further deaminated after more prolonged reaction times, to give an L-($+$)-guanosine analogue [L-($+$)-DXG]. Use of this synthetic procedure for subsequent in-depth studies, for examples those aimed at elucidating the action mechanisms determining antiviral activity of DXG and DADP against drug-resistant HIV RT variants,[113] has been reported.

Access to 3'-oxanucleosides under stereoselective Vorbrüggen conditions was investigated similarly as for 3'-thianucleosides, but with some notable differences. Concurrent with studies on stereoselective *N*-glycosidation of 1,3-oxathiolanes, Liotta et al. discovered an analogous degree of diastereofacial selectivity for 1,3-dioxolanes **73**, even though more oxophilic Ti(IV) catalysts were preferred as Lewis acids[67a,114] (Scheme 18). Treatment of ($\pm$)-**73a** and **73b** with TiCl$_3$(*O*-*i*-Pr) and silylated thymine or other (5-substituted) pyrimidines provided, inter alia, corresponding nucleosides ($\pm$)-**74-76a** and **74-76b** with predominant to complete *cis*-selectivity (typically, *cis/trans* >20:1), as rationalized on the basis of preferential 3'-oxygen/Ti^{IV} interaction (Scheme 18). In the case of 5-fluorocytosine nucleoside, Liotta et al.[114,115] also performed an enzymatic kinetic resolution via a highly enantioselective PLE-mediated hydrolysis of butyrate ester ($\pm$)-**76b**, which acted on the ($+$)-enantiomer only, leading to D-($+$)-FDOC (40%, 95% *ee*). On the other hand, chemical deprotection (MeONa) of unreacted ($-$)-optical isomer brought on formation of L-($-$)-FDOC with even higher purity (90–98% *ee*).[114,115] At first, these highly potent D-($+$)- and L-($-$)-3'-oxanucleosides were both judged too toxic (Scheme 18) to be used for clinical purposes; however, it was later demonstrated[115,116] that cytotoxicity of enantiomerically enriched D-($+$)-FDOC laid exclusively in its antipodean impurity [5% of cytotoxic L-($-$)-FDOC]. As proof of this, a sample of D-($+$)-FDOC obtained as a single enantiomer (>99% *ee*) by chiral chromatography still displayed excellent anti-HIV potency but negligible toxicity (Scheme 18). In an effort to improve its optical purity, 95% *ee* D-($+$)-FDOC underwent formation of diastereomeric L-tartrate salts (L-tartaric acid, then NaOH, 80%), bringing on nucleoside recovery in an excellent 99.7% *ee*.[115] In addition, undesired ($-$)-butyrate ester could be completely racemized [silylated 5-fluorocytosine/TiCl$_3$(*O*-*i*-Pr)/TiCl$_4$] while maintaining very good *cis* selectivity (*cis/trans* >18:1) (Scheme 18). This methodology significantly augmented the chemical efficiency of the entire process, which was therefore judged suitable for large-scale production.[115] Enantiomerically pure material obtained under these conditions was employed either to study intracellular metabolism of D-($+$)-FDOC and for pharmacokinetic studies in rhesus monkeys.[17]

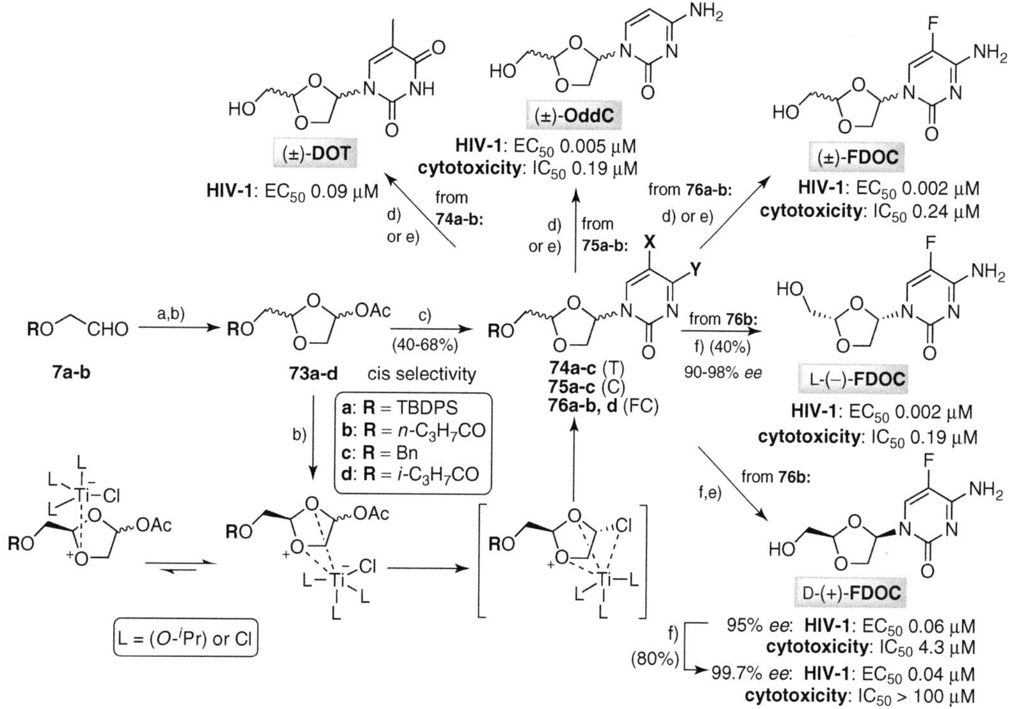

Scheme 18. *Reagents and conditions:* (a) (i) **66**, DCE, PTSA; (ii) Ac$_2$O, DMAP; (c) silylated thymine, cytosine, and 5-fluorocytosine, TiCl$_3$(O-i-Pr), CH$_2$Cl$_2$. Use of other Lewis acids, such as TiCl$_4$, TiCl$_3$(O-i-Pr)$_2$, TMSOTf, SnCl$_4$, and TMSI gave no reaction or caused a drop in anomeric selectivity (see text); (d) TBAF, THF, rt; (e) MeONa, MeOH, rt; (f) PLE, CH$_3$CN-H$_2$O; (g) L-tartaric acid, i-PrOH, rt, 24 h (ii) NaOH, H$_2$O. Anti-HIV-1 activity of all the (+), (−), and (±)-3′-oxanucleosides above was evaluated in acutely infected PBM cells, while cytotoxicity was evaluated in CEM cells.

Experimental Procedures

2-(Butyryl)-methyl-4-(5'-fluorocytosin-1'-yl)-1,3-dioxolane [(±)76b] (Ti^{IV}-Based N-Glycosidation).[115] To a magnetically stirred, clear solution of bistrimethylsilyl-5-fluorocytosine (0.35 g, 1.29 mmol) in anhydrous CH_2Cl_2 (10 mL), a 1 M solution of $TiCl_3(O-i-Pr)$[117] (2.58 mL, 2.58 mmol) in CH_2Cl_2 was slowly added until a reddish-clear solution was obtained. The latter was added dropwise to a solution of (±)-**73b** (0.30 g, 1.29 mmol) in anhydrous CH_2Cl_2 (10 mL). After 3 h at room temperature a mixture of absolute EtOH (40 mL) and concentrated NH_4OH (10 mL) was added to quench the reaction. The resulting mixture was stirred for 30 min at the same temperature, then it was filtered over a 2-inch Celite column. Solvent removal under reduced pressure afforded a light-yellow solid, which was purified on silica gel column chromatography (EtOAc/EtOH, 9:1) affording racemic *cis*-nucleoside (±)-**76b** (dr >20:1) as a white powder (0.27 g, 68% yield). Data for (±)-**76b**[118]: mp 145–146°C. ^{1}H-NMR (600 MHz, CDCl$_3$): δ 0.94 (t, J = 7.2 Hz, 3H), 1.64–1.70 (m, 2H), 2.36 (t, J = 7.2 Hz, 2H), 4.15–4.31 (m, 3H), 4.53 (dd, J = 2.4, and 10.8, Hz, 1H), 5.17 (s, 1H), 6.23 (d, J = 5.4 Hz, 1H), 7.87 (d, J = 6.6 Hz, 1H), 8.60 (br s, 1H). ^{13}C-NMR (150 MHz, CDCl$_3$): ppm 13.7, 18.5, 35.9, 61.6, 72.6, 82.5, 103.2, 125.2, 136.0, 154.3, 158.5, 173.0. HRMS (ESI): expected for $C_{12}H_{16}FN_3O_5$ [M + H]$^+$ found. 302.3; IR (neat): λ_{max} 3324, 3098, 2965, 1740, 1681, 1504, 1100 cm^{-1}.

2-(R)-Hydroxymethyl-4-(R)-(5'-fluorocytosin-1'-yl)-1,3-dioxolane [D-(+)-FDOC] (Enzymatic Resolution).[115] *Step a: Enzymatic hydrolysis*. A solution of racemic *cis*-nucleoside **76b** in a 1:4 mixture of CH_3CN and 0.12 M buffer solution (pH 8) was slightly heated; then, after cooling to room temperature, pig liver esterase (PLE, 1 unit hydrolyzes 1.0 µmol of **76b**) was added slowly. After 20 min, the reaction was stopped ($CHCl_3$) and the organic and aqueous phases were partitioned and separately evaporated. The aqueous layer contained the hydrolyzed desired product [D-(+)-FDOC] isolated in 40% yield and 95% *ee*. On the contrary, the organic phase contained the unreacted L-**76b** in about 90% *ee*.

Step b: Chiral salt crystallization. A 0.22 M solution of L-tartaric acid in *i*-PrOH and a 0.036 M solution of D-(+)-FDOC in the same solvent were prepared by gentle heat. After cooling both mixtures to room temperature, the D-(+)-FDOC solution was added slowly to that containing L-tartaric acid under magnetic stirring. After 10 min, the resulting solution was triturated to produce crystals of tartrate salt, which formation continued over the following 24 h. Then, to a solution of tartrate crystals in water, 1 M NaOH was added until neutral. Subsequent water removal under reduced pressure afforded pure D-(+)-FDOC in 80% o.y. and in 99.7% *ee*. Data for D-(+)-FDOC: ^{1}H-NMR (90 MHz, DMSO-d_6)[91,119]: δ 3.68 (s, 2H, H-5'), 4.05 (dd, $J_{2'a,1'}$ = 4.5, $J_{2'a,2'b}$ = 10.0 Hz, 1H, H-2'a), 4.30 (dd, $J_{2'b,1'}$ = 1.5, $J_{2'b,2'a}$ = 10.0 Hz, 1H, H-2'b), 4.92 (app. t, $J_{4',5'}$ = 1.8, 2.1, 1H, H-4'), 5.32 (app. t, $J_{OH,5'a}$ = 1.2, $J_{OH,5'b}$ = 3.5, 1H, OH), 6.18 (dd, $J_{1',2'b}$ = 1.5, $J_{1',2'a}$ = 4.5 Hz, 1H, H-2'b), 8.23 (d, $J_{6,F}$ = 7.2 Hz, 1H, H-6), 11.80 (s, 2H, NH_2). ^{13}C-NMR (100 MHz, CD_3OD)[115]: ppm 60.15, 71.97, 82.63, 105.70, 125.90, 135.96, 155.54, 158.53. FAB MS [M + Li]$^+$: 238.1. UV[119] (H_2O) λ_{max} 268.0 nm (ε = 7970) (pH 7), 268.0 (ε = 8540) (pH 2), 267.5 (ε = 6340) (pH 11).

Study of 1,3-dioxolane chemistry in Vorbrüggen *N*-glycosidation reactions was widened to other Ti^{IV} catalysts and Lewis acids, with controversial results in some cases. In early studies, Liotta et al.[67a] reported that anomeric selectivity of Ti^{IV}-mediated *N*-glycosidations of (±)-**73a** was closely dependent on Lewis acidity[67a] [TiCl$_4$: *cis/trans* = 7:1; TiCl$_3$(*O-i*-Pr): *cis/trans* = 10:1; TiCl$_2$(*O-i*-Pr)$_2$: *cis* only]. A further dependency on protective groups at 5'-OH was also suggested,[114] since *N*-glycosidation reactions of TBDPS ether **73a** gave relatively lower selectivities in some cases (presumably owing to epimerization of kinetically formed *cis* isomers),[114] while those starting from butyrate esters **73b** always exhibited full *cis* selectivity when using a combination of TiCl$_3$(*O-i*-Pr) and (catalytic or stoichiometric) TiCl$_4$.[114] In more recent reports by the same authors,[115] TiCl$_4$ was found to cause product epimerization if used with butyrate ester **73b**, while TiCl$_2$(*O-i*-Pr)$_2$ lacked suitable Lewis acidity to catalyze *N*-glycosidation. On the other hand, Mansour et al.[120] studied the degree of selectivity as a result of the loss in stereochemical integrity of the 1,3-dioxolane moiety, through use of optically pure[121] benzyl ethers **73c** (Scheme 18). In their hands, from no to moderate *cis*-selectivity was obtained with TiCl$_2$(*O-i*-Pr)$_2$ (*cis/trans* = 1:1) and TiCl$_4$ (*cis/trans* = 73:27), respectively. However, while the former caused no detectable racemization levels, the latter led to complete loss of optical purity.[120] Other Lewis acids, such as

TMSOTf or TMSI, not resulting in any significant *cis*-selectivity (*cis/trans* up to 2:1, according to Chu et al.[122]), caused from low to negligible racemization levels. In the case of TMSOTf, Liotta et al.[115] accurately analyzed the residual optical purity of protected D-(+)-FDOC synthesized starting from enantiomerically pure (>99% *ee*) isobutyrate ester **73d**, depending on the amount of Lewis acid used during the *N*-glycosidation step (0.5 equiv of TMSOTf, 97% *ee*; 1.0 equiv of TMSOTf, 90% *ee*; 2.0 equiv of TMSOTf, 85% *ee*). Eventually, both Liotta et al.[67a] and Mansour et al.[120] independently ascertained that use of $SnCl_4$ proceeded without significant selectivity (*cis/trans* up to 8:7), in contrast to previous results by Chu et al.[123] Mansour et al.[120] eventually highlighted that use of $SnCl_4$ also brought on partial product racemization [residual optical purity of the *cis*-(−)-isomer: 56% *ee*].

Extensive biological evaluation of enantiomerically pure (+) and (−)-3′-oxanucleosides was accomplished by Chu et al.[119,123–125] through seminal "chiron" approaches devised starting from D-mannose (leading to D-3′-oxanucleosides in ≥12 steps) and L-gulose or L-gulonic-1,4-lactone (leading to L-3′-oxanucleosides in ≥9 steps) (Scheme 19). As for 3′-thianucleosides, both synthetic routes basically relied on iterated chemical degradations [$NaIO_4/NaBH_4$, $NaIO_4/RuO_2$, and $Pb(OAc)_4$] of D-**77** and L-**78**, followed by *N*-glycosidation (TMSOTf) of the resulting enantiomeric dioxolanes **73a**, *ent*-**73a**, **79**, and *ent*-**79**. Major synthetic details of these routes have already been reported elsewhere.[1] In the long list of (substituted) purinyl and pyrimidinyl 3′-oxanucleosides, evaluated as tools of SAR studies,[123–125] several enantiomerically pure bioactive compounds, such as D-(−)-DOT,[18,126] L-(−)-OddC (later defined as TRO),[110,127,128] D-(+)-FOddC (later D-FDOC),[24] D-(−)-DXG[129] and its prodrugs D-(−)-DADP,[130] D-(−)-ACDP,[131] and D-(−)-ADP[19,131] were employed for a wide variety of other preclinical studies, involving their uptake, metabolism, pharmacokinetics, and combination studies. Besides anti-HIV and anti-HBV activity, 5-substituted uracil L-3′-oxanucleosides with potent anti-VZV[31,86] and anti-EBV[30] activity in vitro (in some cases, higher than those of anti-HHV drugs ACV and BVDU) were identified (Scheme 19). In recent times, starting from the nucleosides above, Chu et al. engaged in the synthesis and evaluation of a series of 3′-oxanucleoside prodrugs, including 4-*N*-aliphatic amides of TRO,[132] 5′-*O*-aliphatic acid esters and amino acid esters of DOT[133] as well as its 5′-*O*-phosphoramidates, (a)cyclic 5′-*O*-phosphates,[134] and polyamidoamine (PAMAM) dendrimers.[135]

Experimental Procedures

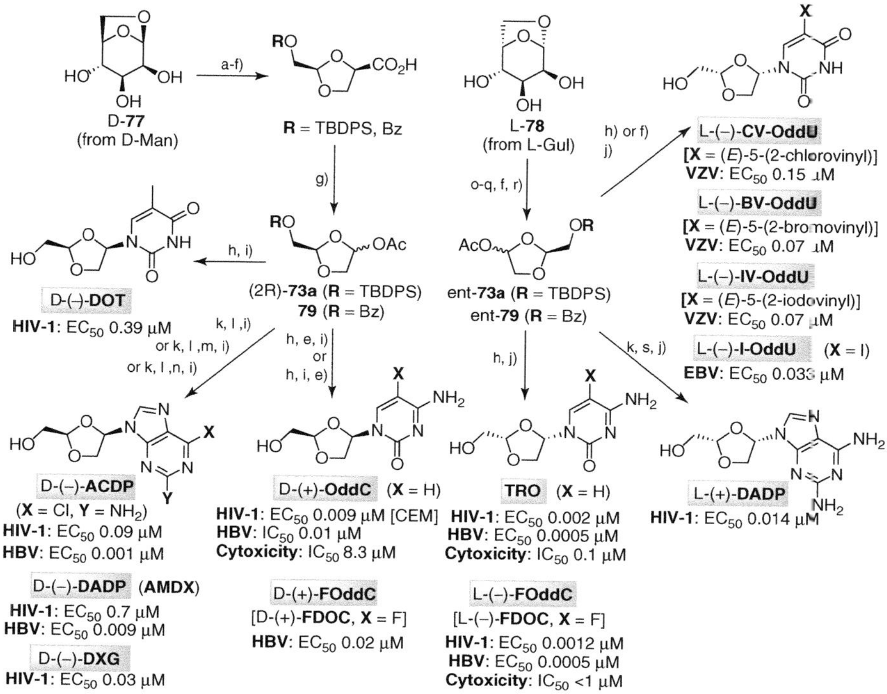

Scheme 19. *Reagents and conditions:* (a) (i) DMP, PTSA, acetone, rt, 24 h; (ii) BzCl, pyridine, 0°C, 45 min; (b) H$_2$SO$_4$, H$_2$O-dioxane, 80°C, 15 h; (c) (i) NaIO$_4$, H$_2$O, EtOH, rt, 1 h; (ii) NaBH$_4$, 10 min; (d) TBDPSCl, imidazole, DMF, rt, 24 h; (e) MeONa, MeOH, rt; (f) NaIO$_4$, RuO$_2$, CH$_3$CN-CCl$_4$, H$_2$O, rt, 5 h; (g) Pb(OAc)$_4$, AcOEt, pyridine, rt, 15 h; (h) (i) N^4-acetylcytosine, N^4-benzoyl-5-fluorocytosine, thymine, or 5-substituted uracil, HMDS, (NH$_4$)$_2$SO$_4$, reflux; (ii) TMSOTf, DCE or CH$_2$Cl$_2$; (i) TBAF, THF, rt; (j) NH$_3$/MeOH; (k) (i) 6-chloropurine, 2,6-dichloropurine; or 2-fluoro-6-chloropurine, HMDS, (NH$_4$)$_2$SO$_4$ reflux; (ii) TMSOTf, CH$_2$Cl$_2$; (l) NH$_3$/DME; (m) 2-mercaptoethanol, NaOMe, MeOH; (n) NH$_3$, EtOH; (o) (i) NaIO$_4$, H$_2$O, MeOH, 0°C, 15 min; (ii) NaBH$_4$, 10 min; (iii) PTSA, acetone, 6 h; (p) BzCl, pyridine-CH$_2$Cl$_2$, rt, 2 h; (q) PTSA, MeOH, rt, 3 h; (r) Pb(OAc)$_4$, THF, pyridine, rt, 45 min; (s) NaN$_3$/EtOH, reflux, 45 min; (t) (i) 5-(*E*)-halovinyluracil, TBDMSOTf, 2,4,6-collidine, CH$_2$Cl$_2$, rt, 30 min; (ii) TMSI, rt, 3 h. For all the (+)- and (−)-3'-oxanucleosides above, cytotoxicity is meant to be >100 μM (unless otherwise specified). Moreover, anti-HIV-1 activity was evaluated in PBM cells (unless otherwise specified), anti-HBV activity in 2.2.15 cells, anti-VZV activity in Ellen strains, and cytotoxicity in Vero cells.

2(R)-(t-Butyl-diphenylsilyloxy)-methyl-4(R)-thymin-1'-yl-1,3-dioxolane (80) (Vorbrüggen N-Glycosidation).[119]

Step a: Silylation of the nucleobase. To a solution of thymine (0.15 g, 1.2 mmol) in hexamethyldisilazane (15 mL), catalytic ammonium sulfate[136] was added. The reaction mixture was refluxed, becoming a clear light-yellow solution after 15 min. Stirring and refluxing were continued under a nitrogen atmosphere for 2.5–4 h. Then HMDS was removed under reduced pressure and anhydrous conditions (coevaporation with anhydrous toluene or xylene improves the process) to give the silylated nucleobase as a white solid.

Step b: N-*Glycosidation.* To a solution of silylated dioxolane (2R)-**73a** (0.24 g, 0.6 mmol) in anhydrous DCE (5 mL) at 5°C, a solution of silylated nucleobase in DCE (5 mL) and TMSOTf (0.23 mL, 1.2 mmol) was subsequently added. The resulting mixture was stirred for about 1 h at room temperature; then, saturated NaHCO$_3$ (10 mL) was slowly added. After an additional 30 min, the mixture was extracted (CH$_2$Cl$_2$) and washed with water until neutral. The organic layer was dried (MgSO$_4$) and evaporated under reduced pressure to give a crude residue from which chromatography over silica gel gave desired nucleosides **80** (*cis/trans* = 1.5:1 dr, 0.21 g, 75% yield) as a white foam.[119] Data for *cis*-**80**: $[a]_D^{21}$ −6.98° (c 0.43, MeOH). ^{1}H-NMR (90 MHz, CDCl$_3$)[91]: δ 1.08 (s, 9H, *t*-Bu), 1.67 (s, 3H, CH$_3$), 3.92 (d, $J_{5'4'}$ = 3.2 Hz, 2H, H-5'), 4.14 (d, $J_{2',1'}$ = 4.0 Hz, 2H, H-2'), 5.06 (d, $J_{4',5'}$ = 3.2 Hz, 1H, H-4'), 6.36 (t, $J_{1',2'}$ = 4.0 Hz, 1H, H-1'), 7.44–7.56 (m, 11H, H-arom), 9.51 (br s, 1H, NH). UV (CH$_3$OH) λ$_{max}$ 265.0 nm (pH 7), 265.0 (pH 2), 264.5 (pH 11).

2-(R)-Hydroxymethyl-4-(R)-(thymin-1'-yl)-1,3-dioxolane [D-(−)-DOT] (TBDPS Group Removal).[119] A solution of nucleoside **80** (93.3 mg, 0.2 mmol) and TBAF (0.24 mL, 0.24 mmol, 1 M in THF) in THF (3 mL) was stirred at room temperature until TLC indicated disappearance of the starting material (ca. 1 h). The reaction mixture was concentrated under reduced pressure, and the resulting residue was purified by silica gel column chromatography to give free nucleoside (42 mg, 92.1% yield) as a white solid. Data for D-(−)-DOT: mp 174–175°C. $[a]_D^{25}$ −18.8° (c 0.17 MeOH). ^{1}H-NMR (90 MHz, DMSO-d_6)[91]: δ 1.75 (d, $J_{CH3,6}$ = 1.2 Hz, 3H, CH$_3$), 3.63 (dd, $J_{5',4'}$ = 2.6 and $J_{5',OH}$ = 6.0 Hz, 2H, H-5'), 4.03 (dd, $J_{2'a,1'}$ = 5.5, $J_{2'a,2'b}$ = 9.9 Hz, 1H, H-2'a), 4.22 (dd, $J_{2'b,1'}$ = 2.0, $J_{2'b,2'a}$ = 9.9 Hz, 1H, H-2'a), 4.90 (t, $J_{4',5'}$ = 2.6, 1H, H-4'), 5.16 (t, $J_{OH,5'}$ = 6.0, 1H, OH), 6.21 (dd, $J_{1',2'b}$ = 2.0, $J_{1',2'a}$ = 5.5 Hz, 1H, H-2'b), 7.67 (d, $J_{6,CH3}$ = 1.2 Hz, 1H, H-6), 11.27 (br s, 1H, NH). UV (H$_2$O) λ$_{max}$ 266.0 nm (ε = 107,600) (pH 7), 266.5 (ε = 9890) (pH 2), 266.3 (ε = 8400) (pH 11).

Almost concurrently to Chu's approach, Mansour et al.[112,121,137] devised an alternative methodology starting from D-mannitol and L-ascorbic acid

(Scheme 20). Whereas oxidative degradation of diol function of protected D-mannitol (RuCl$_3$/NaOCl) gave a diastereomeric pair of enantiomerically pure dioxolanes (2R,4R)-**81** and (2S,4R)-**81** (30%), C–C cleavage of L-ascorbic acid derivative **82** (H$_2$O$_2$, then RuCl$_3$/NaOCl) furnished complementary (2S,4S)-**81** and (2R,4S)-**81** (57%). (2R)-Configured stereoisomers were both converted [Pb(OAc)$_4$] into acetate (2R)-**73d** (80%). Analogously, (2S)-**73d** was obtained from (2S,4S)-**81** and (2S,4R)-**81** (80%). N-Glycosidation (TMSOTf) of (2R)-**73d** and (2S)-**73d** along with final deprotection of the resulting nucleosides, were then accomplished as commonly reported. Biological evaluation data described by the authors were in substantial agreement with those of Chu et al.[119] (Scheme 20). In addition, interesting anti-HSV-1 activity by L-BV-OddU was found,[138] even though this result was later questioned.[86] Optical purity of the synthesized (+) and (−)-3′-oxanucleosides allowed their use for subsequent studies, such as those aimed at elucidating the biochemical mechanisms determining the onset of resistance events to troxacitabine.[139]

Stereoselective synthesis of enantiomerically pure (2R,4S)-**81** was later the subject matter of a chemoenzymatic strategy reported by Kazlauskas et al.[140] First, access to optically active 1,3-dioxolane methyl esters (2S,4S)- and (2R,4S)-**85** (cis/trans = 2:1) from synthetically available 2,3-O-isopropylidene-(S)-glyceric acid methyl ester (**86**) and 2-benzyloxyacetaldehyde (**84**) was performed by a slightly modified synthetic procedure from that described a decade earlier by Norbeck et al.[141] leading to (±)-DOT (Scheme 21). Attention was then focused on discrimination of cis- and trans-**85** (otherwise difficult to separate by common purification techniques) through selective hydrolysis of a methyl ester group mediated by α-chymotrypsin and bovine pancreatic proteases (Scheme 21), screened from 91 commercially available hydrolases.[140] Particularly, a small-scale α-chymotrypsin-catalyzed hydrolysis of the diastereomeric mixture yielded carboxylic acid (2R,4S)-**81** in 56% de, while starting methyl ester (2S,4S)-**85** was recovered in much higher purity (>98% de, 55% yield). The method was employed successively to recruit enantiomerically pure DOT for biochemical studies elucidating its action mechanism against WT and drug-resistant RT strains.[142]

Among synthetic methodologies to enantiomerically pure 3′-oxanucleosides amenable to industrial production, it is worth to mention those reported by Popp et al.[143] (Wacker Chemie) and Sznaidman et al.[111] whose groups independently described large-scale resolution processes of synthetically useful[67a,122] 1,3-dioxolan-4-ones **88** (Scheme 22). After having prepared ton quantities of racemic lactone (R,S)-**88a** bearing an isobutyryl group at the methyleneoxy arm starting from chloroacetaldehyde dimethyl acetal **87** (Scheme 22a), Popp et al.[143] performed chemoenzymatic resolution to (R)-**88a** in high optical purity (22% yield, >98% ee) using Novozym 435 (NZ435), the immobilized formulation of Candida antarctica lipase B (CALB), selected among a variety of 30 commercially available lipases and esterases. Most notably, resolution protocol was even carried out on a pilot-plant scale (1100 kg), displaying same performance. On the other hand, Sznaidman et al.[111] fulfilled a synthesis of enantiomerically pure D-(−)-DADP, starting from lactones (R,S)-**88a**

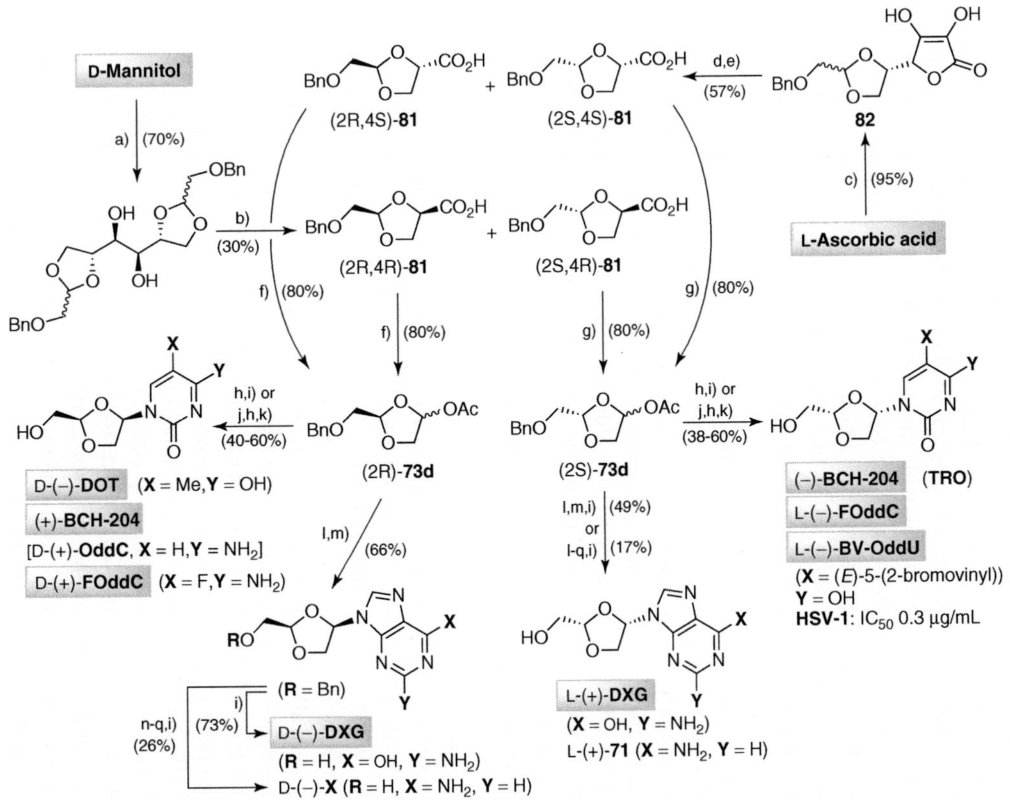

Scheme 20. *Reagents and conditions*: (a) 2-benzyloxyacetaldehyde dimethylacetal, SnCl$_2$, DME, reflux, o.n; (b) RuCl$_3$·H$_2$O, NaOCl, H$_2$O-CH$_3$CN-DCE; (c) 2-benzyloxyacetaldehyde dimethylacetal, PTSA, CH$_3$CN; (d) H$_2$O$_2$, K$_2$CO$_3$, EtOH; (e) (i) RuCl$_3$·H$_2$O, NaOCl, H$_2$O-CH$_3$CN-DCE, Bn(Et)$_3$NCl; (ii) H$^+$; (f) Pb(OAc)$_4$, CH$_3$CN-Py; (g) Pb(OAc)$_4$, CH$_3$CN-CH$_2$Cl$_2$-Py; (h) silylated (N^4-acetyl) pyrimidine, TMSOTf, CH$_2$Cl$_2$, rt, 6 h; (i) 10% PdO, cyclohexene, EtOH, reflux, 3 h; (j) (i) H$_2$, Pd/C, EtOH, rt, 48 h; (ii) Ac$_2$O, pyridine, DMAP, rt; (k) K$_2$CO$_3$, MeOH; (l) (i) N^2-acetyl,O^6-diphenylcarbamoylguanine, BSA, (CH$_2$Cl)$_2$, reflux; (ii) TMSOTf, reflux, 6 h; (m) NH$_2$NH$_2$·H$_2$O, THF, reflux, 3 h; (n) POCl$_3$/PhNMe$_2$/Et$_4$NCl, CH$_3$CN, reflux, 30 min; (o) *t*-BuONO, THF, −20°C; (p) (Me$_3$Si)$_3$SiH, THF, rt; (q) NH$_3$/EtOH, 100°C, 12 h.

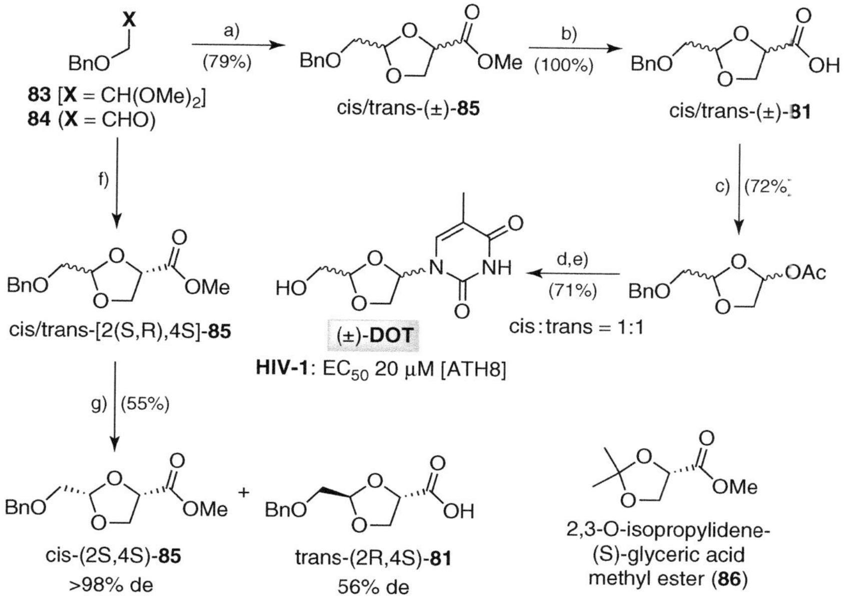

Scheme 21. *Reagents and conditions:* (a) (±)-glyceric acid methyl ester, PTSA, CH$_3$CN, reflux, 2 h; (b) LiOH, THF–H$_2$O, reflux, 1.5 h; (c) Pb(OAc)$_4$, CH$_3$CN, pyridine, rt, 6 h; (d) silylated thymine, TMSOTf, CH$_2$Cl$_2$, rt, 66 h; (e) H$_2$, 20% Pd(OH)$_2$/C, EtOH, rt, 30 min; (f) **86**, PTSA; (g) α-chymotrypsin, BES buffer, pH 7.2.

and **88b** (R = isobutyryl or *p*-methoxybenzoyl), in turn obtained starting from easily available 2,2-dialkoxyethanol **89** (Scheme 22b). At first, a gram-scale separation of (*R,S*)-**88a** and **88b** was carried out by preparative chiral chromatography, while in a subsequent exploratory larger-scale run, enzymatic resolution by chirazyme-L2 (ee not given) was also carried out to afford kilogram quantities of lactone (*R*)-**88a**.[111] Chiral 1,3-dioxolan-4-ones (*R*)-**88a** and (*R*)-**88b** were both used to provide D-(−)-DADP after common transformations (18.7% o.y. from (*R*)-**88a**; 36.4% o.y. from (*R*)-**88b**; 4.5–8.7% o.y. from **89**).

6. CONCLUSIONS

Triggered by their remarkable biological properties discovered over the last two decades, oxathiolane and dioxolane nucleosides belonging to D- and L-series—including 3TC, FTC, RCV, ATC, TRO, AMDX/APD/DXG, DOT, FDOC, and many others—have been at the core of intense synthetic endeavors. Compared with bioactive nucleosides having slight structural changes in the natural ribose (e.g., AZT, ddC, ddI), whose preparation commonly involves routine synthetic strategies, access to 3′-thianucleosides, 3′-oxa-4′-thionucleosides, and

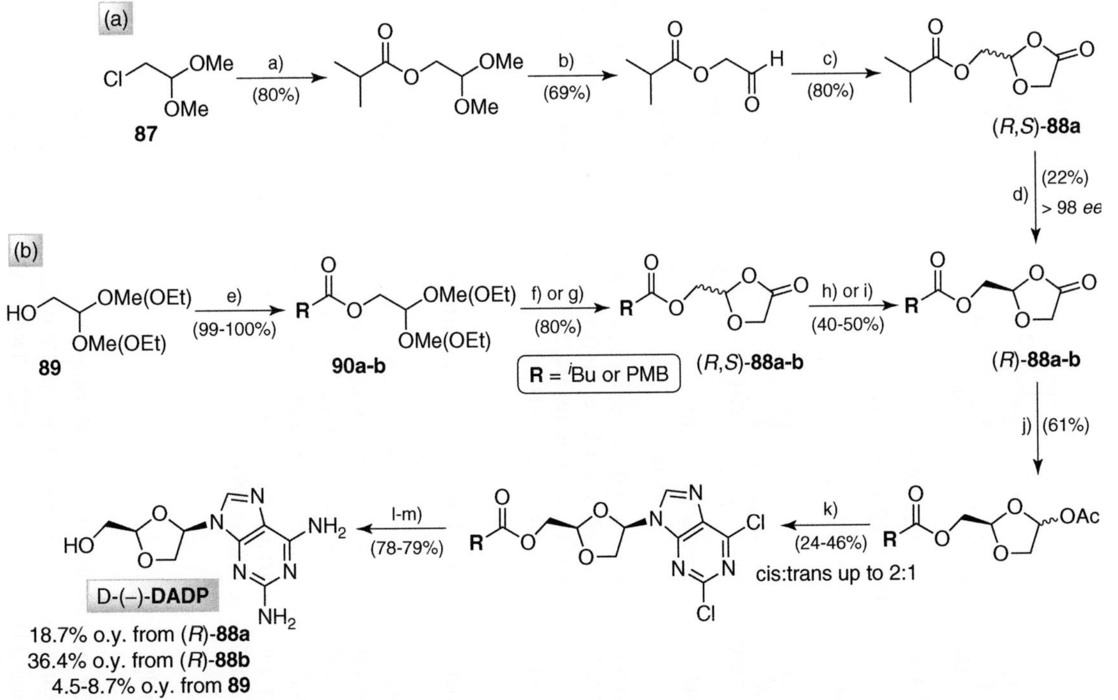

Scheme 22. *Reagents and conditions:* (a) (CH$_3$)$_2$CHCOOK; (b) HCOOH; (c) (i) **66**, HMDS, 93%; (ii) TMSOTf, CH$_2$Cl$_2$, −78°C to rt, 20 h; (d) NZ435, TBME, MeOH, 30°C; (e) 4-methoxybenzoyl chloride or isobutyryl chloride, DMAP, Et$_3$N, TBME, 0°C, 16 h; (f) **66**, BF$_3$·OEt$_2$, CH$_3$CN, 0°C, o.n; (g) from **90b**: (i) TFA, H$_2$O, CH$_2$Cl$_2$, rt, 3.5 h; (ii) **66**, BF$_3$·OEt$_2$, DME, 0°C, o.n; (h) preparative chiral HPLC; (i) from (*R,S*)-**88a**: chirazyme-L2, *i*-ProH, H$_2$O, 0°C; (j) (i) LiAlH(O-*t*-Bu)$_3$, −10°C, 20 min, then rt, 30 min; (ii) Ac$_2$O, DMAP, −15°C, 1 h, then rt, o.n; (k) (i) 2,6-dichloropurine, (NH$_4$)$_2$SO$_4$, HMDS, reflux, 2 h; (ii) TMSOTf, CH$_2$Cl$_2$, −10°C to rt, o.n; (l) (i) NaN$_3$, DMF, rt, 4 h; (ii) H$_2$, Pd/C, rt, o.n; (m) *n*-BuNH$_2$, MeOH, reflux, 4 h.

3′-oxanucleosides has required the study of more creative approaches, especially those leading to enantiomerically pure compounds. Over the years, a wealth of stereoselective routes based on fully chemical (asymmetric de novo syntheses, chiron approaches) or chemoenzymatic strategies have been devised, ranging from small-scale synthesis of enantiomerically enriched mixtures up to pilot-plant-scale production of material in high optical purity. On the one hand, the search for highly enantioselective methodologies turned out to be crucial in those cases (e.g., 3TC, FDOC) where mirror-image nucleosides displayed strikingly different toxicity profiles. On the other hand, stereoselectivity concerns were driven by a need for cost-effective, nonetheless efficient and scalable routes, required while transferring eligible synthetic methods from the academic to the industrial level. For a few nucleosides, the degree of advancement reached in this field nowadays enables large-scale, high-yielding, and fast access to optically pure material, whether dealing with enzymatic resolution processes, chemical asymmetric methods, or a combination of them. In all remaining cases, further efforts are warranted, either in the development of improved synthetic protocols among existing methodologies or in the search for innovative strategies exploiting upcoming synthetic technologies.

Acknowledgments

This chapter is an acknowledgment of the efforts of the large number of eminent scientists working on this challenging topic; particularly, it is dedicated to the memory of Bernard Belleau (1925–1989), pioneering investigator of the medicinal chemistry of oxathiolane and dioxolane nucleosides.

REFERENCES

1. *Antiviral Nucleosides: Chiral Synthesis and Chemotherapy*; Chu, C. K., Ed.; Elsevier: Amsterdam, **2003**.
2. *Modified Nucleosides: In Biochemistry Biotechnology and Medicine*; Herdewijn, P., Ed.; Wiley-VCH: Weinheim, Germany, **2008**.
3. *Chemistry and Biology of Artificial Nucleic Acids*; Egli, M., Herdewijn, P.Eds.; Wiley-VCH: Zurich, Switzerland, **2012**.
4. *Antiviral Drug Strategies*; De Clercq, E., Ed.; Wiley-VCH: Weinheim, Germany, **2011**.
5. (a) De Clercq, E. *Nat. Rev. Drug Discov.* **2002**, *1*, 13–25. (b) De Clercq, E. *Nat. Rev. Microbiol.* **2004**, *2*, 704–720. (c) De Clercq, E. *Nat. Rev. Drug Discov.* **2007**, *6*, 1001–1018.
6. (a) Mansour, T. S.; Storer, R. *Curr. Pharm. Design* **1997**, *3*, 227–264. (b) Rando, R. F.; Nguyen-Ba, N. *Drug Discov. Today* **2000**, *5*, 465–476. (c) Wainberg, M. A. *Antivir. Res.* **2009**, *81*, 1–5.
7. Schinazi, R. F.; Hernandez-Santiago, B. I.; Hurwitz, S. J. *Antivir. Res.* **2006**, *71*, 322–334.
8. Quintas-Cardama, A.; Cortes, J. *Exp. Opin. Investig. Drugs* **2007**, *16*, 547–557.

9. (a) Férir, G.; Kaptein, S.; Neyts, J.; De Clercq, E. *Rev. Med. Virol.* **2008**, *18*, 19–34. (b) Mailliard, M. E.; Gollan, J. L. *Annu. Rev. Med.* **2006**, *57*, 155–166.
10. Mehellou, Y.; De Clercq, E. *J. Med. Chem.* **2010**, *53*, 521–538.
11. (a) Flexner, C. *Nat. Rev. Drug Discov.* **2007**, *6*, 959–966. (b) Stellbrink, H.-J. *Antivir. Chem. Chemother.* **2009**, *19*, 189–200. (c) Ghosh, R. K.; Ghosh, S. M.; Chawla, S. *Expert Opin. Pharmacother.* **2011**, *12*, 31–46.
12. (a) Merino, P. *Curr. Med. Chem.* **2006**, *13*, 539–545. (b) Casu, F.; Chiacchio, M. A.; Romeo, R.; Gumina, G. *Curr. Org. Chem.* **2007**, *11*, 1017–1032. (c) Romeo, G.; Chiacchio, U.; Corsaro, A.; Merino, P. *Chem. Rev.* **2010**, *110*, 3337–3370.
13. The prefix "thio" involves replacement of an oxygen atom of an aldose or ketose by S; on the other hand, "thia" is used to indicate replacement of a CH_2 group by S. See: McNaught, A. D. *Pure Appl. Chem.* **1996**, *68*, 1919–2008.
14. http://www.who.int/hiv/data/2009_global_summary.png.
15. Cihlar, T.; Ray, A. S. *Antivir. Res.* **2010**, *85*, 39–58.
16. Use of a low cost generic fixed-dose combination including d4T, 3TC and the NNRTI Nevirapine has also been explored in the developing world; see ref. 15.
17. Hernandez-Santiago, B. I.; Chen, H.; Asif, G.; Beltran, T.; Mao, S.; Hurwitz, S. J.; Grier, J.; McClure, H. M.; Chu, C. K.; Liotta, D. C.; Schinazi, R. F. *Antimicrob. Agents Chemother.* **2005**, *49*, 2589–2597.
18. Asif, G.; Hurwitz, S. J.; Obikhod, A.; Delinsky, D.; Narayanasamy, J.; Chu, C. K.; McClure, H. M.; Schinazi, R. F. *Antimicrob. Agents Chemother.* **2007**, *51*, 2424–2429.
19. Manouilov, K. K.; Manouilova, L. S.; Boudinot, D. F.; Schinazi, R. F.; Chu, C. K. *Antivir. Res.* **1997**, *35*, 187–193.
20. Corbett, A. H.; Rublein, J. C. *Curr. Opin. Investig. Drugs* **2001**, *2*, 348–353.
21. Herzmann, C.; Arastèh, K.; Murphy, R. L.; Schulbin, H.; Kreckel, P.; Drauz, D.; Schinazi, R. F.; Beard, A.; Cartee, L.; Otto, M. J. *Antimicrob. Agents Chemother.* **2005**, *49*, 2828–2833.
22. http://www.aidsinfo.nih.gov/DrugsNew/DrugDetailT.aspx?int_id=415.
23. Palumbo, E. *Hep. Mon.* **2008**, *8*, 125–127.
24. Schinazi, R. F.; Gosselin, G.; Faraj, A.; Korba, B. E.; Liotta, D. C.; Chu, C. K.; Mathé, C.; Imbach, J.-L.; Sommadossi, J.-P. *Antimicrob. Agents Chemother.* **1994**, *38*, 2172–2174.
25. Furman, P. A.; Davis, M.; Liotta, D. C.; Paff, M.; Frick, L. W.; Nelson, D. J.; Dornsife, R. E.; Wurster, J. A.; Wilson, L. J.; Fyfe, J. A.; Tuttle, J. V.; Miller, W. H.; Condreay, L.; Averett, D. R.; Schinazi, R. F.; Painter, G. R. *Antimicrob. Agents Chemother.* **1992**, *36*, 2686–2692.
26. Gordeau, H.; Jolivet, J. In *Deoxynucleoside Analogs in Cancer Therapy*; Peters, G. J., Ed.; Humana Press: Totowa, NJ, **2006**; pp. 199–214.
27. (a) Lapointe, R.; Létourneau, R.; Steward, W.; Hawkins, R. E.; Batist, G.; Vincent, M.; Whittom, R.; Eatock, M.; Jolivet, J.; Moore, M. *Ann. Oncol.* **2005**, *16*, 289–293. (b) Vose, J. M.; Panwalkar, A.; Belanger, R.; Coiffier, B.; Baccarani, M.; Gregory, S. A.; Facon, T.; Fanin, R.; Caballero, D.; Ben-Yehuda, D.; Giles, F. *Leuk. Lymphoma* **2007**, *48*, 39–45.
28. Seignères, B.; Pichoud, C.; Martin, P.; Furman, P.; Trépo, C.; Zoulim, F. *Hepatology* **2002**, *36*, 710–722.

29. Chu, C. K.; Gadthula, S.; Chen, X.; Choo, H.; Olgen, S.; Barnard, D. L.; Sidwell, R. W. *Antivir. Chem. Chemother.* **2006**, *17*, 285–289.
30. Lin, J.-S.; Kira, T.; Gullen, E.; Choi, Y.; Qu, F.; Chu, C. K.; Cheng, Y.-C. *J. Med. Chem.* **1999**, *42*, 2212–2217.
31. Li, L.; Dutschman, G. E.; Gullen, E. A.; Tsujii, E.; Grill, S. P.; Choi, Y.; Chu, C. K.; Cheng, Y.-C. *Mol. Pharmacol.* **2000**, *58*, 1109–1114.
32. (a) Wang, P.; Hong, J. H.; Cooperwood, J. S.; Chu, C. K. *Antivir. Res.* **1998**, *40*, 19–44. (b) Gumina, G.; Song, G.-Y.; Chu, C. K. *FEMS Microbiol. Lett.* **2001**, *202*, 9–15. (c) Gumina, G.; Chong, Y.; Choo, H.; Song, G.-Y.; Chu, C. K. *Curr. Top. Med. Chem.* **2002**, *2*, 1065–1086.
33. Mathé, C.; Gosselin, G. *Antivir. Res.* **2006**, *71*, 276–281.
34. (a) Maury, G. *Antivir. Chem. Chemother.* **2000**, *11*, 165–189. (b) Zemlicka, J. *Pharmacol. Ther.* **2000**, *85*, 251–266.
35. (a) D'Alonzo, D.; Guaragna, A.; Palumbo, G. *Curr. Org. Chem.* **2009**, *13*, 71–98. (b) D'Alonzo, D.; Guaragna, A.; Palumbo, G. *Curr. Med. Chem.* **2009**, *16*, 473–505. (c) D'Alonzo, D.; Guaragna, A.; Palumbo, G. *Chem. Biodivers.* **2011**, *8*, 373–413.
36. Forsman, J. J.; Leino, R. *Chem. Rev.* **2011**, *111*, 3334–3357.
37. (a) Sabini, E.; Hazra, S.; Konrad, M.; Burley, S. K.; Lavie, A. *Nucleic Acids Res.* **2007**, *35*, 186–192. (b) Sabini, E.; Hazra, S.; Konrad, M.; Lavie, A. *J. Med. Chem.* **2007**, *50*, 3004–3014.
38. Eriksson, S.; Munch-Petersen, B.; Johansson, K.; Eklund, H. *Cell. Mol. Life Sci.* **2002**, *59*, 1327–1346.
39. Sarafianos, S. G.; Marchand, B.; Das, K.; Himmel, D.; Parniak, M. A.; Hughes, S. H.; Arnold, E. *J. Mol. Biol.* **2009**, *385*, 693–713.
40. Kukhanova, M.; Liu, S. H.; Mozzherin, D.; Lin, T. S.; Chu, C. K.; Cheng, Y. C. *J. Biol. Chem.* **1995**, *270*, 23055–23059.
41. Chang, C.-N.; Doong, S.-L.; Zhou, J. H.; Beach, J. W.; Jeong, L. S.; Chu, C. K.; Tsai, C.-H.; Cheng, Y.-C. *J. Biol. Chem.* **1992**, *267*, 13938–13942.
42. Feng, J. Y.; Anderson, K. S. *Biochemistry* **1999**, *38*, 55–63.
43. Lee, K.; Chu, C. K. *Antimicrob. Agents Chemother.* **2001**, *45*, 138–144.
44. Sarafianos, S. G.; Das, K.; Clark, A. D., Jr.; Ding, J.; Boyer, P. L.; Hughes, S. H.; Arnold, E. *Proc. Natl. Acad. Sci. USA* **1999**, *96*, 10027–10032.
45. Chong, Y.; Chu, C. K. *Front. Biosci.* **2004**, *9*, 164–186.
46. Wainberg, M. A.; Cahn, P.; Bethell, R. C.; Sawyer, J.; Cox, S. *Antivir. Chem. Chemother.* **2007**, *18*, 61–70.
47. Chu, C. K.; Yadav, V.; Chong, Y. H.; Schinazi, R. F. *J. Med. Chem.* **2005**, *48*, 3949–3952.
48. Belleau, B.; Dixit, D.; Nguyen-Bu, N.; Kraus, J.-L. International Conference on AIDS. Montreal, Canada, June 4–9, paper T.C.O.1, **1989**.
49. Wilson, L. J.; Hager, M. W.; El-Kattan, Y. A.; Liotta, D. C. *Synthesis* **1995**, 1465–1479.
50. Ray, A. S.; Schinazi, R. F.; Murakami, E.; Basavapathruni, A.; Shi, J.; Zorca, S. M.; Chu, C. K.; Anderson, K. S. *Antivir. Chem. Chemother.* **2003**, *14*, 115–125.
51. (a) Feng, J. Y.; Shi, J.; Schinazi, R. F.; Anderson, K. S. *FASEB J.* **1999**, *13*, 1511–1517. (b) Schinazi, R. F. *J. Acquir. Immune Defic. Syndr.* **2003**, *34*, 243–245.
52. Drogan, D.; Rauch, P.; Hoffmann, D.; Walter, H.; Metzner, K. J. *Antivir. Res.* **2010**, *86*, 312–315.

53. dos Santos Pinheiro, E.; Antunes, O. A. C.; Fortunak, J. M. D. *Antivir. Res.* **2008**, *79*, 143–165.
54. Feng, J. Y.; Murakami, E.; Zorca, S. M.; Johnson, A. A.; Johnson, K. A.; Schinazi, R. F.; Furman, P. A.; Anderson, K. S. *Antimicrob. Agents Chemother.* **2004**, *48*, 1300–1306.
55. Hurwitz, S. J.; Otto, M. J.; Schinazi, R. F. *Antivir. Chem. Chemother.* **2005**, *16*, 117–127.
56. Schinazi, R. F.; McMillan, A.; Lloyd, R. L., Jr.; Schlueter-Wirtz, S.; Liotta, D. C.; Chu, C. K. *Antivir. Res.* **1997**, *34*, A42.
57. Black, P. L.; Ussery, M. A.; Otto, M. J.; Stuyver, L.; Hurwitz, S. J.; Barnett, T.; Mowrey, J. O.; Tharnish, P. M.; Boudinot, F. D.; Schinazi, R. F. *Antivir. Res.* **2001**, *51*, A017.
58. Soudeyns, H.; Yao, X. I.; Gao, Q.; Belleau, B.; Kraus, J. L.; Nguyen-Ba, N.; Spira, B.; Wainberg, M. A. *Antimicrob. Agents Chemother.* **1991**, *35*, 1386–1390.
59. (a) Coates, J. A. V.; Cammack, N.; Jenkinson, H. J.; Mutton, I. M.; Pearson, B. A.; Storer, R.; Cameron, J. M.; Penn, C. R. *Antimicrob. Agents Chemother.* **1992**, *36*, 202–205. (b) Cammack, N.; Rouse, P.; Marr, C. L. P.; Reid, P. J.; Boehme, R. E.; Coates, J. A. V.; Penn, C. R.; Cameron, J. M. *Biochem. Pharmacol.* **1992**, *43*, 2059–2064.
60. Although roughly equipotent in these experiments, (+)-BCH-189 and 3TC revealed different antiretroviral potencies (about an order of magnitude) in subsequent assays. See: Schinazi, R. F.; Chu, C. K.; Peck, A.; McMillan, A.; Mathis, S.; Cannon, D.; Jeong, L. S.; Beach, J. W.; Choi, W. B.; Yeola, S.; Liotta, D. C. *Antimicrob. Agents Chemother.* **1992**, *36*, 672–676.
61. Storer, R.; Clemens, I. R.; Lamont, B.; Noble, S. A.; Williamson, C.; Belleau, B. *Nucleos. Nucleot.*, **1993**, *12*, 225–236.
62. However, it was later pointed out that Yoshikawa's protocol ($POCl_3/P(O)OR_3$) was not well suited if applied to enantiomerically pure 3TC; while use of stochiometric amounts of $POCl_3$ gave unsatisfactory yields, larger excesses even resulted in glycosidic bond cleavage, as reported by detection of free cytosine (Roy, B.; Lefebvre, I.; Puy, J.-Y.; Périgaud, C. *Tetrahedron Lett.* **2011**, *52*, 1250–1252). It has been hypothesized that sulfur atom may take part to stabilization of carbocation species originated from glycosidic bond cleavage, thus increasing chemical instability of 3TC (Crauste, C.; Périgaud, C.; Peyrottes, S. *J. Org. Chem.* **2011**, *76*, 997–1000).
63. Mahmoudian. M.; Baines, B. S.; Drake, C. S.; Hale, R. S.; Jones, P.; Piercey, J. E.; Montgomery, D. S.; Purvis, I. J.; Storer, R.; Dawson, M. J.; Lawrence, G. C. *Enzyme Microb. Technol.* **1993**, *15*, 749–755.
64. Boller, T.; Meier, C.; Menzler, S. *Org. Process Res. Dev.* **2002**, *6*, 509–519.
65. Mahmoudian, M. In *Biocatalysis in the Pharmaceutical and Biotechnology Industries*; Patel, R. N., Ed.; CRC Press: Boca Raton, FL, **2007**; pp. 53–102.
66. Woo, J.-H.; Shin, H.-J.; Kim, T.-H.; Ghim, S.-Y.; Jeong, L.-S.; Kim, J.-G.; Song, B.-H. *Biotechnol. Lett.* **2001**, *23*, 131–135.
67. (a) Choi, W.-B.; Wilson, L. J.; Yeola, S.; Liotta, D. C.; Schinazi, R. F. *J. Am. Chem. Soc.* **1991**, *113*, 9377–9379. (b) Choi, W.-B.; Yeola, S.; Liotta, D. C.; Schinazi, R. F.; Painter, G. R.; Davis, M.; St. Clair, M.; Furman, P. A. *Bioorg. Med. Chem. Lett.* **1993**, *3*, 693–696.

68. Jeong, L. S.; Alves, A. J.; Carrigan, S. W.; Kim, H. O.; Beach, J. W.; Chu, C. K. *Tetrahedron Lett.* **1992**, *33*, 595–598.
69. Hoong, L. K.; Strange, L. E.; Liotta, D. C.; Koszalka, G. W.; Burns, C. L.; Schinazi, R. F. *J. Org. Chem.* **1992**, *57*, 5563–5565.
70. Schinazi, R. F.; McMillan, A.; Cannon, D.; Mathis, R.; Lloyd, R. M.; Peck, A.; Sommadossi, J.-P.; St. Clair, M.; Wilson, J.; Furman, P. A.; Painter, G.; Choi, W.-B.; Liotta, D. C. *Antimicrob. Agents Chemother.* **1992**, *36*, 2423–2431.
71. Doong, S. L.; Tsai, C. H.; Schinazi, R. F.; Liotta, D. C.; Cheng, Y. C. *Proc. Natl. Acad. Sci. USA* **1991**, *88*, 8495–8499.
72. Schinazi, R. F.; Lloyd, R. M., Jr.; Nguyen, M.-H.; Cannon, D. L.; McMillan, A.; Ilksoy, N.; Chu, C. K.; Liotta, D. C.; Bazmi, H. Z.; Mellors, J. W. *Antimicrob. Agents Chemother.* **1993**, *37*, 875–881.
73. Shewach, D. S.; Liotta, D. C.; Schinazi, R. F. *Biochem. Pharmacol.* **1993**, *45*, 1540–1543.
74. Gaede, B. J.; Nardelli, C. A. *Org. Process Res. Dev.* **2005**, *9*, 23–29.
75. Osborne, A. P.; Brick, D.; Ruecroft, G.; Taylor, I. N. *Org. Process Res. Dev.* **2006**, *10*, 670–672.
76. Liotta, D.C.; Schinazi, R. F.; Choi, W.-B. U.S. Patent 5,814,639 (**1998**).
77. Cousins, R. P. C.; Mahmoudian, M.; Youds, P. M. *Tetrahedron: Asymmetry* **1995**, *6*, 393–396.
78. (a) Milton, J.; Brand, S.; Jones, M. F.; Rayner, C. M. *Tetrahedron Lett.* **1995**, *36*, 6961–6964. (b) Milton, J.; Brand, S.; Jones, M. F.; Rayner, C. M. *Tetrahedron: Asymmetry* **1995**, *6*, 1903–1906.
79. Brand, S.; Jones, M. F.; Rayner, C. M. *Tetrahedron Lett.* **1997**, *38*, 3595–3598.
80. (a) Chu, C. K.; Beach, J. W.; Jeong, L. S.; Choi, B. G.; Comer, F. I.; Alves, A. J.; Schinazi, R. F. *J. Org. Chem.* **1991**, *56*, 6503–6505. (b) Jeong, L. S.; Schinazi, R. F.; Beach, J. W.; Kim, H. O.; Shanmuganathan, K.; Nampalli, S.; Chun, M. W.; Chung, W. K.; Choi, B. G.; Chu, C. K. *J. Med. Chem.* **1993**, *36*, 2627–2638.
81. (a) Beach, J. W.; Jeong, L. S.; Alves, A. J.; Pohl, D.; Kim, H. O.; Chang, C. N.; Doong, S. L.; Schinazi, R. F.; Cheng, Y.-C.; Chu, C. K. *J. Org. Chem.* **1992**, *57*, 2217–2219. (b) Jeong, L. S.; Shinazi, R. F.; Beach, J. W.; Kim, H. O.; Nampalli, S.; Shanmuganathan, K.; Alves, A. J.; McMillan, A.; Chu, C. K.; Mathis, R. *J. Med. Chem.* **1993**, *36*, 181–195.
82. Humber, D. C.; Jones, M. F.; Payne, J. J.; Ramsay, M. V. J.; Zacharie, B.; Jin, H.; Siddiqui, A.; Evans, C. A.; Tse, H. L. A.; Mansour, T. S. *Tetrahedron Lett.* **1992**, *33*, 4625–4628.
83. (a) Jin, H.; Siddiqui, A.; Evans, C. A.; Tse, A.; Mansour, T. S.; Goodyear, M. D.; Ravenscroft, P.; Beels, C. D. *J. Org. Chem.* **1995**, *60*, 2621–2623. (b) Siddiqui, M. A.; Jin, H.; Evans, C. A.; DiMarco, M. P.; Tse, H. L. A.; Mansour, T. S. *Chirality* **1994**, *6*, 156–160.
84. Goodyear, M. D.; Hill, M. L.; West, J. P.; Whitehead, A. J. *Tetrahedron Lett.* **2005**, *46*, 8535–8538.
85. Cameron, J. M.; Collis, P.; Daniel, M.; Storer, R.; Wilcox, P. *Drugs Future* **1993**, *18*, 319–323.
86. Choi, Y.; Li, L.; Grill, S.; Gullen, E.; Lee, C. S.; Gumina, G.; Tsujii, E.; Cheng, Y.-C.; Chu, C. K. *J. Med. Chem.* **2000**, *43*, 2538–2546.

87. Turner, K. *Org. Process Res. Dev.* **2009**, *13*, 829–841.
88. Goodyear, M. D.; Dwyer, P. O.; Hill, M. L.; Whitehead, A. J.; Hornby, R.; Hallett, P. *U.S. Patent 6*,051,709 (**2000**).
89. Roy, B. N.; Singh, G. P.; Srivastava, D.; Jadhav, H. S.; Saini, M. B.; Aher, U. P. *Org. Process Res. Dev.* **2009**, *13*, 450–455.
90. Li, J.-Z.; Gao, L.-X.; Ding, M.-X. *Synth. Commun.* **2002**, *32*, 2355–2359.
91. The furanose numbering system was used for NMR spectra interpretation as in ref. 81b.
92. However, it should be noted that (+)-BCH-189 in higher purity could reasonably be obtained through the same method, simply replacing starting (+)-menthyl chloroformate with its (−)-enantiomer.
93. Cox, S.; Deadman, J.; Southby, J.; Coates, J. In *Antiviral Drugs: From Basic Discovery Through Clinical Trials*; Kazmierski, W. M., Ed.;Wiley: Hoboken, NJ, **2011**; pp. 103–116.
94. Cahn, P.; Wainberg, M. A. *J. Antimicrob. Chemother.* **2010**, *65*, 213–217.
95. Cox, S.; Southby, J. *Expert Opin. Investing. Drugs* **2009**, *18*, 199–209.
96. Belleau, B.; Brasili, L.; Chan, L.; DiMarco, M. P.; Zacharie, B.; Nguyen-Ba, N.; Jenkinson, H. J.; Coates, J. A. V.; Cameron, J. M. *Bioorg. Med. Chem. Lett.* **1993**, *3*, 1723–1728.
97. De Muys, J. M.; Gourdeau, H.; Ba-Nguyen, N.; Taylor, D. L.; Ahmed, P. S.; Mansour, T.; Locas, C.; Richard, N.; Wainberg, M. A.; Rando, R. F. *Antimicrob. Agents Chemother.* **1999**, *43*, 1835–1844.
98. (a) Smith, P. F.; Forrest, A.; Ballow, C. H.; Martin, D. E.; Proulx, L. *Antimicrob. Agents Chemother.* **2000**, *44*, 1609–1615. (b) Smith, P. F.; Forrest, A.; Ballow, C. H.; Martin, D. E.; Proulx, L. *Antimicrob. Agents Chemother.* **2000**, *44*, 2816–2823.
99. (a) Wang, W.; Jin, H.; Mansour, T. S. *Tetrahedron Lett.* **1994**, *35*, 4739–4742. (b) Mansour, T. S.; Jin, H.; Wang, W.; Hooker, E. U.; Ashman, C.; Cammack, N.; Salomon, H.; Belmonte, A. R.; Wainberg, M. A. *J. Med. Chem.* **1995**, *38*, 1–4.
100. Bednarski, K.; Dixit, D. M.; Wang, W.; Evans, C. A.; Jin, H.; Yuen, L.; Mansour, T. S. *Bioorg. Med. Chem. Lett.* **1994**, *4*, 2667–2672.
101. Mansour, T. S.; Jin, H.; Wang, W.; Dixit, D. M.; Evans, C. A.; Tse, H. L. A.; Belleau, B.; Gillard, J. W.; Hooker, E.; Ashman, C.; Cammack, N.; Salomon, H.; Belmonte, A. R.; Wainberg, M. A. *Nucleos. Nucleot*, **1995**, *14*, 627–635.
102. (a) Stoddart, C. A.; Moreno, M. E.; Linquist-Stepps, V. D.; Bare, C.; Bogan, M. R.; Gobbi, A.; Buckheit, R. W. Jr.; Bedard, J.; Rando, R. F.; McCune, J. M. *Antimicrob. Agents Chemother.* **2000**, *44*, 783. (b) Gu, Z.; Allard, B.; de Muys, J. M.; Lippens, J.; Rando, R. F.; Nguyen-Ba, N.; Ren, C.; McKenna, P.; Taylor, D. L.; Bethell, R. C. *Antimicrob. Agents Chemother.* **2006**, *50*, 625–631.
103. Taylor, D. L.; Ahmed, P. S.; Tyms, A. S.; Wood, L. J.; Kelly, L. A.; Chambers, P.; Clarke, J.; Bedard, J.; Bowlin, T. L.; Rando, R. F. *Antivir. Chem. Chemother.* **2000**, *11*, 291–301.
104. (a) Caputo, R.; Guaragna, A.; Palumbo, G.; Pedatella, S. *Nucleos. Nucleot*, **1998**, *17*, 1739–1745. (b) Caputo, R.; Guaragna, A.; Palumbo, G.; Pedatella, S. *Eur. J. Org. Chem.* **1999**, 1455–1458.
105. Ramón, D. J.; Yus, M. *Chem. Rev.* **2006**, *106*, 2126–2208.

106. Unsatisfactory enantiomeric excesses obtained hampered larger-scale applications of this method. It's worth mentioning that, when the same protocol was applied to the dithiolane analogue of sulfide 55, a much better 90% ee was accomplished. See Caputo, R.; Guaragna, A.; Palumbo, G.; Pedatella, S. Eur. J. Org. Chem. **2003**, 2, 346–350.
107. Nguyen-Ba, N.; Brown, W.; Lee, N.; Zacharie, B. Synthesis **1998**, 759–762.
108. Marcuccio, S. M.; Epa, R.; White, J. M.; Deadman, J. J. Org. Process Res. Dev. **2011**, 15, 763–773.
109. Mansour, T.S.; Haolun, J. U.S. Patent 6,228,860 B1 (**2001**).
110. Gumina, G.; Chong, Y.; Chu, C. K. In Deoxynucleoside Analogs in Cancer Therapy; Peters, G. J., Ed.; Humana Press: Totowa, NJ, **2006**; pp. 173–198.
111. Sznaidman, M. L.; Du, J.; Pesyan, A.; Cleary, D. G.; Hurley, K. P.; Waligora, F.; Almond, M. R. Nucleos. Nucleot. Nucl, **2004**, 23, 1875–1887.
112. Siddiqui, M. A.; Brown, W.; Nguyen-Ba, N.; Dixit, D. M.; Mansour, T. S.; Hooker, E.; Viner, K. C.; Cameron, J. M. Bioorg. Med. Chem. Lett. **1993**, 3, 1543–1546.
113. Gu, Z.; Wainberg, M. A.; Nguyen-Ba, N.; L'Heureux, L.; de Muys, J.-M.; Bowlin, T. L.; Rando, R. F. Antimicrob. Agents Chemother. **1999**, 43, 2376–2382.
114. Wilson, L. J.; Choi, W.-B.; Spurling, T.; Liotta, D. C.; Schinazi, R. F.; Cannon, D.; Painter, G. R.; St. Clair, M.; Furman, P. A. Bioorg. Med. Chem. Lett. **1993**, 3, 169–174.
115. Mao, S.; Bouygues, M.; Welch, C.; Biba, M.; Chilenski, J.; Schinazi, R. F.; Liotta, D. C. Bioorg. Med. Chem. Lett. **2004**, 14, 4991–4994.
116. Schinazi, R. F.; Mellors, J.; Erickson-Viitanen, S.; Mathew, J.; Parikh, U.; Sharma, P.; Otto, M.; Yang, Z.; Chu, C. K.; Liotta, D. C. Antivir. Ther. **2002**, 7, S15.
117. In a septum-capped 10-mL vial containing 5 mL of anhydrous CH_2Cl_2, 99.9% $TiCl_4$ (0.21 mL, 1.94 mmol, 1.5 equiv) was added dropwise under N_2. After mixing 5 min, $Ti(O-i-Pr)_4$ (0.19 mL, 0.65 mmol, 0.5 equiv) was added dropwise under N_2. Instantaneous mixing occurred to give $TiCl_3(O-i-Pr)$ (1.29 mmol, light yellow solution).
118. Davis, K.B. Enantioselective Synthesis of β-D-Dioxolane-T and β-D-FDOC. Ph.D. dissertation, Emory University, Atlanta, C.A, **2008**.
119. (a) Kim, H. O.; Ahn, S. K.; Alves, A. J.; Beach, J. W.; Jeong, L. S.; Choi, B. G.; Van Roey, P.; Schinazi, R. F.; Chu, C. K. J. Med. Chem. **1992**, 35, 1987–1995. (b) Kim, H. O.; Schinazi, R. F.; Nampalli, S.; Shanmuganathan, K.; Cannon, D. L.; Alves, A. J.; Jeong, L. S.; Beach, J. W.; Chu, C. K. J. Med. Chem. **1993**, 36, 30–37.
120. Jin, H. L.; Tse, H. L. A.; Evans, C. A.; Mansour, T. S.; Beels, C. M.; Ravenscroft, P.; Humber, D. C; Jones, M. F.; Payne, J. J.; Ramsay, M. V. J. Tetrahedron: Asymmetry **1993**, 4, 211–214.
121. Belleau, B.; Evans, C. A.; Tse, H. L. A.; Jin, H. L.; Dixit, D. M.; Mansour, T. S. Tetrahedron Lett. **1992**, 33, 6949–6952.
122. Narayanasamy, J.; Pullagurla, M. R.; Sharon, A.; Wang, J.; Schinazi, R. F.; Chu, C. K. Antivir. Res. **2007**, 75, 198–209.
123. Chu, C. K.; Ahn, S. K.; Kim, H. O.; Beach, J. W.; Alves, A. J.; Jeong, L. S.; Islam, Q.; Van Roey, P.; Schinazi, R. F. Tetrahedron Lett. **1991**, 32, 3791–3794.
124. (a) Kim, H. O.; Shanmuganathan, K. Alves, A. J.; Shanmuganathan, K.; Cannon, D. L.; Alves, A. J.; Jeong, L. S.; Beach, J. W.; Chu, C. K. Tetrahedron Lett. **1992**, 33, 6899–6902. (b) Kim, H. O.; Schinazi, R. F.; Shanmuganathan, K.; Jeong, L. S.; Beach, J. W.; Nampalli, S.; Cannon, D. L.; Chu, C. K. J. Med. Chem. **1993**, 36, 519–528.

125. Lee, M.; Chu, C. K.; Pai, S. B.; Zhu, Y.-L.; Cheng, Y.-C.; Chun, M. W.; Chung, W. K. *Bioorg. Med. Chem. Lett.* **1995**, *5*, 2011–1014.
126. Lennerstrand, J.; Chu, C. K.; Schinazi, R. F. *Antimicrob. Agents Chemother.* **2007**, *51*, 2078–2084.
127. Grove, K. L.; Cheng, Y.-C. *Cancer Res.* **1996**, *56*, 4187–4191.
128. Moore, L. E.; Boudinot, F. D.; Chu, C. K. *Cancer Chemother. Pharmacol.* **1997**, *39*, 532–536.
129. Bazmi, H. Z.; Hammond, J. L.; Cavalcanti, S. C. G.; Chu, C. K.; Schinazi, R. F.; Mellors, J. W. *Antimicrob. Agents Chemother.* **2000**, *44*, 1783–1788.
130. Furman, P. A.; Jeffery, J.; Kiefer, L.; Feng, J. Y.; Anderson, K. S.; Borroto-Esoda, K.; Hill, E.; Copeland, W. C.; Chu, C. K.; Sommadossi, J.-P.; Liberman, I.; Schinazi, R. F.; Painter, G. R. *Antimicrob. Agents Chemother.* **2001**, *45*, 158–165.
131. Chen, H.; Boudinot, F. D.; Chu, C. K.; McClure, H. M.; Schinazi, R. F. *Antivir. Chem. Chemother.* **1996**, *40*, 2332–2336.
132. Radi, M.; Adema, A. D.; Daft, J. R.; Cho, J. H.; Hoebe, E. K.; Alexander, L.-E. M. M.; Peters, G. J.; Chu, C. K. *J. Med. Chem.* **2007**, *50*, 2249–2253.
133. Liang, Y.; Sharon, A.; Grier, J. P.; Rapp, K. L.; Schinazi, R. F.; Chu, C. K. *Bioorg. Med. Chem.* **2009**, *17*, 1404–1409.
134. Liang, Y.; Narayanasamy, J.; Schinazi, R. F.; Chu, C. K. *Bioorg. Med. Chem.* **2006**, *14*, 2178–2189.
135. Liang, Y.; Narayanasamy, J.; Rapp, K. L.; Schinazi, R. F.; Chu, C. K. *Antivir. Chem. Chemother.* **2006**, *17*, 321–329.
136. If the nucleobase does not dissolve promptly after 0.5–2.0 h, addition of a catalytic amount of TMSCl or Py is reported to speed up silylation; cfr: Vorbrüggen, H.; Krolikiewicz, K.; Bennua, B. *Chem. Ber.* **1981**, *114*, 1234–1255.
137. Evans, C. A.; Dixit, D. M.; Siddiqui, M. A.; Jin, H.; Tse, H. L. A.; Cimpoia, A.; Bednarski, K.; Breining, T.; Mansour, T. S. *Tetrahedron: Asymmetry* **1993**, *4*, 2319–2322.
138. Mansour, T. S.; Jin, H. L.; Wang, W.; Dixit, D. M.; Evans, C. A.; Tse, H. L. A.; Belleau, B.; Gillard, J. W.; Hooker, E. U.; Ashman, C.; Cammack, N.; Salomon, H.; Belmonte, A. R.; Wainberg, M. A. *Nucleos. Nucleot.* **1995**, *14*, 627–635.
139. Gourdeau, H.; Clarke, M. L.; Ouellet, F.; Mowles, D.; Selner, M.; Richard, A.; Lee, N.; Mackey, J. R.; Young, J. D.; Jolivet, J.; Lafrenière, R. G.; Cass, C. E. *Cancer Res.* **2001**, *61*, 7217–7224.
140. Janes, L. E.; Cimpoia, A.; Kazlauskas, R. J. *J. Org. Chem.* **1999**, *64*, 9019–9029.
141. Norbeck, D. W.; Spanton, S.; Broder, S.; Mitsuya, H. *Tetrahedron Lett.* **1989**, *30*, 6263–6266.
142. Murakami, E.; Bao, H.; Basavapathruni, A.; Bailey, C. M.; Du, J.; Steuer, H. M.; Niu, C.; Whitaker, T.; Anderson, K. S.; Otto, M. J.; Furman, P. A. *Antivir. Chem. Chemother.* **2007**, *18*, 83–92.
143. Popp, A.; Gilch, A.; Mersier, A.-M.; Petersen, H.; Rockinger-Mechlem, J.; Stohrer, J. *Adv. Synth. Catal.* **2004**, *346*, 682–690.

17 Isoxazolidinyl Nucleosides

UGO CHIACCHIO and ANTONINO CORSARO
Dipartimento di Scienze del Farmaco, Università di Catania, Catania, Italy

SALVATORE GIOFRÈ and GIOVANNI ROMEO
Dipartimento Farmaco-Chimico, Università di Messina, Messina, Italy

1. INTRODUCTION

Interest in nucleoside analogues, in which the furanose ring is replaced by an alternative heterocyclic system, has recently led to a series of modified nucleosides in which an isoxazolidine unit is the mimetic of the sugar moiety of natural nucleosides. Preliminary in vivo tests have shown promising biological activity for some of the members of this new family of potential antiviral drugs. The first systematic review of this topic was published in 1998,[1] and more recently in 2010,[2] including literature reports until 2008. The aim of this chapter is to cover all the synthetic efforts in the construction of this type of nucleoside analogue, extending the literature survey to March 2011.

A variety of nucleoside structures containing an N–O pentatomic ring have been synthesized for exploratory studies to provide strategies for potential drug development. In particular, N- and C-nucleosides have been reported, according to the nucleobase atom involved in the glycosidic bond. In this chapter the isoxazolidine nucleosides have been classified in different groups depending on the substituents present on the pentatomic ring (Figure 1).

Comprehensive literature searches reveal that compounds belonging to classes I to V have been prepared[3]; virtually no syntheses are reported about different types of isoxazolidinyl nucleosides. Consequently, the information known about the ability of isoxazolidine nucleosides to function as bioactive compounds is by no means comprehensive.

2. 4′-AZA-2′,3′-DIDEOXYNUCLEOSIDES I

The first synthetic approach to compounds of class I were reported in 1992 by Tronchet et al.[4] The reaction route is based on the 1,3-dipolar cycloaddition of

Chemical Synthesis of Nucleoside Analogues, First Edition. Edited by Pedro Merino.
© 2013 John Wiley & Sons, Inc. Published 2013 by John Wiley & Sons, Inc.

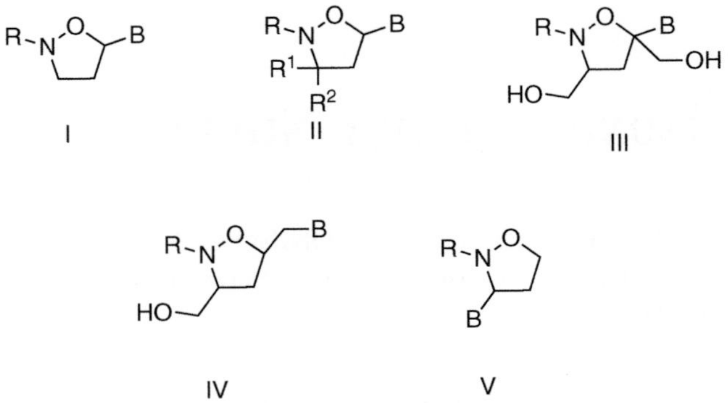

Figure 1.

N-methyl- or *N*-benzylnitrone **1**, prepared in situ from the corresponding hydroxylamine and paraformaldehyde, with vinyl acetate, followed by coupling with silylated thymine, in the presence of trimethyl silyl triflate (TMSOTf), to give *N,O*-nucleosides **3** (Scheme 1). The cycloaddition process shows complete regioselectivity toward isoxazolidine **2**.

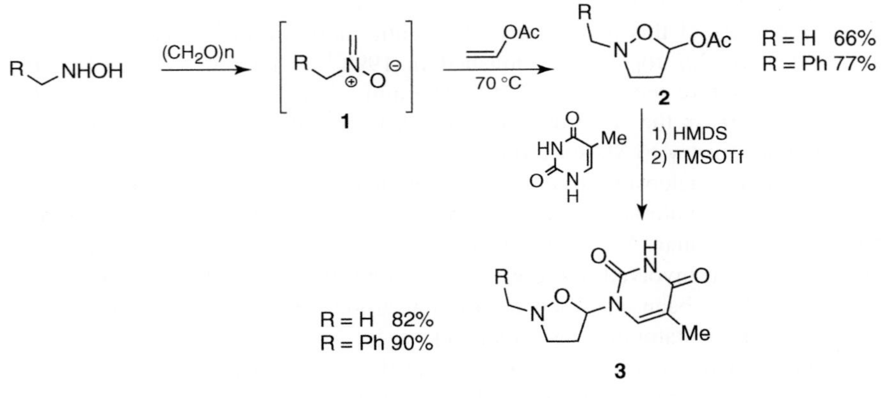

Scheme 1.

Isoxazolidinyl nucleosides, unsubstituted at the nitrogen atom and in racemic form, have been described.[5] These compounds have been prepared by reaction between formyl nitrones bearing an acid-labile protecting group on the nitrogen atom, generated in situ, and vinyl derivatives of the common pyrimidine and purine nucleobases.[6] In a typical procedure, the 4′-aza analogues of 2′,3′-dideoxythymidine and -uridine **8a,b** (AdT and AdU) were prepared by refluxing a chloroform solution of 1-vinyl thymine or 1-vinyl uracil (**4a,b**) and *N*-tetrahydropyranylmethylene nitrone **6**, the latter obtained in situ from the condensation of 5-hydroxypentanal oxime **5** with a molar excess of paraformaldehyde. After removal of the protecting group from **7a, b**, by treatment with *p*-toluenesulfonic acid in a 70:30 chloroform/methanol ratio or a few drops of a diluted solution of $HClO_4$, respectively, AdT and AdU **8a, b** were obtained in racemic form (Scheme 2).[5]

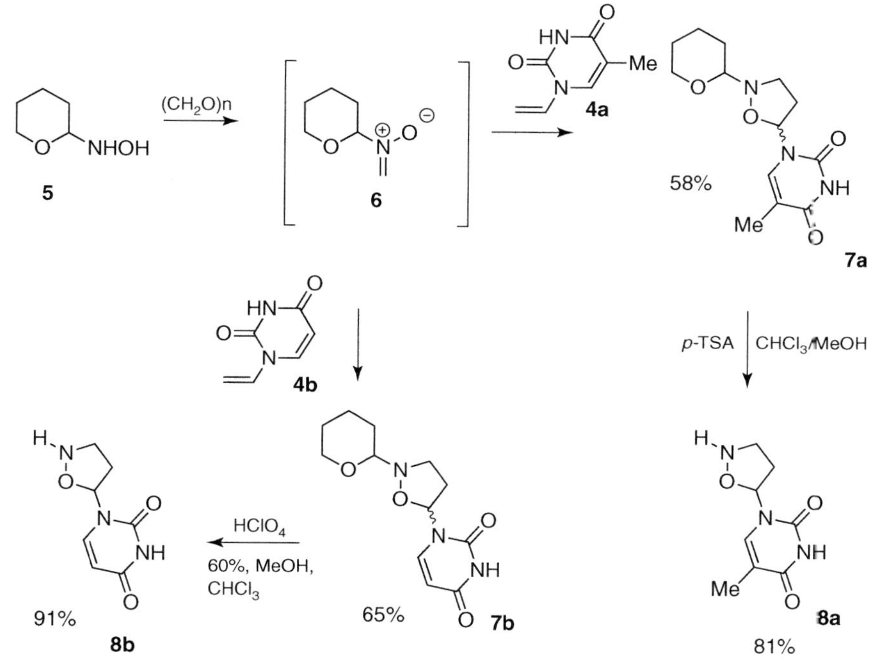

Scheme 2.

AdT is reported to inhibit HIV replication in C 8166 with an activity inversely related to the multiplicity of infection used.[7] A similar protocol has been used for the preparation of the purine analogues of 4′-aza-2′,3′-dideoxynucleosides, AdG **14** and AdA **15**, starting from 5-hydroxypentanal oxime **5** or *N*-trityl hydroxylamine **9** and formaldehyde, respectively. Protection of the vinyl nucleobase as trityl **10** or dimethoxytrityl **11** was required to avoid low yields, due to the formation of side products (Scheme 3).[8]

A convertible nucleoside approach[9] allowed the preparation of 5-methyl cytosine **18** and cytosine derivatives **21** (Schemes 4 and 5).[7]

Both enantiomeric purity and absolute configuration could be key factors in determining the physiological activity of a drug.[10] This consideration has promoted the enantioselective synthesis of (*R*)- and (*S*)-AdT **8a** and (*R*)- and (*S*)-4′aza-2′-3′-dideoxyfluorouridine (AdFU) **8c** (Scheme 6).[11] The key step of the synthetic approach consists of the 1,3-dipolar cycloaddition of a chiral nitrone, containing a stereogenic center at the nitrogen atom, to vinyl acetate in order to construct the isoxazolidine ring stereoselectively. Thus, the Vasella-type nitrone **23**, prepared by the reaction of ribosyl hydroxylamine **22** with paraformaldehyde, was reacted with

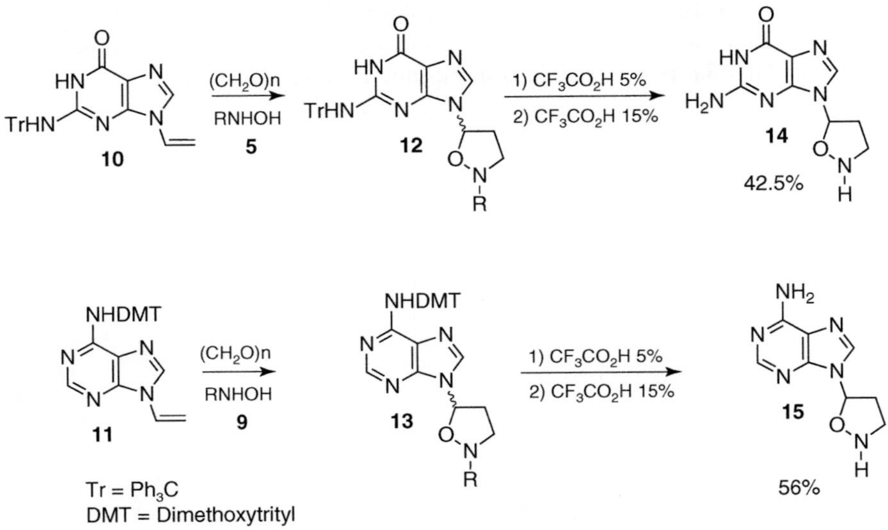

Scheme 3.

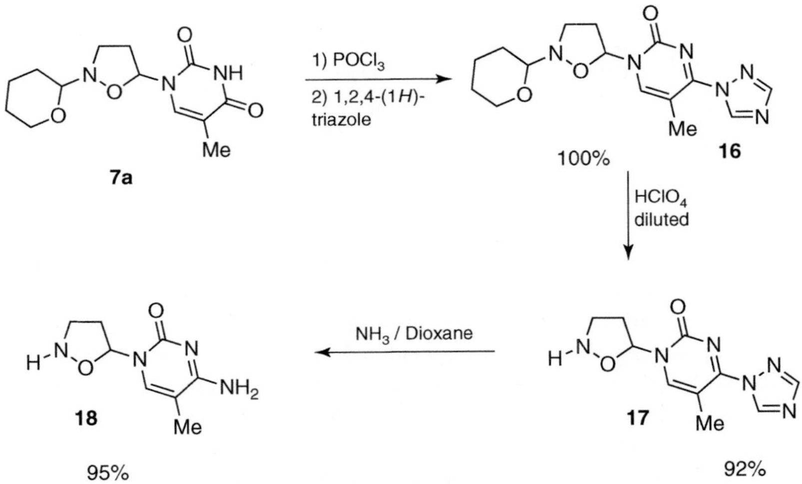

Scheme 4.

vinyl acetate to give a mixture of two homochiral isoxazolidines **24** and **25** (90% global yield), epimeric at C1′, in the ratio 1.4:1.

Isoxazolidines **24** and **25** were then coupled with silylated thymine in the presence of Lewis acids, according to the Vorbrüggen procedure,[12] to afford, with moderate stereoselectivity and in good yields, the isoxazolidinyl nucleosides **26a,c** and **27a,c**, (**26a/27a** ratio 1.4:1, **26c/27c** ratio 1.8:1), which have been separated by

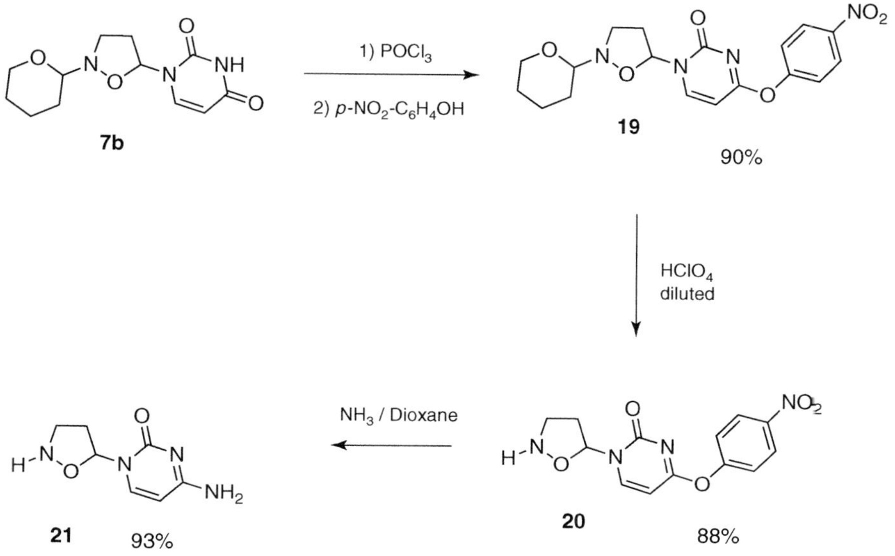

Scheme 5.

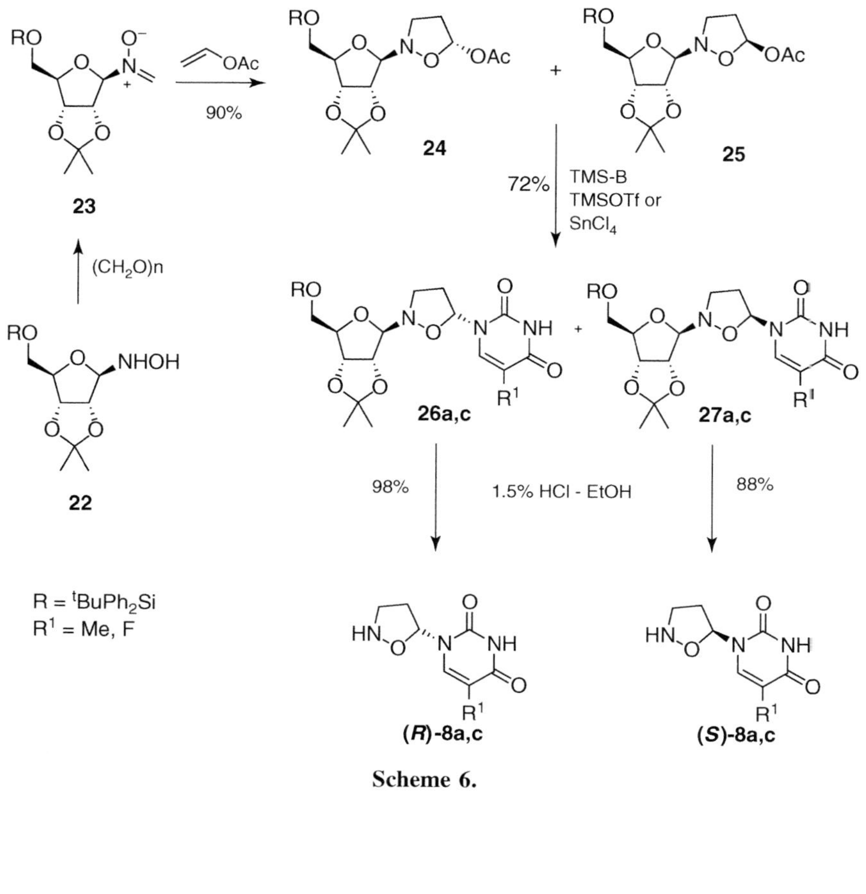

Scheme 6.

R = ^{t}BuPh$_2$Si
R^1 = Me, F

HPLC chromatography. Mixtures of diastereomeric invertomers could be observed for compounds **26** and **27**; however, the NMR data show only one set of resonances ($T = 197-298$ K), suggesting the existence of only one isomer or a nitrogen inversion sufficiently fast to impart time-averaged properties to the compounds observed.

PM3 and AM1 quantomechanical calculations indicate that the major stereoisomers **26a** and **26c** possess the configuration ($1'R$), which is more stable than the configuration ($1'S$) of about 0.8 kcal; this value is in agreement with the experimental **26/27** ratio.

The synthetic scheme toward homochiral 4'-aza-2',3'-dideoxynucleosides has been completed by selective cleavage of the sugar moiety, performed by treatment with 1.5% aqueous HCl; in this way, two separate enantiomers **8a** and **8c** have been obtained.[11]

The same procedure was applied to the synthesis of (*R*)-AdU, (*R*)-AdC, (*R*)-AdFC, (*R*)-AdA, and (*S*)-AdU, (*S*)-AdC, (*S*)-AdFC, (*S*)-AdA. Thus, the coupling reaction of silylated nucleobases (b,d,e,f), in the presence of TMSOTf at 0°C, with the epimeric isoxazolidines **24** and **25** afforded nucleosides **26b,d,e,f** and **27b,d,e,f**. After HPLC separation and selective cleavage of the sugar moiety, pure enantiomers of both **8b,d,e** and **15** have been obtained (Scheme 7).[13]

The use of vinyl nucleobases as dipolarophiles for a one-pot reaction pathway toward enantiomerically pure isoxazolidines **26** and **27** results in lower yields with respect to the two-step procedure based on the Vorbrüggen nucleosidation.[13]

The cytotoxicity of nucleosides **8** and **15** has been tested in cell lines of lymphoid and monocytoid origin. ($5'S$)-5-Fluoro-1-isoxazolidin-5-yl-1*H*-pyrimidine-2,4-dione [(−)-AdFU] **8c**, while showing a low level of cytotoxicity, is a good inductor of apoptosis in both types of cells acting as a strong potentiator of Fas-induced cell death.[13] No relevant antiviral activity has been observed for all nucleosides tested.[13]

N,O-nucleosides, as antiviral agents against HIV, HBV, and HSV, would act as competitive inhibitors of the reverse transcriptase or chain terminators in the DNA synthesis. Since the triphosphorylation is required for incorporation into elongating DNA chains, the need for a hydroxymethyl group or a hydroxymethyl mimic seems to be necessary.

The synthesis of *N*-hydroxymethyl derivatives of isoxazolidinyl nucleosides has not been reported, perhaps due to the unstable aminal functionality. Tronchet et al.[14] have synthesized compounds **30a–c**, where the hydroxyl group is located on the side chain at the nitrogen atom of the isoxazolidine unit (Scheme 8). Compounds **30a–c** exhibit no antiviral activity; however; they represent the first synthetic attempt toward hydroxylated *N,O*-nucleosides.

3. 3'-SUBSTITUTED-4'-AZA-2',3'-DIDEOXY NUCLEOSIDES II

An efficient synthesis of 3-aryl-substituted derivatives **32** in good yields and short reaction times has been reported. The process is based on the microwave-assisted 1,3-dipolar cycloaddition of nitrones **31** with vinyl nucleobases, in the absence of solvents (Scheme 9).[15] A remarkably *cis*-stereoselectivity was observed by tuning the substitutents on the nitrone moiety, and reached a range of 92–98% yield when

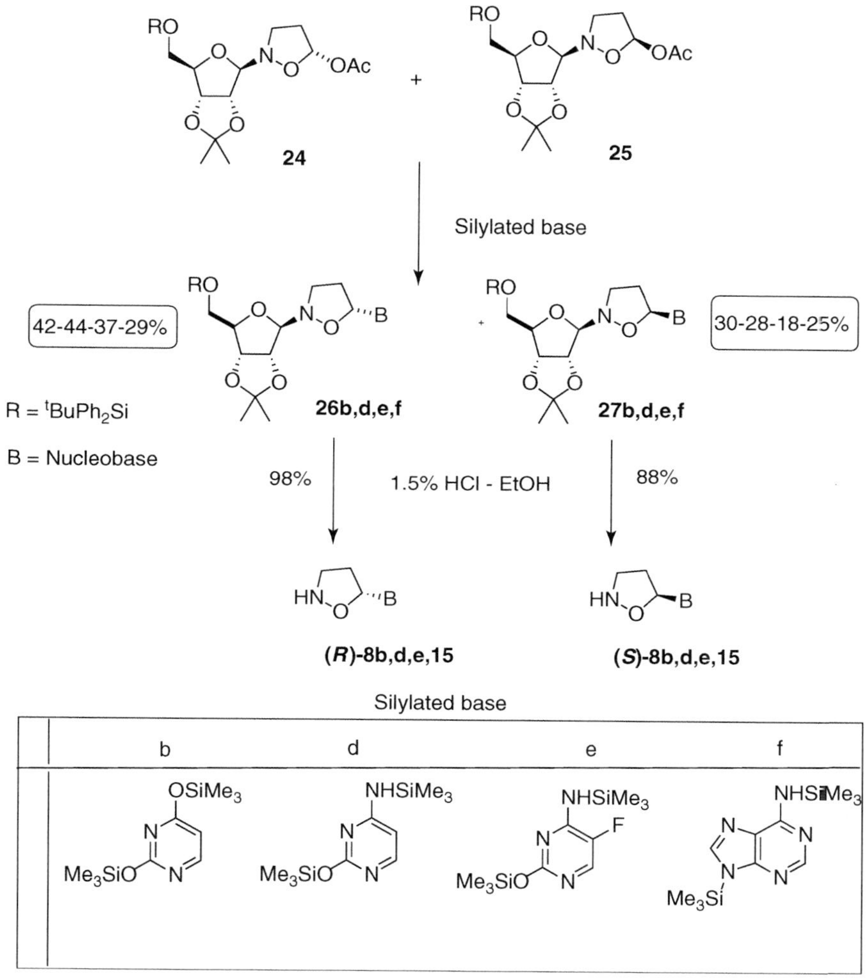

Scheme 7.

a bulky group was present. Regarding their biological activity, the N-methyl-Thy-o-chlorophenyl derivative **32m** and the N-benzyl-Thy-phenyl-substituted compound **32o** have shown an interesting cytotoxic activity in vitro against human lymphoblastoid and JiJoye and Jurkat cell lines [**32m** IC_{50} (μM) 52.6; 28.5; 8.8; **32o** IC_{50} (μM) 55; 36.9; 19.3].

The first isoxazolidinyl nucleosides **36** and **37**, containing an hydroxymethyl group at C3', directly analogous to the ribose-parent structure of nucleosides, were reported by Zhao et al. in 1995.[16] The synthetic scheme involves the Michael addition of N-methylhydroxylamine to α,β-unsaturated ester **33** to form **34**. Subsequent DIBAL reduction, acetyl protection, condensation with silylated thymine, and TBAF treatment give the β- and α-thymine derivatives **36** and **37** (Scheme 10).

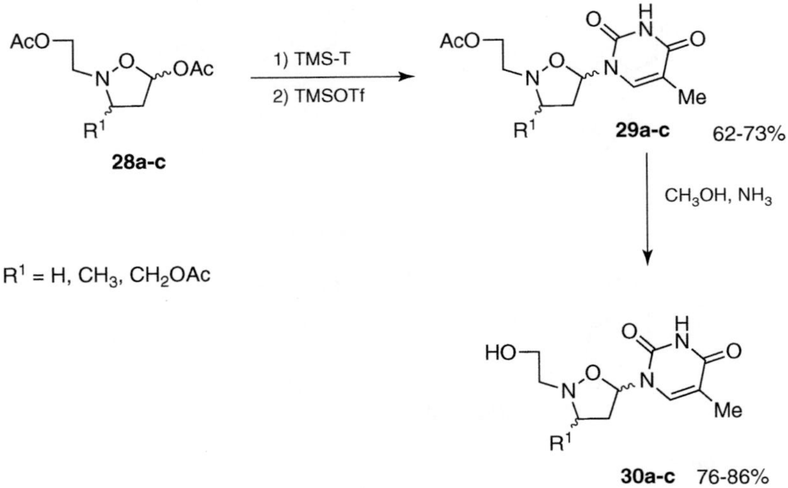

Scheme 8.

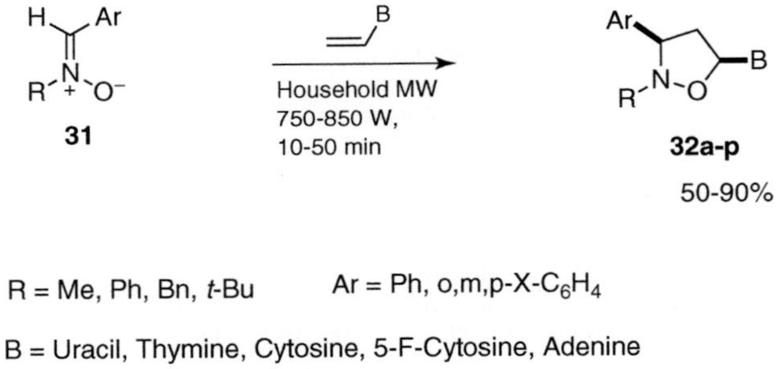

28a–c → (1) TMS-T, 2) TMSOTf) → 29a–c 62-73% → (CH$_3$OH, NH$_3$) → 30a–c 76-86%

R^1 = H, CH$_3$, CH$_2$OAc

Scheme 9.

31 → (Household MW 750-850 W, 10-50 min) → 32a–p 50-90%

R = Me, Ph, Bn, t-Bu   Ar = Ph, o,m,p-X-C$_6$H$_4$

B = Uracil, Thymine, Cytosine, 5-F-Cytosine, Adenine

Isoxazolidinyl thymidine and 5-fluorouridine derivatives, **42a–c** and **43a–c**, have been prepared by the 1,3-dipolar cycloaddition methodology. Nitrones **38a–c**, containing an alkoxycarbonyl or benzoyl group at a nitrone carbon atom, have been reacted with vinyl acetate to give a mixture of diastereomeric products **39a–d**. Coupling with the nucleobase, followed by reduction with NaBH$_4$, produced a mixture of β/α anomers **42a–c** and **43a–c** in about a 2:1 ratio (Scheme 11).[17]

A series of *N*- and *O*-nucleosides **46**, containing an additional hydroxymethyl group at C3′, in racemic form, have been synthesized by cycloaddition of nitrone **44** with vinyl acetate, followed by coupling with silylated nucleobases and TBAF

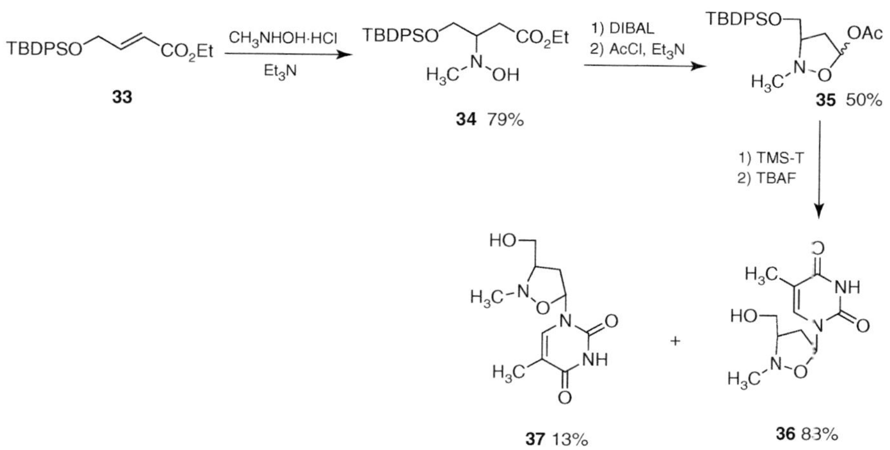

Scheme 10.

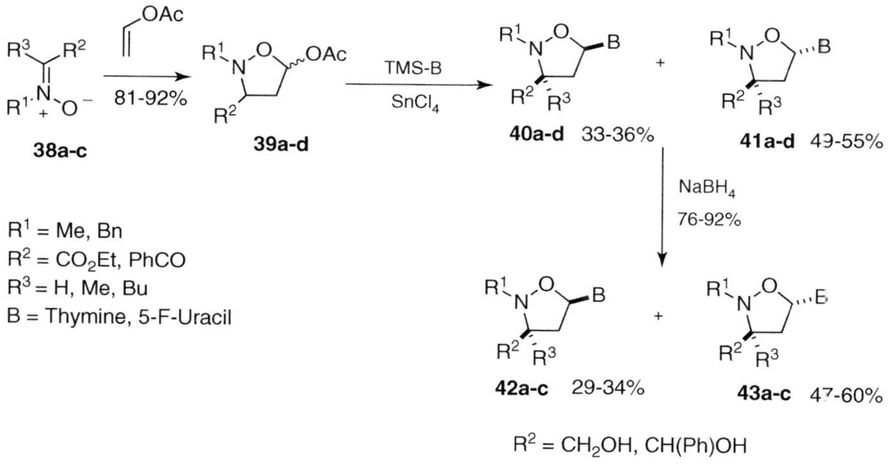

Scheme 11.

R^1 = Me, Bn
R^2 = CO_2Et, PhCO
R^3 = H, Me, Bu
B = Thymine, 5-F-Uracil

R^2 = CH_2OH, CH(Ph)OH

treatment (Scheme 12). Nucleoside **46d** was prepared from **46c** by reaction with K_2CO_3 in methanol.[18]

Following a similar procedure, nitrone **47** has been transformed into a mixture of α- and β-nucleosides **49** and **50** containing a methyl group at C3' (Scheme 13). None of these derivatives showed antiviral activity against HSV-1, HSV-2, and HTLV-1, while a moderate apoptosis against Molt-3-cells was found for compound **50b**.[18]

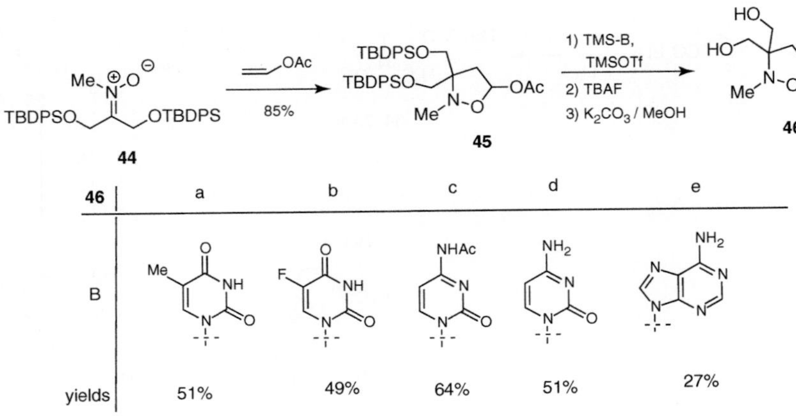

Scheme 12.

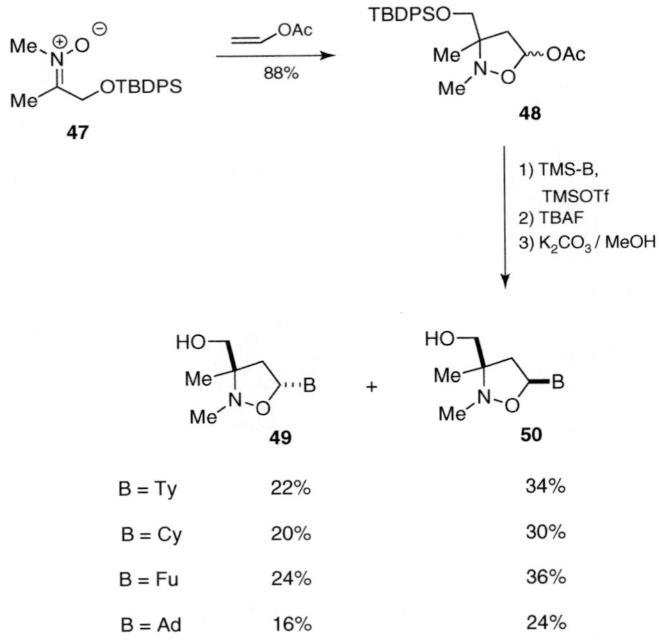

46	a	b	c	d	e
B	Thymine (Me)	5-F uracil	N⁴-Ac cytosine (NHAc)	Cytosine (NH₂)	Adenine (NH₂)
yields	51%	49%	64%	51%	27%

Scheme 13.

	49	50
B = Ty	22%	34%
B = Cy	20%	30%
B = Fu	24%	36%
B = Ad	16%	24%

Several asymmetric routes have been reported for the synthesis of enantiopure N,O-nucleosides containing an hydroxymethyl group ac C3', such as (a) Michael addition of N-methyl hydroxylamine to unsaturated lactones and esters, (b) nucleophilic addition of enolates to α-alkoxy and α-aminonitrones, (c) asymmetric 1,3-dipolar cycloaddition of nitrones, or (d) enantiomeric resolution.

3.1. Michael Addition

Enantiomerically pure (R)-5-(t-butyldimethylsilyloxy)methylfuran-2(5H)-one **51** was treated with N-methylhydroxylamine and TBSCl to give the corresponding protected Michael adduct **52**. This compound was transformed into the key derivative **53** by reaction with DIBAL-H, TBAF, 2,2-dimethoxypropane, and acetyl chloride (27%). Condensation of **53** with silylated thymine or N^4-benzoylated cytosine in CH_3CN in the presence of TMSOTf afforded a single isomer **54a** or **54b** in 53–68 % yield, respectively. Deprotection of **54a,b** to corresponding diols with AcOH at 70°C for 30 min, followed by sodium periodate oxidation, reduction with $NaBH_4$, and in the case of **54b**, debenzoylation reaction, produced the L-3'-hydroxymethyl-4'-azamethyl-2'-3'-dideoxy thymidine **36** and cytidine **55** in 40–19% yield, respectively (Scheme 14).[19]

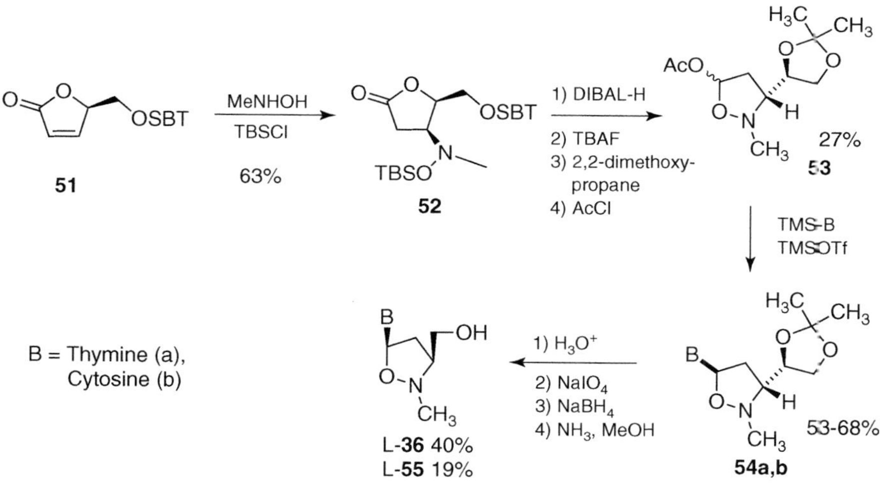

Scheme 14.

Series D of -3'-hydroxymethyl-4'-azamethyl-2'-3'-dideoxy -thimidine, -cytidine, -uridine, and -adenine was obtained starting from α,β-unsaturated esters **56**. Thus, cis- and trans-**56** have been reacted with N-methylhydroxylamine in the presence of $ZnCl_2$ to give the syn-(3R, 4'S)-2-methyl-3(2,2-dimethyl-1,3-dioxolan-4-yl)1,2-isoxazolidin-5-one (76%) **57a** and the anti-(3S,4'S)-2-methyl-3(2,2-dimethyl-1,3-dioxolan-4-yl)1,2-isoxazolidin-5-one (4%) **57b**. The syn-derivative **57a** was treated with DIBAL-H and acetyl chloride to give a mixture of α- and β-anomers **58**. The anomers **58** were condensed with the silylated bases to give, as predominant products, **59a–c**, subsequently transformed into the pyrimidine derivatives D-**36**, D-**55**, and D-**60** by removal of the dioxolane moiety and the benzoyl group. The adenine derivative D-**62** was obtained by coupling of the isoxazolidine acetate **58** with silylated 6-chloropurine in the presence of $Et_2O \cdot BF_3$ at 80°C, followed by hydrolysis of acetonide, oxidation, reduction, and NH_3 treatment (Scheme 15).[20]

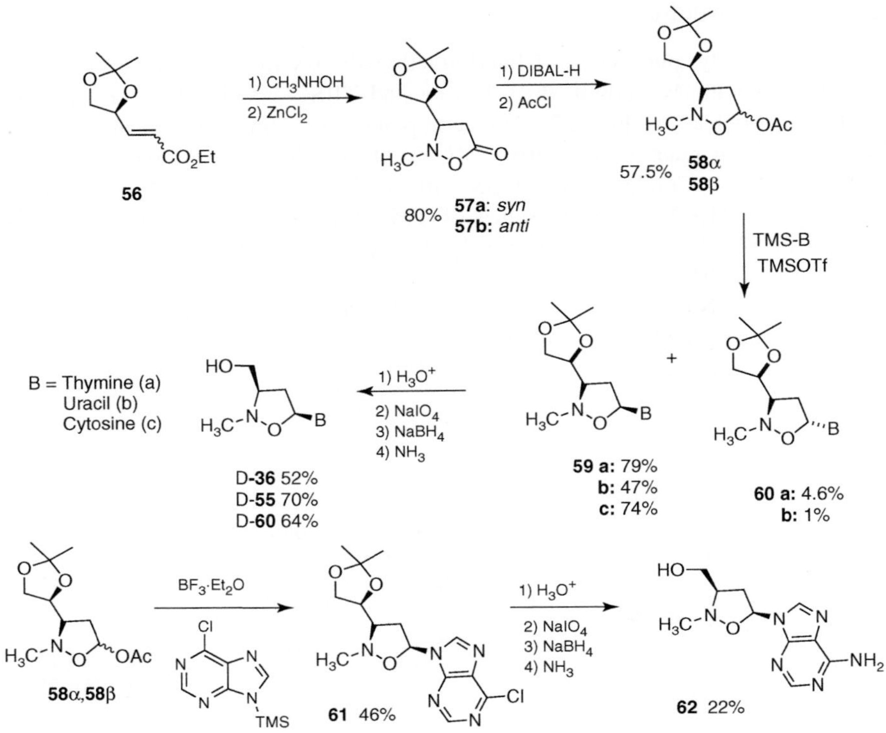

Scheme 15.

α,β-Unsaturated esters **62** and **63**, derived from L-serine, add to *N*-benzylhydroxylamine to give a mixture of β-(hydroxyamino) esters that have been cyclized to *syn*- and *ant*-isoxazolidin-5-ones **64** and **65**, respectively, with sodium methoxide in methanol. Both isoxazolidinones have been then converted, according to Scheme 16, into nucleoside analogues **69a,b**, **70a,b**, and **73**, possessing an opposite configuration at C3'.[21]

Starting from D-serine and following a similar Michael addition approach, the enantiomers **69a** and **70b** have been prepared in 10.36% and 7.17% yields, respectively (Scheme 17).

3.2. Nucleophilic Addition

The isoxazolidine ring of 4'-aza-2',3'-dideoxynucleosides II can be constructed by the addition of ester-derived enolates to nitrones. Thus, the reaction of *N*-methyl- or *N*-benzyl-2,3-*O*-isopropylidene D-Glyceraldehyde nitrones **74** with sodium enolate afforded the *syn*-adduct **75a**, while the same nitrones **74**, in the presence of BF$_3$·Et$_2$O with *O*-methyl-*O*-*tert*-butylidimethylsilyl ketene acetal, afforded the corresponding isoxazolidin-5-ones **75b** in good yields and *anti*-selectivity (Scheme 18). The corresponding derivatives **75a,b** have then been transformed

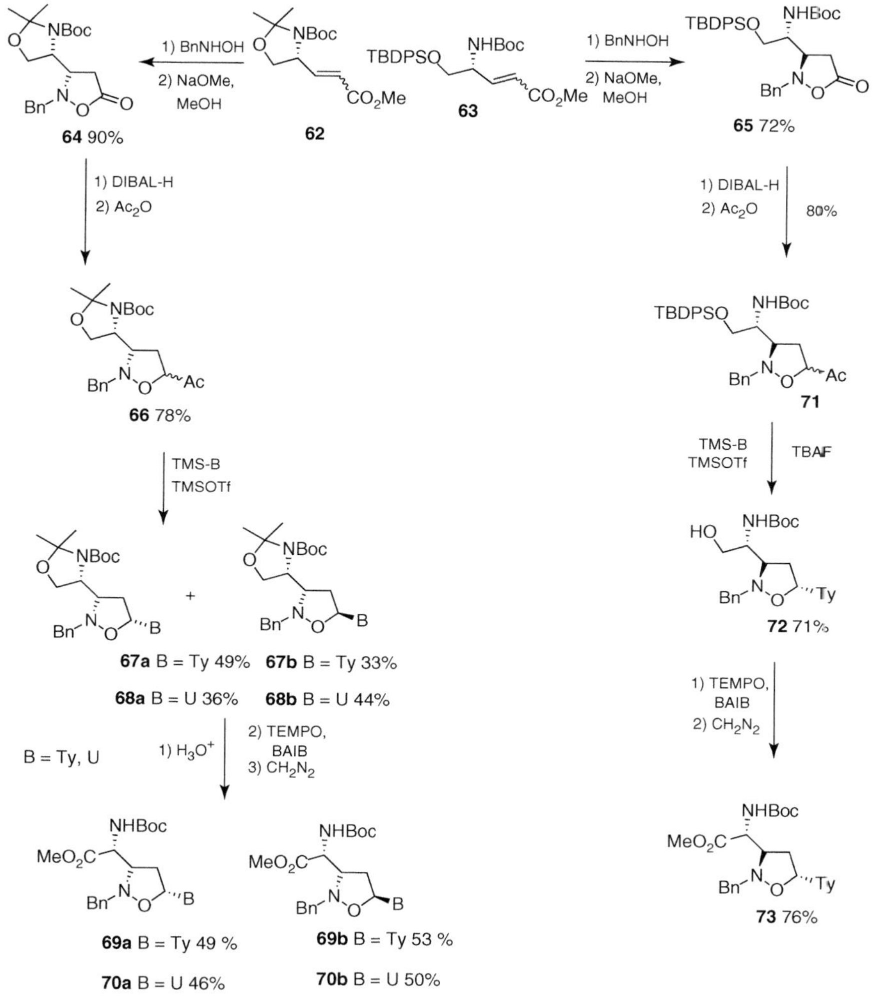

Scheme 16.

into isoxazolidinyl nucleosides D/L-**36** and D/L-**76** as already reported, allowing an enantiodivergent synthesis of both enantiomers.[22]

Using the *N*-benzyl-1,2-di-*O*-ispropilidene-L-glyceraldehyde nitrone **77**, Merino et al. have synthesized, involving the isoxazolidin-5-one **64**, the nucleoside analogues **69a** and **70a** (Scheme 19).[21]

3.3. Asymmetric 1,3-Dipolar Cycloaddition

By using chiral nonracemic nitrones **78**, Romeo and Chiacchio have reported the synthesis of enantiomerically pure branched nuleosides analogues **80a,b** (Scheme 20).[23]

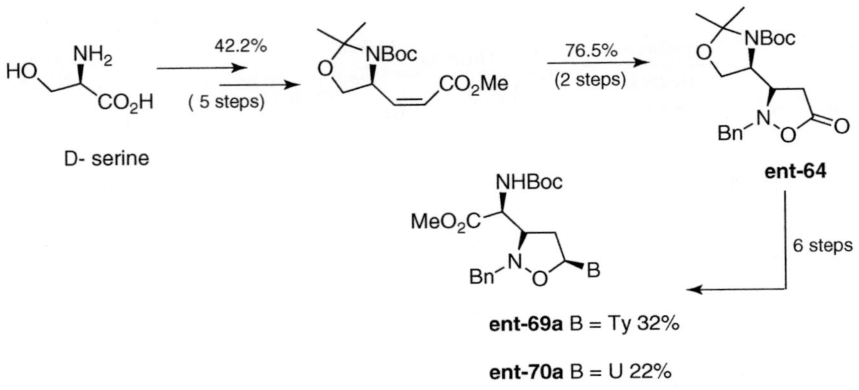

Scheme 17.

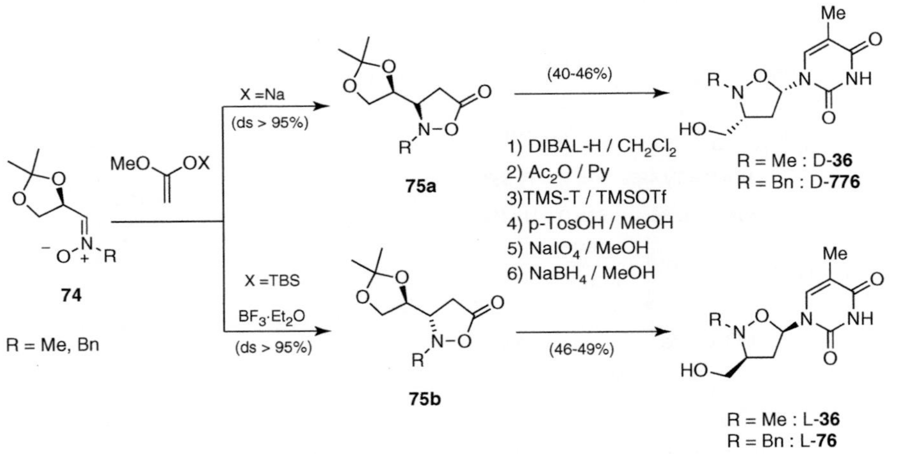

Scheme 18.

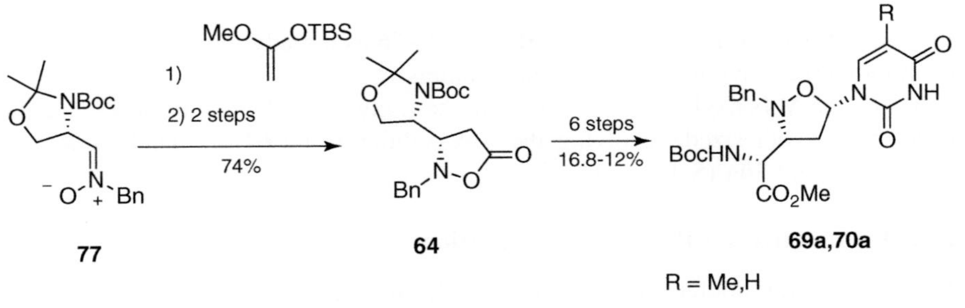

Scheme 19.

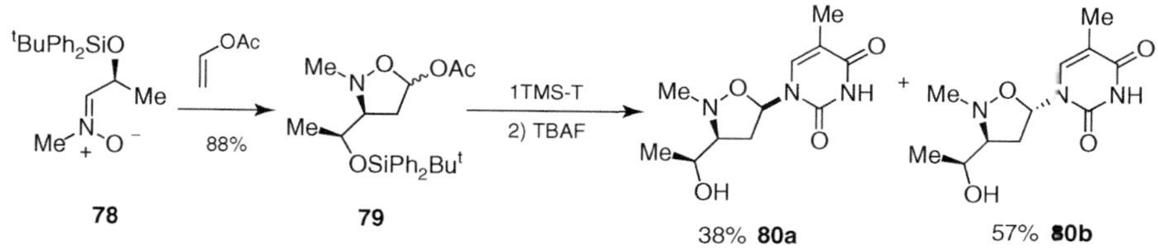

Scheme 20.

The presence of a chiral auxiliary at the C-carbon of the nitrone offers a variable degree of stereochemical control over the cycloaddition process, thus providing a good mean of access to N,O-nucleosides in enantiomerically pure form. Thus, stereoselective construction of the isoxazolidine ring has been achieved through cycloaddition of the chiral nitrone **81**, containing [(1S)-endo]-(−)-borneol as a chiral auxiliary and vinyl nucleobases, followed by a NaBH$_4$ reduction (Scheme 21). The reaction produces the α-anomers (1′S,3′S)-**37** and (1′R,3′R)-**84** as major stereoisomers.[24]

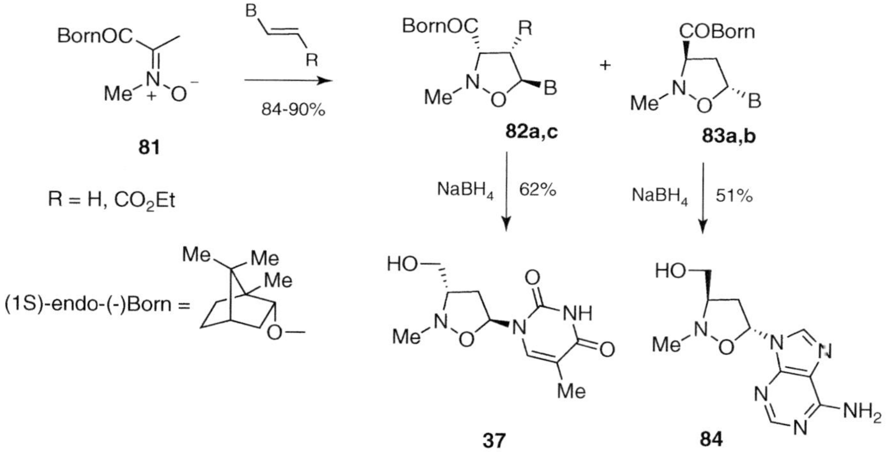

Scheme 21.

Merino et al., using the reaction of nitrones **85** with vinyl acetate followed by the Vorbrüggen procedure, and manipulation of products of nucleosidation, have synthesized the thymidine derivative **87** (Scheme 22).[21,25]

L-Isoxazolidinylthymidine **76** has also been synthesized using as a starting dipole the N-benzyl-1,2-di-O-isopropylidene-D-glyceraldehyde nitrone **74** (Scheme 23).[26]

Isoxazolidinyl nucleosides bearing a polyhydroxylated chain in place of the hydroxymethyl group have been reported by Fisera et al.[27] Cycloaddition reaction of the chiral protected nitrone **90** with vinyl acetate afforded a diasteromeric mixture of cycloadducts from which **92** was separated. Introduction of nucleobases, such

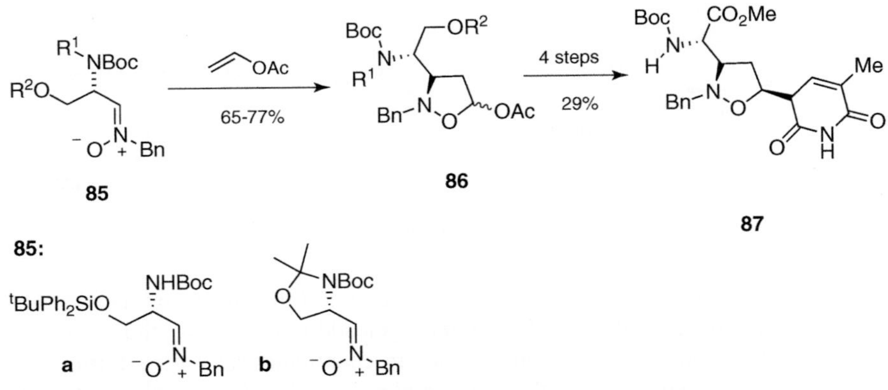

Scheme 22.

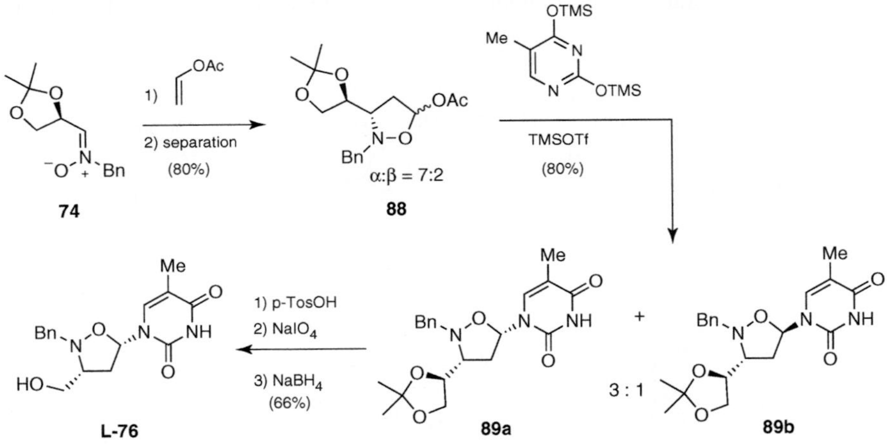

Scheme 23.

as uracil, thymine, acetyl cytosine, and acetyl guanine, afforded D-threonucleoside analogues (Scheme 24). Similarly, by using D-erythronitrone **91**, the nucleoside analogues were obtained.

The same authors have prepared the corresponding D-xyloisoxazolidinyl nucleosides **98** and **99** using the nitrone **96**. The Vorbrüggen nucleosidation was performed at room temperature in methylene chloride, producing the β-anomers **98** as major compounds (Scheme 25).[28]

An improved method of obtaining unprotected enantiomerically pure isoxazolidinyl nucleosides **103a,b** (Scheme 26) involves the use of a chiral auxiliary at the nitrogen atom of the nitrone functionality, using the sugar moiety as an inductor of chirality. In this way, Vasella-like nitrones **100** are versatile building blocks:

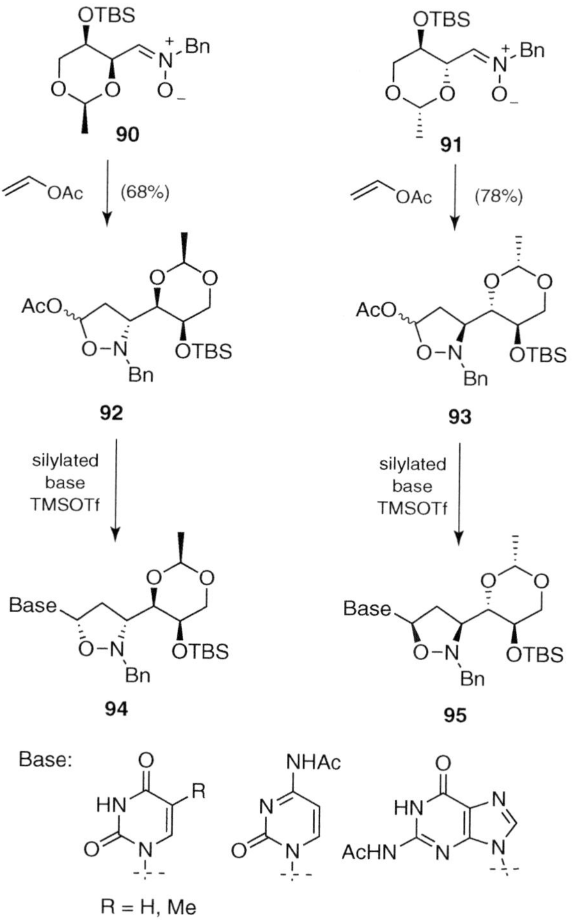

Scheme 24.

the chiral auxiliary is easily introduced before the cycloaddition process and easily removed to give *N,O*-nucleosides unsubstituted at nitrogen atom.[29] The diastereoselectivity of the cycloaddition process is in favor of the α-anomers **103b**. The same authors have found that 1,3-dipolar cycloaddition performed directly with vinyl nucleobases showed better stereoselectivity toward β-nucleosides **103a**.

Similarly, isoxazolidinyl nucleosides, containing an aminomethyl group linked at C3′, have been prepared from nitrone **105**, generated in situ from the *N*-Fmoc-protected glycynal **104** and a D-mannose-derived hydroxylamine. Interestingly, the introduction of a *tert*-butoxycarbonylmethyl group at the nitrogen atom of the isoxazolidine moiety led to the monomer **108**, suitable for the preparation of PNA analogues (Scheme 27).[30]

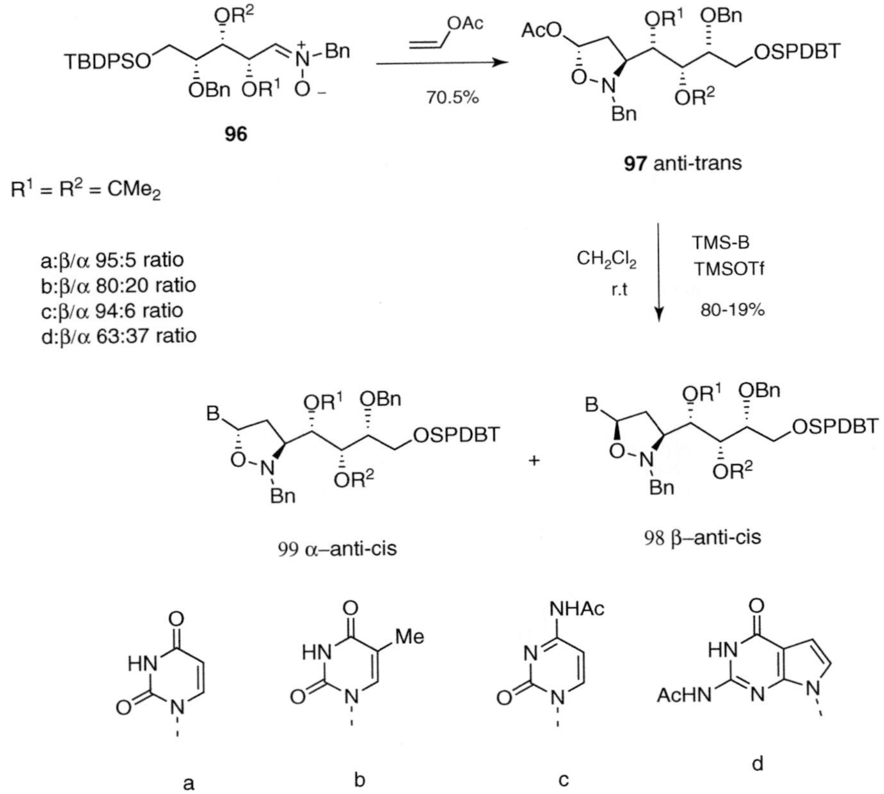

Scheme 25.

R¹ = R² = CMe₂

a: β/α 95:5 ratio
b: β/α 80:20 ratio
c: β/α 94:6 ratio
d: β/α 63:37 ratio

3.4. Enantiomeric Resolution (Pig Liver Esterase and Lipase)

(−)-4′-Benzyl-3-carboxy-2′,3′-dideoxy-4′-azaadenosine **111** in 38.3% overall yield has been synthesized from Sindona et al.[31] by pig liver esterase (PLE) resolution of racemate **110**, obtained from 1,3-dipolar cycloaddition of nitrone **109** with vinyl adenine (Scheme 28). The ester function present at C3′ is indispensable for resolution of the racemate with PLE.

Chiral N,O-nucleoside derivatives have also been obtained by a biocatalytic procedure consisting of lipase-catalyzed resolution of the corresponding racemate in organic solvent. Enantioselective esterification of thymine and cytosine derivatives **36** and **55** has been investigated by comparing the efficiency of different lipases and acyl donors. Since esterification of **36** and **55** occurs with low enantioselectivity, the authors resorted to double-sequential kinetic resolution, using Lipozyme IM as the best catalyst (Schemes 29 and 30). The approach allowed to obtain all the enamtiomers with ee ≥95%.[32]

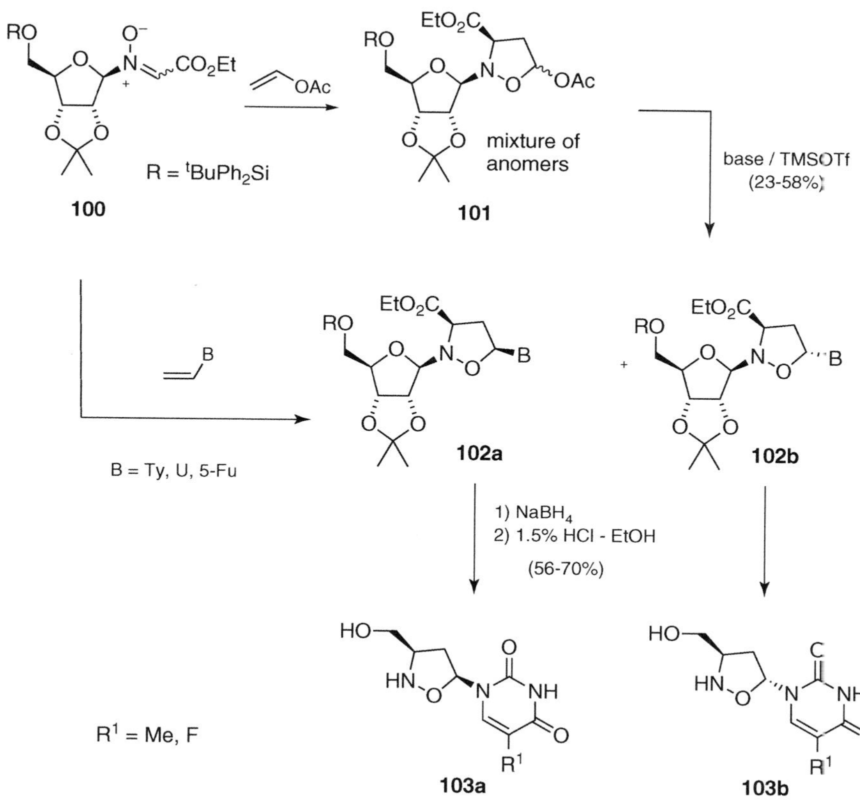

Scheme 26.

4. ISOXAZOLIDINYL C-NUCLEOSIDES

C-Nucleosides, where the typical C–N glycosidic bond is replaced by a non-hydrolyzable C–C bond, have been synthesized according to the 1,3-dipolar cycloaddition methodology. Thus, the new class of modified nucleosides **118**, structurally related to natural pseudouridine, has been approached using thymine-derived nitrones **116**. The reaction with allyl alcohol in a sealed tube for 24 h or MW irradiation for 10 min afforded, in a 1:1 ratio, the corresponding isoxazolidinyl derivatives **117a** and **117b** (Scheme 31). Interestingly, when montmorillonite K-10 was used as an additive, the reaction led directly to the debenzylated derivatives **118a** and **118b** in a 1:1 ratio.[33]

Starting from nitrone **74**, the enantiomerically pure analogue of the antitumoral C-nucleoside, tiazofurin **122**, has been synthesized.[34] The cycloaddition of **74** with acrylonitrile gave a mixture of the four possible stereoisomers, the most important of which, [(3S,5R)] **119**, was transformed into **122**, by condensation with

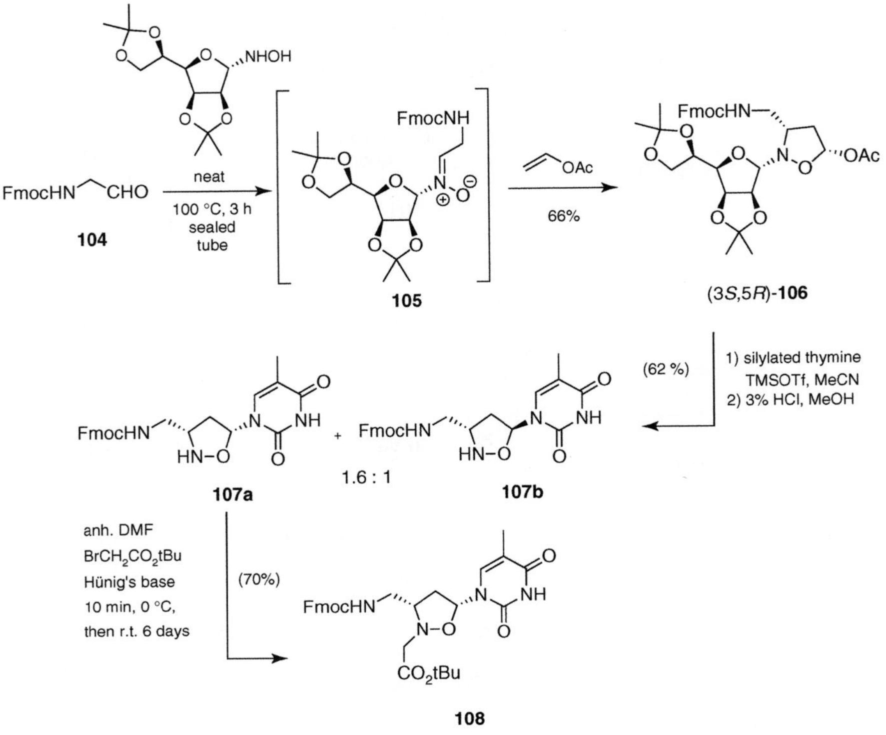

Scheme 27.

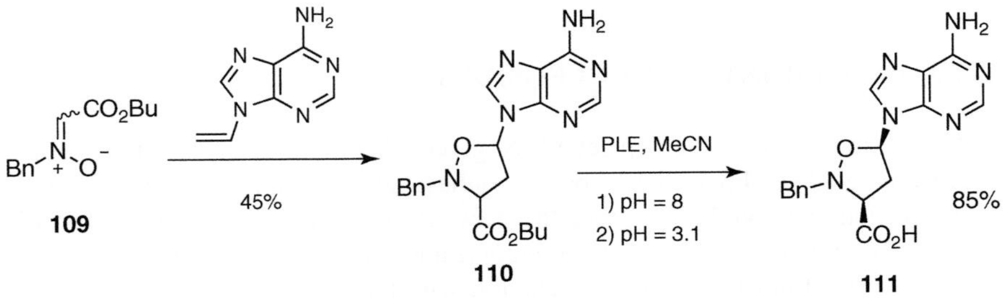

Scheme 28.

L-cysteine, oxidation of the thiazoline ring, hydrolysis of the acetonide moiety, diol cleavage, reduction, and ammonolysis (Scheme 32).

Using the chiral nonracemic allyl alcohol **124** as a dipolarophile, the reverse analogue of benzylated isoxazolidinyl tiazofurin **126** has been prepared. The thiazolyl nitrone **123**, subjected to cycloaddition with (2*S*)-1-[(*tert*-butyl (dimethyl)silyl)oxy]but-3-en-2-ol **124**, in the presence of zinc(II) triflate under

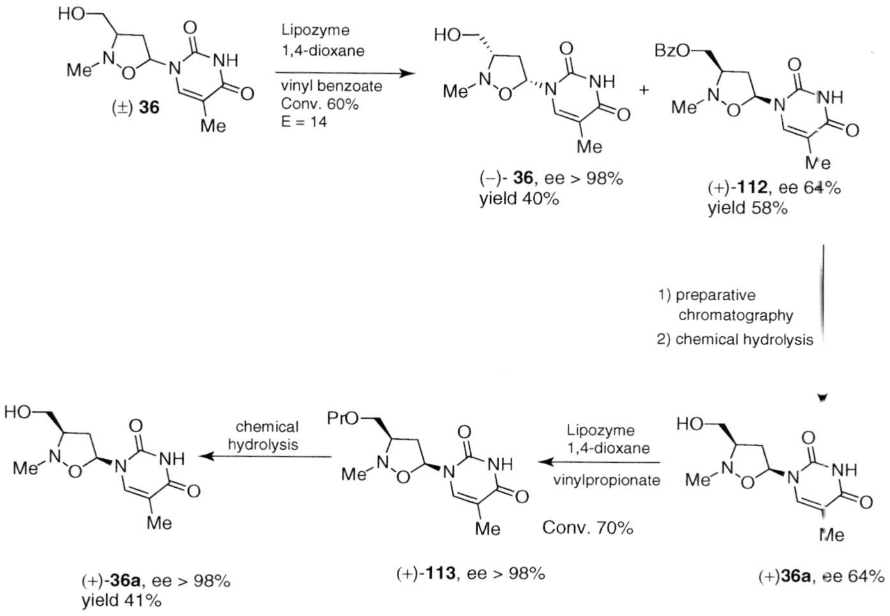

Scheme 29.

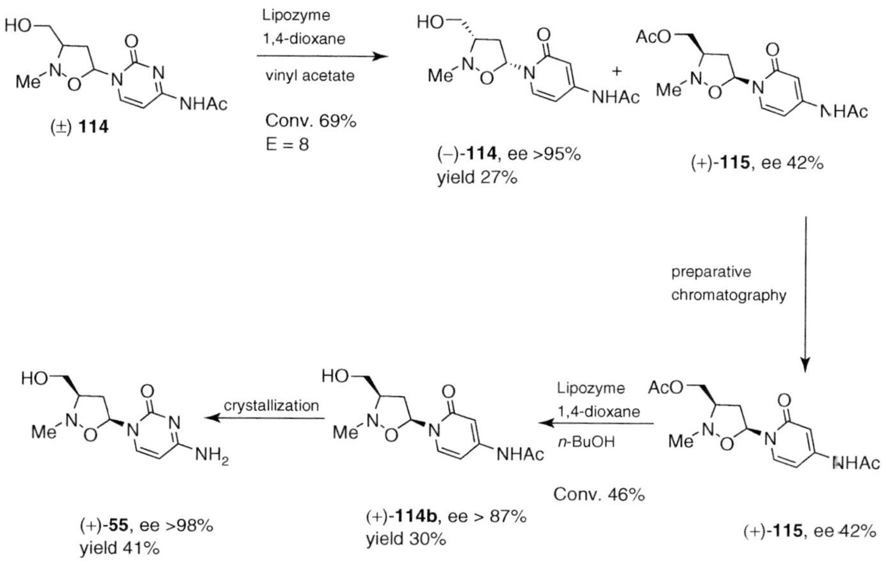

Scheme 30.

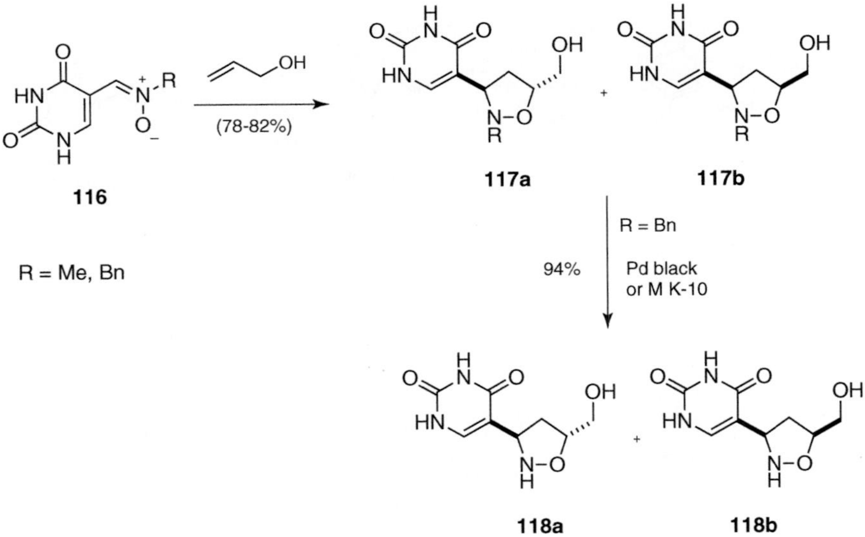

Scheme 31.

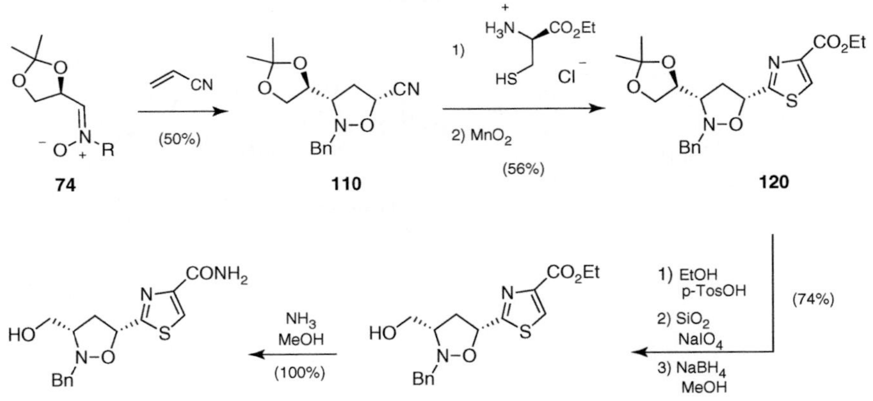

Scheme 32.

microwave irradiation, gave as the major stereoisomer the *syn* cycloadduct **125** in 56% yield. This product was readily transformed into *C*-nucleoside **126** in 45% global yield (Scheme 33). The presence of both Lewis acids and microwave irradiation was crucial for the success of the reaction.[35]

Moreover, starting from nitrone **127**, the same authors have prepared, by treatment of **22** with silylated glycolic aldehyde, the debenzylated isoxazolidinyl tiazofurin **130** (Scheme 34).

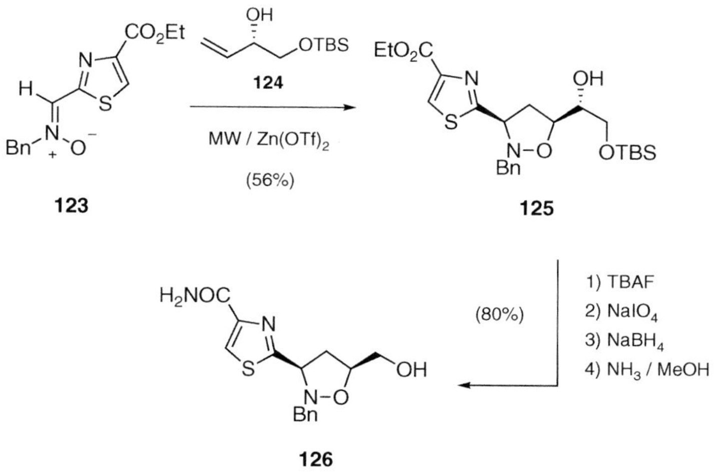

Scheme 33.

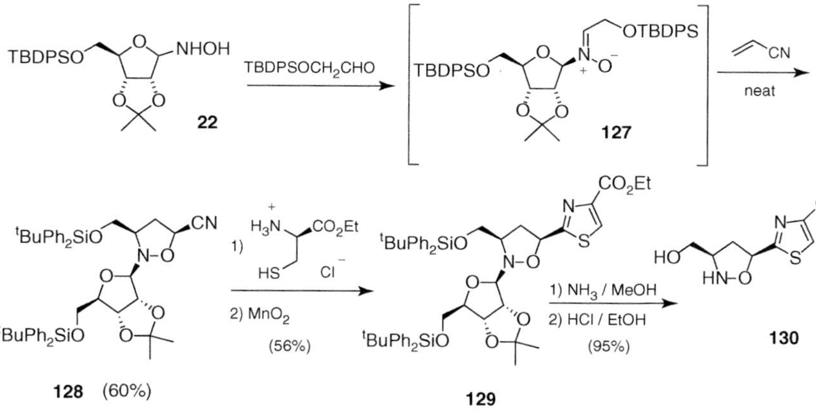

Scheme 34.

In addition, the two approaches depicted in Schemes 33 and 34 have been shown to be complementary, since both D- and L- enantiomeric series were accessible, depending on the starting nitrone employed in the initial step.

5. PHOSPHONATED *N,O*-NUCLEOSIDES

According to the statement that antiviral activity of modified nucleosides is strictly related to their conversion to the triphosphate form,[36] phosphonated *N,O*-nucleosides, as mimetic of monophosphate nucleosides,[37] have been synthesized with the aim to bypass the first limiting step of phosphorylation.[38] In this context the groups of Romeo et al. and Chiacchio et al. have synthesized homo phosphonated-,[39] phosphonated-,[40] and truncated phosphonated-N,O-nucleosides[41] by using nitrones containing a phosphonic group.

In particular, the cycloaddition of nitrone **131** with vinyl acetate afforded a mixture of *trans/cis* adducts (1.85:1 ratio) **132** and **133**, which were converted into α- and β- anomers **134** and **135** in a ratio varying from 1:9 (N-acetylcytosine) to 3:7 (thymine and 5-fluorouracil) (Scheme 35). All the products have been evaluated for their ability to inhibit the reverse transcriptase of avian myeloblastosis retrovirus, and no significant activity was observed.[39]

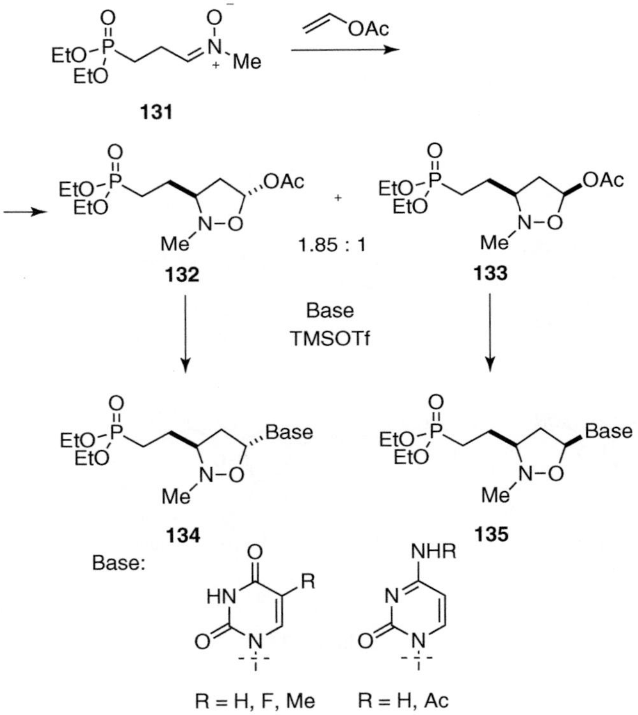

Scheme 35.

A rationalization for the lack of antiviral activity in tyhis class of compounds can be inferred by the data reported by Sigel and Griesser.[42] Chelation with metal ions plays a determinant role: viral polymerases recognize and use triphosphates nucleosides, complexed with metal ions such as a Mg^{2+} and Mn^{2+}, and the type of complexation determines the reaction pattern of nucleotides. Thus, for phosphonated N,O-nucleosides, the proximity of the N- atom to the P-atom could be discriminant with regard to the biological activity. In full-length nucleotides as **135**, the N-atom cannot assure a positive contribution because a seven-membered ring should be formed by chelation, while short length and truncated PCOAN could form six- or five-membered chelates, thus facilitating the bond break between P-α and P-β and allowing transfer of the nucleotide group with release of pyrophosphate.

Scheme 36 describes the route toward short-length PCOANs **138**.[40] The cycloaddition process of phosphonated nitrones **136** conduces, after usual reactions, phosphonated N,O-nucleosides **138** and **139**, containing thymine, 5-fluorouracil, 5-bromouracil, cytosine, adenine, and guanine, in a 2.5:1 ratio, respectively (Scheme 36).

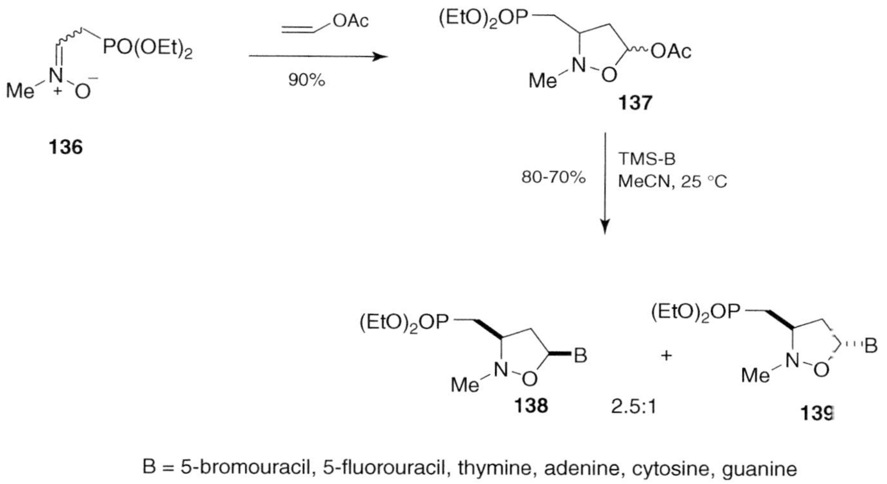

B = 5-bromouracil, 5-fluorouracil, thymine, adenine, cytosine, guanine

Scheme 36.

The cytotoxicity and RT-inhibitory activity of β-anomers **138** were also investigated. Compounds have been shown to present low levels of cytotoxicity assessed by conventional methods to detect viability. Noteworthy, the pyrimidinyl derivatives were as powerful as AZT in inhibiting the RT activity of the human retrovirus T-cell leukemia/lymphotropic virus type 1 and in protecting human peripheral blood mononuclear cells against human retrovirus T-cell leukemia/lymphotropic virus type 1 transmission in vitro.[40]

Experimental Procedure for the Synthesis of PCOAN 138

Synthesis of cis-trans Isoxazolidines 137.[40] A solution of nitrone **136** (5.7 mmol) in vinyl acetate (30 mL) was stirred at 60°C for 24 h. After this period, the reaction mixture was evaporated under reduced pressure and the residue was purified by radial chromatography ($CHCl_3/CH_3OH$ 99.5:0.5 to give a *cis/trans* mixture of isoxazolidines **137**. The first eluted product was **(3RS,5SR)-3-[(diethoxyphosphoryl)methyl]-2-methylisoxazolidin-5-yl**

acetate [(cis) 137]: yield 65%, light yellow oil. ^{1}H-NMR (500 MHz, CDCl$_3$): δ 1.34 (t, 6H, J = 7.0 Hz), 1.92–2.07 (m, 2H), 2.09 (s, 3H, OAc), 2.25 (ddd, 1H, J = 2.5, 8.5, and 14.0 Hz, H$_{4a}$), 2.73 (s, 3H, N-Me), 2.85 (m, 1H, H$_3$), 2.95 (ddd, 1H, J = 6.5, 8.9, and 14.0 Hz, H$_{4b}$), 4.09–4.18 (m, 4H), 6.28 (dd, 1H, J = 2.5 and 6.5 Hz, H$_5$). ^{13}C NMR (125 MHz, CDCl$_3$): δ 16.13, 16.24, 21.02, 28.51 (d, J = 141.0 Hz), 42.74, 43.11, 61.65 (d, J = 12.1 Hz), 62.50, 94.42, 170.26. The second eluted product was **(3RS,5RS)-3-[(diethoxyphosphoryl)methyl]-2-methylisoxazolidin-5-yl acetate [(trans) 137]:** yield 25%, light yellow oil. ^{1}H-NMR (500 MHz, CDCl$_3$): δ 1.33 (t, 6H, J = 7.0 Hz), 1.85–1.92 (m, 1H, H$_{4a}$), 2.05–2.08 (m, 1H), 2.07 (s, 3H, OAc), 2.66 (dd, 1H, J = 5.4 and 13.8 Hz, H$_{4b}$), 2.73 (s, 3H, N-Me), 3.35 (m, 1H, H$_3$), 4.07–4.21 (m, 4H), 6.31 (d, 1H, J = 5.4 Hz, H$_5$). ^{13}C NMR (125 MHz, CDCl$_3$): δ 16.14, 16.23, 21.10, 28.25 (d, J = 115.0 Hz), 42.13, 46.31, 60.65, 60.73 (d, J = 11.7 Hz), 96.44, 170.19.

Diethyl {(1'SR,4'RS)-1'-[[(5-methyl-2,4-dioxo-3,4-dihydropyrimidin-1(2H)-yl)]-3'-methyl-2'-oxa-3'-aza-cyclopent-4'-yl]}methylphosphonate (138c).[40] A suspension of thymine (0.62 mmol) in dry acetonitrile (3 mL) was treated with bis(trimethylsilyl)acetamide (BSA) (2.54 mmol) and refluxed for 15 min under stirring. To the clear solution obtained were added a solution of the epimeric isoxazolidines **137** (0.52 mmol) in dry acetonitrile (3 mL) and trimethylsilyltriflate (TMSOTf) (0.4 mmol) dropwise, and the reaction mixture was stirred at 55°C for 6 h. After being cooled at 0°C, the solution was neutralized by careful addition of aqueous 5% sodium bicarbonate and then concentrated in vacuo. After the addition of dichloromethane (8 mL), the organic phase was separated, washed with water (2 × 10 mL), dried over sodium sulfate, filtered, and evaporated to dryness. The residue was purified by flash chromatography (CHCl$_3$/MeOH, 95:5) to give β-nucleoside **138c**, in 68% yield as sticky oil. ^{1}H-NMR (500 MHz, CDCl$_3$): δ 1.30 (dt, 6H, J = 3.6 and 7.1 Hz), 1.89 (ddd, 1H, J = 10.2, 15.0, and 18.3 Hz, H$_{5'a}$), 1.93 (d, 3H, J = 1.3 Hz), 2.08 (ddd, 1H, J = 3.2, 15.0, and 20.6 Hz, H$_{5'b}$), 2.23 (ddd, 1H, J = 4.6, 10.1, and 13.8 Hz, H$_{6'a}$), 2.75 (s, 3H, N-Me), 2.98 (dddd, 1H, J = 3.2, 7.0, 10.1, and 10.2 Hz, H$_{4'}$), 3.18 (ddd, 1H, J = 7.0, 7.9, and 13.8 Hz, H$_{6'b}$), 4.10–4.17 (m, 4H), 6.20 (dd, 1H, J = 4.6 and 7.9 Hz, H$_{1'}$), 7.66 (q, 1H, J = 1.3 Hz, H$_6$), 9.56 (b s, 1H, NH). ^{13}C-NMR (125 MHz, CDCl$_3$): δ 12.57, 16.33, 16.38, 27.78 (d, J = 143.2 Hz), 42.68, 44.35, 61.93, 61.98, 63.37, 81.94, 110.66, 135.96, 150.56, 164.11.

Truncated phosphonated N,O-nucleosides **142** (TPCOANs),[41] a new class of N,O-nucleosides which contain a diethylphosphonate group directly linked at C3 of the isoxazolidine ring, have been obtained by exploiting the reactivity of phosphonated nitrone **140** as dipole. Thus, the reaction of **140** with vinyl nuleobases, under microwave irradiation, led to the unnatural α-nucleosides **141** as main adducts, while the β-anomers **142** have been obtained in high yield by a two-step procedure involving the vinyl acetate cycloaddition, followed by nucleosidation (Scheme 37).

Preliminary biological assay have shown that the β-anomers **142** completely inhibit the RT of avian Moloney virus (AMV) and human immunodeficiency virus

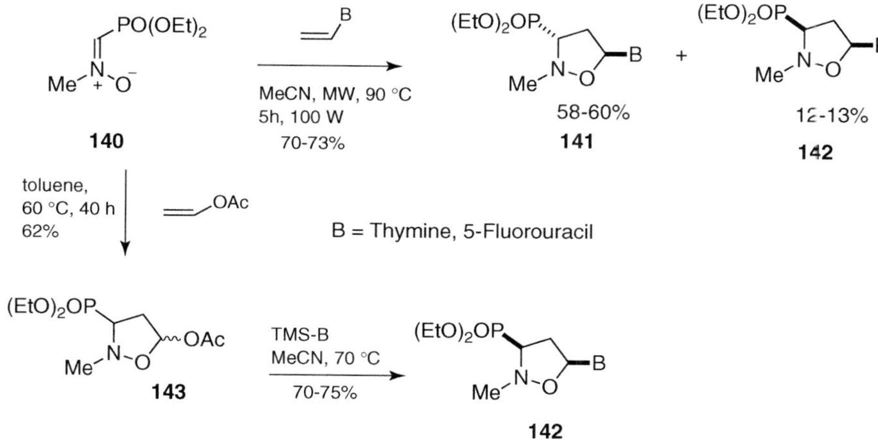

Scheme 37.

(HIV), at concentrations 1 ± 0.1 nM, at a level comparable with that of tenofovir (1 nM) and 10-fold lower than AZT (10 nM). Moreover, the cytotoxicity, evaluated by MTS assay, have indicated a very low toxicity ($CC_{50} > 500$ mM) comparised with AZT (CC_{50} 12.14 mM).[41]

Experimental Procedure for the Synthesis of TPCOAN 142a[41]

1. *One-step procedure*. A solution of **140** (200 mg, 1.02 mmol) in dry acetonitrile (20 mL) and vinyl thymine (156 mg, 1.02 mmol) was put in a sealed tube and irradiated under microwave conditions at 100 W, 90°C, for 5 h. The removal of the solvent in vacuo afforded a crude material which, after MPLC purification by using as eluent a mixture of CHCl$_3$/MeOH 99:1, gives the nucleosides **141a** and **142a**.

2. *Two-step procedure*. A suspension of thymine (252 mg, 2 mmol) in dry acetonitrile (30 mL) was treated with bis(trimethylsilyl)acetamide (1,5 mL, 6 mmol) and left under stirring until the solution was clear. A solution of a mixture of isozaxolidines **143** (282 mg, 1 mmol) in dry acetonitrile (10 mL) and trimethylsilyl triflate (72 µL, 0.4 mmol) was then added, and the reaction mixture was heated at 70°C for 5 h. After being cooled at 0°C, the solution was carefully neutralized by the addition of aqueous 5% sodium bicarbonate and then concentrated in vacuo. After the addition of dichloromethane (20 mL), the organic phase was separated, washed with water (2 × 10 mL), dried over sodium sulfate, filtered, and evaporated to dryness. The ^{1}H-NMR spectrum of the crude reaction mixture shows the presence of β-anomers as nearly exclusive adducts, while the α-anomers are present only in traces. The residue was purified by MPLC on a silica gel column using as eluent a mixture of CHCl$_3$/MeOH 99:1 to afford **142b**.

Diethyl [(1'SR,4'SR)-1'-[5-Methyl-2,4-dioxo-3,4-dihydropyrimid-1(2H)-yl]-3'-methyl-2'-oxa-3'-azacyclopent-4'-yl]phosphonate (142b). Sticky foam (243 mg, 70% yield by the two-step procedure; 40 mg, 12% yield by the one-step procedure). ^{1}H-NMR (700 MHz, CDCl$_3$): δ 9.08 (s, NH, 1H), 7.75 (br q, J = 0.7 Hz, CH=, 1H), 6.23 (dd, J = 7.5 and 3.3 Hz, H$_{1'}$, 1H), 4.23–4.10 (m, 4H), 3.20 (dddd, J = 13.8, 9.0, 7.5, and 4.8 Hz, H$_{5'a}$, 1H), 3.01 (ddd, J = 9.6, 9.0, and 2.7 Hz, H$_{4'}$, 1H), 2.97 (s, 3H), 2.70 (dddd, J = 17.1, 13.8, 9.6, and 3.3Hz, H$_{5'b}$, 1H), 1.95 (d, J = 0.7 Hz, 3H), 1.34 (t, J = 7.0 Hz, 3H), 1.32 (t, J = 7.0 Hz, 3H). ^{13}C-NMR (176 MHz, CDCl$_3$): δ 164.2 (C$_4$), 150.9 (C$_2$), 136.6 (C$_6$), 110.8 (C$_5$), 82.3 (d, J = 10.6 Hz, C$_{1'}$), 64.2 (d, J = 167.3 Hz, C$_{4'}$), 63.6 (d, J = 7.0 Hz), 63.0 (d, J = 7.0 Hz), 46.2 (CH$_3$N), 41.5 (d, J = 3.5 Hz, C$_{5'}$), 16.8 (d, J = 5.2 Hz), 16.7 (d, J = 5.3 Hz), 13.0 (CH$_3$–CH=). ^{31}P-NMR (121.5 MHz, CDCl$_3$): δ 21.44.

Diethyl [(1'SR,4'RS)-1'-[5-Methyl-2,4-dioxo-3,4-dihydropyrimid-1(2H)-yl]-3'-methyl-2'-oxa-3'-azacyclopent-4'-yl]phosphonate (141a). Sticky foam (200 mg, 58% yield by the one-step procedure). ^{1}H-NMR (700 MHz, CDCl$_3$): δ 8.97 (s, NH, 1H), 7.27 (q, J = 0.95 Hz, CH=, 1H), 6.07 (dd, J = 7.7, 3.9 Hz, H$_{1'}$, 1H), 4.27–4.22 (m, 2H), 4.22–4.15 (m, 2H), 3.30–3.25 (brm, H$_{4'}$, 1H), 3.10–3.00 (brm, H$_{5'b}$, 1H), 3.05 (s, 3H), 2.60 (dddd, J = 13.6, 8.0, 6.8, and 3.9 Hz, H$_{5'a}$, 1H), 1.96 (d, J = 0.95 Hz, 3H), 1.39 (t, J = 7.0 Hz, 3H), 1.37 (t, J = 7.0 Hz, 3H). ^{13}C-NMR (176 MHz, CDCl$_3$): δ 163.5 (C$_4$), 150.1 (C$_2$), 135.3 (C$_6$), 111.3 (C$_5$), 83.2 (d, J = 10.2 Hz, C$_{1'}$), 63.1 (d, J = 164.7 Hz, C$_{4'}$), 62.6 (d, J = 7.3 Hz), 62.6 (d, J = 7.3 Hz), 38.7 (CH$_3$N), 29.7 (C$_{5'}$), 16.52 (d, J = 5.6 Hz), 16.50 (d, J = 6.3 Hz), 12.6 (CH$_3$–CH=). ^{31}P-NMR (121.5MHz, CDCl$_3$): δ 20.73.

6. 3'-SUBSTITUTED-4'-AZA-2',3'-DIDEOXYNUCLEOSIDES III (N,O-PSICONUCLEOSIDES)

Natural psicofuranosyl nucleosides, carrying an hydroxymethyl group at the anomeric carbon atom, are endowed with interesting biological activities. On this basis, applying the cycloaddition approach, Chiacchio et al.[43] have designed an easy entry toward *N,O*-psiconucleosides in which the sugar unit was replaced by an isoxazolidine ring (Scheme 38). Condensation between nitrone **38a** and ethyl 2-acetyloxyacrylate **144**, afforded intermediates **145**, which were transformed in two steps into α- and β-anomers **146a** and **146b** in a 1:1 ratio. β-Anomers **146b** (80–50% yield) were obtained as exclusive compounds when the nucleosidation was performed at 45°C.

The reaction of D-glyceraldehyde-derived nitrone **74**, used as a source of chirality, furnishes the β-L and the β-D *N,O*-psiconucleosides **146a** and **146b** (Scheme 39) in enantiomerically pure form.[44]

A new class of truncated phosphonated *N,O*-psiconucleosides has recently been synthesized by reaction of phosphonated nitrone **140** with ethyl 2-acetyloxyacrylate **144**, followed by the usual Vorbrüggen nucleosidation.[45] The cycloaddition approach produces a mixture of *trans/cis* isoxazolidines **148a** and

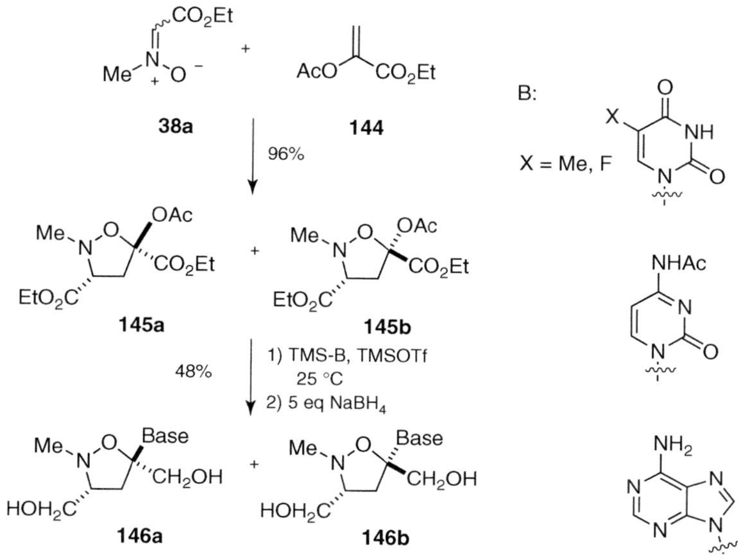

Scheme 38.

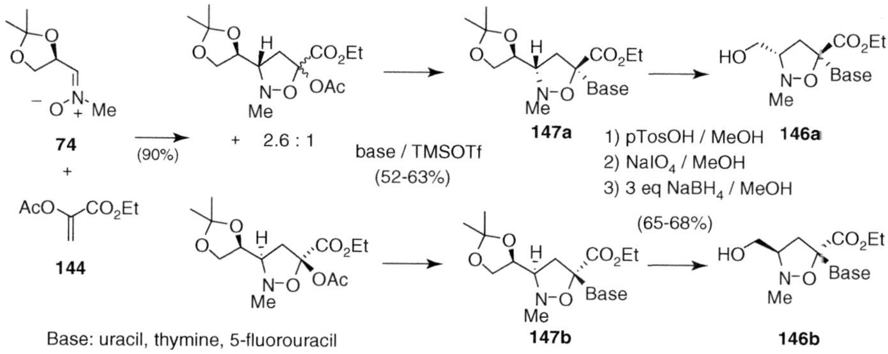

Scheme 39.

148b in an isomeric ratio of 4.5:1 and a combined yield of 80%. Subsequent condensation with silylated nucleobases, performed at 70°C, produces, as expected, the β-anomers **149b** as almost exclusive compounds (Scheme 40).

Preliminary biological assays show that the β-anomers **149b** are able to inhibit the RT of AMV, HTLV-1, and HIV. In particular, the 5-fluorouracil derivative is the more promising compound, acting on AMLV and on HIV at a concentration 1 and 10 nM, respectively. The level of the inhibitory activity of this product toward HTLV-1 and HIV was 10-fold higher than that of tenofovir and similar to that of

AZT. Moreover, as for tenofovir, this compound does not show any cytoxicity, according to MTS assay.[45]

Experimental Procedure Toward 149[45]

Ethyl(3 SR,5 RS)-5-(acetyloxy)-3-(diethoxyphosphoryl)-2-methyl-isoxazolidine-5-carboxylate (148b) and Ethyl(3 SR,5 SR)-5-(acetyloxy)-3-(diethoxy-phosphoryl)-2-methylisoxazolidine-5-carboxylate (148a).[45] A solution of C-diethoxyethylphosphoryl-N-methyl nitrone **140** (3.0 g, 22.9 mmol) and ethyl 2-acetyloxyacrylate **144** (3.7 g, 23 mmol) in THF (100 mL) was stirred at reflux for 24 h. The solvent was evaporated and the residue was purified by flash chromatography (dichloromethane/isopropanol, 98:2). The product eluted first was **148b**; 18% yield; yellow oil. ^{1}H-NMR (500 MHz, CDCl$_3$): δ 4.28–4.12 (m, 6H,), 3.23 (ddd, J = 9.1, 6.9, and 6.5 Hz, H$_3$), 2.93 (ddd, J = 12.5, 12.6, and 9.1 Hz, H$_{4b}$), 2.84 (s, 3 H, CH$_3$N), 2.81 (ddd, J =18.8, 12.5, and 6.9 Hz, H$_{4a}$), 2.07 (s, 3H), 1.31 (t, 6H, J = 6.5 Hz), 1.24 (t, 3H, J = 7.0 Hz). ^{13}C-NMR (125 MHz, CDCl$_3$): δ 170.16 (s, C=O), 165.50 (s, C=O), 102.19 (d, $^3J_{PCCC}$ = 9.5 Hz, C5), 64.57 (d, $^1J_{PC}$ = 163.1, C3), 63.61 (d, J = 6.0 Hz, C–O–P), 62.71 (d, J = 7.8 Hz, C–O–P), 62.61 (s, CH$_2$–O), 46.08 (s, CH$_3$–N), 44.50 (d, $^2J_{PCC}$ = 1.5 Hz, C4), 20.84 (s, CH$_3$ –C=O), 16.38 (d, J = 6.0 Hz), 13.86 (s, CH$_3$). ^{31}P-NMR (121.5 MHz, CDCl$_3$, δ): 22.77.

The fraction eluted second was **148a**; 62% yield; yellow oil. ^{1}H-NMR (500 MHz, CDCl$_3$): δ 4.27–4.09 (m, 6H), 3.27 (ddd, J = 12.0, 6.1, and 1.5 Hz, H$_3$), 2.96 (ddd, J = 16.5, 13.0, and 12.0 Hz, H$_{4b}$), 2.94 (d, J = 0.9 Hz, 3 H, CH$_3$N), 2.85 (ddd, J = 13.0, 6.1, and 4.3 Hz, H$_{4a}$), 2.05 (s, 3H), 1.32 (t, 6H, J = 6.5 Hz), 1.23 (t, 3H, J = 7.0 Hz). ^{13}C-NMR (125 MHz, CDCl$_3$): δ 169.48

(s, C=O), 164.84 (s, C=O), 102.24 (d, $^3J_{PCCC}$ = 13.5 Hz, C5), 64.00 (d, J = 6.0 Hz, C–O–P), 62.78 (d, $^1J_{PC}$ = 171.7, C3), 62.70 (d, J = 8.2 Hz, C–C–P), 62.63 (s, CH$_2$–O), 49.24 (d, J = 4.5 Hz, C–N–C–P), 43.01 (d, $^2J_{PCC}$ = 1.5 Hz, C4), 21.07 (s, CH$_3$–C=O), 16,38 (d, J = 6.0 Hz), 13.86 (s, CH$_3$). ^{31}P-NMR (121.5 MHz, CDCl$_3$): δ 21.66.

Ethyl(3 SR,5 RS)-3-(diethoxyphosphoryl)-5-(5-fluoro-2,4-dioxo-3,4-dihydropyrimidin-1(2 H)-yl)-2-methylisoxazolidine-5-carboxylate (149b) and Ethyl(3 SR, 5 SR)-3-(diethoxyphosphoryl)-5-(5-fluoro-2,4-dioxo-3,4-dihydropyrimi-din-1(2 H)-yl)-2-methylisoxazolidine-5-carboxylate (149a).[45] A suspension of 5-fluorouracil (260 mg, 2 mmol) in dry acetonitrile (30 mL) was treated with bis(trimethylsilyl)acetamide (1,5 mL, 6 mmol) and left under stirring until the solution was clear. A solution of a mixture of isozaxolidines **148a** and **148b** (282 mg, 1 mmol) in dry acetonitrile (10 mL) and trimethylsilyl triflate (72 μL, 0.4 mmol) was then added, and the reaction mixture was heated at 70°C for 5 h. After being cooled at 0°C, the solution was carefully neutralized by the addition of aqueous 5% sodium bicarbonate and then concentrated in vacuo. After addition of dichloromethane (20 mL), the organic phase was separated, washed with water (2 × 10 mL), dried over sodium sulfate, filtered, and evaporated to dryness. The ^{1}H-NMR spectrum of the crude reaction mixture shows the presence of β-anomer as a major adduct, while the α-anomer is present in low amount. The residue was purified by MPLC on a silica gel column using as the eluent a mixture of CH$_2$Cl$_2$/Me$_2$CHOH 98:2 to afford **149b** and **149a** in 80% yield.

Ethyl (3 SR,5 RS)-3-(diethoxyphosphoryl)-5-(5-fluoro-2,4-dioxo-3,4-dihydropyrimidin-1(2 H)-yl)-2-methylisoxazolidine-5-carboxylate (149b). Yield: 68%; white solid, mp 200–201°C. ^{1}H-NMR (500 MHz, CDCl$_3$): δ 8.91 (s, 1H), 7.77 (d, J = 6.5 Hz, 1H), 4.30–4.10 (m, 6H), 3.83 (ddd, J = 14.0, 8.5, and 4.0 Hz, 1H), 3.25 (ddd, J = 10.0, 8.5, and 3.0 Hz, 1H), 2.95 (s, 3H), 2.93 (ddd, J = 16.5, 14.0, and 10.0 Hz, 1H), 1.35 (t, J = 7.5 Hz, 3H), 1.28 (m, 6H).^{13}C-NMR (125 MHz, CDCl$_3$): δ 164.12, 156.93 (d, J = 26.7 Hz), 148.74, 139.90 (d, J = 235.1 Hz), 123.73 (d, J = 35.2 Hz), 92.63 (d, J = 10.6 Hz), 65.13 (d, J = 163.7 Hz), 63.43 (d, J = 7.3 Hz), 63.35, 62.61 (d, J = 6.8 Hz), 46.20, 45.17, 16.41, 16.21, 13.34.

Ethyl (3 SR,5 SR)-3-(diethoxyphosphoryl)-5-(5-fluoro-2,4-dioxo-3,4-dihydropyrimidin-1(2 H)-yl)-2-methylisoxazolidine-5-carboxylate (149a). Yield: 12%; white solid, mp 193–195°C. ^{1}H-NMR (500 MHz, CDCl$_3$): δ 8.65 (b s, 1H), 7.86 (d, J = 6.6 Hz), 4.29–4.14 (m, 6H), 3.81 (ddd, J = 15.0, 14.6, and 1.3 Hz, H$_{5'}$, 1H), 3.18–3.02 (bm, H$_{4'}$ and H$_{5'}$, 2H), 3.13 (s, N-CH$_3$, 3H), 1.38 (t, J = 8.5 Hz, 3H), 1.35 (t, J = 8.2 Hz, 3H), 1.25 (t, J = 8.2 Hz, 3H). ^{13}C-NMR (125 MHz, CDCl$_3$): δ 164.66, 159.82 (d, J = 30.4 Hz), 156.20, 139.95 (d, J = 237.0 Hz), 123.60 (d, J = 35.8 Hz), 92.55 (d, J = 12.5 Hz), 64.07 (d, J = 6.5 Hz), 63.88 (d, J = 175.0 Hz), 63.80, 62.90 (d, J = 6.5 Hz), 53.43, 42.62, 16.48, 16.37, 13.87.

7. HOMOCARBOCYCLIC-2'-OXO-3'AZANUCLEOSIDES IV (HOMO-N,O-NUCLEOSIDES)

Different synthetic approaches toward 1'-homo-C- and N-nucleosides, in which a methylene group has been inserted between the base and the carbohydrate moiety, with the aim to increase the resistance to hydrolytic or enzymatic cleavage or to improve the conformational flexibility and the rotational freedom, compared to commun nucleosides, have been reported in the literature.[46] Thus, by exploiting the cycloaddition process between nitrones **38a, 150** and allyl nucleobases **151**, isoxazolidinyl homonucleosides **152** have been prepared.[47] In particular, the reaction of **38a** with allyl nucleobases has led to a mixture of isoxazolidinyl derivatives **152a** and **152b** in a 1:1 ratio. The reaction of **150** has been found to proceed with better stereoselectivity, affording a mixture of epimeric isoxazolidines with a relative ratio of about 7:1 in favor of *cis*-derivatives **152b**. In the case of allyl adenine, the stereoselectivity was only 2:1 (Scheme 41).[47]

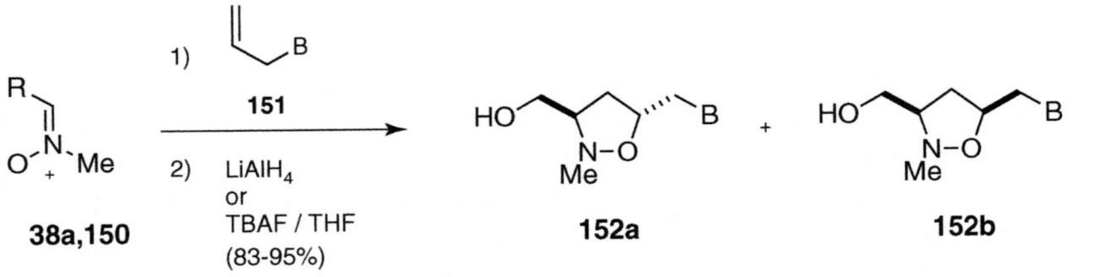

38a R = CO$_2$Et

150 R = CH$_2$OSiPh$_2^t$Bu

R = CO$_2$Et, CH$_2$OSiPh$_2^t$Bu   Base = 5-fluorouracil, thymine, cytosine, adenine

Scheme 41.

The enantioselective version of this reaction was performed using nitrones **100** and **153**, derived from D-ribose and D-mannose, respectively. Both have shown a good degree of enantioselectivity, so allowing easy entry to both enantiomers **155** (Scheme 42).[48]

All the nucleosides were evaluated for antiviral activity against a wide variety of viruses, including HIV-1, HIV-2, KOS, ACV, HSV, and Punta Toro. None of these compounds showed inhibitor activity against any of the viruses tested at concentrations up to 400 μg/mL.[48]

8. 1'-AZA-4'-OXA-2',3'-DIDEOXY NUCLEOSIDES V

Only one report has appeared on the synthesis of 1'-aza-4'-oxa-2',3'-dideoxy nucleosides (truncated reverse isoxazolidinyl nucleosides, TRINs). Some of these

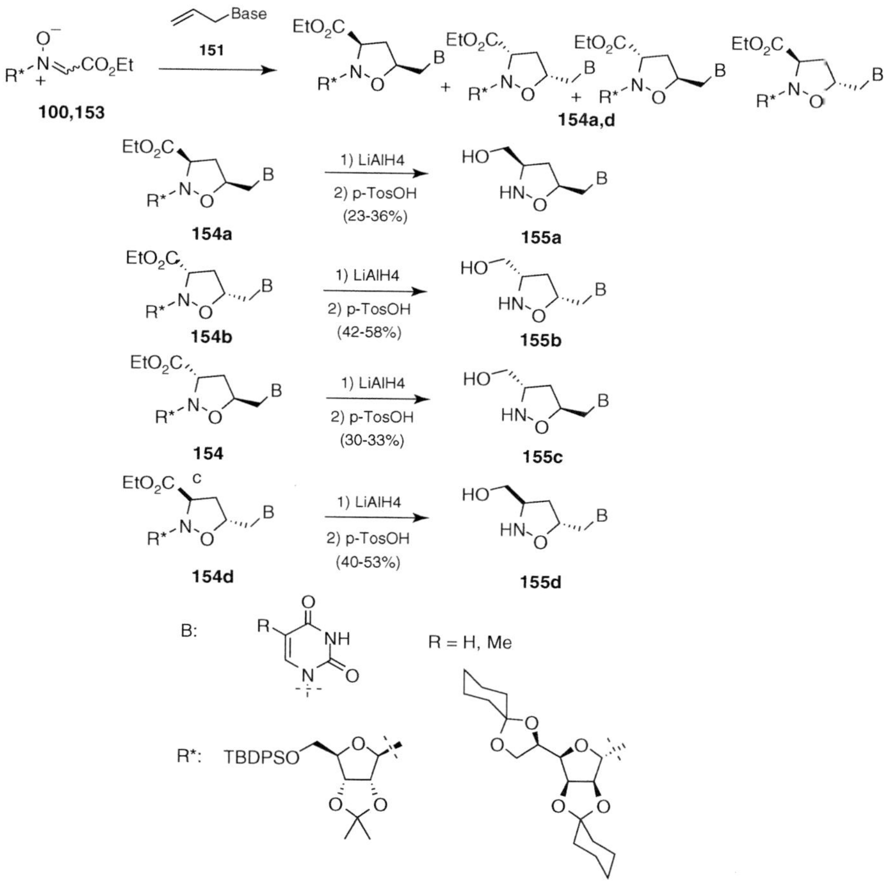

Scheme 42.

compounds have shown a valuable biological activity as nonnucleoside reverse transcriptase inhibitors.[49] The reaction route to TRINs starts from the isoxazolidin-3-one **156** and is based on its reduction to the corresponding azahemiacetal **157**, performed with Schwartz reagent, as the key step. Then, the acetylation of hydroxylic group, with Ac_2O, followed by Vorbrüggen nucleosidation, afforded target compounds **159–167** (Scheme 43).

All these compounds have been tested against AMV-RT and HIV-RT in comparison with the well-known NNRTI nevirapine. In particular, compounds **159** and **160** completely inhibited AMV-RT at the minimum inhibitory concentration of 10 nM, while they inhibited recombinant HIV-RT at concentrations of 20 and 40 nM, respectively. Conversely, compounds **162** and **163** were both unable to inhibit AMV or HIV reverse transcriptase. TRINs **165**, **166**, and **167** were all active toward HIV-RT at a concentration of 10, 1, and 100 nM. This activity

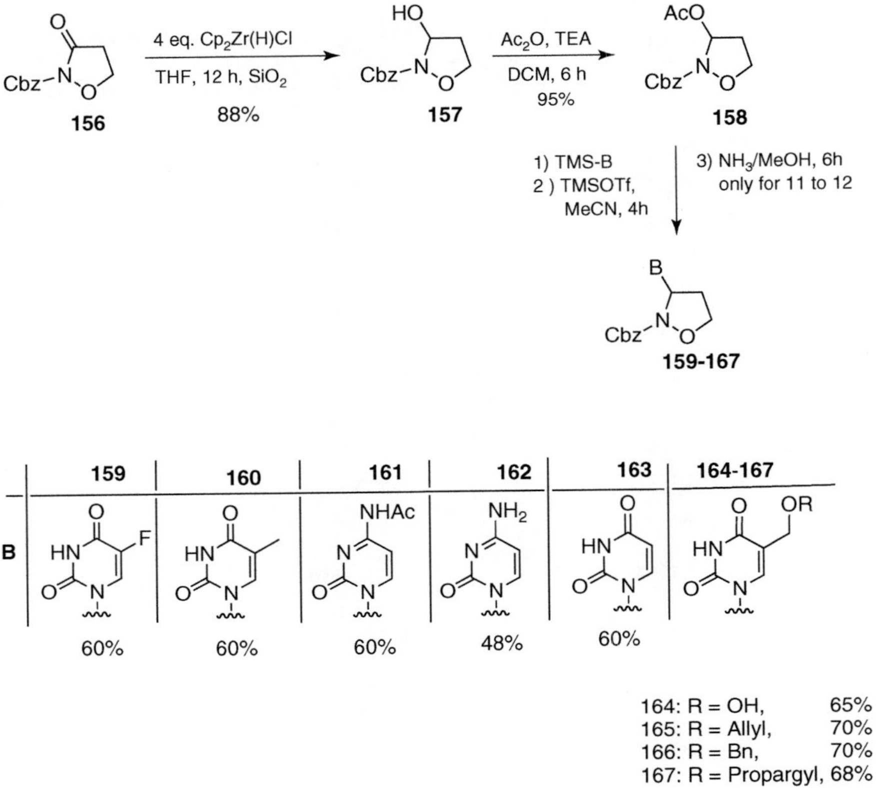

Scheme 43.

was highly specific toward HIV-RT, since, as with nevirapine, they were not able to inhibit AMV-RT. Interestingly, derivative **166** inhibited HIV-RT at a concentration overlapping that of nevirapine. Moreover, it must be emphasized that TRINs **165–167** are less cytotoxic than nevirapine toward MOLT-3 cells.[49]

Computational docking studies confirm the biological results. In fact, when TRINs were inserted into the NNIBP of the TNK651:HIV RT, strong interaction with Tyr188, Tyr181, Phe227, Tyr318, Lys101, Leu234, Leu100, and Trp229 is evident, similarly to TNK-651 that is able to bind at the RT allosteric site, preventing the conformational transition needed for the formation of a productive polymerase–RNA complex.

In conclusion, the SAR within the series of compounds investigated here correlates well with the results of the computational docking studies, indicating that the binding mode suggested by the docking studies is highly plausible.[49]

Procedure for the Preparation of Benzyl 3-(5-((benzyloxy)methyl)-3,4-dihydro-2,4- dioxopyrimidin-1(2H)-yl)isoxazolidine-2-carboxylate (166)

Benzyl 3-Oxoisoxazolidine-2-carboxylate (156). To a solution of isoxazolidin-3-one (1 g, 11.49 mmol) in anhydrous methylene chloride (15 mL), triethylamine (2.40 mL, 17.23 mmol) and benzyl chloroformate (2.42 mL, 17.23 mmol) were added. The resulting mixture was stirred at room temperature for 24 h. The mixture was concentrated under vacuum and extracted with ethyl acetate (10 × 3 mL). Purification by MPLC (5:5 cyclohexane/ethyl acetate) afforded the title compound (2.23 g, 88%) as a white solid mp 88–90°C. ^{1}H-NMR (300 MHz, CDCl$_3$): δ = 2.93 (t, J = 7.5 Hz, 2H, H$_4$), 4.43 (t, J = 7.5 Hz, 2H, H$_5$), 5.35 (s, 2H, CH_2Ph), 7.38–7.48 (m, 5H, H-Ph). ^{13}C-NMR (75 MHz, CDCl$_3$): δ = 34.5, 67.0, 68.1, 128.0, 128.5, 128.6, 131.9, 149.7, 168.5.

Benzyl 3-Hydroxyisoxazolidine-2-carboxylate (157). To a solution of Cp$_2$ZrHCl (13.56 mmol) in dry toluene (5 mL), under nitrogen, benzyl 3-oxoisoxazolidine-2-carboxylate **156** (1 g, 4.52 mmol) in dry toluene (3 mmol) was added. After 1 h of stirring at room temperature, the reaction mixture was loaded directly onto a short plug of silica gel (2 g), eluted with EtOAc and concentrated in vacuo to afford **157** as a white oil (0.959 g, 95%). ^{1}H-NMR (500 MHz, CDCl$_3$): δ = 2.33 (dd, J = 2.2 and 6.4 Hz, 1H, H$_{4a}$), 2.45–2.50 (m, 1H, H$_{4b}$), 3.39 (b s, 1H, OH), 4.00 (dd, J = 6.4 and 7.2 Hz, 1H, H$_{5a}$), 4.23 (dd, J = 7.2 and 14.9 Hz, 1H, H$_{5b}$), 5.22 (s, 2H, CH_2Ph), 5.83 (dd, J = 2.2 and 6.4 Hz, 1H, H$_3$), 7.31–7.40 (m, 5H, H-Ph). ^{13}C-NMR (125 MHz, CDCl$_3$): δ = 36.6, 68.1, 68.6, 81.9, 128.3, 128.5, 128.6, 135.5, 150.1.

Benzyl 3-Acetoxyisoxazolidine-2-carboxylate (158). To a solution of benzyl 3-hydroxyisoxazolidine-2-carboxylate **157** (0.5 g, 2.24 mmol) in pure pyridine (30 mL), acetic anhydride (6.72 mmol) was added. The resulting mixture was stirred at room temperature overnight. After stirring, the mixture was concentrated in vacuum. Purification by MPLC (5:5 cyclohexane/ethyl acetate) afforded **158** (0.475 g, 80%) as a yellow oil. ^{1}H-NMR (300 MHz, CDCl$_3$): δ = 2.02 (s 3H, COCH_3), 2.29–2.37 (m, 1H, H$_{4a}$), 2.59–2.70 (m, 1H, H$_{4b}$), 3.93–4.01 (m 1H, H$_{5a}$), 4.16–4.23 (m, 1H, H$_{5b}$), 5.21 (s, 2H, CH_2-Ph), 6.71 (dd, 1H, J = 2.0 and 6.5 Hz, 1H, H$_3$), 7.30–7.39 (m, 5H, H-Ph). ^{13}C-NMR (75 MHz, CDCl$_3$): δ = 20.5, 35.9, 65.9, 67.9, 82.2, 127.9, 128.3, 135.1, 154.2, 169.3.

Benzyl 3-{5-[(benzyloxy)methyl]-3,4-dihydro-2,4-dioxopyrimidin-1(2H)-yl}isoxazolidine-2-carboxylate (166). A suspension of the corresponding nucleobase (0.84 mmol) in anhydrous acetonitrile (3 mL) was treated with bis(trimethylsilyl)acetamide (BSA) (2.52 mmol) and left under stirring until the solution was clear (15 min). A solution of a mixture of benzyl 3-acetoxyisoxazolidine-2-carboxylate **158** (0.56 mmol) in dry acetonitrile (3 mL) and trimethylsilyl triflate (TMSOTf) (0.11 mmol) was then added and the reaction mixture was heated at 70°C for 5 h. After cooling at 0°C, the solution was carefully neutralized by addition of 5% aqueous sodium bicarbonate and then concentrated in vacuo. After addition of dichloromethane (10 × 3 mL), the organic phase was separated, washed with water (2 × 10 mL), dried over sodium sulfate, filtered, and evaporated to dryness. The residue was purified by MPLC on a silica gel column using as eluent a mixture CH_2Cl_2/MeOH 98:2 to afford **166** in 70% yield. White solid, mp 143–145°C. ^{1}H-NMR (500 MHz, $CDCl_3$): δ = 2.32–2.34 (m, 1H, $H_{4'a}$), 2.98–3.01 (m, 1H, $H_{4'b}$), 3.89–3.94 (m, 1H, $H_{5'a}$), 4.26–4.35 (m, 3H, $H_{5'b}$, CH_2OCH_2Ph), 4.62 (s, 2H, CH_2OCH_2Ph), 5.27 (s, 2H, CH_2Ph), 6.66–6.85 (dd, J = 4.6 and 8.2 Hz, 1H, $H_{3'}$), 7.30-7.40 (m, 10H, H-Ph), 7.55 (s, 1H, H_6), 8.56 (bs, 1H, NH). ^{13}C-NMR (125 MHz, $CDCl_3$): δ 37.5, 64.4, 69.1, 70.1, 70.3, 73.2, 112.7, 127.3, 127.9, 128.1, 128.4, 128.7, 134.8, 137.0, 137.6, 143.2, 149.8, 156.0, 162.0.

REFERENCES

1. Pan, S.F.; Amankulor, N.M.; Zhao, K. *Tetrahedron* **1998**, *54*, 6587–6604.
2. Romeo, G.; Chiacchio, U.; Corsaro, A.; Merino, P. *Chem. Rev.* **2010**, *110*, 3337–3370.
3. Merino, P. *Curr. Med. Anti-Infect. Agents* **2002**, *1*, 389–411; Merino, P. *Curr. Med. Chem.* **2006**, *13*, 359–545; Casu, F.; Chiacchio, M.A.; Romeo, R.; Gumina, G. *Curr. Org. Chem.* **2007**, *11*, 1017–1032; Casu, F.; Chiacchio, M.A.; Romeo, R.; Gumina, G. *Curr. Org. Chem.* **2007**, *11*, 999–1016;
4. Tronchet, J.M.J.; Iznaden, M.; Barbalat-Rey, F.; Dhimane, H.; Ricca, A.; Balzarini, J.; De Clercq, E. *Eur. J. Med. Chem.* **1992**, *27*, 555–560.
5. Leggio, A.; Liguori, A.; Procopio, A.; Siciliano, C.; Sindona, G. *Tetrahedron Lett.* **1996**, *37*, 1277–1280.
6. Dalpozzo, R.; De Nino, A.; Maiuolo, L.; Procopio, A.; Romeo, R. *Synthesis* **2002**, *2*, 172–174.
7. Colacino, E.; Converso, A.; Liguori, A.; Napoli, A.; Siciliano, C.; Sindona, G. *Tetrahedron* **2001**, *57*, 8551–8557.
8. Colacino, E.; Converso, A.; De Nino, A.; Leggio, A.; Liguori, A.; Maiuolo, L.; Napoli, A., Procopio, A.; Siciliano, C.; Sindona, G. *Nucleosides Nucleotides* **1999**, *18*, 581–583; Dalpozzo, R.; De Nino, A.; Maiuolo, L.; De Munno, G.; Sindona, G. *Tetrahedron* **2001**, *57*, 4035–4038.
9. Allerson, C.R.; Chen, S.L.; Verdine, G.L. *J. Am. Chem. Soc.* **1997**, *119*, 7423–7433.
10. Collins, A.N., Sheldrake, G.N.; Crosby, J. Chirality in Industry; Wiley: Chichester, UK, **1992**.
11. Chiacchio, U.; Rescifina, A.; Corsaro, A.; Pistarà, V.; Romeo, G.; Romeo, R. *Tetrahedron: Asymmetry* **2000**, *11*, 2045–2048.
12. Vorbrüggen, J.M.; Kvolikiewicz, K.; Bennua, B. *Chem. Ber.* **1981**, *114*, 1234–1255.
13. Chiacchio, U.; Corsaro, A., Iannazzo, D.; Piperno, A., Pistarà, V., Rescifina, A., Romeo, R., Valveri, V., Mastino, A., Romeo, G. *J. Med. Chem.* **2003**, *46*, 3696–3702.
14. Tronchet, J.M.J.; Iznaden, M.; Barbalat-Rey, Komaroni, I.; Dolotshahi, N.; Bernardinelli, G. *Nucleosides Nucleotides* **1995**, *14*, 1737–1758.

15. Bortolini, O.; D'Agostino, A.; De Nino, A.; Maiuolo, L.; Nardi, M.; Sindona, G. *Tetrahedron* **2008**, *64*, 8078−8081; Bortolini, O.; De Nino, A.; Eliseo, T.; Gavioli, R.; Maiuolo, L.; Russo, B.; Sforza, F. *Bioorg. Med. Chem.* **2010**, *18*, 6970−6976.
16. Xiang, Y.; Chen, J.; Schinazi, R.F.; Zhao, K. *Tetrahedron Lett.* **1995**, *40*, 7193−7196.
17. Chiacchio, U.; Gumina, G.; Rescifina, A.; Romeo, R.; Uccella, N.; Casuscelli, F.; Piperno, A.; Romeo, G. *Tetrahedron* **1996**, *26*, 8889−8898.
18. Chiacchio, U.; Genovese, F.; Iannazzo, D.; Piperno, A.; Quadrelli, P.; Corsaro, A.; Romeo, R.; Valveri, V.; Mastino, A. *Bioorg. Med. Chem.* **2004**, *12*, 3903−3909.
19. Xiang, Y.; Gong, Y.; Zhao, K. *Tetrahedron Lett.* **1996**, *37*, 4877−4880.
20. Xiang, Y.; Gi, H.-J.; Niu, D.; Schinazi, R.F.; Zhao, K. *J. Org. Chem.* **1997**, *62*, 7430−7434.
21. Merino, P.; Franco, S.; Merchan, F.L.; Tejero, T. *Tetrahedron: Asymmetry* **1988**, *9*, 3945−3949; ibid. *J. Org. Chem.* **2000**, *65*, 5575−5589.
22. Merino, P.; Franco, S.; Garces, N.; Merchan, F.L.; Tejero, T. *Chem. Commun.* **1998**, 493−494; Merino, P.; Alamo, E.M.; Franco, S.; Merchan, F.L.; Tejero, T; Vieceli, D. *Tetrahedron Lett.* **2000**, *41*, 9239−9243.
23. Chiacchio, U.; Corsaro, A.; Gumina, G.; Rescifina, A.; Iannazzo, D.; Piperno, A.; Romeo, G.; Romeo, R. *J. Org. Chem.* **1999**, *64*, 9321−9327.
24. Chiacchio, U.; Corsaro, A.; Iannazzo, D.; Piperno, A.; Procopio, A.; Rescifina, A.; Romeo, G.; Romeo, R. *Eur. J. Org. Chem.* **2001**, 1893−1898.
25. Merino, P.; Franco, S.; Merchan, F.L.; Tejero, T. *Tetrahedron Lett.* **1998**, *39*, 6411−6414.
26. Merino, P.; Alamo, E.M.; Franco, S.; Merchan, F.L.; Simon, A.; Tejero, T. *Tetrahedron: Asymmetry* **2000**, *11*, 1543−1554.
27. Hyrosova, E.; Fisera, L.; Kozisek, J.; Fronc, M. *Synthesis* **2008**, *8*, 1233−1238.
28. Hyrosova, E.; Medvecky, M.; Fisera, L.; Hametner, C.; Fröhlich, I.; Marchetti, M.; Allmaier, G. *Tetrahedron* **2008**, *64*, 3111−3118.
29. Chiacchio, U.; Corsaro, A.; Iannazzo, D.; Piperno, A.; Pistarà, V.; A.; Rescifina, A.; Romeo, R.; Sindona, G.; Romeo, G. *Tetrahedron: Asymmetry* **2003**, *14*, 2717−2723.
30. Merino, P.; Tejero, T.; Mates, J.; Chiacchio, U., Corsaro, A.; Romeo, G. *Tetrahedron: Asymmetry* **2007**, *18*, 1517−1520.
31. Leggio, A.; Liguori, A.; Maiuolo, L.; Napoli, A.; Procopio, A.; Siciliano, C.; Sindona, G. *J. Chem. Soc. Perkin Trans. 1* **1997**, 3097−3099.
32. Carnovale, C.; Iannazzo, D.; Nicolosi, G.; Piperno, A.; Sanfilippo, C. *Tetrahedron: Asymmetry* **2009**, *20*, 425−429.
33. Chiacchio, U.; Corsaro, A.; Mates, J. Merino, P.; Piperno, A.; Rescifina, A.; Romeo, G.; Romeo, R.; Tejero, T. *Tetrahedron* **2003**, *59*, 4733−4738.
34. Merino, P.; Tejero, T.; Unzurrunzaga, F.J.; Franco, S.; Chiacchio, U.; Saita, M.G.; Iannazzo, D.; Piperno, A.; Romeo, G. *Tetrahedron: Asymmetry* **2005**, *16*, 3865−3876.
35. Chiacchio, U.; Rescifina, A.; Saita, M.G.; Iannazzo, D.; Romeo, G.; Mates, J.A.; Tejero, T.; Merino, P. *J. Org. Chem.* **2005**, *70*, 8991−9001.
36. Campbell, N.A.; Brad, W.; Robin, J. H.; Pearson Prentice Hall: Boston, **2006**.
37. Perigaud, C.; Girardet, J.L.; Gosselin, G.; Imbach, J.L. Advances in Antiviral Drug Design; De Clercq, R., Ed.; JAI Press: Greenwich, CT, **1995**; Vol. 2, pp. 167−172.
38. Mulato, A.S.; Cherrington, J.M. *Antivir. Res.* **1997**, *36*, 91−97.
39. Chiacchio, U.; Iannazzo, D.; Piperno, A.; Romeo, R.; Romeo, G.; Rescifina, A.; Saglimbeni, M. *Bioorg. Med. Chem.* **2006**, *14*, 955−4738.

40. Chiacchio, U.; Balestrieri, M.; Macchi, B.; Iannazzo, D.; Piperno, A.; Rescifina, A.; Saglimbeni, M.; Sciortino M.T.; Valveri, V.; Mastino, A.; Romeo, G. *J. Med. Chem.* **2005**, *48*, 1389–1394 Chiacchio, U.; Rescifina, A.; Iannazzo, D., Piperno, A.; Romeo, R.; Borrello, L.; Sciortino M.T.; Balestrieri, E.; Macchi, B.; Mastino, A.; Romeo, G. *J. Med. Chem.* **2007**, *50*, 3747–3750.
41. Piperno, A.; Giofrè, S.V.; Iannazzo, D.; Romeo, R.; Romeo, G.; Chiacchio, U.; Rescifina, A.; Piotrowska, D.G. *J. Org. Chem.* **2010**, *75*, 2798–2805.
42. Sigel, H.; Griesser, R. *Chem. Soc. Rev.* **2005**, *34*, 875–900; Sigel, H. *Chem. Soc. Rev.* **2004**, *33*, 191–200
43. Chiacchio, U.; Corsaro, A.; Iannazzo, A.; Piperno, A.; Rescifina, A.; Romeo, R.; Romeo, G. *Tetrahedron Lett.* **2001**, *42*, 1777–1780; Chiacchio, U.; Corsaro, A.; Pistarà, V.; Rescifina, A.; Iannazzo, A.; Piperno, A.; Romeo, R.; Romeo, G; Grassi, G. *Eur. J. Org. Chem.* **2002**, *7*, 1206–1212.
44. Saita, M.G.; Chiacchio, U.; Iannazzo, D.; Corsaro, A.; Merino, P.; Piperno, A.; Previtera, T.; Rescifina, A.; Romeo, G.; Romeo, R. *Nucleosides Nucleotides and Nucleic Acids*, **2003**, *22*, 739–742; Chiacchio, U.; Borrello, L.; Iannazzo, D.; Merino, P.; Piperno, A.; Rescifina, A.; Richichi, B.; Romeo, G. *Tetrahedron: Asymmetry* **2003**, *14*, 2419–2425.
45. Romeo, R.; Carnovale, C.; Giofre, S. V.; Romeo, G.; Macchi, B.; Frezza, C.; Marino-Merlo, F.; Pistarà, V.; Chiacchio, U. *Bioorg. Med. Chem.* **2012**, *20*, 3652–3657.
46. Hossain, N.; Hendrix, C.; Lescrinier, E.; Van Aerschot, A.; Busson, R.; De Clercq, Herdewijn, P. *Bioorg. Med. Chem. Lett.* **1996**, *6*, 1465–1468.
47. Chiacchio, U.; Genovese, F.; Iannazzo, D.; Librando, V; Merino, P.; Rescifina, A.; Romeo, R.; Procopio, A.; Romeo; G. *Tetrahedron* **2004**, *60*, 441–448.
48. Chiacchio, U.; Saita, M.G.; Crispino, L.; Gumina, G.; Mangiafico, S.; Pistarà, V.; Romeo, G.; Piperno, A.; De Clercq, E. *Tetrahedron* **2006**, *62*, 1171–1181.
49. Romeo, R.; Giofrè, S.; Macchi, B.; Balestrieri, E.; Mastino, A.; Merino, P.; Carnovale, C.; Romeo, G.; Chiacchio, U. *ChemMedChem* **2012**, *7*, 565–569.

18 Synthetic Studies on Antifungal Peptidyl Nucleoside Antibiotics

APURBA DATTA
Department of Medicinal Chemistry, University of Kansas, Lawrence, Kansas, USA

1. INTRODUCTION

Despite the renewed interest in the discovery and development of new antifungal agents, the increasing incidence of opportunistic fungal infections during the last 25 years has become a public health concern of serious proportions.[1,2] With a fast-growing population of immune-compromised individuals as a consequence of HIV infection, cancer chemotherapy, organ transplants, and so on, systemic fungal infections are becoming a major cause of morbidity and mortality.[3] A U.S. National Center for Health Statistics (NCHS) study has shown that there is a more than 300% increase in mycotic disease-related fatalities since 1980. Autopsy data indicate that more than 50% of patients who die with malignancies are infected with pathogenic fungi such as *Candida*, *Aspergillus*, and *Cryptococcus*.[4–7] In solid organ transplant recipients, the prevalence of fungal infections ranges from 5% among kidney transplant to 50% among liver transplant patients.[8,9] Similarly, systemic and mucosal fungal infections due to *Candida* and *Cryptococcus* are most common among HIV-infected individuals.[10–14] Over 90% of HIV-positive individuals suffer from fungal infections. Except for the echinocandins and 5-fluorocytosine, the majority of the other commonly used antifungal drugs act by interference with the biosynthesis or functioning of sterols in fungal cell membranes.[15–18] The intensive use of drugs targeting the pathway above (i.e., sterol biosynthesis inhibitors) has resulted in the development of resistance and the consequent reduced efficacy of these drugs. Although the antifungal drugs currently available are potent, the relatively limited choice and narrow spectra of these antifungal regimens, as well as their toxicity, often complicate treatment of deep-seated mycoses. Development of fungal resistance (e.g., to the azoles), unsatisfactory dosage formulations and nephrotoxicity (e.g., for amphotericin B), and emergence of new pathogenic fungal strains also limit the utility of many of the commonly used antifungal drugs.[14,19]

Chemical Synthesis of Nucleoside Analogues, First Edition. Edited by Pedro Merino.
© 2013 John Wiley & Sons, Inc. Published 2013 by John Wiley & Sons, Inc.

Inasmuch as the anatomic and metabolic nature of fungal cells are similar to that of mammalian cells, it has been difficult to devise therapeutic strategies specifically targeted against these pathogens, and yet remain nontoxic to the host. Addressing the issues noted above and development of new and more efficacious antifungal agents to treat life-threatening invasive mycoses is thus a critical need.[20–24]

Streptomyces-derived complex peptidyl nucleoside antibiotics represent a unique class of natural products (Figure 1) with demonstrated antifungal activity against various pathogenic fungi.[25,26] Among the various peptidyl nucleosides, the polyoxins and nikkomycins act by inhibiting the biosynthesis of chitin, an essential component of a fungal cell wall.[27–30] Chitin is responsible for imparting shape and strength to a fungal cell wall, and inhibition of chitin synthesis causes osmotic sensitivity, abnormal morphology, and fungal growth arrest, ultimately leading to cell death.[31,32] Chitin is widely distributed among yeast and mycelial fungi but is absent in mammalian cells. Thus, from a host (human) toxicity perspective, inhibition of chitin biosynthesis is considered a rational, safe, and selective antifungal target.[33,34]

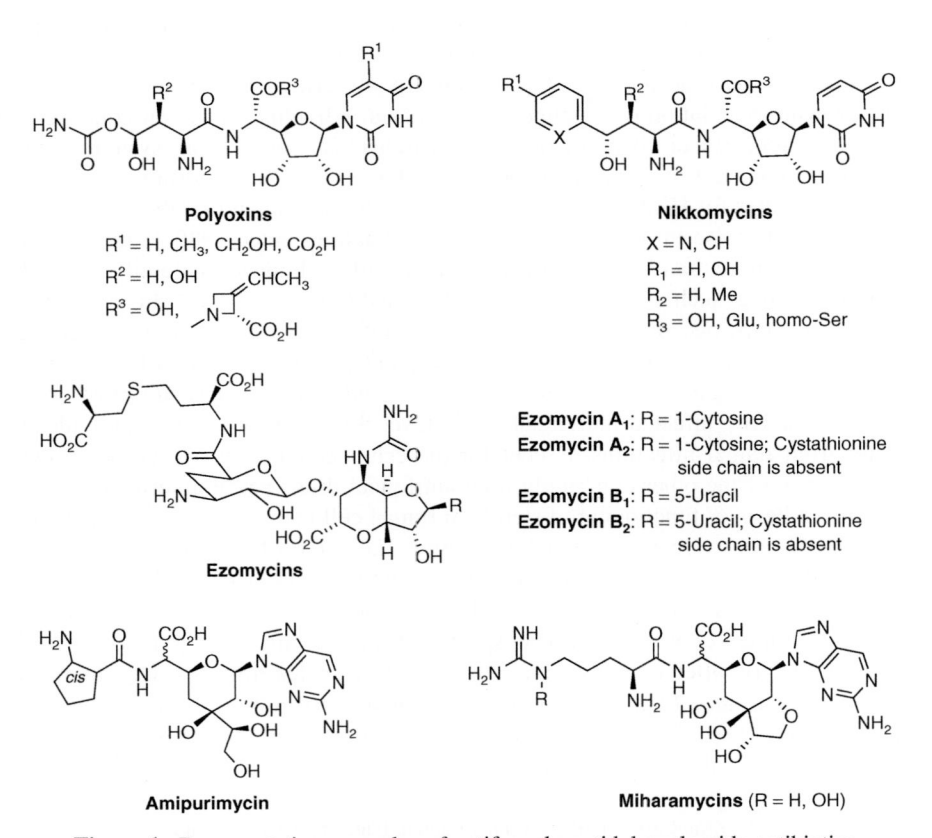

Figure 1. Representative examples of antifungal peptidyl nucleoside antibiotics.

Some of the other natural peptidyl nucleosides, such as the ezomycins, amipurimycin, and miharamycins (Figure 1), also exhibit potent antifungal activity against a broad range of fungal species. However, the exact modes of action of these compounds have yet to be defined.[25,26] The demonstrated antifungal activity, fungal cell specificity, novel mode of action, and unique structural features of the peptidyl nucleosides have generated considerable interest in their potential utility as novel leads for the development of a new generation of antifungal drugs.

Polyoxins isolated (1960) from *Streptomyces cacaoi* var. *asoensis*[27,28] exhibited potent antimicrobial activity against various fungi and have consequently found widespread use as efficient agricultural fungicides, with no adverse side effects.[35] Subsequently, another class of *Streptomyces*-derived peptidyl nucleoside antibiotics, the nikkomycins (neopolyoxins), displayed selective and potent inhibitory activity against fungi such as *Pyricularia oryzae* and *Rhizoctonia solani*.[29,30] The original interest in the polyoxin and nikkomycin class of compounds stemmed from their highly selective activity against various fungi while being nontoxic to bacteria, plants, and animals. Acting via a novel mode of action, these natural products were found to be strong competitive inhibitors of chitin synthases (K_i range: 0.1–1 μM) from fungi and yeasts.[27-30] Chitin, a β-(1 → 4)-linked polymer of N-acetyl glucosamine, is an essential component of the fungal cell wall structure but is not present in green plants or vertebrates.[31,32,36-38] Therefore, chitin synthase and the cellular mechanisms that regulate the activity of this enzyme are excellent targets for pharmaceutical and agricultural pathogen management.[33,34] Importantly, polyoxins and nikkomycins also exhibit impressive in vitro inhibitory activity against various human pathogenic fungi, with nikkomycins being the more potent.[39-41] As these peptidyl nucleosides target chitin synthase, a biosynthetic pathway absent in mammalian cells, these antibiotics represent promising leads for the development of novel, nontoxic antifungal agents.

The structurally more complex ezomycins, amipurimycin and miharamycins (Figure 1), also exhibit antifungal activities against a broad range of fungal species, although the exact modes of action for these compounds have not yet been determined.[42-48] With demonstrated in vitro and in vivo activity against various human pathogenic mycotic infections, such as candidiasis, blastomycosis, and coccidioidomycosis, the complex peptidyl nucleosides represent unique leads in the continuing search for new, effective, and nontoxic antifungal agents. Importantly, as mammalian cells do not have chitin synthase, chitin synthase inhibitors represent safer and potentially useful models toward development of novel antifungal drugs against opportunistic fungal infections. Unfortunately, direct clinical application of these peptidyl nucleosides is compromised by their attenuated in vivo activity, apparently due to their inefficient transport into the fungal cell.[33,34] Consequently, no clinically viable antifungal agent has yet been developed from the peptidyl nucleoside structural leads.

As evident from Figure 1, the peptidyl nucleoside antibiotics shown consists of a purine or pyrimidine base attached to a carbohydrate moiety of varying complexity. A recurring structural motif in these compounds is also the presence of an α-amino acid (1,2-amino alcohol in the ezomycins) functionality at the C4′

or C5′ position of the nucleoside component. However, a major structural difference among these groups of compounds is the presence of a furanosyl nucleosidic core in the polyoxins, nikkomycins, and ezomycins, while the amipurimycin and miharamycin families of compounds consist of a pyranosyl nucleoside core. Among the compounds shown in Figure 1, few total syntheses of the various polyoxins, and nikkomycins, have been reported; however, the total synthesis of amipurimycin, miharamycins and the ezomycins have not yet been achieved.[49–51]

To date, the most commonly used approach for the synthesis of the central carbohydrate amino acid cores of polyoxins and nikkomycins have involved modification of readily available carbohydrate precursors, utilizing the carbon skeleton and resident chirality present in the starting sugar.[52–60] Using such a carbohydrate-based strategy, Kuzuhara et al. achieved the first total synthesis of polyoxin J, a dipeptidyl polyoxin.[52] The length of this synthesis (ca. 40 steps), however, illustrates the usual drawbacks of such an approach. Therefore, a major focus of subsequent studies have been the development of more efficient, flexible, and concise alternatives. In one such effort, Garner's research group utilized D-serine for a stereoselective construction of the glycosyl α-amino acid fragment.[61] In contrast to carbohydrate-based syntheses, Garner's approach utilized an amino acid as a chiral starting material for stereoselective construction of the required carbohydrate framework that obviates the need for a late-stage incorporation of the amine functionality. Similarly, in his formal synthesis of polyoxin J, Mukaiyama utilized L-tartaric acid as a chiral starting material to assemble the central thymine polyoxin C framework.[62] In subsequent studies, Auberson and Vogel[63] and Sethin and Simpkins[64] demonstrated the utility of cycloaddition reactions of furan as attractive ways to access the required carbohydrate framework, although both of the methods required multistep functional group transformations to install the C4′ amino acid functionality.

Unlike the polyoxins and nikkomycins, total syntheses of the structurally more complex peptidyl nucleosides, amipurimycin, miharamycins, and the ezomycins are yet to be accomplished. However, a few approaches to the *trans*-fused furopyran bicyclic core (octosyl acid) of the ezomycins have been developed.[65–70] All of these methods employ a carbohydrate-based strategy to construct the required furopyranose framework and also require multistep reaction sequences to introduce the amine functionality. Similarly, the few synthetic studies toward amipurimycin and the miharamycins were focused on elaboration of the hexopyranoside amino acid core of these compounds. Thus, the research groups of Czernecki and Rauter have utilized carbohydrate starting materials toward developing synthetic routes to various segments of amipurimycin and miharamycin.[48,71–73] In an interesting non-carbohydrate-based approach, a stereocontrolled hetero Diels–Alder reaction between a serine-derived homochiral oxazolidine aldehyde (Garner's aldehyde[74]) and an electron-rich diene was utilized by Garner et al. to assemble a model branched sugar amino acid component of amipurimycin.[75]

In the following sections of this review we describe the results of some of our synthetic studies, detailing strategies and approaches involving stereoselective de novo construction of the nucleoside amino acid cores of the natural products cited above. Starting from a structurally simpler, noncarbohydrate chiral synthon, our approach circumvents some of the disadvantages inherent in the carbohydrate approach, with potential flexibility toward more extensive structure–activity relationship studies.

2. SERINE-BASED SYNTHETIC ROUTE TO THE NUCLEOSIDE AMINO ACID SEGMENT

A common structural feature of the complex peptidyl nucleoside antibiotics is the presence of a pyranosyl or a furanosyl-α-amino acid nucleoside core. Our retrosynthetic plan (Figure 2) for the foregoing family of compounds thus envisaged a unified strategy based on initial stereocontrolled formation of a pivotal furanosyl amino acid precursor **1** (for the polyoxins, nikkomycins, and ezomycins), or the corresponding pyranosyl analogue **2** (for the amipurimycin and miharamycins).

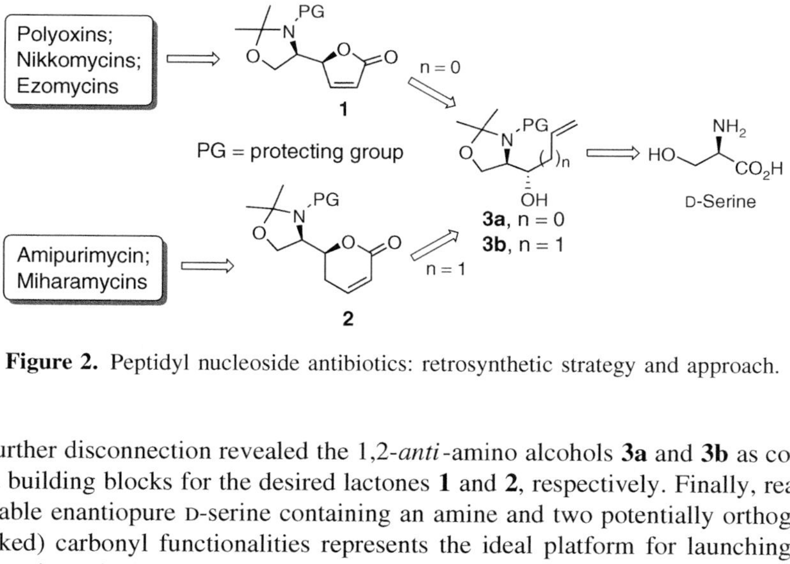

Figure 2. Peptidyl nucleoside antibiotics: retrosynthetic strategy and approach.

Further disconnection revealed the 1,2-*anti*-amino alcohols **3a** and **3b** as convenient building blocks for the desired lactones **1** and **2**, respectively. Finally, readily available enantiopure D-serine containing an amine and two potentially orthogonal (masked) carbonyl functionalities represents the ideal platform for launching our proposed synthetic endeavor. Based on the strategy described above, the results of our synthetic studies toward the various antifungal peptidyl nucleosides are described below.

Our initial attempts at the stereoselective synthesis of the *anti*-1,2-amino alcohol derivatives **3a** and **3b**, via the addition of appropriate Grignard reagents (vinyl-MgBr and allyl-MgBr, respectively) to Garner aldehyde, yielded the products desired with only moderate *anti*-selectivity (ca. 3:1 to 4:1). Exploring alternative approaches, we were ultimately successful in developing a more efficient and stereoselective route to both the desired allylic and homoallylic *anti*-1,2-amino alcohol adducts as required for our proposed synthesis. The syntheses entailed initial conversion of D-serine to the corresponding *N,O*-acetonide-protected Weinreb amide **4** (Scheme 1). Reaction of the amide **4** with allylmagnesium bromide cleanly resulted in the corresponding allylketone **5** in quantitative yield. However, it was observed that prolonged storage or attempted chromatographic purification of **5** caused partial isomerization of the terminal olefin to the corresponding α,β-unsaturated ketone derivative.

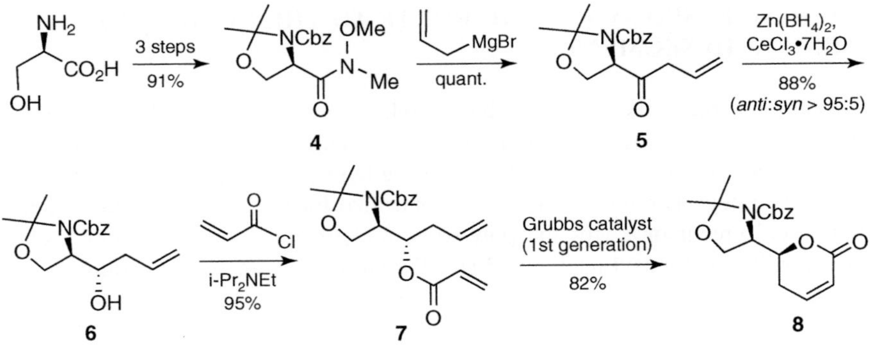

Scheme 1.

Subsequently, instead of purification, the crude ketone was subjected directly to the next reaction. Neighboring *N*-Cbz group-assisted, chelation-controlled reduction of the ketone **5** with zinc borohydride provided the required 1,2-*anti*-amino alcohol derivative **6** with excellent stereocontrol (*anti:syn* >95:5) and high yield. Subsequent conversion of **6** to the corresponding acrylate **7** followed by Grubbs' olefin metathesis protocol resulted in the lactone **8** (Scheme 1) in high overall yield.

The development of a similar route toward the five-membered amino lactone **1** was investigated next. Unlike the allylic ketone **5**, direct conversion of the Weinreb amide **4** to the corresponding vinyl ketone (via addition of vinylmagnesium bromide) was problematic. The reactivity of the product vinylic ketone (a good Michael acceptor) toward adventitious nucleophiles during reaction workup, purification, and subsequent reactions of this intermediate precluded its further use. We overcame this problem by exploiting the allylic ketone **5** itself. Treatment of **5** with neutral alumina resulted in its clean conversion to the corresponding double bond isomerized α,β-unsaturated ketone **9** ($E:Z$ = 93:7) (Scheme 2). Gratifyingly, the reduction of ketone **9** with zinc borohydride proceeded smoothly to form the allylic alcohol derivative **10** with high *anti*-selectivity (*anti/syn* = 92:8).

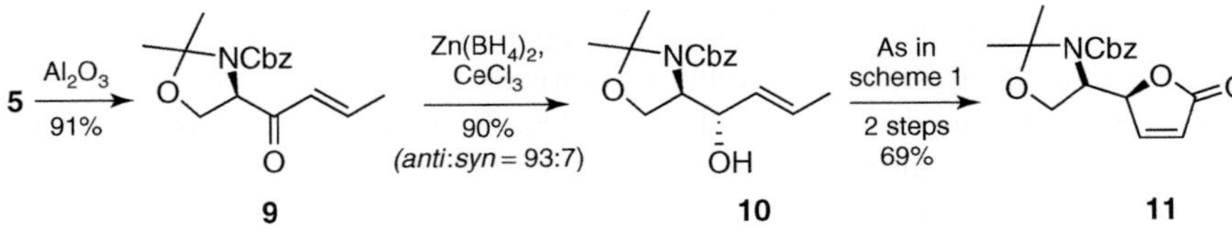

Scheme 2.

Thus, employing the same ketone intermediate 5, the strategy above resulted in an efficient pathway to afford either the homoallylic or the allylic anti-1,2-aminoalcohol adducts 6 and 10, respectively. Following the same sequence of reactions as in scheme 1, the amino alcohol 10 was subsequently converted to the desired aminobutenolide 11 (Scheme 2) in good overall yield.[76] Following the concise, practical, and efficient pathways described in Schemes 1 and 2, we are now able to readily synthesize 12- to 15-g batches of the amino lactones 8 and 11 in consistently good overall yields.

The successful attainment of our initial synthetic goals created the opportunity to further explore the transformation of the pivotal lactones 8 and 11 to the target furanosyl- and pyranosyl nucleoside amino acid structural framework. Accordingly, reaction of the aminopyrones 8 and 11 with potassium osmate resulted in the highly stereoselective (>95%) formation of the corresponding diols 12a and 12b (Scheme 3). Subsequent reduction of the lactone carbonyl to lactol followed by peracetylation afforded the corresponding triacetates 13a and 13b. Deprotection of the acetonide linkage and standard conversion of the resulting primary hydroxy group to the corresponding methyl ester installed the side-chain amino acid functionality of 15a and 15b. N-glycosidation with the corresponding uridine and thymidine nucleosides under Vorbrüggen's conditions[77] provided the corresponding pyranosyl and furanosyl nucleosides 17a-b and 18a-b as the only stereoisomeric products, respectively. Finally, standard removal of the protecting groups culminated in a concise route to the desired nucleoside amino acid derivatives 20a-b and 21a-b (Scheme 3).[76]

The synthetic strategy as described above demonstrates the potential utility of lactones 8 and 11 in the rapid construction of the nucleoside amino acid structural segments of various peptidyl nucleoside antibiotics, and modified analogues thereof. In terms of brevity and overall yields, the routes above compare favorably with the earlier reported syntheses of the foregoing nucleoside fragments. From the perspective of future structure–activity relationship investigations, the versatility of the key lactones 8 and 11 toward easy structural modifications and consequent possible access to a variety of modified nucleoside analogues is also expected to be an added advantage of the present route.

3. EZOMYCINS

The ezomycins are *Streptomyces*-derived antifungal natural products belonging to the complex peptidyl nucleoside superfamily of antibiotics. Isolated and identified during the 1970s, the ezomycins are active against phytopathogenic fungi such as *Sclerotinia* and *Botrytis*.[42–44] The antifungal mechanism of action of the ezomycins has, however, not yet been ascertained. The ezomycins contain an interesting combination of structural features consisting of (1) an octosyl nucleoside core, (2) an ezoaminuroic acid component, and (3) an N-linked pseudopeptide (L-cystathionine) (Figure 3). While some of the ezomycins contain the cytosine nucleobase, others are of the pseudouridine type.

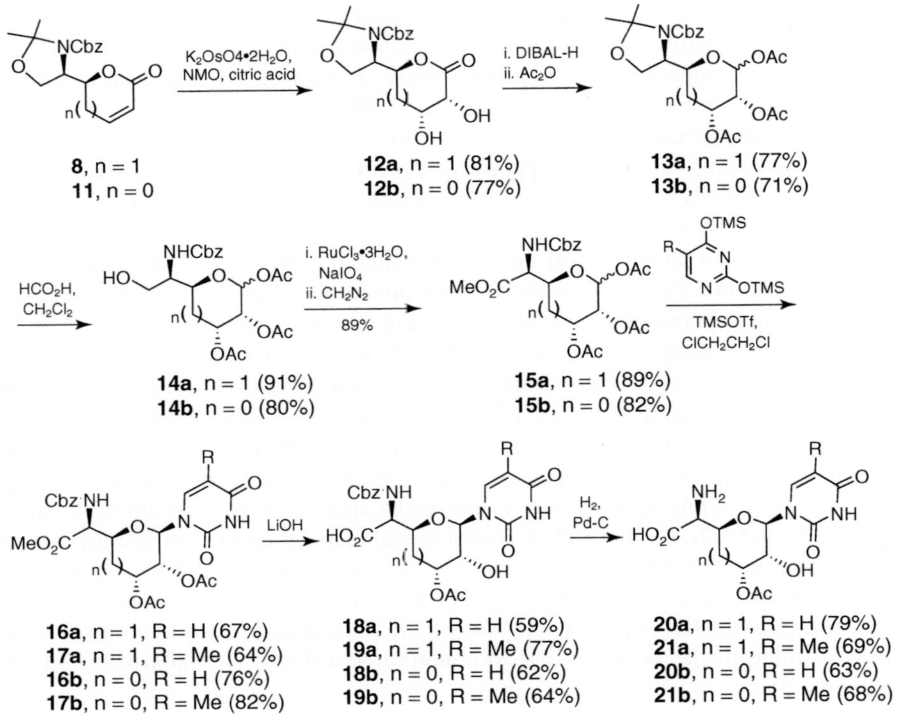

Scheme 3.

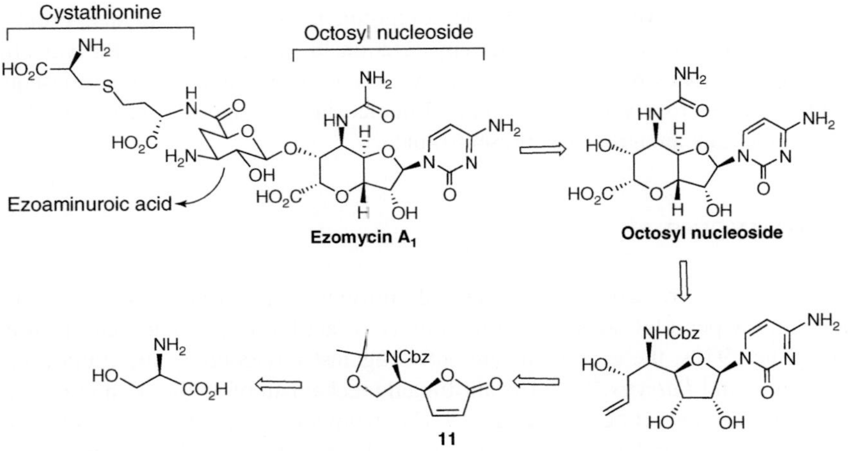

Figure 3. Ezomycin retrosynthetic strategy.

In addition to the ezomycins, differently substituted but otherwise structurally similar bicyclic 3,7-anhydrooctose nucleoside motifs are also found in nucleoside antibiotics, such as the octosyl acids and malayamycin A.[78] Interestingly, in biological studies, the L-cystathionine side chain–containing ezomycins A_1 and B_1 were found to display antifungal activity, whereas those lacking this pseudopeptide (e.g., ezomycins A_2 and B_2) were devoid of activity.[42–44] Although a few syntheses of the ezomycin structural fragments have been reported,[49–51,65–70,78c] the total synthesis, or detailed structure–activity relationship (SAR) studies of these natural products are yet to be accomplished.

In the previously reported syntheses of the octosyl nucleoside core, various carbohydrate starting materials have been employed to construct the bicyclic furopyranyl structural framework. In a strategic deviation from the approaches described above, our plan for the present synthesis involves de novo construction of the target bicyclic nucleoside component, starting from a structurally simpler and more flexible amino acid building block. Our synthetic strategy, utilizing D-serine-derived enantiopure aminobutenolide **11** as an appropriate chiral platform, is shown in Figure 3. Accordingly, initial conversion of **11** to the C4' carbon chain extended furanosyl nucleoside derivative, followed by a stereoselective 6-*exo-trig* cyclization involving the terminal olefin and the C3' oxygen functionality, is expected to lead to the appropriately functionalized octosyl nucleoside structural core of the ezomycins.

In accordance with the strategy above, the D-serine-derived furanosyl triacetate **13b**, as obtained from our previous studies (Scheme 3), was chosen as a suitable starting point for the present endeavor. For the sake of simplicity, we also decided to utilize synthetically more amenable thymine (instead of cytosine) as the nucleobase component. Accordingly, reaction of the glycosyl donor **13b** with bissilylated thymine in the presence of TMSOTf resulted in the formation of the nucleoside derivative **22** (Scheme 4). Interestingly, the acidic reaction conditions of the transformation above also led to clean cleavage of the *N,O*-acetonide protection,[79] thereby eliminating the need of an otherwise necessary additional reaction step. Aiming for the desired two-carbon elongation at the C4' side chain, in a two-step sequence, Dess–Martin periodinane oxidation of the alcohol **22** to the corresponding aldehyde, its subsequent reaction with a stabilized Wittig reagent, and DIBAL reduction of the resulting *E*-α,β-unsaturated ester provided the corresponding allylic alcohol derivative **23** in good overall yield. It is noteworthy that under the reaction conditions employed, the acetate functionalities of **7** were found to be unaffected, and only the allylic ester functionality underwent reduction to the corresponding alcohol. Toward stereoselective installation of the desired C6' hydroxy functionality, epoxidation of the olefin **23** was investigated next. Surprisingly, when subjected to the Sharpless asymmetric epoxidation (SAE),[80] in the presence of either of the chiral ligands (+)-DET or (−)-DET, the allylic alcohol **23** failed to undergo any epoxidation. In an alternative approach, when subjected to reaction with *m*-CPBA, the olefin **23** did form the corresponding epoxide, albeit with poor stereoselectivity (ca. 1:1 mixture of diastereoisomers by ^{1}H-NMR).

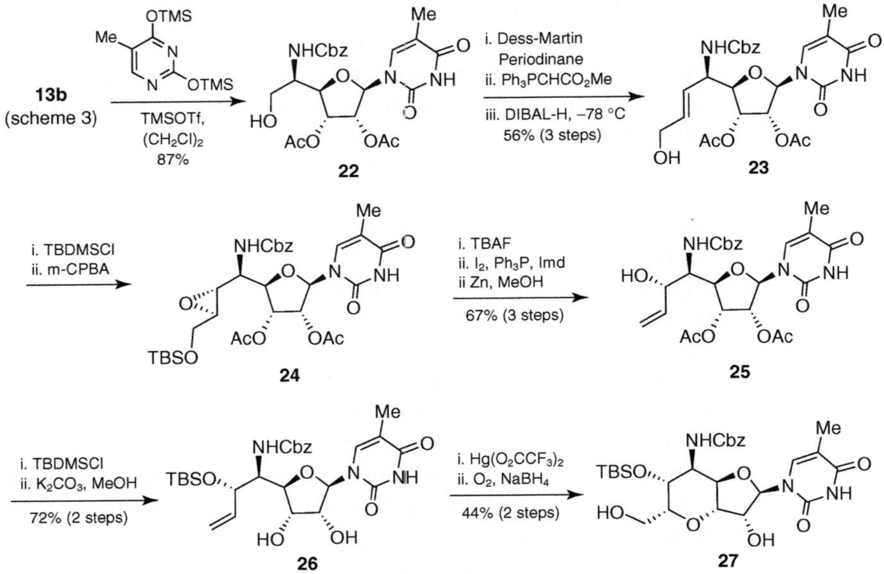

Scheme 4.

The unsatisfactory selectivity in this above epoxidation prompted us to investigate prior protection of the primary hydroxyl group of **23** and study the subsequent effect on *m*-CPBA epoxidation. Gratifyingly, protection of the alcohol **23** to the corresponding TBS-ether derivative, followed by its reaction with *m*-CPBA resulted in the stereoselective formation of the epoxide **24** along with minor quantities of the other diastereoisomer (major/minor = 5:1 by ^{1}H-NMR). At this stage, several attempts to construct the target *trans*-fused bicyclic furopyran framework via acetate deprotection and subsequent 6-endo addition of the resulting C3′ hydroxyl group to the C7′ of the epoxide **24** were, however, not successful. Following an alternative strategy, in a three-step sequence, removal of silyl protection and subsequent conversion of the primary hydroxyl group to iodo, followed by reductive elimination in the presence of Zn, resulted in the clean formation of the rearranged product **25**.[81] The transformation above helped install the secondary hydroxyl bearing the desired chiral center at C6′, and also provided the terminal olefin functionality as a strategic handle toward construction of the desired furopyran bicyclic skeleton. Subsequent TBS protection of the free secondary hydroxy group of **25** followed by cleavage of the acetate functionalities afforded the diol **26** in good yield. Toward performing the desired bicyclic ring formation, an intramolecular 6-exo-*trig* cyclization was investigated next. Accordingly, when **26** was subjected to an oxymercuration–oxidation protocol by initial reaction with mercuric trifluoroacetate in refluxing acetonitrile, followed by treatment of the resulting alkylmercury intermediate with NaBH$_4$ amid continuous bubbling of oxygen, the desired bicyclic furopyranyl nucleoside **27** was obtained in a reasonable yield.

The structure and assigned stereochemistry of the product **27** was confirmed via extensive NMR studies. The stereoselectivity observed in the intramolecular oxymercuration (**26** → **27**) is probably attributable to the sterically more favorable chair-like transition state **II** (Figure 4), leading to the selective formation of the desired *trans*-fused furopyran bicyclic derivative **27**.

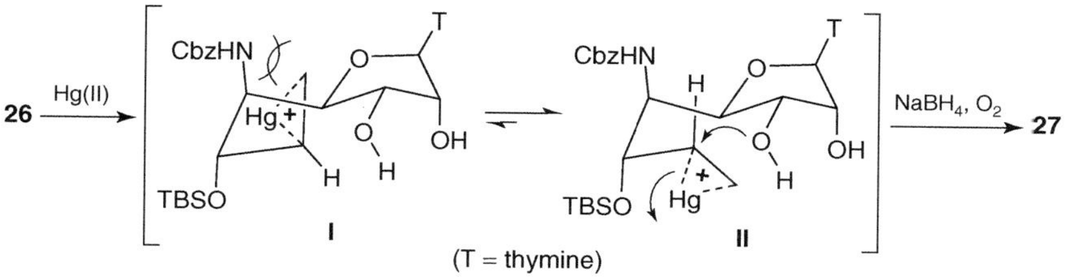

Figure 4. Probable mechanistic pathway toward stereoselective formation of **27**.

Proceeding with the synthesis, in a three-step sequence, the 2′-acetate-protected monohydroxy compound **28** (Scheme 5) was obtained employing a standard protection–deprotection protocol. Finally, oxidation of the hydroxy group to carboxylic acid and subsequent esterification provided the fully functionalized and strategically protected thymine octosyl nucleoside derivative **29** in good overall yield.[82]

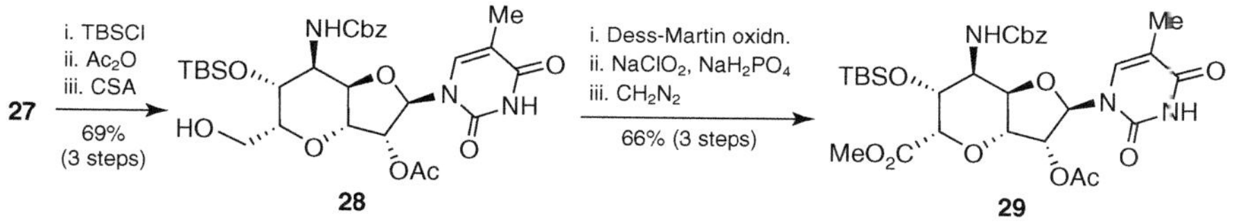

Scheme 5.

The synthesis described represents the first instance wherein the octosyl nucleoside component has been constructed starting from a noncarbohydrate precursor. An advantage of this de novo approach is expected to be the increased synthetic flexibility in terms of potential SAR modifications at the carbohydrate core as well as in the creation of nonnatural stereocenter-containing derivatives. Extension of the present route toward incorporation of cytosine nucleobase and completion of the total syntheses of the various ezomycins are under investigation.

4. AMIPURIMYCIN

The complex peptidyl nucleoside antibiotic amipurimycin displays impressive antifungal activity against several phytopathogenic fungi, such as *Pyricularia oryzae*, *Alternaria kikuchiana*, and *Helminthosporium sigmoideum* var. *irregulare*.[45,46] The mechanism of antifungal action of amipurimycin, however, remains unknown. The proposed structure of amipurimycin (Figure 5) was determined

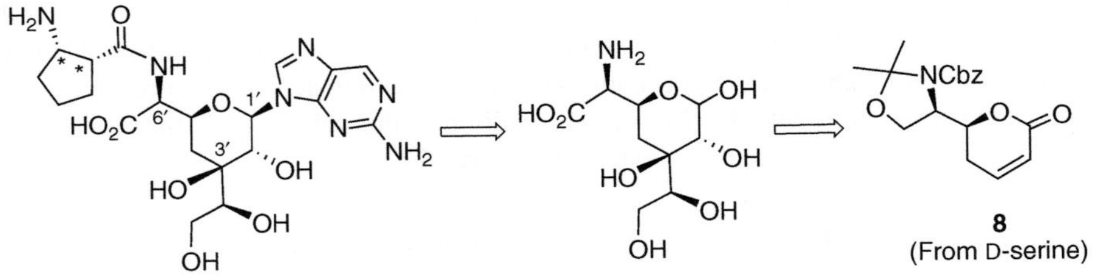

Figure 5. Amipurimycin retrosynthesis.

with the help of extensive spectroscopic and chemical degradation studies. Accordingly, amipurimycin was found to be made up of an unusual C3′ branched pyranose amino acid, appended to an N-terminal amino acid residue at C6′, and a glycosidic purine nucleobase.[83] However, the stereochemistry at C6′, and the absolute configurations at C2″/C3″ for the *cis*-aminocyclopentane carboxylic acid (cispentacin)–derived side chain, could not be ascertained. Interestingly, a recent study has confirmed the C6′ amino acid stereocenter of a structurally related peptidyl nucleoside, miharamycin, to be in the natural S-configuration.[48] As mentioned earlier, a few syntheses of the various structural fragments of amipurimycin have been reported;[71,72,74,84] however, the total synthesis, or structure–activity relationship studies of this natural product are yet to be accomplished.

Continuing with our studies on the peptidyl nucleoside antibiotics, our synthetic strategy for amipurimycin involved utilization of the previously mentioned serine-derived masked amino acid lactone **8** (Figure 5) as an appropriate building block. The enone moiety of **8** provides a strategic handle for incorporation of the required C2′–C3′ *trans*-diol as well as the branched carbon chain at C3′.

Accordingly, stereoselective dihydroxylation of the lactone to the corresponding diol and its acetonide protectiion yielded the lactone **31** (Scheme 6). Partial reduction of the lactone **57** to the corresponding lactol, followed by anomeric *O*-alkylation with 4-methoxybenzylbromide (PMB-Br), yielded the β-glycosidic derivative **32** as the only product. One-pot hydrolytic cleavage of both the acetonide protecting groups of **32** and subsequent selective silyl protection of the resulting terminal primary hydroxy group resulted in the free dihydroxy derivative **33**. Selective, preferential acetylation of the equatorial C2-hydroxy forming the corresponding mono-acetate[85] followed by oxidation of the free C3-hydroxy group resulted in the ketone **34**. Subsequent Wittig olefination under thermodynamic conditions resulted in the selective formation of the *E*-olefin derivative **35**. Exhaustive reduction of the ester functionalities of **35**, followed by acetylation of the resulting diol to its diacetate, and silyldeprotection yielded **36** in good overall yield. Oxidative conversion of the primary hydroxy group to carboxylic acid and its esterification resulted in the side-chain amino acid methyl ester derivative **37**. Toward installation of the C3–C3′ *syn*-diol moiety, stereoselective dihydroxylation of the olefin **37** was investigated. Thus, oxidation of **37** with catalytic quantity of

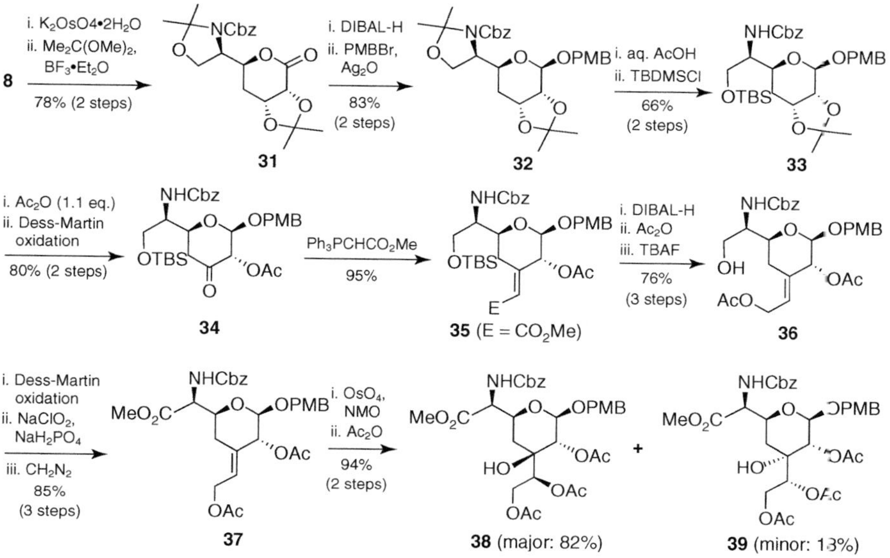

Scheme 6.

osmium tetroxide, followed by acetylation (to facilitate chromatographic separation) of the crude diol resulted in the isolation of the corresponding acetates **38** and **39**, with the expected product **38** being the major diastereoisomer (C3–C3′-syn/anti = 82:18). In the dihydroxylation reaction above, formation of the desired diol **38** as the major diastereoisomer is probably a result of the sterically more favorable approach of the oxidant from the opposite face of the neighboring C2-acetoxy group. Attempts to further improve the selectivity in the reaction above, via employment of the Sharpless asymmetric dihydroxylation (SAD) protocol, however, failed to produce any desired improvement. The failure of SAD in the present substrate is most probably due to steric crowding around the trisubstituted exocyclic olefin **37** and its unfavorable interactions with the bulky chiral catalyst.

Aiming for the purine nucleoside formation, standard deprotection of the PMB-protecting group of **38**, followed by acetylation of the resulting lactol and subsequent reaction with the nucleobase donor bis-TMS-2-(N-acetylamino)purine donor, was undertaken next. Disappointingly, several attempts, involving various combinations of Lewis acid activators, solvents, and various reaction conditions failed to provide the desired nucleoside, resulting in either the recovery, or degradation of the starting materials. Investigating the possibility of the nucleoside formation above with a simpler pyrimidine nucleobase, reaction of **38** with commercially available bis-TMS-thymine was then attempted. Once again the attempted reactions failed to form the desired product.

Failing to achieve N-glycosidation of the pyranoside **38** under traditional Vorbrüggen conditions, we decided to explore alternative methods toward activation of the glycosyl donor above and study its nucleoside-forming reaction.

Formation of activated glycosyl donors via the intermediacy of highly reactive anomeric trichloroacetimidates (Schmidt's trichloroacetimidate protocol)[86] is among the most commonly used methods in contemporary O-glycosidic-bond-forming reactions. Surprisingly, employment of this method in the synthesis of nucleosides remains relatively infrequent. Investigating the trichloroacetimidate protocol in our desired nucleoside synthesis, the anomeric PMB-ether **38** was subjected to standard CAN deprotection, followed by reaction of the resulting crude lactol with trichloroacetonitrile, forming the expected trichloroacetimidate derivative **39** (Scheme 7) in high overall yield. Unfortunately, repeated attempts to couple the trichloroacetimidate **39** with the bis-TMS-2-(N-acetylamino)purine failed to provide the desired nucleoside. Interestingly, reaction of **39** with bis-(TMS)thymine, in the presence of $BF_3 \cdot Et_2O$ as the Lewis acidic activator, did, however, lead to the highly stereoselective formation of the corresponding thymine amipurimycin nucleoside **40**, albeit in a modest yield. Subsequently, peptidic attachment of the cispentacin side chain was accomplished by hydrogenolysis of the N-Cbz protecting group of **40** to unmask the free amine, which was then coupled to (1R,2S)-N-trifluoroacetamido cispentacin derivative to obtain the fully protected peptidyl nucleoside structural core **41**. Global removal of the protecting groups via alkaline hydrolysis completed the total synthesis of the unique "thymine amipurimycin" analogue **42**.[87]

Although attempts to incorporate the aminopurine nucleoside as present in the natural product were unsuccessful, the effort described above represents the first total synthesis of a fully functionalized nucleoside analogue of amipurimycin. In ongoing studies, an alternative strategy of introducing the aminopurine nucleobase on a simpler, early-stage carbohydrate precursor (prior to C3 branching) is being investigated.

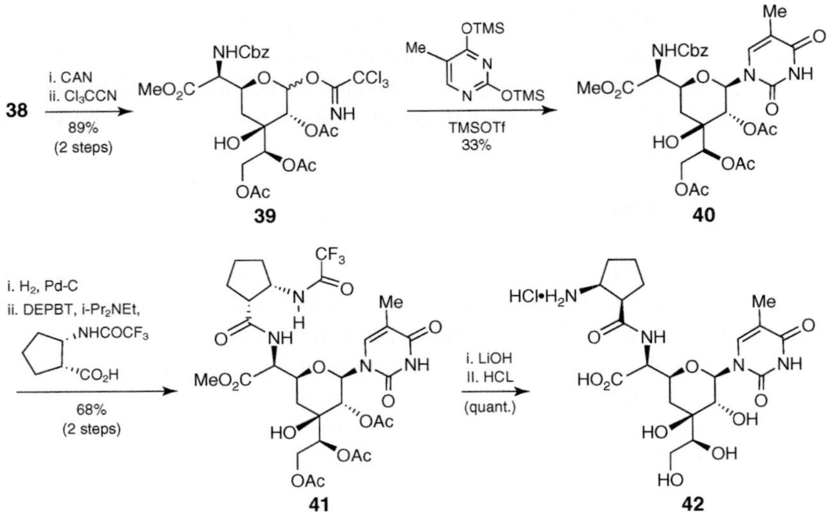

Scheme 7.

5. CONFORMATIONALLY RIGID BICYCLIC PEPTIDYL NUCLEOSIDES

In recent years, conformationally restricted nucleosides and oligonucleotides thereof have attracted considerable attention as biological tools in probing the furanose core conformational preferences, as exhibited by nucleosides and nucleotides in their interactions with the target enzymes.[88] For example, oligonucleotides constructed from a comparatively rigid fused bicyclic carbohydrate core containing nucleosides have been found to display improved recognition of complementary RNA and DNA sequences. In addition to their utility as biological probes, conformationally restrained nonnatural nucleosides and their derivatives are also of potential interest as novel antiviral, anticancer, and antisense agents.[89] Consequently, design, synthesis, and biological evaluation of various conformationally restricted nucleosides, with ring-fused bicyclic sugar backbones, continue to be an active area of research. In continuation of our studies on peptidyl nucleoside antibiotics, in the following section we describe our efforts in the area of conformationally restrained bicyclic nucleosides.

In addition to their utility as convenient platforms for our synthetic studies on natural peptidyl nucleoside antibiotics, the serine-derived lactones **8** (Scheme 1) and **11** (Scheme 2) also proved be useful in the construction of nonnatural peptidyl nucleosides of potential interest. In one such study, employing the masked furanosyl amino acid **11**, we are able to develop a concise synthetic route to a combinatorial library of conformationally rigid 3′,4′-*cis*-fused bicyclic peptidyl nucleosides. Employing the previously mentioned de novo nucleoside synthesis protocol, our strategy and approach involved utilization of an L-serine-derived chiral aminobutenolide toward initial stereoselective formation of a strategic bicyclic furofuranone scaffold **43** (Figure 6). Subsequent incorporation of various nucleobases (first diversity element) via modification of the lactone carbonyl and peptidic attachment of the resident amine functionality with appropriate amino acids (second diversity element) will lead to the desired library of [5,5]-bicyclic ring-fused peptidyl nucleosides **44**.

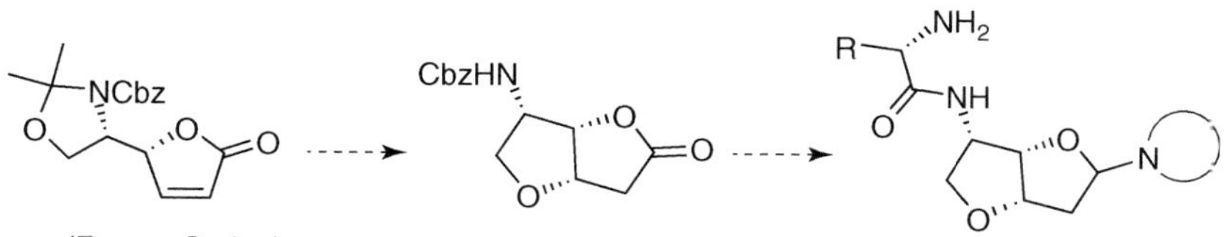

Figure 6. Toward conformationally rigid bicyclic peptidyl nucleosides.

Accordingly, following a previously reported protocol from our laboratory,[90] the L-serine-derived aminobutenolide **45** was converted to the strategic [5,5]-ring-fused bicyclic lactone **43** as shown (Scheme 8). Partial reduction of the lactone **43** to lactol, followed by its treatment with acetic anhydride, provided the corresponding acetate derivative **46** as an appropriate glycosyl donor for nucleoside formation. Subsequent nucleobase introduction via reaction of **46** with bis-silylated uracil in the presence of TMSOTf resulted in the formation of an inseparable anomeric mixture (ca. 2:1 by ^{1}H- NMR) of the corresponding nucleoside derivative **47a** in good yield. In high-resolution NMR studies, NOE correlations observed between the anomeric proton of the minor isomer of **47a** (δ 6.28) and the ring-junction protons (H-3a and H-6a) indicated the minor isomer to be the α-anomer, thereby confirming the major isomer in the nucleoside-forming reaction above to be the expected β-nucleoside. The poor stereoselectivity in the N-glycosidation reaction above can be attributed to the absence of any stereodirecting substituent at the adjacent C2′-position (nucleoside numbering) of the glycosyl donor **46**.

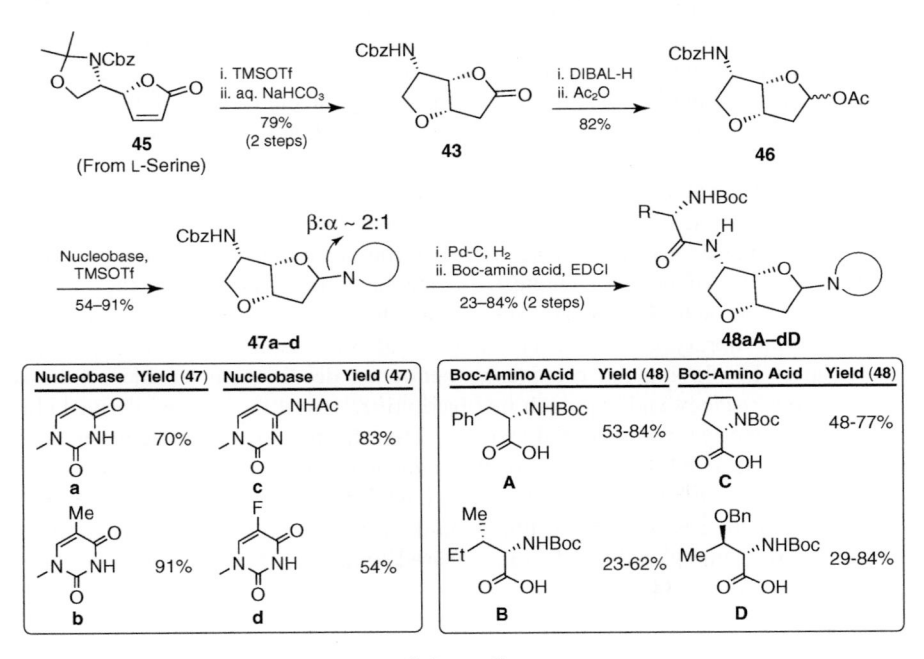

Scheme 8.

To achieve diversification, the nucleoside-forming reaction was subsequently extended to the corresponding thymidine, 5-fluorouridine, and cytidine analogues. Accordingly, reaction of **46** with the bis-silylated nucleobases above resulted in the expected anomeric mixtures of the corresponding nucleoside derivatives **47b–d**, respectively, in moderate to high yields. The final steps toward completion of the synthesis and library construction involved deprotection of the side-chain amine functionality of the nucleosides above, followed by standard peptidic coupling of the resulting free amine with a variety of suitably protected amino acids. Employing the sequence of reactions above, the combination of the four nucleosides **47a–d** with a representative set of four different N-Boc-amino acids (**A–D**) resulted in the construction of a 16-member library of amino acid–linked bicyclic nucleosides **48aA–48aD, 48bA–48bD, 48cA–48cD,** and **48dA–48dD,** as shown (Scheme 8).[91]

6. A NOVEL PYRANOSYL NIKKOMYCIN ANALOGUE

In another structure–activity relationship study investigating the role of the carbohydrate ring size in the antifungal activity of the nikkomycin family of peptidyl nucleoside antibiotics, we undertook the synthesis and biological evaluation of a carbohydrate ring-expanded novel pyranosyl nucleoside analogue of nikkomycin B. Our strategy for the synthesis involved utilization of the D-serine-derived amino pyranone **8** for the initial de novo construction of the pivotal pyranosyl nucleoside amino acid segment, followed by its elaboration to the desired ring-expanded nikkomycin B analogue (Figure 7).

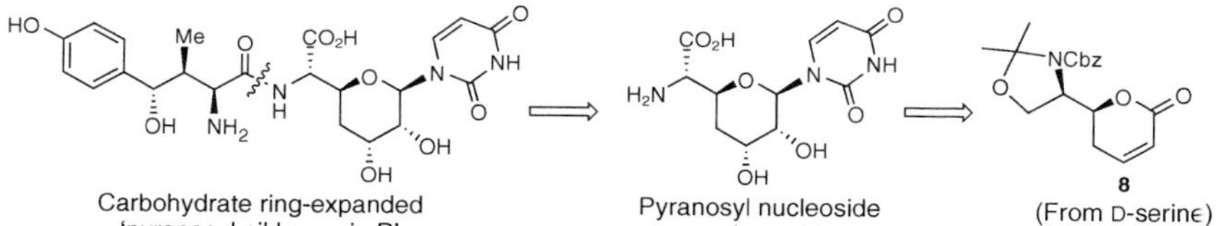

Figure 7. Pyranosyl nikkomycin B: retrosynthetic strategy.

Accordingly, starting from the D-serine-derived amino pyrone **8**, the fully functionalized pyranosyl nucleoside amino acid derivative **16a** was prepared following the procedure described in Scheme 3. Subsequent removal of Cbz protection yielded the free amine **49** (Scheme 9). Standard peptidic coupling of **49** with the nikkomycin B amino acid side-chain derivative **50**[92] resulted in the corresponding dipeptide **51**. Sequential removal of the protecting groups completed the desired synthesis of the novel pyranosyl nikkomycin B analogue **52**.[92]

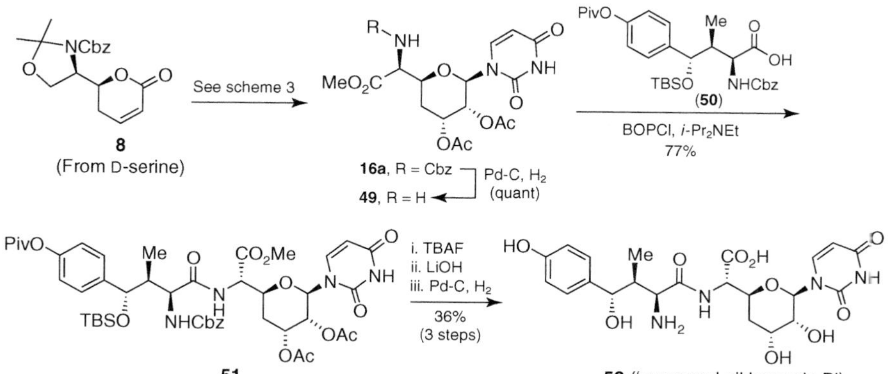

Scheme 9.

Table 1. In Vitro Antifungal Activity (MIC in μg/mL) of Pyranosyl Nikkomycin B (52)

Compound (μg/mL)	Candida albicans	Aspergillus fumigatus	Cryptococcus neoformans	Coccidioides immitis	Blastomyces dermatitidis
Pyranosyl Nikkomycin B (52)	>16	>16	≤0.03	≤0.03	0.5
Nikkomycin Z (ref.)	4	2	≤ 0.03	0.25	0.125
Amphotericin B (ref.)	0.125	0.5	0.25	0.25	0.25

To determine the effect of carbohydrate ring-size enlargement on antifungal activity, the pyranosyl nikkomycin B analogue **52** was screened against clinical isolates of several different human pathogenic fungi. For activity comparison, commercially available nikkomycin Z, the most potent antifungal natural product among the nikkomycin family of peptidyl nucleosides, and the antifungal drug amphotericin B were also included as reference standards in the assay.

As evident (Table 1), the pyranosyl nikkomycin B analogue **52** is inactive against *C. albicans*, and *A. fumigatus*. This is not entirely unexpected, as nikkomycin Z, too, is not a very efficient antifungal agent against most strains of *Candida* and *Aspergillus*. However, against human pathogenic fungal strains of *C. neoformans* and *C. immitis*, the analogue **52** exhibited strong inhibitory activity. Thus, while the antifungal activity of **52** against *C. neoformans* was equipotent to that of nikkomycin Z (and significantly better than that of amphotericin B), against *C immitis*, pyranosyl nikkomycin B (**52**) is more potent than amphotericin B and nikkomycin Z. Although not as active as amphotericin B or nikkomycin Z, the pyranosyl analogue **52** also showed reasonable activity against *B. dermatitidis*.[92]

Results from the studies above are expected to provide a better understanding of the role of the carbohydrate core ring size in the biological activity of the nikkomycin family of antifungal antibiotics. In ongoing studies, we are continuing to explore and optimize the structural features of the nikkomycin family of novel lead compounds toward the potential development of a new class of therapeutically useful antifungal agents.

7. CONCLUDING REMARKS

Carbohydrate starting materials continue to be the mainstay in synthetic endeavors toward various antifungal peptidyl nucleoside antibiotics. However, when the target nucleoside incorporates unusual residues or unnatural stereochemistry, this approach often suffers from lengthy reaction sequences and necessitates extensive protection–deprotection of functional groups. As described herein, we have explored an alternative synthetic strategy utilizing readily available amino acid serine as a simpler and versatile chiral building block toward de novo construction of the nucleoside amino acid cores of the desired peptidyl nucleosides. In addition to circumventing the disadvantages inherent in the carbohydrate approach, the serine-based strategy is more flexible, allowing extensive structural modifications of the nucleoside core. Easy addition or deletion of functionalities, and total control in the creation of new stereocenters, are also some of the additional advantages of the foregoing approach.

8. SELECTED EXPERIMENTAL PROCEDURES

(1S)-1-[(4R)-3-N-Benzyloxycarbonyl-2,2-dimethyl-1,3-oxazolidin-4-yl]-but-3-en-1-ol (6). To a cooled and well-stirred solution ($-10°$C: ice–salt bath) of the allyl ketone **5** (13 g, 42.9 mmol) and $CeCl_3 \cdot 7H_2O$ (5.4 g, 104 mmol) in MeOH (300 mL) was added dropwise (1.5 h) a solution of $Zn(BH_4)_2$ (0.189 M in Et_2O, 550 mL, 104 mmol). After stirring at the same temperature for another 1 h, the reaction was quenched by slow addition of saturated aqueous solution of $NaHCO_3$ (100 mL), allowed to attain room temperature, and then filtered through a sintered funnel. The residual solid was washed thoroughly with EtOAc, organic layer separated, and the aqueous layer extracted with EtOAc (3 × 100 mL). The combined organic extract was washed with brine, dried over anhydrous Na_2SO_4 and concentrated under vacuum. Purification of the crude residue by flash chromatography (hexane/EtOAc, 4:1 to 3:1) yielded the amino alcohol **6** as a colorless oil (12 g, 88%): $[\alpha]_D$ +16.8° (c 1.01, $CHCl_3$); IR (NaCl) 3462, 1695 cm^{-1}. ^{1}H-NMR (125.7 MHz, $CDCl_3$, rotameric mixture): δ 1.27–1.60 (4s, 6H), 2.02–2.40 (m, 2H), 3.14 (br s, 1H, exchangeable with D_2O), 3.86–4.23 (m, 4H), 4.99–5.37 (m, 4H), 5.66–6.0 (m, 1H), 7.38 (s, 5H).

(6S)-6-[(4R)-3-N-Benzyloxycarbonyl-2,2-dimethyl-1,3-oxazolidin-4-yl]-5,6-dihydropyran-2-one (8). A solution of the acrylate **7** (5.68 g, 15.8 mmol) and bis(tricyclohexylphosphine)benzylidene ruthenium(IV) dichloride (Grubbs' first-generation catalyst, 0.65 g, 5 mol%) in anhydrous CH_2Cl_2 (900 mL) was refluxed for 12 h. A second portion of the catalyst (0.65 g, 5 mol%) was then added to the reaction mixture and refluxing continued for another 12 h. After cooling to room temperature, the reaction mixture was treated with DMSO (0.1 mL), activated charcoal powder (10 g), and silica gel (25 g) and stirred for 12 h. After filtering off the solid and thorough washing of the residue with chloroform, the combined filtrate was concentrated under vacuum. The resulting oily liquid was purified by flash chromatography (EtOAc/hexane, 2:3) to afford the lactone **8** as a viscous yellow oil (4.29 g, 82%): $[\alpha]_D$ 29.1° (c 1.05, CH_2Cl_2); IR (NaCl) 1736, 1700 cm^{-1}. ^{1}H-NMR (400 MHz, $CDCl_3$; mixture of rotamers): δ 1.50–1.61 (4s, 6H), 2.26–2.57 (m, 2H), 3.95 (m, 1H), 4.15–4.46 (4m, 3H), 5.08–5.21 (m, 2H), 5.95 and 6.03 (2d, J = 9.4 Hz, 1H), 6.65 and 6.91 (2br s, 1H), 7.37 (s, 5H).

Thymidine Amino Acid Derivative 17b. To a solution of the triacetate **15b** (0.317 g, 0.68 mmol) and bis(trimethylsilyl)thymine (0.65 g, 2.5 mmol) in anhydrous (CH$_2$Cl)$_2$ (3 mL) was added freshly distilled TMSOTf (0.66 mL, 3.0 mmol). The resulting reaction mixture was stirred at room temperature for 5 h and then quenched with aqueous NaHCO$_3$ (2 mL). The organic layer was separated and the aqueous layer extracted with CHCl$_3$ (3 × 5 mL). The combined organic extracts washed with brine (1 × 10 mL), dried (Na$_2$SO$_4$), concentrated under vacuum, and the residue purified by flash chromatography (MeOH/CHCl$_3$, 1:49) to yield the thymine nucleoside amino acid derivative **17b** as a white solid (0.297 g, 82%): mp = 74–77°C; [α]$_D^{25}$ 21.0° (c 0.5, CHCl$_3$). ^{1}H-NMR (500 MHz, CDCl$_3$): δ 1.90 (s, 3H), 2.10 (s, 6H), 3.82 (s, 3H), 4.41 (br t, J = 4.4 Hz, 1H), 4.85 (br s, 1H), 5.16 (dd, J = 5.7 and 11.9 Hz, 2H), 5.31 (t, J = 5.6 Hz, 1H), 5.54 (t, J = 5.9 Hz, 1H), 5.95 (d, J = 5.5 Hz, 2H), 7.07 (s, 1H), 7.33 (s, 5H), 9.18 (s, 1H).

Benzyl(2 R,3 R,3a S,5 R,6 R,7 S,7a R)-6-(tert-butyldimethylsilyloxy)-3-hydroxy-5-(hydro xymethyl)-2-(5-methyl-2,4-dioxo-3,4-dihydropyrimidin-1(2 H)-yl)hexahydro-2 H-furo[3,2-b]pyran-7-ylcarbamate (27). Step 1: A solution of the diol **26** (0.94 g, 1.67 mmol) and mercuric trifluoroacetate (0.912 g, 5.28 mmol) in CH$_3$CN (100 mL) was refluxed overnight. After cooling to room temperature and dilution by addition of EtOAc (50 mL) and brine (50 mL), the resulting mixture was stirred at room temperature for 3 h. The organic layer was separated and the aqueous layer was extracted with EtOAc (3 × 40 mL). The combined organic extract dried over anhydrous Na$_2$SO$_4$ and solvent removed in vacuo to give a white foamy solid that was used as such for the subsequent reaction.

Step 2: To a well-stirred solution of NaBH$_4$ (0.20 g, 5.29 mmol) in DMF (9 mL) at room temperature, oxygen (O$_2$) gas was bubbled for 1 h. To this mixture a DMF solution (9 mL) of the crude mercuric compound (0.828 g) asobtained from the earlier step was added dropwise (2 h) with continuous bubbling of O$_2$. After stirring for 8 h, the reaction mixture was filtered through Celite, the residue washed thoroughly with EtOAc (4 × 20 mL) and the filtrate concentrated under vacuum. Purification of the crude residue by flash chromatography (hexane: MeOH/EtOAc, 18:2:80) yielded the bicyclic diol **27** as a white foamy solid (0.428 g, 44%): mp = 156–158°C; [α]$_D$ = –3.12° (c 0.96, CHCl$_3$). ^{1}H-NMR (400 MHz, CD$_3$OD): δ 0.09 (s, 3H), 0.20 (s, 3H), 0.91 (s, 9H), 1.69 (s, 3H), 3.26–3.28 (m, 2H), 3.57–3.71 (m, 3H), 3.81–3.85 (m, 2H), 4.20 (d, J = 5.1 Hz, 1H), 4.33–4.35 (br m, 1H), 4.24 (br s, 1H), 5.04–5.14 (m, 2H), 5.63 (s, 1H), 7.25–7.36 (m, 6H).

(2 R,3 R,3a R,5 S,6 R,7 S,7a R)-Methyl 7-(benzyloxycarbonylamino)-6-(t-butyldimethyl silyloxy)-3-(ethanoyloxy)-2-(5-methyl-2,4-dioxo-3,4-dihydropyrimidin-1(2 H)-yl)-hexahydro-2 H-furo[3,2-b]pyran-5-carboxylate (29). Step 1: An ice-cooled solution of the alcohol **28** (0.146 g, 0.232 mmol) in CH$_2$Cl$_2$ (10 mL) was treated with Dess–Martin periodinane (15% in CH$_2$Cl$_2$ solution, 0.86 mL, 0.302 mmol), and the reaction mixture was stirred at 0°C for 30 min. The reaction was then stirred at room temperature for another 2 h. After quenching the reaction by addition of a solution of 6 mL of saturated NaHCO$_3$ and 0.3 g of Na$_2$S$_2$O$_3$, the organic layer was separated, the aqueous layer extracted with CH$_2$Cl$_2$ (3 ×

10 mL), and the combined organic extract washed sequentially with 5.0 mL each of saturated aqueous $NaHCO_3$, H_2O, and brine. Drying over anhydrous Na_2SO_4 and removal of solvent under vacuum resulted in a colorless oily residue, which was used directly for the subsequent reaction.

Step 2: To a room-temperature solution of the aldehyde (0.10 g) as obtained from the reaction above and 2-methyl-2-butene (1.20 mL, 11.1 mmol) in *tert*-butanol (16.5 mL) was added dropwise a solution of $NaClO_2$ (0.205 g, 2.27 mmol) and NaH_2PO_4 (0.212 g, 1.76 mmol) in H_2O (4 mL). After stirring at room temperature for 30 min, the reaction mixture was partitioned with 2 mL of H_2O and the layers were separated. The aqueous layer was extracted with EtOAc (3 × 6 mL). The combined organic extract was dried over anhydrous Na_2SO_4, and the solvent was removed in vacuum to provide the product acid as colorless viscous oil and was used as such for the subsequent esterification reaction.

Step 3: [**Caution**: Diazomethane (CH_2N_2) is an explosive and a highly toxic gas. Explosions may occur if the substance is dried and undiluted. All operations involving diazomethane should be carried out in an efficient fume hood following appropriate precautions.] To a biphasic solution of KOH (0.500 g) in H_2O (1 mL) and ether (1.0 mL) at 0°C was added N-methyl-N'-nitro-N-nitrosoguanidine (MNNG, 50% in H_2O, 0.5 g) in one portion. The ethereal layer was decanted into an ice-cooled Erlenmeyer flask containing KOH pellets. The aqueous layer was washed with ether (3 × 5 mL), and the ethereal layers were combined. The CH_2N_2 thus prepared was added to a stirred solution of the crude acid (0.1 g in 2 mL of ether) from step 2 and stirred for 30 min. After removal of excess CH_2N_2 by bubbling nitrogen into the reaction mixture (15 min), the reaction mixture was concentrated under vacuum. Purification of the residue by flash chromatography (hexane/EtOAc, 7:3) yielded the octosyl nucleoside derivative **29** as a white solid (0.095 g, 66% over three steps): mp = 142–144°C; $[\alpha]_D$ +2.83° (c 0.40, $CHCl_3$). ^{1}H-NMR (500 MHz, $CDCl_3$): δ 0.08 (br s, 3H), 0.25 (br s, 3H), 0.89 (s, 9H), 1.89 (s, 3H), 2.22 (s, 3H), 3.77 (s, 3H), 4.24 (br s, 1H), 4.31 (br d, J = 10 Hz, 1H), 4.54 (s, 1H), 4.55–4.58 (m, 2H), 4.99–5.13 (m, 2H), 5.16–5.19 (m, 1H), 5.70 (d, J = 6.0 Hz, 1H), 5.78 (br d, J = 5.2 Hz, 1H), 6.89 (s, 1H), 7.32 (s, 5H), 8.87 and 8.96 (2s, 1H).

Benzyl (R)-2-(tert-butyldimethylsilyloxy)-1-{(2S,5R,6R,E)-5-hydroxy-4-[(E)-methoxycarbonylidene]-6-(4-methoxybenzyloxy)tetrahydro-2H-pyran-2-yl}ethylcarbamate (35). A solution of the ketone **34** (0.186 g, 0.31 mmol) and (carbomethoxymethylene) triphenylphosphorane (0.22 g, 0.62 mmol) in anhydrous benzene (6 mL) was refluxed at 80°C for 12 h. After removal of solvent under vacuum, the resulting residue was purified by flash chromatography (hexane/EtOAc, 4:1) to yield the *E*-alkylidene ester **14** as a viscous oil (0.197 g, 95%): $[\alpha]_D$ −56.3° (c 0.50, $CHCl_3$). ^{1}H-NMR (400 MHz, $CDCl_3$): δ 0.09 (s, 6H), 0.91 (s, 9H), 2.14 (s, 3H), 2.17–2.24 (m, 1H), 3.64 (br t, J = 9.5 Hz, 1H), 3.69 (br s, 4H), 3.81 (s, 3H), 3.89–3.96 (m, 1H), 4.00 (d, 10.0 Hz, 1H), 4.14 (d, J = 14.2, 1H), 4.40 (d, J = 7.7 Hz, 1H), 4.55 (d, J = 11.8 Hz, 1H), 4.81 (d, J = 11.8 Hz, 1H), 5.15 and 5.21 (2d, J = 12.2 Hz, 3H), 5.33 (d, J = 7.5 Hz, 1H), 5.77 (s, 1H), 6.88 (d, J = 8.8 Hz, 2H), 7.23 (d, J = 8.6 Hz, 2H), 7.34–7.39 (m, 5H).

(S)-1-{(3R,4R,6S)-6-[(S)-1-(Benzyloxycarbonylamino)-2-methoxy-2-oxoethyl]-3-(ethanoyloxy)-4-hydroxy-2-(2,2,2-trichloro-1-iminoethoxy)tetrahydro-2H-pyran-4-yl}ethane-1,2-diyl diethanoate (39). Step 1: A stirring room-temperature solution of the PMB-glycoside **38** (0.28 g, 0.42 mmol) in acetonitrile/water (10 mL, 9:1) was treated with ceric ammonium nitrate (0.65 g, 1.19 mmol). After stirring for 1.5 h the reaction mixture was diluted with EtOAc (20 mL) and washed sequentially with water (10 mL), saturated aqueous $NaHSO_3$ (5 mL), and saturated aqueous $NaHCO_3$ (5 mL). The organic layer was dried with Na_2SO_4 and concentrated. The resulting crude lactol (0.22 g) was used for the next reaction without further purification.

Step 2: To a stirred solution of the crude lactol (0.22 g, 0.41 mmol) in anhydrous CH_2Cl_2 (8 mL) was added anhydrous K_2CO_3 (0.2 g, 1.45 mmol), followed by dropwise addition of trichloroacetonitrile (0.4 mL, 4.0 mmol). The reaction was stirred at room temperature for 12 h and then quenched with water (5 mL). The organic layer was separated and the aqueous layer extracted with EtOAc (3 × 10 mL). The combined organic extract was washed with brine (1 × 10 mL), dried (Na_2SO_4), and concentrated. The crude residue was quickly purified by flash chromatography (EtOAc/hexane, 1:1) to afford the trichloroacetimidate **39** as a viscous liquid (0.26 g, 89% over two steps). The product **39** was found to be unstable and was used immediately after purification. ^{1}H-NMR (400 MHz, $CDCl_3$; mixture of anomers): δ 2.05, 2.06, 2.07, and 2.10 (4s, 9H), 2.14−2.30 (m, 2H), 3.73 and 3.76 (2s, 3H), 4.05−4.43 (m, 4H), 4.92−5.17 (m, 4H), 5.41 (s, 1H), 5.61−5.70 (m, 1H), 6.08 and 6.58 (2s, 1H), 7.37 (br s, 5H), 8.58 and 8.68 (2s, 1H).

(S)-1-{(2R,3R,4R,6S)-6-[(S)-1-(Benzyloxycarbonylamino)-2-methoxy-2-oxoethyl]-3-(ethanoyloxy)-4-hydroxy-2-(5-methyl-2,4-dioxo-3,4-dihydropyrimidin-1(2H)-yl)tetrahydro-2H-pyran-4-yl}ethane-1,2-diyldiethanoate (40). A solution of the trichloroacetimidate **39** (0.163 g, 0.24 mmol) dissolved in anhydrous 1,2-dichloroethane (8 mL) was treated with commercially available bis(trimethylsilyl)thymine (0.32 g, 1.2 mmol), followed by freshly distilled TMSOTf (0.17 mL, 0.96 mmol). After stirring at room temperature for 1 h, the reaction was quenched by saturated aqueous $NaHCO_3$ solution (5 mL). The precipitated solid was filtered and washed with $CHCl_3$ (3 × 5 mL). The combined filtrate was transferred to a separating funnel and the organic layer separated. The aqueous layer was extracted with $CHCl_3$ (3 × 5 mL), and the combined organic extracts dried with Na_2SO_4 and concentrated under vacuum. The residue was purified by flash chromatography (EtOAc/hexane, 7:3 to 9:1) to afford the thymine nucleoside **40** as a white solid (0.051 g, 33%): mp = 107−109°C; $[α]_D$ −6.2° (c 1.35, $CHCl_3$). ^{1}H-NMR (400 MHz, $CDCl_3$): δ 1.88 (s, 3H), 2.05, 2.06, and 2.10 (3s, 9H), 2.16−2.29 (m, 2H), 3.80 (s, 3H), 4.07 (s, 1H, exchangeable with D_2O), 4.19 (dd, J = 7.7 and 11.9 Hz, 1H), 4.25 (br s, 1H), 4.55 (dd, J = 2.7 and 12.0 Hz, 1H), 4.61 (d, J = 6.6 Hz, 1H), 5.05 (d, J = 5.2 Hz, 1H), 5.14 (s, 2H), 5.41 (d, J = 4.7 Hz, 1H), 5.80 (d, J = 4.4 Hz, 1H), 6.02 (br s, 1H), 7.27 (s, 1H), 7.36 (br s, 5H), 8.92 (s, 1H).

Synthesis of the Fully Protected Pyranosyl Nikkomycin B Derivative 51. *Step 1*: A solution of the nucleoside amino acid derivative **16a** (0.62 g, 1.16 mmol) in anhydrous EtOAc (10 mL), was treated with 10% palladium on activated carbon (0.7 g) and the mixture was stirred under a hydrogen atmosphere for 3 h. After filtering through Celite, the residue was washed with EtOAc and MeOH. The combined filtrate was concentrated to yield the free amine as a foamy solid (0.45 g, 97% crude), which was taken onto the next step without further purification.

Step 2: To a cooled solution (0°C) of the carboxylic acid **50**[92] (0.75 g, 1.35 mmol) dissolved in anhydrous CH_2Cl_2 (8 mL) was added *N,N*-diisopropylethyl amine (0.23 mL, 1.35 mmol) and BOP-Cl (0.34 g, 1.35 mmol). After stirring at 0°C for 45 min, the crude amine (0.45 g, 1.12 mmol) dissolved in anhydrous CH_2Cl_2 (7 mL) was added to the reaction mixture, followed by *N,N*-diisopropylethyl amine (0.23 mL, 1.35 mmol). After stirring the resulting reaction mixture for 12 h at room temperature, the reaction was quenched with H_2O (10 mL). The two layers were separated, and the aqueous layer was extracted with EtOAc (3 × 10 mL). The combined organic layers were washed with brine (1 × 15 mL), dried (Na_2SO_4), concentrated under vacuum, and the residue was purified by flash chromatography (EtOAc/hexane, 6:4 to 8:2) to yield the fully protected pyranosyl nikkomycin analogue **51** as a white solid (0.83 g, 77%): mp = 134–136°C; $[\alpha]_D$ −18.8° (c 1.05, $CHCl_3$). ^{1}H-NMR (400 MHz, $CDCl_3$): δ −0.28 (s, 3H), 0.03 (s, 3H), 0.80 (d, *J* = 7.1 Hz, 3H), 0.85 (s, 9H), 1.36 (s, 9H), 1.97–2.07 (m, 4H), 2.18–2.25 (m, 4H), 2.33 (br s, 1H), 3.78 (s, 3H), 4.28–4.35 (m, 2H), 4.66–4.73 (m, 2H), 4.90 (d, *J* = 7.2 Hz, 1H), 4.90–5.20 (m, 2H), 5.57 (s, 1H), 5.73 (d, *J* = 6.5 Hz, 1H), 6.03 (d, *J* = 9.6 Hz, 1H), 6.10 (br s, 1H), 6.96 (d, *J* = 8.5 Hz, 2H), 7.18–7.37 (m, 9H), 9.10 (br s, 1H).

Acknowledgment

We thank the Herman Frasch Foundation (American Chemical Society) for financial support.

ABBREVIATIONS

Ac	acetyl
Boc	*tert*-butoxycarbonyl
BOPCl	bis(2-oxo-3-oxazolidinyl)phosphinic chloride
t-Bu	*tert*- butyl
CAN	ammoniun cerium(IV) nitrate
Cbz	carbobenzyloxy
CSA	camphorsulfonic acid
DCE	1,2-dichloroethane
DEPBT	3-(diethoxyphosphoryloxy)1,2,3-benzotriazin-4(3*H*)-one
DET	diethyl tartrate
DIBAL-H	diisobutylaluminum hydride

EDCI	1-(3-dimethylaminopropyl)-3-ethylcarbodiimide hydrochloride
Et	ethyl
EtOAc	ethyl acetate
μM	micromolar
m-CPBA	m-chloroperbenzoic acid
Me	methyl
MeOH	methanol
MIC	minimum inhibitory concentration
NMM	N-methylmorpholine
NMO	N-methylmorpholine oxide
NMR	nuclear magnetic resonance
Piv	pivaloyl
PMB	para-methoxybenzyl
PPTS	pyridinium para-toluenesulfonic acid
Py	pyridine
SAR	structure–activity relationship
TBAF	tetrabutylammonium fluoride
TBDMSCl (or TBSCl)	tert-butyldimethylsilyl chloride
Tf	trifluoromethane sulfonate (triflate)
THF	tetraydrofuran
TLC	thin-layer chromatography
TMS	trimethylsilyl
TMSOTf	trimethylsilyl triflate

REFERENCES

1. For relevant reviews, see (a) Ostrosky-Zeichner, L.; Casadevall, A.; Galgiani, J. N.; Odds, F. C.; Rex, J. H. *Nat. Rev. Drug Discov.* **2010**, *9*, 719–727. (b) Shahid, M.; Shahzad, A.; Tripathi, T.; Sobia, F.; Sahai, S.; Singh, A.; Malik, A.; Shujatullah, F.; Khan, H. M. *Anti-Infect. Agents Med. Chem.* **2009**, *8*, 36–49. (c) Sable, A.; Strohmaier, K. M.; Chodakewitz, J. A. *Annu. Rev. Med.* **2008**, *59*, 361–379. (d) Lorand, T.; Kocsis, B. *Mini-Rev. Med. Chem.* **2007**, *7*, 900–911. (e) Di Santo, R. *Annu. Rep. Med. Chem.* **2006**, *41*, 299–315. (f) Sundriyal, S.; Sharma, R. K.; Jain, R. *Curr. Med. Chem.* **2006**, *13*, 1321–1335. (g) Weimin, Z.; Becker, D.; Cheng, Q. *Recent Patents Anti-infective Drug Disc.* **2006**, *1*, 225–230. (h) Groll, A. H.; Piscitelli, S. C.; Walsh, T.J. *Adv. Pharmacol.* **1998**, *44*, 343–500.

2. (a) Lai, C. C.; Tan, C. K.; Huang, Y. T.; Shao, P. L.; Hsueh, P. R. *J. Infect. Chemother.* **2008**, *14*, 77–85. (b) Shao, P. L.; Huang, L. M.; Hsueh, P. R. *Int. J. Antimicrob. Agents* **2007**, *30*, 487–495. (c) Nucci, M.; Marr, K. A. *Clin. Infect. Dis.* **2005**, *41*, 521–526. (d) Walsh, T. J.; Groll, A.H. *Transpl. Infect. Dis.* **1999**, *1*, 247–261. Walsh, T. J. In *Emerging Targets in Antibacterial and Antifungal Chemotherapy*; Sutcliffe, J.; Georgopapadakou, N. H., Eds.; Chapman & Hall: New York, **1992**, pp. 349–373.

3. Denning, D. W. *J. Antimicrob. Chemother.* **1991**, *28*(*Suppl. B*), 1–6.

4. Bodey, G. P. *Ann. N.Y. Acad. Sci.* **1988**, *544*, 431–442.

5. Anaissie, E. *J. Clin. Infect. Dis.* **1992**, *14*(Suppl. 1), 43–53.
6. Ruhnke, M.; Maschmeyer, G. *Eur. J. Med. Res.* **2002**, *7*, 227–235.
7. Andriole, V. T. *J. Antimicrob. Chemother.* **1999**, *44*, 151–162.
8. Patterson, J. E. *Transpl. Infect. Dis.* **1999**, *1*, 229–236.
9. Paya, C. V. *Clin. Infect. Dis.* **1993**, *16*, 677–688.
10. Ruhnke, M. *Drugs* **2004**, *64*, 1163–1180.
11. Hage, C. A.; Goldman, M.; Wheat, L. J. *Eur. J. Med. Res.* **2002**, *7*, 236–241.
12. Dupont, B.; Crewe Brown, H. H.; Westermann, K.; Martins, M. D.; Rex, J. H.; Lortholary, O.; Kauffmann, C. A. *Med. Mycol.* **2000**, *38*(Suppl. 1), 259–267.
13. Ampel, N. M. *Emerg. Infect. Dis.* **1996**, *2*, 109–116.
14. (a) Hamdan, J. S.; Hahn, R. C. *Anti-Infect. Agents Med. Chem.* **2006**, *5*, 403–412. (b) Diamond, R. D. *Rev. Infect. Dis.* **1991**, *13*, 480–486.
15. For a treatise, see Bennett, J. E. In *Goodman & Gillman's The Pharmacological Basis of Therapeutics*, 11th ed.; Brunton, L. R.; Lazo, J. S.; Parker, K. L., Eds.; McGraw-Hill: New York, **2006**, pp.1225–1241.
16. Dismukes, W. E. *Clin. Infect. Dis.* **2000**, *30*, 653–657.
17. Berg, D.; Plempel, M. *Sterol Biosynthesis Inhibitors: Pharmaceutical and Agrochemical Aspects*; Ellis-Horwood: Chichester, UK, **1988**.
18. Shuter, J. *Cancer Investig.* **1999**, *17*, 145–152.
19. Loeffler, J.; Stevens, D. A. *Clin. Infect. Dis.* **2003**, *36*(Suppl. 1), S31–S41.
20. Anderson, M. B.; Roemer, T.; Fabrey, R. *Annu. Rep. Med. Chem.*; Doherty, A. M., Ed.; Elsevier–Academic Press: New York, **2003**, *38*, 163–172 and references therein.
21. Dolezal, M. *Ceska. Slov. Farm.* **2002**, *51*, 226–235.
22. Watkins, W. J.; Renau, T. E. *Annu. Rep. Med. Chem.*; Doherty, A. M., Ed.; Academic Press: New York, **2000**, *35*, 157–166 and references therein.
23. Andriole, V. T. *Int. J. Antimicrob. Agents* **2000**, *16*, 317–321.
24. Balkovec, J. M. In *Annu. Rep. Med. Chem.*; Bristol, J. A., Ed.; Academic Press: New York, **1998**, *33*, 173–182 and references therein.
25. Isono, K. *J. Antibiot.* **1988**, *41*, 1711–1739.
26. Isono, K. *Pharmacol. Ther.* **1991**, *52*, 269–286.
27. Isono, K.; Ashai, K.; Suzuki, S. *J. Am. Chem. Soc.* **1969**, *91*, 7490–7505.
28. Isono, K.; Suzuki, S. *Heterocycles* **1979**, *13*, 333–351.
29. Dahn, U.; Hagenmeier, H.; Hohne, H.; Konig, W. A.; Wolf, G.; Zahner, H. *Arch. Microbiol.* **1976**, *107*, 143–160.
30. Kobinata, K.; Uramoto, M.; Nishi, M.; Kusakabe, H.; Nakamura, G.; Isono, K. *Agric. Biol. Chem.* **1980**, *44*, 1709–1711.
31. Cabib, E. *Adv. Enzymol.* **1987**, *59*, 59–101.
32. Gooday, G. W. In *Biochemistry of Cell Walls and Membranes in Fungi*; Kuhn, P. J.; Trinci, A. P.; Jung, M. J.; Goosey, M. W.; Copping, L. G., Eds.; Springer-Verlag: Berlin, **1990**, p.61.
33. Behr, J. B. *Curr. Med. Chem. Anti-Infect. Agents* **2003**, *2*, 173–189 and references therein.
34. Ruiz-Herrera, J.; San-Blas, G. *Curr. Drug Targets: Infect. Disord.* **2003**, *3*, 77–91 and references therein.

35. Ko, K. In *Human Welfare and the Environment*; Miyato, J., Ed.; Pergamon Press: Elmsford, NY, **1983**; p. 247.
36. Cabib, E.; Shematek, E. M. In *Biology of Carbohydrates*; Ginsburg, V.; Robius, P. W., Eds.; Wiley: New York, **1981**; Vol. I, p. 51.
37. Bulawa, C. E. *Annu. Rev. Microbiol.* **1993**, *47*, 505.
38. Ruiz-Herrera, J.; Ruiz-Medrano, R. *Mycol. Ser.* **2004**, *20*, 315–330.
39. Becker, J. M.; Covert, N. L.; Shenbagamurthi, P. S.; Steinfeld, A.; Naider, F. *Antimicrob. Agents Chemother.* **1983**, *23*, 926–929.
40. Krainer, E.; Becker, J. M.; Naider, F. *J. Med. Chem.* **1991**, *34*, 174–180 and references therein.
41. Chapman, T.; Kinsman, O.; Houston, J. *Antimicrob. Agents Chemother.* **1992**, *36*, 1909–1914.
42. Sakata, K.; Sakurai, A.; Tamura, S. *Agric. Biol. Chem.* **1974**, *38*, 1883–1890.
43. Sakata, K.; Sakurai, A.; Tamura, S. *Agric. Biol. Chem.* **1975**, *39*, 885–892.
44. Sakata, K.; Sakurai, A.; Tamura, S. *Agric. Biol. Chem.* **1977**, *41*, 2027–2032. For structure determination studies, see pp. 2033–2039.
45. Harada, S.; Kishi, T. *J. Antibiot.* **1977**, *30*, 11–16.
46. Iwasa, T.; Kishi, T.; Matsura, K.; Wakae, O. *J. Antibiot.* **1977**, *30*, 1–10.
47. Seto, H.; Koyama, M.; Ogino, H.; Tsuroka, T. *Tetrahedron Lett.* **1983**, *24*, 1805–1808.
48. Marcelo, F.; Jiménez-Barbero, J.; Marrot, J.; Rauter, A. P.; Sinay, P.; Blériot, Y. *Chem. Eur. J.* **2008**, *14*, 10066–10073.
49. For a review, see Zhang, D.; Miller, M. J. *Curr. Pharm. Des.* **1999**, *5*, 73–99, and references therein.
50. For a review, see Garner, P. In *Studies in Natural Products Chemistry*; Atta-ur-Rahman, Ed.; Elsevier: Amsterdam, **1988**; Stereoselective synthesis (Part A), Vol. 1, pp. 397–435.
51. For a review, see Knapp, S. *Chem. Rev.* **1995**, *95*, 1859–1876 and references therein.
52. Kuzuhara, H.; Ohrui, H.; Emoto, S. *Tetrahedron Lett.* **1973**, 5055–5058 and references therein.
53. Dondoni, A.; Franco, S.; Junquera, F.; Merchan, F. M.; Merino, P.; Tejero, T. *J. Org. Chem.* **1997**, *62*, 5497–5507.
54. Kato, K.; Chen, C. Y.; Akita, H. *Synthesis* **1998**, 1527–1533.
55. (a) Ghosh, A. K.; Wang, Y. *J. Org. Chem.* **1999**, *64*, 2789–2795. (b) Shiro, Y.; Kato, K.; Fujii, M.; Ida, Y.; Akita, H. *Tetrahedron* **2006**, *62*, 8687–8695.
56. Chen, A.; Thomas, E. J.; Wilson, P. D. *J. Chem. Soc. Perkin Trans. 1*, **1999**, 3305–3310.
57. Kutterer, K. M. K.; Just, G. *Heterocycles* **1999**, *51*, 1409–1420.
58. Gonda, J.; Martinkova, M.; Walko, M.; Zavacka, E.; Budesinsky, M.; Cisarova, I. *Tetrahedron Lett.* **2001**, *42*, 4401–4404.
59. Mita, N.; Tamura, O.; Ishibashi, H.; Sakamoto, M. *Org. Lett.* **2002**, *4*, 1111–1114.
60. More, J. D.; Finney, N. S. *Synlett* **2003**, 1307–1310.
61. Garner, P.; Park, J. M. *J. Org. Chem.* **1990**, *55*, 3772–3787.
62. Tabusa, F.; Yamada, T.; Suzuki, K.; Mukaiyama, T. *Chem. Lett.* **1984**, 405–408.
63. Auberson, Y.; Vogel, P. *Tetrahedron* **1990**, *46*, 7019–7032.
64. Gethin, D. M.; Simpkins, N. S. *Tetrahedron* **1997**, *53*, 14417–14436.
65. Knapp, S.; Gore, V. K. *Org. Lett.* **2000**, *2*, 1391–1393 and references therein.

66. Knapp, S.; Jaramillo, C.; Freeman, B. *J. Org. Chem.* **1994**, *59*, 4800–4804.
67. Knapp, S.; Shieh, W.-C.; Jaramillo, C.; Trilles, R. V.; Nandan, S. R. *J. Org. Chem.* **1994**, *59*, 946–948.
68. Maier, S.; Preuss, R.; Schmidt, R. R. *Liebigs Ann. Chem.* **1990**, 483–489.
69. Sakanaka, O.; Ohmori, T.; Kozaki, S.; Tetsuo, S. *Bull. Chem. Soc. Jpn.* **1987**, *60*, 1057–1062.
70. Hanessian, S.; Dixit, D. M.; Liak, T. J. *Pure Appl. Chem.* **1981**, *53*, 129–148.
71. Czernicki, S.; Franco, S.; Valery, J.-M. *J. Org. Chem.* **1997**, *62*, 4845–4847.
72. Rauter, A. P.; Fernandes, A. C.; Czernicki, S.; Valery, J.-M. *J. Org. Chem.* **1996**, *61*, 3594–3598.
73. Rauter, A.; Ferreira, M.; Borges, C.; Durate, T.; Piedade, F.; Silva, M.; Santos, H. *Carbohydr. Res.* **2000**, *325*, 1–15, and references therein.
74. Garner, P. *Tetrahedron Lett.* **1984**, *25*, 5855–5858.
75. Garner, P.; Yoo, J. K.; Sarabu, R.; Kennedy, V. O.; Youngs, W. J. *Tetrahedron* **1998**, *54*, 9303–9316 and references therein.
76. Bhaket, P.; Stauffer, C. S.; Datta, A. *J. Org. Chem.* **2004**, *69*, 8594–860.
77. Vorbrüggen, H.; Ruh-Pohlenz, C. *Org. React.* **2000**, *55*, 1–654.
78. (a) Isono, K.; Crain, R. F.; McKloskey, J. A. *J. Am. Chem. Soc.* **1975**, *97*, 943–945. (b) Benner, J. P.; Boehlendrof, B. G. H.; Kipps, M. R.; Lambert, N. E. P.; Luck, R.; Molleyres, L.-P.; Neff, S.; Schuez, T. C.; Stanley, P. D. WO 03/062242, CAN 139:132519. (c) For a review, see More, J. D. *Org. Prep. Proc. Int.* **2007**, *39*, 107–133.
79. (a) Poon, K. W. C.; Liang, N.; Datta, A. *Nucleosides Nucleotides Nucleic Acids* **2008**, *27*, 389–407. (b) Poon, K. W. C.; Lovell, K. M.; Dresner, K. N.; Datta, A. *J. Org. Chem.* **2008**, *73*, 752–755.
80. For a review, see Johnson, R. A.; Sharpless, K. B. In *Catalytic Asymmetric Synthesis*, 2nd ed.; Ojima, I., Ed.; Wiley-VCH: New York, **2000**, pp. 231–285 and references therein.
81. For an example of a similar functional group transformation, see Yadav, J. S.; Srihari, P. *Tetrahedron: Asymmetry* **2004**, *15*, 81–89 and references therein.
82. Khalaf, J. K., VanderVelde, D. G.; Datta, A. *J. Org. Chem.* **2008**, *73*, 5977–5984.
83. Goto, T.; Toya, Y.; Ohgi, T.; Kondo, T. *Tetrahedron Lett.* **1982**, *23*, 1271–1274.
84. Hara, K.; Fujimoto, H.; Sato, K. I.; Hashimoto, H.; Yoshimura, J. *Carbohydr. Res.* **1987**, *159*, 65–79.
85. For examples of similar selective acylations, see (a) Bhatt, R. K.; Chauhan, K.; Wheelan, P.; Falck, J. R.; Murphy, R. C. *J. Am. Chem. Soc.* **1994**, *116*, 5050–5056. (b) Khalaf, J. K.; Datta, A. *J. Org. Chem.* **2005**, *70*, 6937–6940.
86. (a) Schmidt, R. R.; Michel, J. *J. Carbohydr. Chem.* **1985**, *4*, 141–169. (b) Schmidt, R. R. In *Frontiers in Natural Product Research*, **1996**, *1* (Modern Methods in Carbohydrate Synthesis), 20–54. (c) Schmidt, R. R.; Jung, K.-H. In *Preparative Carbohydrate Chemistry*; Hanessian, S., Ed.; Marcel Dekker: New York, **1997**, pp. 283–312, and references therein.
87. Stauffer, C. S.; Datta, A. *J. Org. Chem.* **2008**, *73*, 4166–4174.
88. For some recent reports, see (a) Choi, Y.; Moon, H. R.; Yoshimura, Y.; Marquez, V. E. *Nucleosides Nucleotides Nucleic Acids* **2003**, *22*, 547–557. (b) Leumann, C. J. *Bioorg. Med. Chem.* **2002**, *10*, 841–854. (c) Imanishi, T.; Obika, S. *Chem. Commun.*

2002, 1653–1659. (d) Meldgaard, M.; Wengel, J. *J. Chem. Soc. Perkin Trans. 1* **2000**, 3539–3554. (e) Herdewijn, P. *Biochim. Biophys. Acta* **1999**, *1489*, 167–179. (f) Kool, E. T. *Chem. Rev.* **1997**, *97*, 1473–1487.

89. For representative publications, see (a) Park, A.-Y.; Moon, H. R.; Kim, K. R.; Chun, M. W.; Jeong, L. S. *Org. Biomol. Chem.* **2006**, *4*, 4065–4067. (b) De Clercq, E. *J. Clin. Virol.* **2004**, *30*, 115–133. (c) Matsuda, A.; Sasaki, T. *Cancer Sci.* **2004**, *95*, 105–111. (d) Vester, B.; Wengel, J. *Biochemistry* **2004**, *43*, 13233–13241. (e) Jepsen, J. S.; Sørensen, M. D.; Wengel, J. *Oligonucleotides* **2004**, *14*, 130–146. (f) Russ, P.; Schelling, P.; Scapozza, L.; Folkers, G.; De Clercq, E.; Marquez, V. E. *J. Med. Chem.* **2003**, *46*, 5045–5054. (g) Shin, K. J.; Moon, H. R.; Georgen, C.; Marquez, V. E. *J. Org. Chem.* **2000**, *65*, 2172–2178. (h) Molas, M. P.; Matheu, M. I.; Castillon, S.; Isac-Garcia, J.; Hernandez-Mateo, F.; Calvo-Flores, F. G.; Santoyo-Gonzalez, F. *Tetrahedron* **1999**, *55*, 14649–14664. (i) Agrofoglio, L. A.; Challand, S. R. *Acyclic, Carbocyclic and L-Nucleosides*; Kluwer Academic: Boston, **1998**. (j) Huryn, D. M.; Okabe, M. *Chem. Rev.* **1992**, *92*, 1745–1768.

90. Bhaket, P.; Morris, K.; Stauffer, C. S.; Datta, A. *Org. Lett.* **2005**, *7*, 875–876.

91. Poon, K. W. C.; Datta, A. *Nucleosides Nucleotides Nucleic Acids*, **2008**, *27*, 914–930.

92. Stauffer, C. S.; Bhaket, P.; Fothergill, A. W.; Rinaldi, M. G.; Datta, A. *J. Org. Chem.* **2007**, *72*, 9991–9997.

19 Chemical Synthesis of Conformationally Constrained PNA Monomers

PEDRO MERINO
Departamento de Quimica Organica, Instituto de Sintesis y Catalisis Homogenea, Universidad de Zaragoza, CSIC, Zaragoza, Aragón, Spain

ROSA MATUTE
Department of Chemical Engineering and Environment Technologies, University of Zaragoza, Zaragoza, Aragón, Spain

1. INTRODUCTION

Peptide nucleic acid (PNA) **1** (Figure 1) is an oligopeptide with nucleobases linked together by an achiral, uncharged aminoethylglycine backbone (*aeg*-PNA), that binds strongly and in a sequence-specific manner to complementary DNA and RNA oligonucleotides.[1] PNA is formed by repetition of a tri-branched unit recalling the sicilian Trinacria.[2]

Since its discovery in 1991 by Nielsen and co-workers,[3] PNAs have became important mimics of natural nucleic acids with application as antisense and antigen agents[4] and as biomedical diagnostic tools.[5,6] Although PNAs were designed initially to recognize dsDNA via triplex formation in the major groove by Hoogsteen pairing,[3] very efficient mimics resulted that bind their sequence-complementary targets in both parallel and antiparallel orientation by Watson–Crick pairing.[7] Moreover, PNAs bind in invasive modes to give PNA/DNA duplexes and PNA_2–DNA or high-order complexes.[8] PNAs bind to DNA and RNA with markedly higher thermal stability relative to any other modified oligonucleotide from synthetic or natural sources.[9,10] Despite these promising features and successful experiments demonstrating that PNAs are capable of inducing transcription and translation inhibition,[11] the therapeutic development of PNAs into drugs is seriously compromised by their limited solubility in physiological media.[12] In pursuit of improved binding affinity of PNAs and to increase their water solubility, different classes of conformationally constrained analogues have been designed.[13] Among these families of

---

Chemical Synthesis of Nucleoside Analogues, First Edition. Edited by Pedro Merino.
© 2013 John Wiley & Sons, Inc. Published 2013 by John Wiley & Sons, Inc.

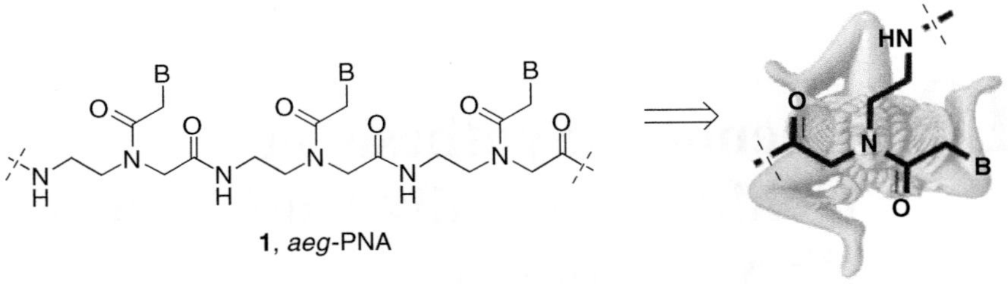

Figure 1. Peptide nucleic acid (PNA). B, heterocyclic base (A, T, C, G).

compounds are those containing a double bond in their structure, which increases the conformational rigidity of the molecule.[14] These analogues are called *olefinic peptide nucleic acids* (OPAs), and Z and E configurations are possible, as in **2** and **3** (Figure 2). Conformational preorganization of the PNA molecule has also been pursued by imposing a carbocyclic ring on the ethylenediamine segment. For structurally distinct carbocycle-containing PNA analogues such as **4–6**, "cyclopropyl-", "cyclopentyl-", and "cyclohexyl-PNA" are common terms.

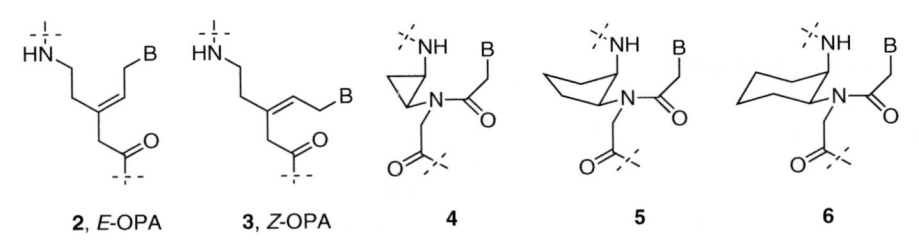

Figure 2. Constrained PNA analogues.

However, the more important and more widely studied constrained analogues of PNA are those formed by connecting two of the three arms of the original PNA with a methylene bridge (Figure 3).[15] Depending on the units connected, PNA analogues **7**, **8** (*ap*-PNA), **9**, and **10** are obtained.[13,16] The connection between the arms can also be done through an ethylene bridge, and pipecolyl-PNA **11** (*pip*-PNA) are then formed. In addition, other analogues are also possible; examples are those in which the lactam moiety has been reduced [e.g., **12** (*aep*-PNA), **13** and **14** (*pyr*-PNA)], compounds with modified functional groups, and those in which the heterocyclic base has been moved to a different position.

The introduction of a pyrrolidine (or piperidine) ring into the backbone of the PNA preorganizes its structure for more favorable and selective nucleic acid recognition and allows incorporation of a basic nitrogen (when lactam moiety is reduced), suitable of being protonated, which contributes to increasing solubility and maintaining the required base orientation.[17,18] The presence of either a carbocyclic ring or a saturated six- or five-membered nitrogen heterocycle gives rise to the

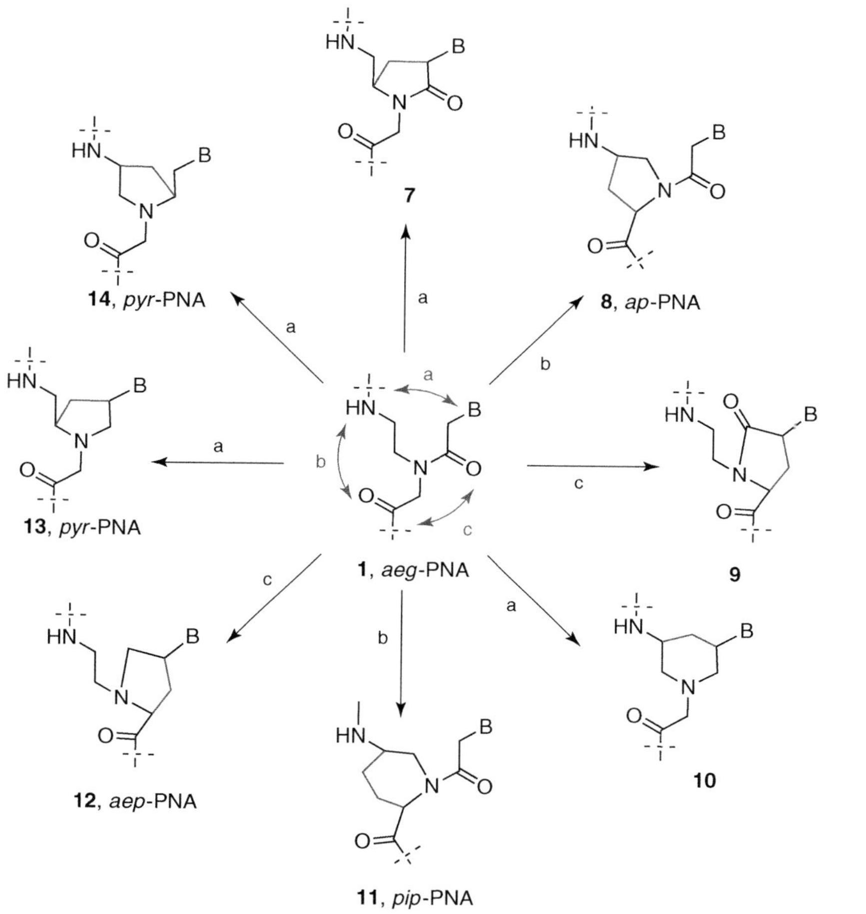

Figure 3. PNA analogues containing saturated nitrogen heterocycles. *(See insert for color representation of the figure.)*

existence of two stereogenic centers. This feature, along with different possible substitution patterns in the ring, lead to a variety of isomers with a wide conformational diversity. To gain direct access to the PNA analogues mentioned above, it is desirable to provide synthetic routes which allow the synthesis of enantiomerically pure monomers starting from easily available precursors and, preferably, leading to both enantiomers. A variety of approaches to the preparation of optically active compounds have been developed so far, and many of them can be applied to the preparation of immediate precursors of PNA monomers either with the aid of starting materials from the chiral pool or through the utilization of chiral auxiliaries. Thus, various advanced methods have been established for some synthetic transformations to constrained analogues, the last step usually being incorporation of the heterocyclic base. Once the monomers are available, the synthesis of the corresponding PNA analogues can be carried out by conventional methods, including automated solid-phase peptide synthesis.[19] In this chapter we highlight the most recent advances in the synthesis of monomers for being incorporated into constrained PNA analogues with a special focus on asymmetric processes leading to enantiomerically pure compounds.

2. OLEFINIC PNA MONOMERS

Oligopeptide arrays of several (up to 10) 3-(2-aminoethyl)-3-pentenoic units bearing a heterocyclic base at C5 were introduced by Leumann et al.[14] and named olefinic peptide nucleic acids (OPAs) (Figure 2, **2** and **3**). On the basis of x-ray crystallography and NMR spectroscopy[20-22] Leumann and co-workers assumed that the double bond efficiently replaced the central amide bond of PNA by fixing rotameric equilibrium of the amide group. Those structural studies indicated that the deleted carbonyl group was not involved in any hydrogen-bond interaction, thus supporting the design of OPAs as suitable PNA analogues. Indeed, these OPAs bind to complementary DNA with affinities similar to those of DNA itself.[14]

The synthesis of the required OPA monomers was designed to be divergent for preparing both (Z) and (E)-OPA monomers from the common precursor **17**. This compound was obtained from intermediate **16** via a Pd-catalyzed coupling with the Reformatsky reagent derived from ethyl α-bromoacetate (Scheme 1).[23] Compound **16** was obtained from commercially available 3-butynol **15** through THP-protection and hydroxymethylation with paraformaldehyde, subsequent conversion into Z-vinyl iodide, and protection of the primary hydroxyl group. Deprotection of the silyl group followed by introduction of protected thymine led to **18**. Removal of the tetrahydropyranyl group and introduction of the azide moiety gave **19**. Finally, ester hydrolysis and hydrogenation over Lindlar catalyst with concomitant Boc introduction furnished the thymine-(Z)-OPA monomer **20**.[23] The adenine (Z)-OPA monomer was also accessible from **17** by a similar route, differing from that of thymine only in the steps in which the base and the terminal amino group were introduced. The corresponding thymine (E)-OPA monomer **24** was prepared by using functional group transformation to reverse the amino acid functionality, thus obtaining **22**.[24] Conversion of **22** into monomer **24** was achieved in five steps and 47% yield (Scheme 1). Compound **21**, protected as monomethoxytrityl at the amino group and having the free carboxylic acid moiety, was suitable for its incorporation into the corresponding PNA analogues through solid-phase peptide synthesis.

The introduction of an additional double bond between the amide group and the nucleobase was reported by Olsen with the aim of designing new PNA analogues able to recognize thymine in a triplex motif through Hoogsteen pairing.[25] Construction of the N-protected formyl heterocycle **28** was achieved from commercially available dimethyl-2-oxoglutarate **25** in four steps and 13.5% overall yield. Wittig-type reaction of **28** afforded **29**, which upon condensation with methyl N-(2-tert-butoxycarbonylaminoethyl)glycinate **30** and basic hydrolysis provided monomer **31** (Scheme 2). Monomer **31** was incorporated into PNA oligomers under

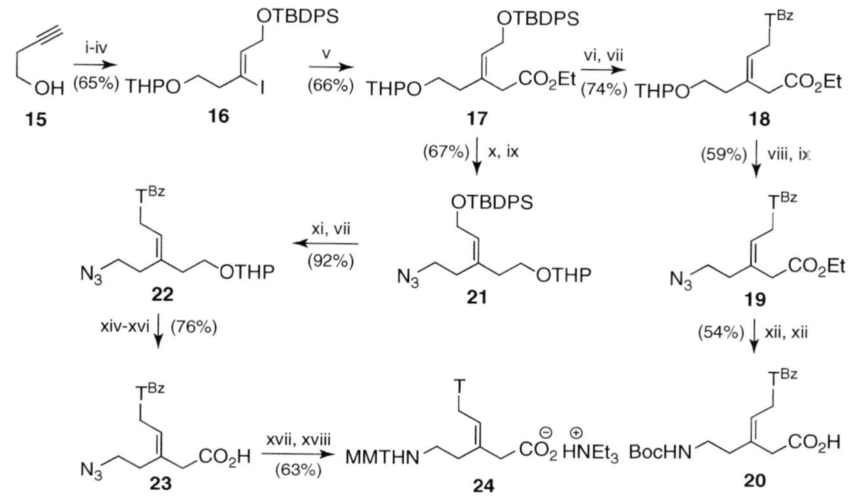

Scheme 1. *Reagents and conditions:* (i) DHP, *p*-TsOH, CH$_2$Cl$_2$. (ii) BuLi, THF, (CH$_2$O)$_n$, rt, 15 h; (iii) Red-Al, THF, 0°C, then NIS, −78°C, 10 min (iv) TBDPSCl, THF, imicazole; (v) BrCH$_2$CO$_2$Et, Zn, CH$_2$(OCH$_3$)$_2$, reflux, 30 min, then 8 mol% Pd(Ph$_3$P)$_4$, DMPU, 4 h, 65°C; (vi) HF, Py, CH$_3$CN; (vii) DEAD, Ph$_3$P, N^{Bz}-thymine, THF, 0°C, 1 h; (viii) *p*-TsOH, EtOH, rt, 3 h; (ix) DIAD, Ph$_3$P, Zn(N$_3$)$_2$·2Py, toluene, THF, 0°C, to rt, 4 h; (x) LiAlH$_4$, THF, 0°C; (xi) TBAF, THF, rt; (xii) LiOH, 0°C, dioxane; (xiii) H$_2$, Lindlar catalyst, MeOH, then Boc$_2$O, Et$_3$N, rt, 40 min; (xiv) *p*-TsOH, MeOH; (xv) Dess–Martin periodinane, CH$_2$Cl$_2$; (xvi) NaOCl$_2$, NaH$_2$PO$_4$, *t*-BuOH; (xvii) Py, Ph$_3$P, then NH$_4$OH (conc.), 2 h; (xviii) MMTCl, DMSO, Et$_3$N.

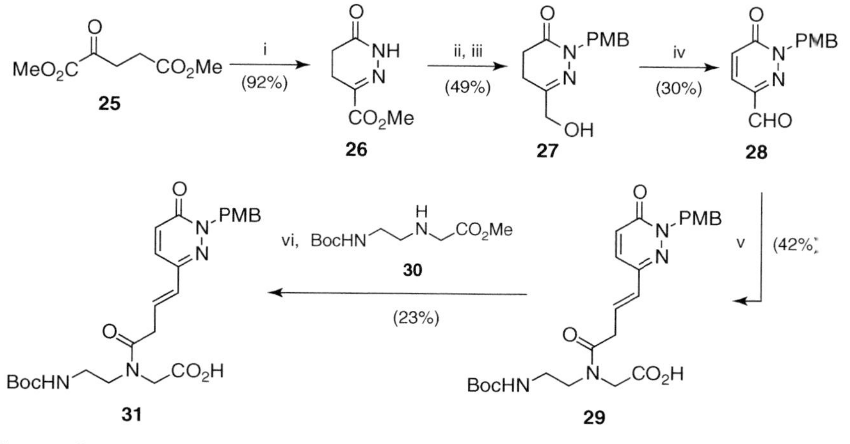

Scheme 2. *Reagents and conditions:* (i) N$_2$H$_4$, AcOH, MeOH, reflux; (ii) *p*-MeOC$_6$H$_4$CH$_2$Cl, NaH, DMF, 0°C; (iii) NaBH$_4$, THF, reflux; (iv) MnO$_2$, toluene, reflux; (v) [Ph$_3$P(CH$_2$)$_2$CO$_2$H]$^+$Br$^-$, NaH, THF, DMSO; (vi) 30, DhbtOH, DCC, DMF, 0°C; then 2 M NaOH, MeOH.

standard solid-phase conditions, which also removed the PMB group.[26] Contrary to the result expected, PNA analogues containing **31** performed more poorly in DNA hybridization assays than did fully flexible analogues, with an almost identical loss of entropy upon triplex formation.

Experimental Procedures

(Z)-Ethyl 5-(tert-butyldiphenylsilyloxy)-3-(2-(tetrahydro-2H-pyran-2-yloxy) ethyl)pent-3-enoate (17).[24] In a dry atmosphere, **16** (16.13 g, 29.3 mmol) was dissolved in dry DMPU (390 mL), then tetrakis(triphenylphosphine)palladium(0) (6.82 g, 5.86 mmol) was added and the mixture was heated to 70°C. In a separate flask, freshly activated zinc powder (10.0 g) was suspended in dry dimethoxymethane (15.0 mL) containing freshly distilled ethyl α-bromoacetate (3.0 mL). The flask was placed in a 50°C oil bath with vigorous stirring until initiation occurred, at which point the reaction was removed from the heat, and with continuous stirring, ethyl α-bromoacetate (12.0 mL) dissolved in dimethoxymethane (80 mL) was added at a rate that maintained a reflux. After addition was complete, the mixture was refluxed for an additional 20 min. The resulting green mixture was then removed from the heat and stirring was stopped. After 20 min the green solution was carefully removed from the zinc deposited and added to the solution of **16** in one batch. After 1 h at 70°C the reaction was quenched by the addition of 100 mL of saturated aqueous NH_4Cl. After a standard aqueous workup, the resulting crude mixture was separated by flash chromatography (SiO_2, 5% ether/hexane to 25% ether/hexane) to give **17** (11.39 g, 22.3 mmol) in 76% yield.

3. CYCLOALKYL PNA MONOMERS

3.1. Cyclohexyl PNA Monomers

The concept of PNA with a conformationally constrained chiral cycloalkyl-derived backbone was pioneered by Nielsen and co-workers.[27] The authors implemented an efficient four-step procedure for the synthesis of cyclohexyl monomers incorporating the four natural nucleobases: thymine, cytosine, adenine, and guanine. The cyclohexane-containing monomers were prepared from 1,2-diaminocyclohexane **32**, commercially available in the (R,R) and (S,S) enantiomeric forms. Monoprotection of **32** followed by alkylation with methyl or ethyl bromoacetate gave rise to **33**, which were coupled with the corresponding N-1-(carboxymethyl)base leading to **34** after base hydrolysis (Scheme 3). Only in the case of thymine was the free base used; for adenine, cytosine, and guanine the Cbz-protected monomers **34** were prepared. Both enantiomeric series were synthesized, and although (S,S)- and (R,R)-PNA bound equally well to complementary achiral PNA strand, only the (S,S)-PNA were found to be well suited to form complexes with DNA or RNA.

The isomeric *cis*-aminocyclohexylglicyl PNAs have been designed by Kumar et al.[28] Synthesis of the required monomers was carried out from a racemic mixture

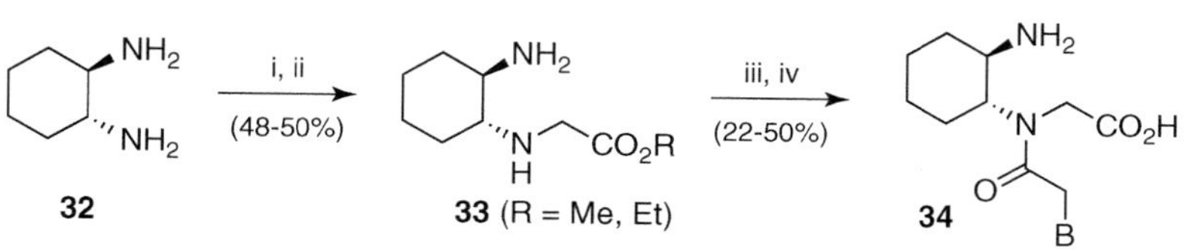

Scheme 3. *Reagents and conditions:* (i) Boc$_2$O, rt, 16 h; (ii) BrCH$_2$CO$_2$R, 0°C, 1 h; (iii) B-CH$_2$CO$_2$H, DCC, rt, 4.5 h; (iv) LiOH, rt, 45 min. B = *N*-Cbz-adenine, *N*-Cbz-cytosine, *N*-Cbz-guanine, thymine.

of *trans*-2-azidocyclohexanoate **36** obtained in two steps from cyclohexene oxide **35** (Scheme 4). Enzymatic resolution with the lipase from *Pseudomonas cepacia* provided enantiomerically pure (1*R*,2*R*)-**37a** and **38**; the latter was hydrolyzed to yield (1*S*,2*S*)-**37b**.

Functional group transformations of **37a** led to **40**, which was transformed into the monoprotected diamine and alkylated with ethyl bromoacetate to provide **41**. Further alkylation with 2-chloroacetyl chloride yielded **42**. Incorporation of the base and basic hydrolysis provided the monomer **43** in eight steps and 29.5% overall yield from **37a**.

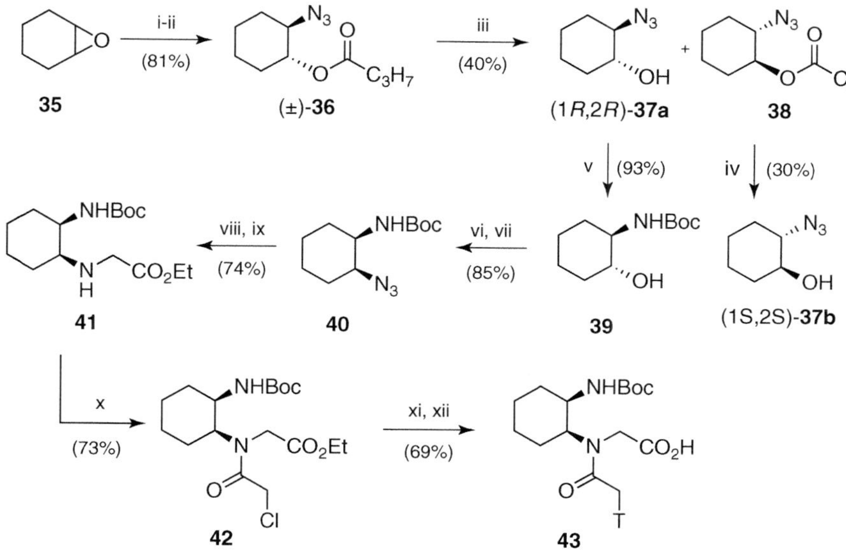

Scheme 4. *Reagents and conditions:* (i) NaN$_3$, NH$_4$Cl, EtOH (aq), reflux; (ii) *n*-butyric anhydride, Py, DMAP, rt; (iii) PCL phosphate buffer, pH 7.2, 2.5 h; (iv) NaOMe, MeOH; (v) H$_2$, PtO$_2$, EtOAc, 35–40 psi, rt; then Boc$_2$O; (vi) MsCl, Py, DMAP, 0°C, 5 h; (vii) NaN$_3$, DMF, 72°C, 18 h; (viii) H$_2$, PtO$_2$, EtOAc, 35–40 psi, rt; (ix) BrCH$_2$CO$_2$Et, CH$_3$CN, rt, 4 h; (x) ClCH$_2$COCl, Na$_2$CO$_3$, dioxane, 0°C, 30 min; (xi) thymine, K$_2$CO$_3$, DMF, 60–65°C, 3.5 h; (xii) LiOH, THF·H$_2$O, 0.5 h.

Following the same reaction sequence, the enantiomer of **43** was obtained from (1*S*,2*S*)-**37b**. Both **43** and *ent*-**43** were incorporated into PNA oligomers by solid-phase synthesis. Measurement of the UV melting temperatures of the modified oligomers with the complementary/mismatched DNA and RNA revealed that whereas **43** preferred to bind RNA, *ent*-**43** showed higher affinity toward DNA. This finding represented a breakthrough in the stereodiscrimination in recognition of DNA and RNA. Indeed, structural x-ray analyses and NMR studies of PNA complexes with DNA and RNA suggested that it might be possible to discriminate between DNA and RNA by tuning the dihedral angle β of the flexible ethylenediamine unit of the PNA.[29] In fact, Kumar et al. demonstrated that PNAs incorporating (1*S*,2*R*)-*cis*-cyclohexanyl thymine monomers such as **43** showed a clear preference to form a duplex with RNA compared with DNA. The differences observed between **43** and other cycloalkyl-containing PNA analogues (see below) were a consequence of the rigidity of the *cis*-cyclohexane ring that resulted in a fixed dihedral angle β of about 65°, more appropriate for RNA selectivity.

3.2. Cyclopentyl PNA Monomers

The same synthetic strategy illustrated in Scheme 4 has also been applied for the preparation of *cis*-cyclopentyl PNA monomers by the same authors.[30] Monoprotected diamine **45** was obtained from cyclopentene oxide **44** in six steps and 53.7% overall yield. Suitable alkylation of the free amino group and deprotection afforded cyclopentyl PNA monomers **46** (Scheme 5). Similarly, *ent*-**46** was also obtained from **44** in 13 steps and 43.5% overall yield.[31]

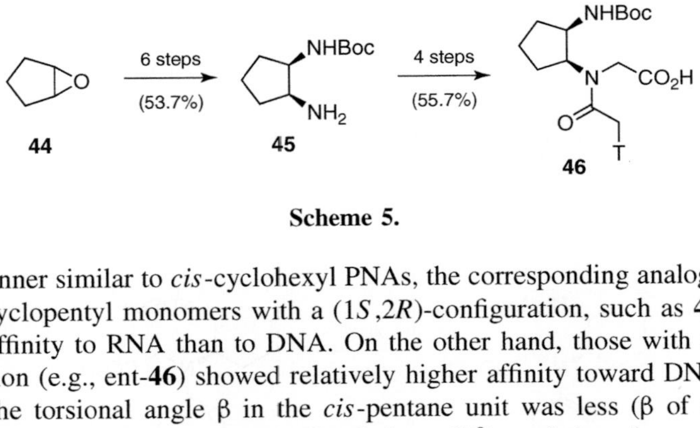

Scheme 5.

In a manner similar to *cis*-cyclohexyl PNAs, the corresponding analogues incorporating cyclopentyl monomers with a (1*S*,2*R*)-configuration, such as **46**, showed a higher affinity to RNA than to DNA. On the other hand, those with a (1*R*,2*S*)-configuration (e.g., ent-**46**) showed relatively higher affinity toward DNA.[31] Interestingly, the torsional angle β in the *cis*-pentane unit was less (β of about 25°) than that in cyclohexane analogues (β of about 65°), and thus the more flexible cyclopentane ring was prone to adopt its conformation through ring puckering complemented by with both DNA and RNA but with stereochemistry-dependent DNA/RNA discrimination. Compared to the cyclopentane ring, the rigidity of the cyclohexane ring resulted in the lower thermal stability of DNA/RNA complexes with cyclohexyl PNA analogues than with parent *aeg*-PNA. On the contrary, the thermal stability of the resulting DNA/RNA complexes with cyclopentyl PNA analogues was increased compared to *aeg*-PNA.[29]

Appella and co-workers reported the synthesis of *trans*-cyclohexyl PNA monomers. The synthesis started by preparing enantiomerically pure **49** through Yamamoto's asymmetric alkylation of (−)-menthyl succinate **47** with 1,3-propanediol ditosylate and further elimination of the chiral auxiliary (Scheme 6).[32] Curtius rearrangement of **49**, followed by trapping the intermediate isocyanate with *tert*-butanol, afforded **50**. Monodeprotection of **50** furnished **51** in only 35% yield, although 60% of **50** was recovered. Attachment of a methylene carboxyethyl unit and coupling with thymine-1-acetic acid after basic hydrolysis provided monomer **53** in eight steps and 16.4% overall yield from **49**. Incorporation of **53** into a typical *aeg*-PNA improved the stability of PNA_2–DNA triplexes and PNA–RNA duplexes for a polyT PNA.[32]

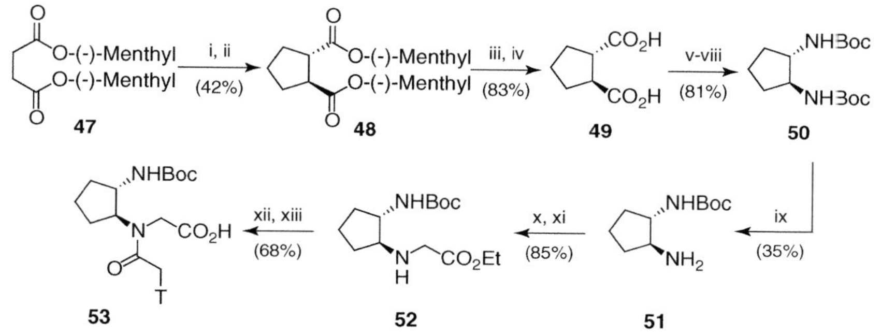

Scheme 6. *Reagents and conditions:* (i) LiTMP, −78°C; (ii) TsO(CH$_2$)$_3$OTs, then terephtalaldehyde; (iii) LiAlH$_4$, THF; (iv) NaOCl, TEMPO; (v) EtOCOCl, Et$_3$N; (vi) NaN$_3$·H$_2$O; (vii) benzene, reflux; (viii) *t*-BuOH, reflux; (ix) TFA (exc.), Et$_2$O; (x) HOC-CO$_2$Et; (xi) H$_2$, Pd-C; (xii) T-CH$_2$CO$_2$H, HATU, *i*-Pr$_2$EtN; (xiii) LiOH, THF, H$_2$O.

The synthetic approach illustrated in Scheme 6 was based on obtaining a chiral nonracemic cyclopentyldiamine with the two amino groups differentiated as a key intermediate. The low yield reactions leading to that intermediate constituted a limitation of the procedure. To increase the availability of cyclopentyl PNA monomers, a new route based on Gellman's synthesis[33] of cyclopentane β-amino acids was developed.[34] By using ethyl 2-oxocyclopentane carboxylate **54**, the orthogonally protected diamine **58** was obtained via enantiomerically pure β-amino acid **56**, prepared through reductive amination, crystallization, and deprotection/protection of the carboxyl and amino groups. Selective deprotection of **58** and consecutive alkylation with methyl α-bromoacetate furnished **59**. Incorporation of the heterocyclic bases followed by basic hydrolysis as described above yielded cyclopentyl PNA monomers **60** bearing thymine and Cbz-protected cytosine and adenine (Scheme 7).

3.3. Cyclopropyl PNA Monomers

An identical synthetic approach to that illustrated in Scheme 6 was employed in preparing *trans*-cyclopropane PNA monomers, the only difference being the formation of the carbocycle, which was carried out with bromochloromethane as an

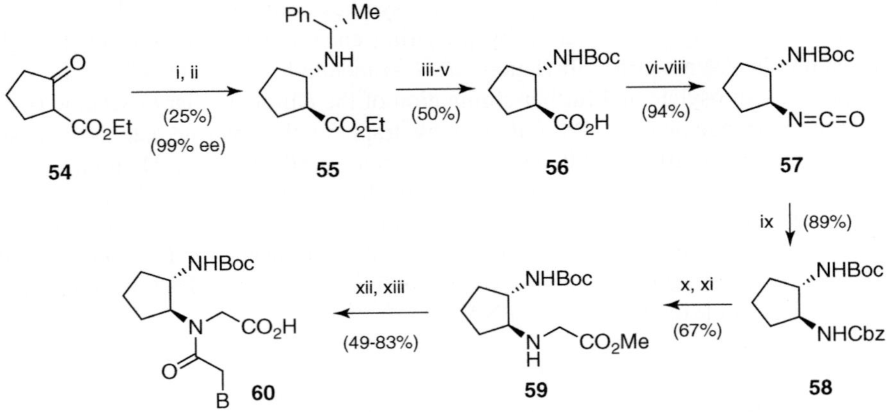

Scheme 7. *Reagents and conditions:* (i) (*R*)-1-phenylethylamine; (ii) NaCNBH$_3$; (iii) LiOH, THF; (iv) H$_2$, Pd-C; (v) Boc$_2$O, dioxane; (vi) ClCO$_2$Et, Et$_3$N, THF, 0°C; (vii) NaN$_3$, H$_2$O; (viii) benzene, reflux; (ix) PhCH$_2$OH, CuCl; (x) H$_2$, Pd-C; (xi) BrCH$_2$CO$_2$Me, Et$_3$N, DMF; (xii) B-CH$_2$CO$_2$H, EDC·HCl, DMAP; (xiii) LiOH, THF·H$_2$O. B = thymine, C^z, A^z.

alkylating agent of (−)-menthyl succinate **47**. Further introduction of a methylene carboxy unit and thymine acetic acid moiety afforded, after ester hydrolysis, the monomer **65** in eight steps and 20.6% overall yield from **62** (Scheme 8).[35] Incorporation of **65** into a PNA oligomer was achieved by employing modified solid-phase peptide synthesis protocols. The *trans*-cyclopropyl PNA formed was shown to be more compatible in the Hoogsteen strand of a PNA$_2$–DNA triplex than was the Watson-Crick strand.

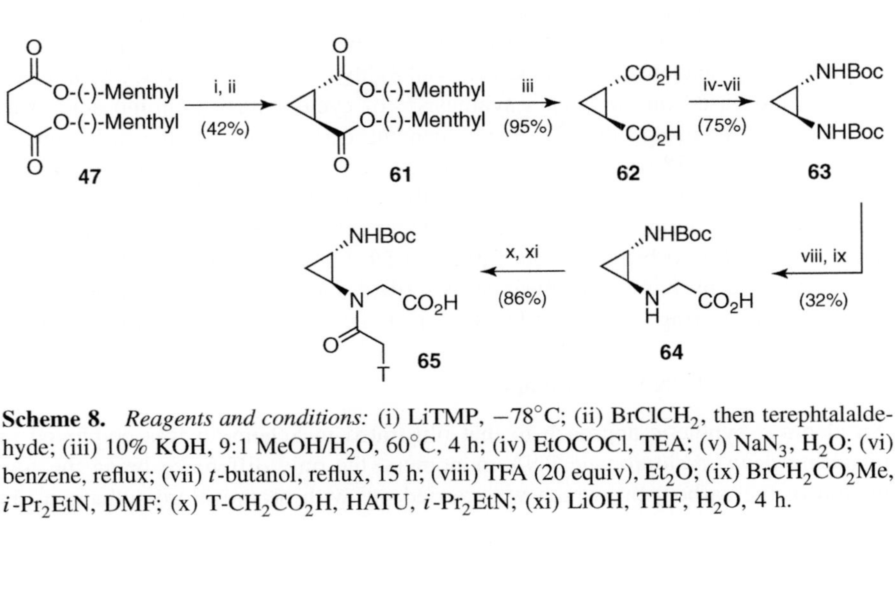

Scheme 8. *Reagents and conditions:* (i) LiTMP, −78°C; (ii) BrClCH$_2$, then terephtalaldehyde; (iii) 10% KOH, 9:1 MeOH/H$_2$O, 60°C, 4 h; (iv) EtOCOCl, TEA; (v) NaN$_3$, H$_2$O; (vi) benzene, reflux; (vii) *t*-butanol, reflux, 15 h; (viii) TFA (20 equiv), Et$_2$O; (ix) BrCH$_2$CO$_2$Me, *i*-Pr$_2$EtN, DMF; (x) T-CH$_2$CO$_2$H, HATU, *i*-Pr$_2$EtN; (xi) LiOH, THF, H$_2$O, 4 h.

Experimental Procedures

N-[(2 R)-tert-*Butyloxycarbonylaminocyclohex-(1 S)-yl]*-N-*(thymin-1-acetyl)* glycine (43).[28] A mixture of chloro compound **42** (1.5 g, 4.0 mmol), thymine (0.55 g, 4.38 mmol), and anhydrous K_2CO_3 (0.66 g, 4.8 mmol) in dry DMF (10 mL) under a nitrogen atmosphere was heated with stirring at 65°C for 3.5 h. After cooling, the solvent was removed under reduced pressure to leave a residue, which was extracted into CH_2Cl_2 (2 × 25 mL) and dried over Na_2SO_4. The solvent was evaporated, and the crude compound was purified by column chromatography (MeOH/CH_2Cl_2) to afford a white solid of thymine monomer ethyl ester, which was suspended in THF (8 mL), to which was added a solution of 0.5 M LiOH (7.7 mL, 3.9 mmol), and the mixture was stirred for 30 min. The mixture was washed with EtOAc (2 × 10 mL). The aqueous layer was acidified to pH 3 and extracted with EtOAc (3 × 50 mL). The EtOAc layer was dried over sodium sulfate and evaporated under reduced pressure to afford monomer **43** as a white solid: yield 0.71 g (69%); mp 172–177°C; $[\alpha]_D^{20}$ −107° (c 1.06, CH_2Cl_2). ^{1}H-NMR (200 MHz, $CDCl_3$): δ 0.5–1.8 (17H, m, *t*-Boc, 4CH_2), 1.88 (3H, s, thymine CH_3), 3.3–5.5 (7H, m, N-CH_2, acyl CH_2, 1-CH, 2-CH, carbamate NH), 7.18 (1H, s, thymine-H), 10.0–11.0 (2H, bd, thymine NH and −COOH).

N-*[(2 S)-Boc-aminocyclopent-(1 S)-yl]*-N-*(thymin-1-yl-acetyl)*-glycine (53).[32] (−)-(1*S*,2*S*)-*N*-Mono-(tert-butoxycarbonyl)cyclopentanediamine **51** (156 mg, 0.78 mmol) was concentrated in a 50-mL round-bottomed flask and dried under vacuum overnight prior to being used in the condensation reaction. Once **51** was dry, EtOAc (10 mL) was added to the reaction flask. Ethylglyoxylate as an 8:1 mixture in toluene (80 μL, 0.78 mmol) was added directly to the reaction solution and stirred for 15 h. Next, the solution was transferred to a Parr reaction vessel, washing the reaction flask with EtOAc (2 mL), and 10% Pd/C (25 mg) was added. The reaction vessel was placed on the Parr, purged three times with H_2(g), and set to shake under 30 psi for 10 h. Once the imine reduction was complete, the reaction solution was filtered through a 50-mL glass fritted funnel, containing 40 mL of Celite, concentrated on a rotary evaporator, and dried under vacuum to obtain a light orange oil. The product was purified by silica gel chromatography, eluting with 100% EtOAc. The fractions containing the desired product, with an R_f of 0.4, were combined, concentrated, and dried under vacuum to give 192 mg (86 % yield) of intermediate glycine ester **52** as a clear viscous oil. Then 102 mg (0.36 mmol) of **52** was dissolved in DMF (12 mL). Next, HATU (220 mg, 0.58 mmol) and thymine-1-acetic acid (131 mg, 0.71 mmol) were added to the reaction mixture. The reaction flask was cooled to 0°C via an ice bath, and *i*-Pr$_2$EtN (186 μL, 1.07 mmol) was added slowly over a 30-min period. The reaction was allowed to warm to room temperature gradually over 8 h. The reaction solution was stirred for a total of 24 h, then transferred to a separatory funnel, washing the reaction flask with EtOAc (50 mL). Next, the organic mixture was washed with a sequence of aqueous solutions: saturated NaCl (50 mL), saturated NaHCO$_3$ (2 × 50 mL), and saturated NaCl (50 mL). The layers were separated and the combined aqueous washes were extracted with EtOAc (4 × 80 mL). All organic aliquots were combined, dried over Na_2SO_4, filtered, via suction filtration, and concentrated on a rotary evaporator to a viscous yellow oil. The product was purified by silica gel chromatography, eluting with 19:1 $MeOH/CH_2Cl_2$. The fractions containing the desired product, with an R_f of 0.3, were combined, concentrated, and dried under vacuum to give 129 mg (80% yield) of the protected monomer as a solid white foam. Then 440 mg (0.97 mmol) of the product above were dissolved in THF (15 mL). The solution was cooled in an ice bath to 0°C. Next, a 1 N aqueous solution of $LiOH·H_2O$ (549 mg, 13.0 mmol) was prepared and added dropwise, via syringe, to the reaction solution over a 5-min period. The ice bath was removed and the solution stirred for 5 h. The basic solution was transferred to a separatory funnel, washing the reaction flask with H_2O (20 mL) and Et_2O (20 ml), and extracted with Et_2O (3 × 40 mL). The remaining aqueous layer was acidified to pH 1 with 3 N HCl and extracted with EtOAc (5 × 50 mL). The organic layers were combined, dried over Na_2SO_4, and filtered by suction filtration. The solution was concentrated on a rotary evaporator and dried under vacuum to give 350 mg of **53** (overall: 58% from **51**) as a brittle white solid: Decomposition occurs at 140°C; ^{1}H-NMR (DMSO-d^6, 500 MHz): major rotamer δ 12.39 (br s, 1H, COOH), 11.29 (s, 1H, imide NH), 7.16 (s, 1H, H_3CCCH), 6.95 (d, J = 8.0 Hz, 1H, carbamate-NH), 4.82 (AB, J = 17 Hz, 1H, thymine-CH_2), 4.60 (AB, J = 17 Hz, 1H, thymine-CH_2), 3.94 (AB, J = 17 Hz, 1H, $COOHCH_2$), 3.68 (AB, J = 17 Hz, 1H, $COOHCH_2$), 1.75 (s, 3H, thymine-CH_3), 1.36 (s, 9H, t-butyl-CH_3), 1.2–2.0 (m, 8H, cyclopentane-CH_2). Minor rotamer δ 11.26 (s, 1H, imide-NH), 6.76 (d, J = 8.0 Hz, 1H, carbamate-NH).

4. PYRROLIDINE-BASED PNA MONOMERS

4.1. Pyrrolidin-2-ones and Pyrrolidines

Pyrrolidinyl PNA analogues are a dominant substance class in the area of conformationally restricted PNAs. The synthesis of the first pyrrolidinone PNA analogue (Figure 3, **7**) has been reported by connecting the aminoethylglycine backbone and the methylene carbonyl linker through a methylene bridge. Structural analyses revealed that the linker-carbonyl points toward the carboxyl end of PNA in PNA–PNA, PNA–DNA, and PNA–RNA duplexes and PNA_2–DNA triplexes.[21,22,36,37] Hence, the formation of a pyrrolidine bearing a heterocyclic base at C3 should prevent rotation around the methylene carbonyl linker and thus facilitating recognition of the single -stranded PNA by the complementary strands (DNA, RNA, or PNA). Nielsen and co-workers reported[38] the synthesis of *cis*- and *trans*-isomers starting from compound **67**, easily available from D-pyroglutamic acid **66** in five steps and 49% overall yield. After enolate oxidation of **67** to give **68**, protection, deprotection, and introduction of the acetate moiety at the nitrogen atom yielded **69**. Desilylation, mesylation, and subsequent azide substitution furnished the intermediate **70**. Double hydrogenation of **70** to deprotect the hydroxyl group and to reduce the azido group afforded **71**, which was transformed into monomer **72** with (3*R*,4*R*)-configuration through introduction of the base and

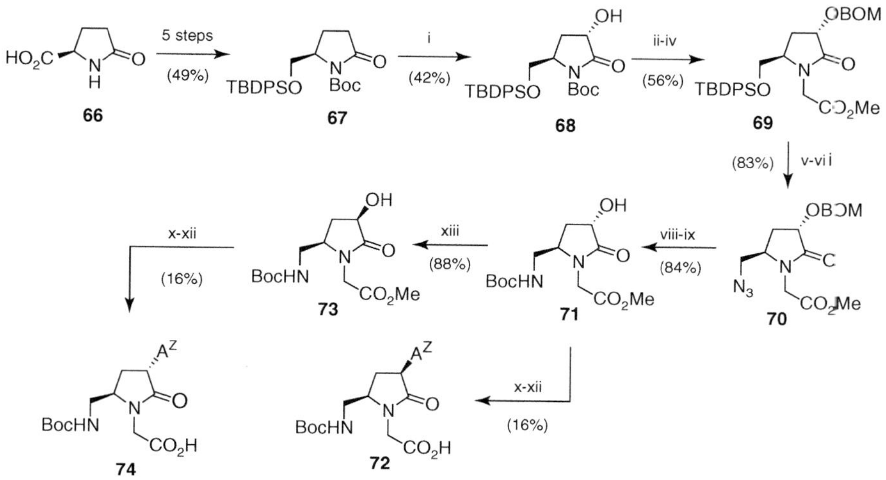

Scheme 9. *Reagents and conditions:* (i) LiHMDS, THF, then MoOPH; (ii) BOMCl, i-Pr$_2$EtN; (iii) TFA, CH$_2$Cl$_2$, 0°C; (iv) NaH, BrCH$_2$CO$_2$Me; (v) HF, Et$_3$N; (vi) MsCl, Py; (vii) NaN$_3$, DMF; (viii) H$_2$, Pd-C, Boc$_2$O; (ix) H$_2$, Pd(OH)$_2$-C, MeOH; (x) adenine, Ph$_3$P, DEAD, dioxane; (xi) N-Cbz-N-Me imidazolium triflate; (xii) LiOH, THF; (xiii) PhCO$_2$H, DEAD, Ph$_3$P, dioxane, then NaOMe, MeOH.

hydrolysis of the ester group. Compound **72** was obtained in 12 steps and 2.62% overall yield from compound **67** (Scheme 9). The (3S,4R)-monomer **74** was prepared by inverting configuration at C3 in **71** and following the same reaction sequence. By this way, compound **74** was obtained in 13 steps and 2.31% overall yield from compound **67**. Similarly, (3S,5S)- and (3R,5S)-isomers were also prepared from L-pyroglutamic acid.

Essentially the same compounds were recently prepared in a more efficient way through an asymmetric 1,3-dipolar cycloaddition.[39] Condensation of Oppolzer's sultam acrylate **78** with the nitrone **77**, generated in situ from commercially available aldehyde **75** and hydroxylamine **76**, afforded **79** in 75% yield as an only isomer (ds > 98%). Hydrogenation of isoxazolidine **79** allowed recovery of the chiral auxiliary, and pyrrolidinone **80** was obtained. Introduction of the base moiety under the usual Mitsunobu-type conditions yielded the PNA monomer (3S,5R)-**81** in only three steps and 55.8% overall yield from **75**. Following Nielsen's methodology,[38] compound **80** was converted into **82** and the PNA monomer (3R,5R)-**83** was obtained in four steps and 30.5% overall yield from **75** (Scheme 10).

In their original paper, Nielsen and co-workers demonstrated that the (3S,5R)-isomer **74** had the highest affinity toward RNA, and it recognized RNA, and PNA better than DNA. A fully modified PNA-decamer prepared with **74** bound to rU$_{10}$ with only a slight decrease in T_m (ca. 1°C) relative to parent *aeg*-PNA.[38] On the other hand, a larger destabilization ($\Delta T_m = -3.5$°C) was observed against complementary RNA.

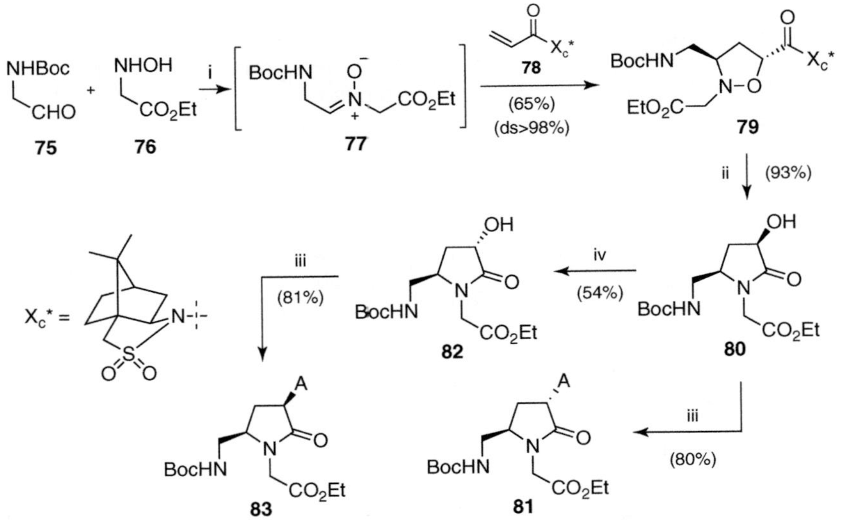

Scheme 10. *Reagents and conditions:* (i) toluene, sealed tube, 60°C, 18 h; (ii) H_2, $Pd(OH)_2$-C, 2000 psi; (iii) adenine, Ph_3P, DIAD, rt, 16 h; (iv) $PhCO_2H$, DEAD, Ph_3P, then NaOEt, EtOH.

In the search for analogues more soluble at physiological pH, Micklefield and co-workers reported the synthesis of a pyrrolidine–amide oligonucleotide mimic in which the lactam moiety connecting the heterocyclic base and the peptide backbone had been reduced completely (Figure 3, **13**). The synthesis of the required monomers is shown in Scheme 11. It started from ester **85**, prepared from *trans*-.4-hydroxy-L-proline **84** in eight steps and 15.3% overall yield, according to the procedure previously reported by Lowe and Vilaivan.[40] Reduction of **85** and introduction of the phtalimido group via Mitsunobu reaction provided intermediate **86**. Selective removal of protecting groups in **86** provided free amines **87** and **88**, which were alkylated to give compounds **89** and **90**, respectively (Scheme 11). Construction of the PNA analogue was carried out through two alternative routes consisting of coupling either **87** with **90** or **88** with **89**. The coupling steps were repeated to give a pentamer, which showed sequence-specific recognition to both DNA and RNA with much higher affinity than native nucleic acids and with faster binding to RNA and DNA.[41] Starting from **86**, the monomer **92** was also prepared to be used in Fmoc solid-phase peptide chemistry.[42]

The corresponding adenine PNA monomer **76** related to **73** was also prepared and employed in the formation of PNA analogues that exhibited higher affinity for complementary DNA and RNA than that of the corresponding adeninyl PNA.[43] The synthesis of the adenine monomer started from **93**, easily accessible from *trans*-4-hydroxy-L-proline **84**. Deprotection of **93**, alkylation to give **94**, and inversion at C4 via Mitsunobu reaction afforded intermediate **95**, which was further transformed into **98** by functional group transformation, Fmoc protection, and acid deprotection (Scheme 12).

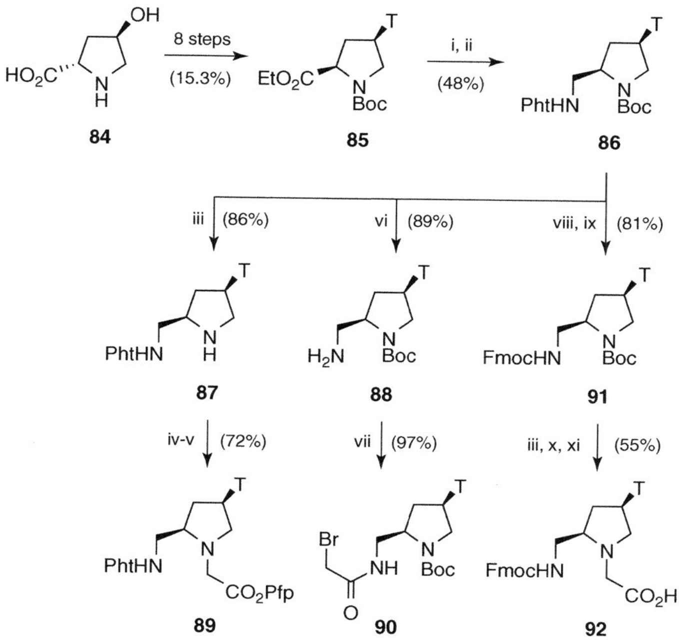

Scheme 11. *Reagents and conditions:* (i) LiBH$_4$, THF, 0°C, 15 h; (ii) phtalimide, Ph$_3$P, DEAD, THF, −15°C to rt, 15 h; (iii) TFA, CH$_2$Cl$_2$, 4 h; (iv) BrCH$_2$CO$_2$-t-Bu, i-Pr$_2$EtN, DMF; (v) TFA, CH$_2$Cl$_2$, then CF$_3$CO$_2$Pfp, Py, DMF, 2 h; (vi) MeNH$_2$, 40°C, 1 h; (vii) (BrCH$_2$CO)$_2$O, CH$_3$CN, CH$_2$Cl$_2$, −8°C to rt, 5 min; (viii) MeNH$_2$, 50°C, 2 h; (ix) FmocCl, Na$_2$CO$_3$, dioxane·H$_2$O; (x) BrCH$_2$CO$_2$-t-Bu, i-Pr$_2$EtN, CH$_2$Cl$_2$; (xi) 4 M HCl, dioxane.

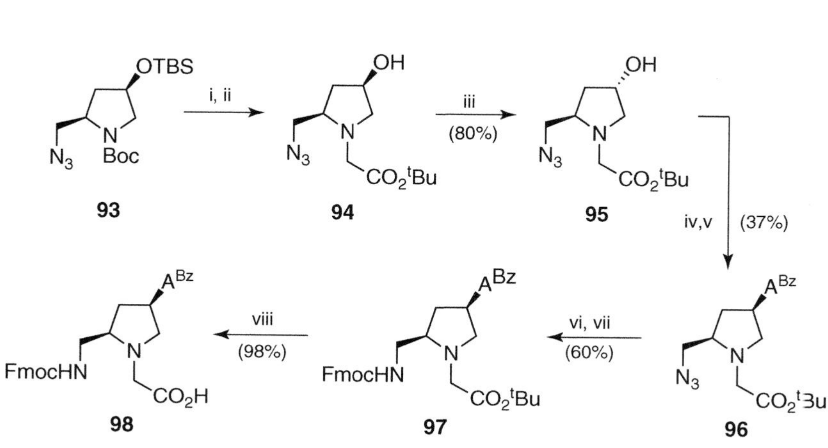

Scheme 12. *Reagents and conditions:* (i) 4 M HCl-dioxane; (ii) BrCH$_2$CO$_2$-t-Bu, i-Pr$_2$EtN, CH$_2$Cl; (iii) HCO$_2$H, Ph$_3$P, DIAD, THF, then NH$_3$, MeOH, rt, 2 h; (iv) TsCl, pyridine, 0°C to rt, 18 h; (v) N^6-benzoyladenine, K$_2$CO$_3$, 18-crown-6, DMF, 80°C; (vi) H$_2$S, Py, H$_2$O, rt, 18 h; (vii) FmocCl, i-Pr$_2$EtN, CH$_2$Cl$_2$; (viii) 4 M HCl-dioxane, CH$_2$Cl$_2$, rt, 24 h.

An approach similar to **92** but protected as *N*-Boc, was reported by Kumar et al. starting from both enantiomers of *trans*-4-hydroxy-L-proline.[44] The corresponding (2*S*,4*S*)- and (2*R*,4*R*)-*cis* monomers were prepared and incorporated into PNA–DNA chimera and PNA analogues.

The reduced form of monomer **72** was prepared from *N*-Boc-*cis*-4-hydroxy-D-proline **99**.[45] After various functional group transformations, compound **99** was transformed into azide **101**. N-Alkylation and transformation of the azido group into the required NHBoc moiety furnished intermediate **103**. Introduction of the heterocyclic base was not carried out through the classical Mitsunobu-type reaction but via tosyl intermediate **104**, which underwent a typical S_N2 reaction. Deprotection of the ester moiety afforded PNA monomer **105** in 11 steps and 2.76% overall yield from **99** (Scheme 13). Similar to compound **72**, incorporation of the pyrrolidine analogue **105** into the PNA strand resulted in destabilization against DNA and RNA as compared to parent *aeg*-PNA. Greater larger destabilization was found for **105** compared to **72**.[45] On the other hand, a fully modified decamer constructed only with **105** formed a 1:2 complex with DNA (DNA$_2$–PNA triplex) in which the PNA is the Watson–Crick strand. Naturally occurring *trans*-4-hydroxy-L-proline **84** was employed to prepare the enantiomer of **105** with thymine as a base. A double Mitsunobu reaction served to preserve the *trans*-configuration of the starting material.[46]

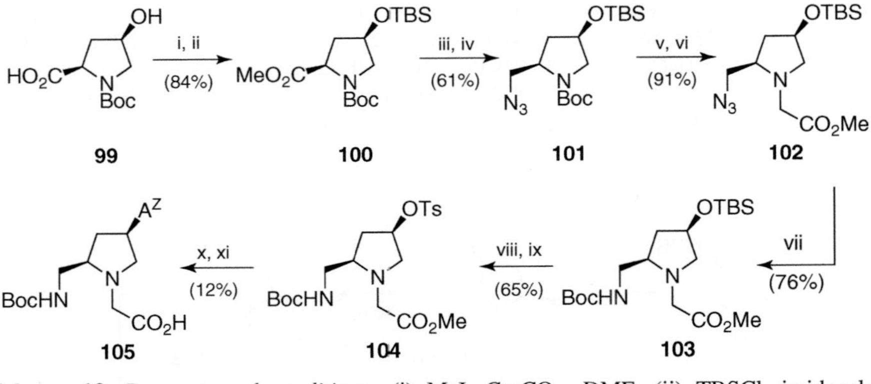

Scheme 13. *Reagents and conditions:* (i) MeI, Cs$_2$CO$_3$, DMF; (ii) TBSCl, imidazole, *i*-Pr$_2$EtN, DMF; (iii) LiBH$_4$, THF; (iv) MsCl, *i*-Pr$_2$EtN, then NaN$_3$, DMF; (v) TFA, CH$_2$Cl$_2$; (vi) BrCH$_2$CO$_2$Me, *i*-Pr$_2$EtN, THF; (vii) H$_2$, 10% Pd-C, Boc$_2$O, EtOAc; (viii) TBAF, THF; (ix) TsCl, Py; (x) A^z, DMF, then Rappoport's reagent; (xi) Ba(OH)$_2$, then H$_2$SO$_4$.

In 2002, Kumar et al. introduced a novel constraint in PNA by connecting the reduced carbonyl group (actually, a methylene) of the nucleobase linker and the β-carbon of the aminomethyl unit of parent *aeg*-PNA.[47] Compound **106**, readily available from *trans*-4-hydroxy-L-proline **84**, was used as the starting material. Reduction of the ester group, introduction of the azide unit, and incorporation of the heterocyclic base afforded intermediate **108**. Further deprotection of the amino group, and alkylation with ethyl 2-bromoacetate furnished **109**, which was converted into PNA monomer **110a** with a (2S,4S)-configuration through conversion of the azide into a NHBoc group and deprotection of the ester. Following the same reaction sequence, the (2R,4S)-isomer **110b** was prepared from protected and mesylated *cis*-4-hydroxy-D-proline **111** (Scheme 14).

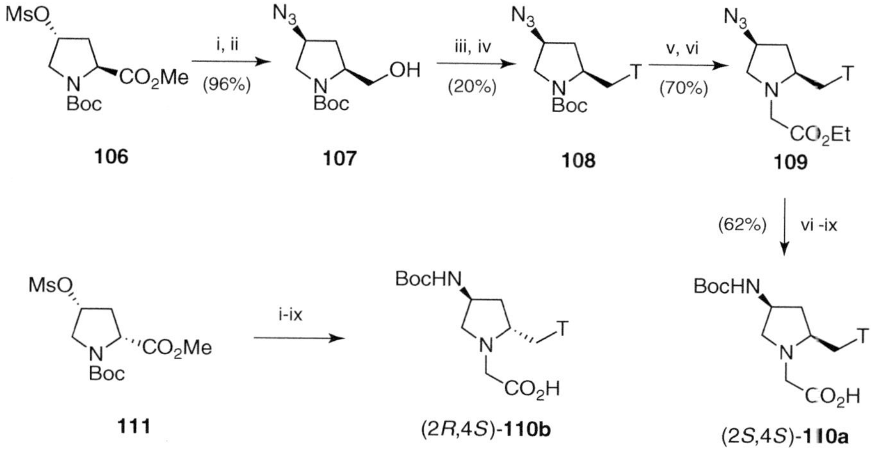

Scheme 14. *Reagents and conditions:* (i) LiBH$_4$, NaBH$_4$; (ii) NaN$_3$, THF; (iii) MsCl, Py; (iv) thymine, K$_2$CO$_3$, 18-crown-6, DMF, 75°C; (v) 50% TFA, CH$_2$Cl$_2$; (vi) BrCH$_2$CO$_2$Et, K$_2$CO$_3$, CH$_3$CN; (vii) H$_2$, Pd-C, MeOH; (viii) BocN$_3$, dioxane-H$_2$O; (ix) 1 N NaOH.

Both **110a** and **110b** were incorporated into *aeg*-PNA sequences. Whereas that containing **110b** showed considerably increased stability in the formation of a PNA$_2$–DNA triplex compared with the control *aeg*-PNA ($\Delta T_m = +16°C$) the triplex melting temperature is much lower ($\Delta T_m = -14°C$) for the PNA oligomer containing **110a**.[48] These results indicated that the binding efficiency of PNA analogues incorporating **110** is dictated by the stereochemistry at C2, in clear contrast with other PNA analogues, in which such an effect was not observed (see below). The methodology outlined in Scheme 14 is also applicable to the preparation of PNA analogues with different bases, including cytosine, adenine, and guanine.[48]

One-carbon-extended PNA monomers have been prepared from *N*-Cbz-*trans*-4-hydroxy-L-proline **112**.[49] Transformation of the carboxyl group into the NBoc aminomethyl group afforded **114**. Deprotection and Micahel addition of **114** to ethyl acrylate served to introduce the extended backbone, and **115** was obtained. Further incorporation of the base and deprotection of the carboxyl group furnished the backbone-extended pyrrolidine PNA (*bep*-PNA) monomer **116** (Scheme 15).[50]

PNA analogues incorporating **116** showed preferred binding selectivity for RNA over DNA in triplex as well as duplex modes.[50] A similar PNA monomer with an extended backbone was used to prepare a PNA–DNA dimer block to be incorporated in DNA sequences at selected positions.[51]

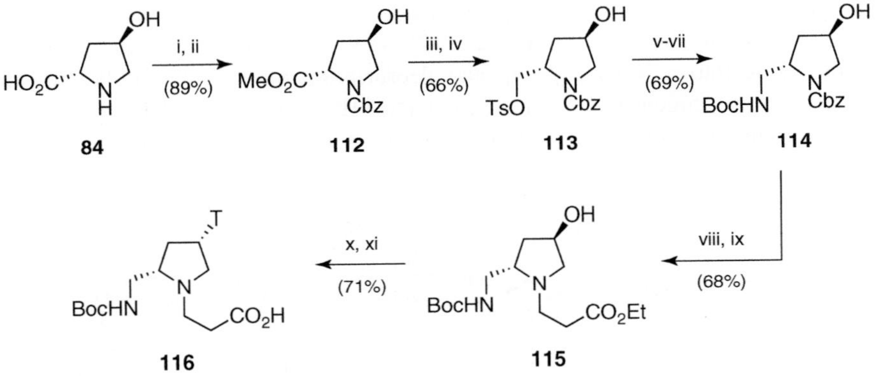

Scheme 15. *Reagents and conditions:* (i) CbzCl (50% toluene soln), Et$_3$N, NaHCO$_3$, H$_2$O, rt, 8 h; (ii) SOCl$_2$, Et$_3$N, MeOH, rt, 7 h; (iii) LiCl, NaBH$_4$, 4:3 EtOH-THF, rt, 7 h; (iv) TsCl, Py, rt, 7 h; (v) NaN$_3$, DMF, 70°C, 8 h; (vi) Raney Ni, H$_2$, MeOH, 35 psi, rt, 3 h; (vii) BocN$_3$, DMSO, 50°C, 5 h; (viii) H$_2$, Pd–C, MeOH, 60 psi, rt, 7 h; (ix) ethyl acrylate, MeOH, rt, 2.5 h; (x) N^3-benzoylthymine, DEAD, Ph$_3$P, benzene, rt, 4 h; (xi) 2 M aq. NaOH, rt, 5 h, Dowex H$^+$.

4.2. 4-Aminoprolines

One of the first approaches to introducing conformational restrictions to PNA was described by Kumar et al. in 1996.[52] These authors reported bridging the ethylenediamine and glycine arms by a methylene group (Figure 3, connection b), leading to PNA analogues with a 4-aminopropyl unit with chirality at C2 and C4 (Figure 3, **8**). To synthesize the corresponding monomer-protected azide **117**, accessible from *trans*-4-hydroxy-L-proline **84** in four steps, it was transformed into *N*-alkylated derivative **118**. Condensation of **118** with a suitable base (thymine, adenine, cytosine, and 2-amino-6-chloropurine), followed by conversion of the azido group into NHBoc and further deprotection of the ester, afforded monomer **120** (Scheme 16). A very similar approach to **120** also starting from **117** has been reported with small variations in the methodology.[53]

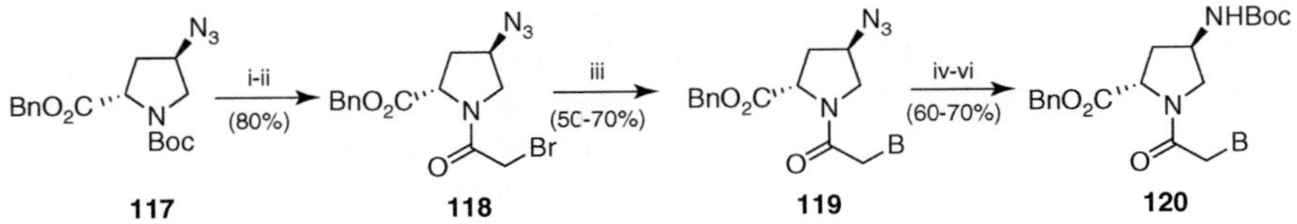

Scheme 16. *Reagents and conditions:* (i) TFA, CH$_2$Cl$_2$; (ii) BrCH$_2$COCl, Na$_2$CO$_3$; (iii) base, K$_2$CO$_3$, or NaH; (iv) HCO$_2$NH$_4$, Pd-C; (v) BocN$_3$, Na$_2$CO$_3$; (vi) 1 M NaOH. B = thymine, adenine, cytosine, 2-amino-6-chloropurine.

The other three diastereomers were also prepared and incorporated into standard PNA chains both at the N-terminus and in the interior to generate chiral nonracemic

PNA analogues.[54] Complementation studies with DNA showed that PNA analogues modified with **120** led to stabilization on the PNA–DNA hybrid.

Novel pyrrolidine carbamate PNA analogues have been described by Meena and Kumar.[55] In these analogues the peptide linkage was replaced with a carbamate linkage, which although longer than the amido linkage, is shorter than a phosphodiester linkage. To prepare carbamate-linked oligomers, compound **124** is required. The synthesis of **124** was done from protected 4-aminoproline **121**, available from protected *trans*-4-hydroxy-L-proline **112** in four steps and an 80% overall yield (Scheme 17). Reduction of the ester and deprotection of the Cbz group followed by alkylation with thymine-1-acetic acid or N^4-Cbz-cytosine-1-acetic acid furnished **123** in good yield. Activation of **123** into **124** was carried out with *p*-nitrophenylcarbonate in the presence of triethylamine. Condensation of **124** with *N*-Boc deprotected **123** in the presence of Hünig's base furnished carbamate–PNA dimers.

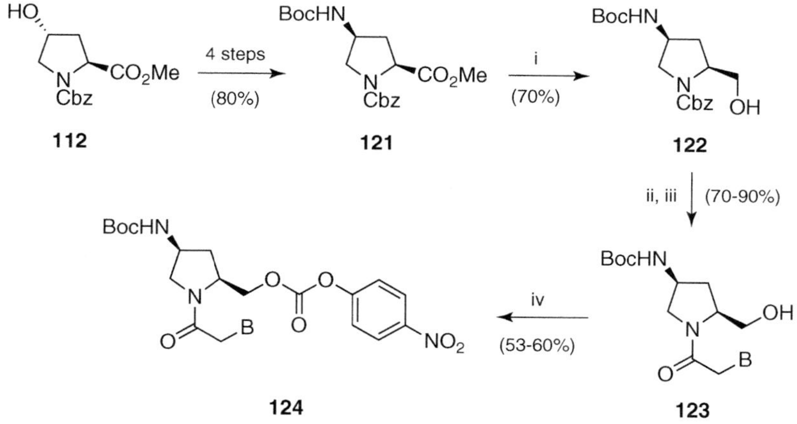

Scheme 17. *Reagents and conditions:* (i) LiBH$_4$, THF; (ii) H$_2$, Pd-C, MeOH; (iii) B-CH$_2$CO$_2$H, DCC, HOBT, DMF; (iv) 4-nitrophenoxycarbonyl chloride. B = thymine, N^4-Cbz-cytosine.

The procedure was amenable to use in solid-phase synthesis methodologies to prepare modified pyrrolidine carbamate PNA analogues **125** incorporating **124** at predetermined positions of *aeg*-PNA (Figure 4). These PNA carbamate analogues were found to have poor water solubility. It was also found that homooligomers and modified PNA in the center of the chain did not bind to DNA. On the other hand, whereas the placement of carbamate linkage at the N-terminus of *aeg*-PNA had no effect on the stability of triplexes ($\Delta T_m = -0.5°C$), modification at the C-terminus caused a large destabilization of the complex ($\Delta T_i = -13°C$). Other attempts to replace the amide linkage were made by Efimov and co-workers, who reported the preparation of phosphonopyrrolidine PNA (*pHyp*-PNA) **126**.[56] Analogues **126** are negatively charged conformationally constrained PNA with a good water solubility and biological stability.[57] They showed high binding affinity to complementary

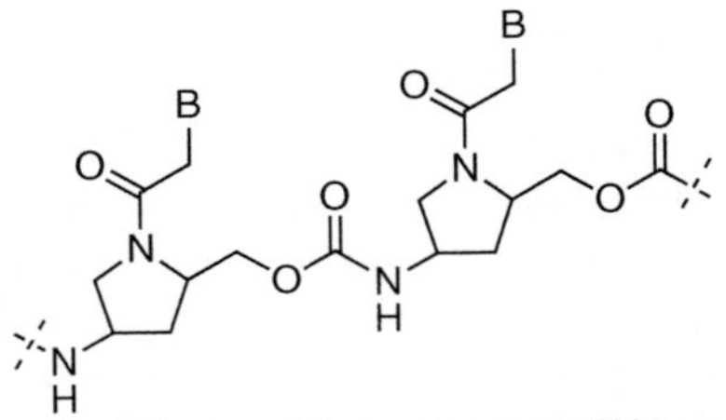

125, pyrrolidine-carbamate PNA

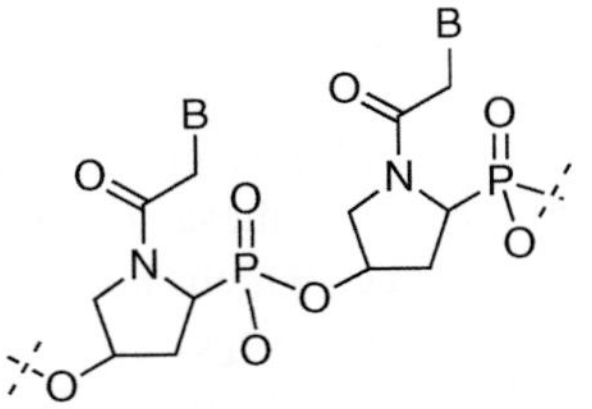

126, pHyp-PNA

Figure 4. PNA analogues with nonamide linkages.

DNA and RNA strands,[58] and their use as in vivo translation inhibitors in zebrafish embryos has been studied.[59] The efficacy of *pHyp*-PNA **126** has also been demonstrated in a cell-free system, in living cells, and in a living model organism.[60]

4.3. *N*-Aminoethylprolines

In 1999, Kumar et al. reported replacement of the glycyl component in PNA by a prolyl unit bearing a base at C4 (Figure 3, connection c). The corresponding aminoethylprolyl PNA analogues (*aep*-PNA, Figure 3, **12**) had a positive charge at physiological pH and form stable triplexes with DNA.[61]

Synthesis of the required monomers for *aep*-PNA was straightforward from the protected *trans*-4-hydroxy-L-proline **127**. Alkylation of the nitrogen, introduction of the base (thymine), and deprotection afforded monomer **129** in 20.4% overall yield (Scheme 18). The *trans*-monomer **131** was obtained in the same way from **130**, which was easily accessible from *trans*-4-hydroxy-L-proline **84**.[52] Both (2*S*,4*S*)-**129** and (2*R*,4*S*)-**131** where used in the preparation of modified PNA oligomers through solid-phase synthesis on Merrifield resin. Interestingly, both diastereomers led to a considerable stabilization of PNA$_2$–DNA triplexes ($\Delta T_m =$ +5–8°C). The introduction of other bases, such as adenine, guanine, and cytosine, was also possible through a typical S$_N$2 reaction from a mesylated intermediate.[62] It was found that incorporation of single (2*S*,4*S*)-isomers with A/T/C/G bases in *aeg*-PNA stabilized the antiparallel duplexes with DNA. Moreover, incorporation of (2*R*,4*S*)-isomer **131** resulted in interesting binding preferences with complementary DNA. A (2*S*,4*S*)-isomer with N^7-substituted guanine influenced the recognition of complementary DNA in an orientation-selective manner during the formation of triplexes.[63] NMR studies demonstrated that the differences observed between thymine and guanine-containing *aeg*-PNA were due to the different pyrrolidine pucker, which was strongly dependent on the effect of pyrimidine or purine substituents.[64] Moreover, guanine containing *aep*-PNA enhanced the stability of G-quadruplexes under various conditions of pH, salt, and metal ions K$^+$ and Na$^+$.[65]

In a novel approach to conformationally constrained pyrrolidine-containing nucleic acid analogues, Lowe and co-workers illustrated the preparation of a new class of chiral PNA **132** in which the sugar phosphate backbone of DNA was

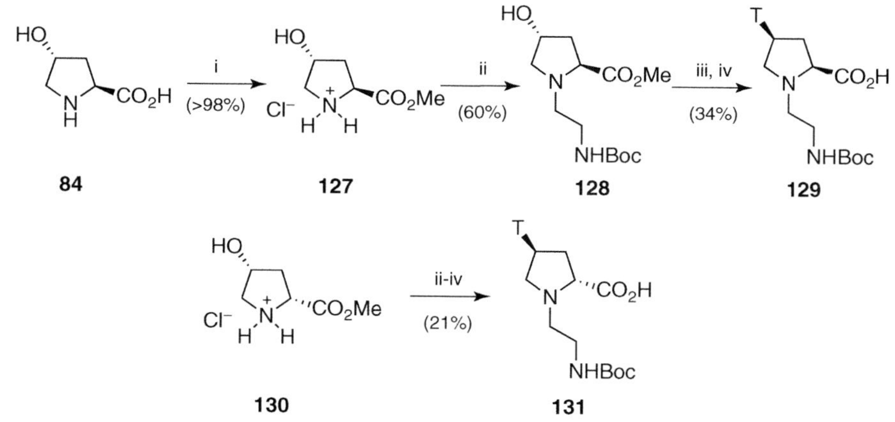

Scheme 18. *Reagents and conditions:* (i) SOCl$_2$, MeOH; (ii) BocHN(CH$_2$)$_2$Br, CH$_3$CN; (iii) N^3-benzoylthymine, DIAD, Ph$_3$P, THF; (iv). 1 N NaOH, MeOH, H$_2$O.

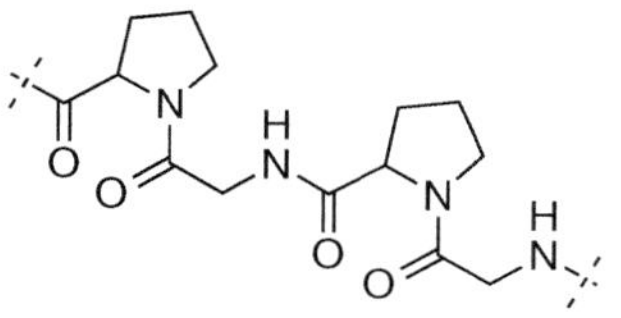

Figure 5. Glycylproline PNA analogue.

replaced by a glycyl proline (Figure 5). Both D- and L- configurations as well as *cis*- and *trans*-geometries were studied.[66]

The synthesis of analogues **132** was made from the corresponding protected dipeptides representing the monomeric unit. In Scheme 19 the synthesis of thymine monomer with (2S,4S)-configuration **136** is illustrated. Introduction of the base was made directly to the protected *trans*-4-hydroxy-L-proline **133** through a Mitsunobu-type reaction.[67] The introduction of guanine (G), cytosine (C), and adenine (A) is also possible, although in the case of G and C, a S$_N$2 reaction was preferred over tosylated **133**. The synthesis of enantiomers with a (2R,4R)-configuration was carried out from the protected form on *trans*-4-hydroxy-D-proline, accessible from the L-enantiomer in seven steps.[40] Deprotection of NBoc in **134** and condensation with Fmoc-protected activated glycine gave **135**, which after deprotection of the carboxyl unit afforded monomer **136** ready to be used in solid-phase synthesis.[67] Decamers with different stereochemistry at the proline ring, including *cis*-L, *trans*-D, and *cis*-D, have been prepared by that method.[68] The polyT$_{10}$ oligomers with *cis*-configuration were found to bind strongly to poly(rA) to form duplexes.[66] Alternation of serine residues improved the solubility of the PNA analogues in aqueous media, but hybridization properties were lost.[69] A modification of compound **136** was introduced by preparing the enantiomeric reduced derivative **137** with a flexible aminoethyl linker from ent-**134** (Scheme 19).[70]

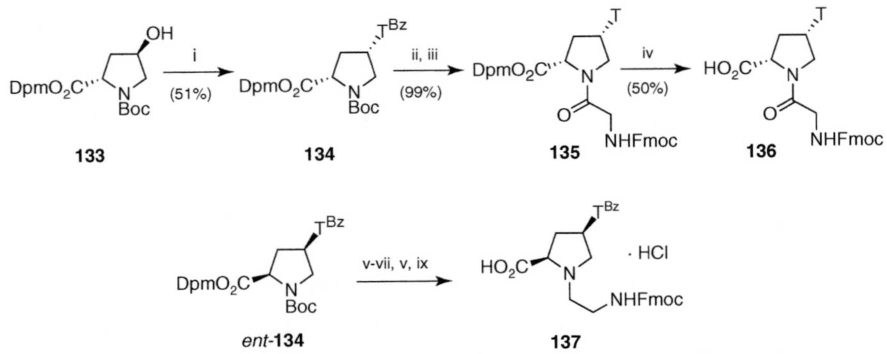

Scheme 19. *Reagents and conditions:* (i) N^3-benzoylthymine, DEAD, Ph$_3$P, THF, $-15°$C to rt, 16 h; (ii) HCl, MeOH, THF, rt, 3 h; (iii) FmocGlyOPfp, i-Pr$_2$EtN, dioxane, rt, 16 h; (iv) HBr, AcOH, rt, 1 h; (v) p-TsOH, CH$_3$CN; (vi) N-nosylaziridine, i-Pr$_2$EtN, CH$_3$CN; (vii) Boc$_2$O, Et$_3$N, DMAP, CH$_2$Cl$_2$; (viii) PhSH, K$_2$CO$_3$, DMF, rt; (ix) FmocCl, i-Pr$_2$EtN, then HCl, dioxane.

Homodecamers prepared from **137** are soluble in aqueous solutions and exhibited strong interactions with oligoribonucleotides but not with oligodeoxyribonucleotides. Modifications at the glycine unit were introduced by the same research group. Additional restrictions were introduced in PNA analogues **132** by replacing the glycine unit with an 1-aminopyrrolidine carboxylic acid. Both 2*R*- and 2*S*-configurations were considered and PNA analogues **138** and **139** were prepared.[71] Whereas no observable binding was found for **139**, decamer **138** exhibited a high level of discrimination toward DNA.[72] The acyclic N-amino-N-methylglycine-derived analogue **140** was also studied, but a complete absence of binding with poly(dA) and poly (rA) was found.[73] Similarly, no binding was observed for homologated **141**, prepared by alkylating ent-**134** with commercially available Fmoc-β-alanine.[71] On the other hand, while constrained homologated PNA **142** did not show evidence of binding to DNA,[71] the *trans*-isomer **143** formed stable duplexes with DNA in both parallel and antiparallel configurations[74] (Figure 6). The introduction of different bases (cytosine, adenine, and guanine) into PNA analogues **143** as well as in those with *trans*-configuration in the proline unit has recently been reported using the same strategy as that illustrated in Scheme 19.[75]

4.4. Other Pyrrolidine-Containing PNA Monomers

Pyrrolidine formation has been a useful tool for introducing constraint in the diethylamino unit, as illustrated by Slaitas and Yeheskiely (Scheme 20).[76] Here, conversion of prolinol **144** into the key amine **146** and subsequent N-alkylation

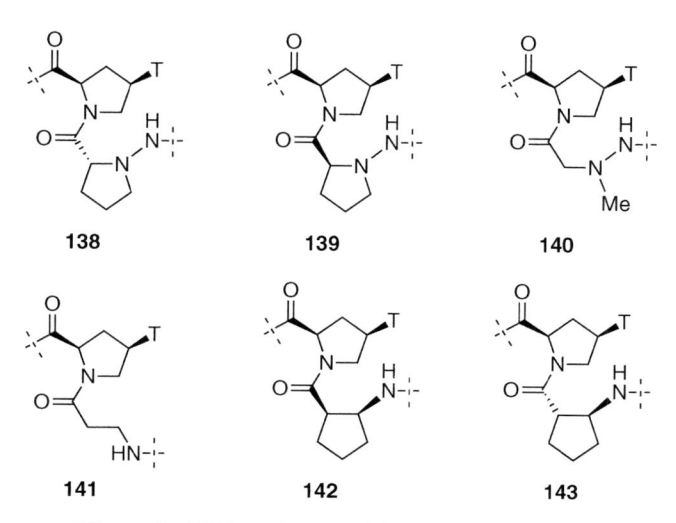

Figure 6. PNA analogues with extended backbones.

with thymine-1-acetic acid and carboxylate deprotection gave monomer **147**. Both *R*- and *S*-monomers (only the isomer *S* is showed in Scheme 20) were prepared and incorporated into individual PNA decamers. In both cases stable duplexes with RNA were formed with some decrease in T_m. The introduction of additional substituents into the pyrrolidine ring has also been studied, and improved binding with both DNA and RNA has been found.[77]

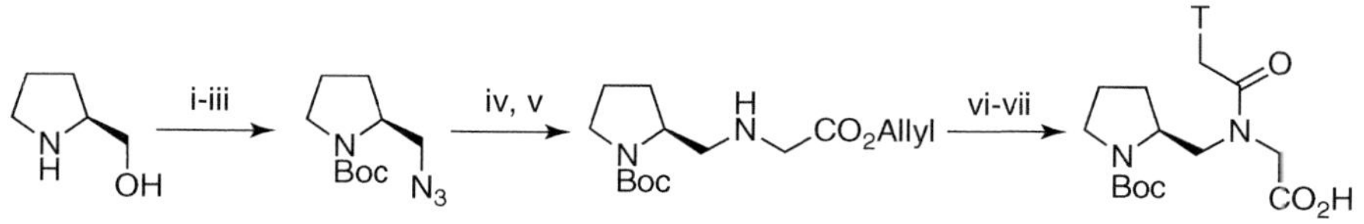

Scheme 20. *Reagents and conditions:* (i) Boc$_2$O, NaOH, H$_2$O; (ii) MsCl, Et$_3$N; (iii) LiN$_3$, DMF; (iv) Ph$_3$P, THF, H$_2$O; (v) allylbromoacetate, Et$_3$N, THF; (vi) HBTU, *i*-Pr$_2$EtN, DMF; (vii) [Pd(Ph$_3$P)$_2$]Cl, Bu$_3$SnH, AcOH, CH$_2$Cl$_2$. (*Note*: No yields were reported in the original publication.)

The combination of ether and amide functionalities led to the oxypeptide nucleic acids (OPNAs), introduced by Sisido and co-workers in 1999.[78,79] The conformationally constrained version of OPNA, pyrrolidine-based OPNA (POPNA), was developed in 2004 by introducing a methylene bridge between the oxyamide chain and the base-containing arm (Figure 7).[80]

To synthesize the prerequisite monomer, *cis*-4-hydroxy-D-proline **150** was converted into intermediate **151**. Introduction of the base (adenine) and deprotection

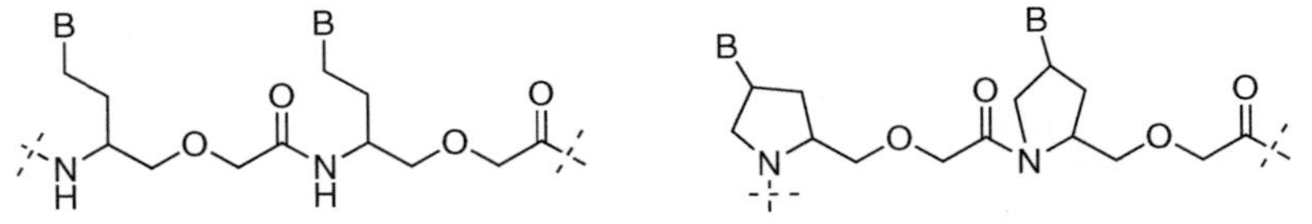

Figure 7. Oxypeptide nucleic acids.

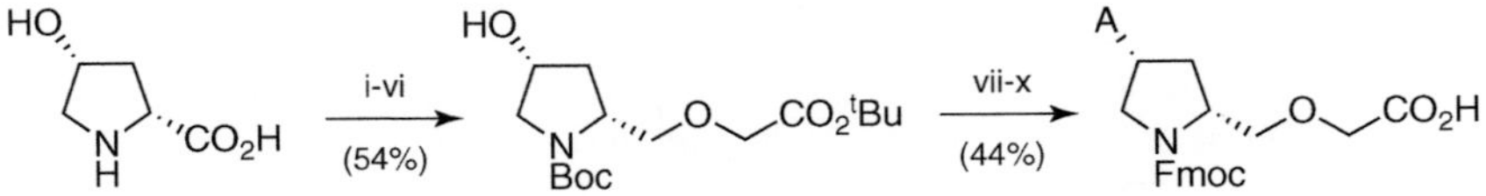

Scheme 21. *Reagents and conditions:* (i) Boc$_2$O, NaHCO$_3$, dioxane, H$_2$O; (ii) EtBr, DMF; (iii) DHP, *p*-TsOH, CH$_2$Cl$_2$, 6 h; (iv) NaBH$_4$, EtOH; (v) BrCH$_2$CO$_2$ *t*-Bu, Bu$_4$NHSO$_4$; (vi) PPTS, EtOH, 3 h; (vii) HCO$_2$H, DEAD, Ph$_3$P, THF, then NH$_3$, MeOH, 6 h; (viii) 6-chloropurine, Ph$_3$P, DEAD, THF, overnight; (ix) NH$_3$, MeOH, 60°C, 24 h; (x) HBr, AcOH, FmocONSu, CH$_3$CN, H$_2$O.

gave rise to **152**, ready for solid-phase synthesis (Scheme 21).[80] By using the same methodology and through the appropriate functional group transformation, the four possible isomers bearing the four natural bases were prepared.[81] Of the four stereoisomers of POPNA bearing adenine, the *cis*-L-isomer showed the highest stability when hybridized with DNA, whereas the *trans*-L-isomer showed the lowest stability.[80] Moreover, with RNA the *trans*-L-isomer formed the most stable hybrid and the *cis*-L-isomer formed the least stable hybrid.[82] Experiments carried out to study the antisense effect of POPNA demonstrated that these analogues have an effect comparable to that of the original *aeg*-PNA.[83]

A variant of oxypeptide **149** consisted of replacing the ether function by an imino group.[84] Synthesis of the required monomer was achieved efficiently from **153**, easily available from *trans*-4-hydroxy-L-proline **84**, as illustrated in Scheme 22. Modified POPNA with **156** acquired a positive charge at physiological pH that facilitated the entry into cells more readily than the oxygen version.

Experimental Procedures

(3R,5R)-5-tert-Butoxycarbonylaminomethyl-3-hydroxy-N-methoxycarbonyl-methyl-2-pyrrolidinone (73).[38] Compound **71** (2.19 g, 7.24 mmol) was dried by evaporation from CH$_3$CN (15 mL) and then redissolved in THF (25 mL). A solution of Ph$_3$P (5.72 g, 21.72 mmol) in THF (25 mL) and a solution of benzoic acid (4.44 g, 36.2 mmol) in toluene (70 mL) was added successively at 0°C, and the mixture was stirred for 2 min before DEAD (5.71 mL, 36.2 mmol) was added dropwise. The clear yellow solution was allowed to warm to room temperature

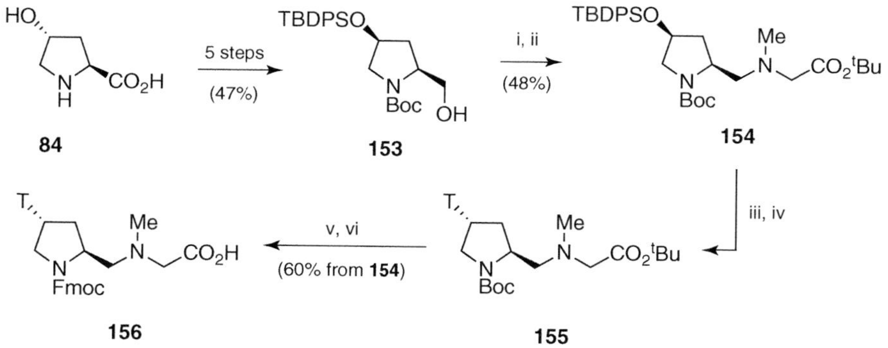

Scheme 22. *Reagents and conditions:* (i) CBr$_4$, Ph$_3$P, THF, 3 h; (ii) MeHNCH$_2$CO$_2$ t-Bu, Na$_2$CO$_3$, DMF, 75°C, 2 days; (iii) TBAF, THF; (iv) N^{Bz}-thymine, Ph$_3$P, DEAD; (v) 30% HBr, AcOH, 30 min; (vi) FmocONSu, NaHCO$_3$, MeOH, H$_2$O. (*Note*: No yields were reported in the original publication.)

and stirred overnight. EtOAc was added, and the mixture was extracted with 0.5 M aqueous citric acid, brine, saturated aqueous NaHCO$_3$, and brine. The organic phase was dried (MgSO$_4$) and evaporated. The crude product was purified by chromatography (EtOAc/hexane 4:1, then pure EtOAc), dissolved in MeOH (50 mL) and cooled to 0°C. NaOMe in methanol (1.05 M, 13.5 mL, 14.18 mmol) was added dropwise, and the solution was stirred at 0°C for 30 min. The reaction was quenched by addition of half-saturated aqueous NH$_4$Cl (100 mL). The aqueous phase was extracted with AcOEt and CH$_2$Cl$_2$. The organic phases were combined, dried over Na$_2$SO$_4$, and evaporated in vacuo. Chromatography (EtOAc, then CH$_2$Cl$_2$/MeOH, 9:1) afforded **73** as a white foam. Yield: 1.88 g (88%). ^{1}H-NMR (300 MHz, DMSO-d^6): δ 6.90 (t, J = 5.7 Hz, 1H), 5.66 (d, J = 5.1 Hz, 1H), 4.19 (d, J = 17.7 Hz, 1H), 4.11 (m, 1H), 3.94 (d, J = 18.0 Hz, 1H), 3.64 (s, 3H), 3.55 (m, 1H), 3.17 (m, 2H), 2.31 (m, 1H), 1.51 (m, 1H), 1.38 (s, 9H).

[(2S,4S)-2-(tert-Butyloxycarbonylaminomethyl)-4-(thymin-1-yl)-pyrrolidin-1-yl] propanoic Acid (116).[50] To a solution of alcohol **115** (1.5 g, 4.74 mmol), N^3-benzoylthymine and triphenylphosphine in dry benzene, cooled to 4°C, was added DIAD dropwise by a syringe under a nitrogen atmosphere. The reaction mixture was stirred for another 5 h at room temperature. The reaction mixture was evaporated to dryness and the residue was purified by column chromatography to obtain monomer ethyl ester as foam (1.8 g). This ethyl ester (1.2 g, 2.8 mmol) was dissolved in methanol (9 mL), 2 M NaOH (9 mL) was added, and the reaction was stirred for 7 h. The aqueous layer was then neutralized with cation-exchange resin (Dowex H$^+$). The reaction mixture was filtered to remove the resin. The aqueous layer was washed with ethyl acetate to remove benzoic acid. The aqueous layer was concentrated to a residue that on coevaporation with dichloromethane (15 mL × 2) afforded monomer **116** as a foam (1.65 g, yield 71%); mp 119–121 °C; $[\alpha]_D^{20}$ −78° (c 0.5, CH$_2$Cl$_2$). ^{1}H-NMR (D$_2$O, 500 MHz):

δ 1.39 (s, 9H, Boc), 1.82 (s, 3H, thy CH$_3$), 2.1–2.3 (m, 1H, C$_3$H′), 2.45–2.75 (m, 2H, N–CH$_2$–CH$_2$–CO), 2.8–2.9 (m, 1H, C$_3$H), 2.95–3.15 (m, 1H, C$_2$H), 3.4–3.8 (m, 5H, COCH$_2$–CH$_2$–N, CH$_2$NH, C$_5$H), 3.95–4.15 (m, 1H, C$_5$H′) 4.8 (m, 1H, C$_4$H), 7.44 (s, 1H, thymine CH).

{(2S, 4S))]-4-tert-butoxycarbonylamino-1-[(thymin-1-yl)methylcarbonyl]pyrrolidin-2-yl}methyl 4-nitrophenyl Carbonate (124, B = T).[55] tert-Butoxycarbonyl-protected thymine monomer **123** (0.53 g, 1.38 mmol) was dried by coevaporation with dry dioxane (3_5 mL) under reduced pressure. The resulting solid was dissolved in a dioxane/Py mixture (11 mL; 10:1) and cooled to 10°C. p-Nitrophenyl chloroformate (0.56 g, 2.78 mmol) was added to the reaction vessel in portions under anhydrous conditions and the mixture was stirred for 4 h after a second addition. The solvent was evaporated, without heating, under reduced pressure to obtain a crude product (1.6 g), which was purified by flash column chromatography using ethyl acetate/petroleum ether (7:3) to obtain the pure product **123** (0.46 g, 60%, R_f = 0.34, EtOAc/petroleum ether 7:3, $[\alpha]_{27}^D$ = −71°; mp = 80°C). ^{1}H-NMR (CDCl$_3$): δ 8.25 (d, 2H, J = 7 Hz), 7.4 (d, 2H, J = 7 Hz), 7.0 (s 1H), 5.2 (bs, 1H), 4.8–4.1 (m, 7H), 4.05–3.7 (m, 1H), 2.6–2.2 (m 1H), 2.15–1.95 (m 1H), 1.9 (s 3H), 1.4 (2s, 9H).

N-tert-Butoxycarbonyl-4-(N^3-benzoylthymin-1-yl)proline Diphenylmethyl Ester (134).[67] N-tert-Butoxycarbonyl-trans-4-hydroxy-L-proline diphenylmethyl ester **133** (0.425 g, 1.07 mmol), triphenylphosphine (0.290 g, 1.10 mmol) and N^3-benzoylthymine (0.250 g, 1.09 mmol) were dissolved in dry THF (10 ml) and the solution was cooled to −15°C. DEAD (180 mL, 1.10 mmol) was then added dropwise to the stirred mixture. The reaction mixture was stirred under an argon atmosphere at room temperature overnight. The solvent was evaporated off and the residue was chromatographed on silica gel with CH$_2$Cl$_2$/acetone (20:1) as eluent to give **134**, which was recrystallized from ethanol to give a fluffy solid (0.310 g, 51%), mp 183–185°C. ^{1}H-NMR (200 MHz; CDCl$_3$); mixture of rotamers: δ 1.30 and 1.49 (9 H, 2 × bs), 1.82 (3 H, bs, thymine CH$_3$), 2.05 and 2.85 [2 H, m, CH$_2$(3′)], 3.65 (1 H, m) and 4.02 (1 H, dd, J = 12.0, 8.0 Hz) [CH$_2$(5′)], 4.54 [1 H, m, CH(2′)], 5.26 [1 H, m, CH(4′)], 6.94 (1 H, s, CHPh$_2$), 7.12 and 7.18 [1 H, 2 × bs, CH(6)], 7.30–7.42 (10 H, br m, phenyl CH), 7.50 (2 H, t, J = 7.0, benzoyl m-H), 7.67 (1 H, t, J = 7.0, benzoyl p-H), and 7.92 (2 H, d, J = 7.0, benzoyl o-H).

5. PIPERIDINYL PNA MONOMERS

The introduction of an extra methylene into the original PNA led to the more flexible aminopropylglycyl PNA (apg-PNA).[85] A conformational constrained analogue of apg-PNA has been constructed by introducing a methylene bridge between the diaminopropyl unit and the glycine arm, thus forming piperidinyl PNA (pip-PNA) analogues (Figure 3, **11**). The thymine monomer **163** was obtained from protected

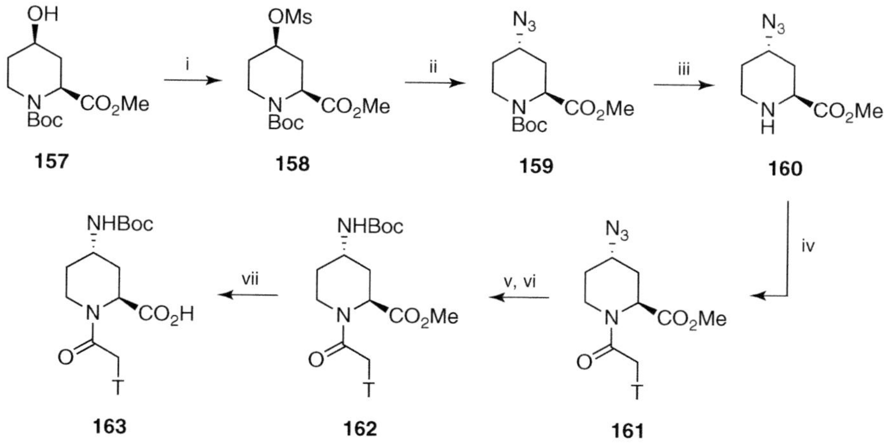

Scheme 23. *Reagents and conditions:* (i) Py, MsCl; (ii) NaN$_3$, DMF; (iii) TFA, CH$_2$Cl$_2$; (iv) T-CH$_2$CO$_2$H, HOBT, DMF; (v) H$_2$, Pd-C; (vi) BocN$_3$; (vii) 2 N NaOH. (Note: No yields were reported in the original publication.)

4-hydroxypipecolic acid **157** in seven steps, as illustrated in Scheme 23.[86] Unfortunately, PNA derived from **163** was not capable of adopting the conformation required for binding to DNA.

A modification to piperidinyl PNA was then introduced by the same authors, who designed a monomer in which the adequate conformation should be frozen.[87] The protected piperidine **165**, obtained in three steps from protected *trans*-4-hydroxy-L-proline **164**, was transformed into azide **166**. The resulting diamine **167** was selectively deprotected, N-alkylated, and desilylated to give **168**. Introduction of the base by a Mitsunobu-type reaction and deprotection of the ester yielded the monomer **169** (Scheme 24), which was incorporated into *aeg*-PNA at the C-terminus. The presence of **169** at that position contributed to increasing the stabilization of PNA$_2$–DNA complexes. On the other hand, installation at the N-terminus destabilized the complexes and in a central position had only a minor effect.

6. ISOXAZOLIDINYL PNA MONOMERS

An isoxazolidine monomer closely related to *pyr*-PNA (Figure 2, **13**) was prepared through the asymmetric 1,3-dipolar cycloaddition of vinyl acetate with nitrone **172** generated in situ from Fmoc glycinal **170** and hydroxylamine **171** (Scheme 25).[88] *N*-Glycosylation of the adduct **173** with silylated thymine, deprotection, and *N*-alkylation furnished the protected monomer **175** in four steps and 28% overall yield from **170**. Compound **175** represents a typical pyrrolidin-2-one monomer in which the carbonyl moiety has been replaced by an oxygen atom.

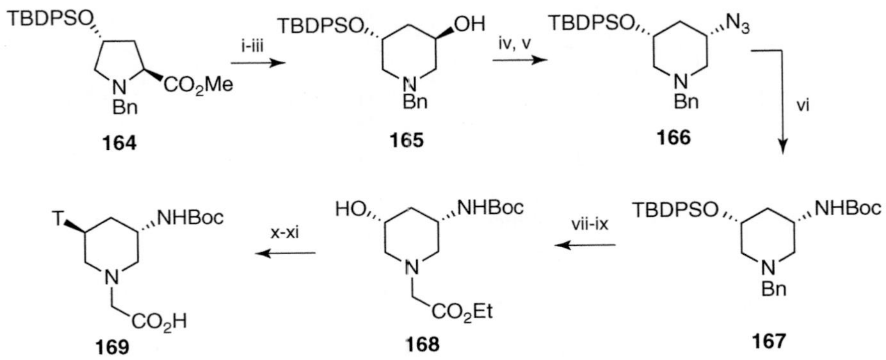

Scheme 24. *Reagents and conditions:* (i) LiBH4/THF; (ii) TFAA; (iii) *i*-Pr$_2$EtN; (iv) MsCl, Py; (v) NaN$_3$, DMF; (vi) Ra-Ni, H$_2$, Boc$_2$O; (vii) H$_2$, Pd-C; (viii) BrCH$_2$COOEt, *i*-Pr$_2$EtN; (ix) TBAF, THF; (x) N^3-benzoylthymine, Ph$_3$P/ DIAD; (xi) NaOH, H$_2$O, MeOH.

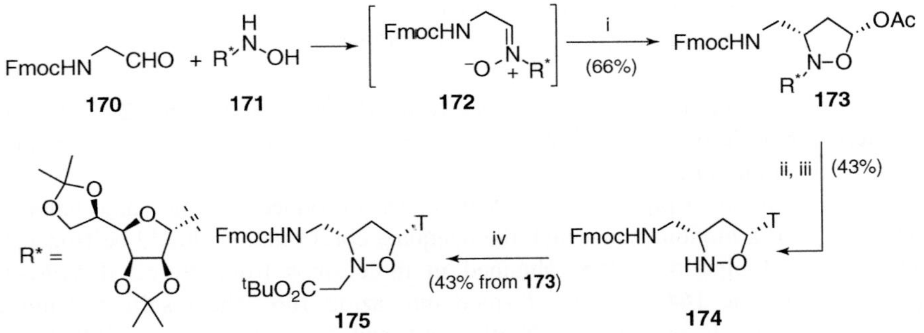

Scheme 25. *Reagents and conditions:* (i) vinyl acetate, neat, sealed tube, 100°C, 3 h; (ii) silylated thymine, TMSOTf, CH$_3$CN; (iii) 3% HCl in MeOH; (iv) BrCH$_2$CO$_2$ *t*-Bu, DMF, *i*-Pr$_2$EtN, rt.

Experimental Procedures

(3 S,5 R)-N-(2,3:5,6-di-O-isopropylidene-D-mannofuranosyl)-3-({[(9H-fluoren-9-yl)methoxy]carbonylamino}methyl)-5-(acetoxy)isoxazolidine (173).[88] A mixture of Fmoc glycinal **170** (1 g, 3.6 mmol), hydroxylamine **171** (0.996 g, 3.6 mmol), and vinyl acetate (30 mL) was placed in a sealed tube and heated at 100°C for 4 h, at which time the reaction mixture was cooled to room temperature and rotatorily evaporated. The residue was purified by column chromatography on silica gel (hexane/EtOac, 4:1) to give pure **173** (1.48 g, 66%) as an oil. $[\alpha]_D^{20} = +50°$ (c 1.28, CHCl$_3$). ^{1}H-NMR (400 MHz, CDCl$_3$, 25°C): δ 1.27 (s, 3H), 1.30 (s, 3H), 1.38 (s, 3H), 1.42 (s, 3H), 1.97 (s, 3H), 2.04 (bd, 1H, $J = 12.6$ Hz), 2.57–2.66 (bddd, 1H, $J = 6.6, 7.6$, and 12.4 Hz), 3.31 (m, 2H), 3.59 (m, 1H), 3.91–3.96 (m, 1H), 3.97–4.03 (m, 2H), 4.10–4.15 (m, 1H), 4.21–4.26 (m, 1H),

4.61 (bs, 2H), 4.67 (bs, 1H), 4.75–4.81 (m, 1H), 4.94 (d, 1H, $J = 5.6$ Hz), 5.13 (bt, 1H, $J = 6.3$ Hz), 6.27 (d, 1H, $J = 6.1$ Hz), 7.25–7.70 (m, 8H).

7. THIAZOLIDINYL PNA MONOMERS

Substitution of a methylene group by a sulfur atom in aminoproline PNA (Figure 2, **8**) led to thiazolidine PNAs. Both *cis*- and *trans*-monomers were prepared from *N*-Boc- glycinal **176**, as depicted in Scheme 26.[89]

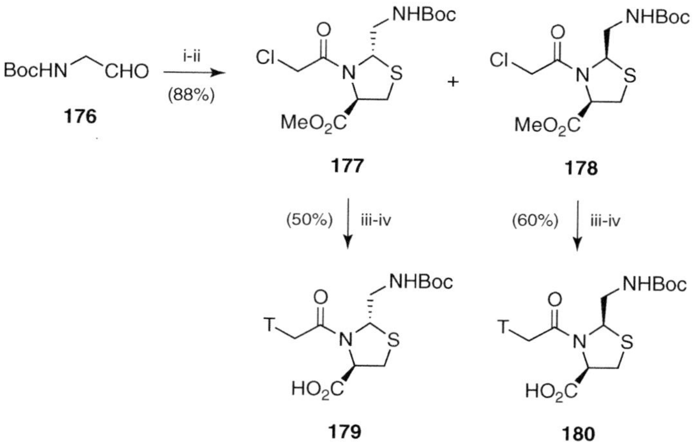

Scheme 26. *Reagents and conditions:* (i) L-cysteine methyl ester, Py, MeOH; (ii) ClCH$_2$CO$_2$H, Py, CH$_2$Cl$_2$, $-78°$C; (iii) thymine, K$_2$CO$_3$, DMF; (iv) LiOH, H$_2$O, dioxane, then KHSO$_4$.

Compounds **179** and **180** were incorporated into a central position in *aeg*-PNA and it was found that they reduced considerably the thermal stability of PNA$_2$–DNA and PNA$_2$–RNA triplexes. A similar effect was observed when monomers with a thiazine ring, instead of a thiazolidine ring, were incorporated into *aeg*-PNA.[90]

Experimental Procedures

(2S,4R)-2-[(tert-butoxycarbonylamino)methyl]-3-(thyminyl-1-acetyl)thiazolidine-4-carboxylic Acid (179).[89] A mixture of thymine (2.84 g, 22.48 mmol), pulverized K$_2$CO$_3$ (12.95 g, 93.8 mmol), and **177** (6.61 g, 18.76 mmol) in anhydrous DMF was stirred at room temperature for 5 h. K$_2$CO$_3$ was removed by filtration through Celite, and DMF was evaporated under reduced pressure. The resulting residue was purified by flash chromatography on silica gel (EtOAc/cyclohexane, 9:1) to give protected **179** as an amorphous white solid which was dissolved in dioxane/H$_2$O (3:2) (20 mL). To this solution, LiOH hydrate (631 mg, 15.05 mmol) was added in portions The reaction mixture was

stirred at room temperature until all the starting material had been consumed. The dioxane was evaporated and water was added. The aqueous layer was washed twice with CH_2Cl_2, neutralized with $KHSO_4$ (2.05 g, 15.05 mmol) and concentrated under vacuum. The resulting residue was triturated with MeOH and the sulfate salts were removed by filtration. The filtrate was concentrated to a small volume and left at 210°C overnight to allow precipitation. The solid was filtered, then washed with ethanol and ether to afford **179** (4.03 g, 75%) as a white solid. ^{1}H-NMR (400 MHz, DMSO-d^6): δ 1.37 (s, 9H, CH_3 Boc), 1.75 and 1.74 (2 s, 3H, CH_3 T), 3.01–3.55 (m, 4H, 5-H, HNCH_2), 4.22–4.91 (m, 2.3 H, 4-H, T-CH_2), 4.96 (t, 0.7H, 4-H), 5.32 (dd, 0.7H, $J = 7.6$ and 3.6 Hz, 2-H), 5.17 (dd, 0.3H, $J = 8.6$ and 3.6 Hz, 2-H), 6.86 and 6.52 (2 m, 1H, Boc-NH), 7.38 and 7.29 (2 s, 1H, CH T), 11.35 and 11.30 (2 s, 1H, NH T).

8. CONCLUDING REMARKS

Synthetic modifications of the original PNA backbone are now possible through the preparation of a series of monomers that feature several kinds of conformational restrictions as well as novel variants of the central backbone (e.g., pyrrolidinyl, isoxazolidinyl, or thiazoldinyl nucleosides). Now, a large pool of structurally diverse linkages, capable of modifying important properties of oligomeric chains, such as binding affinity and water solubility, are available. It is expected that recent promising results with a variety of derivatives qualify conformationally constrained PNA analogues as suitable and convenient structures for testing in antisense research and gene therapy.

Acknowledgments

The authors wish to express their gratitude to the Ministry of Science and Innovation (MICINN, Madrid, Spain, project CTQ2010-19606), the FEDER Program (UE), and the Government of Aragon (Zaragoza, Aragon, Spain, Research Group E-10).

ABBREVIATIONS

A	adenine
A^Z	N-Cbz-adenine
BOM	benzyloxymethyl
C	cytosine
C^Z	N-Cbz-cytosine
DCC	dicyclohexylcarbodiimide
DEAD	diethylazodicarboxylate
DhbtOH	1-oxo-2-hydroxydihydrobenzotriazine
DHP	dihydropyran
DIAD	diisopropylazodicarboxylate
DMAP	4-(N,N-dimethylamino)pyridine
DMF	N,N-dimethylformamide

DMPU	N,N'-dimethyl-N,N'-propylene urea
DMSO	dimethyl sulfoxide
Dpm	diphenylmethyl
G	guanine
G^Z	N-Cbz-guanine
HATU	N-[(dimethylamino)-$1H$-1,2,3-triazolo-[4,5-b]pyridin-1-yl-methylene]-N-methylmethanaminium hexafluorophosphate N-oxide
HBTU	N-[($1H$-benzotriazol-1-yl) (dimethylamino)methylene]-N-methylmethanaminium hexafluorophosphate N-oxide
HMDS	hexamethyldisilazane
HOBT	1-hydroxybenzotriazole
MMT	monomethoxytrityl
MoOPH	oxodiperoxymolybdenum(pyridine)hexamethylphosphoric triamide
Ms	mesyl
NIS	N-iodosuccinimide
OPA	olefinic peptide nucleic acid
OPNA	oxypeptide nucleic acids
PCL	*Pseudomonas cepacia* lipase
Pfp	pentafluorophenyl
Pht	phtalimido
PMB	p-methoxybenzyl
PNA	peptide nucleic acid
POPNA	pyrrolidine oxypeptide nucleic acids
PPTS	pyridine p-toluenesulfonate
Py	pyridine
T	thymine
TBAF	tetrabutylammonium fluoride
TBDPS	*tert*-butyldiphenylsilyl
TBS	*tert*-butyldimethylsilyl
T^{Bz}	N^3-benzoylthymine
TEMPO	2,2,6,6-tetramethylpiperidinyl-1-oxy
TFA	trifluoroacetic acid
TFAA	trifluoroacetic anhydride
THP	tetrahydropyranyl
TMP	2,2,6,6-tetramethylpiperidine
TPP	tetraphenylporphyrin
Ts	tosyl

REFERENCES

1. Nielsen, P. E. In The Chemical Biology of Nucleic Acids; Mayer, G., Ed.; Wiley: Hoboken, NJ, 2010, pp. 103–113.
2. The symbol of Trinacria (mentioned in Homer's *Odyssey*), present in the flag of Sicily, is composed of the head of the Gorgon, with two snakes instead of hair and wings for ears,

from which three folded legs irradiate. The three legs represent the three promontories of Sicily, located in the three corners of the island at Messina (Cape Peloro), Marsala (Cape Lilibeo), and Siracusa (Cape Passero). On August 30, 1302, Sicily became the Kingdom of Trinacria, and its sovereignty was assigned to Federico II of Aragon. The symbol is also present in the flag of the Isle of Man (exported by the Normans in 1072) as well as in the coats of arms of several noble dynasties.

3. Nielsen, P. E.; Egholm, M.; Berg, R. H.; Buchardt, O. *Science* **1991**, *254*, 1497–1500.
4. Nielsen, P. E. *Annu. Rev. Biophys. Biomol. Struct.* **1995**, *24*, 167–183.
5. Ray, A.; Norden, B. *FASEB J.* **2000**, *14*, 1041–1060.
6. Brandt, O.; Hoheisel, J. D. *Trends Biotechnol.* **2004**, *22*, 617–622.
7. Nielsen, P. E. *Curr. Opin. Mol. Ther.* **2010**, *12*, 184–191.
8. Nielsen, P. E. *Chem. Biodivers.* **2010**, *7*, 786–804.
9. Egholm, M.; Buchardt, O.; Christensen, L.; Behrens, C.; Freier, S. M.; Driver, D. A.; Berg, R. H.; Kim, S. K.; Norden, B.; Nielsen, P. E. *Nature* **1993**, *365*, 566–568.
10. Nielsen, P. E. *Acc. Chem. Res.* **1999**, *32*, 624–630.
11. Hanvey, J. C.; Peffer, N. J.; Bisi, J. E.; Thomson, S. A.; Cadilla, R.; Josey, J. A. *Science* **1992**, *258*, 1481–1485.
12. Pensato, S.; Saviano, M.; Romanelli, A. *Expert Opin. Biol. Ther.* **2007**, *7*, 1219–1232.
13. Kumar, V. A.; Ganesh, K. N. *Acc. Chem. Res.* **2005**, *38*, 404–412.
14. Schutz, R.; Cantin, M.; Roberts, C.; Greiner, B.; Uhlmann, E.; Leumann, C. *Angew. Chem. Int. Ed.* **2000**, *39*, 1250–1253.
15. Kumar, V. A. *Eur. J. Org. Chem.* **2002**, 2021–2032.
16. Kumar, V. A. *Nucleosides Nucleotides Nucleic Acids* **2003**, *22*, 1045–1048.
17. Efimov, V. A.; Chakhmakhcheva, O. G. *Collect. Czech. Chem. Commun.* **2006**, *71*, 929–955.
18. Efimov, V. A.; Aralov, A. V.; Chakhmakhcheva, O. G. *Russ. J. Bioorg. Chem.* **2010**, *36*, 663–683.
19. Koch, T. In *Peptide Nucleic Acids: Protocols and Applications*, 2nd ed.; Nielsen, P. E., Ed.; Horizon Biosciences: Essex, UK, **2004**, pp. 37–59.
20. Brown, S. C.; Thomson, S. A. *Science* **1994**, *265*, 777–780.
21. Betts, L.; Josey, J. A.; Veal, J. M.; Jordan, S. R. *Science* **1995**, *270*, 1838–1841.
22. Rasmussen, H.; Kastrup, J. S.; Nielsen, J. N.; Nielsen, J. M.; Nielsen, P. E. *Nat. Struct. Biol.* **1997**, *4*, 98–101.
23. Cantin, M.; Schtitz, R.; Leumann, C. J. *Tetrahedron Lett.* **1997**, *38*, 4211–4214.
24. Roberts, C. D.; Schutz, R.; Leumann, C. *J. Synlett* **1999**, 819–821.
25. Olsen, A. G.; Dahl, O.; Nielsen, P. E. *Nucleosides Nucleotides Nucleic Acids* **2003**, *22*, 1331–1333.
26. Olsen, A. G.; Dahl, O.; Nielsen, P. E. *Bioorg. Med. Chem. Lett.* **2004**, *14*, 1551–1554.
27. Lagriffoule, P.; Wittung, P.; Eriksson, M.; Jensen, K. K.; Norden, B.; Buchardt, O.; Nielsen, P. E. *Chem. Eur. J.* **1997**, *3*, 912–919.
28. Govindaraju, T.; Kumar, V. A.; Ganesh, K. N. *J. Org. Chem.* **2004**, *69*, 1858–1865.
29. Govindaraju, T.; Madhuri, V.; Kumar, V. A.; Ganesh, K. N. *J. Org. Chem.* **2006**, *71*, 14–21.
30. Govindaraju, T.; Kumar, V. A.; Ganesh, K. N. *Chem. Commun.* **2004**, 860–861.

31. Govindaraju, T.; Kumar, V. A.; Ganesh, K. N. *J. Org. Chem.* **2004**, *69*, 5725–5734.
32. Myers, M. C.; Witschi, M. A.; Larionova, N. V.; Franck, J. M.; Haynes, R. D ; Hara, T.; Grajkowski, A.; Appella, D. H. *Org. Lett.* **2003**, *5*, 2695–2698.
33. Leplae, P. R.; Umezawa, N.; Lee, H. S.; Gellman, S. H. *J. Org. Chem.* **2001**, *66*, 5629–5632.
34. Pokorski, J. K.; Witschi, M. A.; Purnell, B. L.; Appella, D. H. *J. Am. Chem. Soc.* **2004**, *126*, 15067–15073.
35. Pokorski, J. K.; Myers, M. C.; Appella, D. H. *Tetrahedron Lett.* **2005**, *46*, 915–917.
36. Eriksson, M.; Nielsen, P. E. *Nat. Struct. Biol.* **1996**, *3*, 410–413.
37. Brown, S. C.; Thomson, S. A.; Veal, J. M.; Davis, D. G. *Science* **1994**, *256*, 777–784.
38. Puschl, A.; Boesen, T.; Zuccarello, G.; Dahl, O.; Pitsch, S.; Nielsen, P. E. *J. Org. Chem.* **2001**, *66*, 707–712.
39. Merino, P.; Greco, G.; Tejero, T.; Chiacchio, U.; Corsaro, A.; Romeo, G. Unpublished results.
40. Lowe, G.; Vilaivan, T. *J. Chem. Soc. Perkin Trans. 1* **1997**, 539–546.
41. Hickman, D. T.; King, P. M.; Cooper, M. A.; Slaterb, J. M.; Micklefield, J. *Chem. Commun.* **2000**, 2251–2252.
42. Tan, T. H. S.; Hickman, D. T.; Morral, J.; Beadham, I. G.; Micklefield, J. *Chem. Commun.* **2004**, 516–517.
43. Tan, T. H. S.; Worthington, R. J.; Pritchard, R. G.; Morral, J.; Micklefield, J. *Org. Biomol. Chem.* **2007**, *5*, 239–248.
44. Kumar, V.; Pallan, P. S.; Meena; Ganesh, K. N. *Org. Lett.* **2001**, *3*, 1269–1272.
45. Pueschl, A.; Tedeschi, T.; Nielsen, P. E. *Org. Lett.* **2000**, *2*, 4161–4163.
46. Kumar, V. A.; Meena. *Nucleosides Nucleotides Nucleic Acids* **2003**, *22*, 1285–1288.
47. D'Costa, M.; Kumar, V.; Ganesh, K. N. *Tetrahedron Lett.* **2002**, *43*, 883–886.
48. Lonkar, P. S.; Ganesh, K. N.; Kumar, V. A. *Org. Biomol. Chem.* **2004**, *2*, 2604–2611.
49. Govindaraju, T.; Kumar, V. A. *Chem. Commun.* **2005**, 495–497.
50. Govindaraju, T.; Kumar, V. A. *Tetrahedron* **2006**, *62*, 2321–2330.
51. Kumar, V. A.; Meena. *Nucleosides Nucleotides Nucleic Acids* **2003**, *22*, 1101–1104.
52. Gangamani, B. P.; Kumar, V. A.; Ganesh, K. N. *Tetrahedron* **1996**, *52*, 15017–15030.
53. Jordan, S.; Schwemler, C.; Kosch, W.; Kretschmer, A.; Schwenner, E.; Stropp, U.; Mielke, B. *Bioorg. Med. Chem. Lett.* **1997**, *7*, 681–686.
54. Gangamani, B. P.; Kumar, V. A.; Ganesh, K. N. *Tetrahedron* **1999**, *55*, 177–192.
55. Meena; Kumar, V. A. *Bioorg. Med. Chem.* **2003**, *11*, 3393–3399.
56. Efimov, V.; Chakhmakhcheva, O. *Collect. Symp. Series* **2002**, *5*, 136–144.
57. Efimov, V. A.; Chakhmakhcheva, O. G.; Wickstrom, E. *Nucleosides Nucleotides Nucleic Acids* **2005**, *24*, 1853–1874.
58. Efimov, V. A.; Buryakova, A. A.; Choob, M. V.; Chakhmakhcheva, O. G. *Nucleosides Nucleotides Nucleic Acids* **1999**, *18*, 1393–1396.
59. Efimov, V. A.; Klykov, V. N.; Chakhmakhcheva, O. G. *Nucleosides Nucleotides Nucleic Acids* **2003**, *22*, 593–599.
60. Efimov, V. A.; Birikh, K. R.; Staroverov, D. B.; Lukyanov, S. A.; Tereshina, M. B.; Zaraisky, A. G.; Chakhmakhcheva, O. G. *Nucleic Acids Res.* **2006**, *34*, 2247–2257.
61. D'Costa, M.; Kumar, V. A.; Ganesh, K. N. *Org. Lett.* **1999**, *1*, 1513–1516.

62. D'Costa, M.; Kumar, V.; Ganesh, K. N. *Org. Lett.* **2001**, *3*, 1281–1284.
63. D'Costa, M.; Kumar, V. A.; Ganesh, K. N. *J. Org. Chem.* **2003**, *68*, 4439–4445.
64. Sharma, N. K.; Ganesh, K. N. *Tetrahedron* **2010**, *66*, 9165–9170.
65. Sharma, N. K.; Ganesh, K. N. *Org. Biomol. Chem.* **2011**, *9*, 725–729.
66. Lowe, G.; Vilaivan, T.; Westwell, M. S. *Bioorg. Chem.* **1997**, *25*, 321–329.
67. Lowe, G.; Vilaivan, T. *J. Chem. Soc. Perkin Trans. 1* **1997**, 547–554.
68. Lowe, G.; Vilaivan, T. *J. Chem. Soc. Perkin Trans. 1* **1997**, 555–560.
69. Vilaivan, T.; Khongdeesameor, C.; Wiriyawaree, W.; Mansawat, W.; Westwell, M. S.; Lowe, G. *Sci. Asia* **2001**, *27*, 113–120.
70. Vilaivan, T.; Khongdeesameor, C.; Harnyuttanakorn, P.; Westwell, M. S.; Lowe, G. *Bioorg. Med. Chem. Lett.* **2000**, *10*, 2541–2545.
71. Vilaivan, T.; Suparpprom, C.; Harnyuttanakorn, P.; Lowe, G. *Tetrahedron Lett.* **2001**, *42*, 5533–5536.
72. Vilaivan, T.; Lowe, G. *J. Am. Chem. Soc.* **2002**, *124*, 9326–9327.
73. Vilaivan, T.; Suparpprom, C.; Duanglaor, P.; Harnyuttanakorn, P.; Lowe, G. *Tetrahedron Lett.* **2003**, *44*, 1663–1666.
74. Siriwong, K.; Chuichay, P.; Saen-oon, S.; Suparpprom, C.; Vilaivan, T.; Hannongbua, S. *Biochem. Biophys. Res. Commun.* **2008**, *372*, 765–771.
75. Taechalertpaisarn, J.; Sriwarom, P.; Boonlua, C.; Yotapan, N.; Chotima Vilaivan; Vilaivan, T. *Tetrahedron Lett.* **2010**, *51*, 5822–5826.
76. Slaitas, A.; Yeheskiely, E. *Nucleosides Nucleotides Nucleic Acids* **2001**, *20*, 1377–1379.
77. Gokhale, S. S.; Kumar, V. A. *Org. Biomol. Chem.* **2010**, *8*, 3742–3750.
78. Kuwahara, M.; Arimitsu, M.; Sisido, M. *J. Am. Chem. Soc.* **1999**, *121*, 256–257.
79. Kuwahara, M.; Arimitsu, M.; Sisido, M. *Tetrahedron* **1999**, *55*, 10067–10078.
80. Kitamatsu, M.; Shigeyasu, M.; Okada, T.; Sisido, M. *Chem. Commun.* **2004**, 1208–1209.
81. Kitamatsu, M.; Takahashi, A.; Ohtsuki, T.; Sisido, M. *Tetrahedron* **2010**, *66*, 9659–9666.
82. Kitamatsu, M.; Shigeyasu, M.; Saitoh, M.; Sisido, M. *Biopolymers* **2006**, *84*, 267–273.
83. Kitamatsu, M.; Kurami, S.; Ohtsuki, T.; Sisido, M. *Bioorg. Med. Chem. Lett.* **2011**, *21*, 225–227.
84. Kitamatsu, M.; Kashiwagi, T.; Matsuzaki, R.; Sisido, M. *Chem. Lett.* **2006**, *35*, 300–301.
85. Hyrup, B.; Egholm, M.; Rolland, M.; Nielsen, P. E.; Berg, R. H.; Buchardt, O. *Chem. Commun.* **1993**, 518–519.
86. Lonkar, P. S.; Kumar, V. A.; Ganesh, K. N. *Nucleosides Nucleotides Nucleic Acids* **2001**, *20*, 1197–1200.
87. Lonkar, P. S.; Kumar, V. A. *Bioorg. Med. Chem. Lett.* **2004**, *14*, 2147–2149.
88. Merino, P.; Tejero, T.; Mates, J.; Chiacchio, U.; Corsaro, A.; Romeo, G. *Tetrahedron: Asymmetry* **2007**, *18*, 1517–1520.
89. Bregant, S.; Burlina, F.; Vaissermann, J.; Chassaing, G. *Eur. J. Org. Chem.* **2001**, 3285–3294.
90. Bregant, S.; Burlina, F.; Chassaing, G. *Bioorg. Med. Chem. Lett.* **2002**, *12*, 1047–1050.

INDEX

1,1′-thiocarbonyldiimidazole, 349
1,2,3,4-tetrahydroisoquinoline, 708
1,2,3-triazole, 437
1,2,4-triazole, 324, 435, 567
1,2,4-triazolines, 488
1,2-diaminobenzene, 267
1,2-diaminocyclohexane, 853
1,3,5-triazine, 8-iodo[1,5-*a*], 298
1,3-dipolar cycloaddition, 64, 152, 509, 546, 782–783, 787, 791, 798–799, 860, 874–875
1,3-oxathiolane, 2-benzoyloxy-4-acetoxy-1,3-oxathiolane, 752
1,4-addition, 482
1,4-Brook rearrangement, 580
1,4-dithiane-2,5-diol, 736
1,6-hydrogen transfer, 396
1.2.4-triazoles, 63
1′,2′-azetine pyrimidine nucleosides, 349
10-camphorsulfonic acid, 751
1-aminothiohypoxanthine, 212
1-cyclohexanyl, 639
1*H*-tetrazole, 222, 232, 235, 336–337, 409
1-hydroxybenzotriazole, 240
1-methylnaphtalene, 302
[2+2]-cycloaddition, 357, 592, 617–618, 621
2,2′-dipyridyl diselenide, 218
2,2-dimethoxypropane, 670
2,3-dihydrofurans, 305–306
2,3-methanohomoserine, 609
(*R*)-2,3-*O*-isopropylideneglyceraldehyde, 520, 607
2,3-*O*-isopropylidene-(*S*)-glyceric acid, 769
2,4,6-triisopropylbenzenesulfonyl chloride, 231, 378
2,4-diaminopyrimidine nucleosides, 201–203
2,6-di-*tert*-butyl-4-methylpyridine, 680
2′,3′-anhydrotiazofurin, 267
2′,3′-didehydro-3′-deoxythymidine, 27, 28, 31–32, 110, 163

2′,3′-dideoxy-3′,3′-difluoropyrimidine nucleosides, 490
2′,3′-dideoxy-3′-thianucleosides, 735
2′,3′-dideoxynucleosides, 299, 488–495
  2′-fluoro, 490
2′,4′-carbocyclic nucleosides, 365
2′-amino-1,2-oxazoline, 476
2′-amino-LNA nucleosides, 364–365
2′-carba-LNA nucleosides, 364–368
2′-*C*-dibranched nucleosides, 361
2′-*C*-methyl, 377
2′-deoxy-*C*-nucleosides, 297
2′-deoxynucleosides, 13, 40
  3′-*O*-dimethoxytrityl, 26
  5′-*O*-levulinyl, 26
2′-deoxypseudouridine, 297
2′-fluoro-2′,3′-unsaturated nucleosides, 490
2′-methylene nucleosides, 429
2′-thio-LNA nucleosides, 364–365
2-aminophenol, 267
2-aminothiophenol, 267
2-benzoyloxy acetaldehyde, 520, 769
2-bromofuran, 302
2-bromopyridine, 273
2-chloro-1,3,2-oxathiaphospholane, 245
2-chlorophenylphosphoro-bistriazolide, 329
2-cyanoethyl *N*,*N*,*N*′,*N*′-tetraisopropylphosphorodiamidite, 338
2-deoxyfuranosyl chloride, 302
2-glycosylfuran, 300
2-iodopyridine, 273
2-mercaptoacetaldehyde, 736
2-mercaptoethanol, 503
2-oxoalkylidentriphenyl phosphorane, 439
[3,3]-sigmatropic rearrangement, 565–566
[3+2]-cycloaddition, 715
3,4-dihydroxypyrrolidine, 710
3,5-ditoluoyl-2-deoxyribofuranosyl chloride, 302
3′,4′-linked bicyclic nucleosides, 373
3′,5′-diaminonucleosides, 4

Chemical Synthesis of Nucleoside Analogues, First Edition. Edited by Pedro Merino.
© 2013 John Wiley & Sons, Inc. Published 2013 by John Wiley & Sons, Inc.

3'-*C*-cyano-3'-deoxyribonucleosides, 488
3'-*C*-dibranched nucleosides, 361
3'-cyanomesyl nucleosides, 434
3'-fluoro-2',3'-dideoxynucleosides, 490
3'-oxa-4'-thionucleosides, 728
3'-oxanucleosides, 728
3'-thianucleosides, 728
3-bromocamphor-8-sulfonic acid, 751
3-butynol, 851
3-chloroperbenzoic acid, 232, 236, 243, 386, 493, 512, 514, 665, 755, 760, 665, 828–829
3-iodopyridine, 273
3-iodoselenophen, 75
3-mercaptolactic acid, 743
3-methoxyacryloyl chloride, 593
3-methoxyacryloyl isocyanates, 544
3TC, 27, 735–736, 750, 759
[4+1] cycloadditions, 305
4,4'-dimethoxytrityl chloride, 334–335, 352, 377
4,6-dichloroformamidopyrimidine, 607
4',5'-didehydronucleosides, 456
4'-aza-nucleosides, 492
(*S*)-4-isopropylthiazolidine-2-thione, 246
4'-phenylselenylation, 379
4'-thiostavudines, 691
4-amino-1,2,3-oxathiazole-2,2-dioxide, 453
4-amino-2-oxazolone, 453
4-aminoprolines, 865–867
4-chlorophenyl phosphorodichloridate, 324
4-deoxypyrazofurin, 586
4-hydroxypipecolic acid, 874
4-hydroxyproline, 701
4-nitrophenylphosphorodichloridate, 236
4-nitrophenylphosphotriester, 236
4-thioarabitol, 683
4-thiofuranose, 661
5,6-dichloroindole-2-thione, 301
5'-(α-*P*-boranotriphosphates), 224
5'-(α-*P*-selenotriphosphates), 224
5'-(α-*P*-thiotriphosphates), 224
5'-cyclic hemiacetal, 447
5'-deoxy-5'-fluorocarba-nucleosides, 589
5'-fluorocarba-nucleosides, 586
5'-homoaristeromycin, 557
5'-levulinate esters, 5
5'-noraristeromycin analogues, 545, 549
5'-phosphorylation, 332
5'-triphosphorylation, 211
5'-α-thiophosphates, 213
5-azacytidine, 488
5-exo-cyclization, 379
5-exo-dig radical cyclization, 577
5-halogenonucleosides, 401
5-hydroxypentanal oxime, 784

6-azathymine, 517
6-azauridine, 9
6-*exo-trig* cyclization, 828–829
7-functionalyzed-7-deazapurines, 488
7-iodo-5-aza-7 deazaguanine, 488
8-quinoyl phosphate, 218

abacavir, 27, 163, 165, 317, 536, 575–576, 728
acetonoxime levulinate, 5, 24
acetyl furanosides, 374
acetyl, 111
acetyl-7-(2-acetoxyethoxymethyl), 107
acidic methanolysis, 683
acyclic aza-nucleosides, 116
acyclic nucleoside phosphonates, 192–201
acyclic nucleosides, 103–156
acyclic thioaminals, 515, 688–691
acyclovir, 10, 103, 106, 109–110, 728
acyl isocyanate, 550
Adams' catalyst, 371
adefovir, 130–131, 198, 728
adenine, 3, 38, 125, 132, 138, 166–167, 175, 179, 181, 188–189, 212, 323, 375, 400–401, 434–435, 476, 497, 501, 512, 537–542, 547, 557, 559, 563, 565–566, 570, 573–575, 578, 581–582, 585, 588–589, 591–593, 610–612, 615, 619, 633, 635, 646, 656, 659–660, 683, 703, 753, 761–762, 789, 792, 806, 813, 851, 853, 856, 860–861, 864–865, 868–869, 871
  3'-hydroxymethyl-4'-azamethyl-2'-3'-dideoxy, 792
  3-hydroxy-4-hydroxymethyl-(*R*)-9-(2-hydroxypropyl)-, 199
  7-deaza, 546
  8-bromo, 674
  9-(1'-methyl-2'-phosphonomethoxyethyl, 136
  9-(4-hydroxybutyl), 111
  allyl, 813
  L-9-(α,β-1-methyl-5-ribofuranosyl), 192
  N^6,N^6-bis(*tert*-butoxycarbonyl), 540
  N^6-benzoyl, 348, 363, 549
  N^6-methyl, 676
  ribocyclohexenyl, 644
adenosine, 5, 7, 24–26, 33, 39, 56, 68, 76–77, 86, 188, 209, 214, 222–223, 348, 363, 373, 377, 401, 474, 476, 657–660, 670, 673–674, 753
  1'2'-oxetane, 348
  2'-deoxy, 329
  2'-deoxy, 476
  2-azido, 68
  2-chloro-N^6-methyl-4'-thio, 667
  2-ehtynyl, 68

INDEX   883

2-hexynyl-4'-thio, 673
2-hexynyl-N^6-methyl, 667
2-iodo-N^6-methyl, 68
3',4'-oxetane, 377
3'-O-methyl, 24
4'-Benzyl-3-carboxy-2',3'-dideoxy-4'-aza, 799
4'-thio, 672–673
5'-alkylidencarbazoyl, 34
6'-Fluoro-6'-methyl-5'-nor-, 189
6-N-benzoylisodeoxy, 335
7,9-dideaza-7-oxa-2'-C-methyl, 279
8,5'-cyclo, 401
8-di(α-hydroxyisopropyl), 86
8-hexenyl-4-thio, 674
8-phenylthio, 401
9-deaza, 279
9-deaza-2'-C-methyl, 279
N^6-methyl-4'-thio, 667
N^6-substituted-5'-hydromethyl-4'-thio, 666–667
    agonists, 659
    antagonists, 670
    deaminase, 38–39, 318, 659
    receptors, 657
adenovirus, 193
ADP, 657
ADP-ribose, 551
AIBN, 321, 366, 379, 389–390, 396, 400, 505, 644
β-L-alanine, 462
alcoholysis, enzymatic, 31
aldol condensation, 380
aldol cyclization, 580
aldol reactions, 276–278
    intramolecular, 398
alkenyl glycosides, 271–275
alkoxycarbonylations, 31–37
alkylation, Rh-catalyzed, 52
alkylazides, 438
alkylcyanoformates, 265
alkynyl glycosides, 271–275
alkynyl stannannes, 444
D-allose, 561
β-allyl C-glycoside, 271, 274
allyl chloroformate, 440
allyl stannannes, 444
allylcarbonates, 293
allylic alkylation, 291, 487, 581–584
allylic oxidation, 392
allylic substitution, 290–293
allylmagnesium bromide, 824
allyloxybis(diisopropylamino)phosphite, 409
Alternaria kikuchiana, 830
altritol, 634

aluminum trichloride, 448
amdoxovir, 731
amidoimidates, 265
amino acid, 241
α-amino acid, 822
α-amino acids, methyl esters, 462
amino cyclization, 718
amino lactone, 826
aminoacetonitrile, 545
aminoacylphosphoramidate, 238, 245
aminocyclopentitol, 570
aminoethylglycine, 848
aminothiadiazole, 265
aminyl radical, 400
amipurimycin, 821–823, 830–833
ammonium fluoride, 678
AMP cyclic, 672
AMP, 657
amphotericin B, 820
anhydrous diazotation, 446
anioninc cyclization, 576–581
Antifungal activity, 837
antigene therapies, 345
antigunfal peptidyl nucleosides, 820–842
antiherpes activity, 630
antiherpetic agents, 608
antisense agents, 345, 834
antisense oligonucleotides, 605
antisense therapies, 345
antitumoral activity, 606
antiviral activity, 318
antiviral therapy, 605
apio sugars, 319
apricitabine, 729, 752
aptamers, 16
D-arabinose, 577–578
L-arabinose, 474–476, 478–480, 491
aracytosine, 10
Arbuzov reaction, 187, 165, 497, 535–536, 554–556, 571–573, 583
aristeromycin, 9-deaza, 545
aryl(aminoacyl)phosphoramidate, 239
aryloxyaminoacylchlorophosphate
aryloxyphosphoramidate, 240
arylphosphodichloridite, 239
arylphosphoramidate diesters, 238
arylphosphoramidate, 239
ascorbic acid, 150, 482, 517, 768
Aspergillus fumigatus, 837
Atherton–Todd reaction, 238, 241
ATP, 657
avian Moloney virus, 807
aza-C-2,3-dideoxynucleosides, 710
aza-C-nucleosides, 707

aza-d4T, 700
aza-ddU, 700
aza-isonucleosides, 702–703
aza-Michael addition, 104, 154
azanucleosides, 504–509, 699–722
azetidine, 377, 457
azetine-locked nucleosides, 349
azide–alkyne cycloaddition, 67
azido-2-deoxyribose, 64
azomethine ylides, 713
AZT, 18, 23, 27, 31–33, 117, 163, 317, 535–536, 728, 760

Bacillus megaterium, 614
Barton-McCombie reaction, 351, 367, 389–390
Barton reaction, 560, 634
base-pairing, 632
Bayer–Villiger rearrangement, 760
Bayer–Williger oxidation, 574
Baylis-Hillman reaction, 557
benzaldehyde dimethyl acetal, 643
benzimidazol C-nucleosides, 267
benzimidazole, 52, 57, 152, 488
benzoic anhydride, 8
Benzoquinone, 271
benzothiazol C-nucleosides, 267
benzothiophene, 302
benzoxazol C-nucleosides, 267
benzoyl chloride, 8
benzoyl cyanide, 351
benzoyl glycolaldehyde, 737
benzoylation, enzymatic, 13
benzylhydrazine, 265
benzyloxycarbonyl chloride, 35
Bestmann–Ohira reagent, 373
bi(aryl)phosphoramidate, 247
bicyclic nucleosides, 346–393, 834–835
  3′,5′-linked, 386–391
  5′,5′-spiro, 391–393
bicyclic peptidyl nucleosides, 834–835
Biginelli reaction, 276–278
binding affinity, 672, 866
binding efficiency, 864
binding modes, 734
(S)-BINOL, 522–523, 751
biocatalysis, 1–42
biocatalytic processes, 2
biological probes, 834
biomolecular recognition, 732
bis(diisopropylamino)chlorophosphine, 234
bis-S-acyl thioethyl esters, 164
Blastomyces dermatitidis, 837
blastomycosis, 822
Bohlmann–Rahtz reaction, 275
borane dimethylsulfide complex, 245, 380

Boranephosphorothioate, 245
Boranophosphates, 244–245
boranophosphoramidate, 245
borneol, 796
Botrytis, 826
bromination, 631, 647
bromocyclobutanones, 620
bromophenylacetylene, 59
bromotrimethylsilane, 199, 519
Buchwald ligands, 303
buciclovir, 149
bulky silyl groups, 440
butyl lithium, 304
BVDU, 49

camphorsulfonic acid, 670
Candida albicans, 837
candida antartica lipase, 2, 4–7, 11, 16–18, 21–23, 35, 40–41, 479, 550, 713, 769
candidiasis, 822
Cannizzaro reaction, 354, 380
carbanucleosides, 593–595
carbathymidine, 219
carbene insertions, intramolecular, 574–575
carbocyclic nucleosides, 14, 495–504, 535–595, 615–625
carbonates, 36, 37
carbonyl functionalities, 824
carbonyl methylenation, 617
carbovir, 165, 495–497, 535–536, 548, 553, 567, 571–572, 575, 582–585
carvone, 591, 641, 643
catabolic enzymes, 318
cell-free translational experiments, 632
ceric ammonium nitrate, 65, 443, 833, 377
cerium (III) chloride, 289
cesium carbonate, 53, 201, 431
C-glycosyl-β-lactams, 278
C-H bond activation, 50–60
chemical resolution, 616
chemoenzymatic resolution, 607
chitin biosynthesis, 821
chitin synthase, 822
chitin, 822
chlorodimethoxytriazine, 708
chloromercury benzamidopurine, 474
chloromethylation, 548
chloromethylpivalate, 231
chlorophosphate, 231
chlorosulfonylisocyanate, 453
chromium carbene, 572, 616
α-chymotrypsin, 769
cidofovir, 104, 145, 197, 728
cis-aminocyclohexylglicyl PNAs, 853
cispentacin, 833

cladribine, 656
Claisen rearrangement, 590
clevudine, 728
Cl-IB-MECA, 661–665
clofarabine, 728
C-nucleosides, 263–310, 486–487
Coccidioides immitis, 837
coccidioidomycosis, 822
codonopsinine, 718
codonopsinol, 718
competitive inhibitors, 317
conformational analysis, 413
conformationally restricted nucleosides, 834–835
conglomerates, 758
constrained nucleosides, 345–417
copper (I) iodide, 53
copper (II) chloride, 218
β-coupling reactions, 374
cross coupling reactions, 295–299
cross-communication, 629
cross-enyne metathesis, 271
cross-metathesis, 825
cross-resistance, 317
Crotalus atrox venom, 737
crotonaldehyde, 567
crotonyl chloride, 7
Cryptococcus neoformans, 837
Curtius rearrangement, 571, 593, 856
cyanide, β-D-ribofuranosyl, 72
β-cyanoethylphosphate, 331
cyanonitrile oxide, 428
cyclaradine, 554, 578
cyclic imine, 708, 712
cyclic phosphoester nucleotides, 404–416
cyclic phosphorochloridate, 235
cyclic phosphotriesters, 235–236
cyclic sulfate, 121, 561
cyclic sulfite, 560
cyclic sulfonate ester, 463
cyclic sulfonate, 461
cyclic thionocarbonates, 497
cycloaddition reactions, 571–574, 274
cyclobut-A, 616, 619
cyclobutanol, 618
cyclobutanone, 572, 592, 616–618, 620–621
cyclobutene anhydride, 621
cyclobut-G, 615, 617
cyclobutyl carbanucleosides, 592–594
cyclobutyl nucleosides, 181–192
cyclohexanyl carbanucleosides, 590–591
cyclohexene oxide, 854
cyclohexenyl carbanucleosides, 590–591
cyclohexyl-PNA, 849
cyclonucleosides, 393–404, 84–90

cyclopentanone, 381, 497, 545, 581
cyclopentene oxide, 538, 855
cyclopentylamine, 542–545
cyclopentyl-PNA, 849
cyclopropanation, 168, 374, 560
cyclopropavir, 613–614
cyclopropyl carbanucleosides, 592–594
cyclopropyl homo-nucleosides, 610
cyclopropyl nucleosides, 165–181
cyclopropyl-PNA, 849
cyclotrimerization proces, 271
cystathionine, 826–827
cytarabine, 10
cytidine deaminase, 22
cytidine deaminase, 738
cytidine, 5, 106, 677
  2'-C-methyl, 677
  2',3'-dideoxy, 27–28, 163, 317
  2'-C-methyl-4'-thio, 677
  2'-deoxy-2'(S)-methyl, 677
  2'-deoxy-2'-fluoro-4'-thio, 687
  2'-fluoro-4'-thio, 493
  5-fluoro, 488
  2'-keto-4'-α-thio, 677
  3-hydroxymethyl-4'-azamethyl-2'-3'-dideoxy, 792
  3',4'-oxetane, 377–378
  3'-oxa-4'-thio(fluoro), 755
  3'-thia(fluoro), 736, 743
  3-deaza-2-deoxy, 79
  4'-C–methyloxa, 518
  4'-C-ethynyl-2',3'-dideoxy, 16
  4-N-acetyl-2',3',5'-tri-O-acetyl, 31
  N-benzoyl-3'-O-acetyl, 330
  N-benzoyl-3'-O-levulinyl, 7
cytomegalovirus, 103, 106, 121, 345s, 613
cytosine, 3, 11, 18, 106, 133, 145, 151, 170, 173, 175, 178.9, 181, 195–196, 198, 200, 219, 324, 335, 352, 355–356, 434, 497, 499, 514, 518, 522, 524, 537, 541, 544, 593, 630, 634, 636, 683, 746–747, 748, 752–753, 756, 758, 784, 792–793, 806, 811, 813, 821, 826, 828, 830, 853, 864, 865, 867, 868–869
5-iodo, 273
5-fluoro, 510, 517, 756, 762
5-methyl, 434, 637
N^4-acetyl, 512, 521, 755
N^4-acetyl-5-fluoro, 755
N^4-benzoyl, 493, 512, 518, 758, 792
N^4-benzoyl-5-fluoro, 518
N^4-Cbz-1-acetic acid, 866
O-alkylation, 335
persilylated, 521

d3T, 317
d4T, 27, 28, 31–32, 110, 163, 728
DAST, 483, 491, 623
Davies oxaziridine, 393, 692
DBU, 360, 365, 374, 688
ddA, 28
ddC, 27–28, 163, 317, 728
ddI, 10, 27–28, 163, 317, 728
deacetylation, 21
debromination, 647
dehydroxylation, 309
density functional theory, 452
deoxyctytidine, 476
deoxycytidine deaminase, 732
deoxygenation, 390
deoxyribose, α-chloro-3,5-di-*p*-toluyl, 476
D-eritadenine, 149
Dess–Martin periodinane, 354, 367, 373, 408, 828
desymmetrization, 487, 581–588, 616
di(2-cyanoethyl)
  N,*N*-diisopropylphosphoramidite, 338
di(alkyl)phosphoramidite, 231
diacetone-D-glucose, 347, 389
diacetoxyiodobenzene, 701
diacylated nucleosides, 8
diamide pronucleotide, 242
diaminopurine, 762
diazocompounds, 305–306
diazomethane, 168, 176
diazotation–reduction sequence, 542
dibenzoyl tartaric acid, 751
dichloroacetic acid, 330
dichloro-*N*,*N*-diisopropylphosphoramidite, 232
dichlorophosphates, 210
dichlorophosphoramidate, 243
dicyclohexylcarbodiimide, 5, 213, 237, 331, 398
dideoxyformycin B, 267
dideoxynucleosides, 267, 317, 318
dideoxyshowdomycin, 267
β-l-2′,3′-dideoxy-3′-thia-5-fluorocytidine, 317
Diels–Alder reaction, 271, 360, 644
  asymmetric, 571
diethoxymethylacetate, 404
diethyl cyanomethylenephosphonate, 486
diethyl tartrate, 828
diethyl(ethoxycarbonyl)(difluoromethyl)
  phosphonate, 492
diethylaluminium chloride, 515
diethylphosphonomethyl triflate, 177
dietylaminosulfur trifluoride, 483
difluorocyclopropanation, 177
difluoromethane, 683
dihydroisothiazole dioxide, 455

dihydroxylation, 293, 382, 393, 487, 548, 550, 552, 582
dihydroxylation, 717
dihydroxylation, 831–832
diiodobenzene, 54–55, 57, 60
diisobutylaluminium hydride, 150, 367, 379, 492, 607, 613, 788, 828
diisopinocampheylborane, 586
diisopropyl phosphite, 170
diisopropylethylamine, 240
dimethoxypropane, 501
dimethoxytrityl, 13, 25
dimethyl-2-oxoglutarate, 851
dimethylethynyldicarboxylate, 271
dioxaphosphorinane constrained nucleosides, 411–416
dioxaphosphorinane, 404, 411
dioxoimidazoline, 448
dioxolane aminopurine, 729
dioxolane fluorocytosine, 729
dioxolane nucleosides, 727–734
dioxolane T, 729
dioxolanyl nucleosides, 517–520
diphenyl carbonate, 87
diphenyl chlorophosphate, 232
diphenyl phosphite, 614
diphenyldisulfide, 302
diphenylphosphite, 238, 241
Dipivoxil, 168
disconnective analysis, 489–490
disobutylaluminium hydride, 393
DNA elongation, 657
DNA methylation, 657
DNA polymerase, 535, 732
DNA stability, 416
docking studies, 815
duplex stabilization, 637
dynamic kinetic resolution, 745

Edman degradation, 569
efavirenz, 104
electronic factors, 438
elvucitabine, 752
emtricitabin
emtricitabine, 104, 727–728, 745
enamine, 442–443
enamine–iminium mechanism, 464
β-enaminones, 275
endocyclic oxygen, 318
enolate oxidation, 859
entecavir, 536, 539, 676, 728
enzymatic acylation, 3, 8
enzymatic desymmetrization, 20, 616
enzymatic hydrolysis, 5
enzymatic polymerization, 148

INDEX **887**

enzymatic resolution, 14, 386, 616, 740, 753, 771, 799–800, 854
enzymatic synthesis, 1–42
epichlorohydrin, 611
epimerization, 619, 765
epoxidation, 290, 361, 548, 621, 630, 641
epoxide opening, 386, 567, 634, 706
epoxide, 538
Epstein-Barr virus, 170, 729
erythronolactone, 556
Escherichia coli, 737
(ethoxycarbonylmethylene)triphenylphosphorane, 281
ethyl (trimethylsilyl)acetate, 557
ethyl 2-acetyloxyacrylate, 809
ethyl 2-oxocyclopentane, 856
ethyl aminoacetate, 545
ethyl bromopiruvate, 519, 267
ethyl cyanoacetate, 545
ethyl α-bromoacetate, 851
ethylene-bridged nucleic acids, 368
ethynyl *C*-glycosides, 271
ethynylmagnesium bromide, 562
eukaryotic phosphotriesterase, 230
ezoaminuroic acid, 826
ezomycins, 821–823, 826–830

famciclovir, 728
feline immunodeficiency virus, 729
Ferrier rearrangement, 356
Fesulphos–copper complex, 715
fetal calf serum, 451
fleximers, 76–79
floxuridine, 9
fludarabine, 656
fluorenylmethanol, 241
fluorescent nucleosides, 284
fluorinated carbanucleosides, 588–590
fluorination, 588
fluoromethylene group, 503
fluoroneplanocin A, 589
fluoronucleosides, 483–486
fluorouridine, 9, 10
Fmoc-β-alanine, 869
formycin A analogue, 567
formycin B, 297
formyl *C*-glycosides, 276
four-membered nucleosides, 614–625
Friedel–Craft reaction, 299
Friedel–Crafts dehydrative glycosylation, 284
D-fructose, 350
TC, 317, 735–736
fungal cell membranes, 820
fungal infections, 820
furan, electrochemical activation, 14

furanoid glycals, 295
furanose, triacetyl, 18
furanose, tributanoyl, 18
furanosid-2′-ulosyl nucleosides, 442
furanosyl-α-amino acid, 824
furyl nucleosides, 14
furyllithium, 701

D-galactose, 480, 564, 704, 743
ganciclovir, 121, 127, 616, 705, 728
Garner's aldehyde, 716–717, 823–825
gemcitabine, 535–536, 684, 728
glucal, 625
gluconolactone, 273
D-glucose, 267, 273, 359, 363, 375, 380, 384, 480, 577, 656, 684, 704
  3-*O*-benzyl-1,2-isopropylidene-α-D-, 281
  2,3,4,6-tetra-*O*-benzyl, 300
  3-deoxydiacetone, 631
  bromoacetyl-α, 636
  diacetone, 347, 388–389, 413, 414, 631, 683
  peracetylated, 27
glucosyl phophoramidite, 27
glycals, 28, 29
glycals, 297–300
glyceraldehyde, 493
glycinate, methyl
  N-(2-*tert*-butoxycarbonylaminoethyl), 851
glycosidic bond, 11
glycosyl cyanides, 264–270
glycosyl epoxides, 290
glycosyl α-amino acid, 823
glycosylazides, 439
glyoxylic acid monohydrate, 521
glyoxylic acid, 745
G-quadruplexes, 867
Grignard addition, 188
Grignard reagents, 286
Grubbs' catalyst I, 387, 556
Grubbs' catalyst II, 371–373, 384, 409, 566
L-gulonic-γ-lactone, 492
L-gulose, 521, 743
guanine, 3, 76, 106, 112, 138, 149, 151, 163, 170, 181, 518, 537, 539–540, 542, 545, 547, 550, 553, 583, 593, 608, 614, 636, 762, 806, 853, 864, 867–869
  1-(2-hydroxymethyl)cyclopropylmethyl, 608
  1-benzyl-9-deaza, 302
  5-hydroxy-4-hydroxymethyl-2-cyclohexenyl, 643
  6-mercapto, 511
  9-(5-hydroxy-4-hydroxymethyl-2-cyclohexen-1-yl), 644
  9-[(1,3-dihydroxy-2-propoxy)methyl, 126
  9-deaza, 545

guanine (*Continued*)
  diacetoxyglyoxal-N^2-diacetyl, 126
  diisobutyroxyglyoxal-N^2-acetyl, 111
  N^2,N^9-diacetyl, 106
  N^2-acetyl-9-(2-acetoxyethoxymethyl), 107
guanosine, 15, 76–77, 87, 109, 164, 170, 219, 222, 503, 730–731, 762
  2'-deoxy, 476
guanosine (*Continued*)
  8-bromo, 87
  N^2-isobutyryl-5'-*O*-levulinyl-2'-deoxy
gulonic γ-lactone, 608

Hairpin loop, 416
haloethylaminophosphodiester, 240
halogenated nucleosides, 31
Hantzsch reaction, 276
Heck reaction, 298
Helminthosporium sigmoideum, 830
Henry reaction, 578
hepatitis B virus, 14, 173, 473, 676, 720, 727
hepatitis C virus, 676
hepatotoxicity, 637
herpes simplex virus, 103, 680
hetero Diels–Alder reaction, 823
hexamethyldisilazane, 296, 375
hexitol nucleic acids, 625
hexose-like nucleosides, 605
Hilbert–Johnson protocol, 701
HIV integrase, 320
Hoffer's chlorosugar, 282–283
hog-liver esterase, 386
homoadenosine, 626
homo-DNA, 626
Hoogsteen strand, 857
Horner–Wadsworth–Emmons reaction, 279, 558–559, 565, 568
H-phosphonate, 210, 216, 232, 237, 239, 241
Hüisgen reaction, 545
human immunodeficiency virus, 1, 14, 31, 103, 184, 317, 429, 473, 521, 691, 727, 807
human leukemia, 517, 680
Hünig's base, 866
hybridization affinity, 637
hydantocidin, 427, 467
hydrazine, 33
hydrazinolysis, 34
hydrobenzoin, 647
hydroboration, 560, 586, 617, 623, 639, 641
  asymmetric, 586–588
hydroboration–oxidation, 274
hydrogen bromide, 348
hydrogen radical, 400

hydrogen-bond, 41, 93, 263, 427, 439, 637, 666, 670, 851
hydrogenation, 107, 168, 176, 182, 188, 309, 354–355, 370, 373, 384, 389, 393, 398, 407, 409, 450, 460, 484, 497, 505, 545, 561, 573, 591, 608, 610, 630, 637, 702, 851, 859–860
hydrolysis, enzymatic, 31
hydrophobic interaction, 263
hydroxycyanation, 361
hydroxylamine, 874–875
hydroxymethyl nucleosides, 359
hypoxanthine, 3, 29, 38, 112, 135, 138, 434–436, 476, 512, 537, 539, 542, 545, 676, 762
  9-deaza, 545, 714
hypoxanthine, 117, 537

imidazo[1,2-*a*]-s-triazines, 488
imidazole, 52, 152
imidazolium triflate, 235
immobilized cholesterol esterase, 740
immucilins, 263, 508, 690, 711–713
immucillin-H, 699, 700, 711–714
immunomodulation, 427
indole, 1-acetyl-2,5,6-trichloro, 301
indole, 1-phenylsulfonyl, 302
inosine, 39, 476
  2',3'-dideoxy, 10, 27–28, 163, 317
intramolecular free-radical ring-closure, 370
intramolecular glycosylation, 395
iodine chloride, 405, 680
iodine, 336
iodomethylphosphonic acid, 192
iodotrimethylsilane, 756–757, 766
ion-exchange chromatography, 133
ionic liquids, 9, 290
iridium catalysis, 152
isoascorbic acid, 556
isobutyryl chloride, 440
isocyanate, β-methoxy-α-methylacrylolyl, 645
isocyanates, 448
isodeoxyadenosine, 329
Isodeoxycytidine, 335
isodeoxyuridine, 335
isodideoxynucleosides, 676
isomeric nucleos(t)ides, 317–340
isonucleos(t)ides, 317–340
isothionucleosides, 657
isoxantosine, 5,9-anhydro-2',3'-*O*-isopropylidene, 404
isoxazolidine, 309, 509, 709, 785–786, 793, 874–875
isoxazolidinyl nucleosides, 509, 782–817
isoxazolidinyl PNA monomers, 874–875
isoxazolidinyl-*C*-nucleosides, 800–804

ketene dimethyl thioacetal, 621
kinetic resolution, 18, 575
Knoevenagel reaction, 301
K-selectride, 170

lactonization, 550, 644
lamivudine, 14, 163, 317, 473, 521, 676, 683, 728, 745, 750, 759
lead tetraacetate, 487, 518, 666, 672
Lindlar catalyst, 851
lipase, 554
lipophilicity, 428
lipozyme, 18
lithiated heterocycles, 506–508, 710
lithium bis(t-butoxymethyl)cuprate, 497
lithium borohydride, 293
lithium hexamethyldisilazane, 393, 396, 493, 666
D-lyxose, 504–505, 509–510, 555
L-lyxose, 474, 656
locked nucleic acids, 362
Luche reduction, 591, 389

macrocyclic diuridine nucleotides, 405–407
Macugen, 25
malayamycin A, 828
maleimide sulfone, 291
malic acid, 701
maltitol, 634
mandelic acid, 391, 522
Mannich reaction, 508, 713–714
D-mannitol, 360–361, 637, 768–770
D-mannose, 386, 556, 569, 670–671, 673–675, 704, 710, 743, 766, 798, 813
Meerwein–Ponndorf–Verley reaction, 393
Meldrum's acid, 582
menthol, 521, 745
menthyl chloroformate, 750
menthyl glyoxylate, 522, 745
menthyl succinate, 856
mercuric trifluoroacetate, 829
Merrifield resin, 867
mesyl chloride, 352
α-mesylcyanohydrins, 442
methanolic ammonia, 348
β-methoxyacryloyl isocyanate, 497
β-methoxy-α-methacryloyl isocyanate, 497
methyl lithium, 678
methyl N-cyanoformimidate, 486
methyl thioglycolate, 741
methyl α-bromoacetate, 857
methylenation, 560, 562, 566
methylenecyclopropylcarbinol, 611
methylidine phosphorane, 676
methylmagnesium iodide, 667
methylmandelic acid, 639

methyloxalyl chloride, 448
methylribonucleotides, 24
methylthioadenosine nucleosidase, 119
Michael addition, 463, 493, 580, 613, 620, 788, 792–793, 702
Michaelis–Arbuzov reaction, 173
microwave heating, 373
miharamycin, 821–823, 831
minor groove, 637
mirror image, 473
Mitsunobu reaction, 84, 152, 170, 182, 188 189, 289–290, 499, 502, 525, 539–542, 549, 557–559, 586, 590–591, 632, 645, 665, 678, 683, 860, 863
Moffat oxidation, 492–493
Moffatt oxidation, 623
molibdenum dioxide acetylacetonate, 479
monomethoxytrityl, 851
mononucleotide prodrugs, 229–257
monophosphorylation, 164
montmorillonite K-10, 800
morpholine, 215
Mucor michel lipase, 522, 741
Mukaiyama's reagent, 265
multicomponent process, 633
myeloid leukemia, 729

N,N-thiocarbonyldiimidazole, 441
N,N-diisopropylethylamine, 65
N,N-dimethylamine, 457
N,N-dimethylethylenediamine, 456, 458
N,N-disubstituted uronamides, 663
N,O-acetonide, 824
N,O-bis(trimethylsilyl)acetamide, 217, 680, 702
N,O-bond reduction, 709
N,O-nucleosides, 18
N^3-selective methylation, 450
N-alkylphosphoramidate monoesters, 242
N-aminomethylprolines, 867–869
N-Aryl Phosphoramidates, 242
N-benzylhydroxylamine, 508
N-Boc-amino acids, 835
N-bromosuccnimide, 391
N-chlorosuccinimide, 239, 503
N-cyaniminooxazolidine, 247
neighboring group participation, 453
neoplacin A, 7, 499–501, 536–537, 574, 557, 560, 580, 583, 728
  7-deaza analogues, 54
  C-linked, 545
neoplanocin C, 559
neoplanocin F, 557
neoplanocin A, 165
neopolyoxins, 822
nephrotoxicity, 820

neutral alumina, 825
nevirapine, 815
N-hydroxypyrrolidines, 286
nifedipine, 276
nikkomycins, 821–823, 835–836
N-iodosuccinimide, 511–512, 691
nitroacetate, 550
nitrobenzoic acid, 60
nitrones, 286–287, 309, 508–509, 571, 709, 716, 783–787, 787, 789–791, 793–794, 796–801, 803–807, 809, 811, 813, 860, 874
   2-alkenyl, 430
   cyclic, 711, 716
   nucleophilic addition, 508–509
   1,3-dipolar cycloaddition, 509
nitroso Diels-Alder reaction, 569–571
N-methylhydroxylamine, 788, 791–792
N-methylimidazole, 232–233, 240
N-methylmorpholine, 708
N-methylmorpholine N-oxide, 382, 487
N-methylpyrrolidinone, 444
Nolan's catalyst, 564
nonafluorobutanesulfonyl fluoride, 637
nopinone, 620
norbornene, 586
norcarbovir, 621
norephedrine, 745
n-pentyl nitrite, 446
N-trityl hydroxylamine, 784
nucleophilic addition, 710, 791, 793–794
L-nucleosides, 14, 473–526
nucleoside antibiotics, 820–842
nucleoside triphosphates, 209–224

O-acetylquinidine, 572
O-benzylhydroxylamine, 367
olefin cyclopropanation, 357
olefinic peptide nucleic acids, 849
oligonucletides, 473
O-mesylcyanohydrins, 431
O-methyl-O-tert-butylidimethylsilyl ketene acetal, 793
Oppolzer's sultam acrylate, 860
origin of life, 625
osmium tetroxide, 293, 382, 393, 487, 832
osmylation, 582
oxaphospholane, 611
oxathioacetals, 743
oxathiolane nucleosides, 521–525, 727–773
oxazinomycin, 545
oxazole, 59
oxazolidinethione, 567
oxazolidinone, 617
oxetane, 348, 359
oxetanocin, 592, 614–615, 623

oxidation–reduction protocol, 482
oxidative amination, 241
oxidative cleavage, 359, 382, 593, 619, 641, 661, 666, 672
oxidative decarboxylation, 517–518
oxime esters, 4
oxone, 219
oxonium intermediate, 287
oxonium-promoted rearrangement, 391
oxonucleosides, 688
oxymercuration, 829
oxypeptide, 871
ozonization, 480

P2X receptor antagonists, 111
P2X3 receptors, 111
P2Y receptor antagonists, 116
Palladium acetate, 297, 53
papilomavirus, 201
paraformaldehyde, 783
Pauson-Khand reaction, 575–576
PCC oxidation, 389, 392
Pearlman's catalyst, 53, 354, 367, 389
penciclovir, 728
peptide nucleic acids, 848–877
perbenzoic acid, 756
perchloric acid, 784
perfluoro-1-butanesulfonyl fluoride, 688
periodate-promoted oxidation, 569
Pfitzner–Moffatt oxidation, 377, 413
pharmacological probes, 660
phase-transfer conditions, 297, 460
phenyl stannannes, 444
phenyldichlorophosphate, 239
phenylphosphoramidate, 241
phenylsulfonylnitromethane, 581
phorphoramidite, 688
phosphate oxyanion, 411
phosphites, 210
phosphitylation, 614, 636
phosphodiester prodrugs, 237–238
phospholipid nucleosides, 237
phosphonate coupling, 4
phosphonated nucleosides, 163–204
   cyclobutyl, 181
   cyclopentyl, 181–192
   cyclopropyl, 165–181
   N,O-nucleosides, 804–806
phosphonopyrrolidine, 866
phosphoramidate, 164, 210, 213
phosphoramidite prodrugs, 238–243
phosphoramidite, 27, 293, 632
phosphorochloridates, 246
phosphorodiamidates, 246
phosphoromorpholidate, 213

INDEX 891

phosphoroselenoate, 238
phosphoroselenoates, 244
phosphorothioate, 238
phosphorothiolates, 244
phosphorous oxychloride, 242
phosphorus chloride, 240
phosphorus oxychloride, 238–240
phosphorus trichloride, 242–243
phosphorylating enzymes, 732
phosphotriester prodrugs, 230–237
photochemical initiation, 395
photocycloaddition, 621
photo-isomerization, 401
photolysis reaction, 572, 616
pig liver esterase, 738, 799–800
pig liver lipase, 799–800
pinene, 620
pinolic acid, 620
pinonic acid, 620
pipecolyl-PNA, 849
piperidine, 457
piperidyl PNA monomers, 873–874
piranosyl-α-amino acid, 824
pivaloyl 4-pentenoate, 567
pivaloyl chloride, 234, 241
PNA monomers, 848–877
PNA-DNA triplex, 857
p-nitrophenylcarbonate resin, 503
polyamidoamine, 766
polycondensed aromatics, 302
polymethylene linkers, 435
polyoxin J, 823
polyoxins, 821–823
   carbocyclic, 552
potassium carbonate, 147
potassium thioacetate, 492, 510, 623
preorganized structure, 632
preparative HPLC, 246, 519
preparative TLC, 525
Prins reaction, 551
D-proline,
   N-Boc-*cis*-4-hydroxy, 863–864, 870
   trans-4-hydroxy, 868
L-proline, 71
   cis-4-hydroxy, 870
   N-Cbz-*trans*-4-hydroxy, 864
   trans-4-hydroxy, 702, 861–863, 865–868, 871, 874
prolinol, 869
propargyl ketones, 265
propylene oxide, 391
pseudomonas cepacia lipase, 2, 4, 5, 7, 15–18, 21–23, 35, 854
Pseudomonas fluorescens lipase, 522

Pseudomonas fluorescens, 552–553
pseudouridine, 49, 309, 709, 826
psiconucleosides, 809
p-toluenesulfonylmethyl isocyanide, 281
Pummerer reaction, 661–663, 678, 757
Pummerer rearrangement, 512, 623–624, 692
purine cyclonucleosides, 400, 402–404
purine nucleoside phosphorilase, 263, 508, 699, 168
purine,
   2,6-dichloro, 476, 515, 539, 662, 666, 670, 675
   2-amino-6-benzyloxy, 176
   2-amino-6-chloro, 540, 548, 581, 608, 613, 865
   2-amino-6-iodo, 636
   6-benzyloxy-2-amino, 586
   6-benzylsulfanyl-9-benzyl, 54
   6-chloro, 16, 166, 170, 357, 360, 435, 491, 519, 521, 540, 583, 590, 626, 672, 675–676, 683, 688, 792
   2-fluoro-6-chloro, 491
   6-cyclohexylamino, 680
   bis-TMS-2-(*N*-acetylamino), 832–833
   C-8 arylation, 52
   N-benzyl-6-chloro, 59
pyran ribonuclosides, 382
pyranoid glycals, 295
pyranosyl nucleosides, 625–647
pyrazole, β-D-ribofuranosyl, 73
pyrazolo[3,4-*d*]pyrimidines, 488
Pyricularia oryzae, 822, 830
pyridazine *C*-nucleosides, 271
pyridinium chloride, 235
pyridinium dichromate, 491, 663
pyridinium phosphoramidate, 216
pyridinium trifluoroacetate, 28, 398
pyrimidine cyclonucleosides, 395–400, 401–402
pyrimidine nucleoside phosphorylase, 657
pyrimidine,
   2-amino-4,6-dichloro, 542
   2-amino-5-nitro-4,6-dichloro, 542
   4,6-dichloro-5-nitro, 621
   5-amino-4,6-dichloro, 617
   6-aza, 680
pyroglutamic acid, 702, 712, 859–860
pyrophosphate, 221
pyrrolidin-2-ones, 859–865
pyrrolidine, 457, 859–865
pyrrolidine-2,5-dicarboxamides, 708
pyrrolidinium phosphoramidates, 215
pyrrolidinyl nucleosides, 504–509
pyrrolo[2,3-*d*]pyrimidines, 488

radical cyclization, 576
radical deamination, 446

radical-induced cyclization, 275
radicamine B, 716–717
reductive cleavage, 634
reductive debenzylation, 390
reductive ring opening, 571
Reformatsky–Claisen rearrangement, 683
Reformatsky reaction, 265
Reformatsky reagent, 851
Reformatsky–Claisen rearrangement, 590
reverse transcriptase inhibitors, 163, 727
reverse transcriptase, 184, 429, 446
(R)-glycidol, 145
Rhizoctonia solani, 822
D-ribose, 718
ribavirin, 10, 49, 70, 217
    5′-O-alanyl, 11
    analogues, 70–73
ribofuranosylethine, 271
β-ribofuranosyl ketoesters, 265
ribonolactone, 273, 295, 555–557
ribonucleoside diphosphate reductase, 429
ribonucleotide reductase, 535
D-ribose, 813
    3-C-hydroxymethyl, 413
    3-O-isopropylidene, 301
L-riboadenosine, 474
L-ribofuranose, 1-O-acetyl-2,3-dideoxy, 489
L-ribose, 1-bromo-2,3-dibenzoyl-5-fosfate, 476
ribosyl bromides, 284
ribozyme, 605
ring contraction, 561, 614
ring-closing metathesis, 371, 384, 387–389, 404, 408–411, 501, 556, 188
RNase H enzymes, 605
ruthenium tetroxide, 393, 607
ruthenium trichloride, 378, 769

S-adenosylmethionine, 657
S-pivaloyl-2-thioethyl, 240
S-adenosylhomocysteine hydrolase, 589
Sakurai cyclization, 309
salicyl alcohol, 235
salicyl chlorophosphite, 241
salicyl phosphorochloridite, 224
savinase, 553
Schramm's aza-C-nucleoside, 714
Schwartz reagent, 814
Sclerotinia, 826
selective hydrolysis, 661
selenazofurin, 74
selenophenfurin, 74, 75
self-pairing, 625–626
semiempirical calculations, 787
semithiocarbazide, 265
L-selectride, 289, 618

L-Serinal, N-(9-phenyl-9-fluorenyl), 578
D-serine, 835–836
serine-derived lactone, 834
Sharpless reagent, 361
showdomycin, 49, 263, 291, 293, 487, 545
silver tetrafluoroborate, 302
silver thiocyanate, 74
silver triflate, 297
simian immunodeficiency virus, 729
Simmons–Smith cyclopropanation, 177, 557, 593
six-membered nucleosides, 625–648
sodim periodate, 380
sodium borohydride, 289
sodium carbonate, 87
sodium cyanoborohydride, 710
sodium hexamethyl disalazane, 382
sodium hexamethyldisilazane, 348
sodium periodate, 192
sodium triacetoxyborohydride, 298
soidum chlorite, 769
solid tumors, 729
solid-phase synthesis, 502, 851, 855, 861, 867–868
solvation free energies, 455
Sonogashira crosscoupling, 62, 90, 273, 444, 673
spirocyclic aglycone, 392
spirocyclic C-nucleosides, 273
spiroisoxazolines, 427–428, 430
spironucleosides, 391, 427–467, 692
stacking interaction, 90
stannic chloride, 521, 738
Staudinger reaction, 278, 439, 349, 365
steric factors, 438
steric hindrance, 452, 267, 444
(S)-stericol, 617
Stille reaction, 303
Streptomyces cacaoi, 822
sugar puckering restriction, 404
sulfamoyl chloride, 453
sulfonylation, 544, 561
Suzuki coupling, 54, 61, 303, 673
Swern oxidation, 383, 617
synadenol, 611
synguanol, 611

tandem Wittig–Michael sequence, 279
tartaric acid, 611
tartrate salts, 762
telbivudine, 728
tellurides, 304
tenofovir, 104, 136, 163, 198, 728, 810
tert-butoxymethylcuprate, 557
tert-butyl hydroperoxide, 409
tert-butyl magnesium bromide, 239

tert-butyl magnesium chloride, 236, 239, 246–247
tert-butylhydroperoxide, 232–233, 240, 401
tert-butylsilyl triflate, 382
tetrahydrofuranylation, enzymatic, 13
tetrahydropyranylation, enzymatic, 13
tetrakis(triphenylphosphine)palladium, 182
tetra-*O*-benzylgalactopyranoside, 564
tetronic acid, 291–293
thermal stability, 637, 656, 855
thiazolidinyl PNA monomers, 877
thienyl stannannes, 444
thietane, 623
thioacetal integrity, 743
thioalkyl group activation, 688
thioamide, 265, 519
thioanisole, 302
thiocarbonyldiimidazole, 680
thioether oxidation, 755
thioglycolaldehyde diethylacetal, 741
thioglycolic acid, 736
thionoester, 489
thionucleosides, 509–517, 655–693
thionyl chloride, 87
thiophenfurin, 75
thiophenyl radical, 400
thioribofuranosyl bromide, 510
thioxonucleosides, 657
three-membered nucleosides, 605–614
threose nucleoside triphosphates, 148
thymidilate synthetase, 535
thymidine kinase, 164
thymidine phosphorylase, 152
thymidine, 21–23, 40, 106, 150, 218, 220, 223, 246, 295, 347, 380, 401, 403, 408–409, 411–412, 476, 521, 587, 700, 835
  1-(2′-deoxy-4′-thio-β-D-erythropentofuranosyl)-(6-aza), 680
  2-deoxy-aza-, 700
  3′-hydroxymethyl-4′-azamethyl-2′-3′-dideoxy, 792
  3′,5′-diamino-3′,5′-dideoxy, 40
  3′-4′-oxetane, 375
  3′-amino-5′-crotonylamino-3′,5′-dideoxy, 7
  3′-*O*-acetyl, 332
  3′-*O*-benzoyl, 334
  3′-*O*-dimethoxytrity, 26
  3′-*O*-levulinyl, 7
  3′-azido-3′-deoxy, 18, 23, 27, 31–33, 117, 163, 317
  5′-*C*hydroxylethyl, 413
  dibenzoyl, 476
  soxazolidinyl, 789
  L-α-arabinofuranosyl, 479

thymidinyl mercury, 474
thymidylate biosynthesis, 93
thymine, 3, 18, 40–42, 117–119, 129, 135, 178–179, 181–182, 256, 296, 327, 347, 359, 361, 364, 367–369, 376, 380–388, 391, 415, 427, 430, 432–435, 439–440, 450, 467, 480, 497, 499, 521, 537, 540, 544, 557, 564, 587, 593, 614, 624, 627, 629–630, 633, 637–640, 691, 702–704, 706, 753, 755, 756, 768, 788, 792–793, 797, 799–800, 805–808, 810–811, 813, 823, 828, 830, 833, 839, 841, 851, 853–854, 856–858, 863–864, 865, 867–868, 873, 876
  (*E*)-OPA monomer, 851
  1-(2,3-dideoxy-β-d-erythrohexopyranosyl), 625
  1-acetic acid, 866
  1-acetic acid, 869
  3′-*C*-cyano-3′-*O*-mesyl-β-D-xylo-, 431
  3-hydroxy-4-hydroxymethyl-5′-*O*-allylphosphoramidite, 409
  arabinofuranosyl, 24
  bis(trimethylsilyl), 691, 832
  iodination, 401
  N^3-benzoyl, 540, 676, 683, 703
  N^4-acetyl, 755
  silylated, 135, 347, 356–357, 362, 381–382, 384, 386, 389, 392–394, 414, 431, 509, 512, 518, 626, 687, 690, 762–763, 771, 783, 785, 788, 792, 801, 828, 874, 375
  xylopiranosyl, 24
tiazofurin, 75, 263, 519, 708, 800–802
tosyl chloride, 352
tosyl hydrazine, 298
transacetalization, 176
transamination, 446, 448–451
transcription inhibition, 848
transesterification, 450
transglycosylation, 478
transient conformations, 405
translation inhibition, 848
translation process, 286
Traube synthesis, 542
triacetoxy sodium borohydride, 296
triazole, 545–546
triazolo nucleosides, 60–70
tributyl amine, 211
tributylammonium pyrophosphate, 212, 214, 217
tributyltin hydride, 321, 366, 379, 380, 389, 400, 476, 505, 631, 644
trichloroacetamidite, 300–301, 833
triethylamine trihydrofluoride, 490
triethylammonium carbonate, 211
triethylorthoformate, 542

triethylphosphate, 242
triethylsilane, 290, 309, 634, 755
triflic anhydride, 365
trifluoroacetic acid, 482
trifluoroacetoxy borohydride, 370
trifluoromethyluridine, 488
triisopropylbenzenesulfonyltetrazolide, 329
trimethyl orthoformate, 499
trimethylaluminium, 678
trimethylphosphate, 210
trimethylphosphine, 349, 365
trimethylphosphonate, 555
trimethylsilyl azide, 582–583
trimethylsilyl bromide, 85, 177
trimethylsilyl chloride, 352
trimethylsilyl cyanide, 265
trimethylsilyl diazomethane, 574, 582
trimethylsilyl iodide, 79, 182
trimethylsilyl silane, 478, 489
trimethylsilyl triflate, 347, 360, 375, 431, 480,
 514–515, 521, 626, 662, 666–668, 670,
 738, 762, 766, 769, 783, 787, 792, 828, 835
trimethylsilyl uridine, 476
triplex formation, 853
triplex melting temperature, 864
tris(imidazolyl)phosphine, 238
troxacitabine, 759, 769
trypanosomal nucleoside hydrolase, 714
TSAO nucleosides, 427–467
Tsuji-Trost reaction, 541–542, 564–566, 548,
 576, 581–584

uracil, 3, 79, 107, 116, 118, 156, 187–189, 219,
 295, 309, 324, 338, 476, 497, 499, 506,
 521, 525, 537–538, 540, 544, 550,
 552–553, 583, 593, 608, 634–636, 647,
 702, 706–707, 789, 793, 797, 821
 1-(3′-deoxy-β-D-psicofuranosyl), 353
 1,3-dimethyl, 58
 1,5-anhydro-5-iodo, 631
 3-benzoyl, 525
 3-benzoyl-5-fluoro, 525
 5-bromo, 434, 709, 719, 806
 3-octil-5-formil, 708
 5-[(E)-bromovinyl], 219
 5-fluoro, 118, 488, 510–511, 520, 524, 535,
 790, 805–806, 808, 810
 5-iodo, 49, 296, 434, 631
 2-thio, 118
 N^3-benzoyl, 540, 676, 703
uridine, 17, 84, 220, 223, 286, 350, 378,
 395–396, 402, 405, 409–410, 499, 503, 826
 (E)-5-(2-bromovinyl)-2′-deoxy, 35
 (E)-5-(2-bromovinyl)-2′-deoxy, 657
 (S)-4′aza-2′-3′-dideoxyfluoro, 784

2,2′-O-anhydro-4′-thio, 687
2′,3′,5′-tri-O-acetyl, 31
2′,3′,5′-tri-O-acetyl-2-C-methyl, 31
L-β-deoxy, 476
2′-deoxy-2′-fluoro-
2′-deoxy-6-(hydroxyalkyl), 428
3′,5′-di-O-TBS-protected, 371
3′-hydroxymethyl-4-azamethyl-2′-3′-dideoxy,
 792
3′-O-acetyl, 405
3-deaza, 78
5′,6-cyclo-5′-deoxy, 395
5′-azido-5-bromo-5′-deoxy-2′,3′-O-
 isopropylidene, 402
5′-deoxy-5′-iodo, 396
5′-O-dimethoxytrityl-2′-O-methyl, 405
5-fluoro, 835
5-iodo, 407
6,5′-cyclo, 398
6-amino-5′,6-N-anhydro, 402
dibenzoyldeoxy, 476
isoxaozlidinyl 5-fluoro, 789
persilylated, 349
etraisopropyldiaminophosphite, 407
uridine–cytidine kinase, 678
U-turn mimics, 404

valaciclovir, 106
valganciclovir, 104, 121
varicella zoster virus, 103, 106, 125, 657
verbenone, 620
vinyl carbonates, 33
vinyl glycosides, 274
vinyl iodide, 851
vinyl magnesium bromide, 271, 387, 408, 556,
 559, 566, 824–825
vinyl nucleobases, 787, 796
vinyl stannannes, 444
vinyl thymine, 509
vinylidenecarbene, 574
viral polymerases, 732
virazole, 28
Vitravene, 345
Vorbrüggen protocol, 347, 358, 361, 386, 392,
 479–481, 486, 489, 504, 515–517, 52 626,
 686, 702, 736, 762, 765, 787, 796–797,
 809, 826, 832

Wang resin, 56
Watson–Crick base-pairing, 605, 848
Weinreb amide, 717, 824
Whistler reaction, 510
Williamson reaction, 441
Wittig homologation, 361
Wittig methylenation, 366, 383

Wittig reaction, 278–282, 380, 564, 593, 607, 618, 621, 676, 831
Wittig-Horner reaction, 386
Wolff rearrangement, 614

Xenopus oocytes, 659
D-xylose, 320–322, 483, 510, 623, 656

L-xylose, 374, 481–482, 486, 614, 623

zalcitabine, 535–536
zinc (II) chloride, 222
zinc borohydride, 825
zinc bromide, 271

Printed and bound by CPI Group (UK) Ltd, Croydon, CR0 4YY
15/09/2022
03148339-0001